# THE SAILING NAVY LIST

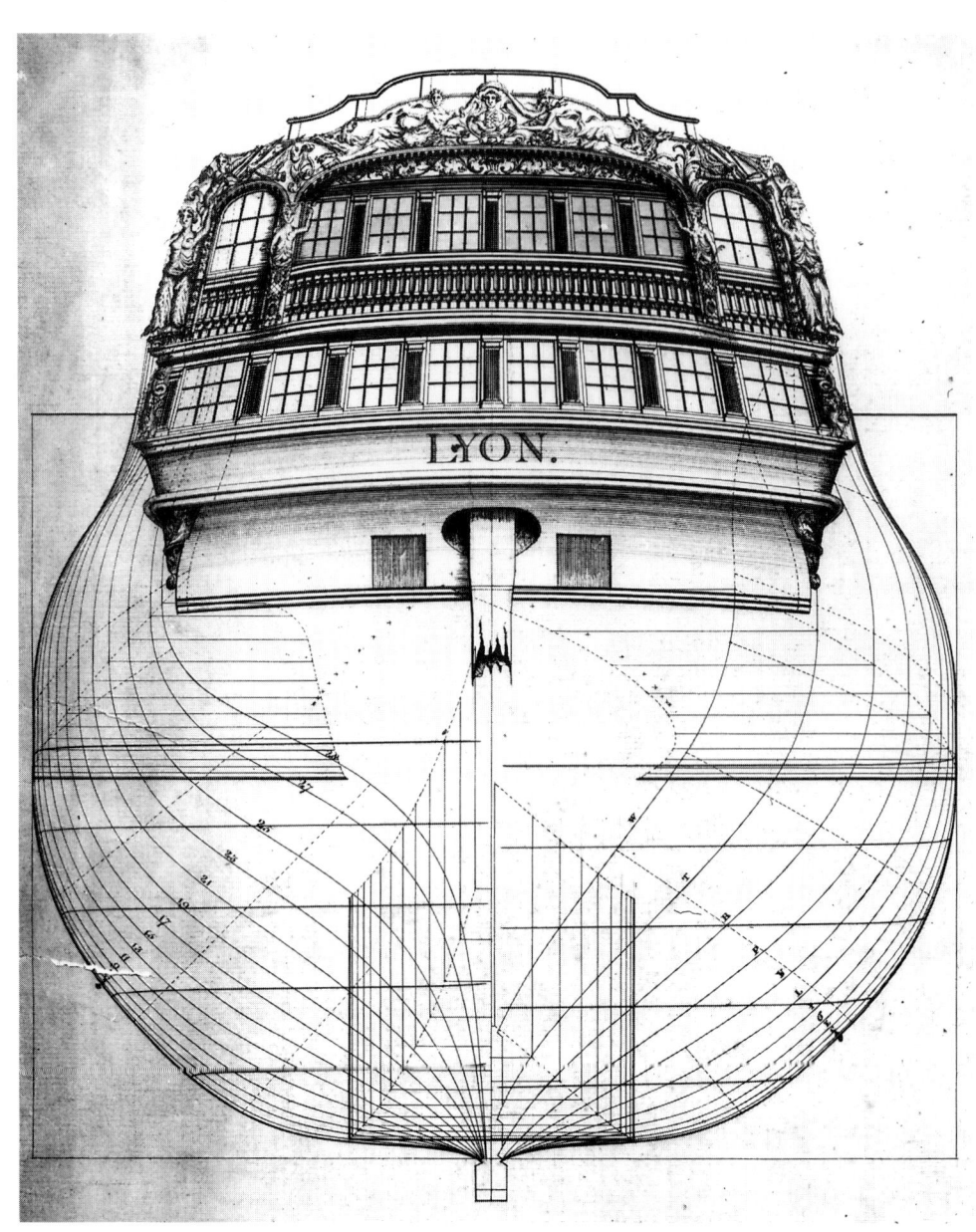

Stern of the *Lyon* (*Lion*) of 1777

# The SAILING NAVY LIST

## All The Ships of the Royal Navy
— Built, Purchased and Captured —
1688-1860

◆

*David Lyon*

# For E.C.L. and J.A.L.

*Illustrated from the collections of the National Maritime Museum, Greenwich*

© David Lyon 1993

First published in Great Britain 1993 by
Conway Maritime Press, an imprint of Brassey's (UK) Ltd,
101 Fleet Street,
London EC4Y 1DE.

**British Library Cataloguing-in Publication Data**

Lyon, David
   Sailing Navy List: All the Ships of the Royal Navy -
   Built, Purchased and Captured, 1688-1855
   I. Title
   623.8
   ISBN 0-85177-617-5 (Limited Edition)
   ISBN 0-85177-864-X (Standard Edition)

All rights reserved. Unauthorised duplication contravenes applicable laws.

Typeset and designed by Eden Valley Press, Kirkby Stephen.
Printed and bound in Great Britain by The Alden Press, Oxford.

# Contents

| | |
|---|---|
| ACKNOWLEDGEMENTS | *vii* |
| PREFACE | *viii* |
| ABBREVIATIONS & DEFINITIONS | *x* |
| GENERAL INTRODUCTION | *1* |

## INTRODUCTION
## SHIPS IN SERVICE IN 1688

| | | |
|---|---|---|
| 1 | Ships Built for the Royal Navy | *11* |
| 2 | Prizes and Purchases | *15* |

## PART I
## SHIPS BUILT FOR THE ROYAL NAVY

| | | |
|---|---|---|
| 1 | Before the Establishments *1689-1705* | *16* |
| 2 | The Preliminary Establishment *1706-1718* | *33* |
| 3 | The Establishment Period *1719-1745* | *39* |
| 4 | The Slade Era *1745-1785* | *62* |
| 5 | Henslow to Seppings *1786-1830* | *104* |
| 6 | Symonds and After *1830-1860* | *170* |

## PART II
## PRIZES AND PURCHASES

| | | |
|---|---|---|
| 1 | King William's War *1689-1697* | *184* |
| 2 | The War of Spanish Succession *1701-1713* | *191* |
| 3 | The Wars of Jenkins' Ear & of the Austrian Succession *1739-1748* | *197* |
| 4 | The Seven Years' War *1756-1763* | *203* |
| 5 | The American War of Independence *1776-1783* | *214* |
| 6 | The French Revolutionary War *1793-1801* | *236* |
| 7 | The Napoleonic War *1802-1815* | *267* |
| 8 | The Period after *1815* | *292* |

## PART III
## AUXILIARIES & ANCILLARIES

| | | |
|---|---|---|
| 1 | Vessels on the Canadian Great Lakes | *297* |
| 2 | Royal and Other Yachts | *303* |
| 3 | Conversions and Reclassifications | *307* |
| 4 | Yard Craft | *315* |
| 5 | Coastguard and Customs Vessels | *331* |
| 6 | Hulks | *336* |

| | |
|---|---|
| INDEX | *351* |

# Acknowledgements

I should like to express my warmest thanks to Robert Gardiner, Patrick Gossett, Andrew Lambert and Brian Lavery; all fellow-researchers who have generouly shared their findings and lists with me. Without that exchange of information this work would be very considerably diminished. Rif Winfield's dedication and kindness at the last minute in checking my proofs against his own listing was immensely useful and much appreciated. Jim Colledge, Fred Dittmar and the late Commander Pitcairn Jones all produced listings which have been of very great use to me. Indeed my battered copies of both editions of 'Colledge' bear witness to just how useful a working tool *Ships of the Royal Navy* is.

Teddy Archibald, Daniel Baugh, David K Brown, Joan Horsley, Roger Knight, Hugh Lyon, Iain MacKenzie, Roger Morriss, the late George Osbon, Alan Pearsall, Stewart Roderick, Tony Sainsbury, David Sambrook, David Syrett, the late Len Tucker and Chris Ware have all given both encouragement and help on particular subjects.

I would also like to thank Martine Acerra, Hubert Berti, Hans-Christian Bjerg, Jean Boudriot, Alec Douglas, Jan Glete, Jose Merino, Jan-Piet Puype, the late Christian de Saint-Hubert, also helpful colleagues at the Prins Hendrik Museum, Rotterdam, and officials at many other institutions in Europe and the United States. All of these have been most helpful in providing information on vessels of non-British origin or fate.

Alan Bax, John Broadwater, Martin Dean, Graeme Henderson, Peter Marsden, Colin Martin, the late Keith Muckelroy, Myra Stanbury, Gordon Watts and many other divers and archaeologists have provided me with valuable information about wrecks.

A number of colleagues and ex-colleagues at the National Maritime Museum have all helped at one time or another in the production of this work, especially Michael Dandridge, George Osbon, Arda Pilkington, David Spence and David Taylor. At the publisher's end of the production process Julian Mannering and Linda Jones have done a grand job.

Many other colleagues, friends and correspondents have wittingly or unwittingly contributed to the book, my apologies for not naming each individually, and my thanks to all.

I should, however, end by thanking the trio whose (sometimes understandably impatient) encouragement finally pushed this work from manuscript to print: Robert Gardiner, Roger Knight and Leo (Eleanor) Sharpston.

Any merit this work possesses is in large part attributable to the generosity and enthusiasm of my many helpers. The remaining errors, omissions and other deficiencies are all my own work.

This book is dedicated in gratitude to my father and mother, John and Kay.

# Preface

This book lists the sailing ships of the British Royal Navy from the late seventeenth century to the age of Victoria. During nearly all this period it was the largest navy in the world. For this reason alone it would be desirable to record its ships in detail. The Royal Navy was also, however, the most industrialised organisation of its time. The role of the Royal Navy as both cause and symptom of the political and economic growth of Britain is only now being fully appreciated by historians. In particular the Royal Dockyards provide a very early example of major factory-style organisation before the rise of factories as such, and in many ways foreshadow the Industrial Revolution. It is therefore important to have an idea of the exact output of those dockyards: the volume and types of ship which they built and fitted out. The ships themselves were the most complex single products of human ingenuity and amongst the most technologically advanced of artifacts. A single ship of the line carried more and heavier guns than would be attached to an entire army, she would support a crew equivalent to the population of a small town for months on end, and had to be able to navigate the seas, survive bad weather and to fight during this time.

I have aimed to provide as complete a listing as possible; first, of the fighting ships designed and built for this navy, then of the ships acquired by capture or purchase, and finally of the auxiliary vessels built or converted for dockyard service, or belonging to ancillary services such as the customs. It may come as a surprise that, even if one discounts the fighting ships and looks only at dockyard craft, the Royal Navy was by far the largest shipowner of the period. It is also noteworthy how very many merchant ships were either purchased or hired for naval use, and for which there is therefore some technical data. A symptom of this is that there are many more plans of later eighteenth-century merchantmen in the Admiralty Collection than survive elsewhere.

At a more restricted level, the aim was to provide basic data on the ships and classes of ship of the British Navy during the 'great age of sail' to serve as an aid to researchers. This came out of my own need for such a work when working on cataloguing the Admiralty Collection of ship plans held at the National Maritime Museum, Greenwich.

This book is compiled from these plans and a series of contemporary listings of the Navy. It follows these sources in giving basic dimensional data and information on armament and complement. It is restricted to constructional data and data relating to major changes in function, structure and major fittings of the vessel. I have not, in the present work, followed up detailed changes in armament or in complement; nor is any attempt made to provide an account of the operational history of each vessel. The detailed origins and rationale of these data are discussed in the Introduction. Several more general observations should also be made. First, the sources themselves are quite often inconsistent, both internally and as between different sources. However, as a matter of policy they have been followed. I should naturally be very pleased to hear from those who have found primary material which amplifies, or corrects, what is set out here. Secondly, where there are inconsistencies between sources and I am unsure which is to be preferred, I have given both versions.

I have attempted to draw a distinction between the period when a ship was in active service, or potentially available for active service, on the one hand, and the time spent in 'harbour service' as a hulk, or other second line duties, on the other. This provides the necessary basis for working out the real strength of the navy at any particular time. As the book is intended as a reference work there is a certain amount of repetition of definitions and other matters of importance between, for example, the general overview of warship development in the Introduction, the introductory notes to the various chapters, and the dictionary of abbreviations and definitions.

As far as possible the categories of ship used are those provided by the original compilers of the navy lists consulted. Thus 'sloop, rig unknown' is a heading taken direct from Admiralty lists of the time, and reflects the uncertainty of the Admiralty clerks themselves.

To avoid confusion the general rule appears to have been to prevent any duplication of names. Inevitably, in wars being fought across the oceans of the world with long, slow and uncertain lines of communication back to London, ships would be purchased on distant stations and be given names already allocated to others. It is significant that such names would usually be changed as soon as the news got back to the Admiralty. This explains the number of name changes in the first year of service of prizes and purchases in the Royal Navy. The major exceptions to the general rule of no duplication of the names of commissioned warships were cutters. These seem to have been regarded as similar to 'tenders' - vessels which were administered on a different level to larger warships (they had a 'progress book' to themselves). Quite often they duplicated the names of larger vessels. Hired ships by their nature were not proper naval vessels; and nearly all retained their mercantile names even when there were several of the same name in service at the same time. Up until the middle of the eighteenth century it was a common usage to retain the name of a captured foreign vessel even if it duplicated a name already in naval service, or to give her the name of the ship which captured her, in both cases differentiating the vessel by the suffix 'Prize' (or in a few cases by giving the nationality, eg. *French Ruby, Sallee Rose*). It is never very clear whether the word Prize was actually part of the name of the ship or just a 'detachable label', though I have normally treated it as the former. Another example of uncertain usage is whether the definite article (*le, la, l'*) is given or omitted with French names. Although there are a number of examples of ship names where the definite article seems usually to have been retained by the Royal Navy, I have, for the avoidance of confusion and to make looking up easier, always omitted it.

The parts of this work covering the late seventeenth century and the opening years of the eighteenth are less securely based on original sources and research than for the period after about 1715 and these should, therefore, be regarded as provisional.

The focus throughout has been on the British Navy and my research has been much less detailed in respect of vessels that came from, or went into, other nations' control. Such information as I do possess relating to their origins and eventual fates is from secondary sources or research done by others: it is therefore more open to error and may be inadequate, incomplete or sometimes simply wrong. I very much hope that works similar to the present compilation are being, or will be, undertaken for those other navies; so that the errors in the present listing which will inevitably emerge if those projects are undertaken can be put right in a later edition. Much the same applies to the origins of the very large number of merchant ships purchased or hired for naval use as the naval sources are not always precise about the origins of such vessels. Also many of the vessels noted as 'sold' will have seen further service as merchantmen; though only rarely has any information at all about such subsequent service been found. Even the naval service of many prizes and purchases, especially those acquired and remaining overseas, is often not well documented. This imperfect documentation explains why what I have chosen to call 'unregistered vessels' are shown in supplementary lists.

Only basic data have been given - it is usually possible to find out more (sometimes a great deal more) about individual ships. My intention has been, however, to provide a basic work of reference where the emphasis is placed on breadth of coverage of the navy, rather than depth of coverage of any particular entry.

I start with a list of ships in service in 1688. The year of the 'Glorious Revolution' in which William and Mary replaced James II on the English throne was chosen partly because it is the point of departure of the earliest primary source I have used, partly because that year marks the beginning of the first round in the 'second Hundred Year's War' between Britain and France, and partly because the navy of the later Stewarts has been very well listed by Frank Fox in his *Great Ships* (although that listing ends with the death of Charles II in 1685, there were few ships added between that date and 1688). The ship listing begins with short notes on the vessels in service at that time. I end in 1860 which was the year of the launch of HMS *Warrior*, the first iron seagoing ironclad and the ship that marked the final, abrupt, demise of the wooden line of battle ship as the dominant warship. Steam propulsion was initially a supplement to the sailing ship of the line. The ironclad completely replaced that type of ship. Steam came in slowly, sail departed equally slowly. Wood was not replaced by iron for hull construction without false starts and a period of overlap, but the *Warrior* provides the clearest marker to the true watershed between the 'sailing navy' and its successor.

NOTE: The original intention had been to publish eight additional appendices with this work; unfortunately considerations of space have meant they have had to be excluded. These appendices consisted of the following lists which presented in a different format material already available under the entries on individual ships:

(1) A chronological list of Royal Naval shipwrecks and founderings starting slightly earlier than the rest of the work, in 1660, and running through to 1860.

(2) A matching chronological list (also starting in 1660) of Royal Naval vessels captured by the enemy.

(3) A list of British shipbuilders (starting with the Royal Dockyards) showing the sailing warships built by them from the early sixteenth century to 1860 (again starting earlier than the ship lists). This is complemented by a Geographical Index of builders. That index also includes references to ships where only the location of the builder is known, and not his (or her) name.

(4) A list of the 'Master Shipwrights' and 'Surveyors of the Navy' from 1660 to 1830, together with brief notes on their careers and a listing of the classes they designed and the ships whose construction they supervised. This is followed by a list of the designs of foreign origin, classified by nationality and the names of the ships whose lines were used as a basis for the British classes concerned.

(5) The index as originally compiled included brief notes of types, classes and conversions to go with the names and dates – thus providing a dictionary of ship names for the sailing navy.

Two further appendices would have provided financial information on a sample of costs of (6) construction of vessels built for the navy and (7) purchase and fitting out prizes and purchases. This information is taken from the 'Progress Books' (and therefore commences with the second quarter of the eighteenth century). Unlike the other appendices, which put information already available in this book into a different and more accessible form, they contain information not previously used. At the time of writing, these appendices cover the sample of ships for which such information is available whose names begin with the letters A to D; but the task of compiling them is continuing.

There are also (8) lists by type of ship in chronological order giving a year-by-year record of additions and deletions, though these lists are in a less developed state than (1) to (7).

If you are interested in obtaining copies of any or all of these, please write to the author:

D J Lyon, c/o the National Maritime Museum, Greenwich, London SE10 9NF, UK.

**David John Lyon, Navarino Day (20 October) 1993**

## Abbreviations & Definitions

**A/F** 'As fitted' (of a plan) showing a ship either 'as built' or at some specific date later in her career. When used of a captured or purchased vessel indicates that the plan shows the vessel as converted for service in the Royal Navy as opposed to her state as purchased or captured (see 'taken off')

**AO** Admiralty Order (date)

**BU** Broken up (ship taken to pieces)

**Class** A group of ships built to the same lines (and therefore to a common design)

**Dyd** Dockyard (ie. Royal Dockyard as opposed to a private shipbuilder)

**Establishment** See section of Introduction dealing with Rebuilds and Establishments for a discussion of the meanings of this word. The 1706, 1719 and 1733 Establishments were a series of more or less prescriptive sets of dimensions for the larger ships. The 1745 Establishment a combination of prescriptive dimensions and designs

**Fc** Forecastle

**Framing** Plan showing side view of the framing

**GD** Gun deck (usually the lowest deck with guns)

**GR** 'Great Repair', a very major repair, virtually a rebuild (in the twentieth century, not the eighteenth century, meaning of the word). There were also 'Middling' and 'Small' Repairs

**K** Keel laid (date)

**L** Launch (date)

**LD** Lower deck

**Lines** Plan showing the shape of a ships hull (like a contour map as seen from four angles)

**MD** Middle deck

**Midships section** (Plan) constructional midships cross-section

**Minion** 4 pounder gun

**NBW** Navy Board Warrant (date)

**NMM** National Maritime Museum

**Ord** Ordered (date)

**Ordinary** Ships laid up in reserve were said to be 'in ordinary'; The ordinary was the organisation that looked after the ships in ordinary

**Orlop** (Plan) The lowest deck in a larger ship; did not stretch the length of the ship and was well below the waterline

**Planking expansion** Plan showing the planking of one side of the vessel as if it had been ironed flat

**Platforms** (Plan) the platforms (short decks) in the hold of a ship which did not possess an orlop deck

**Profile** Plan showing the 'internal works' of the ship as seen from the side; a longitudinal elevation of the hull as sectioned vertically along the centre line. Also usually shows some external features

**QD** Quarterdeck

**RB** 'Rebuild'; see the section in the Introduction on Rebuilds and Establishments for a discussion on the changing meanings iof this word. Basically, after the beginning of the eighteenth century, a 'rebuild' was an administrative fiction for the production of what was essentially a new ship

**RH** 'Round House' (poop)

**Saker** 5¼ pounder gun

**Specfication** A document listing materials to be used, thicknesses of timber, etc, for a ship

**Surveyor** The supervisor of the construction of a vessel who would also, before the middle of the eighteenth century, usually be her designer. For a ship built in a Royal Dockyard would normally be the Master Shipwright of that yard

**Surveyor of the Navy** Chief designer, the equivalent of the more recent DNC (Director of Naval Construction). For most of our period there were usually two at any one time (eg. Williams and Hunt), but at one stage there were three

**T/O** 'taken off' (of a plan) the results of measuring and taking the lines off a ship; usually applied to a purchase or a prize, and indicates that the plan shows the original appearance as captured or bought as opposed to what the ship looked like after conversion for the RN (see 'as fitted')

**UD** Upper deck

**42, 32, 24, 18, 12, 9, 6, 3** (of guns) weight of shot in pounds; with asterisk [*] added means those weapons are carronades

**13", 10"** (of guns) mortars of these calibres

**★** (of gun) carronade

## Note on Pictures

The National Maritime Museum has been generous in allowing me to use the riches of its collections not only as a basis of the researches which make up this work but also, and appropriately, to illustrate it. Faced with the difficult task of choosing 300 illustrations I have decided to concentrate on using the Admiralty plans which were the origin of my researches and which have provided so much of my data. They are also less familiar than the bulk of the oil paintings, more precise than most of the drawings and prints, more firmly identified than all but a few of the models and much more complete in their coverage of the ships of the sailing Royal Navy than all of these others put together. For information on obtaining copies of the plans and other images used please contact the Maritime Information Centre, National Maritiem Museum, Greenwich, London SE10 9NF, UK.

## Bibliography

Anderson, RC, *Lists of Men of War 1650-1700 - Part 1 English Ships*, London 1935 & 1966

Baugh, D, *Naval Administration in the Age of Walpole,* Princeton 1965

Boudriot, J, *The 74 gun ship* (4 vols), Paris 1986-8; *La Vénus* (with H. Berti), Paris Ca 1982; *La Belle Poule - Le Coureur* (with H. Berti), Paris 1985; *Le Cerf,* Paris 1982; *Le Cygne,* Paris Ca 1986

Chapelle, H, *History of the American Sailing Navy,* New York 1949; *The Search for Speed Under Sail,* London 1967

Charnock, J, *History of Marine Architecture* (3 vols), London 1800-2

Colledge, J J, *Ships of the Royal Navy* (2 vols), Newton Abbot 1969; ditto (2nd edition), London 1989-91

Fincham, J, *A History of Naval Architecture*, London 1851

Fox, F, *Great Ships,* London 1980

Gardiner, R, *The First Frigates*, London 1992

Glete, J, *Navies & Nations* (2 vols), Stockholm 1993

Gossett, W P, *The Lost Ships of the Royal Navy,* London 1986

Laird Clowes, W, *The Royal Navy - a History* Vols II-VII, London 1898-1903

Lambert, A, *Battleships in Transition,* London 1984; *The Last Sailing Battlefleet,* London 1991; (ed) *Steam, Steel and Shellfire,* London 1992

Lavery, B, *The Ship of the Line* (2 vols), London 1983-1984; *The Arming and Fitting of English Ships of War*, London, 1987; (ed) *The Line of Battle,* London, 1992

Oppenheim, M, *The Administration of the Royal Navy 1485-1660*, London 1896

Padfield, P, *Tide of Empires*, Vol 2, London 1982

Smythe, Admiral, *The Sailor's Word Book,* London 1867

Syrett, D, *Shipping and the American War,* London University Press 1970

Tanner, J, (ed) *The Naval Manuscripts in the Pepysian Library,* Vol 1, Navy Records Society, London 1903

Vichot, J, (ed) *Repetoire des Navires de Guerre Français,* Musée de la Marine, Paris 1967

Also *The Mariner's Mirror, The Belgian Shiplover, Warship*

## Ship Types Glossary

The definitions given apply to British vessels during the period from the mid-seventeenth to the mid-nineteenth centuries. Even within these bounds it will be noted that the meanings change according to period and to who is using the term. Also, that the same word may be used simultaneously with completely different meanings for the Navy and for the merchant service (sloop and frigate), or according to whether a ship or a boat is being referred to (cutter).

Some names refer to rig (cutter, schooner, ketch), others to hull type (bark as defined here, pink), still others to function or armament (ship of the line, fireship, bomb vessel, gunboat). The larger vessels are usually defined by classified number of guns (from the appearance of the carronade this no longer corresponded to numbers actually carried) and to rate. Other names (frigate, corvette) were names in their own right, whilst 'sloop' in the eighteenth and nineteenth centuries had a sociological significance - referring to the rank of the commanding officer - and could cover vessels of different rigs and types though of a limited range of size and power. Words like 'cruiser' and 'convoy' (used at this time to mean convoy escort as well as the ships being convoyed), 'survey ship' and 'troopship' could mean a ship of almost any type in use for a particular task.

**Advice Boat** Small vessel used to carry despatches and for scouting. Numbers were built in the late seventeenth century, probably fore and aft rigged but exact appearance uncertain. The two built at the beginning of the nineteenth century appear to have been schooner rigged.

**Bark** In the middle of the eighteenth century 'cat bark' as used for Cook's *Endeavour* meant a ship rigged vessel with a plain bluff bow but a full stern with windows, etc. In other words, it was the description of a particular type of hull and not, as it (or its variant form 'barque') had certainly come to mean by the end of the eighteenth century, a particular type of rig, three (or more) masted with square sails on fore and main but only fore and aft sails on the mizzen.

**Bomb Vessel/'Bomb'** Specialised vessel built or converted to carry (usually) two heavy mortars (usually one 13" and one 11") to throw explosive shells at high trajectory in shore bombardments. Needed to be heavily built to take the recoil of these heavy weapons. Invented by the French to bombard Algiers in 1685 and first adopted by the Royal Navy in 1687. The earlier vessels were mostly ketch rigged, hence the usage '**bomb ketch**', but at least one of the early vessels, and, from the end of the 1750s, all bombs built or purchased for the RN, were ship rigged, so 'bomb ketch' for the type is incorrect and not used here. Smaller craft of the gunboat variety fitted with mortars (usually only one) were **mortar vessels**, whilst a number of **mortar brigs** were converted from 10 gun brigs during the Napoleonic War. The mortar craft and floats of the Crimean War were built in considerable numbers but are so close to the end of our period that they have been omitted from this work, apart from a couple of converted dockyard craft. Bomb vessels spent much of their lives being employed as (and often fitted as) sloops. Their sturdy construction also fitted them for use in polar exploration and a number were converted for this purpose.

**Brig** The naval brig appeared in 1779, a type of vessel defined by its square rigged, two masted rig. This was, in fact, almost exactly similar to the rig of most of the earlier two masted sloops, and it is usually thought that the word 'brig' is merely an abbreviation of the older **brigantine**. By the time the Royal Navy adopted the type, brigs had become common in the merchant service. With two masts rather than the three of a ship, brigs were more economical to man, also very manoeuvrable; so it is not surprising that they shortly became the most numerous type of smaller vessel in the Navy. The majority of these vessels were classed as **brig sloops** (with a Commander ['Master and Commander'] in charge). Only the smallest were classed simply as **brigs** (with a Lieutenant as captain), though the numbers of these were increased as the size and seaworthiness of **gunbrigs** (vessels which evolved from the two masted shallow draught sailing **gunboats** of 1794) themselves grew in size and were reclassed. By the second quarter of the nineteenth century, brigs were the predominant smaller vessels of the service and the last active sailing ships belonging to the RN were the Symondsite brigs, *Martin, Sealark,* etc.

**Brigantine** The late seventeenth century brigantines of the RN appear to have been small two masted vessels designed to be rowed as well as to sail. It seems probable that they were rigged with square sails on both masts and that this was the origin of the term 'brig', an abbreviation of 'brigantine'. Some of the two masted sloops of the first half of the eighteenth century were described as 'brigantine rigged'. The word reappeared as a naval usage in the 1820s, by which time it referred to the rig still called 'brigantine', ie. square rigged on the foremast and fore-and-aft rigged on the mizzen. By this time it was being used merely as a description of rig and not as a ship type name. Some packets were brigantine rigged, as were some other small vessels.

**Chain Boat/Lighter** Dockyard craft for laying chains for moorings, etc/**Mooring Lighter**

**Corvette** French name for the small Sixth Rates of 24-20 guns or large ship sloops of 18 guns or so. Increasingly used in the early nineteenth century for the vessels of those types which had a single gun deck and no quarterdeck or forecastle; in other words, flush decked single-deckers. Although used in this work as the French did for the prizes and purchases which come between frigates and sloops in size and power, this is done for convenience and does not represent the Royal Navy's contemporary usage (which was to split these vessels between Sixth Rates and ship sloops). The Royal Navy only started using 'corvette' as a classification during the second quarter of the nineteenth century, when it was in intermittent use for flush decked vessels which had previously been classed as either sloops or Sixth Rates and including a number of razeed frigates (ie. frigates cut down to the single, flush, gun deck).

**Cruiser** Until after the end of our period this did not designate a type of ship, merely the particular duties on which a warship, any warship, was employed. A 'cruiser' was any ship on detached (cruising) duties, not part of a fleet or squadron, and not employed as a 'convoy' (ie. convoy escort for merchantmen)

**Cutter** (1) A single masted fore-and-aft rigged vessel with a jib as well as a foresail and, up to the early nineteenth century, usually also with a square topsail. Appeared in the Royal Navy late in the Seven Years War. Represented one of the commonest types of British coastal craft by this time and seems to have been an English development. Used by the Navy for scouting, despatch carrying and the like. Most revenue vessels were of this kind, but some 'revenue cutters', were brigs or had some other rig. Some early cutters were large enough to be rated as sloops but this was unusual. They were normally Lieutenant's commands. (2) Larger ships boat intended for sailing as

**Dredger** Harbour craft for dredging mud or sand from channels and anchorages. The usual English type used before steam was a barge fitted with a manpower operated scoop on the end of a pole. In 1802 Samuel Bentham brought the first steam dredger into service at Portsmouth. This had a chain of buckets operated by the steam engine, which could be emptied into hopper barges built for the purpose.

**Fifth Rate** At the very beginning of our period this was used for ships of between 40 and 20 guns, but by the eighteenth century had moved up to ships of between 30 and 48 guns. The larger vessels were small two-deckers, the smaller ones the 'one and a half-deckers' of 32 guns or thereabouts, with a few heavy guns on the lower deck and then a continuous upper gun deck of lighter guns. These were ships mostly used as 'cruisers and convoys', for attack on, and particularly defence of, trade. The 'one and a half-deckers' were not particularly successful ships and the last of them were altered to 20 gun Sixth Rates. For the first half of the eighteenth century there would be nothing between the 20 and 24 gun Sixth Rates and the 40 (later 44) gun small two-deckers which were the only Fifth Rates. The 12 pounder frigates (32 and 36 gun ships) altered this picture when they appeared in the 1750s. They, like the later 18 pounder frigates of 36 and 38 guns (42 and 46 respectively after the reclassification of 1816) were Fifth Rates and, after a brief attempt to revive the 44 gun ship in the 1770s completely replaced the older form of Fifth Rate.

**Fireship** Specialised vessel converted or built for the special purpose of the attack of moored or disabled vessels. Fireships would be fitted with an extra 'fire deck', a 'tween deck subdivided to accommodate containers of combustibles and explosives, with channels to take fuses or gunpowder 'trains'. Gunports would be hinged at the bottom to fall open when the retaining ropes burnt through, so as to increase the draught and thereby both the speed and fierceness of combustion. 'Fireworks' of the Roman Candle variety would be placed inside the ports. Special copper or brass chimneys would be fitted on deck to carry the flames up into the rigging, and grappling irons attached to the yardarms. Fireships, like bombs, were specialised attack craft, true weapons systems, and they, even more than bombs, were rarely used as such. The majority of them spent most of their lives 'fitted as sloops'. The last group of purpose built fireships had only been in service for a few months when they were permanently reclassed as ship sloops (1808).

**Fire Vessel** Smaller fireships, all mercantile conversions, and appear to have had little or no armament.

**First Rate** The largest ship of the line, always a three-decker and by the beginning of our period, except for a very few survivals of slightly less force, invariably carried 100 guns. This total was not exceeded until the last quarter of the eighteenth century when, following the French example, first 110 gun ships and then 120s were ordered.

**Floating Battery** An old warship, converted merchantman or (occasionally) purpose built vessel carrying a strong battery of heavy guns for the defence of a port or haven. Usually with only limited powers of movement. Purpose built ones were usually more or less flat bottomed and often with square bows and sterns.

**Fourth Rate** In the 1680s this had included ships of 40 guns (and the galley frigates of even less) but by the beginning of the eighteenth century included ships of between 58 and 50 guns. These had been considered fit for the line of battle, but by this time were coming to be considered too small and weak and were being used for more general tasks from trade protection to being ships of force on distant stations (see 'Small two-deckers' below). These were two-decked ships. By the mid eighteenth century the 60 gun ships had been reassigned to this rate and were no longer felt to be fit for the line of battle. Shortly afterwards, the 60s fell out of use, but 50s were still building in the first decade of the nineteenth century, though in small numbers. The large frigates razeed from 74s or built from new to cope with the big American frigates after the defeat of three British frigates in the opening rounds of the War of 1812 were classed in the Fourth Rate, and the last two-decker 50 building was altered on the stocks to become a 50 gun frigate.

**Frigate** (1) Initially (*ca* 1640) a flush decked, fine-lined type copied from the Dunkirk privateers, usually a two-decker. During the later seventeenth century (2) came to be used in a general and rather vague way to mean any reasonably fine-lined ship. In this meaning it was falling into disuse by the early eighteenth century. Meanwhile (3) merchant ships were being described as 'frigate built' when they had an extra continuous deck (a 'tween deck) combined with a full ship rig, a bow with a proper head, a full stern with quarter galleries as well as stern windows, and both a forecastle and quarterdeck. Just such an extra 'tween deck was to become the standard feature of the new type (4) of Sixth and Fifth Rates which appeared from the 1740s, introduced by the French: two-decked ships without any guns on the lower deck which could therefore be lower in the hull, making the whole ship a snugger and more sea kindly design whilst raising the height of the main gun battery above the waterline. This was the **'true frigate'**, the type for which the name 'frigate' is used as a category in this list, at first 12 pounder armed (ie. the guns in the main battery on the upper deck were of this calibre) 32 and 36 gun ships and 9 pounder armed 28s. By 1780 the inevitable increases in size and power led to the introduction of 18 pounder 36 and 38 gun ships. Within the decade there were a few 24 pounder armed vessels, and by the end of the Napoleonic wars there were some frigates with 50 or more guns; these larger vessels, built in increasing numbers as the nineteenth century continued, are lumped together in this list under the heading **large frigates**. Confusingly, whilst the previous reasonably precise usage was in use by the men who built the ships, sea officers tended from the late seventeenth century to use 'frigate' to describe (5) any cruising ship which was not a ship of the line, and therefore including the small two-deckers of 50 or 44 guns, small Sixth Rates of 24 or 20 guns and even sloops. To confuse matters further, there was a brief period in the 1740s when (6) the first ship sloops were described as 'frigates', probably a side effect of usage (5).

**Galley** (1) The traditional oared warship of the Mediterranean. A couple of these were built for Henry VIII's navy. One (*Margaret*) was purchased in the Mediterranean in 1671, but was given away six years later. No vessels of this type were used by the RN in our period. (2) Term used for oared gunboats serving in North American waters (including on the lakes) during the War of American Independence. Some of these vessels were purpose built and captured from the States navies that built them; others appear to have been conversions of local craft. Most were probably open boats, but the variety of dimensions and armament would tend to indicate a similar diversity in appearance and type, though there is unfortunately little visual evidence for these craft.

**Galley Frigate** Special design for use in the Mediterranean, a small two-decker with sweeps (large oars) on the lower deck. Designed to row as well as to sail, and with fine lines. About 30-40 guns. First appeared in the 1670s, only a few built, the last of which was in service in the early eighteenth century.

**Guard Ship** Usually a ship of the line fitted in peacetime with part of her armament and rig and a nucleus crew, capable of acting as a floating battery for the defence of the port she was at, and also capable of being rapidly fitted out in an emergency before the ships in 'Ordinary' (reserve) became available.

**Gunboat** (1) Small open boats intended mainly for rowing, though often capable of sailing as well. Often lateen rigged with one or two masts. Armed with one, two or possibly three guns, usually fairly large long guns, though sometimes supplemented with carronades, all of which were intended to fire over bow, stern, or both, on a fore and aft line. The earliest seem to have been modified from the 'flat bottomed boats' ('flat bottomed boats for the landing of men') which were the standard landing craft of the time. Such gunboats date back at least as far as the Seven Years War. (2) Shallow draught sailing vessels built for anti-invasion duties in 1794. They were armed with a couple of long guns mounted and firing forward, supplemented by a broadside armament of carronades; the two masted rig was probably that of a brig. The Crimean Gunboats of the 1850s were essentially a steam powered variant of this design. In the more immediate future the type developed in the later 1790s into the deeper draught and more conventional gunbrig, which in its turn was enlarged and made more seaworthy to rejoin the brigs proper. (3) Conversions of hoys, barges, purchased Dutch vessels and the like intended for anti-invasion duties during the Great French War. Usually manned by the 'Sea Fencibles' (the maritime equivalent of the militia and volunteers). Some of the smaller schooners in service in Napoleonic times were classed as 'gun schooners'. The *Azov* and *Kertch* of the Crimean War were referred to as both schooners (which is how they were rigged) and as gunboats (the task they were intended for).

**Gunbrig** Development (deeper draught and more conventional) of the sailing gunboats of 1794, later evolved into vessels classed simply as brigs.

**Gundalow** Flat bottomed sailing barge of the American lakes. Numbers of these vessels with a sloop rig were adapted for use as gunboats.

**Hospital Ship** Conversion of an old warship or of a merchantman to act as a floating hospital, usually to accompany a fleet or to be a hulk. Not purpose built at this period.

**Hoy** Small single masted sailing cargo vessel, gaff rigged with single fore sail. Usually full bodied. Used as dockyard craft.

**Hulk** Dismasted ship, usually old and past active service, used as a **receiving ship**, **sheer hulk**, hospital or accommodation ship or some sort of stationary storeship, etc.

**Ketch** A few small ketches were in naval service late in the seventeenth century, presumably ketch rigged vessels. Later numbers of bomb vessels, also of two masted sloops and yachts, were ketch rigged, but the rig was going out of favour with the Navy by the middle of the eighteenth century. The seventeenth and eighteenth century ketch was square rigged on both masts as well as having fore and aft sails. In the nineteenth century a number of cutters were rigged as fore and aft ketches, or as yawls.

**Launch** Large, open boat intended mainly for rowing. Originally a type of dockyard craft, later adapted for use as the largest type of ships boat.

**Lazaretto** The name, and the concept, came from Italy, where the practice of rigid quarantine to prevent or at least limit the spread of the Plague and other infectious diseases was developed. These were hulks adapted as accommodation for men undergoing quarantine.

**Lighter** A barge-like vessel, in the naval context used for dockyard purposes. Two varieties, sailing lighters and 'dumb' lighters. The latter, as well as having no sails, were usually of simpler shape, often being 'swim' lighters with flat sloping bows and sterns. Strictly defined, a lighter is a vessel which loads or unloads another.

**Line of Battle Ship/Ship of the Line** A ship large enough, powerful enough and strongly built enough to lie in the line of battle and have some chance of standing up to the largest ship the enemy could produce. By the late seventeenth century this meant a ship of at least two complete decks of guns ('two-decker') and at least 50 guns. By the early eighteenth century the 50 was generally considered to be rather too small for this duty (see 'Small two-deckers') though 50s served in the line as late as the battles of the Nile (1798) and Copenhagen (1801). The ship of the line was the capital ship of her day. The majority were two-deckers, but the biggest were the 'three-decker' First and Second Rates of 100 and 90 guns. Up to the middle of the eighteenth century the two-deckers were of 60 and 70 guns. There was also the unsatisfactory group of small three-deckers of 80 guns. From the 1750s the 74 and 64 gun ship types became the standard two-deckers. Up to the 1790s the 100s and 90s (reclassed as 98s from the 1770s) remained the largest vessels, but then the RN followed the French and Spaniards by building 110 and 120 gun First Rates. It also followed those nations in beginning to build 80 (and later 84) gun two-deckers. After the end of the Napoleonic War improvements in construction techniques permitted the building of two-decker 90s.

**Longboat** (1) The largest type of ships boat (later replaced by launches). (2) A decked sailing vessel used by dockyards.

**Lugger** Lug rigged small craft, usually with three masts. Numbers were hired but only a very few purchases or prizes of this type were commissioned into the Navy, whilst only one was ever purchase built.

**Lump** 'Short, heavy lighter used in our Dockyards for carrying anchors, chains or heavy stores to or from vessels....' (Smyth, *Sailors Word Book*).

**Machine** ('Infernal machine'). An explosion vessel, the equivalent of a fireship but intended to blow up after burning. Not specially built, usually a conversion of a small merchantman. Term used around 1700. Later, some fireships were fitted to blow up (for example, the *Mediator* at Basque Roads in 1809) but by then the separate term 'machine' had been dropped.

**Mooring lighter** Dockyard craft fitted to lay or take up moorings, usually with a heavy 'horse' (derrick with sheaves) fitted over the bow and a capstan for raising and lowering the heavy weights involved.

**Mortar vessel/mortar brig** See under **Bomb Vessel**

**Mud Boat** Barge or lighter to take the spoil excavated by a dredger and transport it to the place where it was to be dumped. The later examples were mostly hopper barges.

**Ordnance lighter/hoy/barge** Dockyard craft belonging to the Ordnance Department (at first separate from either the Army or Navy, later under the Army) to carry guns, powder, shot and other ordnance stores.

**Packet** Vessel used to carry mail, often operated by the General Post Office, though from 1837 the Admiralty took over the GPO's packet service for a time. The Admiralty had already been running its own packets, usually converted from brigs.

**Pink** A smaller vessel, usually in the coasting trade as a merchantman. The name seems to have come from a particular type of hull with round bow and stern, but with the upper part of the stern having the characteristic pinched in shape of the 'pink' stern. In the naval context used in the late seventeenth century for small vessels, probably chiefly used as transports.

**Powder hulk** Vessel for storing and issuing gunpowder, usually moored well away from the dockyard to which it was attached. Usually under the control of the Ordnance (which was a separate organisation to the Army or Navy, though later coming under the control of the former).

**Radeau** From the French for 'raft', a floating battery of rectangular plan and flat bottom. In the RN only the *Thunderer* on Lake Champlain in the 1770s bore this type name; probably it was a local mercantile term that had been adopted for this odd vessel.

**Rated Ship/Rates** Originally 'Rate' was a reference to the rate of pay of the Captain (ie. Captain by rank). There were eventually six different rates of pay - and naturally the largest ships got the highest paid captains - and the largest ships carried the most guns. By the middle of the seventeenth century 'Rate' had become associated with the gunpower of ships, and by the 1680s the system was stabilising on the divisions familiar from the eighteenth century: 100 guns = First Rate, etc (see under the entries for the individual rates from First to Sixth in this glossary). All ships commanded by a Captain were 'rated' (which explains apparent anomalies: Royal Yachts which were Fifth or Sixth Rates, the *Sirius* being a Sixth Rate when flagship of the 'First Fleet', etc). Ships which were not commanded by a Captain were 'unrated'. By the early eighteenth century the lightest armed normal rated ships were of 20 guns or more.

**Razee** A ship was razeed (the term comes from the French) by cutting off upperworks or entire decks. 74 and 64 gun ships were razeed into frigates, frigates razeed into corvettes. A razee was usually a bigger, more powerfully armed and heavily built vessel than the commonality of its new class, whilst probably becoming more handy and seaworthy than she had been before being 'cut down'.

**Receiving ship** Hulks intended to receive men, or stores, or both. Accommodation hulks and store hulks and vessels used for both purposes. The Royal Navy did not build barracks for its seamen in home ports until the beginning of the twentieth century; before this hulks were used for accommodating men between commissions, or before they were assigned to ships (particularly pressed men who needed to be kept in some form of confinement to prevent desertion).

**Schooner** Fore and aft rigged small vessel, a speciality of the North American colonies, first in Royal Naval service in the 1760s. Mostly two masted with the foremast equal in size to, or smaller than, the main, though some of the later naval examples were three masted. Used for scouting or communications duties.

**Second Rate** The next to largest ship of the line. Up to the early nineteenth century (when Seppings' improvements in construction permitted the building of longer ships) these were three-deckers, then the two-decker 90 appeared. By the 1680s armament was generally 90 guns. By the extension of the quarterdeck, 90 gun ships were reclassed as 98s just before the outbreak of the War of American Independence. The two-decker 84s of the 1820s were also classed as Second Rates (with a brief period of being considered Third Rates).

**Sheer hulk** A vessel fitted with a pair of 'sheer legs' (two large spars forming an 'A frame') to hoist masts in and out of vessels. In effect, a floating crane.

**Ship of the Line** See **Line of Battle Ship**.

**Sloop** The oldest use for this word (1), and one which remained the main mercantile one throughout our period and beyond (and is still used by yachtsmen), is to describe a small single masted fore-and-aft rigged vessel. The naval sloops of the late seventeenth century appear to have fitted this description, as did the sloops on the American Great Lakes and some sloop rigged vessels acquired on the North American coast during the eighteenth century. From the beginning of the eighteenth century (2) when the term reappears in the lists after a short period of disuse, it was used to describe the small fighting vessels immediately below the smallest rated vessels, the chief distinguishing feature of which was that they were captained by an officer of what would later be called the rank of Commander, at this stage a very senior Lieutenant in the position of 'Master and Commander'. This would continue throughout our period to be the reason for calling a particular vessel a sloop, and individual vessels could come and go from the category depending on who was in charge of them. This also means that various descriptions of craft; ship rigged, brigs, even cutters, could be 'sloops'. The early eighteenth century naval sloops ('**sloops of war**') had two masted rigs, being **snows**, **ketches** and **brigantines** (what would later be called brig rig - square rigged on both masts and not brigantines in the nineteenth century usage which is still current - with fore and aft sails only on the mizzen). In the 1740s the first ship rigged vessels were brought into the Navy as **ship sloops**, and the American War of Independence saw the first **brig sloops**. In the reclassification of 1816 the majority of the larger ship sloops, those with quarterdecks and forecastles, became Sixth Rates (nicknamed **jackass frigates** or **donkey frigates**) and the comparatively few ship sloops left were mostly flush decked (ie. without forecastle or quarterdeck, and therefore could also be called corvettes).

**Smack** Small fore and aft rigged single masted coastal craft, often a fishing boat, but a type also used as dockyard craft.

**Small two-decker** The description used here for the smaller Fourth Rates of 50 guns or so and the Fifth Rates of 44 or 44 guns which carried their guns on two main continuous gun decks, but which, unlike larger two-deckers, were not generally considered fit to lie in the line of battle by the time our period starts. They were much used by the Royal Navy for cruising duties, trade defence, and to provide ships of force for distant stations. The 40 of the early eighteenth century had evolved into the 44 of the middle years of that century. These short ships did not compare in fighting ability or speed with their replacement, the new 'true frigate' type introduced in this period, which carried a similar gun deck armament of 12 pounder guns but carried rather more of them and higher above the waterline. Despite a brief attempt to introduce a rather larger variant of two-decker 44 (also, loosely, called 'frigates' by sea officers but not by those who built them) to suit the exigencies of the colonial uprising phase of the War of American Independence, these were soon relegated to secondary trooping and transport tasks when the war became general. The 50s, which usually had a main armament of 18 pounders, lasted until the end of the Napoleonic Wars; there were clearly a few tasks for them (such as acting as the controlling ship for anti-invasion forces), even though the 18 pounder frigate had taken over the majority of their cruising tasks.

**Snow** See under **sloop**.

**Storeship** Strictly speaking, a ship intended to carry naval stores (spars, timber, cordage, tar, etc - all the material needed to repair sailing warships) whereas a **transport** was intended to carry men. Storeships were auxiliary vessels with a defensive armament. Most were converted from merchantmen but a few were specially built and some were converted from first-line fighting vessels of various kinds.

**Survey ship** Originally 'sloop on survey', usually small, but no particular type of ship.

**Tank vessel** Dockyard craft fitted with iron tanks and pumps to provide water to ships.

**Tender** Any vessel (usually small) attached to another vessel (or establishment) for administrative purposes.

**Third Rates** The Third Rate covered ships of the line of from 80 down to 60 (later 64) guns at the beginning of the eighteenth century. The three-decker (originally 'two and a half-decker') 80s of the 1690s, which were a type built in numbers until the mid-eighteenth century, were of this designation. So were the 80 gun two-deckers of the end of the eighteenth century on. The 70 gun and later 74 gun two-deckers which formed the basis of the line of battle throughout our period also belonged, as did the 64s built in some numbers in the period from 1750 to the mid 1780s but then abandoned as being not powerful enough to sustain the role of ships of the line. The 60 gun ships which were their earlier equivalent began the eighteenth century as Third Rates, but later slid back into the Fourth Rate.

**Transport**  See under Storeship

**Troopship**  Ship converted to carry troops, could be a regular warship or a converted merchantman. Not specially built at this period.

**'Unregistered vessel'**  A term used in this list only to indicate vessels not listed in the main original Admiralty sources consulted, but for whose existence there is reasonable evidence in secondary sources (mostly themselves compiled from other Admiralty sources).

**Victualling lighter/hoy/barge, etc**  Dockyard craft owned by the Victualling Board (a separate naval department under the Admiralty), used for carrying victualling stores (food, etc).

**Yacht**  The word came from the Dutch, who presented the first English example to the restored King Charles II in 1660, originally meaning a scouting craft. By the time the first British examples had appeared in the 1660s, they were used as despatch vessels, for carrying important people, and for racing. Most British examples were used as Royal Yachts, though lesser ones were also used by Commissioners of the Navy and of Dockyards, the Viceroy of Ireland and the Governor of the Isle of Wight. The earlier examples were rigged with what was called 'yacht rig' (like the later cutter rig) on one mast, or were ketches. Later, some of the larger yachts were ship rigged. The larger ones were classed as Fifth Rates, the smaller as Sixth Rates, as they were commanded by captains.

**Xebec/Xebeck/Chebec/Zebec, etc**  A Mediterranean type of vessel capable of proceeding under both sails and oars. Usually fine-lined, with pole masts, and capable of setting either lateen or square sails. The only vessel of this kind built for the RN (at Port Mahon) is shown on her plans to look very like contemporary sloops, and very little like the usual rakish image of a xebec.

# GENERAL INTRODUCTION

This work gives a basic listing of all the ships known to have served with the Royal Navy[1] from the 'Glorious Revolution' of 1688 to the introduction of steam in the 1820s. Thereafter it lists only sailing ships, excluding steamers,[2] terminating at 1860.

It divides into two main listings, both chronological, and also several subsidiary ones. The main listings are of operational fighting ships, divided between those designed and built for the RN on the one hand, and on the other those taken from the enemy or purchased from mercantile sources. Both these sections are divided into chronological chapters: in the first section the chapters coincide with broad eras in the evolution of ship design for the navy, whilst the second has a chapter for each war which caused the captures and necessitated the purchases. In both cases the chapters list the ships by type in descending order of power and size.

The subsidiary listings given as appendices deal with vessels which belonged to the navy or were associated with it but were not fully part of the fighting arm: the dockyard craft, the coastguard cutters, the royal yachts, the older ships which had been turned into hulks, plus an attempt to list the sailing fighting vessels which served on the North American Great Lakes (including Lake Champlain). As the main lists give ships under their original descriptions there is a separate key listing of conversions. The book finishes with a general index of ship names which is the key to following up an individual ship or name.

## The Data

This is given in slightly different forms in the opening chapters of the main listing of ships designed and built for the Navy, in the remainder of those chapters and in the chapters listing purchases and prizes. There are more variations in the appendices where dimensions, etc, are given. Individual variations in the scheme are described in the introductions to the relevant chapters but the general scheme is as follows:

**Class** In this book a 'Class' is taken as being a group of ships built to the same lines (ie. with the hull of the same size and shape). The 'name ship' of the class is normally the first ship of the class to be ordered. Manuscript navy lists often give the name of the ship whose lines were used as the class design, but unfortunately this can be misleading because if the ship concerned was lost the usual procedure seems to have been to substitute the name of the senior surviving member of the class. Thus the *Apollo* Class 36 gun frigates are often referred to as the *Euryalus* Class simply because *Apollo* was lost in 1804, and the second ship of the class, *Blanche,* a year later, with *Euryalus* being third in line. A further complication is trying to judge when a design has been altered enough for the ships built after the alterations to be considered as a separate class or just a sub-group of the first class. This is particularly true of the constant alterations Sir Thomas Slade made in the lines of his 74 gun ship designs; much the same applies to some of the long-drawn out gestations of designs in the second quarter of the nineteenth century. Plans and documentary sources can be interpreted in different ways, and quite how much can a design be altered before it is considered to be a different one? In the less clear-cut cases one may eventually be driven back onto personal judgement between the alternatives of inclusion or separation.[3]

**Name(s)** Particularly in the earlier years of the period, spelling could be somewhat erratic and more than one version of a name can be found in the sources. The variants are given thus: ***Cruiser/Cruizer***. Names were sometimes changed during the course of a ship's naval career. This is indicated as follows: ***Zebra*** 1781-1785 ***Diligence*** 1791 when the name was changed in 1785 and the ship went out of active service in 1791.

In general, the Royal Navy tried to avoid duplicating names: the policy seems to have been to have only one ship with a particular name in service at any particular time. The official exception to this was for cutters, which appear to have been considered to be 'tenders' (ie. vessels attached to other ships and therefore not commissioned in their own right) and which can sometimes be found bearing names also carried by bigger vessels at the same time. Dockyard craft did not count and so might well duplicate names used by fighting ships (eg. *Lion*). Hired ships normally retained their mercantile names and plenty of examples of duplicated names (especially amongst the hired vessels themselves) will be found in the sections devoted to these craft.

In practice, with fighting spreading across the oceans of the world and Admirals commanding on distant stations naming and commissioning prizes and other ships they had purchased, there was bound to be 'double booking' of names. This makes for a considerable source of confusion and uncertainty for the compiler of a list like this, not all of which has been resolved. As one might expect, the problems are worst in the case of the American War of Independence, the war of this period which had the heaviest emphasis on areas beyond Europe. It is, however, striking how often and how soon the Admiralty attempted to remedy such confusions, especially when the ships concerned arrived at a home dockyard. This explains the large number of name changes of prizes and purchases after they had been taken into the Navy.

A seventeenth-century naming habit which died in the course of the following century was to name a prize after the ship which captured her (*Beaver's Prize* is about the last example), whilst the same applied to using the nationality of the prize as part of her name (eg. *Sallee Rose, French Ruby*). It may be worth noting that the large number of geographical names (towns, counties, etc) in use in the first half of our period seem to have been given, as far as the somewhat scanty evidence available shows, to honour noblemen (via their geographical titles) rather than after the places themselves.

**Dates** In the chapters where ships are given by class, the class heading shows the date of the design. When available this is specifically the year

---

[1] Initially this was the English navy, for up to 1707 and the Act of Union, Scotland had its own small navy. After that date the Scots ships were taken into the larger force, and the navy becomes British. To call this force 'The Royal Navy' actually begs many questions: for most of our period of 1688-1860 most navies were Royal, the English/British one merely being the largest (apart from a few years at the beginning of our period when the French was a serious rival) and most successful. It was not until well into the nineteenth century that the usage 'Royal Navy' indicating this service comes into general use. This probably had less to do with the undisputed predominance at this time of the RN than with the fact that from the middle of the nineteenth century the French Navy was either Imperial or republican, the Russian was Imperial, the Austrian Imperial and Royal, and the American republican. When the German Navy began to grow it, too, was Imperial. Only the much diminished Dutch and Scandinavian navies were Royal. In these circumstances the RN really was *the* Royal Navy. However, though it may be anachronistic, the usage 'Royal Navy' (and its abbreviation 'RN') is both convenient and familiar, and therefore is used here.

[2] Except those steamers which were originally designed, ordered or commenced as sailing ships or converted from sail later in their lives. There are also a couple of hulks converted from steamers noted.

[3] It is circumstances of this kind which explain the few and minor differences between the list of classes in this work and that in Brian Lavery's *The Ship of the Line*.

the design lines plan was approved by the Admiralty. This is not always given on the plan or known from other sources, and here the more uncertain method of using the earliest date on the class plans, or the date of the Admiralty order or Navy Board warrant asking for the design, is given.

Following through the class entries, the dates given for individual ships have the following pattern: first the date of ordering (**Ord**) is given. This is normally the date of the Admiralty Order stating that a ship of such and such a type should be built. In cases where more than one order date is given the first may merely specify the type to be built, the second the precise design, or state where the building should take place. In a few cases a second order date long after the first means that the intention to build has been confirmed and work should actually start (the *Royal George* ordered in 1727 and then again in 1746 with work actually starting in 1747 is a case in point). Then follows the date on which the keel was laid [**K**] and then the launch [**L**]; in the case of some of the Dockyard-built ships, this might be the date of floating the ship out of the dock in which she was built, but the distinction has not been drawn here and both means of entering the water are considered as the 'launch'. With sailing ships this date is the most significant. A ship was launched with hull complete and ready for sea. Masts, yards, rigging, fittings (including guns) and stores could be put on board in a finite and fairly short time (under a month, usually) if matters were urgent enough. This process could be done by the crew of the ship herself if necessary. The case is entirely different to the metal hulled, machine driven warships of a later period with their permanent fittings and ever growing complexity. For such ships completion dates matter far more than the launch date, which can happen at a variety of stages in the production process but is far from the end of that process and often quite near to its start. A wooden ship, if launched in time of peace, can be laid up for years before completion[4] but be available to be put in service with the minimum of delay at any time in that period.

These dates are given in numerical form in the following style: Day.Month.Year. When the day is not known the alternative of Month.Year is used. Up to 1751 the dates are old style (Julian Calendar) except when noted as [**NS**] New Style (Gregorian Calendar, already in use on the Continent). However, the earlier English usage of starting the year on Lady Day in late March instead of on New Year's Day is not adhered to, which explains why some dates given here are a year out as compared with contemporary sources or modern works which have not been corrected for this factor. It should also be noted that there are a number of cases for the period up to Ca 1710 where other works (for example, Colledge) give a precise date of launch when I give only the year, or where we differ on launch dates. This is due to the quoting of the date of the Admiralty Order for launching the ship as the actual launching date, which it was not. It will also be noted that my dates for the ordering of a ship sometimes differ from those given by Brian Lavery in his *Ship of the Line* Vol 1. This is due to discrepancies in the sources we have used (plans and progress books in my case, lists of Admiralty Orders and other documentary sources in his). It is quite possible that both dates are part of the ordering process (the first stating that a ship of a particular type should be ordered, the second determining the design or the builder, for example) - in a few cases Lavery appears to have used the date when an order was issued naming a vessel that had already been ordered - whilst other anomalies may be due to clerical error or may be resolved by more detailed research on particular vessels.

In the prizes and purchases section the entries are by individual ship, and the name is followed by the year of entry into service with the Royal Navy and the year of leaving service. The year of build (when known) is given within brackets together with the original name(s), and name and location of the builder.

**Type** Normally indicated at the start of each section in each chapter, but sometimes (usually in the prizes and purchases section) when a section includes a less uniform group of ships; the number of guns or precise designation may be given at the end of the first line of the entry on an individual ship. For details of the different types see the overview of development below, and the glossary.

**Dimensions** are given in imperial feet and inches and fragments of an inch. Those who wish to convert these figures to metric can use the usual conversion factors (1 foot ['] = 0.3048 of a metre, 1 inch ["] = 2.540 centimetres. 1 foot = 12 inches) but should note that the documents, naturally enough, always use imperial measure. Incidentally, it would appear from much evidence including some from archaeological work that British naval shipwrights in the period under discussion did work to close standards, and the recording of dimensions down to the nearest one-eighth of an inch was not just for show.

In this list dimensions are always given in the following form: [i] Feet & inches, [ii] feet & inches x [iii] feet & inches x [iv] feet & inches. The initial two sets of figures are for length, the first [i] being for length on the lower deck (for larger ships) or **on the range of deck** (for smaller vessels which had only the one deck). In either case, this was a rough equivalent of the modern length between perpendiculars, and was usually measured from the further edge of the rabbet cut into the stem and stern posts to take the deck. The second figure [ii] given for length is the **length of keel for tonnage**, a somewhat artificial figure used as its name suggests for working out the tonnage - not the same as the actual length of the keel itself. Because of its importance to the tonnage, and because it is almost always given in the sources, it is quoted here. This is followed by the breadth [iii], usually the moulded breadth at the broadest part of the midships section. 'Moulded' in this context means the width of the frame; in other words, to the inside and not the outside of the planking (the latter would be breadth 'extreme'). The final dimension [iv] is the **depth in hold**, another more or less artificial figure used in working out the tonnage. There are sometimes differences between figures obtained from different sources. Figures taken from the plans have been preferred to those from manuscript lists, dimensions books and the like. For classes and individual ships built for the Navy the **design dimensions** (usually obtained from the plans) are given (except in the earlier years covered) as one set of figures for a class. 'As built' figures are also available for most of the vessels but are not given here as being of insufficient general significance (they do not usually vary much, in any case). However, the purchased or captured vessels need more individual treatment, and so for these the 'as taken off' figures are given (when available).

**Tonnage** This is Builders Old Measurement worked out according to the naval formula: a measure of burthen (carrying capacity, ie. volume of the hull) rather than of displacement or therefore of weight. It follows that this sort of ton has no metric equivalent. The way in which it was calculated changed slightly in the early eighteenth century. It was based on a formula involving length of keel for tonnage, breadth and depth in hold, the result being divided by 94; which explains the fractions being in ninety-fourths. This is the figure given in contemporary documents and does at least give some standard of comparison for different vessels. Like the dimensions the tonnage figures are normally the design ones for ships and classes built for the Navy; for prizes and purchases they are the specific 'as taken off' figures.

**Crew** This is the 'established' figure; the designated complement of officers and men. This might vary during a vessel's career. Especially in wartime, this was more often than not a pious hope. For ships built for the Navy the figure given will usually be the design one. In the cases of major

---

[4] As is the case of the still incomplete frigate *Unicorn* at Dundee, a century and three-quarters after her launch at Chatham.

changes or conversions the altered figures for the complement will be given, when they are known. Otherwise, usually only one figure has been taken from plans, dimensions books or navy lists (in that order of preference) and no attempt made to record the variations.

**Guns** The guns are given by decks, working upwards: LD = Lower deck, MD = Middle deck, UD = Upper deck, QD = Quarter deck, Fc = Forecastle. For each deck the number of guns is given by [x] the weight of their shot in pounds (1 pound [lb] = 453.592 grammes). The weight of the shot was the method initially used by the Dutch to designate their guns in the seventeenth century, and was slowly adopted by the British (and others) who had previously used a somewhat confusing medley of gun type names such as 'culverin', 'falconet', etc. At the beginning of our period armament lists show both systems still in use by the Royal Navy, though by the end of the first decade of the eighteenth century the older names were falling into disuse. Some names (demi-cannon = 32 pounder, culverin = 18 pounder, etc) have direct later equivalents in the new numerical system of designation, and are therefore given under that system here to avoid confusion; this is not the case with '**sakers**' (which fired a shot of $5\frac{1}{4}$ pounds) and '**minions**' (4 pounders) which are given their names on the few occasions they are found in this list.

During the late seventeenth century and into the first few years of the eighteenth the practice was to have three separate 'establishments' of guns, one for peacetime, one (usually similar) for overseas service in wartime, and another, more powerful outfit, for home waters in wartime. As ships grew larger and stronger within each rate this distinction lapsed. This is the explanation for ships being given as, for example, 42/36.

Besides the 'carriage guns' resting on wheeled mountings (the wheels usually being four of the small solid 'trucks' - hence the expression 'truck carriages') ships carried numbers of the smaller anti-personnel weapons usually known as 'swivels' because they were usually fired from a swivel mount rotating on a 'stock' or the gunwale of a ship, and which could also be taken up into the tops and fired from there. The standard weight of shot fired was $\frac{1}{2}$ pound. Other weapons which were so heavy and powerful that they needed to be fired from a swivel or other rest were the 'musketoons', looking like a large musket or blunderbuss.

At the end of the 1770s, during the American War of Independence, the Royal Navy introduced the '**carronade**', named after the Scottish iron foundry that developed and produced it, into service. This was the revival of an old idea of building a short lightweight gun which fired a heavy shot over a short range. It could be accommodated on comparatively lightly built upperworks, both because of its own lightness and because of its less powerful recoil. It also required a smaller crew to operate it than what now came to be called the 'long guns'. Despite initial teething troubles, especially with its pivoted 'slide' mounting, the carronade proved to be very popular with the Royal Navy in its later, modified, form.[5] For the next forty years virtually all Royal Naval ships carried at least some carronades, whilst the majority of the smaller unrated vessels had a main armament of carronades.[6] In this list carronades are indicated by an asterisk [*] marking the weight of shot, thus the following entry: 2 x 6 + 16 x 32* indicates an armament of 2 six-pounder long guns and 16 thirty-two pounder carronades.

Carronades are always given *after* long guns here. This reflects a situation which has produced much confusion. When the new weapons were first introduced they were seen as alternatives, not to some of the main armament of long guns, but to the swivels. On Sixth Rates and above this basically remained the case - there might be a few less long guns carried on quarter deck and forecastle but in general carronades were an addition to rather than a substitute for the long guns. It was primarily for this reason that the new weapons were not counted towards the number of guns used for classifying a ship. This has the convenience for our purposes of keeping the type and class to which a ship belonged clear. However, this did mean that the rating of a ship as (say) a 36 gun Fifth Rate, which up to the introduction of the carronade was actually the number of guns above the size of swivels carried, turned into something which began to seem somewhat fictitious when the actual number of heavier weapons carried (including carronades) was perhaps 42. In the end this led to increasing pressure for change in the rating system. This came in 1816 when a general rerating was introduced for all Royal Naval ships. However, this introduces problems for us, trying to follow the continuity of development and the history of individual designs and classes. It makes little sense from our point of view to see early members of, for example, the *Leda* Class referred to as 38s and later ones as 46s, still less to use this distinction to split them into two classes because of this apparent difference in rating. In this list I have therefore adhered to the original system, not to the post-1816 one, though I have noted the reclassifications introduced then.

A further and fertile source of confusion is that for unrated ships; sloops, brigs and below, carronades were counted from the start as main armament (and are here). The *Cruiser* Class brigs were from the start referred to as being of 18 guns, despite the fact that only two of these guns were long ones, and the other 16 all carronades.

The mortars used aboard bomb vessels were classed by the diameter of their bore, either 13 or 10 inches, and are so described here. Some vessels (mostly prizes) were armed with Gover's lightweight guns towards the end of the Napoleonic War as a half-way house between carronades and long guns. Later, for much the same reason, 'gunnades' were mounted aboard a number of ships. By the 1830s ammunition supply aboard larger warships was being simplified by the adoption of an all-32 pounder armament by fitting guns of different lengths and weights on different decks. The next development was the increasing use of heavier guns (68 pounders) for bow or stern chasers, and the fitting of shell firing guns. Both of these tended to be fitted on slide pivot mountings as they were getting too large to be handled easily on the old truck carriages. The shell guns, like mortars which also fired shells, were classed by bore in inches (usually 8") rather than weight of shot. All these developments are noted here.

**Building** The name and location of builder are given after the name of the ship. In the case of a Royal Dockyard [Dyd] the dockyard is given first followed by the name of the Surveyor in brackets. The Surveyor was initially the man who designed the ship as well as supervising its construction. Later, he fulfilled only the second function whilst the designing was most often done by one of the Surveyors of the Navy. With private shipyards the name of the builder is given first, followed by the location of the yard. When the date, place and identity of the builder of vessels not built for the RN are known these details are given with other details of origin (nationality or mercantile origin and original name) within brackets after the Royal Naval name and dates.

**Alterations** In some cases vessels were lengthened and/or broadened, alterations which necessarily resulted in an increased tonnage. Major alterations in designated complement or armament (when noted in the sources used) are given here, but no attempt has been made at comprehensiveness. The actual complement of a ship was very variable (and usually, especially in wartime, well below the intended one), whilst the actual armament could be varied by the whim of the individual captain as well as by changing official ideas of what the outfit could be. Alterations

---

**5** For further discussion of this, and for illustrations, see Brian Lavery's *The Arming and Fitting of English Ships of War 1600-1815* (Conway Maritime Press 1987) which is a mine of well researched and presented information on this and other matters.

**6** Most brig sloops, for example, had a couple of 6 pounder long guns as bow chasers, and the rest of their guns were carronades. A few larger vessels (*Rainbow* in the early 1780s, *Glatton* at the battle of Copenhagen 1801) had experimental all-carronade armaments.

are noted under the date they took place. When they took place because of an Admiralty Order or Navy Board Warrant they are indicated thus: 'AO.1.1.1801' or 'NBW.1.1.1801'.

**Fates** In the chapters where ships are dealt with by class the main entry on each ship includes only a short note of the way in which the vessel left first-line service. In cases where more details of loss or second-line service are known these are given in a separate 'Fates' section at the end of the entry on the class. In all cases details of subsequent second-line service, or service in another navy, are differentiated by being enclosed within square brackets [thus]. This is to make as clear as possible a distinction between the career of the vessel in first-line service with the Royal Navy and the rest of her career.

**Plans** This listing started from work on the Admiralty Collection of ship plans at the National Maritime Museum, Greenwich, and much of the data and other information contained in it came from that plans collection or associated documents held by the NMM. It therefore seemed useful to have brief notes of what plans existed in that collection for the ships listed here. This is not intended to be a comprehensive and complete listing, but to give some idea of the extent of this splendid source. I have also included notes of some of the more interesting plans in two very important Scandinavian collections. Both contain many British plans and many plans of British ships (not necessarily the same thing) from our period. A number of plans are quite clearly Admiralty plans, others copies of Admiralty originals. The two collections are: that assembled by the Swede af Chapman, now in the Sjöhistoriskas Museum at Stockholm; the other is more important for our purposes and is the work of a number of Danish shipwrights. This assemblage is held by the Rigsarkivet at Copenhagen and includes a number of plans of early eighteenth-century British warships for which there are no plans in the Admiralty Collection.

As can be seen from a glance at the pages that follow there are hardly any plans until about 1715, thereafter the number, coverage, detail and variety of plans grow. The main types of plan listed are as follows:

**Lines** Shows the shape of the ship's hull (or rather half of it) seen from below ('half breadth') the side ('sheer') and either end ('body plan'). Most of the earliest surviving plans are of this description. A lines plan is also part of:

**Lines and profile** A lines plan combined with a 'profile of inboard works' which is merged with the 'sheer'; it is a longitudinal vertical section of the hull, an elevation which shows the decks and other internal features of the hull, and often also includes details of its external features, such as decorations. This sort of plan appears fairly early in the eighteenth century but becomes more common from the middle of that century.

**Profile** (originally 'profile of inboard works') See above, usually found in a set with the deck plans, done at the same time. More common from about 1750.

**Decks** Plan view of each deck, sometimes on separate sheets, sometimes together and usually found in combination with the profile plan (see above).

**Platforms** 'Platforms in hold', short decks for magazines, store rooms, etc, in those ships which did not have an:

**Orlop** A short deck, not running the entire length of the vessel, above the hold. The lowest deck of a larger ship. Well below the waterline.

**Framing** A constructional plan showing the frames which were the basic structure of a ship of this period, seen from the side. Found from the latter part of the eighteenth century onwards, but not as common as profile and deck plans.

**Midships section** A cross-section of the hull structure; not very common.

**Planking expansion** A plan showing the planks from one side of a ship (either internal or, more often, external planking) as if they had been ironed flat. Very rare.

**Sails** A plan showing the outline of the sails above an outline of the hull; not very common until the second quarter of the nineteenth century; not very useful or pictorial, and no substitute for a:

**Rig or Rigging plan** A plan showing details of masts, sails and rigging. Unfortunately almost non-existent for the Royal Navy of this period. It would appear that there was no need for plans on the part of those who dealt with the rigging of warships. This did tend to be fairly standard, at least for ship rigged vessels.[7]

**Specification** Not a plan at all but a document laying down quality and quantity of materials to be used in the construction of a ship. Particularly useful in giving the 'scantlings' (thickness) of the timbers used. By the 1770s the RN was using printed books for ships ordered from private contractors, which had spaces left for individual dimensions, etc. These also had a contract form included as part of the book and are sometimes referred to as 'contracts', though this is misleading as the financial details are not filled in in the ones that form part of the Admiralty Collection at the NMM.

The majority of the plans noted will be at the Admiralty standard scale of $\frac{1}{4}$" to 1' (1:48), though some of the deck plans are at half that scale (1:96 or $\frac{1}{8}$" to 1') and sail plans will usually be at half that scale again ($\frac{1}{16}$" or 1:192). The majority of the plans for ships designed and built for the navy will be design or building plans, mostly for the class rather than the individual ship. However, some plans show a ship 'as built' or 'as fitted' [a/f] at a particular date, usually in more detail and often with decorations (figurehead and other bow decorations with the stern and quarter decorations as well).

The plans for ships purchased or captured by the navy are of their very nature records of the ship as she actually existed (unlike design or building plans) because they were 'taken off' the ship. In a few cases two sets of plans may be found, one as 'taken off' [t/o] - as first brought into the RN - and another 'as fitted', in other words before and after alterations made to convert the vessel for British naval service.

Those plans without a note of origin belong to the Admiralty Collection, and for information about availability of copies, etc, please contact the Maritime Information Centre (Ships Plans), National Maritime Museum, Greenwich, London SE10 9NF, UK.

## Sources

This work is written from a limited number of sources. This stems from a deliberate decision that a working list was needed, rather than a definitive one carefully checked from the enormous number of sources available; that 95 percent accuracy now would be more general use than 98.5 percent in perhaps twenty-five years time. It is intended to be a foundation that others can build on, correct and amend[8] rather than the nearest approach to a definitive list that could be achieved in a lifetime's research, and which might never reach the stage of publication. The basis of the work is the juxtaposition of information from three main official sources. It grew initially from the need to work on the earlier parts of the enormous and unrivalled **collection of Admiralty plans** and associated technical documents held at the National Maritime Museum [NMM]. This was

---

**7** Those who wish to follow this further are referred to Jim Lees' *Masting and Rigging of English Ships of War* (Conway Maritime Press 1974).

**8** If you find errors, or have further information based on original material which could improve a subsequent edition of this work, the author would be very glad to hear from you.

supplemented by reference to the copies of the **Admiralty Progress Books** held at the same Museum (the originals of these invaluable documents are held at the Public Record Office [PRO]). Further information was obtained from assorted **Admiralty Dimensions Books** and other similar lists at the NMM, some of them copies of documents whose originals are now held at the PRO, others original documents in the Museum's collections. All three sources have been abstracted in much more detail than given here in manuscript and note form. This gives variants of dimensions obtained from plans and dimension books (often there are discrepancies), more detail on armament variations from the same sources, and much fuller versions of constructional histories (from the Progress Books). It is on this fully referenced foundation that the present simpler and less detailed list is based. The Admiralty plans have been used as the ultimate authority on dimensions, design armament, and the division into classes, when such information can be obtained from them. Though there are not plans for all the ships listed here, they do cover the vast majority of the classes built for the Navy from the second decade of the eighteenth century, and a sizeable proportion of the prizes and purchases. The Progress Books have provided most of the dates of acquisition, building, subsequent alterations and disposal or loss given here. However, they do not start until the end of the War of Spanish Succession (1713). For earlier years these details have had to be obtained from other sources (see below). The Progress Books are essentially abstracts of expenditure in the English Royal Dockyards (Deptford, Woolwich, Chatham, Sheerness, Portsmouth, Plymouth and Pembroke), but usually also include details of building or acquisition for ships built elsewhere, and often other information as well. They are particularly detailed for the period from the 1740s to the 1770s, and also for the later 1780s and early 1790s. They are at their worst for the period of intense activity and stress of the Napoleonic War.

The chapters of this list which cover the years up to 1713 depend mostly on a large dimensions book (Dimensions Book B). This exists at the NMM as a copy of an original now in the PRO. The evidence from this has been supplemented by copies of a yearly distribution list covering the period 1699 to 1713, the original of which is also in the PRO. Though Dimensions Book B covers ships in service from the 1680s it would appear that its compilation began in the early eighteenth century. It was added to until well after the middle of that century.

Further dimensions books compiled at intervals in the latter half of that century carry the listing of the Navy up to 1800. The originals of these are also at the PRO, but the NMM copies have been used here. There is also a series of Admiralty dimensions books which were held by the Royal United Services Institute (RUSI) Museum but came to the NMM when the former museum closed down. This series of sources has been used for dimensions, armament and dates when these have not been available from plans or Progress Books, and has been invaluable for some of the more obscure prizes, purchases and dockyard craft. They have also helped on the identification of classes, though again information from the plans and Progress Books has always been preferred when available. The dimensions books themselves seem to have been compiled for internal reference within the Admiralty (or possibly Navy Board) office.

Three published works have proved invaluable in building up information on the earliest ships listed here. Oppenheim's *Administration of the Royal Navy*, the listing in the first volume of Tanner's Navy Record Society series on the Pepysian Papers, and Frank Fox's *Great Ships* have all been plundered, and the reader interested in further details on the ships listed in

Chapter 1 is referred to these excellent works. There has been a considerable amount of exchange of information and discussion between the author and two friends who have now published. Andrew Lambert provided much information on those line of battle ships which were converted to steam some time before his *Battleships in Transition* appeared. The exchange of listings and other information with Brian Lavery both when he was preparing his *The Ship of the Line* and since has been very considerable. Another friend, Jan Glete of Stockholm University, has checked through much of the work from his own mammoth compilation of published sources and work on a number of foreign archives. His contribution has been especially valuable in the first chapter and on the origins of foreign prizes.[9]

The origins of prizes captured by the RN, as well as the fates of those ships taken from that Navy by others, are sometimes given in notes in the Dimensions Books mentioned earlier. More often they have been taken from a variety of sources, mostly secondary.

**Ships in French Service** I have used the *Repertoire des Navires de Guerre Français* but preferred to it original material communicated by Dr Martine Acerra (based on her researches on Rochefort-built ships) and information taken from the works of M Jean Boudriot when they are in conflict. M Boudriot and his collaborator M Hubert Berti have also very kindly communicated more information in the form of corrections to lists I had sent them. M Boudriot also very kindly provided a copy of a document from the Archives Nationales listing all the French losses of the 'Great War' (1793-1815), which has been most useful.

**Ships in Spanish Service** I have found copies of two lists compiled by M Christian de Saint Hubert, one of Spanish ships of the line (subsequently published in *Warship* Magazine), the other of frigates of great use. These were communicated to me by José Merino. I subsequently corresponded with the very helpful M de Saint Hubert and was shocked recently to hear of his murder whilst in the service of the Belgian Foreign Office.

I have been unable to find a source on Spanish ships smaller than frigates.

**Ships in Danish Service** I utilised a list held in the Danish Rigsarkivet whilst working on the magnificent collection of plans there. I was greatly helped by the compiler of that list, Hans-Christian Bjerg, who has subsequently published it.

**Ships in Dutch Service** I was sent copies of lists of warships built at Rotterdam and Amsterdam by a colleague at the Prins Hendrik Museum, Rotterdam, whose name I have unfortunately forgotten. Further details of Dutch ships have been provided through the generosity of Jan Glete. Unfortunately, there are still gaps left in the records.

**Ships in American Service** I have relied on the classic works of Howard Chapelle here: the *History of the American Sailing Navy* and, for most of the vessels of mercantile origin, his *Search for Speed under Sail* and *History of American Sailing Ships*.

Other sources have been used for some of the appendices, and are discussed in the introductions to those sections. Finally, three major works, two published, one only available in typescript, have proved immensely useful in sorting out many of the uncertainties (some of which, however, still remain) about so many of the smaller prizes and purchases, particularly those which remained on overseas stations for all their service lives, and some of which may have been only hired rather than purchased and commissioned. The late Commander Pitcairn Jones' ship histories (now held at the NMM in typescript and microfiche form after considerable revision by Alan Pearsall and Christopher Ware) have been the main source for the vessels listed as 'unregistered' in this work. These ship histories are mostly based on the Distribution Lists now in the PRO. J J Colledge's magnificent dictionary of *Ships of the Royal Navy* has been an invaluable

---

[9] He has now published the results of his enquiries into the fleets of all the navies of the modern world from 1500 to 1860 in his *Nations and Navies* (Stockholm University 1993), a copy of which he sent me just in time for me to read and benefit from before writing this introduction. This wide ranging and thought provoking work is very strongly recommended.

'backstop', always to hand as a readily available check. Though the research for this work has uncovered a few errors, some corrected in the second edition of 1987, these do not detract from the usefulness of this pioneering work which is still the standard dictionary of ships of the Royal Navy. There are a very few ships listed in 'Colledge' and not in any other source I have used for whose existence there does not seem to me to be enough evidence to justify including their names here. When in doubt, I have erred on the side of including rather than excluding these problematic vessels.

Patrick Gosset's abstract of the Admiralty Court Martial Records, now published as *The Lost Ships of the Royal Navy*, came to the author for comment when still a computer list, and resulted in another fertile exchange of information. In this case, of course, Mr Gosset's source is both more detailed and liable to be more accurate than those used by the author, and has usually been followed when there is a discrepancy. Those readers wanting more information on losses by wreck or capture are referred to this most useful work. The old and typescript ship lists by Commander Rupert Jones and the (generally inferior) one by Captain Manning held at the NMM have, though occasionally referred to, not been found particularly useful for the present work. Both tend to be imprecise about ship types, origins and fates; Colledge is generally much better in these respects. They are also much inferior in the quantity and quality of operational history details given to that in the Pitcairn Jones' ship histories.

When I was preparing this work for publication Rif Winfield, who has been drawing up his own list of ships of the Royal Navy mainly from listings in the Public Record Office, got in touch with me and we have since exchanged much information. Mr Winfield has looked through my lists and drew my attention to some errors and omissions that were there from the start and rather more that had crept in during the long drawn out process of copying and developing the work from a manuscript list, via several typescripts to first one and then another word processor. Mr Winfield has been very generous with his information and I very much hope that his own excellent work (rather different from the present one in organisation, data presented, and coverage) will find a publisher. Meanwhile I am extremely grateful to him for his unselfish assistance.

## Ship Types - An Overview of Development

At the beginning of our period there was already a clearly recognisable division of ship types in Royal Naval service. The biggest vessels were the 'three-deckers', usually used as flagships. They had more than three decks, of course, in the form of platforms, orlops, fore and after castles etc, but they were identified by the three continuous decks with guns all along their length (Lower or Gun deck, Middle deck and Upper deck). They were further designated by the maximum number of guns carried as 100, 96 (an obsolete type) or 90 gun ships. The 100s and 96s were First Rates, the 90s Second Rates. This rating system had originally referred to the rate of pay of the captain (and continued to do so) but became so closely linked to the number of guns carried that it had long since been taken to be chiefly related to that factor. The actual number of guns (and men) carried varied at this time and, until the early years of the eighteenth century, continued to do so according to whether the Navy was at peace or at war and whether the vessel was to serve in home or distant waters. In the latter case, and in peacetime, the number of guns and men carried was less than the possible maximum; so, for example, a 100 gun ship would only carry 90 guns in these circumstances. This is indicated in the list in the following manner: 100/90.

The three-deckers were also, of course, 'ships of the line' or 'line of battle ships', a designation which came from the period of the First Anglo-Dutch War[10] when opposing fleets started to fight in the long lines which became the basis of all battle tactics and of major warship design from then on. Only ships of a certain size and power - in effect those large enough and with enough guns to have some chance of standing up to a three-decker - were fit to lie in the line of battle. At this period that meant a ship of 50 guns or more - a ship of at least two continuous gun decks ('two-decker'), and one capable of carrying a reasonable number of larger cannon (from 18 pounders upwards). The largest of the two-deckers which formed the majority of the ships of the line were Third Rates of 70 (in a couple of cases 74) guns. Next down the scale were the 60s, also Third Rates (though later classed amongst the Fourth Rates) then the 56, 50 or 48 gun Fourth Rates.

The next type down, usually classed as 40 gun ships of the Fifth Rate, were small two-deckers, not normally powerful enough to lie in the line of battle and usually used as convoy escorts, 'cruisers' (a word which throughout the sailing navy period was used for any ship on detached service and did not come to mean a particular type of ship until late in the nineteenth century) or flagships on distant stations. The Fifth Rate also included smaller ships of 36, 32, 30 or 28 guns, usually carrying most of their guns on a single gun deck, though sometimes with a few heavier guns on a lower deck (what the French called a 'demie batterie'). In the 1670s a few examples of a specialised type of Fifth or Sixth Rate, intended for service 'in the Straights' (Mediterranean) and evolved from experience in wars against the Barbary Corsairs, had appeared. These were the 'galley frigates', ships with fine lines and fewer guns for their size in order to make use of an outfit of large oars or 'sweeps' on their lower decks when circumstances made these useful. Although the word 'frigate' had earlier had a specific use to describe a fine lined, flush decked two-decker taken from prototypes used by Dunkirk privateers it had, by the 1680s, declined into a general term of dubious utility for any comparatively fine lined vessel, and is best ignored until it came to designate a particular type of warship again in the 1750s (the 'true frigate').

The next type down were single decked vessels of 24 or 20 guns and of the Sixth Rate. A ragbag of smaller vessels were still designated Sixth Rates, but by this time were a disappearing breed. These were the smallest and least powerful vessels to be commanded by a Captain, and therefore to be rated. It seems probable that all, or nearly all, were ship rigged (three masted, with square sails on all masts), and throughout the rest of the history of the sailing navy all rated vessels were to remain ship rigged. The Royal Yachts were an exception to this; as prestige vessels they, too, were commanded by Captains. The larger ones were usually considered as Fifth Rates, the smaller as Sixth Rates. By the eighteenth century the largest were ship rigged, but most were one or two masted, and predominantly fore and aft rigged (some as ketches, some as sloops). During the Dutch wars of the 1660s and 70s these vessels were used as scouting craft and despatch vessels for the fleet, but as the end of the century approached this practice stopped. Thereafter they were not used for warlike purposes, and are therefore to be found in part III.

During the last quarter of the seventeenth century a variety of small vessels filled the place later taken by sloops. These would be commanded by a Lieutenant, or, perhaps, a Master. The names of these types might refer to rig, hull type or function. Unfortunately, there is very little in the way of clearly identified visual material to enable one to be precise about definitions. It seems probable that Ketches actually had a ketch rig as it was later known: two masted, the after mast shorter than the other and fore and aft sails less predominant than later. At this stage a ketch resembled a ship rigged vessel missing its foremast. However, the name may originally

---

[10] Peter Padfield in the first volume of his excellent general maritime history of the age of sail *Tides of Empire* shows that ships had already, on occasion, fought in line ahead, a point repeated by Jan Glete in his recent *Nations and Navies*. However, the point remains that the true beginning of fleet actions fought as a matter of course in this formation, the beginning of building ships optimised to fight in this way and the first steps in creating the regulatory system set up to organise this were during the war of the English Commonwealth against the United Provinces.

have referred to the hull type rather than the rig. Brigantines were quite probably smallish vessels fitted to row as well as sail.[11] The rig may have been two masted, with some resemblance to that later known as 'brigantine' (square sails on the foremast, fore and aft on the main). However, it must be remembered that the classic definitions of ship rig types crystallise in the early nineteenth century, usually on the basis of late eighteenth-century usage. Earlier usage appears to have been more flexible, less certain and more varied. The word 'Pink' appears to have identified a type of hull – with a 'pinched in' poop above a rounded stern. This was the form associated with the Dutch 'flute', which also had a rounded bow. It seems likely that 'pink' and 'flute' were similar, if not even identical types. Other small craft were defined by function: the 'Advice Boats' were despatch vessels and scouts, though most of the smaller vessels and not just the Advice Boats were used for such tasks from time to time. It is not clear what hull form or rig was used for the Advice Boats, if indeed these did not vary. At least one small British warship of the last years of the seventeenth century, the *Royal Transport*, had a rig that a later generation would describe as 'schooner' and it is not impossible that at least some of the Advice Boats may have had some form of fore and aft rig. It rather looks as if the earliest sloops (there had been some in service in the 1670s though none lasted until 1688) may well have had the single masted fore and aft rig associated with their name. They were certainly small enough for this.

At this period, however, the type of small ship of which there were most in service was the fireship. Many were conversions from merchantmen or from warships of the lower rates. However, there was also a fair number of purpose-built fireships. Whether a conversion or designed as such, the fireship was intended as a weapon in her own right, to be 'expended' in setting a larger enemy on fire. Targets would be vessels disabled in battle, grounded or trapped in a port or anchorage. A vessel still fully able to manoeuvre with a reasonably competent crew had little to fear from such a weapon. Fireship crews had the very tricky problem of ensuring contact with the enemy at just the right moment whilst preserving their own skins both from enemy fire and the combustion of their own vessel. The physical features that distinguished fireships from other warships is that they were fitted with an extra deck, gunports that hinged at the bottom rather than the top so that they would fall open when the retaining ropes burnt through and therefore increase the draught, and grappling hooks on the yardarms. The 'firedeck' was subdivided by channels to take fuzes so that the barrels and bundles of combustibles placed in the subdivision could be set on fire as rapidly and comprehensively as possible. Fireworks of the 'Roman candle' variety would be placed immediately inside the ports to be set off as the fire took hold. Special chimneys would be shipped to carry the flames up the rigging. Fireships had been used to considerable effect in the Anglo-Dutch wars, but perhaps their most effective use was in the aftermath of the Anglo-Dutch victory of Barfleur (1692), in the series of inshore actions against the scattered elements of the French fleet known as the Battle of La Hogue. Thereafter, the use of fireships for their primary purpose declined, though ships continued to be built or converted for the purpose for over a century. However, they were mostly used, as they had been earlier, as general-purpose small craft; only occasionally would one fulfil its name and purpose in a final blaze. In the 1690s a variant of the fireship saw service under the title 'machine' (as in 'infernal machine'). Machines were in general smaller craft than fireships, always conversions rather than purpose-built, and their purpose was to be floating bombs. Whilst fire (often backed by the potential threat of an explosion) was the chief weapon of the fireship, that of a machine was an explosion. Packed with gunpowder they were the successors of the 'hellburners of Antwerp'

used against the Spaniards a century earlier. A century later the 'fireships' used by Cochrane against the French in Basque Roads (1809) would be similar explosion vessels.

In 1687 the first of a new kind of vessel was built for the Royal Navy, a bomb vessel (or 'bomb'), a specialised shore bombardment vessel which, like the fireship, can be considered a 'weapons system' in the modern phrase – a vessel built round a particular weapon and purpose in a way the other warships of the time were not.[12] The chief weapon of the bomb was the mortar firing an explosive shell at a high trajectory. These vessels had to be stoutly built and specially reinforced under the mortar 'beds'. The type had been invented by the French and first employed in the bombardment of Algiers (1685). Though the first English examples seem to have been copies of the French vessels, with the mortars fixed in both traverse and elevation (ie. pointing forward at a fixed angle of elevation) in a side-by-side installation, by the early eighteenth century the English 'bombs' had surpassed their French prototypes in introducing mortars capable of both traverse and elevation mounted on rotating turntables in pairs one behind the other on the centre line. Besides introducing a much greater flexibility in the use of these weapons it is arguable that this is the origin of the turret at sea. Most, though not all, of the early Royal Naval vessels of this type were ketch rigged; the term 'bomb ketch' therefore acquired the general currency it still holds. However, as some early bombs were ship rigged, as were all those taken into service after early in the Seven Years War, 'ketch' is wrong for the type as a whole.

The latter part of the 1690s produced a greater emphasis on the building of the smaller types of ship because of the French shift from *Guerre d'Escadre* (war based on the battlefleet) to *Guerre de Course* (commerce raiding). In particular, numbers of the somewhat unsatisfactory 'one and a half-decker' Fifth Rate ships of 32 guns or so were built. However, the battlefleet could not be neglected and a new type of large battleship was produced to fill the gap between First and Second Rate three-deckers (now respectively standardising on 100 and 90 guns as their maximum armament) and the two-deckers of around 70 to 60 guns (the 50 gun ships were beginning to slip down into the category of ships no longer powerful enough to stand in the line of battle but extremely useful for trade protection and other miscellaneous 'sea control' and policing tasks). The result of this was the 80 gun Third Rate. The first group of these were large two-deckers with long quarter decks. These proved structurally somewhat weak, overgunned and unsatisfactory so their successors and replacements were built with quarter decks and forecastles connected and thereby converted to short three-deckers. These ships have been universally condemned by subsequent commentators, beginning with Charnock (in his *History of Naval Architecture*) and with much justification. The three-decker 80s were poor sailers, often unable to utilise their heaviest guns which were too close to the waterline. However, there were some points in their favour, principally that they were comparatively cheap to build and man, whilst having the very real advantage in battle of being a three-decker with the higher command of the uppermost guns and the protection of the lower batteries that brought with it.[13] Certainly, the type remained a mainstay of the British fleet for over half a century until the revival of the larger two-decker in the shape of the '74' underlined its obsolescence.

From 1700 no more pinks, ketches, advice boats or brigantines were built; instead, the term 'sloop' was revived to cover the small non-rated craft being built for the Navy. Those built between 1700 and 1715 were probably two masted (unfortunately, the only surviving plan known does not give any indication of the number or position of the mast or masts); certainly those built between 1715 and 1739 all were, though the rig varied among 'brigantine' (what would later be called 'brig'), 'snow'(probably then, as

---

11 See Robert Gardiner's contribution on the smaller types of warship in Lavery (ed), *The Line of Battle* (Conway Maritime Press's History of the Ship 1982).

12 In the same way that the galley of classical times had been built round its ram.
13 I owe this point to Andrew Lambert who makes it with force in his *Last Sailing Battlefleet*.

later, differing in rigging a square mainsail as well as a spanker, probably on a trymast) and ketch. These were the commands of senior lieutenants serving as 'masters and commanders', who would usually, confusingly, be given the courtesy title (but not the rank!) of 'captain'. Apparently as a result of wartime experience, the old 32 gun Fifth Rate was transformed into the 20 gun Sixth Rate of the 1719 Establishment, some by conversion but mostly by new building. This attractive, almost flush decked design foreshadowed the 'true frigate' in having a 'tween deck below the single complete gun deck and thus carrying her guns high. It was a thoroughly sensible, seaworthy and fast-sailing design.[14]

## A Note on 'Rebuilding' and on 'Establishments'

At this stage in the story of the development of the sailing warship it is as well to clarify the way in which the two words 'rebuilding' and 'Establishment' are used here. The full story of the change of sense in which the former was used has been clarified by Brian Lavery,[15] and the present author's researches have certainly confirmed the picture he has drawn. In outline form this is as follows. Up to the beginning of the War of Spanish Succession (1702) ships that were getting old, worn out, rotten or old-fashioned would be rebuilt by being put in dry dock, taken to pieces to a greater or lesser extent, and then the good timbers reused in the rebuilt vessel. Rotten timbers would be stripped out and replaced, and the opportunity might well be taken to modify the lines, insert more frames (and therefore lengthen the ship) or to make other alterations to the shape, size or structure of the ship. The resemblance of the result to what had been there before was a variable quality, but basically the term 'rebuild' was an accurate description of what had been done. It is therefore fair to presume a degree of continuity between the ship before and after its rebuild.

From 1702 this is no longer the case. Because of the need to use the comparatively small numbers of docks available for wartime repairs and refits it was undesirable for them to be occupied by a vessel undergoing rebuild - a necessarily long drawn out process. Instead, the operation was reallocated to slipways in place of docks. Increasingly, the ship might be taken to pieces on one slip and the 'rebuild' on another, not always even in the same dockyard. Some timber from the old vessel might be used in the 'rebuilt' one, but in effect the operation became an administrative fiction for building a new ship. Ships continued to be described as being 'rebuilt' into the 1740s when the pressures of a new war caused the term to be abandoned.[16] It should be made clear that contemporary documents then (and later) make a clear distinction between 'rebuilds' on the one hand and 'Great', 'Middling' and 'Small' repairs on the other. Normally, the 'repair' did not involve a major change in the ship, though there are a few exceptional cases in which dimensions and/or appearance might change, and a very few when some sources use the word 'repair' when the others use 'rebuild'. In this work, when there is doubt, the operation has been treated as a rebuild, and rebuilds (designated in the list by the abbreviation **RB**) are treated as new ships by being given separate entries, though it should be borne in mind that pre-1702 (and therefore for all the ships in Chapter 1 and the majority noted in Chapter 2) there is a strong though variable element of continuity which diminishes rapidly after that date.

The word 'establishment' was the sailing navy's equivalent of 'outfit', or 'allocation', usually with the overtone of this being the officially laid down scale of allowances. It could be used to describe the outfit of guns, or of victualling stores, or even of men. In the early eighteenth century (usually when used with a capital letter) 'Establishment' came to be used of ship types in the sense of laid down dimensions and even design. The impulse to standardisation goes back at least as far as Pepys, and by the early eighteenth century resulted in sets of rules which became increasingly rigid. There were obvious advantages in building ships of similar size and gunpower in as standardised a form as possible, particularly in the stockpiling of structural timber, spars and rigging for repairs in the dockyards. In 1706 a set of guidelines for the dimensions of larger warships appeared. Subsequent lists refer to ships built under this as '1706 Establishment', though it appears not to have been adhered to too closely. In the aftermath of the end of the War of Spanish Succession under the dominating and excessively conservative influence of Ackworth, the Surveyor of the Navy,[17] the rigid and detailed '1719 Establishment' laid down dimensions, 'scantlings' (thickness of timber), spar dimensions, armament and much else. This was generally closely adhered to, though it did not cover the unrated sloops, which were small and cheap enough to allow for some adventurousness, as a variety of dimensions and designs show, and there was some experimenting with increased dimensions in the 60 gun ships. In general the dimensions were too small for satisfactory performance at sea with the armament carried (though the 60s sailed fast and were copied by the discriminating Danes, which gives the lie to the suggestion that all British designs at this date were slow). In 1733 dimensions were increased in a modification of the 1719 Establishment. This was described in contemporary lists (as it is here) as the '1733 Establishment', though strictly speaking it never was officially established. The same comment applies to the next and somewhat more drastic increase in size brought on by the first impact of renewed war, the '1741 Establishment'. All these Establishments cramped design by insisting on dimensions which were to be strictly adhered to. This considerably limited the ability of the shipwrights of the individual dockyards to experiment with different lines, disposition of armament and so on. What happened next was that pressure from the Admiralty to produce bigger and more powerful designs to counter French and Spanish developments led to an even more rigid 'Establishment' - that of 1745. This specified the actual lines to be used, removing all influence on the designs by the Master Shipwrights of the yards in favour of ones prepared by the Surveyor of the Navy. However, what appeared at first sight to be a hardening of the Establishment system into total rigidity was, in fact, the precursor of the breaking of this particular fetter on development. Further pressure from the Admiralty, seemingly engineered by Anson, led to the appointment of a new pair of Surveyors of the Navy. The more important of the pair was [Sir] Thomas Slade who can fairly claim to have been the most successful warship designer of the eighteenth century.[18]

The reforms apparently masterminded by Anson and Slade[19] led to a whole series of new types of warship appearing during the middle years of the eighteenth century. The 74 gun ship with 32 pounders on the lower deck became the standard two-decker line of battle ship, with the 64 gun ship as a smaller and cheaper version (and a replacement of the old 70s). The 9 and 12 pounder armed 28 and 32 gun frigates filled the gap between the small Sixth Rates and the small two-deckers, and were more successful as cruisers than either. These frigates were in effect two-deckers with the lower deck not fitted with guns. This 'tween deck (initially, and very

---

14  For this and much which follows see the excellent summary of Robert Gardiner's researches and views in his *The First Frigates* (Conway Maritime Press 1992).

15  See his *Ship of the Line*, also his earlier article on the subject in *The Mariner's Mirror*.

16  There was the exceptional case of the fir frigate *Clyde* which was rebuilt at the beginning of the nineteenth century, but this is a 'one off' recurrence of the term of no general significance.

17  In other words, the Navy's chief designer.

18  This is judged on results. His masterpiece, the *Victory*, was no accident; Slade designed a whole series of very successful three-deckers, two-deckers, frigates and smaller craft. There can be no doubt he was by some way the most successful British warship designer. The Swede Af Chapman was probably the most distinguished naval architect (a title which it would be difficult to justify for Slade who was still very much in the shipwright tradition, pragmatic rather than overtly scientific) of the century, but as a practical designer of ships he was more original and innovative than practical and successful. The Frenchman Sané produced numbers of excellent designs but neither in the volume nor showing the innovation or flexibiltiy of which Slade was capable. The nearest to Slade in practical genius was probably the earlier Frenchman, Blaise Ollivier, so important in the development of the 74 and the frigate, but even his designs do not seem to have been so consistently successful.

confusingly, known as the 'gun deck') below the waterline simultaneously provided better accommodation for crew and stores, and enabled the one continuous deck of guns to be placed well above the waterline, giving plenty of clearance between the lower 'cills' of the gunports and the waterline. This meant the guns could be fought in rougher conditions, and made the frigates into the ideal type of cruising warship.[20] Both 74s and frigates had already appeared in the navies of France and Spain, though what was arguably the first 'true' frigate was designed by a Frenchman for the Danes. The British felt that they had to emulate these types, and often used the lines of French prizes; the designs which resulted were not simple copies, and were adapted to British needs.[21]

At the same time, the first ship sloops (larger sloops with a three masted ship rig rather than the old two masted ones) were built. The ship rig also replaced the ketch rig for bomb vessels. These developments occurred during the Seven Years War (1757-1763) at the end of which the Royal Navy also made a large purchase of cutters. These single masted fore and aft rigged small craft had already been hired in some numbers. They were a standard type of local craft in British waters, whilst in North America the equivalent rig was the schooner. As tension in the American Colonies increased in the next decade, so numbers of these schooners were acquired in North America for local use by the Navy.

With both cutters and schooners the fore and aft rig permitted smaller crews and greater handiness inshore, and both types were used until the end of sail in the Royal Navy. The success of the 74s meant that the old three-decker 80s ceased building and went out of service rapidly. The medium sized three-deckers of 90 guns would last for longer, but reclassed as 98 gun ships with weapons added on the quarterdeck. The 70 had changed into the 64, whilst the 60, no longer big enough for the line of battle, soon vanished into limbo. The small two-deckers of 50 guns were still considered useful on distant stations and as convoy escorts, and continued to be built, but the smallest two-deckers, the 44s, were so obviously outshone by the new frigates that none was built for a while. A brief revival of an enlarged version of the type in the 1770s would be doomed to rapid relegation to second-line duties because of the enlargement of the frigate.

As colonial unrest boiled up into armed revolt in the 1770s there was a temporary reaction in favour of building small two-deckers and Sixth Rates, which were considered the most useful types for the situation. New and somewhat larger 50s, 44s, 24s and 20s were ordered in some numbers, as were sloops. Once France and then Spain came into the war, however, the reversion to the main types useful in an all-out war with rival navies was swift. 74s and frigates came into their own.

The American War of Independence witnessed the appearance of two new types in the Royal Navy. It was inevitable that the very successful frigate would grow in size and power of armament. For once it was not the French who took the lead, for the British were first to put 18 pounder guns in the main battery of an enlarged frigate, an example shortly followed by the French and other navies.[22] The 38 gun ships which followed were often built to French lines (though with modified structure, upperworks and layout to conform with British needs and practice) but nearly all the 36 gun frigates were built to wholly British designs.

It was also inevitable that the growth in size and complexity of the ship sloop would create a need for a somewhat smaller vessel which required a smaller crew. Brigs had been hired earlier, but the first ones ordered for the Navy were building in 1778. They could be seen as a reversion to the old two masted sloops, and the majority of the new brigs were also classed as sloops ('brig sloops'), though the smallest were usually just 'brigs'; this could change with the change of commanding officers, the distinction between brig sloop and brig being simply whether he was a Commander or just a Lieutenant. The purpose-built fireship was revived at the same time, built to a fine lined and speedy model to provide the fleet with an answer to the danger of the Franco-Spanish combined fleet which threatened to dominate the Channel.[23] At the same time the Royal Navy introduced a new weapon which had the effect of confusing the issue of the classification of sailing warships more than any other factor.

This weapon was the carronade, a stubby, short range weapon firing a heavy ball with a light charge. Developed by the Carron Iron Works in Scotland, it was normally carried on a pivoted mount.[24] The experimental arming of the *Rainbow* (44) with a very heavy, albeit very short ranged, carronade armament was exceptional. The carronade was first generally introduced as an armament for the light upper decks of warships, and as a substitute for the half-pound swivel guns which were mounted on posts in the bulwarks of these decks. They were not, initially, a substitute, or even really a supplement, for the long guns which were considered the main armament. They were, therefore, not counted as part of that main armament in rated ships. In unrated ships, sloops, brigs and other lesser craft, carronades soon became the majority of the armament and were counted in the official total of the guns of that vessel. For example, brig sloops carrying a standard armament of 2 'long guns' and 16 carronades were called 18 gun vessels. However, rated vessels were referred to by the originally designed long gun armament only, even though the number of carronades carried on forecastle and quarterdeck might considerably increase the broadside of the ship, and sometimes reduce the number of long guns aboard. Whilst this complicates matters when working out what the armament of a particular ship was at a given time, the use of the old designations ('74', '38', '24' etc) does give a clear indication of what type the ship belonged to. It is for this reason - the clear indication of type - that I have continued to use these designations even after the great reclassification of 1816 when it was decided to give a better indication of fighting power by including at least some of the carronades carried in the title. This means that 38s became 46s, 36s were renumbered as 42s, whilst some ship sloops became 28s or 26s. In this work I have noted these new designations but continued to use the old ones as the main type 'names'. To do otherwise would be to introduce a totally artificial discontinuity, an arbitrary fault line across a continuous development. This is reinforced by

---

19   The decisions and consequences of this revolution are easy to find, the detail documentation of the reasoning and parts played by individuals almost non-existent. See Robert Gardiner's *The First Frigates* and Brian Lavery's *Ship of the Line*.

20   See Robert Gardiner's *The First Frigates* for an elegant and cogent analysis of this development.

21   Again, Robert Gardiner has much penetrating comment on this matter in his *First Frigates* and the reader interested in a sophisticated and detailed analysis of the differences between British and French designs should refer to that work. It might be worth adding that the pragmatic policy of copying a good design wherever it comes from seems to be both sensible and perhaps more genuinely scientific than the French eighteenth-century over-reliance on an as yet incompletely developed science of naval architecture. Even in the twentieth century experience with earlier designs has been the best way of designing a successful sailing ship as there are so many variables. The legend of French superiority in warship design is at best a half-truth of diminishing validity as the eighteenth century progressed, and has too often been used as a substitute for careful analysis of an extremely complicated and changing situation.

22   I owe this point to the kindness of Robert Gardiner who documents this in his forthcoming monograph on heavy frigates, which I have had the privilege of reading in manuscript. Gardiner hints, almost certainly rightly, that this upward step in power must have owed much to the unaccustomed postion of temporary numerical inferiority that the RN found itself in after France and Spain both came into the American War of Independence against Britain.

23   I owe this point to Robert Gardiner drawing my attention to Kempenfelt's 5.12.1779 letter to Middleton suggesting fast fireships which could keep up with the fleet (J K Laughton [ed], *The Barham Papers*, Vol 1, pp 296-7, Navy Records Society 1907).

24   See Brian Lavery's *Arming and Fitting of the Sailing Man of War* ( a work of more general historical interest than might be apparent from the title, and written with the clarity and readability one has come to expect from this author) for more detail. It is worth noting that the design of gun and of mounting were altered after its first introduction and before the Great French War; in particular the gun became longer and the mounting less liable to oversetting. The form we are used to from illustrations is the later version.

the ironic fact that the 1816 change was made just at the time that the Navy was beginning to move away from its heavy use of carronades.

The last 64 and 44 gun two-decker ships to be completed for the Royal Navy were already building at the end of the American War of Independence, whilst the same also applied to the 9 pounder 28 gun frigates. Indeed the 44s were so completely eclipsed by the new large frigates (both carried the same main battery of 18 pounders) that the last ones were completed as troopships; the others were converted to transports, hospital ships, or for other second-line tasks. Already the next stage in the rise of the frigate was preparing in the 1780s. First the Swede af Chapman and then the Americans built 24 pounder-armed frigates. The initial British reaction was to 'razee' (cut down) three 64 gun ships into heavy frigates. One of these was Pellew's very successful *Indefatigable*. Eventually, and with the impact of the victories of the American heavy frigates in 1812, the Royal Navy, rather reluctantly, built numbers of heavy frigates itself. The Royal Navy at this time also began to increase the size of both two- and three-decker line of battle ships, in both cases by following French examples, with 80 and 120 gun ships respectively. As with the large frigates, the initial prototypes would not be followed until the war with Napoleon was drawing to its close a quarter of a century later.

The earlier years of the Great War with France (a pre-1914 usage which simplifies referring to the Revolutionary and Napoleonic Wars of 1793 to 1815) were marked by the building of larger 74s and 18 pounder frigates in numbers. From the start, British 74s had divided into two main groups: the standard ('common class', not a class as we use the term but a group) and the occasional larger vessels which usually had a couple more guns in their main battery. The policy of distant blockade adopted in the early years of the war seemed to require more of the bigger vessels, but later on the emphasis shifted to close blockade and the building of 'common class' vessels. In the same years the last 98 gun three-deckers were built for the Navy. It is an indication of the growth in relative sizes that some of these were built to the lines of Slade's *Victory*, a 100 gun First Rate. The same growth was demonstrated in two-decker ships of the line, with the 80 gun *Caesar* built for the Navy whilst other vessels of this size and armament were captured from the French. The standard frigates now were 36s and 38s; some smaller ones were still built and armed with 12 pounders but these were either built of fir for speed of building or were ordered from small yards which could not build anything larger[25] (one thinks of the 'Weapons' Class destroyers as a parallel).

The threat of invasion by France and a more general need for small, heavily armed ships for inshore work was met by the building of a series of classes of vessel with (usually) brig rig and 10 or 12 guns. The initial classes of 1794 were shallow draught and armed with heavy long guns in the bow. These were described as '**gunboats**'. Their successors of 1797 were more conventional in both draught and armament and generally referred to as '**gunbrigs**', having the appearance of small flush decked brigs, as did subsequent classes built in the early years of the new century. Later vessels were classed simply as '**brigs**' and some were even brig sloops. By this time they had merged into the main group of brigs, which became the main type of vessel below the rated level. Meanwhile, the larger type of ship sloop, equipped with quarterdeck and forecastle, was continuing to grow in size and gunpower and merge with the Sixth Rates. The use of 'gunboat' here was confusing as, from the American War of Independence, the Navy had been using large rowing boats with one or two guns in the bow and/or stern as **gunboats**, and large numbers of vessels of this type were to be built during the Great French War. They could be sailed as well as rowed (often with a lateen rig – betraying Mediterranean influence; it seems likely that the type was invented in the Mediterranean, perhaps by the Neapolitan Navy, as a cheaper replacement for the galley). Also large numbers of mercantile coastal craft, barges, hoys, Dutch galliots, etc, were converted, some of them with considerable ingenuity, to anti-invasion gunboats. These were usually to be manned by the Sea Fencibles (a sort of maritime militia for local defence).

In 1816 the great reclassing took place, with 38s becoming 46s, the larger ship sloops being reclassed as Sixth Rates of 24, 26 or even 28 guns (to be referred to as 'jackass frigates' by seamen) and so on. However, most of the types so reclassed were already becoming obsolete, the tendency being to build larger vessels. The structural reforms associated with the name of Seppings made this possible. What was in effect a form of girder construction and very solid structures made it possible to build longer wooden ships. 120 and 110 gun three-deckers, two-decker 80s, 84s and then even 90s and 50 gun frigates became the order of the day.[26] A few medium sized frigates, small Sixth Rates and ship sloops were built, the vast majority of new construction of smaller vessels being of brigs. In the 1820s steamships intended chiefly for towing the sailing fighting ships began to come into service with the Royal Navy. By the 1830s these were becoming more heavily armed and bigger, but paddle steamers were essentially still auxiliaries to the main fighting vessels which were still sail-powered. In the 1840s screw engines made it possible for these fighting ships, ships of the line and frigates to be converted to, or built with, auxiliary steam power. It was not until the ironclad appeared, at the beginning of the 1860s, that change became radical[27] and the essential continuity of the development of what were still basically sailing warships was broken.

---

**25** I owe this point to Robert Gardiner (private communication).
**26** For the reasons behind the Royal Navy's building policy after 1815 and for its details see Andrew Lambert's *The Last Sailing Battlefleet* (Conway Maritime Press 1990). Those wishing for more information should refer to this deeply researched and carefully reasoned account.
**27** On this change see Andrew Lambert's three chapters on the development of the steam fighting ship in the volume in Conway's *History of the Ship* series, entitled *Steam, Steel and Shellfire* (1993, edited Lambert), for a clear and coherent account of the slow triumph of steam.

# INTRODUCTION
# SHIPS IN SERVICE 1688

## Chapter 1: Ships Built For The Royal Navy

This chapter and the following one give basic details of those fighting vessels which were in service with the Royal Navy at the beginning of our period. The year 1688 witnessed the 'Glorious Revolution' in which James II was driven out of his kingdom and succeeded by William of Orange and his wife (James' daughter) Mary. This was the signal for the outbreak of war with Louis XIV of France, the first of a series of naval wars between Great Britain and France. For convenience, this section of two chapters includes both those ships built for the Royal Navy (given by type in descending order of size and gunpower, and in chronological order by date of build within each type) and, in the second chapter, ones which came into service through capture or purchase (these are again grouped by type and placed together at the end of the chapter, again grouped by type and within each type by date of acquisition). For further information about these vessels see the lists in Tanner's first volume of *The Naval Manuscripts in the Pepysian Library* (Navy Records Society 1903), also (except for ships built during the reign of James II and fireships) Frank Fox's *Great Ships* (Conway Maritime Press 1980) and (large vessels only) Brian Lavery's *The Ship of the Line* Vol 1 (Conway Maritime Press 1983). This section is a minimal listing based on the sources noted above and not on original material. (Although the majority of the ships mentioned are in Dimensions Book B, this would appear to have been compiled at a later date.) For these reasons I have not attempted to unravel the sometimes contradictory information available in these sources on dimensions, tonnage, detail of armaments, etc, and give merely a brief outline of the history of the ships concerned. Nor have I attempted to include precise dates of ordering, launching, etc.

In these two chapters *only*, individual entries are given in the following order:

**(1)** Name/Alternative form(s) of name. [RB] given after the name indicates that this is a 'Rebuild'. **(2)** Number of guns (war - home waters)/number of guns (war - overseas *or* peacetime; sometimes the latter is less than the former and is given separately). **(3 when relevant)** Previous names are given in brackets thus: (ex Name renamed *date*). **(4)** Year of launch (the calendar used up to 1753 is Old Style, but without using the convention in force at the time which began the year on Lady Day in late March rather than on New Year's Day. This may lead to the figure given here being a year out compared to contemporary sources or certain modern works which have not been corrected for this factor). **(5)** Built by and at: *either* Royal Dockyard (abbreviated to 'Dyd') with Surveyor (responsible for design at this stage as well as supervising the ship's construction) given in brackets thus: ..... Dyd (surveyor), **or** commercial builder and location of yard. **(6 when relevant)** entry(ies) for subsequent changes in function or name, etc. RB = 'Rebuild'. **(7)** Fate (BU = broken up, RB = 'Rebuild') with date. **(8 when relevant)** brief notes of plans in the Admiralty Collection (and a few other small collections) at the National Maritime Museum, Greenwich. In exceptional cases, the existence of plans in other collections (eg. the Danish Rigsarkivet Collection, or the Swedish Chapman Collection) may be noted.

### First Rates (three-deckers)

***Royal Sovereign*** 100, (1637 as *Sovereign Of The Seas/Royal Sovereign*) Woolwich Dyd (Peter Pett). Renamed *Sovereign* during the Civil War; 'rebuilt' 1659 at Chatham Dyd (Captain Taylor); 1660 renamed *Royal Sovereign*; RB 1685 Chatham Dyd (Lee); 1696 burnt by accident (in 1701 a new *Royal Sovereign* was built at Woolwich and referred to as a 'rebuild' but cannot be accepted as such in any meaningful sense of the word).
PLANS: Early nineteenth century reconstructions of Lines.

***Saint George*** 96/90/80, (ex *Charles* renamed 1687), 1668 Deptford Dyd (Jonas Shish). 1691 reclassed as a Second Rate; 1701 RB at Portsmouth.

***London*** 96/86, 1670 Deptford Dyd (Jonas Shish). 1679 RB at Chatham (Sir Phineas Pett); 1701 BU for RB at Chatham.

***Saint Andrew*** 96/86, 1670 Woolwich Dyd (Byland). 1701 RB at Chatham as *Royal Anne*.

***Prince*** 100/90, 1670 Chatham Dyd (Phineas Pett). 1692 RB (or Great Repair) as *Royal William* at Chatham.

***Royal Charles/Charles Royall*** 100/90, 1673 Portsmouth Dyd (Deane). 1693 RB (or Great Repair) as *Queen* at Woolwich.

***Royal James*** 100/90, 1675 Portsmouth Dyd (Deane). 1691 renamed *Victory*; 1695 RB (or Great Repair) at Chatham.

***Britannia/Britania*** 100/90, 1682 Chatham Dyd (Phineas Pett). 1691 'girdled' (beam increased); (1701? Great Repair; according to Frank Fox but no evidence found in sources consulted); 1716 BU.

***Saint Michael*** 90/80, 1669 Portsmouth Dyd (Tippetts) as a Second Rate. 1672 reclassed as a First Rate; 1689 reclassed as a Second Rate; 1706 RB at Blackwall as *Marlborough*.

### Second Rates (three-deckers)

***Katherine/Royal Katherine*** 84/82/74, 1664 Woolwich Dyd (Christopher Pett). 1701/2 rebuilding.

***Victory*** 82/72, [1620 Deptford (Burrill)]. 1665 RB Chatham Dyd (Phineas Pett); 1690 BU.

***Vanguard*** 90/82, 1678 Portsmouth Dyd (Furzer). 1705/6 rebuilding at Chatham.

***Windsor Castle*** 90/82, 1678 Woolwich Dyd (Jonas Shish). 28.4.1693 wrecked on the Southsand Head.

***Duchess/Dutchesse*** 90/82, 1679 Deptford Dyd (John Shish). 1703 renamed *Princess Anne*; 1702 renamed *Windsor Castle*; 1706 renamed *Blenheim*; 1709 RB.

***Sandwich*** 90/82, 1679 Harwich Dyd (Betts). 1709 rebuilding at Chatham.

***Albermarle*** 90/82, 1680 Harwich Dyd (Betts). 1701 rebuilding at Chatham.

***Duke*** 90/82, 1682 Woolwich Dyd (Thomas Shish). 1701 rebuilding as *Prince George*.

***Ossory*** 90/82, 1682 Portsmouth Dyd (Furzer). By 1707 rebuilding at Deptford.

***Neptune*** 90/82, 1683 Deptford Dyd (John Shish). 1708? rebuilding at Blackwall.

***Coronation*** 90/82, 1685 Portsmouth Dyd (Betts). 3.9.1691 foundered off Rame Head, Plymouth Sound. (Divers have found remains both immediately under the Head and also some distance out in the Sound which seem all be from this vessel, which could have come apart after sinking. So far no final solution has been found to this problem, though investigations have continued and it now looks as if the above explanation is winning general acceptance.)

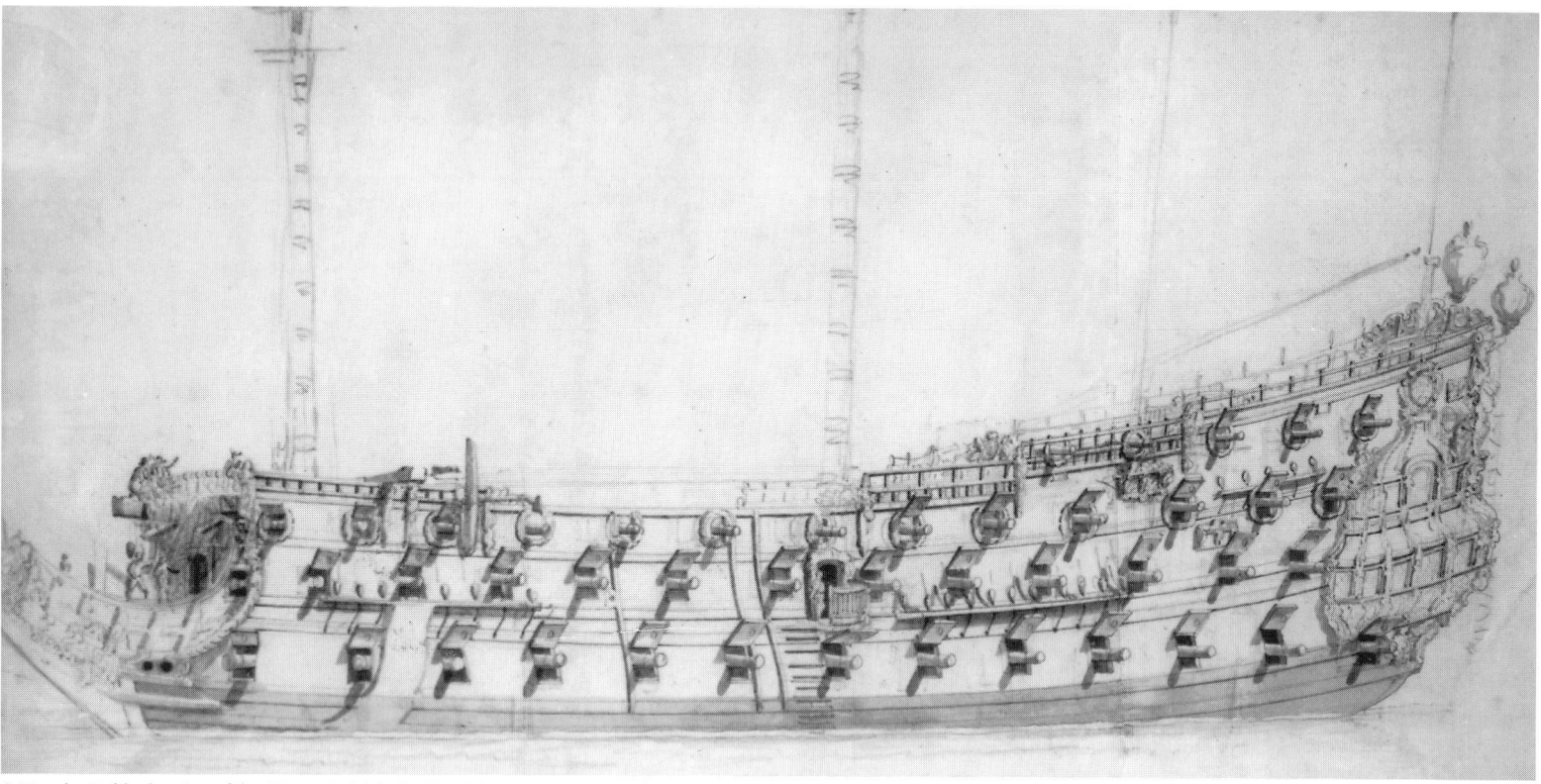

A Van der Velde drawing of the 90 gun *St Michael* of 1669 (MM Negative number V.537)

## Third Rates (two-deckers)

***Lion/Lyon*** 60/52/48, [1640 Chatham Dyd (Apsley)]. RB 1658 Chatham (Taylor); 1696/7? reclassed as a Fourth Rate; 1698 sold.

NOTE: Oppenheim alleges that the 1640 *Lion* was a RB of the 1609 *Red Lion*, which in her turn was a RB of the 1582 *Red Lion*, supposedly a RB of the *Lion* of 1557, herself a RB of the *Lion* of 1536 (which was only 120 tons!). This seems a little unlikely, though not impossible. It is probably simpler for our purposes if the 1640 vessel is regarded as a new one, for which point of view there is some evidence.

***Mary*** 64/54 or 62/54 [originally 56/50?], (ex *Speaker* renamed 1660), 1658 Woolwich Dyd (Christopher Pett). 'Girdled' (breadth increased) soon afterwards; probably RB, or at least subject to a large repair some time before 1688; 1696/7? reclassed as Fourth Rate; 26-27.11.1703 wrecked on the Goodwins during the 'Great Gale'. (It is possible, though not probable, that remains of this vessel were found in 1980 by a team of divers including the present author. The remains are more likely to be those of one of the 70s lost at the same time, *Restoration* or *Northumberland*.)

NOTE: Was built as a 'frigate' in the style of the Dunkirk privateers and the *Constant Warwick*. She can be considered the prototype of the two-decker line of battle ship. However she has little direct connection with the 'true' frigates of the later eighteenth century.

***Dunkirk*** 60/52, (ex *Worcester* renamed 1660), 1651 Woolwich Dyd (Burrell). (1692? Great Repair according to Frank Fox); 1696/7 reclassed as Fourth Rate; 1704 RB at Blackwall.

***Plymouth*** 60/52 (originally 52/54), 1653 John Taylor, Wapping. Probably RB early in her career; 1696/7? reclassed as Fourth Rate; 1703 to be RB.

***Henrietta*** 62/54 [originally 50], (ex *Langport* renamed 1660), 1654 John Bright, Horsleydown. 25.12.1689 wrecked on Mount Batten, Plymouth.

***Montague/Mountague*** 62/54 [originally 52], (ex *Lyme* 1654 at Portsmouth Dyd (Tippetts). 1660 renamed); RB 1675 Chatham Dyd (Phineas Pett); 1696/7? reclassed as Fourth Rate; 1698 RB.

***Dreadnought*** 62/54 [originally 52], (ex *Torrington* renamed 1660), 1650 Henry Johnson, Blackwall; 16.10.1690 foundered off the North Foreland.

***York*** 60/52 [originally 52], (ex *Marston Moor* renamed 1660), 1654 Henry Johnson, Blackwall. 1696/7? reclassed as Fourth Rate; 12.11.1703 wrecked on the Shipwash (off Harwich).

***Monck*** 60/52 [originally 52], 1659 Portsmouth Dyd (Tippetts). 1696/7? reclassed as a Fourth Rate; 1701 rebuilding at Rotherhithe.

***Cambridge*** 70/60, 1666 Deptford Dyd (Jonas Shish). 19.2.1694 wrecked: 'cast away at the back of Gibraltar'.

***Warspite/Warspight*** 70/60, 1666 Henry Johnson, Blackwall. 1702 rebuilding at Blackwall.

***Rupert*** 66/58, 1666 Harwich Dyd (Deane). 1701 rebuilding at Plymouth.

***Resolution*** 70/60, 1667 Harwich Dyd (Deane). 1698 RB at Chatham.

***Monmouth*** 66/58, 1667 Chatham Dyd (Phineas Pett). 1700 RB at Woolwich.

***Edgar*** 70/62, 1668 Francis Bayley/Baly/Boly, Bristol. 1701 rebuilding at Portsmouth (according to some lists actually rebuilt in 1700, but as she is definitely recorded as 'rebuilding' in 10.1701, a 1702 launch date seems more likely).

***Swiftsure*** 70/60, 1673 Harwich Dyd (Deane). 1696 RB at Deptford.

***Harwich*** 70/60, 1674 Harwich Dyd (Deane). 3.9.1691 wrecked at Plymouth.

***Royal Oak/Oake Royall*** 74/64, 1674 Deptford Dyd (Jonas Shish). 1690 RB at Chatham.

***Defiance*** 64/56, 1676 Chatham Dyd (Phineas Pett). 1695 RB at Woolwich.

***Anne*** 70/62, 1678 Chatham Dyd (Phineas Pett). 6.7.1690 run ashore and burnt at Pett Levels, Rye Bay - after being heavily damaged at the battle of Beachy Head - to avoid capture by the French. (Remains still visible at extreme low tide; investigated some years ago by Peter Marsden, Bill Wilkes and the present author. The site is protected under the Historic Wrecks Act, and now belongs to the Nautical Museums Trust which runs the new Hastings Shipwreck Heritage Centre.)

***Captain*** 70/62, 1678 Woolwich Dyd (Thomas Shish). 1705 rebuilding at Portsmouth.

***Hampton Court*** 70/62, 1678 Deptford Dyd (John Shish). 1701 RB at Blackwall.

***Hope*** 70/62, 1678 Robert Castle, Deptford. 6.4.1695 taken by the French.

***Lenox*** 70/62, 1678 Deptford Dyd (John Shish). 1700 rebuilding at Deptford.

***Restoration/Restauracon*** 70/62, 1678 Harwich Dyd (Betts). 1701 rebuilding at Portsmouth.

***Bredah*** 70/62, 1679 Harwich Dyd (Betts). 12.10.1690 blown up by accident at Cork.

***Berwick/Barwick*** 70/62, 1679 Chatham Dyd (Phineas Pett). 1699 rebuilding at Deptford.

***Burford*** 70/62, 1679 Woolwich Dyd (Thomas Shish). 1698 RB at Deptford.

***Eagle*** 70/62, 1679 Portsmouth Dyd (Furzer). 1699 RB at Chatham.

***Elizabeth*** 70/62, 1679 Robert Castle, Deptford. 1704 RB at Portsmouth.

***Essex*** 70/62, 1679 Henry Johnson, Blackwall. 1699 rebuilding at Rotherhithe.

SHIPS BUILT FOR THE ROYAL NAVY

*Expedition* 70/62, 1679 Portsmouth Dyd (Furzer). 1699 RB at Chatham.
*Grafton* 70/62, 1679 Woolwich Dyd (Thomas Shish). 1699 rebuilding at Rotherhithe.
*Kent* 70/62, 1679 Henry Johnson, Blackwall. 1698 RB at Rotherhithe.
*Northumberland* 70/62, 1679 Francis Bayley/Bailey, Bristol. 1701 rebuilding at Chatham.
*Pendennis* 70/62, 1679 Chatham Dyd (Phineas Pett). 26.10.1689 wrecked on the Kentish Knock.
*Sterling Castle/Stirling Castle* 70/62, 1679 Deptford Dyd (John Shish). 1699 RB at Chatham.
*Exeter* 70/62, 1680 Henry Johnson, Blackwall. 1691 burnt by accident at Plymouth; afterwards hulked at Portsmouth; [1717 BU].
*Suffolk* 70/62, 1680 Henry Johnson, Blackwall. 1698 RB at Blackwall.

## Fourth Rates (two-deckers)

*Assurance* 42/36 [originally 32], 1646 Deptford (Peter Pett, senior). Probably reclassed as a Fifth Rate 1691; 07.1.1698 (or 1699); 27.1. sold.
*Constant Warwick* 42/36 [originally 30, then 32], (1645 Peter Pett, Ratcliffe, for the Earl of Warwick as a privateer. 1649 purchased for the Navy); RB 1666 Portsmouth Dyd (Tippetts); 12.7.1691 taken by the French (Tourville's Fleet).
NOTE: Built to the lines of Dunkirk privateers and the first 'frigate' built in England. However this does not mean that she was similar to the 'true' frigates of the later eighteenth century, which were a much later and rather different development.
*Tiger/Tyger* 46/40 [originally 38, later 44/38], (1647 Deptford Dyd (Peter Pett)). RB 1681 Deptford Dyd (John Shish); 1702 rebuilding at Rotherhithe.
*Adventure* 44/38 [originally 40], 1646 Woolwich Dyd (Commissioner Pett). 1692 RB.
*Dragon* 46/40 [originally 38], 1647 Chatham Dyd (Goddard). 1690 RB at Deptford.
*Assistance* 48/42 [originally 40], 1650 Henry Johnson, Deptford. 1699 RB at Deptford.
*Reserve* 48/42 [originally 40], 1650 Peter Pett, Woodbridge. 1701 RB at Deptford.
*Advice* 48/42 [originally 40], 1650 Peter Pett, Woodbridge. Probably RB at some time before 1688; 1698 RB at Woolwich.
*Portsmouth* 46/40 [originally 38], 1650 Portsmouth Dyd (Eastwood). 9.8.1689 taken by the French and blew up.
*Foresight* 48/42 [originally 40], 1650 Deptford Dyd (Jonas Shish). 4.7.1698 wrecked in the West Indies.
*Centurion* 48/42 [originally 40], 1650 Peter Pett, Ratcliffe. 25.12.1689 wrecked on Mount Batten.
*Bonaventure* 48/42, (ex *President* 42), 1650 Deptford Dyd (Peter Pett). 1660 renamed; 1663 RB at Chatham Dyd (Phineas Pett); RB 1683 Portsmouth Dyd (Betts); 1699 RB at Woolwich.
*Diamond* 48/42 [originally 40], 1651/2 Deptford Dyd (Peter Pett). 20.9.1693 taken by the French; [renamed *Le Diamant* and in French service until 1697].
*Ruby* 48/42 [originally 40], 1651 Deptford Dyd (Peter Pett). 1705 rebuilding.
*Portland* 50/44 [originally 44], 1653 John Taylor, Wapping. 12.4.1692 burnt to avoid capture off Malaga.
*Bristol* 48/42 [originally 44], 1653 Portsmouth Dyd (Tippetts). 1693 RB at Deptford.
*Swallow* 48/42 [originally 40], (ex *Gainsborough* renamed 1660), 1653 Thomas Taylor, 'Pitch House' (Thames). 8.2.1692 wrecked at Kinsale.
*Antelope/Anthelope* 48/42, (ex *Preston* renamed 1660), 1653 Carey, Woodbridge. 11.6.1693 sold.
*Hampshire* 46/40 [originally 38], 1653 Deptford Dyd (Phineas Pett). 26.8.1697 sunk by the French in Hudson's Bay.
*Newcastle* 54/46 [originally 44 then 46], 1653 Phineas Pett, Ratcliffe. 1692 RB at Rotherhithe.
*Happy Return* 54/46 [originally 44], (ex *Winsby* renamed 1660), 1654 Edgar, Yarmouth. 4.11.1691 taken by the French.
*Jersey* 48/42 [originally 40], 1654 Starling, Maldon. 18.12.1691 taken by the French; [renamed *Le Jersey* and in French service until 1716].
*Mary Rose* 48/42 [originally 40], (ex *Maidstone* renamed 1660), 1654 Monday, Woodbridge. 12.7.1691 taken by the French; [renamed *Le Vaillant* and in French service until 1716].
*Dover* 48/42 [originally 40], 1654 William Castle/Castell, Shoreham. 1695 RB.
*Crown* 48/42 [originally 40], (ex *Taunton* renamed 1660), 1654 William Castle, Rotherhithe. (1689? RB noted by Frank Fox, but no mention in Dimensions Book); 1704 RB.
*Greenwich* 54/46, 1666 Woolwich Dyd (Christopher Pett). 1699 RB.
*Falcon* 42/36 [originally 36/30], (Fifth Rate, 1668 reclassed as a Fourth Rate), 1666 Woolwich Dyd (Christopher Pett). Probably reclassed as Fifth Rate in 1691; 1.5.1694 taken by the French in the West Indies. (renamed *Le Falcon/Le Faucon/Le Faucon Anglais* and in French service until 1708).
*Sweepstakes* 42/36, (Fifth Rate [36/30], 1668 reclassed as a Fourth Rate), 1666 Edgar, Yarmouth. Probably reclassed as a Fifth Rate 1691; 24.5.1698 sold.
*Saint David* 54/46, 1667 Daniel Furzer, Conpill. ('Forest', ie of Dean). 11.11.1698 sank at her moorings, raised and hulked; [at Kinsale, later to Woolwich; 1713 sold].
*Nonsuch* 42/36 [originally 36/30], (Fifth Rate, 1669 reclassed as a Fourth Rate), 1668 Portsmouth Dyd (Deane). Probably reclassed as a Fifth Rate 1691; 1695 taken by the *French Le Français*; [renamed *Le Sans Pareil* and out of French service 1697].
*Phoenix* 42/36, (Fifth Rate. 1672 reclassed as a Fourth Rate), 1671 Portsmouth Dyd (Deane). Probably reclassed as a Fifth Rate in 1691; 12.4.1692 burnt near Malaga to avoid capture.
*Oxford* 48/42, 1674 Francis Bayley, Bristol. 1702 RB (or Great Repair; but whichever it was it altered her dimensions).
*Woolwich* 54/46, 1675 Woolwich Dyd (Phineas Pett). 1702 RB (or Great Repair; but whichever it was it altered her dimensions).
*Kingfisher* 46/40, 1675 Phineas Pett, Woodbridge. 1699 RB.
*Charles Galley* 32/28 [originally 32], ('galley frigate'), 1676 Woolwich Dyd (Phineas Pett). 1693 RB.
NOTE: The prototype of the 'galley frigates' intended for use against the Barbary Corsairs in the Western Mediterranean and therefore built with very fine lines and designed to be rowed as well as sailed. The lower deck was mainly occupied by the large sweeps required for rowing so the number of guns carried was less than for an ordinary warship of the same size.
PLANS: Lines.
*James Galley* 30, ('galley frigate'). 1676 Anthony Deane Jnr, Blackwall. 25.11.1694 wrecked on Longsand Head.
NOTE: A slightly smaller version of the *Charles Galley*.
*Mary Galley* 34, ('galley frigate'). 1687 Cuckold's Point, John Deane, Rotherhithe. 1708 RB (or Great Repair?).
NOTE: 'To be built to the same form as the *James Galley*'.
PLANS: (Contract PRO Adm 106/3070).
*Deptford* 50/44 or 48/42, 1687 Woolwich Dyd (Thomas Shish). 1700 RB.
*Saint Albans* 50/48/44, 1687 Deptford Dyd (John Shish or Harding). 9.12.1693 lost at Kinsale.
*Sedgemore* 50/44, 1687 Chatham Dyd (Lee); 2.1.1689 lost at St Margaret's Bay near Dover.

## Fifth Rates

*Pearl* 30/28 [originally 22], 1651 Peter Pett, Ratcliffe. (1688-1689 fireship); 1697 foundation at Harwich.
*Garland/Guardland* 30/28 [originally 22], (ex *Grantham* renamed 1660), 1654 Furzer, Southampton. 1688-1689 fireship; 1698 sold.
*Guernsey* 30/28 [originally 22], (ex *Basing* renamed 1660). 1654 Shish, Walderswick. (1688-1689 fireship); 26.10.1693 sold.
*Dartmouth* 32/30/28 [originally 22], 1655 Portsmouth Dyd (Tippetts). (1688-1689 fireship); 9.10.1690 wrecked in the Sound of Mull; remains recently excavated by divers from Bristol under the direction of Colin Martin.
*Richmond* 28/26 [originally 22 or 26], (ex *Wakefield* renamed 1660), 1656 Portsmouth Dyd (Tippetts). (1688-1689 fireship); 1698 sold.
*Rose* 28/26, 1674 Edgar, Yarmouth. 1689 fireship [10]. 20.9.1698 sold.
*Saphire* 32/28, 1675 Harwich Dyd (Deane). 11.9.1696 burnt at Bay Bulls, Newfoundland to prevent capture by the French; [remains recently

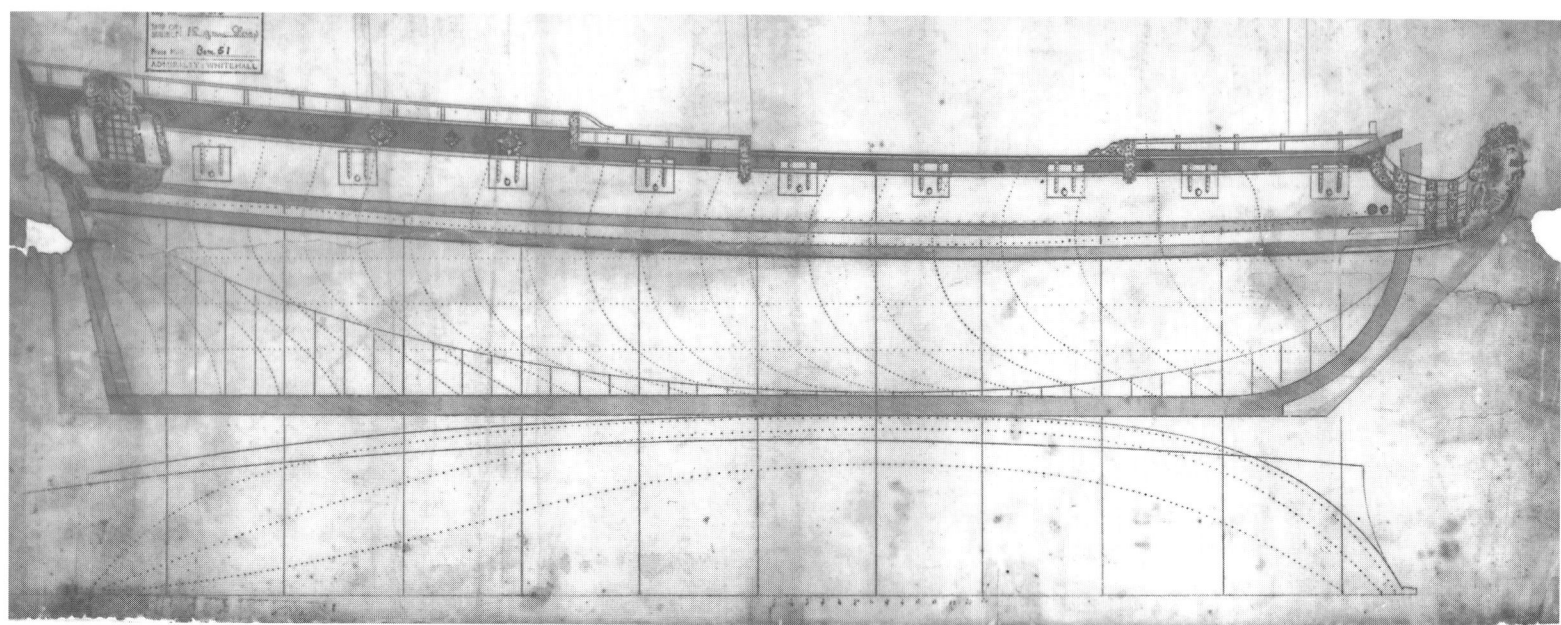

A plan showing an unidentified 18 gun ship of the 1670s (Admiralty plan number 6815)

located and excavations conducted by local divers and Parks Canada].

## Sixth Rates

*Drake* 16/14 [originally 12 or 14], 1652/3 Deptford Dyd (Phineas Pett). 1690/1691 'cast and condemned' at Jamaica.

*Fanfan* 4, 1665/6 Harwich Dyd (Deane) as yacht. 1692 pitch boat at Portsmouth; [no further record?]

*Saudadoes* 16/14/10, [1669/1670 as a yacht (8/6) at Portsmouth Dyd (Deane)]. RB 1673 Deptford Dyd (Jonas Shish). 1696 ( _____ ) taken by the French.

*Greyhound* 16/14, 1672 Portsmouth Dyd (Deane). 24.5.1698 sold.

*Larke* 18/16, 1675 Anthony Deane, Blackwall. 3.5.1698 sold.

## Ketches

*Deptford* 10, 1665 Deptford Dyd (Jonas Shish). 26.8.1689 wrecked off Virginia.

## Fireships

(see also under Fifth Rates for temporary conversions)

*Eagle* 12, (ex *Selby* renamed 1660), 1654 Taylor, Wapping. As a Fifth Rate 22, later 32/28; 1674 fireship; 1694 foundation at Sheerness.

*Mermaid* _____ 1655 Graves, Limehouse. As a Fifth Rate 22; later 32/28; 1688 fireship; 1689 BU.

## Bombs

(see under Part I Chapter 1)

## Yachts, Hulks and Harbour Craft

(see Part III)

# Chapter 2: Prizes and Purchases

A Van der Velde drawing of the fireship *Sophia*, captured from the Duke of Argyll in 1684 (NMM negative number V.611)

## Fourth Rates

*Tiger Prize* 48/42, (Algerine ['Turkish'] ____ ). Taken 1678; 1696 foundation at Sheerness.
*Mordaunt* 46/40, (Privateer *Mordaunt* 1681 William Castle. Deptford for Lord Mordaunt). Purchased from Lord Mordaunt 1683; 21.11.1693 wrecked off Cuba.

## Fifth Rates

*Swan/Swann* 32/28, (Dutch [?] ____ ). Purchased from Captain Anthony Young 1673; (1688-1688 fireship); 7.6.1692 wrecked in the earthquake at Port Royal, Jamaica (as a fireship?).
*Saint Paul* 32/28, (Algerine ____ , Dutch built). Taken 1678; 1688 fireship [10]; 3.5.1688 sold.
*Helderenberg* 30/28,( Duke of Monmouth, Dutch built ____ ). Given by the Dutch to the Duke of Monmouth and taken during his rebellion 1686; 10.1688 made a hospital ship; 17.12.1688 run down by *Bonaventure* off the Isle of Wight.

## Sixth Rates

*Sally Rose/Rose* 16/14, (Salee ____ ). Taken 1684; (1688/1689 fireship); 1696 sold.
*Dumbarton/Dunbarton* 20/18, (Earl of Argyle's ship). Taken 1685 during the Argyle rebellion; 1691, 'cast and condemned (6.6.1691) in Virginia'.

## Ketches

*Quaker* 10, ( ____ ). Purchased from Mr Moore 1671; 20.5.1698 sold.
*Kingfisher* 4, ( ____ ). Purchased 1684; 25.3.1690 (New Style) taken by the French.

## Fireships

*Castle* 8, ( ____ ). Purchased 1671/2; Pepys suggests she was sold 12.1683 but by Admiralty Order of 26.8.1692 she was to be sunk as a breakwater at Harwich.
*Ann And Christopher* 8, ( ____ ). Purchased 1671/2; ordered to be sold 1686; 1692/3 sold as a breakwater at Portsmouth.
*Young Sprag/Spragge* 6, ( ____ ). Purchased 1673 from Admiral Sir Edward Spragge (Sprag) as Sixth Rate (10); 1677 fireship; 1693 sunk as a foundation at Portsmouth.
*Sophia* ____ (Duke of Argyle's ____ ). Captured 1685 by *Kingfisher* from the Duke of Argyle during the latter's rebellion; taken into the navy as a Sixth Rate (10); 1688 fireship; 1690 hoy; 1713 BU.
*Half Moon* ____ . (Salee ____ 32 guns?) Captured 1685 and became a Sixth Rate [18/16]; 1688 fireship; 24.5.1692 expended at La Hogue.

## Unlisted Vessels

ie. not mentioned in original sources consulted but only by secondary sources (eg. Colledge, Pitcairn Jones)

*Little London*, smack built at Chatham 1672 and sold 1697.
*Sampson*, fireship (8/12). Purchased in 1679 and expended 14.3.1689.

# PART I
# SHIPS BUILT FOR THE ROYAL NAVY

## Chapter 1: Before the Establishments 1689-1705

This chapter deals with a period before design had stabilised into what became known as the 'Establishments' (a system of fixed dimensions and, later, designs to which nearly all larger vessels were built - see the next two chapters). However, there was already a fair degree of similarity amongst the vessels built, as there had been in the previous decades of expansion under the later Stuart kings. Expansion continued, though the impact of the wars with the French (1689-1698 and 1702-1713) produced a rather different emphasis. Initially, the French seemed to be launching a successful challenge to the combined battlefleets of England and the United Provinces with their victory at Beachy Head (1690). This challenge was the more serious because the French appeared to be outbuilding the two older-established battlefleet powers in the largest two-decker and three-decker ships of the line. So far the situation called for the same building response as the three wars with Holland had done, to build up the battlefleet with ships of the line. However, in 1692 the French received a severe check at Barfleur and La Hogue, and abandoned the attempt to build up a powerful battlefleet just when it seemed they were succeeding[1] (probably because of financial worries). Instead, they posed a new challenge to the Allies by a vigorously prosecuted raiding war against the commerce which was vital to Holland and not much less so to Britain. The battlefleet could not be abandoned, and line of battle ships had still to be built; but the emphasis now shifted to 'convoys and cruisers'. The contemporary escort ships were the small two-deckers and other lesser vessels of the Fourth, Fifth and Sixth Rates. The increase in numbers of these craft is obvious in the following pages. Something else to notice in this chapter (and both the following one and Chapters 1 and 2 of Part II) is the number of smaller vessels that changed hands, often more than once, between the combatants. This was clear evidence of a ding-dong struggle in which there was, and could be, no easy victory for either side.

The individual ships are grouped by type (generally indicated by rate and number of guns) and by date of launch within those groups. Information is given in the following order for each vessel: **(1)** Name/Variants of name. The abbreviation [RB] = 'Rebuild'. **(2)** Year of launch-year of leaving first line service (ie. year of loss, of disposal or of hulking). **(3)** Where built and by whom (Dyd = [Royal] Dockyard); the name given after a dockyard is that of the Surveyor (at this time usually the designer as well as the man who supervised the building of the vessel). Private builders have their names given first, then the location of the shipyard. **(4)** Detailed designation - this is usually the number of guns (at this period usually given as two numbers: the first wartime home waters complement; the other being wartime overseas or peacetime complement). **(5)** Dimensions (normally 'as built') given in the following sequence: Length on the range of deck (the lower deck except for single-decked vessels), length of keel for tonnage (not, it should be noted, the actual length of the keel itself) x breadth x depth in hold. The next figure is the tonnage, the result of a formula involving these dimensions and a divisor of 94. **(6)** The complement of men, like the number of guns, is given as the (higher) wartime home waters figure followed by the lower figure, which was the establishment for overseas wartime service and for peacetime. It should be noted that this was the official allocation, rarely if ever achieved in the reality of the manning problems of a growing Navy in the age of the press gang. **(7)** Detailed gun establishment given in the following form, deck by deck: for each deck the number of guns is given followed by their weight of shot in pounds, the usual designation of guns by now - in a few cases the older English names [eg. 'saker', 'minion'] - are used. The decks are given in ascending order (the heaviest guns normally being on the lowest decks): G(un) D(eck), the lowest deck to carry guns, sometimes the 'lower deck'; M(iddle or Main) D(eck), in three-deckers only; U(pper) D(eck), the uppermost continuous deck, Q(uarter) D(eck) and Fc (= forecastle). In the largest ships the R(ound) H(ouse) or poop would often carry guns as well. Besides the main carriage guns there would normally be a number of 'swivels', small weapons mounted on swivels upon the gunwales, bulwarks or in the tops. The most frequent calibre by this time for a swivel was that which would fire a half-pound shot. Both this and the following chapters contain particular detail on armament as I have some information on the actual guns placed aboard particular vessels at specific times from a certain amount of research on the Ordnance returns which had been preserved at the Naval Ordnance Museum at Priddy's Hard when I was researching there in the mid 1980s.[2] These are noted under the year of the return, and are a record of what was actually done, as opposed to the 'Establishment' figures which are also given here, which represent merely what was intended, or hoped, to be put aboard. The proposed or 'Establishment' figures are given first and the figures from armament returns are given afterwards in brackets, prefixed by the year of the return in question. Some older types of gun were still in use, as was a mixture of the old gun names and the newer designation by weight of shot which had been taken over from the Dutch. Whilst the weight of shot is the system used here in almost all cases, there were some older types of gun still in use for which it is more appropriate to give the old names: 'saker' ($5\frac{1}{4}$ pound shot) and 'minion' (4 pound shot). **(8)** The next entry is a brief summary of the front-line service of the ship in question, and whether there was a 'Rebuild' [**RB**]. At this period the operation was changing in its nature. Up to just after the beginning of the new century (about 1703) was an actual rebuilding of the

---

[1] On this and much else see the very wide ranging work of the Swedish economic historian Jan Glete, *Nations and Navies* (Stockholm University 1993), a copy of which the author very kindly sent to me just in time to read it before writing the introduction. Jan Glete covers all the navies of Europe and America from 1500 to 1860, and has made use of an earlier version of the present work as well as an impressive variety of sources in many languages.

[2] I also had, thanks to the generosity of the bookseller Antony Simmonds, the opportunity of transcribing the contents of a number of armament returns which he had purchased at an auction after they had been taken from this collection to be put up for sale in aid of naval charities. Presumably, the grounds for this sale were that there was so much material in the collection that a few being removed would make little difference. This from a collection whose chief importance was its range and comprehensiveness! The very substantial and important collection of returns of late seventeenth- and eighteenth-century dates which remain are now (1992), apparently, boxed in storage as both the Naval armament depot and its attached museum have been closed. It is to be hoped that it will eventually be found a good and permanent home where it can again be made available to researchers and where it will no longer be exposed to the dangers of plundering for the sake of fund raising activities. [The latest news in mid 1993 on this collection is that it is being catalogued by Hampshire Record Office, so the prospects now seem better.]

vessel, which would be stripped down in a dry dock and possibly completely reassembled with unsound timbers replaced, and alterations made to the shape and the scantlings as required. The ship would usually be substantially the same vessel, though often altered to a greater or lesser extent. With the beginning of war against France in 1702, this altered. Dry docks were needed for wartime repairs and the usual procedure became to break up the vessel ('take to pieces'); some of the old timbers might be reused in the new vessel (on at least one occasion a small hoy was quite adequate to carry all the reusable timbers of a large two-decker from the yard where she was broken up to the one in which she was to be 'rebuilt'). Increasingly, the 'Rebuild' was becoming an administrative fiction. The 'Rebuilt' vessel might contain some material from the old ship but was, in effect, a new warship, usually built to a new design.[3] Next come (when available) the dates of the Order [**Ord**], of laying down [**K** for 'Keel laid'], and launching [**L**]. In a few cases the ship might have been built in a dry dock and 'floated out' rather than launched, but the distinction is not worth drawing here. In the days of wooden sailing warships the actual fitting out would normally be done mostly by the crew with a greater or lesser degree of assistance from the Dockyards. Guns and masts would be hoisted aboard, rigging and gun tackles would be set up, carpenters would make and fit partitions and so on. In time of war or emergency this would take less than a month (an average three weeks or so). Ships completed in peacetime would often be left as empty shells for some years (and in the nineteenth century might have a roof built over them and used for harbour service throughout their lives). With metal hulled steamships fitting out became a much more complex, technical and long-drawn out process, which might commence well before the vessel was launched. So from the middle of the nineteenth century the launch date has become progressively less significant. In our period, and for our purposes, however, it is the most important date of all. It signifies that the ship was almost ready to put into service after a fitting out process of limited and fairly standard length. These construction dates are followed by any details of changes in name, designation or function. Each stage in the ship's career is separated from the next by a semicolon (;), as is the final fate which is given at the end of the entry. This fate is as a front-line, serving vessel in the Royal Navy [BU = broken up, 'taken to pieces'], whilst subsequent second-line harbour service as a hulk of some kind, or service in another navy after capture, are given within square brackets [....] at the end of the entry. **(9)** The final entry is a brief description of any plans that might exist in the National Maritime Museum's Admiralty Collection (or in a number of other plans collections).

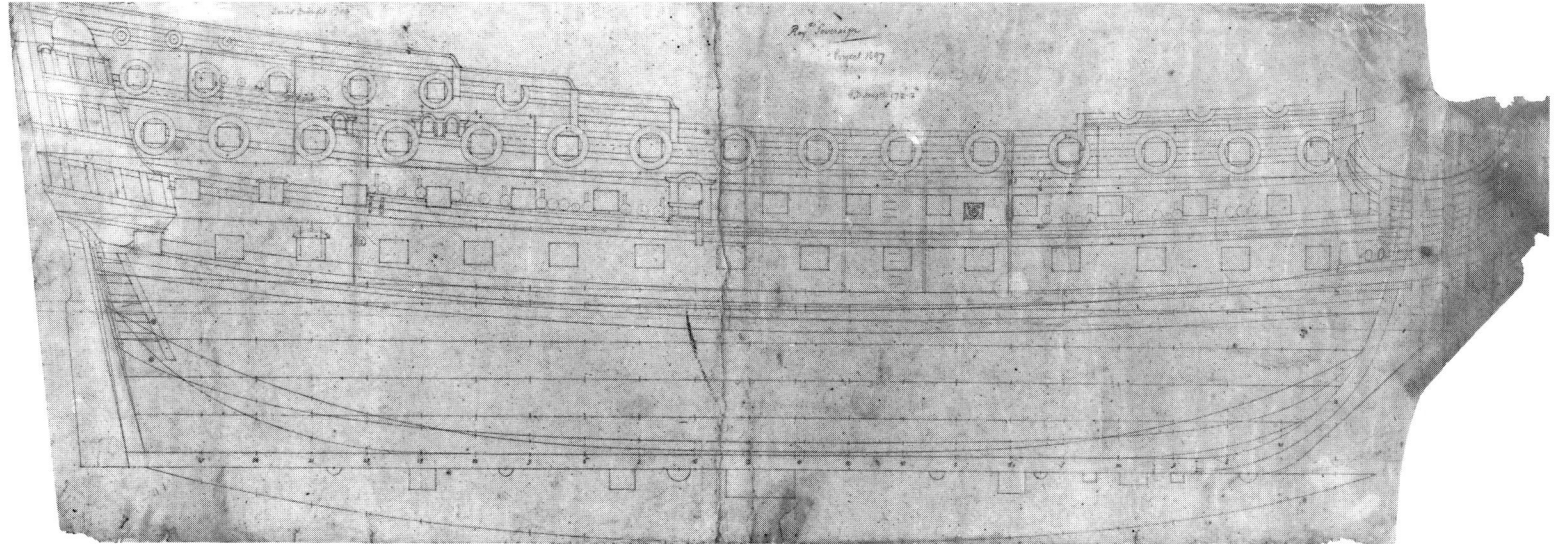

Admiralty sheer plan (the side-on view part of a lines plan) of the 100 gun *Royal Sovereign* of 1701 (plan X 123)

# First Rates, 100 Guns (three-deckers)

***Royal William*** 100/90, 1692-1714. Chatham Dyd (Surveyor: Lee).
   Dimensions & tons: 1672˝3", 132'6" x 47'2" x 18'. 1568
   Men: 780/580. Guns: GD 28 x 32, MD 28 x 18, UD 28 x 9, QD 12 x 6, Fc 4 x 6.
   RB of *Prince* (same dimensions) L _____ 1692; by 1714 BU.

***Queen*** 100/90, 1693-1709. Woolwich Dyd (Surveyor: Lawrence).
   Dimensions & tons: 170'6", 137'8" x 47' 7" x 18'. 1658
   Men: 780/580. Guns: GD 26 x 32, MD 28 x 18, UD 28 x 9, QD 12 x 6, Fc 4 x 6, RH 2 x 6.
   RB of *Royal Charles* L _____ 1693; 1709 BU.

***Victory*** 100/90, 1695-1714 *Royal George*-1715 *Victory*-1716. Chatham Dyd (Surveyor: Lee).
   Dimensions & tons: 163'1", 136" x 45'4" x 18'6". 1486.
   Men: 780/580. Guns: GD 26 x 32, MD 28 x 18, 28 x 9, QD 12 x 6, Fc 4 x 6, RH 2 x 6.
   RB L _____ 1695; 1714 renamed *Royal George*; 1715 renamed *Victory*; 1716 cut down to the middle deck at Cork; 1721 BU.

PLANS: Lines (although labelled, probably later, as *Victory* may well instead be *Royal George* of 1715).

***Royal Sovereign*** 100/90, 1701-1766. Woolwich Dyd (Surveyor: Harding).
   Dimensions & tons: 174'6", 141'7" x 50'3½" x 20'1". 1883.
   Men: 780/580. Guns: GD 28 x 32, MD 28 x 18, UD 28 x 9, QD 12 x 6, Fc 4 x 6, RH 2 x 6.
   Theoretical RB of previous *Royal Sovereign* (burnt 1696); L 25.7.1701; 1723/9 Great Repair at Chatham; 1756 reclassed as 90 gun Second Rate; 1759 guard ship in the Downs; 1766 BU.

NOTE: As a ship that was 'genrally [sic] approved of' her dimensions were taken for the 100 gun ships of the 1719 Establishment. Her Great Repair of the 1720s is counted as producing a new ship by Lavery [1983]. In 1763: 'The [Dockyard] Officers remark upon this ship that her frame appears sounder than could have been expected for a ship of her age and that she has, a great character for sailing well, but as her scantling is too small for the weight of metal lately established for ships of 100 guns they proposed the reducing of her to an 80 gun ship'.

PLANS: Sheer [identification with this *Royal Sovereign* not entirely certain].

---

**3** For further information on this subject see Brian Lavery's article in *The Mariner's Mirror* (Vol 66) and his *Ship of the Line*.

***Royal Anne*** 100, 1704-1727. Woolwich Dyd (Surveyor: Lee).
   Dimensions & tons: 170', 140'6" x 48' x 19'4". 1721 82/94.
   Men: 780/580. Guns: GD 26 x 32, MD 28 x 18, UD 28 x 9, QD 18 x 6.
   RB of *Saint Andrew*; L 05.1704; 1727 BU.
***London*** 100, 1706-1747. Chatham Dyd (Surveyor: Rosewell).
   Dimensions & tons: 168', 137'6" x 48' x 19'2". 1685.
   Men: 780/580. Guns: ____ .
   L ____ 1706; 1718-1721 Great Repair (almost a RB) at Chatham; 1747 BU.
PLANS: Lines.

## Second Rates, 90 Guns (three-deckers)

***Association*** 90/82, 1696-1707. Portsmouth Dyd (Surveyor: Bagwell).
   Dimensions & tons: 165', 133'6" x 45'4" x 18'3". 1459.
   Men: 680/500. Guns: GD 26 x 32, MD 26 x 18, UD 26 x 9, QD 12 x 6, Fc 4 x 6, RH 2 x 6.
   Ord ____ 1695; L ____ 1696; 22.10.1707 wrecked off the Scilly Islands (Sir Cloudesley Shovell's flagship). [The wreck was located and became the scene of operations by rival teams of divers in the 1960s. Excavations have continued intermittently since.]
***Barfleur*** 90/82, 1697-1713. Deptford Dyd (Surveyor: Harding).
   Dimensions & tons: 162'10 ½", 129' 4 ½" x 46'4" x 18'2 ½". 1476.
   Men: 680/500. Guns: GD 26 x 32, MD 26 x 18, UD 26 x 9, QD 12 x 6, Fc 4 x 6, RH 2 x 6.
   Ord ____ 1695; L 10. 8. 1697; 1713 BU.
***Namure*** 90/82, 1697-1716. Woolwich Dyd (Surveyor: Lawrence).
   Dimensions & tons: 160'9", 130' x 45'8" x 18'6". 1442 .
   Men: 680/540. Guns: GD 26 x 32, MD 26 x 18, UD 26 x 9, QD 12 x 6, Fc 4 x 6, RH 2 x 6.
   Ord ____ 1695; L ____ 1697; 1716 cut down to the middle deck; 1723 BU.
***Triumph/Tryumph*** 90/82, 1698-1714 *Prince* 1738. Chatham Dyd (Surveyor: Lee).
   Dimensions & tons: 160'1", 131" x 46'1½" x 18'3". 1482.
   Men: 680/500. Guns: GD 26 x 32, MD 26 x 18, UD 26 x 9, QD 12 x 6, Fc 4 x 6, RH 2 x 6.
   Ord ____ 1695; L early 1698; 1714 renamed *Prince*; 1738 BU.
PLANS: Lines.
***Prince George*** 90, 1701-1719. Chatham Dyd (Surveyor: Shortiss).
   Dimensions & tons: 162'10", 131'6" x 45'1" x 18'7". 1421.
   Men: 680/500. Guns: GD 26 x 32, MD 26 x 18, UD 26 x 9, QD 12 x 6, Fc 4 x 6, RH 2 x 6.
   RB of *Duke*; L ____ 1701; 1719 BU.

PLANS: Sheer [possibly of the 1723 rebuild, but more likely to be this ship].
***Saint George*** 90, 1701-1726. Portsmouth Dyd (Surveyor: Waffe).
   Dimensions & tons: 160', 132'6" x 44'6" x 18'6". 1395 62/94.
   Men: 680/500. Guns: GD 26 x 32, MD 26 x 18, UD 26 x 9, QD 12 x 6, Fc 4 x 6, RH 2 x 6.
   RB L ____ 1701; 1726 BU.
***Royal Katherine*** 90, 1703-1706 *Ramillies* 1741. Portsmouth Dyd (Surveyor: Waffe).
   Dimensions & tons: 160', 132'6" x 44'6" x 18'6". 1395 62/94.
   Men: 680/500. Guns: GD 26 x 32, MD 26 x 18, UD 26 x 9, QD 12 x 6, Fc 4 x 6, RH 2 x 6 (1703: GD 26 x 24, MD 28 x 12, UD & QD & Fc: 8 x 6 + 28 sakers).
   RB L 23.2.1703; 1706 renamed *Ramillies*; 1741 BU.
***Albemarle*** 90, 1704-1709 *Union* 1718. Chatham Dyd (Surveyor: Shortiss).
   Dimensions & tons: 163'6", 135' x 45' x 18'4". 1398.
   Men: 680/500 [?]. Guns: GD 26 x 32, MD 26 x 18, UD 26 x 9, UD 12 x 6, Fc 4 x 6, RH 2 x 6.
   RB L early 1704; 1709 renamed *Union*; 1718 BU.
***Marlborough*** 90, 1706-1725. Johnson, Blackwall.
   Dimensions & tons: 162'8", 132'6" x 47'4" x 18'6". 1579.
   Men: 680/500. Guns: GD 26 x 32, MD 26 x 18, UD 26 x 9, QD 12 x 6, Fc 4 x 6, RH 2 x 6.
   RB of *Saint Michael*; L ____ 1706; 1725 BU.

## Third Rates, 80 Guns

(Earlier vessels are two deckers, those later ones marked with an asterisk [*] are three deckers.)

***Cornwall*** 80/72, 1692-1704. Winter, Southampton.
   Dimensions & tons: 156'4", 129'6" x 41'6" x 17'3". 1186.
   Men: 500/355. Guns: GD 26 x 24, UD 28 x 12, QD 16 x 6, Fc 6 x 6, RH 4 x 3.
   Ord 12.3.1691; L 23.4.1692; 1704 BU.
***Devonshire*** 80/72, 1692-1701. Wyatt, Bursledon.
   Dimensions & tons: 154', 126'11" x 41'5" x 17'4". 1158.
   Men: 500/355. Guns: GD 26 x 24, UD 28 x 12, QD 16 x 6, Fc 6 x 6, RH 4 x 3.
   Ord 12.3.1691; L 5.4.1692; 1701 BU.
***Boyne*** 80/72, 1692-1708. Deptford Dyd (Surveyor: Harding).
   Dimensions & tons: 157', 128'2" x 41'3" x 17'3". 1160.
   Men: 500/355. Guns: GD 26 x 24, D 28 x 12, QD 16 x 6, Fc 6 x 6, RH 4 x 3.
   Ord ____ 1691; L 21.5.1692; 1708 BU.
***Russell*** 80/72, 1692-1707. Portsmouth Dyd (Surveyor: Stigant).

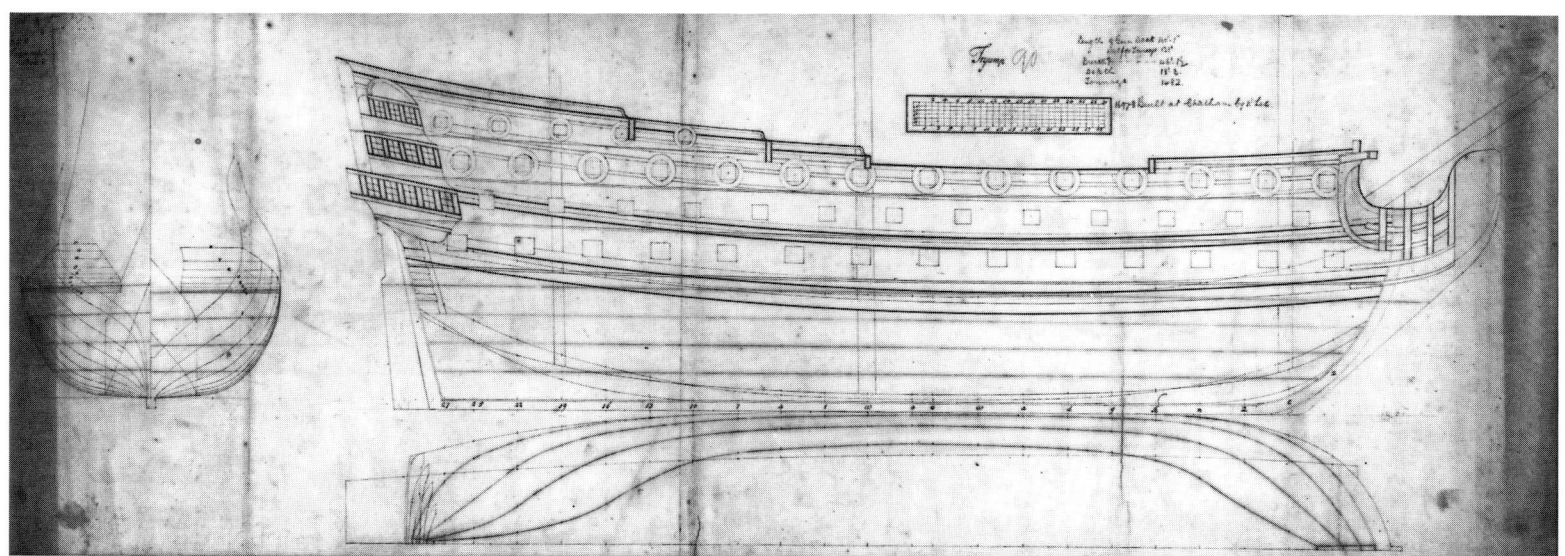

Lines plan of the *Triumph* of 1698, 90 guns (Admiralty plan number 328)

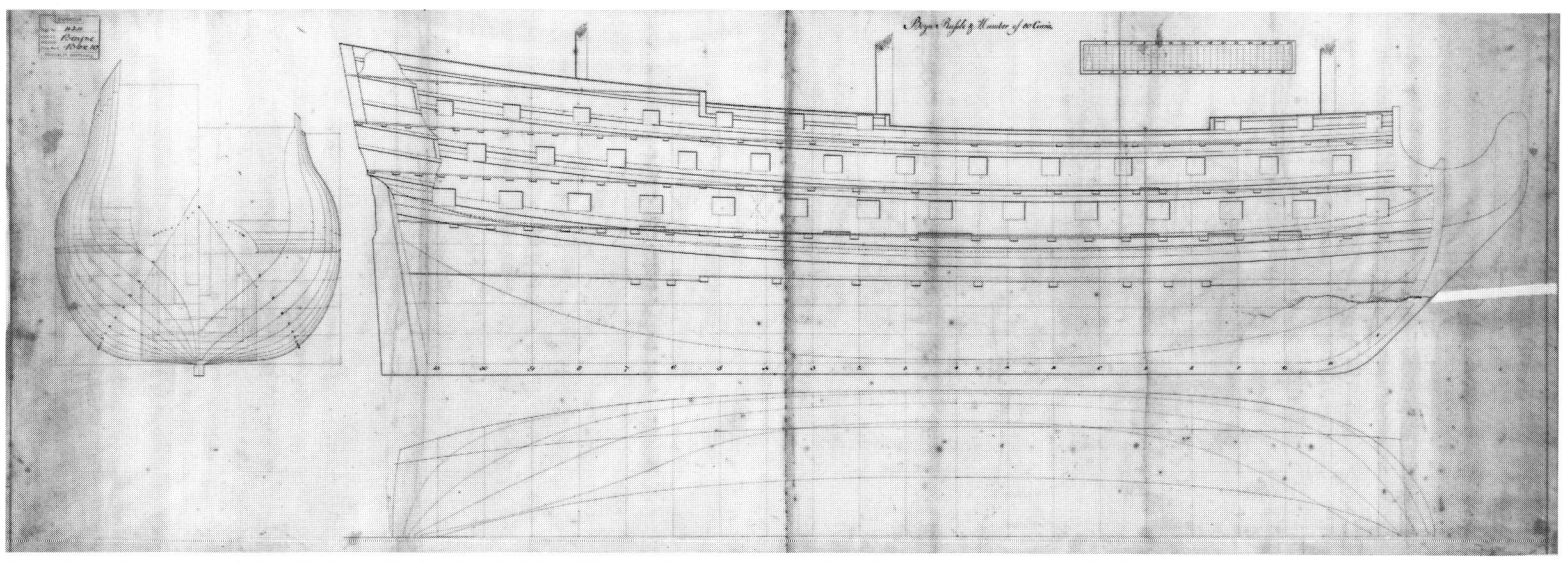

Lines plan of the early two-decker 80s *Boyne, Humber, Russel* of the 1690s, a weak design soon transformed into the stronger though far from satisfactory three-decker 80s whose continuation in Royal Naval use for over half a century is one of the more puzzling aspects of early eighteenth-century warship history

Dimensions & tons: 155'6", 128'6" x 41'6" x 17'2". 1177.
Men: 500/355. Guns: GD 26 x 24, UD 28 x 12, QD 16 x 6, Fc 6 x 6, RH 4 x 3.
Ord ____ 1691; L 3.6.1692; 1707 BU.

*Norfolk* 80/72, 1693-1718. Winter, Southampton.
Dimensions & tons: 156'6", 129'3¼" x 41'6" x 17'4". 1184.
Men: 500/355. Guns: GD 26 x 24, UD 28 x 12, QD 16 x 6, Fc 6 x 6, RH 4 x 3.
Ord 21.12.1691; L 28.3.1693; 1718 BU.

*Humber* 80/72, 1693-1707. Frame, Hull.
Dimensions & tons: 156'3", 129'9" x 42'1½" x 17'4". 1223.
Men: 500/355. Guns: GD 26 x 24, UD 28 x 12, QD 16 x 6, RH 4 x 3 (1702: GD 24 x 32, D 28 x 9, D & Fc 20 x 6 + 2 x 3).
Ord 12.3.1691; L 30.3.1693; 1707 BU.

*Sussex* 80/72, 1693-1694. Chatham Dyd (Surveyor: Lee).
Dimensions & tons: 157'2", 132'5" x 41'4" x 17'1½". 1203 26/94.
Men: 500/355. Guns: GD 26 x 24, UD 28 x 12, QD 16 x 6, Fc 6 x 6, RH 4 x 3.
Ord ____ 1691; L 11.4.1693; 17.2.1694 foundered off Gibraltar.

*Torbay* 80/72, 1693-1716. Deptford Dyd (Surveyor: Harding).
Dimensions & tons: 156', 128'9" x 41'11" x 17'4". 1202.
Men: 500/355. Guns: GD 26 x 24, UD 28 x 12, QD 16 x 6, Fc 6 x 6, RH 4 x 3 (1708: GD 26 x 32, UD 28 'x 18, QD & Fc 26 x 6).
Ord ____ 1691; L 16.12.1693; 1716 BU.

*Lancaster* 80/72, 1694-1710. Wyatt, Bursledon.
Dimensions & tons: 156'1", 128'9" x 41'10" x 18'6". 1198.
Men: 500/355. Guns: GD 26 x 24, UD 28 x 12, QD 16 x 6, Fc 6 x 6, RH 4 x 3 (1703 & 1706: GD 24 x 32, D 30 x 18, QD & Fc 26 x 6).
Ord 3.3.1693; L 3.4.1694; 1710 BU.

*Dorsetshire* 80/72, 1694-1706. Winter, Southampton.
Dimensions & tons: 153'4½", 125'2" x 42' x 18'. 1176.
Men: 500/355. Guns: GD 26 x 24, UD 28 x 12, QD 16 x 6, Fc 6 x 6, RH 4 x 3 (1706: 24 x 32, 30 x 18, 26 x 6).
Ord 1.6.1693; L 8.12.1694; 1706 BU.

*Shrewsbury*★ 80/72, 1695-1711. Portsmouth Dyd (Surveyor: Stigant).
Dimensions & tons: 158', 130'5" x 42'6" x 17'5". 1257.
Men: 520/360. Guns: GD 26 x 32, MD 26 x 18, UD 22 x 6, QD 6 x 6.
Ord ____ 1691; L 6.2.1695; 1711 BU.

*Cambridge* 80/72, 1695-1713. Deptford Dyd (Surveyor: Harding).
Dimensions & tons: 156', 127'9" x 41'11½" x 17'. 1194.
Men: 500/355. Guns: GD 26 x 24, UD 28 x 12, QD 16 x 6, Fc 6 x 6, RH 4 x 3.
Ord ____ 1691; L 21.2.1695; 1713 BU.

*Chichester* 80/72, 1695-1705. Chatham Dyd (Surveyor: Lee).
Dimensions & tons: 157'3", 130'5" x 41'9½" x 17'. 1210.
Men: 500/355. Guns: GD 26 x 24, UD 28 x 12, QD 16 x 6, Fc 6 x 6, RH 4 x 3 (1704: 24 x 32, 30 x 12, 22 sakers, 4 x 3).
Ord ____ 1691; L 6 or 8.3.1695; 1705 BU.

*Newark* 80/72, 1695-1713. Frame, Hull.
Dimensions & tons: 157'1½", 130'5½" x 41'10½" x 18'. 1216 75/94.
Men: 500/355. Guns: GD 26 x 24, D 28 x 12, QD 16 x 6, Fc 6 x 6, RH 4 x 3.
Ord 17.3 or 5.1693; L 3.6.1695; 1713 BU.

*Cumberland*★ 80/72, 1695-1707. Bolton & Wyatt, Bursledon (completed by Mrs Wyatt).
Dimensions & tons: 156', 130' x 42'(43'4" after girdling) x 18'. 1219 74/94.
Men: 520/360. Guns: GD 26 x 32, MD 26 x 18, UD 22 x 6, QD 6 x 6.
Ord 4.5.1694; L 12.11.1695; 10.10.1707 taken by the French *Le Lys* of Duguay-Trouin's squadron off Ushant. [1715 sold by the French to Genoa; 1717 sold by Genoa to Spain as *San Carlos*; 1718 recaptured by the RN at Cape Passaro and laid up at Port Mahon until; 1720 sold to Austria for the Neapolitan Navy; 1737? BU at Trieste after foundering in that harbour because of her poor state of repair.]

*Ranelagh*★ 80/72, 1697-1723. Deptford Dyd (Surveyor: Harding).
Dimensions & tons: 158'8", 129'3" x 41'8¾" x 17'4". 1198 89/94.
Men: 520/360. Guns: GD 26 x 32, MD 26 x 18, UD 22 x 6, QD 6 x 6.
Ord ____ ; L 25.6.1697; 1723 BU.

*Somerset*★ 80/72, 1698-1715. Chatham Dyd (Surveyor: Lee).
Dimensions & tons: 158', 130' x 42'9" x 17'. 1262 ⅔.
Men: 520/360. Guns: GD 26 x 32, MD 26 x 18, UD 22 x 6, QD 6 x 6.
Ord ____ ; L 31.5.1698; 1715 hulked at Woolwich; [1740 BU.]

*Devonshire*★ 80, 1704-1707. Woolwich Dyd (Surveyor: Lee).
Dimensions & tons: 156', ____ x 42'1½" x 17'. 1220.
Men: 520/360? Guns: GD 26 x 24, MD 26 x 12, UD 22 x 6, QD 6 x 6 (1707: 20 x 24, 28 x 12, 22 sakers, 4 x 3).
RB L ____ 1704; 10.10.1707 blew up in action with the French.

*Cornwall*★ 80, 1706-1722. Burchett, Rotherhithe.

Dimensions & tons: 156'7½", 127'11" x 42'8¼" x 17'7". 1241 21/94.
Men: 520/360? Guns: GD 26 x 24, MD 26 x 12, UD 22 x 6, QD 6 x 6.
RB L early 1706; 1722 BU.

***Chichester**** 80, 1706-1749. Woolwich Dyd (Surveyor: Stacey).
Dimensions & tons: 155'6", 127'6" x 43'5" x 17'10". 1278⅓.
Men: 520/360? Guns: GD 26 x 24, MD 26 x 12, UD 22 x 6, QD 6 x 6.
RB begun 1705; L ____ 1706; 1749 BU.

## Third Rates, 70 Guns, etc (two-deckers)

***Royal Oak*** 74, 1690-1710. Chatham Dyd (Surveyor: Lee).
Dimensions & tons: 157'6", 127' x 41' 4" x 18'6". 1154.
Men: 500/355. Guns: GD 24 x 32, UD 30 x 18, QD & Fc 16 x 9, RH 4 x 3 (1703 establishment: GD 26 x 24, UD 28 x 12, QD 16 x 6, Fc 6 x 6, RH 4 x 3 [?1].
RB L ____ 1690; 1710 BU.

NOTE: Arguably this vessel was a predecessor/prototype of the two-decker 80s listed above.

***Bredah*** 70, 1692-1730. Woolwich Dyd (Surveyor: Lawrence).
Dimensions & tons: 150', 126' x 40'5" x 16'8". 1094.
Men: 440/320. Guns: GD 24 x 32 + 2 x 18, UD 26 x 12, QD & Fc 14 x 6, RH 4 x 3 (1703 establishment: 24 x 24, 26 x 12, 12 x 6, 4 x 6, 4 x 3)
Ord ____ 1690; L 23.4.1692; 1730 BU.

***Yarmouth*** 70, 1691-1707. Barrett, Harwich.
Dimensions & tons: 150'10", 123'11" x 40'1" x 16'11". 1059.
Men: 440/320. Guns: GD 26 x 24, UD 26 x 9, QD 12 x 6, Fc 4 x 6, RH 4 x 3 (1703 establishment: 24 x 24, 26 x 12, 12 x 6, 4 x 6, 4 x 3. 1704: 26 x 24, 26 x 18, 14 sakers, 4 x 3).
Ord 23.1.1691; L 7.1.1695; 1707 BU.

***Ipswich*** 70, 1694-1727. Barrett, Harwich.
Dimensions & tons: 148', 123'5" x 40' x 16'8". 1049.
Men: 440/320. Guns: GD 26 x 24, UD 26 x 9, QD 12 x 6, Fc 4 x 6, RH 4 x 3 (1703 establishment: 24 x 24, 26 x 12, 4 x 6, 4 x 3. 1696: 22 x 32, 4 x 18, 26 x 12, 14 sakers, 4 minions. 1703: 26 x 9 instead of 12s).
Ord ____ 1690; L 19.4.1694; 1727 BU.

PLANS: 1712: Lines.

***Swiftsure*** 70, 1696-1716 *Revenge* 1716. Snellgrove, Deptford.
Dimensions & tons: 148', 122' x 39' x 14'10", 987.
Men: 440/320. Guns: GD 26 x 24, UD 26 x 9, QD 12 x 6, Fc 4 x 6, RH 4 x 3 (1703 establishment: 24 x 24, 26 x 12, 12 x 6, 4 x 6, 4 x 3. 1707 & 1708- 24 x 18, 26 x 9, 16 x 6).
RB L ____ 1696; 1716 renamed *Revenge*; 1716 BU.

***Bedford*** 70, 1698-1736. Woolwich Dyd (Surveyor: Harding).
Dimensions & tons: 151', 124'1" x 40'4" x 16'9". 1073.
Men: 440/320. Guns: (1703 establishment: GD: 24 x 24, UD 26 x 12, QD 12 x 6 Fc 4 x 6 RH 9 x 3 . 1703: 22 x 24, 4 x 18, 26 x 9, 14 sakers, 4 x 3. 1717: 26 x 24, 24 x 9, 10 sakers).
L 12.9.1699; 1713 to 1718? Fitted for the South Sea Company; 1736 BU.

***Burford*** 70, 1698-1719. Snellgrove, Deptford.
Dimensions & tons: 152'9", 126'2"x 40'8¾" x 16' 4¼", 1113.
Men: 440/320. Guns: GD 26 x 24, UD 26 x 9, QD 12 x 6, Fc 4 x 6, RH 4 x 3. (1703 establishment: 24 x 24, 26 x 12, 12 x 6, 4 x 6, 4 x 3).
RB L 12.9.1698; 14.2.1719 wrecked 'near Pentimiglia on the coast of Italy'.

***Kent*** 70, 1698-1722. Wells, Rotherhithe.
Dimensions & tons: 151'6", 123'7½" x 40' 3" x 16'7", 1064.
Men: 440/320. Guns: GD 26 x 24, UD 26 x 9, QD 12 x 6, Fc 4 x 6, RH 4 x 3. (1703 establishment: 24 x 24, 26 x 12, 12 x 6, 4 x 6, 4 x 3, 1703: 22 x 32, 4 x 18, 26 x 12, 26 x 6).
RB L ____ 1698; 1722 BU.

***Resolution*** 70, 1698-1703. Chatham Dyd (Surveyor: Lee/Furzer).
Dimensions & tons: 148'2", 120'9" x 37'6" x 15'9". 902.
Men: 440/320. Guns: GD 26 x 24, UD 26 x 9, QD 12 x 6, Fc 4 x 6, RH 4 x 3. (1703 establishment: 24 x 24, 26 x 12, 12 x 6, 4 x 6, 4 x 3).
RB L ____ 1698; 26/27.11.1703 wrecked off 'Pemsey' (Pevensey) during the 'Great Gale'.

***Orford*** 70, 1698-1709. Snellgrove, Deptford.
Dimensions & tons: 150'5", 122'6" x 40'6" x 17'1". 1051.
Men: 440/320. Guns: GD 26 x 24, UD 26 x 9, QD 12 x 6, Fc 4 x 6, RH 4 x 3. (1703 establishment: 24 x 24, 26 x 12, 12 x 6, 4 x 6, 4 x 3).
Ord ____ 1695; L ____ 1698; 1709 BU.

***Nassau*** 70, 1698-1706. Portsmouth Dyd (Surveyor: Waffe).
Dimensions & tons: 150'9", 127' x 40' x 17'2". 1080.
Men: 440/320. Guns: GD 26 x 24, UD 26 x 9, QD 12 x 6, Fc 4 x 6, RH 4 x 3. (1703 establishment: 24 x 24, 26 x 12, 12 x 6, 4 x 6, 4 x 3. 1699: 22 x 24, 4 x 18, 26 x 12, 14 sakers, 4 x 3).
Ord ____ 1695; L 27.4.1698; 30.10.1706 grounded between the Dean Sand and the Horse near Chichester Harbour.

***Suffolk*** 70, 1698-1735. Johnson, Blackwall.
Dimensions & tons: 151'4", 124' x 40'4½" x 16'7½". 1078.
Men: 440/320. Guns: GD 26 x 24, UD 26 x 9, QD 12 x 6, Fc 4 x 6, RH 4 x 3 (1703 establishment: 24 x 24, 26 x 12, 12 x 6, 4 x 6, 4 x 3).
RB L ____ 1698; 1717-1718 Great Repair at Chatham Dyd; 1735 BU.

***Revenge*** 70, 1699-1711 *Buckingham*-1727. Miller, Deptford.
Dimensions & tons: 150'3", 123'8" x 40'3" x 17'1½". 1065.
Men: 440/320. Guns: GD 26 x 24, UD 26 x 9, QD 12 x 6, Fc 4 x 6, RH 4 x 3. (1703 establishment: 24 x 24, 26 x 12, 12 x 6, 4 x 6, 4 x 3. 1704: 22 x 24, 4 x 18, 26 x 9, 14 sakers, 4 x 3).
Ord ____ 1695; L ____ 1699; 1711 renamed *Buckingham*; 1727 hulked.

***Expedition*** 70, 1699-1709. Chatham Dyd (Surveyor: Furzer/Shortiss).
Dimensions & tons: 152'1", 125'10" x 40'10" x 17'1½". 1116.
Men: 440/320. Guns: GD 26 x 24, UD 26 x 9, QD 12 x 6, Fc 4 x 6, RH 4 x 3. (1703 establishment: 24 x 24, 26 x 12, 12 x 6, 4 x 6, 4 x 3).
RB L ____ 1699; 1709 BU.

***Sterling Castle*** 70, 1699-1703. Chatham Dyd (Surveyor: Furzer/Shortiss).
Dimensions & tons: 151'2", 124'8" x 40'6" x 17'8". 1087.
Men: 440/320. Guns: GD 26 x 24, UD 26 x 9, QD 12 x 6, Fc 4 x 6, RH 4 x 3. (1703 establishment: 24 x 24, 26 x 12, 12 x 6, 4 x 6, 4 x 3).
RB L ____ 1699; 26/27.11.1703 wrecked on the Goodwins during the 'Great Gale'. [The semi-intact wreck - apparently protected by being buried in the sand until shortly before its discovery - was located and dived on by a team including the present author in 1979. In 1980 the wreck was covered the sand again, though there have been subsequent reports that it has re-emerged. It is protected under the Historic Wrecks Act.]

***Eagle*** 70, 1699-1707. Chatham Dyd (Surveyor: Furzer).
Dimensions & tons: 151'6", 125' x 40'8" x 17'3". 1099.
Men: 440/320. Guns: GD 26 x 24, UD 26 x 9, QD 12 x 6, Fc 4 x 6, RH 4 x 3. (1703 establishment: 24 x 24, 26 x 12, 12 x 6, 4 x 6, 4 x 3. 1704: 22 x 32, 4 x 18, 26 x 12, 16 sakers, 4 x 3).
RB L ____ 1699; 22.10.1707 wrecked off the Scilly Islands.

***Essex*** 70, 1700-1736. Wells, Rotherhithe.
Dimensions & tons: 150'4", 124' x 40'7½" x 16'6". 1090.
Men: 440/320. Guns: GD 26 x 24, UD 26 x 9, QD 12 x 6, Fc 4 x 6, RH 4 x 3. (1703 establishment: 24 x 24, 26 x 12, 12 x 6, 4 x 6, 4 x 3).
RB L ____ 1700; 1736 BU.

***Grafton*** 70, 1700-1707. Wells, Rotherhithe.
Dimensions & tons: 150'8½", 124'5" x 40'10" x 16'8". 1102.
Men: 440/320. Guns: GD 26 x 24, UD 26 x 9, QD 12 x 6, Fc 4 x 6, RH 4 x 3. (1703 establishment: 24 x 24, 26 x 12, 12 x 6, 4 x 6, 4 x 3).
RB L ____ 1700; 1.5.1707 taken by the 9 ships of Forbin's Dunkirk squadron off Beachy Head. [In French service until 1744.]

***Berwick/Barwick*** 70, 1700-1715. Snellgrove, Deptford.
Dimensions & tons: 150'9", 125'3" x 40'5½" x 16'10". 1090.
Men: 440/320. Guns: GD 26 x 24, D 26 x 9, QD 12 x 6, Fc 4 x 6, RH 4 x 3. (1703 establishment: 24 x 24, 26 x 12, 12 x 6, 4 x 6, 4 x 3. 1708: 26 x 24, 26 x 12, 14 sakers, 4 x 3).
RB L ____ 1700; 1715 hulked [at Portsmouth; 1742 BU].

***Hampton Court*** 70, 1701-1707. Johnson, Blackwall.
Dimensions & tons: 150'6", 123'10" x 40'4½" x 16'11". 1073.
Men: 440/320. Guns: GD 26 x 24, UD 26 x 9, QD 12 x 6, Fc 4 x 6,

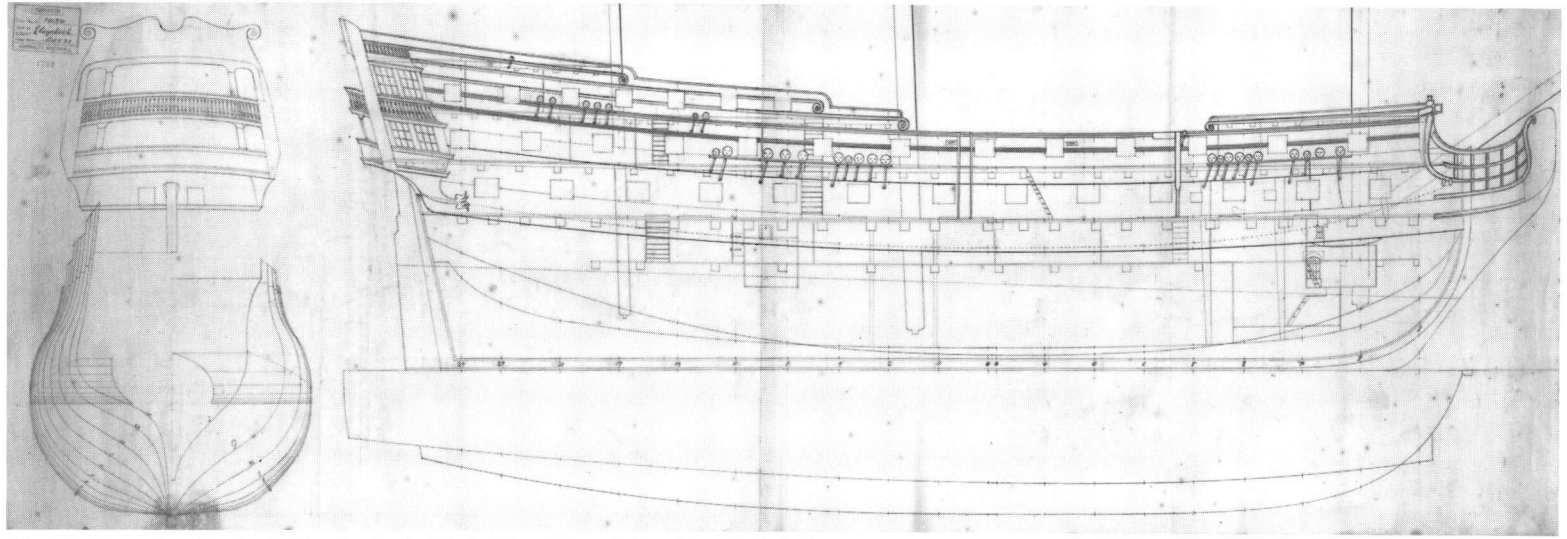

Admiralty lines and profile plan (lacking the detail of the half breadth), probably of the *Elizabeth* of 1706, 70 guns (plan number 1050), the plan probably having been drawn some time after the ship was built

RH 4 x 3. (1703 establishment: 24 x 24, 26 x 12, 12 x 6, 4 x 6, 4 x 3).
RB Ord 27.9.1699; L ____ 1701; 1.5.1707 taken by the 9 ships of Forbin's Dunkirk squadron. [In French service until sold to Spain in 1711 as *Nuestra Senora Del Carmen Y San Antonio*; 1715 wrecked off Florida; the wreck site has been located and plundered.]

***Lenox*** 70, 1701-1721. Popely, Deptford.
Dimensions & tons: 152'7½", 126'1½" x 40'3½" x 17'1". 1089.
Men: 440/320. Guns: GD 26 x 24, UD 26 x 9, QD 12 x 6, Fc 4 x 6, RH 4 x 3. (1703 establishment: 24 x 24, 26 x 12, 12 x 6, 4 x 6, 4 x 3).
RB L ____ 1701; 1721 BU.

***Edgar*** 70, 1702-1708. Portsmouth Dyd (Surveyor: Waffe).
Dimensions & tons: 153'6", 124' x 37'9" x 15'6". 1199.
Men: 440/320. Guns: GD 26 x 24, UD 26 x 9, QD 12 x 6, Fc 4 x 6, RH 4 x 3. (1703 establishment: 24 x 24, 26 x 12, 12 x 6, 4 x 6, 4 x 3).
RB L ____ 1702; 1708 BU.

***Northumberland*** 70, 1702-1703. Chatham Dyd (Surveyor: Shortiss).
Dimensions & tons: 152', 126'8" x 40'4" x 17'3". 1096.
Men: 440/320. Guns: GD 26 x 24, UD 26 x 9, QD 12 x 6, Fc 4 x 6, RH 4 x 3. (1703 establishment: 24 x 24, 26 x 12, 12 x 6, 4 x 6, 4 x 3).
RB L ____ 1702; 26/27.11.1703 wrecked during the 'Great Gale' on the Goodwins. [A site located during 1980 by a team of divers which included the present author was probably this wreck, though it could have been the *Restoration*.]

***Restoration*** 70, 1702-1703. Portsmouth Dyd (Surveyor: Waffe).
Dimensions & tons: 150'9", 122'9" x 40' x 17'. 1045.
Men: 440/320. Guns: GD 26 x 24, UD 26 x 9, QD 12 x 6, Fc 4 x 6, RH 4 x 3. (1703 establishment: 24 x 24, 26 x 12, 12 x 6, 4 x 6, 4 x 3. 1703: 22 x 32, 4 x 18, 28 x 9, 12 sakers, 4 x 3).
RB L ____ 1702; 26/27.11.1703 wrecked during the 'Great Gale' on the Goodwins. [See under the entry on *Northumberland* above.]

***Warspite*** 70, 1703-1716 *Edinburgh*-1717. J & R Burchett, Rotherhithe.
Dimensions & tons: 147'7", 120'6" x 38'6½" x 15'8". 952 3/94.
Men: 400. Guns: ____ .
RB Ord 1.12.1701; L 20.2.1703; 1716? renamed *Edinburgh*; 1717 BU.

PLANS etc: Contract.

***Elizabeth*** 70, 1704-1704. Portsmouth Dyd (Surveyor: Podd).
Dimensions & tons: 153'2", 126'2" x 41'5½" x 17'1½". 1152.
Men: 440/320. Guns: GD 26 x 24, UD 26 x 9, QD 12 x 6, Fc 4 x 6, RH 4 x 3 (1703 establishment: 24 x 24, 26 x 12, 12 x 6, 4 x 6, 4 x 3).

RB L ____ 1704; 12.11.1704 taken by the French *Le Jason* 12 leagues to the South West of Scilly. [1714 repaired by the French but out of service by 1718.]

***Northumberland*** 70, 1705-1719. Deptford Dyd (Surveyor: Harding).
Dimensions & tons: 150'8", 123'8" x 41' x 17'6". 1106.
Men: 440/320. Guns: GD 26 x 24, UD 26 x 9, QD 12 x 6, Fc 4 x 6, RH 4 x 3.
L 29.3.1705; 1719 BU.

***Resolution*** 70, 1705-1707. Woolwich Dyd (Surveyor: Lee).
Dimensions & tons: 150'10", 123'10" x 40'11" x 17'1". 1103.
Men: 440/320. Guns: GD 26 x 24, UD 26 x 9, QD 12 x 6, Fc 4 x 6, RH 4 x 3.
L 31.3.1705; 11.3.1707 burnt near 'Ventimiglia .....in the Straights' (Mediterranean) to prevent capture by the French.

***Sterling Castle*** 70, 1705-1720. Chatham Dyd (Surveyor: Rosewell).
Dimensions & tons: 151', 125'6" x 41' x 17'6". 1122.
Men: 440/320. Guns: GD 26 x 24, UD 26 x 9, QD 12 x 6, Fc 4 x 6, RH 4 x 3. (1713: 24 x 24, 26 x 9, 16 x 6, 4 x 3).
L 21.9.1705; 1720 BU.

***Restoration*** 70, 1706-171_. Deptford Dyd (Surveyor: Allin).
Dimensions & tons: 151', 123'9" x 41' x 17'6". 1106.
Men: 440/320. Guns: GD 26 x 24, UD 26 x 9, QD 12 x 6, Fc 4 x 6, RH 4 x 3.
L 15.8.1706; 9.11.1711 wrecked 'in the Straights' (Mediterranean, off Leghorn/Livorno?).

***Elizabeth*** 70, 1706-1732. Woolwich Dyd (Surveyor: Stacey).
Dimensions & tons: 150'6", 124'4" x 40'11¾" x 17'4" (1720: 150'6", 125'6" x 40'8" x 17'2"). 1110 (1720: 1103 42/94).
Men: 440/320. Guns: GD 26 x 24, UD 26 x 9, QD 12 x 6, Fc 4 x 6, RH 4 x 3.
L 14.4.1706 (or just after 5.8.1706); 1718-1720 Great Repair at Woolwich Dockyard (Surveyor: Hayward); L 15.2.1720; 1731/1732 BU.

PLANS: Lines. Danish Rigsarkivet Collection: Body plan & midsection/1728 masts, spars & tops.

# Third Rates 64-60 Guns and Fourth Rates 64-60 Guns (two-deckers)

Note that these vessels were all eventually classed as Fourth Rate 60s.

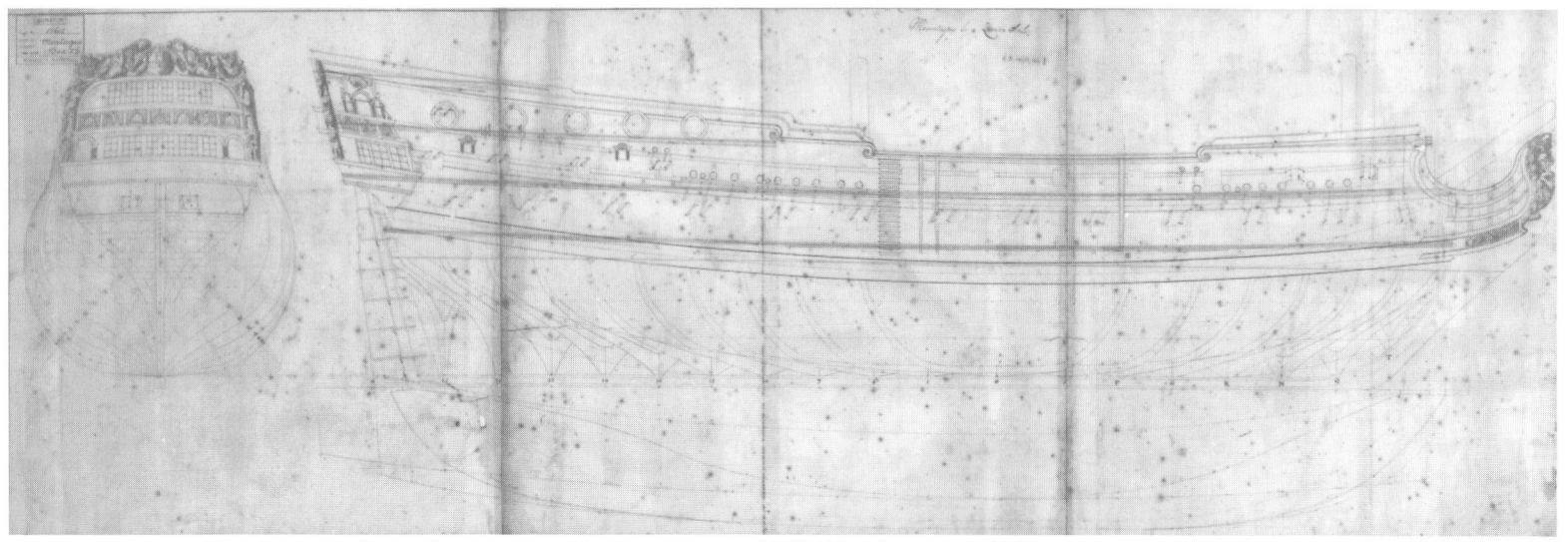

A rather archaic form of lines plan showing bow and stern carvings of the 60 gun *Montague* of 1698 (Admiralty number 1162)

**Dreadnought** 64/54, 1691-1705. Johnson, Blackwall.
 Dimensions & tons: 142', 120'6" x 36'5½" x 17'6". 852.
 Men: 365/240? Guns: GD 24 x 24, UD 26 x 12, QD & Fc 14 x 6, RH 2 x 3. (1703 establishment: 24 x 24, 26 x 9, 10 x 6, 4 x 6).
 L ____ 1691; 1697 reclassed as Fourth Rate; 1705 BU.

**Defiance** 64, 1695-1705. Woolwich Dyd (Surveyor: Lawrence).
 Dimensions & tons: 143'10", 118' x 37'11" x 15'8". 902.
 Men: 365/240? Guns: (1703 establishment: GD 24 x 24, UD 26 x 9, QD 10 x 6, Fc 4 x 6).
 RB L ____ 1695; 1705 BU.

**Monmouth** 66, 1700-1716. Woolwich Dyd (Surveyor: Harding).
 Dimensions & tons: 147'9", 123' x 38' x 16'. 944.
 Men: 400. Guns: (1703 establishment: GD 24 x 24, UD 26 x 9, QD 10 x 6, Fc 4 x 6).
 RB L ____ 1700; 1716? BU.

**Rupert** 66/58, 1703-1736. Plymouth Dyd (Surveyor: Rosewell).
 Dimensions & tons: 143'4", 119' x 38'4" x 15'2". 8930 1/94.
 Men: 365/240? Guns: (1703 establishment: GD 24 x 24, UD 26 x 9, QD 10 x 6, Fc 4 x 6).
 RB L 10.1703; 1716 reclassed as a Fourth Rate of 60 guns; 1736 BU.

**Mary** 64/60, 1704-1728 *Princess Mary* -1737. Chatham Dyd (Surveyor: Shortiss).
 Dimensions & tons: 145', 122'3" x 37'6" x 15'8". 914.
 Men: 365/240? Guns: (1703 establishment: GD 24 x 24, UD 26 x 9, QD 10 x 6, Fc 4 x 6. 1714: 24 x 18, 24 x 9, 8 x 6).
 L 12.5.1704?; 1728 renamed *Princess Mary*; 1737 BU.

**Sunderland** 60, 1694-1715. Winter, Southampton.
 Dimensions & tons: 145'2", 120'2" x 37'10" x 15'10". 915.
 Men: 365/240. Guns: GD 22 x 24, UD 24 x 9, QD 10 x 6, RH 4 x 3 (1703 establishment: 24 x 24, 26 x 9, 10 x 6, 4 x 6. 1703: 22 x 24, 22 x 9, 14 sakers, 2 x 3).
 Ord 20.4.1693; L 17.3.1694; 1715 hulked. [At Chatham; 1741 to Port Mahon as hospital ship; 1744 condemned.]

**Carlisle** 60, 1693-1696. Snellgrove, Redhouse.
 Dimensions & tons: 145', 118'10" x 38' x 15'7½". 912.
 Men: 365/240. Guns: GD 22 x 24, UD 24 x 9, QD 10 x 6, RH 4 x 3.
 Ord 30.12.1691; L 11.2.1693; 28.1.1696 wrecked on the Shipwash (off Harwich).

**Medway** 60, 1693-1716. Sheerness Dyd (Surveyor: Furzer).
 Dimensions & tons: 145'3", 119' x 38' x 15'7". 914.
 Men: 365/240. Guns: GD 22 x 24, UD 24 x 9, QD 10 x 6, RH 4 x 3. (1703 establishment: 24 x 24, 26 x 9, 10 x 6, 4 x 6).
 Ord ____ 1691; L 20.9.1693; 1716 BU.

**Canterbury** 60/54, 1693-1720. Snellgrove, Redhouse.
 Dimensions & tons: 144'9", 119'3" x 38'1½" x 15'7". 903.
 Men: 365/240. Guns: GD 22 x 24, UD 24 x 9, QD 10 x 6, RH 4 x 3. (1703 establishment: 24 x 24, 26 x 9, 10 x 6, 4 x 6. 1714: 24 x 24, 24 x 9, 8 sakers).
 Ord 24.2.1693; L 18.12.1693; 1720 BU.

**Winchester** 60, 1693-1695. Wyatt, Burlesdon.
 Dimensions & tons: 146'2½", 121'7" x 38'2" 15'11". 941.
 Men: ____ . Guns: ____ .
 Ord 20.1.1692; L 11.4.1693; 24.9.1695 foundered off Cape Florida.

**Pembroke** 60, 1694-1709. Snellgrove, Redhouse.
 Dimensions & tons: 145', 120'6" x 37'7¾" x 15'9½". 908 28/94.
 Men 365/240. Guns: GD 22 x 24, UD 24 x 9, QD 10 x 6, RH 4 x 3. (1703 establishment 24 x 24, 26 x 9, 10 x 6, 4 x 6).
 Ord 8.1.1694; L 22.11.1694; 29.12.1709 taken by the French; 23.3.1711 retaken and foundered.

**Gloucester** 60, 1695-1708. Clements, Bristol.
 Dimensions & tons: 145'2", 120'4" x 37'5" x 15'8". 896 9/94.
 Men 365/240. Guns: GD 22 x 24, UD 24 x 9, QD 10 x 6, RH 4 x 3. (1703 establishment: 24 x 24, 26 x 9, 10 x 6, 4 x 6).
 Ord 27.3.1693; L 5.2.1695; 1708 hulked. [At Deptford; 1731? BU.]

**Windsor** 60, 1695-1725. Snellgrove, Redhouse.
 Dimensions & tons: 146'2½", 120' x 37'9" x 15'8½". 910.
 Men: 365/240. Guns: GD 22 x 24, UD 24 x 9, QD 10 x 6, RH 4 x 3. (1703 establishment: 24 x 24, 26 x 9, 10 x 6, 4 x 6).
 Ord 22.10.1694; L 31.10.1695; 1725 BU.

**Kingston** 60, 1697-1716. Frame, Hessle (near Hull).
 Dimensions & tons: 145', 120'10" x 37'11" x 15'9". 923 82/94.
 Men: 365/240. Guns: GD 22 x 24, UD 24 x 9, QD 10 x 6, RH 4 x 3. (1703 establishment: 24 x 24, 26 x 9, 10 x 6, 4 x 6).
 Ord 10.8.1694?; L 13.3.1697; 1716 BU.

**Exeter** 60, 1697-1740. Portsmouth Dyd (Surveyor: Bagwell).
 Dimensions & tons: 148', 122'5" x 38'2" x 15'9". 948 50/94.
 Men: 365/240. Guns: (1703 establishment: GD 24 x 24, UD 26 x 9, QD 10 x 6, Fc 4 x 6. 1702: 22 x 24, 22 x 9, 14 sakers, 8 minions).
 Ord ____ 1691; L 26.5.1697; 1740 BU.

**Montague/Mountague/Montagu** 60, 1698-1714. Woolwich Dyd (Surveyor: Harding).
 Dimensions & tons: 143'10", 119'11" x 37'8" x 15'4". 905.
 Men: 365/240. Guns: GD 22 x 24, UD 24 x 9, QD 10 x 6, RH 4 x 3. (1703 establishment: 24 x 24, 26 x 9, 10 x 6, 4 x 6).
 RB L ____ 1698; 1714 BU.

PLANS: Lines.

**Monck** 60, 1702-1720. Burchett, Rotherhithe.
 Dimensions & tons: 137'6½", 114'3" x 36'5½" x 14'5½". 807¾.

BEFORE THE ESTABLISHMENTS 1689-1705

(1703 establishment: 24 x 24, 26 x 9, 10 x 6, 4 x 6).
RB L early 1702; 1720 50 guns; 24.11.1720 lost in Yarmouth Road.

*Nottingham* 60/64, 1703-1716 Deptford Dyd (Surveyor: Harding).
Dimensions & tons: 145'9¾", 120'4" x 38' x 15'11". 920.
Men: 365/240. Guns: GD 22 x 24, UD 24 x 9, QD 10 x 6, RH 4 x 3.
(1703 establishment: 24 x 24, 26 x 9, 10 x 6, 4 x 6).
L 10.6.1703; 1716 BU?

*Dunkirk* 60, 1704-1729. Johnson, Blackwall.
Dimensions & tons: 141'6", 116'6" x 38'3" x 15'7½". 906.
Men: 365/240. Guns: GD 22 x 24, UD 24 x 9, QD 10 x 6, RH 4 x 3.
RB L 12.1704; 1729 BU.

*Plymouth* 60, 1705-1705. Johnson, Blackwall.
Dimensions & tons: 140'5", 115'3" x 38'3" x 15'7". 896 84/94.
Men: 365. Guns: ? GD 22 x 24, UD 24 x 9, QD 10 x 6, Fc 4 x 6?
RB L ___ 1705; 11.8.1705 supposed foundered with all hands in a violent storm 'In the Soundings'.

*Dreadnought* 60, 1706-1721. Johnson, Blackwall.
Dimensions & tons: 142'10½", 118'0½" x 38'1" x 15'9". 911.
Men: 365. Guns: GD 22 x 24, UD 24 x 9, QD 10 x 6, RH 4 x 3.
RB (or Great Repair?); L 02.1706; 1721 BU (for 'Great Repair' rather than 'Rebuild' - but such a radical one that it resulted in virtually a new ship - see underneath the 1719 Establishment 60s for subsequent history).

*York/Yorke* 64/60, 1706-1737. Plymouth Dyd (Surveyor: Rosewell/Lock?).
Dimensions & tons: 146'6", 122' x 39' x 16'. 987 3/94.
Men: 365. Guns: GD 24 x 18, D 26 x 9, QD 10 x 6, Fc 4 x 6.
L 8.4.1706; 1737 (?) BU or Great Repair?

## Fourth Rates, 50 Guns, etc (two-deckers)

*Centurion* 50/48/42, 1690-1729. Deptford Dyd (Surveyor: Harding).
Dimensions & tons: 125'8½", 105' x 33'2" x 13'5". 614 35/94.
Men: 280/185. Guns: GD 20 x 12, UD 22 x 8, QD 8 x 6.
L ___ 1690; 1729 BU.

*Dragon* 48/40, 1690-1707. Deptford Dyd (Surveyor: Harding).
Dimensions & tons: 118'11", 99' x 31'9" x 12'2". 479.
Men: 230/160. Guns: GD 18 x 12, UD 22 x 8, QD 8 x 6. (1703 establishment: 20 x 12, 20 x 6, 6 x 6. 1703: 18 x 12, 28 sakers).
RB L ___ 1690; 1707 BU.

*Chester* 50/48/42, 1691-1707. Woolwich Dyd (Surveyor: Lawrence).
Dimensions & tons: 125', 105'10" x 34'4" x 13'10½". 663.
Men: 230/160. Guns: GD 22 x 12, UD 22 x 6, QD 6 x 4. (1703 establishment: 20 x 12, 20 x 6, 6 x 6.).
L 21.3.1691; 10.10.1707 taken by *Le Jason* and the rest of Duguay-Trouin's French squadron off Ushant. [Sold by the French in 1709.]

*Chatham* 50/48/42, 1691-1718. Chatham Dyd (Surveyor: Lee).
Dimensions & tons: 126', 109'6" x 34'4" x 13'4". 686.
Men: 230/160. Guns: GD 22 x 18, UD 20 x 8, QD 10 x 4 (1703 establishment: 20 x 12, 20 x 6, 6 x 6).
L 20.4.1691; 1718 BU.

*Norwich* 48/42, 1691-1692. Portsmouth Dyd (Surveyor: Stigant).
Dimensions & tons: 125'7", 101'6" x 33'8" x 13'4". 616.
Men: 230/160. Guns: GD 20 x 18, UD 20 x 6, QD 8 x 4.
L ___ 1691; 6.10.1692 wrecked in the West Indies.

*Rochester* 48/42, 1692-1714. Chatham Dyd (Surveyor: Lee).
Dimensions & tons: 125'5", 107' x 32'8" x 13'6". 607.
Men: 230/160. Guns: GD 18 x 12, UD 20 x 9, QD 8 x 6. (1703 establishment: 20 x 12, 20 x 6, 6 x 6.).
L ___ 1692; 1714 BU.

*Portland* 48/42, 1693-1719. Woolwich Dyd (Surveyor: Lawrence).
Dimensions & tons: 125'6", 103'6" x 34' x 14'. 636.
Men: 230/160. Guns: (1703 establishment: 20 x 12, 20 x 6, 6 x 6.).
L 28.3.1693; 1719 BU.

*Falmouth* 48/42, 1693-1704. Snellgrove, Redhouse.
Dimensions & tons: 124', 101'6½" x 33'7½" x 13'9". 610.
Men: 280/185? Guns: (1703 establishment: GD 22 x 12, UD 22 x 6, QD 8 x 6, Fc 4 x 6).
Ord ___ 1693; L 26.6.1693; 1702 possibly rebuilt at Chatham, dimensions not changed, but reclassed as a 54 gun ship; 4.8.1704 taken about 15 leagues south from Scilly by 6 French warships. [1705 sold by the French for mercantile use.]

*Southampton* 48/42, 1693-1699. Winter, Southampton.
Dimensions & tons: 121'9", 100' x 33'10" x 13'9". 609.
Men: 230/160. Guns: ___ .
Ord ___ 1693; L 10.6.1693; 1699 BU.

*Bristol* 50/48/42, 1693-1709. Castle, Deptford.
Dimensions & tons: 130'2", 109' x 35'2" x 13'. 710.
Men: 230/160. Guns: (1703 establishment: 20 x 12, 20 x 6, 6 x 6).
RB L ___ 1693; 25.4.1709 taken 'In the Soundings' by Du Guay Trouin's French squadron; then sunk.

*Dartmouth* 48/42, 1693-1695/1702 *Vigo*-1703. Shish, Rotherhithe.
Dimensions & tons: 122', 100' x 33'8" x 13'7". 602 84/94.
Men: 230/160. Guns: (1703 establishment: 20 x 12, 20 x 6, 6 x 6).
L 24.7.1693; 4.2.1695 taken by the French *Saint Esprit*; 1702 retaken by *Barfleur* at Vigo, reclassed as 50 guns and renamed *Vigo*; 26/27.11.1703 wrecked off Helvoetsluis, Holland, during the 'Great Gale'.

*Severn* 50/48/42, 1693-1734. Johnson, Blackwall.
Dimensions & tons: 131'3", 109' x 34'4" x 13'6". 683.
Men: 230/160. Guns: (1703 establishment: 20 x 12, 20 x 6, 6 x 6. 1717: 22 x 12, 22 x 9, 8 minions - alternatively 22 x 18, 6 x 6?).
L ___ 1693; 1734 BU.

*Norwich* 48/42, 1693-1712. Castle, Deptford.
Dimensions & tons: 123'8", 101'6" x 33'10" x 13'6½". 618.
Men: 230/160. Guns: (1703 establishment: 20 x 12, 20 x 6, 6 x 6).
Ord ___ 1693; L ___ 1693; 1712 BU.

*Weymouth* 48/42, 1693-1717. Portsmouth Dyd (Surveyor: Stigant).
Dimensions & tons: 132'4", 107'10" x 34'3" x 13'10". 673.
Men: 230/160. Guns: GD 18 x 12, UD 18 x 6, QD 8 x 4. (1703 establishment: 20 x 12, 20 x 6, 6 x 6. 1707: 20 x 12, 20 x 6, 8 minions).
Ord ___ 1693; L ___ 1693; 1717 BU.

*Anglesea* 48/42, 1694-1719. Plymouth Dyd (Surveyor: Waffe) or Flint, Plymouth?
Dimensions & tons: 125', 106' x 33'2" x 14'. 620.
Men: 230/160. Guns: (1703 establishment: 20 x 12, 20 x 6, 6 x 6).
Ord ___ 1693; L ___ 1694; 1719 BU.

*Colchester* 50/48/42, 1694-1704. Johnson, Blackwall.
Dimensions & tons: 131'4", 111'8" x 34'3" x 13'7". 696.
Men: 230/160. Guns: (1703 establishment: 20 x 12, 20 x 6, 6 x 6).
Ord ___ 1693; L ___ 1694; 16.1.1704 foundered at Whitsand Bay, Cornwall.

*Coventry* 50/48/12, 1694-1704/1709-1709. Deptford Dyd (Surveyor: Harding).
Dimensions & tons: ___ , 106' x 34'5" x 13'6". 670.
Men: 230/160. Guns: (1703 establishment: 20 x 12, 20 x 6, 6 x 6).
Ord ___ 1693; L ___ 1694; 24.7.1704 taken by the French *Le Jason* of Duguay-Trouin's squadron; 1709 retaken by *Portland* & BU.

*Romney/Rumney* 54/50/48/42, 1694-1707. Johnson, Blackwall.
Dimensions & tons: 131'0½", 109' x 34'4" x 13'7". 683.
Men: 280/185. Guns: (1703 Establishment: GD 22 x 12, UD 22 x 6, QD 8 x 6, Fc 4 x 6).
L ___ 1694; 22.10.1707 wrecked off the Scilly Islands (Cloudesley Shovell's squadron). [There are reports that her wreck has been found.]

*Litchfield* 50/48/42, 1694-1720. Portsmouth Dyd (Surveyor: Stigant).
Dimensions & tons: 130'3", 107'7" x 34'7½" x 13'6". 682.
Men: 230/160 Guns: (1703 establishment: 20 x 12, 20 x 6, 6 x 6. 1703: 22 x 12, 20 x 9, 8 minions).
L ___ 1694; 1720 BU.

*Lincoln* 48/42, 1695-1703. Woolwich Dyd (Surveyor: Lawrence).
Dimensions & tons: 130'7", 108'4" x 34'3½" x 13'6½". 675 91/94.
Men: 230/160. Guns: (1703 establishment: 20 x 12, 20 x 6, 6 x 6).
Ord 1693; L 19.2.1695; 26.1.1703 parted from *Ipswich* in a storm and not seen since, presumed foundered.

*Burlington* 50/48/42, 1695-1733. Johnson, Blackwall.
Dimensions & tons: 131'3", 109' x 34'3" x 13'7". 680.
Men: 230/160. Guns: (1703 establishment: 20 x 12, 20 x 6, 6 x 6. 1703:

20 x 12, 22 x 9, 14 sakes, 8 minions).
L ____ 1695; 1733 BU.

***Dover*** 48/42, 1695-1730. Portsmouth Dyd (Surveyor: Bagwell).
Dimensions & tons: 118′, 98'6" x 34'4" x 12'7". 604.
Men: 230/160. Guns: (1703 establishment: 20 x 12, 20 x 6, 6 x 6. 1707; 20 x 18, 30 x 6).
RB L ____ 1695; 1716 reclassed as a Fifth Rate of 40 guns; 1730 BU.

***Falkland/Faulklannd*** 48/42, 1696-1702/1702-1718. Holland, New England (built on speculation).
Dimensions & tons: 128'6", 109' x 33'2" x 13'9". 638.
Men: 230/160. Guns: (1703 establishment: 20 x 12, 20 x 6, 6 x 6).
L ____ 1695; purchased 1696; 1702 RB at Chatham Dyd, but no changes in dimensions or tonnage; 1718 BU.

***Harwich*** 48/42, 1695-1700. Castle, Deptford.
Dimensions & tons: 130'2", 109' x 34'4" x 13'8". 683.
Men: 230/160. Guns: (1703 establishment: 20 x 12, 20 x 6, 6 x 6).
L ____ 1695; 5.10.1700 'Lost at Island of Cullinshaw at Amoy in China heaving off from the place where she was lay'd ashore to be clean'd'.

***Pendennis*** 50/48/42, 1695-1705. Castle, Deptford.
Dimensions & tons: 130'2½", 109' x 34'3½" x 13'6½". 681.
Men: 230/160. Guns: (1703 establishment: 20 ´x 12, 20 x 6, 6 x 6).
L ____ 1695; 20.10.1705 taken by the French *Le Protée*. [Sold by the French in 1706.]

***Blackwall*** 50/48/42, 1696-1705. Johnson, Blackwall.
Dimensions & tons: 131'1½", 109' x 34'2½" x 13'7½". 678.
Men: 230/160. Guns: (1703 establishment: 20 x 12, 20 x 6, 6 x 6).
L ____ 1696; 20.10.1705 taken by four French men of war. [In French service until 1719.]

***Nonsuch*** 50/48/42, 1696-1716. Castle, Deptford.
Dimensions & tons: 130'5" , 109' x 34'2" x 13'9". 676 77/94.
Men: 230/160. Guns: (1703 establishment: 20 x 12, 20 x 6, 6 x 6. 1703: 20 x 18, 22 x 9).
L ____ 1696; 1716 BU.

***Warwick*** 50/48/42, 1696-1709. Castle, Deptford.
Dimensions & tons: 130'5", 109' x 34'5" x 13'9". 686 71/94.
Men: 230/160. Guns: (1703 establishment: 20 x 12, 20 x 6, 6 x 6).
L ____ 1696; 1709 BU.

***Guernsey*** 50/48/42, 1696-1716. Johnson, Blackwall.
Dimensions & tons: 131'9", 109' x 34'3" x 13'6". 680.
Men: 230/160. Guns: (1703 establishment: 20 x 12, 20 x 6, 6 x 6).
L ____ 1696; 1716 BU.

***Dartmouth*** 48/42, 1698-1714. Parker, Southampton.
Dimensions & tons: 131'3¾", 108' 10½" x 34'3½" x 13'6½". 681 47/94.
Men: 230/160. Guns: (1703 establishment: 20 x 12, 20 x 6, 6 x 6).
Ord ____ 1695; L 3.3.1698; 1714 BU.

***Winchester*** 48/42, 1698-1716. Wells, Rotherhithe.
Dimensions & tons: 130', 107'5" x 34'4" x 13'7". 673.
Men: 230/160. Guns: GD 18 x 18?, UD 20 x 8, QD 8 x 4. (1703 establishment: 20 x 12, 20 x 6, 6 x 6).
Ord ____ 1695; L 17.3.1698; 1716 BU.

***Advice*** 48/42, 1698-1711. Woolwich Dyd (Surveyor: Harding).
Dimensions & tons: 118', 99'3" x 32'4" x 12'1". 551.
Men: 230/160. Guns: (1703 establishment: 20 x 12, 20 x 6, 6 x 6).
RB L ____ 1698; 27.6.1711 taken by 8 French privateers off Yarmouth. [Acquired by the French East India Company?]

***Hampshire*** 48/42, 1698-1739. Taylor, Rotherhithe.
Dimensions & tons: 132', 110'7" x 34'3" x 13'8". 690 1/94.
Men: 230/160. Guns: (1703 establishment: 20 x 12, 20 x 6, 6 x 6. 1701: 18 x 12, 20 x 6, 6 minions. 1704: 20 x 12, 22 x 6, 8 minions).
Ord ____ 1695; L 3.3.1698; 1739 BU.

***Salisbury*** 50/48/42, 1698-1703/1708 *Salisbury Prize*-1716 *Preston*-1739. Herring, Beaulieu.
Dimensions & tons: 134', 109'9½" x 34'2" x 13'6". 681.
Men: 230/160. Guns: (1703 establishment: 20 x 12, 20 x 6, 6 x 6).
Ord ____ 1695; L 3.3.1698; 10.4.1703 taken by the French *L'Adriot* (Captain St. Pol); 15.3.1708 retaken by *Leopard*, renamed *Salisbury Prize*; 1716 renamed *Preston*; 1739 BU.

***Worcester*** 48/42, 1698-1713. Winter, Southampton.
Dimensions & tons: 131'8½", 109'7" x 34'4¾" x 13'6¼". 689 50/94.
Men: 230/160. Guns: (1703 establishment: 20 x 12, 20 x 6, 6 x 6).
Ord ____ 1695; L 31.5.1698; 1713 BU.

***Jersey*** 50/48, 1698-173_. Nye & Moor, Cowes.
Dimensions & tons: 132'1", 109' x 34'2", 13'8". 677.
Men: 280/185. Guns: (1703 Establishment: GD 22 x 12, UD 22 x 6, QD 8 x 6, Fc 4 x 6).
Ord ____ 1695; L 4.11.1698; 1731 hulked at Plymouth [1763 BU].

***Bonadventure*** 48/42, 1699-1711. Woolwich Dyd (Surveyor: Harding).
Dimensions & tons: 125'5", 102'5" x 33'1½" x 12'5". 597.
Men: 230/160. Guns: (1703 establishment: 20 x 12, 20 x 6, 6 x 6).
RB L ____ 1699; 1711 BU.

***Carlisle*** 1699-1700 48/42, Plymouth Dyd (Surveyor: Waffe)
Dimensions & tons: ____ , 112' x 34'6" x 13'2". 709.
Men: 230/160. Guns: (1703 establishment: 20 x 12, 20 x 6, 6 x 6).
Ord ____ 1695; L ____ 1699; 19.9.1700 blew up by accident in the Downs.

***Assistance*** 50, 1699-1710. Deptford Dyd (Surveyor: Miller).
Dimensions & tons: 119'7", 103'4" x 33'3" x 12'. 607.
Men: 230/160. Guns: 1702: 20 x 12, 22 x 9, 8 minions.
RB L ____ 1699; 1710 BU.

***Southampton*** 48/42, 1699-1728. Deptford Dyd (Surveyor: Miller).
Dimensions & tons: 122'3", 102'2" x 34'2½" x 13'2". 636.
Men: 230/160. Guns: (1703 establishment: 20 x 12, 20 x 6, 6 x 6).
RB L ____ 1699; 1716 reclassed as a Fifth Rate; 1726 40 guns; 1728 hulked. [At Jamaica; 1771 sank and BU after salvage.]

***Tilbury*** 48, 1699-1726. Chatham Dyd (Surveyor: Furzer) 48.
Dimensions & tons: 130'1½", 110'3" x 34'4" x 13'7½". 691.
Men: 230/160. Guns: (1703 establishment: 20 x 12, 20 x 6, 6 x 6).
Ord ____ 1695; L ____ 1699; 1726 BU.

***Kingfisher*** 46/40, 1699-1706. Woolwich Dyd (Surveyor: Harding).
Dimensions & tons: 125'8", 110' x 34'4½" x 12'9". 691⅓.
Men: 230/160. Guns: (1703 establishment: 20 x 12, 20 x 6, 6 x 6).
RB L ____ 1699; 1706 hulked. [At Deptford; 1708 to Harwich; 1709 to Sheerness; 1728 BU.]

***Greenwich*** 54/46, 1699-1724. Portsmouth Dyd (Surveyor: Waffe).
Dimensions & tons: 135'10", 110' x 36' x 13'6½". 785.
Men: 280/185. Guns: (1703 Establishment: GD 22 x 12, UD 22 x 6, QD 8 x 6, Fc 4 x 6).
RB L ____ 1699; 1724 BU.

***Deptford*** 48/42, 1700-1717. Woolwich Dyd (Surveyor: Harding).
Dimensions & tons: 128'4", 106'9" x 34'4" x 13'5". 669.
Men: 230/160. Guns: (1703 establishment: 20 x 12, 20 x 6, 6 x 6. 1701: 20 x 18, 22 x 6, 8 perriers).
RB L ____ 1700; 1717 BU.

***Reserve*** 48, 1701-1703. Deptford Dyd (Surveyor: ____ ).
Dimensions & tons: 117'6", 96'5" x 33'7½" x 13'. 579 80/94.
Men: 230/160. Guns: (1703 establishment: 20 x 12, 20 x 6, 6 x 6).
RB L ____ 1701; 27.11.1703 'Sunk down right in Yarmouth Roads ....in the Great Storm and all of men on board her lost'.

***Woolwich*** 54/46, 1702-1736. Woolwich Dyd (Surveyor: Lee).
Dimensions & tons: 139', 110' x 36'1" x 14'11". 761½.
Men: 280/185. Guns: (1703 Establishment: GD 22 x 12, UD 22 x 6, QD 8 x 6, Fc 4 x 6).
RB L 01.1702; 1736 BU.

***Tiger/Tyger*** 50, 1702-1718. Wells, Rotherhithe.
Dimensions & tons: 124'8½", 103'6" x 33'4½" x 13'9". 613 7/94.
Men: 230/160? Guns: (1703 establishment: 20 x 12, 20 x 6, 6 x 6 ?).
RB early 1702; 1718 BU.

***Oxford*** 54/46, 1702-1723. Deptford Dyd (Surveyor: Harding).
Dimensions & tons: 126'8½", 103'5" x 35'0½" 1" x 14'9½". 675.
Men: 280/185. Guns: (1703 Establishment: GD 22 x 12, UD 22 x 6, QD 8 x 6, Fc 4 x 6. 1717: 20 x 12, 26 x 6).
RB? L ____ 1702; 1723 BU.

***Falkland***. RB 1702; see earlier entry under 1695.

***Swallow*** 54/50, 1703-1717. Deptford Dyd (Surveyor: Harding).
Dimensions & tons: 130', 106'3" x 34'6" x 13'6". 672 64/94.

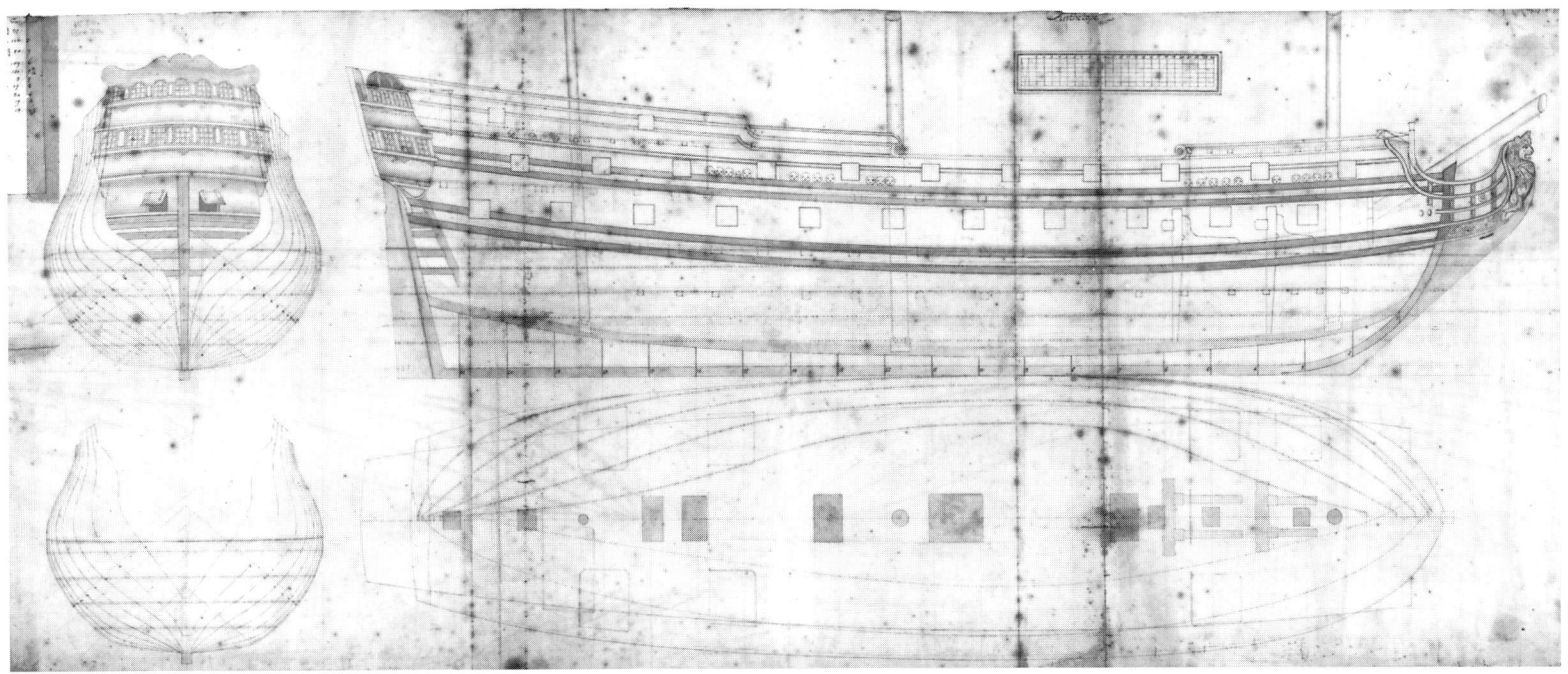

Another early example of an Admiralty lines plan showing the *Anthelope*, 50, of 1703, with also some decorations and a number of lower deck fittings shown on the half breadth (Admiralty number 1580)

Men: 280/185. Guns: (1703 Establishment: GD 22 x 12, UD 22 x 6, QD 8 x 6, Fc 4 x 6).
L 22.2.1703; 1717 BU.

**Leopard** 54/50, 1703-1719. Swallow, Rotherhithe.
Dimensions & tons: 131'1", 108'9" x 34'4½" x 13'6½". 683 12/94.
Men: 280/185. Guns: (1703 Establishment: GD 22 x 12, D 22 x 6, QD 8 x 6, Fc 4 x 6).
L 15.3.1703; 1719 BU.

**Anthelope** 50, 1703-1738. Taylor, Rotherhithe.
Dimensions & tons: 131'5", 108'11½" x 34'4½" x 13'9". 684.
Men: 280/185. Guns: 1703: 20 x 12, 30 x 6.
L 13.3.1703; 1738 BU.
PLANS: Danish Rigsarkivet: 1727: Lines.

**Panther** 50, 1703-1713. Popely, Deptford.
Dimensions & tons: 131'3½", 108'9½" x 34'4" x 13'8¼". 683 27/94.
Men: 280/185. Guns: (1703 Establishment: GD 22 x 12, UD 22 x 6, QD 8 x 6, Fc 4 x 6).
Ord 8.7.1702?; L 15.3.1703; 1713 BU.

**Newcastle** 54/50, 1704-1728. Sheerness Dyd (Surveyor: Allin).
Dimensions & tons: 130'2", 109' x 34'2" x 13'7". 676 77/94.
Men: 280/185. Guns: ___ .
L 24.3.1704; 1728 BU.

**Reserve** 54/50, 1704-1716 *Sutherland*-1741. Deptford Dyd (Surveyor: Harding?).
Dimensions & tons: 130', 107' x 34'5½" x 13'6½". 676.
Men: 280/185. Guns: 1714: 20 x 12, 26 x 6.
L 18.3.1704; 1716 renamed *Sutherland*; 1741 hulked. [Hospital ship at Port Mahon; 1754 BU.]

**Crown** 54/46, 1704-1719. Deptford Dyd. (Surveyor: Harding).
Dimensions & tons: 126'8", 103'4" x 34'5½" x 13'6". 652 59/94.
Men: 280/185. Guns: ___ .
RB L ___ 1704; 29.1.1719 lost under Saint Julian's Fort at the entrance to the Tagus

**Ruby** 50, 1706-1707. Deptford Dyd (Surveyor: Allin).
Dimensions & tons: 128'4", 105'7" x 34'8" x 13'7". 674¾.
Men: 280/185. Guns: ___ .

RB L 18.2.1706; 10.10.1707 taken by the French *Le Maure* of Duguay-Trouin's squadron off Ushant. [In French service as *Le Rubis* in 1708.]

**Saint Albans** 54/50, 1706-1717. Burchett, Rotherhithe.
Dimensions & tons: 130'8", 109'7" x 34'4" x 13'7½". 687 8/94.
Men: 280/185. Guns: ___ .
L 27.8.1706; 1717 BU.

**Colchester** 54/50, 1707-1718. Deptford Dyd (Surveyor: Allin).
Dimensions & tons: 130'6", 108'3" x 34'5" x 13'6½". 682 3/94.
Men: 280/185. Guns: 1708: 22 x 12, 32 x 6.
L 22.2.1707; 1718 BU.

## Fifth Rates, 40 Guns, etc (two-deckers)

**Adventure** 42/36, 1692-1709. Chatham Dyd (Surveyor: Lee).
Dimensions & tons : 117', 98' x 29" x 11'8½". 438.
Men: 190/130. Guns: 1703 establishment: GD 18 x 9, UD 20 x 6, QD 4 x 6. 1702: 18 x 9, 20 sakers, 6 x 3.
RB L ___ 1692; 1.3.1709 taken off Martinique by a French 36 gun ship.

**Hector** 42/40. 1703-1718. Burchett, Rotherhithe.
Dimensions & tons: 116'4½", 95'7" x 31'2" x 12'11". 493 74/94.
Men: 190/130 . Guns: GD 18 x 9, UD 20 x 6, QD 4 x 6. 1703: 18 (possibly 20) x 9, 14 sakers, 6 x 3.
Ord 1702?; L 20.2.1703-1718 BU.

**Greyhound** 42/40, 1703-1711. Hubbard, Ipswich (completed by Mrs Hubbard?).
Dimensions & tons: 114'3", 95'3" x 31'3" x 12'10½". 494 71/94.
Men: 190/130. Guns: GD 18 x 9, UD 20 x 6, QD 4 x 6.
L 03.1703; 26.8.1711 wrecked on the Hind Sand near Teignmouth Bar.

**Lark** 42/40, 1703-1723. Wells, Rotherhithe.
Dimensions & tons: 115'2", 95' x 31'2½" x 12'10". 492 5/94.
Men: 190/130. Guns: GD 18 x 9, UD 20 x 6, QD 4 x 6.
Ord 1702?; L 02 or 03.1703.

**Garland/Guarland** 42/40, 1703-1709. Woolwich Dyd (Surveyor: Lee).
Dimensions & tons: 115'6", 96' x 31'2" x 12'11½". 496 1/94.
Men: 190/130. Guns: GD 18 x 9, UD 20 x 6, QD 4 x 6.
L 05.1703; 29.11.1709 wrecked near Cape Henry, Virginia.

**Folkestone** 42/40, 1703-1728. Deptford Dyd (Surveyor: Harding).

Dimensions & tons: 115'8", 95' x 31'4" x 13'. 496 10/94.
Men: 190/130. Guns: GD 18 x 9, UD 20 x 6, QD 4 x 6.
L 15.10.1703; 1728 BU.

***Roebuck*** 40, 1704-1725. Portsmouth Dyd (Surveyor: Podd).
Dimensions & tons: 115', 95'2" x 31'3" x 13'. 494 18/94.
Men: 190/130. Guns: Two less than preceeding vessels.
L 5.4.1704; 1725 BU.

***Sorlings*** 40, 1706-1717. Sheerness Dyd (Surveyor: Acworth).
Dimensions & tons: 116'6", 94'7" x 31'8¾" x 13'1½". 506 45/94.
Men: 190/130. Guns: GD 18 x 9, UD 20 x 6, QD 4 x 6.
L 18.2.1706; 17.12.1717 cast away on the coast of Holland.

## Small Fifth Rates (two- or one and a half-deckers)

***Experiment*** 32/28, 1689-1724. Chatham Dyd (Surveyor: Lee).
Dimensions & tons: 105', 92' x 27'6" x 10'6". 370.
Men: 145/100 (115/85 as 20). Guns: 1703 establishment: GD 4 x 9, UD 22 x 6, QD 6 x 4 (as 20: UD 20 x 6).
L 17.12.1689; 1717 20 gun Sixth Rate; 1724 BU.

***Mermaid*** 28/24, 1689-1707. Woolwich Dyd (Surveyor: Lawrence).
Dimensions & tons: 106', 86' x 27'4¾" x 9'6". 34⅓.
Men: 125/90. Guns: 1703 establishment: GD 4 x 9, UD 20 x 6, QD 4 x 4.
RB L ___ 1689; 1707 BU.

***Milford*** 32/28, 1689-1694. Woolwich Dyd (Surveyor: Lawrence).
Dimensions & tons: 105'2", 88'5" x 27'6" x 10'. 355.
Men: 145/100. Guns: GD 4 x 9, UD 22 x 6, QD 6 x 4.
L ___ 1689; 4.12.1694 taken by the French. [In French service until 1709.]

***Pembroke*** 32/28, 1690-1694. Deptford Dyd (Surveyor: Harding).
Dimensions & tons: 105'6", 96'6" x 27'2½" x 10'2". 356.
Men: 145/100. Guns: GD 4 x 9, UD 22 x 6, QD 6 x 4.
L 3.3.1690; 23.2.1694 taken by the French.

***Dolphin*** 28/24, 1690-1709. Chatham Dyd. (Surveyor: Lee).
Dimensions & tons: 93'6", 82' x 24'9½" x 9'8". 267.
Men: 125/90. Guns: 1703 establishment: GD 4 x 9, UD 20 x 6, QD 4 x 4. Begun building as a fireship?
L 29.3.1690 as a Fifth Rate? or converted 1691? 1709 BU.

***Portsmouth*** 32/28, 1690-1696. Portsmouth Dyd (Surveyor: Stigant).
Dimensions & tons: 106'3", 89' x 29'6" x 10'. 412.
Men: 145/100. Guns: GD 4 x 9, UD 22 x 6, QD 6 x 4.
L 13.5.1690; 1.10.1696 taken by the French off Beachy Head.

***Roebuck*** 26, 1690-1701. Snellgrove, Wapping. 24/26/24.
Dimensions & tons: 96', 84'5" x 25'6" x 9'9¾". 292.
Men: 125/90. Guns: ___ .
Began building as a fireship; L 17.4.1690 as a Fifth Rate(?) or converted 1694(?); 24.2.1701 foundered off Ascension Island by springing a leak in the bow.

NOTE: Dampier's ship [searches have been made for this wreck, so far without success].

***Sheerness*** 32/28, 1690-1730. Sheerness Dyd (Surveyor: Furzer).
Dimensions & tons: 105'9", 89'3" x 27'6" x 10'. 359.
Men: 145/100 (as 20: 115/85). Guns: 1703 establishment: GD 4 x 9, UD 22 x 6, QD 6 x 4 (as 20: UD 20 x 6).
L ___ 1690; 1717 20 gun Sixth Rate; 1730 BU.

***Speedwell*** 28/24, 1690-1702. Gressingham, Rotherhithe.
Dimensions & tons: 94', 78'6" x 24'11" x 9'8". 259.
Men: 125/90. Guns: ___ .
Begun as a fireship; L 3.4.1690 as a Fifth Rate (?) or converted 1695 (?); 1702 BU.

***Charles Galley*** 36/30-32/28 (Galley Frigate), 1693-1710. Woolwich Dyd (Surveyor: Lawrence).
Dimensions & tons: 130'4", 124' x 28'10" x 9'. 548¼.
Men: 155/110. Guns: 1703 establishment: GD 8 x 12, UD 22 x 6, QD 6 x 4.
RB L ___ 1693; 1710 BU.

***Shoreham/Shoram*** 32/28, 1693-1719. Ellis, Shoreham.
Dimensions & tons: 103', 85'7" x 28'1½" x 10'8". 359.
Men: 145/100. Guns: 1703 establishment: GD 4 x 9, UD 22 x 6, QD 6 x 4.
L 16.1.1694; 1719 BU.

***Scarborough/Scarboro'*** 36/32 - 32/28, 1694-1694/1697 *Milford*-1705. Woolwich Dyd (Surveyor: Lawrence).
Dimensions & tons: 104'10", 84'5" x 28'10½" x 11'7". 374.
Men: 155/110. Guns: 1703 establishment: GD 8 x 12, UD 22 x 6, QD 6 x 4.
L 15.2.1694; 18.7.1694 taken by the French off the north of Ireland [renamed *Le Duc De Chaulne*]; 15.2.1697 retaken and renamed *Milford*; 1705 BU.

***Lyme/Lime*** 32/28, 1694-1720. Plymouth Dyd (Surveyor: Waffe) or Flint, Plymouth?
Dimensions & tons: 109', 88' x 28'8" x 10'6". 384.
Men: 145/100. Guns: GD 4 x 9, UD 22 x 6, QD 6 x 4.
L ___ 1694; 1717 reclassed as a 24 gun Sixth Rate; 1720 BU.

***Sorlings*** 32/28, 1694-1705. Barrett, Shoreham.
Dimensions & tons: 102'8½", 85'8" x 28'2½" x 10'9". 362.
Men: 145/100. Guns: GD 4 x 9, UD 22 x 6, QD 6 x 4.
L 19.3.1694; 20.10.1705 taken by the French *Le Jersey* and three other French men of war. [In French service until 1710; ?1711 recaptured and sold?]

***Terrible*** 28/26/24, 1694-1710. Ellis, Shoreham.
Dimensions & tons: 92'3", 76' x 25' x 9'. 253.
Men: 125/90. Guns: 1703 establishment: GD 4 x 9, UD 20 x 6, QD 4 x 4.
L ___ 1694; 1701 fireship, 14 guns; 1710 reclassed as a Fifth Rate 20.9.1710 taken by the Spanish off Cape Saint Marys.

***Winchelsea*** 36/30 - 32/28, 1694-1706 Wyatt, Redbridge.
Dimensions & tons: 103'5", 85'4½" x 28'4" x 10'7¼". 364½.
Men: 155/110. Guns: (?) GD 4 x 9, UD 22 x 6, QD 6 x 4, 1703 Establishment. 1703 establishment: GD 8 x 12, UD 22 x 6, QD 6 x 4.
L ___ 1694; 6.6.1706 taken by four French privateers off Hastings.

***Arundell*** 32/28, 1695-1713. Ellis, Shoreham.
Dimensions & tons: 108'7", 90'7" x 28' x 10'6". 378.
Men: 145/100. Guns: 1703 establishment: GD 4 x 9, UD 22 x 6, QD 6 x 4.
L 10.9.1695; 1713 sold.

***Hastings*** 32/28? 1695-1697. Ellis, Shoreham.
Dimensions & tons: 108'8", 90'9" x 28'2½"' x 10'7½". 383 90/94.
Men: 145/100. Guns: 1703 establishment: GD 4 x 9, D 22 x 6, D 6 x 4.
L 5.2.1695; 10.12.1697 'cast away by Dublin' (off Waterford?).

***Milford*** 32/28, 1695-1697. Hubbard, Ipswich.
Dimensions & tons: 107'10", 90'2¼" x 28'4½" x 11'3". 385.
Men: 145/100. Guns: 1703 establishment: GD 4 x 9, UD 22 x 6, QD 6 x 4.
L ___ 1695; 01.1697 taken by the French. [Served as *Le Milford* until 1720.]

***Scarborough*** 36/30 - 32/28, 1695-1710/1712 *Garland*-1721. Parker, Southampton.
Dimensions & tons: 108', 90' x 28'7"' x 10'9". 391.
Men: 155/110. Guns: 1703 establishment: GD 8 x 12, UD 22 x 6, QD 6 x 4.
L ___ 1695; 1.11.1710 taken by the French; 1712 retaken by *Weymouth* and *Fowey* renamed *Garland/Guardland* ; 1717 fireship; 1721 BU.

***Fowey/Fowy/Fowye*** 32/28, 1696-1704. Burgiss & Briggs, Shoreham.
Dimensions & tons: 108', 89'5½" x 28'2" x 10'6½". 377½.
Men: 145/100. Guns: 1703 establishment: GD 4 x 9, UD 22 x 6, QD 6 x 4.

L 7.5.1696; 1.8.1704 taken when a league off Scilly by seven sail of French men of war. [In French service until 1709.]

*Faversham/Feversham* 36/30 - 32/28, 1696-1711. Ellis & Collins, Shoreham.
Dimensions & tons: 107', 88'5½" x 28'1½" x 10'8". 372.
Men: 155/110. Guns: 1703 establishment: GD 8 x 12, UD 22 x 6, QD 6 x 4.
L 1.10.1696. 7.10.1711 'Cast away on the Island of Cape Briton (Cape Breton) at New England'.

*Gosport* 32/28, 1696-1706. Collins & Chatfield, Shoreham.
Dimensions & tons: 107'9", 89'10" x 28'1" x 11'. 376.
Men: 145/100. Guns: 1703 establishment: GD 4 x 9, UD 22 x 6, QD 6 x 4.
L 3.9.1696; 28.8.1706 taken going to Jamaica by a French 54.

*Looe* 32/28, 1696-1697. Plymouth Dyd (Surveyor: Waffe).
Dimensions & tons: 110', 93' x 28' x 11'. 384 80/94.
Men: 145/100. Guns: ____.
L ____ 1696; 30.4.1697 wrecked near Baltimore, Ireland.

*Lynn* 32/28, 1696-1713. Ellis, Shoreham.
Dimensions & tons: 107'9½", 88' x 28'3" x 10'8½". 380.
Men: 145/100. Guns: 1703 establishment: GD 4 x 9, UD 22 x 6, QD 6 x 4.
L ____ 1696; 1713 sold.

*Poole* 32/28, 1696-1737 Nye, Cowes.
Dimensions & tons: 108'6", 90'1½" x 28'2½" x 10'7". 381½.
Men: 145/100. Guns: 1703 establishment: GD 4 x 9, UD 22 x 6, QD 6 x 4.
L ____ 1696; 1719 fireship; 1737 breakwater.

*Rye* 32/28, 1696-1727 Sheerness Dyd (Surveyor: Shortiss).
Dimensions & tons: 109'6½", 90' x 28'4" x 11'3". 384⅛.
Men: 145/100. Guns: 1703 establishment: GD 4 x 9, UD 22 x 6, QD 6 x 4. 1703: 6 x 9, 20 sakers, 6 x 3.
L ____ 1696; 1718 20 gun Sixth Rate; 1727 breakwater.

*Southsea Castle* 32/28, 1696-1697. Knowles, Redbridge.
Dimensions & tons: 106'6", 88'8" x 28'1½" x 10'8½". 373.
Men: 145/100. Guns: 1703 establishment: GD 4 x 9, UD 22 x 6, QD 6 x 4.
L 1.8.1696; 16.9.1697 'cast away at Highlake'.

*Bedford Gally* 34/28 (galley frigate), 1697-1709. Holland, New England.
Dimensions & tons: 103'3½", 85'2" x 28'8" x 10'7½". 372.
Men: 145/100. Guns: 1703 establishment: GD 4 x 9, UD 22 x 6, QD 6 x 4.
L ____ 1697; purchased just after completion for the RN from a Mr Taylor (probably built for the RN from the start); 1709 BU.

*Looe* 36/30 - 32/28, 1697-1705. Portsmouth Dyd (Surveyor: Bagwell).
Dimensions & tons: 108'1", 89'8" x 28'7" x 11'1". 390.
Men: 155/110 Guns: 1703 establishment: GD 8 x 12, UD 22 x 6, QD 6 x 4.
L ____ 1697; 12.12.1705 'Lost in Scratchwell Bay on the back of the Isle of Wight a little to the Eastward of the Needles'.

*Lowestoff/Lastoff/Lowestaffe/Laystaffe* 32/28, 1697-1722. Chatham Dyd (Surveyor: Lee).
Dimensions & tons: 104'4", 89'9" x 27'8" x 10'4". 357.
Men: 145/100. Guns: 1703: 6 x 32 [?], 22 sakers, 4 x 3.
L ____ 1697; 1722 BU.

*Southsea Castle* 32/28, 1697-1699 Deptford Dyd (Surveyor: Harding).
Dimensions & tons: 108', 89'9" x 28'6" x 10'9". 387.
Men: 145/100. Guns: ____.
L ____ 1697; 12.11.1699 wrecked on the 'Isle of Ash' (Point Basque).

*Bridgwater* 36/30 - 32/28, 1698-1738. Sheerness Dyd (Surveyor: Shortiss).
Dimensions & tons: 110'5", 90'11" x 29'2" x 11'5½". 411.
Men: 155/110. Guns: 1703 establishment: GD 8 x 12, UD 22 x 6, QD 6 x 4.
L ____ 1698; 1727 fireship; 1738 BU.

*Hastings* 32/28, 1698-1707. Betts, Woodbridge.
Dimensions & tons: 108'4", 89'10" x 28'3" x 10'6". 381.
Men: 145/100. Guns: 1703 establishment: GD 4 x 9, UD 22 x 6, QD 6 x 4. 1703: 6 x 9, 20 sakers, 6 x 3.

L ____ 1698; 9.2.1707 cast away (or capsized?) off Yarmouth.

*Ludlow* 32/28, 1698-1703. Munday, Woodbridge.
Dimensions & tons: 108', 90' x 28'3" x 10'7". 381.
Men: 145/100. Guns: 1703 establishment: GD 4 x 9, UD 22 x 6, QD 6 x 4.
L 12.9.1698; 16.1.1703 taken by *L'Adroit* and one other French man of war 'about 4 leagues off the Goree'. [In French service until 1717.]

*Kinsale* 36/30, 1700-1723. Kinsale Dyd (Surveyor: Stacey).
Dimensions & tons: 117'6", 99'6" x 31'9" x 12'. 533.
Men: 155/110. Guns: 1703 establishment: GD 8 x 12, UD 22 x 6, QD 6 x 4.
L 22.5.1700; 1723 BU.

*Speedwell* 28/24, 1702-1715. Newman & Graves, Limehouse.
Dimensions & tons: 94'9", 78'10½" x 25'4" x 9'8½". 269 22/94.
Men: 125/90. Guns: 1703 establishment: GD 4 x 9, UD 20 x 6, QD 4 x 4.
RB L 05.1702?; 1715 BU.

*Tartar* 32, 1702-1732. Woolwich Dyd (Surveyor: Harding or Lee?).
Dimensions & tons: 108', 90'9" x 29'6" x 13'. 420.
Men: 145/100. Guns: 1704: 22 sakers, 10 minions.
L 12.9.1702; 1716/20? converted to Sixth Rate of 20 guns; 1732 BU.

NOTE: Possibly a 'pink'?

*Falcon/Faulkon/Faulcon* 32/28, 1704-1709. Deptford Dyd (Surveyor: Harding).
Dimensions & tons: 106'5", 88'5" x 29'7" x 13'. 411 53/94.
Men: 145/100. Guns: GD 4 x 9, UD 22 x 6, QD 6 x 4.
L 12.1704; 29.12.1709 taken by the French *Le Sérieux* in the Mediterranean.

*Fowey* 32, 1705-1709. Chatham Dyd (Surveyor: Shortiss). 32/28.
Dimensions & tons: 108', 89' x 29'6" x 13'. 411 92/94.
Men: 145/100. Guns: GD 4 x 9, UD 22 x 6, QD 6 x 4.
L 14.3.1705; 14.4.1709 taken by a French 44 and a 40 on passage from Alicante to Lisbon.

NOTE: Possibly a 'pink'.

*Milford* 32, 1705-1728. Deptford Dyd (Surveyor: Allin). 36/30.
Dimensions & tons: 108'7½", 88'11" x 29'10" x 11'10½". 420 88/94.
Men: 145/100. Guns: GD 4 x 9, UD 22 x 6, QD 6 x 4.
RB L ____ 1705; 18.6.1728 wrecked on the Island of Cuba.

*Winchelsea* 36, 1706-1707. Johnson, Blackwall. 36/30.
Dimensions & tons: 105'6", 86' x 30'4½" x 12'7". 422.
Men: 145/100. Guns: GD 4 x 9, UD 22 x 6, QD 6 x 4.
L 09.1706; 2.9.1707 lost in a hurricane 'near St. Christopher' (St. Kitts) or 'off Antego' (Antigua).

## Large Sixth Rates, 20-24 Guns

*Jersey* 24/22, 1693-1698 *Margate*-1707. Deptford Dyd (Surveyor: Harding).
Dimensions & tons: 94'6", 81' x 24'8" x 10'8". 262.
Men: 115/85. Guns: 1703 establishment: UD 20 x 6, QD 4 x 4.
L 17.1.1694; 1698 renamed *Margate*; 1.12.1707 wrecked near Carthagena, 'New Spain'.

*Lizard* 24/22, 1693-1696. Chatham Dyd (Surveyor: Lee).
Dimensions & tons: 94'3", 79'6" x 24'4" x 10'8". 250.
Men: ____. Guns: ____.
L ____ 1693; 31.5.1696 wrecked 'in the Streights' (Mediterranean off Toulon?).

*Maidstone* 24/22, 1693-1714. Chatham Dyd (Surveyor: Lee).
Dimensions & tons: 94'3", 79'6" x 24'4" x 10'8". 250.
Men: 115/85. Guns: 1703 establishment: UD 20 x 6, QD 4 x 4. 1705: 20 sakers, 4 x 3.
L ____ 1693; 1714 sold.

*Falcon/Faulcon* 24/22, 1694-1695. Barrett, Shoreham.
Dimensions & tons: 91'6", 77'6" x 24'6½" x 10'8". 240.
Men: ____. Guns: ____.
Ord 28.3.1694; L ____ 1694; 3.1.1695 taken by the French; 1703 retaken by *Romney* and BU.

*Drake* 24/22, 1694-1694. Fowler, Rotherhithe.

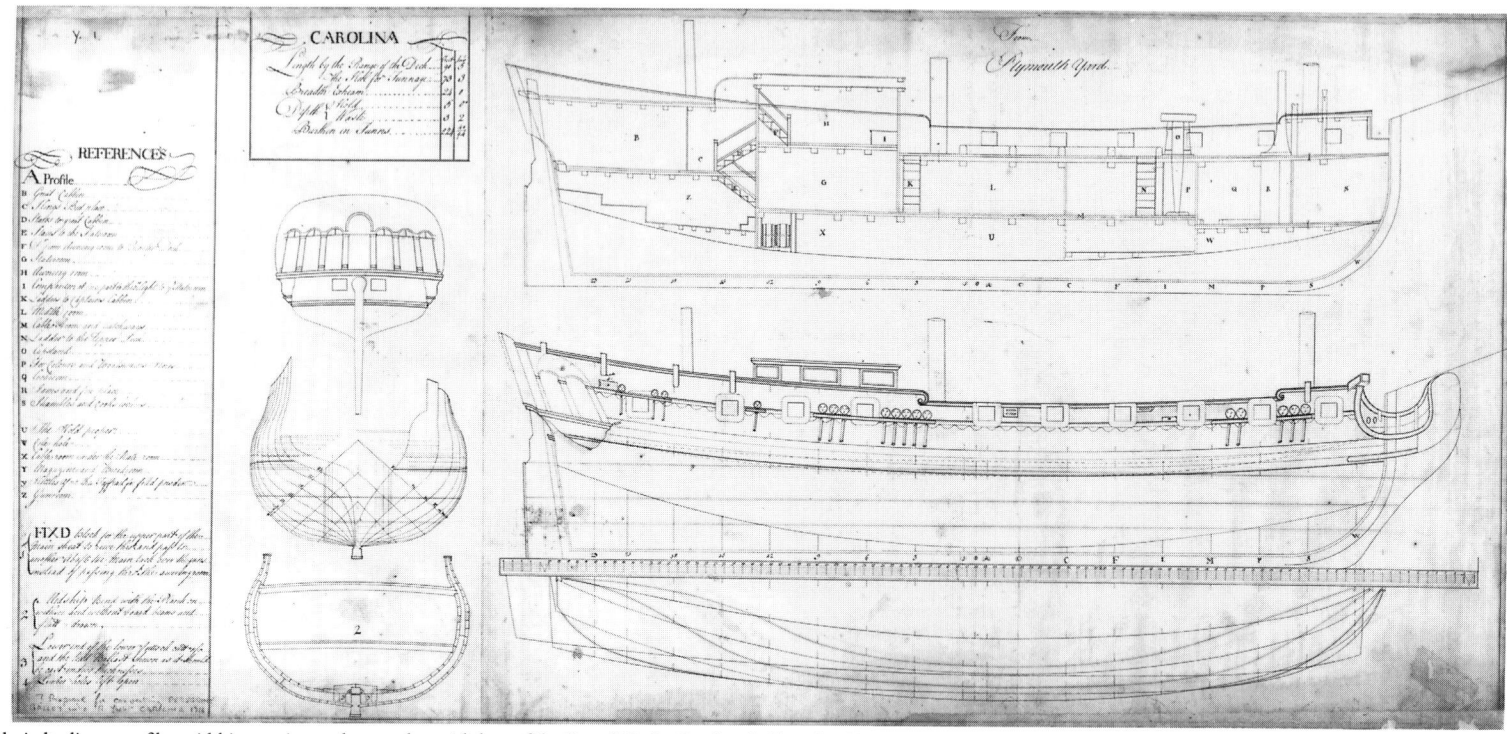

Admiralty lines, profile, midships section and stern plan with key of the Royal Yacht *Carolina*, built as the 20 gun *Peregrine Galley* in 1700

Dimensions & tons: 93', 77'8" x 24'9" x 10'8". 253.
Men: ____ . Guns: ____ .
Ord 2.5.1694; L ____ 1694; 20.12.1694 wrecked off Ireland.

*Swann* 24/22, 1694-1707. Castle, Deptford.
Dimensions & tons: 93'3", 78'1" x 24'6" x 10'8". 249.
Men: 115/85. Guns: 1703 establishment: UD 20 x 6, QD 4 x 4.
Ord 2.5.1694; L ____ 1694; 17.8.1707 foundered in a violent storm coming from Jamaica.

*Newport* 24/22, 1694-1696. Portsmouth Dyd (Surveyor: Stigant).
Dimensions & tons: 94'3", 78'9" x 24'7" x 10'11". 253.
Men: ____ . Guns: ____ .
L ____ 1694; 6.7.1696 taken by the French.

*Queenborough/Quinbrough* 24/22, 1694-1719. Sheerness Dyd (Surveyor: Bagwell).
Dimensions & tons: 96'4", 80'4½" x 24'9" x 10'10". 261.
Men: 115/85. Guns: 1703 establishment: UD 20 x 6, QD 4 x 4. 1717: 2 sakers, 18 minions, 4 x 3.
L ____ 1694; 1719 sold.

*Seahorse* 24/22, 1694-1704. Haydon, Limehouse.
Dimensions & tons: 93'10", 78'8" x 24'9" x 10'11". 256.
Men: 115/85. Guns: 1703 establishment: UD 20 x 6, QD 4 x 4.
Ord 6.5.1694; L ____ 1694; 14.3.1704 wrecked on rocks off the north east side of Jamaica.

*Solebay* 24/22, 1694-1709. Snelgrove, Redhouse.
Dimensions & tons: 92'1", 77'8" x 24'11" x 10'8". 256.
Men: 115/85. Guns: 1703 establishment: UD 20 x 6, QD 4 x 4.
L ____ 1694; 25.12.1709 wrecked on the Boston Rock near Lyme.

*Penzance* 24/22, 1694/1695-1713. Ellis, Shoreham.
Dimensions & tons: 94'3", 77'10½" x 24'9" x 10'8". 246.
Men: 115/85. Guns: 1703 establishment: UD 20 x 6, QD 4 x 4.
Ord 3.10.1694; L 22.4.1695 or ____ 1695; 1713 sold.

*Biddeford/Biddiford* 24/22, 1695-1699. Barrett, Harwich.
Dimensions & tons: 93'1", 78'6" x 24'9" x 10'9". 255¾.
Men: ____ . Guns: ____ .
Ord 11.5.1694; L ____ 1695; 12.11.1699 wrecked on the 'Isle of Ash' (Point Basque).

*Dunwich* 24/22, 1695-1714. Collins & Chatfield, Shoreham.
Dimensions & tons: 93'7", 78'4" x 24'6½" x 10'8". 250.
Men: 115/85. Guns: 1703 establishment: UD 20 x 6, QD 4 x 4. 1701: 18 sakers, 4 x 3.
L ____ 1695; 1714 breakwater.

*Seaford* 24/22, 1695-1697. Herring, Bursledon.
Dimensions & tons: 98'5", 81'2" x 26'1" x 10'10". 293 60/94.
Men: ____ . Guns: ____ .
L 27.12.1695; 5.5.1697 taken and burnt by the French off the Scilly Islands.

*Orford* 24/22, 1695-1698, *Newport*-1714. Ellis, Shoreham.
Dimensions & tons: 93'8", 77'6" x 24'8" x 10'8". 244.
Men: 115/85. Guns: 1703 establishment: UD 20 x 6, QD 4 x 4.
L 19.11.1695; 1698 renamed *Newport*; 1714 sold.

*Lizard* 24/22, 1697-1714. Sheerness Dyd (Surveyor: Shortiss).
Dimensions & tons: 95', 79'4" x 25' x 10'10". 263.
Men: 115/85. Guns: 1703 establishment: UD 20 x 6, QD 4 x 4.
L early 1697; 1714 sold.

*Deal Castle* 24/22, 1697-1706. Deptford Dyd (Surveyor: Harding).
Dimensions & tons: 91'11", 77'6" x 24'1¼" x 10'9½". 240.
Men: 115/85. Guns: 1703 establishment: UD 20 x 6, QD 4 x 4.
L ____ 1697; 3.7.1706 taken by three French ships off Dunkirk.

*Flamborough* 24/22, 1697-1705. Chatham Dyd (Surveyor: Lee).
Dimensions & tons: 94', 78' x 24'8" x 10'4". 252.
Men: 115/85. Guns: 1703 establishment: UD 20 x 6, QD 4 x 4.
L 10.7.1697; 10.10.1705 taken by the French *Le Jason* 54 and sank 20 leagues north west by west of Cape Spartel.

*Seaford* 24/22, 1697-1722. Portsmouth Dyd (Surveyor: Bagwell).
Dimensions & tons: 93'2", 77'2" x 24'7" x 10'10". 248.
Men: 115/85. Guns: 1703 establishment: UD 20 x 6, QD 4 x 4.
L 15.10.1697; 1722 BU.

*Peregrine Galley* 20, 1700-1716, *Carolina*-1733. Sheerness Dyd (Surveyor: Lee).
Dimensions & tons: 86'6", 71' x 22'10" x 10'. 196 84/94.
Men: 50. Guns: 1703: 16 x 6 (?), 4 x 3.
L ____ 1700; 1716 became royal yacht *Carolina*; 1733 BU & RB.

*Squirrel* 24, 1703-1703. Portsmouth Dyd (Surveyor: Podd).
Dimensions & tons: 93'6", 80' x 24'8" x 10'8". 258.
Men: ____ . Guns: 1703: 20 sakers, 4 x 3.
L 14.6.1703?; 2.10.1703 taken by two French privateers off Dungeness.

*Nightingale* 24, 1703-1707/1708, *Fox*-1724. Chatham Dyd (Surveyor: Shortiss).
Dimensions & tons: 93', 78'9" x 24'6" x 10'8". 251 40/94.
Men: 115/85. Guns: GD 20 x 6, QD 4 x 4.
L 19.12.1702; 1707 taken by six French galleys off Harwich; 01.1708 taken by *Ludlow Castle* and renamed *Fox*; 1724 BU.

*Squirrell/Squerrell* 24, 1704-1706. Portsmouth Dyd (Surveyor: Podd).
Dimensions & tons: 93'6", 80' x 24'8" x 10'8". 258 85/94.

Men: ____ . Guns: 1705: 20 x 6, 4 minions.
L 11.1704 (?); 7.7.1706 taken by three French ships 'on the back of the Goodwin'. [The French renamed her *L'Ecureil*]; 1708 retaken and sank.
*Alborough/Aldbrough* 24, 1706-1727. Johnson, Blackwall.
Dimensions & tons: 94'3", 74'11" x 26'10½" x 11'1½". 287¾.
Men: 115/85. Guns: GD 20 x 6, QD 4 x 4.
L 6.3.1706; 1727 BU.
*Deal Castle* 24, 1706-1722. Burchett, Rotherhithe.
Dimensions & tons: 98'2", 74'6" x 26'2½" x 11'. 272.
Men: 115/85. Guns: GD 20 x 6, QD 4 x 4.
L 9.9.1706.

## Smaller Sixth Rates

*Royal Transport* 18, 1695-1698. Chatham Dyd (Surveyor: Lee).
Dimensions & tons: 90', 75' x 23'6" x 7'9". 220.
Men: 100. Guns: 18. x ____ .
L 11.12.1695; 1698 given to the 'Czar of Muscovy' (Peter the Great).
NOTE: Experimental design by the Earl of Denbigh (Peregrine Butler).

## Fireships

NOTE: *Roebuck, Speedwell* and, probably, *Dolphin* were begun and possibly completed as fireships, but before or soon after completion converted to Fifth Rates; see under Small Fifth Rates for details.

*Fox* 8, 1690-1692. Barrett, Shoreham.
Dimensions & tons: 93'4", 79'1" x 25'1" x 9'8". 264 63/94.
Men: 45. Guns: 8 x ____ .
Ord 6.12.1689; L 16.4.1690; 19.5.1692 expended at the battle of Barfleur.
*Griffin* 8, 1690-1701. Castle & Rolf, Rotherhithe.
Dimensions & tons: 94'9½", 82'4" x 24'8½" x 9'9¼". 266.
Men: 45. Guns: 8 x ____ .
Ord 6.12.1689; L 17.4.1690; 1701 BU.
*Hawk* 8, 1690-1712. Frame, Wapping.
Dimensions & tons: 94'10½", 86'8" x 25' x 9'7". 288.
Men: 45. Guns: 8 x ____ .
Ord 6.11.1689; L 17.4.1690; 1712 breakwater.
*Hopewell* 8, 1690-1690. Ellis, Shoreham.
Dimensions & tons: 93'8½". 79'4" x 24'10½" x 9'8½". 261.
Men: 45. Guns: 8 x ____ .
L 15.4.1690; 3.6.1690 burnt by accident in the Downs.
*Hound* 8, 1690-1690. Graves, Limehouse.
Dimensions & tons: 93'8", 81'8½" x 25'x 9'8". 271 60/94.
Men: 45. Guns: 8 x ____ .
L ____ 1690; 22.5.1692 expended at Cherbourg.
*Hunter* 8, 1690-1710. Shish, Rotherhithe.
Dimensions & tons: 93'6", 84'7" x 24'10 x 9'11". 277.
Men: 45. Guns: 1707: 6 minions, 2 falconets.
Ord 6.12.1689; L 29.4.1690; 1710 reclassed as 24 gun Fifth Rate; 20.10.1710 taken by the Spaniards off Cape Saint Mary.
*Spy* 8, 1690-1693. Taylor, Rotherhithe.
Dimensions & tons: 91'6", 80' x 25'3½" x 9'6½". 272.
Men: 45. Guns: 8 x ____ .
Ord 6.12.1689; L 6.4.1690; 12.1.1693 blown up by accident at Portsmouth.
*Vulture* 8, 1690-1708. Castle, Rotherhithe.
Dimensions & tons: 93'1", 82'2" x 24'10¼" x 9'7½". 270 30/94.
Men: 45. Guns: 8 x ____ .
Ord 6.12.1689; L 18.4.1690; 10.2.1708 taken coming from the West Indies by the French (and sank?).
*Wolf/Woolf* 8, 1690-1692. Castle, Deptford.
Dimensions & tons: 93'1", 81'10" x 24'11" x 9'9". 272 23/94.
Men: 45. Guns: 8 x ____ .
Ord 6.12.1689; L 18.4.1690; 22.5.1692 expended at Cherbourg.
*Flame* 8, 1690-1697. Gressingham, Rotherhithe.
Dimensions & tons: 91'7", 80' x 25'4½" x 9'8". 273.
Men: 45. Guns: 8 x ____ .
Ord 31.10.1690; L 6.3.1690; 15.9.1697 foundered in the Atlantic coming from Barbados.
*Aetna/Etna* 8, 1691-1697. Frame, Hull.
Dimensions & tons: 90'11¼", 81'3" x 25'7½" x 9'6". 283.
Men: 45. Guns: 8 x ____ .
Ord 20.10.1690; L 19.3.1691; 18.4.1697 taken by the French. [Became French *L'Aetna*.]
*Blaze* 8, 1691-1692. Snellgrove, Deptford.
Dimensions & tons: ____ , 76'1" x 25'4" x ____ . 259 68/94
Men: 45. Guns: 8 x ____ .
Ord 31.10.1690; L 5.3.1691; 22.5.1692 expended at Cherbourg. [The contract in the PRO for this ship is dated 31.10.1690.]
*Lightning* 8, 1691-1705. Taylor, Rotherhithe.
Dimensions & tons: 91'2", 80'x 25'2"x 9'8½". 270.
Men: 45. Guns: 8 x ____ .
Ord 31.10.1690; L 20.3.1691; 24.11.1705 taken by the French.
*Phaeton* 8, 1691-1692. Castle, Deptford.
Dimensions & tons: 91'5", 79' x 25'7½" x 9'6". 263.
Men: 45. Guns: 8 x ____ .
Ord 31.10.1690; L 19.3.1691?; 19.5.1692 expended at the Battle of Barfleur.
*Strombolo* 8 later 24, 1691-1709. Johnson, Blackwall.
Dimensions & tons: 91'6", 78'1½" x 25'4" x 9'6¾". 266.
Men: 45. 1704: ____ . Guns: 8 x ____ . 1704: ____ .
L 7.3.1691?; 1704 converted to a 28/24 gun Fifth Rate; 1713 sold.
*Vesuvius* 8, 1691-1692/1693. Taylor, Rotherhithe.
Dimensions & tons: 92', 80' x 25'2" x 9'8½". 248
Ord 31.0.1690; L 30.3.1691; 19.11.1693 expended at Saint Malo.
*Vulcan* 8, 1691-1708. Shish, Rotherhithe.
Dimensions & tons: 91'2", 80' x 25'4½" x 9'6". 273.
Men: 45. Guns: 8 x ____ .
Ord 31.10.1690; L 21.2.1691?; 1708 breakwater.
*Blaze* 8, 1694-1697. Johnson, Blackwall.
Dimensions & tons: 92', 77'8" x 24'9" x 10'. 253.
Men: 45. Guns: 8 x ____ .
L ____ 1694 [may have been purchased from Johnson in 04.1694 whilst building]; 05.1697 taken by the French off Scilly.
*Firebrand* 8, 1694-1707. Haydon, Limehouse.
Dimensions & tons: 92'3", 78' x 25'5" x 9'7". 268.
Men: 45. Guns: 1703: 6 minions, 2 falconets.
Ord 13.12.1693; L 31.3.1694; 22.10.1707 wrecked off Scilly (Cloudesley Shovell's Fleet). [There are reports that this wreck has been found recently.]
*Phoenix/Phenix* 8, 1694-1708. Gardner & Dalton, Rotherhithe.
Dimensions & tons: 91', 76' x 25'2½" x 9'10½". 256.
Men: 45. Guns: 1703: 6 minions, 2 falconets.
Ord 8.12.1693; L ____ 1694; 22.10.1707 stranded off Scilly (Cloudesley Shovell's Fleet) but got off again; 1708 BU.
*Griffin* 8, 1702-1737. Sheerness Dyd (Surveyor: Allin).
Dimensions & tons: 94'7", 78'9½" x 25'1½" x 9'6". 263 63/94.
Men: 45. Guns: 8 x ____ .
RB L early 1702; 1737 sold.
PLANS: ? Lines & profile.

## Bomb Vessels

As this type of vessel first appeared in the Royal Navy a year before our period begins, and had only been invented by the French a year before that (for an attack on the corsair power of Algiers), it seems logical to begin listing the type in this chapter, rather than putting the first example in the introductory chapter.

It would appear that the Royal Navy rapidly modified the original French concept to include mortars mounted on traversing turntables rather than fixed to fire forwards. Unfortunately, the available information is fragmentary so the details of this development are difficult to sort out. This is a pity as the bomb, with the fireship, was a very early example of a true 'weapons system', and it can be argued that the traversing turntable for the mortar was the ancestor of the rotating turret and other traversing gun mountings. After the initial prototypes craft of this type appear to have been built in groups (or classes) to common dimensions and even, possibly, to common designs. So, for the first time in this work, a group of ships is listed under a common head, with the complement and armament being given at the start of the entry rather than for each individual vessel, dimensions placed beneath the entry or entries to which they refer, whilst extended details of the vessels' respective fates and notes on the plans are given at the end of the entry.

*Salamander* 10, 1687-1703. Chatham Dyd (Surveyor: Lee).
    Dimensions & tons: 64'4", 54'6" x 21'6" x 8'4". 134.
    Men: 35. Guns: 1 or 2 mortars + 10 x 2.
    L ____ 1687; 1703 BU.

NOTE: The first English bomb vessel. Probably a copy of the original French craft of this kind.

*Firedrake* 12, 1688-1689. Deptford Dyd (Surveyor: Harding).
    Dimensions & tons: 85'9", 68' x 27' x 9'10". 279.
    Men: ____. Guns: 1 mortar? + 12 x ____ (as sloop 16 x ____ ).
    L —— 1688; 12.11.1689 taken by the French; [renamed *La Tempête* as a 'frégate' and in French service until 1726].

*Portsmouth* 2, 1694-1703. Deptford Dyd (Surveyor: Harding).
    Dimensions & tons: 59', ____ x 21'4" x 9'. 142.
    Men: 30. Guns: 2 mortars + 2 guns.
    RB/conversion of 1679 yacht; L ____ 1688; 27.11.1703 foundered at anchor at the Nore during the 'Great Gale'.

*Kitchen/Kitchin* 8. 1692?-1698 Castle, Rotherhithe.
    Dimensions & tons: 59', 49'6" x 19'6" x 8'. 100.
    Men: 30. Guns: 8 x ____.
    RB/conversion of 1670 yacht; sold 1698.

## 1693 Group

Men: 65. Guns: 1 or 2 mortars? + 12 x 6 + 6 swivels ('patteroes').

NOTE: *Mortar* is known to have been a two-decked ship-rigged vessel. *Serpent* built at the same yard by the same constructor was presumably a sister, and the two others are likely to have been similar, though dimensions varied as can be seen below.

*Firedrake* Deptford Dyd (Surveyor: Harding).
    L 06.1693; 1703 foundered.
    85'2½", 66' x 24'1" x 9'10". 202 86/94.

*Granadoe* Fowler, Rotherhithe.
    L 26.6.1693; 1694 blown up.
    87', 73' x 26'10" x ____ . 279.

*Mortar* Chatham Dyd (Surveyor: Lee).
    L ____ 1693; 1703 wrecked.
    85', 66" x 24' x 9'10". 279.

*Serpent* Chatham Dyd (Surveyor: Lee).
    L ____ 1693; 1694 lost.
    86', 69'9" x 26'6" x 9'9". 260 46/94.

FATES: *Firedrake* foundered 'in the Streights' (Mediterranean) on 12.10.1703. *Granadoe* blew up before Dieppe (or Le Havre?) on 16.7.1694. *Mortar* was wrecked on the coast of Holland on 2.12.1703. *Serpent* was lost 'coming into the Streights' (near Gibraltar?) on 19.2.1694.

PLANS: Danish Rigsarkivet: *Mortar*?: Profile & midsection.

*Star/Starr* 8, 1694-1712. Johnson, Blackwall.
    Dimensions & tons: ____ , 53'11" x 26'2½" x 9'. 117.
    Men: 35. Guns: 8 x ____ (as sloop?)
    L ____ 1694 (may well have been purchased - in 04.1694 - whilst building); 29.5.1712 'cast away on a shoal of rocks off the Island Heneago to windward off Jamaica'.

## 1695 Group

Men: 30. Guns: 1 or 2 mortars? + 4 x ____ . Clearly much smaller vessels than the 1693 ones, and very probably ketch rigged.

*Blast* Johnson, Blackwall.
    L ____ 1695; 1721 yard craft.
    66', 50'6" x 23'2" x 10'. 143.

*Carcass* Taylor, Rotherhithe.
    L ____ 1695; 1713 sold.
    66'6", 50'6" x 23'2" x 10'. 143 14/94.

*Comet* Johnson, Blackwall.
    L ____ 1695; 1706 taken.
    66'2", 50'6" x 23'2" x 10'. 144.

*Dreadfull* Graves, Limehouse.
    L 6.5.1695; 1695 burnt.
    66'10½", 50'6" x 23'6" x 10'1". 147 23/94.

*Furnace* Wells, Horsleydown.
    L 18.4.1695; 1725 BU.
    65'6", 50'6" x 23'4" x 10'. 144 21/94.

*Granadoe* Castle, Deptford.
    L ____ 1695; 1718 BU.
    64'5", 50'6" x 23'8½" x 10'. 147 75/94.

*Serpent* Chatham Dyd (Surveyor: Lee).
    L ____ 1695; 1703 taken.
    65'6", 49'8" x 23' x 10'. 139 71/94.

*Terror/Terrour* Davis, Limehouse.
    L 11.1.1696; 1705 taken & blown up.
    65'6", 50'6" x 23'6½" x 10'2". 149.

*Thunder* Snellgrove, Limehouse.
    L ____ 1695; 1696 taken.
    65'6", 50'6" x 23'6" x 10'. 147 88/94.

FATES: *Blast* was a yard craft at Port Mahon until deleted in 1724. *Comet* was taken by the French on 10.10.1706; served as *La Comète* until 1708. *Dreadfull* was burnt off Saint Malo on 5.7.1695. *Serpent* was taken in the Atlantic on 15.10.1703 by French privateers. *Terror* was taken by the French at Gibraltar on 17.10.1705 and then blown up. *Thunder* was taken off the Dutch coast on 6.4.1696 by a French privateer.

*Basilisk* 4?, 1695-1729. Redding, Wapping. 4?
    Dimensions & tons: 72'2", 57'4" x 23'2" x 10'2". 163 ¼.
    Men: 30. Guns: 1 or 2 mortars? + 4 x ____ (minions in 1714).
    L ____ 1695; 1729 BU.

*Salamander* 4?, 1703-1713. Woolwich Dyd (Surveyor: Lee).
    Dimensions & tons: 66', 51'4" x 21'2" x 8'8½". 122.
    Men: 30. Guns: 1 or 2 mortars? + 4 x ____ ?
    RBT ____ 1703; 1713 sold.

## Ketches

Probably ketch-rigged (ie. two masts, with the forward one larger than the after one, and referred to as the mainmast. At this stage would carry square sails as well as fore and aft ones, and can be visualised as a ship-rigged vessel with the foremast removed and the other two moved forward to compensate.

*Albrough* 10, 1691-1696. Johnson, Aldborough.
    Dimensions & tons: ____ . 52'9" x 18'10" x 9'1". 100.
    Men: 50. Guns: 10 x ____ .
    Ord 31.10.1690; L ____ 1691; 17.8.1696 accidentally blown up.

*Eaglet* 10, 1691-1693. Shish, Rotherhithe.
    Dimensions & tons: 62'6", 52' x 18'10" x 9'0½". ____ .
    Men: 50. Guns: 10 x ____ .
    Ord 31.10.1690; L 7.4.1691; 05.1693 taken by the French off Arran.

*Harpe* 10, 1691-1693. Frame, Scarborough.
    Dimensions & tons: ____ , ____ x ____ x ____ . ____ .
    Men: 50. Guns: 10 x ____ .
    Ord 20.10.1690; L ____ 1691; (17.6.1693 new style) taken by the French.

*Hart* 10, 1691-1692. Castle, Rotherhithe.
    Dimensions & tons: 62'6", 52'8" x 18'11¼" x 9'1". 96.
    Men: 50. Guns: 10 x ____ .

Ord 31.10.1690; L 23.3.1691; 05.1692 taken by two French privateers off Saint Ives.

*Hind* 10, 1691-1697. Snellgrove, Wapping.
Dimensions & tons: 63', 52'6" x 18'8½" x 9'1". 96.
Men: 50. Guns: 10 x ____ .
Ord 31.10.1690; L. 2.4.1691; 7.1.1697 taken by the French in the North Sea.

*Roe* 10, 1691-1697. Haydon, Limehouse.
Dimensions & tons: 62'2", 48'8" x 18'10" x 9'0½". 92.
Men: 50. Guns: 10 x ____ .
L 8.4.1691; 08.1697 or 12.10.1697 stranded Virginia (York River?).

*Scarborough* 10, 1691-1694. Frame, Scarborough.
Dimensions & tons: ____ , ____ x ____ x ____ . ____ .
Men: 50. Guns: 10 x ____ .
Ord 20.10.1690; L ____ 1691; 12.1.1694 taken by the French.

*Martin* 10, 1694-1702. Parker, Southampton.
Dimensions & tons: ____ , 52'7" x 19' x 8'4". 99 80/94.
Men: 50. Guns: 10 x ____ .
Ord 2.7.1694; L ____ 1694; 29.8.1702 taken off Jersey by three French privateers.

## Pinks

The distinguishing feature of the pink appears to be its stern, rounded up to the upper deck and above that the gunwales 'pinched in' to form a a very small transom and a platform to take the rudder head. Similar to the Dutch 'Flute' in appearance and (probably) in construction. Probably ship rigged; and much used as armed transports.

*Talbott* 10, 1691-1691/1693-1694. Taylor, Rotherhithe.
Dimensions & tons: 62', 51'2" x 18'9¼" x 9'2". 94.
Men: 50. Guns: 10 x ____ .
Ord 31.10.1690 (as ketch?); L ____ 1691; 19.6.1691 (new style) taken by the French; 10.11.1693 (new style) retaken; 15.10.1694 wrecked.

*Paramour* 10/6, 1694-1706. Deptford Dyd (Surveyor: Harding).
Dimensions & tons: 64', 52' x 18' x 9'7". 89.
Men: 50/25/10. Guns: 10 x ____ or 6 x ____ + 2 swivels.
L ____ 1694; 1706 sold.

NOTE: This was the vessel commanded by the astronomer Halley; not an altogether successful experiment!

*Wrenn* 10, 1694-1697. Stigant, Redbridge.
Dimensions & tons: ____ , 53'7" x 19'2" x 8'4". ____ .
Men: 50. Guns: 10 x ____ .
Ord 2.11.1694; L ____ 1694; 04.1697 taken by the French.

## Brigantines

This type of vessel appears to have been a small, light, fine-lined vessel intended to row with sweeps as well as to sail, see Robert Gardiner on the type in Lavery (ed), *The Line of Battle*. Used as a despatch vessel and scout.

*Shark* 8?, 1691-1698. Deptford Dyd (Surveyor: Harding).
Dimensions & tons: 57'10", 48' x 15'1" x 5'2". 58.
Men: 30. Guns: 8 x ____ (including swivels?).
L ____ 1691; 1698 sold.

*Diligence* 6, 1692-1708. Deptford Dyd (Surveyor: Harding).
Dimensions & tons: 63'3", 53' x 16'9" x 6'2¼". 80.
Men: 35. Guns: 6 x ____ + 4 falconets.
L ____ 1692; 1708 sold.

*Dispatch* 6, 1692-1712. Deptford Dyd (Surveyor: Harding).
Data as *Diligence* above.
L ____ 1692; sold 1712.

*Discovery* 6, 1692-1705. Woolwich Dyd (Surveyor: Lawrence).
Dimensions & tons: 64', 52' x 16' x 5'7½". 75.
Men: 35. Guns: 6 x ____ + 4 falconets.
L ____ 1692; 1705 BU.

*Spy/Spye* 6, 1693-1706. Woolwich Dyd (Surveyor: Lawrence).
Data as *Discovery* above.
L ____ 1693; 1706 BU.

*Fly/Flye* 6, 1695- 1712. Portsmouth Dyd (Surveyor: Bagwell).
Dimensions & tons: 61'6", 52'x 20' x 6'. 70.
Men: 35. Guns: 6 + 2 swivels.
L ____ 1695; 1712 sold.

*Intelligence* 6, 1695-1700. Woolwich Dyd (Surveyor: Lawrence).
Dimensions & tons: ____ , 52' x 16'6" x 6'. 75.
Men: 35. Guns: 6 x ____ + 2 swivels.
L ____ 1695; 3.3.1700 wrecked on the Isle of Man (Douglas Bay?).

*Post Boy* 6, 1695-1702. Deptford Dyd (Surveyor: Harding).
Dimensions & tons: ____ , 51'6" ´x 16'8½" x 6'5". 76 40/94.
Men: 35. Guns: 6 x ____ + 2 swivels.
L late 1695; 30.5.1702 taken by the French off Beachy Head.

*Swift* 6, 1695-1696. Chatham Dyd (Surveyor: Lee).
Dimensions & tons: 63', 52' x 17' x 6'. 79 88/94.
Men: 35. Guns: 6 x ____ + 2 swivels.
L ____ 1695; 17.8.1696 foundered.

## Sloops

Like the bombs these vessels appear to have been built in groups, and are treated as such here with the design data given at the start of the entry and details of fates and plans grouped at the end. These early sloops were very small and lightly armed vessels indeed, and there are some grounds for suspecting that they were 'sloop rigged'; in other words fitted the mercantile definition of a sloop - a vessel with one mast rigged with fore and aft sails.

### Two Gun Group

For dimensions and tons see individual entries.
Men: 35/50. Guns: 2 falconets + 2 pattaroes (swivels).

*Bonetta* Deptford Dyd (Surveyor: Miller).
L ____ 1699; 1712 sold.
Dimensions & tons: 58'2", 48' x 16'1" x 6'1". 66 4/94.

*Fox* Sheerness Dyd (Surveyor: Lee).
L ____ 1699; 1699 wrecked.
Dimensions & tons: 58'6", 47'6" x 16' x 6'. 64 64/94.

*Merlin* Chatham Dyd (Surveyor: Shortiss).
L ____ 1699; 1712 sold.
Dimensions & tons: 59'2", 48'10" x 16' x 6'2½". 66 46/94.

*Prohibition* Woolwich Dyd (Surveyor: Harding).
L ____ 1699; 1702 taken.
Dimensions & tons: ____ .

*Shark* Deptford Dyd (Surveyor: Miller).
L ____ 1699; 1703 taken.
Dimensions & tons: ____ , 48' x 16'1" x 6'1". 66.

*Swallow* Chatham Dyd (Surveyor: Shortiss).
L ____ 1699; 1703 taken.
Dimensions & tons: ____ , 48'10" x 16' x 6'2½". 66 46/94.

*Swift* Portsmouth Dyd (Surveyor: Waffe).
L ____ 1699; 1702 taken.
Dimensions & tons: ____ , 48' x 16'6" x 6'. 65

*Woolfe* Portsmouth Dyd (Surveyor: Waffe).
L ____ 1699; 1704 taken; 1704? retaken?; 1708 taken; 1708 retaken; 1712 sold.
Dimensions & tons: ____ .

FATES: *Fox* cast away on the coast of Ireland, 12.1699. *Prohibition* taken by the French off Land's End on 14.8.1702. *Shark* taken by the French 29.3.1703. *Swallow* taken by a French privateer 31.3.1703. *Swift* taken whilst on passage to the West Indies by a French privateer 18.18.1702. *Woolfe* taken 24.6.1704 by a French privateer of 10 guns; retaken (12.1704?) by the Dutch and handed back to the RN ('salvage paid' to the Dutch); 19.6.1708 taken by the French between Milford and Dublin; 21.6.1708 retaken by *Speedwell*.

### Four Gun Group

Men: 50/35. For individual dimensions and tonnage see under each sloop. Guns: 4 falconets + 4 pattaroes (swivels).

***Hound*** Deptford Dyd (Surveyor: Miller).
　　L ____ 1700; 1714 sold.
　　Dimensions & tons: 61', 50' x 17'8" x 7'8". 83.
***Otter*** Deptford Dyd (Surveyor: Miller).
　　L 02.1701?; 1702 taken.
　　Dimensions & tons: 61'1", 50' x 17'8" x 7'8", 83.
FATES: *Otter* taken 28.7.1702 by two French vessels off Land's End.
***Ferret*** 10, 1704-1706. Dummer, Blackwall.
　　Dimensions & tons: 72'. 60'0½" x 20'0½" x 7'3". 128 26/94.
　　Men: 80. Guns: 8 minions + 2 falconets.
　　L. 9.9.1704; 25.5.1706 taken by the French.
***Swift*** 12/10, 1704-1719. Woolwich Dyd (Surveyor: Lee).
　　Dimensions & tons: 73'5", 60'6" x 19'7" x 7'6". 123 39/94.
　　Men: 80. Guns: 10 x 3 (7' length) + 2 falconets + 10 swivels?
　　L 25.10. or 1.11.1704.; 1719 sold.
***Weazle/Weasell*** 10, 1704-1712. Dummer, Blackwall.
　　Dimensions & tons: 71'11", 59'11½" x 20'0¼" x 7'1½". 128 8/94.
　　Men: 80. Guns: 8 minions + 2 falconets.
　　L 9.9.1704; 1712 sold.
***Drake*** 14, 1705-1715. Woolwich Dyd (Surveyor: Poulter).
　　Dimensions & tons: 83'8", 70'2½" x 21'8¼" x 6'6". 175 61/94.
　　Men: 85. Guns: LD 8 x 3 ('usual weight'), UD 6 x 3 ('lighter').
　　L 7 or 8.11.1705; still in service in 1715, probably disposed of soon after.

## Advice Boats

These were vessels used for scouting and for carrying messages and important people. Rig is uncertain; there are some indications that it may even have been some early form of the schooner rig; but more evidence is needed.

### Group I: Stigant design
　　Dimensions & tons: 61'6", 53'4" x 16'1" x 6'. 73.
　　Men: 40. Guns: 4 x ____ + 6 swivels.
***Fly*** Portsmouth Dyd (Surveyor: Stigant).
　　L ____ 1694; 1695 lost.
***Mercury*** Portsmouth Dyd (Surveyor: Stigant).
　　L ____ 1694; 1697 taken.
FATES: *Fly* wrecked 22.8.1695. *Mercury* taken by a French privateer on 19.6.1697 (new style).

### Group II: Waffe design
　　Dimensions & tons: ____ , 50'11" x 16'5" x 6'. 73.
　　Men: 40. guns: 4 x ____ + 6 swivels.
***Messenger*** Plymouth Dyd (Surveyor: Waffe).
　　L ____ 1694; 1701 lost.
***Post Boy*** Plymouth Dyd (Surveyor: Waffe).
　　L ____ 1694; 1694 taken.
FATES: *Messenger* supposed foundered on her passage back from Maryland during 1701. *Post Boy* taken 1.10.1694 by the French off Calais.

### Group III: Stigant design
　　Dimensions & tons: 65'6", 56'6" x 16' x 6'10". 76 88/94.
　　Men: 40. Guns: 4 x ____ + 6 swivels.
***Express*** Portsmouth Dyd (Surveyor: Stigant).
　　L ____ 1695; 1713 sold.
***Post Boy*** Portsmouth Dyd (Surveyor: Stigant).
　　L ____ 1695; 1695 taken.
FATES: *Post Boy* taken by French privateer *Facteur de Bristol* off Plymouth 3.7.1695.

### Group IV: Stigant design
　　Considerably smaller than the others:
　　Dimensions & tons: ____ , 38' x 13'8" x 6'4". 38.
　　Men: ____ . Guns: 4 x ____ .
***Scout Boat*** Portsmouth Dyd (Surveyor: Stigant).
　　L ____ 1695; 1703 sold.

### Group V
　　Two much larger and more powerfully armed vessels which differed slightly from one another in dimensions.
　　Men: 55. Guns: 10 x ____ .
***Eagle*** Fugar, Arundel.
　　L ____ 1696; 1703 wrecked.
　　Dimensions & tons: 76', 62'5" x 21'1¼" x 8'6". 152 70/94.
***Swift*** Moore, Arundel.
　　L ____ 1696; 1698 wrecked.
　　Dimensions & tons: 76'11", 63'10" x 21'4" x 8'7½". 154 49/94.
FATES: *Eagle* wrecked on the Sussex coast during the 'Great Gale' 27.11.1703. *Swift* was 'cast away on the coast of Virginia' (actually North Carolina?) 24.1.1698.

## Transports, etc

***Greenfish*** 2, 1690-1705. Sheldon & Green, Ireland.
　　Dimensions & tons: 61', 51' x 15'8" x 7'9". 67.
　　Men: 8. Guns: 2 x ____ .
　　L ____ 1690 (Fisher suggests purchased 1693); 1705 sold.

# Chapter 2: The Preliminary Establishment 1706-1718

In 1706 there was an attempt (not apparently over-successful) to standardise the dimensions, if not the design, of the larger warships. This is probably more significant in the retrospect of the 'Establishment' of 1719 and its successors than at the time, but it did mean some degree of uniformity amongst the larger ships. For the first part of the period covered by this chapter (the Act of Union with Scotland was passed in 1707 and with it the small Scots Navy was absorbed into the Royal Navy proper - see the chapter on the Prizes and Purchases of the War of Spanish Succession for the ships concerned) war continued with Louis XIV's France. The trend seen in the previous chapter towards building small two-deckers, Sixth Rates and assorted smaller craft for trade defence continued; there were no significant fleet actions in this period, though the battlefleet itself was maintained, both now and later. The not very successful 'one and a half-deckers' of 30 guns or so continued to be built for a few years, but did not, as a type, continue beyond the end of the war (in 1713). There was to be nothing between the small two-deckers of 40 guns and the Sixth Rates of 20 and 24 guns until the frigate type was copied from the French in the middle of the eighteenth century. By the end of the war, however, one new type had emerged, the two masted sloop (rigged as snows, ketches or 'brigantines'[1]) armed at this stage with 10 or so light guns, and presumably evolving from a smaller, earlier type of 'sloop'. It was to be this type which replaced the mixture of ketches, pinks, brigantines, advice boats and smaller Sixth Rates of less than 20 guns which had formed the lower ranks of fighting ships for the last few decades.

The entries in this chapter are given by ship type, and in order of building within that type. The designation of the type is followed by basic data (except for the Sixth Rates and lesser vessels where the varying dimensions and tonnage are given under the individual ships) as the 'Establishment' meant that the design data for each ship of each type would be similar; individual small variations 'as built' still existed (and continued into the twentieth century, though of progressively less range and importance). The data is given in the form laid down in the previous chapter: length on deck; length on keel for tonnage x breadth x depth in hold; tonnage; complement of men (by this time the earlier variation between overseas and home service, and war and peace, had gone). Designed armament is given deck by deck (and with known alterations indicated in brackets with the relevant date) in the form: number of guns x calibre in weight of shot (pounds - by this time the old English gun names were disappearing from use, as were the weapons themselves) fired. The group entry is then followed by entries for each ship, with name (and variants thereof), year of launch, and builder (Royal Dockyard [and Surveyor] or commercial builder and location). When known, dates of Ord(er), K(eel laying), and L(aunch) are given, followed by alterations of name, classification or function, ending with the date of removal from front-line service. The individual entries are followed by more extended notes on fates [including second-line (harbour) service, or service after capture in foreign navies within square brackets], and finally a brief noting of relevant plans.

## First Rates, 100 Guns (three-deckers)

NOTE: There was not actually an establishment for 100s at this time. *Royal William* and *Britannia* were later listed with the 1719 Establishment (and are placed there in this work) but really belong to this period instead.

Dimensions & tons: 171'9", 139'7" x 49'3" x 19'6". 1801.
Men: 800/780/580 (1745 as 90: 750). Guns: (to 1716: 100 war 90 peace) GD 26 x 32, MD 28 x 18, UD 28 x 9, QD 12 x 6, Fc 4 x 6, RH 2 x 6 (From 1716: 28 x 32, 28 x 24, 28 x 12, 12 x 6, 4 x 6. 1745 as 90: 26 x 32, 26 x 24, 226 x 12, 10 x 6, 2 x 6).

**Royal George** 1715. Woolwich Dyd (Surveyor: Acworth) [RB of *Queen*].
Ord ____ ; K 3.5.1709; L 3.9.1715; 1745 90 gun Second Rate; 1756 renamed *Royal Anne*; 1767 BU.
PLANS: Lines.

## Second Rates, 90 Guns (three-deckers)

Dimensions & tons: 162', 132' x 47' x 18'6". 1551.
Men: 680/500, later 750 (1754 as 80s: 700). Guns: (up to 1716 96 war 86 peace) GD 26 x 32, MD 28 x 18, UD 28 x 9, QD 12 x 6, Fc 4 x 6, RH 2 x 6 (1716: 26 x 32, 26 x 18, 26 x 9, QD & Fc 12 x 6. 1754 as 80s: 26 x 32, 26 x 18, 24 x 9, 4 x 6).

**Blenheim** 1709. Woolwich Dyd (Surveyor: Stacey) [RB of *Duchess*].
Ord ____ ; K ____ 1708; L 15.4.1709; 1740 hulked.
**Neptune** 1710. Johnson, Blackwall [RB].
Ord 12.5.1708; K ____ ; L.6.5.1710; 1724 BU.
**Ossory** 1711 Deptford Dyd (Surveyor: Allin) [RB].
Ord ____ ; K ____ 1708; L 21.7.1711; 1716 renamed *Princess*; 1728 renamed *Princess Royal*; 1763 BU.
**Vanguard** 1710. Chatham Dyd (Surveyor: Rosewell) [RB].
Ord ____ ; K ____ ; L 2.8.1710; 1728 renamed *Duke*; 1733 BU.
**Sandwich** 1715. Chatham Dyd (Surveyor: Rosewell) [RB].
Ord 20.7.1709; K ____ ; L 21.4.1715; 1749 hulked.
**Barfleur** 1716. Deptford Dyd (Surveyor: Stacey) [RB].
Ord ____ ; K 9.12.1713; L 27.6.1716; 1753 Third Rate 80 guns; 1764 hulked.

FATES: *Blenheim* hulked at Portsmouth as hospital ship; 1763 BU. *Sandwich* hulked at Chatham; 1752 converted to lazaretto and transferred to Sheerness; 1770 BU. *Barfleur* hulked at Plymouth; BU 1783.

PLANS: *Ossory*: Lines/decks. 1728 as *Princess Royal*: Lines. *Blenheim*: Danish Rigsarkivet: Lines. As hospital ship 1743 [Admiralty]:decks.

## Third Rates, 80 Guns (three-deckers)

Dimensions & tons: 156', 127'6" x 43'6" x 17'8". 1283.
Men: 520/360. From 1716: 500 or 476. Guns: (up to 1716 80 war 72 peace) GD 26 x 32, MD 26 x 18, UD 22 x 6, QD 6 x 6 (1716: 26 x 32, 26 x 12, 24 x 6, 4 x 6).

**Boyne** 1708. Johnson, Blackwall [RB].
Ord ____ ; K ____ ; L 26.3.1708; 1732 BU.
**Humber** 1708. Wicker, Deptford [RB].
Ord ____ ; K ____ ; L 26.3.1708; 1723 BU.
**Dorsetshire** 1712. Portsmouth Dyd (Surveyor: Stacey) [RB].
Ord ____ ; K 24.11.1706; L 20.9.1712; 1749 BU.
**Russell** 1709. Wells, Rotherhithe [RB].
Ord 3.10.1707; K ——; L 16.3.1709; 1726 BU.
**Cumberland** 1710. Deptford Dyd (Surveyor: Allin).
Ord ____ ; K ____ ; L 27.12.1710; 1732 BU.
**Devonshire** 1710. Woolwich Dyd (Surveyor: Acworth).
Ord ____ ; K ____ ; L 12.12.1710; 1740 hulked.
**Shrewsbury** 1713. Deptford Dyd (Surveyor: Allin) [RB].
Ord ____ ; K 15.11.1711; L 12.8.1713; 1750 BU.
**Cambridge** 1715. Woolwich Dyd (Surveyor: Acworth) [RB].

---

[1] Probably what would later be called 'brig' rig, as used by the type known as the 'brigantine' mentioned in the previous chapter ('brig' almost certainly being an abbreviation of 'brigantine'), with square sails on both of the masts, and not the type of rig with square sails only on the fore and fore-and-aft sails on the mainmast, which was what became known as brigantine rig by the early nineteenth century and later.

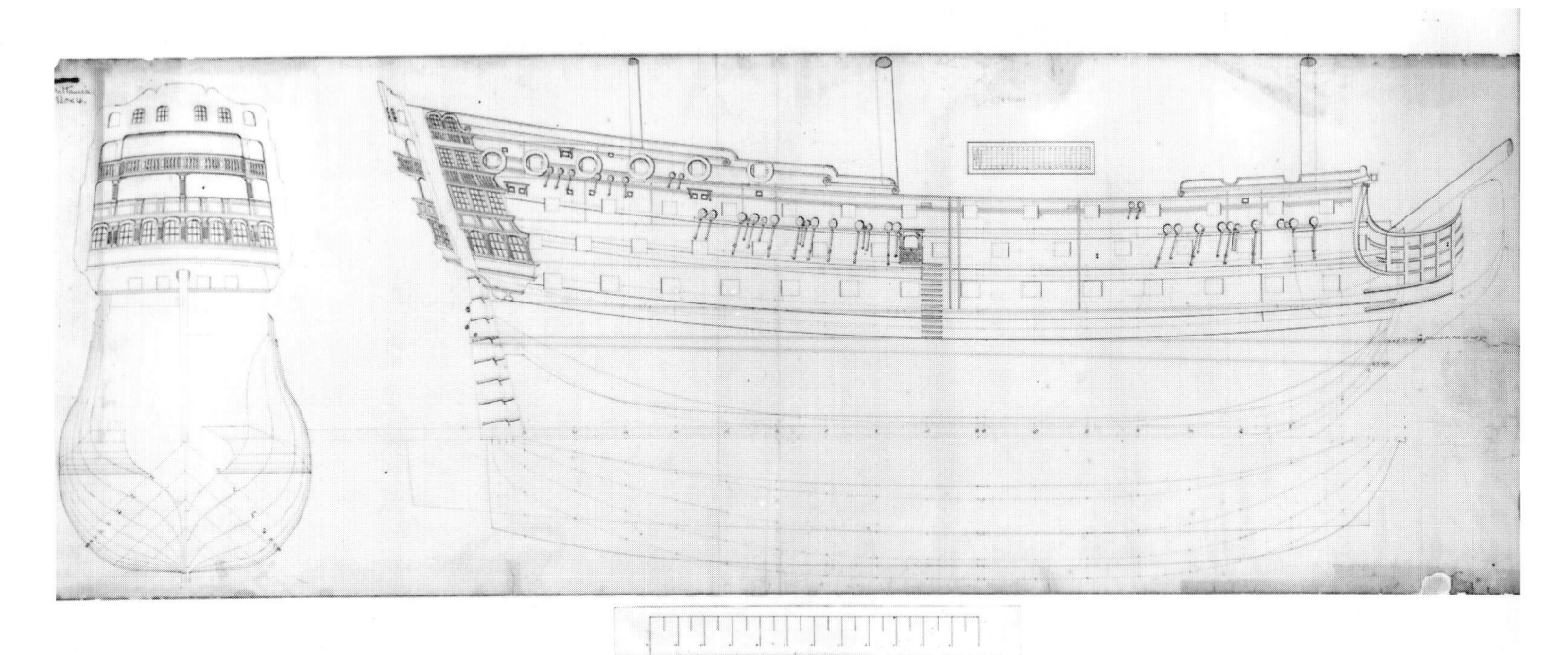

This Admiralty plan (6317) shows the 100 gun ship *Britannia*, probably the 'rebuild' of 1719, but it could be of her 1682 predecessor, identification not being entirely certain, though the plan itself is of a later rather than earlier style. As the 1719 ship was usually listed under the 1719 Establishment the entry on her is in the next chapter, but her design was earlier.

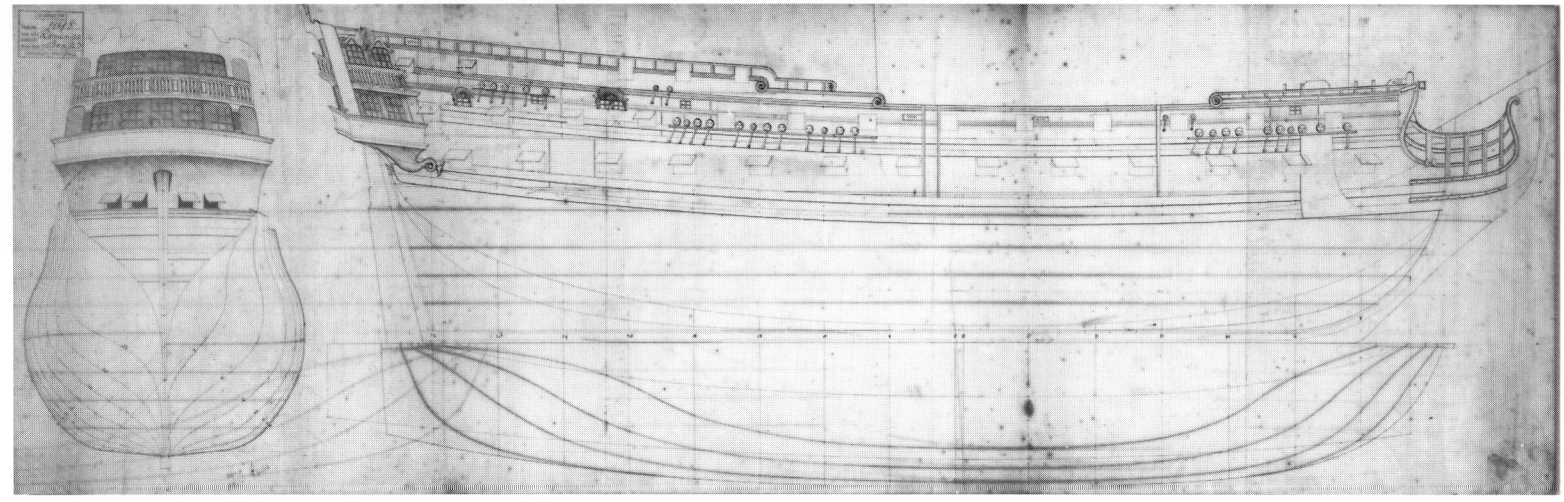

Admiralty plan number 1198 shows the 70 gun ship *Revenge* of 1718

Ord ____ ; K 30.8.1713; L 17.9.1715; 1750 BU.
**Newark** 1717. Chatham Dyd (Surveyor: Rosewell) [RB].
　　Ord ____ ; K 12.5.1713; L 29.7.1717; 1741 BU.
**Torbay** 1719. Deptford Dyd (Surveyor: Stacey) [RB].
　　Ord ____ ; K 23.7.1716; L 23.5.1719; 1750 BU.
FATES: *Devonshire* hulked at Woolwich; 1760 sold.
PLANS: *Boyne, Humber & Russel*: Lines. *Cumberland*: Lines/decks. *Devonshire*: as hulk: Profile. *Torbay*: Lines & profile.

## Third Rates, 70 Guns (two-deckers)

Dimensions & tons: 150', 123'4" x 41'1¾" x 17'4". 1110.
Men: 440/320. Guns: (up to 1716 70 war, 62 peace) GD 24 x 24, UD 26 x 12, QD 12 x 6, Fc 4 x 6, RH 4 x 3 (1716: 26 x 24, 26 x 12, 14 x 6, 4 x 3, no RH guns – examples of actual armaments issued: *Captain* in 1714 was issued with 22 x 32, 24 x 12, 12 sakers, 4 x 3. *Hampton Court* in 1714 had 22 x 24, 24 x 9, 12 x 6, 4 x 3. *Yarmouth* in 1713 had 22 x 24, 24 x 9, 12 sakers, 4 x 3, then 26 x 24, 26 x 12, 18 x 6.

**Nassau** 1707. Portsmouth Dyd (Surveyor: Podd).
　　Ord ____ ; K ____ ; L 9.1.1707; 1736 BU.
**Captain** 1708. Portsmouth Dyd (Surveyor: Podd).
　　Ord ____ ; K ____ 1705; L 6.7.1708; 1720 BU.
**Resolution** 1708. Deptford Dyd (Surveyor: Allin).
　　Ord ____ ; K ____ ; L 26.3.1708; 1711 wrecked.
**Edgar** 1709. Burchett, Rotherhithe [RB].
　　Ord ____ ; K ____ ; L 31.8.1709; 1711 blown up.
**Grafton** 1709. Swallow, Limehouse.
　　Ord ____ ; K ____ ; L 12.8.1709; 1722 BU.
**Hampton Court** 1709. Taylor, Rotherhithe.
　　Ord ____ ; K ____ ; L 27.8.170; 1741 BU.
**Yarmouth** 1709. Wicker, Deptford [RB].
　　Ord ____ ; K ____ 1707; L 9.9.1709; 1740 hulked.
**Orford** 1713. Fowler [&/or Johnson?], Limehouse [RB].
　　Ord 26.2.1709 or 1.8.1709; K 23.8.1709; L.17.3.1713; 1745 wrecked.
**Royal Oak** 1713. Woolwich Dyd (Surveyor: Acworth) [RB].
　　Ord ____ ; K 05.1710; L 14.5.1713; 1737 BU.
**Expedition** 1714. Portsmouth Dyd (Surveyor: Stacey) [RB].

# THE PRELIMINARY ESTABLISHMENT 1706-1718

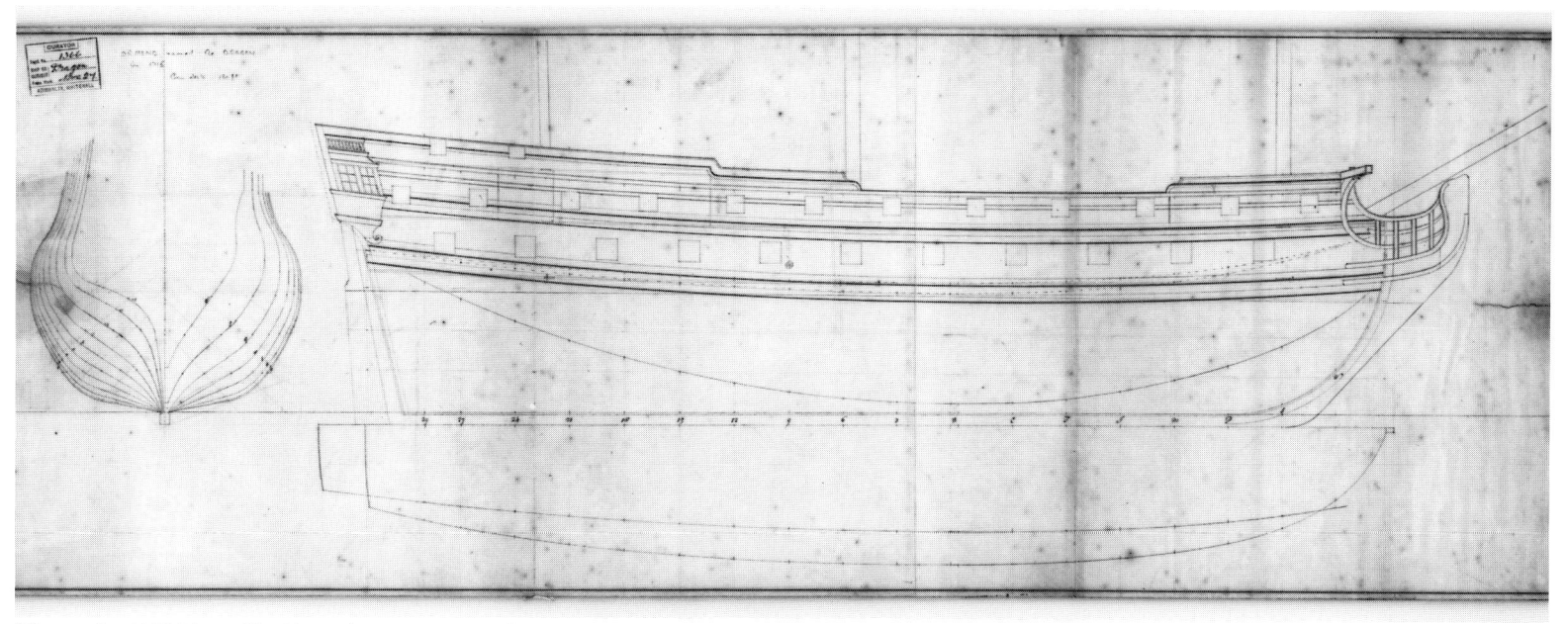

(Plan number 1366) Lines of the 50 gun ship *Dragon* (originally named *Ormond*). She has a single level stern gallery like the later frigates

Ord ____ ; K 11.3.1709; L 16.8.1714; 1715 renamed *Prince Frederick*; 1736 BU.
**Monmouth** 1718. Portsmouth Dyd (Surveyor: Naish) [RB].
Ord ____ ; K 04.1716; L 3.6.1718; 1744 BU.
**Revenge** 1718. Woolwich Dyd (Surveyor: Hayward) [RB].
Ord ____ ; K 10.1716; L 30.10.1718; 1740 BU.

NOTE: *Suffolk* (1698) had a Great Repair in 1718 which changed her dimensions (see under the original entry) to the extent that she could be considered virtually a new ship.

FATES: *Resolution* was wrecked in a storm at Barcelona on 10.1.1711. *Edgar* blew up by accident at Spithead on 15.10.1711. *Yarmouth* was hulked at Portsmouth; 1768 BU. *Orford* was wrecked on 'the Island Heneago' on 14.2.1745.

PLANS: *Resolution*: Lines/decks. *Yarmouth*: Spar dimensions. *Revenge*: Lines.

## Third Rates later Fourth Rates 60 Guns (two-deckers)

Dimensions & tons: 144', 119' x 38' x 15'8". 914.
Men: 365/240 later 355. Guns: (up to 1716 64 war 56 peace) GD 24 x 18, UD 26 x 9, QD 10 x 6; Fc 4 x 6 (1716: 24 x 24, 26 x 9, 8 x 6, 2 x 6).
**Defiance** Wicker, Deptford [RB?].
Ord 18.8.1704; K ____ ; L ____ 1707; 1729 BU (or Great Repair?).
**Plymouth** Plymouth Dyd (Surveyor: Lock).
Ord 7.7.1705; K ____ ; L 25.5.1708; 1720 BU.
**Gloucester** Burchett, Rotherhithe.
Ord ____ ; K ____ ; L 25.7.1709; 1709 taken.
**Lion** Chatham Dyd (Surveyor: Rosewell).
Ord ____ ; K ____ ; L 20.1.1710; 1735 BU.
**Rippon** Deptford Dyd (Surveyor: Allin).
Ord ____ ; K ____ ; L 23.8.1712; 1729 BU.
**Montague/Mountague** Portsmouth Dyd (Surveyor: Naish) [RB].
Ord ____ ; K 13.5.1714; L 24.7.1716; 1749 BU.
**Kingston** Portsmouth Dyd (Surveyor: Naish) [RB].
Ord ____ ; K 22. 7. 1716; L 9. 5. 1719; 1736 BU.
**Medway** Deptford Dyd (Surveyor: Stacey) [RB].
Ord ____ ; K 9.7.1716; L 1.8.1718; 1742 hulked.
**Nottingham** Dept ford Dyd (Surveyor: Stacey) [RB].
Ord ____ ; K 8.4.1717; L 5.10.1719; 1739 BU.

FATES: *Gloucester* taken on 26.10.1709 by the French *Lys* off Cape Clear; 1711 sold by the French to Genoa; 1720 sold by the Genoese to Spain as the *Conquistador*; out of service 1738. *Medway* was hulked as a Receiving Ship at Portsmouth; 1749 BU.

PLANS: *Rippon*: Lines/decks. *Montague*: Lines. *Kingston*: Lines. *Medway*: Lines.

## Fourth Rates 50 Guns (two-deckers)

Dimensions & tons: 130', 108' x 35' x 14'. 704.
Men: 280/185 later 300. Guns: (up to 1716 54 war 46 peace) GD 22 x 12, UD 22 x 6, QD 8 x 6, Fc 2 x 6 (1716: 22 x 18, 22 x 9, 4 x 6, 2 x 6. Those reclassed as 44s in 1744 were then armed as follows: 20 x 18, 20 x 6, QD 4 x 6).
**Salisbury** Chatham Dyd (Surveyor: Rosewell).
Ord ____ ; K ____ ; L 3.7.1707; 1716 BU.
**Dragon** Taylor, Rotherhithe [RB].
Ord ____ ; K ____ ; L ____ 1707; 1712 wrecked.
**Falmouth** Woolwich Dyd (Surveyor: Stacey).
Ord ____ ; K ____ ; L 26.2.1708; 1724 BU.
**Ruby** Deptford Dyd (Surveyor: Allin).
Ord ____ ; K ____ ; L 25.3.1708; 1744 renamed *Mermaid* as Fifth Rate 44 guns; 1748 sold.
**Chester** Chatham Dyd (Surveyor: Rosewell).
Ord ____ ; K ____ ; L 18.10.1708; 1742 reclassed as Fifth Rate 44 guns; 1743 hospital ship; 1750 BU.
**Romney** Deptford Dyd (Surveyor: Allin).
Ord ____ ; K ____ ; L 2.12.1708; 1723 BU.
**Warwick** Burchett Rotherhithe [RB].
Ord 9.5.1709; K ____ ; L. 9.11.1711; 1726 BU.
**Pembroke** Plymouth Dyd (Surveyor: Lock).
Ord ____ ; K ____ ; L 18.5.1710; 1726 BU.
**Bristol** Plymouth Dyd (Surveyor:Lock).
Ord ____ ; K ____ ; L 8.5.1711; 1743 BU.
**Bonadventure** Chatham Dyd (Surveyor: Rosewell) [RB].
Ord ____ ; K ____ ; L 18.9.1711; 1716 renamed *Argyle*; 1719 BU.
**Gloucester** Deptford Dyd (Surveyor: Allin).
Ord ____ ; K ____ ; L 4.10.1711; 1725 BU.
**Ormond** Woolwich Dyd (Surveyor: Acworth).
Ord ____ ; K ____ ; L 18.10.1711; 1715 renamed *Dragon*; 1733 BU.
**Advice** Deptford Dyd (Surveyor: Allin).
Ord ____ ; K ____ ; L 8.7.1712; 1744 renamed *Milford* & reclassed as a Fifth Rate 44; 1749 sold.
**Assistance** Johnson, Limehouse [RB].
Ord 23.6.1710; K 3.10. 1710; L 16.1.1713; 1720 BU.
**Worcester** Deptford Dyd (Surveyor: Allin) [RB].
Ord ____ ; K 2.5.1713; 31.8.1714; 1733 BU.
**Strafford** Plymouth Dyd (Surveyor: Phillips).
Ord ____ ; K ____ ; L 18.7.1714; 1726 BU.
**Panther** Woolwich Dyd (Surveyor: Hayward) [RB].

Ord ____ ; K 11.12.1713; L 26.4.1716; 1745 hulked.
**Dartmouth** Woolwich Dyd (Surveyor: Hayward) [RB].
Ord ____ ; K 7.5.1714; L 7.8.1716; 1733 BU.
**Rochester** Deptford Dyd (Surveyor: Stacey) [RB].
Ord ____ ; K 30.7.1714; L 13.3.1716; 1744 hulked.
**Salisbury** Deptford Dyd (Surveyor: Stacey) [RB].
Ord ____ ; K 25.5.1716; L 10.10.1717; 1724 BU (or docked for a Great Repair?).
**Guernsey** Woolwich Dyd (Surveyor: Hayward) [RB].
Ord ____ ; K 7.9.1716; L 24.10.1717; 1737 BU.
**Nonsuch** Portsmouth Dyd (Surveyor: Naish) [RB].
Ord ____ ; K ____ ; L 29.4.1717; 1740 hulked.
**Saint Albans** Plymouth Dyd (Surveyor: Phillips) [RB].
Ord ____ ; K 3.4.1717; L 6.3.1718; 1734 BU.
**Winchester** Plymouth Dyd (Surveyor: Phillips) [RB].
Ord ____ ; K 10.7.1716; L 12.11.1717; 1744 hulked.
**Norwich** Chatham Dyd (Surveyor: Rosewell) [RB].
Ord ____ ; K ____ ; L 20.5.1718; 1744 renamed *Enterprise* as Fifth Rate 44; 1771 BU.
**Deptford** Plymouth Dyd (Surveyor: Phillips) [RB].
Ord ____ ; K 21.3.1717; L 19.6.1719; 1724? BU or sold 1726?
**Weymouth** Woolwich Dyd (Surveyor: Hayward) [RB].
Ord ____ ; K 13.6.1717; L 26.2.1719; 1733 BU.
**Swallow** Chatham Dyd (Surveyor: Rosewell) [RB].
Ord ____ ; K 11.9.1717; L 25.3.1719; 1728 BU.
**Tiger** Sheerness Dyd (Surveyor: Ward) [RB].
Ord ____ ; K 1.5.1718; L 9.11.1722; 1743 lost.

FATES: *Dragon* 'run upon the Caskett [Casquet] Rocks and lost' off Alderney 15.3.1712. *Panther* hulked at Deptford; 1768 sold. *Rochester* hulked as hospital ship at Port Mahon; 1744 renamed *Maidstone*; 1748 BU. *Nonsuch* hulked at Portsmouth; 1745 BU. *Winchester* hulked at Harwich; 1745 to Sheerness; 1748 to Chatham; 1781 BU. *Deptford* sold per A.O. dated 3.5.1726. *Tiger* 12.1.1743 wrecked on a Key (Island) near Tortuga.

PLANS: *Ormond* (as *Dragon*): Lines.

## Fifth Rates 40 Guns (two-deckers)

Dimensions & tons: 118', 97'6" x 32' x 13'6". 531.
Men: 190/130 later 190. Guns: (up to 1716 42 war 36 peace) GD 18 x 9, UD 20 x 6, QD 4 x 6 (1716: 20 x 12, 20 x 6, no QD guns). In 1707 at least one of these vessels was armed with 16 x 9 and 20 x 6. In 1708 *Gosport* had 20 x 9 + 20 x 6; as did *Pearl* in 1710 but in 1717 the latter had 20 x 12 + 20 x 6. In 1714 *Saphire* had 18 x 9 + 24 sakers, whilst *Adventure* had 18 x 9 + 24 x 6.

**Gosport** Woolwich Dyd (Surveyor: Stacey).
Ord ____ ; K ____ ; L 8.3.1707; 1735 BU.
**Looe** Johnson, Blackwall.
Ord 5.5.1706?; K ____ ; L 7.4.1707; (1717-1724 hospital ship); 1735 hospital ship; 1737 breakwater.
**Ludlow Castle** Sheerness Dyd (Surveyor: Acworth).
Ord ____ ; K ____ ; L 10.4.1707; 1722 BU.
**Portsmouth** Deptford Dyd (Surveyor: Allin).
Ord ____ ; K ____ ; L 1.4.1707; 1721 hospital ship; 1728 BU.
**Hastings** Portsmouth Dyd (Surveyor: Podd).
Ord ____ ; K ____ ; L 2.10.1707; 1740 hulked.
**Pearl** Burchett, Rotherhithe.
Ord ____ ; K ____ ; L 5.8.1708; 1723 BU.
**Mary Gally** Deptford Dyd (Surveyor: Allin) [RB].
Ord ____ ; K ____ ; L 6.8.1708; 1721 BU.
**Saphire** Portsmouth Dyd (Surveyor: Podd).
Ord ____ ; K ____ ; L 08.1708; 1740 hulked.
**Diamond** Johnson, Blackwall.
Ord 8.10.1708?; K ____ ; L ____ 1708/1709?; 1721 BU.
**Southsea Castle** Swallow, Rotherhithe.
Ord 15.3.1708; K ____ ; L 18.11.1708; 1723 BU (or Great Repair?).
**Enterprise** Plymouth Dyd (Surveyor: Lock).
Ord ____ ; K ____ ; L 28.4.1709; 1740 hulked at Plymouth.
**Adventure** Sheerness Dyd (Surveyor: Acworth).
Ord ____ ; K ____ ; L 17.6.1709?; 1724 BU.
**Fowey** Portsmouth Dyd (Surveyor: Stacey).
Ord ____ ; K ____ ; L 7.12.1709; 1744 Sixth Rate 24; 1746 sold.
**Launceston** Portsmouth Dyd (Surveyor: Stacey).
Ord ____ ; K ____ ; L 17.10.1711; 1726 BU.
**Faversham** Plymouth Dyd (Surveyor: Phillips).
Ord ____ ; K ____ ; L 22.8.1712; 1730 BU.
**Charles Gally** Deptford Dyd (Surveyor: Allin) [RB].
Ord ____ ; K ____ ; L 29.8.1710; 1726 breakwater.
**Lynn** Sheerness Dyd (Surveyor: Hayward).
Ord ____ ; K ____ ; L 8.4.1715; 1732 BU.
**Southsea Castle** Portsmouth Dyd (Surveyor: Naish) [RB or Great Repair].
Ord ____ ; K 26.2.1723; L 10.7.1724; 1744 sold.

FATES: *Hastings* hulked at Plymouth; 1745 sold. *Saphire* hulked at Plymouth; 1745 sold. *Enterprise* hulked at Plymouth; 1745 hospital ship; 1749 BU.

**Galley Frigate 40/42 guns** Appears to have been designed by the Marquis of Carmarthen (Lord Dursley):
Dimensions & tons: 127', 100' x 31' x 13', 511 16/94.
Men and guns as the 40 gun ships listed above (?).

**Royal Anne Gally** Woolwich Dyd (Surveyor: Stacey).
Ord ____ ; K ____ ; L 18.6.1709; 10.11.1721 wrecked 'on the Staggs' off the Lizard.

## Fifth Rates 32 Guns ('one and a half'-deckers)

NOTE: These were not built to an Establishment of dimensions and so their 'as built' dimensions are given here; other data as follows: *Sweepstakes, Scarborough, Bedford Gally*: Men: 145/100. Guns: (before 1716 32 war 28 peace) GD 4 x 9, UD 22 x 6, QD 6 x 4. *Mermaid, Dolphin*: Men: 155/110. Guns: (till 1716 36 war 30 peace) GD 8 x 12, UD 22 x 6, QD 6 x 4. In 1716 all the survivors of the above were to be armed as 30 gun ships as follows: GD: 8 x 9, UD 20 x 6, QD 2 x 4.

**Mermaid** Chatham Dyd (Surveyor: Rosewell) [RB].
Dimensions & tons: 108', 90' x 29'8" x 12'. 421.
Ord ____ ; K ____ ; L ____ 1707; 1725 hulked [?]
**Sweepstakes** Woolwich Dyd (Surveyor: Stacey).
Ord ____ ; K ____ ; L 20.9.1708; 1709 taken.
Dimenstions & tons: 108'5", 90' x 29'6" x 12'. 416 57/94.
**Scarborough** Sheerness Dyd (Surveyor: Poulter?).
Ord ____ ; K ____ ; L 23.5.1711; 1720 BU.
Dimensions & tons: 108', 90' x 29'6" x 12'. 416 57/94.
**Dolphin** Portsmouth Dyd (Surveyor: Stacey) [RB].
Ord ____ ; K ____ ; L 7.6.1711; later re-rated as a 20 gun Sixth Rate; 1730 BU.
Dimensions & tons: 110', 90'6" x 29'8" x 12'10". 423 62/94.

FATES: *Mermaid* moorings hulk at Deptford?; 1734 BU. *Sweepstakes* taken by French 40 and a 26 on 16.4.1709; retaken 05.1709 by Lord Dursley's squadron in the Soundings, but apparently not put back into service.

PLANS: *Sweepstakes*: Danish Rigsarkivet: Lines & profile. *Dolphin*? [shown pierced for 40 guns]: Lines & profile.

**Also RB of 'Galley frigate'** of similar armament but different dimensions & tonnage: 103'3½", 91'8" x 29' x 11'2". 410.
**Bedford Gally** Portsmouth Dyd (Surveyor: Podd) [RB].
Ord ____ ; K ____ ; L 24.8.1709; 1717 fireship; 1725 breakwater.

## Large Sixth Rates (24 or 20 Guns)

These do not seem to have been built to a standard design or 'Establishment' and so their 'as built' dimensions are given here also their classified number of guns: 24 gun ships normally carried: Men: 115/85, 20s later 100. Guns: (24s were 24 war 22 peace up to 1716) UD 20 x 6, 24s were established with extra 4 x 6 on the QD until 1716 when the QD guns were deleted. In 1717 *Hind* carried 20 x 6. In 1714 *Biddeford* had 18 sakers. In 1716 *Greyhound* was armed with 18 x 6.

**Flamborough/Flambrough** 24, Woolwich Dyd (Surveyor: Stacey).

Dimensions & tons: 94', 79'8" x 25' x 10'8". 261 49/94.
Ord ____ ; K ____ ; L 02.1707?; 1727 BU.

***Nightingale*** 24, Johnson, Blackwall.
Dimensions & tons: 90'3¾", 74'0½" x 25'4½" x 10'6". 253 55/94.
Ord ____ ; K ____ ; L 15.10.1707; 1716 sold.

***Squirrell*** 24, Woolwich Dyd (Surveyor: Stacey).
Dimensions & tons: 94', 79' x 25' x 10'8". 262 59/94.
Ord ____ ; K ____ ; L 29.12.1707; 1727 BU.

***Phoenix*** 24, Plymouth Dyd (Surveyor: ____ ).
Built as a fireship (which see) in 1709; 02.1711 20 gun Sixth Rate; 07.1711 24 guns; 1727 BU for RB.

***Solebay*** 24, Portsmouth Dyd (Surveyor: Stacey).
Dimensions & tons: 95'10", 80'10" x 25'2" x 10'8". 272 30/94.
Ord ____ ; K ____ ; L 21.8.1711; 1726 bomb vessel (3 mortars); 1734 fireship; 1735 Sixth Rate (20?); 1742 hulked.

***Gibraltar*** 20, Deptford Dyd (Surveyor: Allin).
Dimensions & tons: 94'1",76'7" x 26'2¾" x 11'6". 280 23/94.
Ord ____ ; K ____ ; L 18.10.1711; 1725 BU.

***Port Mahon*** 20, Deptford Dyd (Surveyor: Allin).
Dimensions & tons: 94'4½", 76'10" x 26'3¼" x 11'6". 282 5/94.
Ord ____ ; K ____ ; L 18.10.1711; 1740 BU.

***Blandford*** 20, Woolwich Dyd (Surveyor: Acworth).
Dimensions & tons: 94', 76'9" x 26' x 11'6". 276.
Ord ____ ; K ____ ; L 29.10.1711; 1719 lost.

***Seahorse*** 20, Portsmouth Dyd (Surveyor: Stacey).
Dimensions & tons: 94', 78' x 26' x 11'6". 280 44/94.
Ord ____ ; K ____ ; L 13.2.1712; 1727 BU.

***Hind*** 20, Woolwich Dyd (Surveyor: Acworth).
Dimensions & tons: 94', 76'9" x 26' x 11'7". 276.
Ord ____ ; K ____ ; L 31.10.1711; 1721 wrecked.

***Biddeford*** 20, Deptford Dyd (Surveyor: Allin).
Dimensions & tons: 94'3", 76'8" x 26'3" x 11'6". 281.
Ord ____ ; K ____ ; L 14.3.1712; 1726 or 1727 BU.

***Rose*** 20, Chatham Dyd (Surveyor: Rosewell).
Dimensions & tons: 94', 76' x 26' x 11'6". 273.
Ord ____ ; K ____ ; L 25.4.1712; 1722 BU.

***Success*** 20, Portsmouth Dyd (Surveyor: Stacey).
Dimensions & tons: 94', 76'6" x 26'6½" x 11'10". 275 7/94.
Ord ____ ; K ____ ; L 30.4.1712; 1739 fireship; 1743 BU.

***Greyhound*** 20, Woolwich Dyd (Surveyor: Acworth).
Dimensions & tons: 94', 76'9" x 26' x 11'6". 276.
Ord —— ; K —— ; L 21.6.1712; 1718 taken

***Lively*** 20, Plymouth Dyd (Surveyor: Phillips).
Dimensions & tons: 94'10", 76'7" x 26'2" x 11'6". 278 73/94 .
Ord ____ ; K ____ ; L 28.5.1713; 1738 BU.

***Speedwell*** 20, Deptford Dyd (Surveyor: Stacey) [RB].
Dimensions & tons: 95'5½", 78'9¾" x 25'6½" x 11'6". 273 69/94.
Ord ____ ; K 16.11.1715; L 27.3.1716; 1719 bomb vessel; 1720 wrecked.

***Dolphin*** de-rated from a 32 gun Fifth Rate (which see). Ca 1715.

FATES: *Solebay* became a hospital hulk at the Tower of London; 1748 sold. *Blandford* was 'cast away at St. Jean de Luz' on 28.3.1719 *Hind* was 'run upon a rock going into Jersey' on 7.12.1720. *Greyhound* was taken by Spain in Saint Jeremy's (Jerome's) Bay on 1.9.1718.

PLANS: *Success* [not certain it is this vessel of the name, but probable]: Lines.

Also **'Galley frigate'** of larger dimensions but similar armament, presumably associated with Lord Dursley (the Marquis of Camarthen) and likely to be one of his designs:

***Dursley Galley*** 20, Deptford Dyd (Surveyor: Stacey).
Dimensions & tons: 105', 86'9" x 28'4½" x 10'. 371 48/94.
Ord ____ ; K ____ ; L 13.2.1719; 1745 sold [or BU; but there is also a note that she became a privateer after her sale and taken by the French 8.5.1746]

## Small Sixth Rates (14 or 12 Guns)

These were not 'Established' and were built to a variety of designs. Therefore their individual 'as built' dimensions and tonnage are given here. These vessels seem to have been the equivalent of the slightly later sloops which are listed immediately below, except in that the Sixth Rates were probably ship rigged, and the sloops probably two-masted. Men: 60/ 70. Guns: ____ ____ ____ .

***Lively*** 12, Wicker, Deptford.
Dimensions & tons: ____ , 56'5" x 20'5¾" x 9'1¾". 125 80/94.
Ord —— ; K —— ; L 06.1709?; 1712 sold [Possibly purchased whilst building?]

***Hind*** 12, Dummer, Thames.
Dimensions & tons: 78'4", 63'5½" x 21'10" x 9'11¾". 160 88/94.
Ord ____ ; K ____ ; L 13.7.1709; 1709 wrecked.

***Delight*** 14, Woolwich Dyd (Surveyor: Acworth).
Dimensions & tons: 77'8", 63' x 22'1" x 9'3". 163 2/5.
Ord ____ ; K ____ ; L 18.10.1709; 1713 sold.

***Margate*** 14, Deptford Dyd (Surveyor: Allin).
Dimensions & tons: 77'2½", 63' x 22'0¼" x 9'3½". 162 75/94.
Ord ____ ; K ____ ; L 11.1709; 1712 sold.

***Otter*** 14, Smith, Rotherhithe.
Dimensions & tons: 76'3¼", 62'3" x 22'6" x 9'3". 167 57/94.
Ord ____ ; K ____ ; L 14.2.1710; 1713 sold.

***Seahorse*** 14, Yeames, Limehouse.
Dimensions & tons: 76'0½", 62'1" x 22'1½" x 9'3". 161 61/94.
Ord ____ ; K ____ ; L early 1710; 1711 wrecked.

***Swann*** 12, Dummer, Thames.
Dimensions & tons: 78'4", 63'6½" x 21'10¾" x 10'. 162 3/94.
Ord ____ ; K ____ ; L early 1710 [or 17.9.1709]; 1713 sold.

FATES: *Hind* was wrecked off Hurst Castle on 16.7.1709. *Seahorse* was wrecked on the rocks at the West side of Dartmouth Harbour on 26.12.1711.

## Sloops

These were also built to a number of individual designs and dimensions. They appear to have been the direct successors of the earlier small Sixth Rates in dimensions and armament if not in rig (see above). Certainly they were much bigger and more powerful vessels than the earlier craft listed as sloops. It would seem that these later vessels were two-masted (ketch or 'brigantine' - what would later be called brig - or 'snow' rigged). They also seem to have replaced the earlier collection of ketches pinks and advice boats which constituted the lowest rank of fighting vessels in the previous period.

### Allin design (all sisters?). 10, all built by Deptford Dyd (Surveyor: Allin).

Dimensions & tons: 64'6", 50' x 20'8" x 9'1". 113.
Men: 100. Guns: 10 x 3 (or 8 x 4 + 2 x 3) + 4 swivels. In 1714 *Sharke* had 10 x 3 + 4 falconets.

***Jamaica*** Ord 20.6.1709?; L 30.5.1710; 1715 wrecked.
***Tryall*** Ord 20.6.1709?; L 30.9.1710; 1719 BU for RB.
***Ferrett*** Ord ____ ; L 20.4.1711; 1718 taken.
***Sharke*** Ord ____ ; L 20.4.1711; 1722 BU for RB.

FATES: *Jamaica* was wrecked on 'the rock of the Great Caimans' on 2.10.1715. *Ferrett* was taken by Spain in Cadiz Bay on 1.9.1718.

PLANS: Lines. (Contract for *Jamaica* & *Tryall*?; Public Record Office ADM 106/3071.)

### Acworth design (both sisters?). 10, both built by Woolwich Dyd (Surveyor: Acworth).

Dimensions & tons: 64'6", 50'2" x 20'8" x 9'. 114.
Men: 80. Guns: 10 x 5¼ (sakers) + 4 swivels.

***Hazard*** Ord ____ ; L 19.4.1711; 1714 wrecked.
***Happy*** Ord ____ ; L 17.4.1711; 1735 sold.

FATES: *Hazard* was 'wrecked on Cony Hazard Rock off Boston, New England' on 12.11.1714.

***Tryall*** 10, Deptford Dyd (Surveyor Stacey) [RB].
Dimensions & tons: 76", 59'3" x 21'3" x 9'6". 142 39/94.
Men & guns as before? (see above) L 16.5.1719; 1731 BU.

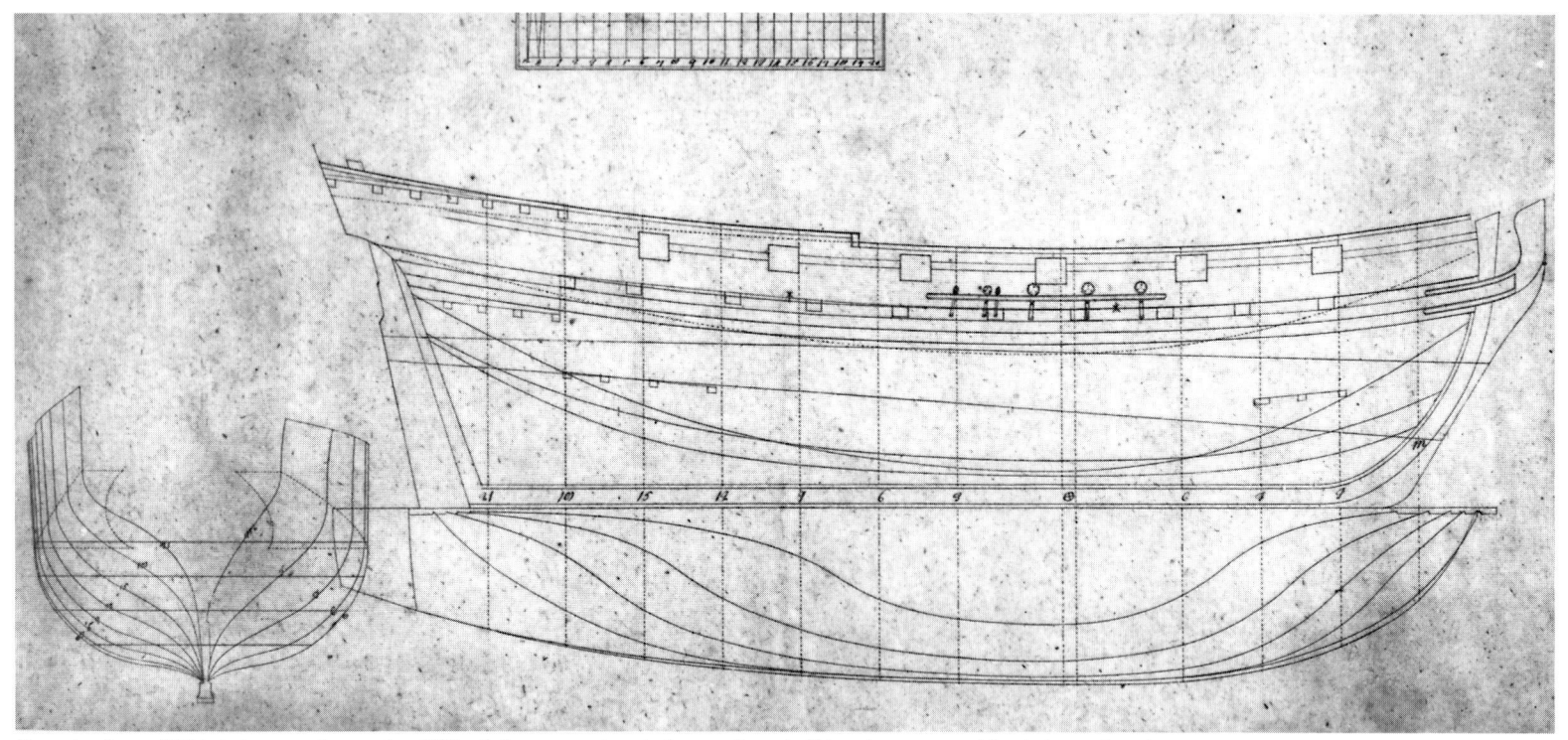

*Ferret*, the sloop of 1711, is show on plan number 3779 as being apparently a single masted vessel. Although no mast position is shown there is only one set of channels marked on the plan

## Fireships

***Phoenix*** RB at Portsmouth Dyd.
> Dimensions & tons: 93'6", 76'9" x 25'10½" x 9'8". 273 21/94.
> Men: 100 (as 20), 115 (as 24).
> Ord ____ , K ____ , L ____ 1709; 02.1711 20 gun Sixth Rate; 07.1711 24 gun; 1727 BU for RB.

## Transports etc

**Allin design.** Both built at Deptford Dyd (Surveyor: Allin).
> Dimensions & tons: 126'3", 105'7" x 31'2" x 13'6". 545 (approximately).
> Men: 90. Guns: 24 x ____ .

***Fortune*** Ord ____ ; L 31.5.1709; 1713 sold.
***Success*** Ord ____ ; L 10.9.1709; 1733 hulked at Deptford [1748 sold].

# Chapter 3: The Establishment Period 1719-1745

This was a period of peace, broken by episodes of quasi-war with Spain (1717-21 and a more muted struggle in 1727-29), in the first of which a fleet action was fought against the newly established Spanish national navy. The Battle of Cape Passaro (1718, fought off Sicily) resulted in a defeat for the Spaniards and the taking of several prizes, some of which were retained at Port Mahon (Minorca) for some years, but which were not commissioned into the navy. In 1739 war broke out with Spain ('The War of Jenkins' Ear') which merged into a more general war with France in 1744, and the central European struggle of the War of the Austrian Succession; peace was signed at Aix-la-Chapelle (Aachen) in 1748.

It is significant that the long period of effective peace between 1713 and 1739 did not result in the neglect of the Royal Navy, whose numbers of ships were well maintained throughout. The deficiency came to be one of quality, not quantity, design not procurement, and it was a deficiency which was clearly revealed by the test of war with Spain and then France. The Surveyor of the Navy of this period was Sir Jacob Ackworth, who had a deeply conservative belief that things had been done better in his youth, during the late seventeenth century. The very detailed Establishment of dimensions and scantlings of 1719 set the pattern. The subsequent 'Establishments' of 1733 and of 1741 were merely modified versions of this prescriptive document. These Establishments laid down specific dimensions and scantlings (thicknesses of timber) for the construction of rated ships, down to and including 20 gun Sixth Rates. They did not prescribe the dimensions of lesser vessels, sloops and below, nor did they actually lay down one design to be followed. Provided the very tight dimensional requirements were met, the individual surveyor responsible for the design and building of a particular vessel could vary the lines (ie. the shape) of the vessel, though obviously his options were severely restricted. The next Establishment, that of 1745, did prescribe the actual design to be used, but it very rapidly produced a reaction in exactly the opposite direction, and is therefore relegated to the next chapter in this work.

Despite the wish to standardise and to retain an unchanging set of types, the pressures for change were irresistible. The two revisions of the 1719 Establishment (1733 and 1741) were both in an upwards direction. Size and strength were increased in each, although not adequately to meet the challenge of new types produced abroad.

Ships continued to be 'rebuilt' [indicated here by the abbreviation **[RB]** until the outbreak of war, though it is clear that most, if not all, of these vessels were, to all intents and purposes, new. The onset of war brought with it urgent needs for the use of dry docks for repairs, thereby removing the possibility of leisurely dismantling a vessel for subsequent 'rebuilding', and with this the administrative fiction of 'rebuilding' finally disappeared.[1]

The same basic format is used here as in the previous chapter: ships are given by type in descending order of power, and within the type in approximate chronological order of build. However, there is a further refinement at this stage as the types are divided into the groupings by 'Establishment' for all the rated vessels. Smaller types, with their greater variety and freedom from design restraints, are treated by class (ie. of vessels built to the same lines) whilst those which do not fall into these groups, or do not appear to do so, are treated individually. The classes are given in chronological order of design (when known). It will be noted that a few of the larger ships (such as Anson's *Centurion* of 60 guns) were not 'built to the Establishment' and are therefore listed separately in the appropriate chronological place.

As is fairly obvious from looking at the entries under the Plans heading here, the real starting point for the Admiralty Ship Plans Collection is the years just after 1713; from the end of the War of Spanish Succession on plans seem to have been kept fairly systematically, though both coverage of ships and variety of types of plan increase as the eighteenth century progresses.

## First Rates, 100 Guns (three-deckers)

**1719 Establishment** (listed as 1719 Establishment, but in fact from the previous period; built to the dimensions of the *Royal Sovereign* of 1701, which were adopted as the guide lines for the actual 1719 Establishment for this type of vessel).

Dimensions & tons: 174', 140'7", x 50' x 20'. 1869 42/94. (*Britannia* as built 1895 tons).
Men: 800 (*Royal William* as 84: 750. *Royal William* in 1782: 650). Guns: GD 28 x 32 or 42, MD 28 x 24, UD 28 x 12, QD 12 x 6, Fc 4 x 6 . (*Royal William* as 84: GD 28 x 32, MD 28 x 18, UD 28 x 9. *Royal William* in 1782: GD 28 x 24, MD 28 ́x 18, UD 28 x 9, Fc 2 x 9).

**Royal William** Portsmouth Dyd (Surveyor Naish) [RB].
Ord ____ ; K 31.7.1714; L 3.9.1719; never fitted for sea as a 100 gun ship; 1756 cut down to a Second Rate of 84 guns; 1790 hulked.
**Britannia** Woolwich Dyd (Surveyor Hayward) [RB].
Ord ____ ; K ____ 1716?; L 3.10.1719; 1745 hulked.
**Royal Sovereign** See entry in Chapter 2 for Great Repair/Rebuild of 1724-1728.
FATES: *Royal William* as cut down to a 3- decked 84 'answers well and is a very useful ship'; hulked as a Receiving Ship at Portsmouth; 1813 BU. In view of the number of times she is dragged into discussions on the longevity or otherwise of wooden warships it should be remembered that she was a fictitious 'Rebuild', in other words she was a new ship to all intents and purposes when she was launched in 1719. Furthermore she was laid up all the time between her launch and her cutting down in 1756 and never was fitted for sea, still less sent to sea, during all that time. So her actual active life was restricted to the years between 1756 and 1790, rather longer than most of her contemporaries, but hardly as impressive as legend would have it, and partly due to the extraneous factor that King George III was reputed to have had a fondness for her which meant she was retained in the Navy List for longer than her real effectiveness warranted (and she was not in commission for much of this time). *Britannia* was a hospital hulk at Chatham; 1749 BU.

PLANS: *Royal William*: Lines. As 84 gun ship in 1757: Lines. *Britannia*: Lines. [Danish Rigsarkivet: Lines]

### 1733 Establishment

Dimensions & tons: 174'9", 141'7" x 50'6" x 20'6". 1869.
Men: 850. Guns: GD 28 x 32 or 42, MD 28 x 24, UD 28 x 12, QD 12 x 6, Fc 4 x 6.
**Victory** Portsmouth Dyd (Surveyor: Allin) [RB].
Ord 11.9.1733?; K 6.3.1727?; L 23.2.1738; 1744 lost.

FATE: Night of 5.10.1744 lost without trace; thought to have been wrecked on the Casquets.

NOTE: Often known as 'Balchin's *Victory*' after the Admiral lost with her. Last First Rate in the RN to be armed entirely with brass guns. Despite a number of

---

[1] There is an isolated example of rebuilding in the Napoleonic War, the fir built 38 gun frigate *Clyde*, built in 1795 and rebuilt in 1805/6, but this is an exceptional case. This is ignoring the very special case of the Royal Yachts.

searches for this wreck it has not yet been located (1992).
PLANS: Lines/decks.

**No 1741 Establishment First Rates built.** However the following particulars were proposed:
>Dimensions & tons: 175', 142'4" x 50' x 21'. 1892 69/94.
>Men: 850. Guns: GD 28 x 32 or 42, MD 28 x 24, UD 28 x 12, QD 12 x 6, Fc 4 x 6.

# Second Rates, 90 Guns (three-deckers)

### 1719 Establishment
>Dimensions & tons: 164'. 132'2" x 47'3" x 18'10". 1566 89/94.
>Men: 750 (1755 as 80s: 700). Guns: GD 26 x 32, MD 26 x 18, UD 26 x 9, QD 10 x 6, Fc 2 x 6 (1755 as 80s: GD 26 x 32, MD 26 x 18, UD 24 x 9, QD 4 x 6).

***Prince George*** Deptford Dyd (Surveyor: Stacey) [RB].
>Ord 3.9.1719; K ____ 1719; L 4.9.1723; 1755 Third Rate 80; 1758 burnt by accident and foundered.

***Union*** Chatham Dyd (Surveyor: Rosewell) [RB].
>Ord 13.1.1722?; K 5.3.1718?; L 8.2.1726; 1749 BU.

***Namure*** Deptford Dyd (Surveyor: Stacey) [RB].
>Ord 1.6.1723; K 08.1723; L 13.9.1729; 1746 Third Rate (two-decker) 74; 1749 wrecked.

***Neptune*** Woolwich Dyd (Surveyor: Hayward) [RB].
>Ord 11.8.1724; K 19.1.1725; L 18.10.1730; 1749 Third Rate (two-decker) 74; 1750 renamed *Torbay*; 1784 hulked.

***Marlborough*** Chatham Dyd (Surveyor: Rosewell) [RB].
>Ord 5.4.1725; K 1.9.1725; L 26.9.1732; 1756 Third Rate 80; 1761 Third Rate (two-decker) 68; 1762 foundered.

FATES: *Prince George* was burnt by accident at sea on 13.4.1758. *Namure* was lost in 'a hurricane of wind' in the Road of Fort Saint David, India 12/14.4.1749. *Neptune* became a sheer hulk at Portsmouth; 1816 BU. *Marlborough* foundered returning from Manilla in 12.1762.

PLANS: *Prince George* & *Namure*: Lines. *Union*?: Lines. *Namure* as 74: Lines & profile/decks. *Neptune* as *Torbay* 74: Lines & profile. *Marlborough*: Lines. As 68, 1761: Profile/decks.

### 1733 Establishment
>Dimensions & tons: 166', 134'1" x 47'9" x 19'6". 1623.
>Men: 750 Guns: GD 26 x 32, MD 26 x 18, UD 26 x 9, QD 10 x 6, Fc 2 x 6

***Duke*** Woolwich Dyd (Surveyor: Hayward) [RB].
>Ord 25.9.1733; K 25.5.1734; L 28.4.1739; 1769 BU.

***Saint George*** Portsmouth Dyd (Surveyor: Allin) [RB].
>Ord 4.9.1733; K 4.2.1739; L 3.4.1740; 1774 BU.

PLANS: *Duke*: Lines. *St George*: Lines & profile/orlop, gun & middle decks/upper deck, quarter deck & forecastle.

### 1741 Establishment
>Dimensions & tons: 168', 137' x 48' x 20'2". 1679.
>Men: 750. Guns: GD 26 x 32, MD 26 x 18, UD 26 x 12, QD 10 x 6, Fc 2 x 6.

***Ramillies*** Portsmouth Dyd (Surveyor: Lock).
>Ord 29.7.1739/30.11.1742; K 22.2.1743; L 8.2.1749; 1760 wrecked.

***Prince*** Chatham Dyd (Surveyor: Ward).
>Ord 13.12.1742; K 10.11.1743; L 8.8.1750; 1776 BU.

FATES: *Ramillies* was wrecked 'off the Start' (Bolt Head). The wreck was found in the 1950s and has been intermittently plundered since.

PLANS: *Ramillies*: Lines/lines & profile. *Prince*: Lines.

# Third Rates 80 Guns (three-deckers)

### 1719 Establishment
>Dimensions & tons: 158', 128'2" x 44'6" x 18'2". 1350.
>Men: 600. Guns: GD 26 x 32, MD 26 x 12, UD 24 x 6, QD 4 x 6.

***Lancaster*** Portsmouth Dyd (Surveyor: Naish) [RB].
>Ord ____ ; K 25.5.1719; L 1.9.1722; 1743 BU.

***Cornwall*** Deptford Dyd (Surveyor: Stacey) [RB].
>Ord 16.1.1722; K 27.3.1723; L 7.10.1726; 1755 hulked.

***Humber*** Portsmouth Dyd (Surveyor: Naish) [RB].
>Ord 30.10.1723?; K 7.10.1723; L 4.10.1726; 1728 renamed *Princess Amelia*; 1752 BU.

***Norfolk*** Plymouth Dyd (Surveyor: Lock) [RB].
>Ord ____ ; K 6.2.1717; L 21.9.1728; 1749 BU.

***Princess Caroline*** Woolwich Dyd (Surveyor: Hayward) [RB of *Ranelagh*].
>Ord 20.8.1723; K 04.1724; L 15.3.1731; 1755 hulked.

***Somerset*** Woolwich Dyd (Surveyor: Hayward).
>Ord 23.12.1725; K ____ ; L 21.10.1731; 1746 BU.

***Russell*** Deptford Dyd (Surveyor: Stacey) [RB].
>Ord 4.2.1729; K 16.4.1729; L 8.9.1735; 1761 Breakwater.

FATES: *Cornwall* hulked as a prison ship at Chatham; 1761 BU. *Princess Caroline* hulked at Sheerness as a hospital ship; 1764 BU.

PLANS: *Humber*: Lines & profile. *Norfolk*: Lines. *Princess Caroline* & *Somerset*:

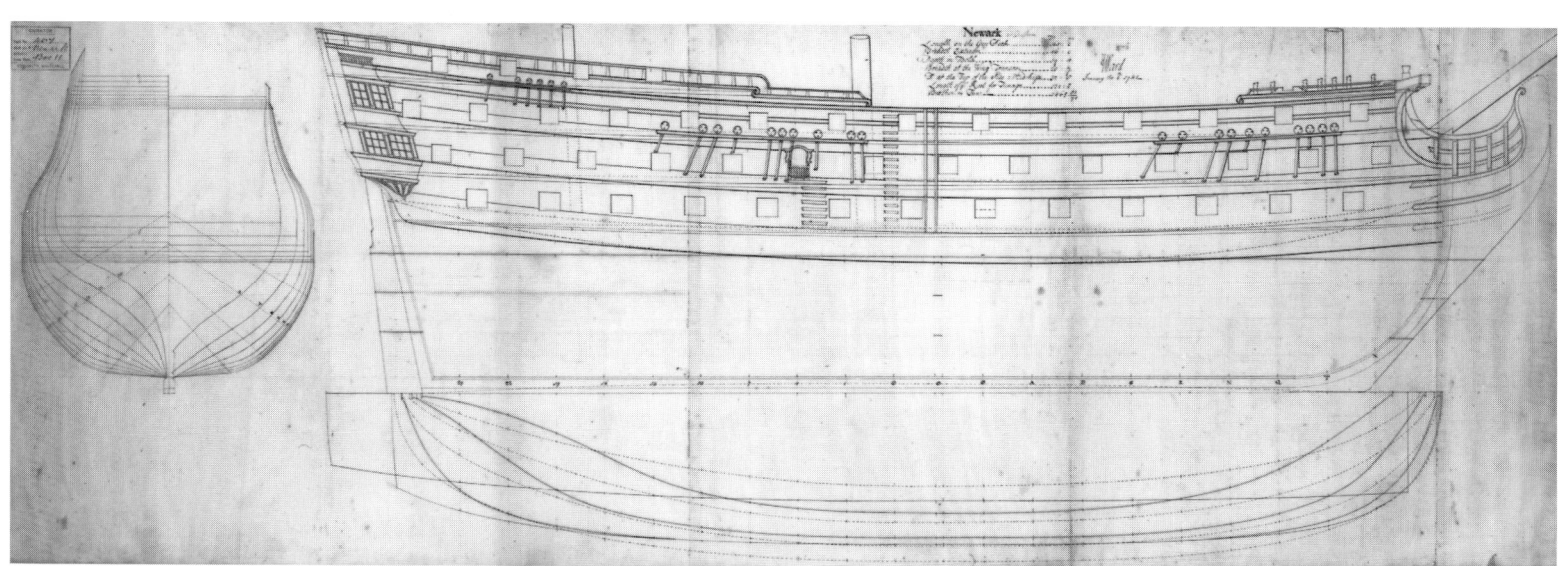

Lines plan of the 1733 Establishment 80 gun ship *Newark*

THE ESTABLISHMENT PERIOD 1719-1745

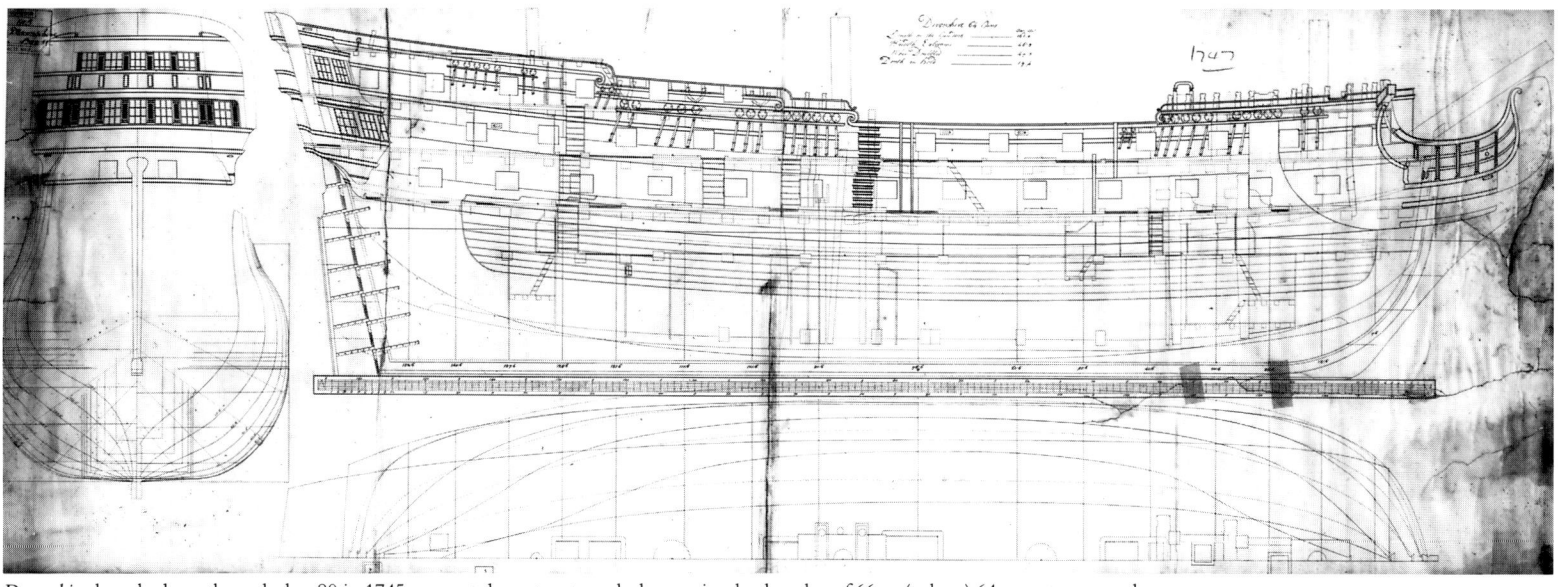

*Devonshire*, launched as a three-decker 80 in 1745, was cut down to a two-decker, variously classed as of 66 or (as here) 64 guns, two years later as shown in this lines and profile plan (number 1115). This plan also shows what appears to be extra planking ('girdling') at around waterline level on each side. This would increase the breadth and make the ship more stable

Body plan. *Somerset*?: Lines.

## 1733 Establishment

Dimensions & tons: 158', 127'8" x 45'5" x 18'7". 1400.
Men: 600. Guns: GD 26 x 32, MD 26 x 12, UD 24 x 6, QD 4 x 6.

**Cumberland** Woolwich Dyd (Surveyor: Hayward) [RB].
Ord 6.7.1733; K ____ ; L 11.7.1739; 1748 Third Rate (two-decker) 66 guns; 1760 foundered.

**Boyne** Deptford Dyd (Surveyor: Stacey) [RB].
Ord 4.11.1733; K ____ ; L 28.5.1739; 1763 BU.

FATES: *Cumberland* foundered at anchor near Goa on 2.11.1760.
PLANS: *Cumberland*: Lines. Danish Rigsarkivet: profile.

## 1741 Establishment

Dimensions & tons: 161', 130'3" x 46 " x 19'4". 1466. (*Devonshire* was broadened to 47'3" after 'doubling').
Men: 600. Guns: GD 26 x 32, MD 26 x 18, UD 24/28 x 9, QD 4 x 6.

**Devonshire** Woolwich Dyd (Surveyor: Holland).
Ord 25.4.1741; K ____ ; L 19.7.1745; 1747 Third Rate (two-decker) 66; 1772 BU.

**Newark** Chatham Dyd (Surveyor: Ward).
Ord 24.4.1741; K ____ ; L 27.8.1747; 1770 hulked

**Lancaster** Woolwich Dyd (Surveyor: Holland).
Ord 15.2.1743; K ____ ; L as a 66 gun two-decker in 1749 (see under 70 gun ships).

FATES: *Newark* hulked as a lazaretto at Sheerness; 1787 BU.
PLANS: *Devonshire*: Lines/lines & profile. As 64: Lines & profile. *Newark*: Lines. Also design lines plan 'as proposed by Wm. Gantlett' showing dimensions of this establishment 'to be lengthened to 165 feet' [the length of the 1745 Establishment].

## Third Rates 70 Guns (two-deckers)

### 1719 Establishment

Dimensions & tons: 151', 123'2" x 41'6" x 17'4". 1128 9/94.
Men: 480. Guns: GD 26 x 24, UD 26 x 12, QD 14 x 6, UD 4 x 6.

**Edinburgh** Chatham Dyd (Surveyor: Rosewell) [RB].
Ord ____ ; K 31.7.1717; L 30.6.1721; 1742 BU.

**Northumberland** Woolwich Dyd (Surveyor: Hayward) [RB].
Ord ____ ; K 1.4.1719; L 13.7.1721; 1739 BU.

**Captain** Portsmouth Dyd (Surveyor: Naish) [RB].
Ord ____ ; K ____ ; L 21.5.1722; 1740 hulked.

**Burford** Deptford Dyd (Surveyor: Stacey).
Ord 11.10.1720; K ____ ; L 19.7.1722; 1752 BU.

**Sterling Castle** Woolwich Dyd (Surveyor: Hayward) [RB].
Ord 12.3.1720; K 12.7.1721; L 23.4.1723; 1740 hulked.

**Berwick** Deptford Dyd (Surveyor: Stacey).
Ord 31.3.1721; K ____ ; L 22.7.1723; 1743 hulked.

**Lenox** Chatham Dyd (Surveyor: Rosewell) [RB].
Ord 2.5.1721; K 1.2.1722; L 29.11.1723; 1756 breakwater.

**Kent** Woolwich Dyd (Surveyor: Hayward) [RB].
Ord 14.2.1722; K 7.3.1722; L 27.4.1724; 1744 BU.

**Grafton** Woolwich Dyd (Surveyor: Hayward) [RB].
Ord 21.9.1722; K 09.1722; L 25.11.1725; 1744 BU.

**Ipswich** Portsmouth Dyd (Surveyor: Allin) [RB].
Ord 24.8.1727?; K 24.4.1727; L 30.10.1730; 1755 64 guns; 1757 hulked.

**Buckingham** Deptford Dyd (Surveyor: Stacey).
Ord 22.2.1727; K ____ ; L 13.4.1731; 1745 BU.

**Prince of Orange** (ex *Bredah*?) Deptford Dyd (Surveyor: Stacey).
Ord 5.5.1729; K ____ ; L 5.9.1734; 1772 hulked.

FATES: *Captain* hulked at Portsmouth; 1762 BU. *Sterling Castle* hulked at Chatham; 1742 to Sheerness; 1771 BU. *Berwick* hulked at Plymouth; 1783 BU. *Ipswich* hulked at Gibraltar; Ca 1754 BU. *Prince Of Orange* hulked at Sheerness; 1810 sold.

PLANS: *Northumberland*: Lines & profile. *Burford & Berwick*: Lines, profile, gun deck. [Danish Rigsarkivet: Lines]. *Sterling Castle*: Lines & profile. *Berwick?*: Lines & profile/decks. *Kent & Grafton*: Lines & profile. *Ipswich*: Lines & profile. As 64?: Lines & profile/decks.

### 1733 Establishment

Dimensions & tons: 151', 122'2" x 43'5" x 17'9". 1224.
Men: 480, 1743: 470 (*Captain* in 1778: 350). Guns: as 70s: GD 26 x 24, UD 26 x 12, QD 14 x 6, Fc 4 x 6 1743 as 64s GD 26 x 24, UD 26 x 18, QD 10 x 6, Fc 2 x 6 (*Captain* in 1778: GD 22 x 24, UD 22 x 12, QD 4 x 6, Fc 2 x 6).

NOTE: Originally classed as 70s, later as 64s.

**Elizabeth** Chatham Dyd (Surveyor: Ward) [RB].
Ord 4.9.1733; K 15.3.1736; L 29.11.1737; 1766 BU.

1719 Establishment rig & lines for a 70 gun ship (Admiralty plan number 6914), one of the very few eighteenth-century rigging plans to be found in the Admiralty Collection

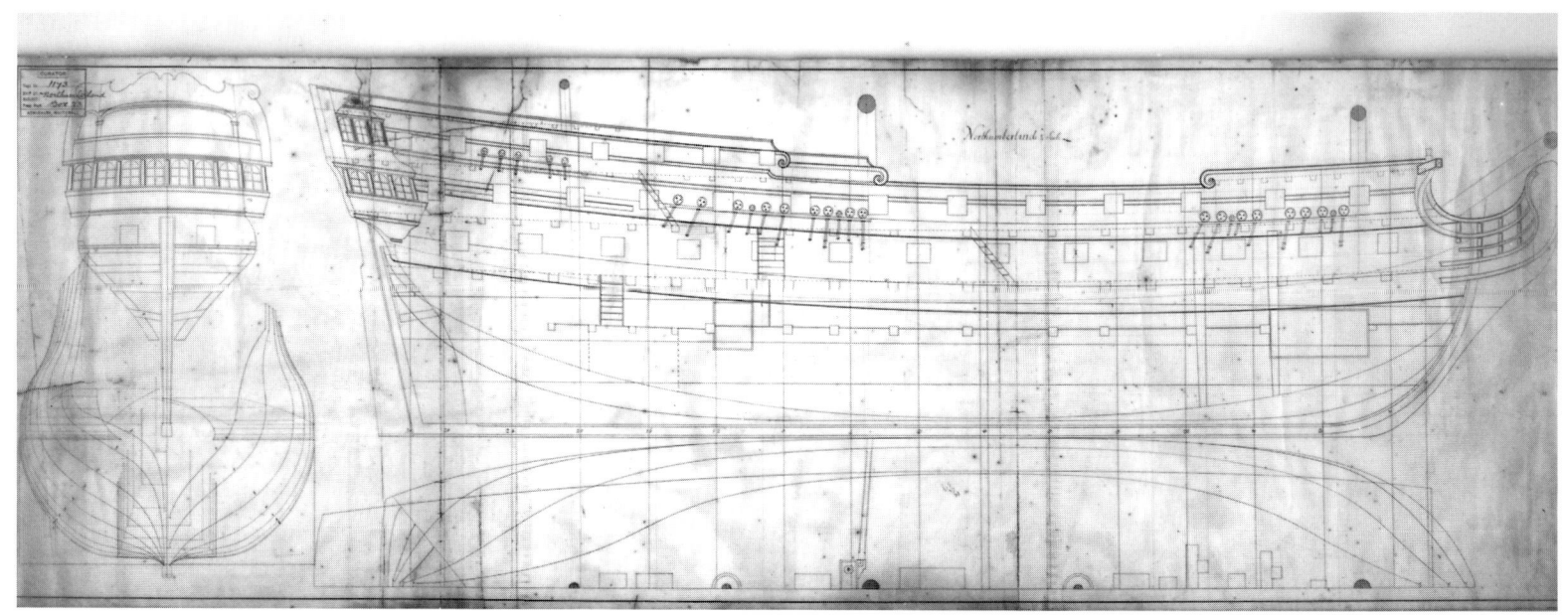

*Northumberland*, 70 gun ship of the 1719 Establishment. Lines and profile plan number 1173

**Prince Frederick** Deptford Dyd (Surveyor: Stacey) [RB of *Expedition*].
  Ord ____ ; K 10.1736; L 18.3.1740; 1764 hulked.

**Nassau** Chatham Dyd (Surveyor: Ward) [RB].
  Ord 25.5.1736; K 23.5.1737; L 25.4.1740; 1770 sold.

**Suffolk** Woolwich Dyd (Holland) [RB].
  Ord 3.12.1735; K ____ ; L ____ 1740; 1770 sold.

**Essex** Woolwich Dyd (Surveyor: Holland) [RB].
  Ord 20.5.1736; K 20.8.1736; L 21.2.1741; 1759 wrecked.

**Bedford** Portsmouth Dyd (Surveyor: Allin) [RB].
  Ord 8.10.1736; K 1.4.1737; L 9.3.1741; 1767 hulked.

**Royal Oak** Plymouth Dyd (Surveyor: Lock) [RB].
  Ord 8.3.1737; K 10.10.1737; L 29.8.1741; 1764 BU.

**Monmouth** Deptford Dyd (Surveyor: Allin) [RB].
  Ord 9.9.1739; K 12.1739; L 6.9.1742; 1767 BU.

**Revenge** Deptford Dyd (Surveyor: Stacey) [RB].

# THE ESTABLISHMENT PERIOD 1719-1745

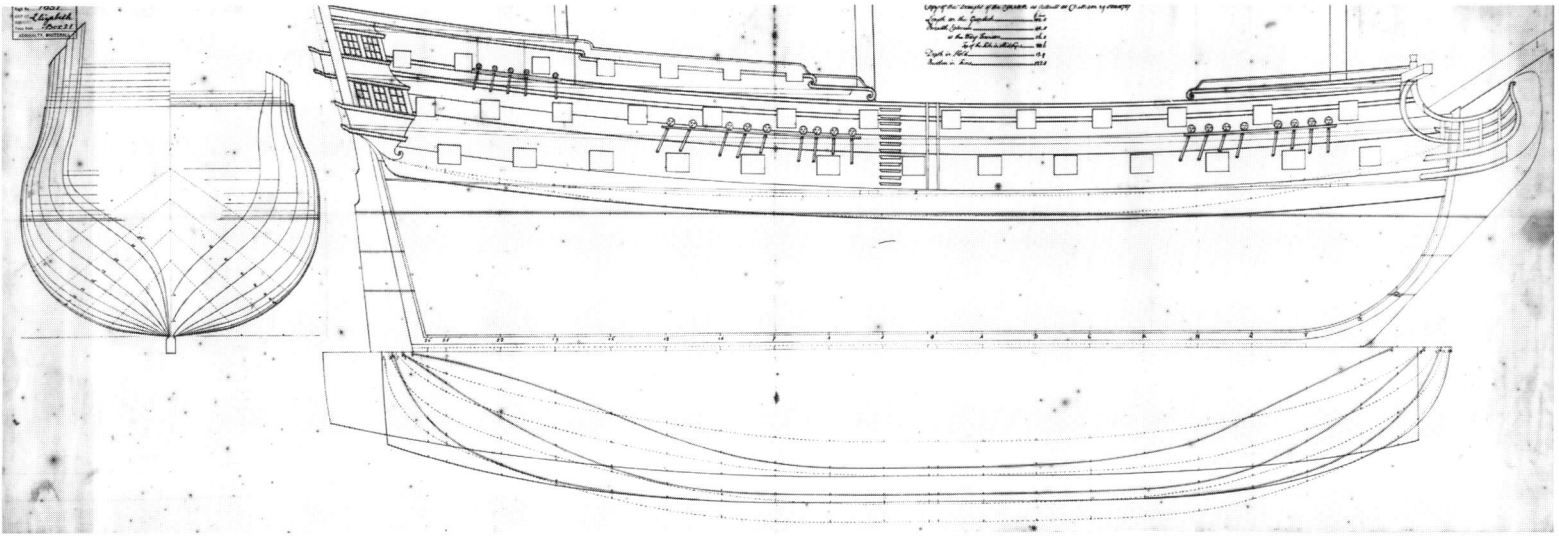

Lines plan of the *Elizabeth* of 1737 (Admiralty number 1051)

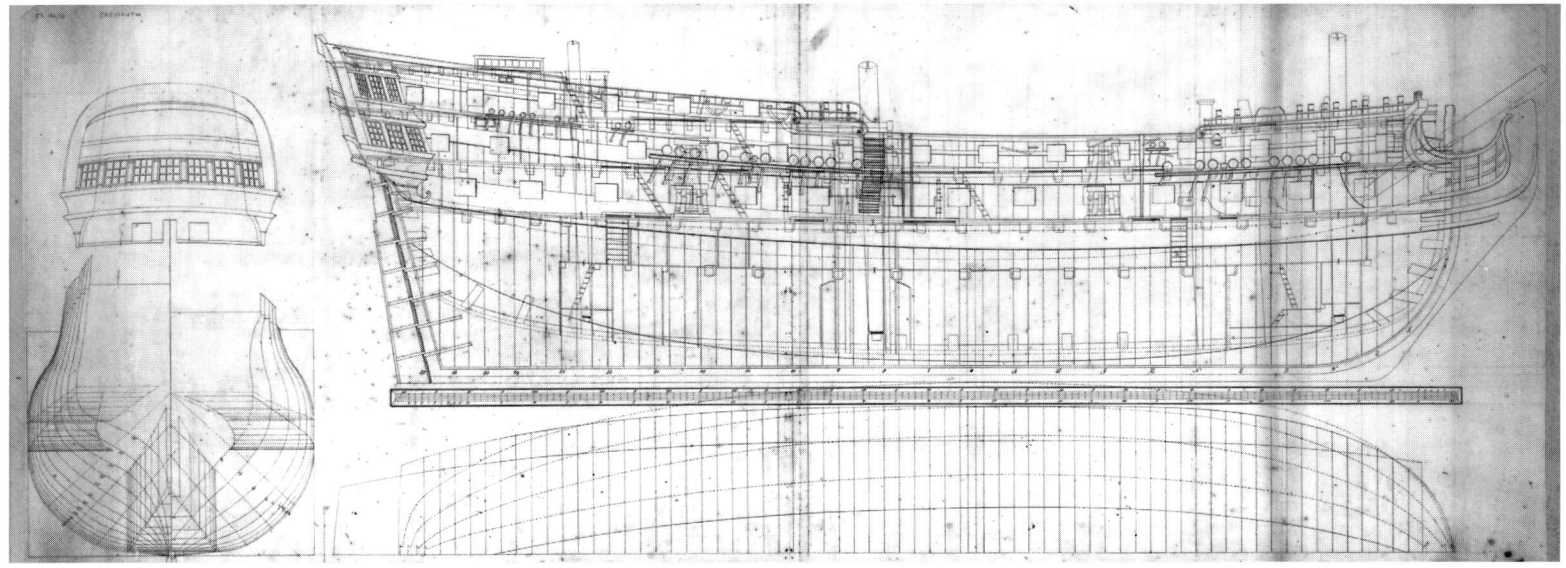

Lines and profile plan of the *Yarmouth* of 1742 (Charnock Collection, NMM, plan Ch. 136)

Ord 3.5.1739; K 24.6.1740; L 25.5.1742; 1787 sold.

**Sterling Castle** Chatham Dyd (Surveyor: Ward).
Ord 6.8.1739; K 25.8.1740; L 24.4.1742; 1762 scuttled.

**Captain** Woolwich Dyd (Surveyor: Holland).
Ord 7.9.1739; K 20.2.1740; L 14.4.1743; 1777 storeship (Sixth Rate) 30, renamed *Buffalo*; 1778 Fourth Rate 50; 1783 BU.

**Berwick** Deptford Dyd (Surveyor: Allin).
Ord 15.12.1740; K 1.1.1741; L 13.6.1743; 1760 BU.

FATES: *Prince Frederick* hulked at Sheerness as a temporary lazaretto; 1784 sold. *Essex* was wrecked on the Four Bank during the Battle of Quiberon Bay 20/21.11.1759. *Bedford* hulked at Deptford; 1787 sold. *Sterling Castle* was sunk by Admiral Pocock's orders in the harbour at Havanna.

PLANS: *Elizabeth*: Lines/lines & profile. *Prince Frederick*: Danish Rigsarkivet: Lines. *Nassau*: Lines. *Suffolk*: Lines/Lines & profile. *Essex*: Lines & profile. *Bedford*: Lines. *Royal Oak*: Lines/lines & profile/outline decks. *Monmouth*:? Lines/Lines & profile. *Captain*: Stern. *Berwick*: Lines/decks.

**Northumberland** Woolwich Dyd (Surveyor: Holland).
Ord 20.9.1739; K 16.10.1740; L 7.10.1743; 1744 taken.

**Hampton Court** Deptford Dyd (Surveyor: Allin).
Ord 28.10.1741; K 20.7.1742; L 3.4.1744; 1774 BU.

**Edinburgh** Chatham Dyd (Surveyor: Ward).
Ord 5.12.1740; K 2.12.1741; L 31.5.1744; 1771 BU.

**Kent** Deptford Dyd (Surveyor: Allin).
Ord 10.5.1743; K 8.12.1743; L 10.5.1746; 1757 (or 1760?) hulked.

**Grafton** Portsmouth Dyd (Surveyor: Lock); begun as 1741 Establishment but completed as 1745 Establishment (which see).

FATES: *Northumberland* taken by the French *Content* off Ushant served as *Le Northumberland*; 1779 renamed *L'Atlas* as a 'flûte' 1781 wrecked off Ushant.

**Kent** Hulked in the East Indies; fate uncertain.

PLANS: *Northumberland*: Lines & Profile. *Hampton Court*: Lines/Lines & profile/decks. *Kent*: Lines. *Grafton*: Lines.

## 1741 Establishment

Dimensions & tons: 154', 125'8" x 44' x 18'11", 1291 49/94.
Men: 480. Guns: GD 26 x 32, UD 26 x 18, QD 10 x 9, Fc 2 x 9.
Originally classed as 70s but later as 64s.

## *Yarmouth* Class 1740.
1741 Establishment design, lengthened by 6 feet.
Dimensions & tons: 160', 131'8" x 44' x 18'11". 1359.
Men: 520. Guns: GD 26 x 32, UD 26 x 18, QD 10 x 9, Fc 2 x 9.
Classed as a 64.

**Yarmouth** Deptford Dyd (Surveyor: Allin).

Ord 25.4.1740; K 25.11.1742; L 8.3.1745; 1781 classed as a 74 for local service; 1783 hulked.
FATE: Receiving ship at Plymouth; 1801 sold.
PLANS: Lines.

**Lancaster Class 1743.** Begun as a 1741 Establishment 80 gun three-decker, but in 1747 decided to complete her as a 66 gun two-decker. At the same time her sister *Devonshire,* already completed as an 80, was given a similar conversion. (see under 80s).
Dimensions & tons: 161', 130'10" x 46'1" x 19'4". 1478.
Men: 520. Guns: GD 26 x 32, UD 26 x 18, QD 12 x 9, Fc 2 x 9.
*Lancaster* Woolwich Dyd (Surveyor: Holland).
Ord 15.2.1743; K 4.2.1744; L 22.4.1749; 1773 BU.
*Devonshire* Woolwich Dyd (Surveyor: Holland) See under 80s.
PLANS: See under 80s.

# Third Rates, later Fourth Rates, 60 Guns (two-deckers)

## 1719 Establishment
Dimensions & tons: 144', 117'7" x 39' x 16'5". 951 27/94.
Men: 400. Guns: GD 24 x 24, UD 26 x 9, QD 8 x 6, Fc 2 x 6.

NOTE: *Centurion* and *Rippon* varied somewhat from the Establishment dimensions (see below). *Dreadnought* underwent a Great Repair rather than a rebuild, but was so altered that it is more convenient to consider her here as a new ship of this group rather than as a continuation of her previous existence. The same probably applies to *Defiance*.

*Canterbury* Portsmouth Dyd (Surveyor: Naish) [RB].
Ord 23.1.1721?; K 19.11.1720; L 15.9.1722; 1741 BU.
*Dreadnought* Portsmouth Dyd (Surveyor: Naish) [Great Repair].
Ord ____ ; K 30.6.1721; L 11.3.1723; 1742 hulked .
*Plymouth* Chatham Dyd (Surveyor: Rosewell) [RB].
Ord 26.5.1720?; K 13.4.1720; L 2.8.1722; 1764 BU.
*Sunderland* Chatham Dyd (Surveyor: Rosewell).
Ord 31.3.1721; K ____ ; L 30.4.1724; 1742 BU.
*Windsor* Deptford Dyd (Surveyor: Stacey) [RB].
Ord 18.11.1725?; K 10.1725; L 27.10.1729; 1742 BU.
*Defiance* Chatham Dyd (Surveyor: Rosewell). [RB or Great Repair].
Ord ____ ; K 1.7.1729; L 1.9.1730; 1742 hulked.
*Deptford* Deptford Dyd (Surveyor: Stacey).
Ord 3.5.1726; K 12.12.1729; L 26.9.1732; 1752 Fourth Rate 50; 1767 sold.
*Swallow* Plymouth Dyd (Surveyor: Lock).
Ord 7.1.1729; K 1.12.1729; L 6.10.1732; 1737 renamed *Princess Louisa*; 1742 BU.
*Pembroke* Woolwich Dyd (Surveyor: Hayward).
Ord 8.9.1726; K 09.1729; L 27.11.1733; 1749 wrecked.
*Tilbury* Chatham Dyd (Surveyor: Ward).
Ord 15.12.1726; K 25.3.1731; L 2.6.1733; 1742 burnt.
*Dunkirk* Portsmouth Dyd (Surveyor: Allin) [RB].
Ord 12.9.1729; K 01.1731; L 3.9.1734; 1749 BU.
FATES: *Dreadnought* hulked as a receiving ship at Portsmouth; 1749 BU. *Pembroke* 'lost in a hurricane of wind' in the Road of Fort Saint David, India 12/14.4.1749. *Tilbury* burnt by accident off Hispaniola 21.9.1744.
PLANS: *Windsor*: Lines. *Deptford*: Lines. *Swallow*: Lines. *Pembroke*: Lines. *Plymouth*: Lines.

## Modified 1719 Establishment
Dimensions & tons: [*Centurion*] 144', 177'7" x 40' x 16'5". ____ . [*Rippon*] 145', 118'6" x 40' x 16'5". The first being 1 foot broader than the Establishment, the second both this and 1 foot longer.
*Centurion* Portsmouth Dyd (Surveyor: Allin).
Ord 17.10.1729; K 9.9.1729; L 6.1.1733; 1744 Fourth Rate 50 renamed *Eagle*; 1745 renamed *Centurion*; 1769 BU.
NOTE: Anson's *Centurion*.
*Rippon* Woolwich Dyd (Surveyor: Hayward) [RB].
Ord 23.6.1730; K 19.10.1730; L 29.3.1735; 1751 BU.
PLANS: *Centurion*: Lines. *Rippon*: Lines.

## 1733 Establishment
Dimensions & tons: 144', 117' x 41'5" x 16'11". 1068.
Men: 400. Guns: GD 24 x 24, UD 26 x 9, QD 8 x 6, Fc 2 x 6.
*Warwick* Plymouth Dyd (Surveyor: Lock).
Ord 14.3.1727; K 04.1730; L 25.10.1733; 1756 taken.
*Strafford* Chatham Dyd (Surveyor: Ward).
Ord 22.5.1733; K 15.9.1733; L 24.7.1735; 1756 breakwater at Sheerness.
*Worcester* Portsmouth Dyd (Surveyor: Allin).
Ord 4.9.1733; K 13.11.1733; L 20.12.1735; 1765 BU.
*Augusta* Deptford Dyd (Surveyor: Stacey).
Ord 22.5.1733; K 7.11.1733; L 1.7.1736; 1765 BU.
*Dragon* Woolwich Dyd (Surveyor: Hayward).
Ord 19.10.1733; K 12.11.1733; L 11.9.1736; 1757 breakwater at Sheerness.
*Jersey* Plymouth Dyd (Surveyor: Lock).
Ord ____ ; K. 11.1733; L 14.6.1736; 1771 hospital ship; Ca 1780 hulked.
*Superb* Woolwich Dyd (Surveyor: Hayward).
Ord 4.9.1733; K 29.9.1733; L 27.8.1736; 1757 BU.
*Weymouth* Plymouth Dyd (Surveyor: Lock)
Ord 19.4.1735; K 09.1733; L 31.3.1736; 1745 wrecked.
*Lion/Lyon* Deptford Dyd (Surveyor: Stacey) [RB].
Ord 9.12.1735; K 23.2.1736; L 25.4.1738; 1765 sold.
*York* Plymouth Dyd (Surveyor: Lock) [RB or Great Repair].
Ord ____ ; K 04.1737; L 31.1.1739; 1751 breakwater at Sheerness.

Anson's *Centurion* of 60 guns shown on plan number 1278, one of two Admiralty lines plans of this successful ship

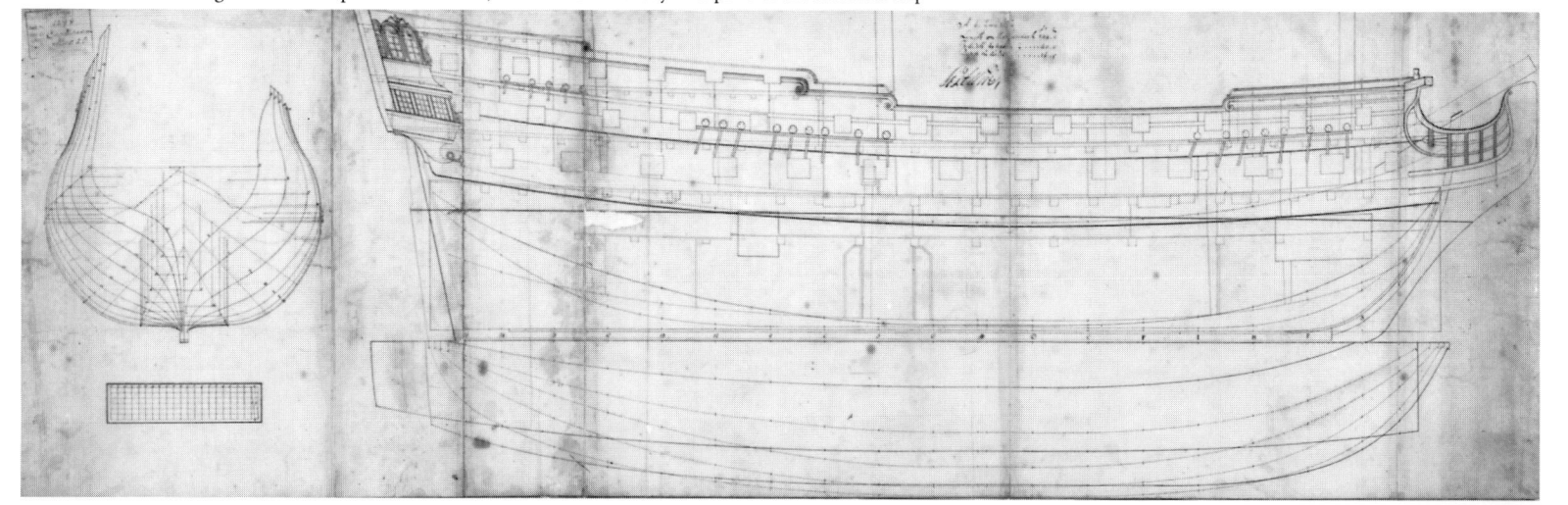

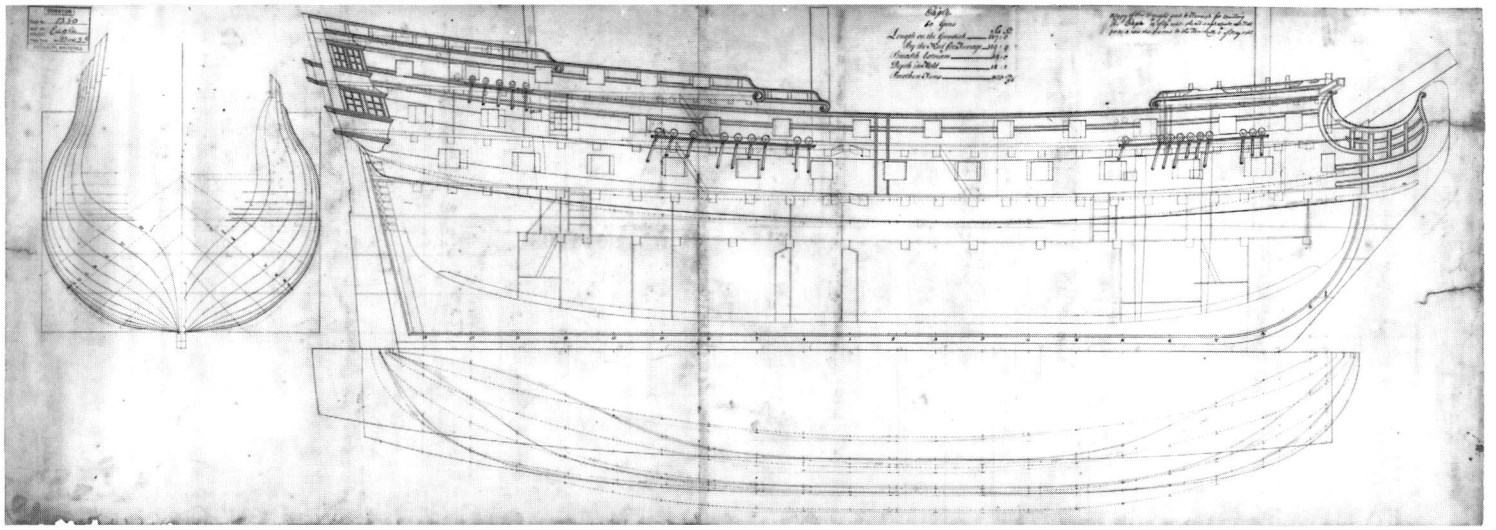

*Eagle*, 60 gun ship of 1745 (plan number 1280)

Ord ____ ; K 04.1737; L 31.1.1739; 1751 breakwater at Sheerness.
**Kingston** Plymouth Dyd (Surveyor: Lock) [RB].
Ord ____ ; K 28.2.1737; L 8.10.1740; 1762 sold.
**Rupert** Sheerness Dyd (Surveyor: Rosewell) [RB].
Ord 16.8.1736; K 27.6.1737; L 27.10.1740; 1769 BU.
**Dreadnought** Bronsdon & Wells, Deptford.
Ord 5.12.1740; K 7.1.1741; L 23.6.1742; 1784 sold.
**Medway** Bird, Rotherhithe.
Ord 5.12.1740; K 7.1.1741; L 26.5.1742; 1748 hulked.
**Exeter** Plymouth Dyd (Surveyor: Fellows) [RB].
Ord ____ ; K 27.10.1740; L 19.3.1744; 1763 BU.
**Nottingham** Sheerness Dyd (Surveyor: Pool).
Ord 18.5.1739; K 12.12.1740; L 17.8.1745; 1773 breakwater at Sheerness.
**Princess Mary** Portsmouth Dyd (Surveyor: Lock) [RB].
Ord 15.12.1736; K 01.1738; L 5.10.1742; 1766 sold.

FATES: *Warwick* taken by the French *L'Atalante* at Martinique; 22.1.1761 retaken by *Minerva* and BU. *Jersey* hulked at New York as a prison hospital ship; 1783 sold. *Weymouth* wrecked at Antigua 16.2.1745. *Medway* hulked at Portsmouth; 1749 BU.

PLANS: *Worcester*: Danish Rigsarkivet: Lines. *Augusta*: Danish Rigsarkivet: Lines. *Weymouth*: Danish Rigsarkivet: Lines. *Lion*: Danish Rigsarkivet: Lines. *Kingston*: Lines. *Worcester*: Lines/lines & profile. *Augusta*: Lines. *Exeter*: Lines. *Warwick*: Lines. *Superb*: Lines. *Princess Mary*: Lines. *Jersey*: Lines/Lines & profile. *Dragon, Weymouth, Nottingham, Medway, Dreadnought*: Lines. *Strafford*: Lines. *Rupert*: Lines. *Nottingham*: Lines.

### 1741 Establishment. Classed as 58s not 60s, and Fourth Rates.
Dimensions & tons: 147', 119'9" x 42' x 18'1". 1,123 57/94.
Men: 400. Guns: GD 24 x 24, UD 24 x 12, QD 8 x 6, Fc 2 x 6.
**Canterbury** Plymouth Dyd (Surveyor: Fellows).
Ord 5.3.1741; K 1.3.1742; L 5.2.1745; 1770 BU.
**Defiance** West, Deptford.
Ord ____ ; K 22.3.1743; L 12.10.1744; 1766 sold.
**Princess Louisa** Carter, Limehouse.
Ord 23.12.1742; K 20.3.1743; L 1.7.1744; 1766 BU.
**Sunderland** Portsmouth Dyd (Surveyor: Lock) [RB].
Ord 13.11.1742; K 25.1.1743; L 4.4.1744; 1761 foundered.
**Eagle** (ex *Centurion*) Barnard, Harwich.
Ord 10.4.1744; K 24.7.1744; L 2.12.1745; 1767 sold.
**Tilbury** Portsmouth Dyd (Surveyor: Lock).
Ord 10.12.1742; K 8.2.1743; L 20.7.1745; 1757 wrecked.

FATES: *Sunderland* foundered off Pondicherry 1.1.1761. *Tilbury* foundered in a hurricane off Louisbourg 24.9.1757.

PLANS: Lines/Lines & profile. *Princess Louisa*: Lines & profile/orlop/gun deck/upper deck/quarter deck & forecastle. *Eagle*: Lines & profile. *Tilbury*: Lines & profile. *Sunderland*: Lines. *Canterbury*: Lines.

**Enlarged design.** 'To Master Shipwright's [of Woolwich?] Draught'.
Dimensions & tons: 152', 126' x 42' x 17'10". 1182.
Men & Guns as above.
**Windsor** Woolwich Dyd (Surveyor: Holland) [RB].
Ord 1.11.1742; K 28.2.1743; L 26.2.1745; 1777 BU.
PLANS: Lines/Lines & profile. Danish Rigsarkivet: Lines.

## Fourth Rates 50 Guns (two-deckers)

### 1719 Establishment
Dimensions & tons: 134', 109'8" x 36' x 15'2". 755 89/94.
Men: 300. Guns: GD 22 x 18, UD 22 x 9, QD 4 x 6, Fc 2 x 6.
**Falkland** Deptford Dyd (Surveyor: Stacey) [RB].
Ord ____ ; K 8.4.1718; L 23.8.1720; 1743 BU.
**Chatham** Deptford Dyd. (Surveyor: Stacey) [RB].
Ord ____ ; K 4.7.1718; L 15.8.1721; 1749 breakwater at Sheerness.
**Colchester** Chatham Dyd (Surveyor: Rosewell) [RB].
Ord ____ ; K 1.4.1719; L 26.10.1721; 1742 BU.
**Leopard** Woolwich Dyd (Surveyor: Hayward) [RB].
Ord ____ ; K 1.8.1719; L 10.4.1721; 1740 BU.
**Argyle** Woolwich Dyd (Surveyor: Hayward) [RB].
Ord 27.1.1720?; K 20.10.1719; L 5.7.1722; 1748 breakwater at Harwich.
**Portland** Portsmouth Dyd (Surveyor: Naish) [RB].
Ord ____ ; K 6.2.1719; L 25.2.1723; 1743 BU.
**Assistance** Woolwich Dyd (Surveyor: Naish?) [RB].
Ord 21.3.1720?; K 01.1720; L 25.11.1725; 1746 breakwater.
**Romney** Deptford Dyd (Surveyor: Stacey) [RB].
Ord 11.6.1723; K 21.8.1723; L 17.10.1726; 1747 Fifth Rate 44; 1757 sold.
**Salisbury** Portsmouth Dyd (Surveyor: Naish) [RB or Great Repair].
Ord 9.4.1725?; K 23.12.1724; L 30.10.1726; 1745 hulked.
**Oxford** Portsmouth Dyd (Surveyor: Allin) [RB].
Ord 29.6.1723; K 13.3.1724; L 10.7.1727; 1758 BU.
**Falmouth** Woolwich Dyd (Surveyor: Hayward) [RB].
Ord 14.5.1724; K 24.6.1724; L 30.4.1729; 1747 BU.
**Litchfield** Plymouth Dyd (Surveyor: Lock) [RB].
Ord ____ ; K 11.1727; L 25.3.1730; 1744 BU.
**Greenwich** Chatham Dyd (Surveyor: Rosewell) [RB].
Ord 16.4.1724; K 1.7.1724; L 10.2.1731; 1744 lost.
**Newcastle** Woolwich Dyd (Surveyor: Hayward) [RB].
Ord 31.5.1728; 03.1729; L 6.1.1733; 1746 BU.

FATES: *Salisbury* hulked at Kinsale; 1749 sold. *Greenwich* wrecked in a hurricane at Jamaica 20.10.1744.

PLANS: *Falkland*: Lines & profile. *Chatham*: Danish Rigsarkivet: Lines. *Colchester*: Lines & Profile. *Oxford*: Lines & profile.

### 1733 Establishment
Dimensions & tons: 134', 108'8" x 38'6" x 15'9". 853.

Plan number 6196 shows both rigging and lines of a 1719 Establishment 50 gun ship and is noted as representing the *Oxford*

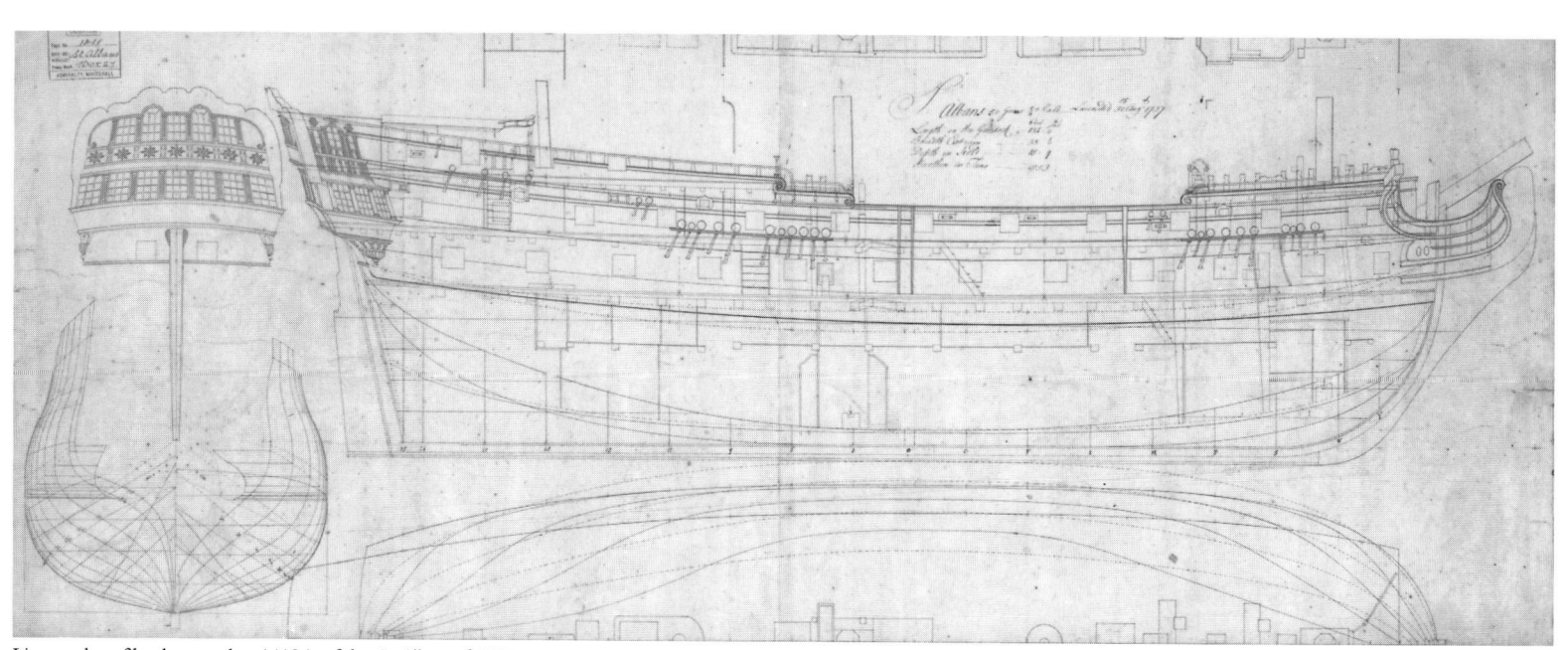

Lines and profile plan number 1418 is of the *St Albans* of 1737

Men: 300 Guns: GD 22 x 18, UD 22 ´x 9, QD 4 x 6, Fc 2 x 6.
**Gloucester** Sheerness Dyd (Surveyor: Rosewell) [RB].
   Ord 22.5.1733; K 19.8.1734; L 22.3.1737; 1742 scuttled.
**Saint Albans** Plymouth Dyd (Surveyor: Lock) [RB].
   Ord 10.9.1734; K 12.1.1736; L 30.8.1737; 1744 lost.
**Severn** Plymouth Dyd (Surveyor: Lock) [RB].
   Ord 13.5.1734; K 10.2.1735; L 28.3.1739; 1746 taken.
**Guernsey** Chatham Dyd (Surveyor: Ward) [RB].
   Ord 23.2.1737; K 03.1738; L 11.8.1740; 1769 hulked.
**Dartmouth** Woolwich Dyd (Surveyor: Hayward).
   Ord 8.10.1736; K 19.10.1736; L 22.4.1741; 1747 blown up.
**Woolwich** Deptford Dyd (Surveyor: Stacey) [RB].
   Ord 10.6.1736; K 08.1738; L 6.4.1741; 1747 B.U.
**Antelope/Anthelope** Woolwich Dyd (Surveyor: Hayward) [RB].
   Ord 9.1.1738/19.7.1738; K 28.9.1738; L 27.2.1742; 1783 sold.
**Preston** Plymouth Dyd (Surveyor: Fellows) [RB].
   Ord 8.5.1739/25.7.1739; K 12.1739; L 18.9.1742; 1748 hulked.
**Hampshire** Barnard, Ipswich [RB].
   Ord 28.4.1740; K 11.6.1740; L 13.11.1741; 1766 BU.
**Leopard** Perry, Blackwall.
   Ord 28.4.1740; K 20.6.1740; L 30.9.1741; 1761 BU.
**Nonsuch** Quallett, Rotherhithe.
   Ord 28.4.1740; K 19.10.1740; L 29.12.1741; 1766 BU.
**Sutherland** Taylor, Rotherhithe.

THE ESTABLISHMENT PERIOD 1719-1745

Ord 28.4.1740;  1.7.1740;[1] 14.10.1741; 1770 sold.

FATES: *Gloucester* burnt and sunk off the Ladrone Islands 'because of lack of hands and leakiness;' on Anson's voyage 15.8.1742. *Saint Albans* wrecked in a hurricane at Jamaica 20.10.1744. *Severn* taken by Conflans with two French men of war 19.10.1746 ; retaken by Hawke 14.10.1747 but not recommissioned. *Guernsey* hulked at Woolwich; 1786 sold. *Dartmouth* blown up off Cape Saint Vincent in action with the Spanish *Glorioso* 3 or 7.10.1747. *Preston* hulked at Trincomalee; 1749 scuttled.

PLANS: *Saint Albans*: Lines/Lines & profile. *Severn*: Lines. *Guernsey*: Lines. As hulk; 1769: Profile & external profile/orlop/upper deck/rig [pencil, very faint]. *Dartmouth*: Lines. *Woolwich*: Lines. *Antelope*: Lines. *Preston*: Lines. *Leopard*: Danish Rigsarkivet: Lines & profile. *Sutherland*: Lines. *Hampshire, Sutherland, Leopard, Nonsuch*: Lines. *Nonsuch*: Lines.

## 1741 Establishment

Dimensions & tons: 140', 113'9" x 40'x 17'2½". 968 8/94.
Men: 350. Guns: GD 22 x 24, UD 22 x 12, QD 4 x 6, Fc 2 x 6.

**Harwich** (ex *Tyger*) Barnard, Harwich.
   Ord 21.8.1742; K 11.1742; L 22.12.1743; 1760 wrecked.
**Colchester** [I] Barnard, Harwich.
   Ord 6.9.1742; K 14.12.1742; L 14.8.1744; 1744 wrecked.
**Falkland** Ewer, Bursledon.
   Ord 15.11.1742; K 20.1.1743; L 17.3.1744; 1768 hulked.
**Chester** Bronsdon, Deptford.
   Ord 24.1.1743; K 02.1743; L 18.2.1744; (1757-1762 floating battery at Milford); 1767 sold.
**Winchester** Bird, Rotherhithe.
   Ord 28.3.1743; K 7.5.1743; L 3.5.1744; 1769 sold.
**Portland** Snellgrove, Limehouse.
   Ord 26.4.1743; K 29.4.1743; L 11.10.1744; 1763 sold.
**Maidstone** (ex *Rochester*) Bronsdon, Deptford.
   Ord 16.5.1743; K 30.5.1743; L 12.10.1744; 1747 wrecked.
**Panther** Plymouth Dyd (Surveyor: Fellows).
   Ord 16.5.1743; K 27.6.1743; L 24.6.1746; 1756 BU.
**Gloucester** Whetstone, Rotherhithe.
   Ord 15.6.1743; K 12.7.1743; L 23.3.1745; 1758 hulked.
**Norwich** Perry, Blackwall.
   Ord 30.9.1743; K 23.11.1743; L 4.7.1745; 1768 sold.
**Ruby** Ewer, Bursledon.
   Ord 30.9.1743; K 18.2.1744; L 3.8.1745; 1765 BU.
**Advice** Rowcliffe, Southampton.
   Ord 30.3.1744; K 06.1744; L 26.2.1746; 1756 BU.
**Salisbury** Ewer, Bursledon.
   Ord 2.5.1744; K 23.5.1744; L 29.1.1746; 1761 condemned.
**Litchfield** Barnard, Harwich.
   Ord 1.6.1744; K 24.7.1744; L 26.6.1746; 1758 wrecked.
**Colchester** [II] Carter, Southampton.
   Ord 6.11.1744; K ____ 1744; L 20.9.1746; 1773 BU.

FATES: *Harwich* wrecked on the Isle of Pines, Cuba 4.10.1760. *Colchester* [I] wrecked on the Kentish Knock 21.9.1744 (less than a month after completion). *Falkland* hulked at Chatham for the Commissioners of Victualling; fate uncertain. *Maidstone* wrecked near Saint Malo 27.6.1747; the site has recently been located and excavated. *Gloucester* hulked at Chatham; 1759 receiving ship at Sheerness; 1764 BU. *Litchfield* wrecked on the African coast 28.11.1758.

PLANS: Lines & profile/orlop/gun deck/upper deck/quarter deck & forecastle. *Harwich & Colchester* [I]: Lines & profile. *Falkland*: Lines & profile. *Panther*: Lines. *Litchfield & Colchester* [II]: Lines & profile.

### *Bristol* Class 1742.
1741 Establishment design lengthened by 6 feet.
Dimensions & tons: 146', 120' x 40' x 16'10". 1021.
Men: 350. Guns: GD 22 x 24, UD 22 x 12, QD 4 x 6, Fc 2 x 6.

**Bristol** Woolwich Dyd (Surveyor: Fellows/Holland).
   Ord 22.11.1742; K 24.6.1743; L 9.7.1746; 1768 BU.
**Rochester** Deptford Dyd (Surveyor: Holland).
   Ord 8.3.1747; K 24.9.1747; L 3.8.1749; 1770 sold.

PLANS: Lines & profile. *Bristol*: Lines (incomplete). *Rochester*: Lines & profile.

## Fifth Rates 40/44 Guns (two-deckers)

### 1719 Establishment [40s]

Dimensions & tons: 124', 101'8" x 33'2" x 14'. 594 55/94 .
Men: 250. Guns: GD 20 x 12, UD 20 x 6.

**Hector** Plymouth Dyd (Surveyor: Powell/Phillips) [RB].
   Ord ____ ; K 24.3.1718; L 16.2.1721; 1742 BU.
**Anglesea** Chatham Dyd (Surveyor: Rosewell) [RB or Great Repair].
   Ord ____ ; K 01.1719; L 19.5.1725; 1742 breakwater.
**Mary Gally** Plymouth Dyd (Surveyor: Lock) [RB].
   Ord ____ ; K 01.1721; L 13.2.1727; 1743 BU.
**Diamond** Deptford Dyd (Surveyor: Stacey) [RB].
   Ord ____ ; K 05.1720?; L 13.3.1723; 1744 sold.
**Ludlow Castle** Woolwich Dyd (Surveyor: Hayward) [RB].
   Ord ____ ; K 3.2.1722; L 1.2.1724; 1744 hulked.
**Kinsale** Portsmouth Dyd (Surveyor: Naish) [RB].
   Ord 20.12.1722; K ____ 1723; L 13.4.1724; 1741 BU.
**Pearl** Deptford Dyd (Surveyor: Stacey) [RB].
   Ord —— ; K 01.1723; L 7.10.1726; 1744 sold.
**Lark** Woolwich Dyd (Surveyor: Hayward) [RB].
   Ord ____ ; K 17.10.1723; L 2.8.1726; 1742 hulked.
**Adventure** Portsmouth Dyd (Surveyor: Naish) [RB or Great Repair].

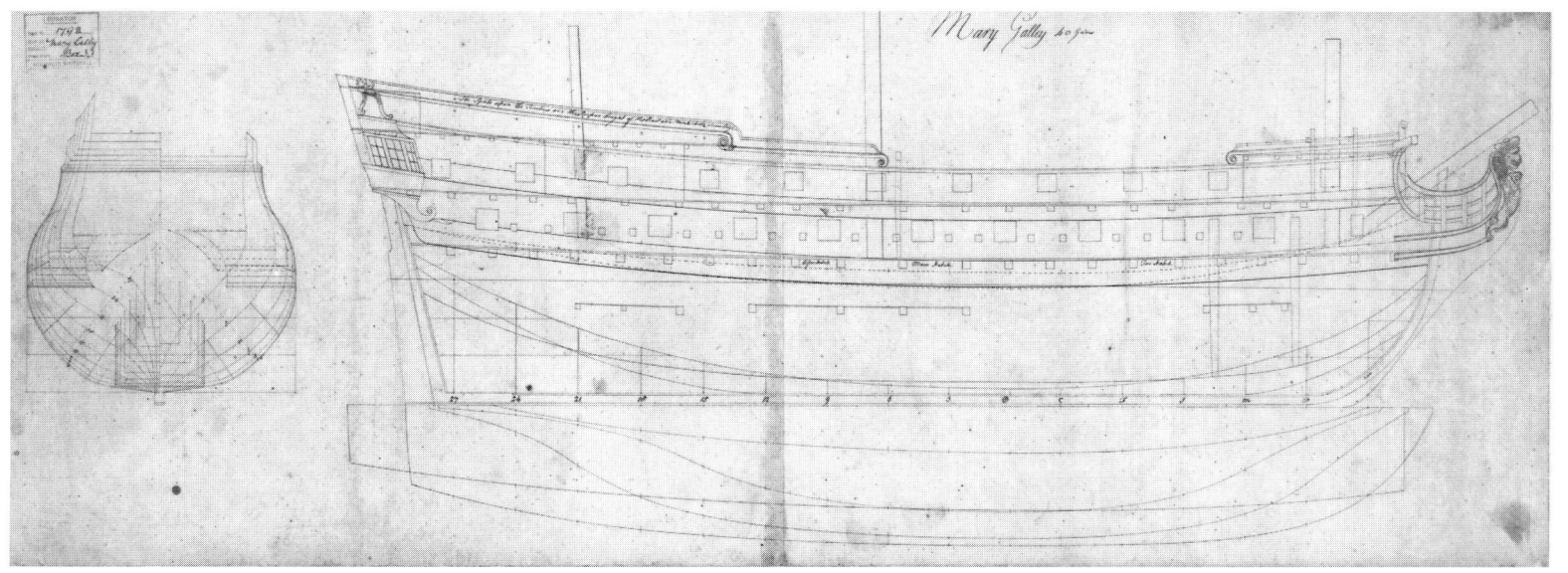

This plan (number 1792) is particularly interesting and unusual in showing 'sweep' (oar) ports on the lower deck of a two-decker warship. This is the *Mary Gally* of the 1719 Establishment, a 40 gun ship and one of the last 'galley frigates', as her name indicates. Notice her one-level stern, like the later frigates

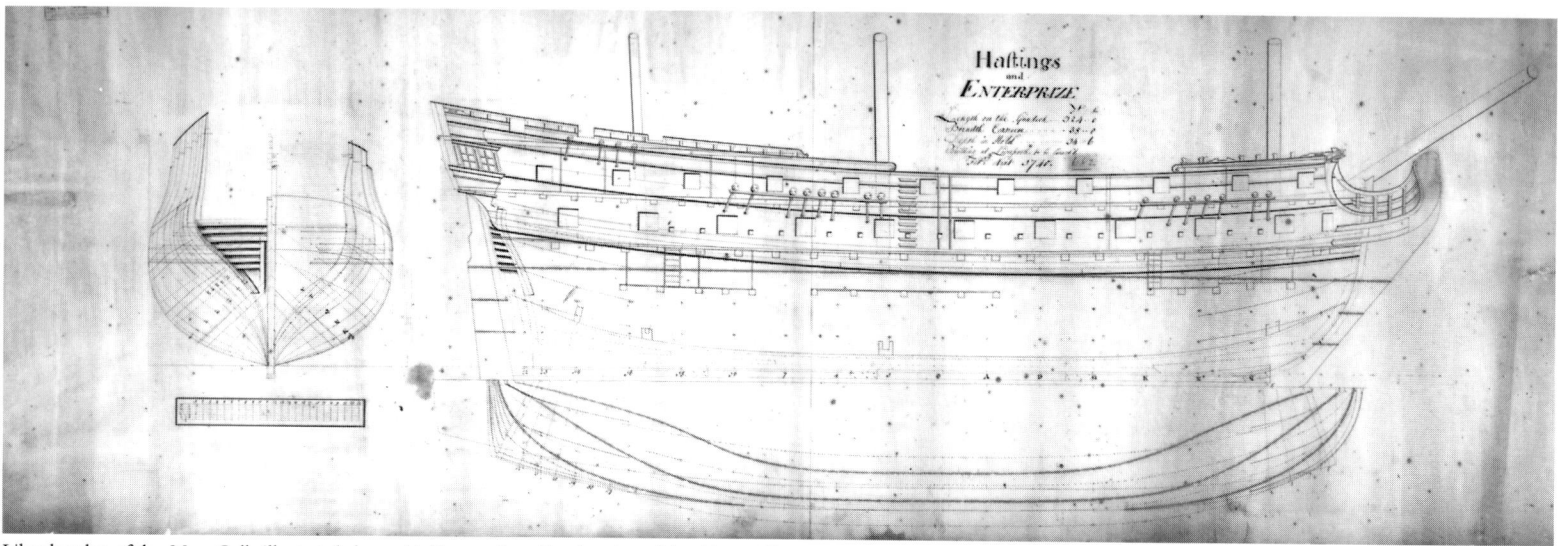

Like the plan of the *Mary Gally* illustrated above this lines and profiles drawing of the 44 gun *Hastings* and *Enterprize* of 1741 is unusual in showing oar ports on the lower deck. These vessels were at the upper limit of size for rowing. Like in the previous plan the stern galleries are on one level only

Ord ____ ; K 20.11.1724; L 4.6.1726; 1740 hulked.
**Princess Louisa** Woolwich Dyd (Surveyor: Hayward) [RB of *Launceston*].
Ord ____ ; K ____ 1726; L 8.8.1728; 1736 wrecked.
**Torrington** Deptford Dyd (Surveyor: Stacey) [RB of *Charles Gally*].
Ord ____ ; K ____ 1726; L 17.7.1729; 1744 sold.
**Roebuck** Woolwich Dyd (Surveyor: Hayward) [RB].
Ord ____ ; K 02.1729; L 16.6.1733; 1743 breakwater.

FATES: *Ludlow Castle* hulked at Antigua; 1750 sold. *Lark* hulked at Jamaica; 20.10.1744 wrecked. *Adventure* hulked at Deptford; 1741 BU. *Princess Louisa* 'Lost .... in a violent storm by beating her keel coming over the Hinder she beached, bilged upon the flatts about a mile from the Helvoet Sluijce' 29.12.1736.

PLANS: *Diamond*: Lines. *Diamond, Pearl & Torrington*: Danish Rigsarkivet: Lines. *Pearl*: Lines (?). Danish Rigsarkivet: Lines/lines & profile. *Ludlow Castle*: Lines. *Mary Gally*: Lines.

## 1733 Establishment 40/later 44s

Dimensions & tons: 124', 100'3" x 35'8" x 14'6". 678.
Men: 250 (as 32s: 210). Guns: GD 20 x 12, UD 20 x 9, QD 4 x 6. *Eltham* only: UD 18 x 9, QD 2 x 6 (as 32s: UD 26 x 12, QD 4 x 6, Fc 2 x 6).

**Eltham** Deptford Dyd (Surveyor: Stacey) [RB of *Portsmouth*].
Ord ____ ; K 14.10.1734; L 30.6.1736; 1763 BU.
**Dover** Bronsdon, Deptford [RB].
Ord 31.10.1739; K 23.1.1740; L 7.1.1741; 1763 sold
**Folkestone** Bird, Rotherhithe [RB].
Ord 31.10.1739; K 23.1.1740; L 8.1.1741; 1749 sold.
**Lynn** West, Deptford [RB].
Ord 4.12.1739; K 24.12.1739; L 7.3.1741; 1763 sold.
**Feversham** Perry, Blackwall [RB].
Ord 14.12.1739; K 24.1.1740; L 7.1.1741; 1749 sold.
**Gosport** Snelgrove, Limehouse [RB].
Ord 14.12.1739; K 10.1.1740; L 20.2.1741; 1768 BU.
**Hastings** Okill, Liverpool.
Ord 8.2.1740; K 6.3.1740; L 7.3.1741; 1763 BU.
**Liverpool** (ex *Enterprise*) Okill, Liverpool.
Ord 8.2.1740; K 2.6.1740; L 19.7.1741; 1756 sold; 1759 repurchased as *Looe* (30)?; 1763 sold?
**Saphire** Carter, Limehouse.
Ord 8.2.1740; K 20.3.1740; L 21.2.1741; 1757 32 guns (one-decker); 1784 sold.
**Kinsale** Bird, Rotherhithe.
Ord 10.6.1740; K 14.1.1741; L 27.11.1741; 1758 hulked .
**Adventure** Blades, Hull.
Ord 25.7.1740; K 11.1740; L 1.10.1741; 1757 32 guns (one-decker); 1770 sold.
**Diamond** Carter, Limehouse.
Ord 5.12.1740; K 26.1.1741; L 30.10.1741; 1756 sold.

**Launceston** Buxton, Rotherhithe.
Ord 22.12.1740; K 26.1.1741; L 29.12.1741; (1752-1759 hulked at Sheerness); 1784 sold.
**Looe** Snelgrove, Limehouse.
Ord 22.12.1740; K 26.1.1741; L 29.12.1741; 1744 wrecked.

FATES: *Kinsale* hulked at Antigua; 1763 sold. *Launceston* served as a quarantine hulk at Sheerness 1752-1759. *Looe* wrecked 'off the Bahama Islands' (and near the Florida coast); the wreck has been found and excavated.

PLANS: *Dover*: Danish Rigsarkivet: Lines. *Gosport*: Danish Rigsarkivet: Lines. *Launceston*: Sunderland Museum: Lines & profile. *Hastings & Liverpool*: Lines & profile. *Hastings*: Lines. *Kinsale*: Lines. *Gosport, Feversham, Saphire*: Lines.

## 1741 Establishment 44s

Dimensions & tons: 126', 102'6" x 36' x 15'5½". 706 36/94.
Men: 250 (as 24: 180). Guns: GD 20 x 18, UD 20 x 9, QD 4 x 6 ( as 24: UD 20 x 18, QD 2 x 6, Fc 2 x 6).

**Anglesea** Blades, Hull.
Ord 28.9.1741; K 11.1741; L 3.11.1742; 1745 taken.
**Torrington** Rowcliffe, Southampton.
Ord 17.11.1741; K 02.1742; L 15.1.1743; 1763 sold.
**Hector** Blades, Hull.
Ord 1/13.10.1742; K 2.1.1743; L 24.10.1743; 1762 sold.
**Roebuck** Rowcliffe, Southampton.
Ord 1.12.1742; K 12.1742; L 21.12.1743; (1763-1764 lent as a privateer); 1764 returned and sold.

NOTE: This would appear to be the only time in the eighteenth century that the Royal Navy lent or hired a warship for privateering - unlike the French who regularly did this. The *Roebuck* was lent to an association calling themselves 'The Antigallican Private Ship of War'.

**Lark** Golightly, Liverpool.
Ord 14.4.1743; K 04.1743; L 30.6.1744; 1757 sold.
**Pearl** Okill, Liverpool.
Ord 22.4.1743; K 04.1743; L 29.6.1744; 1759 sold.
**Mary Gally** Bird, Rotherhithe.
Ord 26.6.1743; K 18.5.1743; L 16.6.1744; 1764 breakwater.
**Ludlow Castle** Taylor, Rotherhithe.
Ord 16.5.1743; K 06.1743; L 31.7.1744; 1762 Sixth Rate 24 (one decker); 1771 BU.
**Fowey** Blades, Hull.
Ord 3.9.1743; K ____ ; L 14.8.1744; 1748 wrecked.
**Looe** Gorill, Liverpool.
Ord 18.4.1744; K 04.1744; L 17.8.1745; 1750 hulked.
**Chesterfield** Quallett, Rotherhithe.
Ord 23.5.1744; K 2.6.1744; L 31.10.1745; 1762 wrecked.
**Poole** Blades, Hull.
Ord 23.5.1744; K 06.1744; L 5.6.1745; 1765 BU.
**Southsea Castle** Okill, Liverpool.

# THE ESTABLISHMENT PERIOD 1719-1745

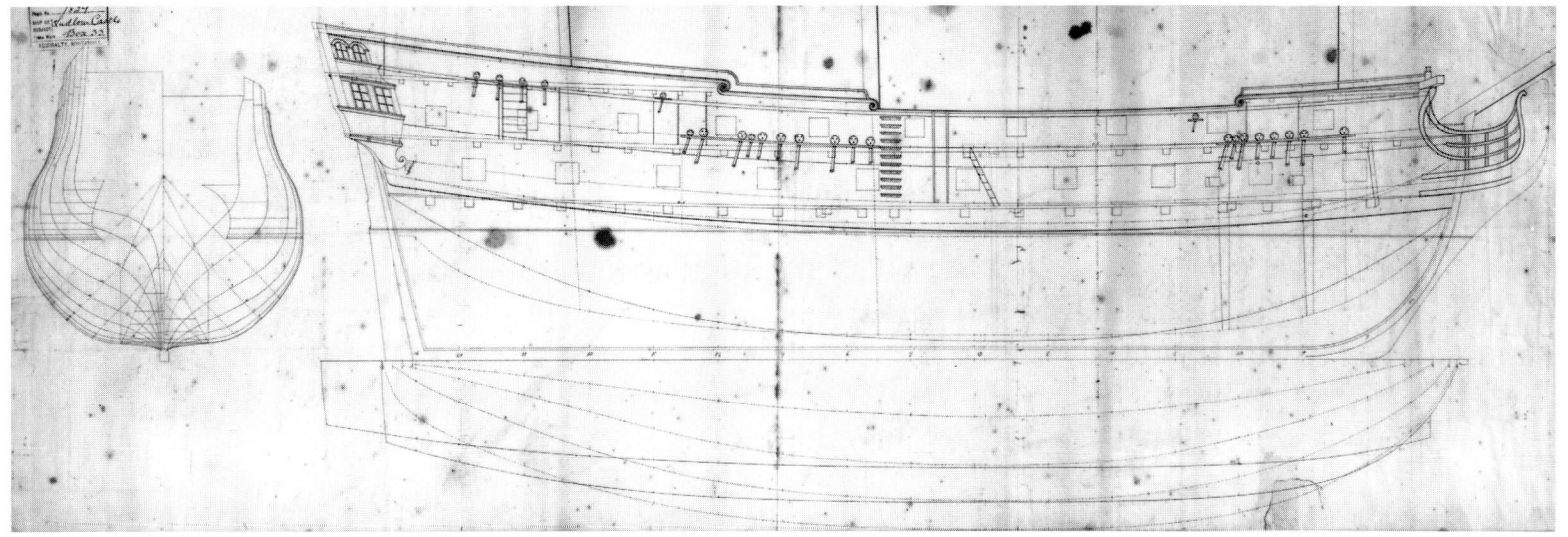

Lines plan (number 1827) of the 44 gun *Ludlow Castle* of 1744. In this plan the stern is galleried on two levels unlike the earlier 44s and 40s with their single-level sterns

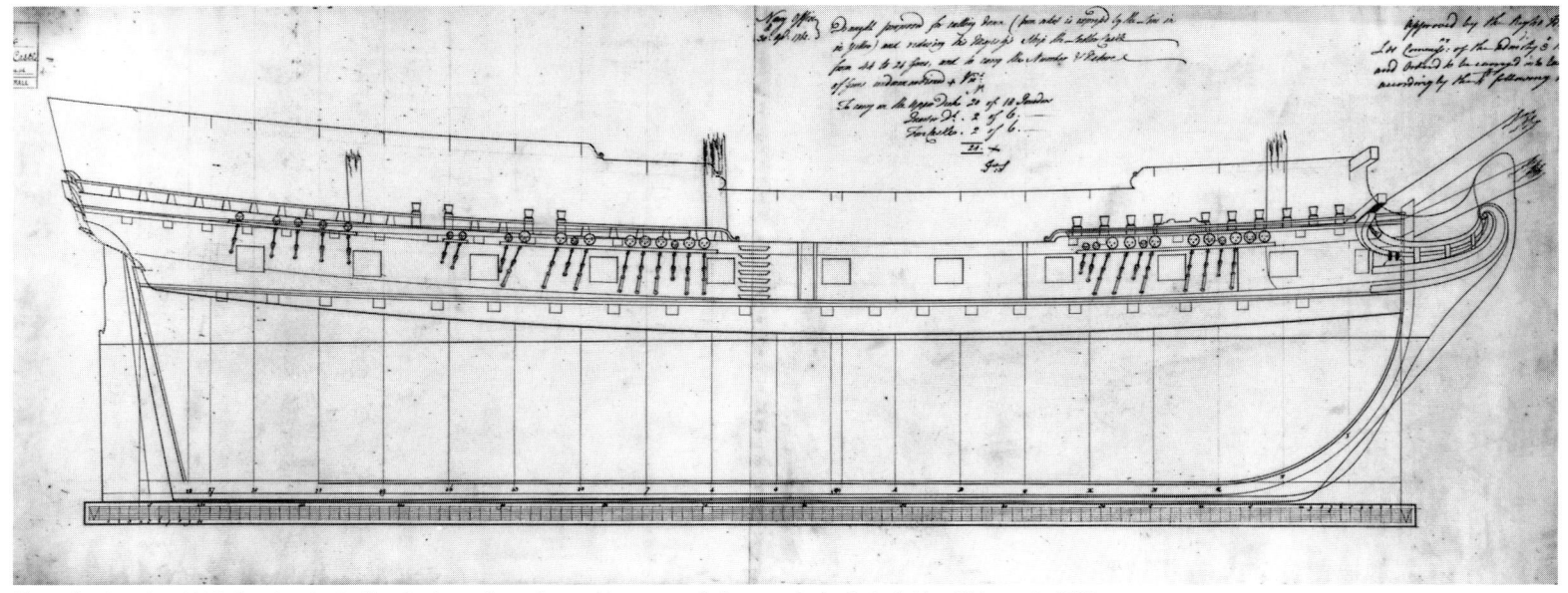

Sheer plan (number 2848) showing the *Ludlow Castle* cut down from a 44 gun two-decker to a single-decked ship of 24 guns in 1762

Ord 23.5.1744; K 29.5.1744;
10.8.1745; 1760 storeship 24; 1762 lost.
**Prince Edward** Bird, Rotherhithe.
Ord 21.6.1744; K 12.7.1744; L 2.9.1745; 1766 sold.
**Anglesea** Gorill, Liverpool.
Ord 19.6.1745; K 07.1745; L 3.12.1746; 1764 breakwater.
**Thetis** Okill, Liverpool.
Ord 19.6.1745; K 07.1745; L 15.4.1747; 1757 hospital ship 22; 1767 sold.

FATES: *Anglesea* taken by the French *Apollon* 50 on 9.3.1745; in French service until 1753. *Fowey* was wrecked in the Gulf of Florida 27.6.1748; there are reports that this wreck has been found. *Looe* was hulked at Sheerness; 1759 breakwater at that Yard. *Chesterfield* lost in the entrance to the Old Strait of Bahama 24.7.1762. *Southsea Castle* was lost off Manilla in 08.1762.

PLANS: *Ludlow Castle* as 24; 1762: profile/platforms/lower deck/quarter deck & forecastle. *Torrington*: Lines & profile. *Anglesea*: Lines & profile/orlop/lower deck/upper deck/quarter deck & forecastle. *Anglesea, Torrington, Hector, Roebuck, Lark, Pearl, Mary Gally, Ludlow Castle, Poole*: Lines & profile. *Roebuck, Lark, Pearl, Mary Gally, Southsea Castle, Chesterfield, Fowey, Looe*: Lines & profile. *Prince Edward*: Lines. *Prince Edward, Looe, Southsea Castle, Chesterfield, Poole, Looe*: Lines & profile.

## Sixth Rates 20/24 Guns (one-deckers)

### 1719 Establishment 20s

Dimensions & tons: 106', 87'9" x 28'4" x 9'2". 374 49/94.

Men: 140. Guns: UD 20 x 6.
**Blandford** Deptford Dyd (Surveyor: Stacey).
Ord ____ ; K ____ ; L 13.2.1720; 1742 sold.
**Greyhound** Deptford Dyd (Surveyor: Stacey).
Ord ____ ; K ____ ; L 13.2.1720; 1741 BU.
**Shoreham** Woolwich Dyd (Surveyor: Hayward) [RB].
Ord ____ ; K 10.1719; L 5.8.1720; 1737 bomb vessel; 1744 sold.
**Lyme** Deptford Dyd (Surveyor: Stacey) [RB].
Ord ____ ; K 02.1720; L 8.11.1720; 1739 BU.
**Scarborough** Deptford Dyd (Surveyor: Stacey) [RB].
Ord ____ ; K 10.1720; L 19.7.1722; 1737 sold.
**Garland/Guardland** Sheerness Dyd (Surveyor: Ward) [RB].
Ord ____ ; K 10.1721; L 1.5.1724; 1744 sold.
**Lowestoffe** Portsmouth Dyd (Surveyor: Naish) [RB].
Ord ____ ; K 07.1722; L 18.12.1723; 1744 sold.
**Seaford** Deptford Dyd (Surveyor: Stacey) [RB].
Ord ____ ; K 08.1722; L 22.10.1724; 1727 bomb vessel; 1740 BU.
**Rose** Woolwich Dyd (Surveyor: Hayward) [RB].

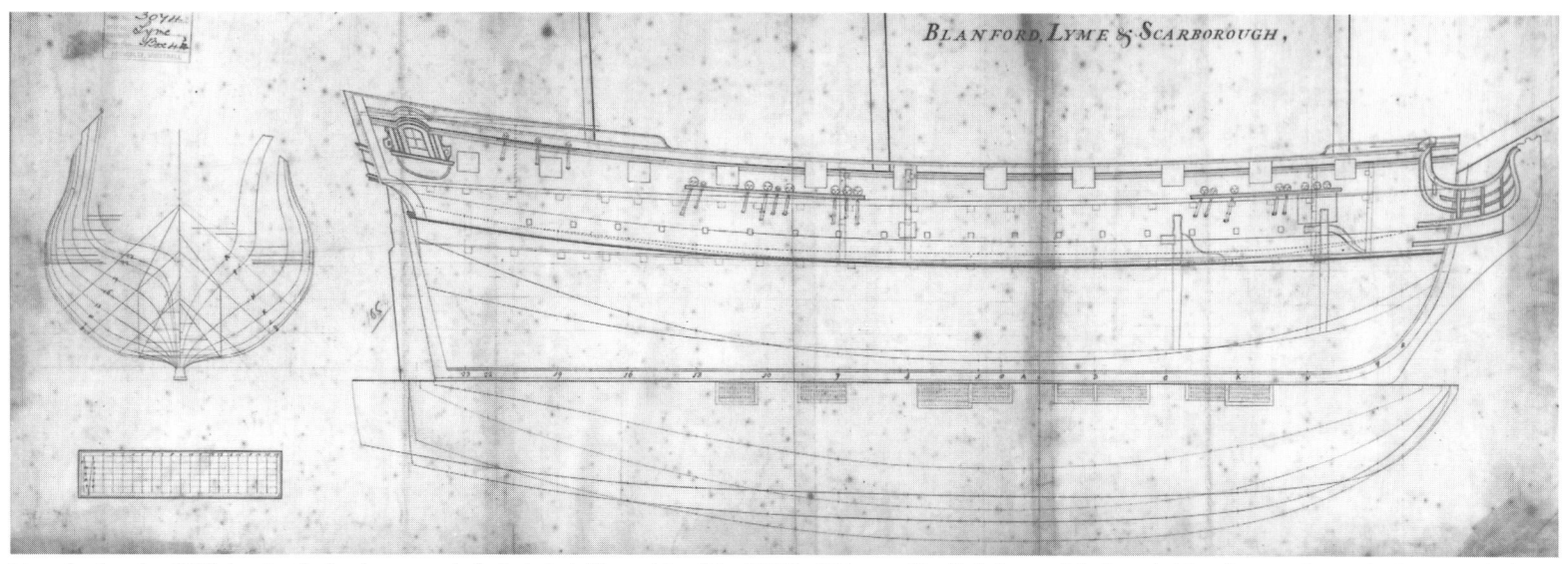

Lines plan (number 3074) showing the handsome, nearly flush-decked, 20 gun ships of the 1719 Establishment, *Blandford*, *Lyme* and *Scarborough*. Note the row of sweep (oar) ports on the lower deck, broken by the single small loading port amidships

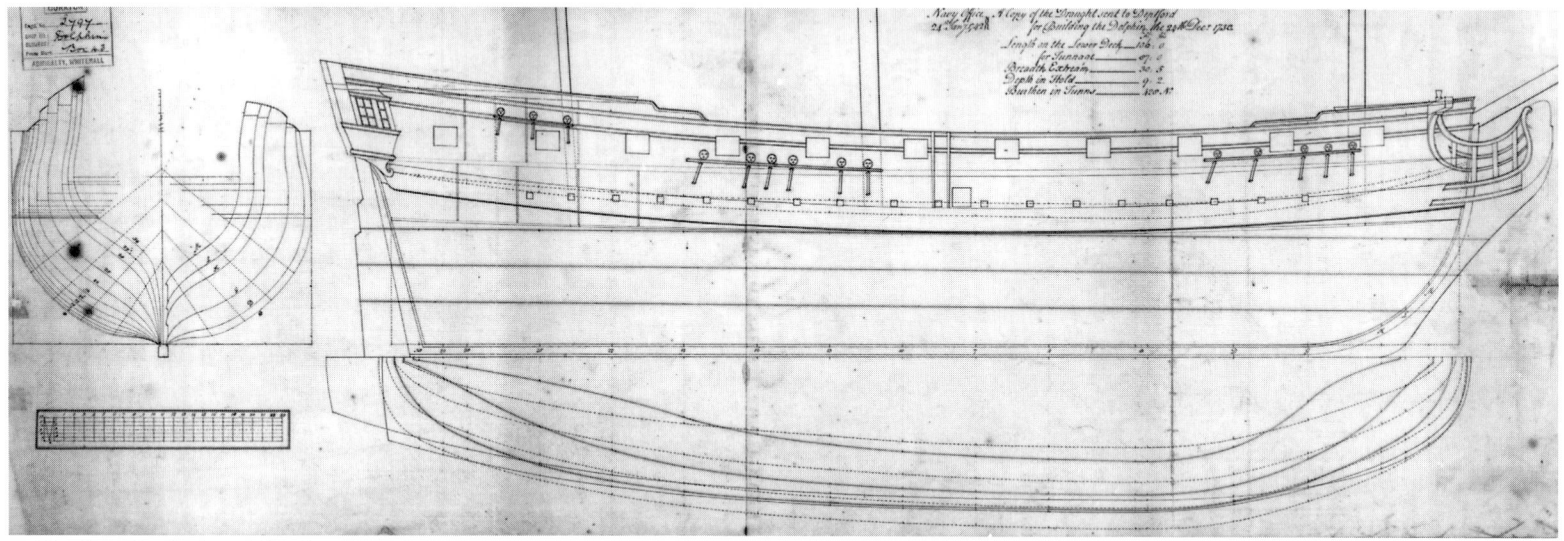

Building lines plan for the *Dolphin* of 1732 (number 2797)

Ord ____ ; K 12.11.1722; L 8.9.1724; 1739 hulked.
**Deal Castle** Sheerness Dyd (Surveyor: Ward) [RB].
    Ord ____ ; K 12.1720?; L 6.4.1727; 1747 sold.
**Fox** Deptford Dyd (Surveyor: Stacey) [RB].
    Ord ____ ; K 02.1724?; L 18.11.1727; 1738 BU.
**Experiment** Plymouth Dyd (Surveyor: Lock) [RB].
    Ord ____ ; K 03.1727?; L 1.11.1727; 1738 BU.
**Gibraltar** Deptford Dyd (Surveyor: Stacey) [RB].
    Ord ____ ; K 26.9.1724; L 8.8.1727; 1749 sold.
**Bideford** Chatham Dyd (Surveyor: Rosewell) [RB].
    Ord ____ ; K 02.1726; L 2.10.1727; 1736 lost.
**Alborough** Portsmouth Dyd (Surveyor: Allin) [RB].
    Ord ____ ; K 29.3.1727; L 21.10.1727; (1737; Ca 1740 fireship); 1742 BU.
**Flamborough** Portsmouth Dyd (Surveyor: Allin) [RB].
    Ord ____ ; K 29.3.1727; L 21.10.1727; 1749 sold.
**Seahorse** Deptford Dyd (Surveyor: Stacey) [RB].
    Ord ____ ; K 11.3.1727; L 7.10.1727; 1748 sold.
**Squirell** Woolwich Dyd (Surveyor: Hayward) [RB].
    Ord ____ ; K 12.4.1727; L 19.10.1727; 1749 sold.
**Phoenix** Woolwich Dyd (Surveyor: Hayward) [RB].
    Ord ____ ; K 04.1727; L 16.1.1728; 1742 hulked.
**Rye** Chatham Dyd (Surveyor: Rosewell).
    Ord ____ ; K ____ ; L 21.10.1727; 1735 BU.

FATES: *Rose* hulked at Deptford; 1744 sold. *Bideford* lost off Flamborough Head 18.3.1736. *Phoenix* became moorings hulk at Woolwich; 1744 sold.

PLANS: *Blandford, Lyme & Scarborough*: Lines & profile. *Greyhound*: Danish Rigsarkivet: Lines/platforms. *Lowestoff*: Lines. *Gibraltar*: Danish Rigsarkivet: Lines, profile & lower deck/platforms & upper deck. *Garland*: Lines & profile. *Experiment, Biddeford, Flamborough, Squirrel*: Lines. *Biddeford*: Sheer, half breadth & profile/body plan.

## Modified 1719 Establishment 20s.
Appear to have been a preliminary trial of the increase in size introduced in the 1733 Establishment. In the lists they are referred to as '1733 Establishment'; despite the fact that they were completed before it was issued. However breadth and depth seem to have been slightly less than those given in that Establishment, so they are treated as a separate group here.

    Dimensions & tons: 106", 87' x 30'5" x 9'2". 428.
    Men: 140. Guns: UD 20 x 6.

**Dolphin** Deptford Dyd (Surveyor: Stacey).
    Ord ____ ; K 12.1730; L 6.1.1732; 1747 fireship; 1755 renamed *Firebrand*; 1757 20 renamed *Penguin*; 1760 taken.
**Sheerness** Deptford Dyd (Surveyor: Stacey).
    Ord ____ ; K 12.1730; L 4.1.1732; 1744 sold.

FATES: *Dolphin* (as *Penguin*) taken by two French frigates off Oporto, then burnt 28.3.1760.

PLANS: *Dolphin*: Lines. *Sheerness*: Lines.

## 1733 Establishment 20s

    Dimensions & tons: 106', 87' x 30'6" x 9'5". 442 4/94.

# THE ESTABLISHMENT PERIOD 1719-1745

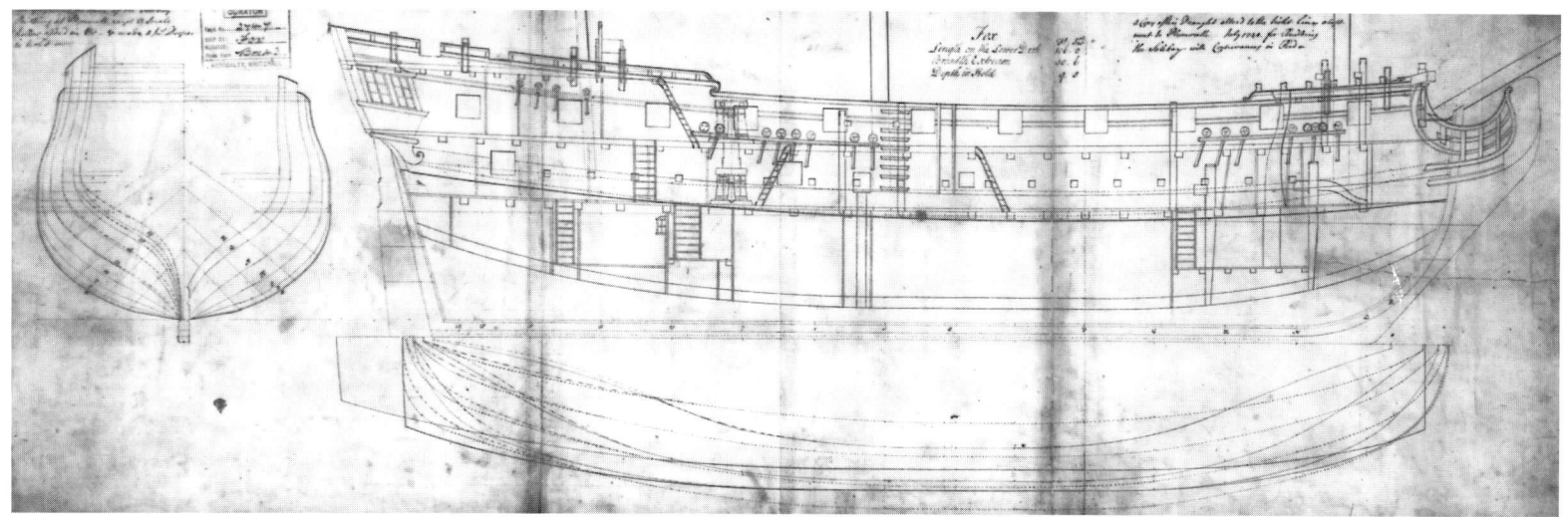

*Fox*, like the earlier 20s, has oar ports and a loading port on the lower deck. As shown in the lines and profile plan (number 2747) she also had two, widely separated, gunports a side on the after part of the lower deck. This plan was also used for building the *Solebay*

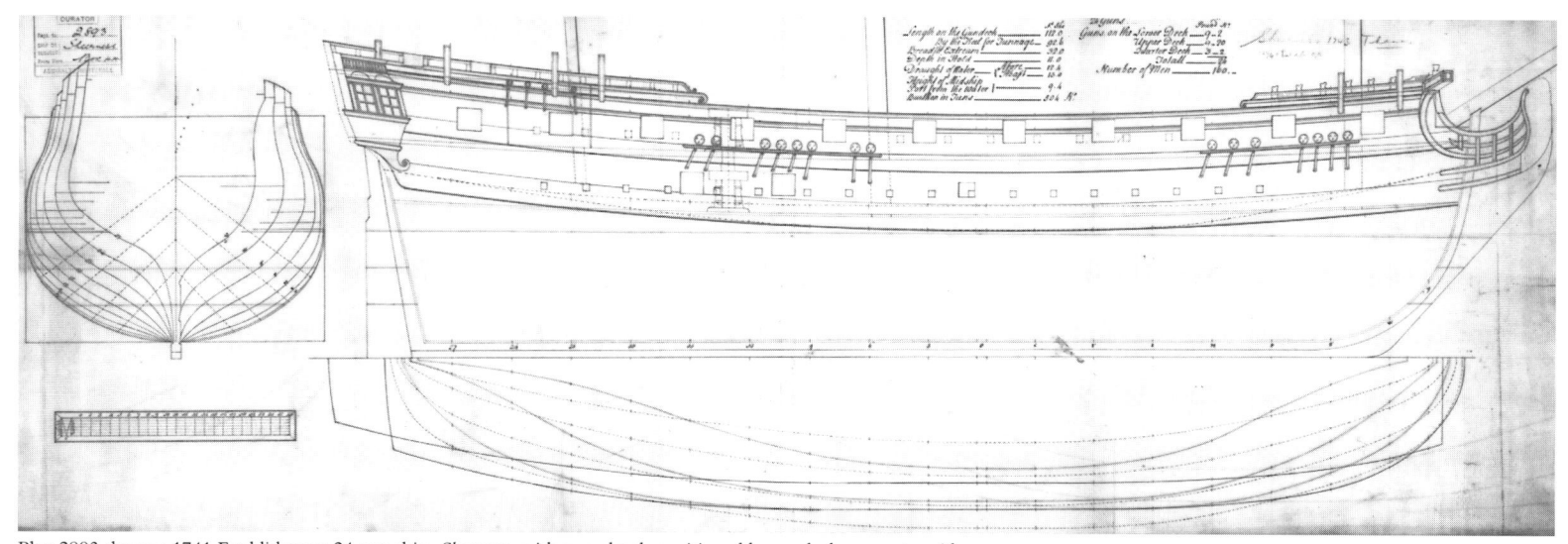

Plan 2893 shows a 1741 Establishment 24 gun ship, *Sheerness*, with two closely positioned lower deck gunports a side

Men: 150. Guns: UD 20 x 9. Towards the end of their careers some (eg. *Success* and *Greyhound*) were classed as 24s.

**Tartar** Deptford Dyd (Surveyor: Stacey) [RB].
Ord ____ ; K 3.8.1733; L 28.3.1734; 1755 BU.

**Kennington** (ex *Mermaid*) Deptford Dyd (Surveyor: Stacey).
Ord 1.2.1735; K 23.4.1735; L 30.6.1736; 1749 BU.

**Fox** Buxton, Rotherhithe.
Ord 08.1739?; K 16.9.1739; L 1.5.1740; 1745 wrecked.

**Rye** Bird, Rotherhithe.
Ord 13.8.1739?; K 24.9.1739; L 1.4.1740; 1744 wrecked.

**Lyme** Taylor, Rotherhithe.
Ord 13.8.1739; K 22.9.1739; L 17.5.1740; 1747 foundered.

**Winchelsea** Carter, Rotherhithe.
Ord 13.8.1739; K 22.9.1739; L 3.5.1740; (1758-1758 in French hands); 1761 BU.

**Experiment** Bird, Rotherhithe.
Ord 3.9.1739; K 21.9.1739; L 18.4.1740; 1763 sold.

**Lively** Quallett, Rotherhithe.
Ord 3.9.1739; K 22.1.1740; L 10.6.1740; 1750 sold.

**Scarborough** Perry, Blackwall.
Ord 14.10.1739; K 20.10.1739; L 31.5.1740; 1749 sold.

**Port Mahon** Buxton, Deptford.
Ord 4.9.1739; K 17.10.1739; L 26.8.1740; 1763 sold.

**Rose** Bird, Rotherhithe.
Ord 4.10.1739; K ____ ; L 14.8.1740; 1755 sold.

**Biddeford** Barnard, Ipswich. Ord 14.10.1739; K 6.11.1739; L 15.6.1740; 1754 BU.

**Success** Blades, Hull.
Ord 1.10.1739; K 30.10.1739; L 14.8.1740; 1779 BU.

**Bridgwater** Pearson, Lynn.
Ord 1.12.1739?; K 22.1.1740; L 11.12.1740; 1743 lost.

**Seaford** Stow, Shoreham.
Ord ____ ; K 30.6.1740; L 6.4.1741; 1754 BU.

**Solebay** Veale (begun), completed by Plymouth Dyd (Surveyor: Lock).
Ord 30.6.1740; K 17.7.1740; L 20.7.1742; (1744-1746 in French hands); 1763 sold.

**Greyhound** Snelgrove, Limehouse.
Ord 5.12.1740; K 26.1.1741; L 19.9.1741; 1768 sold.

**Blandford** West, Deptford.
Ord 27.11.1740?; K 3.12.1740; L 2.10.1741; 1767 sold.

FATES: *Fox* wrecked near Dunbar 14.11.1745. *Rye* wrecked on the coast of Norfolk 27.11.1744. *Lyme* foundered in the Atlantic 15.9.1747. *Winchelsea* taken by the French *Le Bizarre* and *La Mignonne* 10.10.1758; but recaptured 27.10.1758. *Bridgwater* lost off Newfoundland, 18.9.1743. *Solebay* was taken by four French warships 5.9.1744; 1746 cut out of Saint Martin's Road by the British privateer *Alexander*.

PLANS: *Fox, Solebay*: Lines & profile. *Lyme*: Lines & profile. *Experiment, Lively, Success, Biddeford, Bridgwater*: Lines. *Scarborough*: Lines.

## 1741 Establishment 24s

Dimensions & tons: 112', 91'6" x 32' x 11'. 498 34/94.
Men: 160. Guns: LD 2 x 9 (right aft), UD 20 x 9, QD 2 x 3.

**Lowestoff** Buxton, Deptford.
Ord 24.8.1741; K 28.10.1741; L 8.7.1742; 1749 sold.

*Alborough* Okill, Liverpool.
   Ord 17.12.1741; K 6.1.1742; L 16.3.1743; 1749 sold.
*Alderney* (ex *Squirrel*) Reed, Hull.
   Ord ____ ; K 29.3.1742; L 18.3.1743; 1749 sold.
*Phoenix* Graves, Limehouse.
   Ord 7.11.1742; K 3.12.1742?; L 27.7.1743; 1757 hulked.
*Sheerness* Buxton, Rotherhithe.
   Ord 7.1.1743?; K 24.1.1743; L 8.10.1743; 1768 sold.
*Wager* Quallett, Rotherhithe.
   Ord 30.4.1743; K 19.5.1743; L 2.6.1744; 1763 sold.
*Shoreham* Reed, Hull.
   Ord 16.5.1743; K 17.7.1743; L 31.5.1744; 1758 sold.
*Bridgwater* Rowcliffe, Southampton.
   Ord 13.12.1743; K 20.12.1743; L 13.10.1744; 1758 wrecked.
*Glasgow* Reed, Hull.
   Ord 1.5.1744?; K 05.1744; L 22.5.1745; 1756 sold.
*Tryton* Heather, Bursledon.
   Ord 23.5.1744; K 06.1744; L 17.8.1745; 1758 wrecked.
*Mercury* Golightly, Liverpool.
   Ord 11.8.1744; K 09.1744; L 13.10.1745; 1753 BU.
*Surprize* Wyatt, Beaulieu.
   Ord 11.8.1744; K 09.1745; L 27.1.1746; 1770 sold.
*Fox* Horn, Bursledon.
   Ord 27.8.1744; K 09.1744; L 26.4.1746; 1751 wrecked.
*Siren* Snelgrove, Limehouse.
   Ord 27.8.1744; K 09.1744; L 3.9.1745; 1764 sold.
*Rye* Carter, Southampton.
   Ord 30.11.1744; K 29.12.1744; L 11.2.1746; 1763 sold.
*Centaur* Blades, Hull.
   Ord 19.6.1745; K 22.7.1745; L 11.6.1746; 1761 sold.
*Deal Castle* Golightly, Liverpool.
   Ord 19.6.1745; K 5.7.1745; L 2.12.1746; 1756 BU.
FATES: *Phoenix* hulked at the Tower of London; 1760 sold. *Bridgwater* and *Tryton* both run ashore and burnt at Cuddalore, India, to avoid capture by the French on 28.4.1758. *Fox* wrecked in a hurricane at Jamaica on 11.9.1751.
PLANS: *Lowestoft*: Lines & profile. *Lowestoff, Aldborough, Alderney*: Lines & profile. *Phoenix*: Lines & profile. *Sheerness* : Lines/Lines Profile. *Wager*: Lines/stern. *Bridgwater*: Lines, profile & decks/platforms. *Tryton*: Lines. *Centaur*: Upper deck. Danish Rigsarkivet: Lines *Centaur* & *Deal Castle*: Lines & Profile/platforms/lower deck/upper deck. *Phoenix*: Lines.

## Modified 1741 Establishment 24 Gun design: Poole
produced this design which then became the basis for the 1745 Establishment vessels of this type.
   Dimensions & tons: 113'1½", 94' x 32'2" x 11'. 517 12/94.
*Garland* Sheerness Dyd (Surveyor: Poole).
   Ord 23.4.1744; K 18.11.1745; L 13.8.1748; 1783 sold.
PLANS: Lines & profile.

## Sloops

All probably two-masted; snows, brigantines or ketches (rig is indicated when known). Sloops were not 'Established' and appear to have been built to an assortment of designs. Some of the later vessels can be grouped in what appear to be classes; where numbers were built to the same lines. These groups have been listed as such here. It is quite possible that there were more such groups than it has been possible to identify here. For those vessels not grouped in this fashion the 'as built' dimensions and number of guns (when known) are given in each case. It should be noted that in some 'classes' the same lines could be used for vessels with different rigs. The 14 gun sloops seem to have been intended to carry 110 men, the smaller vessels perhaps 50 or so. The guns would be 3 or 4 pounders, and would be supplemented with swivels.

NOTE: Some extra information on dimensions, etc., was provided through the kindness of Rif Winfield from his researches at the PRO and elsewhere; and this is noted between brackets - (thus).

*Bonetta* 1721-1731. Deptford Dyd (Surveyor: Stacey).
   Dimensions & tons: 55'2", 42'11¼" x 17' x 7'6". 66 .
   Guns: 4?
   Ord ____ ; K ____ ; L 18.4.1721; 1731 sold.
*Ferret* 1721-1731. Woolwich Dyd (Surveyor: Hayward).
   Dimensions & tons: 55'5½", 43'1" x 17'1½" x 7'6". 67 19/94.
   Guns: 4?
   Ord ____ ; K ____ ; L 6.5.1721; 1731 sold.
*Otter* 1721-1742. Deptford Dyd (Surveyor: Stacey).
   Dimensions & tons: 64'6", 51'5" x 18'3" x 8'3". 91 8/94.
   Guns: 4.
   Ord ____ ; K ____ ; L 8.8.1721; 13.1.1742 wrecked off Aldborough.
*Swift* 1721-1741. Woolwich Dyd (Surveyor: Hayward) snow.
   Dimensions & tons: 60', 47' x 19'2" x 8'3". 90½.
   Guns: 4?
   Ord ____ ; K ____ ; L 19.8.1721; 1741 sold.
PLANS: Lines.
*Cruizer* 1721-1732. Deptford Dyd (Surveyor: Stacey).
   Dimensions & tons: 62', 47' x 19'10" x 9'. 100 7/94.
   Guns: 4?
   Ord ____ ; K ____ ; L 24.10.1721; 1732 sold.
*Hawk* 1721-1742. Chatham Dyd (Surveyor: Rosewell).
   Dimensions & tons: (62', 47'9½" x 19'10" x 9', 100).
   Guns: 4?
   Ord ____ ; K ____ ; L 23.11.1721; 1740 'supposed to have been lost'.
*Spye* 1721-1731. Portsmouth Dyd (Surveyor: Naish).
   Dimensions & tons: 62'1", 48'6" x 20' x 9'. 103 18/94.
   Guns: 4?
   Ord ____ ; K ____ ; L 9.12.1721; 1731 sold.
*Weazell* 1723-1732. Woolwich Dyd (Surveyor: Hayward).
   Dimensions & tons: (61'6", 47'11" x 20' x 9', 102) .
   Guns: 4?
   Ord ____ ; K ____ ; L 7.11.1723; 1732 sold.
*Sharke* 1723-1732. Deptford Dyd (Surveyor: Stacey) [RB].
   Dimensions & tons: 69'2", 54'8" x 20'8" x 9'6". 124 18/94.
   Guns: 4/6?
   Ord ____ ; K 09.1722; L 3.9.1723; 1732 sold.
NOTE: With *Happy* (see below) the only case of a sloop being 'rebuilt' in this period. These vessels were really too small and cheap to justify such a proceeding.
*Spence* 1723-1729. Deptford Dyd (Surveyor: Stacey ).
   Dimensions & tons: 64'6", 50'1½" x 20'8" x 9'6". 113 82/94.
   Guns: 4? (or 8).
   Ord ____ ; K ____ ; L 3.3.1723; 1729 BU.
*Happy* 1724-1735. Deptford Dyd (Surveyor: Stacey) [RB].
   Dimensions & tons:(? as *Sharke*?).
   Guns: 14.
   Ord ____ ; K 09.1724; L 10.7.1725; 1735 sold.
NOTE: See remarks under *Sharke* (above).
*Drake* 1729-1740. Deptford Dyd (Surveyor: Stacey).
   Dimensions & tons: (87', 73'5" x 23' x 6' 207).
   Men: (100). Guns: 14. (or 8 x 4pdrs + 12 swivels) .
   Ord ____ ; K 09.1728; L 3.4.1729; 1740 BU.
*Spence* 1730-1749. Deptford Dyd (Surveyor: Stacey).
   Dimensions & tons: 87', 73'5" x 23' x 6'. 206 54/94.
   Men: 110/100. Guns: 8 (x 4 + 12 swivels?).
   Ord ____ ; K 14.6.1729; L 24.6.1730; 1749 sold.
PLANS: Lines & decks.
*Grampus* 1731-1742. Woolwich Dyd (Surveyor: Hayward).
   Dimensions & tons: (70', 56'1" x 23'2" x 6'. 160).
   Men: (80). Guns: 6 (x 4 + 10 swivels). Men: (80).
   Ord ____ ; K ____ ; L 21.10.1731; 10.10.1742 lost whilst cruising off the French coast.
*Wolf* 1731-1741. Deptford Dyd (Surveyor: Stacey) snow.
   Dimensions & tons: 87', 73'6" x 25' x 6'. 244 32/94.
   Men: (100). Guns: 8 (x 4 + 12 swivels).
   Ord ____ ; K 04.1731; L 20.11.1731; 2.3.1741 wrecked on the Caicos Bank.

THE ESTABLISHMENT PERIOD 1719-1745

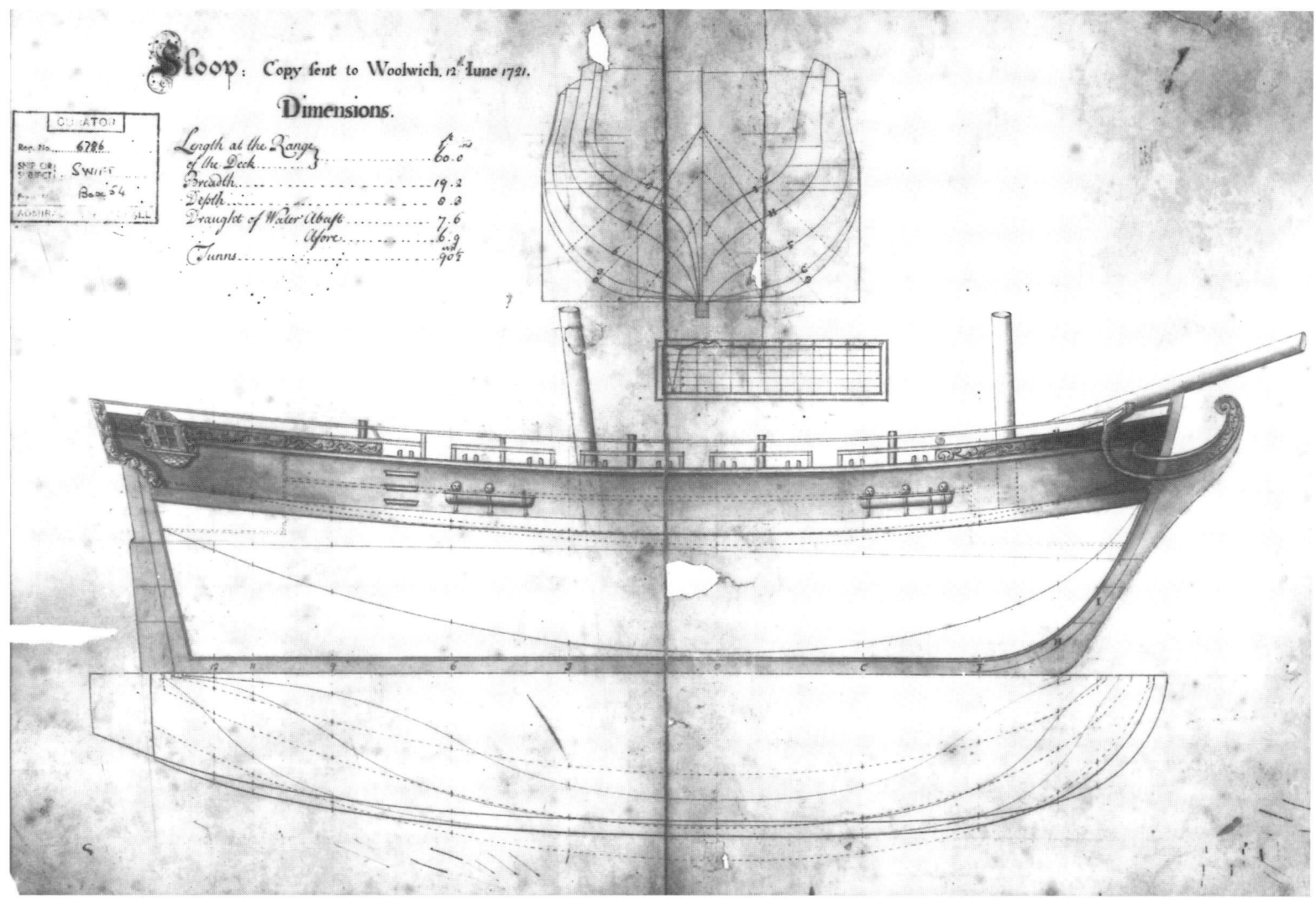

*An attractive and unusual lines plan of the sloop* Swift *of 1721 (number 6786)*

PLANS: Lines.

**Bonetta** 1732-1744. Woolwich Dyd (Surveyor: Hayward).
   Dimensions & tons: (81'4", 65'6" x 24' x 10'. 201.)
   Men: (80). Guns: 8 x 4 + 12 swivels.
   Ord ____ ; K 5.7.1732; L 24.8.1732; 20.10.1744 wrecked in a hurricane at Jamaica.

**Spy** 1732-1745. Chatham Dyd (Surveyor: Rosewell) snow.
   Dimensions & tons: 85'7", 69'5" x 23'4" x 10'6". 200 91/94.
   Men (80). Guns: 8 (x 4 + 12 swivels).
   Ord ____ ; K 10.7.1732; L 25.8.1732; 1745 sold.

PLANS: Lines & decks.

***Cruizer* Group.** 8 guns 1732. Stacey design.
   Dimensions & tons: 84', 71'1" x 23' x 9'. 200 (*Druid* and *Lynx* of the 1760s were built to the same lines but are listed separately; see next Chapter).
   Men: (80). Guns: 8 (x 4 + 12 swivels).
**Cruizer** Deptford Dyd (Surveyor: Stacey) snow.
   Ord 5.5.1732; K 11.7.1732; L 6.9.1732; 1745 sold.
**Hound** Deptford Dyd (Surveyor: Stacey).
   Ord 5.5.1732?; K 11.7.1732; L 6.9.1732; 1745 BU.
**Tryall** Deptford Dyd (Surveyor: Stacey).
   Ord 5.5.1732?; K 11.7.1732; L 6.9.1732; 1741 foundered.
FATES: *Tryall* scuttled in the Pacific (Anson's voyage) 4.10.1741.
PLANS: Lines.

**Sharke** 1732-1755. Portsmouth Dyd (Surveyor: Allin) ketch.
   Dimensions & tons: 80', 63' x 24'6" x 9'11¼". 201 13/94.
   Men: (80). Guns: 8/10 (x 4 + 12 swivels).
   Ord ____ ; K 4.7.1732; L 7.9.1732; 1755 sold.

PLANS: Lines.

**Saltash** 1732-1741. Plymouth Dyd (Surveyor: Lock) snow.
   Dimensions & tons: 85'7", 69'3" x 23'4" x 9'6". 200 37/94.
   Men: (80). Guns: 8 (x 4 + 12 swivels).
   Ord ____ ; K 10.7.1732; L 7.9.1732; 1741 sold.

NOTE: According to her plan she had the figurehead of a sea nymph and not the usual lion used for smaller vessels at the time.

PLANS: Lines.

**Fly** 1732-1751. Sheerness Dyd (Surveyor: Ward) snow.
   Dimensions & tons: 86'6", 69'7" x 23'3" x 10'6". 200.
   Men: (80). Guns: 8/10 (x 4 + 12 swivels).
   Ord ____ ; K 7.7.1732; L 15.9.1732; 1751 sold.

PLANS: Lines & decks.

***Drake* Group.** 8 guns 1740. Designer uncertain, but probably the surveyor.
   Dimensions & tons: 85', 68'8" x 23'6" x 9'6". 204.
   Men: (80). Guns: 8 (later 10?) x 4 (+ 12 swivels).
**Drake** West, Wapping. Snow.
   Ord 24.6.1740; K 25.9.1740; L 19.2.1741; 1742 lost.
**Hawk** Greville & Whetstone, Limehouse. Snow.
   Ord 25.8.1740; K 20.10.1740; L 10.3.1741; 1747 BU.
**Swift** Carter, Limehouse. Snow.
   Ord 6.12.1740; K 26.1.1741; L 30.5.1741; 1756 foundered.
FATES: *Drake* Lost in Gibraltar Bay 22.11.1742. *Swift* 'supposed cast away in the North Seas' 31.10.1756.

PLANS: Lines & decks.

**Otter** 1742-1763. Buxton, Rotherhithe. Snow.
   Dimensions & tons: 87'6", 73'4¾" x 25' x 6'6". 243 74/94.

54　　　　　　　　　　　　　　　　　　　　　　　　　　　　　　　　　　　THE ESTABLISHMENT PERIOD 1719-1745

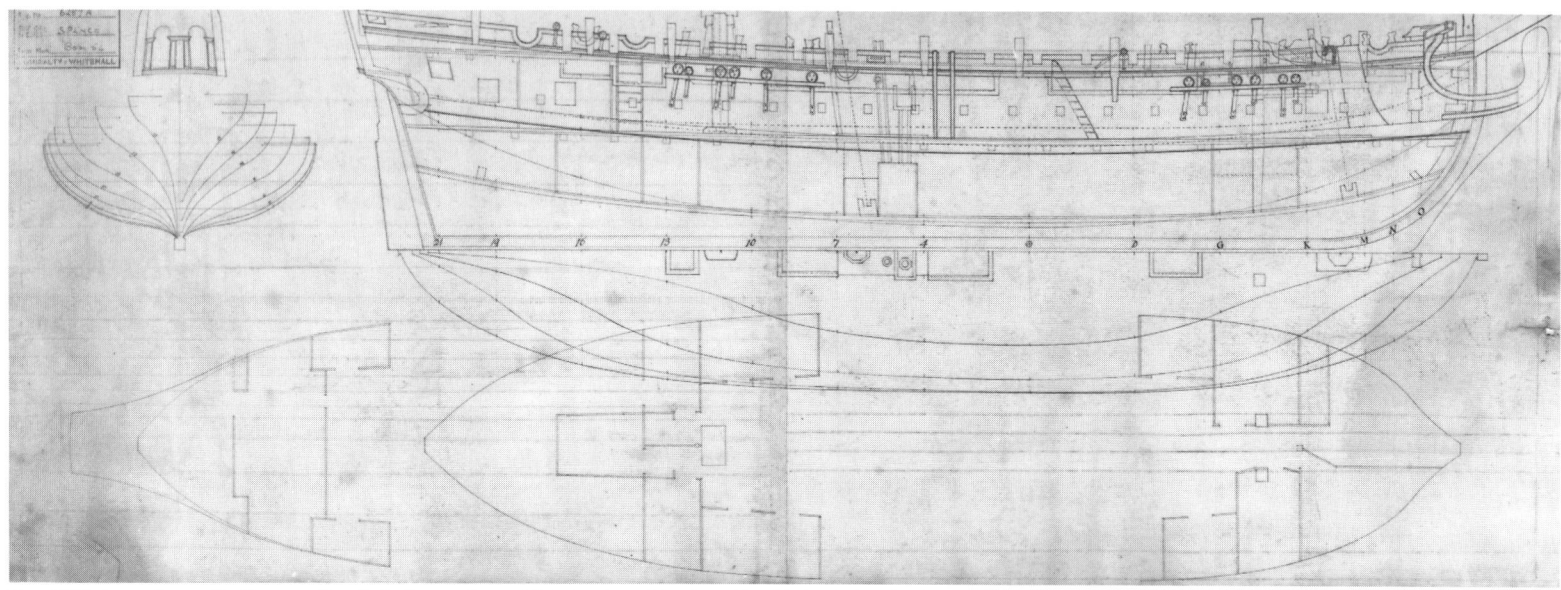

Plan 6289 shows the profile, lines, outline deck plans and a partial body plan of the sloop identified as being the *Spence* of 1730. It is not certain if the vessel was actually built to this design. Certainly the arrangement shown here, of a vessel capable of being rowed from both upper and lower decks, is most unusual

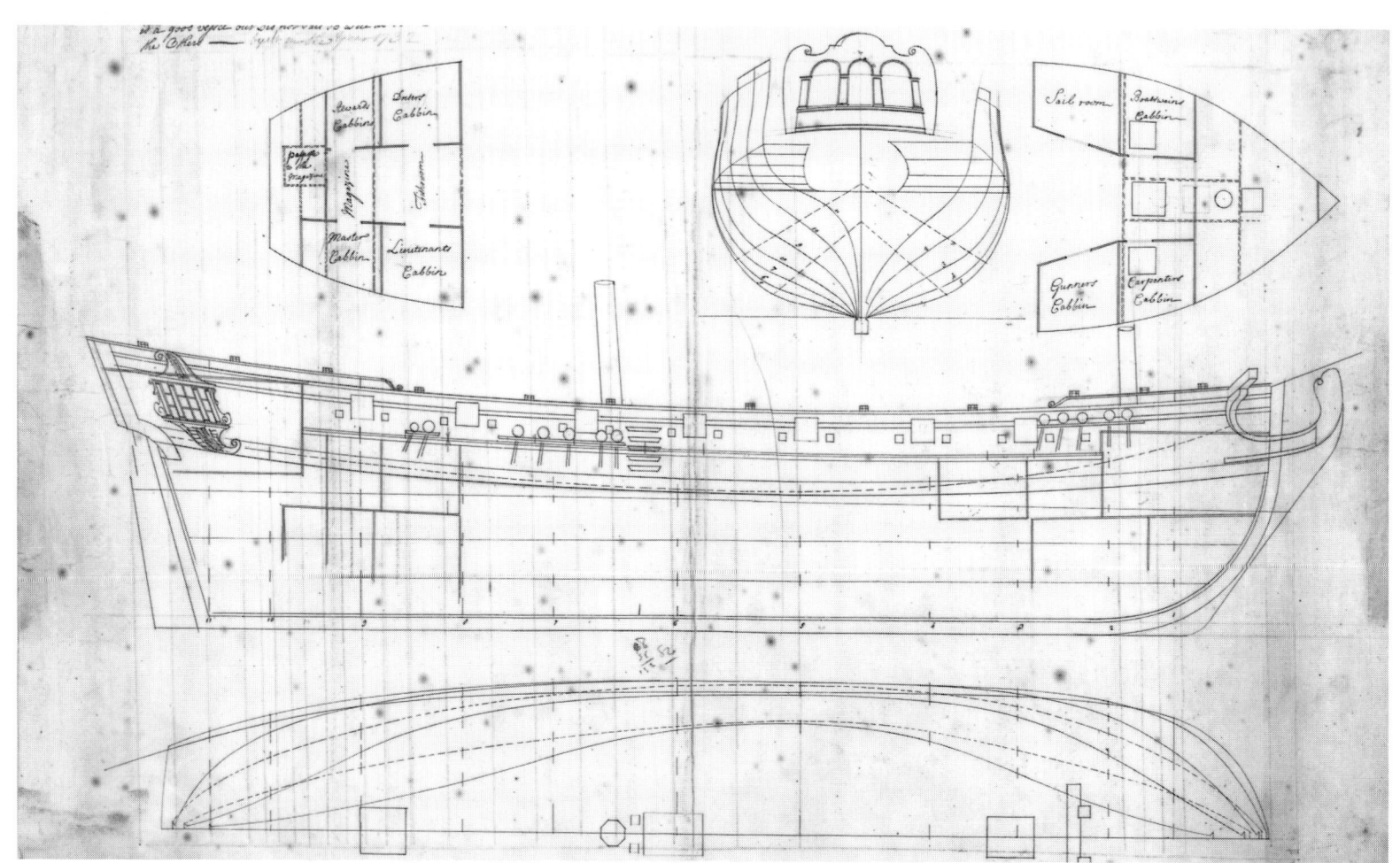

This lines, profile and decks plan (number 3345) of the sloop *Spy* of 1732 was drawn, as were quite a few other plans of the time, on the back of printed Admiralty forms - both print and lines can be seem showing through the paper

Men: (110). Guns: 14 x 4 + 12 swivels.
Ord 23.1.1742; K 20.2.1742; L 19.8.1742; 1763 sold.
PLANS: Lines.

**Baltimore** 1742-1762. West, Deptford, 'Projected' by Lord Baltimore.
Dimensions & tons: (89', 73'10" x 25'3¼" x 10'6". 251).
Men: 110. Guns: 14 x 4. + 12 swivels.
Ord ____ ; K 12.8.1742; L 30.12.1742; 1758 bomb vessel; 1762 sold.
PLANS: As bomb 1758: profile, decks & midships section.

**Saltash** 1742-1746. Quallett, Rotherhithe. Snow.
Dimensions & tons: 87'10½", 74'1" . x 25'1½" x 10'6½". 248.
Men: 110. Guns: 14 x 4 + 14 swivels.
Ord 19.7.1742; K 6.8.1742; L 30.12.1742; 24.6.1746 'oversett off Beachy Head'.

**Wolf** 1742-1749. West, Deptford. Snow.
Dimensions & tons: 87'6", 73'0½" x 25' x 6'. 244.
Men: 110. Guns: 14 x 4 + 14 swivels.
Ord 21.7.1741; K 31.7.1741; L 27.2.1742; 29.10.1745 taken by a French

THE ESTABLISHMENT PERIOD 1719-1745

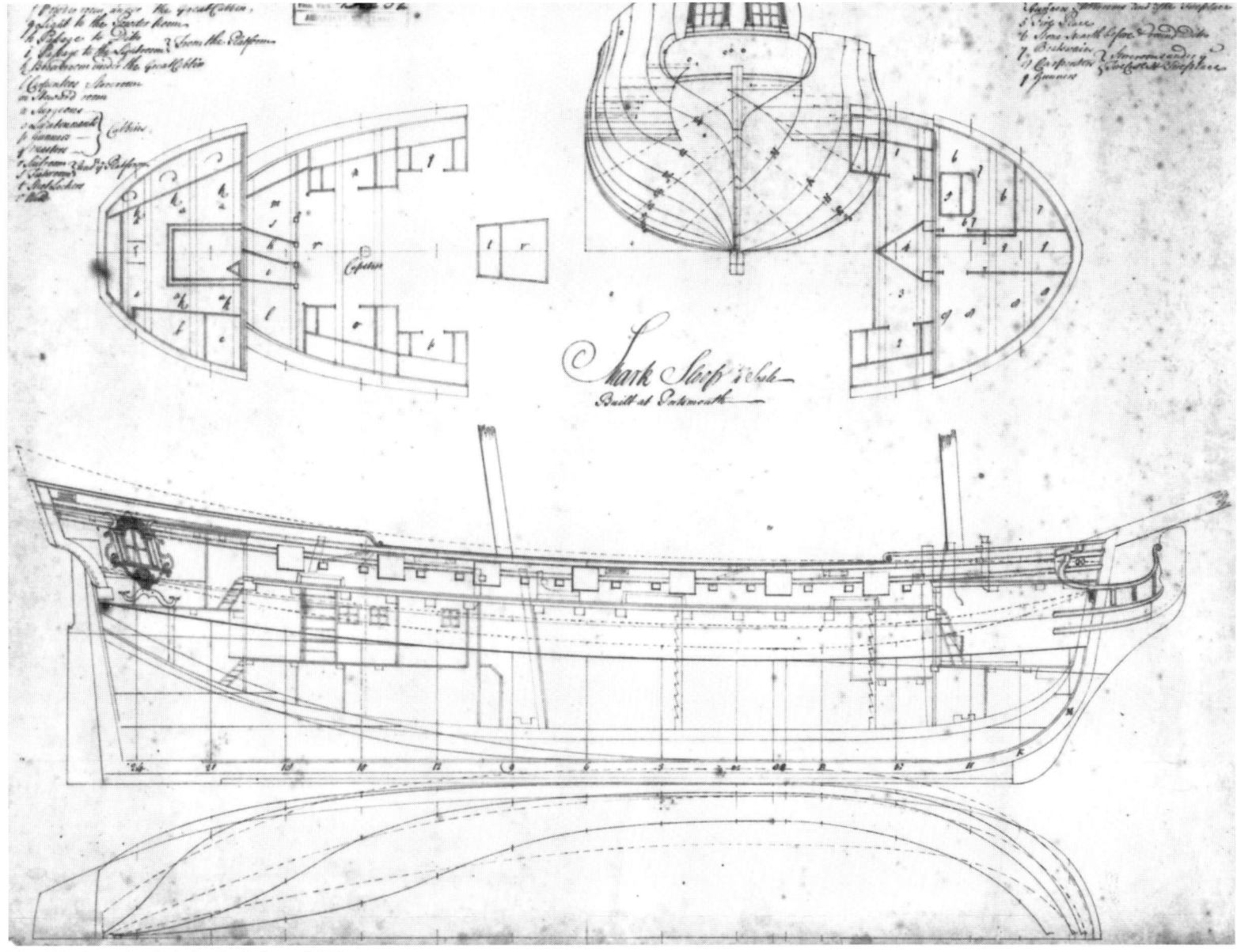

Plan 3654 shows the *Sharke* (or *Shark*) sloop of 1732

privateer of 32 guns; 1.3.1747 retaken by *Grand Turk* & *Amazon*; 31.12.1748 wrecked off the coast of Ireland.

PLANS: Lines.

**Drake** 1743-1748. Buxton, Deptford. Snow.
   Dimensions & tons: 88'3½", 74'4½" x 25'1¼" x 11'0¼". 249 30/94.
   Men: (110). Guns: 14 (x 4 + 12 or 14 swivels).
   Ord 25.2.1743; K 11.2.1743; L 28.7.1743; 1748 sold.

**Grampus** 1743-1744. Perry, Blackwall. Snow.
   Dimensions & tons: (87'10", 74'4" x 25'1¼" x 11'. 249.)
   Guns: 14.
   Ord 5.2.1743; K 15.2.1743; L 27.7.1743; 30.9.1744 foundered in the Channel.

NOTE: Built to the same sheer as the *Drake* and the same body as the *Otter*, and probably to the same dimensions as the latter.

**Merlin** 1744-1748. Greville & Whetstone, Limehouse. Snow.
   Dimensions & tons: 91'10¼", 75'0⅞" x 26'0¾" x 6'10". 271.
   Men: 110. Guns: 10 x 6 + 14 swivels.
   Ord ____ ; K ____ ; L 20.2.1744; 1748 sold.

**Swallow class 10 guns 1743.** Acworth design. Though dimensions and general lines did not vary there appear to have been two separate subclasses; the second of which is indicated by asterisking [★] the names. Some vessels later carried 14 guns and were re-rigged as ships.
Dimensions & tons: 91', 74'9" x 26' x 12'. 268 77/94.

Men: 110 (as bombs: 60). Guns: 10 (later 12) x 6 . 18 swivel; in some cases 14 x 6 + 14 swivels. (as bombs: 1 x 13" + 1 x 10" mortars + 8 x 4 + 12 swivels).

**Swallow** (?ex *Galgo*? ) Buxton, Deptford.
   Ord ____ ; K ____ ; L 17.2.1744; 1744 wrecked.

**Speedwell★** Buxton, Deptford. Ketch.
   Ord 30.3.1744; K 04.1744; L 9.11.1744; 1750 sold.

**Hazard★** Buxton, Rotherhithe. Snow?
   Ord 4.4.1744; K 26.4.1744; L 11.12.1744; 1745 taken by the Jacobites and handed to the French who renamed her *Le Hasard*; became French privateer *Le Prince Charles*; 1746 retaken; 1749 sold.

**Tavistock★** Darly, Gosport. Snow.
   Ord 23.5.1744; K ____ ; L 22.3.1745; 1747 renamed *Albany*; 1754 ship rig; 1763 sold.

**Hornet** Chitty & Quallett, Chichester. Snow.
   Ord 11.8.1744; K 09.1744; L 3.8.1745; (1747 briefly in French hands); 1770 sold.

**Raven** Blades, Hull. Snow.
   Ord 27.8.1744; K 09.1744; L 4.7.1745; 1753 ship rig; 1762 fireship; 1763 sold.

**Swan** Hinks, Chester. Snow?
   Ord 6.10.1744; K 11.1744; L 14.12.1745; 1755 ship rig; 1763 sold.

**Badger** Janvrin, Bursledon . Snow.
   Ord 10.10.1744; K 20.12.1744; L 5.8.1745; 1762 lost.

**Dispatch** Stow & Bartlett, Shoreham. Snow.

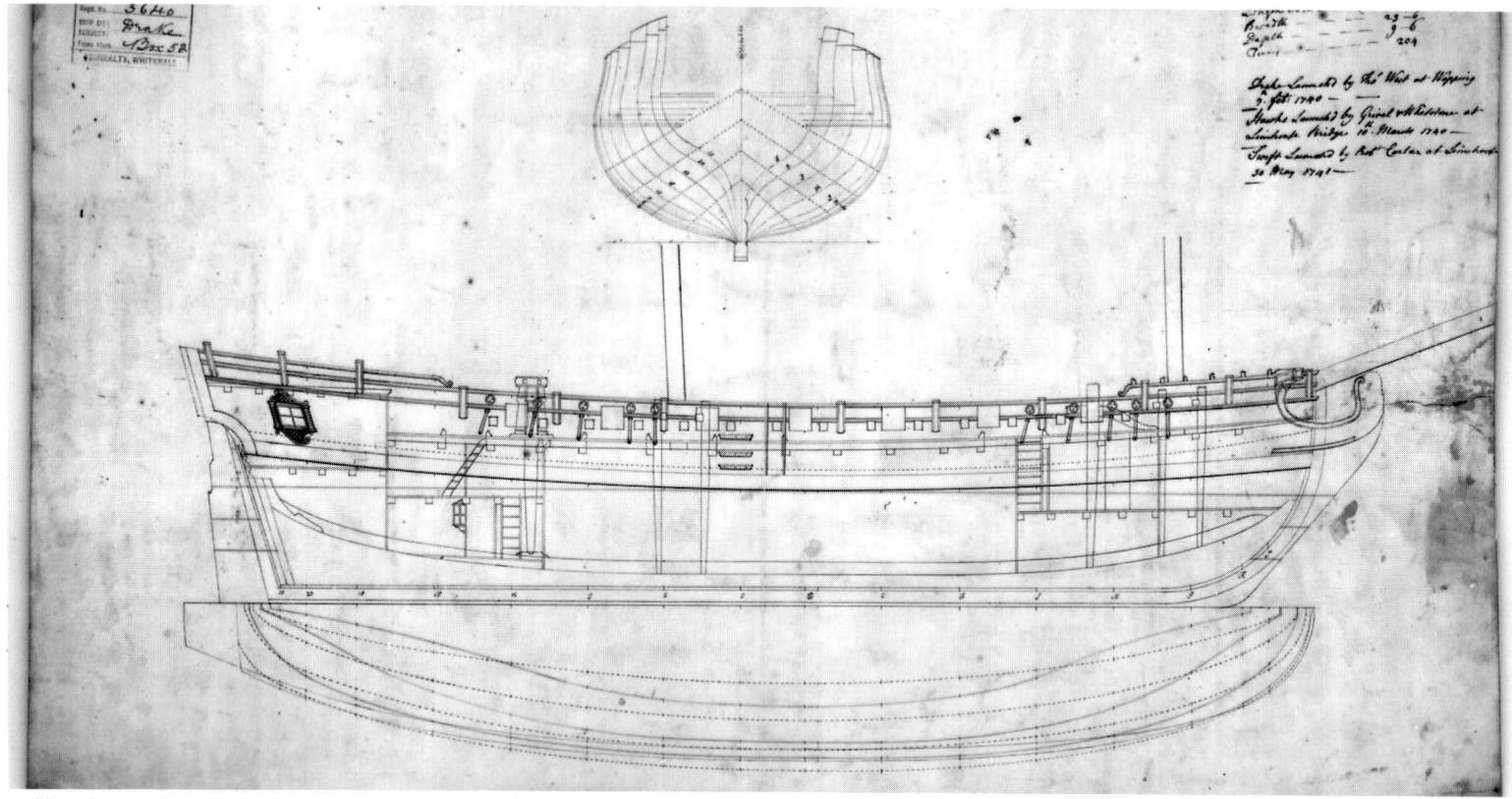

Building plan (number 3640) for the group – *Drake*, *Hawk* and *Swift* all being built from this

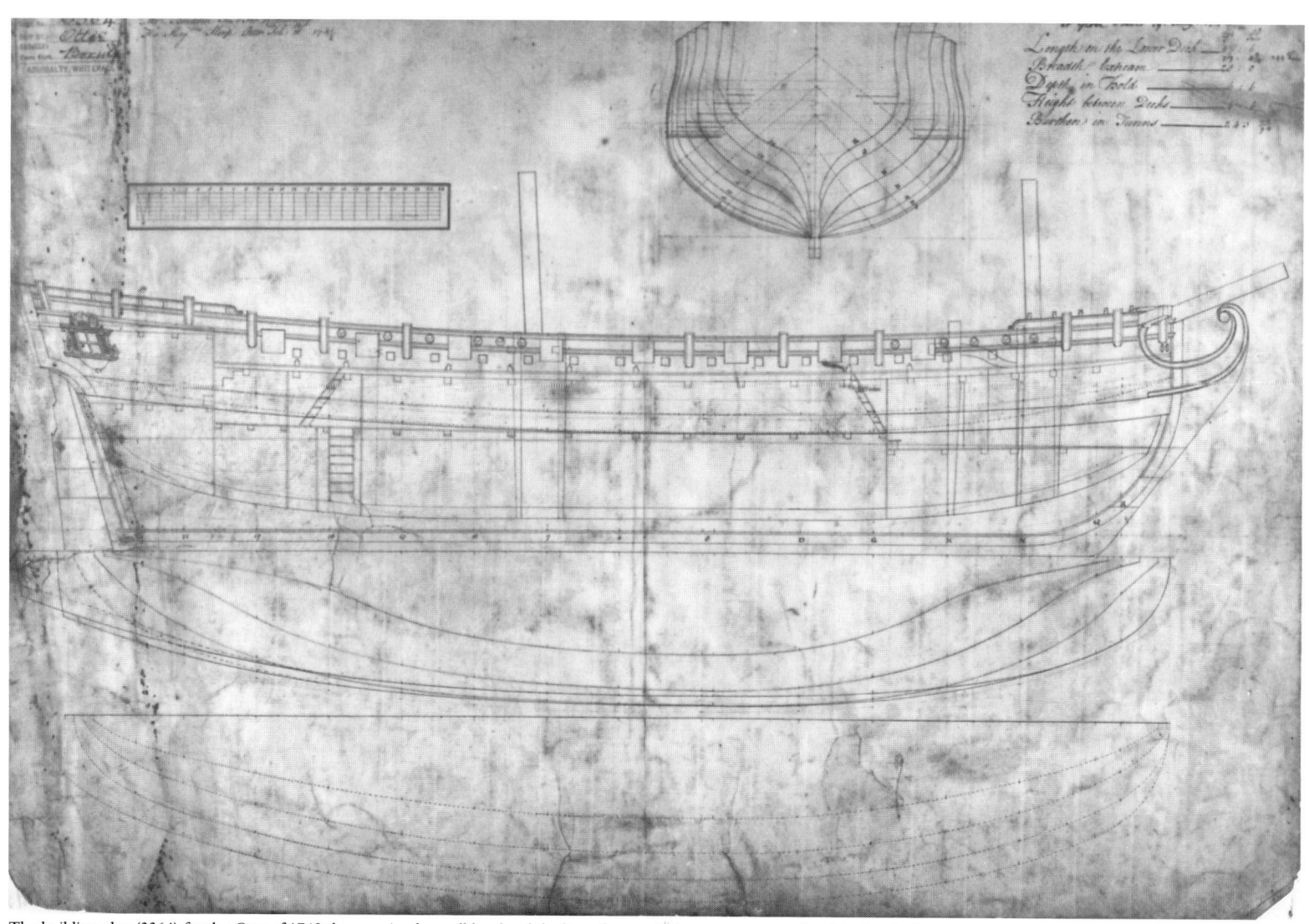

The building plan (3364) for the *Otter* of 1742 shows a simple scroll head and the line of upper deck sweep ports usual at this time in sloops and other smaller warships

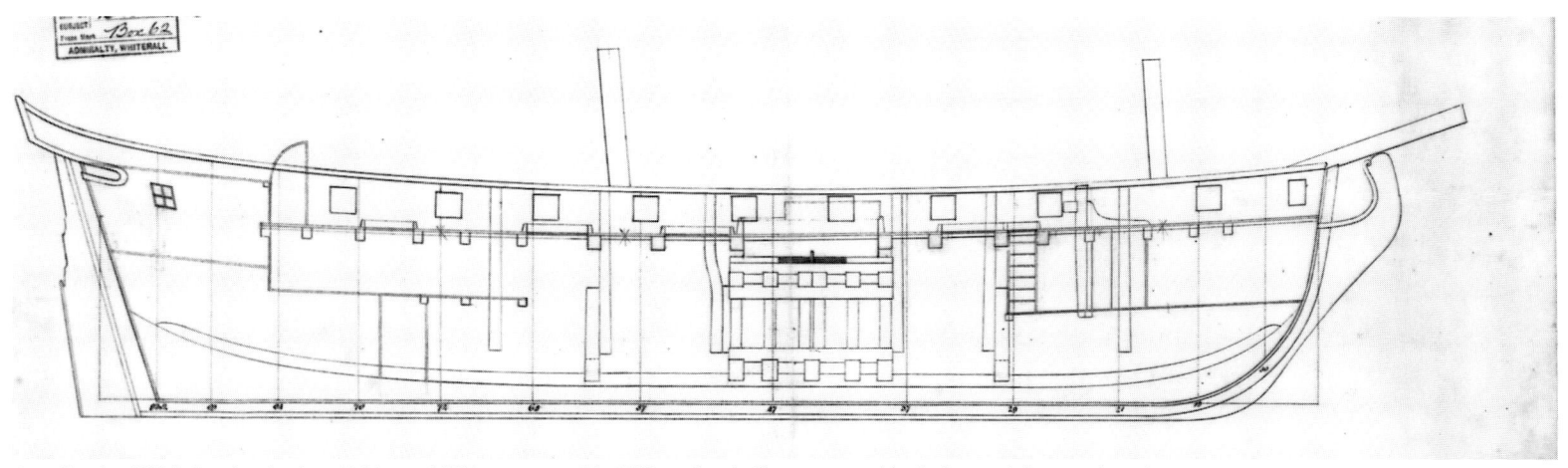

A profile plan (4374) showing the sloop *Baltimore* of 1742 as converted in 1758 to a bomb. She was unusual in the latter role because she only carried one mortar instead of the usual two. The plan slso shows that she had a 'pink' stern

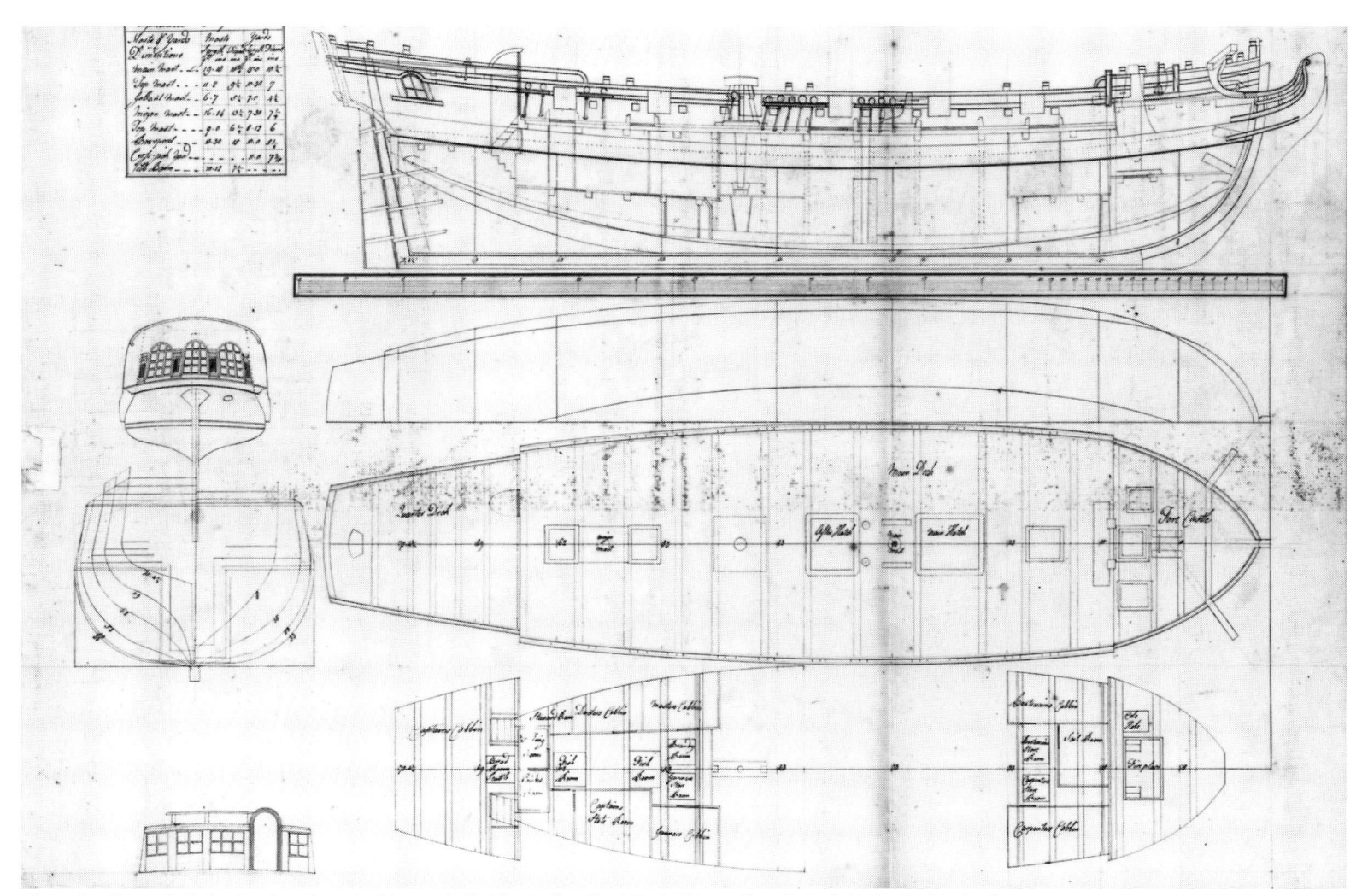

Charnock collection hull plan (number Ch. 166) of the *Speedwell* of the *Swallow* Class (1744). This excellent plan includes details of the break of the poop, and a set of spar dimensions

Ord 5.4.1745; K 05.1745; L 30.12.1745; 1773 sold.

**Falcon** Alexander, Thames. Ketch (?).

Ord 5.4.1745; K 04.1745; L 30.11.1745; 1758 bomb vessel; 1759 wrecked.

**Kingfisher** Darly, Gosport. Snow.

Ord 5.4.1745; K 05.1745; L 12.12.1745; (1758-1760 bomb vessel); 1763 sold.

**Hound** Stow & Bartlett, Shoreham. Snow.

Ord 15.4.1745?; K ___ ; L 22.5.1745; 1773 sold.

**Scorpion** Wyatt & Major, Beaulieu. Snow.

Ord 5.4.1745; K 04.1745; L 8.7.1746 ; 1762 lost.

FATES: *Swallow* wrecked off Bahama Islands 24.12.1744. *Hazard* taken in Montrose Harbour by the Jacobites 24.11.1745; 25.3.1746 retaken. *Hornet* taken by a French privateer 'off ye Berryhead'; 10.1747 retaken by *Tryton* in the Channel. *Badger* lost at some time in 1762 (? foundered at sea?). *Falcon* wrecked on the Saintes, Guadeloupe 19.4.1759. *Scorpion* cast away off the Isle of Man 23.9.1762.

PLANS: Lines. *Swan*: Lines & decks. *Speedwell* type: Lines.

**Hind** 1744-1747. Perry, Blackwall.

Dimensions & tons: 91'6", 75'2½" x 26'1½" x 12'2". 272 57/94.

Men: 110. Guns: 10 x 6 + 14 swivels.

Ord 6.8.1743; K 11.9.1743; L 19.4.1744;1.9.1747 foundered off Louisbourg Harbour.

**Vulture** 1744-1761. Greaves, Limehouse.

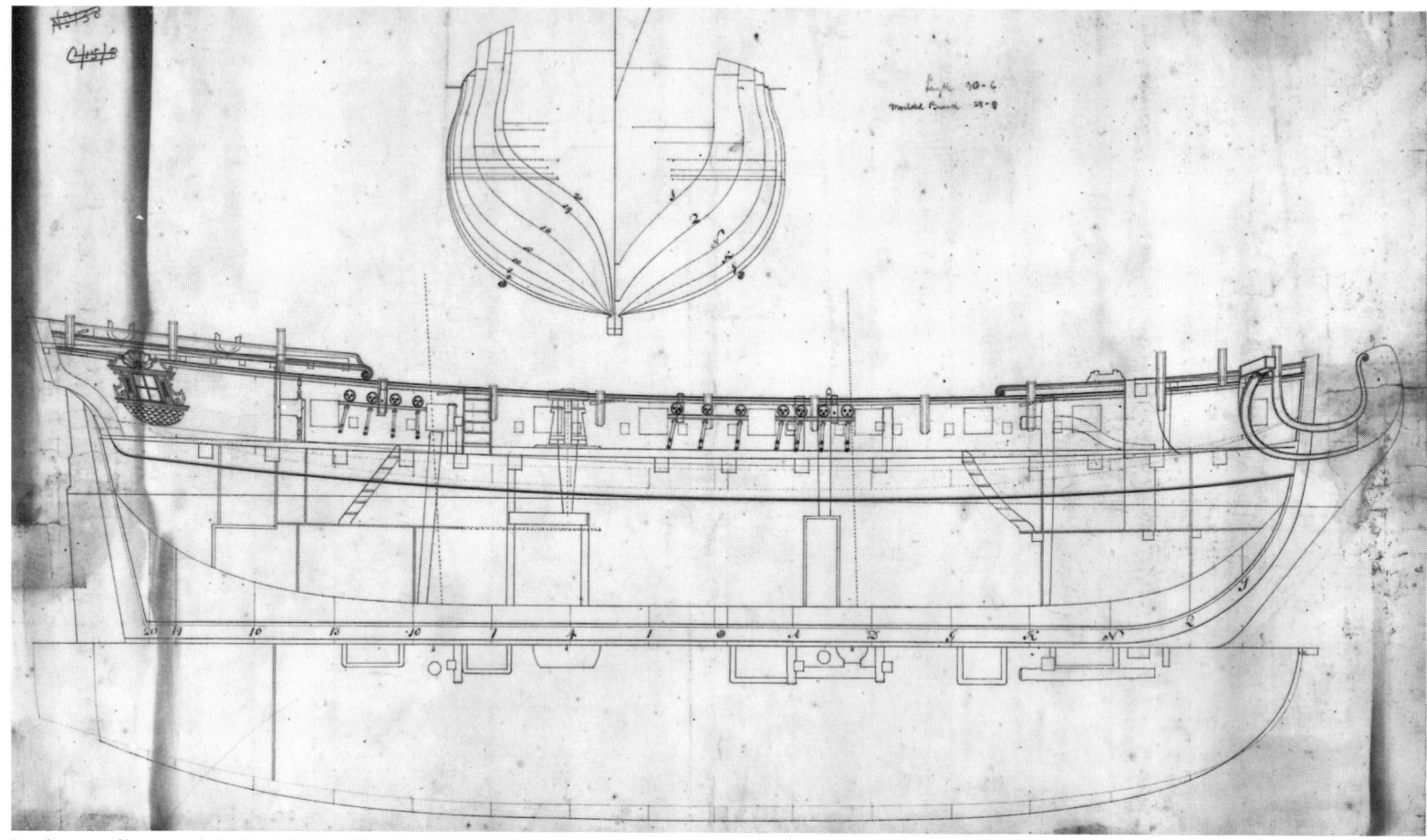

Part lines, profile & part deck plan (6433A) of the *Falcon* of the *Swallow* Class. Note the dotted lines that show the position of the masts, indicating that this vessel was ketch-rigged

Dimensions & tons: 91'3", 73'10⅞" x 26'0¾" x 12'1¾". 267.
Men: 110. Guns: 10 x 6 + 14 swivels (later classed as a 14 gun vessel).
Ord 6.8.1743; K 16.9.1743; L 4.5.1744; 1761 sold.

### *Jamaica* Class 10/14 guns 1743. Allin design. Snows?
Dimensions & tons: 91'3", 75'x 26'1¾" x 12'. 272.
Men: 110. Guns: 10 x 6 + 14 swivels.
*Jamaica* Deptford Dyd (Surveyor: Allin).
  Ord 18.8.1743; K 15.9.1743; L 17.7.1744; 1770 wrecked.
*Trial/Tryall* Deptford Dyd (Surveyor: Allin).
  Ord 18.8.1743; K 15.9.1743; L 17.7.1744; 1754 ship rig; 1776 BU.
FATES: *Jamaica* wrecked on the 'cobradoes' at Jamaica 27.1.1770.

### *Falcon* Class 10 guns 1744. Designer uncertain. Snows.
Dimensions & tons: 91'3", 74'9" x 26'2" x 12'. 272.
Men: 110. Guns: 10 x 6 + 14 swivels.
*Falcon* Barnard, Harwich.
  Ord 30.3.1744; K 15.5.1744; L 12.11.1744; (1745-1746 in French hands); 1746 renamed *Fortune*; 1770 sold.
*Lizard* Ewer, Bursledon.
  Ord 4.4.1744; K ____ ; L 22.12.1744; 1748 wrecked.
FATES: *Falcon* taken in the Channel by a Saint Malo privateer of 22 guns 10.1745; 6.3.1746 retaken. *Lizard* wrecked 'on the Rocks of Scilly' 27.2.1748.
PLANS: Lines.

*Hinchinbrook* 1745-1746. Janvrin, Bursledon.
  Dimensions & tons: 91'4", 74'6" x 26'1¾" x 12'0½". 270 84/94.
  Men: ____ . Guns: 10 x 6 + 14 swivels.
  Ord 23.5.1744; K ____ ; L 8.3.1745; 10.11.1746 taken 'coming from Cape Breton' by a French privateer.
*Swallow* 1745-1769. Bird, Rotherhithe. Snow.
  Dimensions & tons: 92', ____ x 26'6" x ____ . 278.
  Men: 110 Guns: 14 x 4 + 14 swivels.
  Ord ____ ; K ____ ; L 14.12.1745; 1755 ship rig; 1769 sold.
PLANS: Lines/decks.

*Viper* 1746-1755. *Lightning*-1762 Durrell, Poole.
  Dimensions & tons: 91', 74'9" x 26'1" x 12'1". 270 48/94.
  Men: 110 (as fireship: 45). Guns: 14 x 4 + 14 swivels (as fireship: 8 x 4 + 6 swivels).
  Ord 11.4.1745; K 06.1745; L 11.6.1746; 1755 fireship, renamed *Lightning*; 1762 sold.

## Bomb Vessels

### 1730 Group. Designer uncertain.
Dimensions & tons: 83', 65'4" x 27'6" x 11'. 262 76/94.
Men: 60. Guns: 2 mortars, 8 guns, 14 swivels.
*Salamander* Woolwich Dyd (Surveyor: Hayward).
  Ord ____ ; K ____ ; L 7.7.1730; (1730-1735 sloop); 1744 sold.
*Terrible* Deptford Dyd (Surveyor: Stacey).
  Ord ____ ; K ____ ; L 4.8.1730; 1749 sold.
PLANS: Lines & decks.

### *Alderney* class 'bomb sloop' 1734. 'From a draught drawn by HRH the Duke of Cumberland'.
Dimensions & tons: 90'6", 72'10" x 26'1" x 11'. 262.
Men: 60. Guns: 2 mortars, 10 guns, 14 swivels.
*Alderney* Woolwich Dyd (Surveyor: Hayward).
  Ord 27.3.1734; K 1.4.1734; L 29.3.1735; (1735-1738 sloop); 1742 hulked at Jamaica; fate uncertain.
PLANS: Lines & profile/decks & midships section.

### 1740 Group. 'Same as *Alderney*' (of 1734).
Dimensions & tons: 90'6", 73'9" x 26'x 11'. 265 17/94.
Men: 60. Guns: 1 x 13" + 1 x 10" mortars, 8 x 4 + 10 swivels (as sloops: 10 x 4 + 14 swivels. In 1742 *Furnace* was armed as follows: GD 4 x 6, UD 8 x 3 + 12 swivels).
*Basilisk* Snelgrove, Limehouse.
  Ord 11.3.1740; K 3.4.1740; L 30.8.1740; (1740-1741 sloop); 1750 sold.

# THE ESTABLISHMENT PERIOD 1719-1745

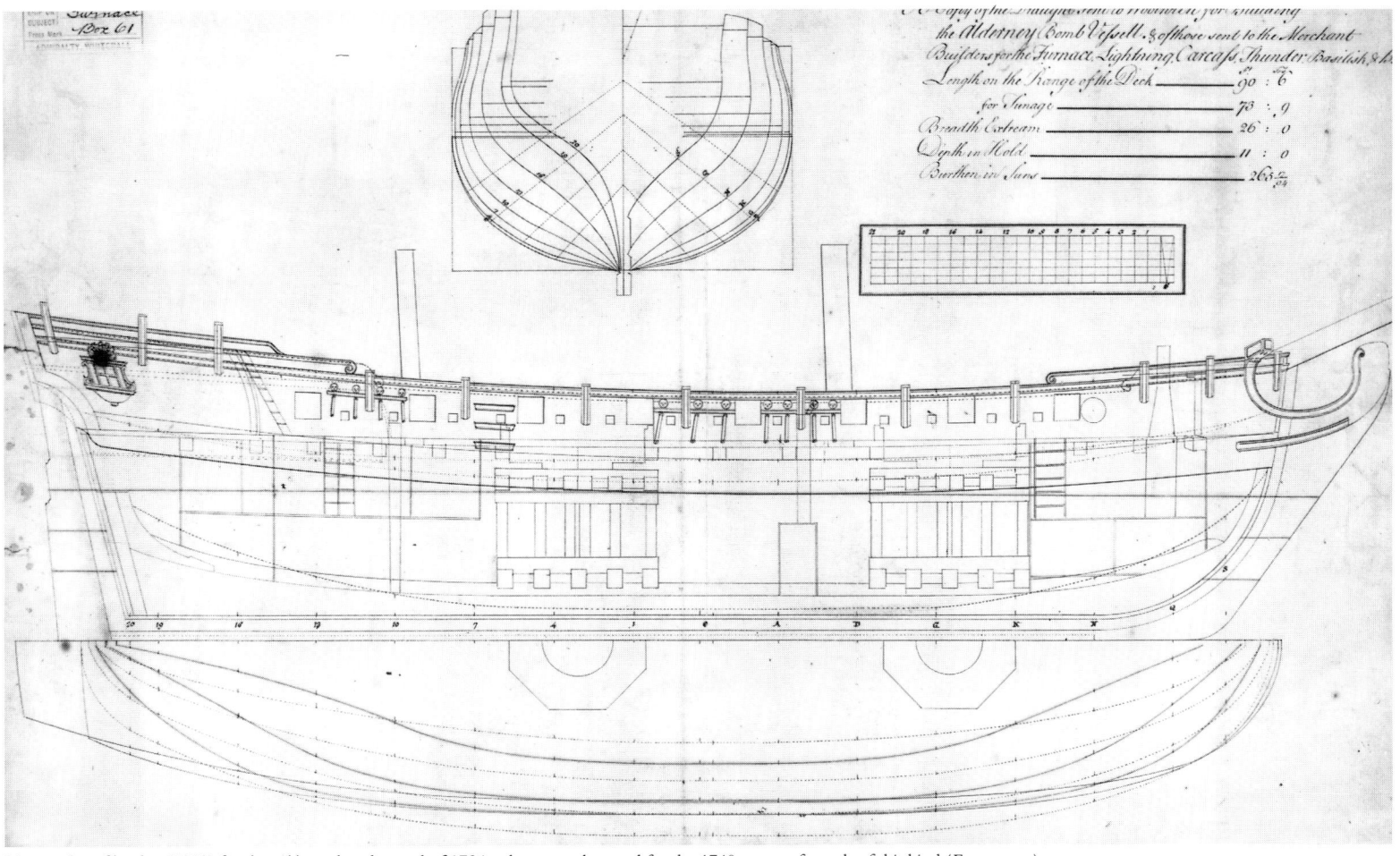

Lines and profile plan (4312) for the *Alderney* bomb vessel of 1734 – the same plan used for the 1740 group of vessels of this kind (*Furnace*, etc)

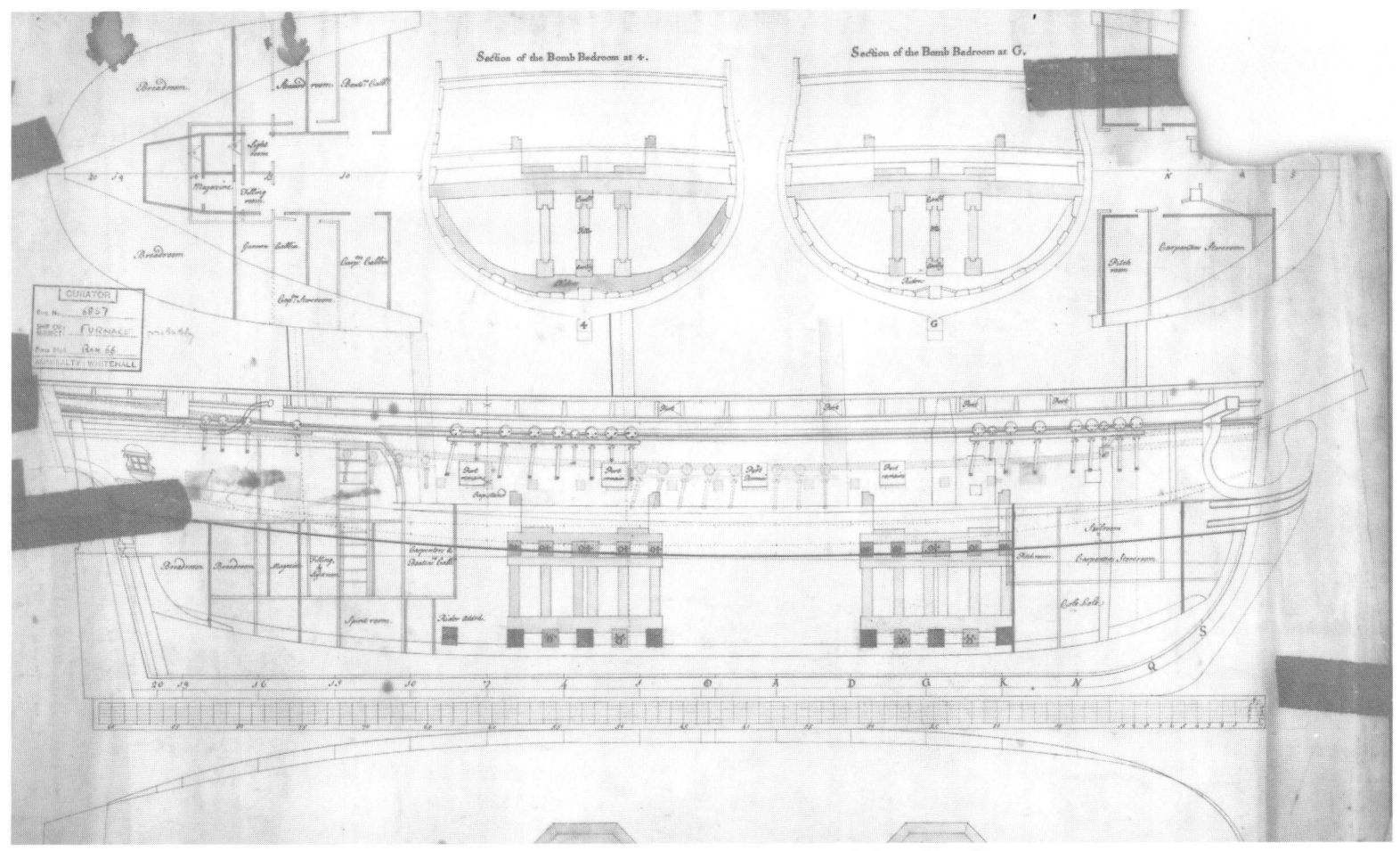

This profile deck & sections plan (6857) is believed to show the *Furnace* of the 1740 group of bomb vessels built to the lines of the *Alderney* (see above). The arrangements were clearly different from the 1734 vessel, but the resemblance of this plan to the one used to illustrate Part III Chapter 3 clearly labelled as *Furnace* as fitted for Arctic exploration makes the identification likely

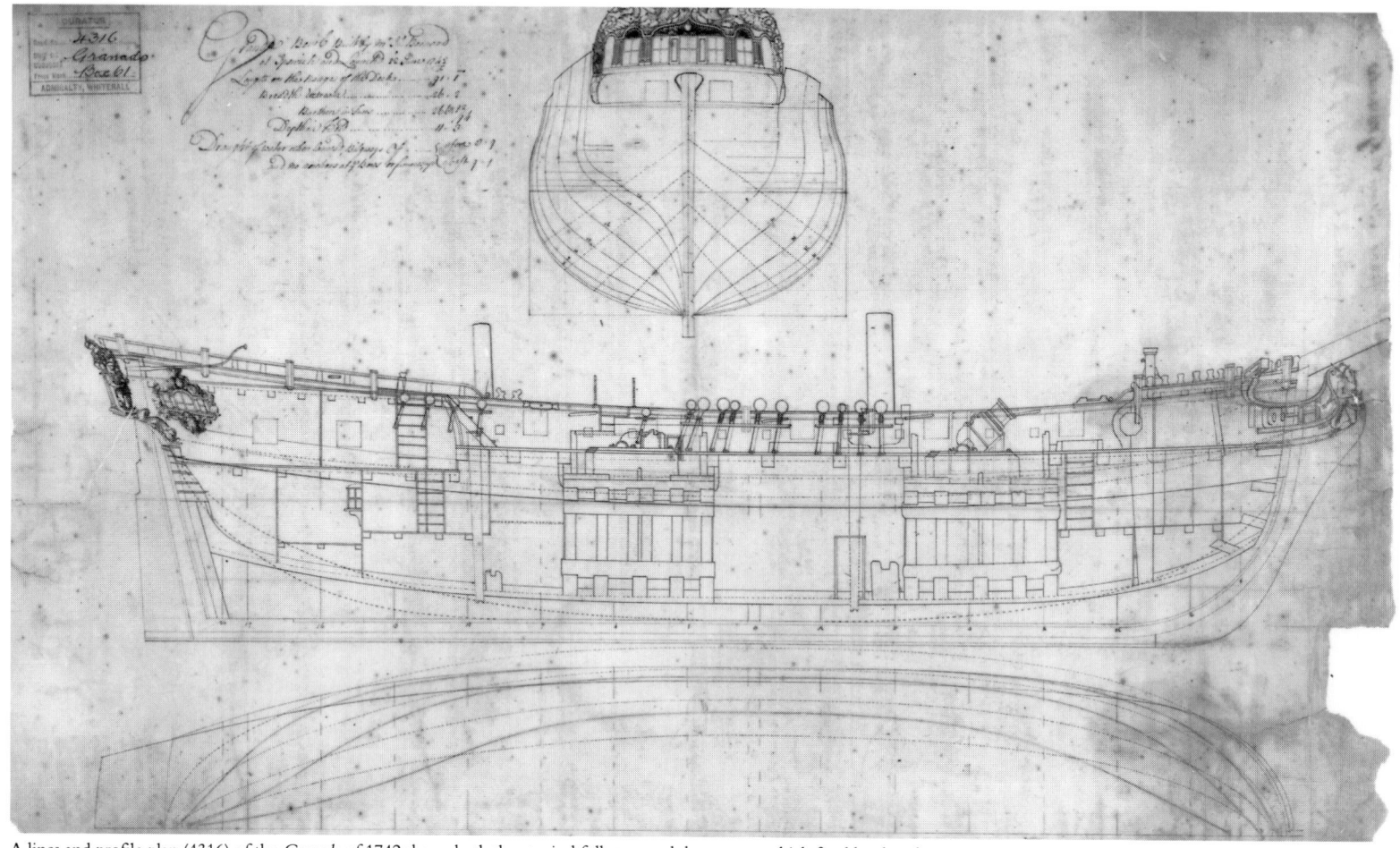

A lines and profile plan (4316) of the *Granado* of 1742 shows both the atypical full stern and the mortars which fired her bombs

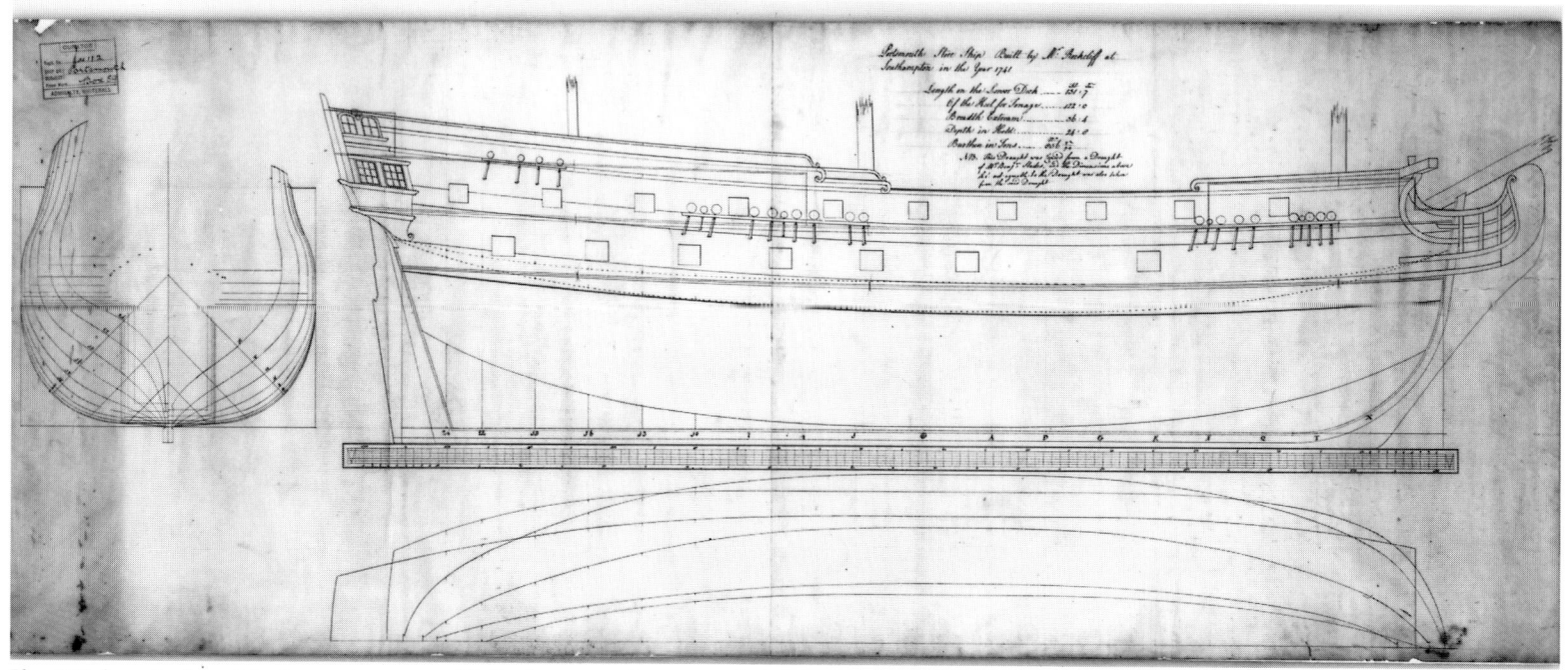

Plan 4112 shows the lines of the *Portsmouth* transport of 1742 – the pattern of comparatively few gunports on the lower deck and those mostly aft was similar to contemporary East Indiamen

**Blast** West, Deptford.
  Ord 11.3.1740; K 19.3.1740; L 28.8.1740; (1740-1741 sloop); 1745 taken.
**Carcass** Taylor, Rotherhithe.
  Ord 11.3.1740; K 23.4.1740; L 27.9.1740; (1740-1741 sloop); 1749 sold.
**Furnace** Quallett, Rotherhithe.
  Ord 11.3.1740; K 23.4.1740; L 25.10.1740; (1740-1741 sloop); 1741 converted to ship rig as an exploration ship; 1756 bomb; 1763 sold.
**Lightning** Bird, Globe Stairs, Rotherhithe.
  Ord 11.3.1740; K 23.4.1740; L 24.10.1740; (1740-1741 sloop); 1746 foundered.
**Thunder** Bird, Great Wet Dock, Rotherhithe.
  Ord 11.3.1740; K 23.4.1740; L 30.8.1740; (1740-1741 sloop); 1744 foundered.

FATES: *Blast* taken by two Spanish 'polacras' near Jamaica 19.10.1745. *Lightning* capsized 'off the island Gorgona' near Leghorn (Livorno) 16.6.1746. *Thunder* wrecked in a hurricane at Jamaica 20.10.1744.

PLANS: As *Alderney*.

**1741 Group.** Apparently a slightly modified version of the 1740 group.
   Dimensions & tons: 91', 75'6" x 26' x 11'3". 270.
   Men: 60. Guns: 1 x 13" + 1 x 10" mortar. 8 x 4 + 12 swivels (as sloops: 10 guns + 14 swivels).
*Comet* Taylor, Rotherhithe.
   Ord 14.9.1741; K 8.10.1741; L 29.3.1742; 1743 sloop; 1749 sold.
*Firedrake* Perry, Blackwall.
   Ord 14.9.1741; K 7.10.1741; L 20.2.1742; (1755-1758 sloop); 1763 sold.
*Mortar* Perry, Blackwall.
   Ord 14.9.1741; K 8.10.1741; L 25.2.1742; (1742- 1745 sloop); 1749 sold.
*Serpent* Snellgrove, Limehouse.
   Ord 14.9.1741; K 9.10.1741; L 15.3.1742; 1742 sloop; 1748 wrecked.
*Terror* Greville, Limehouse.
   Ord 14.9.1741; K 9.10.1741; L 13.3.1742; (1742-1743 sloop); 1754 sold.

FATES: *Serpent* wrecked near the Barbadoes 1.9.1748.

PLANS: As *Alderney*?

***Granado* Class 1741.** Designer uncertain. Differed from her otherwise similar contemporaries in having a full stern with windows instead of the pink stern normal in this type of ship at the time. Possibly she had originally been intended as a sloop or a fireship; though there is no proof for this apart from her stern.
   Dimensions & tons: 91'1", 73'10¼" x 26'2" x 11'4". 268 92/94.
   Men: 60 (as sloop: 110). Guns: 1 x 13″ + 1 x 10" mortar + 8 x 4 + 12 swivels (as sloop: 10 x 4).
*Granado* Barnard, Ipswich.
   Ord 14.9.1741; K 18.11.1741; L 22.6.1742; (1742-1756 sloop); (1760-1761 sloop); 1763 sold.

PLANS: Lines, profile & decoration.

## Storeships, etc

*Deptford* 1735-1756. Deptford Dyd (Surveyor: Stacey).
   Dimensions & tons: 124', 110'3" ′x 35'8" x 15'6". 678 8/94.
   Men: 120. Guns: LD 4 x 12, UD 16 x 9, QD 4 x 6. 16 swivels.
   Ord ____ ; K 5.4.1734; L 29.4.1735; 1756 BU.
PLANS: Lines.

*Portsmouth* 1742-1747. Rowcliffe, Southampton. Acworth design.
   Dimensions & tons: 124'1", 100'5" x 35'8" x 14'8". 693 75/94.
   Men: 120. Guns: 24 (probably similar in type and distribution to *Deptford*'s).
   Ord 28.11.1740; K 03.1741; L 12.1.1742; 1747 foundered. [3.12.1747 between the Galloper and Longsand Head].
PLANS: Lines.

# Chapter 4: The Slade Era 1745-1785

By 1745 war with Spain and then France had revealed that British warship design had been left behind by developments in the other two navies. Specifically, there was a need for larger and more powerful two-decker line of battle ships, and to follow the French development of a new type of warship of middling size and power, the frigate. Pressure from sea officers and from the Admiralty upon the Navy Board, which was responsible for designing and building ships for the Navy, continued to mount. In the first instance, it met an uncompromisingly conservative reaction. The 1745 Establishment did increase the size and gunpower of rated vessels, but it stuck to previously existing types of ship, and it not only specified both dimensions and scantlings but also the actual lines to be used. For the first time, the design itself was standardised. This step towards total rigidity was very rapidly reversed, however.

In the early 1750s there appears to have been a total revolution in ship design, but a very quiet one which is very difficult to document. The results are clear as the arguments and administrative steps by which they were produced are obscure. The large three-deckers were not particularly affected, new designs were produced (notably the classic one for *Victory* by Slade) but the basic size, shape and power of these designs were still very much in the pattern of preceding vessels. However, the small 80 gun three-deckers were discontinued; instead, the size and power of the largest two-deckers was increased to follow the French and Spanish in producing the type which was to become the standard ship of the line for the next 60 years. This was the 74 gun Third Rate. This, from the start, appeared in two main types, the smaller 'common class' built in very large numbers, and the larger type copied from the French *Invincible* of which the first examples were the *Valiant* and *Triumph*. Except in the 1790s, this second, larger, type was never built in any numbers by the Royal Navy, whose main requirement was for numbers of competent vessels of moderate size rather than fewer, more powerful, ships. The need for numbers and for economy produced a requirement for a cheaper version of the two-decker line of battle ship as a replacement for the older 70s and 60s. This appeared in the form of the '64'. Built in numbers for the next three decades, this proved in the long run to be too small and weak compared to the 74s and the last ones were being built as the period covered by this chapter draws to a close. However, the surviving examples of the type would continue to give useful service until the end of the Napoleonic War.

To revert to the middle of the eighteenth century, war experience had shown that, just as the '74' was the battle ship of the future, the French had produced the ideal cruising warship in the form of the 'true frigate', a vessel initially armed with 9 or 12 pounder guns as the main battery and carrying these well above the waterline thanks to a lower deck well below the waterline and not armed with guns at all. With gun ports well above the waterline, plenty of room for stores, and a comparatively long hull, these vessels were well suited to the multifarious duties hitherto dealt with by a far from ideal combination of small two-deckers and small Sixth Rates. The pioneer British 28 gun 9 pounder frigates were built in 1748 (though originally classed as 24s). The first 32 gun 12 pounder frigates built for the Royal Navy appeared in 1757.

At much the same time, the first ship (ie. ship rigged) sloops appeared, and rapidly these larger vessels replaced the two masted sloops of previous years. However, this created a gap to be filled by a smaller and cheaper type of vessel, a gap which would be filled by the appearance of the brig in the late 1770s. By that time two still smaller types had already (1760s) been added to the naval inventory. These were the standard fast sailing small ship of British coastal waters, the single masted cutter, and also the North American equivalent, the schooner. The Seven Years War had also seen the building of the last ketch rigged bomb vessels, some of their sisters having been fitted from the start with the ship rig, which was henceforward to be standard for this type of vessel.

During the early 1770s increasing colonial unrest produced orders for numbers of new versions of older types of warship, small two-deckers of 50 and 44 tons, and 20 and 24 gun Sixth Rates. Presumably, they were thought to be more suitable for employment as 'colonial gunboats'. However, as soon as major European naval powers joined in the conflict against Britain the emphasis shifted back towards the newer fighting ships. There was a place for small numbers of 50s and of small Sixth Rates, and these continued to be built for another generation; but the 44 was so clearly less satisfactory than the frigates that the type was soon relegated to second-line duties and no more were built.

Towards the end of the 1770s Britain found herself confronted with an alliance of most other European navies. It was no doubt this position of temporary numerical inferiority that caused her to take the lead in developing a more powerful type of frigate, with 18 pounders as the main battery (the 36 and 38 gun frigates). The French and Spanish had already been building larger two-deckers, and the first British two-decker 80 was building by the end of the 1780s.

## First Rates, 100 Guns (three-deckers)

### 1745 Establishment/*Royal George* Class. *Britannia* was originally to be built 'by the Establishment draught with some variations' (Admiralty Order 28.3.1751) but this was changed (AO 21.5.1757) to building her by the draught of the *Royal George*.

> Dimensions & tons: 178', 144'6½" x 51'10" x 21'6". 2065 58/94.
> (Establishment design had breadth of 51' and was of 2000 tons).
> Men: 850. Guns: GD 28 x 42 (later 32), MD 28 x 24, UD 28 x 12, QD 12 x 6, Fc 4 x 6.

**Royal George** (ex *Royal Anne*) Woolwich Dyd.
> Ord 21.3.1727/27.8.1746; K 8.1.1747; L 18.2.1756; 1782 foundered.

**Britannia** Portsmouth Dyd.
> Ord 28.3.1751/21.2.1757; K _____ ; L 19.10.1762; 1813 hulked.

FATES: *Royal George* foundered at Spithead 'she being heeled to come at the pipe that leads to the well'. The story that her bottom dropped out of her is most unlikely, she was merely heeled over too far, flooded and sank; 'a land breeze shook her shrouds and she was overset' as the poet Cowper put it. The wreck was dived on and dispersed in the 1830s. *Britannia* was hulked at Plymouth as a prison ship; 1815 renamed *Saint George* as a harbour flag and receiving ship; 1825 renamed *Barfleur* and then BU.

PLANS: Spar plan (1745 Establishment). *Royal George:* Lines/lines & profile/profile/orlop/gun deck/main deck/upper deck/quarter deck/forecastle/roundhouse/decks/above waterline profile with weather deck fittings. *Britannia*: Lines & profile/gun deck/main deck/upper deck/quarter deck & forecastle.

### *Victory* Class 1759. Slade design. The classic 100 gun ship. She sailed as well as a two-decker which helps to explain her long active career, and her frequent appearance as a flagship. The years spent weathering in frame before her launch also contributed to her longevity. The *Boyne* class of 1801, though originally classed as 98s, were built to her lines, and could be considered as sisters.

> Dimensions & tons: 186", 151'3⅝" x 51'10" x 21'6". 2162 22/94.
> Men: 850. Guns: GD 30 x 42 (later 32), MD 28 x 24, UD 30 x 12, QD 10 x 6, Fc 2 x 6.

**Victory** Chatham Dyd.
> Ord 13.12.1758/14.6.1759; K 2.7.1759; L 7.5.1765; 1808 Second Rate 98;

THE SLADE ERA 1745-1785

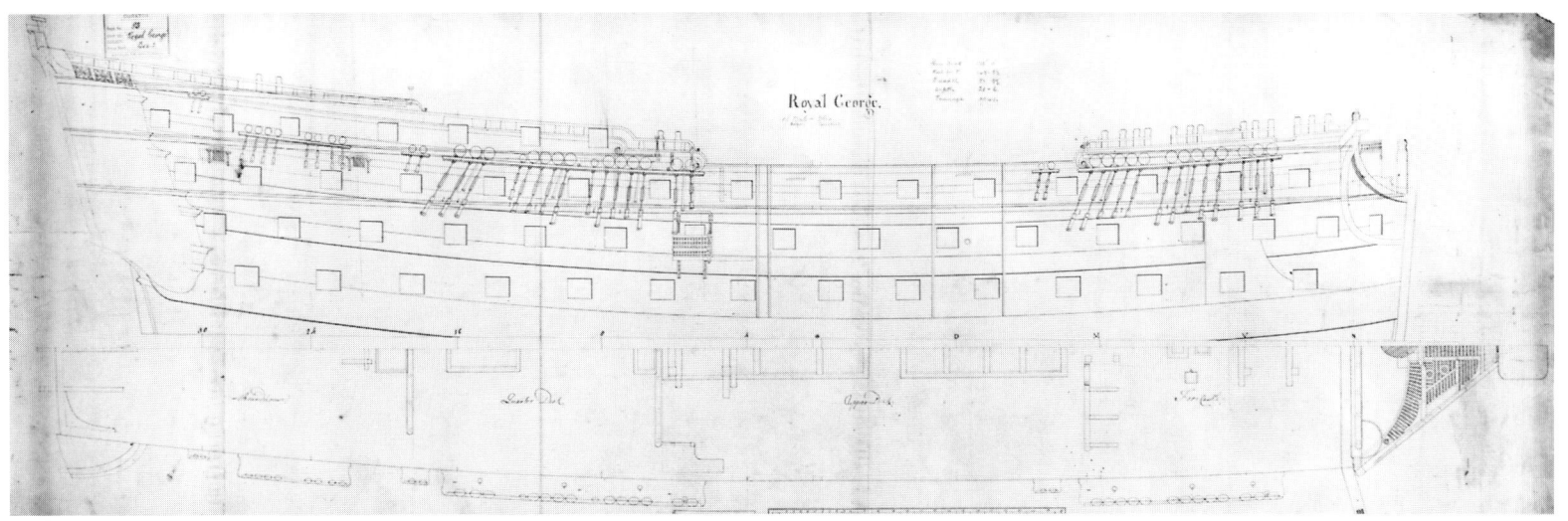

*The Royal George of 1756 is illustrated in this very rare type of plan showing the above waterline external profile and outline of the upper decks with a plan of the heads (plan number 83)*

later reclassed as 100 guns again; 1824 hulked at Portsmouth as a harbour flag and receiving ship; converted during the 1920s to her appearance at the time of Trafalgar still in commission and open to the public today at Portsmouth.

PLANS: Lines. Ca 1800: lines & profile/profile/orlop/gun deck/main deck/upper deck. As in 1805 (later plan): Lines & profile & decoration. 1830: Upper deck.

NOTE: None of these plans are particularly good for modelmaking; the prospective builder of this much-modelled and indeed hackneyed subject can be referred to E N Longridge's *Anatomy of Nelson's Ships* (latest editions by Model & Allied Publications) which includes good plans, and to the much less easy to obtain *H.M.S. VICTORY, building, restoration and repair* by A R Bugler (HMSO).

***Royal Sovereign* Class 1772.** Williams design.
Dimensions & tons: 186', 152'6" x 52' x 22'3". 2193 38/94.
Men: 850. Guns: GD 28 x 32 (design: 42), MD 28 x 24, UD 30 x 12, QD 10 x 12 (design: 6), Fc 4 x 12 (design: 6).

***Royal Sovereign*** Plymouth Dyd.
Ord 3.2.1772; K 01.1774; L 11.9.1786; 1825 hulked.

FATE: Hulked at Plymouth and renamed *Captain*; 1841 BU.

PLANS: Lines/profile/orlop/gun deck/main deck/upper deck/quarter deck & forecastle/stern decoration/capstan. As fitted: Lines & profile & decoration.

***Umpire* Class 1772.** Hunt design. The design was lengthened by 3' from the original concept.
Dimensions & tons: 190', 156'5" x 52'4" x 22'4". 2278 62/94.
Men: 850. Guns: First pair: GD 30 x 32 (design: 42), MD 28 x 24, UD 30 x 12, QD 10 x 12 (design: 6), Fc 2 x 12 (design: 6), Second *Queen Charlotte*: GD 30 x 32, MD 30 x 24, UD 30 x 12, QD 2 x 12 + 12 x 32 ____, Fc 2 x 12 + 2 x 32 ____, RH 6 x 18 ____; classed as a 104 gun ship.

***Royal George*** (ex *Umpire*) Chatham Dyd.
Ord 3.2.1772; K 06.1784; L 16.9.1788; 1822 BU.

***Queen Charlotte*** [I] Chatham Dyd.
Ord 12.12.1782; K 09.1785; L 15.4.1790; 1800 blown up.

***Queen Charlotte*** [II] Deptford Dyd.
Ord ____ ; K 10.1805; L 17.5.1810; 1859 hulked.

FATES: *Queen Charlotte* [I] caught fire by accident and blew up off Leghorn (Livorno) on 16.3.1800. *Queen Charlotte* [II] was hulked at Portsmouth as the gunnery training ship *Excellent*; 1892 sold for BU.

PLANS: Lines/profile/orlop/gun deck/main deck/upper deck/quarter deck & forecastle/midships section. *Queen Charlotte* [I]: figurehead (port)/(starboard). *Queen Charlotte* [II] planking expansion (external)/(internal).

## Second Rates, 90, later 98, Guns (three-deckers)

Originally 90s, these ships were reclassed as 98s in 1778 with the addition of 8 guns to the quarter deck.

**No 1745 Establishment 90s,** though a design was prepared of:
170', 138'4" x 48'6" x 26'6". 1730 tons.

PLANS: Lines/spar plan.

***Namure* Class 1750.** 'Admiralty' (Bately) design, 1745 Establishment amended.
Dimensions & tons: 175', 142'2½" x 40'6" x 20'6". 1779 30/94.
Men 750. Guns: GD 26 x 32, MD 26 x 18, UD 26 x 12, QD 10 x 6, Fc 2 x 6.

***Namure*** Chatham Dyd.
Ord 12.7.1750; K 18.12.1750; L 3.3.1756; 1805 cut down to Third Rate two-decker 74; 1807 hulked.

FATE: Hulked at Chatham as a receiving ship (for the Nore?); 1833 BU.

PLANS: Lines. 1802 proposed cutting down to 74: Sheer.

***Neptune* Class 1750.** Allin design, 1745 Establishment amended.
Dimensions & tons: 171', 143'3" x 48'6" x 20'6". 1792 31/94.
Men: 750. Guns: GD 26 x 32, MD 26 x 18, UD 26 x 12, QD 10 x 6, Fc 2 x 6.

***Neptune*** Portsmouth Dyd.
Ord 12.7.1750; K 20.7.1750; L 8.12.1756; 1784 hulked.

***Union*** Chatham Dyd.
Ord 12.7.1750; K 5.6.1751; L 25.9.1756; 1788 hulked.

FATES: *Neptune* became a sheer hulk at Portsmouth; 1816 BU. *Union* was hulked at Chatham as a hospital ship; 1790 renamed *Sussex*; 1816 BU.

PLANS: Lines. *Union* as hospital ship 1788: Decks. As hospital ship 1813: Orlop & forecastle/other decks.

***Sandwich* Class 1755.** Slade design.
Dimensions & tons: 176', 142'7½" x 49' x 21'. 1821 30/94.
Men: 750. Guns: GD 28 x 32, MD 28 x 18, UD 30 x 12, QD 2 x 6, Fc 2 x 6.

***Sandwich*** Chatham Dyd.
Ord 12.11.1755; K 14.4.1756; L 15.4.1759; 1790 hulked.

***Blenheim*** Woolwich Dyd.
Ord 12.11.1755; K 1.5.1756; L 5.7.1761; 1801 cut down to Third Rate two-decker 74; 1807 foundered.

FATES: *Sandwich* became a receiving ship at Chatham; 1794 prison ship (rated as a sloop); 1816 BU. *Blenheim* foundered in a storm in the Indian Ocean off Rodriguez in 02.1807.

PLANS: Lines/profile/stern. *Sandwich* as receiving ship 1787: orlop, gun deck,

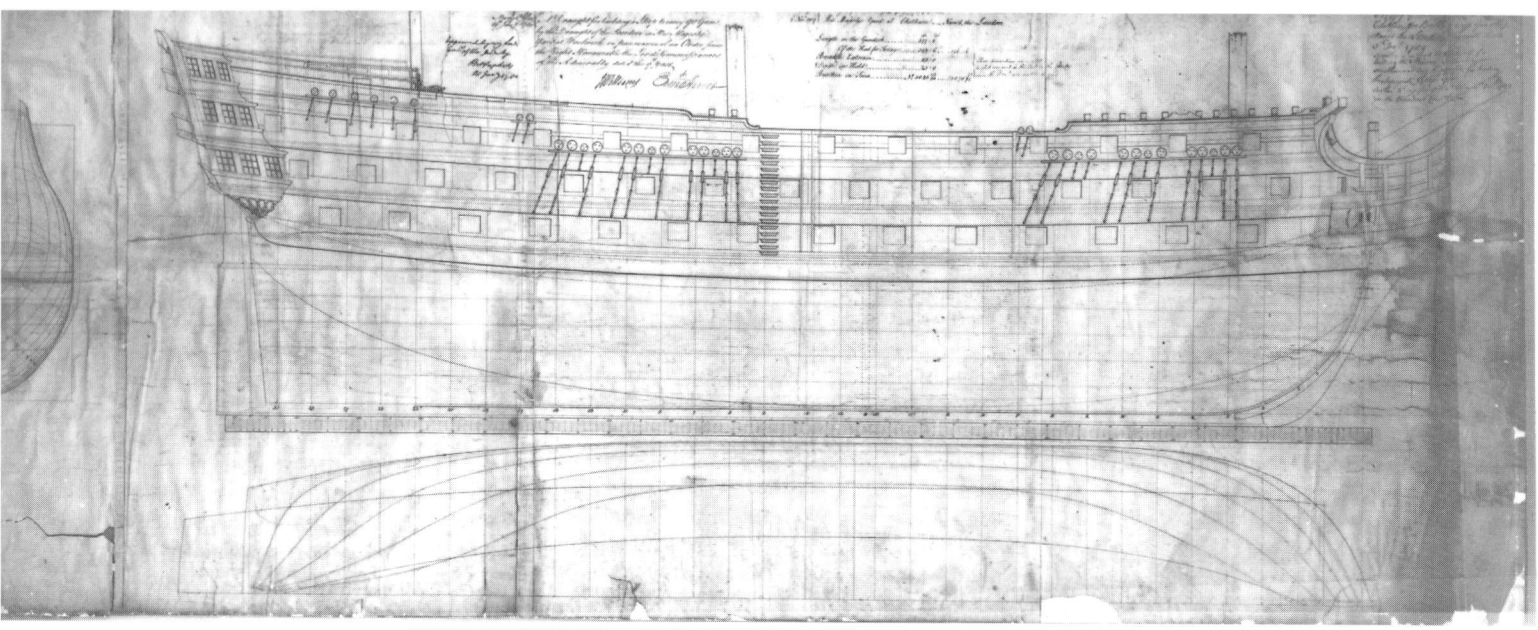

Design lines plan for the *London* Class 90 gun ships (243)

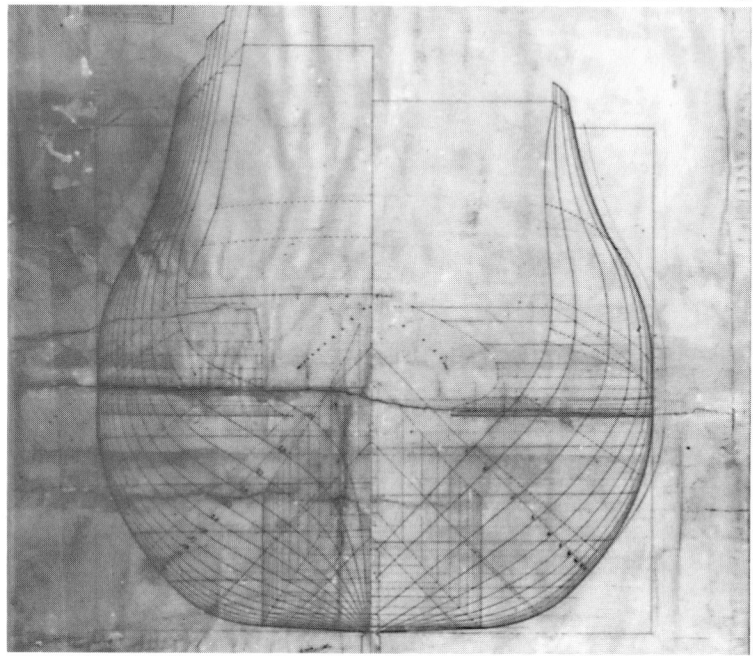

main deck/upper deck, quarter deck & forecastle. *Blenheim* as built: Lines & profile. As 74 in 1801: sheer.

### Ocean Class 1758. Slade design, modification of 1755 *Sandwich* Class.
Dimensions & tons: 176', 143'x 49'x 21'. 1826 89/94.
Men: 750 Guns: GD 28 x 32, MD 30 x 18, UD 30 x 12, Fc 2 x 6.
**Ocean** Chatham Dyd.
Ord 22.4.1758; K 4.8.1758; L 21.4.1761; 1791 BU.
PLANS: Sheer & half breadth/body plan/profile/orlop/gun deck/main deck/upper deck/quarter deck & forecastle.

### London Class 1759. Slade design.
Dimensions & tons: 177'6", 146'6" x 49'x 21'. 1870 93/94.
Men: 750 Guns: GD 28 x 32, MD 30 x 18, UD 30 x 12, Fc 2 x 6.
**London** Chatham Dyd.
Ord 13.12.1758/6.11.1759; K 4.11.1759; L 24.5.1766; 1811 BU.
**Prince** Woolwich Dyd.
Ord 9.12.1779; K 1.1.1782; L 4.7.1788; 1796 lengthened by 17' (increased to 2088 tons); 1816 hulked.
**Impregnable** Deptford Dyd.
Ord 13.9.1780; K 10.1781; L 15.4.1786; 1799 wrecked.
**Windsor Castle** Deptford Dyd.
Ord 19.8.1782/10.12.1782; K 19.8.1784; L 21.5.1790; 1814 cut down to a Third Rate two-decker 74; 1839 BU.

FATES: *Prince* was hulked as a victualling vessel and as accommodation for officers (dockyard officials); 1837 BU. *Impregnable* was wrecked near Langstone, Sussex, on 19.10.1799; there are reports that this wreck has been located.

PLANS: Lines/profile/orlop/gun deck/main deck/upper deck/quarter deck & forecastle/midships section. *Prince*: 1796 for lengthening: sheer. *Windsor Castle*: 1814 as a 74: sheer.

### Barfleur Class 1761. Slade design, based on the *Royal William* 100/84 of 1719..
Dimensions & tons: 177'6", 144'0¾" x 50'3" x 21'. 1934 87/94.
Men: 750 Guns: GD 28 x 32, MD 30 x 18, UD 30 x 12, Fc 2 x 6.
**Barfleur** Chatham Dyd.
Ord 19.10.1761/1.3.1762; K 22.11.1762; L 30.7.1768; 1819 BU.
**Prince George** Chatham Dyd.
Ord 11.6.1766; K 18.5.1767; L 31.8.1772; 1817 hulked.
**Princess Royal** Portsmouth Dyd.
Ord 10.9.1767; K 31.10.1767; L 18.10.1773; 1807 BU.
**Formidable** Chatham Dyd.
Ord 17.8.1768/21.9.1768; K 12.1769; L 20.8.1777; 1813 BU.

FATES: *Prince George* became a sheer hulk at Portsmouth; 1839 BU.

PLANS: Lines/profile/orlop/gun deck/main deck. *Barfleur* as built: Lines & profile/gun deck, main deck, upper deck. *Prince George*: 1839 as target: Profile & midsection (before)/(after). *Princess Royal*: as built: Lines & profile & decoration.

# THE SLADE ERA 1745-1785

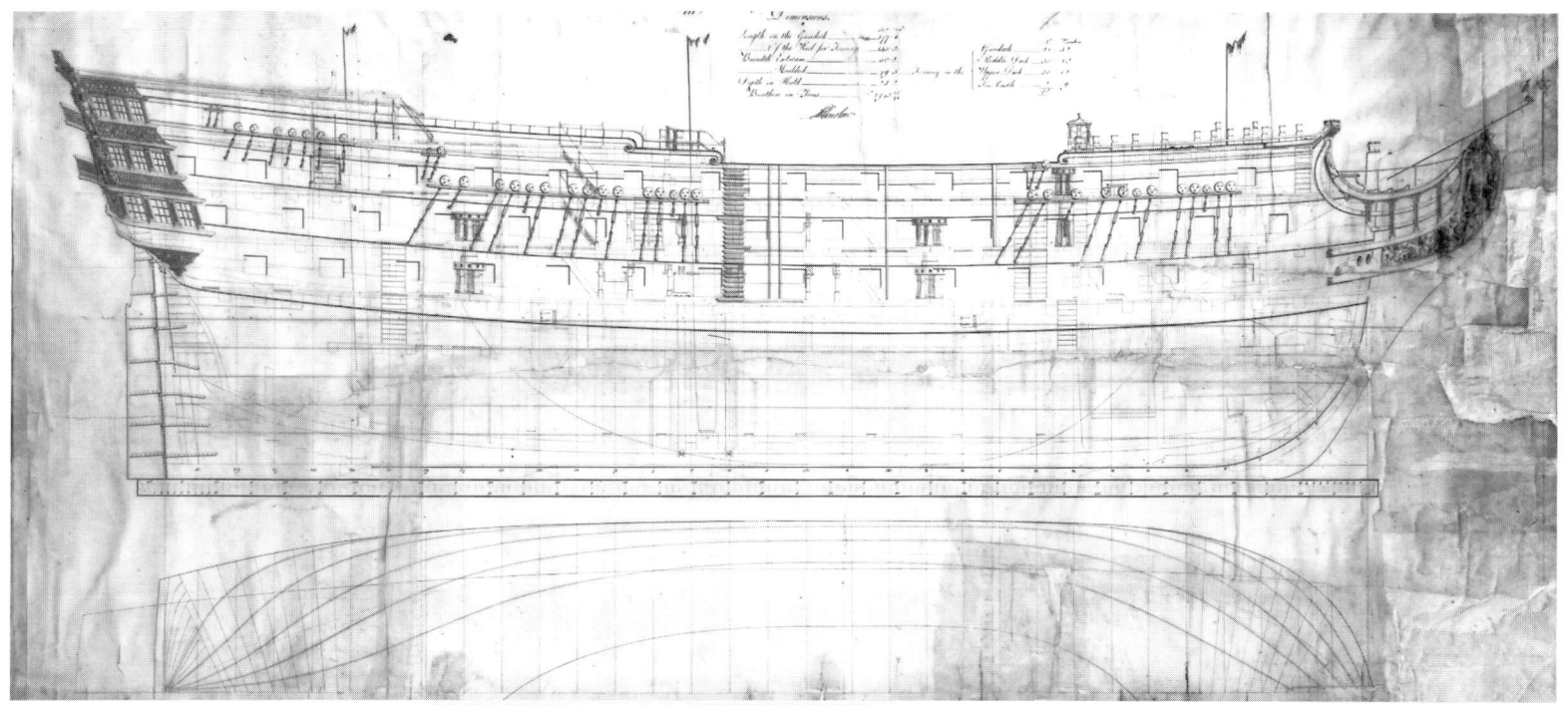

During the 1770s and early 1780s numbers of 'as built' plans combining lines and profile plans with a greater or lesser amount of decoration were produced. This plan of the 98 gun *Duke* is an example

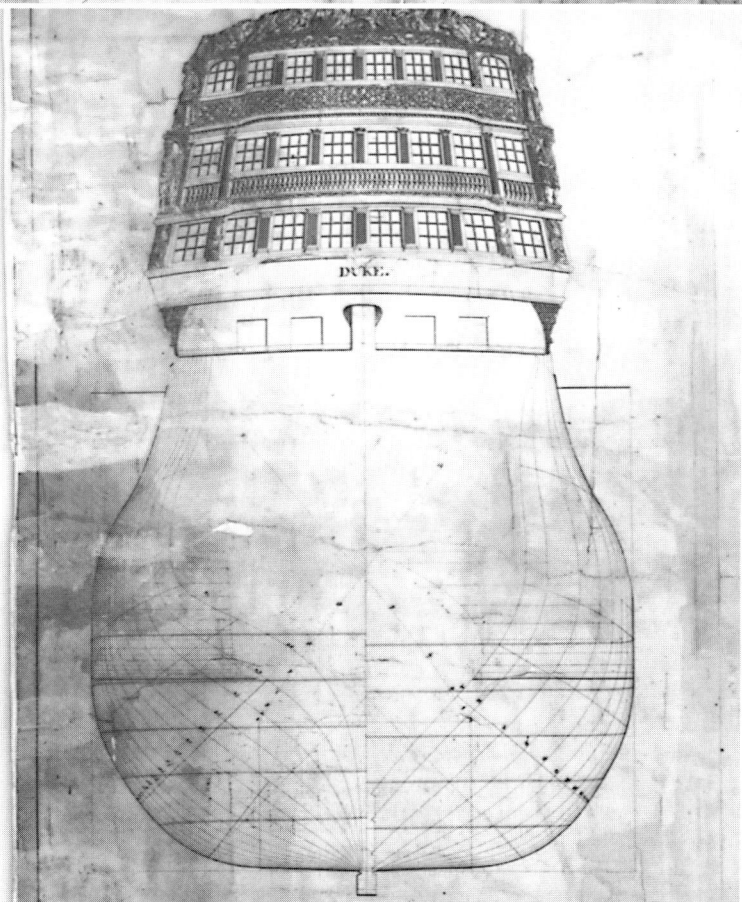

Proposed as 74: sheer.

### Queen Class 1762. Bately design.
Dimensions & tons: 177'6", 143'7" x 49'6" x 21'9", 1871 33/94.
Men: 750. Guns: GD 28 x 32, MD 30 x 18, UD 30 x 12, Fc 2 x 6.
**Queen** Woolwich Dyd.
  Ord 19.10.1761/21.1.1762; K 1.4.1762; L 18.9.1769; 1811 cut down to a Third Rate two-decker 74; 1821 BU.

PLANS: Lines & profile/orlop/gun deck/main deck/upper deck/quarter deck & forecastle. 1784: after part of gun & upper decks.

### Duke Class 1771. Williams design.
Dimensions & tons: 177'6", 145'3" x 50' x 21'2", 1931 45/94.
Men: 750. Guns: GD 28 x 32, MD 30 x 18, UD 30 x 12, Fc 2 x 6.
**Duke** Plymouth Dyd.
  Ord 18.6.1771; K 10.1772; L 18.10.1777; 1798 hulked.
**Saint George** Portsmouth Dyd.
  Ord 16.7.1774; K 08.1774; L 4.10.1785; 1811 wrecked.
**Glory** Plymouth Dyd.
  Ord 16.7.1774; K 7.4.1775; L 5.7.1788; 1825 BU.
**Atlas** Chatham Dyd.
  Ord 5.8.1771; K 1.10.1777; L 13.2.1782; 1804 cut down to a Third Rate two-decker 74; 1814 hulked.

FATES: *Duke* became a lazaretto at Portsmouth; 1799 to Sheerness; 1843 BU.
*Saint George* was wrecked on the coast of Jutland on 24.12.1811; the wreck has

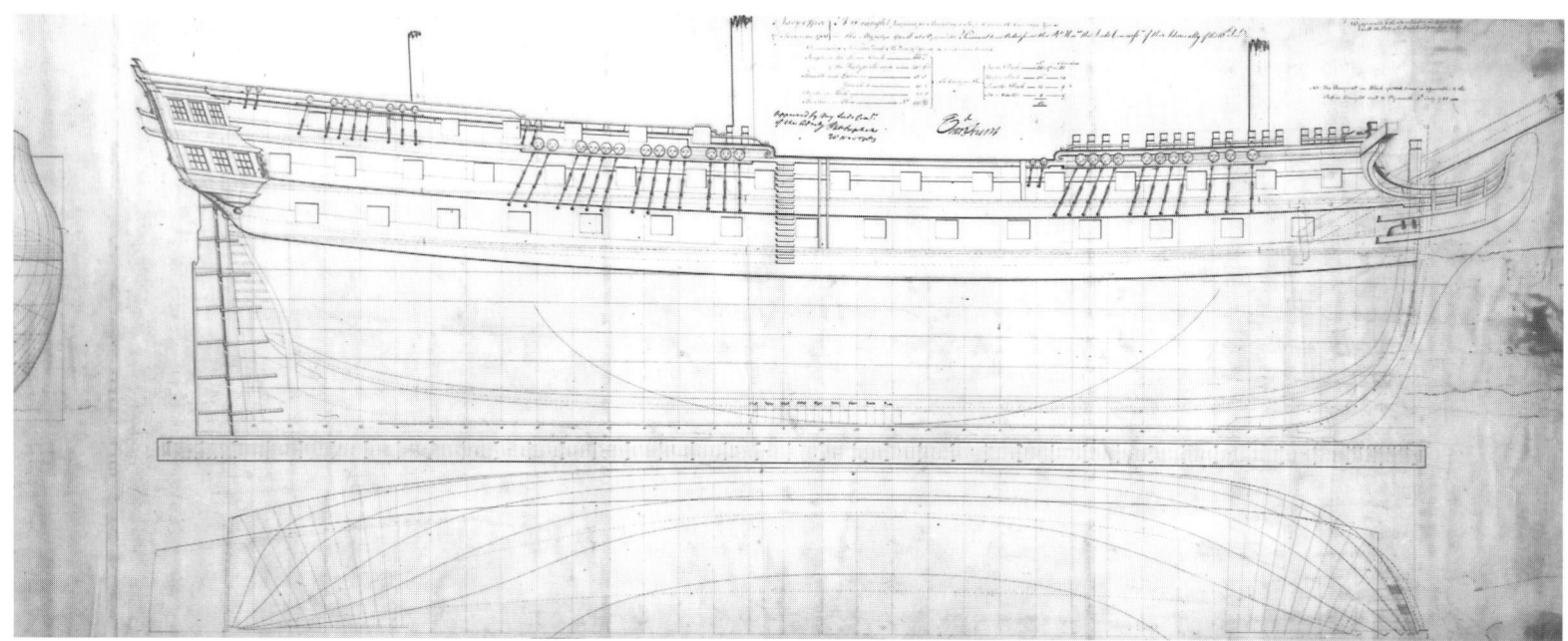

The design lines plan for the first two-decker 80 built for the Royal Navy for nearly a century: the *Caesar* of 1793

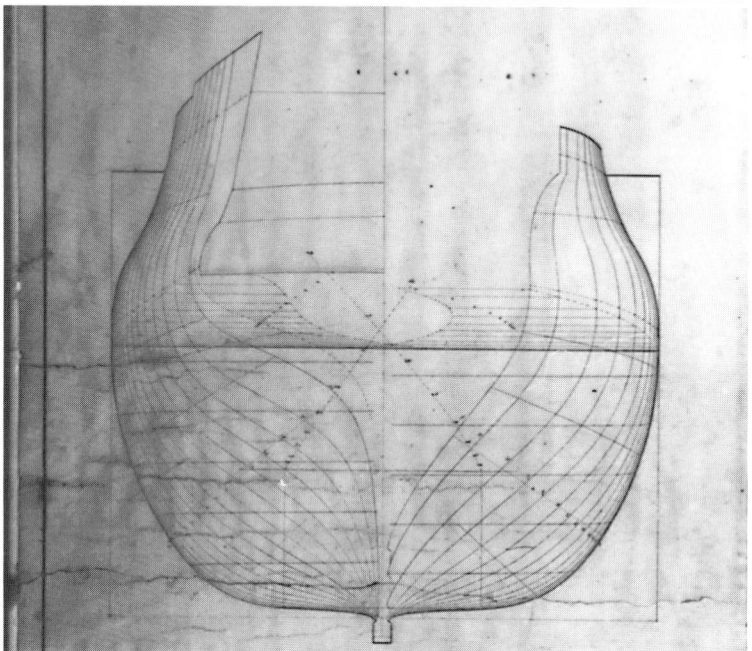

been located and is being excavated. *Atlas* was hulked as a temporary prison ship at Portsmouth; 1815 powder magazine; 1821 BU.

PLANS: Lines/profile/orlop/gun deck/upper deck/quarter deck & forecastle. *Duke:* as built: Lines & profile & decoration. 1794 as lazaretto: Profile & main & upper decks/midships section. *Saint George:* as built: lines & profile. *Glory:* Roundhouse. As built: Lines & profile & decoration. *Atlas:* as built: Lines & profile & decoration.

### *Boyne* Class 1783. Hunt design.
Dimensions & tons: 182', 149'8" x 50'3" x 21'9". 2010 18/94.
Men: 750. Guns: GD 28 x 32, MD 30 x 18 (originally to be 24s), UD 30 x 12, QD 8 x 6, Fc 2 x 6.

**Boyne** Woolwich Dyd.
Ord 21.1.1783; K 4.11.1783; L 27.6.1790; 1795 burnt.

**Prince of Wales** Portsmouth Dyd.
Ord 13. & 29.11.1783; K 05.1784; L 28.6.1794; 1822 BU.

FATES: *Boyne* caught fire by accident and blew up at Spithead on 1.5.1795.

Plans: Lines/profile/orlop/gun deck/main deck/upper deck/quarter deck & forecastle.

## Third Rate 80 Guns (three-deckers)

### 1745 Establishment
Design for an 80, but no ships were built to it inunmodified form.
PLANS: Spar plan.

### *Cambridge* Class 1750. Allin design, modified 1745 Establishment.
Dimensions & tons: 166', 139'3" x 47' x 20'. 1636 17/94.
Men: 650. Guns: GD 26 x 32, MD 26 x 18, UD 24 x 9, QD 4 x 6.

**Cambridge** Deptford Dyd.
Ord 12.7.1750; K 29.8.1750; L 21.10.1755; (1777-1782 hulked); 1790 hulked.

FATE: 1777-1782 prison ship at Plymouth; 1790 receiving ship at Plymouth; 1808 BU.

PLANS: Lines. As built: Lines. 1779 as prison ship: Decks & midships section. 1781 as cut down: Sheer.

### *Princess Amelia* Class 1751. Modified 1745 Establishment design.
Dimensions & tons: 165', 134'10¾" x 47 x 20'. 1585 5/94.
Men: 650. Guns: GD 26 x 32, MD 26 x 18, UD 24 x 9, QD 4 x 6.

**Princess Amelia** (ex *Norfolk*) Woolwich Dyd.
Ord 28.3.1751/26.4.1751; K 15.8.1751; L 7.3.1757; 1788 hulked.

FATE: Lazaretto at Sheerness; 1818 sold.

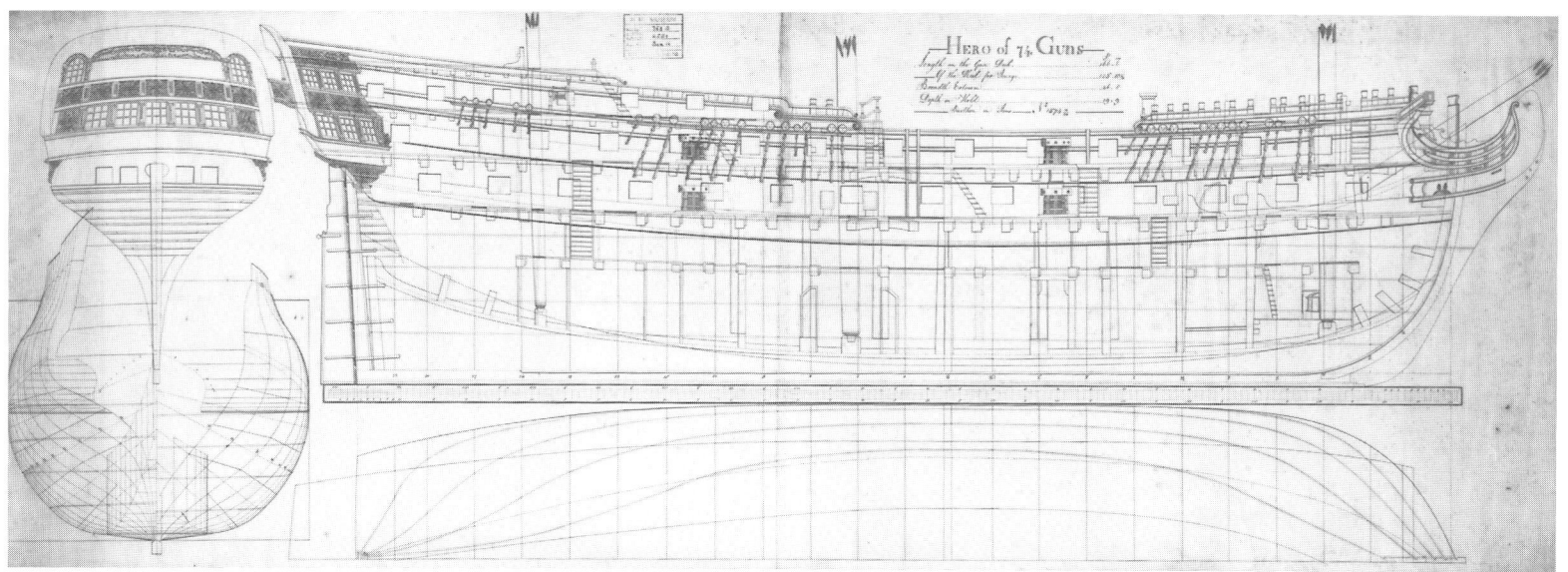

Admiralty plan 758B is an early 'as built' lines and profile of the 74 gun *Hero* of 1759

PLANS: Lines.

## Third Rate 80 Guns (two-deckers)

***Caesar* Class 1783.** Hunt design, originally ordered as a 74 of the *Alfred* Class in 1777. The first British built two-decker 80 since the first 80s of the 1690s.

Dimensions & tons: 181', 148'3⅛" x 50'3" x 22'11". 1991 68/94.
Men: 650. Guns: GD 30 x 32, UD 32 x 24, QD 14 x 9, Fc 4 x 9.
***Caesar/Cesar*** Plymouth Dyd.
Ord 13.11.1783; K 24.1.1786; L 16.11.1793; 1814 hulked.

FATE: Army depot at Plymouth; 1821 BU.

PLANS: Lines/profile/orlop/gun deck/upper deck/quarter deck & forecastle/specification.

## Third Rates 74 Guns (two-deckers)

***Culloden* Class 1744.** Design by Acworth modified from the 1741 Establishment 80 gun ship design. Therefore not the true prototype of the British 74 gun ship; though a step in that direction.

Dimensions & tons: 161', 132'3" x 46' x 19'4". 1487.
Men: 580/650 Guns: GD 28 x 32, UD 28 x 18, QD 14 x 9, Fc 4 x 9.
***Culloden*** Deptford Dyd.
Ord 31.12.1744; K 23.5.1745; L 9.9.1747; 1770 sold.

PLANS: As built: Lines & profile & decoration.

***Dublin* Class 1755.** Slade design. Originally classed as 70s but reclassed as 74s before completion; and therefore the first true British 74s.

Dimensions & tons: 165'6", 134'6" x 46'6" x 19'9". 1546 87/94.
Men: 600/550. Guns: GD 28 x 32, UD 28 x 18, QD 14 x 9, Fc 4 x 9.
***Dublin*** Deptford Dyd.
Ord 26.8.1755; K 18.11.1755; L 6.5.1757; 1784 BU.
***Norfolk*** (ex *Princess Amelia*) Deptford Dyd.
Ord 26.8.1755; K 18.11.1755; L 28.12.1757; 1774 BU.
***Lenox*** Chatham Dyd.
Ord 28.10.1755; K 8.4.1756; L 25.2.1758; (1778-1781 hulked); 1789 BU.
***Mars*** Woolwich Dyd.
Ord 28.10.1755; K 1.5.1756; L 15.3.1759; 1778 hulked.
***Shrewsbury*** Wells, Deptford.
Ord 28.10.1755; K 14.1.1756; L 23.2.1758; 1783 scuttled.
***Warspight*** West, Deptford.
Ord 14.11.1755; K 11.1755; L 8.4.1758; 1778 hulked.
***Resolution*** Bird, Northam.
Ord 24.11.1755; K 12.1755; L 14.12.1758; 1759 wrecked.

FATES: *Lenox* was a receiving ship at Portsmouth from 1779-1781. *Mars* was a prison and receiving ship at Portsmouth; 1784 sold. *Shrewsbury* was condemned and scuttled at Jamaica ___ 1783. *Warspight* was a receiving ship at Portsmouth; 1800 renamed *Arundel* ; 1802 sold. *Resolution* was wrecked during the Battle of Quiberon Bay on the Four Bank 20/21.11.1759.

PLANS: Lines/profile/orlop/gun deck/upper deck/quarter deck & forecastle. *Resolution*: as built: Lines & profile.

***Fame* Class 1756.** Bately design.

Dimensions & tons: 165'6", 135'8" x 46'7" x 19'10½". 1565 88/94.
Men: 550. Guns: GD 28 x 32, UD 28 x 18, QD 14 x 9, Fc 4 x 9.
***Fame*** Bird, Deptford.
Ord 13.4.1756; K 28.5.1756; L 1.1.1759; (1764-1771 troopship); 1799 hulked.

FATE: Prison ship at Portsmouth renamed *Guildford*; 1814 sold.

PLANS: No plans.

***Hero* Class 1756.** Slade design. Originally *Hercules* and *Thunderer* were to be built to this design, but they were altered whilst building (see below).

Dimensions & tons: 166'6", 136' x 46'6" x 19'9". 1564 17/94.
Men 550. Guns: GD 28 x 32, UD 28 x 18, QD 14 x 9, Fc 4 x 9.
***Hero*** Plymouth Dyd.
Ord 25.5.1756?/7.7.1756; K 08.1756; L 28.3.1759; 1787 hulked.

FATE: Receiving ship at Plymouth; 1788 to Chatham as receiving ship; 1793 prison ship; 1800 renamed *Rochester*; 1810 BU.

PLANS: Lines/profile/orlop/gun deck/upper deck/quarter deck & forecastle. As built: Lines & profile. 1793 as prison ship: Decks.

***Hercules* Class 1756.** Slade design. Originally to be built to *Hero*'s design, but during construction Slade altered their plans.

Dimensions & tons: 166'6" x 136' x 47' x 19'9". 1598.
Men: 550. Guns: GD 28 x 32, UD 28 x 18, QD 14 x 9, Fc 4 x 9.
***Hercules*** Deptford Dyd.
Ord 15.7.1756; K 30.3.1757; L 25.2.1759; 1784 sold.
***Thunderer*** Woolwich Dyd.
Ord 15.7.1756; K 17.9.1756; L 19.3.1760; 1780 lost.

FATES: *Thunderer* presumed lost in hurricane in the West Indies by 31.10.1780.

PLANS: Lines/profile/orlop/gun deck/upper deck/quarter deck & forecastle/steering wheel.

***Bellona* Class 1757.** Slade design, and a very successful one. Later to be adapted for the *Arrogant* class, which in turn was altered to become the *Monarch* class.

Dimensions & tons: 168', 138' x 46'9" x 19'9". 1603 89/94.
Men: 550. Guns: GD 28 x 32, UD 28 x 18, QD 14 x 9, Fc 4 x 9.
***Bellona*** Chatham Dyd.
Ord 28.12.1757/31.1.1758; K 05.1758; L 19.2.1760; 1814 BU.

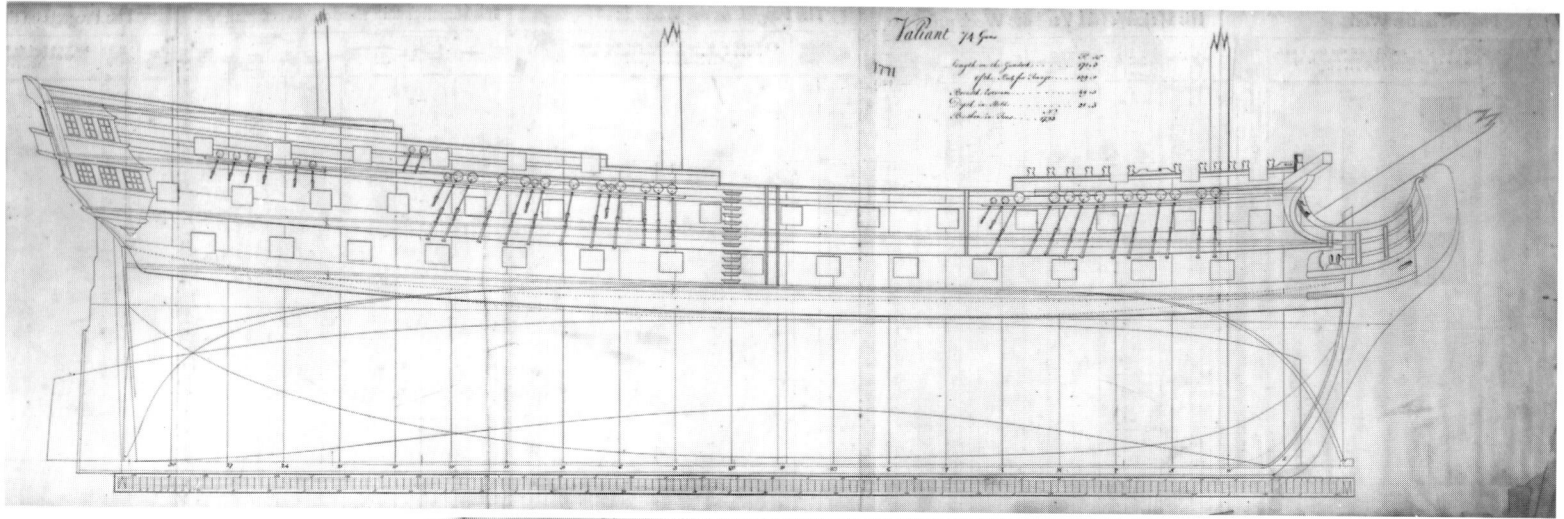

The first of the larger type of 74s built for the Royal Navy, the *Valiant* of 1759 shown in a partial lines plan (number 1001)

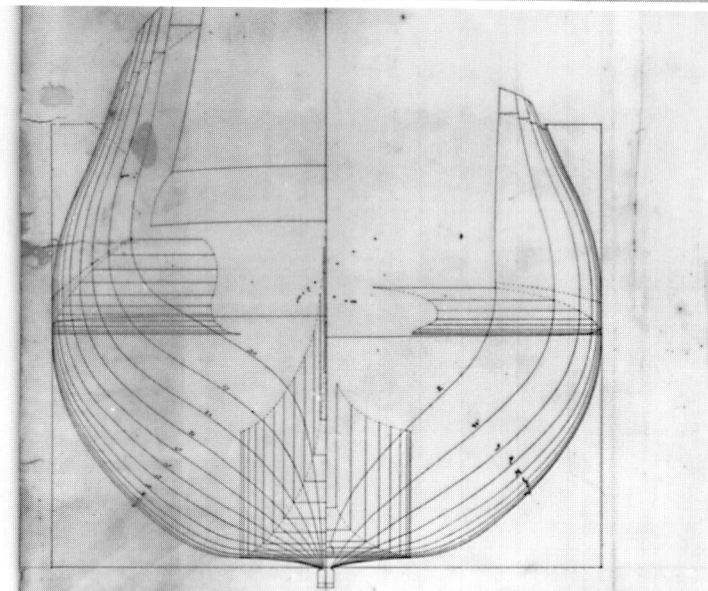

**Dragon** Deptford Dyd.
Ord 28.12.1757/31.1.1758; K 28.3.1758; L 4.3.1760; 1769 troopship; 1780 hulked.

**Superb** Deptford Dyd.
Ord 28.12.1757/31.1.1758; K 12.4.1758; L 27.10.1760; (1768-1770 troopship); 1783 wrecked.

FATES: *Dragon* became a receiving ship at Portsmouth; 1784 sold. *Superb* was run ashore at Telicherry Roads, India, 7.11.1783.

PLANS: Lines/profile/orlop/gun deck/upper deck/quarter deck & forecastle.

## Valiant Class 1757.
Originally ordered as two extra members of the *Dublin* class, but the order was then altered to one for two larger vessels copied from the design of the French *Invincible* (captured 1747). There seem to have been minor differences in detail design between the two ships, so they should perhaps be considered as half-sisters. In the 1790s the design was revived and modified for the *Ajax* and *Kent*.

Dimensions & tons: 171'3", 139'x 49'3" x 21'3". 1799 41/94. (*Valiant*), 1793 22/94 (*Triumph*).

Men: 650. Guns: GD 28 x 32, UD 30 x 24, QD 10 x 9, Fc 2 x 9.

**Valiant** Chatham Dyd.
Ord 11.1.1757/21.5.1757; K 1.2.1758; L 10.8.1759; 1799 hulked.

**Triumph** Woolwich Dyd.
Ord 11.1.1757/21.5.1757; K 2.1.1758; L 3.3.1764; 1813 hulked.

FATES: *Valiant* was a lazaretto at Sheerness; 1826 BU. *Triumph* was a lazaretto at Pembroke; 1850 BU.

PLANS: Lines/profile/orlop/gun Deck/upper Deck. *Valiant*: as built: Decks. *Triumph*: 1814 as lazaretto: Profile & midships section.

## Arrogant Class 1758.
Slade design, slightly modified from the *Bellona* Class. *Edgar* was originally to have been built to the *Alfred* Class design.

Dimensions & tons: 168', 138'x 46'9" x 19'9". 1604 27/94.

Men: 550 (as frigates: 450). Guns: GD 28 x 32, UD 28 x 18, QD 14 x 9, Fc 4 x 9. (*Elephant* & *Saturn* as frigates: UD 28 x 18, Spar dk. 28 x 42★ + 2 x 12, *Saturn* added a single 12 on the Fc).

**Arrogant** Barnard, Harwich.
Ord 13.12.1758/26.1. & 6.2.1759; K 03.1760; L 22.1.1761; by 1806 hulked.

**Cornwall** Wells, Deptford (or Rotherhithe).
Ord 13.12.1758/26.1.1759; K 19.2.1759; L 19.5.1761; 1780 condemned and sunk.

**Defence** Plymouth Dyd.
Ord 15.12.1758/26.1.1759; K 14.5.1759; L 31.3.1763; 1811 wrecked.

**Kent** Deptford Dyd.
Ord 13.12.1758/26.1.1759; K 24.4.1759; L 26.3.1762; 1784 sold.

**Edgar** Woolwich Dyd.
Ord 16.7. & 25.8.1774; K 26.8.1776; L 30.6.1779; 1813 hulked.

**Goliath** Deptford Dyd.
Ord 21.2.1778; K 10.4.1779; L 19.10.1781; 1813 cut down to frigate (Fourth Rate) 58; 1815 BU.

**Vanguard** Deptford Dyd.
Ord 9.12.1779; K 16.10.1782; L 6.3.1787; 1812 hulked.

**Excellent** Graham, Harwich.
Ord 9.8.1781; K 03.1782; L 27.11.1787; 1825 cut down to a frigate (Fourth Rate) 58, but conversion never completed?; 1830 hulked.

**Saturn** Raymond, Northam.
Ord 27.12.1781; K 08.1782; L 22.11.1786; 1813 cut down to a frigate (Fourth Rate) 58; 1825 hulked.

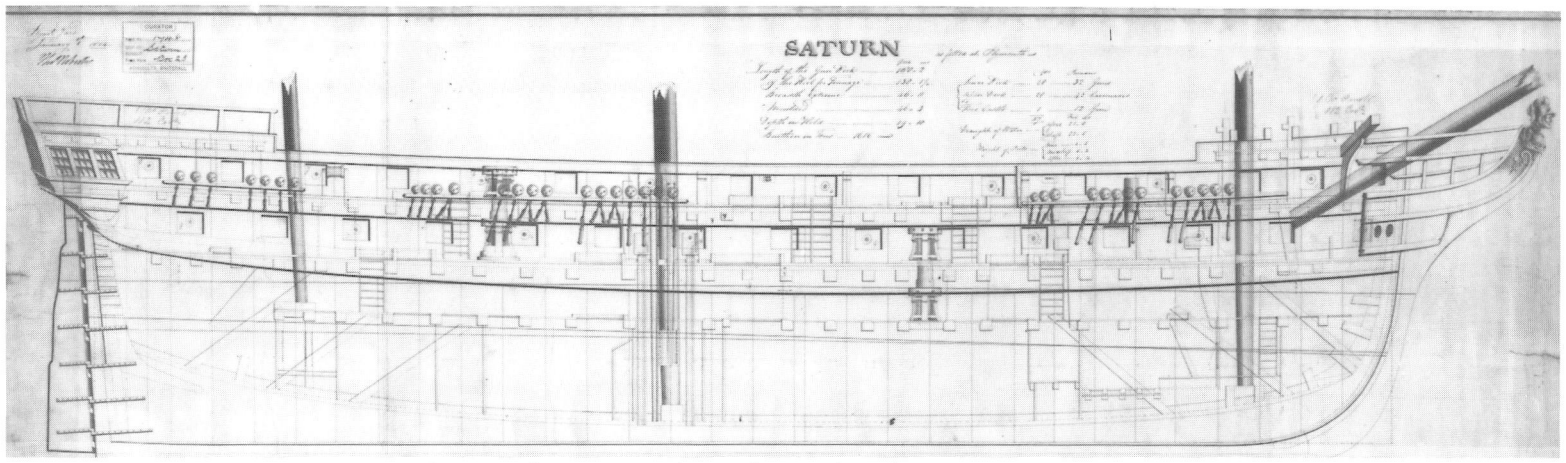

*This profile plan (number 1748) shows the 74 gun ship Saturn of 1781 (Arrogant Class) cut down to a heavy frigate in 1813*

**Zealous** Barnard, Deptford.
Ord 19.6.1782; K 12.1782; L 25.6.1785; 1816 BU.

**Elephant** Parsons, Bursledon.
Ord 27.12.1781; K 02.1783; L 24.8.1786; 1818 cut down to a frigate (Fourth Rate) 58; 1830 BU.

**Bellerophon** Greaves, Frindsbury.
Ord 11.1.1782; K 05.1782; L 17.10.1786; 1816 hulked.

**Audacious** Randall, Rotherhithe.
Ord ____ ; K 08.1783; L 23.7.1785; 1815 BU.

**Illustrious** Adams, Bucklers Hard.
Ord 31.12.1781; K 09.1784; L 7.7.1789; 1795 wrecked.

FATES: *Arrogant* became a receiving ship at Bombay; 1810 sold for BU. *Cornwall* scuttled at Saint Lucia, being unserviceable, 30.6.1780. *Defence* wrecked on the coast of Jutland in a storm 24.12.1811; it is reported that this wreck has been found, but that not much seems to survive. *Edgar* became a receiving ship at Chatham; 1814 renamed *Retribution*; 1835 BU. *Vanguard* became a receiving ship at Plymouth; 1814 powder ship; 1821 BU. *Excellent* became the gunnery training hulk at Portsmouth; 1835 BU. *Saturn* became a lazaretto at Pembroke; 1845 harbour flag and receiving ship; 1850 guard ship; 1868 BU. *Bellerophon* became a convict hulk at Sheerness; 1824 renamed *Captivity*; 1836 sold. *Illustrious* was wrecked near Avenza on 17.3.1795.

PLANS: Lines/profile/orlop/gun deck/upper deck/quarter deck & forecastle/internal planking/specification. *Cornwall*: as built: Decks. *Edgar*: as built: Lines & profile & decoration. Goliath: as built: Lines & profile. *Saturn*: as frigate 1814: Profile/orlop/lower deck/upper deck. *Elephant*: as built: Lines & profile. As frigate 1816: Profile/orlop/lower deck/upper deck/quarter deck & forecastle.

### Canada Class 1759. Bately design.

Dimensions & tons: 170', 140'5" x 46'7" x 20'6". 1632 35/94.
Men: 550. Guns: GD 28 x 32, UD 28 x 18, QD 14 x 9, Fc 4 x 9 (later an extra pair of 9s were added to *Canada*'s QD and she was classed as a 76).

**Canada** Woolwich Dyd.
Ord 1.12.1759/24.4.1760; K 1.7.1760; L 17.9.1765; 1780 reclassed as a 76; 1810 hulked.

**Majestic** Adams & Barnard, Deptford.
Ord 23.7.1781; K 07.1782; L 11.2.1785; 1813 cut down to a frigate (Fourth Rate) 58; 1816 BU.

**Orion** Adams & Barnard, Deptford.
Ord 2.10.1782; K 02.1783; L 1.6.1787; 1814 BU.

**Captain** Batson, Limehouse.
Ord 14.11.1782; K 05.1784; L 26.11.1787; 1809 hulked.

FATES: *Canada* became a receiving ship at Chatham; 1814 powder magazine; 1826 convict ship; 1834 BU. *Captain* became a receiving ship at Plymouth; 1813 burnt by accident and BU.

PLANS: Lines/lines & profile/orlop/gun deck/upper deck/quarter deck & forecastle. *Majestic*: as built: Lines & profile & decoration.

### Albion Class 1759. Design adapted by Slade from the lines of the

*Torbay*, the old 90 gun *Neptune* of the 1719 Establishment, which had been cut down to a 74.

Dimensions & tons: 168', 139'1" x 47' x 18'10". 1634 21/94 (contract built ships) / 168', 139'1¼" x 47'3" x 18'10". 1651 75/94 (dockyard built ships).
Men: 550. Guns: GD 28 x 32, UD 28 x 18, QD 14 x 9, Fc 4 x 9.

**Albion** Deptford Dyd.
Ord 1.12.1759/24.4.1760; K 26.6.1760; L 16.5.1763; 1794 floating battery, 60 guns; 1797 wrecked.

**Grafton** Deptford Dyd.
Ord 10.9 & 22.10.1767; K 1.7.1768; L 26.9.1771; 1792 hulked.

**Alcide** Deptford Dyd.
Ord 31.8.1774; K 14.6.1776; L 30.7.1779; 1794 hulked.

**Fortitude** Randall, Rotherhithe.
Ord 22.2.1778; K 4.3.1778; L 22.3.1780; 1795 hulked.

**Irresistible** Barnard, Harwich.
Ord 5.2.1777/3.7.1778; K 10.1778; L 6.12.1782; 1803? hulked.

FATES: *Albion* was wrecked in the Swin, 26.4.1797. *Grafton* became a receiving ship at Portsmouth; 1816 BU. *Alcide* became a receiving ship at Portsmouth; 1817 BU. *Fortitude* was a powder ship at Plymouth; 1820 BU. *Irresistible* became a powder hulk (?); 1806 BU.

PLANS: Lines/profile/orlop/gun deck/upper deck/quarter deck & forecastle/framing. *Alcide*: as built: Lines & profile/decks. *Fortitude*: as powder hulk 1801: Decks.

### Monarch/Ramillies Class 1760. Slade design 'very nearly the same as *Arrogant*', therefore a further modified version of the *Bellona* Class.

Dimensions & tons: 168'6", 138'3⅞"x 46'9" x 19'9". 1607 50/94.
Men: 550/600. Guns: GD 28 x 32, UD 28 x 18, QD 14 x 9, Fc 4 x 9.

**Monarch** Deptford Dyd.
Ord 22.11.1760; K 2.6.1761; L 21.7.1765; 1813 BU.

**Ramillies** Chatham Dyd.
Ord 1.12.1759/25.4.1760; K 25.8.1760; L 15.4.1763; 1782 scuttled.

**Terrible** Barnard, Harwich.
Ord 1 & 13.1.1761; K 02.1761; L 4.9.1762; 1781 scuttled.

**Russell** West, Deptford.
Ord 13.1.1761; K 06.1761; L 12.11.1764; 1811 sold.

**Invincible** Wells, Deptford.
Ord 12.10.1761; K 12.1761; L 9.3.1765; 1801 wrecked.

**Magnificent** Deptford Dyd.
Ord 16.12.1760/7.1.1761; K 15.4.1762; L 20.9.1766; 1804 wrecked.

**Prince of Wales** (ex *Hibernia*) Bird, Milford.
Ord 16.12.1761/7.1.1762; K 03.1762; L 4.6.1765; 1783 BU.

**Marlborough** Deptford Dyd.
Ord 4.12.1762; K 3.6.1763; L 26.8.1767; 1800 wrecked.

**Robust** Barnard, Harwich.
Ord 16.12.1761/7.1.1762; K 02.1762; L 25.10.1764; 1802? hulked.

FATES: *Ramillies* burnt and scuttled after storm damage off the Banks of Newfoundland 21.9.1782. *Terrible* burnt and scuttled after battle damage suffered at the Battle of the Capes of the Chesapeake 11.9.1781. *Invincible* wrecked on

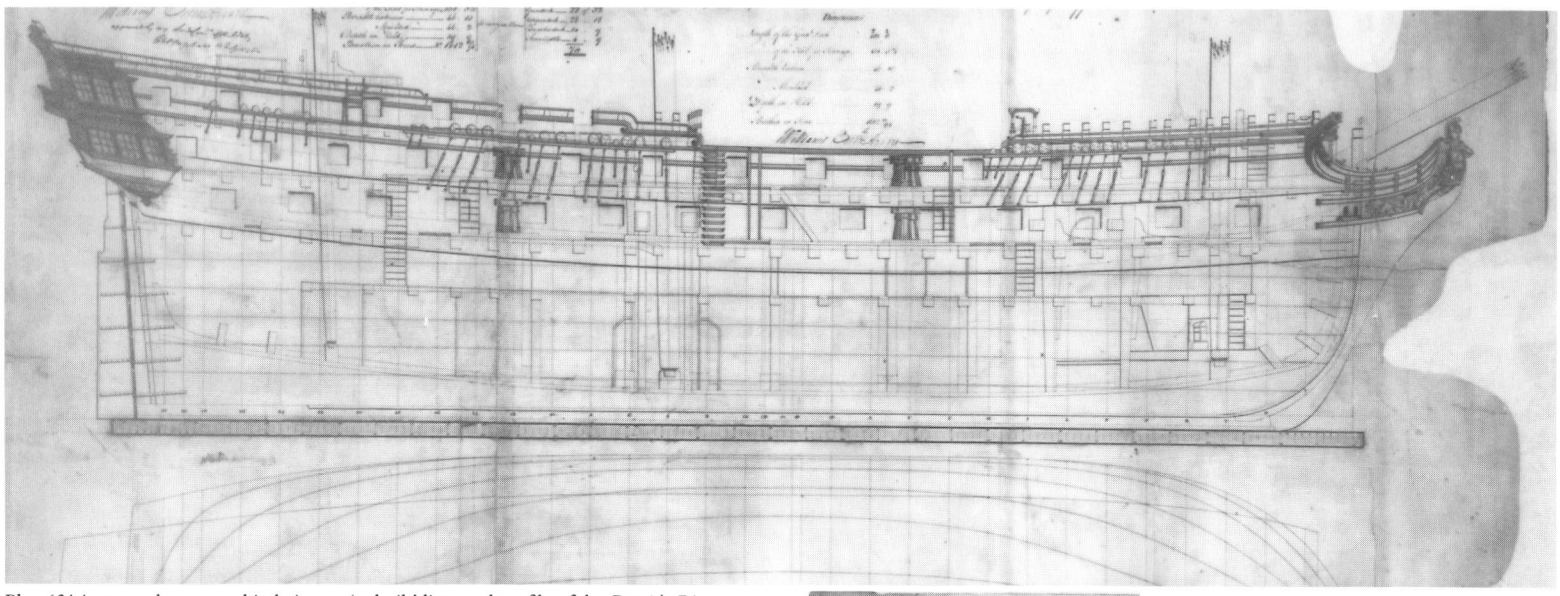

Plan 631 is somewhat unusual in being an 'as built' lines and profile of the *Berwick*, 74 (1775 - of the *Elizabeth* Class) which has subsequently been adapted as the building plan for later ships of the class (*Bombay Castle, Powerful, Defiance* - all of whose builders were sent copies of this plan)

Haseborough Sands, Yarmouth, 16.3.1801. *Magnificent* wrecked on the Pierres Noires off Brest 25.3.1804; this wreck has apparently been found and plundered. *Marlborough* wrecked on a sunken rock near Belleisle 4.11.1800. *Robust* became a receiving ship at Portsmouth; 1817 BU.

PLANS: Lines/profile/orlop/gun deck/upper deck/quarter deck & forecastle/specification. *Marlborough*: 1768: lead sheathing (experimental).

### Suffolk Class 1761. 
Bately design. It is possible that these two ships were actually built to two separate designs, though more likely that they were the same.
  Dimensions & tons: 168', 138'9" x 46'9" x 20'3". 1613 1/94.
  Men: 550/600. Guns: GD 28 x 32, UD 28 x 18, QD 14 x 9, Fc 4 x 9.
**Suffolk** Randall, Rotherhithe.
  Ord 1.1. & 13.1.1761; K 4.3.1761; L 22.2.1765; 1803 BU.
**Ajax** Portsmouth Dyd.
  Ord 4.12.1762; K 6.9.1763; L 23.12.1767; 1785 sold.

PLANS: Lines/decks/midships section.

### Royal Oak Class 1765. 
Williams design.
  Dimensions & tons: 168'6", 138'2" x 46'9" x 20'. 1606 21/94.
  Men 550/600. Guns: GD 28 x 32, UD 28 x 18, QD 14 x 9, Fc 4 x 9.
**Royal Oak** Plymouth Dyd.
  Ord 16.11.1765; K 05.1766; L 13.11.1769; 1796 hulked.
**Bedford** Woolwich Dyd.
  Ord 12.10.1768; K 10.1769; L 27.10.1775; 1815 hulked.
**Conqueror** Plymouth Dyd.
  Ord 12.10.1768; K 10.1769; L 18.10.1773; 1794 BU.
**Vengeance** Randall, Rotherhithe.
  Ord 14.1.1771; K 04.1771; L 25.6.1774; 1808 hulked.
**Sultan** Barnard, Harwich.
  Ord 14.1. & 23.3.1771; K 03.1771; L 23.12.1775; 1797 hulked.
**Hector** Adams & Barnard, Deptford.
  Ord 14.1. & 23.3.1771; K 04.1771; L 27.5.1774; 1808? hulked.

FATES: *Royal Oak* became a prison ship at Portsmouth; 1805 renamed *Assistance*; 1815 BU. *Bedford* was hulked at Portsmouth; 1817 BU. *Vengeance* became a prison ship at Portsmouth; 1816 BU. *Sultan* became a prison ship at Portsmouth; 1805 renamed *Suffolk*; 1816 BU. *Hector* was hulked, probably as a prison ship, at Plymouth; 1816 BU.

PLANS: Lines/profile/orlop/gun deck/upper deck/quarter deck & forecastle/framing. *Bedford*: as built: Lines & profile & decoration. *Conqueror*: as built: Lines & profile & decoration. *Hector*: as built: Lines & profile & decoration.

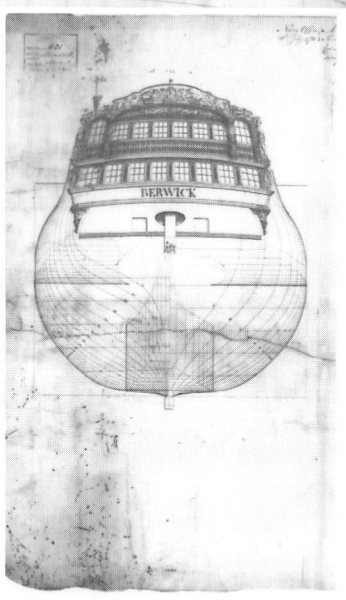

### Egmont Class 1765. 
Slade design.
  Dimensions & tons: 168'6", 140'0¾" x 46'10" x 19'9". 1607 53/94.
  Men: 550/600. Guns: GD 28 x 32, UD 28 x 18, QD 14 x 9, Fc 4 x 9.
**Egmont** Deptford Dyd.
  Ord 6.6.1765; K 10.1766; L 29.8.1768; 1799 BU.

PLANS: Lines/profile/orlop/gun deck/upper deck/quarter deck & forecastle.

### Elizabeth Class 1766. 
Slade design.
  Dimensions & tons: 168'6", 138'3⅛" x 46'10" x 19'9". 1612 88/94.
  Men: 550/600. Guns: GD 28 x 32, UD 28 x 18, QD 14 x 9, Fc 4 x 9.
**Elizabeth** Portsmouth Dyd.
  Ord 6.11.1765; K 6.5.1766; L 17.10.1769; 1797 BU.
**Resolution** Deptford Dyd.
  Ord 16.9.1766; K 07.1767; L 12.4.1770; 1813 BU.
**Cumberland** Deptford Dyd.
  Ord 8.6.1768; K 01.1769; L 9.3.1774; 1804? BU.
**Berwick** Portsmouth Dyd.
  Ord 12.10.1768; K 05.1769; L 18.4.1775; 1795 taken.
**Bombay Castle** (ex *Bombay*) Perry Blackwall.
  Ord 14.7.1779; K 06.1780; L 14.6.1782; 1796 wrecked.

NOTE: This ship was paid for by the East India Company; but she was not, as is sometimes stated, a conversion from an East Indiaman.

**Powerful** Perry, Blackwall.
  Ord 8.7.1780; K 04.1781; L 3.4.1783; 1812 BU.
**Defiance** Randall, Rotherhithe.
  Ord 11.7.1780; K 04.1782; L 10.12.1783; 1813 hulked.

THE SLADE ERA 1745-1785

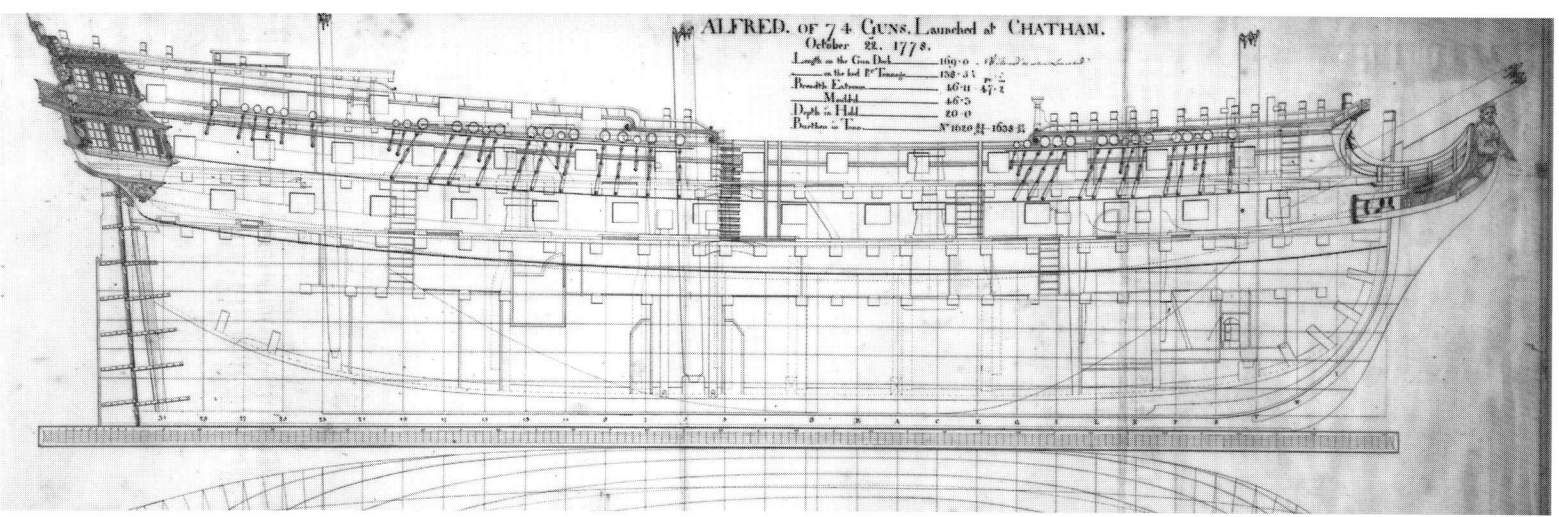

*Alfred* of 74 guns – an 'as built' lines and profile plan (546) showing the name ship of her class, launched in 1778

**Swiftsure** Wells, Deptford (or Rotherhithe?).
  Ord 19.6.1782; K 05.1784; L 4.4.1787; (1801-1805 in French hands); 1805 renamed *Irresistible*; 1816 BU.

FATES: *Berwick* taken by the French fleet in the Mediterranean 7.3.1795; 21.10.1805 retaken at Trafalgar but lost on the 23rd off San Lucar in the subsequent storm. *Bombay Castle* wrecked in the Tagus whilst attempting to enter the river 21.12.1796. *Defiance* became a temporary prison hulk at Chatham; 1817 BU. *Swiftsure* was taken in the Mediterranean by Gantheaume's squadron 24.6.1801; 21.10.1805 retaken at Trafalgar.

PLANS: Lines/profile/orlop/gun deck/upper deck/quarter deck & forecastle/specification. *Cumberland* as built: Lines & profile & decoration. *Berwick* as built: Lines & profile & decoration. *Bombay Castle*: as built?: Lines & profile. *Powerful:* as built: Lines & profile. *Defiance*: decks/capstan.

### Culloden/Thunderer Class 1769. Slade design.

Dimensions & tons: 170', 140'1⅞" x 47' x 19'11". 1658 52/94 (dockyard built ships) / 170', 139'8" x 46'6" x 19'11". 1652 65/94 (contract-built ships).
Men: 550/600. Guns: GD 28 x 32, UD 28 x 18, QD 14 x 9, Fc 4 x 9.
**Culloden** Deptford Dyd.
  Ord 30.11.1769; K 07.1770; L 18.5.1775; 1781 wrecked.
**Thunderer** Wells, Deptford.
  Ord 23.8.1781; K 03.1783; L 13.11.1783; 1814 BU.
**Venerable** Perry, Blackwall.
  Ord 9.8.1781; K 04.1782; L 19.4.1784; 1804 wrecked.
**Victorious** Perry, Blackwall.
  Ord 28.12.1781; K 11.1782; L 27.4.1785; 1803 BU.
**Ramillies** Randall, Rotherhithe.
  Ord 19.6.1782; K 12.1782; L 12.7.1785; 1831 hulked.
**Terrible** Wells, Deptford.
  Ord 13.12.1781; K 01.1783; L 28.3.1785; 1823 hulked.
**Hannibal** Perry, Blackwall.
  Ord 19.6.1782; K 04.1783; L 15.4.1786; 1801 taken.
**Theseus** Perry, Blackwall.
  Ord 11.7.1780; K 09.1783; L 25.9.1786; 1814 BU.

FATES: *Culloden* wrecked on the East end of Long Island (New York) 23.1.1781 this wreck has been found and is being excavated. *Venerable* wrecked in Torbay 25.11.1804; this wreck has apparently been located by divers. *Ramillies* became a lazaretto at Sheerness; 1850 BU. *Terrible* became a receiving ship at Sheerness; 1819 coal depot; 1836 BU. *Hannibal* was taken by the French at the Battle of Algeciras; in French service as *L'Hannibal* until 1823.

PLANS: Lines/profile/orlop/gun deck/upper deck/quarter deck & forecastle/steering wheel/specification. *Culloden*: as built: Lines & profile & decoration. *Ramillies*: as built: Lines & profile.

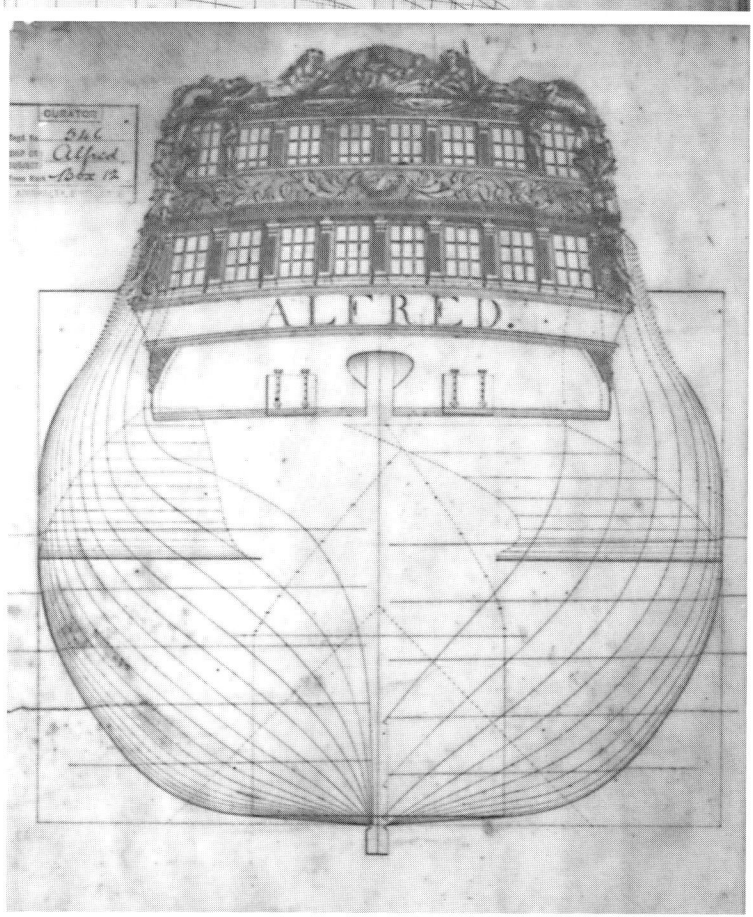

### Alfred Class 1772. Williams design. *Edgar* was ordered as one of this class, but was completed to the design of *Arrogant* instead.

Dimensions & tons: 169', 138'5¼'" x 46'11" x 20'. 1620 82/94.
Men: 550/600. Guns: GD 28 x 32, UD 28 x 18, QD 14 x 9, Fc 4 x 9.
**Alfred** Chatham Dyd.
  Ord 13.8.1772; K 11.1772; L 22.10.1778; 1814 BU.
**Alexander** Deptford Dyd.
  Ord 21.7.1773; K 6.4.1774; L 8.10.1778; (1794-1795 in French hands); 1805 hulked.
**Warrior** Portsmouth Dyd.
  Ord 13.7.1773; K 11.1773; L 18.10.1781; 1819 hulked.
**Montague** Chatham Dyd.
  Ord 16.7.1774; K 30.1.1775; L 28.8.1779; 1818 BU.

### Cancelled 1783. **Bulwark** Portsmouth Dyd.
  Ord 5.2.1777/11.6.1778; keel never laid.
**Caesar** Plymouth Dyd.
  Ord 5.2.1777/11.6.1778; keel never laid; 80 gun ship built instead.

FATES: *Alexander* taken by a French squadron off Sicily 6.11.1794; 23.6.1795 retaken at Groix. Became a lazaretto at Portsmouth (on the Motherbank); 1819

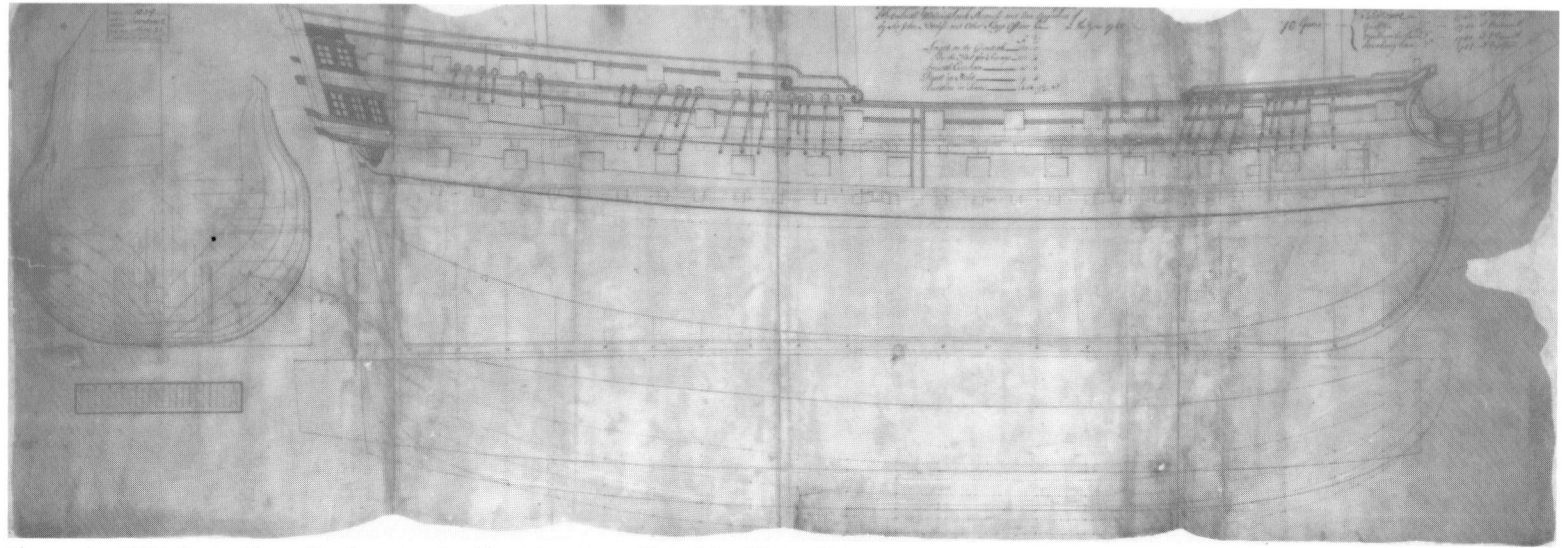

Plan number 1039 is for the 70 gun ships (later classed as 66 and then 64 gun ships) of the 1745 Establishment. Though usually identified as the lines for the *Somerset* the inscription on the plan makes it quite clear that it applies to the entire class

BU. *Warrior* became a receiving ship at Chatham; 1831 temporary quarantine ship; 1832 receiving ship at Woolwich; 1857 BU.

PLANS: Lines/profile/orlop/gun deck/upper deck/quarter deck & forecastle/framing. *Alfred*: as built: Lines & profile & decoration. *Alexander*: as built: Lines & profile & decoration. Lazaretto 1805: outline decks. *Warrior*: as built: Lines & profile & decoration. *Montague*: as built: Lines & profile & decoration.

### *Ganges/Culloden* Class 1778. Hunt design.

Dimensions & tons: 169'6", 138'11¼" x 47'4" x 20'3". 1656 64/94.
(*Tremendous* lengthened whilst building to 170', tonnage increased to 1712 51/94).
Men: 550/600. Guns: GD 28 x 32, UD 28 x 18, QD 14 x 9, Fc 4 x 9.
(Those built in the nineteenth century had instead:
QD 2 x 18 + 12 x 32★, Fc 2 x 18 + 2 x 32★. *Tremendous* as a frigate:
500 men. UD 6 x 8" + 22 x 32, QD & Fc 4 x 8" + 18 x 32.)

**Ganges** Randall, Rotherhithe.
  Ord 6.2.1778; K 04.1780; L 30.3.1782; 1811 hulked.
**Culloden** Randall, Rotherhithe.
  Ord 12.7.1779; K 02.1782; L 16.6.1783; 1813 BU.
**Tremendous** Barnard, Deptford.
  Ord 30.6.1779; K 13.8.1782; L 30.10.1784; 1845 cut down to frigate (Fourth Rate) 50 & renamed *Grampus*; 1866 hulked.
**Invincible** Woolwich Dyd.
  Ord 9. & 16.7.1801; K 1.1.1806; 15.3.1808; by 1857 hulked.
**Minden** Bombay Dyd. Teak built.
  Ord 9.7.1801; K ____ ; L 19.6.1810; 1842 hulked.
**Minotaur** Chatham Dyd.
  Ord 3.12.1811; K 12.1812; L 15.4.1816; 1842 hulked.

FATES: *Ganges* became a prison ship at Plymouth; 1814 sold. *Tremendous*, as *Grampus*, became a powder depot at Portsmouth; 1883 War Department powder depot; 1897 sold. *Invincible* became a receiving ship and coal depot at Plymouth; 1861 BU. *Minden* became a seamens' hospital at Hong Kong; 1861 sold. *Minotaur* became a receiving ship at Sheerness; 1859 guard ship; 1866 renamed *Hermes*; 1869 BU.

PLANS: Lines/profile/orlop/gun deck/upper deck/quarter deck & forecastle/specification. *Ganges*: as built: Lines & profile. As prison ship 1811: decks. *Tremendous*: as built: Lines & profile & decoration. 1810: Lines.

### *Leviathan* Class 1779. Built to the lines of the French *Courageux* (1761). The 1805 *Blake* Class was a lengthened version of this design.

Dimensions & tons: 172'3", 140'5¼" x 47'9" x 20'9". 1703 21/94.
Men: 640/600. Guns: GD 28 x 32, UD 28 x 18, QD 14 x 9, Fc 4 x 9.

**Leviathan** Chatham Dyd.
  Ord 9.12.1779; K 05.1782; L 9.10.1790; 1816 hulked.
**Carnatic** Adams & Barnard, Deptford
  Ord 14.7.1779; K 03.1780; L 21.1.1783;
  1805 hulked

NOTE: Paid for by the East India Company, but not, as has sometimes been stated, a converted East Indiaman.

**Colossus** Cleverley, Gravesend.
  Ord 13.12.1781; K 10.1782; L 4.4.1787; 1798 wrecked.
**Minotaur** Woolwich Dyd.
  Ord 3.12.1782; K 01.1788; L 6.11.1793; 1810 wrecked.
**Aboukir** Brindley Frindsbury.
  Ord 24.11.1802; K 06.1804; L 18.11.1807; 1824 hulked.
**Bombay** Deptford Dyd.
  Ord 23.7.1805; K 10.1805; L 28.3.1808; 1819 renamed *Blake*; 1823 hulked.

FATES: *Leviathan* became a convict ship at Portsmouth; 1846 target; 1848 sold. *Carnatic* became a receiving ship at Plymouth; 1815 renamed *Captain*; 1825 BU. *Colossus* was wrecked on the Scillies 10.12.1798; the wreck has been excavated under the auspices of the British Museum but with the aim of recovering as much as possible of Sir William Hamilton's collection of antique vases rather than to record the ship herself, or her other contents. *Minotaur* was wrecked on the Haak Sand, Texel 22.12.1810. *Aboukir* became a receiving ship at Chatham; 1831 hospital ship; 1838 sold. *Bombay*, as *Blake*, was hulked as a receiving ship at Portsmouth; 1855 BU.

PLANS: Lines/profile/orlop/gun deck/upper deck/quarter deck & forecastle/specification. *Carnatic*: as receiving ship 1806: Profile & decks. *Colossus*: as built: Lines & profile.

## Third Rates, 70 & 64 Guns (two-deckers)

### 1745 Establishment, 68s (later classed as 64s). Also see *Temple* Class.

Dimensions & tons: 160', 131'4" x 45'x 19'4". 1414 56/94.
Men: 520. Guns: GD 26 x 32, UD 28 x 18, QD 12 x 9, Fc 2 x 9.
Note that *Grafton* was originally ordered as a 1741 Establishment ship.

**Grafton** Portsmouth Dyd.
  Ord 28.8.1744; K 11.9.1745; L 29.8.1750; 1767 sold.
**Somerset** Chatham Dyd.
  Ord 8.11.1744, K 5.5.1746; L 18.7.1748; 1778 wrecked.
**Northumberland** Plymouth Dyd.
  Ord 22.6.1744; K 14.8.1744; L 1.12.1750; 1777 storeship, 30, renamed *Leviathan*; 1779 Fourth Rate, 50; 1780 foundered.
**Orford** Woolwich Dyd.
  Ord 31.8.1745; K 24.2.1746; L 15.11.1749; 1777 hulked.
**Swiftsure** Deptford Dyd.
  Ord 31.8.1745; K 26.1.1747; L 25.5.1750; 1773 sold.
**Vanguard** Ewer, Cowes.
  Ord 3.10.1745; K 11.1745; L 16.4.1748; 1774 sold.
**Buckingham** Deptford Dyd.

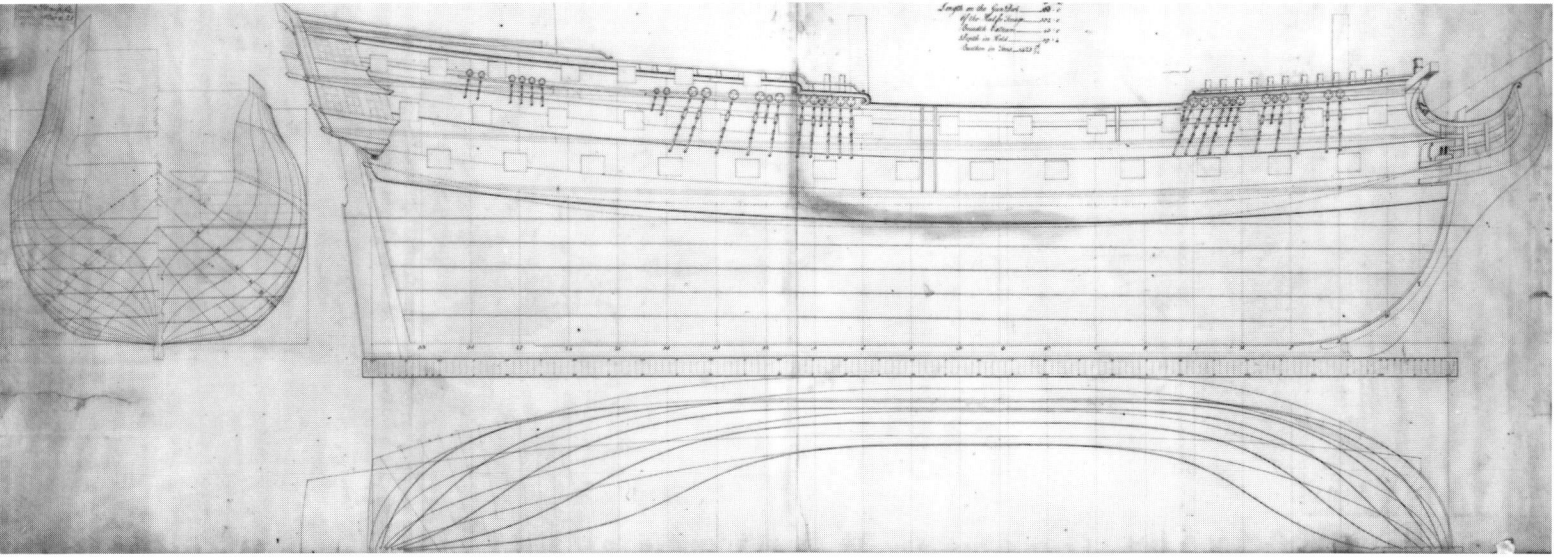

Lines plan (number 1067) for the *Temple* Class 70s (later 66s or 64s)

Ord 28.10.1745; K 26.1.1747; L 30.4.1751; 1777 storeship, 30, renamed *Grampus;* 1779 foundered.

FATES: *Somerset* wrecked near Cape Cod on 12.10.1778. The wreck is still visible at low tide. *Northumberland*, as *Leviathan,* foundered at sea 27.2.1780. *Orford* became a hospital ship at Sheerness; 1783 BU. *Buckingham*, as *Grampus*, foundered at sea in the West Indies 11.11.1779.

PLANS: Lines. *Grafton:* ? Lines & profile. *Orford*: Danish Rigsarkivet: Lines & profile.

### *Chichester* Class 1750.
Allin design, 1745 Establishment, amended, classed as 68 gun ship (later 64).

Dimensions & tons: 160', 131'6½" x 44'9" x 19'6". 1401 33/94.
Men: 520. Guns: GD 26 x 32, UD 28 x 18, QD 12 x 9, Fc 2 x 9.

*Chichester* Portsmouth Dyd.
Ord 12.7.1750; K 28.7.1750; L 4.6.1753; 1779 hulked.

FATE: Became a temporary hulk at Plymouth; 1783 receiving ship; 1803 BU.
PLANS: Lines.

### *Burford* Class 1754.
Allin design, modified from the 1745 Establishment; classed as 70s.

Dimensions & tons: 162', 134'6" x 44'8" x 19'8". 1426 87/94.
Men: 520. Guns: GD 26 x 32, UD 28 x 18, QD 12 x 9, Fc 2 x 9.

*Burford* Chatham Dyd.
Ord 15.1.1754; K 30.10.1754; L 5.5.1757; (1764-1769 troopship); 1785 sold.

*Dorsetshire* Portsmouth Dyd.
Ord 13.1.1754; K 22.6.1754; L 13.12.1757; 1768 troopship; 1775 BU.

*Boyne* Plymouth Dyd.
Ord 13.5.1758; K 9.8.1758; L 31.5.1766; 1783 BU.

PLANS: Lines/profile/decks. *Dorsetshire*: as built: decks.

### *Temple* Class 1756.
Copied from the 1745 Establishment *Vanguard*; classed as 68s (later 64s).

Dimensions & tons: 160', 132'x 45'x 19'4". 1421 76/94.
Men: 520. Guns: GD 26 x 32, UD 28 x 18, QD 12 x 9, Fc 2 x 9.

*Temple* Blades, Hull.
Ord 9.9.1756; K 17.11.1756; L 3.11.1758; 1762 foundered.

*Conqueror* Barnard, Harwich.
Ord 11.1.1757; K 9.2.1757; L 24.5.1758; 1760 wrecked.

FATES: *Temple* foundered, returning from Manilla, 19.12.1762. *Conqueror* wrecked on Saint Nicholas Island, Plymouth Sound 26.10.1760.

PLANS: Lines/profile/orlop/gun deck/upper deck/quarter deck & forecastle.

### *Africa/Asia* Class 1758.
Slade design; the first true 64s.

Dimensions & tons: 158', 129'9" x 44'4" x 18'10". 1346 27/94.
Men: 500. Guns: GD 26 x 24, UD 26 x 18, QD 10 x 9, Fc 2 x 9.

*Asia* Portsmouth Dyd.
Ord 4.3.1758; K 18.4.1758; L 15.6.1764; 1804 BU.

*Essex* Stanton, Rotherhithe.
Ord 13.12.1758; K 31.1.1759; L 28.8.1760; 1777 hulked.

*Africa* Perry, Blackwall.
Ord 13.12.1758; K 7.5.1759; L 1.8.1761; 1777 sold.

FATES: *Essex* became a receiving ship at Portsmouth; 1799 sold.

PLANS: Lines/profile/orlop/gun deck/upper deck/quarter deck & forecastle.

### *Saint Albans* Class 1761.
Slade design 64s.

Dimensions & tons: 159', 130'7½" x 44'4" x 18'10". 1365 57/94.
Men: 500. Guns: GD 26 x 24, UD 26 x 18, QD 10 x 9, Fc 2 x 9.

*Saint Albans* Perry, Blackwall.
Ord 1.1.1761; K 08.1761; L 13.8.1764; (1803-1807 floating battery); 1814 BU.

*Augusta* Stanton, Rotherhithe.
Ord 1.1.1761; K 28.2.1761; L 13.7.1763; 1777 burnt.

*Director* Cleverley, Gravesend.
Ord 2.8.1780; K 11.1780; L 9.3.1784; (1794-1797 hulked); 1801 BU.

FATES: *Augusta* grounded on Mud Island in the River Delaware then burnt by hostile artillery fire (red hot shot from shore batteries) 23.10.1777; there have been reports of people working on this wreck recently. *Director* was a lazaretto at Chatham 1794-1797.

PLANS: Lines/profile/orlop/gun deck/upper deck/quarter deck & forecastle/framing.

### *Exeter* Class 1761.
Bately design 64s.

Dimensions & tons: 159', 130'9" x 44'4" x 19'4". 1366 86/94.
Men: 500 Guns: GD 26 x 24, UD 26 x 18, QD 10 x 9, Fc 2 x 9.

*Exeter* Henniker, Chatham.
Ord 1.1.1761; K 28.1.1761; L 26.7.1763; 1784 destroyed.

*Europa/Europe* Adams, Bucklers Hard.
Ord 16.12.1761; K 02.1762; L 21.4.1765; 1796 hulked.

*Prudent* Woolwich Dyd.
Ord 16.12.1761; K 04.1765; L 28.9.1768; 1794 hulked.

*Trident* Plymouth Dyd.
Ord 4.12.1762/28.2.1763; K 10.1763; L 20.4.1768; 1808 hulked.

FATES: *Exeter* destroyed at the Cape of Good Hope 'being in a very defective state' 12.12.1784. *Europa* became a prison ship at Plymouth; 1814 sold. *Prudent* became a prison ship at Plymouth 1802 powder hulk; 1814 sold. *Trident* became a receiving ship at Malta; 1816 sold.

PLANS: Lines & profile. *Prudent*: as built: Lines & profile/quarterdeck & forecastle. 1794 as prison ship: decks.

### *Ardent* Class 1761.
Slade design 64s.

Dimensions & tons: 160', 131'8" x 44'4" x 18'. 1376 47/94
Men: 500 Guns: GD 26 x 24, UD 26 x 18, QD 10 x 9, Fc 2 x 9.

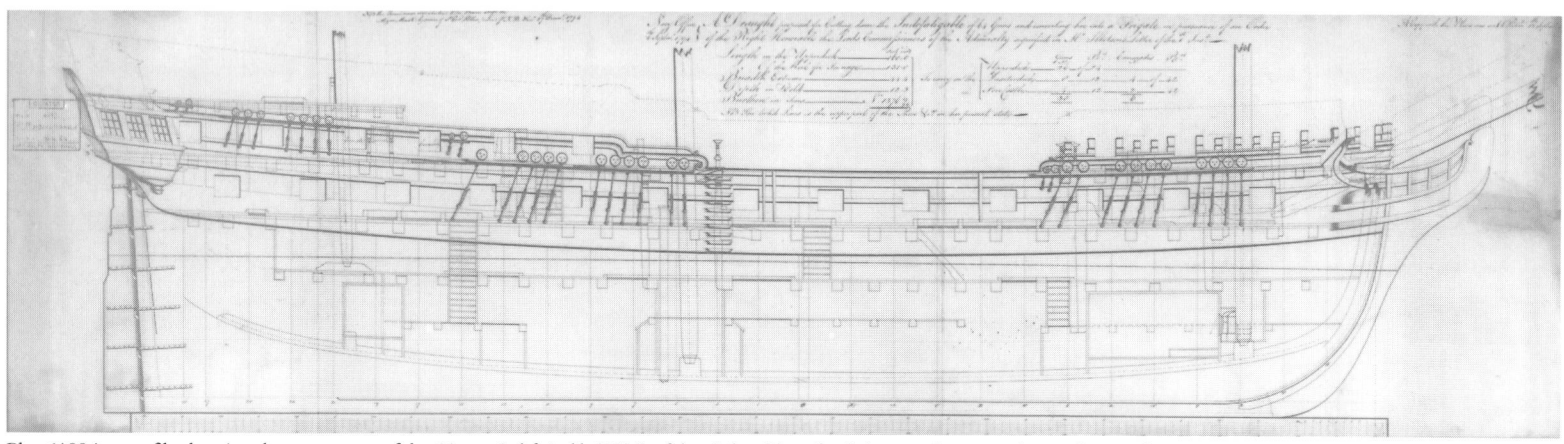

Plan 6199 is a profile showing the appearance of the 64 gun *Indefatigable* (1784) of the *Ardent* Class after being cut down to a heavy frigate of 38 guns – in which guise she achieved fame under the command of Edward Pellew. The original outline of her upperworks is shown in a dotted line.

(*Indefatigable* as a frigate: UD 26 x 24, QD 8 x 12 + 4 x 42★, Fc 4 x 12 + 4 x 42★).

**Ardent** Blades, Hull.
 Ord 16.12.1761/7.1.1762; K 15.1.1762; L 13.8.1764; 1779 taken; 1782 retaken and renamed *Tiger*; 1784 sold.
**Raisonable** Chatham Dyd.
 Ord 4.12.1762; K 25.11.1765; L 10.12.1768; 1810 hulked.
**Agamemnon** Adams, Bucklers Hard.
 Ord 5.2.1777; K 05.1777; L 10.4.1781; 1809 wrecked.
**Belliquex** Perry, Blackwall.
 Ord 11.2.1778; K 06.1778; L 5.6.1780; 1814 hulked.
**Stately** Raymond, Northam.
 Ord 5.2.1777/10.2.1778; K 25.5.1779; L 22.12.1784; (1799-1805 troopship); 1814 BU.
**Indefatigable** Adams, Bucklers Hard.
 Ord 3.8.1780; K 05.1781; L 07.1784; 1795 cut down to a frigate (Fifth Rate) 38; 1816 BU.
**Nassau** Hillhouse, Bristol.
 Ord 14.11.1782; K 03.1783; L 20.9.1785?; 1799 troopship; 1799 wrecked.

FATES: *Ardent* taken by the Franco-Spanish fleet in the Channel 17.8.1779; 1782 retaken from the French. *Raisonable* hulked at Sheerness; 1815 BU. *Agamemnon* was wrecked in the River Plate 20.6.1809. There are recent reports of divers working on her wreck (1992). *Belliquex* became a prison ship at Chatham; 1816 BU. *Nassau* was wrecked on the coast of Holland 14.10.1799.

PLANS: Lines/profile/orlop/gun deck/upper deck/quarter deck & forecastle/specification. *Indefatigable*: as built: Lines & profile. 1794 as frigate: profile/orlop/lower deck/upper deck/quarter deck & forecastle.

## Worcester Class 1765. Slade design 64s.
 Dimensions & tons: 159', 130'7½" x 44'6" x 19'. 1373 73/94.
 Men: 500. Guns: GD 26 x 24, UD 26 x 18, QD 10 x 9, Fc 2 x 9.
**Worcester** Portsmouth Dyd.
 Ord 16.11.1765/7.5.1766; K 6.5.1766; L 17.10.1769; 1788 hulked.
**Lion/Lyon** Portsmouth Dyd.
 Ord 12 & 27.10.1768; K 05.1769; L 3.9.1777; 1814 hulked.
**Stirling Castle** Chatham Dyd.
 Ord 12 & 27.10.1768; K 10.1769; L 28.6.1775; 1780 wrecked.

FATES: *Worcester* hulked at Deptford; 1816 BU. *Lion* sheer hulk at Plymouth; 1816 to Sheerness; 1837 sold. *Stirling Castle* wrecked on the Silver Keys during the great West Indian hurricane of 1780 (on 9.10.1780).

PLANS: Lines/profile/orlop/gun deck/upper deck/quarter deck & forecastle. *Lion* as built: Lines & profile and decorations/decks. 1822 as sheer hulk at Sheerness (?): sheer legs & details [2 differing plans]. *Stirling Castle* as built: Lines, profile & decorations/orlop & lower decks/upper deck, quarter deck & forecastle.

## Intrepid/Magnanime Class 1765. Williams design 64s.
 Dimensions & tons: 159'6", 131' x 44'4" x 19'. 1369 50/94.
 Men: 500. Guns: GD 26 x 24, UD 26 x 18, QD 10 x 9, Fc 2 x 9. (*Anson* & *Magnanime* as frigates: UD 26 x 24, QD 8 x 12 + 4 x 42★, Fc 4 x 12 + 2 x 42★).

**Intrepid** Woolwich Dyd.
 Ord 18.11.1765; K 01.1767; L 4.12.1770; 1810 hulked.
**Monmouth** Plymouth Dyd.
 Ord 10.9.1767; K 05.1768; L 18.4.1772; 1796 hulked.
**Defiance** Woolwich Dyd.
 Ord 9.6.1768; K 10.1768; L 31.8.1772; 1780 lost.
**Nonsuch** Plymouth Dyd.
 Ord 30.11.1769; K 01.1772; L 17.2.1774; 1794 floating battery; 1802 BU.
**Ruby** Woolwich Dyd.
 Ord 30.11.1768; K 9.9.1772; L 26.11.1776; 1811 hulked.
**Vigilant** Adams, Bucklers Hard.
 Ord 14.1. & 23.3.1771?; K 02.1771; L 6.10.1774; 1795 hulked.
**Eagle** Wells, Rotherhithe.
 Ord 14.1.1771; K 04.1771; L 12.5.1774; 1794 hulked.
**America** Deptford Dyd.
 Ord 18.6.1771; K 10.1771; L 5.8.1777; 1800 hulked.
**Anson** Plymouth Dyd.
 Ord 24.4.1773; K 01.1774; L 4.1.1781; 1794 cut down to a frigate (Fiftth Rate) 38; 1807 wrecked.
**Polyphemus** Sheerness Dyd.
 Ord 1.12.1773; K 01.1776; L 7.4.1782; 1813 hulked.
**Magnanime** Deptford Dyd.
 Ord 16.10.1775; K 23.8.1777; L 14.10.1780; 1795 cut down to a frigate (Fifth Rate) 38; 1813 BU.
**Sampson** Woolwich Dyd.
 Ord 25.7.1776; K 20.10.1777; L 8.5.1781; 1808 hulked.
**Repulse** Fabian, Cowes.
 Ord 5.2.1777/16.5.1777; K 12.1.1778; L 28.11.1780; 1800 lost.
**Diadem** Chatham Dyd.
 Ord 5.2.1777/11.6.1778; K 2.11.1778; L 19.12.1782; (1798-1799 troopship); 1815 hulked.
**Standard** Deptford Dyd.
 Ord 5.8.1779; K 05.1780; L 8.10.1782; (1799-1801 hulked); 1816 BU.

FATES: *Intrepid* receiving ship at Plymouth; 1828 BU. *Monmouth* prison ship at Portsmouth renamed *Captivity*; 1818 BU. *Defiance* wrecked on Savannah Bar, Georgia 2.2.1780. *Ruby* became a depot ship at Bermuda; 1821 BU. *Vigilant* became a prison ship at Portsmouth 1816 BU. *Eagle* became a lazaretto at Chatham; 1800 renamed *Buckingham*; 1812 BU. *America* was hulked (as a prison ship?), after stranding, but where is not certain; lent to the Transport Board (?) 1804; 1807 BU. *Anson* was wrecked off Mounts Bay 29.12.1807. *Polyphemus* became a powder magazine at Chatham; 1826 fitted for the Lieutenants of the Ordinary; 1827 BU. *Sampson* became a prison ship at Chatham for Danish prisoners; 1814 to Woolwich as a sheer hulk; 1832 sold. *Repulse* was wrecked on a rock off Ushant 10.3.1800. *Diadem* became a temporary receiving ship at Plymouth; 1832 BU. *Standard* between 1799 and 1801 was a convalescent ship.

PLANS: Lines/profile/orlop/gun deck/upper deck/quarter deck & forecastle/framing. *Monmouth*: as convict ship: Profile & decks/decks. 1812; Decks. *Defiance*: Capstans & firehearth. *Ruby*: as built: Lines & profile &

# THE SLADE ERA 1745-1785

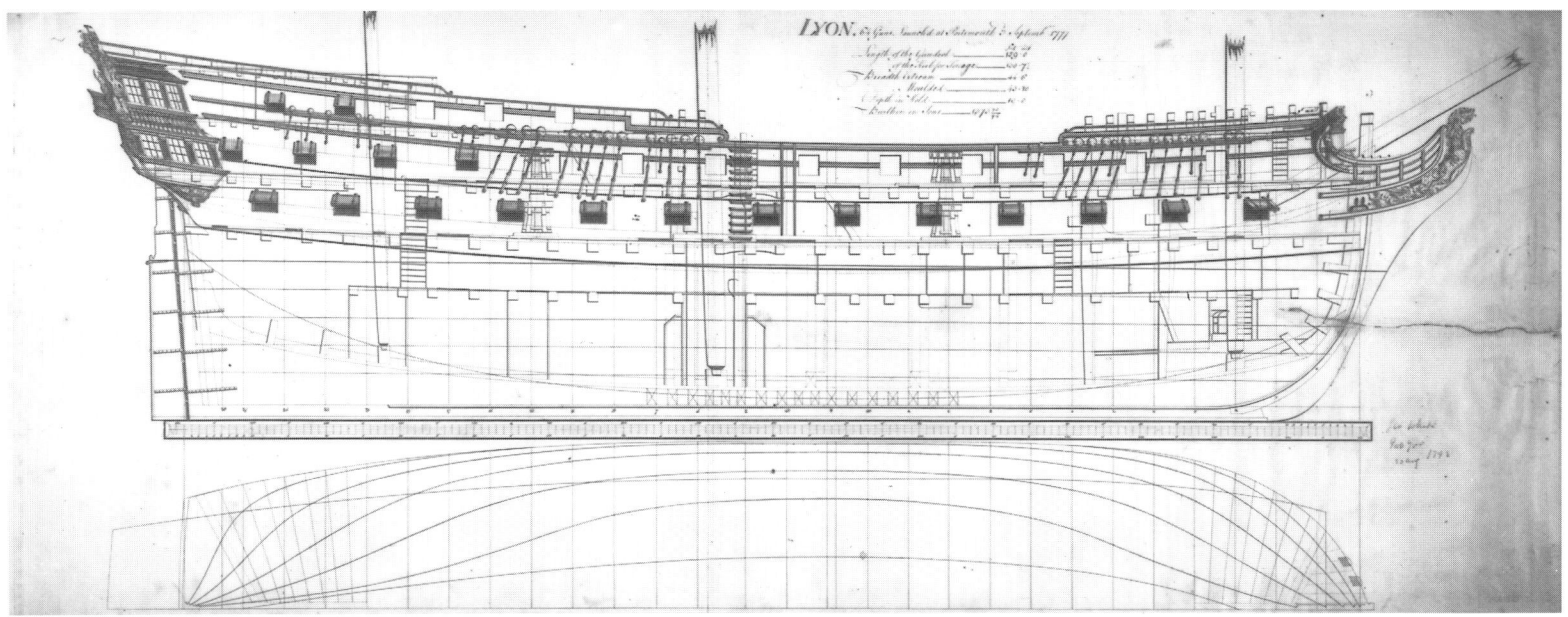

This 'as built' lines and profile plan (x1160) of the 64 gun *Lion* of the *Worcester* Class, launched in 1777, is a particularly fine example of the type - the archaic spelling of the name with a Y instead of an I was still common in the eighteenth century

decoration. *Vigilant*: as built: Lines & profile & decoration. *Anson*: as built: Lines & profile & decoration. 1794 as frigate: sheer/profile/orlop/lower deck/upper deck/quarter deck & forecastle. *Diadem* as troopship 1810: Profile/orlop/upper deck/quarter deck & forecastle. *Standard*: as built: Lines & profile. 1814: orlop.

## *Inflexible* Class 1777.
Williams design 64s 'similar to *Albion*' (Slade's 74 gun ship design).

Dimensions & tons: 159'6", 131' x 44'6" x 18'. 1379 8/94.
Men: 500. Guns: GD 26 x 24, UD 26 x 18, QD 10 x 9, Fc 2 x 9.

*Inflexible* Barnard, Harwich.
Ord 5.2.1777; K 04.1777; L 7.3.1780; (1793-1795 storeship); (1800-1807 troopship); 1809 hulked.

*Africa* Adams & Barnard, Deptford.
Ord 11.2.1778; K 2.3.1778; L 11.4.1781; (1795-1805 hospital ship); 1814 BU.

*Dictator* Batson, Limehouse.
Ord 21.10.1778; K 05.1780; L 6.1.1783; (1798-1803 troopship); (1803-1805 floating battery); 1813 troopship; 1817 BU.

*Sceptre* Randall, Rotherhithe.
Ord 16.1.1779; K 05.1780; L 8.6.1781; 1799 lost.

FATES: *Inflexible* was a floating magazine at Halifax, Nova Scotia; 1820 BU. *Sceptre* was wrecked in Table Bay 5.12.1799.

PLANS: Lines/profile/orlop/gun deck/upper deck/quarter deck & forecastle/framing. *Sceptre*: as built: Lines & profile.

## *Crown* Class 1779.
Hunt design, 64s.

Dimensions & tons: 160'6", 131'8½" x 44'6" x 19'5". 1387 29/94.
Men: 500. Guns: GD 26 x 24, UD 26 x 18, QD 10 x 9, Fc 2 x 9.

*Crown* Perry, Blackwall.
Ord ____ ; K 09.1779; L 15.3.1782; 1798 hulked.

*Ardent* Stares & Parsons, Bursledon.
Ord 15.10.1779; K 10.1780; L 15.3.1782; 1794 blew up.

*Scipio* Barnard, Deptford.
Ord 11.11.1779; K 01.1780; L 22.10.1782; 1798 BU.

*Veteran* Fabian, Cowes.
Ord 3.9.1780; K 07.1781; L 14.8.1787; 1809 hulked.

FATES: *Crown* became a prison ship at Portsmouth; 1802 powder hulk; 1806 prison ship; 1816 BU. *Ardent* accidentally caught fire and blew up off Corsica 11.4.1794. *Veteran* became a prison ship at Portsmouth; 1816 BU.

PLANS: Lines/profile/orlop/gun deck/upper deck/quarter deck & forecastle/framing/specification. *Crown*: as built: Lines & profile/orlop/gun deck/upper deck/quarter deck & forecastle. *Ardent* as built: Lines & profile. *Scipio*: as built: Lines & profile.

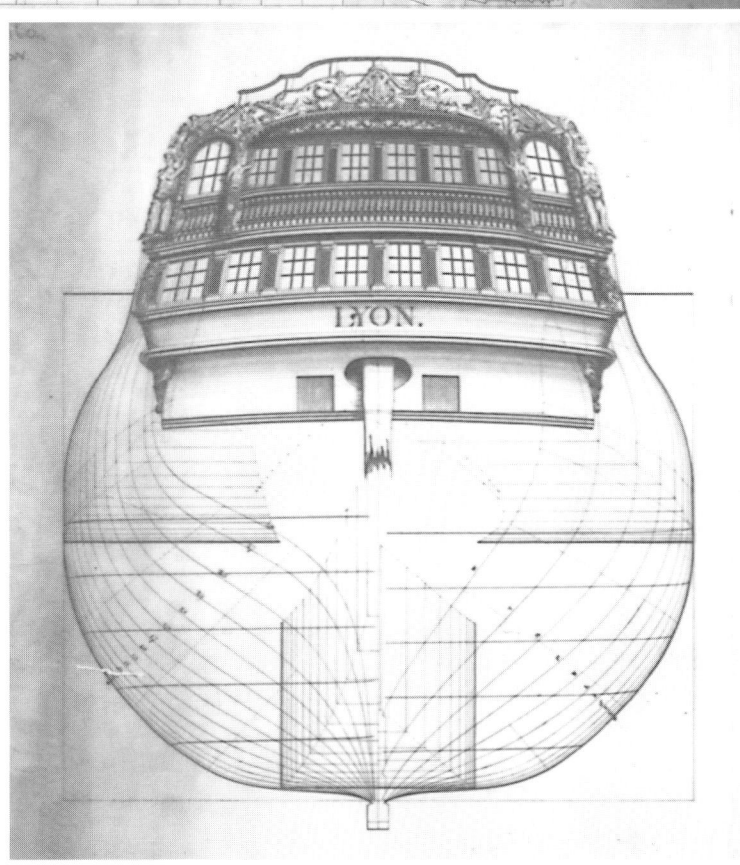

# Fourth Rates 60 Guns (two-deckers)

## 1745 Establishment.
*Anson* appears to have been designed by Acworth. *Tiger* by Lock.

Dimensions & tons: 150', 123'0½" x 42'8" x 18'6". 1191.
Men: 420. Guns: GD 24 x 24, UD 24 x 12, QD 8 x 6, Fc 2 x 6.

*Anson* Ewer, Bursledon.
Ord 6.8.1745; K 09.1745; L 10.10.1747; 1773 sold.

*Saint Albans* West, Deptford.
Ord 6.8.1745; K 09.1745; L 23.12.1747; 1765 sold.

*Tiger/Tyger* Wells, Deptford.
Ord 17.3.1746; K 04.1746; L 23.11.1747; 1761 hulked.

*Weymouth* Plymouth Dyd.
Ord 8.11.1744; K 26.9.1748; L 18.2.1752; 1772 BU.

FATES: *Tiger* hulked at Bombay; 1765 sold.

PLANS: Lines & profile. *Anson*: as built: Lines & profile. *Anson* & *Saint Albans*: Orlop/gun deck/upper deck/quarter deck & forecastle. *Saint Albans*: as built?:

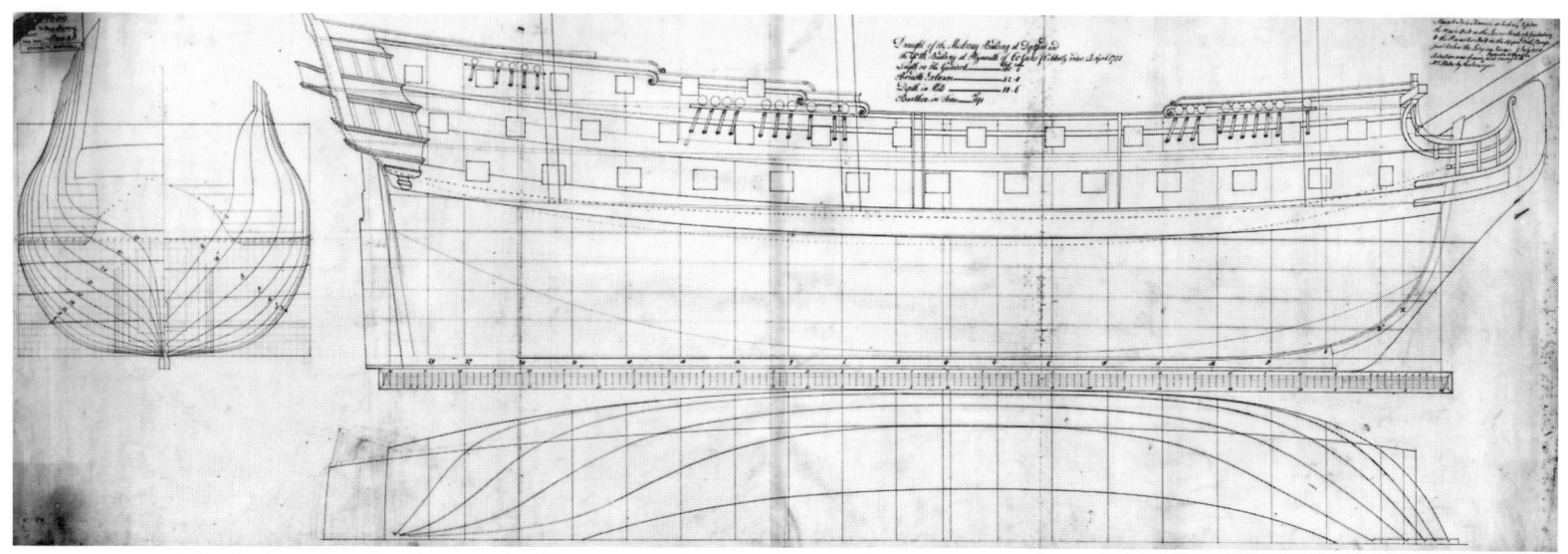

Lines plan number 1300 is for the *Medway* Class 60s ordered in 1751, amongst the last two-decker 60s built

lines & profile. *Tiger*: as built: lines & profile & decoration/orlop/gun deck/upper deck/quarter deck & forecastle. *Weymouth*: Lines. As built: Decks.

### Dunkirk Class 1751. Allin design. *Dunkirk* appears originally to have been ordered to the 1745 Establishment design; but in 1750 was 'to be built agreeable to the Draught and scheme of dimensions and scantlings proposed by the Surveyor of the Navy'.

Dimensions & tons: 153'6", 127'1¾" x 42'5" x 18'9". 1216 64/94.
Men: 420. Guns: GD 24 x 24, UD 26 x 12, QD 8 x 6, Fc 2 x 6.
**Dunkirk** Woolwich Dyd.
Ord 15.11.1745/12.7.1750; K 1.4.1746; L 22.7.1757; 1778 hulked.
**Achilles** Barnard, Harwich.
Ord 14.11.1755; K 12.1755; L 6.2.1757; 1780 hulked?; 1784 sold.
**America** Wells, Rotherhithe.
Ord 24.11.1755; K 12.1755; L 21.2.1757; 1771 BU.
FATES: *Dunkirk* became a receiving ship at Chatham; 1779 to the Downs and then to Plymouth as a receiving ship for convicts; 1785 receiving ship; 1792 sold. If *Achilles* was hulked in 1780 there is no record of where this was done.
PLANS: Lines & profile/orlop/gun deck/upper deck.

### Montague Class 1750. 'Admiralty' design; amendment of 1745 Establishment design.
Dimensions & tons: 157'3", 132'6½" x 42' x 18'6". 1245 20/94.
Men: 420. Guns: GD 24 x 24, UD 26 x 12, QD 8 x 6, Fc 2 x 6.
**Montague** Sheerness Dyd.
Ord 12.7.1750; K 16.12.1751; L 15.9.1757; 1774 breakwater.
PLANS: Lines/profile.

### Medway Class 1751. Allin design, modified from 1745 Establishment. *Rippon* (1752) seems to have been modified from this design.
Dimensions & tons: 150', 123'0½" x 42'8" x 18'6". 1191.
Men: 420. Guns: GD 24 x 24, UD 25 x 12, QD 8 x 6, Fc 2 x 6.
**Medway** Deptford Dyd.
Ord 28.3.1751/25.4.1751; K 13.6.1751; L 14.2.1753; 1789 hulked.
**York** Plymouth Dyd.
Ord 28.3.1751/25.4.1751; K 18.6.1751; L 10.11.1753; 1772 BU.
FATES: *Medway* was a receiving ship at Plymouth; 1802 renamed *Arundel*; 1811 BU.
PLANS: Lines/decks.

### Rippon Class 1752. Allin design, appears to have been modified from that of *Medway*; therefore a further modification of the 1745 Establishment.
Dimensions & tons: 155', 129' x 42'5" x 18'7". 1234 47/94.
Men: 420. Guns: GD 24 x 24, UD 25 x 12, QD 8 x 6, Fc 2 x 6.
**Rippon** Woolwich Dyd.
Ord 8.11.1752; K 23.11.1752; L 26.1.1758; 1782 fitted as a 54 gun ship with an all-carronade armament and the masts and spars of a 50 gun ship; 1788 hulked.
FATES: Receiving ship at Plymouth; 1808 BU.
PLANS: Lines.

### Pembroke Class 1752. 'The Draught is similar to the *Monarch* [French prize, 74 guns, captured 1747] and scantlings of the frames and beams are to be agreeable to those proposed for the *Rippon*.'
Dimensions & tons: 156', 132'6" x 42'1" x 18'. 1247 52/94.
Men: 420. Guns: GD 24 x 24, UD 26 x 12, QD 8 x 6, Fc 2 x 6.
**Pembroke** Plymouth Dyd.
Ord 8.11.1752; K 1.1.1753; L.2.6.1757; 1776 hulked.
FATE: Hulked at Halifax, Nova Scotia; 1793 BU.
PLANS: Lines.

### Edgar/Firm Class 1756. Slade design.
Dimensions & tons: 154', 126'0¼" x 43'6" x 18'4". 1268 38/94.
Men 420. Guns: GD 24 x 24, UD 26 x 12, QD 8 x 6, Fc 2 x 6.
**Edgar** Randall, Rotherhithe.
Ord 13.4.1756; K. 08? 1756; L 16.11.1758; 1774 breakwater at Sheerness.
**Panther** Martin, Chatham.
Ord 25.5.1756; K 06.1756; L 22.6.1758; 1787 hulked.
**Firm** Perry, Blackwall.
Ord 15.7.1756; K ____ 1756; L 15.1.1759; 1778 hulked.
FATES: *Panther* became a temporary hulk at Plymouth; 1813 BU. *Firm* hulked at Plymouth; 1784 receiving ship; 1791 sold.
PLANS: Lines/profile/orlop/gun deck/upper deck/quarter deck & forecastle.

## Fourth Rates, 50 Guns (two-deckers)

### 1745 Establishment
Dimensions & tons: 144', 117'8½" x 41' x 17'8". 1052 47/94.
Men: 350. Guns: GD 22 x 24, UD 22 x 12, QD 4 x 6, Fc 2 x 6.
**Severn** Barnard, Harwich.
Ord 17.3.1746; K 04.1746; L 10.7.1747; 1759 sold.
**Assistance** Sedger, Chatham.
Ord 6.8.1745; K 08.1745; L 22.12.1747; 1770 hulked.
**Greenwich** Janvrin, Beaulieu.
Ord 3.10.1745; K 11.1745; L 19.3.1747; 1757 taken.
**Tavistock** Blades, Hull.
Ord 18.10.1745; K 11.1746; L 26.8.1747; 1760 hulked.
**Falmouth** Woolwich Dyd.
Ord 15.11.1745; K 22.8.1746; L 7.12.1752; 1765 beached.
**Newcastle** Portsmouth Dyd.
Ord 11.2.1745; K 17.6.1746; L 4.12.1750; 1761 wrecked.

### Cancelled 1748. *Woolwich* Chatham Dyd.

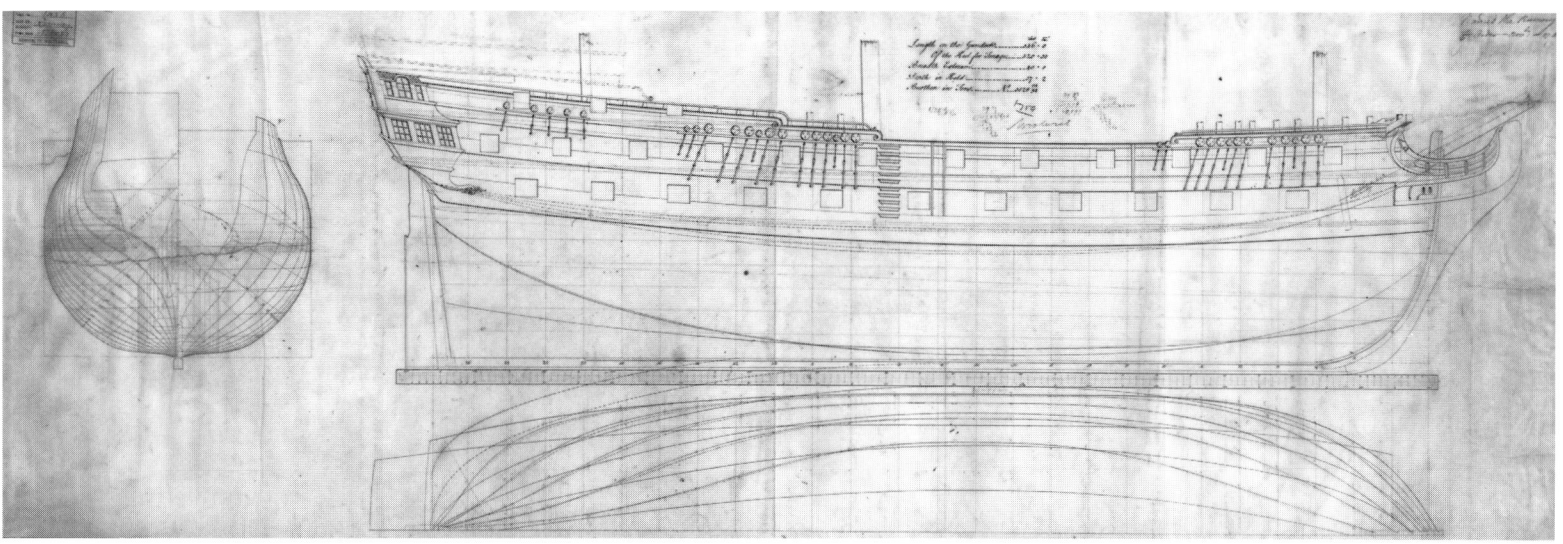

Design lines plan number 1425 for the 50 gun *Romney* of 1762 shows what is probably the later addition of a 'roundhouse' (poop deck) in dotted lines

Ord 6.5.1747; not begun?; name transferred to a 44 building y Janvrin at Beaulieu.

**Cancelled 1749.** *Dartmouth* Plymouth Dyd.
Ord 10.1.1746; not begun?

FATES: *Assistance* became a temporary hulk at Portsmouth; 1773 sold. *Greenwich* taken by the French in the West Indies; 14.1.1758 retaken but no further record, presumably sold or BU. *Tavistock* hulked at Woolwich; 1768 sold. *Falmouth* was beached and abandoned at Batavia on 16.1.1765 as unseaworthy. *Newcastle* was run ashore a little to the southward of Ariancopang near Pondicherry during the 'Pondicherry Hurricane' on 1.1.1761.

PLANS: Lines/spar plan. *Assistance*: Lines & profile/orlop/gun deck/upper deck/quarter deck & forecastle. *Assistance*, *Tavistock* & *Greenwich*: Lines & profile. *Greenwich*: Lines & profile.

## Preston Class 1751. Allin's modification of the 1745 Establishment.
Dimensions & tons: 144', 117'8½" x 41' x 17'8". 1052 47/94.
Men: 350. Guns: GD 22 x 24, UD 22 x 12, QD 4 x 6, Fc 2 x 6.
*Preston* Deptford Dyd.
Ord 28.3.1751; K 13.6.1751; L 7.2.1757; 1785 hulked.
FATE: Hulked at Woolwich; 1715 BU.
PLANS: Lines/stern.

## Chatham Class 1752. Allin design, modified from the 1745 Establishment.
Dimensions & tons: 147', 124' x 40' x 17'8". 1055 30/94.
Men: 350. Guns: GD 22 x 24, UD 22 x 12, QD 4 x 6, Fc 2 x 6.
*Chatham* Portsmouth Dyd.
Ord 26.10. & 8.11.1752; K 14.12.1752; L 25.4.1758; 1793 hulked.
FATE: Became a convalescent ship at Plymouth; 1797 to Falmouth; 1805 floating magazine at Chatham; 1810 renamed *Tilbury*; 1814 BU.
PLANS: Lines.

## Warwick Class 1758. Bately design.
Dimensions & tons: 151', 122'9" x 40'2" x 18'3". 1053 36/94.
Men: 350. Guns: GD 22 x 24, UD 22 x 12, QD 4 x 6, Fc 2 x 6.
*Warwick* Portsmouth Dyd.
Ord 13.12.1758/20.7.1759; K 27.8.1762; L 28.2.1767; 1783 hulked.
FATE: Receiving ship at Chatham; 1802 sold.
PLANS: Lines & profile.

## Romney Class 1759. Slade design.
Dimensions & tons: 146', 120'10" x 40' x 17'2". 1028 34/94.
Men: 350. Guns: GD 22 x 24, UD 22 x 12, QD 4 x 6, Fc 2 x 6.
*Romney* Woolwich Dyd.
Ord 20.7.1759; K 1.10.1759; L 8.7.1762; 1804 wrecked.
FATE: Wrecked near the Texel on 19.11.1804.
PLANS: Lines/profile/specification.

## Salisbury Class 1766. Slade design, modification of the *Romney* design (and possibly just to be considered as a sub-division of that class?).
Dimensions & tons: 146', 120'7⅝" x 40'4" x 17'4". 1043 77/94.
Men: 350. Guns: GD 22 x 24, UD 22 x 12, QD 4 x 6, Fc 2 x 6.
*Salisbury* Chatham Dyd.
Ord 8. & 18.1.1766; K 19.8.1766; L 2.10.1769; 1796 wrecked.
*Centurion* Barnard, Harwich.
Ord 25.12.1770; K 05.1771; L 27.5.1774; 1808 hulked.
FATES: *Salisbury* was wrecked on the Isle of Vache 13.5.1796. *Centurion* became a hospital ship at Halifax, Nova Scotia; 1824 sank & BU.
PLANS: Lines/profile/orlop/gun deck/upper deck/quarter deck & forecastle/specification. *Salisbury* as built: Decks. *Centurion* as built: Lines & profile/decks.

## Portland Class 1767. Williams design.
Dimensions & tons: 146', 119'9" x 40'6" x 17'6". 1044 73/94.
Men: 350. Guns: GD 22 x 24, UD 22 x 12, QD 4 x 6, Fc 2 x 6 (*Leander* carried an extra pair of 6 pounders on the QD and was classed as a 52, but was otherwise identical to the rest of the class).
*Portland* Sheerness Dyd.
Ord ____ ; K 01.1767; L 1.4.1770; 1801 storeship; 1802 hulked.
*Bristol* Sheerness Dyd.
Ord 12.10.1768; K 05.1771; L 25.10.1775; 1794 hulked.
*Renown* Fabian, Northam.
Ord 25.12.1770; K 05.1771; L 4.12.1774; 1794 hulked.
*Isis* Henniker, Cowes.
Ord ____ ; K 12.1772; L 19.11.1774; 1810 BU.
*Leopard* Portsmouth Dyd.
Ord 16.10.1775; K 01.1776 but not completed, instead frames taken to: Sheerness Dyd K 7.5.1785; L 29.4.1790; 1811 troopship; 1814 wrecked.
*Hannibal* Adams, Bucklers Hard.
K 24.5.1776; K 07.1776; L 26.12.1779; 1782 taken.
*Jupiter* Randall, Rotherhithe.
Ord 21.6.1776; K 07.1776; L 13.5.1778; (1805-1807 hospital ship); 1808 wrecked.
*Leander* Chatham Dyd.
Ord ____ ; K 1.3.1777; L 1.7.1780; 1798 taken; 1799 retaken; 1806 hulked.
*Adamant* Baker, Liverpool.
Ord 13.11.1776; K 6.9.1777; L 24.1.1780; 1809 hulked.
*Europa* Woolwich Dyd.
Ord ____ ; K 26.9.1778; L 19.4.1783; 1798 troopship; 1814 sold.
*Assistance* Baker, Liverpool.
Ord ____ ; K 4.7.1778; L 18.3.1781; 1802 wrecked.
FATES: *Portland* became a convict ship at Portsmouth (in Langstone Harbour); 1817 sold. *Bristol* became a lazaretto at Sheerness; 1794 BU. *Leopard* was wrecked off Anticosti, Saint Lawrence River on 28.6.1814. *Hannibal* was taken

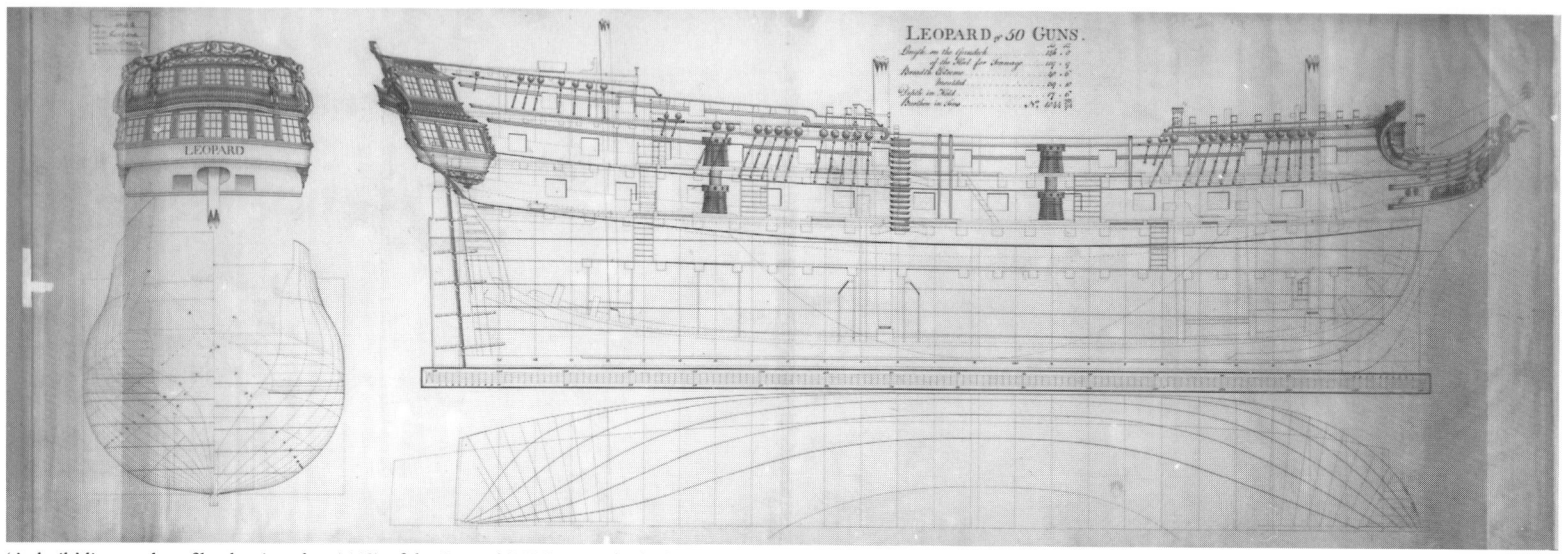

'As built' lines and profile plan (number 1462) of the *Leopard* (1790), a *Portland* Class 50 gun ship. Admirers of Patrick O'Brian's novels will be interested to know that this plan was the basis of his description of 'the horrible old *Leopard*' in his *Desolation Island*

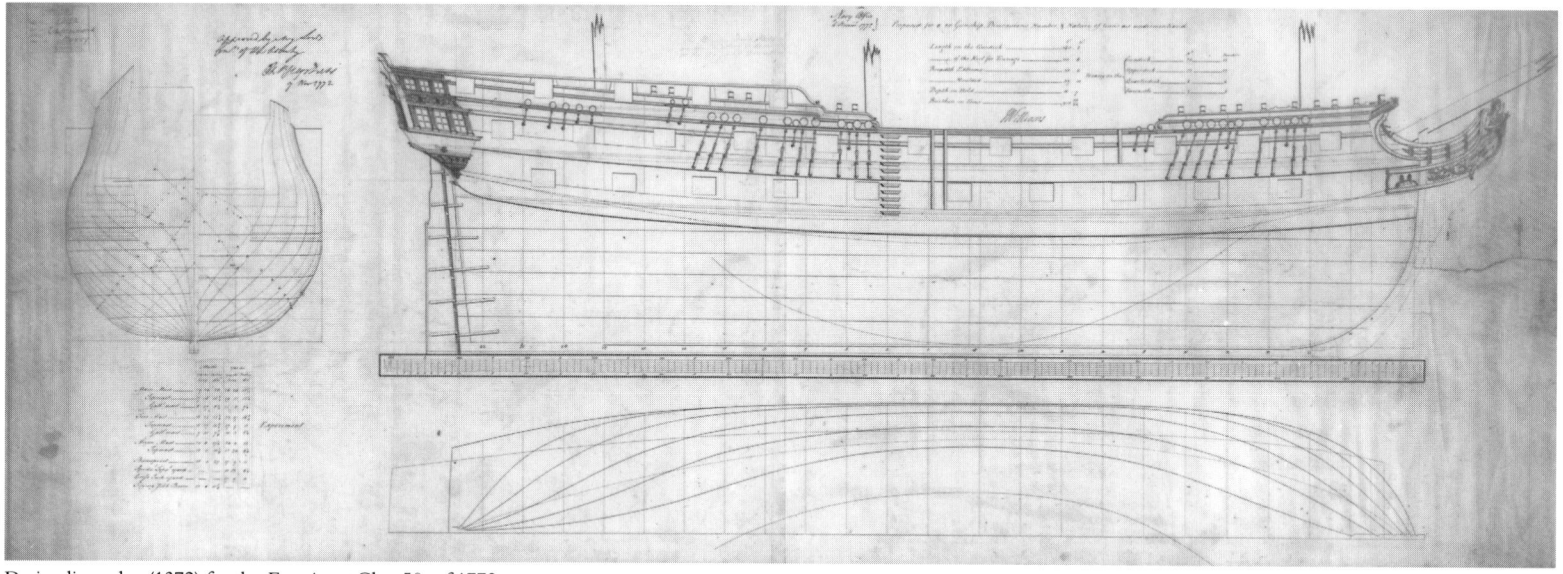

Design lines plan (1372) for the *Experiment* Class 50s of 1772

by Suffren off Sumatra 21.1.1782; in French service until 1782. *Jupiter* was wrecked in Vigo Bay 10.12.1808. *Leander* was taken by the French *Généreux* in the Mediterranean on 17.8.1798; 3.3.1799 was taken by the Russians at the fall of Corfu and handed back to the RN; 1806 became the medical depot ship *Hygeia* at Portsmouth; 1817 sold. *Adamant* became a receiving ship at Chatham; possibly later to Sheerness?; 1814 BU. *Assistance* was wrecked near Dunkirk on 29.3.1802.

PLANS: Lines/framing/profile. *Bristol* as built: Lines & profile & decorations/orlop & gun deck/upper deck, quarter deck & forecastle. As prison ship 1794: Orlop/gun deck/upper deck & quarter deck. *Isis* as built: Lines & profile & decorations. *Leopard* as built: Lines & profile & decorations. As troopship 1812: Orlop & gun deck/upper & quarter decks & forecastle. *Jupiter* as hospital ship 1805: Profile/orlop/gun deck/upper deck/quarter deck & forecastle/midships section.

### Experiment Class 1772. 
Williams design. Appears to have been an attempt to produce a light, fast two decker with a main armament of 12 pdrs. instead of the more usual mixture of 24s and 12s. Possibly an attempt to produce an alternative to the new frigates. If so, it was a failure, and not repeated.

Dimensions & tons: 140'6", 115'6" x 38'6" x 16'7". 910 59/94.
Men: 350. Guns: GD 20 x 12, UD 22 x 12, QD 6 x 6, Fc 2 x 6.

**Experiment** Adams & Barnard, Deptford.
Ord ____ ; K 12.1772; L 23.8.1774; 1779 taken.
**Medusa** Plymouth Dyd.
Ord 1.8.1775; K 03.1776; L 23.7.1785; (1793-1797 hulked); 1797 hospital ship; 1798 wrecked.

FATES: *Experiment* taken by the French fleet under D'Estaing off Georgia after being dismasted, 22.9.1779. *Medusa* was a receiving ship at Plymouth 1793-1797, then wrecked in Rosas Bay 22.11.1798.

PLANS: Lines/profile/orlop/gun deck/upper deck/quarter deck & forecastle/framing. *Medusa* as built: Lines, profile & decoration/orlop/gun, upper & quarter decks & forecastle.

### Grampus Class 1780. 
Hunt design.
Dimensions & tons: 148', 121'9½" x 40'6" x 17'9". 1062 58/94.
Men: 350. Guns: GD 22 x 24, UD 22 x 12, QD 4 x 6, Fc 2 x 6.

**Grampus** Fisher, Liverpool.
Ord 15.2.1780; K 03.1781; L 8.10.1782; 1794 BU.
**Cato** Cleverley, Gravesend.
Ord 17.2.1780; K 06.1780; L 29.5.1782; 1782 foundered.

FATES: *Cato* was supposed to have foundered going to the East Indies about 31.12.1782.

PLANS: Lines/profile/orlop/gun deck/upper deck/quarter deck & forecastle/specification.

### Trusty Class 1780. 
Hunt design.
Dimensions & tons: 150', 123'9½" x 40'6" x 17'9". 1080 63/94.
Men: 350. Guns: GD 22 x 24, UD 22 x 12, QD 4 x 6, Fc 2 x 6.

**Trusty** Hilhouse, Bristol.

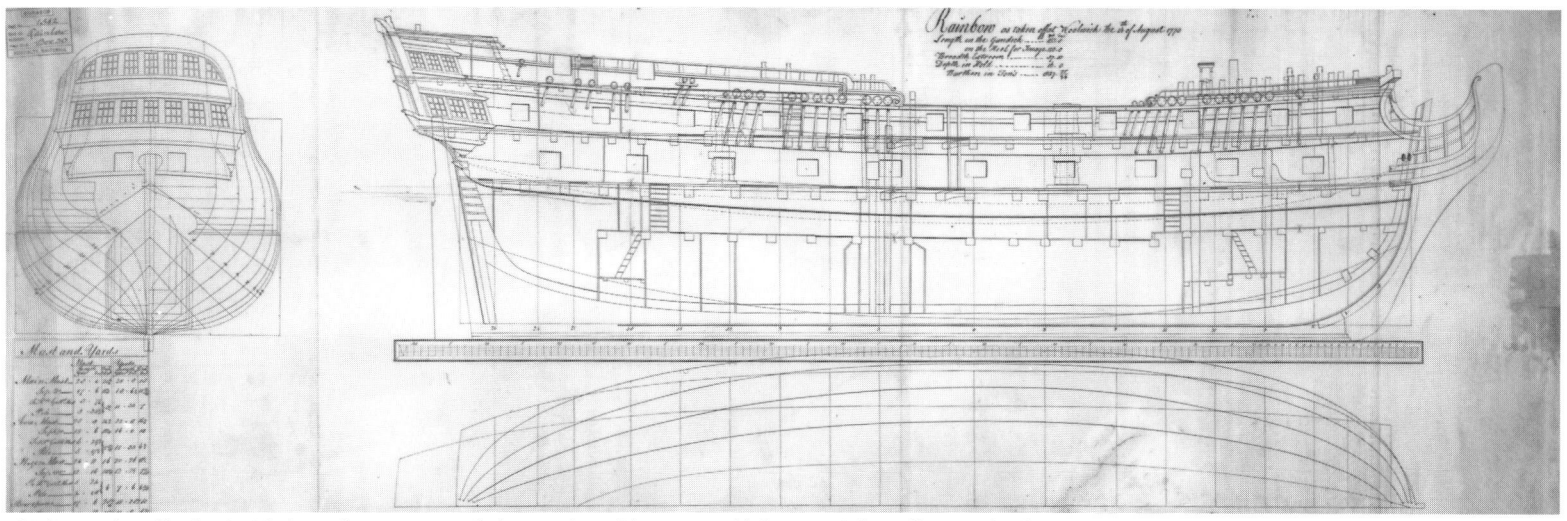

This lines and profile plan (1582) shows the 44 gun two-decker *Rainbow* of the 1745 Establishment as 'taken off' at Woolwich in 1770, with spar dimensions given

## Fifth Rates, 44 Guns (two-deckers)

**1745 Establishment.** There are some indications that *Prince Henry* was a separate design by Allin, though all the rest were certainly to the 1745 Committee design.

Dimensions & tons: 133', 108'10" x 37'6" x 16'. 814 54/94.
Men: 280. Guns: GD 20 x 18, UD 20 x 9, QD 4 x 6.

*Prince Henry* (ex *Culloden*) Gorill, Liverpool.
Ord 3.10.1745; K 11.1745; L 12.3.1747; 1764 BU.

*Assurance* Heather, Bursledon.
Ord 11.12.1745; K 01.1746; L 26.9.1747; 1753 wrecked.

*Expedition* Okill, Liverpool.
Ord 6.1.1746; K 01.1746; L 11.7.1747; 1764 BU.

*Penzance* Chitty, Chichester.
Ord 6.1.1746; K 01.1746; L 7.11.1747; 1766 sold.

*Crown* Taylor, Rotherhithe.
Ord 4.3.1746; K 04.1746; L 13.7.1747; 1757 storeship, 24; 1770 sold.

*Rainbow* Carter, Limehouse.
Ord 4.3.1746; K 03.1746; L 30.5.1747; 1782 all-carronade armament; 1784 hulked.

NOTE: The experimental all-carronade armament consisted of 68 pdrs on the GD, 42 pdrs on the UD, 32 pdrs on the QD. The moral effect of this formidable, if short-ranged, armament caused the surrender of the fine new French frigate *Hébé* and was therefore indirectly responsible for introducing the Royal Navy to the design of the largest class of frigates (the *Leda*s) which it built.

*Humber* Smith, Bursledon.
Ord 5.4.1746; K 04.1746; L 5.3.1748; (1759-1762 hulked.); 1762 wrecked.

*Woolwich* Janvrin, Beaulieu.
Ord 5.4.1746; K 04.1746; L 7.3.1749; 1762 sold.

FATES: *Assurance* wrecked on the Needles, Isle of Wight 24.4.1753; this wreck has been reported found, but there is some confusion with the remains of the later wreck of the *Pomone* which is very close, and most, if not all, the artifacts so far recovered appear to belong to the latter. *Rainbow* became a receiving ship at Woolwich; 1802 sold. *Humber* was a receiving ship at Sheerness 1759-1762 and was wrecked on the 16th or 18th of September 1762 on Hazeboro' Sands off Yarmouth.

PLANS: Lines/orlop/gun deck/upper deck/quarter deck & forecastle/spar plan. *Assurance*: Danish Rigsarkivet: Lines & profile. *Prince Henry*: Lines & profile/orlop/gun deck/upper deck/quarter deck & forecastle. *Crown* as storeship 1757: Profile/orlop & gun deck/upper deck & quarter deck & forecastle. *Rainbow* 1770: Lines & profile/decks/lines & profile/orlop/gun deck/upper deck/quarter deck & forecastle.

**Lengthened 1745 Establishment.** 6' longer, otherwise similar.

*America* Messerwe, Portsmouth, New Hampshire, N. America.
Ord 1.9.1746; K 7.4.1747; L 4.5.1749; 1756 renamed *Boston*; 1757 sold.

PLANS: No plans.

**Phoenix Class 1758.** Slade design.

Dimensions & tons: 140', 117'2" x 36'8" x 16'. 837 83/94.
Men: 280. Guns: GD 20 x 18, UD 22 x 9, Fc 2 x 6.

*Phoenix* Batson, Limehouse.
Ord 7.2.1758; K 02.1758; L 25.6.1759; 1780 wrecked.

FATE: Wrecked on Cuba during the Great West Indian Hurricane on 4.10.1780.

PLANS: Lines/profile/orlop/gun deck/upper deck/quarter deck & forecastle/specification. As built: Lines & profile.

**Roebuck Class 1769.** Slade design. The early members of this class had what appeared to be a 'two level' stern, like larger two-deckers, with windows on two separate levels. (This was an illusion, the windows only actually illuminated the one stern cabin.) These ships are indicated below by a single asterisk [★]. The later ships had a 'single level' frigate-type stern - and are denoted by two asterisks [★★]. Those ships marked by three asterisks [★★★] are those for which the stern arrangement is uncertain, but are more likely to be 'single level' than not. The first *Serapis* is in this group; though pictures showing her engagement with the *Bonhomme Richard* show the frigate-style stern. These pictures cannot be presumed to be accurate, however, as they are unlikely to be drawn from the ship herself. Though this *Serapis* is often referred to a 'frigate' this is not, of course, the correct description of a small two decker of this type.

Dimensions & tons: 140', 115'9" x 37'9½" x 16'4". 879 20/94.
Men: 300 Guns: GD 20 x 18, UD 22 x 9 (short 12s in later vessels), Fc 2 x 6.

*Roebuck*★ Chatham Dyd.
Ord 20.11.1769; K 10.1770; L 28.4.1774; 1790 hospital ship; 1799 troopship; 1803 floating battery; 1811 BU.

*Romulus*★ Adams, Bucklers Hard.
Ord 14.5.1776; K 07.1776; L 17.12.1777; 1781 taken.

*Acteon/Actaeon*★ Randall, Rotherhithe.
Ord 3.7.1776; K 07.1776; L 29.1.1778; 1793 hulked.

*Janus*★ Batson, Limehouse.
Ord 24.7.1776; K 9.8.1776; L 15.5.1778; 1789 troopship; 1800 wrecked.

*Charon*★ [I] Barnard, Harwich.
Ord 9.10.1776; K 01.1777; L 8.10.1778; 1781 sunk.

*Dolphin*★★ Chatham Dyd.
Ord 8.1.1777; K 1.5.1777; L 10.3.1781; 1781 hospital ship; 1800 troopship; 1804 storeship; 1817 BU.

*Ulysses*★★★ Fisher, Liverpool.
Ord 16.4.1777; K 28.6.1777; L 14.7.1779; (1790-1802 troopship); 1816 sold.

*Serapis*★★★ [I] Randall, Rotherhithe.

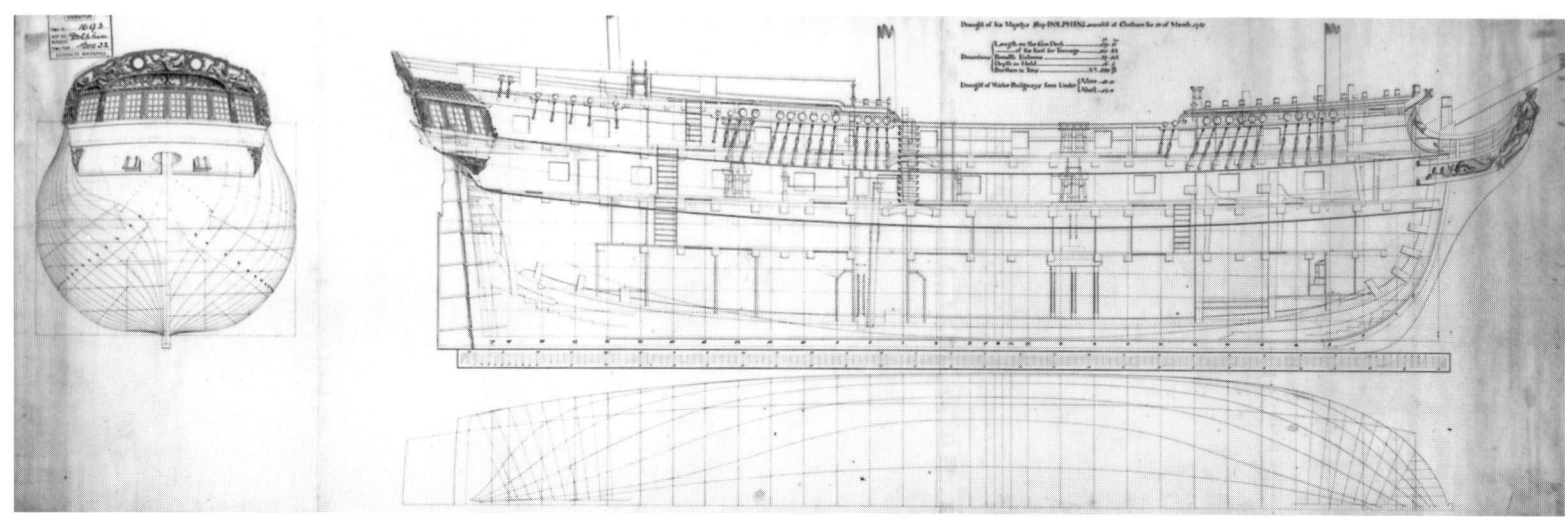

The *Dolphin* of 1781, like most of her sisters of the large *Roebuck* Class of small two-decker 44s, had a single-level 'frigate style' stern gallery as is shown in this 'as built' lines and profile plan (1693)

Ord 11.2.1778; K 3.3.1778; L 4.3.1779; 1779 taken.
**Endymion**★★★ Greaves, Limehouse.
Ord 2.2.1778; K 18.3.1778; L 28.8.1779; 1790 lost.
**Assurance**★★★ Randall, Rotherhithe.
Ord 28.5.1778; K 11.6.1778; L 20.4.1780; 1791 troopship; 1796 transport; 1799 hulked.
**Argo**★★★ Baker, Newcastle.
Ord ____ ; K 18.8.1779; L 8.6.1781; 1791 troopship; 1816 sold.
**Diomede**★★★ Hillhouse, Bristol.
Ord ____ ; K 03.1780; L 18.10.1781; 1795 lost.
**Guardian**★ Batson, Limehouse.
Ord ____ ; K 12.1780; L 23.3.1784; 1791 sold after damage.
**Mediator**★★ Raymond, Northam.
Ord ____ ; K 07.1780; L 30.3.1782; 1788 storeship, renamed *Camel*; 1810 BU.
**Regulus**★★ Raymond, Northam.
Ord ____ ; K 06.1781; L 10.2.1785; 1793 troopship; 1816 BU.
**Resistance**★★ Greaves, Deptford (or Limehouse?).
Ord ____ ; K 04.1781; L 11.7.1782; 1798 blown up.
**Serapis**★★ [II] Hillhouse, Bristol.
Ord 13.7.1780; K 05.1781; L 7.11.1782; 1795 storeship; (1801-1803 floating battery); 1819 hulked.
**Gladiator**★★ Adams, Bucklers Hard.
Ord 13.7.1780; K 04.1781; L 20.1.1783; by 1807 hulked.
**Experiment**★★ Fabian, Cowes.
Ord 13.7.1780; K 06.1781; L 27.11.1784; 1793 troopship; 1805 hulked.
**Charon**★★[II] Hillhouse, Bristol.
Ord ____ ; K 05.1782; L 17.5.1783; 1794 hospital ship; 1800 troopship; 1805 BU.

FATES: *Romulus* was taken off the Capes of Virginia by a French 64 and two frigates; the French renamed her *La Resolution* in 1784; in service until 1789. *Acteon* became a receiving ship at Liverpool; 1802 sold. *Janus* was wrecked (as the *Dromedary*) near Trinidad on 10.8.1800. The 1778 *Charon* was set on fire by French and American shore batteries firing red-hot shot and sank at Yorktown on 10.10.1781; the wreck was found and looted in the 1930s but the remains have more recently been properly excavated. The 1779 *Serapis* was taken off Flamborough Head by John Paul Jones in the *Bonhomme Richard* and the rest of a Franco-American squadron on 23.9.1779 and taken into the French Navy; 1781 wrecked off Madagascar. *Endymion* was lost off Turks Island on 28.9.1790. *Assurance* became a receiving ship at Woolwich; 1815 BU. *Diomede* was wrecked near Trincomalee on 2.8.1795; there are reports of the wreck having been found by divers. *Guardian* was sold at Capetown because of severe damage sustained in striking an iceberg in the Southern Indian Ocean and after an epic voyage to safety under Captain Riou. *Resistance* was accidentally set on fire and then blew up in the Straits of Banca on 24.7.1798. The 1782 *Serapis* became a convalescent ship at Jamaica; 1826 sold. *Gladiator* was hulked at Portsmouth and eventually became a convalescent ship; 1817 BU. *Experiment* became a storeship at Falmouth; 1815 to Liverpool as a lazaretto; 1836 sold.

PLANS: Lines/profile/orlop/gun deck/upper deck/quarter deck & forecastle/specification. *Roebuck*: as built: Lines & profile & decoration. As hospital ship 1780: profile/orlop/gun deck/upper deck/quarter deck & forecastle. *Janus*: as built: Lines & profile/orlop/gun deck/upper deck/quarter deck & forecastle. *Charon*: as built: Lines & profile & decoration/orlop & gun deck/upper deck & quarter deck & forecastle. *Dolphin*: as built: Lines & profile & decoration. *Guardian*: as built: Lines & profile. *Mediator:* as built: Lines & profile. *Regulus*: as built: Lines & profile. 1810 as troopship: profile/orlop/gun deck/upper deck/quarter deck & forecastle. *Resistance*: as built: Lines & profile & decoration. *Serapis* [II] : as built: Lines & profile & decoration. As storeship 1813: Profile & decks/orlop/gun deck/upper deck/quarter deck & forecastle. *Gladiator:* as built: Lines & profile. *Charon* [II]: 1790: gun deck. As hospital ship 1793: profile/orlop/gun deck/upper deck/quarter deck & forecastle.

### *Adventure* Class 1782. Hunt design.

Dimensions & tons: 140', 115'2½" x 38'3" x 16'10". 896 54/94.
Men: 300. Guns: GD 20 x 18, UD 22 x 12, Fc 2 x 6. (later 4 x 12★ were added to the QD).
**Adventure** Perry, Blackwall.
Ord ____ ; K ____ ; L 19.7.1784; 1799 troopship; 1801 hulked.
**Chichester** Smith & Co/Taylor (Crookenden, Taylor & Smith), Itchenor.
Ord ____ ; K 08.1782; L 10.3.1785; 1787 troopship; 1794 storeship; 1810 to the West India Dock Co.
**Expedition** Randall, Rotherhithe.
Ord ____ ; K ____ ; L 29.10.1784; 1798 troopship; 1810 hulked.
**Gorgon** Perry, Blackwall.
Ord ____ ; K ____ ; L 27.1.1785; (1787-1789 troopship); 1793 storeship; 1805 floating battery; 1817 BU.
**Woolwich** Calhoun, Bursledon.
Ord ____ ; K 01.1783; L 15.12.1785; 1793 troopship; 1798 storeship; 1813 troopship; 1813 wrecked.
**Severn** Hillhouse, Bristol.
Ord ____ ; K 06.1783; L 04.1786; 1804 wrecked.
**Dover** Parsons, Bursledon.
Ord ____ ; K 08.1783; L 05.1786; 1795? transport; 1806 burnt.
**Sheerness** Adams, Bucklers Hard.
Ord ____ ; K 12.1783; L 16.7.1787; completed as a troopship; 1805 wrecked.

FATES: *Adventure* became a receiving ship at Sheerness; 1816 BU. *Chichester* became a hulk for the West India Dock Co; 1815 BU. *Expedition* became a ballast ship at Chatham; 1817 BU. *Woolwich* was wrecked on Barbuda Island on 6.11.1813. *Severn* was wrecked in Granville Bay, Jersey on 21.12.1804. *Dover* was accidentally burnt off Woolwich on 20.8.1806 and then BU. *Sheerness* was wrecked on York Island in the inner harbour of Trincomalee on 8.1.1805.

PLANS: Lines/profile/orlop/gun deck/upper deck/quarter deck & forecastle /framing/planking expansion (internal)/(external)/specification. *Chichester*: as built: Lines & profile. 1788: orlop/gundeck. As troopship 1789: Profile/orlop/upper deck. *Gorgon*: As troopship 1790: Gun deck. As hospital

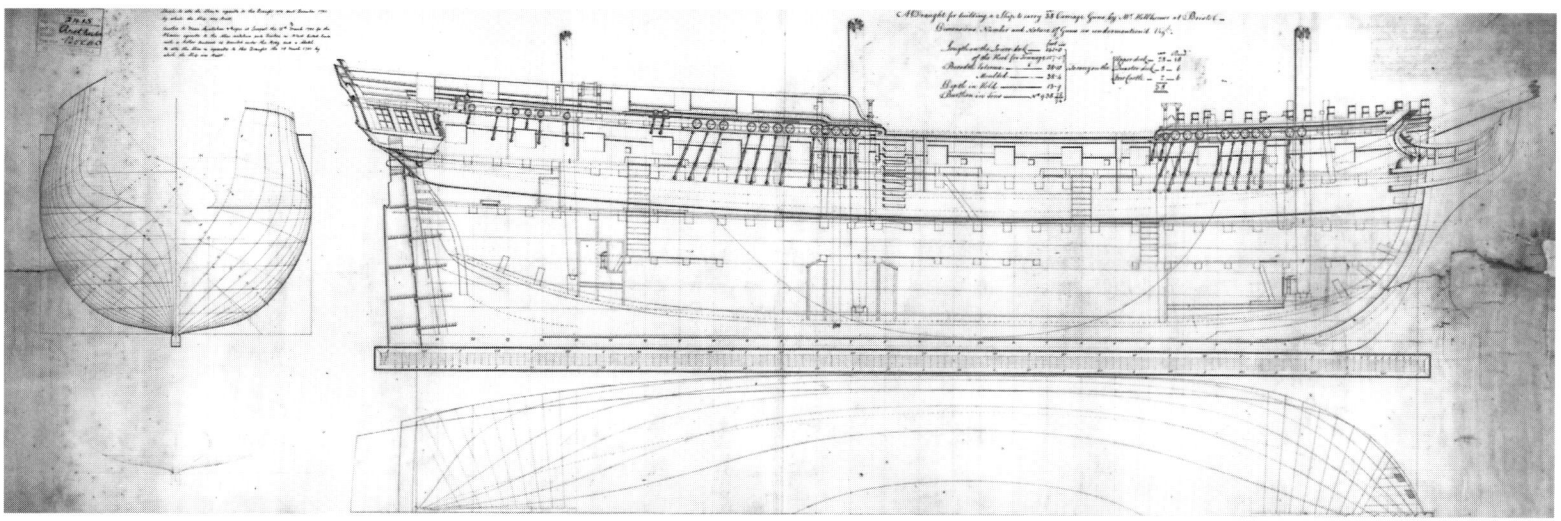

Lines and profile plan (2415) for building the 38 gun frigate *Arethusa* of 1781

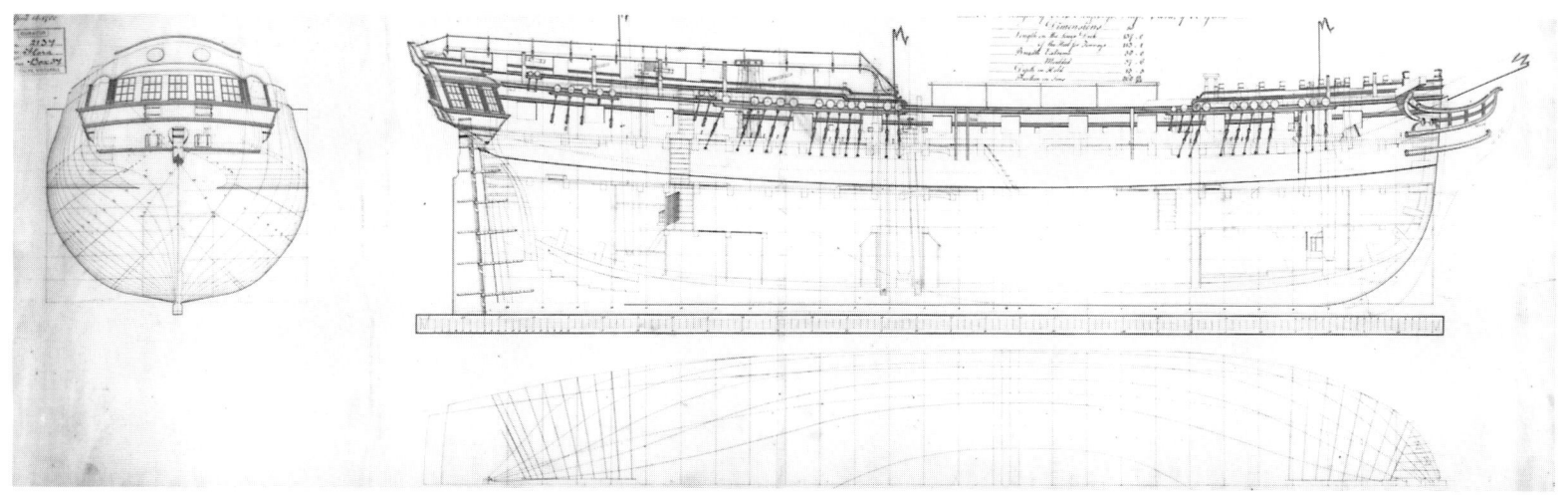

Lines plan (2137) for the 36 gun frigate *Flora* of 1780, the first 18 pounder frigate to be completed

ship 1808: profile/orlop/gun deck/upper deck/quarter deck & forecastle. *Sheerness*: as built: Lines & profile.

# 18 pounder Frigates, 38 Guns (Fifth Rates)

### *Arethusa* Class 1778. Hunt design.
Dimensions & tons: 141', 117' x 38'10" x 13'9". 928 72/94.
Men: 270. Guns: UD 28 x 18, QD 8 x 6, Fc 2 x 6. Also 14 swivels.

*Arethusa* Hillhouse, Bristol.
Ord ____ ; K 23.8.1779; L 10.4.1781; 1814 BU.

*Minerva* Woolwich Dyd.
Ord 6.11.1778; K 11.1778; L 3.6.1780; 1803 BU.

*Phaeton* Smallshaw & Co, Liverpool.
Ord 08.1780; K 06.1780; L 12.6.1782; 1827 sold; 1828 resold.

PLANS: Gun deck/upper deck/quarter deck & forecastle/specification. *Arethusa*: Lines & profile. *Minerva*: as built: Lines & profile. *Phaeton*: as built: Lines & profile.

### *Latona* Class 1779. Williams design.
Dimensions & tons: 141', 116'7" x 38'10" x 13'6". 935.
Men: 270. Guns: UD 28 x 18, QD 8 x 6, Fc 2 x 6. Also 14 swivels (later 10 x 18★ instead).

*Latona* Greaves & Purnell, Limehouse.
Ord ____ ; K 10.1779; L 13.3.1781; 1810 troopship; 1813 hulked.

FATE: Became receiving ship at Leith; 1816 sold.

PLANS: Lines/profile/orlop/gun deck/upper deck/quarter deck & forecastle/framing.

### *Thetis* Class 1781. Hunt design, modified *Arethusa*.
Dimensions & tons: 141'3", 117'0⅜" x 38'10" x ____ 938 72/94.
Men: 270. Guns: UD 28 x 18, QD 8 x 6, Fc 2 x 6. Also 14 swivels.

*Thetis* Randall, Grey & Brent, Rotherhithe.
Ord ____ ; K 12.1781; L 23.9.1782; (1800-1805 troopship); 1814 sold.

PLANS: Lines/profile/orlop/gun deck/upper deck/quarter deck & forecastle/framing.

# 18 pounder Frigates, 36 Guns (Fifth Rates)

### *Flora* Class 1778. Williams design.
Dimensions & tons: 137', 113'1" x 38' x 13'3". 868 53/96.
Men: 260. Guns: UD 26 x 18, QD 8 x 9, Fc 2 x 9. 12 swivels.

*Flora* Deptford Dyd.
Ord 6.11.1778; K 21.11.1778; L 6.5.1780; (1801-1806 registered as a sloop); 1808 wrecked.

*Thalia* (ex *Unicorn*) Calhoun, Bursledon.
Ord ____ ; K 03.1781; L 7.11.1782; (1801-1805 troopship); 1814 BU.

*Crescent* Calhoun, Bursledon.
Ord ____ ; K 11.1781; L 28.10.1784; 1808 wrecked.

*Romulus* Greaves, Deptford.
Ord ____ ; K 11.1782; L 21.9.1785; 1799 troopship; 1803 floating battery?; 1810 troopship; 1813 hulked.

FATES: *Flora* wrecked and then destroyed on the Dutch coast 19.1.1808. *Crescent* was wrecked on the coast of Jutland 6.12.1808 the wreck has recently been located and excavated. *Romulus* was hulked as a hospital ship at Bermuda; 1816 BU.

PLANS: Lines/profile/orlop/gun deck/upper deck/quarter deck & forecastle/framing/stern. *Flora*: as built: Lines & profile/gun deck/upper deck/quarter

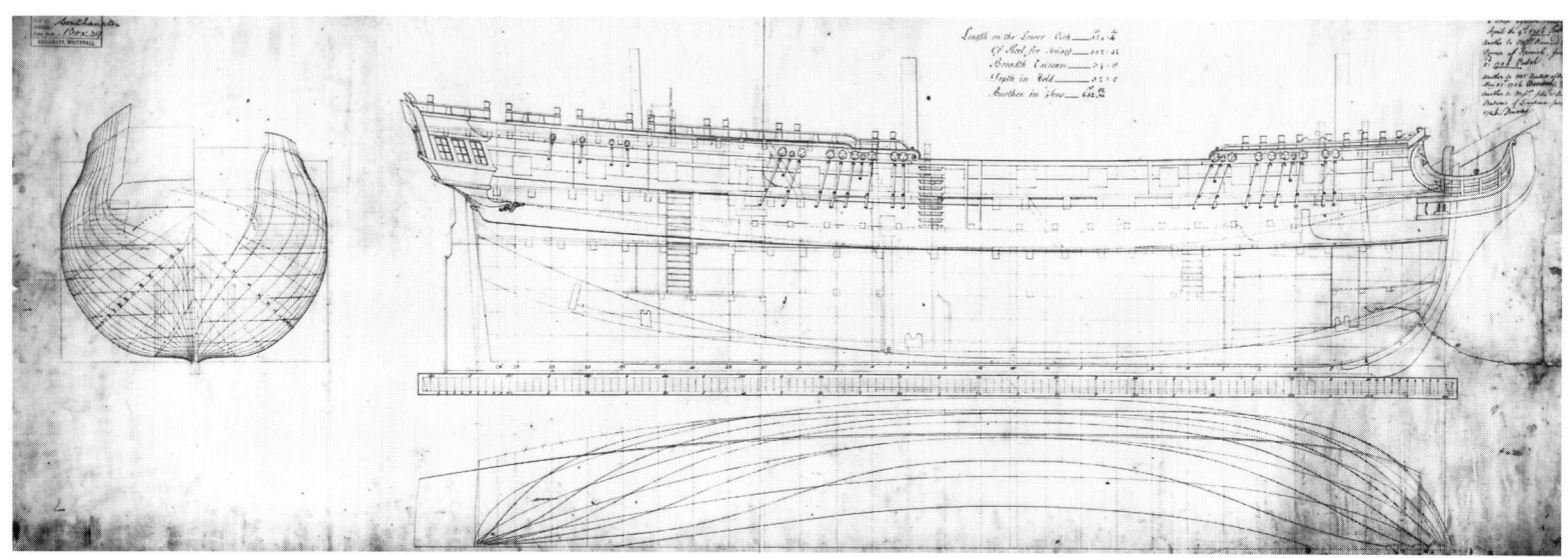

The lines and profile building plan (2355) for the first British 12 pounder frigates, the 32 gun *Southampton* Class of 1756

deck & forecastle. *Crescent:* as built: Lines & profile.

### Cassandra Class 1780. Lines of the French *Prudente* (captured 1779).
Dimensions & tons: ____ ____ .
Men: 260. Guns: UD 26 x 18, QD 8 x 9, Fc 2 x 9. 12 swivels.

**Cancelled 1782.** *Cassandra* Sheerness Dyd. Not begun?
PLANS: No plans. It seems unlikely that any plans were actually drawn (information from Robert Gardiner).

### Perseverance/Inconstant Class 1780. Hunt design. The 1794 *Phoebe* Class was a lengthened version of this design.
Dimensions & tons: 137', 113'5½" x 38' x 13'5". 871 42/94.
Men: 260. Guns: UD 26 x 18, QD 8 x 9, Fc 2 x 9. 12 swivels.
***Perseverance*** Randall, Rotherhithe.
Ord ____ ; K 08.1780; L 10.4.1781; Ca 1806 hulked.
***Phenix/Phoenix*** Parsons, Bursledon.
Ord ____ ; K 08.1781; L 15.7.1783; 1816 wrecked.
***Inconstant*** Barnard, Deptford.
Ord ____ ; K 12.1782; L 28.10.1783; (1798-1806 troopship); 1817 BU.
***Leda*** Randall, Rotherhithe.
Ord ____ ; K 01.1783; L 12.9.1783; 1795 foundered.
***Tribune*** Parsons, Bursledon.
Ord ____ 1801? K 07.1801; L 5.7.1803; 1833 cut down to a corvette 24 (Sixth Rate); 1839 lost.
***Shannon*** (ex *Pallas*) Brindley, Frindsbury.
Ord ____ 1801? K 08.1801; L 2.9.1803; 1803 run ashore & burnt.
***Meleager*** Chatham Dyd.
Ord 21.9.1801/9.7.1802; K 06.1804; L 25.11.1806; 1808 wrecked.
***Iphigenia*** Chatham Dyd.
Ord 21.9.1801/9.7.1802; K 02.1806; L 26.4.1808; 1833 to the Marine Society.
***Orlando*** Chatham Dyd.
Ord ____ 1808?; K 03.1809; L 20.6.1811; 1819 hulked.

### Teak-built version. *Salsette* Bombay Dyd.
Ord 1803?; K 16.4.1806; L 24.3.1807; 1831 hulked.
NOTE: Purchased after completion; teak-built; uncertain whether built to this class design or not.
***Doris*** (*Pitt* until 1807?) Bombay Dyd (built for the Bombay Marine?).
Ord ____ 1803?; K 9.7.1803; L 17.1.1805; 1807 name changed from *Pitt* and purchased by RN (or 1808?); 1829 sold.

**Cancelled 1805.** *Lowestoffe* Woolwich Dyd.
Ord 9.3.1802; not begun?
FATES: *Perseverance* became a receiving ship; 1823 sold. *Phenix* was wrecked near Smyrna on 20.2.1816. *Leda* capsized in a squall off Madeira 11.12.1795. *Tribune* was wrecked near Tarragona on 28.11.1839. *Shannon* was run ashore near La Hogue and burnt to avoid capture on 10.12.1803. *Meleager* was wrecked on Bare Bush Key, Jamaica on 30.7.1808. *Salsette* became a lazaretto at Pembroke; 1835 to Woolwich as a receiving ship; 1874 BU. *Iphigenia* became a training ship for the Marine Society; 1851 BU. *Orlando* became a hospital ship at Trincomalee; 1824 sold. *Doris* was sold at Valparaiso because of her decayed state in 1829.
PLANS: Lines/profile/orlop/gun deck/upper deck/quarter deck & forecastle/specification.

### Melampus Class 1782. Hunt design 'originally intended for a 38'.
Dimensions & tons: 141', 117'0⅜" x 38'10" x 13'9". 938 73/94.
Men: 260. Guns: UD 26 x 18, QD 8 x 9, Fc 2 x 9. 12 swivels.
***Melampus*** Hillhouse, Bristol.
Ord ____ ; K 12.1782; L 8.6.1785; 1815 sold to the Dutch Navy.
PLANS: Lines/orlop/gun deck/upper deck/quarter deck & forecastle/specification.

## 12 pounder Frigates, 36 Guns (Fifth Rates)

### Venus Class 1756. Slade design.
Dimensions & tons: 128'4", 106'2⅝" x 35'8" x 12'4". 718 18/94.
Men: 240. Guns: UD 24 x 12, QD 6 x 6, Fc 2 x 6 (1793: QD 6 x 18★).
***Venus*** Okill, Liverpool.
Ord 13.6. & 26.7.1756; K 16.8.1756; L 11.3.1758; 1792 reclassed as a 32; 1809 renamed *Heroine*; 1824 hulked.
***Pallas*** Wells, Deptford.
Ord 13.7. & 23.7.1756; K 07.1756; L 30.8.1757; 1783 run aground.
***Brilliant*** Plymouth Dyd.
Ord 29.7.1756; K 28.8.1756; L 27.10.1759; 1776 sold.
FATES: *Venus* (as *Heroine*) became a temporary convict ship at Woolwich; 1828 sold. *Pallas* was run ashore on Saint George's Isle 'being so leaky it was impossible to keep her above water'.
PLANS: Lines/profile/orlop/gun deck/upper deck/quarter deck & forecastle. *Brilliant:* as built: Lines & profile/decks.

## 12 pounder Frigates, 32 Guns (Fifth Rates)

### Southampton Class 1756. Slade design.
Dimensions & tons: 124'4", 102'1½" x 34'8" x 12'. 652 52/94.
Men: 220. Guns: UD 26 x 12, QD 4 x 6, Fc 2 x 6. 12 swivels.
***Southampton*** Inwood, Rotherhithe.
Ord 12.3.1756; K 04.1756; L 5.5.1757; 1812 wrecked.
***Minerva*** Quallett, Rotherhithe.
Ord 25.5.1756; K 1.6.1756; L 17.1.1759; (1778-1781 in French hands); 1781 renamed *Recovery*; 1784 sold.
***Vestal*** Barnard, Harwich.

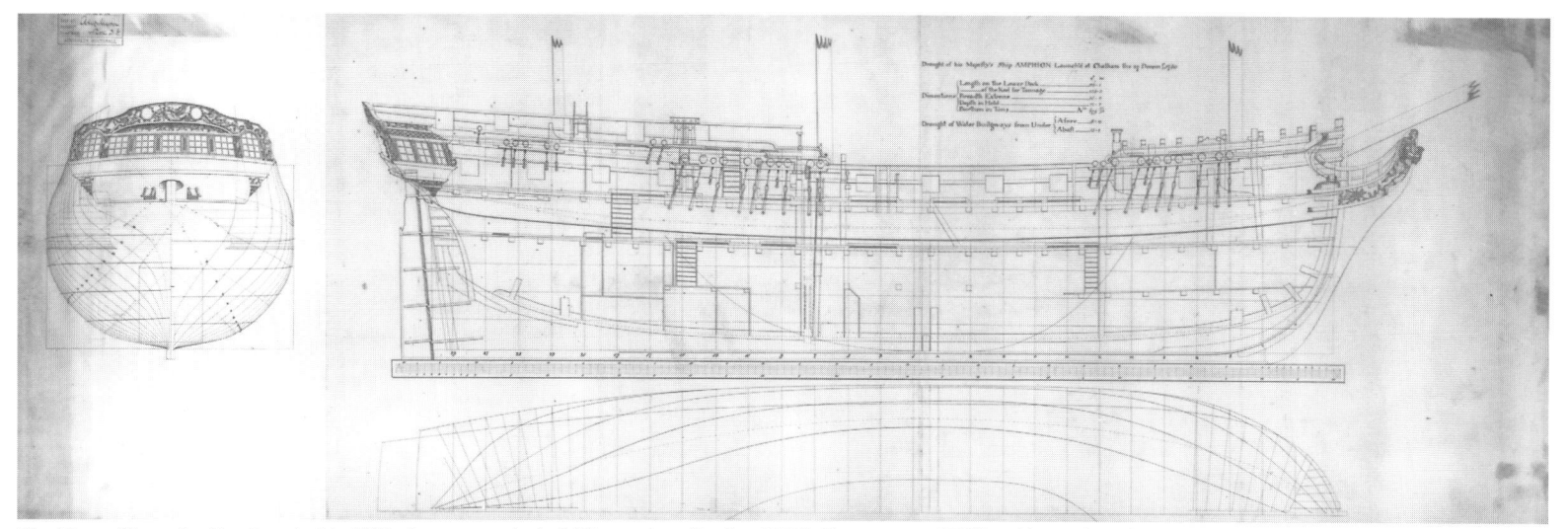

The 32 gun frigate *Amphion,* launched in 1780, shown on an 'as built' lines and profile plan (2262). She was one of Williams' large *Thetis* (later called *Amazon*) Class

Ord 25.5.1756; K 06.1756; L 17.6.1757; 1775 BU.
**Diana** Batson, Limehouse.
Ord 1.6.1756; K 06.1756; L 30.8.1757; 1793 sold.
FATES: *Southampton* was wrecked off Conception Island, Bahamas on 27.11.1812. *Minerva* was taken by the French *La Concorde* in the West Indies on 22.8.1778; 4.1.1781 retaken by *Courageux*.
PLANS: Lines & profile/decks. *Southampton*: Ca 1790: Decks. *Minerva*: 1770: Decks. *Diana*: 1774: Decks.

### *Richmond* Class 1756.
Bately. For the *Thames* Class modified from this design see the next chapter.
Dimensions & tons: 127', 105'1" x 34' x 11'9". 646 12/94.
Men: 220. Guns: UD 26 x 12, QD 4 x 6, Fc 2 x 6. 12 swivels.
**Richmond** Buxton, Deptford.
Ord 12.3.1756; K 04.1756; L 12.11.1757; 1781 taken.
**Juno** Alexander, Rotherhithe.
Ord 1.6.1756; K 06.1756; L 29.9.1757; 1778 scuttled.
**Thames** Adams, Bucklers Hard.
Ord 11.1.1757; K 02.1757; L 10.4.1758; (1793-1796 in French hands); 1803 BU.
**Boston** Inwood, Rotherhithe.
Ord 24.3.1761; K 5.5.1761; L 11.5.1762; (1777-1779 classed as a 28 gun Sixth Rate); 1811 BU.
**Lark** Bird, Rotherhithe.
Ord 24.3.1761; K 5.5.1761; L 10.5.1762; 1778 scuttled.
**Jason** Batson, Limehouse.
Ord 30.1. & 13.2.1762; K 1.4.1762; L 13.2.1763; (1777-1779 classed as a 28 gun Sixth Rate); 1785 sold.
FATES: *Richmond* was taken by De Grasse in the Chesapeake; in French service until burnt at Sardinia in 1795. *Juno* was abandoned and burnt at Rhode Island on 5.8.1778 when blockaded by the French fleet. *Thames* was taken by *La Carmagnole* when going into Gibraltar after an action with *L'Uranie*; 9.6.1796 retaken by *Santa Margarita*. *Lark* was abandoned and burnt at Rhode Island on 5.8.1778, when blockaded by the French.
PLANS: Lines & profile/decks. *Richmond*: 1771: orlop & gun deck. *Lark*, 1776: Decks.

### *Alarm* Class 1757.
Slade design.
Dimensions & tons: 125', 103'4" x 35'2" x 12'. 679 67/94.
Men 220. Guns: UD 26 x 12, QD 4 x 6, Fc 2 x 6. 12 swivels.
**Alarm** Barnard, Harwich.
Ord 19.9.1757; K 26.9.1757; L 19.9.1758; 1812 BU.
**Eolus** West, Deptford.
Ord 19.9.1757; K 09.1757; L 29.11.1758; 1796 hulked.
**Stag** Stanton, Rotherhithe.
Ord 19.9.1757; K 26.9.1757; L 4.9.1758; (1777-1779 classed as a 28 gun Sixth Rate); 1783 BU.
**Pearl** Chatham Dyd.

Ord 24.3.1761; K 6.5.1761; L 27.3.1762; 1803 hulked.
**Glory** Hodgson, Hull.
Ord 30.1. & 13.2.1762; K 03.1762; L 24.10.1763; 1774 renamed *Apollo*; 1786 BU.
**Emerald** Blades, Hull.
Ord 24.3.1761; K 13.5.1761; L 8.6.1762; 1793 BU.
**Aurora** Chatham Dyd.
Ord 8.12.1762; K 10.10.1763; L 13.1.1766; 1769 foundered.
FATES: *Eolus* became a receiving ship at Sheerness; 1800 renamed *Guernsey;* 1801 BU. *Pearl* became a slop ship at Portsmouth; 1814 receiving ship; 1825 renamed *Prothee* 1832 sold. *Aurora* supposed foundered going to the West Indies in 1769.
PLANS: Lines/profile/orlop/gun deck/upper deck/quarter deck & forecastle/stern/specification. *Glory:* 1769: Decks.

### *Niger* Class 1757.
Slade design very similar to the *Alarm* Class.
Dimensions & tons: 125', 103'4" x 35'2" x 12'. 679 62/94.
Men: 220. Guns: UD 26 x 12, QD 4 x 6, Fc 2 x 6. 12 swivels.
**Niger** Sheerness Dyd.
Ord 19.9.1757; K 7.2.1758; L 25.9.1759; 1799 troopship; 1804 Sixth Rate 28 guns, renamed *Negro*?; 1814 sold.
**Montreal** Sheerness Dyd.
Ord 6.6.1759; K 21.4.1760; L 15.9.1761; 1779 taken.
**Quebec** Barnard, Harwich.
Ord 16.7.1759; K 07.1759; L 14.7.1760; (1778-1779 Sixth Rate 28 guns); 1779 blown up.
**Winchelsea** Sheerness Dyd.
Ord 8 & 11.8.1761; K 29.3.1762; L 31.5.1764; 1800 troopship; 1803 hulked.
FATES: *Montreal* taken by French ships of the line off Malaga 29.4.1779; in French service until; 18.12.1793 burnt at Toulon during the British evacuation. *Quebec* blew up whilst engaging the French *Surveillante* 5.10.1779. *Winchelsea* became a mooring hulk (?) at Sheerness; 1814 sold.
PLANS: Lines/profile/orlop/gun deck/upper deck/quarter deck & forecastle/decks/stern/specification. *Niger:* 1769: Orlop & gun deck & upper deck 1777: Orlop & gun deck & upper deck. *Montreal:* 1768: Gun deck, upper deck & quarter deck & forecastle. 1773: Decks.

### *Tweed* Class 1757.
Built to the lengthened lines of the *Tartar* (28); the Slade modification of the *Tyger Prize* design. Built to lighter scantlings than the other 32s, after the French style, and does not seem to have been a success.
Dimensions & tons: 128'4½", 107'9" x 33'11½" x 10'4". 660 86/94.
Men: 220. Guns: UD 26 x 12, QD 4 x 6, Fc 2 x 6. 12 swivels.
**Tweed** Blades, Hull.
Ord 26.11.1757; K 19.1.1758; L 28.4.1759; 1776 sold.
PLANS: Orlop, lower deck & upper deck.

### *Lowestoffe* Class 1760. Adapted by Slade from the lines of the French

*Aurora* (*L'Abénakise* captured 1757).

Dimensions & tons: 130', 107' x 35' x 12'6". 701 35/94. *Diamond* and *Orpheus* differed slightly with 108'2½" keel for tonnage and 705 7/94 tons.

Men: 220. Guns: UD 26 x 12, QD 4 x 6, Fc 2 x 6. 12 swivels.

**Lowestoffe** West, Deptford.
Ord 24.4.1760; K 9.5.1760; L 5.6.1761; 1801 foundered.

**Diamond** Hodgson, Hull.
Ord 25.12.1770; K 05.1771; L 28.5.1774; 1784 sold.

**Orpheus** Barnard, Harwich.
Ord ____ ; K 05.1771; L 7.5.1774; 1778 scuttled.

FATES: *Lowestoffe* was wrecked off Ingua in the West Indies on 10.8.1801. *Orpheus* was abandoned and burnt at Rhode Island when blockaded there by the French on 5.8.1778.

PLANS: Lines/profile/orlop/gun deck/upper deck/quarter deck & forecastle/framing/stern. *Lowestoffe*: 1769: Decks. *Orpheus*: as built: Lines & profile/decks.

### *Thetis*/*Amazon* Class 1771. Williams design.

Dimensions & tons: 126', 104' x 35' x 12'2". 677 62/94.
Men: 220. Guns: UD 26 x 12, QD 4 x 6, Fc 2 x 6. 12 swivels.

**Thetis** Adams, Bucklers Hard.
Ord 25.12.1770; K 02.1771; L 2.11.1773; 1781 wrecked.

**Amazon** Wells, Deptford.
Ord 25.12.1770; K 04.1771; L 24.5.1773; 1794 BU.

**Ambuscade** Adams & Barnard, Deptford.
Ord 02.1771?; K 04.1771; L 17.9.1773; (1798-1803 in French hands); 1810 BU.

**Cleopatra** Hillhouse, Bristol.
Ord 13.5.1778; K 6.7.1778; L 26.11.1779; 1814 BU.

**Amphion** Chatham Dyd.
Ord 11.6.1778; K 1.10.1778; L 27.12.1780; 1796 blown up.

**Iphigenia** Betts, Mistleythorn.
Ord 22. & 26.2.1779; K 25.5.1779; L 27.12.1780; 1798 hulked; 1801 troopship; 1801 burnt.

**Success** Sutton, Liverpool.
Ord 22. & 26.2.1779; K 8.5.1779; L 10.4.1781; (1801 briefly in French hands); 1812 troopship; 1813 hulked.

**Juno** Bateson, Limehouse.
Ord 21.10.1778; K 12.1778?; L 30.9.1780; 1811 BU.

**Andromache** Adams & Barnard, Deptford.
Ord 1.2.1780; K 06.1780; L 17.11.1781; 1811 BU.

**Siren** Betts, Mistleythorn.
Ord 1781?; K 02.1781; L 24.9.1782; 1805 hulked.

**Greyhound** Betts, Mistleythorn.
Ord 1781?; K 01.1782; L 11.12.1783; 1808 wrecked.

**Iris** Barnard, Deptford.
Ord 5.10.1781; K 01.1782; L 2.5.1783; (1803-1805 to Trinity House); 1811 hulked.

**Terpsichore** Betts, Mistleythorn.
Ord 1782?; K 11.1782; L 17.12.1785; 1811 hulked.

**Meleager** Greaves, Frindsbury.
Ord 1782?; K 12.1782; L 28.2.1785; 1801 wrecked.

**Castor** Graham, Harwich.
Ord 1782?; K 01.1783; L 26.5.1785; 1819 sold.

**Orpheus** Adams & Barnard, Deptford.
Ord 2.10.1778; K 7.7.1779; L 3.6.1780; 1807 wrecked.

**Solebay** Adams & Barnard, Deptford.
Ord 1.12.1780?; K 05.1783; L 26.3.1785; (1803-1806 to Trinity House); 1809 wrecked.

**Blonde** Calhoun, Bursledon.
Ord ____ ; K 09.1783; L 22.1.1787; 1798 troopship; 1803 hulked.

FATES: *Thetis* struck on a rock and sank at Saint Lucia on 12.5.1781. *Ambuscade* was taken by the French corvette *Bayonnaise* on 14.12.1798; renamed *Embuscade* in French service; 1803 retaken by *Victory*. *Amphion* accidentally caught fire and blew up in the Hamoaze, Plymouth on 22.9.1796. *Iphigenia* became a temporary prison hospital ship at Plymouth 1798-1801. *Success* taken by Gantheaume's squadron in the Mediterranean 13.2.1801; in French service as *Le Succès* 2.9.1801 recaptured by *Pomone*; 1813 prison ship at Halifax, Nova Scotia; 1820 BU. *Siren* became a lazaretto at Pembroke; 1822 BU. *Greyhound* wrecked on the coast of 'Luconia' (Luzon) in the Philippines 4.10.1808. *Iris* became a receiving ship at Yarmouth; 1809 renamed *Solebay* (?); 1815 to the Marine Society as a training ship; 1833 BU. *Terpsichore* became a receiving ship at Chatham; 1830 BU. *Meleager* was wrecked on the Triangle Bank, Gulf of Mexico 9.6.1801. *Orpheus* was wrecked on a coral reef in the West Indies 23.1.1807. *Solebay* was wrecked in action with a fort at Senegal 11.6.1809. *Blonde* was placed in 'stationary service' at Portsmouth; 1805 sold.

PLANS: Lines/profile/orlop/gun deck/upper deck/quarter deck & forecastle /framing. 1810 alterations for trooping shown on profile & deck plans. *Amphion*: as built: Lines & profile & decoration. *Juno*: as built: Lines & profile & decoration. 1798: Profile/orlop & lower deck/upper deck & quarter deck & forecastle. *Andromache*: as built: Lines & profile & decoration. *Orpheus*: as built: Lines & profile & decoration.

### *Dedalus*/*Active* Class 1778. Hunt design.

Dimensions & tons: 126', 103'9⅜" x 35'4" x 12'2". 689 25/94.
Men: 220. Guns: UD 26 x 12, QD 4 x 6, Fc 2 x 6. 12 swivels.

**Dedalus** Fisher, Liverpool.
Ord 25.6.1778; K 07.1778; L 20.5.1780; (1803-1806 to Trinity House); 1811 BU.

**Mermaid** Woolwich Dyd.
Ord 27.8.1778; K 09.1778 then transferred to: Sheerness Dyd.
Ord 27.8.1778; K 29.7.1782; L 29.11.1784; 1811 troopship 1815 BU.

**Cerberus** Randall, Rotherhithe.
Ord 14.10.1778; K 24.11.1778; L 15.7.1779; 1783 wrecked.

**Active** Raymond, Northam.
Ord 10.12.1778; K 02.1779; L 30.8.1780; 1796 wrecked.

**Fox** Parsons, Bursledon.
Ord 10.12.1778; K 02.1779; L 2.6.1780; 1812 troopship; 1816 BU.

**Ceres** Fearon, Liverpool.
Ord 7.5.1779; K 4.9.1779; L 19.9.1781; 1803 hulked.

**Astrea** Fabian, East Cowes.
Ord 7.5.1779; K 09.1779; L 24.7.1781; (1800-1805 troopship); 1808 wrecked.

**Quebec** Stares, Bursledon.
Ord ____ ; K 06.1780; L 24.5.1781; (1803-1805 hulked); 1813 hulked.

FATES: *Cerberus* wrecked near Bermuda 30.4.1783. *Active* wrecked in the Saint Lawrence River 15.7.1796. *Ceres* became a receiving ship at Sheerness; 1812 to Chatham as harbour flagship; 1816 victualling depot; 1830 BU. *Astrea* wrecked on rocks off Anegada, West Indies 24.5.1808. *Quebec* was hulked at Woolwich 1803-1805; 1813 receiving ship at Sheerness; 1816 BU.

PLANS: Lines/orlop/gun deck/upper deck/quarter deck & forecastle/framing/specification. *Daedalus*: 1792: Lines & profile.

### *Hermione*/*Andromeda* Class 1780. Hunt design.

Dimensions & tons: 129', 107' x 35'4" x 12'8". 710 43/94.
Men: 220. Guns: UD 26 x 12, QD 4 x 6, Fc 2 x 6. 12 swivels.

**Hermione** Teast, Tombs & Blaming, Bristol.
Ord 20.3.1780; K 06.1780; L 9.9.1782; (1797-1799 in Spanish hands); 1799 renamed *Retaliation*; 1800 renamed *Retribution*; 1803 to Trinity House.

**Druid** Teast, Tombs & Blaming, Bristol.
Ord 20.3.1780; K 08.1780; L 16.6.1783; (1798-1805 troopship); 1813 BU.

**Andromeda** Sutton, Liverpool.
Ord ____ ; K 05.1781; L 04.1784; 1811 BU.

**Penelope** Barton, Liverpool.
Ord ____ ; K 02.1782; L 27.10.1783; 1797 BU.

**Aquilon** Young & Woolcombe, Thames.
Ord ____ ; K 11.1782; L 23.11.1786; 1816 BU.

**Blanche** Calhoun, Bursledon.
Ord ____ ; K 07.1783; L 10.7.1786; 1799 troopship; 1799 wrecked.

FATES: *Hermione* was seized by mutineers 22.9.1797; they handed her to the Spaniards at La Guaira who retained her name and classed her as a 34; cut out of harbour and retaken 25.10.1799. 1803 transfer to Trinity House; 1805 BU.

THE SLADE ERA 1745-1785

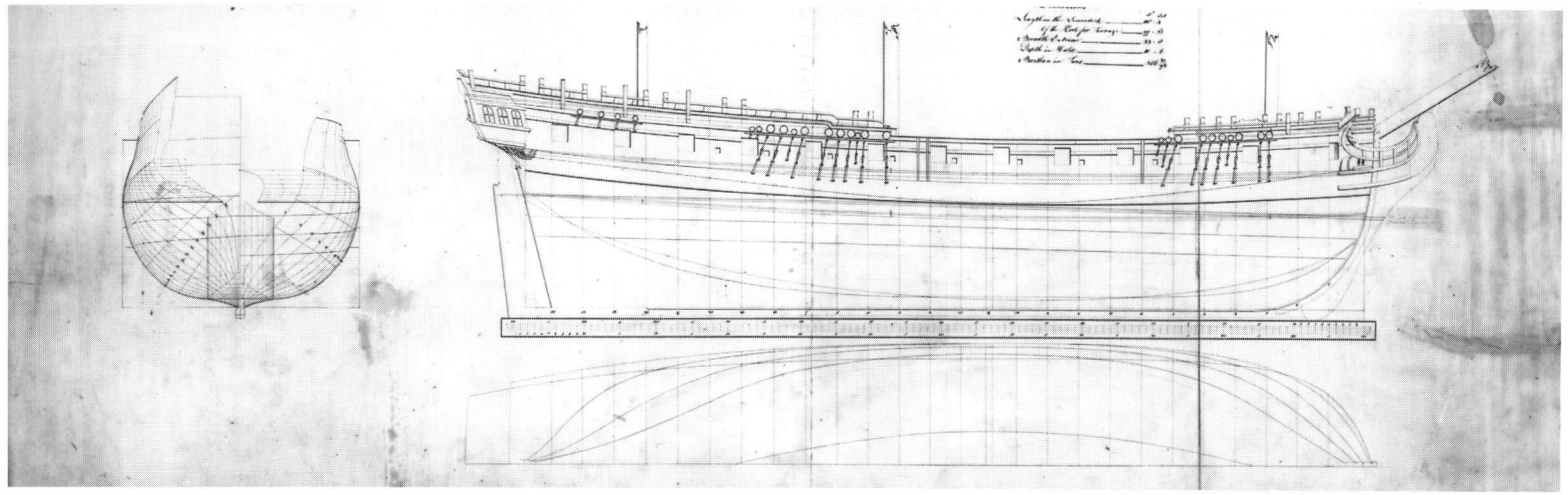

Lines plan 2460 shows the *Carysfort*, a 28 gun frigate of 1766, a modified version of the *Coventry* Class designed a decade earlier

*Blanche* wrecked in the entrance to the Texel.
PLANS: Lines/profile/orlop/gun deck/upper deck/quarter deck & forecastle/framing/stern /specification. *Andromeda*: as built: Lines & profile.

## Frigates, 28 Guns, 9 pounder (Sixth Rates)

***Lyme* Class 1747.** Built to the lines of the *Tyger*, French prize (a privateer which was not taken into RN service). Originally classed as 24s, not reclassed as 28s until 1756. Nonetheless they were the prototypes of the 28 gun frigates, and the first true frigates built for the RN. There were slight differences between the two, particularly in the structure of the bow. Both the *Lowestoff* and *Coventry* classes were developed from this design, as was the enlarged *Tweed* (32 guns).

 Dimensions & tons: 117'10", 96'5½" x 33'8" x 9'10½". 597 92/94.
 Men: 160/180. Guns: UD 22 x 9, QD 2 x 3. 12 swivels (1756: UD 24 x 9, QD 4 x 3).
***Lyme*** Deptford Dyd.
 Ord 29.4.1747; K 24.9.1747; L 10.12.1748; 1760 wrecked.
***Unicorn*** Plymouth Dyd.
 Ord 29.4. & 1.5.1747; K 3.7.1747; L 7.12.1748; 1771 BU.
FATES: *Lyme* wrecked on the Baltic coast of Sweden 18.10.1760.
PLANS: Lines & profile. *Lyme*: as built: Lines & profile & decoration/orlop & gun deck & upper deck/gun deck/upper deck. *Unicorn*: (As built: Lines & profile & decoration in the Chapman Collection at Stockholm; reproduced in Chapman's 'Architectura Navalis Mercatoria', this plan looks very much as if it was an original Admiralty plan, removed by Chapman during his visit to Britain; a visit which was interrupted by his being gaoled for espionage.)

***Lowestoff/Tartar* Class 1753.** Slade design based on the *Lyme* 'with such alterations as may tend to the better stowing of men and carrying for guns'. There is some suggestion that *Argo* and *Guadeloupe* should be included in the next (*Coventry*) Class rather than this one.

 Dimensions & tons: 117'10", 96'8½" x 33'8" x 10'2". 583 3/94.
 Men: 200. Guns: UD 24 x 9, QD 4 x 3. 12 swivels.
***Lowestoff*** Greaves, Limehouse.
 Ord 20.5.1755; K 06.1755; L 17.5.1756; 1760 wrecked.
***Tartar*** Randall, Rotherhithe.
 Ord 12.6.1755; K 4.7.1755; L 3.4.1756; 1797 wrecked.
***Argo*** Bird, Rotherhithe.
 Ord 19.9.1757; K 27.9.1757; L 20.7.1758; 1776 BU.
***Guadeloupe*** Williams, Milford.
 Ord 19.9.1757; K 29.6.1758 builder failed & order transferred to: Plymouth Dyd.
 Ord ____ ; K 3.8.1759; L 5.12.1763; 1781 scuttled.
FATES: *Lowestoff* wrecked in the Saint Lawrence at the raising of the siege of Quebec 16.5.1760. *Tartar* wrecked at San Domingo 27.4.1797. *Guadeloupe* sunk at Yorktown to avoid capture 10.10.1781.
PLANS: Lines & profile/orlop/gun deck/upper deck/quarter deck & forecastle/specification. *Tartar*: 1763: Gun & upper decks. 1770: sheer. 1790: sheer. 1792: orlop. *Argo*: 1775: Decks. *Guadeloupe*: as built ?: Lines & profile. 1770: Decks.

***Coventry* Class 1756.** Slade design 'by the draught of the *Tartar* with such alterations withinboard as may be judged necessary'. In other words a further development of the *Lyme* design.

 Dimensions & tons: 118'4", 97'3½" x 33'8" x 10'6". 586 76/94.
 Men: 200. Guns: UD 24 x 9, QD 4 x 3. 12 swivels.
***Coventry*** Adams, Beaulieu.
 Ord 15.4.1756; K 31.5.1756; L 30.5.1757; 1783 taken .
***Lizard*** Bird, Rotherhithe.
 Ord 15.4.1756; K 5.5.1756; L 7.4.1757; 1800 hulked.
***Liverpool*** Gorill, Liverpool.
 Ord 3.9.1756; K 1.10.1756; L 10.2.1758; 1778 wrecked.
***Maidstone*** Sewards, Rochester.
 Ord 3.9.1756; K 1.10.1756; L 9.2.1758; 1794 BU.
***Active*** Stanton, Rotherhithe.
 Ord 6.5.1757; K 13.6.1757; L 11.1.1758; 1778 taken.
***Aquilon*** Inwood, Rotherhithe.
 Ord 6.5.1757; K 15.6.1757; L 24.5.1758; 1776 sold.
***Cerberus*** Fen, Cowes.
 Ord 6.5.1757; K 13.6.1757; L 5.9.1758; 1778 destroyed.
***Griffin*** Janverin, Bursledon.
 Ord 6.5.1757; K 06.1757; L 18.10.1758; 1761 wrecked.
***Levant*** Adams, Beaulieu .
 Ord 6.5.1757; K 06.1757; L 6.7.1758; 1780 BU.

**Fir Built**. ***Actaeon*** Chatham Dyd.
 Ord 5.5./6.5.1757; K 26.5.1757; L 30.9.1757; 1766 sold.
***Hussar*** Chatham Dyd.
 Ord 18.4.1757; K 3.5.1757; L 23.7.1757; 1762 taken.
***Shannon*** Deptford Dyd.
 Ord 18.4.1757; K 11.5.1757; L 17.8.1757; 1765 BU.
***Trent*** Woolwich Dyd.
 Ord 5.5.1757; K 19.5.1757; L 31.10.1757; 1764 sold.
***Boreas*** Woolwich Dyd.
 Ord 18.4.1757; K 21.4.1757; L 29.7.1757; 1770 sold.

**Modified Version?** ***Milford*** Chitty, Milford Haven.
 Ord 19.9.1757; K 11.1757; L 20.9.1759; 1785 sold.
***Carysfort*** Sheerness Dyd.
 Ord 20.2.1764; K 06.1764; L 23.8.1766; 1813 sold.
***Hind*** Clayton & Wilson, Sandgate.
 Ord 1783?; K 02.1783; L 22.7.1785; 1811 BU.

**Cancelled 7.10.1783.** *Laurel* Jacobs, Sandgate. Ord 1783? Not begun (builder failed).

FATES: *Coventry* taken by the French off Ganjam in the Bay of Bengal; French service until 1785. *Lizard* became a hospital ship for the lazarettos at Sheerness;

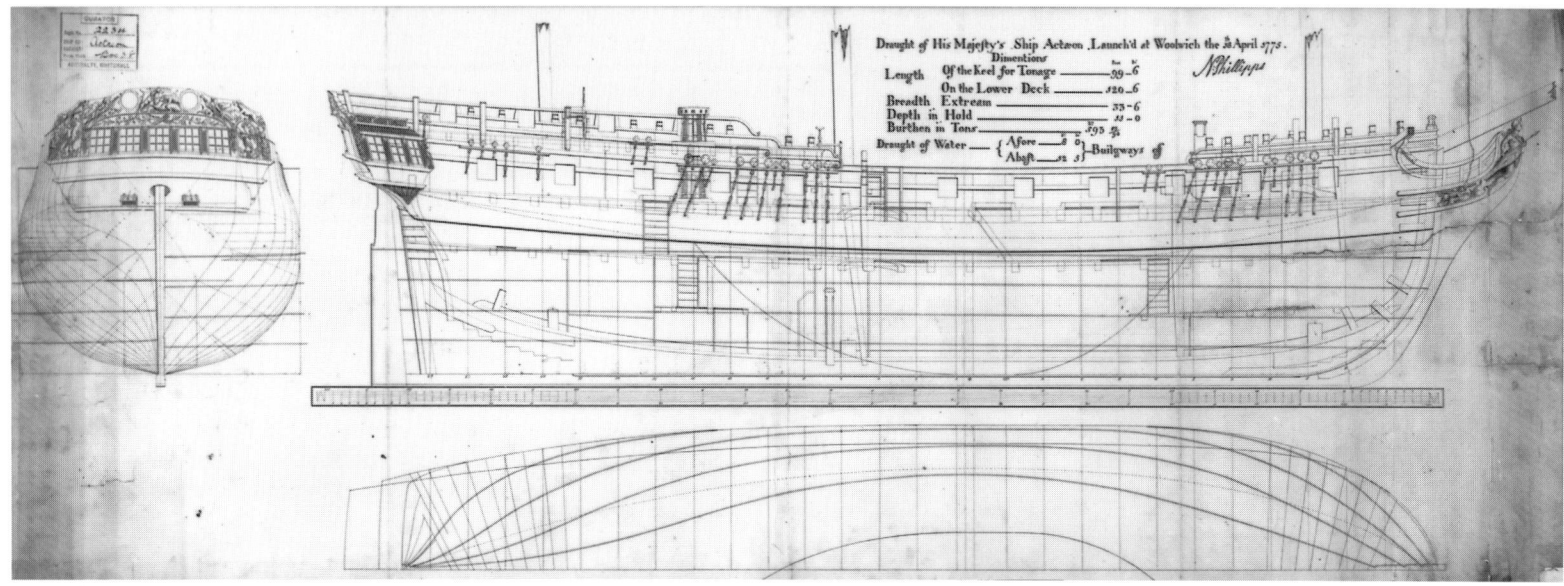

As so often this detailed 'as built' lines and profile plan (2234) illustrates a member (*Actaeon/Acteon* 1775) of one of Williams' classes (*Enterprise* Class 28 gun frigates) built in numbers at the time of the American War of Independence

1828 sold. *Liverpool* was wrecked on Long Island 11.2.1778. *Active* taken by the French *Dédaigneuse* and *Charmante* off San Domingo 1.9.1778; in French service until 1788. *Cerberus* abandoned and burnt at Rhode Island when blockaded there by the French 5.8.1778; wreck recently excavated. *Griffin* wrecked on the North East shoals off Barbuda 4.10.1761. *Hussar* stranded on Cape François (or on the East end of Cuba) and taken by the French 22.5.1762 (but not salvaged?).

PLANS: Lines/profile/orlop/gun deck/upper deck/quarter deck & forecastle/framing. *Coventry*: 1776: Decks. *Liverpool*: as built?: Lines & profile/orlop, gun & upper decks. *Active*: as built ?: Lines & profile. 1771: Decks. *Cerberus*: as built?: Decks. *Levant*: as built: Lines & profile. Fir built: Lines & profile/profile/orlop/gun deck/upper deck/quarter deck & forecastle/framing. Modified version: Lines/profile/orlop/gun deck/upper deck/quarter deck & forecastle/framing. *Milford*: 1775: Decks. *Carysfort*: as built?: Lines & profile/decks.

### *Mermaid* Class 1760.
Slade design. Reduced and adapted from the lines of the French *Aurora* 36 (*L'Abénakise* taken 1757) 'nearly similar to the *Aurora* prize.'

Dimensions & tons: 124', 102'8⅛" x 33'6" x 11'. 612 72/94. The last three of the class differed in being 103'4¾" keel for tonnage and 617 22/94 tons.

Men: 200. Guns: UD 24 x 9, QD 4 x 3. 12 swivels.

**Mermaid** Blades, Hull.
Ord 24.4.1760; K 27.5.1760; L 6.5.1761; 1778 wrecked.
**Hussar** Inwood, Rotherhithe.
Ord 30.1. & 13.2.1762; K 1.4.1762; L 26.8.1763; 1780 wrecked.
**Solebay** Airey & Co, Newcastle.
Ord 30.1.1762; K 10.5.1762; L 9.9.1763; 1782 wrecked.
**Greyhound** Adams, Bucklers Hard.
Ord ____ ; K 02.1771; L 20.7.1773; 1781 wrecked.
**Triton** Adams, Bucklers Hard.
Ord ____ ; K 02.1771; L 1.10.1773; 1796 BU.
**Boreas** Hodgson, Hull.
Ord ____ ; K 05.1771; L 23.8.1774; 1797 hulked.

FATES: *Mermaid* was driven ashore by D'Estaing's fleet in the Delaware 8.7.1778 (or 08.1778?); there are reports that this wreck has been located. *Hussar* was wrecked on a rock at Hell Gate, New York 29.11.1780; it is believed that this wreck was located and salvaged some time ago. *Solebay* was wrecked 'by running ashore to avoid the French fleet on their attacking the fleet under Sir Samuel Hood' off Nevis Point, and then burnt 25.1.1782. *Greyhound* was wrecked on the South Sand near Deal in August 1781. *Boreas* becamea slop ship at Sheerness; 1802 sold.

PLANS: Lines/profile/orlop/gun deck/upper deck/quarter deck & forecastle/framing. *Mermaid*: as built?: Decks. *Hussar*: as built ?: Decks. *Solebay*: 1769: Orlop, gun & upper decks. *Boreas*: as built?: Decks.

### *Enterprize* Class 1770.
Williams design.

Dimensions & tons: 120'6", 99'6" x 33'6" x 11'. 593 89/94.
Men: 200. Guns: UD 24 x 9, QD 4 x 3. 12 swivels.

**Enterprize** Deptford Dyd.
Ord 01.1771; K 9.9.1771; L 24.8.1774; 1791 hulked.
**Siren** Henniker, Chatham.
Ord 25.12.1770; K 04.1771; L 2.11.1773; 1777 wrecked.
**Fox** Calhoun, Northam.
Ord 25.12.1770; K 05.1771; L 2.9.1773; (1777 briefly in American hands); 1778 taken.
**Surprize** Woolwich Dyd.
Ord 01.1771; K 5.9.1771; L 13.4.1774; 1783 sold.
**Acteon** Woolwich Dyd.
Ord 5.11.1771; K 10.1772; L 18.4.1775; 1776 burnt.
**Medea** Hillhouse, Bristol.
Ord 14.5.1776; K ____ 1776; L 28.4.1778; 1801 hulked.
**Proserpine** Barnard, Harwich.
Ord 14.5.1776; K 06.1776; L 7.7.1777; 1799 wrecked.
**Andromeda** Fabian, Cowes.
Ord 24.5.1776; K 07.1776; L 18.1.1777; 1780 foundered.
**Aurora** Perry, Blackwall.
Ord 3.7.1776; K 07.1776; L 7.6.1777; 1814 sold.
**Sibyl** Adams, Bucklers Hard.
Ord 24.7.1776; K 10.12.1776; L 2.1.1779; 1795 renamed *Garland*; 1798 lost.
**Brilliant** Adams, Bucklers Hard.
Ord 9.10.1776; K 02.1777; L 15.7.1779; 1811 BU.
**Pomona** Raymond, Southampton.
Ord 02 or 03.1777; K 8.5.1777; L 22.9.1778; 1795 renamed *Amphitrite*; 1811 BU.
**Crescent** Hillhouse, Bristol.
Ord 19.7.1777; K 19.8.1777; L 03.1779; 1781 taken.
**Nemesis** Jolly, Liverpool.
Ord [originally] 30.9.1777; K 11.777; builder failed, transferred to: Smanshall, Liverpool.
Ord ____ ; L 23.1.1780; (1795-1796 in French hands); 1812 troopship; 1814 sold.
**Resource** Randall, Rotherhithe.
Ord 30.9.1777; K 11.1777; L 10.8.1778; 1799 troopship; 1803 hulked.
**Mercury** Mestaers, Thames.
Ord 22.1.1778; K 25.3.1778; L 9.12.1779; 1803 floating battery; 1810 troopship 1814 BU.
**Cyclops** Menetone, Thames.
Ord 6.3.1778; K 3.4.1778; L 31.7.1779; 1800 troopship; 1807 hulked.

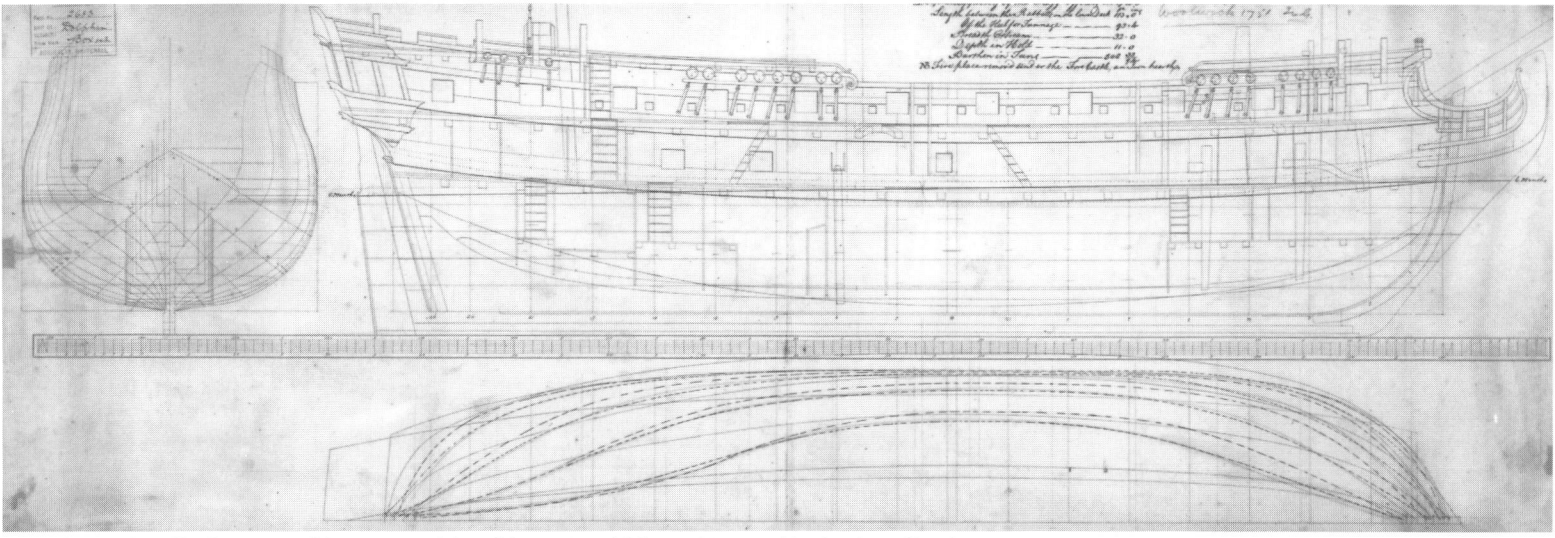

Design lines and profile plan (2683) of the 24 gun *Dolphin* of the 1745 Establishment 'Approved by the Flag Officers'

*Vestal* Batsons, Thames.
  Ord 18.3.1778; K 1.5.1778; L 24.12.1779; 1800 troopship; (1803-1810 to Trinity House); 1810 troopship; 1814 hulked.
*Laurel* Raymond, Southampton.
  Ord 30.4.1778; K 3.6.1778; L 27.10.1779; 1780 wrecked.
*Pegasus* Deptford Dyd.
  Ord 21.2.1778; K 20.6.1778; L 1.6.1779; 1800 troopship; 1814 hulked.
*Hussar* Wilson, Sandgate.
  Ord 1782?; K 06.1782; L 1.9.1784; 1796 wrecked.
*Rose* Stewart & Hall, Sandgate.
  Ord 1782?; K 06.1782; L 1.7.1783; 1794 wrecked.
*Dido* Stewart & Hall, Sandgate.
  Ord 1782?; K 09.1782; L 27.11.1784; 1800 troopship; 1804 hulked.
*Thisbe* King, Dover.
  Ord 1782? K 09.1782; L 25.11.1783; 1800 troopship; 1815 sold.
*Alligator* Jacobs, Sandgate.
  Ord 1782?; K 12.1782; L 18.4.1787; 1810 hulked.
*Circe* Ladd, Dover.
  Ord 1782?; K 12.1782; L 30.9.1785; 1803 wrecked.
*Lapwing* King, Dover.
  Ord 02.1783?; K 02.1783; L 21.9.1785; 1810 hulked.

NOTE: Thanks to what appears to be a clerical error in the list of ships in Charnock's *History of Naval Architecture* subsequent historians have often listed a vessel called *Ariel* launched in 1785 at Dover as the last of this class and type, an error repeated and reinforced by the 'identification' of a model in the Science Museum as this vessel. In fact there was never such a vessel; there are no plans, no mention in Admiralty lists, no recorded history, etc. In short this is a classical illustration of the danger of using near-contemporary printed works as 'authorities'.

FATES: *Enterprize* became a receiving ship for impressed men at the Tower of London; 1806 to Deptford; 1807 BU. *Siren* was wrecked on the coast of Connecticut 6.11.1777. *Fox* was taken by American 'privateers' (actually USS *Hancock* and consort) 7.6.1777; 8.7.1777 retaken by *Flora* and *Rainbow*; 10.9.1778 taken by the French *Junon* off Brest; 1778 or 1779 wrecked. *Acteon* grounded at Charleston and was burnt to avoid capture 28.6.1776. *Medea* went aground (1801) and was then hulked as a hospital ship at Portsmouth; 1805 sold. *Proserpine* was wrecked off Heligoland 7.1.1799. *Andromeda* was overset and foundered during the Great West Indian Hurricane about 6 leagues to the Windward of Martinique on 11/12.10.1780. *Sibyl* (as *Garland*) was lost off Madagascar on 26.7.1798. *Crescent* was taken by the French *Gloire* and *Friponne* (together with the Dutch *Castor* which had just been taken by *Crescent* and *Flora*) on 20.6.1781 served the French until; 1786 wrecked. *Nemesis* was taken by the French *Sensible* and *Sardoine* 9.12.1795; 9.3.1796 retaken by *Egmont* near Tunis. *Resource* became a receiving ship at the Tower of London and was renamed (?) *Enterprize* 1803; 1816 BU. *Cyclops* became a receiving ship at Portsmouth; 1814 sold. *Vestal* became a prison ship at Barbados; 1816 sold. *Laurel* was driven ashore at Martinique 'and very soon went to pieces' during the Great West Indian Hurricane 11/12.10.1780. *Pegasus* became a receiving ship at Deptford or Chatham; 1816 sold. *Hussar* was wrecked near Isle Bas 24.12.1796. *Rose* was wrecked on Rocky Point, Jamaica 28.6.1794. *Dido* became an Army prison ship at Portsmouth; 1817 sold. *Alligator* was used as a salvage ship at Cork in 1810; then laid up at Plymouth?; but sold at Cork 1814. *Circe* was wrecked on the Lemon and Ower (shoal) near Yarmouth 16.11.1803. *Lapwing* became a salvage vessel at Cork 1810 to raise *Britannia*; 1813 to Pembroke as accommodation for clerks etc; 1828 BU.

PLANS: Lines/profile/orlop/gun deck/upper deck/quarter deck & forecastle/midships section/framing/planking expansion (internal)/(external). *Siren*: as built: Lines & profile & decoration /decks. *Surprize*: as built: Lines & profile & decoration. 1776: Decks. *Acteon*: as built: Lines & profile & decoration. *Rose*: as built: Lines & profile/orlop/lower deck/upper deck/quarter deck & forecastle. *Dido*: as built: Lines & profile/orlop/lower deck/upper deck/quarter deck & forecastle. *Circe* as built: Lines & profile/decks.

## Sixth Rates, 24 Guns

### 1745 Establishment.
  Dimensions & tons: 113', 93'4" x 32' x 11'. 508 32/94.
  Men: 160. Guns: LD 2 x 9 (aft), UD 20 x 9, QD 2 x 3.
*Arundel* Chitty, Chichester.
  Ord 3.10.1745; K 10.1745; L 23.11.1746; 1765 sold.
*Queenborough* Sparrow, Rotherhithe.
  Ord 9.8.1746; K 26.8.1746; L 2.1.1748; 1761 wrecked.
*Fowey* Janvrin, Beaulieu.
  Ord 26.9.1747; K 10.1747; L 4.7.1749; 1781 sunk.
*Hind* Chitty, Chichester.
  Ord 26.9.1747; K 10.1747; L 29.11.1749; 1782 transport; 1784 sold.
*Sphinx* Allen, Rotherhithe.
  Ord 26.9.1747; K 11.1747; L 10.12.1748; 1770 sold.
*Dolphin* Sparrow, Rotherhithe.
  Ord 26.9.1747; not begun? transferred to:
  Woolwich Dyd.
  Ord 10.6.1748; K 3.8.1748; L 1.5.1751; 1777 BU.

FATES: *Queenborough* was driven ashore a little to the Southward of Ariancopang near Pondicherry during a hurricane 1.1.1761. *Fowey* was sunk by shore batteries at Yorktown 10.10.1781; the wreck has apparently been found.

PLANS: Lines/lines & profile/lower deck/upper deck/quarter deck & forecastle/spar plan. *Queenborough*: Gun deck/upper deck. As built?: Lines & profile. *Fowey*: 1772: Decks. *Dolphin*: Lines & profile. As built: Sheer & profile. 1770 as expedition ship: Decks.

### Modified 1745 Establishment (American built).
  Dimensions & tons: 118', 97'7½" x 32'8½" x 11'1¾". 555 51/94.
  Men: 160. Guns: LD 2 x 9 (aft), UD 20 x 9, QD 2 x 3?
*Boston* Hallowell, Boston, Massachusetts.
  Ord 09.1746; K late 1746; L 3.5.1748; 1752 BU.

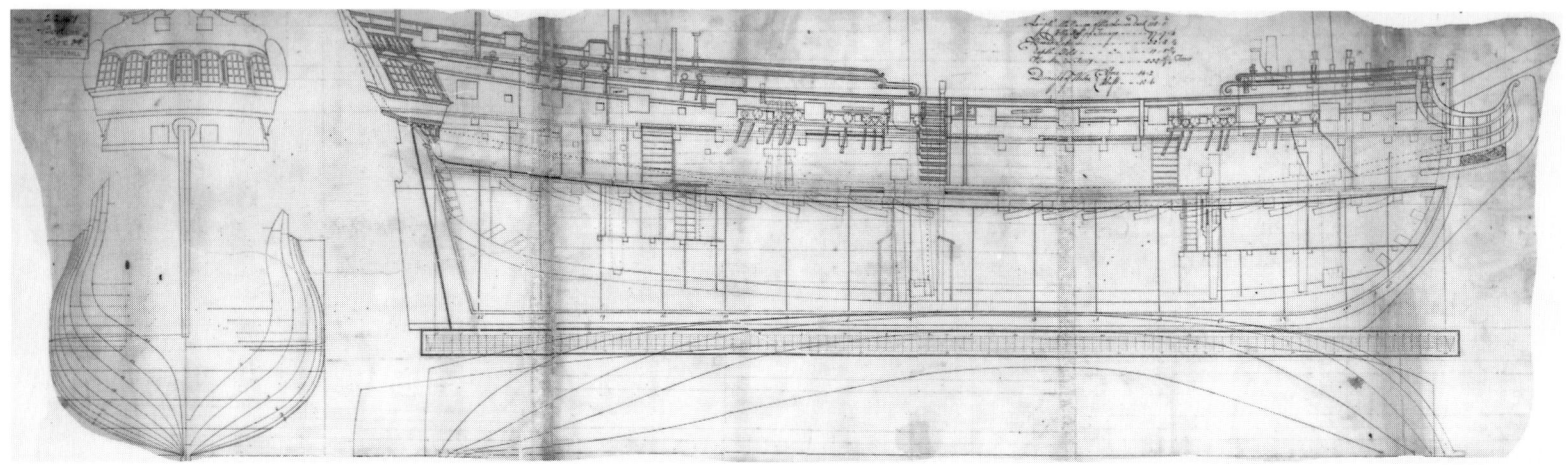

This lines and profile plan (2247) of the American-built *Boston* makes an interesting comparison with the plan illustrated on the previous page of the *Dolphin* – the former being modified from that design

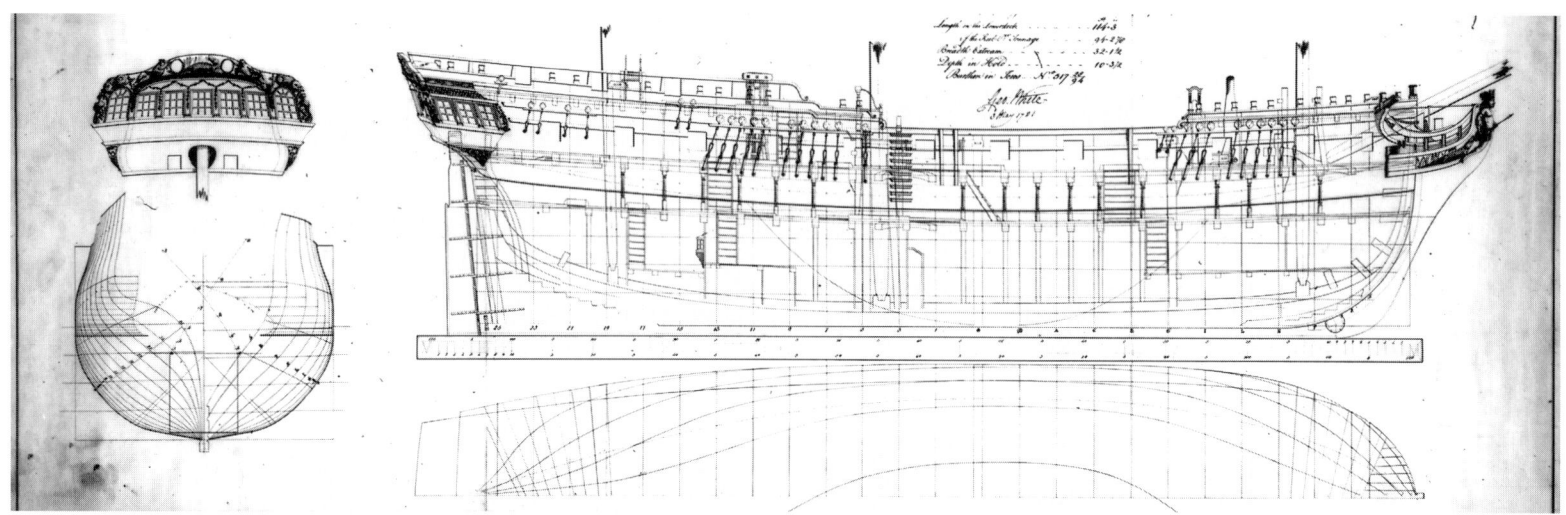

Plan 2868 is the lines and profile of the *Crocodile* of 1781 'as built'. She was a 24 gun ship of the *Porcupine* Class

PLANS: No plans.

**Lyme Class 1747.** Though at first classed as 24 gun ships, they were the prototypes of the 28 gun frigates, and are listed in full under this type (see above).

**Mermaid Class 1748.** Allin design, an intermediate stage in the introduction of the frigate.
  Dimensions & tons: 115', 96'8" x 32' x 10'2". 526 45/94.
  Men: 160. Guns: LD 2 x 9 (aft), UD 20 x 9, QD 2 x 3, also 6 x ½pdr. swivels.
*Mermaid* Adams, Beaulieu.
  Ord 4.2. & 5.3.1748; K 2.4.1748; L 22.5.1749; 1760 lost.
FATE: Wrecked off the coast of South Carolina on 6.1.1760.
PLANS: Lines/lines & profile.

**Seahorse Class 1748.** Acworth design, like *Mermaid* (above), an intermediate stage in the development of the frigate.
  Dimensions & tons: 112', 92'6" x 32' x 10'. 504.
  Men: 160. Guns: UD 22 x 9, QD 2 x 4.
*Seahorse* Barnard, Harwich.
  Ord 4.2.1748; K 23.2.1748; L 13.9.1748; 1784 sold.
PLANS: Lines/orlop/gun deck/upper deck/quarter deck & forecastle. As built: Lines & profile. 1770: Sheer/orlop & quarter deck & forecastle/lower & upper decks.

**Porcupine Class 1776.** Williams design
  Dimensions & tons: 114'3", 94'3½" x 32' x 10'3". 513 55/94.
  Men: 160 (*Hyena* 1798:100). Guns: UD 22 x 9, QD 2 x 6 (in 1790 *Pandora* was given 4 x 18★ in place of her QD guns. In 1798 *Hyena* was armed with 20 x 32★ on the UD and no guns on the QD).
*Porcupine* Graves, Limehouse.
  Ord 21.6.1776; K 07.1776; L 17.12.1777; 1805 BU.
*Pelican* Adams & Barnard, Deptford.
  Ord 24.7.1776; K 08.1776; L 24.4.1777; 1781 wrecked.
*Eurydice* Portsmouth Dyd.
  Ord 24.7.1776; K 8.10.1776?; L 26.3.1781; 1824 hulked.
*Hyena* Fisher, Liverpool.
  Ord 9.10.1776; K 05.1777; L 1.2.1778; (1793-1797 in French hands); 1798 reclassed as 20; 1802 sold.
*Amphitrite* Deptford Dyd.
  Ord 8.1.1777; K 2.7.1777; L 28.5.1778; 1794 wrecked.
*Crocodile* Portsmouth Dyd.
  Ord 8.1.1777; K 02.1777; L 25.4.1781; 1784 wrecked.
*Penelope* Baker, Liverpool.
  Ord ____ ; K 28.6.1777; L 25.6.1778; 1779 foundered.
*Siren* Baker, Newcastle.
  Ord ____ ; K 21.1.1778; L 30.7.1779; 1781 wrecked.
*Pandora* Adams & Barnard, Deptford.
  Ord 11.2.1778?; K 2.3.1778; L 17.5.1779; 1791 wrecked.
*Champion* Barnard, Harwich.
  Ord 11.2.1778?; K 04.1778; L 17.5.1779; 1809 hulked.
FATES: *Pelican* was wrecked in a hurricane at Jamaica 18.8.1781. *Eurydice* became a receiving ship at Woolwich; 1834 BU. *Hyena* was taken by the French *La Concorde* in the West Indies 27.5.1793 renamed *L'Hyène* in French service; 27.10.1797 retaken by *Indefatigable*. *Amphitrite* was wrecked on the rocks off Leghorn (Livorno) on 30.1.1794. *Crocodile* was wrecked on the Scilly Rocks off Prawle Point on 9.5.1784; the wreck has been found and is being excavated by local divers. *Penelope* was supposed cast away (or foundered) in the West Indies during November 1779. *Siren* was wrecked near Beachy Head in January 1781

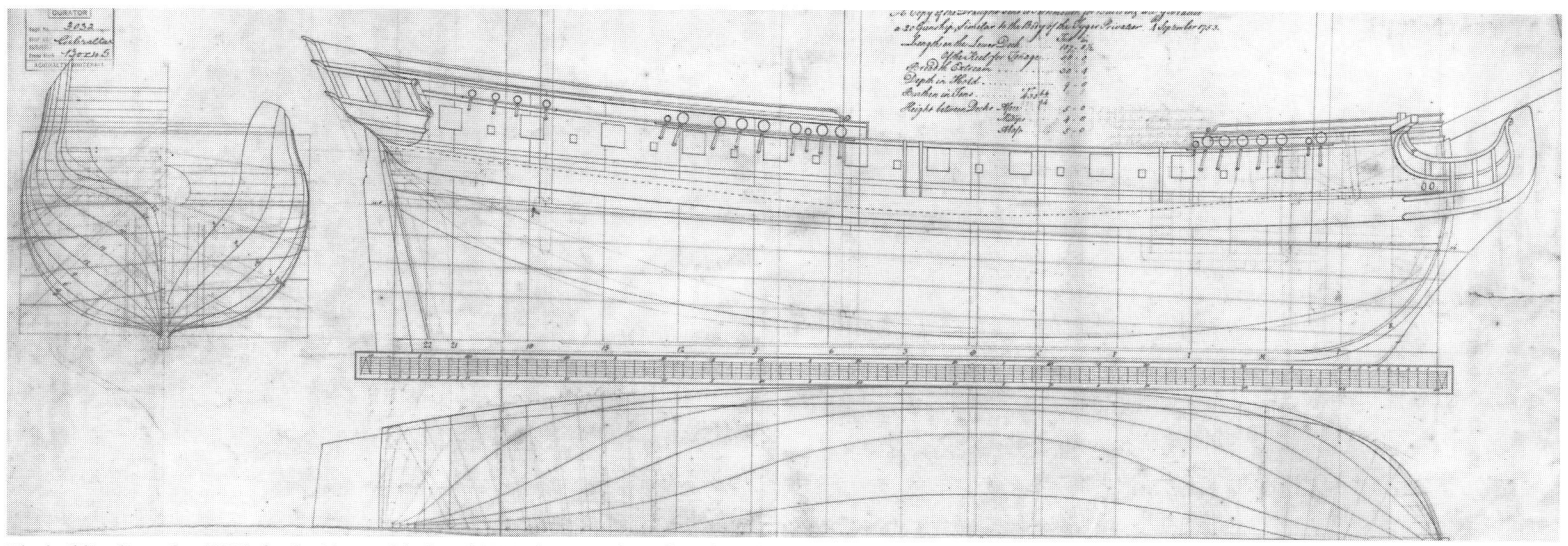

The building lines plan (3032) for the 20 gun *Gibraltar* of 1754, the name ship of her class, shows the stem post curved back at the top - a characteristic of French construction which will have been adopted because her lines were a reduced copy of those of the French privateer *Tyger*

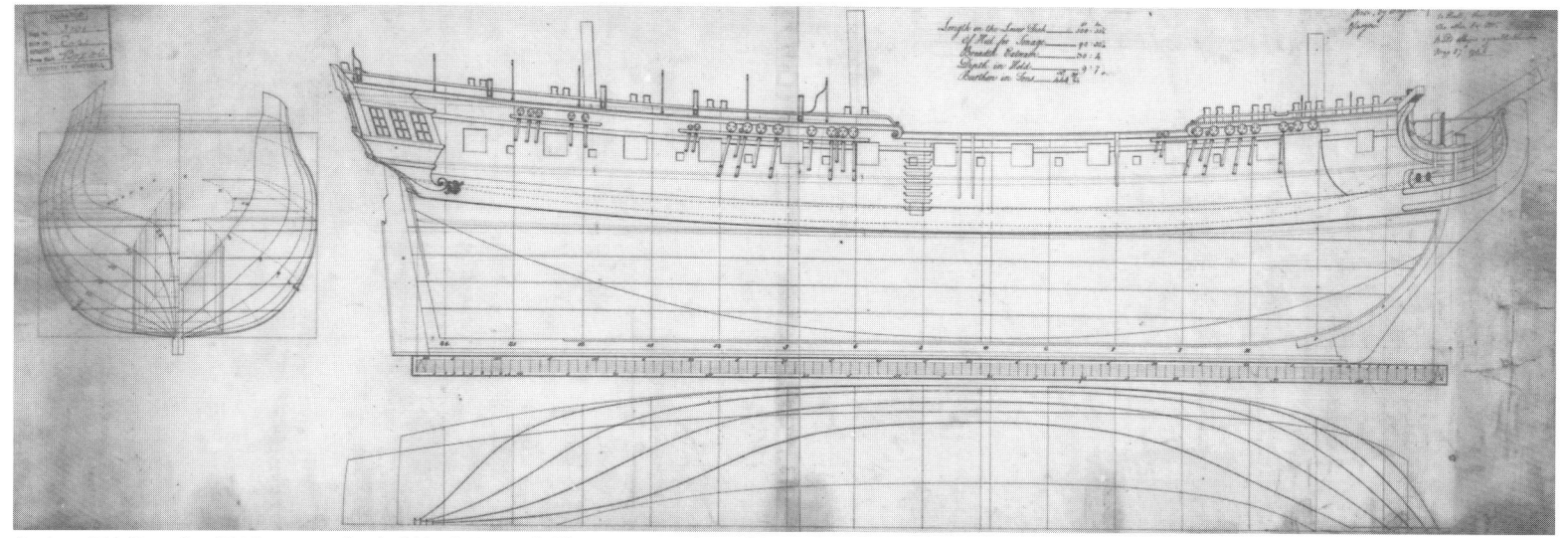

Copies of this lines plan (3101) were used to build both *Rose* and *Glasgow,* 20 guns ships of the *Seaford* Class, in 1757

and the wreck was sold for BU. *Pandora* was wrecked in the entrance to Endeavour Straits (off the Northern tip of Queensland) whilst taking captured *Bounty* mutineers back for trial; the wreck has recently been discovered and is being excavated under the auspices of the Queensland Museum. *Champion* became a receiving ship at Sheerness; 1816 sold.

PLANS: Lines/profile/orlop/gun deck/upper deck/quarter deck & forecastle/framing/planking expansion (internal). *Eurydice:* as built: Lines & profile & decoration. *Crocodile*: as built: Lines & profile & decoration.

### *Myrmidon* Class 1779.
'In all respects similar to the *Amazon* which was taken from the French in the War preceding the last.' (Privateer taken 1745.) Classed as a ship of 22 guns.
  Dimensions & tons: 113'9½", 94'2" x 31' x 10'2". 481 15/94.
  Men: 160. Guns: UD 20 x 6, QD 2 x 3.
***Myrmidon*** Deptford Dyd.
  Ord 10.6.1779; K 19.11.1779; L 9.6.1781; 1798 hulked?
FATE: Harbour service at Plymouth from 1798?; 1811 BU.
PLANS: Lines & profile/orlop/gun deck/upper deck/quarter deck & forecastle.

### *Squirrel* Class 1782.
Hunt design.
  Dimensions & tons: 119', 99' x 32'5" x 10'3". 553 34/94.
  Men: 200. Guns: UD 22 x 9, QD 2 x 6.
***Squirrel*** Barton, Liverpool.
  Ord ____ ; K 03.1783; L 9.5.1785; 1812 hulked.
FATE: Receiving ship at Portsmouth; 1817 sold.
PLANS: Lines/profile/orlop/gun deck/upper deck/quarter deck & forecastle/framing/stern/planking expansion.

## Sixth Rates, 20 Guns

### *Gibraltar* Class 1753.
Built to the reduced lines of the French privateer *Tyger* (Compare the *Lyme* Class of 24/28 gun frigates). *Biddeford* was originally to be built at Portsmouth then at Deptford to the lines of the *Royal Caroline* (*Seaford* Class) but this was soon altered to the lines of the *Tyger*. The contract built vessels had more height between decks than the dockyard built pair.
  Dimensions & tons: 107'8½", 88' x 30'4" x 9'8". 430 64/94. (Dockyard built.) 107'8", 88'6" x 30'4" x 9'8". 433 12/94 (Contract built.)
  Men: 160. Guns: UD 20 x 9. (In 1756 *Aldborough* had 2 x 3 added to the Fc).
***Gibraltar*** Portsmouth Dyd.
  Ord 29.8.1753; K 11.9.1753; L 9.5.1754; 1773 BU.
***Biddeford*** Deptford Dyd.
  Ord 30.10.1754; K 9.1.1755; L 2.3.1756; 1761 wrecked.
***Aldborough*** Perry, Blackwall.
  Ord 11.4.1755; K 7.5.1755; L 15.5.1756; 1756 reclassed as a 22; 1777 BU.
***Flamborough*** Batson, Limehouse.
  Ord 11.4.1755; K 27.5.1755; L 14.5.1756; 1772 sold.
***Kennington*** Adams, Bucklers Hard.
  Ord 20.5.1755; K 06.1755?; L 1.5.1756; 1774 BU.
***Lively*** Janvrin, Hamble.
  Ord 20.5.1755; K 06.1755?; L 10.8.1756; (1778-1781 in French hands);

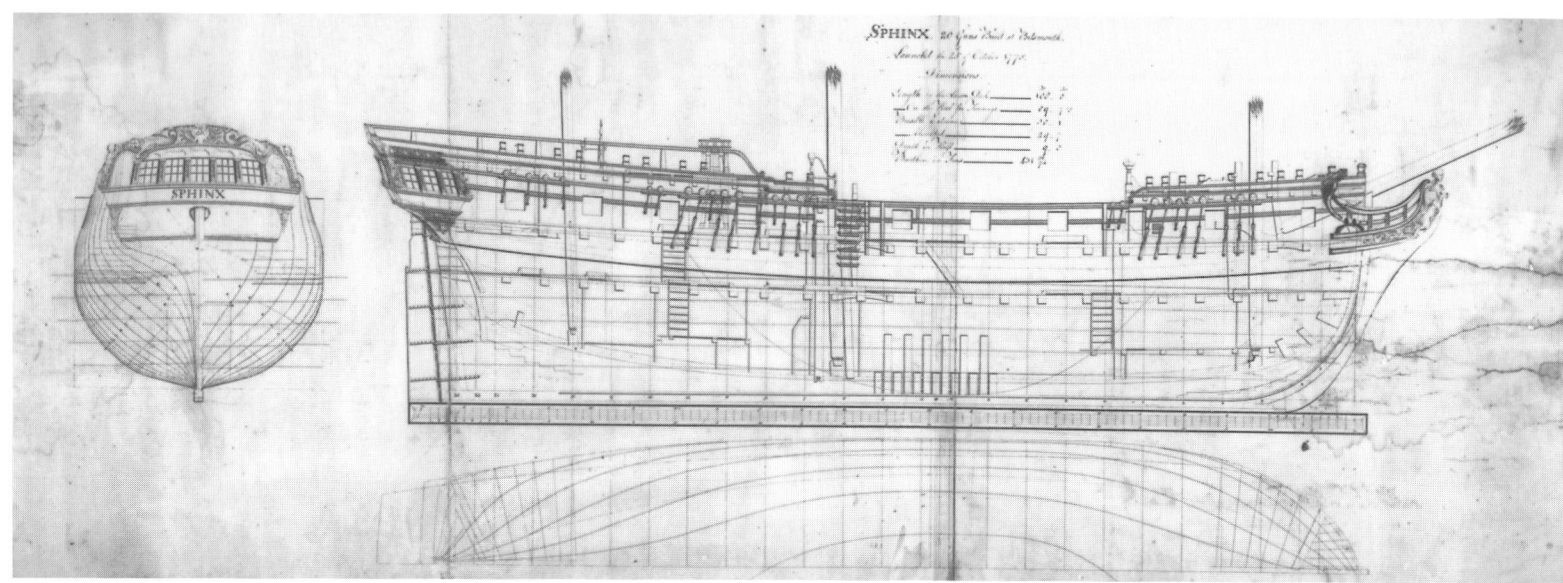

The 20 gun ship *Sphinx*, depicted in this 'as built' lines and profile plan (2998), was the name ship of her class

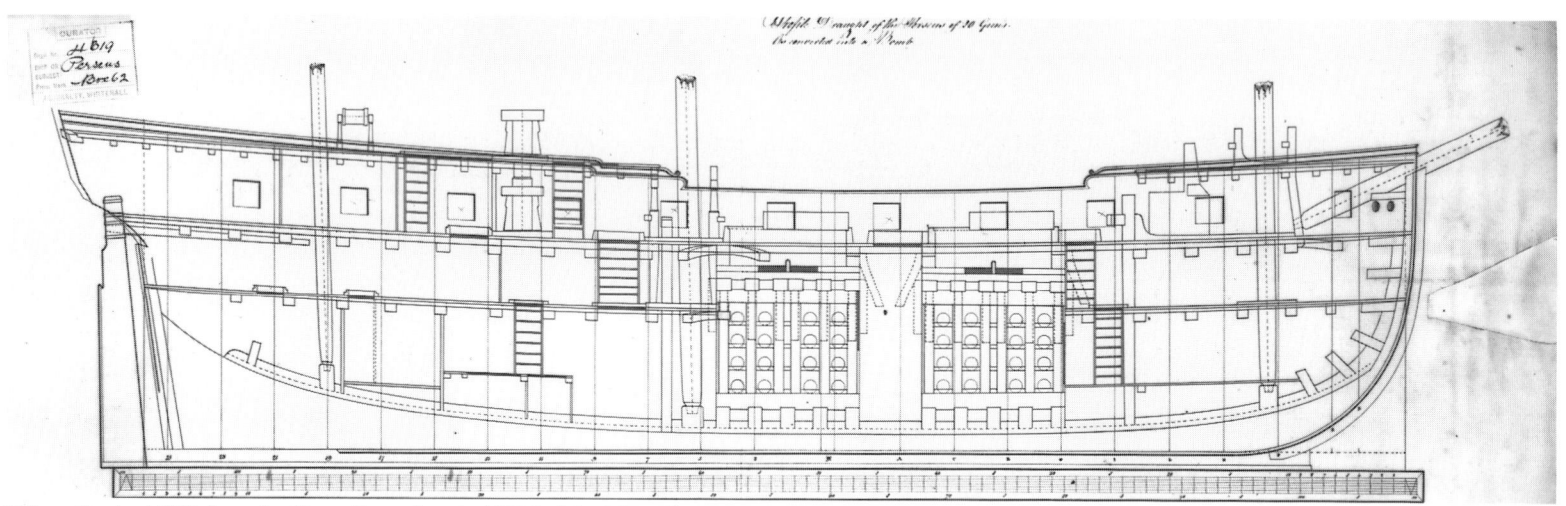

This profile plan (4319) shows the 20 gun *Perseus* (launched 1776) of the *Sphinx* Class (compare the plan of that ship illustrated above) fitting as bomb in 1798

1784 sold.
**Scarborough** Blades, Hull.
   Ord 12.6.1755; K 21.7.1755; L 17.4.1756; 1780 foundered.
**Mercury** Barnard, Harwich.
   Ord 12.6.1755; K 1.7.1755; L 2.3.1756; 1777 wrecked.
FATES: *Biddeford* was cast away off Yarmouth on 30.10.1761. *Lively* was taken by the French *Iphigénie* 10.7.1778; 29.7.1781 retaken by *Perseverance*. *Scarborough* was presumed lost in the Great West Indian Hurricane (by 31.10.1780). *Mercury* was wrecked near New York in 1777.
PLANS: Lines/lines & profile/orlop/gun deck/upper deck/quarter deck & forecastle /specification. *Gibraltar*: as built: Lines & profile/decks. 1768: Decks. *Lively*: 1769: Decks. *Mercury*: as built: Decks.

### Seaford Class 1753. Built to the lines of the yacht *Royal Caroline*.
*Biddeford* was originally to be built to this design. (See *Gibraltar* Class above).
   Dimensions & tons: 108', 89'4⅞" x 30'1" x 9'6". 430 37/94 (Dockyard built.) 108'11½", 90'10¼" x 30'4" x 9'7". (Contract built.)
   Men: 160. Guns: UD 20 x 9.
**Seaford** Deptford Dyd.
   Ord 29.8.1753; K 25.9.1753; L 3.9.1754; 1780 reclassed as a 22; 1784 sold.
**Rose** Blades, Hull.
   Ord 13.4.1756; K 5.6.1756; L 8.3.1757; 1779 scuttled.
**Glasgow** Reed, Hull.
   Ord 13.4.1756; K 5.6.1756; L 31.8.1757; 1779 burnt.
FATES: *Rose* was sunk to block Savannah Bar to prevent the entrance of the French 19.9.1779 [A rather inaccurate 'replica' was constructed some years ago; on the grounds that the *Rose*'s depredations had been the direct cause of the foundation of the US Navy; and is still in existence at the time of writing (1992), persistently and inaccurately referred to as 'the frigate *Rose*'.] *Glasgow* was accidentally burnt by the carelessness of the Steward in Montego Bay, Jamaica 1.6.1779
PLANS: Lines/lines & profile/profile. *Seaford*: 1769: Decks. 1770: Decks. 1771: Decks.

### Squirrel Class 1754. Lines of the yacht *Royal Caroline*. It is not clear whether these two ships should be counted as a sub-group of the previous class, or a separate, though similar, class.
   Dimensions & tons: 107', 89'7½" x 29' x 9'2". 400 87/94.
   Men: 160. Guns: UD 20 x 9.
**Squirrel** Woolwich Dyd.
   Ord 30.10.1754; K 19.3.1755; L 23.10.1755; 1783 sold.
**Deal Castle** Deptford Dyd.
   Ord 30.10.1754; K 9.1.1755; L 20.1.1756; 1780 lost.
FATES: *Deal Castle* was lost off Puerto Rico having been driven from Saint Lucia by the Great West Indian Hurricane on 15.10.1780.
PLANS: Lines & profile/specification. *Deal Castle*: as built: Decks/quarter deck & forecastle. 1768: Decks. 1775: Decks.

### Sphinx Class 1773. Williams design.
   Dimensions & tons: 108', 89'8" x 30' x 9'8". 429 23/94.
   Men: 160. Guns: 20 x 9. (Later those reclassed as 24s were given 4 x 4 on QD & Fc. *Perseus* as a bomb carried 1 x 13" + 1 x 10" mortars & 10 x 6.)
**Sphinx** Portsmouth Dyd.
   Ord 15.4.1773; K 11.1773; L 25.10.1775; (?1778 briefly fitted

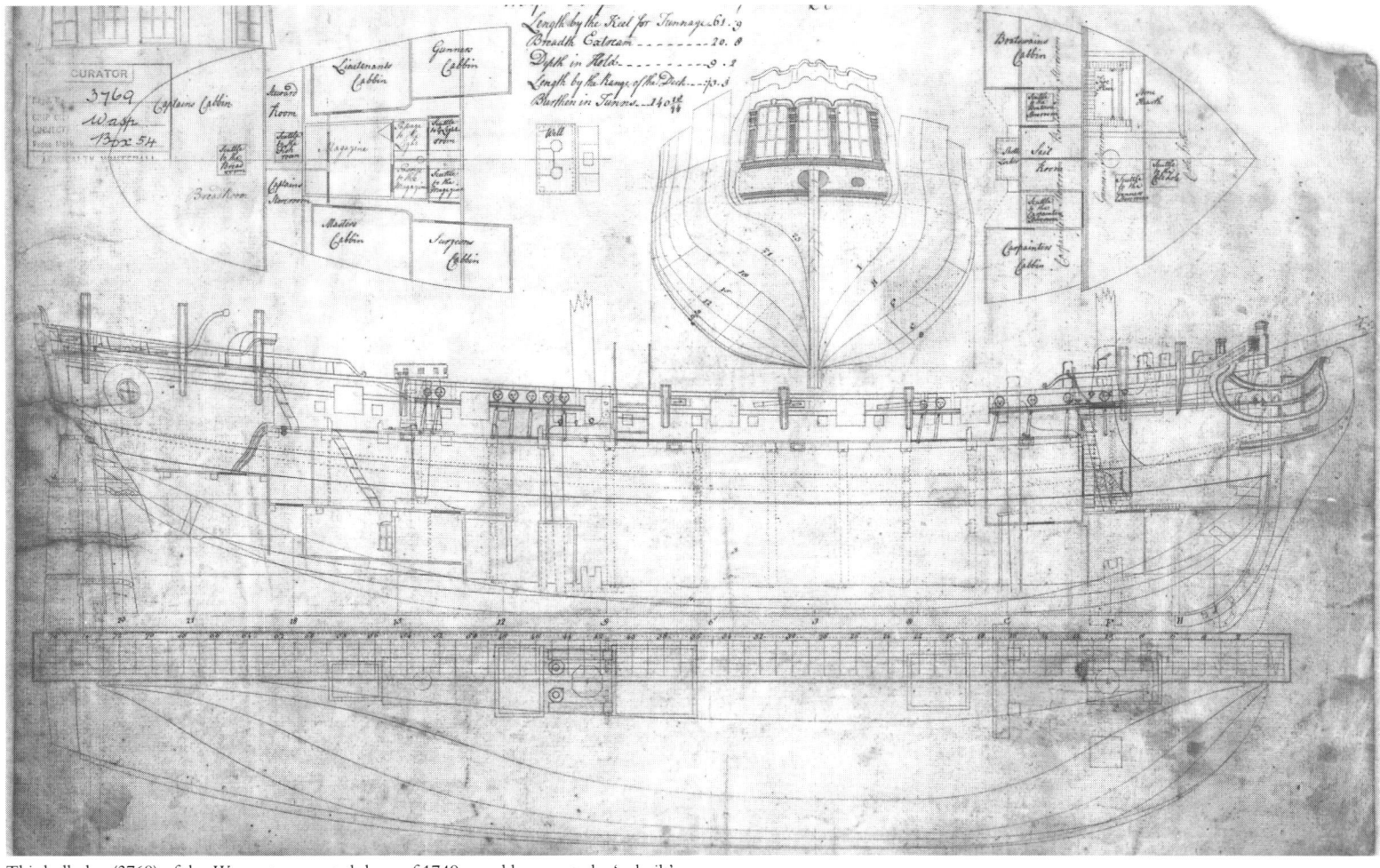

This hull plan (3769) of the *Wasp*, a two masted sloop of 1749, would appear to be 'as built'

as a bomb vessel?); 1811 BU.
**Camilla** Chatham Dyd.
Ord 1.12.1773; K 05.1774; L 20.4.1776; 1809 hulked.
**Daphne** Woolwich Dyd.
Ord 1.12.1773; K 08.1774; L 21.3.1776; 1793 reclassed as a 24; (1795-1797 in French hands); 1802 sold.
**Galatea** Deptford Dyd.
Ord 1.12.1773; K 10.1774; L 21.3.1776; 1783 BU.
**Ariadne** Chatham Dyd.
Ord 10.6.1775?; K 05.1775?; L 27.12.1776; 1793 reclassed as a 24; 1814 sold.
**Vestal** Plymouth Dyd.
Ord 1.8.1775; K 02.1776; L 23.5.1777; 1777 foundered.
**Perseus** Randall, Rotherhithe.
Ord 30.10.1775; K 11.1775; L 20.3.1776; 1798 converted to a bomb vessel; 1805 BU.
**Unicorn** Randall, Rotherhithe.
Ord 30.10.1775; K 11.1775; L 23.3.1776; (1780-1781 in French hands); 1787 BU.
**Ariel** Perry & Co, Blackwall.
Ord 3.7.1776; K 07.1776; L 7.7.1777; 1779 taken.
**Narcissus** Plymouth Dyd.
Ord ___ ; K 13.6.1777; L 9.5.1781; 1796 wrecked.

FATES: *Camilla* became a receiving ship at Sheerness; later floating barracks; 1825 breakwater; 1831 sold. *Daphne* was taken by the French *Tamise*, *Méduse* etc; 28.12.1797 retaken by *Anson*. *Vestal* was supposed to have foundered off Newfoundland about 3.10.1777. *Unicorn* was taken by the French *Andromaque* off Hispaniola on 4.9.1780; renamed *Licorne* in French service; 20.4.1781 retaken by *Resource*. *Ariel* taken by the French *Aréthuse* off South Carolina; lent to the Americans; 1783 returned by them to the French; 1793 lost. *Narcissus* wrecked off New Providence 3.10.1796.

PLANS: Lines/profile/orlop/gun deck/upper deck/quarter deck & forecastle/framing/planking expansion (internal)/(external). *Sphinx*: as built:

Lines & profile & decoration. *Camilla*: as built: Lines & profile & decoration/orlop & lower deck/upper deck & quarterdeck & forecastle. *Daphne*: as built: Lines & profile & decoration/decks. *Vestal*: as built: Lines & profile & decoration. *Perseus*: 1798 as bomb: Profile/lower deck/upper deck/midships section.

## Early Two-masted Sloops

Type of rig (snow, ketch) indicated where known.

### Grampus Class 1746. Snows. Origin of design uncertain.
Dimensions & tons: 90', 74'6" x 26' x 12'. 268.
Men: 110. Guns: 14 x 4 + 14 swivels.
**Grampus** Reed, Hull.
Ord 9.1.1746; K 02.1746; L 3.12.1746; 1762 fireship; 1772 renamed *Strombolo*; 1780 hulked.
**Saltash** Quallett & Allen, Rotherhithe.
Ord 9.1.1746; K 02.1746; L 19.12.1746; 1753 ship rig; 1763 sold.

FATES: *Grampus* (as *Strombolo*) became a prison ship at New York; 1780 sold.

PLANS: Lines & profile/orlop. *Grampus*: as fireship: Lines & profile/above waterline profile/decks.

### Peggy Class 1749. Allin design. Joint winner of the design competition held by Admiralty order of 30.1.1749 for building two sloops 'to draw a small draught of water to cruize against the smugglers'.
Dimensions & tons: 74'6", 61'6½" x 20'9" x 9'6". 140 90/94.
Men: 60 Guns: 8 x 3 + 10 swivels.
**Peggy** Deptford Dyd.
Ord 30.1.1749; K 11.5.1749; L 26.7.1749; out of service soon after 1769.

PLANS: Lines & profile/lines & profile & midships section.

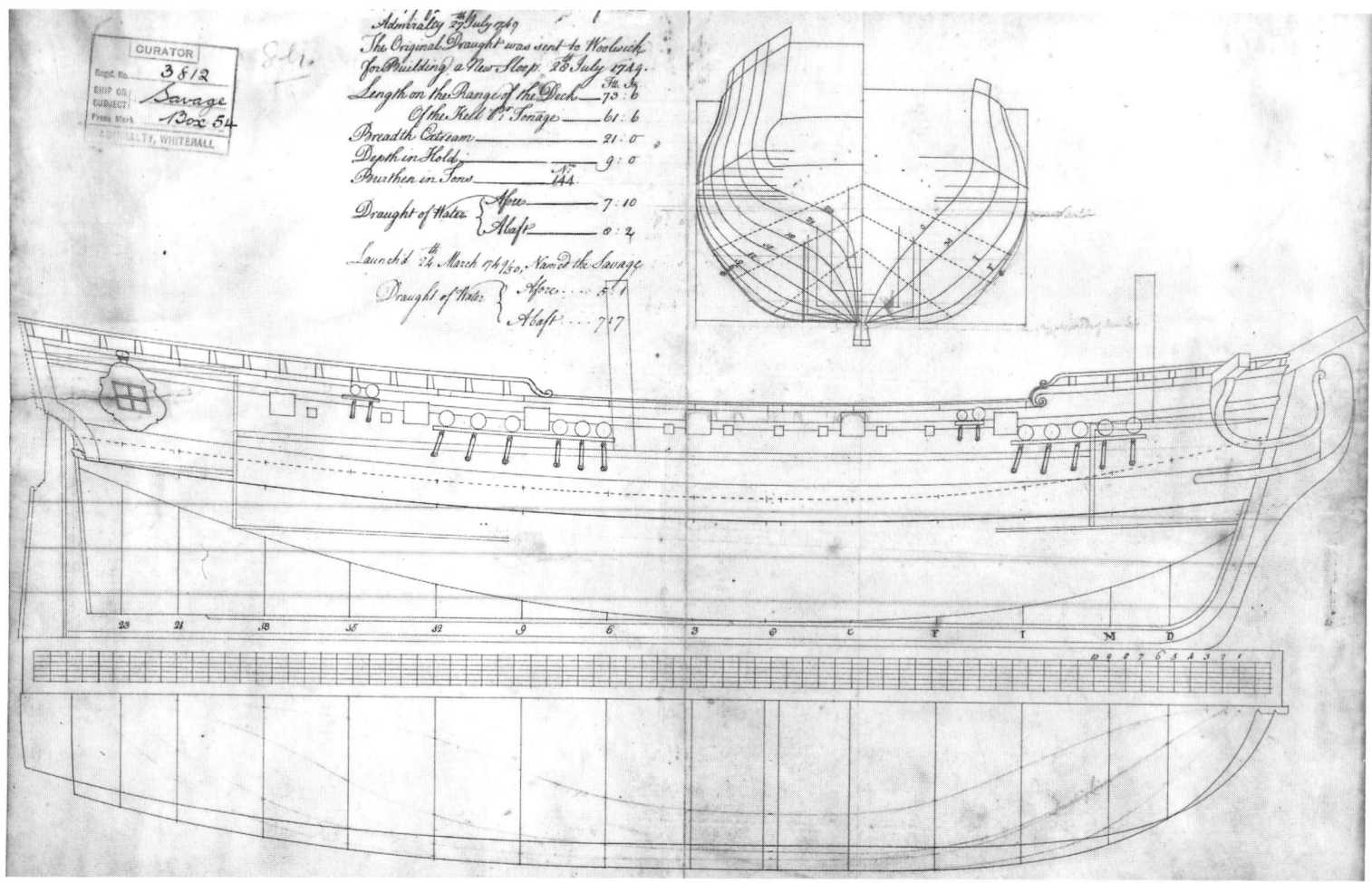

The building lines plan (3812) for the two masted sloop *Savage*, 1750

**Wasp Class 1749.** Lock design, joint winner with *Peggy* of the design competition for an anti-smuggler vessel.
Dimensions & tons: 73'9", 61'9" x 20'8" x 9'2". 140 18/94.
Men: 60. Guns: 8 x 3 + 10 swivels.
*Wasp* Portsmouth Dyd.
Ord 30.1.1749; K 18.4.1749; L 4.7.1749; 1781 sold.
PLANS: Lines. As built: Lines & profile. 1769: Decks. 1770: Decks.

**Savage Class 1749.** 'Admiralty' design.
Dimensions & tons: 73'6", 61'6" x 21' x 9'. 144 24/94.
Men: 50. Guns: 8 x 3 + 10 swivels.
*Savage* Woolwich Dyd.
Ord 12.7.1749; K 1.8.1749; L 24.3.1750; 1776 wrecked.
FATE: Wrecked at Louisbourg, Nova Scotia 16.9.1776.
PLANS: Lines.

**Hazard Class 1749.** See under Ship Sloops.

**Cruizer Class 1752.** Allin design? Modified from the yacht *Royal Caroline*. ('Similar to the body of the *Royal Caroline*'.)
Dimensions & tons: 75'6", 62'3" x 28'7½" x 9'4". 140 79/94.
Men: 50. (*Speedwell* as fireship: 40). Guns: 8 x 3 + 10 swivels. (*Speedwell* as fireship: 8 x ____ ).
*Cruizer* Deptford Dyd, Snow.
Ord 14.1.1752; K 2.3.1752; L 31.8.1752; 1753 ship rig; 1776 burnt.
*Speedwell* Chatham Dyd, Ketch.
Ord 14.1.1752; K 11.2.1752; L 21.10.1752; 1779 fireship renamed *Spitfire*; 1780 sold.
*Happy* Woolwich Dyd, Ketch.
Ord 29.8.1753; K 26.9.1753; L 22.7.1754; 1766 wrecked.
*Wolf* Chatham Dyd, Snow.
Ord 25.8.1753; K 10.10.1753; L 24.5.1754; 1781 sold.
FATES: *Cruizer* condemned and burnt at Cape Fear by order of Lord Howe 2.10.1776. *Happy* wrecked off Winterton, near Yarmouth 14.9.1766.
PLANS: Lines/lines & profile/profile. Swedish Chapman Collection: rig (as snow).

**Fly Class 1752.** To build 'by a draught similar to the *Monarch*, French prize', adapted by Lock.
Dimensions & tons: 75', 64'3" x 20'3" x 9'10". 140 13/94.
Men: 80. Guns: 8 x 3 + 10 swivels.
*Fly* Portsmouth Dyd. Ketch.
Ord 14.1.1752; K 1.2.1752; L 9.4.1752; 1772 sold.
*Ranger* Woolwich Dyd, Snow.
Ord 14.1.1752; K 27.1.1752; L 7.10.1752; 1783 sold.
PLANS: Lines. *Fly*: as built: Lines & profile & upper deck/orlop & gun deck. *Ranger*: as built: Lines & profile.

**Hawk Class 1755.** Allin design.
Dimensions & tons: 88', 72'1½" x 24'10" x 10'7½". 220 86/94.
Men: 80. Guns: 8 x 3 + 10 swivels.
*Hawk* Batson, Limehouse. Ketch.
Ord 9.1.1755; K 23.7.1755; L 1.4.1756; (1759-1761 in French hands); 1781 sold.
FATE: Taken off Cape Clear by a French privateer 11.1759; renamed *Le Faucon* in French service; 01.1761 retaken.
PLANS: Lines & profile. 1769: Lines & profile.

**Hunter Class 1755.** Slade design, Pink.
Dimensions & tons: 88'8", 75'11⅜" x 24'3" x 7'. 238.
Men: 100. Guns: 10 x 6 (short, light guns) + 12 swivels.
*Hunter* Stanton & Wells, Rotherhithe. Pink.
Ord 5.8.1755; K 08.1755; L 28.2.1756; 1779 hulked.
FATE: Became a prison ship at New York; 1780 sold.
PLANS: No plans.

THE SLADE ERA 1745-1785

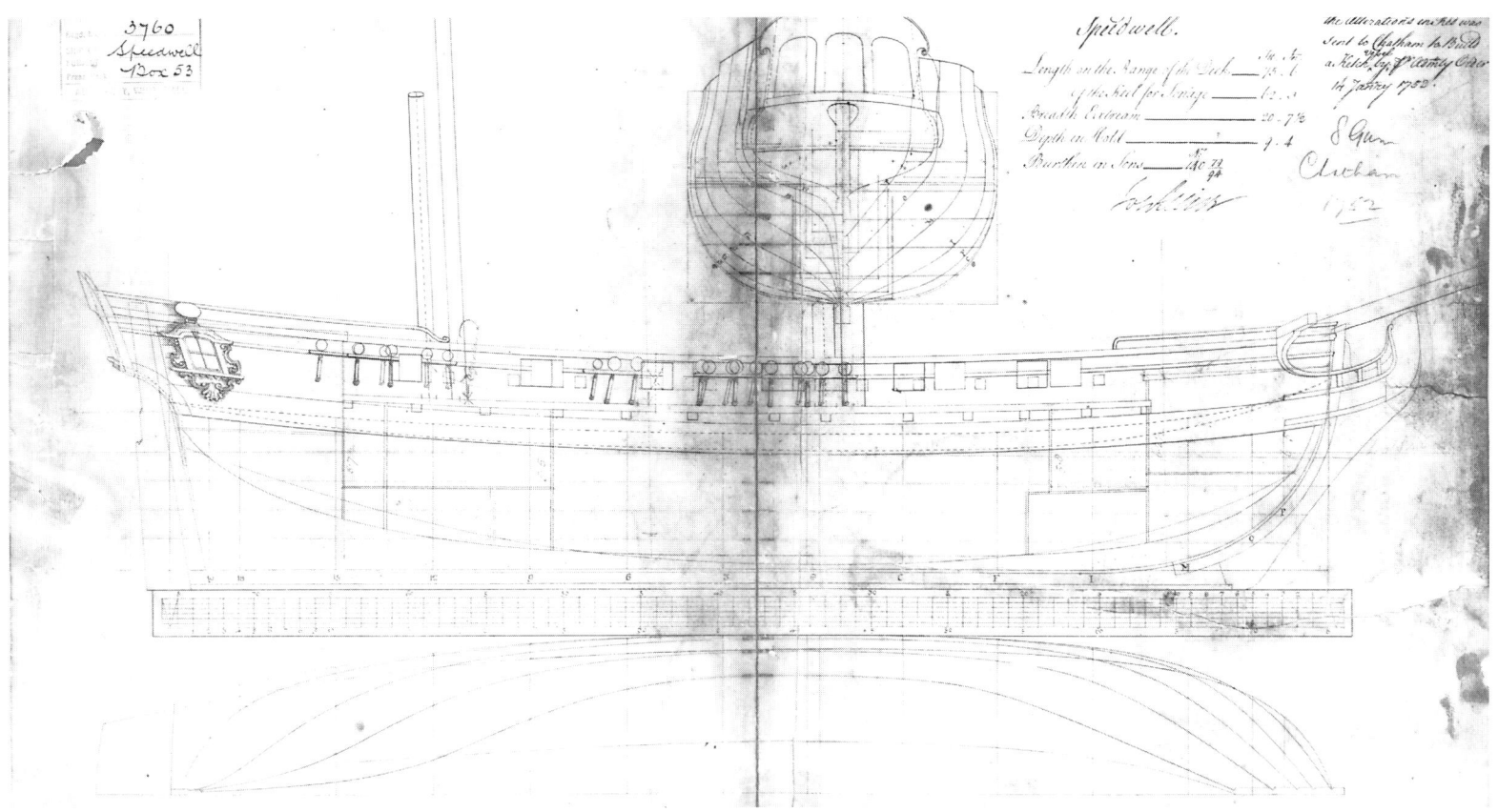

This building plan (3760) of the *Speedwell* (launched 1752), of the *Cruizer* Class sloops, shows this ketch-rigged vessel with altered mast positions

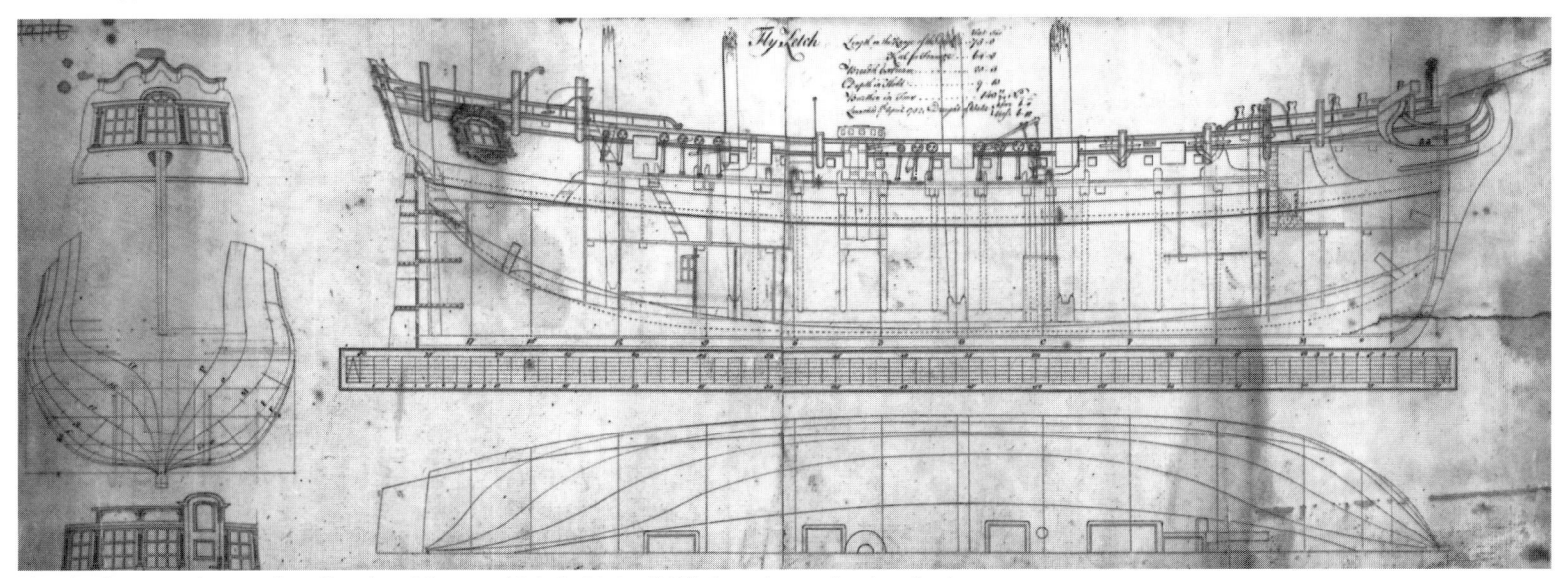

The *Fly* of 1752 was the name ship of her class of sloops; as this 'as built' plan (6419) shows she was rigged as a ketch

**Bonetta Class 1755.** Slade design 'by the draught similar to the *Lyme*' on the reduced lines of the pioneer 28 gun frigate, adapted from the French *Tyger*. All rigged as snows.
    Dimensions & tons: 85'10", 70' x 24'4" x 10'10". 220 43/94.
    Men: 100. Guns: 10 x 6 (short, light guns) + 12 swivels.

**Bonetta** Bird, Rotherhithe.
    Ord 9.7.1755; K 16.7.1755; L 4.2.1756; 1776 sold.

**Spy** Inwood, Rotherhithe.
    Ord 9.7.1755; K 25.7.1755; L 3.2.1756; 1773 sold.

**Merlin** Quallett, Rotherhithe.
    Ord 9.7.1755; K 18.7.1755; L 20.3.1756; (1757 briefly in French hands); 1757 renamed *Zephyr*; 1778 taken.

FATES: *Merlin* taken by the French *Le Machault* in 04.1757; 08.1757 retaken; 25.8.1778 (as *Zephyr*) taken by the French *La Gracieuse*; 1780 retaken and burnt.

PLANS: Lines & profile & upper deck/orlop.

**Viper Class 1755.** Designer uncertain.
    Dimensions & tons: 88'3", 73' x 24' x 7'. 223 62/94.
    Men: 100. Guns: 10 x 6 (short, light guns) + 12 swivels

**Viper** West, Deptford. Snow.
    Ord 5.8.1755; K 16.8.1755; L 31.3.1756; 1779 wrecked.

FATE: Wrecked in the Gulf of Saint Lawrence 15.11.1779.
PLANS: Lines & profile & decks.

**Alderney Class 1755.** See under ship sloops.

**Diligence Class 1756.** 'Admiralty' design. (Dimensions are those of the *Alderney* Class; is she therefore of the same design?)
    Dimensions & tons: 88'4", 72'3" x 24'6" x 10'10". 230 64/94.
    Men: 100 (as a fireship: 45). Guns: 10 (later 12) x 4 + 12 swivels (as a fireship: 8 guns).

**Diligence** Wells, Deptford. Snow.
    Ord 23.2.1756; K 18.3.1756; L 29.7.1756; 1779 to be hulked

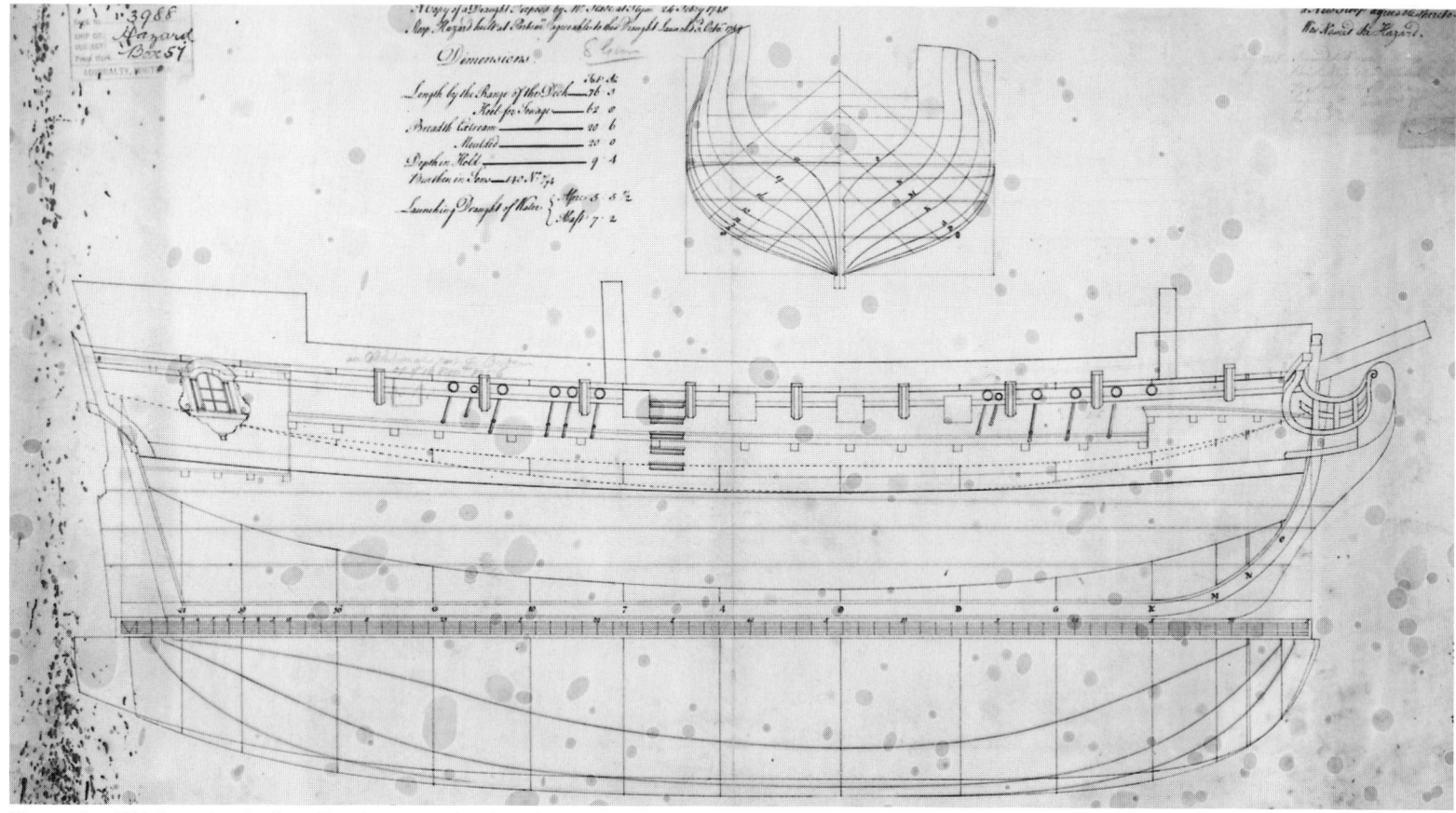

Plan number 3988 shows that the sloop *Hazard* appears to have been built in 1749 as a two-master. The built up upperworks shown in outline on the plan seem to have been added in 1755 and it would seem likely that it was at, or around, this date that this sloop was converted to the ship rig she featured for most of her long career. The row of posts shown along the side are the 'stocks' for mounting swivel guns

at Sheerness, but instead converted to a fireship & renamed *Comet*; 1780 sold.

PLANS: Lines & profile. 1778: Decks.

**Druid Class 1760.** 'Of the same dimensions and by the same Draught as the *Cruizer* sloop built at Deptford in 1732'. Both rigged as snows.
Dimensions & tons: 86'7", 72'11" x 23'2" x 9'5". 208 14/94.
Men: 100. Guns: 10 x 4 + 12 swivels.
*Druid* Barnard, Harwich.
Ord 19.8.1760; K 09.1760; L 21.3.1761; 1773 breakwater.
*Lynx* Stanton & Wells, Rotherhithe.
Ord 19.8.1760; K 4.9.1760; L 11.3.1761; 1777 sold.

PLANS: Lines & profile/decks.

**Vulture Class 1762.** See under ship sloops.

## Ship Sloops

**Hazard Class 1749.** Slade design.
Dimensions & tons: 76'3", 62'8" x 20'6" x 9'4". 140 7/94.
Men: 50. Guns: 8 x 3 + 10 swivels.
*Hazard* Portsmouth Dyd.
Ord 12.7.1749; K 25.7.1749; L 3.10.1749; 1783 sold.

NOTE: It seems likely this vessel was built as a two masted sloop and not converted to ship rig until about 1755 (see the plan illustrated here).

PLANS: Lines/lines & profile & orlop.

**Alderney Class 1755.** Bately design. Begun as snows but completed as ships.
Dimensions & tons: 88'4", 72'3" x 24'6" x 10'10". 230 64/94.
Men: 100. Guns: 10 (later 12) x 4 + 12 swivels (in 1780 2 x 12★ were added to *Alderney*).
*Alderney* Snook, Saltash.
Ord 14.11.1755; K 12.1.1756; L 5.2.1757; 1783 sold.

*Stork* Stow, Shoreham.
Ord 14.11.1755; K 11.1755?; L 8.11.1756; 1758 taken.

FATES: *Stork* was taken by the French in the West Indies 16.8.1758.

PLANS: Lines & profile/decks.

**Favourite Class 1757.** Slade design.
Dimensions & tons: 96'4", 79'10" x 27' x 8'6". 360 51/94.
Men: 125. Guns: 16 x 6 + 14 swivels.
*Favourite* Sparrow, Shoreham.
Ord 11.1.1757; K 18.3.1757; L 15.12.1757; 1784 sold.
*Tamar* Snook, Saltash.
Ord 11.1.1757; K 14.3.1757; L 23.1.1758; 1778 fireship, renamed *Pluto*; 1780 taken.

**Cancelled 1757?** Possibly, but not certainly, this class.
*Flora* Deptford Dyd.
Ord ____ ; K ____ .

FATES: *Tamar* (as *Pluto*) taken on 30.11.1780.

PLANS: Lines. *Favourite*: as built ?: Lines & profile/decks.

**Ferret Class 1760.** Slade design (not certain if originally ship rigged).
Dimensions & tons: 95'3", 78'8" x 26'2" x 7'2". 286 47/94.
Men: 125. Guns: 14 x 6 + 12 swivels.
*Ferret* Stanton, Rotherhithe.
Ord 24.4.1760; K 9.5.1760; L 6.12.1760; 1775 lost.

FATE: 'Supposed to be lost' in the West Indies 31.8.1775.

PLANS: Lines & profile/decks/orlop/forecastle/upperworks etc. 1763: Decks. 1771: Decks.

**Beaver Class 1760.** Bately design, based on the French *Aurora*. (*L'Abénakise*, taken 1757.)
Dimensions & tons: 96'7", 79'9" x 25'10" x 12'6". 283 7/94.
Men: 125. Guns: 14 x 6 + 12 swivels.
*Beaver* Inwood, Rotherhithe.
Ord 24.4.1760; K 19.5.1760; L 3.2.1761; 1783 sold.

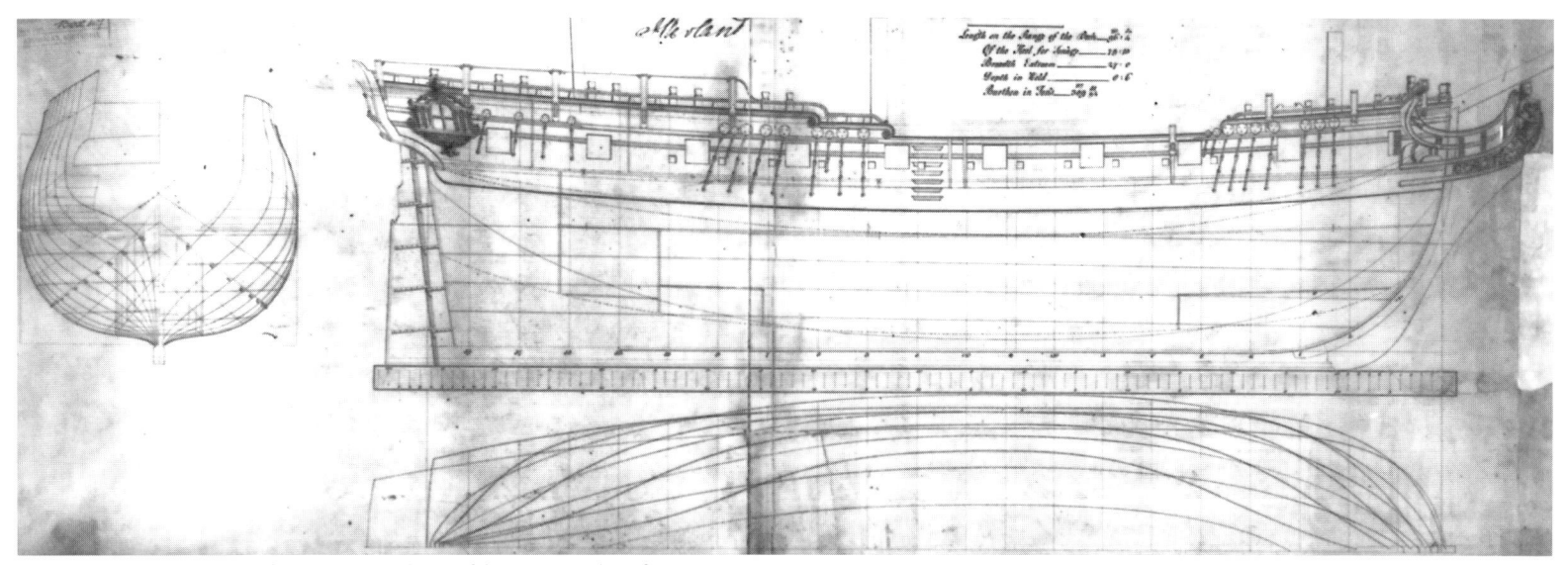

The building lines plan (3306) for the two ship sloops of the *Favourite* Class of 1757

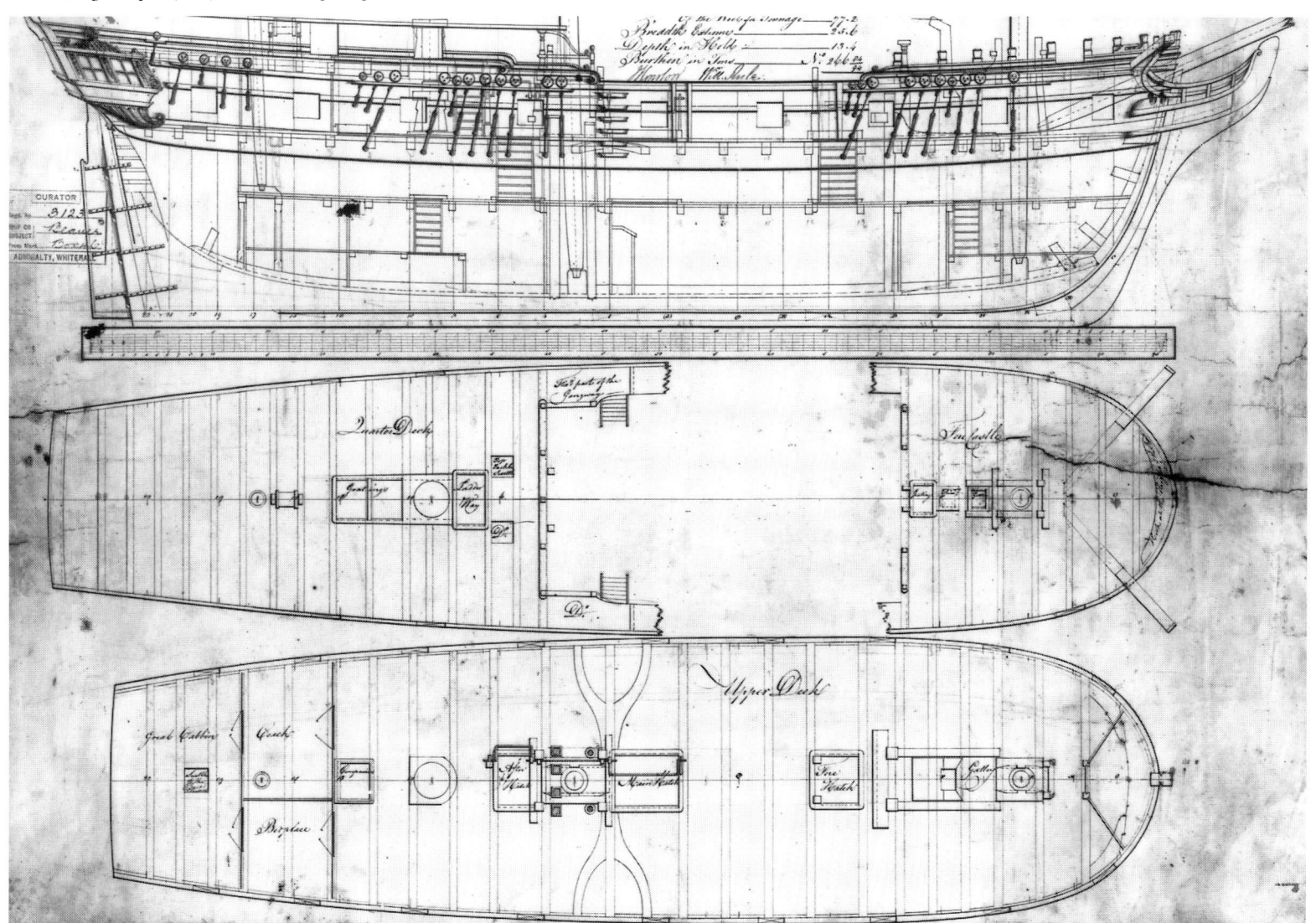

Combined profile and decks plan (3123) of the ship sloop *Beaver*, built in 1761 as the name ship of her class

**Martin** Randall, Rotherhithe.
Ord 24.4.1760; K 29.5.1760; L 7.2.1761; 1784 sold.

**Senegal** Bird, Rotherhithe.
Ord 24.4.1760; K 05.1760; L 24.12.1760; (1778-1780 in French hands); 1780 blown up.

FATES: *Senegal* taken by the French fleet under D'Estaing 14.8.1780; 2.11.1780 retaken; 22.11.1780 blown up at Goree 'by some unknown accident as they were refitting her'.

PLANS: Lines & profile/decks. Beaver: Lines & profile. *Martin:* Lines & profile. *Senegal:* Lines & profile.

### Nautilus Class 1761.
Slade design, probably a variant of the *Favourite* class. Originally (1757) to be built at Deptford Dyd.
Dimensions & tons: 98′, 80′9″ x 27′ x 7′6″. 314 17/94.
Men: 125. Guns: 16 x 6 + 14 swivels?

**Nautilus** Hodgson, Hull.
Ord 17.4.1761; K 05.1761; L 24.5.1762; 1780 out of commission;

eventual fate unknown.
PLANS: Lines & profile/profile/decks. As built: Lines & profile.

**Vulture Class 1762.** 'Of the same dimensions and as near as may be to the Draught of the *Epreuve*'. (Taken from the French 1760.) Designed as snows, but completed with ship rig?
Dimensions & tons: 91'4", 73'7½" x 25'11" x 13'5½". 263 3/94.
Men: 125 Guns: 14 x 6 + 12 swivels.
*Vulture* Davis, Northam.
Ord 15.2.1762; K 02.1762?; L 12.1.1763; 1771 BU.
*Swift* Greaves, Limehouse.
Ord 15.2.1762; K 02.1762?; L 1.3.1763; 1770 wrecked.
FATES: *Swift* wrecked on the coast of Patagonia in 03.1770, site now located and excavation proceeding.
PLANS: Lines & profile/decks.

**Otter Class 1766.** Slade design. Some differences between individual members of the class.
Dimensions & tons: 95', 78' x 27' x 12'6". 302 43/94.
Men: 125. Guns: 14 x 6 + 12 swivels.
*Otter* Deptford Dyd.
Ord 8.5.1766; K 14.10.1766; L 26.10.1767; 1778 wrecked.
*Swallow* Deptford Dyd.
Ord 8.5.1768; K 8.9.1768; L 30.12.1769; 1777 foundered.
*Falcon* Portsmouth Dyd.
Ord 27.10.1768; K 04.1769; L 15.6.1771; 1778 scuttled later raised; 1779 foundered.
FATES: *Otter* wrecked 'on the South Coast of America' (Florida) on 25.8.1778. *Swallow* foundered returning from the Cape of Good Hope about 31.12.1777. *Falcon* was scuttled at Narrangansett Bay as a blockship 5.8.1778; later raised; 20.9.1779 'supposed to be lost'.
PLANS: *Otter*: Lines & profile/profile/decks. As built: Lines & profile. *Swallow*: Lines/profile & decks/orlop & lower deck. *Falcon*: Lines/orlop & lower deck/profile & upper deck & quarter deck & forecastle.

**Swan Class 1766.** Williams design. It is possible that the design was modified for the later members of the class.
Dimensions & tons: 96'7", 78'10" x 26'9" x 12'10". 300 4/94.
Men: 125/121. Guns: 14 (later 16) x 6 + 14 swivels.
*Swan* Plymouth Dyd.
Ord 8.5.1766; K 06.1766; L 21.11.1767; 1779 fireship, renamed *Explosion*; 1783 reconverted to sloop and again renamed *Swan*; 1814 sold.
*Kingsfisher* Chatham Dyd.
Ord ____ ; K 01.1769; L 13.7.1770; 1778 burnt (scuttled).
*Cygnet* Portsmouth Dyd.
Ord ____ ; K 11.1773; L 24.1.1776; 1790s hulked.
*Atalanta* Sheerness Dyd.
Ord 1.12.1773; K 9.4.1774; L 12.8.1775; (1781 briefly in American hands); 1781 renamed *Helena*; 1802 sold.
*Pegasus* Chatham Dyd.
Ord 10.4.1775; K 05.1775; L 27.12.1776; 1777 lost.
*Fly* Sheerness Dyd.
Ord 1.8.1775; K 01.1776; L 14.9.1776; 1802 foundered.
*Fortune* Woolwich Dyd.
Ord 16.10.1775; K 19.4.1777; L 28.7.1778; 1780 taken.
*Swift* Portsmouth Dyd.
Ord 16.10.1775; K 01.1776; L 9.1.1777; 1778 wrecked.
*Vulture* Wells, Deptford.
Ord 30.10.1775; K 11.1775; L 18.3.1776; 1792 hulked.
*Hound* Adams & Barnard, Deptford.
Ord 30.10.1775; K 11.1775; L 8.3.1776; 1784 BU.
*Spy* Graves, Limehouse.
Ord 30.10.1775; K 11.1775; L 6.4.1776; 1778 wrecked.
*Hornet* Perry, Blackwall.
Ord 30.10.1775; K 11.1775; L 19.3.1776; 1791 sold.
*Cormorant* Barnard, Ipswich.
Ord 30.10.1775; K 11.1775; L 21.5.1776; 1781 taken.
*Dispatch* Deptford Dyd.
Ord ____ ; K 7.4.1776; L 10.2.1777; 1778 lost.
*Zebra* Barnard, Ipswich.
Ord 24.5.1776; K 07.1776; L 8.4.1777; 1778 wrecked.
*Cameleon* Randall, Rotherhithe.
Ord 21.6.1776; K 07.1776; L 26.3.1777; 1780 lost.
*Nymph* Chatham Dyd.
Ord ____ ; K 04.1777; L 27.5.1778; 1783 burnt.
*Savage* Barnard, Ipswich.
Ord 12.3.1777; K 1.6.1777; L 28.4.1778; 1803 hulked.
*Fairy* Sheerness Dyd.
Ord ____ ; K 9.6.1777; L 24.10.1778; (1781 briefly in French hands); 1811 BU.
*Fury* Sime & Mackenzie, Leith.
Ord ____ ; K 08.1777; L 18.3.1779; 1787 BU.
*Thorn* Betts, Mistleythorn.
Ord ____ ; K 12.1777; L 17.2.1779; (1779-1782 in American and then French hands); 1799 to the Marine Society.
*Delight* Graves, Limehouse.
Ord ____ ; K 24.12.1777; L 7.11.1778; 1781 foundered.
*Bonetta* Perry & Hankey, Blackwall.
Ord ____ ; K 06.1778; L 29.4.1779; (1781-1782 in French hands); 1797 BU.
*Shark* Walton, Hull.
Ord ____ ; K 26.12.1778; L 23.11.1779; by 1807 hulked.
*Alligator* Fisher, Liverpool.
Ord ____ ; K 10.1779; L 11.11.1780; 1782 taken.
FATES: *Kingsfisher* burnt to avoid capture at Rhode Island 7.8.1778. *Cygnet* became a temporary receiving ship at Portsmouth (?); 1802 sold. *Pegasus* foundered off Newfoundland about 31.10.1777. *Fly* foundered off Newfoundland during 1802. *Fortune* was taken by the French (Guichen) in the West Indies 26.4.1780; 1783 to the French East India Co as *Le Courier De L'Orient*. *Swift* run ashore and burnt off Cape Henry to prevent capture 23.11.1778. *Vulture* became a slop ship at Portsmouth; 1802 sold. *Spy* was wrecked at night on the rocks near Cape Race, Newfoundland 16.6.1778. *Cormorant* was taken by De Grasse's fleet off Charleston Bar on 24.8.1781; in French service until 1798. *Dispatch* capsized in the Gulf of Saint Lawrence about 8.12.1778. *Zebra* was wrecked on the bar at Egg Island Harbour, River Delaware on 5.11.1778. *Cameleon* was lost during the Great West Indian Hurricane on 11.10.1780. *Nymph* was burnt at the island of Tortola through the carelessness of the Purser's steward 28.6.1783. *Savage* became a hospital ship for convicts at Woolwich; 1815 sold. *Thorn* was taken 'by two rebel (American) frigates' 25.8.1779; 20.8.1782 retaken from the French by *Arethusa*; 1799 to the Marine Society as a training ship; 1816 sold. *Delight* sailed from Spithead going to America 3.9.1781 presumed to have foundered. *Shark* became a receiving ship at Jamaica; 1818 wrecked. *Alligator* taken by the French *La Fée* at the mouth of the Channel; 1783 became the French mercantile packet *Le Courrier De New York* and served as such until about 1794.
PLANS: Lines & profile/framing/decks /planking expansion (external). *Kingsfisher*: as built: Lines & profile & decorations/decks. *Atalanta*: as built: Lines & profile & decorations/orlop & lower deck/upper deck & quarterdeck & forecastle. *Pegasus*: as built: Lines & profile & decorations/orlop & lower deck/upper deck, quarter deck & forecastle. *Fly*: as built: Lines & profile & decorations/orlop & lower deck/upper deck, quarter deck & forecastle. *Swift*: as built: Lines & profile & decorations/decks. *Zebra*: as built: Lines & profile & decorations/decks.

**Ceres Class 1774.** 'Agreeable to the *Pomona* Sloop [ex French *Chevert* taken 1761] when taken from the French except the upper deck which is proposed to be carried quite aft & a quarter deck over it to come as far forward as the aft part of the main channel.'
Dimensions & tons: 108', 90'11¼" x 27'4" x 12'6". 361 36/94.
Men: 125. Guns: 18 x 6 + 12 swivels.
*Ceres* Woolwich Dyd.
Ord 16.7.1774; K 27.5.1776; L 25.3.1777; (1778-1782 in French hands); 1782 renamed *Raven*; 1783 taken.
FATE: Taken by the French *Iphigénie* off Saint Lucia; French service as *La Ceres*; 19.4.1782 retaken by Hood; 7.1.1783 (as *Raven*) taken by the French frigates *Nymphe* and *Concorde* in the West Indies; in French service as *La Ceres* until

# THE SLADE ERA 1745-1785

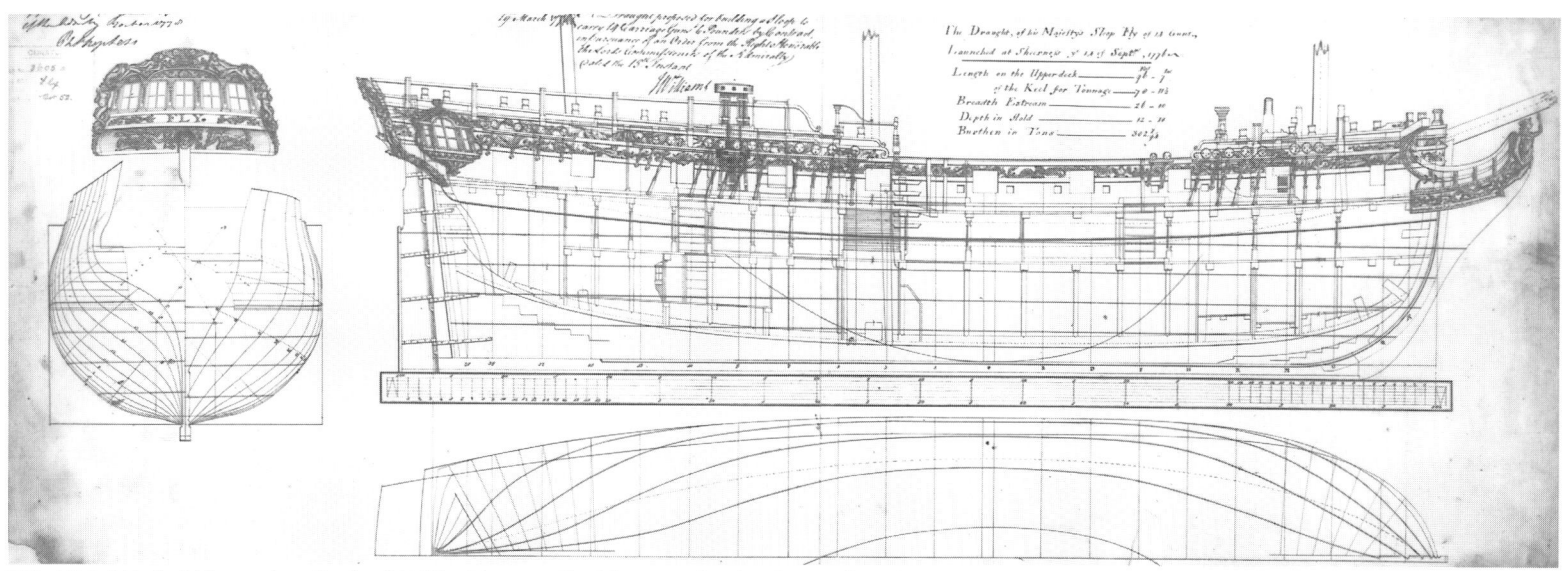

This delightful 'as built' lines and profile plan (3605A) of the 1776 *Fly* of the *Swan* Class ship sloops is the only such plan to include representation of not only all the carved work on head, quarters and stern, but also the painted figures along the upperworks. This plan was also copied for building a later vessel of this large class

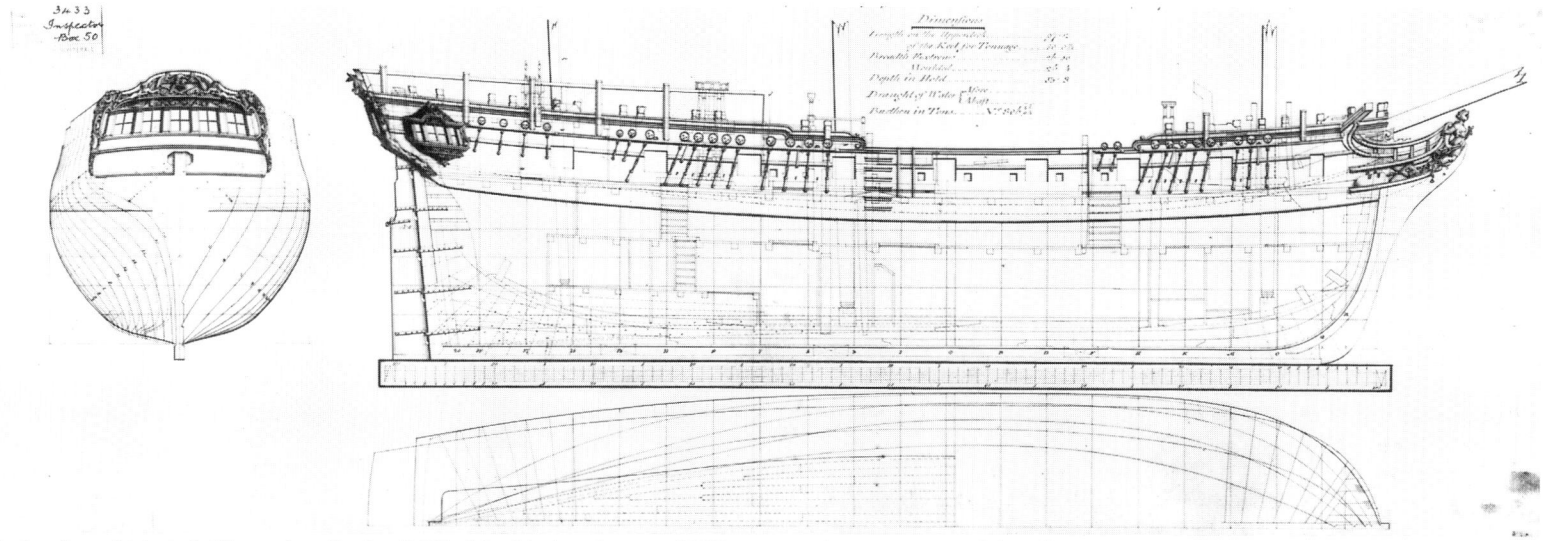

At first glance this 'as built' lines and profile plan (3433) of the ship sloop *Inspector* of 1782 appears to show a conventional ship of her type and period. However, a closer inspection of the lower part of the hull shows an exceptionally fine run aft, combined with an extraordinary bulged fin-like structure growing out of the bottom of the deadwood on either side, presumably intended to provide extra buoyancy to compensate for that fine run (or just possibly as a hydrodynamic experiment to decrease resistance?). This does not seem to have been a satisfactory experiment; in any event the form was never repeated

1791.

PLANS: Lines.

### Zebra Class 1779. Hunt design.
Dimensions & tons: 98', 80'1¼" x 27'2" x 13'4". 314 42/94.
Men: 125/110. Guns: 14 (design)/16 (on completion) x 6 + 12 swivels.

*Zebra* Cleverley, Gravesend.
Ord 16.9.1779; K 10.1779; L 31.8.1780; 1798 converted to bomb vessel; 1812 sold.

*Ariel* Baker, Liverpool.
Ord 14.4.1780; K 06.1780; L 18.10.1781; 1802 sold.

*Bulldog* Ladd, Dover.
Ord ____ ; K 10.1781; L 10.11.1782; 1798 bomb vessel; 1801 hulked.

### Cancelled 1783. (builder failed) *Serpent* Jacobs, Sandgate.
Ord ____ ; K 02.1783.

FATES: *Bulldog* became a powder hulk at Portsmouth; 1829 BU.
PLANS: Lines/lines & profile/profile /decks/framing/specification. *Bulldog*: As bomb 1798: Profile & midships section.

### Inspector Class 1780. 'Admiralty' design.
The run of the lines aft was exceptionally finely tapered, but with a bulge at the bottom of the after deadwood, just in front of the rudder, presumably to give extra buoyancy rightaft

Dimensions & tons: 97'2", 80'0¼" x 26'10" x 13'3". 306 46/94.
Men: 125. Guns: 16 x 6 + 12 swivels.

*Inspector* Game, Wivenhoe.
Ord ____ ; K 06.1780; L 29.4.1782; 1802 sold.

PLANS: Lines & profile/orlop & lower deck/upper deck & quarter deck & forecastle/stern frames. As built: Lines & profile & decorations/decks.

### Echo Class 1781. Hunt design 'to carry 16 carriage guns'.
The first sloop class to be intended from the start to carry carronades. Bore a distinct resemblance to French ships of the earlier part of the century. The upper end of the stempost was bent backwards, there was a very French midships section with extreme tumble-home and quarter badges and stern gallery layout of typically French shape. Despite this the design does not seem to have been based on any particular French vessel. Every other known class based on such borrowing of French designs has a particular French prototype quoted, but no such prototype is given in any of the sources in this particular case.

Dimensions & tons: 101'4", 83'4½" x 27'7" x 12'10". 337 37/94.
Men: 125. Guns: UD 16 x 6, QD 6 x 12★, Fc 2 x 12★.

*Echo* Barton, Liverpool.

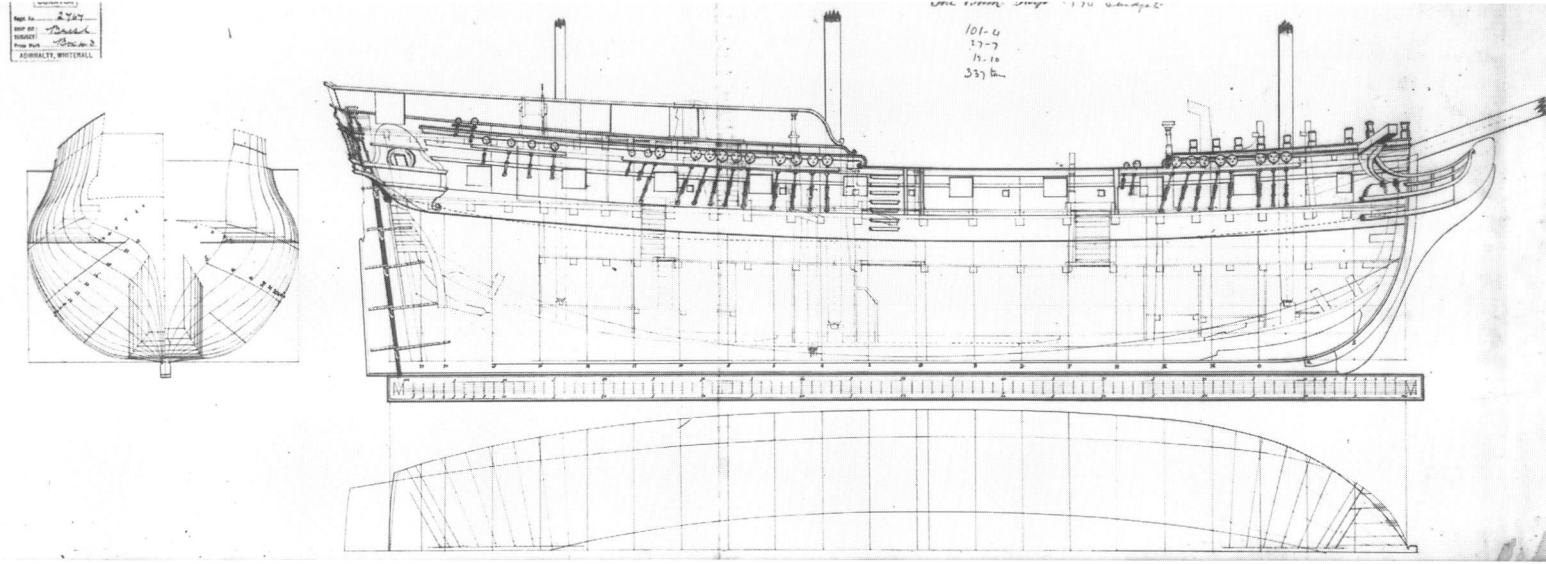

The *Brisk* (built 1784) of the *Echo* Class of ship sloops is shown in this lines and profile plan (number 2767)

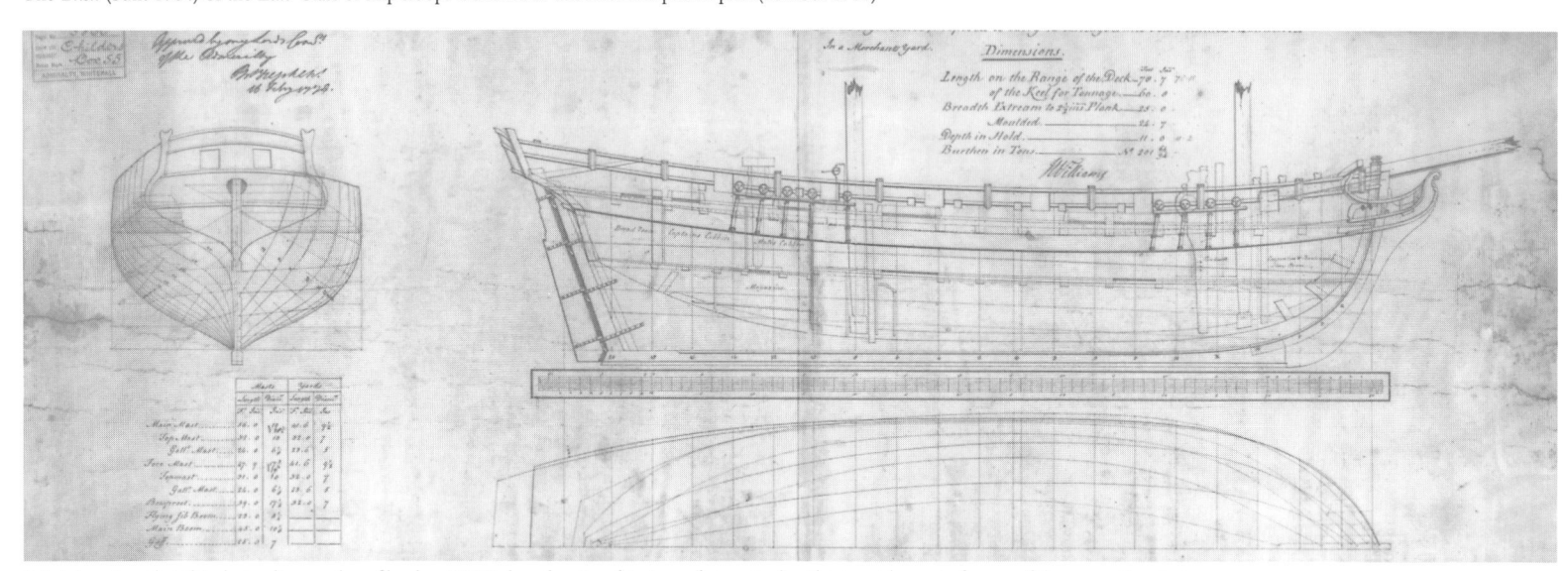

It is interesting that this design lines and profile plan (3866) describes its subject as a 'brigantine' as this is evidence in favour of the suggestion that the word 'brig' did originate as an abbreviation of this term. What is actually shown is the hull form of the first class of brigs (later classed as brig sloops) built for the Royal Navy, the *Childers* Class of 1778

Ord ____ ; K 11.1781; L 8.10.1782; 1797 BU.
**Rattler** Wilson, Sandgate.
Ord ____ ; K 03.1782; L 22.2.1783; 1792 sold.
**Calypso** Graves, Limehouse.
Ord ____ ; K 05.1782; L 27.9.1783; 1803 sunk.
**Nautilus** Crookenden, Taylor & Smith/Smith & Co, Itchenor.
Ord ____ ; K 08.1782; L 9.1.1784; 1799 wrecked.
**Brisk** Jacobs, Sandgate.
Ord ____ ; K 05.1782; L 6.5.1784; 1805 sold.
**Scorpion** Ashman & Son, Shoreham.
Ord ____ ; K 11.1782; L 26.3.1785; 1802 sold.
FATES: *Calypso* run down by a merchantman in the Atlantic, 08.1803. *Nautilus* wrecked off Flamborough Head 2.2.1799. *Brisk* may have been a receiving ship at Portsmouth from 1790 on.
PLANS: Lines/orlop/lower deck/upper deck/quarter deck & forecastle/stern/framing/specification. *Echo:* as built: Lines & profile /lower deck, upper deck, quarter deck & forecastle. *Nautilus:* as built: Lines & profile. *Brisk:* as built: Lines & profile/orlop/lower deck, upper deck, quarter deck & forecastle. *Scorpion:* as built: Lines & profile.

## Brigs and Brig Sloops

**Childers Class 1778.** Williams design or builders designs to common dimensions? Originally classed as brigs, but seem to have been reclassed as brig sloops shortly afterwards.
Dimensions & tons: (approx.) 78'7", 60'8" x 25' x 11'. 202 (some minor variations).
Men: 80/90. Guns: 10 x 4 + 12 swivels (later 14 x 4 + 12 swivels. Those converted to fireships had: 10 x 4).
**Childers** Menetone & Son, Thames.
Ord ____ ; K 3.4.1778; L 7.9.1778; 1811 BU.
**Alert** King, Dover.
Ord ____ ; K 10.1778; L 1.10.1779; 1792 sold.
**Falcon** Hills, Sandwich.
Ord ____ ; K 08.1781; L 23.9.1782; 1800 fireship; 1800 expended.
**Otter** Hills, Sandwich.
Ord ____ ; K 08.1781; L 17.3.1782; 1799 fireship; 1801 sold.
**Weazle** Hills, Sandwich.
Ord ____ ; K 09.1782; L 18.4.1783; 1799 wrecked.
**Ferret** Hills, Sandwich.
Ord ____ ; K 02.1783; L 17.8.1784; 1801 sold.
FATES: *Falcon* expended in Dunkirk Road 7.7.1800. *Weazle* wrecked in Barnstaple Bay 11.1.1799; apparently the wreck has been discovered by divers.
PLANS: Lines/profile & decks/decks.

**Speedy Class 1781.** Builder's design?
Dimensions & tons: 78'3", 59'0½" x 25'8¼" x 10'10". 287 12/94.
Men: 80/90. Guns: 14 x 4 + 12 swivels.
**Speedy** King, Dover.

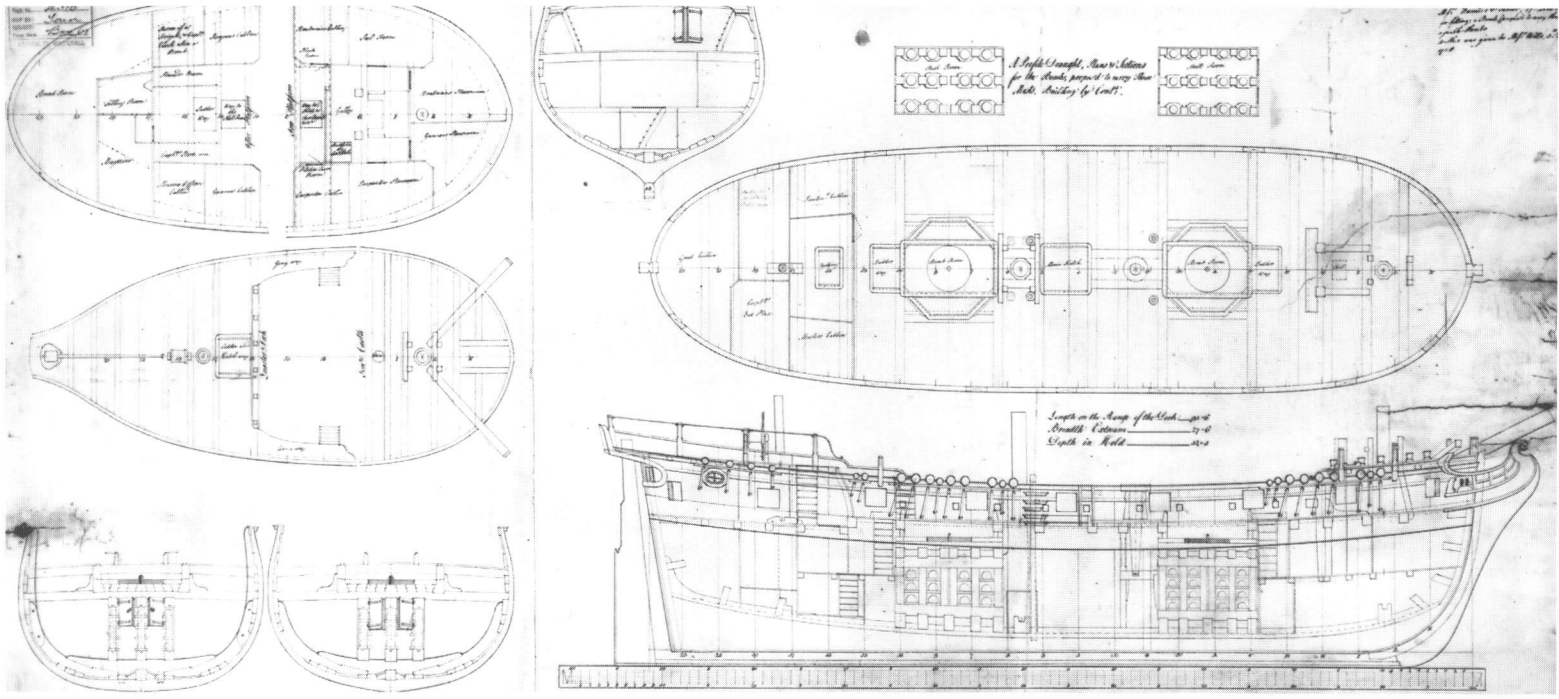

Profile, decks and sections building plan (4310) for the ship-rigged version of the *Infernal* Class bomb vessels of 1756, for *Basilisk* and *Carcass* as well as the *Terror* whose name appears on the (much later) Admiralty stamp on the plan

Ord ____ ; K 06.1781; L 29.6.1782; (1794-1795 in French hands); 1801 taken.

NOTE: This was Cochrane's *Speedy* which took the *Gamo*.

**Flirt** King, Dover.
 Ord ____ ; K 08.1781; L 4.3.1782; 1795 sold.

FATES: *Speedy* was taken by French frigates off Nice 28.6.1794 03.1795 retaken; 2.3.1801 taken by Linois' squadron off the Spanish coast; 1802 Napoleon gave her (as *Le Saint Pierre*) to the Pope.

PLANS: Lines & profile/lower deck/upper deck.

## Bomb Vessels

***Infernal* Class 1756.** Slade design. Four of the first seven (marked with an asterisk [*] here) were rigged as ketches, like all previous eighteenth century bombs. The remainder, like all subsequent RN bombs, were ship rigged. All had the usual pink stern.

 Dimensions & tons: 91'6", 74'1¾" x 27'6" x 12'1". 298 22/94.
 Men 60 (110 as sloops). Guns 1 x 13" + 1 x 10" mortar + 8 x 6 + 14 swivels (14 x 6 fitted as sloops; only 12 x 6 for ketch-rigged version).

**Infernal*** West, Northam.
 Ord 5.10.1756; K 11.1756; L 4.7.1757; (1757- 1758 sloop); 1760 sloop; 1774 sold.
**Blast*** Bird, Northam.
 Ord 21.9.1758; K 10.1758; L 27.2.1759; (1760-1760 sloop); 1771 BU.
**Basilisk** Wells, Deptford.
 Ord 21.9.1758; K 2.10.1758; L 10.2.1759; (1760-1761 sloop); 1762 taken.
**Carcass** Stanton, Rotherhithe.
 Ord 21.9.1758; K 28.9.1758; L 27.1.1759; (1759-1762 sloop); (1773-1775 exploration ship); 1784 sold.
**Mortar*** Wells, Deptford.
 Ord 21.9.1758; K 10.1758; L 14.3.1759; 1760 sloop; 1774 sold.
**Terror** [I] Barnard, Harwich.
 Ord 21.9.1758; K 10.1758; L 16.1.1759; 1774 sold.
**Thunder*** [I] Henniker, Chatham.
 Ord 21.9.1758; K 16.10.1758; L 15.3.1759; (1760-1761 sloop); 1774 sold.
**Etna** Randall, Rotherhithe.
 Ord 14.2.1776; K 03.1776; L 20.6.1776; 1784 BU.
**Vesuvius** Perry, Blackwall.
 Ord 14.2.1776; K 03.1776; L 3.7.1776; 1812 sold.
**Terror** [II] Randall, Rotherhithe.
 Ord 13.11.1778; K 12.1778; L 2.6.1779; 1812 sold.
**Thunder** [II] Randall, Rotherhithe.
 Ord 13.11.1778; K 12.1778; L 18.5.1779; 1781 foundered.

FATES: *Basilisk* taken by the French privateer *L'Audacieux* from Bayonne 01.1763. *Thunder* sailed from Spithead 2.1.1781 and not heard of since.

PLANS: Lines & profile/decks & midships section. *Carcass* 1768: decks. 1773: Profile & midships section/orlop & upper deck.

## Fireships

***Incendiary* Class 1778.** Design of uncertain origin. It is quite possible, indeed probable, that this vessel was purchased on the stocks whilst already building, and therefore should be included under the purchases rather than here, but she has been given the benefit of the doubt.

 Dimensions & tons: 110'8", 87'4" x 29'2⅝" x 12'0½". 396 55/94.
 Men: 45. Guns: uncertain.

**Incendiary** Mestaers, Thames.
 Ord ____ ; K ____ ; L 6.11.1778; 1781 wrecked.

FATE: Lost on 'the Needle Rocks', Isle of Wight in 1781.

PLANS: Lines & profile/orlop & main deck/lower & upper decks.

***Tisiphone/Pluto* Class 1780.** Built to the lines of the French *Amazon* of the 1740s. In some lists *Pluto, Comet, Vulcan* and *Megaera* are stated to be similar to the French *Aurora* (*L'Abénakise*) captured in 1757; but dimensions, plans and other evidence seem to suggest that this is an error. The design was adapted by Henslow for similar vessels built in 1805. (See the next chapter for the 'Modified *Tisiphones*').

 Dimensions & tons: 108'9", 90'7" x 29'7" x 9'. 421 63/94.
 Men: 55. Guns: 14 x 18* (originally 12 long guns + 18 swivels?).

**Tisiphone** Ladd, Dover.
 Ord ____ ; K 03.1780; L 9.5.1781; 1803 floating battery; 1816 sold.
**Alecto** King, Dover.
 Ord ____ ; K 03.1780; L 26.5.1781; 1802 sold.
**Spitfire** Teague, Ipswich.
 Ord ____ ; K 12.1780; L 19.3.1782; 1825 sold.
**Incendiary** King, Dover.
 Ord ____ ; K 05.1781; L 12.8.1782; 1801 taken.
**Conflagration** Junot & Pelham, Shoreham.
 Ord ____ ; K 04.1782; L 28.10.1783; 1790 partially fitted as a hospital ship but then refitted as a fireship; 1793 expended.

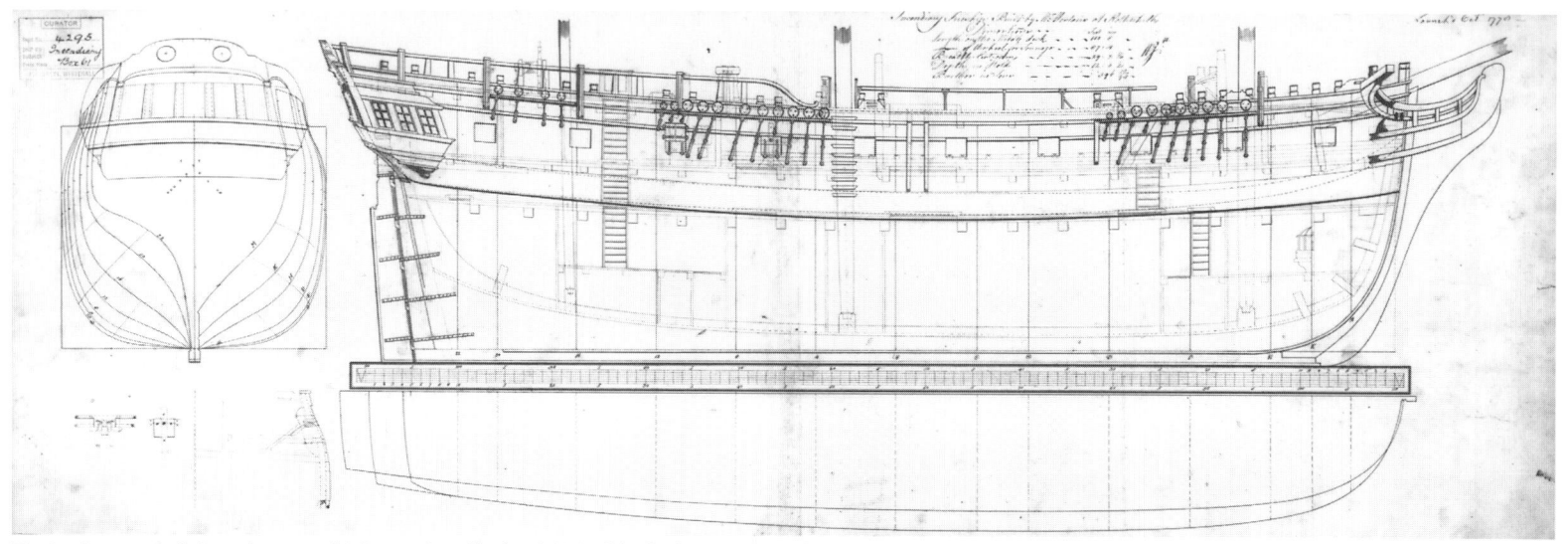

The burdensome hull shape shown on this lines and profile plan (4295) of the fireship *Incendiary* of 1778 would tend to reinforce the possibility that she was of mercantile origin

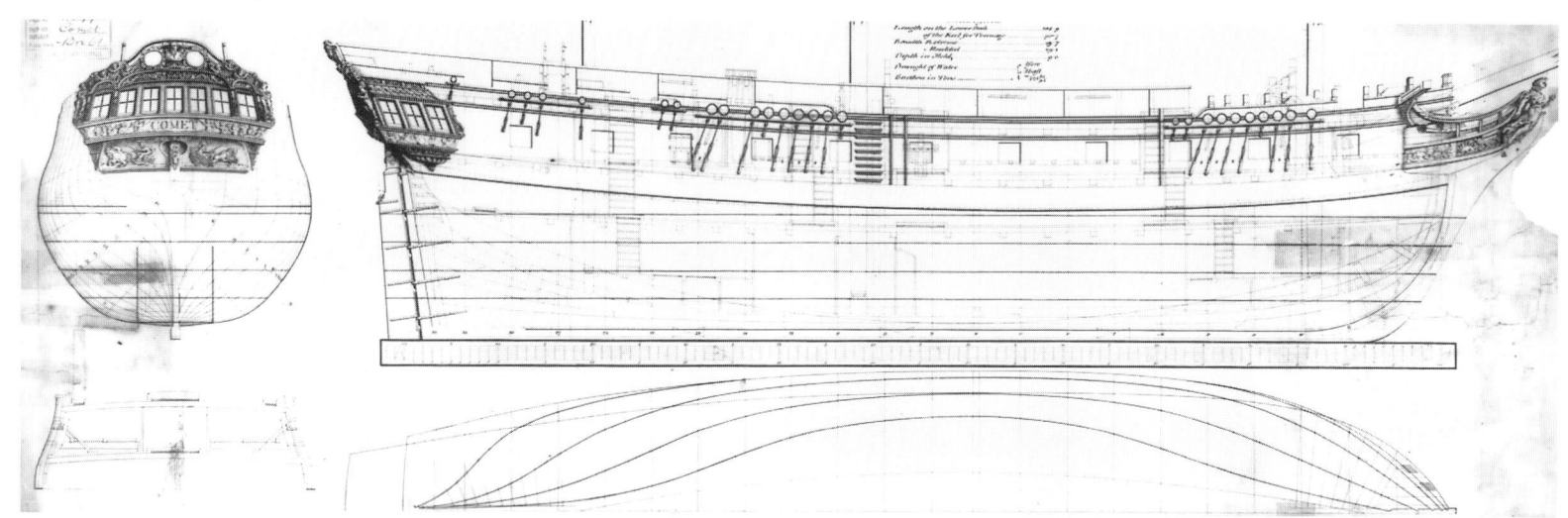

This 'as built' lines and profile plan of the fireship *Comet* of the 1780 *Tisiphone* Class clearly shows the very fine lines of this design

*Pluto* Stewart, Sandgate.
    Ord ____ ; K 01.1781; L 1.2.1782; 1793 fitted as a sloop (but not registered as one); 1809 hulked.
*Comet* Game, Wivenhoe.
    Ord ____ ; K 05.1781; L 11.11.1783; 1800 expended.
*Vulcan* Edwards, Shoreham.
    Ord ____ ; K 04.1782; L 12.9.1783; 1793 expended.
*Megaera* Teague, Ipswich.
    Ord ____ ; K ____ ; L 05.1783; 1817 sold.

FATES: *Incendiary* taken by Gantheaume's squadron in the Mediterranean 10.2.1801. *Conflagration* expended at the evacuation of Toulon 19.12.1793. *Pluto* became a receiving ship at Portsmouth; 1817 sold. *Comet* expended in Dunkirk Road 7.7.1800. *Vulcan* expended at the evacuation of Toulon 18.12.1793.
PLANS: Lines & profile/orlop/lower deck/main deck/upper deck/framing. *Spitfire* 1781: as built: Lines & profile. *Pluto* as built: Lines & profile. *Conflagration* as built: Lines & profile. *Comet* as built: Lines, profile & decorations. *Vulcan* as built: Lines & profile.

## Cutters, etc

### Sherbourne Class 1763. Slade design.
    Dimensions & tons: 54'6", 44'7" x 19' x 8'1". 85 60/94.
    Men: 30. Guns: 6 x 3 + 8 swivels.
*Sherbourne* Woolwich Dyd.
    Ord ____ ; K 27.6.1763; L 3.12.1763; 1784 sold.
PLANS: Lines & profile & decks.

### Ferret Class 1763. Slade design?
    Dimensions & tons: 50', 39' x 20' x 7'10". 82 92/94.
    Men: 30. Guns: 6 x 3 + 8 swivels.
*Ferret* Chatham Dyd.
    Ord ____ ; K 8.7.1763; L 8.10.1763; 1781 sold.
*Lurcher* Deptford Dyd.
    Ord ____ ; K 3.6.1763; L 26.9.1763; 1778 sold.
PLANS: Lines & profile & decks/lower deck & midships section.

### Builder's designs? 1763 (Wells Class?). Differing dimensions and tonnages (given below) but common to all :
    Men: 30. Guns: 6 x 3 + 8 swivels.
*Lapwing* White, Broadstairs (47'9", 34'10½" x 21'1¾" x 8'1½". 82¾).
    Ord ____ ; K 08.1763; L 21.1.1764; 1765 sold.
*Folkestone* Hall, Folkestone (48'2", 36'5" x 21'5" x 7'1". 88 82/94).
    Ord ____ ; K 08.1763; L 13.10.1764; 1778 taken.
*Wells* Jacobs, Folkestone (47'8", 35'2" x 21'3" x 8'2". 84½).
    Ord ____ ; K 08.1763; L 30.6.1764; 1780 sold.

FATES: *Folkestone* taken by the French *La Surveillante* off the French coast 24.6.1778; in French service until 1783.
PLANS: Lines & Profile/decks/midships section. *Wells*: profile & cabins.

### Kite Class 1764. Bately design.
    Dimensions & tons: 55'10", 44'8½" x 18'8" x 6'. 82 81/94.
    Men: 30. Guns: 4 x 3 + ____ swivels.
*Kite* Deptford Dyd.
    Ord ____ ; K 8.3.1764; L 7.9.1764; 1771 sold.
PLANS: No plans?

# THE SLADE ERA 1745-1785

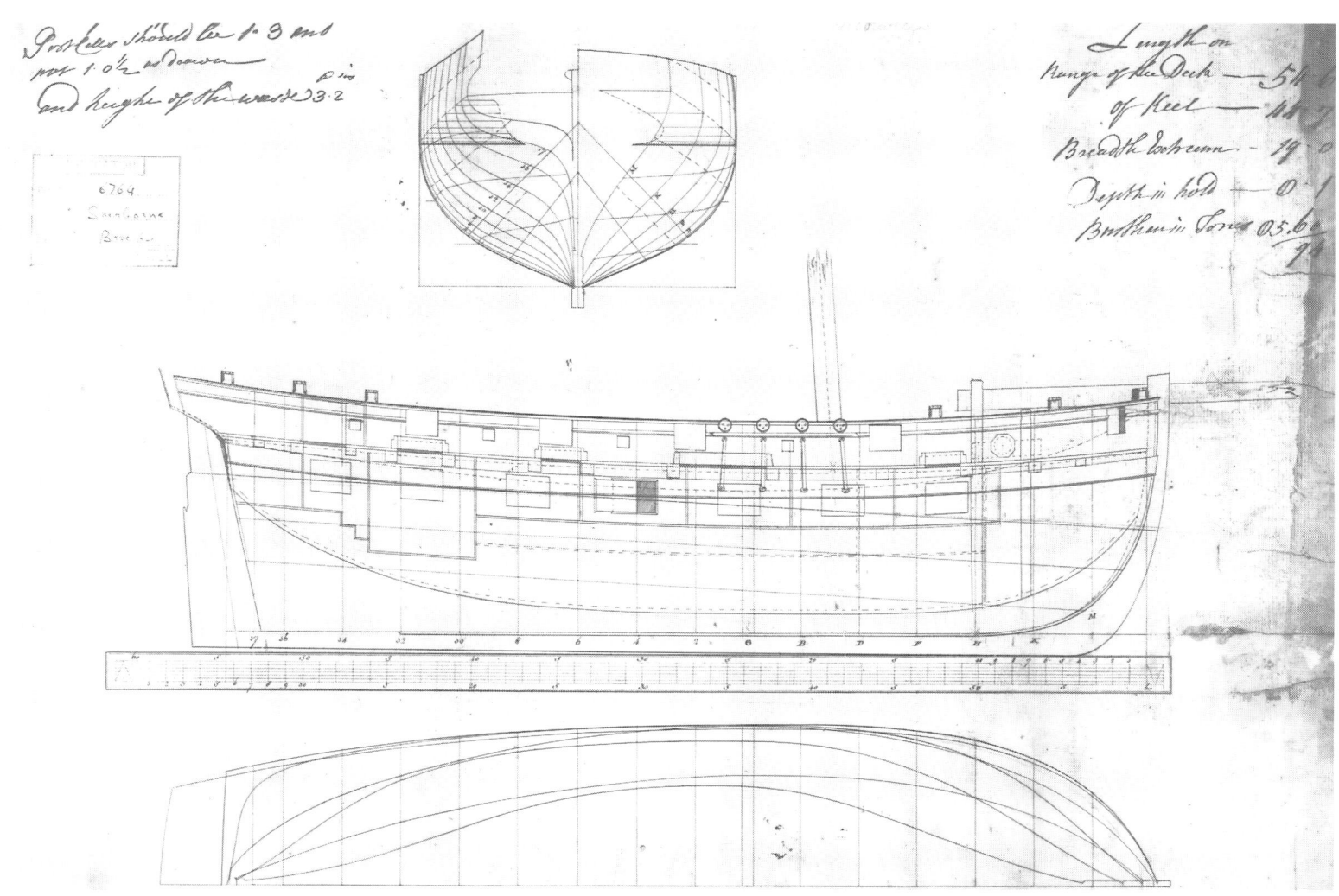

Lines and profile plan (number 6764) shows the cutter *Sherbourne* of 1763

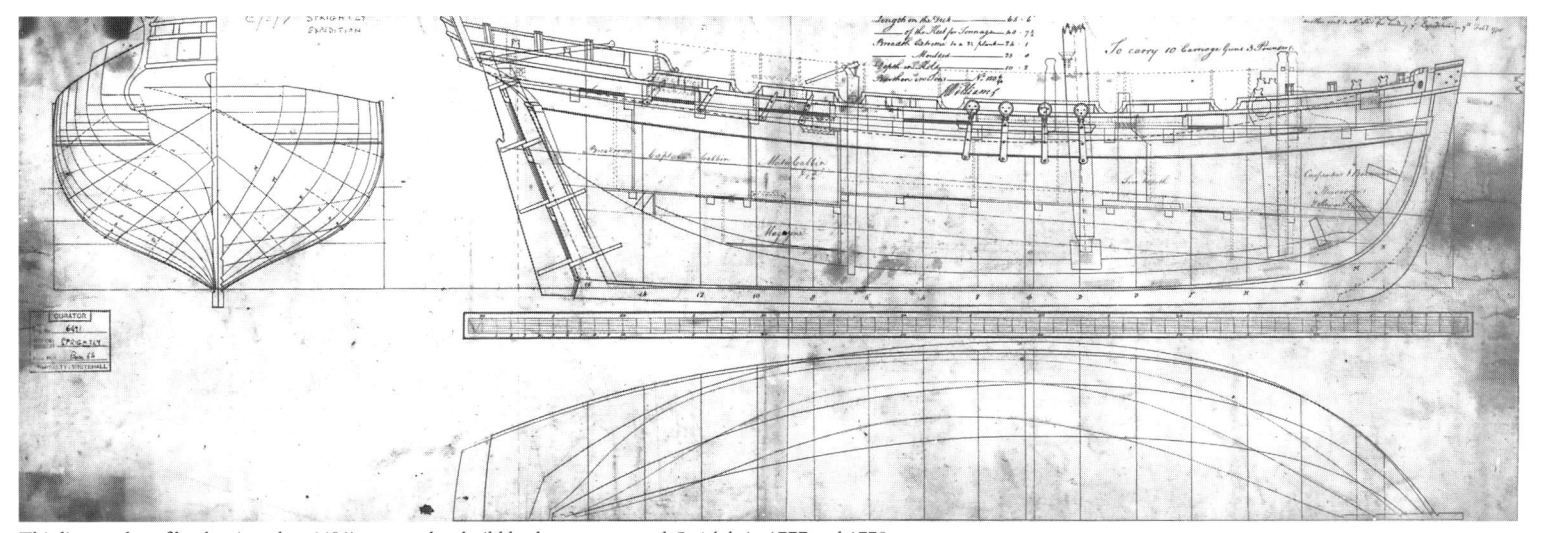

This lines and profile plan (number 6491) was used to build both cutters named *Sprightly* in 1777 and 1778

**Sprightly Class 1777.** Williams design?
    Dimensions & tons: 65'6", 48'6" x 24'1" x 10'2". 150 6/94.
    Men: 45/55. Guns: 10 x 3 + 12 swivels. (*Expedition* had 2 x 12★ temporarily added. *Sprightly* No. II later carried 16 x 12★ + 8 swivels.)

**Sprightly** [I] King, Dover.
    Ord ____ ; K 02.1777; L 4.8.1777; 1777 foundered.

**Expedition** Ladd, Dover.
    Ord ____ ; K 24.3.1778; L 3.8.1778; still listed in 1816 (clerical error?), ultimate fate uncertain.

**Sprightly** [II] King, Dover.
    Ord ____ ; K 28.3.1778; L 4.8.1778; 1801 taken.

FATES: [I] Capsized off Guernsey 22.12.1777; wreck found by divers. [II] Taken in the Mediterranean by Gantheaume's squadron who then sank her 10.2.1801.

PLANS: Lines & profile/decks.

**Rattlesnake Class 1776.** Builder's design.
    Dimensions & tons: 69'4", 52' x 25'7" x 10'9". 180.
    Men: 45. Guns: 10 x 4 + 12 swivels (*Alert* later 10 x 4)

**Rattlesnake** Farley, Folkestone.
    Ord ____ ; K 12.1776; L 7.6.1777; 1779 designated sloop (though still cutter-rigged); 1781 wrecked.

**Alert** Ladd, Dover.
    Ord ____ ; K 01.1777; L 24.6.1777; classed on completion as a sloop (though still cutter-rigged); 1778 taken.

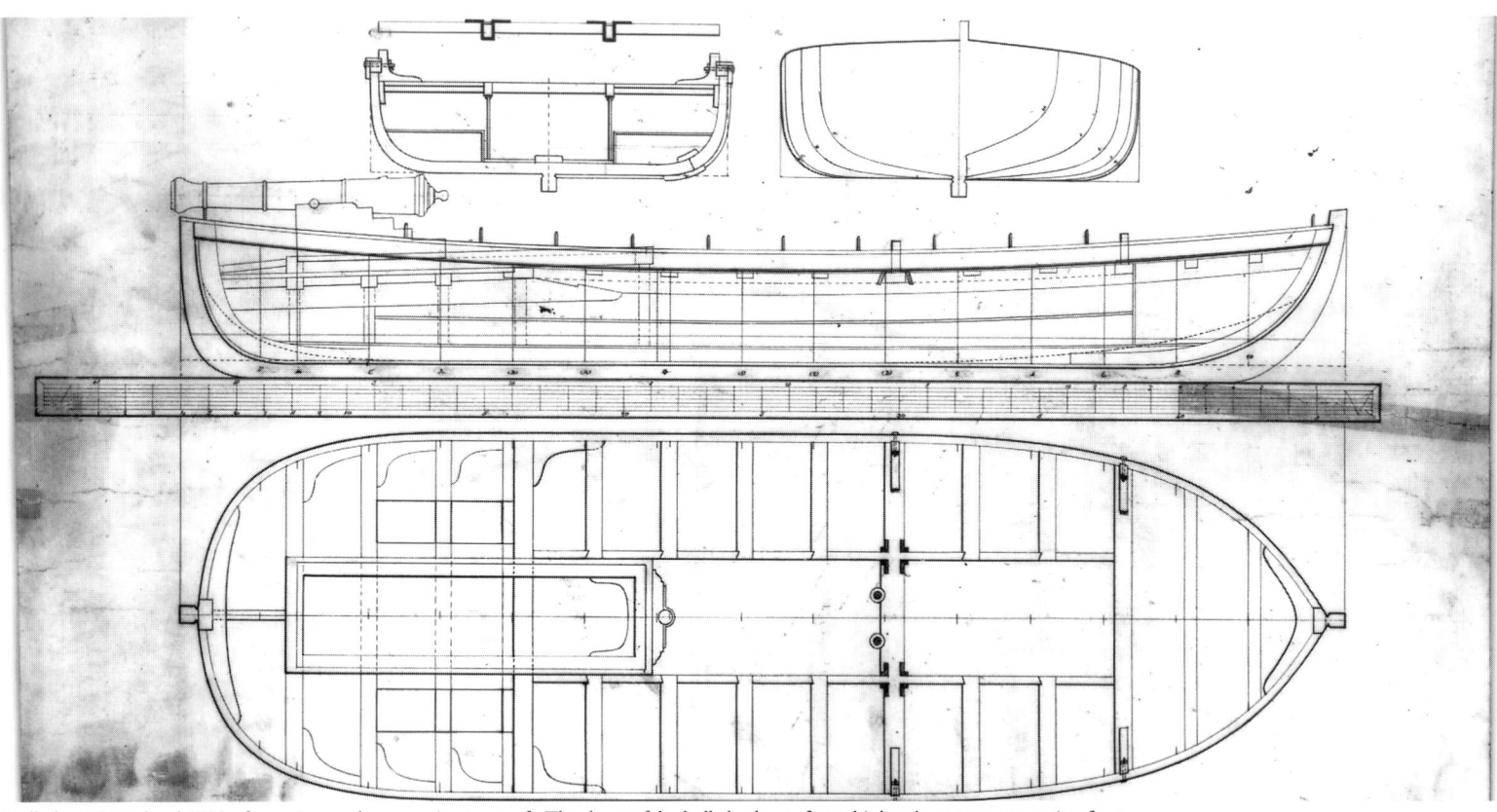

Hull plan (4069) dated 1782 of a rowing gunboat carrying a gun aft. The shape of the hull clearly confirms this boat's ancestry as coming from the 'flat bottomed boats for the landing of men' - the landing craft of the day. Like these it would appear that the gunboat design was intended for prefabrication

FATES: *Rattlesnake* wrecked on the island of Trinidad 11.10.1781. *Alert* taken by the French *Junon* 17.7.1778; 15.12.1779 lost without trace.
PLANS: As built: Lines & profile.

**Pigmy Class 1780.** Williams design *or* Builder's design?
Dimensions & tons: 69'4", 52' x 25'7" x 10'9". 181.
Men: 55. Guns: 10 x 4 + 12 swivels.
*Pigmy* King, Dover.
Ord ____ ; K 02.1781; L 05.1781; (1781-1782 in French hands); 1783 renamed *Lurcher*, 1783 renamed *Pigmy*; 1793 wrecked.
*Cockatrice* King, Dover.
Ord ____ ; K ____ ; L 3.7.1781; 1802 sold.
FATES: *Pigmy* wrecked on the Motherbank 16.12.1793.
PLANS: Lines, profile & decks.

## Xebeck

**Minorca Class 1777.** Designer unknown, but probably the Master Shipwright at Port Mahon Dockyard. Adapted version of standard Mediterranean rig (which normally employed square sails before the wind and lateens off the wind; three masted) combined with a hull very similar to the standard British ship sloop of the time.
Dimensions & tons: 96'9", 78'6" x 30'6" x 10'. 388 40/94.
Men: ____ . Guns: 18 x 6 + 20 swivels.
*Minorca* Port Mahon, Minorca.
Ord ____ ; K 02.1778; L 27.8.1779; 1781 scuttled.
FATE: Sunk at Port Mahon at the 'key' to block the passage during the French siege.
PLANS: Lines.

## Survey ships, etc

**Penguin Class 1772.** Survey ship ('sloop on survey') designer uncertain.
Dimensions & tons: 44', 31' x 14'6" x ____ . 35.
Men: 50. Guns: 10 swivels.

**Cancelled 1783.** *Penguin* Woolwich Dyd.
Ord ____ ; K ____ ; 1772; kept in frame at Woolwich till 1783.

**Conversion from Yard Craft.** *Lion/Lyon* 1776. Survey ship, built as a hoy (transport) at Portsmouth, 1753 and had been a tender at the Tower of London.
Dimensions & tons: 72', 58'6" x 22'6" x 10'7". 157½.
Men: ____ . Guns: 4 swivels (1783: 12 swivels).
FATE: 1776 conversion to survey ship; 1777 tender for press gang; 1783 survey ship for Newfoundland; 1786 sold.

## Floating Battery: Conversion from Yard Craft

*Goree* Sloop/floating battery. Conversion of *Hayling*, Portsmouth Dockyard hoy (built 1729) for service as a floating battery on the coast of Africa.
Dimensions & tons: 66'8", 51'5" x 21'6" x 10'. 126.
Men: 50. Guns: 10 x 12.
FATE: 1759 converted; 1763 BU.

## Small Gunboats

Small one or two gun gunboats, usually adapted for both rowing and sailing. They were not listed in contemporary Navy Lists (with a very few exceptions) or the Progress Books. However it is possible to trace at least some of these un-named (though often numbered) craft from other sources - especially the Admiralty plans. The earliest of these gunboats appear to have evolved from the landing craft of the time; the 'flat bottomed boats for the landing of men' many of which were themselves converted to act as gunboats (for example during the landing at Havanna during the Seven Years War or on the American Coast during the War of Independence). Known examples of these craft for this period are:

**Rowing gunboat.** Gun forward 1776.
Dimensions & tons: 37', ____ x 12' x 3'4". ____ .
Guns: one on truck carriage recoiling on inclined plane forward.
FATE: In 1776 [?] 10 were sent out to Canada in frame and assembled there. In

1780 another 14 were built and 'then taken asunder' for easier transport.
PLANS: Lines, profile & deck.

### Rowing gunboat. Gun aft 1779.
Dimensions & tons: 39'6", ____ x 12' x 3'4". ____ .
Guns: one carriage gun (18 or 24 pdr?) on inclined plane firing over stern. Steering by oar over stern.
FATES: 20 were to be built in 1779 for the defence of the River Medway.

PLANS: Lines, profile & deck.

### Rowing gunboat. Gun aft 1782.
Dimensions & tons: 42', ____ x 12'11" x 3'10". ____ .
Guns: one long gun (18 pdr?) on slide aft. 'Made use of in the defence of Gibraltar 1782 ____ and found to answer the purpose in every respect'.

PLANS: Lines. Profile & deck.

# Chapter 5: Henslow to Seppings 1786-1830

In the first half of our period, wooden warship design was reaching the limits of its effectiveness with traditional methods of construction. Most of the possible variants had been or were being tried; designs half a century old were revived but without the chance of longer, stronger hulls the possibilities were limited. The structural innovations of Robert Seppings (preceded by less radical developments by Gabriel Snodgrass, who chiefly worked for the East India Company), who eventually became Surveyor of the Navy, introduced from the early years of the nineteenth century, gave warship design a new lease of life. Braces, iron knees and other developments resulted virtually in a form of girder construction; various other structural improvements, none major in themselves but producing an overall strengthening of the structure, permitted ships of greater length, and therefore of greater speed and power, to be built.

In the first part of this period, the opening years of the conflict with revolutionary France, there was a tendency to build larger ships within each type, especially with both 74s and frigates. The reason behind this appears to have been the policy of distant blockade, of letting the enemy break out from his harbours and then trying to run him down. Many of these larger vessels were built to French designs, some of them copied from ships captured up to half a century earlier. Later, there was a reversion to building standardised designs of a medium size, usually to the designs of the Surveyors of the Navy, the sort of ships required in large numbers for a policy of close blockade. This was the period of the Napoleonic War and of very large classes, the 'Forty Thieves' for the 74s, the *Leda* Class for the frigates, above all the *Cruiser* and *Cherokee* Class brigs. These brigs illustrate a major alteration in the pattern of building smaller craft. Hitherto, the usual tendency with most sloops and similar vessels (though not all, as some much earlier designs illustrate) was to build miniature versions of bigger warships complete with quarterdecks and forecastles. By the 1790s it was becoming normal for the smaller sloops and nearly all the brigs to be built as flush decked vessels (though sometimes fitted with small 'topgallant' forecastles and possibly poops). Many of these flush-deckers also had a different shape of hull, with a 'vee' shaped bottom (ie. a lot of 'deadrise') and often with a raked keel considerably deeper aft than forward. This 'peg top' section, in an extreme form, came to be associated with the designs of William Symonds, but had appeared long before. It probably originated in North America, though this is not certain.

Radical experiments were tried in hull shape and structure by General Samuel Bentham, who built vessels constructed with strong hulls by the use of watertight bulkheads rather than conventional framing. Carronade armaments were mounted on new principles and the rigging of these vessels was also experimental. Bentham's period at Portsmouth also led to experiments with steam power for dredging. As will be seen, Schanck's 'sliding keels' (a form of dagger-board which could be hoisted through a trunk in the hull, giving variable keel area to shallow draught vessels) were tried in a variety of smaller vessels, including one sloop (*Cynthia*). Captain Erasmus Gower's experimental hull form and rig were tried in the despatch vessel *Transit*, whilst a whole series of remarkable experimental gunboats were built or converted under the aegis of Sir Sidney Smith at Plymouth in the 1790s. The Royal Navy was far from unwilling to make limited experiments with radical new ideas. After the end of the Napoleonic War a steamer was building for African exploration (the unsuccessful *Congo*).

This chapter has benefited from Andrew Lambert kindly casting his eye over it shortly before publication.

## First Rates, 120 Guns (three-deckers)

***Caledonia* Class 1794.** Rule design.
Dimensions & tons: 205', 170'11" x 53'6" x 23'2". 2602 14/94.
Men: 875. Guns: GD 32 x 32, MD 34 x 24, UD 34 x 18, QD 16 x 12, Fc 4 x 12 (later: QD 6 x 12 + 10 x 32★, Fc 2 x 12 + 2 x 32★. *Prince Regent*: GD 34 x 32, MD 34 x 24, UD 34 x 24, QD 6 x 12 + 10 x 32★, Fc 2 x 12 + 2 x 32★)

***Caledonia*** Plymouth Dyd.
Ord 6.11.1794/15.12.1795; K 01.1805; L 25.6.1808; 1856 hulked.

***Britannia*** Plymouth Dyd.
Ord 6.11.1812; K 12.1813; L 20.10.1820; 1855 hulked.

***Prince Regent*** Chatham Dyd.
Ord by 01.1815; K 17.7.1815; L 12.4.1823; 1847 cut down to a Second Rate 92 (two-decker); 1861 converted to a screw line of battle ship; 1873 BU.

***Royal George*** Chatham Dyd.
Ord 2.9.1819; K 06.1823; L 22.9.1827; 1854 converted to a screw line of battle ship; 1875 sold.

FATES: *Caledonia* to the Greenwich Seamens' Hospital in 1856; name changed to *Dreadnought*; 1870 returned to the RN; offered as a church ship but not accepted; 1871 to Metropolitan Asylums as a hospital ship for male convalescent smallpox patients; 1875 BU. *Britannia* hospital ship at Portsmouth 1855; 1859 to Dartmouth as a cadet training ship; 1869 BU.

PLANS: Lines/profile/orlop/gun deck/main deck/upper deck/quarter deck & forecastle/framing/midships section. *Caledonia*: as built: Lines & profile & decoration. *Britannia*: 1811: profile. 1843: profile/orlop/gun deck/main deck/upper deck/quarter deck & forecastle. 1859: orlop/gun deck/main deck/upper deck/quarter deck & forecastle. *Prince Regent*: as built: Lines & profile & decoration. 1843: Profile/hold & hold sections/orlop/gun deck/main deck/upper deck/quarter deck & forecastle. *Royal George*: 1852: Profile/orlop/midships section. 1858: hold & hold sections/orlop/gun deck/main deck/quarter deck & forecastle. 1860: Profile/orlop/gun deck/main deck/quarter deck & forecastle.

***Nelson* Class 1806.** 'Surveyors' (Rule & Peake) design, modified from *Caledonia*.
Dimensions & tons: 205'. 170'10⅛" x 53'6" x 24'. 2601 4/94.
Men: 875. Guns: GD 32 x 32, MD 34 x 24, UD 34 x 18, QD 6 x 12 + 10 x 32★, Fc 2 x 12 + 2 x 32★.

***Nelson*** Woolwich Dyd.
Ord 10.5. & 10.8. & 24.9. & 1.10.1806; K 12.1809; L 4.7.1814; 1858 converted to a screw line of battle ship; 1867 given to the New South Wales Government; 1898 sold; 1928 BU.

***Howe*** Chatham Dyd.
Ord 10.5. & 10.8. & 24.9. & 1.10.1806; K 06.1808; L 28.3.1815; 1854 BU.

***Saint Vincent*** Plymouth Dyd.
Ord 10.5. & 10.8. & 24.9. & 1.10.1806; K 05.1810; L 11.3.1815; by 1862 hulked.

FATES: *Saint Vincent* became a training ship for boys; 1906 sold for BU.
PLANS: Lines/profile/orlop/gun deck/main deck/upper deck/quarter deck & forecastle/framing/specification. *Nelson*: as built: Lines & profile & decoration. (small scale). 1867 as screw ship: Profile/hold & hold section/orlop/gun deck/main deck/upper deck/quarter deck & forecastle. *Howe*: 1838: Orlop/gun deck/main deck/upper deck/quarter deck & forecastle. *Saint Vincent*: 1852: Profile/hold & sections/orlop/main deck/upper deck/quarter deck & forecastle.

***Hibernia*.** See under 110 gun ships.

***Saint George* Class 1819.** Broadened version of the *Caledonia*. The *Royal Frederick* was originally ordered to this design in 1827 though later modified into what became the *Queen*.
Dimensions & tons: 205'5½", 170'6" x 54'6" x 23'2". 2693 71/94.
Men: 1000. Guns: GD 2 x 68 + 30 x 32, MD 2 x 68 + 32 x 32, UD 2 x 68 + 32 x 32, QD 16 x 32 (short), Fc 2 x 32 + 2 x 32 (short).

***Saint George*** Plymouth Dyd.

# HENSLOW TO SEPPINGS 1786-1830

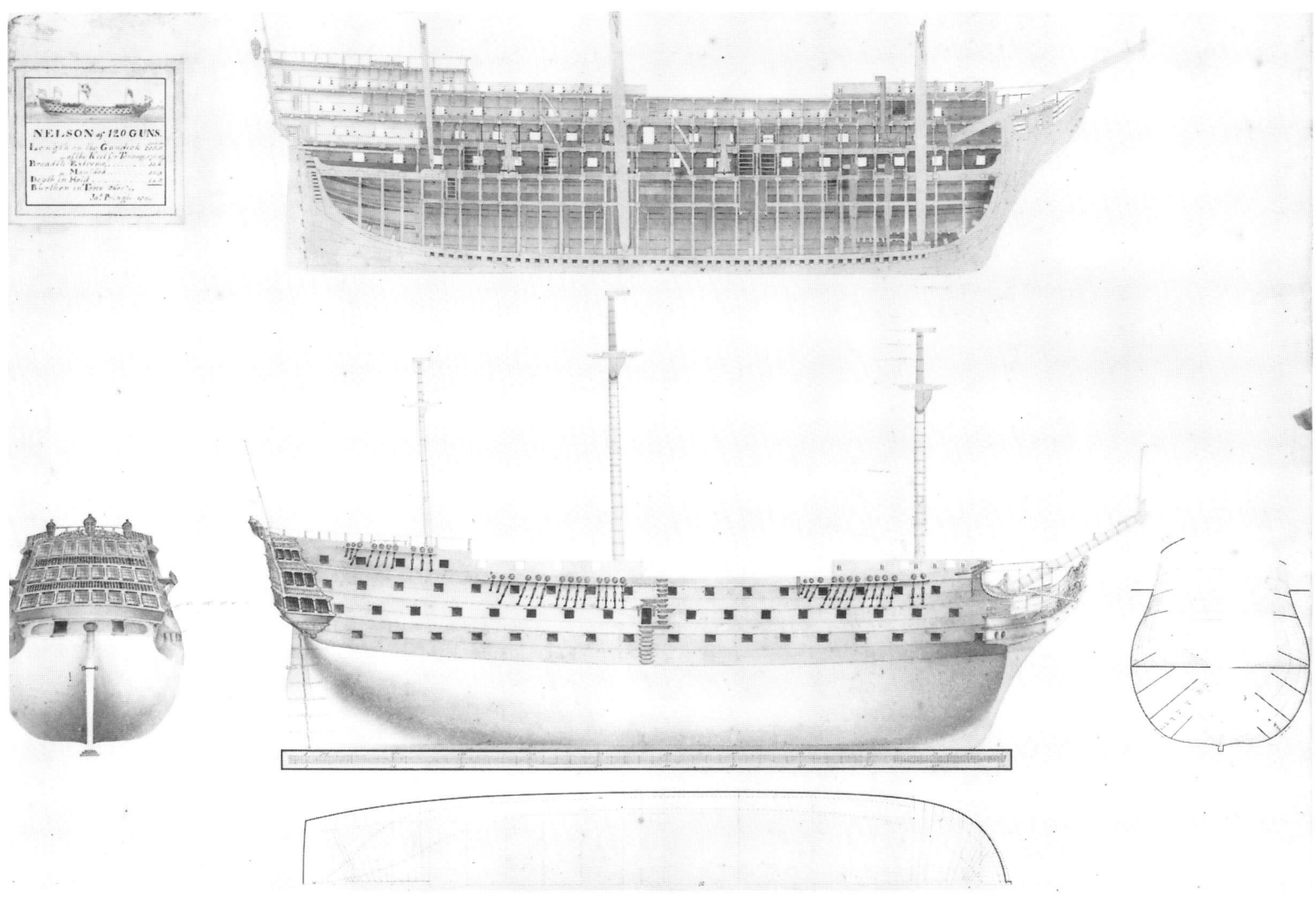

*This beautifully detailed small scale hull plan (number x22) shows the Nelson of 120 guns, launched in 1814 and the name ship of her class*

Ord 25.7.1819; K 05.1827; L 27.8.1840; 1859 converted to a screw line of battle ship; 1883 sold.

**Royal William** Pembroke Dyd.
Ord 30.12.1823; K 10.1825; L 2.4.1833; 1860 cut down and converted to a screw line of battle ship of 72 guns; 1885 hulked.

**Neptune** Portsmouth Dyd.
Ord ____ 1822?; K 01.1827; L 27.9.1832; 1859 cut down and converted to a screw line of battle ship of 72 guns; 1875 BU.

**Waterloo** Chatham Dyd.
Ord 9.9.1823; K 03.1827; L 18.6.1833; 1859 converted to a screw line of battle ship; 1862 renamed *Conqueror*; 1876 to the Marine Society.

**Trafalgar** Woolwich Dyd.
Ord 19.2.1825; K 12.1829; L 21.6.1841; 1859 converted to a screw line of battle ship; 1873 renamed *Boscawen*; 1906 sold.

FATES: *Royal William* hulked at Chatham 1885 as the training ship *Clarence*; 1889 burnt by accident. *Waterloo* training ship for the Marine Society from 1876; 1918 burnt.

PLANS: Lines/profile/orlop/gun deck/main deck/upper deck/quarter deck & forecastle/rig/framing. *Saint George* as screw two-decker 90: Lines/profile/hold & hold sections/orlop/gun deck/main deck/upper deck/rig.

## First Rates, 110 Guns (three-deckers)

**Ville de Paris Class 1788.** Henslow design, originally intended to be a 100 gun ship, but changed to a 110 shortly after the order was placed.
Dimensions & tons: 190', 156'1⅛" x 53' x 22'4". 2332 24/94.
Men: 837. Guns: GD 30 x 32, MD 30 x 24, UD 32 x 18, QD 14 x 12, Fc 4 x 12.

**Ville De Paris** Chatham Dyd.
Ord 10.1. & 5.8.1788; K 1.7.1789; L 7.7.1795; 1825 hulked.

**Hibernia** see below.

FATES: *Ville de Paris* became a lazaretto at Pembroke; 1845 BU.

PLANS: Lines/profile/orlop/gun deck/main deck/upper deck/quarter deck & forecastle.

**Hibernia Class 1790.** Lengthened version of the *Ville de Paris*. The lengthening was done during building; she was begun to the same dimensions as the *Ville de Paris*. Later classed as a 120 gun ship.
Dimensions & tons: 201'2", 167'3⅛" x 53' x 22'4". 2499.
Men: 850. Guns: GD 32 x 32, MD 32 x 24, UD 34 x 18, QD 12 x 32★, Fc 2 x 18 + 4 x 32★.

**Hibernia** Plymouth Dyd.
Ord 9.12.1790; K 11.1792; L 17.11.1804; 1855 hulked.

FATE: Became a guard and receiving ship at Malta; 1902 sold.

PLANS: Profile/orlop/gun deck/main deck/upper deck/quarter deck & forecastle/framing/specification. 1821: profile. 1833: wales.

**Ocean Class 1797.** Lengthened version of Henslow's 98/104 gun *Dreadnought/Neptune* class (and begun as one of that class).
Dimensions & tons: 196', 164'0⅜" x 51' x 21'6". 2269 63/94.
Men: 850. Guns: GD 30 x 32, MD 32 x 18, UD 32 x 18, QD 2 x 18 + 10 x 32★, Fc 2 x 18 + 2 x 32★, RH 6 x 18★. Later as an 80 had a full 32 pounder battery.

**Ocean** Woolwich Dyd. (as 98): Ord 9.12.1790; K 1.10.1792.
(as 110): Ord 4.6.1797; L 24.10.1805; 1821 cut down to Third Rate 80 (two-decker); 1831 hulked.

FATE: Became a lazaretto at Sheerness; 1832? reverted to an 80? later flagship for the Captain of the Ordinary; 1852 coal depot then to Chatham; 1875 BU.

PLANS: Sheer. 1850 as coal depot: Profile/gun deck/upper deck/quarter deck &

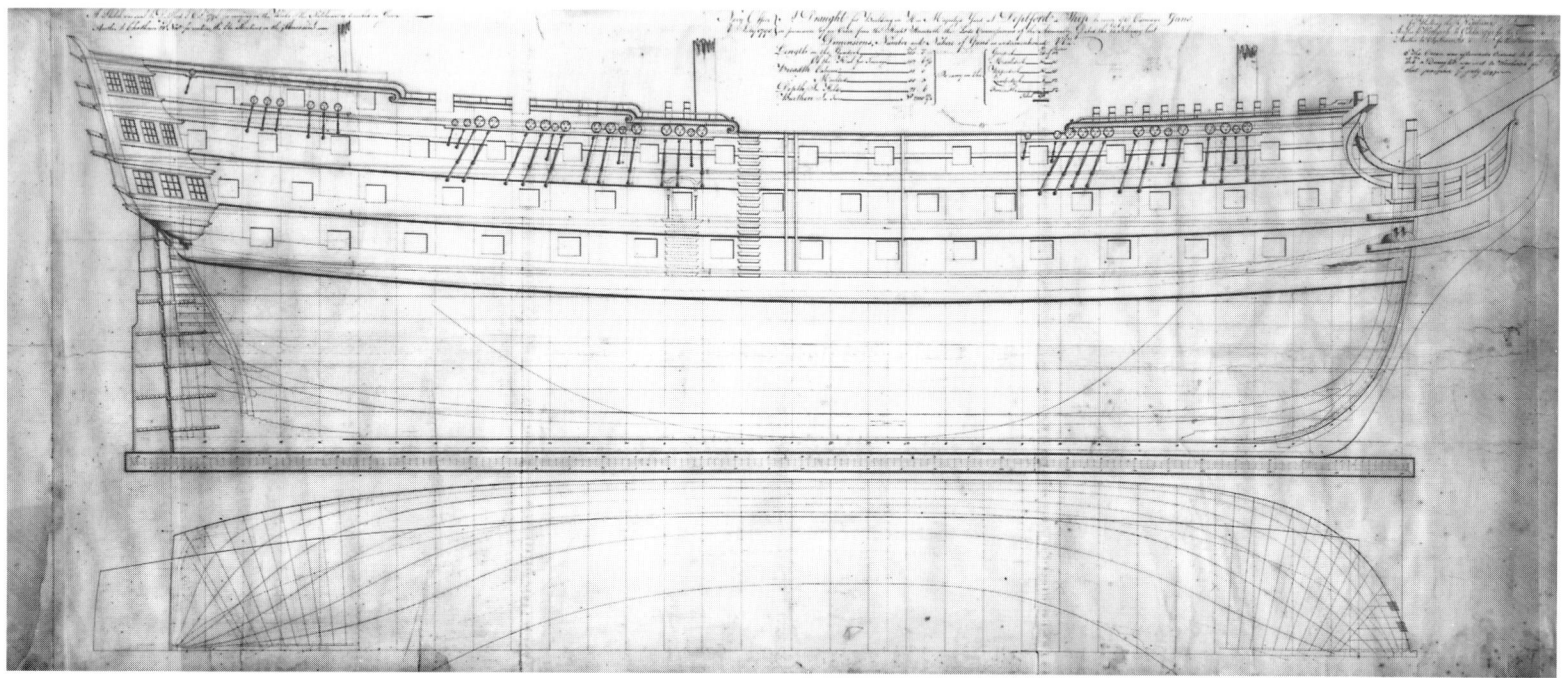

Part of a lines plan (sheer and half breadth - plan number x163) for building the *Dreadnought* Class 98 gun three-deckers of 1788

forecastle. 1853 as coal depot: cabins.

## First Rates, 100/104 Guns (three-deckers)

[later classed as 110s].

### *Impregnable* Class 1797.
Rule design. *Trafalgar* was a modified form of this design, and the *Princess Charlotte*'s a lengthened version.

Dimensions & tons: 196', 163'7⅜" x 52'8" (originally 51') x 22'. 2404 (originally 2263 58/94).
Men: 738. Guns: GD 28 x 32, MD 30 x 24, UD 30 x 12, QD 14 x 32★, Fc 2 x 12 + 2 x 32★.

*Impregnable* Chatham Dyd.
Ord 23.11.1797/13.12.1798; K 22.2.1802; L 1.8.1810; 1862 hulked.

FATE: 1839 harbour flagship at Plymouth; 1862 training ship; 1883 renamed *Kent*; 1891 renamed *Caledonia*; 1906 sold.

PLANS: Lines/profile/orlop/gun deck/main deck/upper deck/quarter deck & forecastle/framing/midships section. 1817: Orlop. 1848: Orlop/cabins.

### *Boyne* Class 1801.
Built to the lines of Slade's *Victory*. Originally classed as 98s.

Dimensions & tons: 186', 153'0¾" x 51'3" x 21'6". 2138 42/94.
Men: 738. Guns: GD 28 x 32, MD 30 x 24, UD 30 x 18, QD 4 x 12 + 8 x 32★, Fc 2 x 12 + 2 x 32★.

*Boyne* Portsmouth Dyd.
Ord 9. & 16.7.1801; K 04.1806; L 3.7.1810; 1834 hulked.
*Union* Plymouth Dyd.
Ord 9. & 16.7.1801; K 10.1805; L 16.11.1811; 1830 cut down to a Third Rate 80/76 (two-decker); 1833 BU.

FATES: Both were intended to be razeed into 80 gun ships following the example of *Ocean*, with a full 32 pounder battery. *Boyne* became the gunnery training ship *Excellent* at Portsmouth in 1834; 1859 renamed *Queen Charlotte*; 1861 BU.

PLANS: Lines/profile/orlop/gun deck/main deck/upper deck/quarter deck & forecastle/framing/sheer/midships section.

### *Trafalgar* Class 1806.
Rule design, modified from *Impregnable*. In 1813 breadth was increased by 1½ feet for better stability.

Dimensions & tons: 196', 162'9" x 52'6" (originally 51') x 22'8". 2386 5/94.
Men: 738. Guns: GD 28 x 32, MD 30 x 18, UD 30 x 12, QD 6 x 12 + 8 x 32★, Fc 2 x 12 + 2 x 32★.

*Trafalgar* Chatham Dyd.
Ord 17.8. & 18.9 & 24.9.1806; K 05.1813; L 26.7.1820; 1825 renamed *Camperdown*; 1860 hulked.

FATE: Became coal depot at Portsmouth; 1882 renamed *Pitt*; 1906 sold.

PLANS: Lines/Profile/orlop/gun deck/main deck/upper deck/quarter deck & forecastle/framing/specification.

### *Princess Charlotte* Class 1812.
Enlarged version of design of 1801 *Boyne* Class, therefore a development of Slade's *Victory*. Extra section added amidships to get the extra length and 'additional breadth to be obtained by cross framing'.

Dimensions & tons: 197'7", 163'4" x 52'9" x 22'6". 2417 44/94.
Men: 738. Guns: GD 28 x 32, MD 30 x 24, UD 30 x 18, QD 4 x 12 + 8 x 32★, Fc 2 x 12 + 2 x 32★.

*Princess Charlotte* Portsmouth Dyd.
Ord 6.1.1812; K 11.1818; L 14.9.1825; 1857 hulked.
*Royal Adelaide* (ex *London*) Plymouth Dyd.
Ord 6.1.1812; K 05.1819; L 28.7.1828; 1869 hulked.

FATES: *Princess Charlotte* became floating barracks at Hong Kong 1875 sold. *Royal Adelaide* became the flag and receiving ship at Plymouth; 1891 to Chatham; 1905 sold.

PLANS: Lines/profile/orlop/gun deck/main deck/upper deck/quarter deck & forecastle/midships section. *Royal Adelaide*: 1835: sails.

### *Dreadnought*. See under 98s.

## Second Rates, 98 Guns (three-deckers)

### *Dreadnought/Neptune* Class 1788.
Henslow design. *Ocean* was begun as one of this class, but then lengthened and reclassed as a 110 gun ship. *Dreadnought* was later reclassed as a First Rate of 104 guns.

Dimensions & tons: 185', 152'6⅝" x 51' x 21'6". 2110 53/94.
Men: 750. Guns: GD 28 x 32, MD 28/30 x 18, UD 30 x 18, QD 8 x 12, Fc 2 x 12.

*Dreadnought* Portsmouth Dyd.
Ord 17.1. & 22.3.1788; K 07.1788; L 13.6.1801; 1825 hulked.
*Neptune* Deptford Dyd.
Ord 15.2.1790; K 04.1791; L 28.1.1797; 1813 hulked.
*Temeraire* Chatham Dyd.
Ord 9.12.1790; K 07.1793; L 11.9.1798; 1813 hulked.
*Ocean* Woolwich Dyd.
Ord 9.12.1790; K 1.10.1792; see under 110s.

FATES: *Dreadnought* became a lazaretto at Pembroke; 1831 to Greenwich as the

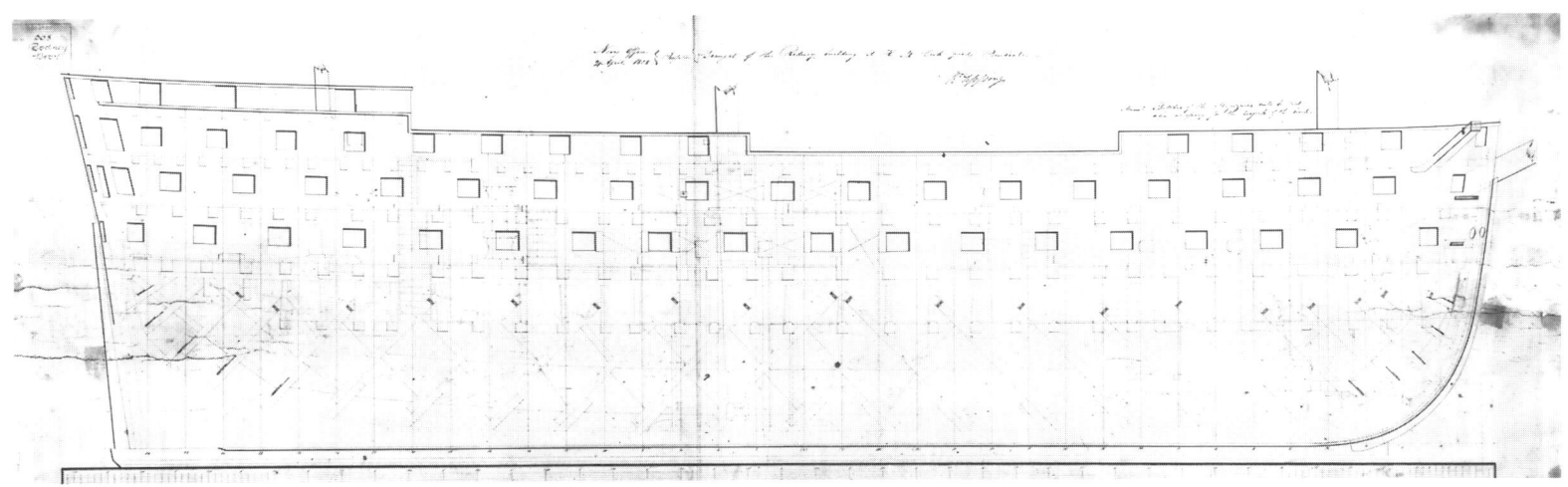

Profile plan (number 308) for building the *Rodney* of 1833, the name ship of her class. The pattern of diagonal bracing which permitted the building of these very big and long two-deckers can be seen low in the hull. The round, built up, bow and stern were other sources of strength

seamens' hospital; 1857 BU. *Neptune* became temporary prison ship at Plymouth; 1818 BU. *Temeraire* became a prison ship at Plymouth; 1820 to Sheerness as receiving ship; 1838 sold for BU; it was when she was being towed up the Thames to be broken up at Charlton that Turner saw her and was inspired to paint his picture of 'The fighting *Temeraire*'.

PLANS: Lines/profile/orlop/gun deck/main deck/upper deck/quarter deck & forecastle/wales. *Dreadnought*: as built: Lines & profile & decoration. *Neptune*: capstan.

## Second Rates, 90/92 Guns (two-deckers)

**Rodney Class 1826.** Seppings design. Originally intended to be 100 gun ships. Response to French *Royal Charles* with a full spar deck.
Dimensions & tons: 205'6", 170'4" x 53'6" x 23'. 2598 1/94.
Men: 720/830. Guns: GD 2 x 68★ + 30 x 32, UD 34 x 32, QD & Fc 26 x 32.

**Rodney** Pembroke Dyd.
 Ord ____ 1827; K 07.1827; L 06.1833; 1860 converted to a screw line of battle ship; 1882 BU.

**London** Chatham Dyd.
 Ord by 11.1826; K 10.1827; L 28.9.1840; 1858 converted to a screw line of battle ship; 1874 hulked.

**Nile** Plymouth Dyd.
 Ord ____ 1827?; K 10.1827; L 28.6.1839; 1854 converted to a screw line of battle ship; 1876 hulked.

FATES: *London* became harbour storeship at Zanzibar; 1884 sold for BU. *Nile* became the training ship *Conway* for the Mercantile Marine Association of Liverpool; later to the Menai Strait; 1953 stranded; 1956 wreck burnt.

PLANS: Lines/profile/orlop/gun deck/upper deck/quarter deck & forecastle/midships section/framing/sails. *Rodney*: as built: Profile. 1851 as screw: Lines/profile/hold & hold sections/orlop/gun deck/upper deck/quarter deck & forecastle. 1859: sails/machinery arrangement. *London*: 1857 as screw: Lines/profile/hold & hold sections/orlop/gun deck/upper deck/quarter deck & forecastle. 1874: Profile/hold & hold sections/orlop/gun deck/upper deck/quarter deck & forecastle/rig. 1875: sails. *Nile*: 1853: Profile/hold & hold sections/gun deck/upper deck/quarter deck & forecastle/machinery arrangement.

## Second/Third Rates, 84 Guns (two-deckers)

**Formidable Class 1815.** Lines of French *Canopus* (taken 1798) modified by Seppings. Originally to be classed as an 80. Timber captured from the French at Genoa was used in the construction of *Formidable*. Originally ordered with an open stern.
Dimensions & tons: 193'10", 160'2⅝" x 51'5¼" x 22'6". 2254 69/94.
Men: 700. Guns: GD 32 x 32, UD 32 x 24, QD 4 x 24 + 14 x 32★, Fc 2 x 24 + 2 x 32★.

**Formidable** Chatham Dyd.
 Ord 8.5. & 30.8.1815; K ____ ; L 19.5.1825; 1869 hulked.

### Modified Version, Teak Built in India.
Dimensions & tons: 196'1½", 161'11½" x 51'5¼" x 22'6". 2279 29/94.

**Ganges** Bombay Dyd.
 Ord 4.6.1816; K 05.1819; L 10.11.1821; 1857 hulked.

**Asia** Bombay Dyd.
 Ord 22.4.1819; K 01.1822; L 19.1.1824; 1859 hulked.

**Bombay** Bombay Dyd.
 Ord 26.1.1825; K 05.1826; L 17.3.1828; 1861 converted to a screw line of battle ship; 1864 burnt.

### Further Modified Design to be Built in India, to carry 32 pounders throughout, her waterlines fore and aft were filled out.

**Calcutta** Bombay Dyd.
 Ord 6.4.1827; K 03.1828; L 14.3.1831; 1863 hulked.

### Modified Version
Dimensions & tons: 195'4½". 161'11½" x 51'5¼" x 22'6". 2279 29/94.

**Monarch** Deptford Dyd.
 Ord 18. & 23.7.1817; not begun? transferred to: Chatham Dyd.
 Ord 22.2.1825; K 08.1825; L 8.12.1832; 1866 BU.

**Vengeance** Pembroke Dyd.
 Ord 24.1.1817; K 07.1819; L 27.7.1824; 1861 hulked.

**Thunderer** Woolwich Dyd.
 Ord 30.8.1817; K 04.1823; L 22.9.1831; 1863 hulked.

**Powerful** Chatham Dyd.
 Ord 23.1.1817; K 08.1820; L 21.6.1826; 1862 hulked.

**Clarence** (ex *Goliath*) Pembroke Dyd.
 Ord 27.5.1819; K 08.1824; L 25.7.1827; 1864 hulked.

FATES: *Formidable* transferred to the Bristol Training Ship Association; 1906 sold. *Ganges* became a training ship at Portsmouth; 1866 to Plymouth; 1906 renamed *Tenedos* III; 1910 renamed *Indus* V; 1922 renamed *Impregnable* III; 1929 sold for BU. *Asia* became Ordinary guard ship at Portsmouth; 1906 sold for BU. *Bombay* burnt by accident at Montevideo 14.12.1864. *Calcutta* became gunnery training ship at Portsmouth; 1908 sold. *Monarch* became a receiving ship at Plymouth; 1864 hulk; 1879 masting hulk; 1897 sold. *Thunderer* became a target ship at Portsmouth; 1869 renamed *Comet*; 1870 renamed *Nettle*; 1901 sold. *Powerful* became a target ship at Chatham; 1864 BU. *Clarence* was lent to the Liverpool Roman Catholic Reformatory Society; 1884 burnt and BU.

PLANS: Lines/gun deck/upper deck/quarter deck & forecastle/framing/midships section. Teak built version: Lines/profile/orlop/gun deck/upper deck/quarter deck & forecastle/framing/midships section. *Bombay* as screw 1860: Lines/profile/hold & hold sections/orlop /gun deck/upper deck/quarter deck & forecastle. 1864 as fitted: Profile/hold & hold sections/orlop/gun deck/upper deck/quarter deck & forecastle. *Calcutta* in 1830: profile/orlop/gun deck/upper deck/quarter deck & forecastle. In 1858: orlop/gun deck/upper deck/quarter deck & forecastle. Modified version: Lines/profile/orlop/gun deck/upper deck/quarter deck & forecastle/framing/midships section/specification.

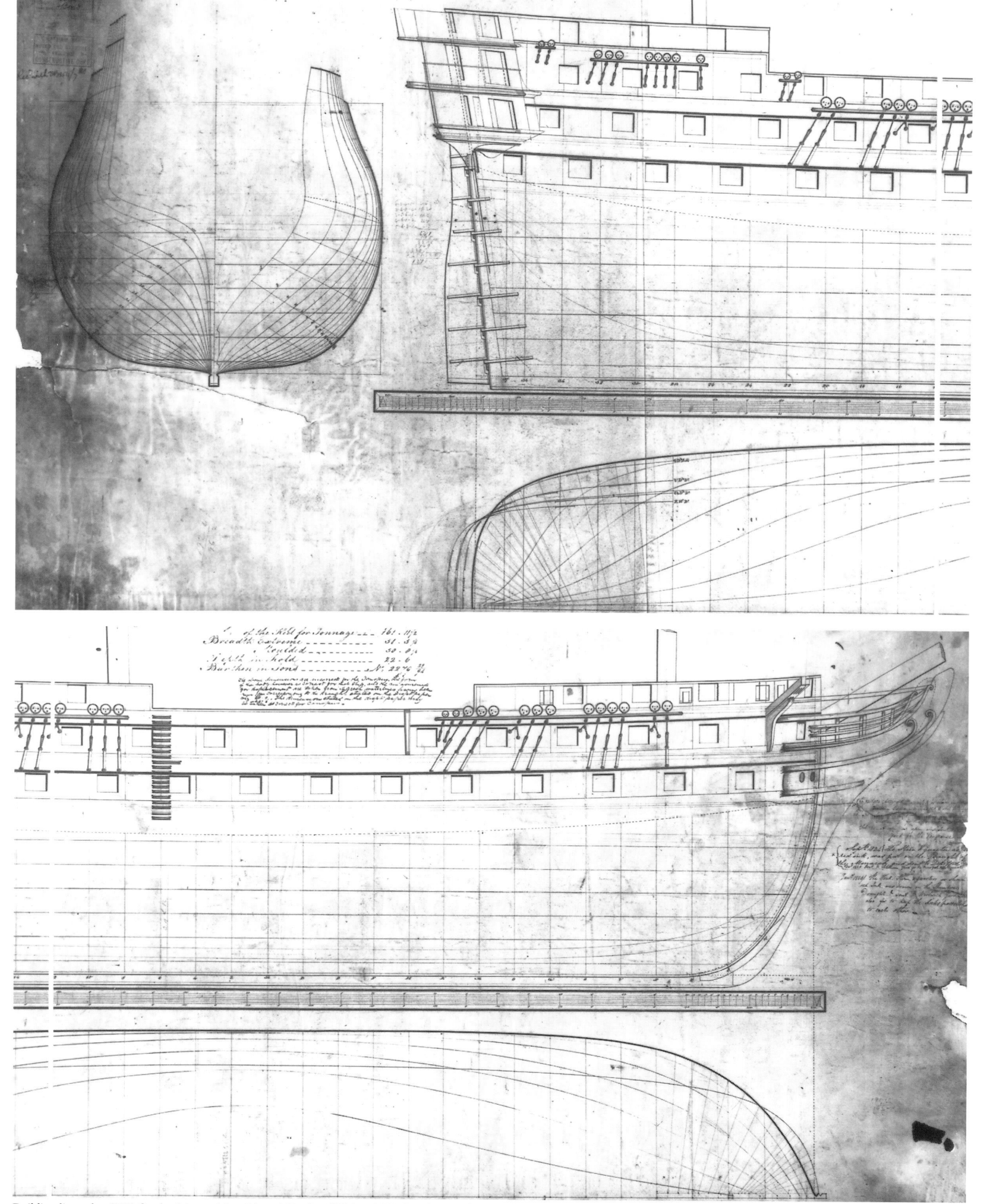

Building lines plan (332) for the 84 gun ships *Asia* and *Bombay* (1824 & 1828, Bombay Dockyard) of the *Formidable* Class. These plans were also used for those ships of the modified design built in home yards

NOTE: (provided by Andrew Lambert). There was a Seppings design for a new 84 gun ship, based on *Canopus*, but with a full two deck battery of long 32s in 1831-2, which Symonds destroyed.

## Second/Third Rates, 80 Guns (two-deckers)

### *Foudroyant* Class 1788. Henslow design.
Dimensions & tons: 184', 151'5⅝" x 50'6" x 22'6". 2054 65/94.
Men: 650. Guns: GD 30 x 32, UD 32 x 24, QD 14 x 12, Fc 4 x 12.
*Foudroyant* (ex *Superb*) Plymouth Dyd.
Ord 17.1.1788; K 05.1789; L 31.8.1798; 1861 hulked.

FATE: Became gunnery training ship at Plymouth; 1890 sold to Wheatley Cobb as a boys' training ship; 1897 wrecked on Blackpool Sands whilst on a fund raising and propaganda cruise; replaced by the frigate *Trincomalee* who took her name and still retains it at her berth in Portsmouth harbour (though in West Hartlepool for repairs in 1991-3 and apparently due to revert to her original name of *Trincomalee* if restored to her original state).
PLANS: Lines/profile/orlop/gun deck/upper deck/quarter deck & forecastle/framing.

### *Waterloo* Class 1809. Peake design.
Dimensions & tons: 192', 159'10" x 49' x 21'. 2041 45/94.
Men: 650. Guns: GD 30 x 32, UD 32 x 18, QD 4 x 12 + 10 x 32★, Fc 2 x 12 + 2 x 32★.
*Waterloo* (ex *Talavera*) Portsmouth Dyd.
Ord ____ 1809; K 11.1813; L 16.10.1818; 1824 renamed *Bellerophon*; 1856 hulked.

FATE: As *Bellerophon* became a receiving ship at Portsmouth; 1892 sold.
PLANS: Lines/profile/framing. As receiving ship, 1855: orlop/gun deck/upper deck/quarter deck & forecastle. 1856: profile/orlop/gun deck/upper deck/quarter deck & forecastle/round house.

### *Cambridge* Class 1810. Lines of Danish *Christian* VII (taken 1807). Later classed as an 82.
Dimensions & tons: 187'2", 154'10½" x 50'10½" x 21'7". 2125 27/94.
Men: 700. Guns: GD 30 x 32, UD 32 x 18, QD 4 x 12 + 10 x 32★, Fc 2 x 12 + 2 x 32★.
*Cambridge* Deptford Dyd.
Ord 8. & 26.7.1810; K 12.1811; L 23.6.1815; 1856 hulked.
FATE: Became gunnery training ship at Plymouth; 1869 BU.
PLANS: Lines/profile/orlop/gun deck/upper deck/quarter deck & forecastle/framing/midships section. 1871: Cabins (orlop)/(gun deck)/(upper deck). 1883: accommodation (orlop)/(upper deck).

### *Boscawen* Class 1817. Seppings design 'on enlarged lines of the old *Minotaur*'.
Dimensions & tons: 187'4½", 153'8" x 50' x 21'6"'. 2043 41/94.
Men: 700. Guns: probably as *Cambridge* above.

### Cancelled after 1822 (in 1835?). *Boscawen* Woolwich Dyd.
Ord ____ 20.11.1817; not begun?
PLANS: Lines/profile/orlop/gun deck/upper deck/quarter deck & forecastle/framing/midships section.
See Lambert's *Last Sailing Battlefleet* for the full complicated story of the *Boscawen* as a 74, as a stretched 80, as a Vernon class frigate (named *Indefatigable*) and finally as Symonds powerful 70 gun ship.

### *Indus* Class 1817. Enlarged lines of the Danish *Christian* VII (taken 1807); probably via an enlargement of the *Black Prince* (74) class design; as this ship was originally ordered as a 74, but then by Admiralty Order dated 18.5.1820 to be enlarged and classed as an 80. She was built using teak frames brought from Bombay aboard Bombay-built warships.
Dimensions & tons: 188'6", 155" x 50'5" x 22'6". 2095 62/94.
Men: 700. Guns: GD 28 x 32, UD 32 x 24, QD 4 x 12 + 10 x 32★, Fc 2 x 12 + 2 x 32★
*Indus* Portsmouth Dyd (teak frames from Bombay).
Ord 11.11.1817/18.5.1820; K 07.1824; L 16.3.1839; 1860 hulked.

FATE: Became harbour flagship at Plymouth; 1898 sold for BU.
PLANS: Lines/profile/orlop/gun deck/upper deck/quarter deck & forecastle/round house. 1842: orlop/gun deck/upper deck/quarter deck & forecastle.

### *Hindostan* Class 1819. 'Lines of the *Repulse* [Rule 1800 design 74] enlarged'; using teak frames brought from Bombay in the *Malabar*.
Dimensions & tons: 185'8", 153'0¾" x 50'8" x 21'. 2035 37/94.
Men: 650/590. Guns: GD 28 x 32 + 2 x 68★, UD 32 x 24, QD 4 x 12 + 10 x 32★, Fc 2 x 12 + 2 x 32★.
*Hindostan* Plymouth Dyd (teak frames from Bombay).
Ord 24.9.1819; K 09.1828; L 2.8.1841; in ordinary (reserve) from then on; 1884 hulked.

FATE: Cadet training ship at Dartmouth; 1905 training ship for boy artificers at Portsmouth; renamed *Fisgard* III; 1920 renamed *Hindostan*; 1921 sold for BU.
PLANS: Lines/profile/orlop/gun deck/upper deck/quarter deck & forecastle/midships section.

## Third Rates, 74 Guns (two-deckers)

### *Brunswick* Class 1785. 'Admiralty' design.
Dimensions & tons: 176', 145'2" x 48'8" x 19'6". 1828 72/94.
Men: 600. Guns: GD 28 x 32, UD 30 x 18, QD 12 x 9, Fc 4 x 9.
*Brunswick* Deptford Dyd.
Ord 10.1.1785; K 05.1786; L 30.4.1790; 1812 hulked.

FATE: Became prison ship at Chatham; 1814 powder magazine; 1825 lazaretto at Sheerness; 1826 BU.
PLANS: Lines/profile/orlop/gun deck/upper deck/quarter deck & forecastle/specification. As built: Lines & profile.

### *Mars* Class 1788. Henslow design. *Conqueror* appears to have been a modified version of this design.
Dimensions & tons: 176', 144'3" x 49' x 20'. 1842 24/94.
Men: 640 Guns: GD 28 x 32, UD 30 x 24, QD 12 x 9, Fc 4 x 9.
*Mars* Deptford Dyd.
Ord 17.1.1788; K 10.10.1789; L 25.10.1794; 1814 hulked.
*Centaur* Woolwich Dyd.
Ord 17.1.1788; K 11.1790; L 14.3.1797; 1819 BU.

FATES: *Mars* became a receiving ship at Portsmouth; 1823 BU.
PLANS: Lines/profile/orlop/gun deck/upper deck/quarter deck & forecastle/round house/specification.

### *Courageux* Class 1792. Henslow design. It may have been intended to build another ship to this design at Portsmouth which was never proceeded with.
Dimensions & tons: 181', 150'10⅛" x 47' x 19'9". 1772 38/94.
Men: 590. Guns: GD 28 x 32, UD 28 x 18, QD 2 x 18 + 12 x 32★, Fc 2 x 9 (later 18?) + 2 x 32★.
*Courageux* Deptford Dyd.
Ord 6.11.1794; K 10.1797; L 26.3.1800; 1814 hulked.

FATE: Became a lazaretto at Chatham; 1832 BU.
PLANS: Lines/profile/orlop/gun deck/upper deck/quarter deck & forecastle.

### *Plantagenet* Class 1794. Rule design.
Dimensions & tons: 181', 151'3⅛" x 47'x 19'9". 1777 28/94.
Men: 590. Guns: GD 28 x 32, UD 28 x 18, QD 14 x 9, Fc 4 x 9.
*Plantagenet* Woolwich Dyd.
Ord 6.11.1794/9.9.1796; K 11.1798; L 22.10.1801; 1817 BU.
PLANS: Lines/profile/orlop/gun deck/upper deck/quarter deck & forecastle.

### *Bulwark* Class 1795. Rule design. *Valiant* to be built to *Bulwark*'s lines but with some variations in layout.
Dimensions & tons: 182', 150'8" x 49' x 20'6". 1940. (*Valiant* depth in hold 20'10", tons 1925 12/94.)
Men: 640. Guns: GD 28 x 32, UD 30 x 24, QD 12 x 9 Fc 4 x 9. (*Valiant* to be QD 4 x 12 + 10 x 32★, Fc 2 x 12 + 2 x 32★).
*Bulwark* (ex *Scipio*) Portsmouth Dyd.

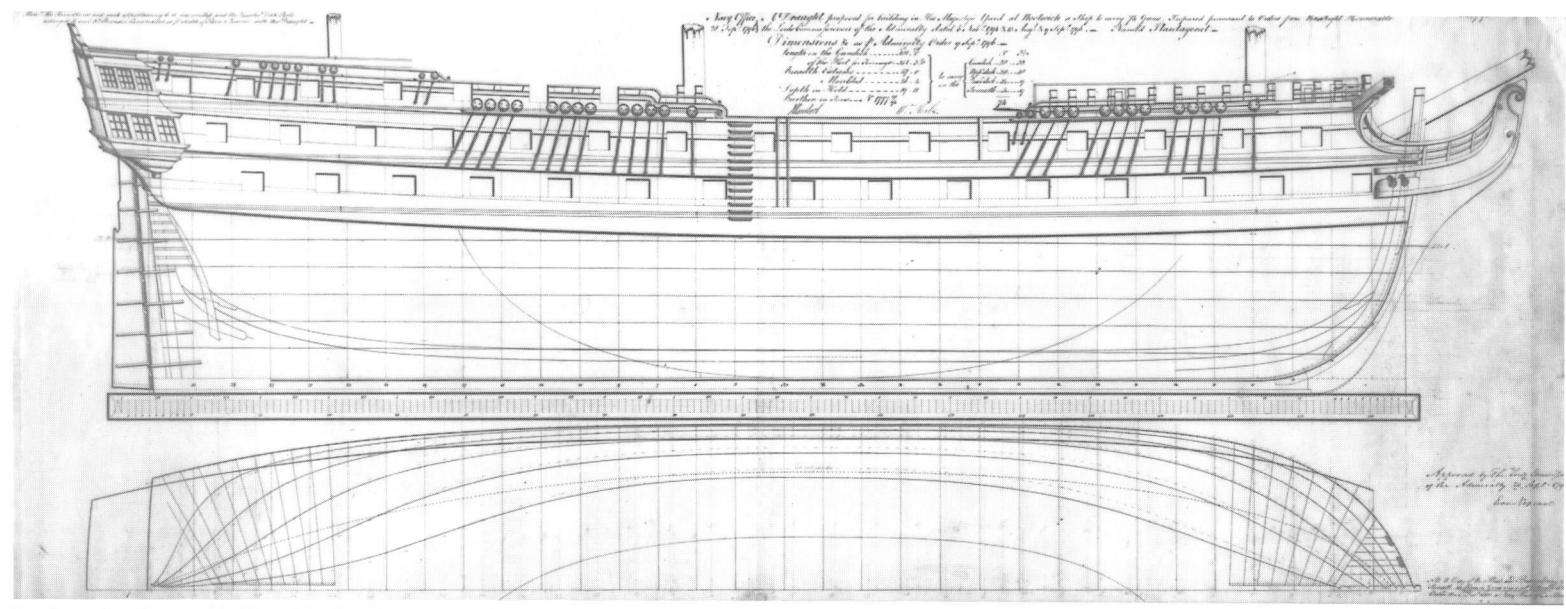

Part lines plan (sheer and half breadth of number 854 with body plan omitted) showing the design of the 74 gun ship *Plantagenet* of 1801

Ord 6.11.1794; K 04.1804; L 23.4.1807; 1826 BU.

**Cancelled 1832.** *Valiant* Plymouth Dyd.
Ord ___ 1826 (not begun).

PLANS: Lines/profile/orlop/gun deck/upper deck/quarter deck & forecastle/specification. *Valiant*: Lines/profile/orlop/gun deck/upper deck/quarter deck & forecastle/midships section/framing.

**Kent Class 1795.** Modified version of the 1757 *Valiant* Class (itself copied from the lines of the French *Invincible* taken in 1747). Originally intended to be of the same dimensions as the *Valiant* Class, but with the decks raised 6". During construction they were lengthened by 11'.
Dimensions & tons: 182'3", 141'8⅝" x 49'3" x 21'3". 1931 62/94.
Men: 690 Guns: GD 28 x 24, UD 28 x 24, QD 14 x 9
(later 4 x 9 + 8 x 32★) , Fc 4 x 9.

*Kent* Perry, Blackwall.
Ord 10.6.1795; K 10.1795; L 17.1.1798; 1857 hulked.
*Ajax* Randall, Rotherhithe.
Ord 10.6.1795; K 09.1795; L 3.3.1798; 1807 burnt.

FATES: *Kent* became a sheer hulk at Plymouth; 1880 BU. *Ajax* accidentally burnt off Tenedos 11.2.1807.

PLANS: Lines/profile/orlop/gun deck/upper deck/quarter deck & forecastle/specification.

**Conqueror Class 1795.** Henslow design, modified from the *Mars* Class. It is not certain whether she should be considered as a new class, or just a sub-group of the earlier one.
Dimensions & tons: 176', 144'3" x 49' x 20'9". 1842 24/94.
Men: 590. Guns: GD 28 x 32, UD 30 x 18, QD 14 x 9,
Fc 2 x 9 + 2 x 32★, RH 6 x 18★.

*Conqueror* Graham, Harwich.
Ord 30.4.1795; K 10.1795; L 23.11.1801; 1822 BU.

PLANS: As *Mars* class.

**Dragon Class 1795.** Rule design.
Dimensions & tons: 178', 146'9" x 48' x 20'6". 1798 44/94.
Men: 640. Guns: GD 28 x 32, UD 28 x 18, QD 4 x 18 + 10 x 32★,
Fc 2 x 18 + 2 x 32★, RH 6 x 18★.

*Dragon* Wells, Deptford.
Ord 30.4.1795; K 08.1795; L 2.4.1798; 1824 hulked.

FATE: Became a lazaretto at Pembroke; 1832 receiving ship and marine barracks; 1842 renamed *Fame*; 1850 BU.

PLANS: Lines/profile/orlop/gun deck/upper deck/quarter deck & forecastle/framing.

**Northumberland Class 1795.** Lines of the French *Impetueux* (taken 1794).
Dimensions & tons: 182', 150'1¼" x 48'7½" x 21'7". 1887 74/94.
Men: 640. Guns: GD 30 x 32, UD 30 x 18, QD 4 x 18 + 10 x 32★,
Fc 2 x 18 + 2 x 32★, RH 6 x 18★.

*Northumberland* Barnard, Deptford.
Ord 10.6.1795; K 10.1795; L 2.2.1798; 1827 hulked.
*Renown* (ex *Royal Oak*) Dudman, Deptford.
Ord 10.6.1795; K 11.1795; L 2.5.1798; 1814 hulked.

FATES: *Northumberland* became a lazaretto at Sheerness; 1850 BU. *Renown* became a hospital ship at Plymouth; later to Deptford?; 1835 BU?

PLANS: Lines/profile/orlop/gun deck/upper deck/quarter deck & forecastle.

**Spencer Class 1795.** Barralier design (Barralier was a French emigré naval architect). This design was modified for the *Milford* Class.
Dimensions & tons: 181', 148'10¼" x 49' x 21'10". 1901 5/94.
Men: 640. Guns: GD 30 x 32, UD 30 x 18, QD 4 x 18 + 10 x 32★,
Fc 2 x 18 + 2 x 32★, RH 6 x 18★.

*Spencer* Adams, Bucklers Hard.
Ord 19.9.1795; K 09.1795; L 10.5.1800; 1822 BU.

PLANS: Lines/profile/orlop/gun deck/upper deck/quarter deck & forecastle.

**Achille Class 1795.** Lines of the French *Pompée* (taken 1793).
Dimensions & tons: 182'2", 149'9¾" x 49'0½" x 21'10½". 1916.
Men: 640. Guns: GD 30 x 32, UD 30 x 18, QD 4 x 18 + 10 x 32★,
Fc 2 x 18 + 2 x 32★, RH 6 x 18★.

*Achille* Cleverley, Gravesend.
Ord 10.6.1795; K 10.1795; L 16.4.1798; 1865 sold for BU.
*Superb* Pitcher, Northfleet.
Ord 10.6.1795; K 08.1795; L 19.3.1798; 1826 BU.

PLANS: Lines/profile/orlop/gun deck/upper deck/quarter deck & forecastle/framing/midships section. *Achille* as built: Midships section.

**Revenge Class 1796.** Henslow design. Originally intended to be built as an 80. Though completed as a 74, she was later reclassed as a 78, then as a 76.
Dimensions & tons: 182', 150'3" x 49' x 20'9". 1918 83/94.
Men: 590. Guns: GD 28 x 32, UD 30 x 24, QD 14 x 9,
Fc 2 x 9 + 2 x 32★, RH 6 x 18★.

*Revenge* Chatham Dyd.
Ord 29.9.1796; K 08.1800; L 13.4.1805; 1849 BU.

PLANS: Lines/profile/orlop. 1827: hold. 1834: sails.

**Milford Class 1797.** Barralier design, modified from that of the *Spencer*.
Dimensions & tons: 181', 149'3¼" x 49' x 21'. 1906 35/94.

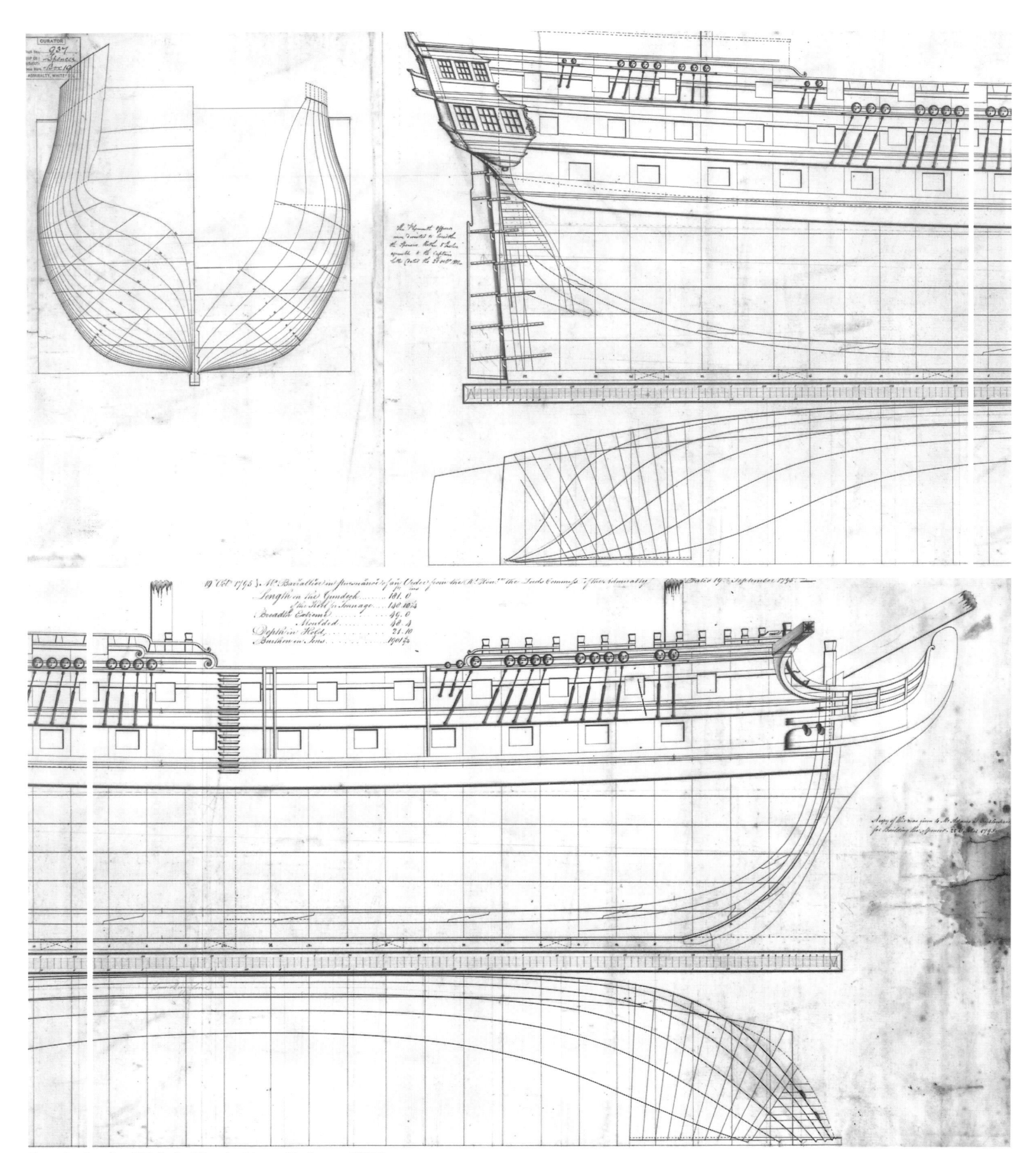

Lines plan (number 937) for building the 74 gun ship *Spencer* of 1800

Men: 590. Guns: GD 28 x 32, UD 30 x 24, QD 14 x 9, Fc 2 x 9 + 2 x 32★, RH 6 x 18★.

*Milford* Jacobs, Milford Haven.
Ord 6.12.1796; K 06.1798; builder failed so completed by 'Government' (this was the beginning of the establishment of Pembroke Dyd): *Milford* (Pembroke Dyd) L 1.4.1809; 1825 hulked.

**Cancelled 1811.** *Sandwich* Milford (Pembroke Dyd).
Ord 1809?; K 12.1809.

FATE: *Milford* became a lazaretto at Pembroke; 1846 BU.

PLANS: Lines/profile/orlop/gun deck/upper deck/quarter deck & forecastle/specification.

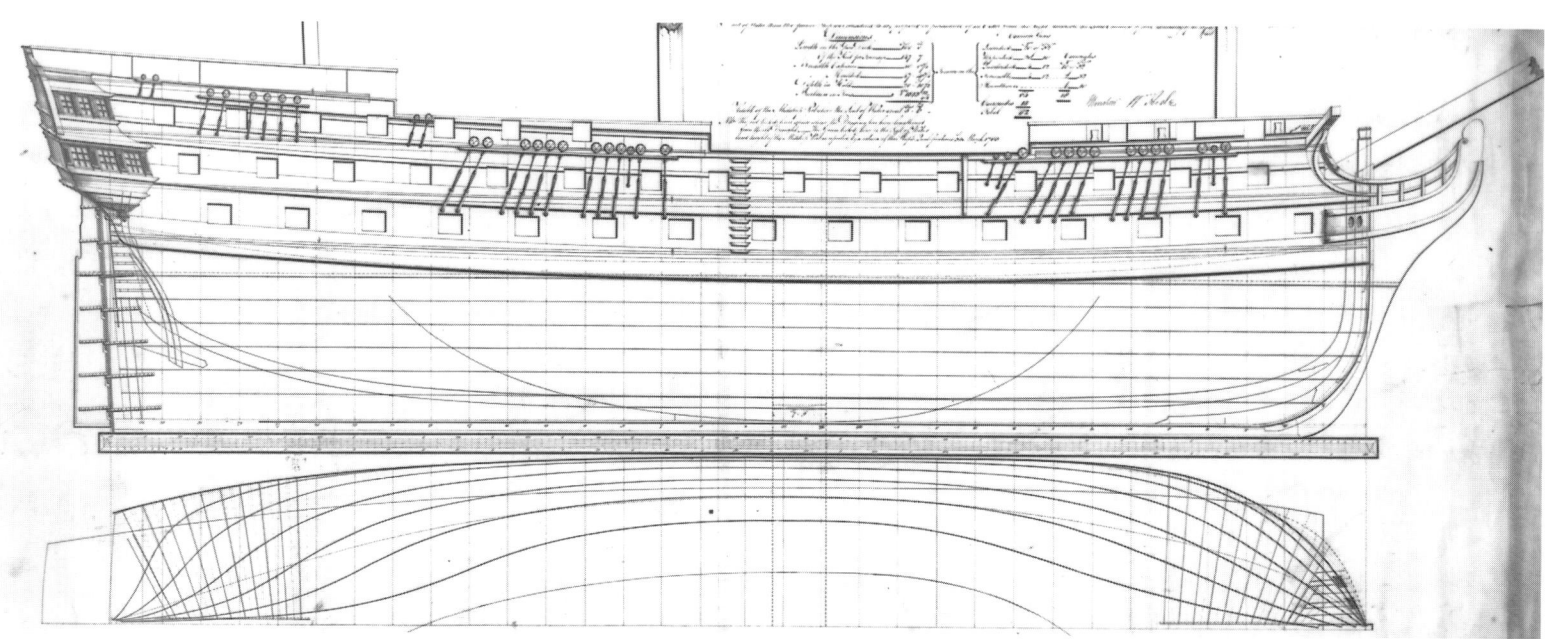

Lines plan (636) for building the *Blake* Class 74s of 1805

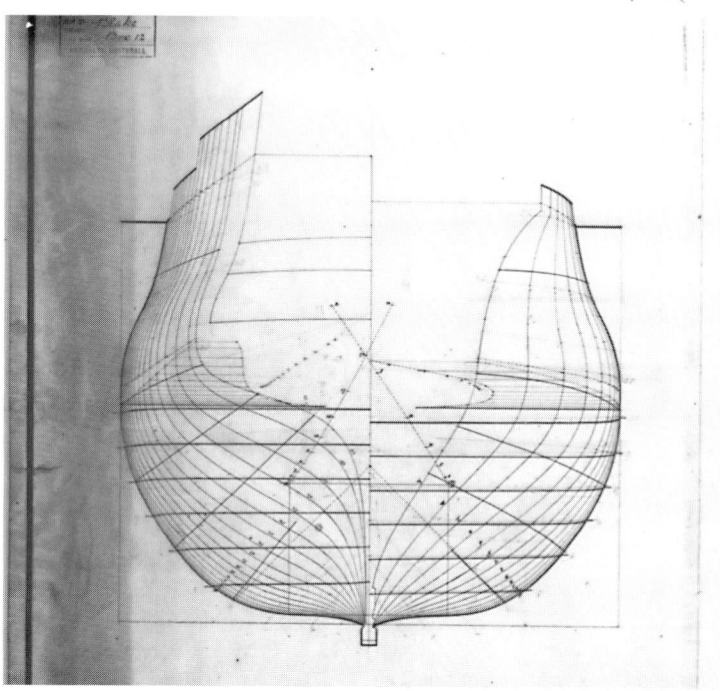

***Colossus* Class 1798.** Henslow design, later reclassed as 76s.
Dimensions & tons: 180', 148'3½" x 48'10" x 21'. 1880 88/94.
Men: 590. Guns: GD 28 x 32, UD 30 x 24, QD 4 x 24 + 10 x 32★, Fc 2 x 24 + 2 x 24★, RH 6 x 18★. [*Warspite* as frigate: Men 500, Guns: UD 6 x 8" + 22 x 32, QD & Fc 4 x 8" + 18 x 32].

***Colossus*** Deptford Dyd.
Ord 23.11.1797; K 05.1799; L 23.4.1803; 1826 BU.

***Warspite*** Chatham Dyd.
Ord 13.1.1798; K 3.12.1805; L 16.11.1807; 1840 cut down to frigate of 50 guns (Fourth rate); 1862 hulked.

FATES: *Warspite* lent to the Marine Society as a training ship; 1876 burnt.

PLANS: Lines/framing. *Warspite* as frigate 1838: profile/orlop/gun deck/upper deck/quarter deck & forecastle. 1842: profile/hold/orlop/gun deck/upper deck/quarter deck & forecastle.

***Fame/Hero* Class 1799.** Henslow design.
Dimensions & tons: 175', 144' x 47'8" x 20'6". 1740 27/94. (Dockyard built). 175', 144'1⅝" x 47'6" x 20'6". 1729 70/94 (Contract built).
Men: 640. Guns: GD 28 x 32, UD 28 x 18, QD 2 or 4 x 18 + 12 x 32★, Fc 2 x 18 + 2 x 32★, RH 6 x 18★.

***Fame*** Deptford Dyd.
Ord 15.10.1799; K 22.1.1802; L 8.10.1805; 1817 BU.

***Albion*** Perry, Blackwall.
Ord 24.6.1800; K 06.1800; L 17.6.1802; 1831 hulked.

***Hero*** Perry, Blackwall.
Ord 24.6.1800; K 08.1800; L 18.8.1803; 1811 wrecked.

***Illustrious*** Randall, Rotherhithe.
Ord 4.2.1800?; K 01.1800?; L 3.9.1803; 1848 hulked.

***Marlborough*** Barnard, Deptford.
Ord 31.1.1805; K 08.1805; L 22.6.1807; 1835 BU.

***York*** Brent, Rotherhithe.
Ord 31.1.1805; K 08.1805; L 7.7.1807; 1819 hulked.

***Hannibal*** Adams, Bucklers Hard.
Ord 31.1.1805; K 12.1805; L 05.1810; 1825 hulked.

***Sultan*** Dudman, Deptford.
Ord 31.1.1805; K 12.1805; L 19.9.1807; 1861 hulked.

***Royal Oak*** Dudman, Deptford.
Ord 1804?; K 06.1806; L 4.3.1809; 1825 hulked.

FATES: *Albion* became a lazaretto at Portsmouth; 1836 BU. *Hero* was wrecked on the Haak Islands 25.12.1811. *Illustrious* became an Ordinary guard ship at Portsmouth; 1853 hospital ship; 1859 reverted to Ordinary guard ship; 1868 BU. *York* became a convict ship at Portsmouth; 1853 BU. *Hannibal* became a lazaretto at Plymouth; later to Pembroke?; 1834 BU. *Sultan* became a receiving ship at Portsmouth; 1862 target ship; 1864 BU. *Royal Oak* became a receiving ship at Bermuda; 1850 BU.

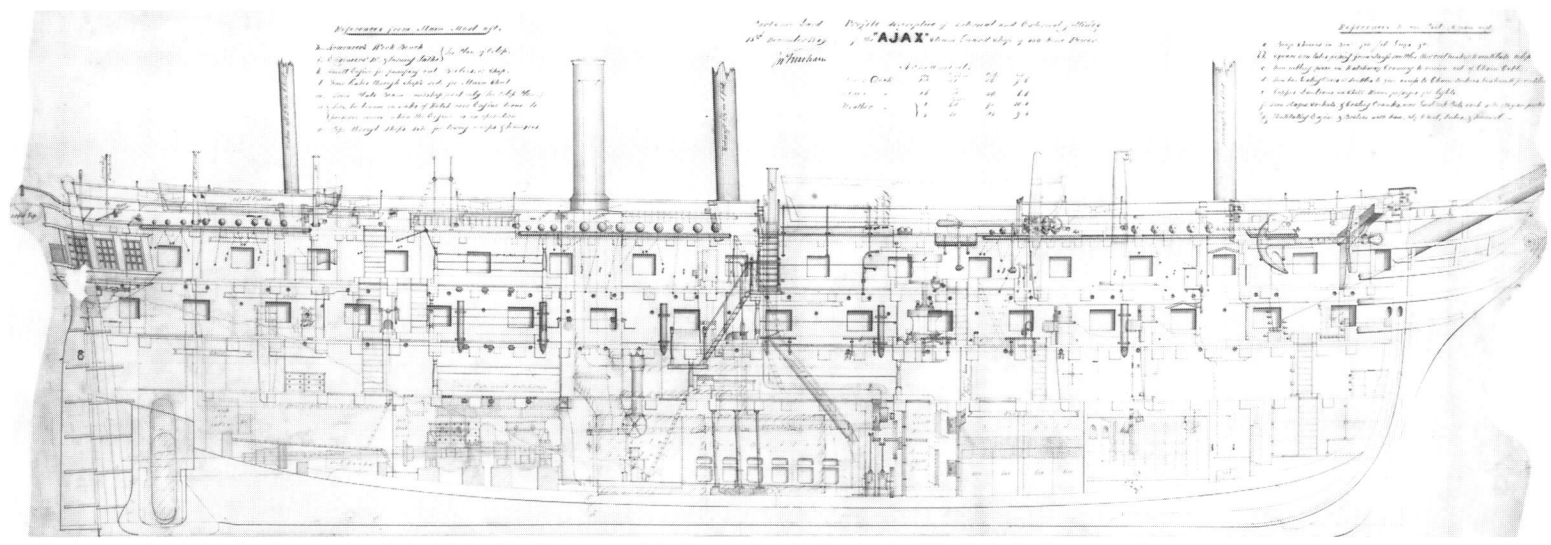

Profile plan (number 8237) showing the *Ajax* (built 1809) of the 'Surveyors of the Navy' (*Armada*) Class 'as fitted' in 1849 after her conversion to a 'screw guard (or block) ship'; in effect one of the Royal Navy's first screw battleships

PLANS: Lines/profile/orlop/gun deck/upper deck/quarter deck & forecastle.

### Swiftsure Class 1800. Henslow design.

Dimensions & tons: 173', 142' x 47'6" x 20'9". 1704 17/94.
Men: 590. Guns: GD 28 x 32, UD 28 x 18, QD 4 x 18 + 10 x 32★,
Fc 2 x 18 + 2 x 32★, RH 6 x 18★.

**Swiftsure** Adams, Bucklers Hard.
Ord ____ 1800 (or 1801?); K 08.1802; L 23.7.1804; 1819 hulked.
**Victorious** Adams, Bucklers Hard.
Ord ____ 1804?; K 02.1805; L 20.10.1808; 1826 hulked.

FATES: *Swiftsure* became a receiving ship at Portsmouth; 1845 BU. *Victorious* became a receiving ship at Portsmouth; 1861 BU.

PLANS: Lines/profile/specification. *Victorious* as receiving ship 1841: Orlop/gun deck/upper deck/quarter deck & forecastle.

### Repulse Class 1800.
Rule design The lines were enlarged for the *Hindostan* Class 80s of 1819. *Talavera* was timbered according to Seppings' principle, using smaller timbers than usual.

Dimensions & tons: 174', 143'2" x 47'4" x 20'. 1706 8/94.
Men: 590. Guns: GD 28 x 32, UD 28 x 18, QD 2 x 18 + 12 x 32★,
Fc 2 x 18 + 2 x 32★, RH 6 x 18★.
[*Eagle* as frigate: Men: 500. Guns: UD 6 x 8" + 22 x 32, QD & Fc 4 x 8" + 18 x 32.]

**Repulse** Barnard, Deptford.
Ord 4.2.1800; K 09.1800; L 21.7.1803; 1820 BU.
**Eagle** Pitcher, Northfleet.
Ord 4.2.1800; K 08.1800; L 27.2.1804; 1831 cut down to frigate of 50 guns (Fourth Rate); 1857 hulked.
**Sceptre** Dudman, Deptford.
Ord 4.2.1800; K 12.1800; L 11.12.1802; 1821 BU.
**Magnificent** Perry, Blackwall.
Ord 31.1.1805; K 04.1805; L 30.8.1806; 1823 hulked.
**Valiant** Perry, Blackwall.
Ord ____ 1804; K 04.1805; L 24.1.1807; 1823 BU.
**Elizabeth** Perry, Blackwall.
Ord ____ 1803; K 08.1805; L 23.5.1807; 1820 BU.
**Cumberland** Pitcher, Northfleet.
Ord 31.1.1805; K 08.1805; L 19.8.1807; 1830 hulked.
**Venerable** Pitcher, Northfleet.
Ord 31.1.1805; K 12.1805; L 12.4.1808; 1825 hulked.
**Talavera** (ex *Thunderer*) Woolwich Dyd. (Seppings' timbering.)
Ord 28.2.1814; K 07.1814; L 15.10.1818; 1840 burnt.
**Belleisle** Pembroke Dyd.
Ord 18.12.1812; K 02.1816; L 26.4.1819; 1841 troopship; 1854 hulked.
**Malabar** Bombay Dyd (Teak built).
Ord ____ 1815?; K 04.1817; L 28.12.1818; 1848 hulked.

FATES: *Eagle* hulked at Falmouth for the Coastguard; 1860 training ship in Southampton Water; 1862 to Liverpool; 1910 Mersey Division RNVR; 1918 renamed *Eaglet*; 1926 burnt; 1927 wreck sold for BU. *Magnificent* became a receiving ship at Jamaica; 1843 sold. *Cumberland* became a convict ship and coal depot at Chatham; 1833 renamed *Fortitude*; by 1856 to Sheerness as a coal depot; 1870 sold. *Venerable* became a church ship at Portsmouth; 1838 BU. *Talavera* was accidentally burnt by accident at Plymouth in 10.1840, then BU. *Belleisle* became a hospital ship at Sheerness 1866-1868 lent to the Seamens Hospital at Greenwich; 1872 BU. *Malabar* became a coal depot at Portsmouth; 1883 renamed *Myrtle*; 1905 sold.

PLANS: Lines/profile/orlop/gun deck/upper deck/quarter deck & forecastle/framing/midships section.[Note: deck plans show appearance after the end of the Napoleonic War]. *Eagle* as drill ship 1874: iron roof (2 sheets). *Belleisle* as troopship 1843: Orlop/gun deck/upper deck/quarter deck & forecastle/round house. As hospital ship 1857: Orlop/gun deck/upper deck/quarter deck & forecastle. *Malabar* as coal depot 1849: Profile/gun deck/upper deck/quarter deck & forecastle. 1868: upper deck/midships section.

### Blake Class 1805.
'Constructed on the exact plan of the old French *Courageux* [taken 1761] except the length on the Gun Deck is made 7'9" longer to obtain an increase in capacity to enable her to carry her midship port about 5½" higher out of the water .....'. Therefore can be regarded as a lengthened version of the *Leviathan* Class: Note the very short active careers of both these vessels.

Dimensions & tons: 180', 147'7" x 48'0¾" x 20'10½". 1813.
Men: 640? Guns: GD 28 x 32, UD 28 x 18, QD 4 x 12 + 10 x 32★,
Fc 4 x 12 + 2 x 32★, RH 6 x 18★.

**Blake** (ex *Bombay*) Deptford Dyd.
Ord 23.7. & 30.10.1805; K 04.1806; L 23.8.1808; 1814 hulked.
**Saint Domingo** Woolwich Dyd.
Ord 30.10.1805; K 06.1806; L 3.3.1809; 1816 sold.

FATES: *Blake* became a temporary prison ship at Portsmouth; 1816 sold.

PLANS: Lines/profile/orlop/gun deck/upper deck/quarter deck & forecastle/framing.

### Armada / Conquestadore / Vengeur Class 1806.
'Surveyors of the Navy' design. An attempt to keep the best features of previous classes, as well as to combine Henslow's and Rule's designs. It illustrates how the traditional methods of construction and design had been taken just about as far as they could go. Only improvements in constructional techniques would enable design to go forward – and this was to come with the developments associated with Seppings. These ships later became objects of dislike, and, because of the poor construction of numbers of the class, were nicknamed 'The forty thieves'. In fact the universal condemnation of the class has probably gone too far; the design, though uninspired, seems to have been adequate. The *Cornwallis* Class was a modification of this design, numerically the largest class of sailing line of battle ships ever built.

Dimensions & tons: 176'. 145'1" x 47'6" x 21'. 1741 17/94.

Men: 590. Guns: GD 28 x 32, UD 28 x 18, QD 4 x 12 + 10 x 32★, Fc 2 x 12 + 2 x 32★

[By the 1830s those ships razeed as frigates were armed as follows: Men: 500. Guns: UD 6 x 8" + 22 x 32, QD & x 4 x 8" + 18 x 32]

**Armada** Blackburn, Turnchapel.
Ord 21.10.1806; K 02.1807; L 22.3.1810; 1863 sold.

**Cressy** Brindley, Frindsbury.
Ord 1.10.1806; K 03.1807; L 22.3.1810; 1827 to be converted to 50 gun frigate but instead; 1832 BU.

**Vigo** Ross, Rochester.
Ord 20.10.1806; K 04.1807; L 21.2.1810; 1827 hulked.

**Vengeur** Graham, Harwich.
Ord 20.10.1806; K 07.1807; L 19.6.1810; 1824 hulked.

**Ajax** Perry, Blackwall.
Ord 1.7.1807; K 08.1807; L 2.5.1809; 1847 converted to screw 'block ship'; 1864 BU.

**Conquestadore** Guillaume, Northam.
Ord 20.10.1806; K 08.1807; L 1.8.1810; 1831 cut down to a frigate of 50 guns (Fourth Rate); 1856 hulked.

**Poictiers** King, Upnor.
Ord 1.10.1806; K 08.1807; L 9.12.1809; 1857 BU.

**Berwick** Perry, Blackwall.
Ord 1.7.1807; K 10.1807; L 11.9.1809; 1821 BU.

**Egmont** Pitcher, Northfleet.
Ord 13.7.1807; K 10.1807; L 7.3.1810; 1863 hulked.

**Clarence** Blackburn, Turnchapel.
Ord 13.7.1807; K 11.1807; L 11.4.1812; 1827 renamed *Centurion* and to be converted to a 50 gun frigate, but instead; 1828 BU.

**Edinburgh** Brent, Rotherhithe.
Ord 13.7.1807; K 11.1807; L 26.1.1811; 1852 converted to a screw 'block ship'; 1866 sold.

**America** Perry, Blackwall.
Ord 22.8.1807; K 01.1808; L 21.4.1810; 1835 cut down to a Frigate of 50 guns (Fourth Rate); 1864 hulked.

**Scarborough** Graham, Harwich.
Ord 13.7.1807; K 01.1808; L 29.3.1812; 1836 sold.

**Asia** Brindley, Frindsbury.
Ord 13.7.1807; K 02.1808; L 2.12.1811; 1819 renamed *Alfred*; 1828 cut down to a frigate of 50 guns (Fourth rate); 1858 hulked.

**Mulgrave** King, Upnor.
Ord 23.6.1807; K 02.1808; L 1.1.1812; 1836 hulked.

**Anson** Steemson, Hull.
Ord 2.11.1807; K 03.1808; L 1.5.1812; 1831 hulked.

**Gloucester** Pitcher, Northfleet.
Ord 11.6.1808?; K 03.1808?; L 27.2.1812; 1832 cut down to a frigate of 50 guns (Fourth Rate); 1861 hulked.

**Rodney** Barnard, Deptford.
Ord 28.5.1808; K 03.1808; L 8.12.1809; 1827 renamed *Greenwich* and cut down to frigate of 50 guns (Fourth Rate); conversion probably never completed, instead; 1836 sold.

**Hogue** Deptford Dyd.
Ord 1.10.1806; K 04.1808; L 3.10.1811; 1848 converted to screw 'block ship'; 1865 BU.

**Dublin** Brent, Rotherhithe.
Ord 31.7.1807; K 05.1808; L 13.2.1812; 1826 cut down to a frigate of 50 guns (Fourth Rate); 1845 laid up until; 1885 sold.

**Barham** Perry, Blackwall.
Ord 2.11.1807; K 06.1808; L 8.7.1811; 1826 cut down to a frigate of 50 guns (Fourth Rate); 1840 BU.

**Benbow** Brent, Rotherhithe.
Ord 11.6.1808; K 07.1808; L 3.2.1813; 1848 hulked.

**Stirling Castle** Ross, Rochester.
Ord 12.8.1807; K 07.1808; L 31.12.1811; 1839 hulked.

**Vindictive** Portsmouth Dyd.
Ord 15.1.1806; K 07.1808; L 23.11.1813; 1833 cut down to frigate of 50 guns (Fourth Rate); 1862 hulked.

**Blenheim** Deptford Dyd.
Ord 4.1.1808; K 08.1808; L 31.5.1813; 1847 converted to screw 'block ship'; 1858 hulked.

**Duncan** Dudman, Deptford.
Ord 13.7.1807; K 08.1808; L 2.12.1811; 1826 hulked.

**Rippon** Blake, Bursledon.
Ord 1.1.1808; K 10.1808; L 8.8.1812; 1821 BU.

**Medway** Pitcher, Northfleet.
Ord 19.8.1807; K 12.1808; L 19.11.1812; 1847 hulked.

**Cornwall** Barnard, Deptford.
Ord 13.7.1807; K 03.1809; L 16.1.1812; 1830 cut down to a frigate of 50 guns (Fourth Rate); 1859 hulked.

**Pembroke** Wells, Blackwall.
Ord 17.5.1808; K 03.1809; L 27.6.1812; 1855 converted to a screw line of battle ship; 1873 hulked.

**Indus** Dudman, Deptford.
Ord 31.7.1807; K 04.1809; L 19.12.1812; 1818 renamed *Bellona*; 1842 hulked.

**Redoubtable** Woolwich Dyd.
Ord 29.12.1806; K 04.1808; L 26.1.1815; 1841 BU.

**Devonshire** Barnard, Deptford.
Ord 28.5.1808; K 02.1810; L 23.9.1812; 1849 hulked.

**Defence** (ex *Marathon*) Chatham Dyd.
Ord 23.3.1808; K 05.1812; L 25.4.1815; 1848 hulked.

**Hercules** Chatham Dyd.
Ord 16.5.1809; K 08.1812; L 5.9.1815; 1838 troopship; 1852 emigrant ship; after 1853 hulked.

**Agincourt** Plymouth Dyd.
Ord 6.1.1812; K 05.1813; L 19.3.1817; after 1848 hulked.

**Pitt** Portsmouth Dyd.
Ord ____ 1807?; K 05.1813; L 13.4.1816; 1853 hulked.

**Wellington** (ex *Hero*) Deptford Dyd.
Ord ____ 1812?; K 07.1813; L 21.9.1816; 1848 hulked.

**Russell** Deptford Dyd.
Ord ____ 1811?; K 08.1814; L 22.5.1822; 1855 converted to a screw line of battle ship; 1858 Coastguard guard ship at Sheerness; 1865 BU.

## Cancelled 1809. Un-named Rio de Janeiro, Brazil.

Ord ____ ; never begun. [Was this the *Orford*; supposedly cancelled 1809?]

**Akbar** 'Prince of Wales Island' (India).
Ord ____ K 4.4.1807; cancelled 12.10.1809; never begun? (uncertain whether she was of this particular class).

FATES: *Vigo* became a receiving ship at Plymouth; 1865 BU. *Conquestadore* was lent as a powder depot to the War Office at Purfleet; 1863 powder depot at Plymouth; 1897 sold. *Egmont* became a storeship at Rio de Janeiro; 1875 sold. *America* became a target ship at Portsmouth; 1869 BU. *Asia* (as *Alfred*) became a gunnery trials ship at Portsmouth; 1865 BU. *Mulgrave* became a lazaretto at Pembroke; 1844 powder ship; 1854 BU. *Anson* became a temporary lazaretto at Portsmouth; by 1843 to Chatham and then to Tasmania as a convict ship; 1851 BU. *Gloucester* became a receiving ship at Chatham; 1884 sold. *Benbow* became a marine barracks at Sheerness; 1854 prison ship for Russians; 1859 coal depot; 1894 sold for BU. *Stirling Castle* became a convict ship at Plymouth; 1844 to Portsmouth; 1861 BU. *Vindictive* became a depot ship at Fernando Po; 1871 sold. *Blenheim* was hulked at Portsmouth; 1865 BU. *Duncan* became a lazaretto at Portsmouth; 1831 to Sheerness; 1863 BU. *Medway* became a convict ship at Bermuda; 1865 sold. *Cornwall* was lent to the London School Ship Society as a reformatory; 1868 to the Tyne as the *Wellesley* hulk; 1875 BU. *Pembroke* became the base ship at Chatham; 1890 renamed *Forte* as a receiving hulk; 1905 sold. *Indus* (as *Bellona*) became a receiving ship at Plymouth; 1868 BU. *Devonshire* was lent to the Greenwich Seamens Hospital as a temporary hospital ship; 1854 to Sheerness as a prison ship for Russians; 1860 school ship in 'Queenborough Swale' 1869 BU. *Defence* became a convict ship at Woolwich; 1857 burnt and BU. *Hercules* became an army depot ship at Hong Kong; 1865 sold. *Agincourt* became a training ship at Plymouth; 1866 cholera hospital ship; 1870 receiving ship; 1884 sold for BU. *Pitt* became a coal depot and receiving ship at Portsmouth; 1860 to Portland; later back to Portsmouth; 1877 BU. *Wellington* became a receiving and depot ship at Sheerness; 1857 to the Coastguard at Sheerness; 1862 to the Liverpool Juvenile Reformatory Association Ltd as a training ship & renamed *Akbar*; 1908 sold for BU.

PLANS: Lines/profile/orlop/gun deck/upper deck/quarter deck & forecastle/specification /framing. *Ajax* as screw ship as fitted 1849: Profile/hold & hold sections/orlop/gun deck/upper deck/quarter deck & forecastle. 1850 sails. *Egmont* 1861: Profile/hold & hold sections/orlop/gun deck/upper deck/quarter deck & forecastle. As storeship 1863: Profile/orlop/gun deck/upper deck/quarter deck & forecastle. *Edinburgh* 1829: Orlop. As screw ship 1852 as fitted: Profile/hold & hold sections/orlop/gun deck/upper deck/quarter deck & forecastle. *Gloucester* as frigate 1832: Profile/orlop/gun deck/upper deck/quarter deck & forecastle. *Hogue* as screw ship as fitted 1849: Profile/hold & hold sections/orlop/gun deck/upper deck/quarter deck & forecastle. *Dublin* as frigate 1831: quarter deck & forecastle. *Barham* as frigate 1827: Profile/orlop/gun deck/upper deck /quarter deck & forecastle. *Vindictive* as frigate 1842: Profile/hold & hold sections/orlop/gun deck/upper deck/quarter deck & forecastle. As depot ship: Profile/orlop/gun deck/upper deck. *Blenheim* as screw ship 1847: Profile/hold & hold sections & orlop /gun deck/upper deck/quarter deck & forecastle. 1849: Profile/gun deck/upper deck/quarter deck & forecastle. *Hercules* as troopship 1838: gun deck/upper deck. As emigrant ship 1853: gun deck/upper deck/quarter deck & forecastle.

### *Rochfort* Class 1809.
Barrallier design, later classed as an 80.
Dimensions & tons: 192'9½", 160'6½" x 49'4½" x 21'10". 2082.
Men: 640. Guns: GD 30 x 32, UD 30 x 18, QD 12 x 32★, Fc 2 x 18 + 6 x 32★, RH 6 x 18★.

*Rochfort/Rochefort* Pembroke Dyd.
Ord 1.6.1809; K 08.1809; L 6.4.1814; 1826 BU.

PLANS: Lines/profile/orlop/gun deck/upper deck/quarter deck & forecastle/framing.

### *Black Prince* Class 1810.
Reduced lines of the Danish *Christian* VII (taken 1807). Built on Seppings' principle.
Dimensions & tons: 176', 104'9⅞" x 47'6" x 21'. 1738 6/94.
Men: 590. Guns: GD 28 x 32, UD 28 x 18, QD 4 x 12 + 10 x 32★, Fc 2 x 12 + 2 x 32★, RH 6 x 18★.

*Black Prince* Woolwich Dyd.
Ord 14.8.1810; K 07.1814; L 30.3.1816; 1855 BU.

*Melville* Bombay Dyd (teak built).
Ord 6.9.1813; K 07.1815; L 17.2.1817; 1857 hulked.

*Hawke* Woolwich Dyd.
Ord ____ 1812?; K 04.1815; L 16.3.1820; 1855 converted to a screw line of battle ship; 1865 BU.

FATES: *Melville* became a hospital ship at Hong Kong; 1873 sold.

PLANS: Lines/profile/orlop/gun deck/upper deck/quarter deck & forecastle/framing/midships section. *Hawke* 1837 as fitted: profile/orlop/gun deck/upper deck/quarter deck & forecastle.

### *Chatham* Class 1810.
Design using captured frames of Franco-Dutch *Royal Hollondais* building at Flushing (Vlissingen) when that port was captured in 1809. The frames were taken down and shipped to England.
Dimensions and tons: 177'7", 146'7⅞" x 48'10" x 21'6½". 1860.
Men: 590. Guns: GD 28 x 32, UD 28 x 24, QD 4 x 12 + 10 x 32★, Fc 2 x 12 + 2 x 32★, RH 6 x 18★.

*Chatham* Woolwich Dyd (Frames from Flushing).
Ord 1810?; K ____ ; L 14.2.1812; 1817 sold.

PLANS: Lines.

### *Cornwallis* Class 1810?
Teak built version of the Surveyors' Class (*Armada* Class) design. There is some confusion about the *Wellesley* which may have been begun as a member of the *Black Prince* Class. *Carnatic*'s frames were brought to Portsmouth from Bombay.
Dimensions & tons: 177'1", 145'11" x 48' x 21'1¾". 1788 24/94.
Men: 590. Guns: GD 28 x 32, UD 28 x 18, QD 4 x 12 + 10 x 32★, Fc 2 x 12 + 2 x 32★, RH 6 x 18★.

*Cornwallis* Bombay Dyd.
Ord 25.7.1810; K ____ 1812; L 05.1813; 1854/5 screw line of battle ship; 1865 hulked.

*Wellesley* Bombay Dyd.
Ord 3.9.1812; K 05.1813; L 24.2.1815; 1862 hulked.

*Carnatic* Portsmouth Dyd (frames from Bombay).
Ord 30.9.1814; K 01.1818; L 21.10.1823; 1860 hulked.

FATES: *Cornwallis* hulked in 1865 as a jetty at Sheerness; 1916 renamed *Wildfire* as base ship; 1957 BU. *Wellesley* became harbour flag ship and receiving ship at Chatham; 1868 to Purfleet for the London School Ship Society as a reformatory and renamed *Cornwall*; 1940 sunk by the Luftwaffe. *Carnatic* became a coal depot at Portsmouth; 1886 floating magazine for the War Office; 1891 returned to the Admiralty.

PLANS: Lines & profile/sails (see *Armada* Class for decks). *Cornwallis* as screw ship: Profile/ hold & hold sections/orlop/gun deck/upper deck/quarter deck & forecastle.

### Class Uncertain: 74 Gun Ship Cancelled March 1800.
*Princess Amelia* Chatham Dyd. (182' x 49', 1906 tons?)
Ord ____ ; K 1.1.1799.

### 74 Cancelled 1810? *Augusta?* Portsmouth Dyd.
Ord 1806? K 1806?

### 74 Cancelled 1810, 1812 or 1815. (1750 tons?) *Julius?* Chatham Dyd.
Ord ____ 1807?

## Fourth Rates, 50 Guns (two-deckers)

### *Antelope* Class 1790.
Henslow design. The *Diomede* Class seem to have been a variant.
Dimensions & tons: 150', 123'8½" x 41'x 17'8". 1106 7/94.
Men: 350. Guns: GD 22 x 24, UD 22 x 12, QD 4 x 6 , Fc 2 x 6.

*Antelope* Sheerness Dyd.
Ord 15.2.1790; K 06.1790; L 10.11.1802; 1823 hulked,

FATE: Became a convict ship at Bermuda; 1848 (or 1845?) BU.

PLANS: Lines/profile/orlop/gun deck/upper deck/quarter deck & forecastle/planking expansion (internal)/(external). As built: Lines & profile.

### *Diomede* Class 1790.
Henslow design, probably a variant of the *Antelope* design.
Dimensions & tons: 151', 124'7½"x 41' x 17'8". 1114 31/94.
Men: 350. Guns: GD 22 x 24, UD 22 x 12, QD 4 x 6, Fc 2 x 6.

*Diomede* (ex *Firm*) Deptford Dyd.
Ord 9.12.1790; K 08.1792; L 17.1.1798; 1812 troopship; 1815 BU.

*Grampus* (ex *Tiger*) Portsmouth Dyd.
Ord 9.12.1790; K 10.1791; L 20.3.1802; 1820 hulked.

FATE: *Grampus* became a hospital ship in the Thames for the Committee for Distressed Seamen.

PLANS: Lines (see *Antelope* for decks).

### *Jupiter* Class 1811.
Rule design modified from the reduced lines of the Danish 80 *Christian* VII (taken 1807).
Dimensions & tons: 154', 127'5⅜" x 41'6" x 18'. 1167 49/94.
Men: 350. Guns: GD 22 x 24, UD 24 x 12, QD 8 x 24★, Fc 2 x 6 + 2 x 24★.

*Jupiter* Plymouth Dyd.
Ord 30.6.1810; K 07.1810; L 22.11.1813; 1831 hulked; 1832 troopship; 1846 hulked.

### Modified Version of the Design.
Dimensions & tons: 154', 127'3¾" x 41'11" x 17'6", 1189 76/94.

*Romney* Pelham, Frindsbury.
Ord ____ 1811?; K ____ ; L 24.2.1815; 1822 troopship; 1837 hulked.

*Salisbury* Deptford Dyd.
Ord 13. & 17.7.1810; K ____ ; L 21.6.1814; 1837 sold for BU.

*Isis* Woolwich Dyd.
Ord ____ 1811?; K 02.1816; altered whilst building to a 50 gun frigate (see under Large Frigates).

FATES: *Jupiter* became a coal depot; 1870 BU. *Romney* became a receiving ship for freed slaves; 1845 sold.

PLANS: Lines/profile/orlop/gun deck/upper deck/quarter deck & forecastle/framing/midships section/specification. *Romney* as troopship 1824:

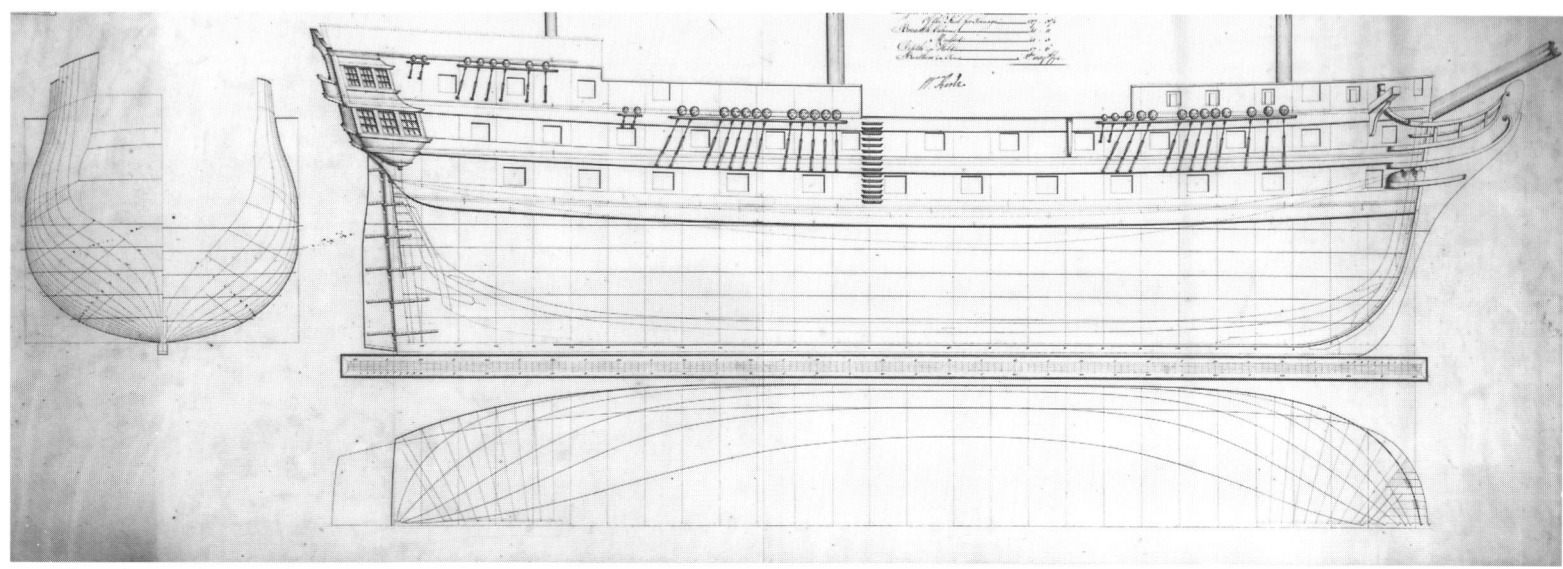

Design lines plan (number 1435) for building the 50 gun ship *Salisbury* of 1814 to a modified form of the *Jupiter* Class design

cabins. 1826: Gun deck/upper deck.

## Large 24 pounder Frigates, 40/60 Guns (Fourth Rates)

### *Endymion* Class 1795. 40/44/50. Lines of the French *Pomone* (taken 1794).

Dimensions & tons: 159'2⅜", 132'4½" x 41'11⅜" x 12'4", 1238 67/94.
Men: 320. Guns: UD 26 x 24, QD 16 x 32★,
Fc 2 x 9 + 4 x 32★ (or 6 x 32★).

*Endymion* Randall, Rotherhithe.
Ord 30.4. & 11.8.1795; K 11.1795; L 29.3.1797; 1868 BU.

### 'Fir built' (actually pitch pine) Version. Classed as 40s, then 50s.

Dimensions & tons: as above.
Men: 340. Guns: UD 28 x 24, QD 16 x 32★, Fc 2 x 9 + 4 x 32★.

*Forth* Wigram, Blackwall.
Ord ___ 1812?; K 02.1813; L 14.6.1813; 1819 BU.
*Severn* Wigram, Blackwall.
Ord ___ 1812?; K 01.1813; L 14.6.1813; 1825 sold.
*Liffey* Wigram, Blackwall.
Ord ___ 1812?; K 02.1813; L 25.9.1813; 1827 BU.
*Glasgow* Wigram, Blackwall.
Ord ___ 1812?; K 05.1813; L 21.2.1814; 1828 BU.
*Liverpool* Wigram, Blackwall.
Ord ___ 1812?; K 05.1813; L 21.2.1814; 1822 sold.

PLANS: Lines/profile/orlop/gun deck/upper deck/quarter deck & forecastle/specification. Fir-built version: Lines/framing.

### *Cambrian* Class 1795. 40. Henslow design.

Dimensions & tons: 154', 128'5¼" x 41' x 14'. 1148 39/94.
Men: 320. Guns: UD 28 x 24, QD 8 x 9 + 6 x 32★, Fc 4 x 9 + 2 x 32★.

*Cambrian* Parsons, Bursledon.
Ord 30.4.1795; K 09.1795; L 13.2.1797; 1828 lost.

FATE: Wrecked off Grabusa in the Mediterranean 31.1.1828.

PLANS: Lines/profile/orlop/gun deck/upper deck/quarter deck & forecastle/framing/specification.

### *Leander* Class 1813. 50/60. Rule design, fir built.

Dimensions & tons: 174', 145'3⅝" x 44'10½" x 14'4". 1556 38/94.
Men: 450. Guns: UD 30 x 24, Spar dk. 2 x 24 + 28 x 42★ + 2 x 24 chase guns.

*Leander* Wigram, Blackwall.
Ord before 05.1813; K 06.1813; L 10.11.1813; 1830 BU.

PLANS: Lines/profile/orlop/gun deck/upper deck/forecastle/framing/small scale lines & profile.

### *Isis* Class 1813. 50/60. Rule design on the reduced lines of the Danish 80 *Christian* VII (taken 1807). Begun as a two-decker 50 of the *Jupiter* class, then lengthened by 11' and converted to a frigate just before her launch.

Dimensions & tons: 164', 138'4" x 41'11" x 13'3". 1292 88/94.
Men: 350. Guns: UD 22 x 24, QD 24 x 12 + 8 x 24★, 2 x 6 + 2 x 24★

*Isis* Woolwich Dyd.
Ord ___ (1811?); K 02.1816; L 5.10.1819; 1861 hulked.

FATE: Became a coal depot at Sierra Leone; 1867 sold.

PLANS: Lines/profile/orlop/gun deck/upper deck/quarter deck & forecastle. As receiving ship 1860: profile/gun deck/upper deck/quarter deck & forecastle.
1861: profile/orlop/gun deck/upper deck/quarter deck & forecastle.

### *Newcastle* Class 1813. 50/60. Barrallier design, fir built.

Dimensions & tons: 177, 150'2¾" x 44'4½" x 14'11". 1573 45/94.
Men: 450. Guns: UD 30 x 24, Spar dk. 4 x 24 + 26 x 32★

*Newcastle* Wigram, Blackwall.
Ord ___ 1813?; K 06.1813; L 10.11.1813; 1824 hulked.

FATE: Hulked at Pembroke; 1827 to Liverpool; ___ .

PLANS: profile/orlop/gun deck/upper deck/quarter deck & forecastle. As built: Lines & profile.

### *Java* Class 1813. 50. 'Surveyors of the Navy' design (Peake, Tucker & Seppings).

Dimensions & tons: 172', 145'1¼" x 42'8" x 14'3". 1450.
Men: 450. Guns: UD 30 x 24, Spar dk. 6 x 24 + 16 (or 26?) x 42★.

*Java* Plymouth Dyd.
Ord ___ 1813?; K 03.1814; L 16.11.1815; 1861 hulked.

### Modified design

Dimensions & tons: 172', 144'9" x 43'8" x 14'6". 1468 11/94.

*Southampton* Deptford Dyd.
Ord ___ 1816?; K 03.1817; L 7.11.1820; 1857 to the Coastguard.
*Portland* (ex *Kingston*) Plymouth Dyd.
Ord ___ 1816?; K 08.1817; L 8.5.1822; 1846 hulked.
*Lancaster* Plymouth Dyd.
Ord ___ 1816?; K 06.1818; L 23.8.1823; 1847 hulked.
*Winchester* Woolwich Dyd.
Ord ___ 1816?; K 11.1818; L 21.6.1822; 1861 hulked.
*Worcester* Deptford Dyd.
Ord ___ 1818?; K 12.1820; L 10.10.1843; 1862 hulked
*Chichester* Woolwich Dyd.
Ord ___ 1817?; K 07.1827; L 12.7.1843; 1866 hulked.

**Cancelled 1829**. (probably, but not certainly, this class) *Liverpool*

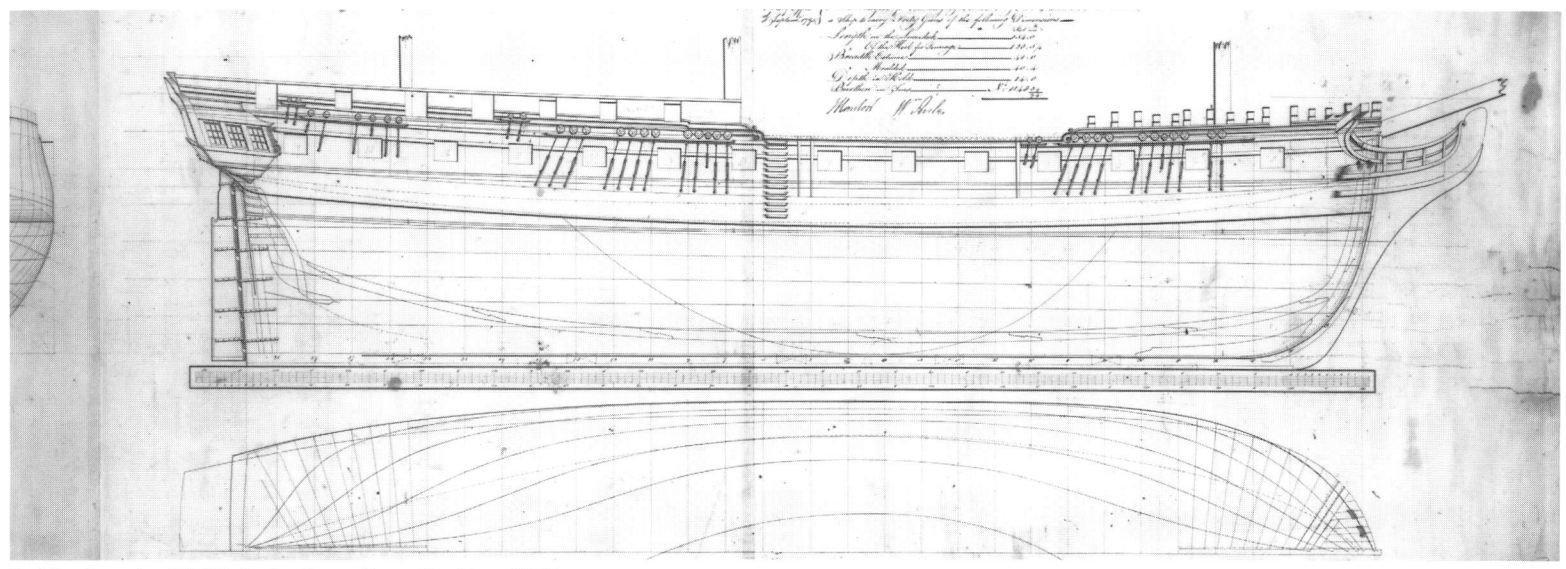

Building lines plan (1828A) for the 40 gun frigate *Cambrian* of 1797

Plymouth Dyd.
    Ord 7.1.1826; never begun?
(also see *Jamaica* under 'Class Uncertain' below).

FATES: *Java* became a target at Portsmouth; 1862 BU. *Southampton* became a Coastguard guard ship at Sheerness; 1868 to the Humber as a training ship; 1912 sold for BU. *Portland* became a provision depot at Plymouth 1846-1850; 1850 fitted as a frigate again?; 1862 sold. *Lancaster* became a hospital ship at Plymouth; later to Ordinary; 1864 sold. *Winchester* became the *Conway* training ship at Liverpool; 1876 training ship *Mount Edgcumbe* for the Devon & Cornwall training ship Society; 1921 sold. *Worcester* became the training ship of that name at Greenhithe; 1885 sold to BU. *Chichester* lent to the National Refuge Society (Greenhithe); 1889 sold.

PLANS: Lines/profile/orlop/gun deck/upper deck/quarter deck & forecastle/framing/midships section. Modified (*Southampton*) design: Lines/profile/orlop/gun deck/upper deck/quarter deck & forecastle/framing/midships section. *Winchester* 1847: Lines. *Worcester* 1847: profile/orlop/gun deck/upper deck/quarter deck & forecastle. 1859 as screw: Lines/profile/hold & hold sections/gun deck/upper deck. *Chichester* 1859: Lines/profile/hold & hold sections/gun deck/upper deck.

### *President* Class 1818. 52/60.
Lines of American *President* (taken 1814).
    Dimensions & tons: 174'10", 143'6" x 44'6" x 13'4". 1511 49/94.
    Men: 450. Guns: UD 30 x 24, Spar dk. 6 x 24 + 16 x 24★.
**President** Portsmouth Dyd.

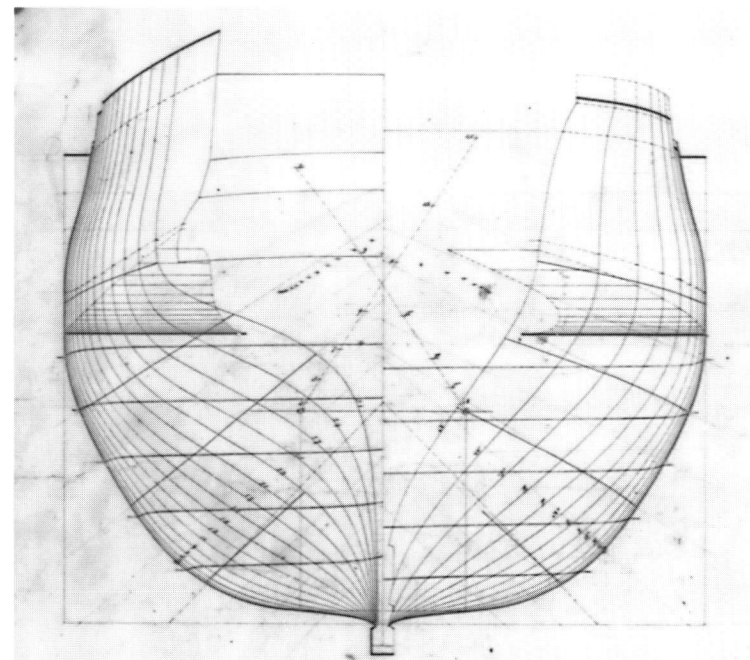

Ord 25.5.1818/7.10.1820; K 06.1824; L 20.4.1829; 1861 hulked.

FATE: Became RNR training ship in the City Canal (London); 1871 in the West India Docks, Poplar; 1903 renamed *Old President*; 1903 sold.

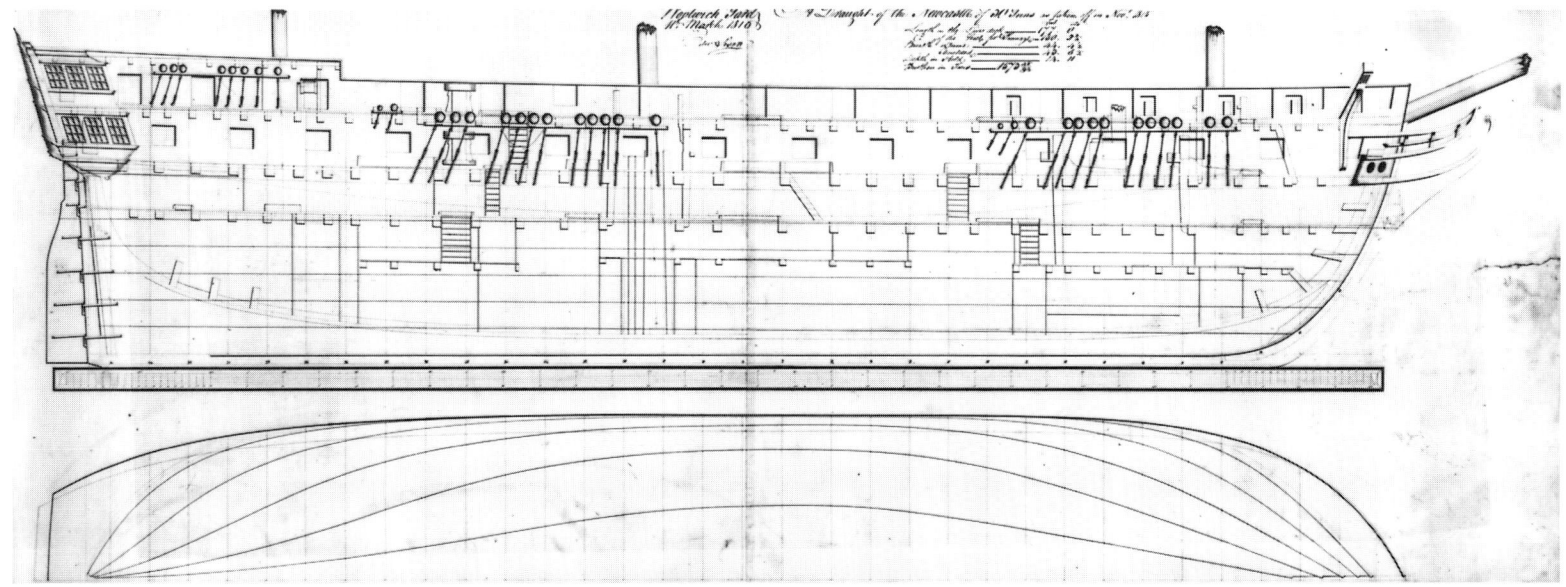

'As taken off' lines and profile plan (1448) of the large spar-decked frigate *Newcastle* of 1813

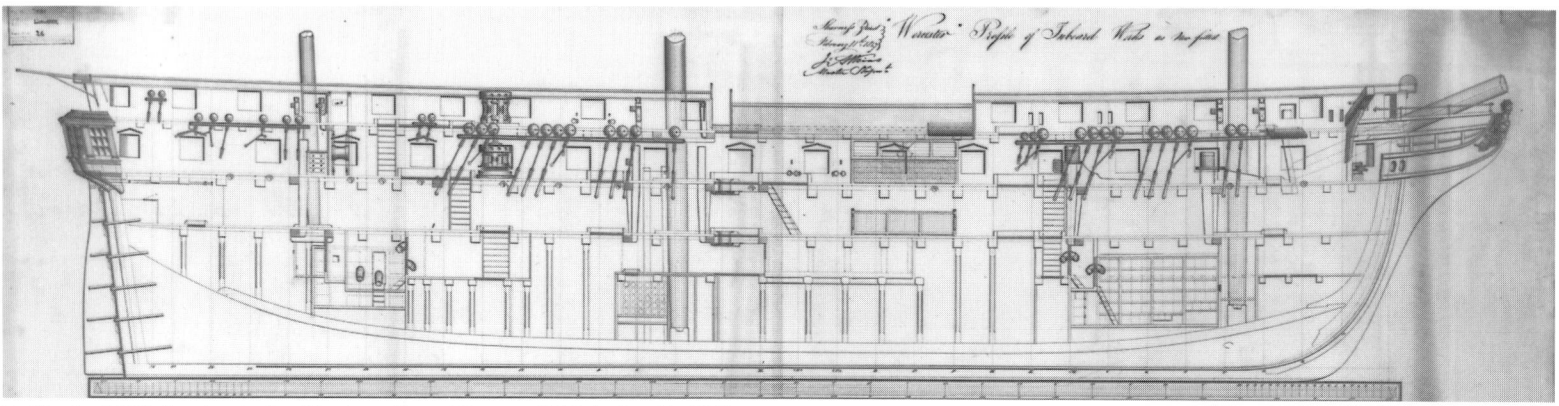

Profile plan (7880) showing the *Java* Class large frigate *Worcester* of 1822 'as fitted' in 1847

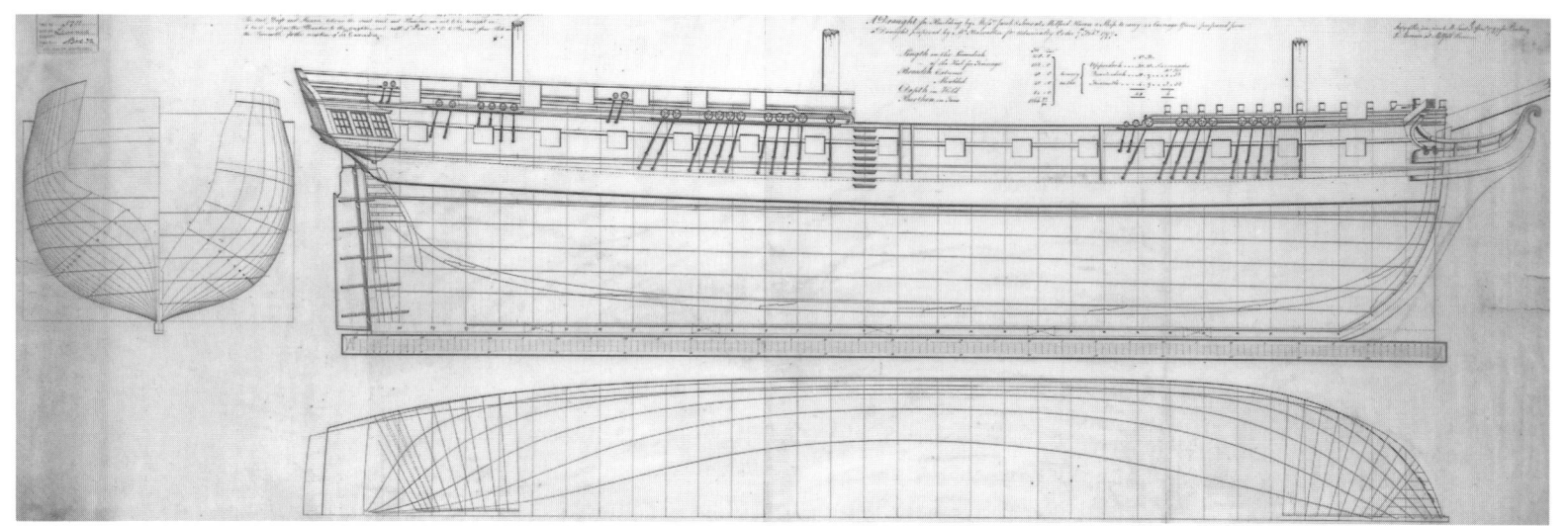

Design lines plan (1711) for the large 18 pounder frigate *Lavinia* of 1806

PLANS: Lines/profile/orlop/gun deck/upper deck/quarter deck & forecastle.

**Class Uncertain: Cancelled 5.3.1829.** [possibly *Java* class]
*Jamaica*? Plymouth Dyd.
   Ord 1.7.1825.

## 18 pounder Frigates, 40/50 Guns (Fifth Rate)

***Acasta* Class 1795.** Rule design.
   Dimensions & tons: 154', 129'0¼" x 40'6" x 14'3". 1127 22/94.
   Men: 320. Guns: UD 30 x 18, QD 8 x 9 + 4 x 32★, Fc 2 x 9 + 4 x 32★.

*Acasta* Randall, Rotherhithe.
   Ord 30.4.1795; K 09.1795; L 14.3.1797; 1821 BU.

PLANS: Lines/profile/orlop/gun deck/upper deck/quarter deck & forecastle.

***Lavinia* Class 1797.** 44/50. Barrallier design.
   Dimensions & tons: 158', 132'8" x 40'8" x 14'. 1166 92/94.
   Men: 340. Guns: UD 30 x 18, QD 10 x 9 + 4 x 32★, Fc 4 x 9 + 2 x 32★.

*Lavinia* Jacobs, Milford Haven.
   Ord 7.2.1797; K 05.1798 (builder failed, completed by 'Government'; at Milford).

'Government' Milford Haven.
   Ord ____ ; L 6.3.1806; 1836 hulked.

FATE: Lazaretto at Liverpool; 1852 coal depot at Plymouth; 1870 sunk in Plymouth Sound by collision with HAPAG (Hamburg Amerika) *Cimbria*; wreck sold as it lay for salvage.

PLANS: Lines/profile/orlop/gun deck/upper deck/quarter deck & forecastle/framing. As coal depot 1852: Profile & quarter deck forecastle & midships section/gun deck & upper deck.

## 18 pounder Frigates, 38 Guns (Fifth Rate)

***Artois/Apollo* Class 1793.** Henslow design.
   Dimensions & tons: 146'. 121'7⅛" x 39' x 13'9". 983 70/94.
   Men: 270. Guns: UD 28 x 18, QD 8 x 9 + 6 x 32★, Fc 2 x 9 + 2 x 32★.

*Artois* Wells, Deptford.
   Ord 2.3.1793; K 03.1793; L 3.1.1794; 1797 wrecked.
*Apollo* Perry, Blackwall.
   Ord 2.3.1793; K 03.1793; L 18.3.1794; 1799 wrecked.
*Diana* Randall, Rotherhithe.
   Ord 2.3.1793; K 03.1793; L 3.3.1794; 1815 Dutch Navy.
*Seahorse* Stalkartt, Rotherhithe.
   Ord 2.3.1793; K 03.1793; L 11.6.1794; 1819 BU.
*Diamond* Barnard, Deptford.
   Ord 2.3.1793; K 04.1793; L 17.3.1794; 1812 BU.
*Jason* Dudman, Deptford.
   Ord 2.3.1793; K 04.1793; L 3.4.1794; 1798 wrecked.
*Ethalion* Graham, Harwich.
   Ord 30.4.1795; K 10.1795; L 14.3.1797; 1799 wrecked.

### Fir Built

*Clyde* [I] Chatham Dyd.
   Ord 4. & 20.2.1795; K 06.1793; L 26.3.1796; 1805 BU for RB.
*Tamar/Tamer* Chatham Dyd.
   Ord 4. & 20.2.1795; K 06.1795; L 26.3.1796; 1810 BU.
*Clyde* [II] Woolwich Dyd [RB].
   Ord ____ 1804?; K 06.1805; L 20.2.1806; laid up since 1810; 1814 sold.

NOTE: The only recorded example since the 1740s of a 'rebuild'; presumably required because of the use of fir and its comparatively rapid deterioration.
FATES: *Artois* wrecked near La Rochelle 31.7.1797. *Apollo* wrecked on the Haak Sand off Holland 7.1.1799. *Diana* sold to the Dutch Navy and served in that navy until the 1830s. *Jason* wrecked near Brest 13.10.1798. *Ethalion* wrecked on the Penmarks 19.12.1799.

# HENSLOW TO SEPPINGS 1786-1830

*The 38 gun frigate* Active *of 1799 shown in 1819 with Lieutenant Burton's man-powered wheels in a pencil sketch by Schetky showing her leaving Portsmouth harbour (NMM photo negative x.1350)*

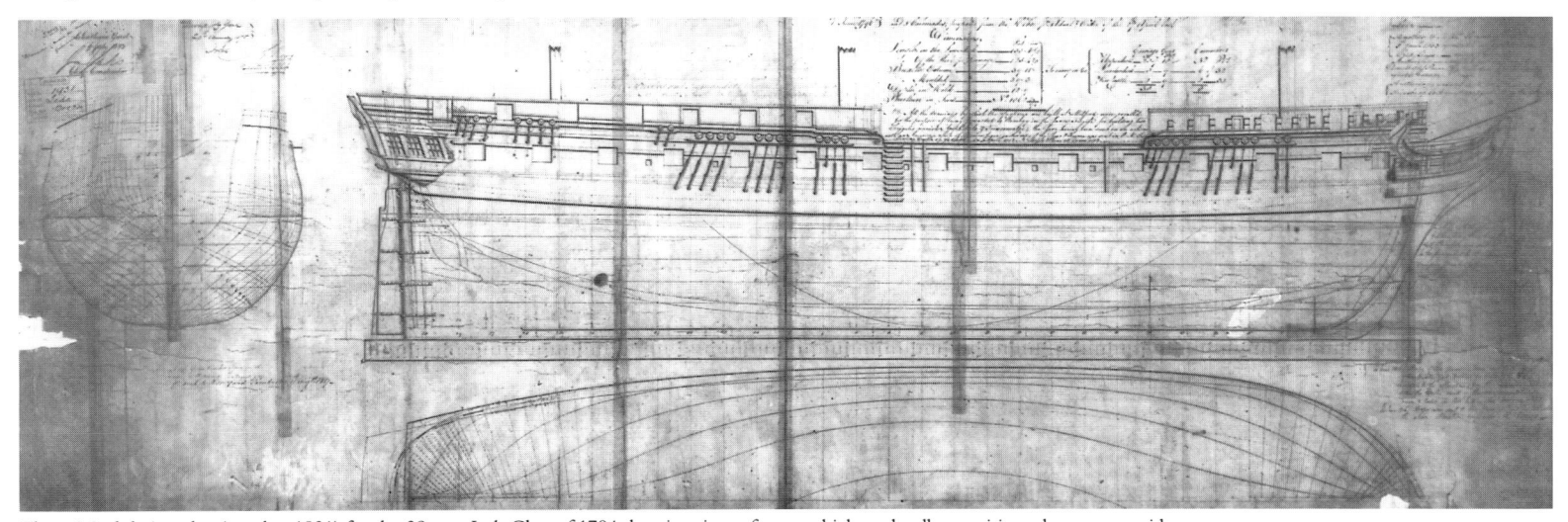

*The original design plan (number 1931) for the 38 gun* Leda *Class of 1794 showing signs of wear which are hardly surprising when one considers that it was used for the numerically largest class of sailing frigate ever built*

PLANS: Lines/profile/orlop/gun deck/upper deck/quarter deck & forecastle/framing/planking expansion (inner)/(outer). Fir built version: Lines.

### Active Class 1794. Henslow design.
Dimensions & tons: 150', 125'2⅞" x 39'9" x 13'9". 1052 55/94.
Men: 284. Guns: UD 28 x 18, QD 8 x 9 + 6 x 32★, Fc 2 x 9 + 2 x 32★.
**Active** Chatham Dyd.
   Ord 6.11.1794/6.5.1797; K 07.1798; L 14.12.1799; 1825 hulked.
FATE: Became receiving ship at Plymouth; 1833 renamed *Argo*; 1860 BU.
PLANS: Lines/profile/orlop/gun deck/upper deck/quarter deck & forecastle.

### Leda Class 1794.
Built to the lines of the French *Hebe* (taken 1782). The largest class of sailing frigates ever built (even larger if one counts the other French frigates built to the lines of the *Hebe*). Two of the class are still in existence, the only remaining British sailing frigates; *Unicorn* at Dundee and *Trincomalee* ('*Foudroyant*') at Portsmouth (in early 1993 at West Hartlepool where she had been for some time awaiting funds for restoration and eventual return to Portsmouth).
Dimensions & tons: 150'1½", 125'2⅞" x 39'9" x 13'9". 1052 55/94
Men: 284. Guns: UD 28 x 18, QD 8 x 9 + 6 x 32★, Fc 2 x 9 + 2 x 32★.
**Leda** Chatham Dyd.
   Ord 27.4.1796; K 1.5.1799; L 18.11.1800; 1808 wrecked.
**Pomone** Brindley, Frindsbury.
   Ord ____ 1802?; K 12.1803; L 17.1.1805; 1811 wrecked.
**Shannon** Brindley, Frindsbury. [Broke's ship.]
   Ord ____ 1803?; K 08.1804; L 5.5.1806; 1831 hulked.
**Leonidas** Pelham, Frindsbury.
   Ord ____ 1805?; K 11.1805; L 4.9.1807; 1872 hulked.
**Surprise** Milford (Pembroke Dyd.)
   Ord ____ 1808?; K 01.1810; L 25.7.1812; 1822 hulked.
**Briton** Chatham Dyd.
   Ord ____ 1808?; K 02.1810; L 11.4.1812; 1841 hulked.
**Lacedemonian** Portsmouth Dyd.
   Ord ____ 1808?; K 05.1810; L 21.12.1812; 1822 BU.
**Tenedos** Chatham Dyd.
   Ord ____ 1808?; K 05.1810; L 11.4.1812; 1843 hulked.
**Lively** (ex *Scamander*) Chatham Dyd.
   Ord 26 & 28.9.1808; K 07.1810; L 14.7.1813; 1831 hulked.
**Amphitrite** Bombay Dyd. Teak built.
   Ord ____ 1812?; K 08.1814; L 14.4.1816; 1846 cut down to a 26 gun Corvette (Sixth Rate); 1857 to the Coastguard.
**Diamond** Chatham Dyd.
   Ord ____ 1812?; K 08.1813; L 16.1.1816; 1827 burnt.
**Thetis** Pembroke Dyd.
   Ord ____ 1812?; K 12.1814; L 1.2.1817; 1830 wrecked.
**Arethusa** Pembroke Dyd.
   Ord ____ 1814?; K 02.1813; L 29.7.1817; 1836 hulked.
**Blanche** Chatham Dyd.
   Ord ____ 1815?; K 02.1816; L 26.5.1819; 1833 hulked.
**Trincomalee** Bombay Dyd. Teak built.
   Ord ____ 1812?; K 05.1816; L 12.10.1817; 1847 cut down to a 26 gun corvette (Sixth Rate); 1861 hulked; still afloat in Portsmouth Harbour as the training ship '*Foudroyant*' [1991-3 at West Hartlepool].
**Fisguard** Pembroke Dyd.
   Ord 24.8.1815; K 02.1817; L 8.7.1819; 1847 hulked.
**Venus** Deptford Dyd.

Ord ____ 1816?; K 03.1817; L 10.8.1820; 1848 hulked.
*Melampus* Pembroke Dyd.
Ord 23.4.1817; K 08.1817; L 10.8.1820; 1855 hulked.
*Amazon* Deptford Dyd.
Ord ____ 1817?; K 10.1817; L 15.8.1821; 1845 cut down to a 26 gun corvette (Sixth Rate); (1848-1852 lent to Liberia?); 1863 sold.
*Minerva* Portsmouth Dyd.
Ord ____ 1816?; K 10.1817; L 13.6.1820; 1895 BU.
*Latona* Chatham Dyd.
Ord ____ 1816?; K 10.1818; L 16.6.1821; 1868 hulked.
*Nereus* Pembroke Dyd.
Ord 6.5.1817; K 01.1819; L 30.7.1821; 1843 hulked.
*Diana* Chatham Dyd.
Ord ____ 1816?; K 02.1819; L 8.1.1822; 1868 hulked.
*Hamadryad* Pembroke Dyd.
Ord ____ 1817?; K 09.1819; L 25.7.1823; 1866 hulked.
*Aeolus/Eolus* Deptford Dyd.
Ord ____ 1817?; K 10.1818; L 17.6.1825; 1846 hulked.
*Thisbe* Pembroke Dyd.
Ord ____ 1817?; K 08.1820; L 9.9.1824; 1863 hulked.
*Hebe* Woolwich Dyd.
Ord 30.8.1817; K 05.1820; L 14.12.1826; 1839 hulked.
*Cerberus* Plymouth Dyd.
Ord ____ 1817?; K 11.1820; L 30.3.1827; 1866 BU.
*Circe* Plymouth Dyd.
Ord ____ 1817?; K 11.1820; L 22.9.1827; 1866 hulked.
*Clyde* Woolwich Dyd.
Ord ____ 1817?; K 01.1821; L 9.10.1828; 1870 hulked.
*Thames* Chatham Dyd.
Ord ____ 1817?; K 06.1821; L 21.8.1823; 1841 hulked.
*Fox* Portsmouth Dyd.
Ord ____ 1817?; K 06.1821; L 17.8.1829; 1856 converted to a screw frigate/transport; 1882 BU.
*Unicorn* Chatham Dyd.
Ord ____ 1817?; K 02.1822; L 30.3.1824; never fitted for sea; 1860 hulked; still afloat at Dundee and undergoing 'restoration' to a seagoing state she has never experienced before. The period of over a century and a half between being launched and being fully fitted out must be some sort of record! However it is the excellent condition of the hull which is the result of this long 'lay up' which makes the 'restoration' such a practical possibility. [1992 note - there are now indications that the ship will stay as she is, as a genuine hulk.]
*Daedalus* Deptford Dyd. Originally to have been built at Sheerness.
Ord ( ____ 1817?) 18 & 23.7.1821; K 10.1822; L 22.5.1826; 1844 cut down to a 19 gun corvette (Sixth Rate); 1861 hulked.
*Proserpine* Plymouth Dyd.
Ord ____ 1817?; K 11.1822; L 1.12.1830; 1864 sold.
*Mermaid* Chatham Dyd.
Ord ____ 1817?; K 09.1823; L 30.7.1826; 1858 hulked.
*Mercury* Chatham Dyd.
Ord ____ 1817?; K 04.1824; L 16.11.1826; 1862 hulked.
*Penelope* Pembroke Dyd.
Ord ____ 1817?; not begun?; transferred to:
Chatham Dyd.
Ord 26.5.1827; K 11.1827; L 13.10.1829; 1843 converted to a paddle frigate; 1864 sold.
*Thalia* Portsmouth Dyd.
Ord ____ ; not begun?; transferred to:
Chatham Dyd.
Ord 26.5.1827; K 02.1828; L 12.1.1830; 1855 hulked.

**Cancelled 1831.** *Medusa* Woolwich Dyd.
Ord ____ 1818; K ____ .
*Pegasus* Deptford Dyd.
Ord 18.7.1818; K ____ transferred to:
Sheerness Dyd.
Ord 17.2.1825; K 03.1828.

**Fir built Variant.** Could be considered a separate class. Modified by Peake from the design of the *Lively* (ex *Scamander*), and so sometimes referred to as the *Scamander* Class - though this is most confusing as the actual ship in service at the time completed with the name *Scamander* was a 36, not a 38, and of another class entirely. One could call these vessels the *Cydnus* Class to avoid that confusion, but it seems better to regard them as a sub-group of the *Leda*s.

*Cydnus* Wigram Blackwall.
Ord ____ 1812?; K 12.1812; L 17.4.1813; 1816 BU.
*Eurotas* Wigram, Blackwall.
Ord ____ 1812?; K 12.1812; L 17.4.1813; 1817 BU.
*Niger* Wigram, Blackwall.
Ord ____ 1812?; K ____ ; L 29.5.1813; 1820 BU.
*Araxes* Pitcher, Northfleet.
Ord ____ 1812?; K 01.1813; L 13.9.1813; 1817 sold.
*Meander* Pitcher, Northfleet.
Ord ____ 1812?; K 01.1813; L 13.8.1813; 1817 BU.
*Pactolus* Barnard, Deptford.
Ord ____ 1812?; K 01.1813; L 14.8.1813; 1818 sold.
*Tanais* Mrs.Ross, Rochester.
Ord ____ 1812?; K 02.1813; L 27.10.1813; 1819 sold.
*Tiber* List, Fishbourne, Isle of Wight.
Ord ____ 1812?; K 02.1813; L 10.11.1813; 1820 sold.

FATES: *Leda* was wrecked at the mouth of Milford Haven 31.1.1808. *Pomone* was wrecked on the Needles 14.10.1811; the wreck has recently been found and excavated. *Shannon* became a receiving ship and temporary hulk at Sheerness; 1844 renamed *Saint Lawrence*; 1859 BU. *Leonidas* became a powder hulk at Sheerness; 1894 sold. *Surprise* became a convict ship at Cork; 1837 sold. *Tenedos* became a convict ship at Bermuda; 1863 accommodation ship; 1875 BU. *Lively* became a receiving ship; 1863 sold. *Amphitrite* lent from the Coastguard in the 1860s to the War Office contractor for the forts at Portsmouth; 1875 BU. *Diamond* accidentally burnt at Portsmouth 18.4.1827 then BU. *Thetis* was wrecked off Cape Frio, Brazil 5.12.1830; the cargo of bullion was recovered at the time in a brilliant feat of improvised salvage; the Belgian diver Robert Sténuit has worked on the site in the late 1980s. *Arethusa* was hulked at Liverpool as a lazaretto (?); 1844 renamed *Bacchus*; 1850 to Plymouth as a lazaretto; 1852 coal depot; 1883 sold to BU. *Blanche* became a receiving ship and temporary hulk at Portsmouth; 1865 sold to BU. *Trincomalee* was hulked at Sunderland as a training ship for Volunteers; 1897 to Wheatley Cobb at Falmouth as a training ship and renamed *Foudroyant*; later to Portsmouth Harbour where she is still afloat as a youth training ship (in late 1987 was towed up to West Hartlepool for restoration and was still there, awaiting further funds, in 1992). *Fisguard* became a harbour flagship (fitted for a Commodore) at Woolwich; 1879 BU. *Venus* was lent to the Marine Society; 1865 BU. *Melampus* was to be hulked at Constantinople, though it is uncertain whether she was ever sent there; 1857 to the Coastguard at Southampton; 1866 returned to the RN at Portsmouth then to the War Office there as an ordnance store until 1891; 1895 returned to the Admiralty; 1906 sold. *Latona* became a mooring vessel at Sheerness; 1872 to Portsmouth as a powder depot; 1874 intended to become a training ship but this was not put into effect and instead; 1875 BU. *Nereus* became a coal depot at Valparaiso, Chile; 1879 sold. *Diana* became a receiving ship at Sheerness; 1874 BU. *Hamadryad* became a hospital ship for sick seamen at Cardiff; 1905 sold. *Aeolus* became a store depot at Sheerness; 1855 to Portsmouth as an accommodation ship; 1861 lazaretto; 1886 BU. *Thisbe* became a floating church at Cardiff; 1892 sold. *Hebe* became a receiving ship at Woolwich 1872 to Sheerness & BU there. *Circe* became an accommodation hulk at Plymouth as tender to *Indefatigable*; 1885 swimming bath; 1916 renamed *Impregnable* IV; 1922 sold for BU. *Clyde* became an RNR training ship at Aberdeen; 1904 sold. *Thames* became a convict ship at Deptford; 1844 to Bermuda as convict ship; 1863 sunk and wreck sold for BU. *Unicorn* hulked at Chatham; 1868 lent to the War Office as a powder hulk; 1873 to Dundee as RNR training ship 1941-1959 renamed *Cressy*; now belongs to the *Unicorn* trust who are fitting her out as a museum ship in Dundee Harbour. *Daedalus* training ship for the RNR at Bristol; 1911 sold for BU. *Mermaid* became an Army powder ship at Purfleet; 1863 returned to the RN and to Dublin as a powder depot; 1875 BU. *Mercury* became a coal depot at Woolwich; by 1873 to Sheerness as a coal depot; 1906 sold. *Thalia* became a Roman Catholic chapel ship at Portsmouth; 1867 BU.

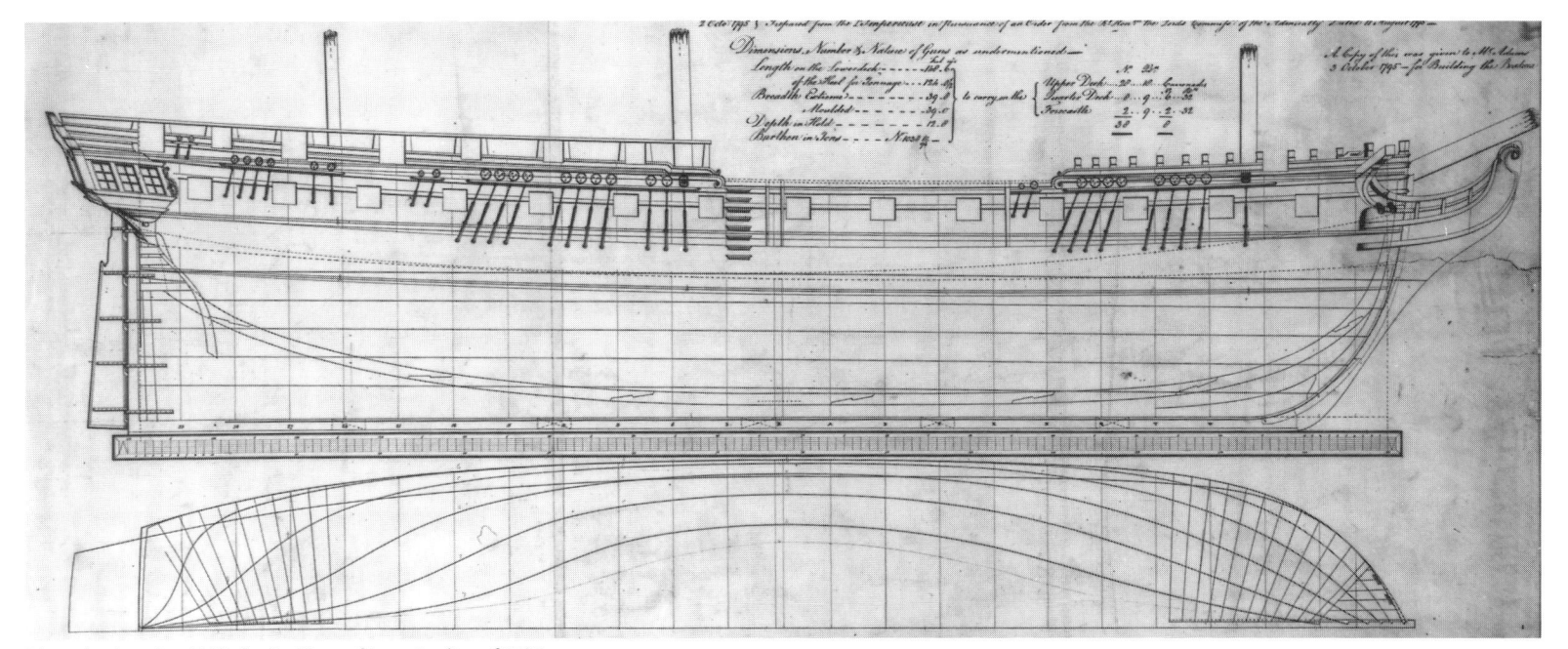

HENSLOW TO SEPPINGS 1786-1830

Lines plan (number 1869) for the 38 gun frigate *Boadicea* of 1797

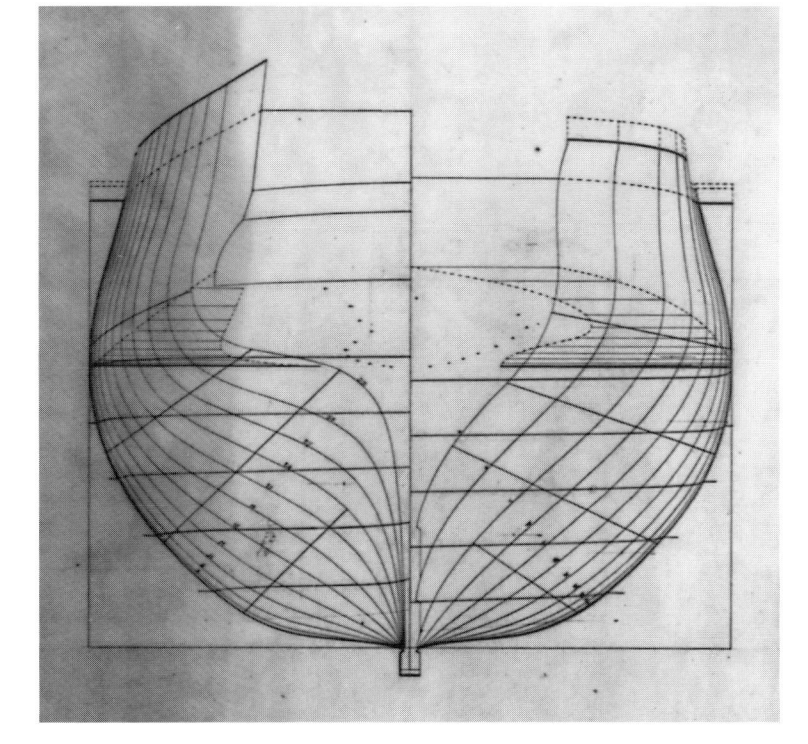

PLANS: Lines/profile/orlop/gun deck/upper deck/quarter deck & forecastle/framing/midships section. *Surprise* as convict ship 1823: profile & decks. *Fox* 1843: Lines. as screw ship 1855: Profile/hold & hold sections/upper deck. *Penelope* as paddle ship 1842: Lines/sails. 1850: Orlop & gun deck/upper deck/quarter deck & forecastle. *Cydnus* (fir-built) group: Lines/profile/orlop/gun deck/upper deck/quarter deck & forecastle/framing/specification.

### Naiad Class 1795.
Rule design. One of the group of 8 large frigates ordered at this time.
  Dimensions & tons: 147', 122'8⅜" x 39'5" x 13'9". 1013 90/94.
  Men: 284. Guns: UD 28 x 18, QD 8 x 9 + 6 x 32★, Fc 2 x 9 + 4 x 32★.
**Naiad** Hill & Co, Limehouse.
  Ord 30.4.1795; K 09.1795; L 27.2.1797; 1847 hulked.
FATE: Coal depot at Callao, Peru; 1866 sold.
PLANS: Lines/profile/framing.

### Hydra Class 1795.
Lines of the French *Melpomene* (taken 1794). Another of the large frigates of 1795.
  Dimensions & tons: 148'2", 123'7⅜" x 39'4" x 12'8". 1017 20/94.
  Men: 284. Guns: UD 28 x 18, QD 8 x 9 + 6 x 32★, Fc 2 x 9 + 2 x 32★.
**Hydra** Cleverley, Gravesend.
  Ord 30.4.1795; K 11.1795; L 13.3.1797; 1813 troopship; 1820 sold.
PLANS: Lines/profile/orlop/gun deck/upper deck/quarter deck & forecastle/framing.

### Boadicea Class 1795.
Lines of the French *Imperieuse* (taken 1793). Another of the large frigates of 1795.
  Dimensions & tons: 148'6", 124'0½" x 39'8" x 12'8". 1038 10/94.
  Men: 284. Guns: UD 28 x 18, QD 8 x 9 + 6 x 32★, Fc 2 x 9 + 2 x 32★.
**Boadicea** Adams, Bucklers Hard.
  Ord 30.4.1795; K 09.1795; L 12.4.1797; 1858 BU.
PLANS: Lines/profile/orlop/gun deck/upper deck/quarter deck & forecastle/framing.

### Amazon Class 1796.
Rule design.
  Dimensions & tons: 150', 125'7¾" x 39'5" x 13'9". 1038 6/94.
  Men 284. Guns: UD 28 x 18, QD 8 x 9 + 6 x 32★, Fc 2 x 9 + 2 x 32★.
**Amazon** Woolwich Dyd.
  Ord ____ 1795?; K 04.1796; L 18.5.1799; 1817 BU.
**Hussar** (ex *Hyena*) Woolwich Dyd.
  Ord 15.2.1797; K 08.1798; L 1.10.1799; 1804 wrecked.
FATES: *Hussar* wrecked on the Saintes, Bay of Biscay 02.1804.
PLANS: Lines/profile/orlop/gun deck/upper deck/quarter deck & forecastle/framing.

### Lively Class 1799.
Rule design.
  Dimensions & tons: 154', 129'8" x 39'5" x 13'6". 1071 51/94.
  Men: 284/300. Guns: UD 28 x 18, QD 2 x 9 + 12 x 32★, Fc 2 x 9 + 2 x 32★.
**Lively** Woolwich Dyd.
  Ord ____ 1799?; K 11.1801; L 23.7.1804; 1810 lost.
**Resistance** Ross, Rochester.
  Ord ____ 1803?; K 03.1804; L 10.8.1805; 1842 troopship; 1858 BU.
**Apollo** Parsons, Bursledon.
  Ord ____ 1803?; K 04.1804; L 27.6.1805; 1822 to be converted to a royal yacht; but conversion not completed; 1838 troopship; 1853 storeship; 1856 BU.
**Horatio** Parsons, Bursledon.
  Ord ____ 1805?; K 07.1805; L 23.4.1807; 1851 converted to a screw frigate; 1855 steam mortar frigate; 1865 sold.
**Spartan** Ross, Rochester.
  Ord ____ 1805?; K 10.1805; L 16.8.1806; 1822 BU.
**Statira** Guillaume, Northam. Ord ____ 1805?; K 12.1805; L 7.7.1807; 1815 wrecked.

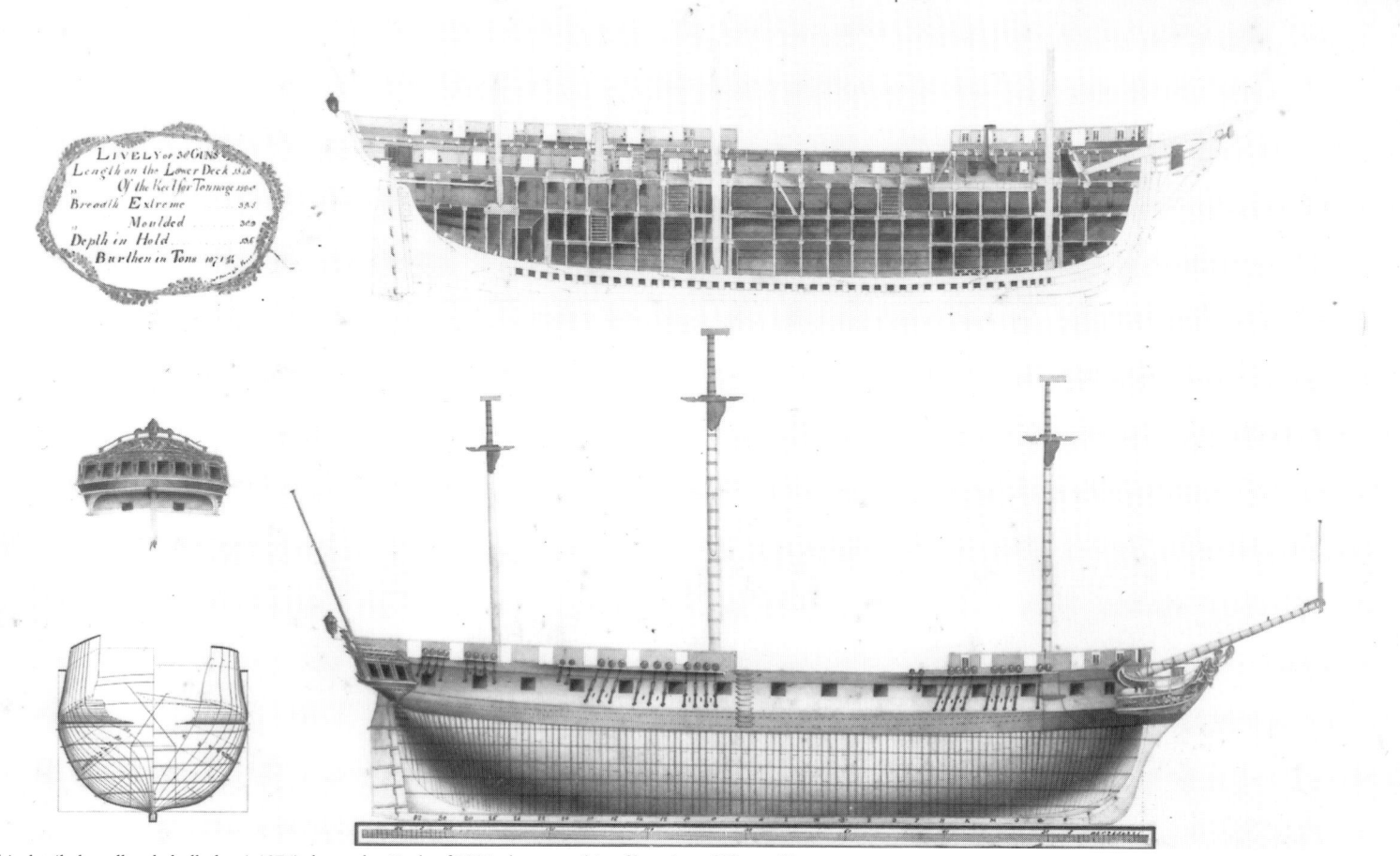

This detailed small scale hull plan (x1976) shows the *Lively* of 1804, the name ship of her class of 38 gun frigates

*Hussar* Adams, Bucklers Hard.
  Ord ____ 1805?; K 03.1806; L 23.4.1807; 1833 hulked.
*Undaunted* Graham, Harwich.
  Ord ____ 1803?; K ____ ; transferred to:
  Woolwich Dyd.
  Ord ____ ; K 04.1806; L 17.10.1807; 1860 hulked.
*Menelaus* Plymouth Dyd.
  Ord ____ 1808?; K 11.1808; L 17.4.1810; 1836 hulked.
*Nisus* Plymouth Dyd.
  Ord ____ 1808?; K 12.1808; L 3.4.1810; 1822 BU.
*Macedonian* Woolwich Dyd.
  Ord ____ 1808?; K 05.1809; L 2.6.1810; 1812 taken.
*Crescent* Woolwich Dyd.
  Ord ____ 1808?; K 09.1809; L 11.12.1810; 1840 hulked.
*Bacchante* Deptford Dyd.
  Ord ____ 1808?; K 07.1810; L 16.11.1811; 1837 hulked.
*Nymphe* (ex *Nereide*) Parsons, Warsash.
  Ord ____ 1811?; K 01.1811; L 13.4.1812; 1837 hulked.
*Sirius* Tyson, Bursledon.
  Ord ____ 1810?; K 09.1811; L 11.9.1813; 1860 hulked.
*Laurel* Parsons, Warsash.
  Ord ____ 1812?; K 07.1812; L 31.3.1813; 1864 hulked.

FATES: *Lively* wrecked at Malta 26.8.1810. *Statira* wrecked on a sunken rock off Cuba 27.2.1815. *Hussar* became a receiving ship at Chatham; 1861 target ship at Shoeburyness & burnt. *Undaunted* became a target ship at Portsmouth; 1860 BU. *Menelaus* became a lazaretto at Sheerness; 1841 to Portsmouth as a lazaretto at the Motherbank; 1897 sold. *Macedonian* taken by the USS *United States* off Madeira 25.10.1812; in American service until BU 1835. *Crescent* became a hulk for freed slaves at Rio de Janeiro; 1854 sold. *Bacchante* became a lazaretto at Sheerness; briefly to Deptford (for Greenwich) as a cholera ship; 1858 BU. *Nymphe* became a receiving ship at Sheerness; 1861 to water police; 1863 Roman Catholic chapel ship; 1871 renamed *Handy*; 1875 BU. *Sirius* became a target ship at Portsmouth; 1862 BU. *Laurel* became a 'lavatory' at Portsmouth; 1885 sold to BU.

PLANS: Lines/profile/orlop/gun deck/upper deck/quarter deck & forecastle/framing/specification. *Resistance* as troopship 1841: orlop/gun deck/upper deck/quarter deck & forecastle. 1845: orlop/gun deck/upper deck/quarter deck & forecastle. *Apollo* as troopship 1838: profile/orlop/gun deck/upper deck/quarter deck & forecastle. 1841: orlop/gun deck/upper deck/quarter deck & forecastle. *Horatio* 1818: framing. Screw frigate 1849: profile/orlop/gun deck/upper deck/quarter deck & forecastle. Screw mortar frigate 1855: profile/hold & hold sections/orlop/gun deck/upper deck/quarter deck & forecastle. *Hussar* as receiving ship 1834: profile/orlop/gun deck/upper deck/quarter deck & forecastle. *Crescent* as depot ship for freed slaves: Profile & quarter deck & forecastle/upper deck. *Nymphe* as receiving ship: upper deck. As guard ship 1837: upper deck. *Lively* as built: small scale & detailed lines & profile

### Forte Class 1801. Lines of French *Revolutionnaire* (taken 1794).
  Dimensions & tons: 157'2", 131'9⅞" x 40'5½" x 12'5". 1147 68/94.
  Men: 284. Guns: UD 28 x 18, QD 2 x 18 + 10 x 32★,
  Fc 2 x 18 + 4 x 32★.
*Forte* Woolwich Dyd. Possibly originally ordered to be built at Sheerness Dyd?
  Ord 9. & 16.7.1801; K 03.1811; L 21.5.1814; 1844 BU.

PLANS: Lines/profile/orlop/gun deck/upper deck/quarter deck & forecastle/specification/framing.

### Seringapatam Class 1813. Lines of French *President* (taken 1806, later known as *Piedmontaise*). These lines were modified for the later members of the class (I would like to thank Rif Winfield for helping to elucidate this slightly tangled story), with an extra foot of breadth added and depth increased by 9 inches. These modified lines were also to be the basis of the 1830 *Castor* Class of 32 pounder 36s.
  Dimensions & tons: 159', 133'9¾" x 40'5" x 12'9". 1162 34/94.
  Men: 315. Guns: UD 28 x 18, QD 14 x 32★, Fc 2 x 9 + 2 x 32★ (later modified in most cases to UD 2 x 8" + 26 x 32, QD 10 or 12 x 32, Fc 4 x 32).
*Seringapatam* Bombay Dyd. Teak built.
  Ord 21.8.1813; K 11.1817; L 5.9.1819; 1847 hulked.
*Druid* Pembroke Dyd.

Ord ____ 1817?; K 08.1821; L 1.7.1825; 1846 hulked.
**Madagascar** Bombay Dyd. Teak built.
Ord ____ 1819?; K 10.1821; L 15.11.1822; 1846 hulked.
**Nemesis** Pembroke Dyd.
Ord ____ 1817?; K 08.1823; L 19.8.1826; 1866 BU.
**Leda** Pembroke Dyd.
Ord ____ 1822?; K 10.1824; L 15.4.1828; 1865 hulked.
**Hotspur** Pembroke Dyd.
Ord ____ 1822?; K 07.1825; L 9.10.1828; 1859 hulked.
**Africaine** Chatham Dyd.
Ord 8.1.1822; K 09.1825; L 20.12.1827; 1867 sold.
**Eurotas** Chatham Dyd.
Ord ____ 1824?; K 02.1827; L 19.2.1829; 1848 converted to a screw frigate; 1856 steam mortar frigate; 1865 sold.

### Modified Version:
159', 133'2¼" x 41'5" x 13'6". 1215 16/94. (It is probable that the cancelled vessels listed beneath this section would have been completed to these dimentions had they been continued with.)
**Seahorse** Pembroke Dyd.
Ord ____ 1826?; K 11.1826; L 22.7.1830; 1847 converted to a screw frigate; 1856 steam mortar frigate; 1870 hulked.
**Andromeda** Bombay Dyd. Teak built.
Ord ____ 1827?; K 08.1827; L 03.1828; 1836 hulked.
**Stag** Pembroke Dyd.
Ord ____ 1820?; K 04.1828; L 2.10.1830; 1866 BU.
**Forth** Pembroke Dyd.
Ord ____ 1825?; K 11.1828; L 1.8.1833; 1847 converted to a screw frigate; 1856 steam mortar frigate; 1869 hulked.
**Meander** Chatham Dyd.
Ord ____ 1824?; K 02.1829; L 5.5.1840; 1857 hulked.

### Cancelled 1831. **Euphrates** Plymouth Dyd (teak frames).
Ord 22.10.1822; K. 30.6.1828?
**Spartan** Portsmouth Dyd.
Ord 13.9.1824? 1825; K ____ .
**Tiber** Portsmouth Dyd.
Ord 9.6.1825; K ____ .
**Jason** Woolwich Dyd.
Ord 18.7.1818? 1818; K ____ .
**Manilla** Bombay Dyd.
Ord 5.4.1819? 1820; K ____ .
**Orpheus** Chatham Dyd.
Ord 9.6.1825? 1825?; K ____ .
**Severn** Plymouth Dyd.
Ord 9.6.1825? 1825?; K ____ .
**Theban** Portsmouth Dyd.
Ord 13.9.1824? 1824?; K ____ .

### Cancelled 1832. **Statira** Plymouth Dyd.
Ord 12.1823; K 12.1823?
**Tigris** Plymouth Dyd.
Ord 1819?; K 06.1822?
**Inconstant** Sheerness Dyd.
Ord 9.6.1825; K ____ .
**Pique** Plymouth Dyd.
Ord 25.10.1820; K 06.1822.

FATES: *Seringapatam* became a coal depot and receiving ship at the Cape of Good Hope; between 1873-1883 BU and remains sold. *Druid* became a lazaretto at Liverpool; 1863 sold. *Madagascar* became a provision depot at Plymouth; 1853 to Rio de Janeiro as a receiving ship; 1863 sold. *Leda* became a water police ship at Plymouth; 1906 sold. *Hotspur* became a Roman Catholic chapel ship at Plymouth; 1868 renamed *Monmouth*; 1902 sold. *Seahorse* became a coal hulk at Plymouth, renamed *Lavinia*; 1902 sold. *Andromeda* was hulked at Liverpool; 1863 sold. *Meander* hulked at ____ for the Coastguard; 1859 to Ascension Island as a coal and store ship; 1870 wrecked. In 1985 this wreck was investigated by RAF divers.

PLANS: Lines/profile/orlop/gun deck/upper deck/quarter deck & forecastle/framing/specification. *Seahorse* as screw mortar frigate 1855: hold & hold sections. As coal depot 1868: profile/hold & hold sections/gun deck/upper deck, quarter deck & forecastle. *Eurotas* as screw mortar frigate 1855: Profile/hold & hold sections/gun deck/upper deck/quarter deck & forecastle. As coal depot 1868: Profile/gun deck/upper deck, quarter deck & forecastle. *Forth* as screw mortar frigate 1855: Profile/hold & hold sections/gun deck/upper deck/quarter deck & forecastle. As coal depot 1868: Profile/gun deck/upper deck & quarter deck & forecastle.

**Blonde Class 1816.** Enlarged lines of the *Euryalus* (ie. Rule's *Apollo* Class 36s).
Dimensions & tons: 155', 143'2" x 39'8" x 13'6". 1103.
Men: 315 Guns: UD 28 x 18, QD 14 x 32★, Fc 2 x 9 + 2 x 32★.
**Blonde** Deptford Dyd.
Ord ____ 1815? or ____ 1816; K 03.1816; L 12.1.1819; 1850 hulked.
FATE: Became a receiving ship (and coal depot?) at Portsmouth; 1866 temporary hospital ship; 1870 renamed *Calypso*; 1895 sold.
PLANS: profile/orlop/gun deck/upper deck/quarter deck & forecastle.

## 18 pounder Frigates, 36 Guns (Fifth Rates)

**Amazon Class 1794.** Rule design.
Dimensions & tons: 143', 119'6" x 38'2" x 13'6". 925 84/94.
Men: 264. Guns: UD 26 x 18, QD 8 x 9 + 6 x 32★, Fc 2 x 12 + 2 x 32★.
**Amazon** Wells, Deptford.
Ord 24.5.1794; K 06.1794; L 4.7.1795; 1797 wrecked.
**Emerald** Pitcher, Northfleet.
Ord 24.5.1794; K 06.1794; L 31.7.1795; 1822 hulked.

**Fir Built Version.** **Glenmore** (ex *Tweed*) Woolwich Dyd.
Ord 24.1. & 4. & 20.2.1795; K 03.1795; L 24.3.1796; 1805 hulked.
**Trent** Woolwich Dyd.
Ord 24.1. & 4. & 20.2.1795; K 03.1795; L 24.2.1796; 1803 hospital ship; 1815 hulked.

FATES: *Amazon* wrecked near Isle Bas during the *Droits de l'Homme* action. *Emerald* became a receiving ship at Portsmouth; 1836 BU. *Glenmore* became a receiving ship at Portsmouth; 1814 sold. *Trent* became a hospital ship laid up at Cork; 1823 BU.
PLANS: Lines/profile/orlop/gun deck/upper deck/quarter deck & forecastle/framing. Fir built version: Lines.

**Phoebe Class 1794.** 'Lengthened *Inconstant*' (ie. Hunt's *Perseverance* Class of 1780 with added length).
Dimensions & tons: 142'6", 118'10½" x 38' x 13'5". 913 13/94.
Men: 264. Guns: UD 26 x 18, QD 8 x 9, Fc 2 x 9.
**Phoebe** Dudman, Deptford.
Ord 24.5.1794; K 06.1794; L 24.9.1795; 1826 hulked.
**Caroline** Randall, Rotherhithe.
Ord 24.5.1794; K 06.1794; L 17.6.1795; 1813 hulked.
**Doris** Cleverley, Gravesend.
Ord 24.5.1794; K 06.1794; L 31.8.1795; 1805 lost.
**Dryad** Barnard, Deptford.
Ord 24.5.1794; K 06.1794; L 4.6.1795; 1838 hulked.
**Fortunee** Perry, Blackwall.
Ord ____ 1799?; K 04.1800; L 17.11.1800; 1818 sold.

FATES: *Phoebe* became a receiving ship and slop ship at Plymouth 1841 sold. *Caroline* became a salvage vessel at Portsmouth; 1815, by then at Deptford, BU. *Doris* wrecked in Quiberon Bay 21.1.1805. *Dryad* became a receiving ship at Portsmouth; 1860 BU.
PLANS: Lines/profile/orlop/gun deck/upper deck/quarter deck & forecastle/framing/specification.

**Sirius Class 1795.** Lines of the French *San Fiorenzo* (taken 1794).
Dimensions & tons: 148'10", 124'0⅜" x 39'7" x 13'3". 1033 66/94.
Men: 274. Guns: UD 26 x 18, QD 6 x 9, Fc 2 x 9.
**Sirius** Dudman, Deptford.
Ord 30.4.1795; K 09.1795; L 12.4.1797; 1810 destroyed.

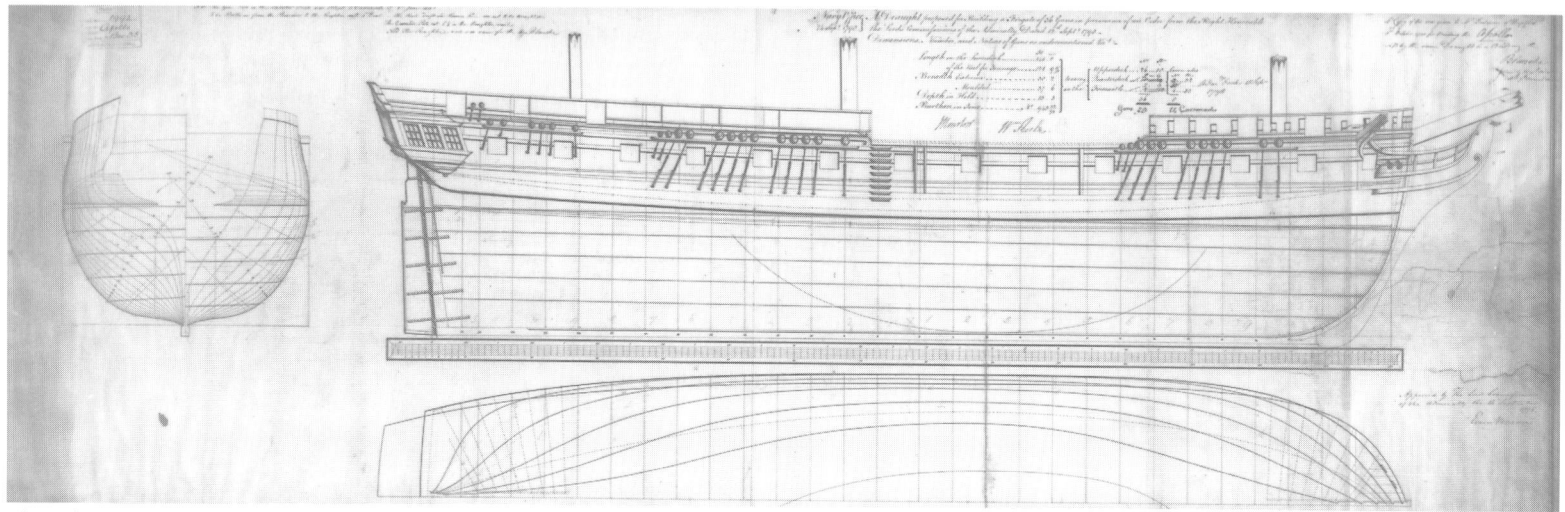

This is the original design lines plan (number 1992) for the 36 gun frigates of the *Apollo* Class of 1798 - initially used to build *Apollo* and *Blanche*, later copied for the rest of the class

FATE: Grounded and destroyed to avoid capture at Grand Port, Mauritius; the wreck has been worked by salvors and divers.

PLANS: Lines/profile/orlop/gun deck/upper deck/quarter deck & forecastle/framing.

## Penelope Class 1797. Henslow design.

Dimensions & tons: 150', 125'6¾" x 39'6" x 13'. 1042.
Men: 274. Guns: UD 26 x 18, QD 6 x 9 + 6 x 32★, Fc 4 x 9 + 2 x 32★.

**Penelope** Parsons, Bursledon.
Ord 4.5.1797; K 06.1797; L 26.9.1798; 1814 troopship; 1815 wrecked.

**Amethyst** Deptford Dyd.
Ord _____ 1797?; K 08.1798; L 23.4.1799; 1811 wrecked & BU.

**Jason** Parsons, Bursledon.
Ord 15.1.1798; K 10.1798; L 27.1.1800; 1801 wrecked.

FATES: *Penelope* wrecked in the River Saint Lawrence 30.4.1815.
PLANS: No plans.

## Aigle Class 1798. Henslow design.

Dimensions & tons: 146', 122'1½" x 38'6" x 13'. 962 81/94.
Men: 264. Guns: UD 26 x 18, QD 4 (later 6) x 9 + 8 (later 10) x 32★, Fc 4 x 9 + 2 x 32★.

**Aigle** Adams, Bucklers Hard.
Ord 8.11.1798; K 11.1798; L 23.9.1801; 1831 cut down to a 24 gun corvette (Sixth Rate); 1852 hulked.

**Resistance** Parsons, Bursledon.
Ord _____ 1800?; K 03.1800; L 29.4.1801; 1803 wrecked.

FATES: *Aigle* became a receiving ship and coal depot at Woolwich; 1869 to Sheerness; 1870 used for torpedo experiments, sunk, raised and remains sold. *Resistance* wrecked off Cape Saint Vincent 31.5.1803.

PLANS: Lines/profile/orlop/gun deck/upper deck/quarter deck & forecastle/framing/specification.

## Apollo/Euryalus Class 1798. Rule design.

Dimensions & tons: 145', 129'9⅜" x 38'2" x 13'3". 943 53/94.
Men: 264. Guns: UD 26 x 18, QD 6 x 9 + 6 x 32★, Fc 2 x 9 + 2 x 32★ (later alternatives: QD 8 x 9 + 4 x 32★ or 2 x 9 + 10 x 32★, Fc 2 x 9 + 4 x 32★ or 2 x 12 + 4 x 32★).

**Apollo** Dudman, Deptford.
Ord 15.9.1798; K 11.1798; L 16.8.1799; 1804 lost.

**Blanche** Dudman, Deptford.
Ord _____ 1799?; K 02.1800; L 2.10.1800; 1805 taken and burnt.

**Euryalus** Adams, Bucklers Hard.
Ord _____ 1800?; K 10.1801; L 6.6.1803; 1825 hulked.

**Dartmouth** Tanner, Dartmouth.
Ord _____ 1803?; K 07.1804. Firm failed, transferred to:
Cook, Dartmouth.
Ord _____ ; L 28.8.1813; 1831 hulked.

**Owen Glendower** Steemson, Hull.
Ord _____ 1805?; K 01.1807; L 21.11.1808; 1842 hulked.

**Semiramis** Deptford Dyd.
Ord _____ 1806?; K 04.1807; L 25.7.1808; 1821 cut down to a 24 gun corvette (Sixth. Rate); 1844 BU.

**Hotspur** Parsons, Bursledon.
Ord _____ 1807?; K 08.1807; L 13.10.1810; 1821 BU.

**Manilla** Woolwich Dyd.
Ord _____ 1805?; K 10.1807; L 11.9.1809; 1812 wrecked.

**Malacca** (ex *Penang*) 'Prince of Wales Island' (Penang).
Ord _____ 1808?; K _____ ; L 6.3.1809?; 1816 BU.

**Curacoa** Kidwell, Itchenor.
Ord _____ 1805?; K 01.1808; L 23.9.1809; 1831 cut down to a 26/24 gun corvette (Sixth Rate); 1849 BU.

**Saldanha** Temple, South Shields.
Ord _____ 1806?; K _____ ; L 8.12.1809; 1811 wrecked.

**Havannah** Wilson, Liverpool.
Ord _____ 1808?; K _____ ; L 26.3.1811; 1845 cut down to a 19 gun corvette (Sixth Rate); 1860 hulked.

**Theban** Parsons, Warsash.
Ord _____ 1808?; K 06.1808; L 22.12.1809; 1817 BU.

**Orpheus** Deptford Dyd.
Ord _____ 1808?; K 08.1808; L 12.8.1809; 1819 BU.

**Leda** Woolwich Dyd.
Ord _____ 1808?; K 10.1808; L 9.11.1809; 1817 sold.

**Belvidera** Deptford Dyd.
Ord _____ 1808?; K 12.1808; L 23.12.1809; 1846 hulked.

**Astrea** Guillaume, Northam.
Ord _____ 1808?; K 12.1808; L 05.1810; 1823 hulked.

**Galatea** Deptford Dyd.
Ord _____ 1809?; K 08.1809; L 31.8.1810; 1829 fitted with experimental propelling machinery; 1836 hulked.

**Maidstone** Deptford Dyd.
Ord _____ 1810?; K 09.1810; L 18.10.1811; 1838 hulked.

**Stag** Deptford Dyd.
Ord _____ 1810?; K 01.1811; L 26.9.1812; 1821 BU.

**Magicienne** List, Fishbourne.
Ord _____ 1810?; K 04.1811; L 8.8.1812; 1831 cut down to a 24 gun corvette (Sixth Rate); 1845 BU.

**Pallas** Guillaume, Northam.
Ord _____ 1811?; K 05.1811; transferred to:
Portsmouth Dyd.
Ord _____ ; K 04.1814; L 13.4.1816; 1836 hulked.

**Creole** Tanner, Dartmouth.
Ord 1805?; never begun?; firm failed, transferred to:
Plymouth Dyd.
Ord _____ ; K 09.1811; L 1.5.1813; 1833 hulked.

**Barrosa** Deptford Dyd.
Ord _____ 1811?; K 10.1811; L 21.10.1812; 1823 hulked.

# HENSLOW TO SEPPINGS 1786-1830

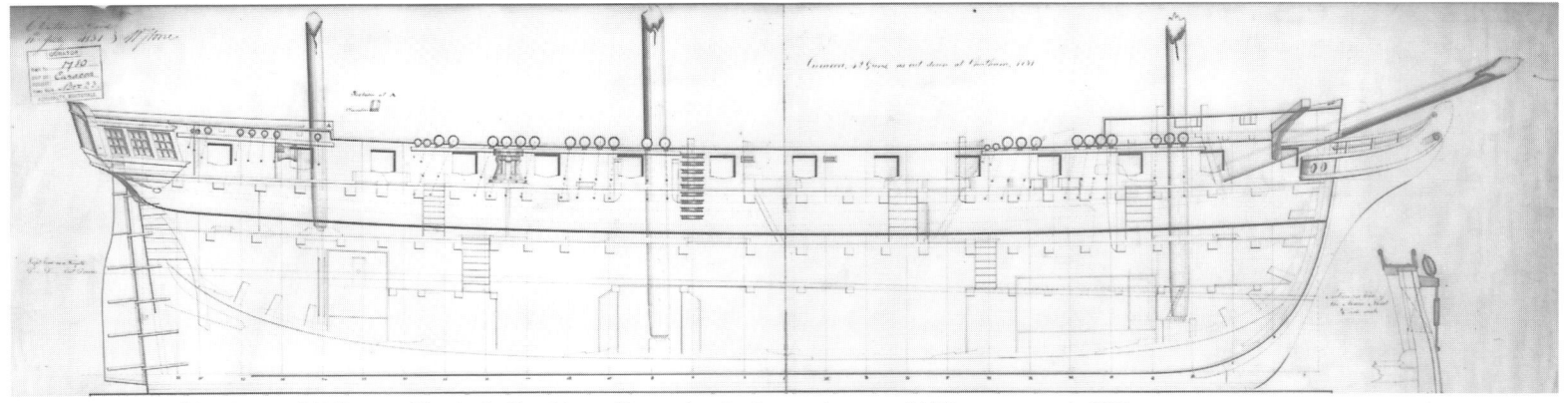

*Profile plan (1780) of the* Curacoa *of 1809, one of the* Apollo *Class 36 gun frigates, showing her cut down to a 24/26 gun corvette in 1831*

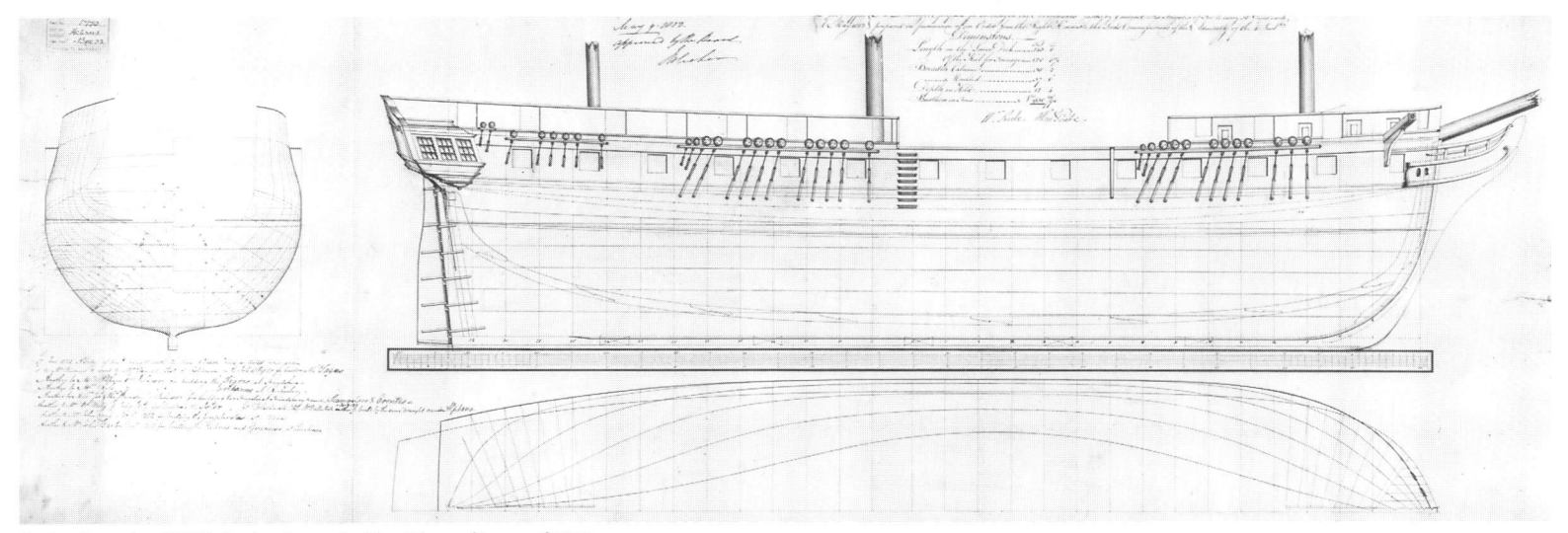

*Design lines plan (1770) for the* Scamander *Class 36 gun frigates of 1812*

**Tartar** Deptford Dyd.
  Ord ____ 1811?; K 10.1812; L 6.4.1814; 1827 hulked.
**Brilliant** Deptford Dyd.
  Ord ____ 1812?; K 11.1813; L 28.12.1814; 1846 cut down to 22 gun corvette (Sixth Rate); 1860 hulked.

FATES: *Apollo* wrecked near Cape Mondego on the coast of Portugal 2.4.1804. *Blanche* taken by the French in the West Indies 19.7.1805, then burnt. *Euryalus* became a convict ship at Chatham; 1847 to Gibraltar as a convict ship; 1859 renamed *Africa;* 1860 sold. *Dartmouth* to quarantine service at Chatham; 1854 BU at Deptford. *Owen Glendower* became a convict ship at Gibraltar; 1884 sold. *Manilla* wrecked on the Haak Sand 28.1.1812. *Saldanha* wrecked off Lough Swilly 4.12.1811. *Havannah* to the Ragged School at Cardiff as a training ship; 1905 sold. *Belvidera* became a store depot at Portsmouth; 1852 receiving ship; 1906 sold: *Astree* became a depot ship at Falmouth; 1851 BU. *Galatea* became a receiving ship and coal depot at Jamaica; 1849 BU. *Maidstone* became a coal depot at Portsmouth; 1867 BU. *Pallas* became a coal depot at Plymouth; 1862 sold. *Creole* to harbour service (unspecified) at Chatham; 1833 BU. *Barrosa* became a slop ship at Portsmouth; later receiving ship and ordnance depot; 1841 sold. *Tartar* became a receiving ship and hulk at Chatham (this may have occurred in 1830 and not 1827); 1859 BU. *Brilliant* became a training ship for Coast Volunteers at Dundee; 1875 to Inverness as an RNR training ship; 1889 renamed *Briton;* 1906 lent to the War Office as sleeping accommodation for militia; 1908 sold.

PLANS: lines/profile/orlop/gun deck/upper deck/quarter deck & forecastle/framing/midships section/specification. *Euryalus* as convict hulk 1847: Lines/profile/orlop/gun deck/upper deck/quarter deck & profile.

## Ethalion Class 1800. 'Admiralty design'.

Dimensions & tons: 152', 129'2¼" x 38' x 13'. 992 25/94 .
Men: 264. Guns: UD 26 x 18, QD 14 x 32★, Fc 2 x 9 + 4 x 32★.
**Ethalion** Woolwich Dyd.
  Ord ____ 1799?; K 05.1800; L 29.7.1802; 1823 hulked.
FATE: Became convict ship at Woolwich; 1824 temporary receiving ship; 1835 became a breakwater at Harwich.
PLANS.: Lines/profile/orlop/gun deck/upper deck/quarter deck & forecastle/framing.

## Pyramus Class 1805. 'Lines of the *Belle Poule* [French, taken 1780] with topside of *Nymphe* [French, taken 1780]'.

Dimensions & tons: 140', 115'11⅝" x 38'2" x 11'11". 898 50/94.
Men: 264. Guns: UD 26 x 18, QD 2 x 9 + 10 x 32★, Fc 2 x 9 + 2 x 32★.
**Pyramus** Greensword & Kidwell, Itchenor.
  Ord 29.6.1805; K 04.1806; transferred to:
  Portsmouth Dyd.
  Ord ____ ; K 11.1808; L 22.1.1810; 1832 hulked.
FATE: Became a receiving and convict ship at Halifax Nova Scotia; 1879 sold.
PLANS: Lines/profile/orlop /gun deck/upper deck/quarter deck & forecastle/framing/trussed frames (starboard)/(port)/specification.

## Scamander Class 1812. Rule design, all 'fir' built (*Alpheus, Hebrus* and *Granicus* of yellow pine, all the others of red pine).

Dimensions & tons: 143', 120'0¾" x 38'2" x 12'4'. 930 25/94.
Men: 274. Guns: UD 26 x 18, QD 12 x 32★, Fc 2 x 9 + 2 x 32★.
**Scamander** (ex *Lively*?) Brindley, Frindsbury.
  Ord 4.5.1812; K 08.1812; L 13.7.1813; 1819 sold.
**Eridanus** (ex *Liffey*) Mrs. Ross, Rochester.
  Ord 4.5.1812; K 08.1812; L 1.5.1813; 1818 sold.
**Ister** (ex *Blonde*) Wallis, Blackwall.
  Ord 4.5.1812; K 08.1812; L 14.7.1813; 1819 sold.
**Orontes** (ex *Brilliant*) Brindley, Frindsbury.
  Ord 4.5.1812; K 08.1812; L 29.6.1813; 1817 BU.
**Tagus** (ex *Severn*) List, Fishbourne.
  Ord 4.5.1812; K 08.1812; L 14.7.1813; 1822 sold.
**Tigris** (ex *Forth*) Pelham, Frindsbury.
  Ord 4.5.1812; K 09.1812; L 26.6.1813; 1818 sold.
**Euphrates** (ex *Greyhound*) King, Upnor.

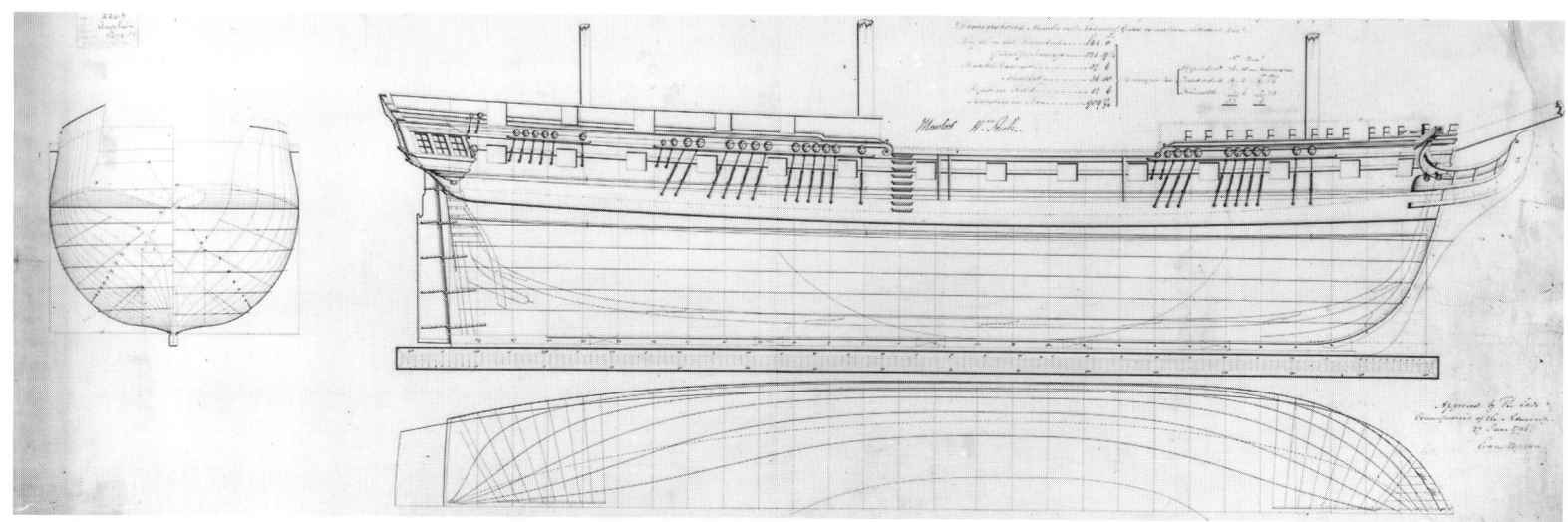

Design lines plan (2263) for the *Amphion* Class 32 gun frigates

Ord 4.5.1812; K 01.1813; L 8.11.1813; 1818 sold.
**Granicus** Barton, Limehouse.
Ord 4.5.1812; K 01.1813; L 25.10.1813; 1817 sold.
**Hebrus** Barton, Limehouse.
Ord 4.5.1812; K 01.1813; L 13.9.1813; 1817 sold.
**Alpheus** Barton, Limehouse.
Ord 4.5.1812; K 07.1813; L 6.4.1814; 1817 sold.
PLANS: Lines/profile/orlop/gun deck/upper deck/quarter deck & forecastle/framing/midships section /specification.
NOTE: The original names of the red pine vessels were changed for 'foreign river' names on 7.12.1812, according to Rif Winfield.

## 32 pounder Frigates 36 gun (Fifth Rate)

### *Castor* Class 1828.
Seppings design, 'similar dimensions to those building after the *Piedmontaise* [*Seringapatam* Class of 38s] but with additional foot in breadth'.
Dimensions & tons: 159', 133'7⅜" x 42'6" x 13'6". 1283 68/94. (*Amphion* as screw frigate: 175'4", 152'3¾" x 43'2" x 13'4½". 1474 77/94.)
Men: ____ . Guns: UD 22 x 32. QD 10 x 18, Fc 4 x 18.
**Castor** Chatham Dyd.
Ord ____ 1828?; K 01.1830; L 2.5.1832; 1860 hulked.
**Amphion** (ex *Ambuscade*) Woolwich Dyd.
Ord ____ 1828?; K 04.1830; L 14.1.1846 as a screw frigate; 1862 sold.
FATES: *Castor* became a training ship for Coast Volunteers at North Shields; 1895 to Sheerness in reserve; 1902 sold.
PLANS: Lines/framing/profile/orlop/lower deck/upper deck/quarter deck & forecastle/midships section. *Amphion* as screw frigate: lines. As fitted: Profile/hold/lower deck/upper deck/quarter deck & forecastle/screw well/screw hoisting apparatus.

## 18 pounder Frigates 32 Guns (Fifth Rate)

### *Pallas* Class 1790.
Henslow design.
Dimensions & tons: 135', 112'8¾" x 36' x 12'6". 776 72/94.
Men: 220. Guns: UD 26 x 18, QD 4 x 6, Fc 2 x 6.
**Pallas** Woolwich Dyd.
Ord 9.12.1790; K 05.1792; L 19.12.1793; 1798 wrecked.
**Stag** Chatham Dyd.
Ord 9.12.1790; K 03.1792; L 28.6.1794; 1800 wrecked.
**Unicorn** Chatham Dyd.
Ord 9.12.1790; K 03.1792; L 12.7.1794; 1815 BU.
FATES: *Pallas* wrecked on Mount Batten Point, Plymouth 4.4.1798. *Stag* wrecked in Vigo Bay 6.9.1800.
PLANS: Lines/profile/orlop/gun deck/upper deck/quarter deck & forecastle/framing/stern.

### *Cerberus* Class 1793.
Henslow design.
Dimensions & tons: 135', 112'4¼" x 36'6" x 12'6". 796 17/94.
Men: 254. Guns: UD 26 x 18, QD 4 x 6 + 6 x 24★, Fc 2 x 6 + 2 x 24★.
**Cerberus** Adams, Bucklers Hard.
Ord 14.2.1793; K 04.1793; L 25.4.1794; 1814 sold.
**Alcmene** Graham, Harwich.
Ord 14.2.1793; K 04.1793; L 8.11.1794; 1809 wrecked.
**Lively** Nowlan, Northam.
Ord 14.2.1793; K 04.1793; L 23.10.1794; 1798 wrecked.
**Galatea** Parsons, Bursledon.
Ord 14.2.1793; K 04.1793; L 17.5.1794; 1809 BU.

### Fir Built 12 Pounder Version.
Guns: UD 26 x 12, QD 4 x 6 + 4 x 24★, Fc 2 x 6 + 2 x 24★
**Maidstone** Deptford Dyd.
Ord 4.2.1795; K 03.1795; L 12.12.1795; 1810 BU.
**Shannon** Deptford Dyd.
Ord 4.2.1795?; K 04.1795; L 9.2.1796; 1802 sold.
FATES: *Alcmene* wrecked off Nantes 29.4.1809. *Lively* wrecked near Rota Point, Cadiz 12.4.1798.
PLANS: Lines/profile/orlop/gun deck/upper deck/quarter deck & forecastle/framing/stern/specification. Fir built: Lines.

### *Triton* Class.
See under 12 pounder 32s.

### *Amphion* Class 1796.
Rule design.
Dimensions & tons: 144', 121'7½" x 37'6" x 12'6". 909 71/94.
Men: 254? Guns: UD 26 x 18, QD 4 x 6 + 4 x 24★, Fc 2 x 6 + 2 x 24★.
**Amphion** Betts, Mistleythorn.
Ord 11.6.1796; K 07.1796; L 19.3.1798; 1820 breakwater.
**Eolus** Barnard, Deptford.
Ord ____ 1800?; K 04.1800; L 28.2.1801; 1817 BU.
**Medusa** Pitcher, Northfleet.
Ord 28.1.1800; K 04.1800; L 14.4.1801; 1813 hulked.
**Proserpine** Steemson, Hull.
Ord ____ 1805?; K 09.1805; L 6.8.1807; 1809 taken.
**Nereus** Temple, South Shields.
Ord ____ 1805?; K 11.1806; L 4.3.1809; 1817 BU.
FATES: *Medusa* became a temporary hospital ship at Plymouth; 1816 BU. *Proserpine* taken by the French *Pénélope* and *Pauline* off Toulon 28.2.1809; in service with the French until 1840 when she became a 'corvette de charge'.
PLANS: Lines/profile/orlop/gun deck/upper deck/quarter deck & forecastle/framing/specification.

### *Narcissus* Class 1797.
Henslow design.
Dimensions & tons: 142', 118'5" x 37'6" x 12'6". 885 90/94.

Men: 254? Guns: UD 26 x 18, QD 4 (or 2) x 6 + 6 (or 8) x 24★, Fc 2 x 6 + 2 x 24★

*Narcissus* Deptford Dyd.
Ord 25.11.1797; K 02.1800; L 12.5.1801; 1823 hulked.

*Tartar* Brindley, Frindsbury.
Ord ____ 1800?; K 08.1800; L 27.6.1801; 1811 wrecked.

*Cornelia* Temple, South Shields.
Ord ____ 1805?; K 06.1806; L 26.7.1808; 1814 BU.

**Cancelled 1806.** *Doris* Record, Appledore.
Ord 6.1.1806; never begun.

*Syren* Record, Appledore.
Ord 16.7.1805; never begun?

FATES: *Narcissus* became a temporary convict ship at Woolwich; 1824 convict hospital ship; 1837 sold. *Tartar* was wrecked in the Baltic 18.8.1811.

PLANS: Lines/profile/orlop/gun deck/upper deck/quarter deck & forecastle/framing/stern.

## *Bucephalus* Class 1805.
Rule design, modified from the lines of the French *Topaze* (taken 1793).

Dimensions & tons: 150', 126'6⅛" x 38' x 12'1". 971 66/94.
Men: 254? Guns: UD 26 x 18, QD 2 x 9 + 10 x 24★, Fc 2 x 9 + 2 x 24★.

*Bucephalus* Row, Newcastle.
Ord 9.8.1805; K 08.1806; L 3.11.1808; 1814 troopship; 1822 hulked.

FATE: Became a receiving ship at Portsmouth; 1834 BU.

PLANS: Lines/profile/orlop/gun deck/upper deck/quarter deck & forecastle/framing/midships section/specification.

## *Hyperion* Class 1805.
Henslow design, modified from the lines of the French *Magicienne* (taken 1781).

Dimensions & tons: 143'9", 118'8⅛" x 39'2½" x 12'4½". 970 39/94.
Men: 254?. Guns: UD 26 x 18, QD 2 x 9 + 10 (or 2) x 24★, Fc 2 x 9 + 2 x 24★ + 2 x 9 chase guns.

*Hyperion* Gibson, Hull.
Ord ____ 1806?; K 02.1806; L 3.11.1807; 1825 to the Coast Blockade; 1833 BU.

PLANS: Lines/profile/orlop/gun deck/upper deck/quarter deck & forecastle/framing/lines & profile/specification.

## *Prompte* Class 1813.
Designer uncertain. Fir built. Frames constructed at Chatham, then taken down to be shipped to Canada for assembly on the Great Lakes, but both sold incomplete at Quebec instead. The upper deck armament was to have been short 24 pounders, but as they were to have been classed as 32 gun ships they have been included in this section.

Dimensions & tons: 130', 107'10¾" x 36' x 10'3". 743 70/94.
Men: ____ . Guns: UD ____ x short 24 , QD ____ , Fc ____ .

*Prompte* Chatham Dyd.
Ord ____ ; K 7.1.1814; taken down 02.1814; 1814 sold.

*Psyche* Chatham Dyd.
Ord ____ ; K 01.1814; taken down 02.1814; 1814 sold.

PLANS: Lines/profile/orlop/gun deck/upper deck/quarter deck & forecastle.

# 12 pounder Frigates, 32 Guns (Fifth Rates)

**Fir Built** *Cerberus* **Class.** See under 18 pounder 32s.

## *Triton* Class 1796.
'Admiralty' design by James Gambier who was a member of the Admiralty Board (information from Robert Gardiner). Fir built. The Upper Deck armament was originally intended to be 18 pounders.

Dimensions & tons: 142', 123'1⅜" x 36' x 11'10". 848 66/94.
Men: 220. Guns: UD 26 x 12, QD 4 x 6 + 6 x 24★, Fc 2 x 6 + 2 x 24★ (later UD 12 x 9 + 10 x 24★, QD & Fc no guns).

*Triton* Barnard, Deptford.
Ord ____ 1795?; K 04.1796; L 5.9.1796; 1803 hulked [?].

FATE: Hulk at Woolwich (?) in 1803; 1810 to Plymouth as a receiving ship; 1814 sold.

PLANS: Lines/profile/orlop/gun deck/upper deck/quarter deck & forecastle/framing. As prison ship 1803: decks.

## *Thames* Class 1804.
All fir built. Modified from Bately's 1756 *Richmond* Class.

Dimensions & tons: 127', 107' x 34' x 11'9". 657 88/94.
Men: 220. Guns: UD 26 x 12, QD 2 x 6 + 6 x 24★, Fc 2 x 6 + 2 x 24★.

*Thames* Chatham Dyd.

NMM Photo Negative 6687 - Oil painting of the 32 gun frigate *Triton* of 1796 by Pocock

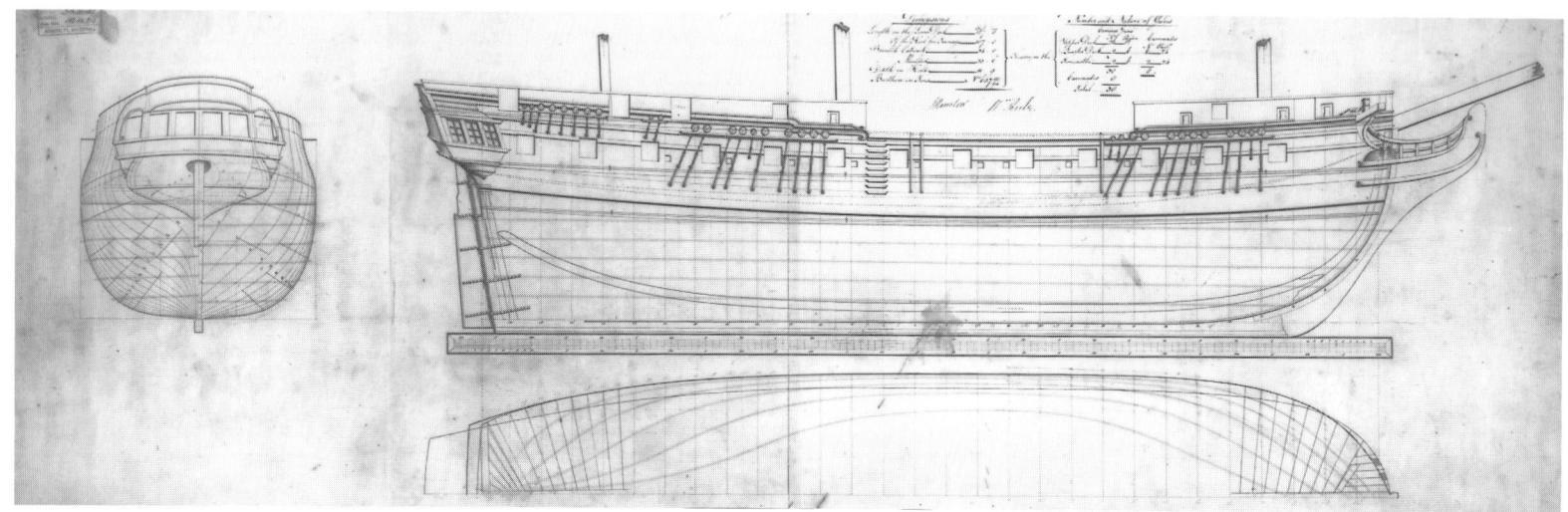

Design lines plan (2357) for building the *Thames* of 1805, the name ship of a class of fir built 32 frigates

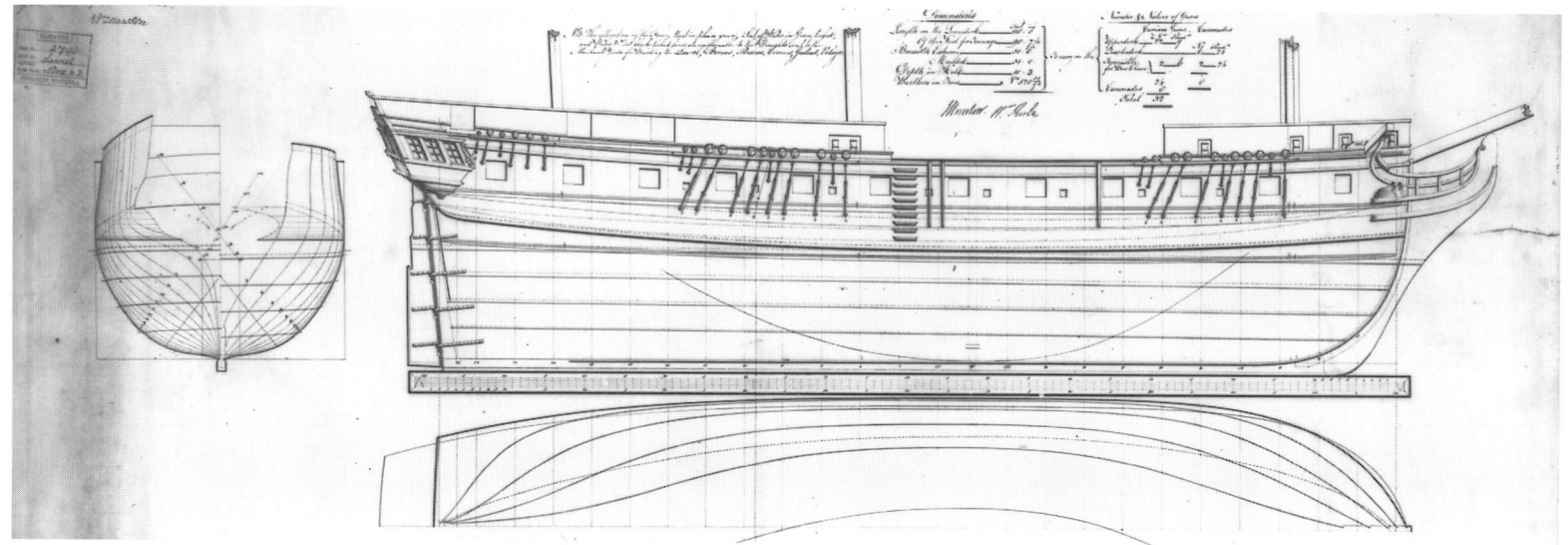

Design lines plan (2796) for the *Laurel* Class of 22 gun ships (1805)

Ord 1.5.1804; K 07.1804; L 24.10.1805; 1814 troopship; 1816 BU.
**Circe** Plymouth Dyd.
Ord 16.3.1804; K 06.1804; L 17.11.1804; 1814 sold.
**Pallas** Plymouth Dyd.
Ord 16.3.1804; K 06.1804; L 17.11.1804; 1810 wrecked.
**Hebe** Deptford Dyd.
Ord 12.7.1804?; K 08.1804; L 31.12.1804; 1813 sold.
**Jason** Woolwich Dyd.
Ord 12.7.1804; K 08.1804; L 21.11.1804; 1815 BU.
**Minerva** Deptford Dyd.
Ord 12.7.1804; K 08.1804; L 25.10.1805; 1816 BU.
**Alexandria** Portsmouth Dyd.
Ord 12.7.1804; K 10.1804; L 18.2.1806; 1817 hulked.

FATES: *Pallas* was wrecked in the Firth of Forth 18.12.1810. *Alexandria* became a receiving ship at Sheerness; 1818 BU.

PLANS: Lines/profile/orlop/gun deck/upper deck/quarter deck & forecastle/framing. Fir built version: profile/orlop/gun deck/upper deck/quarter deck & forecastle/specification/midships section.

**Cancelled 32:** *Medea* Woolwich Dyd.
Ordered 1800 then cancelled. Gardiner gives the same or another Medea as ordered 12.7.1804 from Chatham Dyd and then cancelled – as one of the ships of the above class.

## Sixth Rates Classed as 22 Guns

**Laurel Class 1805.** Henslow design 'to carry 22 carriage guns on the upper deck'.

Dimensions & tons: 118', 98'7¼" x 31'6" x 10'3". 526 39/94.
Men: 155. Guns: UD 22 x 9, QD 6 x 24★, Fc 2 x 24★ + 2 x 6 chase guns.
**Laurel** Bools & Good, Bridport.
Ord 30.1.1805; K 06.1805; L 2.6.1806; (1808-1810 in French hands); 1810 renamed *Laurestinus*; 1813 wrecked.
**Boreas** Stone, Yarmouth.
Ord 30.1.1805?; K 06.1805; L 19.4.1806; 1807 wrecked.
**Comus** Custance, Yarmouth.
Ord 30.1.1805?; K 08.1805; L 28.8.1806; 1816 wrecked.
**Garland** Chapman, Biddeford.
Ord 30.1.1805?; K 08.1805; L 25.4.1807; 1817 sold.
**Perseus** Sutton, Ringmere.
Ord 30.1.1805?; K 11.1805; L 30.11.1812; 1816 hulked.
**Volage** Chapman, Biddeford.
Ord 30.1.1805?; K 01.1806; L 23.3.1807; 1818 sold.

FATES: *Laurel* taken by the French *Cannonière* off Mauritius 15.9.1808; renamed *L'Esperance* in French service; 12.4.1810 retaken; 22.10.1813 (as *Laurestinus*) wrecked on the Silver Keys ('one of the Gallipagos') in the Bahamas. *Boreas* wrecked on Guernsey 7.12.1807. *Comus* was wrecked at Saint Shots, Newfoundland 24.10.1816. *Perseus* became the receiving ship for distressed seamen at Deptford; 1818 to the Tower of London for the same purpose; 1850 BU.

PLANS: Lines/profile/orlop/gun deck/upper deck/quarter deck & forecastle/framing/specification.

**Banterer Class 1805.** Rule design 'to carry 22 carriage guns on the upper deck'.

Dimensions & tons: 118', 98'7⅛" x 32' x 10'6". 537 2/94.

Men: 155. Guns: UD 22 x 9, QD 6 x 24★, Fc 2 x 6 + 2 x 24....
*Banterer* Temple, South Shields.
    Ord 30.1.1805; K 08.1805; L 24.2.1807; 1808 wrecked.
*Crocodile* Temple, South Shields.
    Ord 30.1.1805; K 06.1805; L 19.4.1806; 1816 BU.
*Cossack* (ex *Pandour*) Temple, South Shields.
    Ord 30.1.1805; K 07.1805; L 24.1.1806; 1816 BU.
*Daphne* Davy, Topsham.
    Ord 30.1.1805; K 07.1805; L 2.7.1806; 1816 sold.
*Cyane* Bass, Topsham/Lympstone.
    Ord 30.1.1805; K 08.1805; L 14.10.1806; 1815 taken.
*Porcupine* Owen, Topsham.
    Ord 30.1.1805; K 09.1805; L 26.1.1807; 1816 sold.

FATES: *Banterer* wrecked in the Saint Lawrence 4.12.1808. *Cyane* taken by USS *Constitution* in the Atlantic 20.2.1815; in American service until BU 1836.
PLANS: Lines/profile/orlop/gun deck/upper deck/quarter deck & forecastle/framing/stern.

## Ship Sloops with Quarter Deck & Forecastle

(majority remaining in 1816 reclassed, and those built after that date classed as Sixth Rates). Classified number of guns noted after class headings.

### *Hound* Class 1788. 16. Henslow design. The class was originally intended to carry 16 x 6 + 14 swivels.

    Dimensions & tons: 100', 82'9¾" x 27' x 13'. 320 89/94.
    Men: 125. Guns: UD 16 x 6, QD 4 x 12★, Fc 2 x 12★ (*Serpent* carried 6 QD carronades. *Fury* as a bomb: 2? mortars + 10 x 6 + 12 swivels).
*Hound* (ex *Hornet*) Deptford Dyd.
    Ord 17.1.1788; K 09.1788; L 31.3.1790; 1794 taken.
*Martin* Woolwich Dyd.
    Ord 17.1.1788; K 15.7.1789; L 8.10.1790; 1800 lost.
*Rattlesnake* Chatham Dyd.
    Ord 17.1.1788; K 1.7.1789; L 7.1.1791; 1811 hulked.
*Fury* Portsmouth Dyd.
    Ord 17.1.1788; K 07.1788; L 2.3.1790; 1798 converted to a bomb vessel; 1811 BU.
*Serpent* (ex *Porcupine*) Plymouth Dyd.
    Ord 17.1.1788; K 11.1788; L 3.12.1789; 1806 foundered.

FATES: *Hound* taken by the French *Seine* and *Galathée* when coming from the West Indies 14.7.1794. *Martin* supposed foundered in the North Sea October 1800. *Rattlesnake* became a receiving ship at Plymouth 1814 sold. *Serpent* foundered on the Jamaica Station 09.1806.
PLANS: Lines/profile, upper deck & quarter deck & forecastle/orlop & gun deck/framing. *Fury* as bomb 1798: Profile/profile & midships section/decks/midships section. *Serpent* as built: Lines & profile & decoration/profile & upper deck & quarter deck & forecastle/orlop & gun deck.

### *Hawk* Class 1790. 16. Henslow design.

    Dimensions & tons: 100', 81'11¾" x 27'6" x 13'6". 329 72/94.
    Men: 121. Guns: 16 x 6 + 12 swivels.
*Hawk/Hawke* Deptford Dyd.
    Ord 9.12.1790; K 09.1791; L 24.7.1793; 1803 BU.
*Swift* Portsmouth Dyd.
    Ord ____ 1790?; K 08.1791; L 5.10.1793; 1798 foundered.

FATES: *Swift* presumed foundered when caught in typhoon escorting a convoy in the China Sea; 04.1798.
PLANS: Lines/profile/orlop & gun deck/upper deck & quarter deck & forecastle/framing/planking expansion.

### *Pylades* Class 1793. 16. Henslow design. The 1795 *Brazen* Class was an enlarged version of this design.

    Dimensions & tons: 105', 86'7½" x 28' x 13'6". 361 14/94.
    Men: 125. Guns: 16 x 6 + 4 swivels.
*Pylades* Mestaers, Thames.
    Ord 18.2.1793; K 03.1793; L 04.1794; 1794 wrecked but then salvaged; 1798 re-purchased; 1815 sold.
*Alert* Randall, Rotherhithe.
    Ord 18.2.1793; K 04.1793; L 8.10.1793; 1794 taken.
*Albecore* Randall, Rotherhithe.
    Ord 18.2.1793; K 04.1793; L 19.11.1793; 1802 sold.
*Peterell* Wilson, Frindsbury.
    Ord 18.2.1793; K 05.1793; L 4.3.1794; 1811 hulked.
*Ranger* Hill & Mellish, Thames.
    Ord 18.2.1793; K 05.1793; L 19.5.1794; 1805 taken.
*Rattler* Raymond (or Parsons?), Southampton.
    Ord 18.2.1793; K 05.1793; L 21.2.1795; 1815 sold.

FATES: *Pylades* wrecked in 'Harlesswick' Bay, Shetland 26.11.1794 wreck sold, salved, repurchased in 1798 and returned to service. *Alert* taken by the French *Unité* off Ireland 05.1794; renamed *L'Alerte* in French service; 23.8.1794 foundered off Douarnenez. *Peterell* became a receiving ship at Plymouth; 1827 sold. *Ranger* taken and burnt by a French squadron in the West Indies 19.7.1805.
PLANS: Lines/profile/orlop/gun deck/upper deck/quarter deck & forecastle/framing/stern/specification.

### *Cormorant* Class 1793. 16 etc. Lines of the French *Amazon* of 1745, modified by Henslow and Rule. The exact relationship of this design to the *Tisiphone* Class of fireships, to the larger but similar *Coquette* Class of sloops, and to the *Rosamond* Class, is not altogether clear. It is also not clear whether the last few vessels listed here as a subgroup (*Ranger* etc.) should be considered as such, or as an entirely separate class; or exactly how many of the later vessels should be listed with them. Those built in 1805 and later would eventually (1816) be classed as Sixth Rates of either 26 or 24 guns (22 for *Anacreon*), although originally still classed as 16 gun sloops. Their final classification is given in brackets - ( )

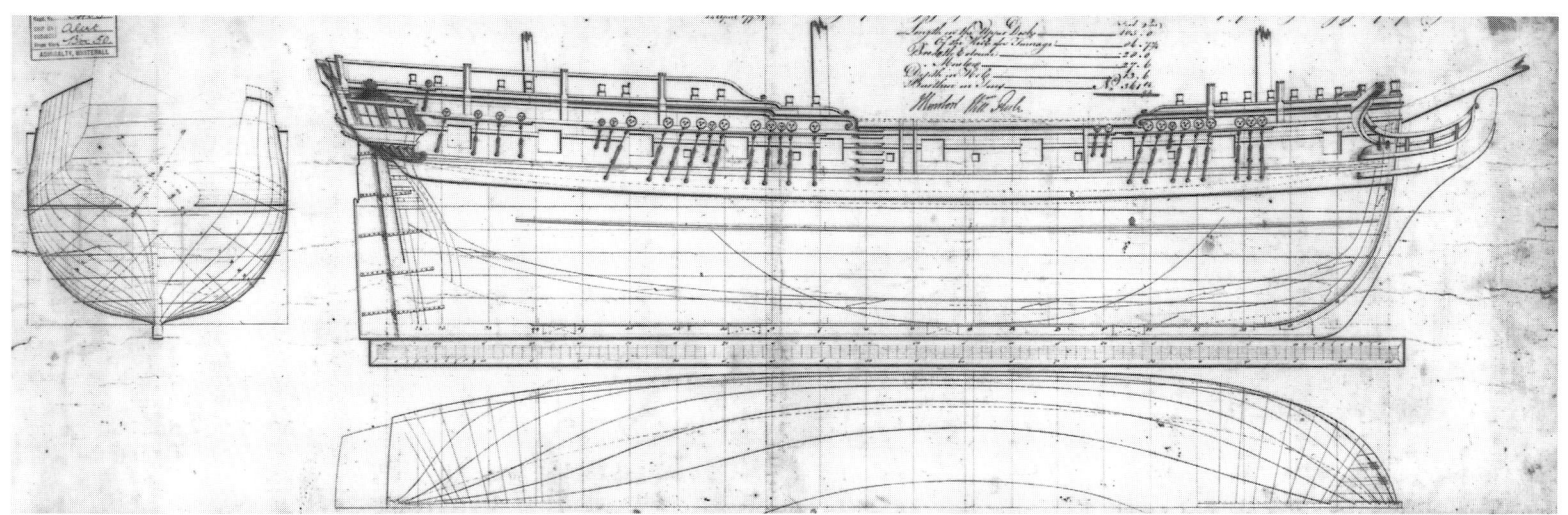

Lines plan (3405) for building *Alert* in 1793 to the *Pylades* Class ship sloop design

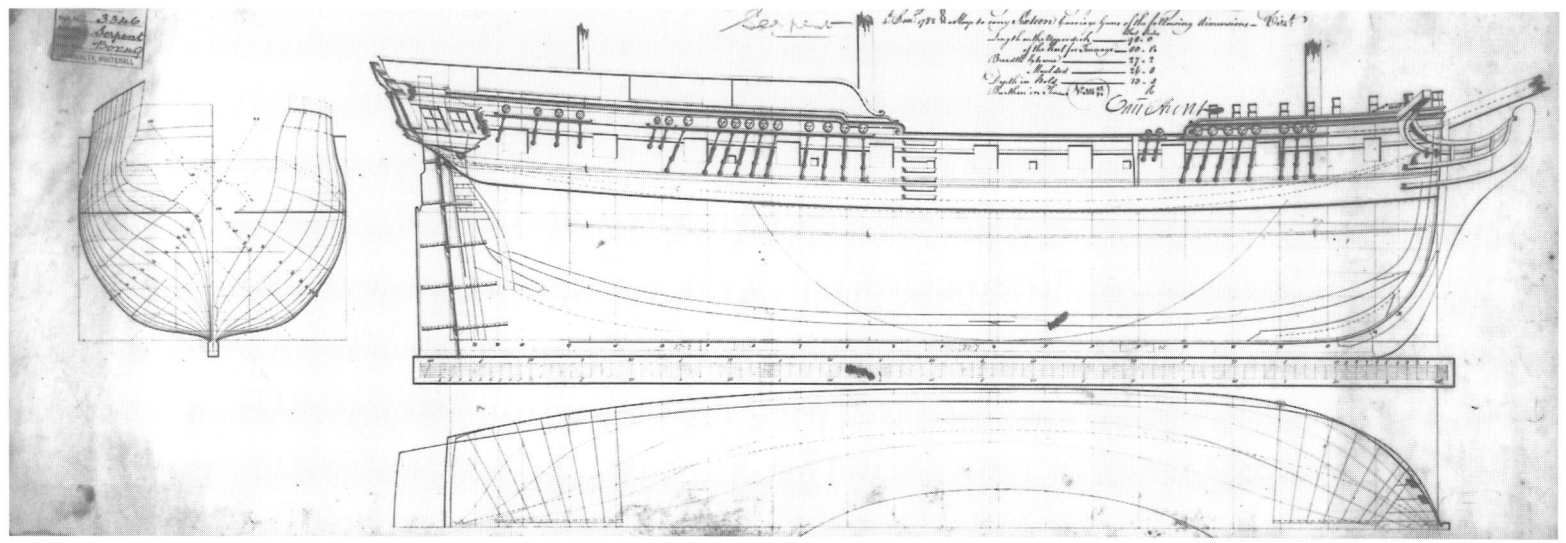

Building lines plan (3346) for the ship sloop *Serpent* of 1789 of the *Hound* Class

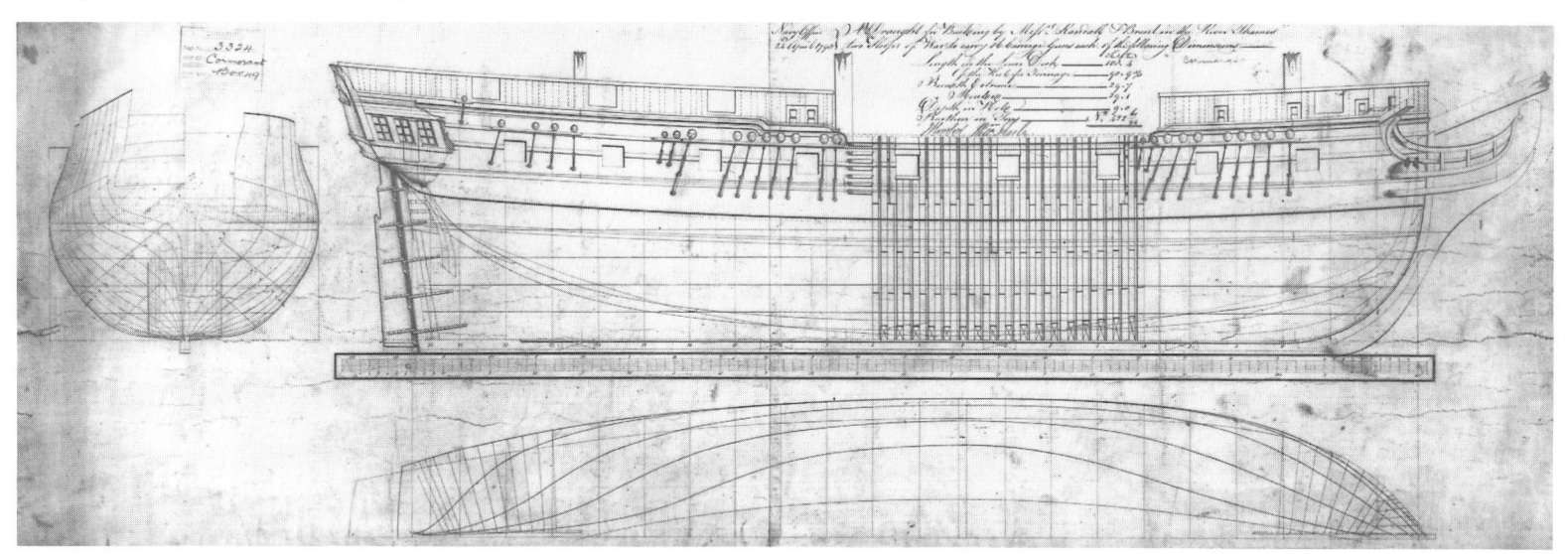

Design lines plan for the 1794 *Cormorant*, the name ship of her class of ship sloops

- after the builder.
  Dimensions & tons: 108'4", 90'9⅝" x 29'7" x 9'. 422 61/94.
  Men: 125. Guns: UD 16 x 6 + 14 swivels (1805 group: UD 16 x 32★, QD 6 x 18★, Fc 2 x 6 + 2 x 18★).
**Cormorant** Randall & Brent, Rotherhithe.
  Ord ___ 1793?; K 04.1793; L 2.1.1794; 1796 blown up.
**Favourite** Randall & Brent, Rotherhithe.
  Ord ___ 1793?; K 04.1793; L 1.2.1794; (1806-1807 in French hands); 1807 renamed *Goree*; 1817 BU.
**Hornet** Stalkartt, Thames.
  Ord ___ 1793?; 04.K 1793; L 3.2.1794; 1805 hulked.
**Hazard** Brindley, Frindsbury.
  Ord ___ 1793?; K 05.1793; L 3.3.1794; 1817 sold.
**Lark** Pitcher, Northfleet.
  Ord ___ 1793?; K 05.1793; L 15.2.1794; 1809 foundered.
**Lynx** Cleverley, Gravesend.
  Ord ___ 1793?; K 05.1793; L 14.2.1794; 1813 sold.
**Stork** Deptford Dyd.
  Ord ___ 1794?; K 12.1795; L 29.11.1796; 1816 sold.
**Hyacinth** Preston, Yarmouth. (later 24).
  Ord ___ 1805?; K 11.1805; L 30.8.1806; 1820 BU.
**Herald** Carver & Corney, Littlehampton. (later 24).
  Ord ___ 1805?; K 12.1805; L 27.12.1806; 1817 BU.
**Sabrina** Adams, Southampton. (later 24).
  Ord ___ 1805?; K 12.1805; L 1.9.1806; 1816 sold.
**Cherub** King, Dover. (later 26).
  Ord ___ 1805?; K 01.1806; L 27.12.1806; 1820 sold.
**Minstrel** Bools, Bridport. (later 24).
  Ord ___ 1805?; K 01.1806; L 25.3.1807; 1817 sold.
**Blossom** Guillaume, Northam. (later 24).
  Ord ___ 1805?; K 02.1806; L 10.12.1806; 1825 exploration ship; 1829 survey ship; 1833 hulked.
**Favourite** Bailey, Ipswich. (later 26).
  Ord ___ 1805?; K 02.1806; L 13.9.1806; 1821 BU.
**Sapphire** Brindley, Lynn. (later 26).
  Ord ___ 1805?; K 02.1806; L 11.11.1806; 1822 sold.
**Wanderer** Betts, Mistleythorn. (later 24).
  Ord ___ 1805?; K 02.1806; L 29.9.1806; 1817 sold.
**Partridge** Avery, Dartmouth. (later 26).
  Ord ___ 1806?; K 03.1806; L 15.7.1809; 1816 BU.
**Tweed** Iremonger, Littlehampton (later 24).
  Ord ___ 1805?; K 03.1806; L 10.1.1807; 1813 wrecked.
**Egeria** Bools & Good, Bridport. (later 26).
  Ord ___ 1805?; K 06.1806; L 31.10.1807; 1826 hulked.

## Modified Design?

  Dimensions & tons: 111'3", 90'9¾" x 29'8" x 9'1". 426?
**Ranger** Thorn, Fremington. (later 26).
  Ord ___ 1805?; K 08.1806; L 5.9.1807; 1814 BU.
**Dauntless** Deptford Dyd. (later 26).
  Ord ___ 1806?; K 11.1807; L 20.12.1808; 1825 sold.
**Jalouse** Plymouth Dyd. (later 26).
  Ord 15.1.1806; K 07.1808; L 13.7.1809; 1819 sold.
**Anacreon** Owen, Ringmore. (22).
  Ord ___ 1806?; K 07.1809; builder failed, transferred to: Plymouth Dyd.

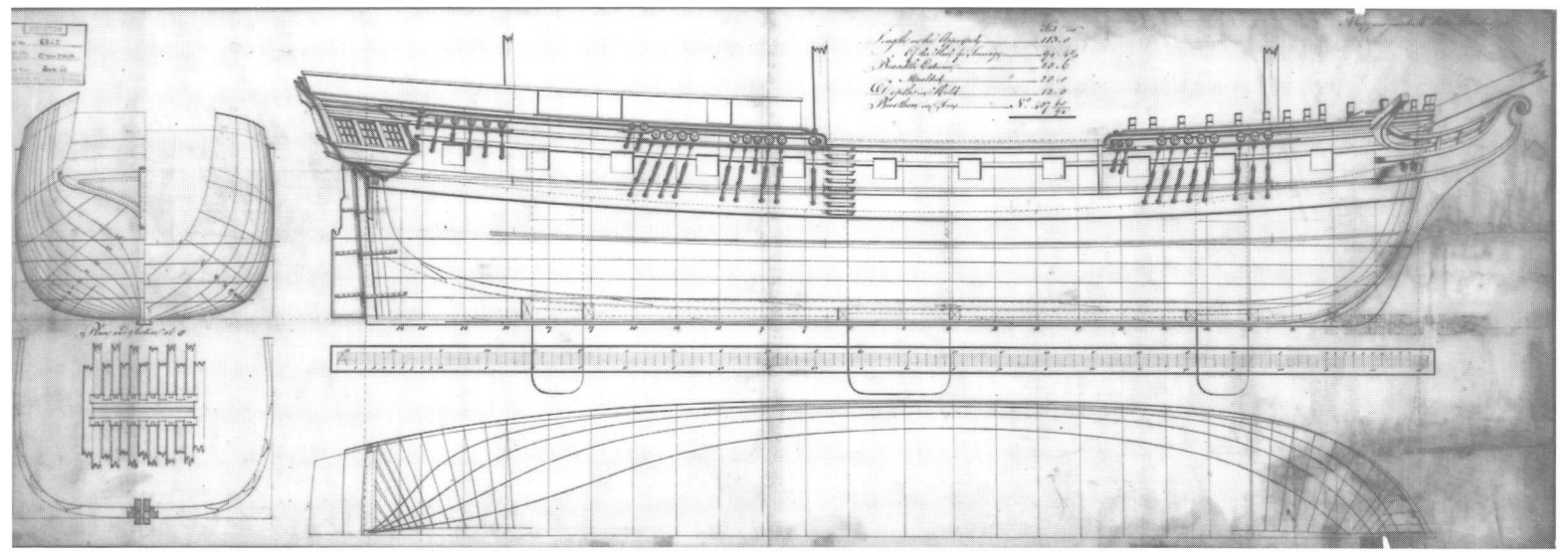

The distinctive feature of the sloop *Cynthia* of 1796, clearly shown in her design lines plan (6352), were her shallow draught and the Schanck 'sliding keels' (dagger boards) with which she was fitted to compensate for the missing keel area. This was, apart from the *Dart* and *Arrow* (see below), the biggest ship built with Schanck keels – though the design for a large frigate with these devices is shown by a model now in the collections of the NMM, Greenwich

    Ord ____ ; K 09.1811; L 1.5.1813; 1814 foundered.

**Cancelled 8.9.1810.** *Serpent* Sheerness Dyd.
    Ord ____ 1809?; not begun?

FATES: *Cormorant* accidentally blown up at Port Au Prince, Haiti 13.9.1796. *Favourite* taken off the Cape Verde Islands by a French squadron 6.1.1806; 1807 retaken. *Hornet* was hulked at Plymouth for the military medical staff; 1817 sold. *Lark* foundered off San Domingo 8.8.1809. *Blossom* became a lazaretto at Sheerness; 1848 BU. *Tweed* was wrecked in the Bay of Shoals, Newfoundland 5.12.1813. *Egeria* became a receiving ship at Plymouth; 1843 to the breakwater department; later police ship; 1865 BU. *Anacreon* foundered on her voyage from Lisbon 28.2.1814.

PLANS: Lines/profile/orlop/gun deck/upper deck/quarter deck & forecastle/framing /planking expansion/specification. Modified design: Lines/profile/orlop/gun deck/upper deck/quarter deck & forecastle/midships section/specification.

### *Cynthia* Class 1795. 18/16. Schanck sliding keel design. (3 daggerboards).
    Dimensions & tons: 113', 94'4⅜" x 28'6" x 12'. 407 65/94.
    Men: 121. Guns: 16 x 6 + 14 swivels.
***Cynthia*** Wells, Rotherhithe.
    Ord ____ 1795?; K 10.1795; L 23.2.1796; 1809 BU.

PLANS: Lines/gun deck/upper deck & quarter deck & forecastle/framing/masts/wells for keels/specification.

### *Brazen/Termagant* Class 1795. 18. Henslow design, enlarged version of the 1793 *Pylades* Class.
    Dimensions & tons: 110', 90'8⅜" x 29'6" x 8'6". 419 76/94.
    Men: 121. Guns: UD 18 x 6, QD 6 x 12★, Fc 2 x 12★
***Bittern*** Adams, Bucklers Hard.
    Ord 2.1.1795; K 06.1795; L 7.4.1796; 1812 tender; 1833 sold.
***Cyane*** Wilson, Frindsbury.
    Ord 2.1.1795; K 05.1795; L 9.4.1796; (1805 in French hands); 1805 renamed *Cerf*; 1809 sold.
***Plover*** Betts, Mistleythorn.
    Ord 2.1.1795; K 05.1795; L 23.4.1796; 1816 hulked.
***Termagant*** Dudman, Deptford.
    Ord 2.1.1795; K 05.1795; L 23.4.1796; 1819 sold.
***Brazen*** Portsmouth Dyd.
    Ord ____ K 15.6.1807; L 26.5.1808; 1827 hulked.

**Cancelled 1799.** *Brazen* Portsmouth Dyd.
    Ord 2.1.1795?; K ____ .

FATES: *Cyane* taken by the French *Hortense* and *Hermione* in the West Indies 12.5.1805; renamed *Cerf* in French service; 5.10.1805 retaken off Tobago and retained the French name. *Plover* to the Committee for Distressed Seamen at Deptford; 1818 returned to the RN; 1819 sold. *Brazen* (presumably a re-order of the vessel cancelled in 1799) became a floating church at Deptford; 1848 BU.

PLANS: Lines/profile/orlop/gun deck/upper deck/quarter deck & forecastle/framing/specification.

### *Merlin* Class 1795. 16. Rule design.
    Dimensions & tons: 106', 87'7" x 28' x 13'9". 365 32/94.
    Men: 121. Guns: UD 16 x 6, QD 4 x 12★, Fc 2 x 12★
***Merlin*** Dudman, Deptford.
    Ord 2.1.1795; K 06.1795; L 25.3.1796; 1803 BU.
***Pheasant*** Edwards, Shoreham.
    Ord ____ 1795?; K 10.1795; L 17.4.1798; 1824 hulked.
***Martin*** Tanner, Dartmouth.
    Ord 27.11.1802; K 09.1803; L 1.1.1805; 1806 foundered.
***Brisk*** Tanner, Dartmouth.
    Ord 27.11.1802; K 03.1804; L 04.1805; 1816 sold.
***Star*** Tanner, Dartmouth.
    Ord 27.11.1802; K 07.1802; L 26.7.1805; 1812 bomb vessel and renamed *Meteor*; 1816 sold.
***Cygnet*** Palmer, Yarmouth.
    Ord 27.11.1802; K 02.1803; L 6.9.1804; 1815 wrecked.
***Ariel*** Palmer, Yarmouth; completed by 'Government' after Palmer failed: 'Government', Yarmouth.
    Ord 27.11.1802; K 02.1805; L 19.4.1806; 1816 sold.
***Helena*** Preston, Yarmouth.
    Ord 27.11.1802; K 03.1803; L 26.4.1804; 1814 sold.
***Kangaroo*** Brindley, Lynn.
    Ord 27.11.1802; K 08.1803; L 12.9.1805; 1815 sold.
***Albacore*** Hillhouse, Bristol.
    Ord 27.11.1802; K 03.1803; L 10.5.1804; 1815 sold.
***Fly*** Parsons, Bursledon.
    Ord 27.11.1802; K 05.1803; L 26.3.1804; 1805 wrecked.
***Kingfisher*** King, Dover.
    Ord 27.11.1802; K 03.1803; L 10.3.1804; 1816 BU.
***Otter*** Atkinson, Hull.
    Ord 27.11.1802; K 07.1803; L 2.3.1805; 1814 hulked.
***Rose*** Hamilton & Breeds, Hastings.
    Ord 27.11.1802; K 10.1803; L 18.5.1805; 1817 sold.
***Halifax*** Halifax Dyd., Nova Scotia.
    Ord 27.11.1802; K ____ ; L 11.10.1806; 1814 BU.

**Fir Built Version.** *Wolf* Tanner, Dartmouth.
    Ord ____ 1802?; K 04.1803; L 4.8.1804; 1806 wrecked.

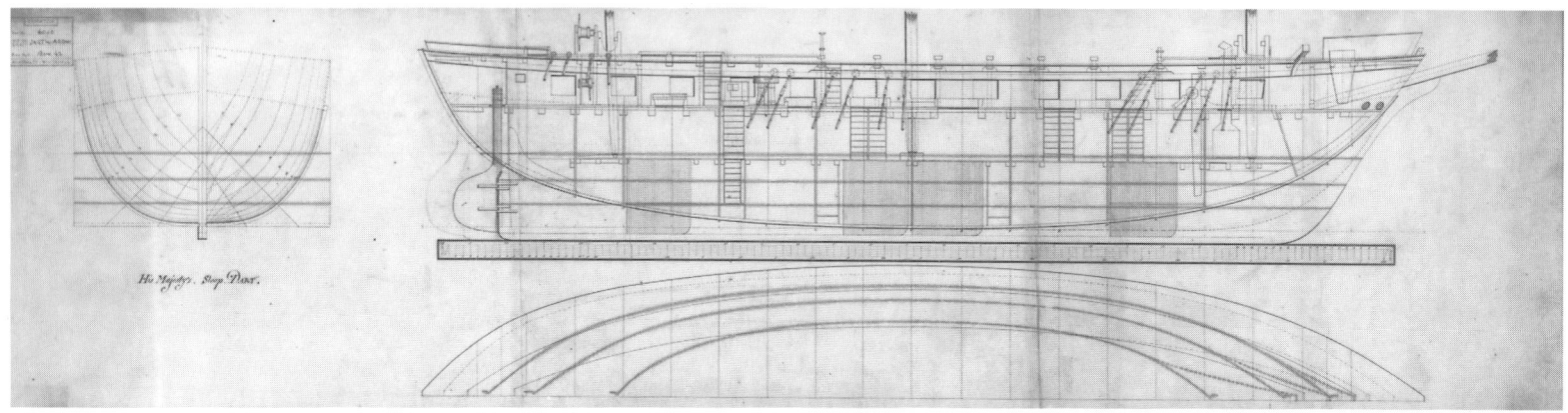

The totally unconventional shape, structure and fittings of the powerful *Dart* of 1796, classed for want of anything better as a 'sloop', shows up well in this lines and profile plan (6060)

FATES: *Pheasant* became a temporary receiving ship at Woolwich; 1827 sold. *Martin* foundered in the Atlantic during 1806. *Cygnet* wrecked off the River Courantyn (Guiana) 7.3.1815. *Fly* lost off the coast of Florida 18.3.1805. *Otter* became a lazaretto at Pembroke; 1828 sold. *Wolf* wrecked amongst the Bahamas 5.9.1806.

PLANS: Lines/profile/orlop/gun deck/upper deck/quarter deck & forecastle/framing/stern/specification. *Star* as bomb 1812: profile/orlop/upper deck/quarter deck & forecastle/midships section.

### Dart Class 1795. 20/28.
Design by Samuel Bentham, who also supervised the construction of these experimental unconventionally built vessels. The structure made use of bulkheads, they were almost double-ended, and were fitted with sliding keels (daggerboards). Nearly everything about these powerful vessels, from lines to armament, was unconventional; which presumably prevented them being classed as Sixth Rates despite their number of guns.

Dimensions & tons: 128'8", 80'8" x 33' (30') x 7'11". 386 116/94.
Men: 121 (later 140). Guns: UD 24 x 32★, QD 2 x 32★ (4 x 32★ for *Dart*, Fc 2 x 32★ - all these guns were Sadler's experimental 24 cwt guns). Later variations included only 18 x 32★ on the UD, nothing on the QD and Fc guns changed to 6 pounders.

**Dart** Hobbs, Redbridge.
Ord ____ 1795?; K ____ ; L ____ 1796; 1809 BU.
**Arrow** Hobbs, Redbridge.
Ord ____ 1795?; K ____ ; L ____ 1796; 1805 taken.

FATES: *Arrow* was taken in the Mediterranean by the French *Hortense* and *Incorruptible* and then sank 4.2.1805.
PLANS: Lines/profile/upper deck/quarter deck & forecastle/midships section. *Dart* as built: Lines & profile/profile & midships section/lower & upper decks.

### Nautilus Class 1796. 18.
Barrallier design.
Dimensions & tons: 112', 94'8¼" x 29'6" x 9'. 438 28/94.
Men: 121. Guns: UD 18 x 9, QD 6 x 12★, Fc 2 x 12★ (originally to have been QD 8 swivels, Fc 6 swivels).

**Nautilus** Jacobs, Milford Haven.
Ord 16.12.1796; K 04.1798 firm failed, transferred to 'Government' Milford Haven (the beginning of Pembroke Dyd). L 12.8.1804; 1807 wrecked.

FATE: Wrecked on Cerrigotto in the Eastern Mediterranean 4.1.1807.
PLANS: Lines/profile/orlop & gun deck/upper deck & quarter deck & forecastle/framing.

### Rosamond Class 1805. 16 (later 26/24 Sixth Rates).
Similar to the fireship *Pluto* of 1782, and therefore based on the French prize *Amazon* of 1745. There is some confusion over just how close in design these vessels were to the *Cormorant* Class of 1793, which had the same origins.

Dimensions & tons: 108'4", 90'9⅝" x 29'7" x 9'. 423.
Men: 121 Guns: UD 16 x 32★, QD 6 x 18★, Fc 2 x 6 + 2 x 18★.

**Rosamond** Temple, South Shields.
Ord ____ 1805?; K 12.1805; L 27.1.1807; 1815 sold.
**Acorn** Crocker, Biddeford. (later 26).
Ord ____ 1805?; K 03.1806; L 30.10.1807; 1819 BU.
**Fawn** Owen, Topsham. (later 26).
Ord ____ 1805?; K 03.1806; L 22.4.1807; 1818 sold.
**Myrtle** Chapman, Biddeford. (later 24).
Ord ____ 1805?; K 06.1806; L 2.10.1807; 1818 BU.
**Raccoon** Preston, Yarmouth. (later 26).
Ord ____ 1805?; K 03.1806; L 30.3.1808; 1819 hulked.
**Hesper** Tanner, Dartmouth. (later 26).
Ord ____ 1805?; K 03.1806; L 3.7.1809; 1817 sold.
**North Star** Tanner, Dartmouth. (later 26).
Ord ____ 1805?; K 05.1806; L 21.4.1810; 1817 sold.

FATES: *Racoon* became a convict hospital ship at Portsmouth; 1838 sold.
PLANS: orlop/gun deck/specification; ? other plans as *Cormorant* Class?

### Coquette Class 1805. 18 (later 28 Gun Sixth Rate).
'Lines of the old French *Amazon* [1745]'; enlarged version of the *Cormorant* Class of 1793. Originally intended to carry 18 x 6 pounders, but altered as below.

Dimensions & tons: 113'3", 94'2" x 31' x 9'4". 481 33/94.
Men: 121. Guns: UD 16 x 32★, QD 6 x 18★, Fc 2 x 6 + 2 x 18★

**Coquette** (ex *Queen Mab*) Temple, North Shields?
Ord 4.10.1805; K 02.1806; L 25.4.1807; 1817 sold.
**Talbot** Heath, East Teignmouth.
Ord 4.10.1805; K 03.1806; L 22.7.1807; 1815 sold.

PLANS: Lines/framing/stern/specification.

### Valorous, Ariadne.
See under the *Hermes* Class of 1810 in the flush-decked sloops category.

### Conway Class 1813. 26/28 sloops/Sixth Rates.
Rule design.
Dimensions & tons: 108', 89'9⅝" x 30'6" x 9'. 444 33/94.
Men: 150. Guns: UD 18 x 32★, QD 6 x 12★. Fc 2 x 6 + 2 x 12★ (in 1817 the carronades were removed from the Fc).

**Conway** Pelham, Frindsbury.
Ord 18.1.1813; K 05.1813; L 10.3.1814; 1825 sold.
**Dee** Bailey, Ipswich.
Ord 18.1.1813; K 05.1813; L 5.5.1814; 1819 sold.
**Eden** Courtney, Chester.
Ord 18.1.1813; K 03.1813; L 19.5.1814; 1833 BU.
**Menai** Brindley, Frindsbury.
Ord 18.1.1813; K 06.1813; L 5.4.1814; 1829 hulked.
**Mersey** Courtney, Chester.
Ord 18.1.1813; K 03.1813; L 03.1814; 1831 hulked.
**Tamar** Brindley, Frindsbury.
Ord 18.1.1813; K 05.1813; L 23.3.1814; 1831 hulked.
**Tees** Taylor, Biddeford.
Ord 18.1.1813; K 10.1813; L 17.5.1817; 1826 hulked.
**Towey** Adams, Bucklers Hard.
Ord 18.1.1813; K 05.1813; L 6.5.1814; 1822 BU.
**Tyne** Davy, Topsham.
Ord 18.1.1813; K 08.1813; L 20.5.1814; 1825 sold.
**Wye** Hobbs, Redbridge.

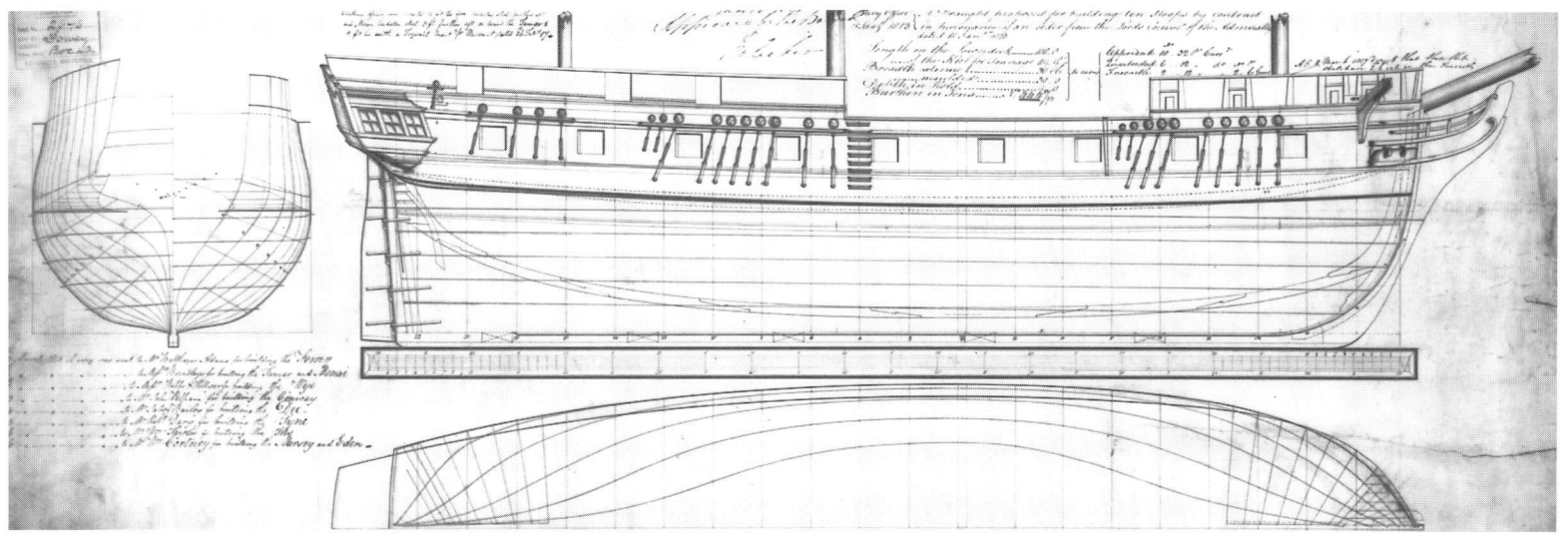

Despite being noted on the Admiralty stamp as being for the 'Fowey' (which should actually be Towey) this lines plan (2676) is actually the design plan to which all the *Conway* Class ship sloops/Sixth Rates were built

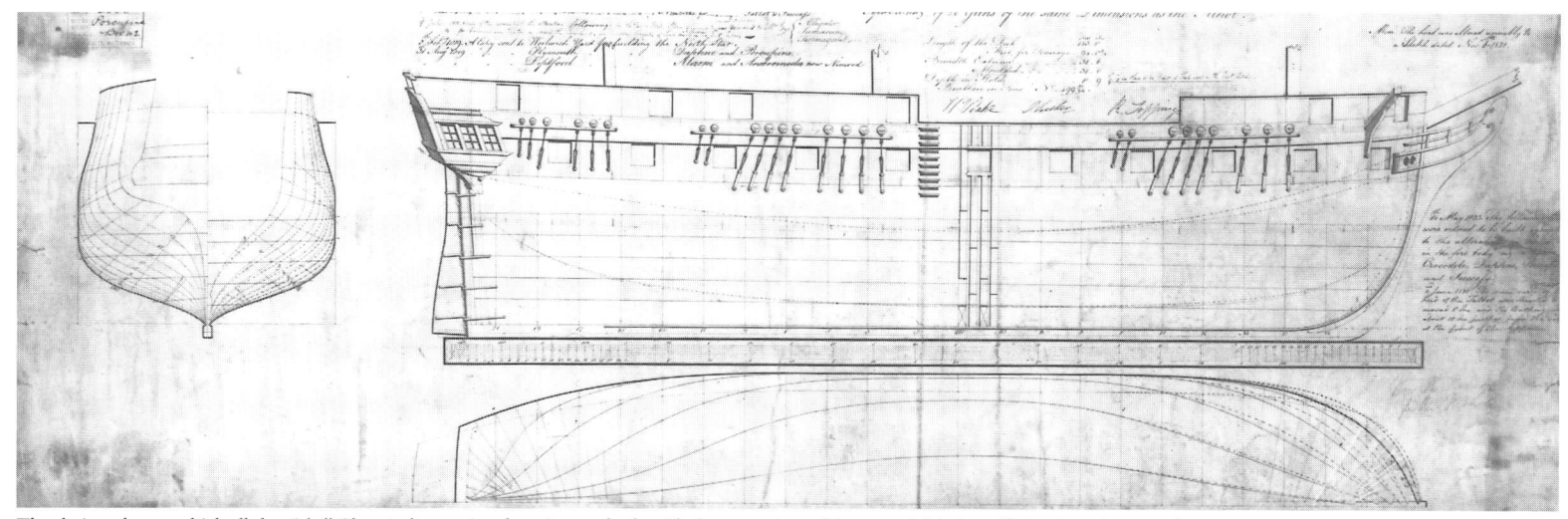

The design plan to which all the *Atholl* Class sixth rates (ex sloops) were built with the exception of the named ship herself (the *Porcupine* named on the stamp being only one of them). Lines plan number 2635

Ord 18.1.1813; K 09.1813; L 17.8.1814; 1834 hulked.

FATES: *Menai* became receiving ship at Portsmouth; 1852 target ship; 1853 BU. *Mersey* became receiving ship at Portsmouth; 1852 BU. *Tamar* became a coal depot at Plymouth; 1837 sold. *Tees* to the Church Society at Liverpool as a church ship; 1872 sold. *Wye* became a convict hospital ship and floating breakwater at Sheerness; by 1850 convict hospital ship at Chatham; 1852 BU.

PLANS: Lines/profile/orlop/gun deck/upper deck/quarter deck & forecastle/framing/stern/midships section/specification.

### *Atholl* Class 1816. 28 Sloop/Sixth Rate.

Design by the Surveyors of the Navy. Several were built of exotic and unusual types of timber; this is indicated after the name of the builder.

Dimensions & tons: 113'8", 94'8¾" x 31'6" x 8'9". 499 91/94.
Men: 175. Guns: UD 20 x 32★, QD 6 x 18★, Fc 2 x 9.

**Atholl** Woolwich Dyd. (larch).
Ord 27.11.1816; K 11.1818; L 23.11.1820; 1832 troopship 1851 hulked.
**Ranger** Portsmouth Dyd.
Ord ____ 1819??; K 01.1819; L 7.12.1820; 1832 sold.
**Niemen** Woolwich Dyd. (Baltic fir).
Ord ____ 1819?; K 07.1819; L 23.11.1820; 1828 BU.
**Rattlesnake** Chatham Dyd.
Ord ____ 1818?; K 08.1819; L 26.3.1822; 1839 troopship; 1846 survey vessel; 1860 BU.
**Alligator** Cochin (East India Co). (teak).
Ord ____ 1819?; K 11.1819; L 29.3.1821; 1854 hulked.
**Herald** (ex *Termagant* renamed 1824?) Cochin (East India Co). (teak).
Ord ____ ; K 03.1820; L 15.11.1822; 1845 survey vessel; 1861 hulked.
**North Star** Woolwich Dyd.

Ord ____ 1818?; K 04.1820; L 7.2.1824; 1849 storeship for the Arctic; 1860 BU.
**Tweed** Portsmouth Dyd.
Ord ____ 1818?; K 12.1820; L 14.4.1823; 1831 cut down to a 20 gun corvette; 1852 sold.
**Talbot** Pembroke Dyd.
Ord ____ 1818?; K 03.1821; L 9.10.1824; 1854 storeship; 1855 hulked.
**Samarang** Cochin (East India Co) (teak).
Ord ____ 1819?; K 03.1821; L 1.1.1822; 1847 hulked.
**Rainbow** Chatham Dyd.
Ord ____ 1818?; K 04.1822; L 20.11.1823; 1838 sold.
**Success** Pembroke Dyd.
Ord ____ 1819?; K 07.1823; L 30.8.1825; 1833 hulked.
**Crocodile** Chatham Dyd.
Ord ____ 1819?; K 12.1823; L 28.10.1825; 1842 troopship; 1850 hulked.
**Nimrod** (ex *Andromeda*) Deptford Dyd. (African Timber).
Ord ____ (1819? as 28?) 12.1.1821; K 10.1821; L 26.8.1828; had been razeed whilst building and completed as 20 gun sloop or corvette; 1853 hulked.

### Cancelled 1820? (or 1826?). *Alarm*? Deptford Dyd?
Ord ____ 1819?; K ____ 1819?

### Cancelled 1.11.1832. *Daphne* Plymouth Dyd.
Ord ____ 1820; K ____ .
**Porcupine** Plymouth Dyd.
Ord 1.6.1819; K ____ .

FATES: *Atholl* became a storeship at Sheerness; 1854 to Greenock as a depot ship

NMM Photo Negative x.1448 *Alligator* (1821, *Atholl* Class) with the line of battle ship *Canopus* (French prize) in the background: painting by J C Joy

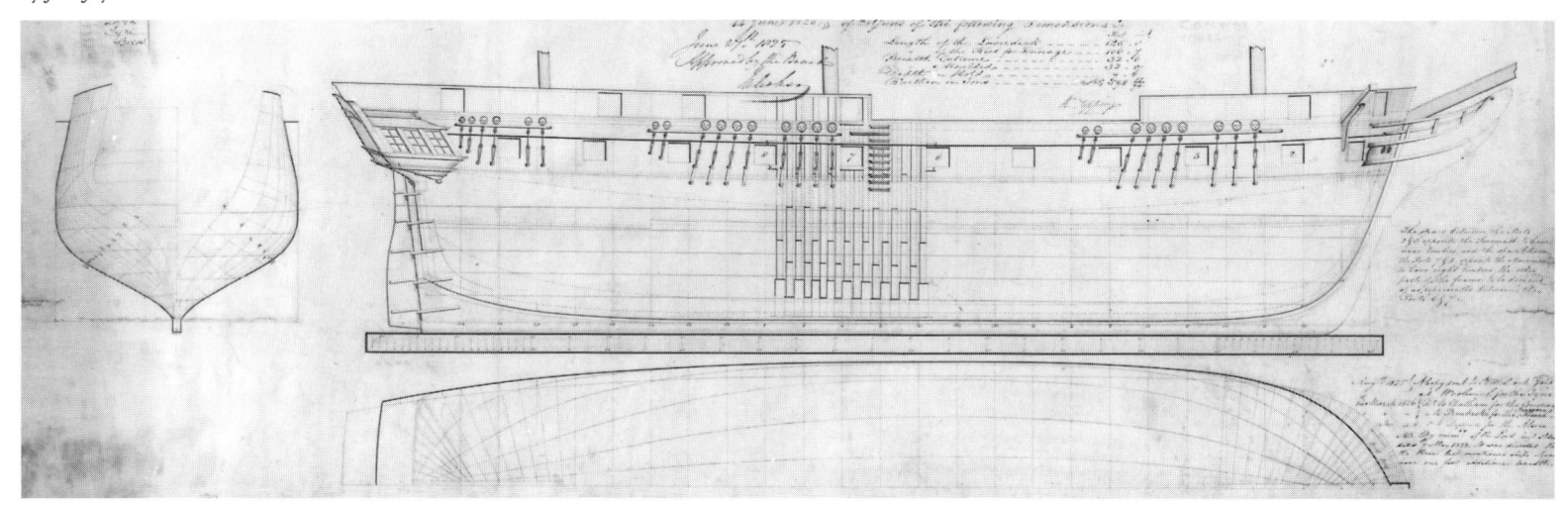

Design lines plan (2492) for the *Tyne* Class 28 gun ships of 1825

and 'rendezvous'; 1861 to Plymouth in reserve; 1863 BU. *Alligator* became a seamens hospital at Hong Kong; 1854 allocated to the use of the Canton Consulate; 1865 sold. *Herald* became a chapel ship at Shoreham; 1862 sold. *Talbot* became an ordnance depot at Woolwich; 1896 sold. *Samarang* became a guard ship at Gibraltar; 1883 BU. *Success* became a receiving ship at Portsmouth; 1849 BU. *Crocodile* became the Tower of London hulk; later to Deptford; before 1861 to Dover; 1861 sold. *Nimrod* became a coal depot (and receiving ship?) at Plymouth; later yard craft C.1; later C.76?; 1907 sold.

Plans: Lines/profile/orlop/gun deck/upper deck/quarter deck & forecastle/framing/midships section/specification. *Atholl* as troopship 1832-1845: profile/orlop/gun deck/upper deck/quarter deck & forecastle. As storeship 1851: profile/hold & hold sections/orlop/gun deck/upper deck/quarter deck and forecastle. *Rattlesnake* as troopship 1839: gun deck/upper deck/quarter deck & forecastle. As survey ship 1846: Profile/gun deck/upper deck/quarter deck & forecastle. *Crocodile* as troopship: profile/gun deck/upper deck/quarter deck & forecastle.

### *Volage* Class 1818. 28.
Sixth Rate. Designed and built by the 'Superior class of Portsmouth shipwright apprentices'.
 Dimensions & tons: 113'10", 96'8½" x 31'10" x 8'9". 521 5/100.
 Men: 175. Guns: UD 20 x 32★, QD 6 x 18★, Fc 2 x 9.
***Volage*** Portsmouth Dyd.

Ord 29.10.1818; K 08.1819; L 20.2.1825; 1847 survey ship; 1855 hulked.
FATE: To the War Department at Upnor (opposite Chatham on the River Medway) as a powder depot; 1874 sold.
PLANS: Lines/profile, hold & hold sections, gun deck, upper deck.

### *Tyne* Class 1825. 28.
Sixth Rate. Seppings design, which was increased by one foot in breadth over that originally proposed.
 Dimensions & tons: 125', 106' x 33'6" x 9'6". 632 71/94.
 Men: 175. Guns: UD 20 x 32★, QD 6 x 18★, Fc 2 x 9.
***Tyne*** Woolwich Dyd.
 Ord ____ 1825?; K 11.1825; L 30.11.1826; 1847 hulked.
***Imogene*** (ex *Pearl*) Pembroke Dyd.
 Ord ____ 1825?; K 11.1829; L 24.6.1831; 1840 burnt.
***Conway*** Chatham Dyd.
 Ord ____ 1825?; K 12.1829; L 2.2.1832; 1859 hulked.

### Cancelled 1832. *Alarm* Deptford Dyd.
 Ord ____ 1826?; K 11.1826; then to:
Pembroke Dyd.
 K 01.1832; completed except for her stern timbers when she began to be taken to pieces in 10.1832.

FATES: *Tyne* became a provision depot at Plymouth; 1850 store ship; 1862 sold.

*Imogene* accidentally burnt at Plymouth 27.9.1840. *Conway* lent as a training ship to the Mercantile Marine Association of Liverpool; 1861 to the Aberdeen RNR, renamed *Worcester;* 1871 BU.

PLANS: Lines/profile/hold & hold sections/orlop & gun deck/upper deck/quarter deck & forecastle/midships section/keels /expansion of coppering/stern/specification. *Tyne* as built: Profile/orlop & gun deck/upper deck/quarter deck & forecastle. 1827: hold & hold sections. 1828: hold & hold sections.

## *Sapphire* Class 1825. 28. Sixth Rate. Designed and built by the superior class of shipwright apprentices at Portsmouth.

Dimensions & tons: 120', 100'6¾" x 33'10" x 8'. 604 30/94.
Men: 175. Guns: UD 20 x 32★, QD 6 x 18★, Fc 2 x 9.

*Sapphire* Portsmouth Dyd.
Ord ____ 1825?; K 11.1825; L 30.1.1827; 1839 troopship; 1847 hulked.

FATE: Became a receiving ship at Trincomalee; 1864 sold.

PLANS: As troopship 1839: Lines/profile/orlop & gun deck/upper deck/quarter deck & forecastle/midships section.

## *Challenger* Class 1825. 28. Sixth Rate. Design by Captain Hayes.

Dimensions & tons: 125'7½", 105'11½" x 32'8½" x 9'3¼". 603.
Men: 175. Guns: UD 20 x 32★, QD 6 x 18★, Fc 2 x 9.

*Challenger* Portsmouth Dyd.
Ord ____ 1825?; K 11.1825; L 14.11.1826; 1835 wrecked.

FATE: Wrecked on the coast of Chile 19.5.1835.

PLANS: No plans.

## *Acteon* Class 1828. 26. Sixth Rate. School of Naval Architecture design.

Dimensions & tons: 121'6", 100'4" x 34' x 9'7". 620.
Men: 175. Guns: (design) UD 20 x 32 ('Gunnades'), QD 4 x 32★, Fc 2 x 9 or 2 x 32★.

*Acteon* Portsmouth Dyd.
Ord ____ 1828?; K 09.1828; L 31.1.1831; completed as an 18 gun survey ship; 1866 hulked.

FATE: Became a hospital ship at Portsmouth; 1874 used for torpedo experiments and attached to HMS *Vernon;* 1889 sold.

PLANS: Lines/profile/hold/orlop/gun deck/upper deck/quarter deck & forecastle/framing.

## *Calliope* Class 1831. 28. Sixth Rate. Seppings design.

Dimensions & tons: 130', 108'10" x 35'5" x 10'6". 709 16/94.
Men: 175. Guns: UD 20 x 32, QD 6 x 32, Fc 2 x 32 (all Dixon's 25 cwt. 32 pounders).

*Calliope* Sheerness Dyd.
Ord ____ 1830?; K 01.1831; L 5.10.1837; 1855 hulked.

*Andromache* Pembroke Dyd.
Ord ____ 1830?; K 08.1831; L 27.8.1832; 1846 hulked.

**Cancelled 1831.** *Vestal* Chatham Dyd.
Ord 29.10.1830; not begun?

FATES: *Calliope* became a chapel ship at Plymouth; later 'floating factory' (workshop); 1883 BU. *Andromache* became a provision depot at Plymouth; 1854 powder depot; later at Pembroke but then returned to Plymouth; 1875 BU.

PLANS: Lines/profile/gun deck/upper deck/quarter deck & forecastle/framing/midships section/specification.

# Flush Decked Ship Sloops (and later Sixth Rates and Corvettes)

## *Snake* Class 1796. 18. Rule design. The ship-rigged version of the *Cruiser* Class brigs, identical except for their rig. Several of the brigs were converted to ships; and often converted back again (see under the *Cruiser* Class brigs) just as some of this class were converted to brigs. The 1825 *Favourite* Class was a lengthened version of this design.

Dimensions & tons: 100', 77'3½" x 30'6" x 12'9", 382 41/94.
Men: 121. Guns: 2 x 6 + 16 x 32★.

*Snake* Adams, Bucklers Hard.
Ord ____ 1797??; K 01.1797; L 18.12.1797; 1809 converted to a brig; 1816 sold.

*Victor* Brindley, Lynn.
Ord ____ 1797?; K 04.1797; L 19.3.1798; 1808 sold.

*Childers* Chatham Dyd.
Ord 10.6.1823; K 11.1825; L. 23.8.1827; 1834 converted to a brig; 1865 sold.

*Cruiser* Chatham Dyd.
Ord 10.6.1823; K 01.1826; L 19.1.1828; by 1840s a brig; 1849 sold.

PLANS: Lines/profile/decks/framing/midships section. *Cruiser* as brig 1844: Hold & hold sections.

## *Echo* Class 1796. 18. Henslow design, same lines and dimensions as the brig *Busy*.

Dimensions & tons: 96', 75'1¾" x 29'6" x 12'9". 355 91/94.
Men: 121. Guns: 16 x 32★ + 2 x 6 chase guns.

*Echo* King, Dover.
Ord 19. & 26.12.1796; K ____ ; L 09.1797; 1809 sold.

PLANS: Lines/profile & decks/framing/specification.

## *Osprey* Class 1797. 18. Rule design, originally intended to be a brig.

Dimensions & tons: 102', 80'5⅝" x 29'11" x 12'9". 383 7/94.
Men: 121. Guns: 2 x 6 + 16 x 32★.

*Osprey* Pitcher, Northfleet.
Ord 15.3.1797; K 05.1797; L 7.10.1797; 1813 BU.

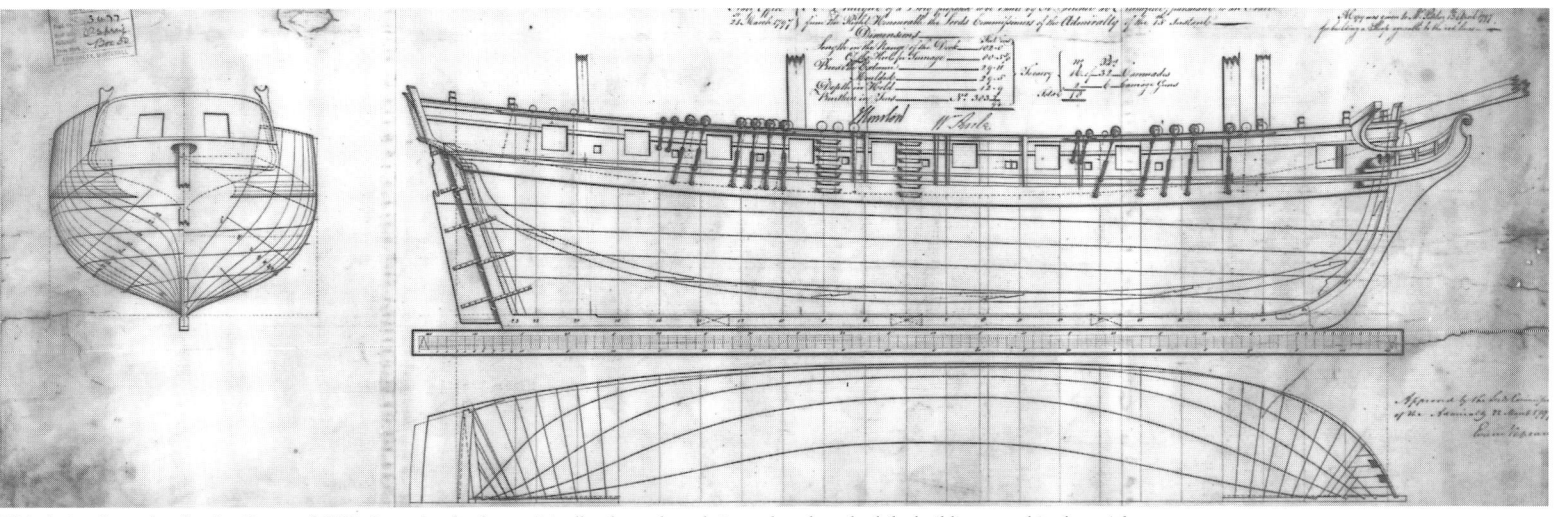

This design lines plan for the *Osprey* of 1797 shows her both as originally planned as a brig, and as altered whilst building to a ship sloop (plan number 3477)

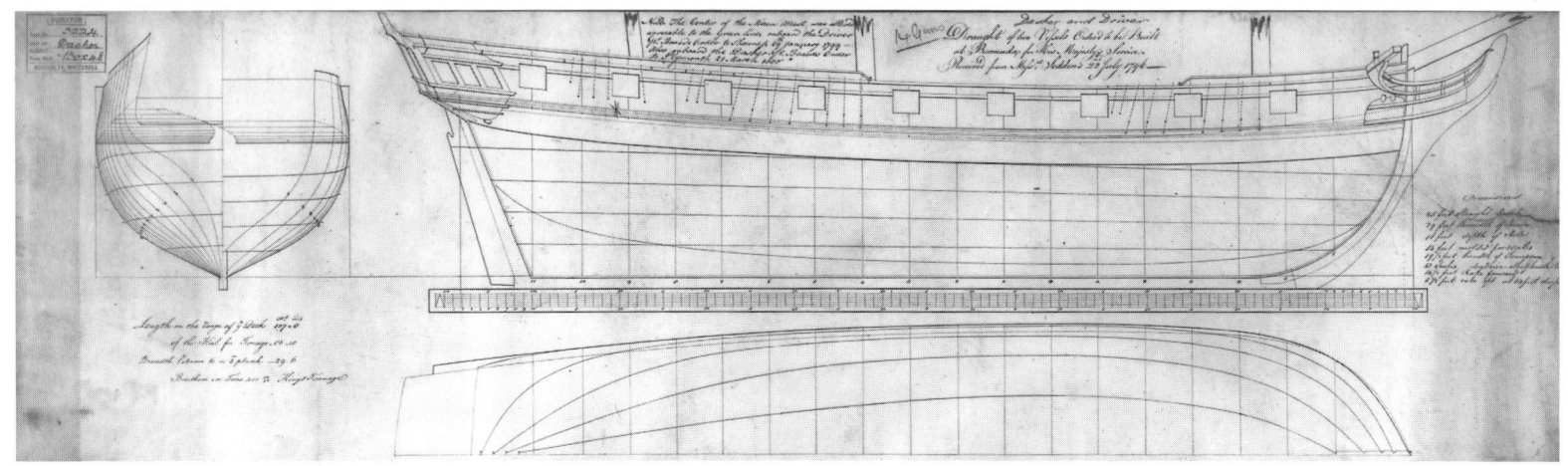

The lines plan (number 3224) of the *Dasher* Class Bermudan-built ship sloops of 1797 clearly shows their origin as brigs (they were converted whilst building)

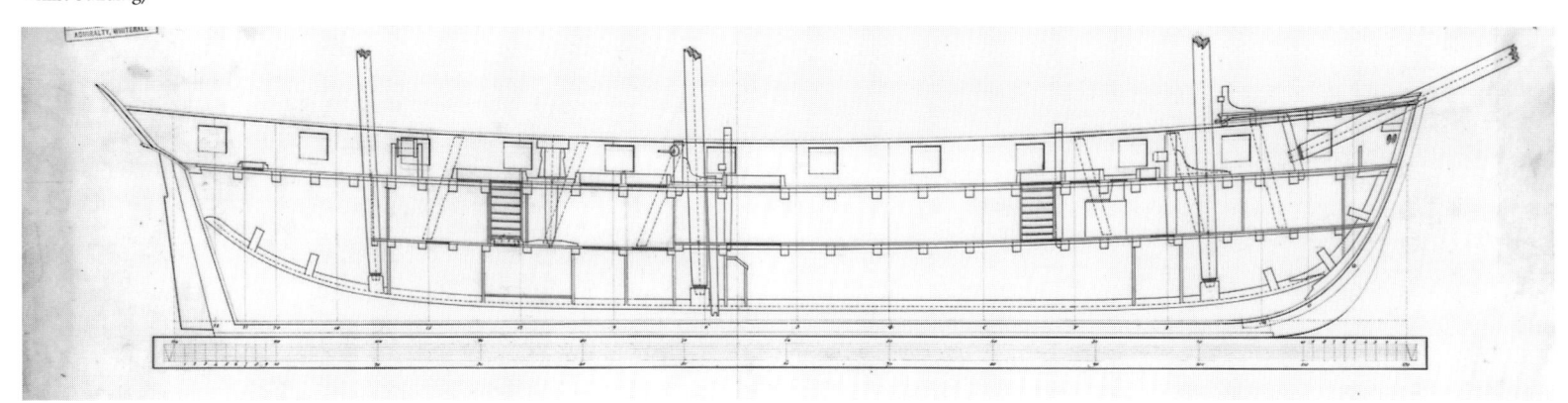

The profile (number 2968) for building the *Combatant* Class ship sloops of 1804 shows their shallow draught

PLANS : Lines.

### *Dasher* Class 1797 (1795?). 16.
Originally designed as, and began building as, brigs, but completed as ship sloops. Possibly designed by Goodrick (or Goodrich), who certainly acted as the contractor for these vessels (which explains why some sources list the builder as 'Goodrich & Co'). The later *Bermuda* Class was based on this design.

Dimensions & tons: 107´. 86'10" x 29'6" x ____ . 401 83/94.
Men: 80 Guns: 16 x 24★.

***Dasher*** Outerbridge & McCallan, Bermuda (cedar built).
Ord ____ 1795?; K ____ ; L ____ 1797; 1818 hulked.

***Driver*** Tynes, Bermuda (cedar built).
Ord ____ 1795?; K ____ ; L ____ 1797; 1824 hulked.

FATE: *Dasher* lent to the Committee for Distressed Seamen; 1820 to be fitted for the Army in the West Indies; but not sent; 1822 laid up at Deptford; 1826 convict ship at Deptford; 1826 BU. *Driver* became coal depot at Deptford; 1834 BU.

PLANS: Lines. *Driver* as built: lower & upper decks.

### *Bermuda* Class 1803. 18 (later Vessels 16).
A modification of the *Dasher* Class of 1797. Will have been built of Bermudan cedar.

Dimensions & tons: 107´, 83'10⅝" x 29'11" x 14'8". 399 31/94.
Men: 121. Guns: UD 16 x 24★ ( the 1803 vessels also carried 2 x 3 on the Fc).

***Bermuda*** Shedden, Bermuda.
Ord 23.6.1803; K ____ ; L ____ 1806; 1808 wrecked.

***Indian*** Shedden?, Bermuda.
Ord 23.6.1803; K ____ ; L ____ 1805; 1817 sold.

***Atalante*** ____ , Bermuda.
Ord 26.4.1806; K ____ ; L 08.1808; 1813 wrecked.

***Martin*** ____ , Bermuda.
Ord 26.4.1806; K ____ 1806; L 05.1809; 1817 wrecked.

***Morgiana*** Hill, Bermuda.
Ord ____ 1809; K ____ 1811; L 12.1811; 1825 sold.

***Sylph*** Tynes, Bermuda.
Ord ____ 1809; K ____ 1811; L ____ 1812; 1815 wrecked.

FATES: *Bermuda* wrecked on the Bahamas 22.4.1808. *Atalante* wrecked on Sister Rocks, Sambro Island off Halifax 10.11.1813. *Martin* wrecked on the West coast of Ireland 8.12.1817. *Sylph* wrecked on Southampton Bar, Long Island 17.1.1815.

PLANS: Lines/profile/decks/midships section.

### *Combatant* Class 1804. 20/22 Sloops.
Designed by John Stainforth, Member of Parliament. Originally intended to be completely flush-decked, but a small forecastle was added whilst building. The first proposal was that they should carry 24 guns (with alternative outfits of 6 pounders or of carronades) but to be registered as 20 gun ships. They were to draw 11' with all stores on board. Some sources give their original classification as gun vessels, but they were normally classed as sloops.

Dimensions & tons: 120'. 99'6" x 28' x 11'3". 416 40/94.
Men: 121. Guns: UD 20 or 22 x 24★.

***Combatant*** Blunt, Thorn. 22.
Ord 05.1804; K ____ ; L 31.11.1804; 1808 fitted for the defence of Gibraltar Bay; 1808 hulked.

***Dauntless*** Blunt, Hull. 20.
Ord 05.1804; K ____ ; L 11.1804; 1807 taken.

***Valorous*** Blunt, Hull. 22.
Ord 05.1804; K ____ ; L 11.1804; 1814 hulked.

FATES: *Combatant* became a receiving ship at Sheerness (never having been sent out to Gibraltar?); 1816 BU. *Dauntless* was taken by the French Army 'on the Holm' at the surrender of Danzig (Gdansk) 19.5.1807. *Valorous* became an army depot at Cowes, Isle of Wight; 1817 sold.

PLANS: Lines/profile/lower deck/upper deck/sails.

### *Hermes* Class 1810. 20 sloops/Sixth Rates.
'Similar to *Bonne Citoyenne*' (French prize taken 1796). The last two of the class were later given quarter decks and forecastles as 'post ships'. The *Hermes* design was used in a slightly reduced version for the *Cyrus* Class of 1812.

Dimensions & tons: 119'10". 99'10⅝" x 30'11" x 8'7". 507 76/94.
Men: 135. Guns: 2 x 9 + 18 x 32★.

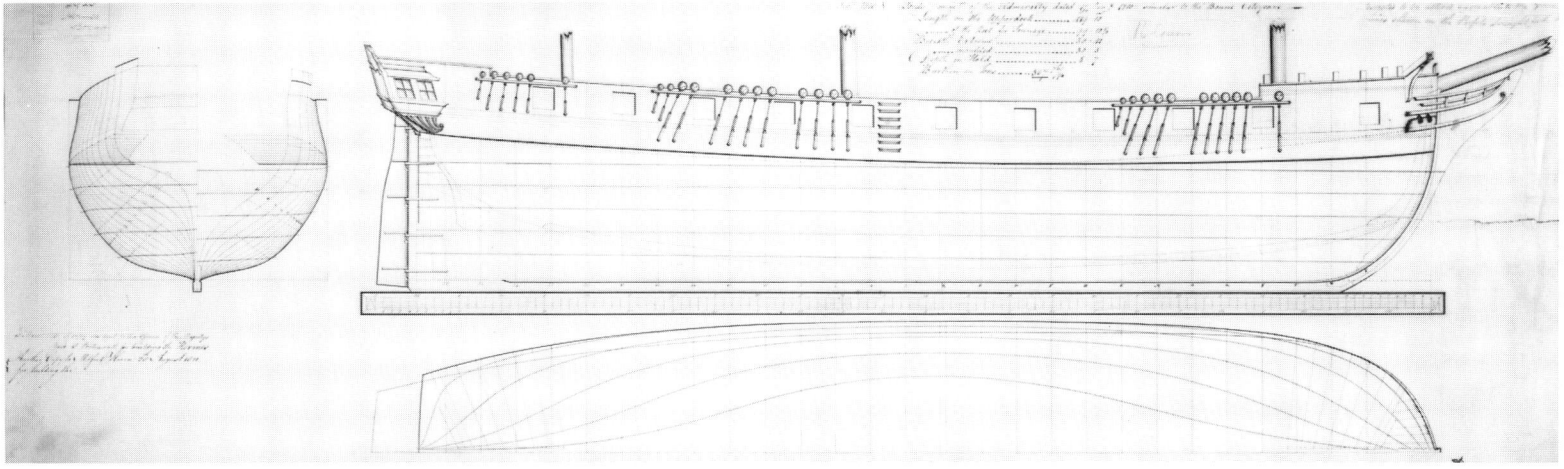

The design lines plan of the *Hermes* Class ship sloops of 1810 clearly shows French influence in both shape and general appearance – not surprising since the class was based on the prize *Bonne Citoyenne*. Plan number 2926

*Hermes* Portsmouth Dyd.
  Ord ____ 1810?; K 05.1810; L 22.7.1811; 1814 wrecked.
*Myrmidon* Pater (Pembroke Dyd.).
  Ord ____ 1811?; K 07.1812; L 18.6.1813; 1823 BU.
*Valorous* Pater (Pembroke Dyd).
  Ord ____ 1813?; K 03.1815; L 10.2.1816; 1821 post ship (26 guns); 1829 BU.
*Ariadne* Pater (Pembroke Dyd).
  Ord ____ 1813?; K ____ ; L 10.2.1816; 1820 post ship (26/28 guns); 1837 hulked.
FATES: *Hermes* grounded and burnt taking a fort near Mobile, Alabama 15.9.1814; it is possible that this wreck has been located by divers. *Ariadne* became a coal depot at Alexandria; 1841 sold.
PLANS: Lines/profile/lower deck/upper deck/framing/midships section/stern/specification. *Hermes* as built: Profile/gun deck/upper deck. *Ariadne* as coal depot 1837: profile & decks.

## *Cyrus* Class 1812. 20 Sloops/Sixth Rates.
'From the lines of the *Myrmidon* sloop' in other words a further development, by Rule, of the *Hermes* Class design, but with slightly reduced dimensions. Therefore a development of the French *Bonne Citoyenne* of 1796. There are some indications that the two vessels built by Adams (*Medina* and *Carron*) were to a slightly altered design.
  Dimensions & tons: 115'6". 97'2" x 29'8" x 8'6". 454 80/94.
  Men: 135. Guns: 2 x 6 + 20 x 32★.
*Cyrus* Courtney, Chester.
  Ord 18.11.1812; K 01.1813; L 26.8.1813; 1823 sold.
*Levant* Courtney, Chester.
  Ord 18.11.1812; K 01.1813; L 8.12.1813; (1815 briefly in American hands) 1820 BU.
*Medina* Adams, Bucklers Hard.
  Ord 18.11.1812; K 01.1813; L 13.8.1813; 1832 sold.
*Carron* Adams, Bucklers Hard.
  Ord 18.11.1812; K 03.1813; L 9.11.1813; 1820 lost.
*Cyrene* Chapman, Biddeford.
  Ord 18.11.1812; K 04.1813; L 4.6.1814; 1828 sold.
*Falmouth* Chapman or Taylor?, Biddeford.
  Ord 18.11.1812; K 04.1813; L 8.1.1814; 1825 sold.
*Hind* (ex *Barbadoes*) Davy, Topsham.
  Ord 18.11.1812; K 05.1813; L 8.3.1814; 1829 sold.
*Esk* Bailey, Ipswich.
  Ord 18.11.1812; K 03.1813; L 11.10.1813; 1827 sold.
*Leven* Bailey, Ipswich.
  Ord 18.11.1812; K 03.1813; L 23.12.1813; 1833 hulked.
*Erne* Newman, Dartmouth.
  Ord 18.11.1812; K 03.1813; L 10.12.1813; 1819 lost.
*Slaney* Brindley, Frindsbury.
  Ord 18.11.1812; K 04.1813; L 9.12.1813; Ca 1830 hulked.
*Lee* Brindley, Frindsbury.
  Ord 18.11.1812; K 05.1813; L 24.1.1814; 1822 BU.
*Spey* Warwick, Eling (Southampton).
  Ord 18.11.1812; K 05.1813; L 8.1.1814; 1822 sold.
*Larne* Bottomley, Lynn.
  Ord 18.11.1812; K 07.1813; L 8.3.1814; 1820 sold.
*Tay* Adams, Bucklers Hard.
  Ord 18.11.1812; K 04.1813; L 26.11.1813; 1816 wrecked.
*Bann* King, Dover.
  Ord 18.11.1812; K 05.1813; L 8.6.1814; 1829 sold.
FATES: *Levant* taken by the USS *Constitution* in the Atlantic 20.2.1815; recaptured shortly afterwards. *Carron* wrecked near Puri, India 6.7.1820. *Leven* became a convict and receiving ship at Woolwich; 1841 to Limehouse as a hulk; 1848 BU. *Slaney* laid up at Bermuda because defective; 1838 BU. *Tay* was wrecked on the Alacreanes Islands, Gulf of Mexico 11.11.1816.
PLANS: Lines/profile/lower deck/upper deck/midships section/framing/specification. *Cyrus* in 1821: cabins.

## *Rose* Class 1819. 18 Sloop.
Designed and built by the superior class of shipwright apprentices at Portsmouth.
  Dimensions & tons: 104'2½". 85'10" x 29'6" x 13'4". 397 28/94.
  Men: 125. Guns: 2 x 6 + 16 x 32★.
*Rose* Portsmouth Dyd.
  Ord 27.2.1819; K 3.4.1820; L 1.6.1821; 1851 BU.
PLANS: Lines/profile/orlop & lower deck/upper deck/midships section. As built: profile/orlop & lower deck/upper deck.

## *Martin* Class 1820 (or 1818?) 20, Sloop.
Surveyors of the Navy design.
  Dimensions & tons: 108'5", 90'7¼" x 28'10½" x 7'9". 401.
  Men: 125. Guns: 2 x 6 + 18 x 32★.
*Martin* Portsmouth Dyd.
  Ord ____ 1818?; K 07.1820; L 18.5.1821; 1826 foundered.
FATE: Supposed foundered off the Cape about 02.1826.
PLANS: Lines/profile/orlop & lower deck/upper deck/midships section.

## *Pylades* Class 1822. Sloop/Corvette 18.
Seppings design.
  Dimensions & tons: 110', 90'1⅜" x 30' x 8'2". 431 37/94.
  Men: 125. Guns: 2 x 9 + 16 x 32★.
*Pylades* Woolwich Dyd.
  Ord 5.6.1822; K 03.1823; L 29.6.1824; 1845 BU.
PLANS: Lines/profile/lower deck/upper deck/midships section.

## *Orestes* Class 1823. 18. Sloop/Corvette.
School of Naval Architecture design (by Professor Inman). There may have been some alterations in the design of *Comet* and *Lightning*. See also *Fly*.
  Dimensions & tons: 113'3", 92'10⅛" x 30'6" x 8'. 459 37/90.
  Men: 125. Guns: 2 x 9 + 16 x 32★.
*Orestes* Portsmouth Dyd.

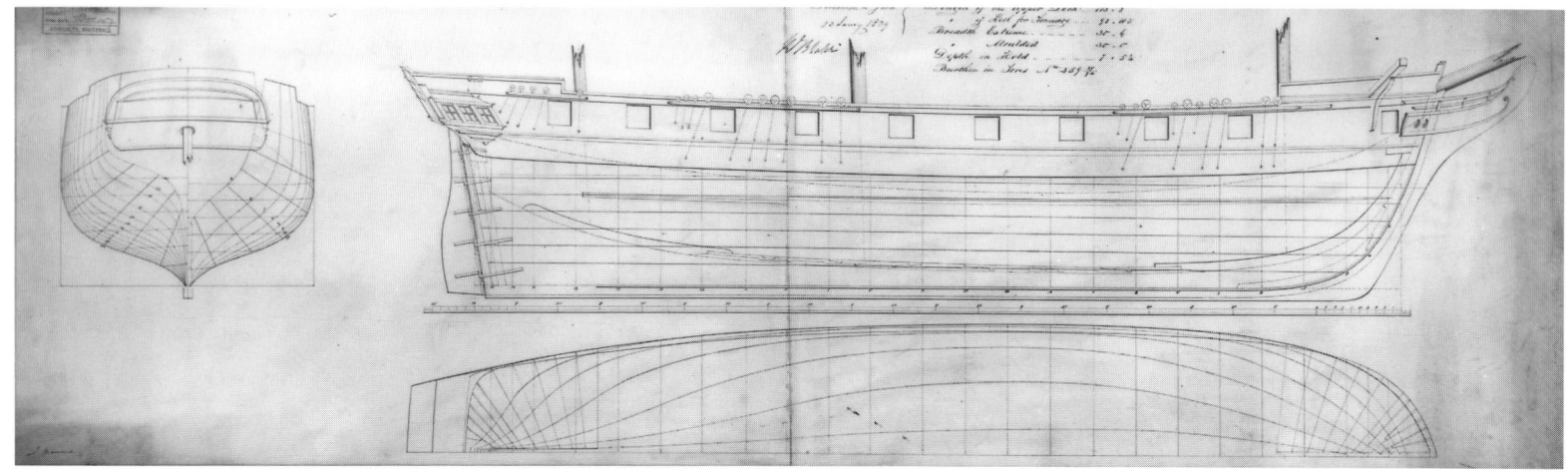

Lines plan for the *Orestes* of 1824, the name ship of her class of ship sloops. Plan number 3201

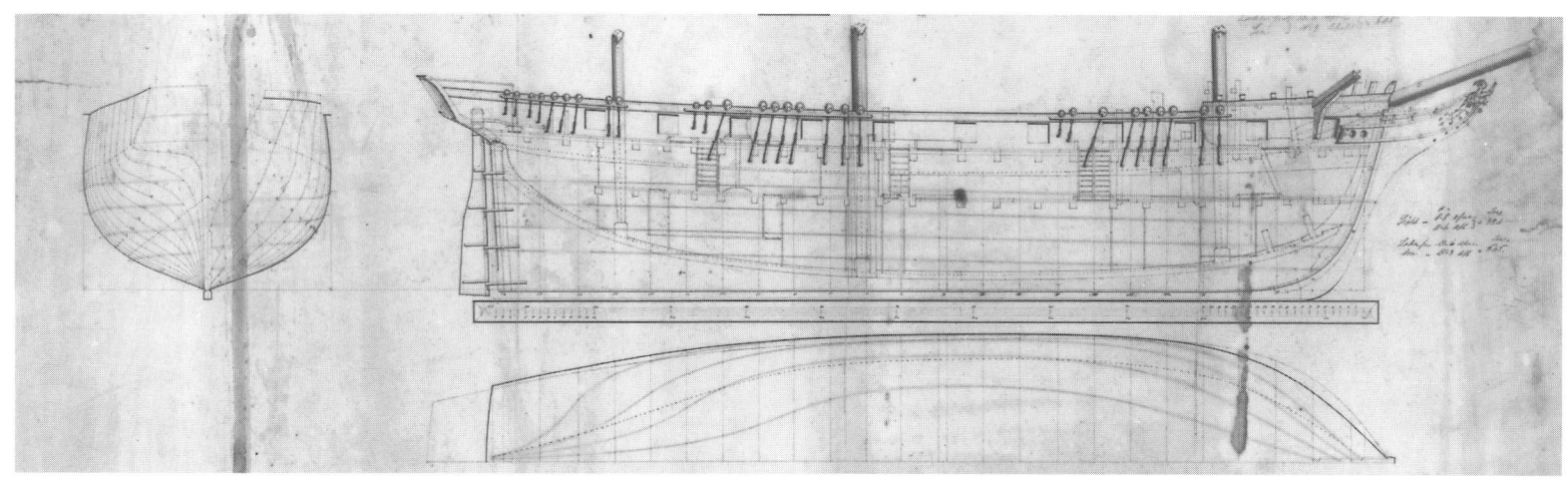

Lines and profile plan (3174) of the ship sloop *Pearl* of 1828

Ord ____ 1823?; K 04.1823; L 31.5.1824; 1852 hulked.
**Comet** Pembroke Dyd.
Ord ____ 1823?; K 10.1826; L 14.8.1828; 1832 renamed *Comus*; 1862 BU.
**Lightning** Pembroke Dyd.
Ord ____ 1823?; K 07.1828; L 2.6.1829; 1832 renamed *Larne*; 1866 BU.
FATES: *Orestes* became a coal depot at Portsmouth; later yard craft C.28; by 1890 to Plymouth; Ca 1905 sold.
PLANS: Lines/profile/lower deck/upper deck/midships section/framing. *Orestes* as built: lines. 1839: profile/lower deck/upper deck. 1851 as coal depot: profile/lower deck/upper deck.

### *Champion* Class 1823. 18, Sloop. Designed by Captain Hayes.
Dimensions & tons: 109'6", 91'10⅛" x 30'10½" x 7'8¾". 456.
Men: 125. Guns: 2 x 9 + 16 x 32★.
**Champion** Portsmouth Dyd.
Ord ____ 1824?; K 11.1823; L 31.5.1824; 1859 hulked.
FATE: Became harbour police ship at Portsmouth; 1864 lent to Committee on floating obstructions and used for explosive experiments; 1867 BU.
PLANS: 1835: Sails.

### *Wolf* Class 1825. 18, Sloop. Designed by Captain Hayes.
Dimensions & tons: 113'4½", 91'8¼" x 30'10¼" x 7'10¾". 454 71/94.
Men: 125. Guns: 2 x 6 + 16 x 32★.
**Wolf** Portsmouth Dyd.
Ord ____ 1825?; K 11.1825; L 1.12.1826; 1848 hulked.
FATE: 10.3.1830 stranded on the Isle of Wight but salved; 1848 became coal depot at Kingstown, Ireland; 1878 BU.
PLANS: Lines & profile.

### *Pearl* Class 1825. 18/20, Sloop. Sainty design. On 14.3.1825 the
Admiralty ordered she was 'to be built on such lines and principles as Mr. Sainty may think proper'. On 11.7.1825 this was cancelled; 'not to be built', but this in turn was reversed by an order of 23.2.1826 to build her.
Dimensions & tons: 118'9", 94'9" x 33'7⅜" x 9'6". 558.
Men: 125. Guns: 2 x 9 + 18 x 32★.
**Pearl** Sainty, Colchester.
Ord 14.3.1825/3.2.1826; K 07.1826; L 17.3.1828; 1851 BU.
PLANS: Lines. As built: lines & profile. 1835: profile/lower deck/upper deck. Original design, not as built?: specification.

### *Favorite* Class 1825 (or 1823?). 18, Sloop. Lengthened
version of the *Snake* Class ship sloops, and hence of Rule's *Cruiser* Class brigs. The lengthening was done by inserting an extra 9'6" long midships section.
Dimensions & tons: 109'6", 86'9½" x 30'6" x 12'9". 429 40/94.
Men: 125. Guns: 2 x 9 + 16 x 32★.
**Favorite** Portsmouth Dyd.
Ord ____ 1823?; K 03.1825; L 21.4.1829; 1859 hulked.
**Hazard** Plymouth Dyd.
Ord ____ 1823?; K 05.1829?; L 21.4.1837; 1866 BU.
**Hyacinth** Plymouth Dyd.
Ord ____ 1823?; K 03.1826; L 6.5.1829; 1860 hulked.
**Racehorse** Plymouth Dyd.
Ord ____ 1823?; K 05.1829; L 24.5.1830; 1861 hulked.
FATES: *Favorite* became a coal depot at Plymouth; Ca 1890 yard craft C.3; later C.77; 1905 sold. *Hyacinth* became a hulk at Portland; 1871 BU. *Racehorse* became a coal depot at Plymouth; 1901 sold. PLANS: Lines/profile/lower deck/upper deck/framing/midships section.

### *Satellite* Class 1825. 18, Sloop. Seppings design. The *Scout* Class
was an enlarged version.

# HENSLOW TO SEPPINGS 1786-1830

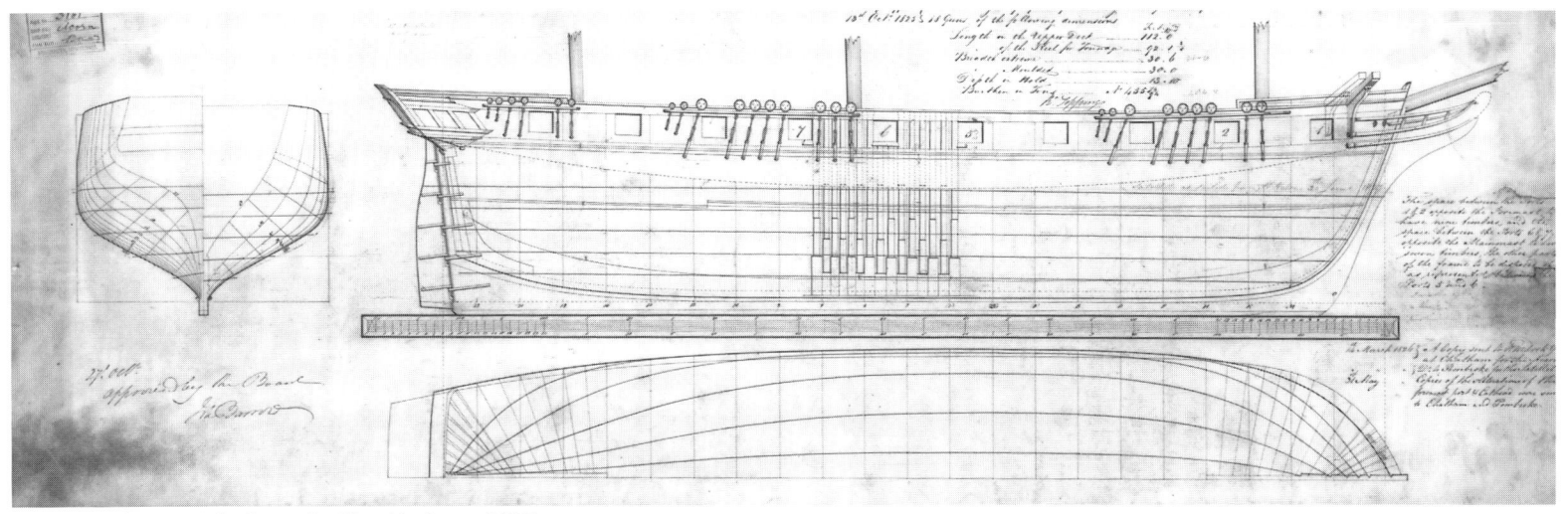

Design lines plan (3138) for the *Satellite* Class ship sloops of 1825

Dimensions & tons: 112', 92'1⅜" x 30'6" x 13'10". 455 74/94.
Men: 125. Guns: 2 x 6 + 16 x 32★.
**Satellite** Pembroke Dyd.
  Ord ____ 1825?; K 06.1826; L 3.10.1826; 1849 BU.
**Acorn** Chatham Dyd.
  Ord ____ 1825?; K 06.1826; L 16.11.1826; 1828 foundered.
FATES: *Acorn* foundered off Halifax 14.4.1828.
PLANS: Lines/profile/lower deck/upper deck/midships section.

## *Scout* Class 1829. 18, sloop. 
Seppings design, 'as *Satellite*' but larger.
Dimensions & tons: 115', 94'0⅜" x 31' x 14'3". 480 81/94.
Men: 120/125. Guns: 2 x 9 + 16 x 32★.
**Scout** Chatham Dyd.
  Ord ____ 1829?; K 10.1831; L 15.6.1832; 1852 BU.

**Cancelled 1831.** **Rover** Chatham Dyd.
  Ord ____ ; 'to be built by another Draught' see Symonds' *Rover* Class.
**Pheasant** Plymouth Dyd.
  Ord 30.1.1829 not laid down and cancelled 1.11.1831.
**Redwing** Plymouth Dyd.
  Ord 30.1.1829 not laid down and cancelled 1.11.1831.
PLANS: Lines/profile/lower deck/upper deck/midships section/stern. As built: lower deck/upper deck.

## *Fly* Class 1829. 18, Sloop/Corvette.
Development of Professor Inman's (School of Naval Architecture) *Orestes* Class design.
Dimensions & tons: 114'4", 93'8" x 31'4" x 14'5". 479 79/94.
Men: 120. Guns: 2 x 9 + 16 x 32★.
**Fly** Pembroke Dyd.
  Ord ____ 1829?; K ____ ; L 25.8.1831; 1854 hulked.
**Harrier** Pembroke Dyd.
  Ord ____ 1829?; K 11.1830; L 8.11.1831; 1840 BU.

**Cancelled 1831.** **Electra** Portsmouth Dyd.
  Ord ____ ; K 02.1830; completed to Symonds' *Rover* design instead.
**Argus** Portsmouth Dyd.
  Ord 29.4.1831; not begun?
**Acorn** Portsmouth Dyd.
  Ord 29.4.1831; not begun?
FATES: *Fly* became a coal depot at Plymouth; later yard craft C.2; 1903 BU.
PLANS: Lines/profile & decks/framing/midships section.

## Sloop, 18, class uncertain; 
cancelled 30.9.1820: **Samarang** Portsmouth Dyd.
  Ord 6.9.1815.

# Brig Sloops & Brigs
(see also gunboats & gunbrigs)

## *Diligence* Class 1795. 16, later 18 Brig Sloops.
Henslow design 'to carry 16 carriage guns'.
Dimensions & tons: 95', 75'2½" x 28'6" x 12'. 313 59/94.
Men: 121. Guns: designed for 16 x 6 + 12 swivels, but completed with 16 x 32★, later 2 x 6 added as chase guns.
**Diligence** Parsons, Bursledon.
  Ord 4.3. & 22.4.1795; K 04.1795; L 24.11.1795; 1800 wrecked.
**Harpy** King, Dover.
  Ord 4.3. & 22.4.1795; K 05.1795; L 02.1796; 1817 sold.
**Hound** Hill, Sandwich.
  Ord 4.3. & 22.4.1795; K 05.1795; L 24.3.1796; 1800 foundered.

**Fir Built Version.** **Curlew** Randall, Rotherhithe.
  Ord 4.3? & 22.4.1795; K 05.1795; L 16.7.1795; 1796 lost.
**Seagull** Wells, Deptford.
  Ord 4.3? & 22.4.1795; K 05.1795; L 07.1795; 1804 (or 1805?) foundered.
**Kangaroo** Wells, Deptford.
  Ord ____ 1795?; K 07.1795; L 30.9.1795; 1802 sold.
**Racoon** Randall, Rotherhithe.
  Ord ____ 1795?; K 07.1795; L 14.10.1795; 1806 BU.
**Camelion** Randall, Rotherhithe.
  Ord ____ 1795?; K 07.1795; L 14.10.1795; 1811 BU.
FATES: *Diligence* wrecked on a shoal on the North West side of Cuba 09.1800. *Hound* presumed foundered off Shetland in a storm 26.9.1800. *Curlew* foundered 'in the North Seas' 31.12.1796. *Seagull* presumed to have foundered in the Channel 02.1805?
PLANS: Lines/profile & decks/framing/specification.

## *Albatross* Class 1795. 16, later 18, Brig Sloops.
Rule design.
Dimensions & tons: 96', 73'9½" x 30'6" x 12'9". 365. 12/94.
Men: 121. Guns: designed for 16 x 6 + 12 swivels, but completed with 16 x 32★, later 2 x 6 chase guns added.
**Albatross** Ross, Rochester.
  Ord ____ 1795?; K 05.1795; L 30.12.1795; 1807 sold.
**Dispatch** Nicholson, Chatham.
  Ord ____ 1795?; K 05.1795; L 15.12.1795; 1796 Russian Navy.

**Fir Built Version.** **Kite** Barnard, Deptford.
  Ord ____ 1795?; K 05.1795; L 17.7.1795; 1805 sold.
**Pelican** Perry, Blackwall.
  Ord 1795?; K 05.1795?; L 17.6.1795; 1806 sold.
**Raven** Wallis, Blackwall.
  Ord ____ 1795?; K 05.1795; L 11.1.1796; 1798 wrecked.
**Star** Perry, Blackwall.

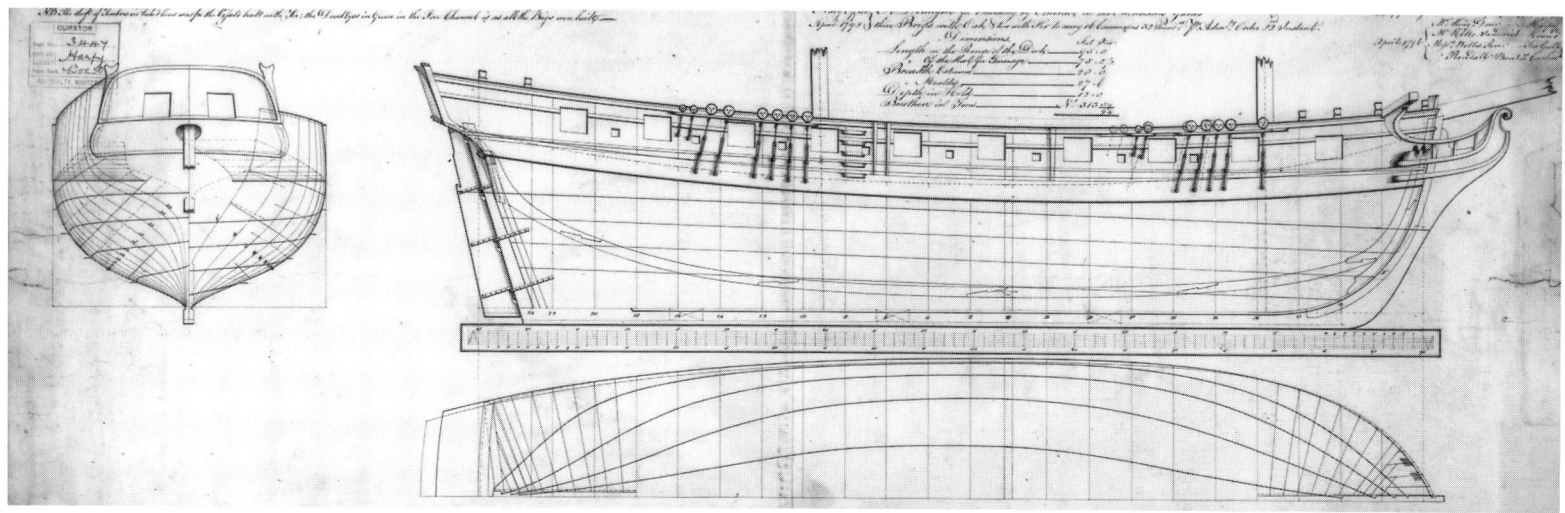

The lines plan (3447) to which the *Diligence* Class brigs of 1795 were built

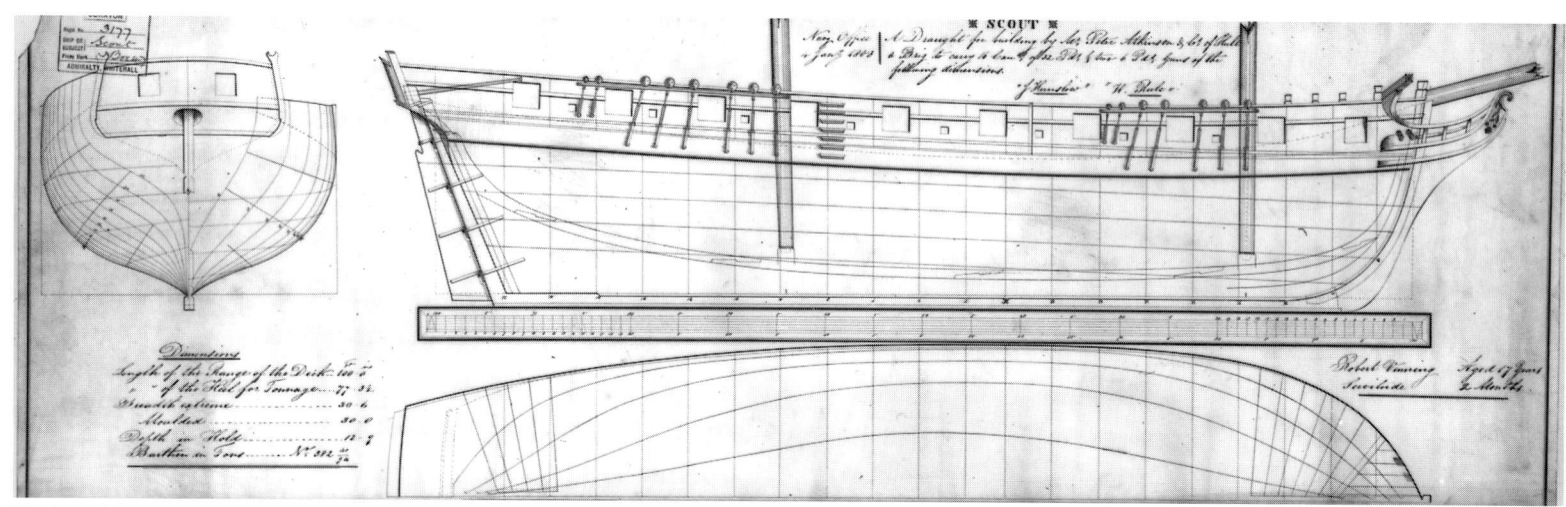

The lines plan (3177) to which the *Cruiser* Class brig *Scout* was built in 1804

Ord 13.7.1795; K 07.1795; L 29.8.1795; 1802 sold.
**Swallow** Perry, Blackwall.
Ord 13.7.1795; K 07.1795; L 10.9.1795; 1802 sold.
**Sylph** Barnard, Deptford.
Ord 13.7.1795; K 07.1795; L 8.9.1795?; 1811 BU.
FATES: *Dispatch* given to the Russian Navy. *Raven* wrecked at the mouth of the Elbe 3.2.1798.
PLANS: Lines/profile/decks/framing. *Dispatch* as built: lines & profile.

### Busy Class 1796. 18, Brig Sloop.
Henslow design. Built to the same lines and dimensions as the ship sloop *Echo*.
Dimensions & tons: 96', 75'1¼" x 29' x 12'9". 335 91/94.
Men: 121. Guns: 16 x 32★ + 2 x 6 x chase guns.
**Busy** Graham, Harwich.
Ord 19. & 26.12.1796; K 03.1797; L 20.11.1797; 1807 lost.
FATE: Foundered on the Halifax Station 02.1807.
PLANS: Lines/profile & decks/specification.

### Cruiser/Cruizer Class 1796. 18, Brig Sloops.
Rule design. Apart from their masts the *Snake* Class of sloop were identical, as were their derivatives; the *Favorite* Class, apart from an extra section inserted amidships to give extra length. With these extra vessels counted the *Cruiser* Class was the most numerous class of warships built in the age of sail; other than their nearest rivals in this respect (the *Cherokee* Class), no other class was anywhere near as large.
Dimensions & tons: 100', 77'3½" x 30'6" x 12'9". 382 41/94.
Men: 121. Guns: 2 x 6 + 16 x 32★.
**Cruiser** Teague, Ipswich.
Ord ____ 1796?; K 02.1797; L 20.12.1797; 1819 sold.

**Scorpion** King, Dover.
Ord 27.11.1802; K 01.1803; L 17.10.1803; 1819 sold.
**Despatch** Symons, Falmouth.
Ord 27.11.1802; K 04.1803; L 26.5.1804; 1811 BU.
**Musquito** Preston, Yarmouth.
Ord 27.11.1802; K 05.1803; L 4.9.1804; 1822 sold.
**Scout** Atkinson, Hull.
Ord 27.11.1802; K 05.1803; L 7.8.1804; 1827 sold.
**Minorca** Brindley, Lynn.
Ord 27.11.1802; K 05.1804; L 06.1805; 1814 BU.
**Amaranthe** Dudman, Deptford.
Ord 27.11.1802?; K 12.1803; L 20.11.1804; 1815 sold.
**Espoir** King, Dover.
Ord ____ 1803?; K 02.1804; L 22.9.1804; 1821 BU.
**Surinam** Ayles, Topsham.
Ord ____ 1803?; K 02.1804; L 01.1805; 1825 sold.
**Wolverene** Owen, Topsham.
Ord ____ 1803?; K 02.1804; L 1.3.1805; 1816 sold.
**Calypso** Dudman, Deptford.
Ord ____ 1803?; K 03.1804; L 2.2.1805; 1821 BU.
**Moselle** King, Dover.
Ord ____ 1803?; K 03.1804; L 10.1804; 1815 sold.
**Swallow** Tanner, Dartmouth.
Ord ____ 1802?; K 05.1804; L 24.12.1805; 1815 BU.
**Weazle** Owen, Topsham.
Ord ____ 1803?; K 05.1804; L 2.3.1805; 1815 sold.
**Ferret** Tanner, Dartmouth.
Ord ____ 1802?; K 08.1805; L 4.1.1806; 1813 wrecked.
**Racehorse** Hamilton, Hastings
Ord ____ 1803?; K 06.1804; L 17.2.1806; 1822 wrecked.

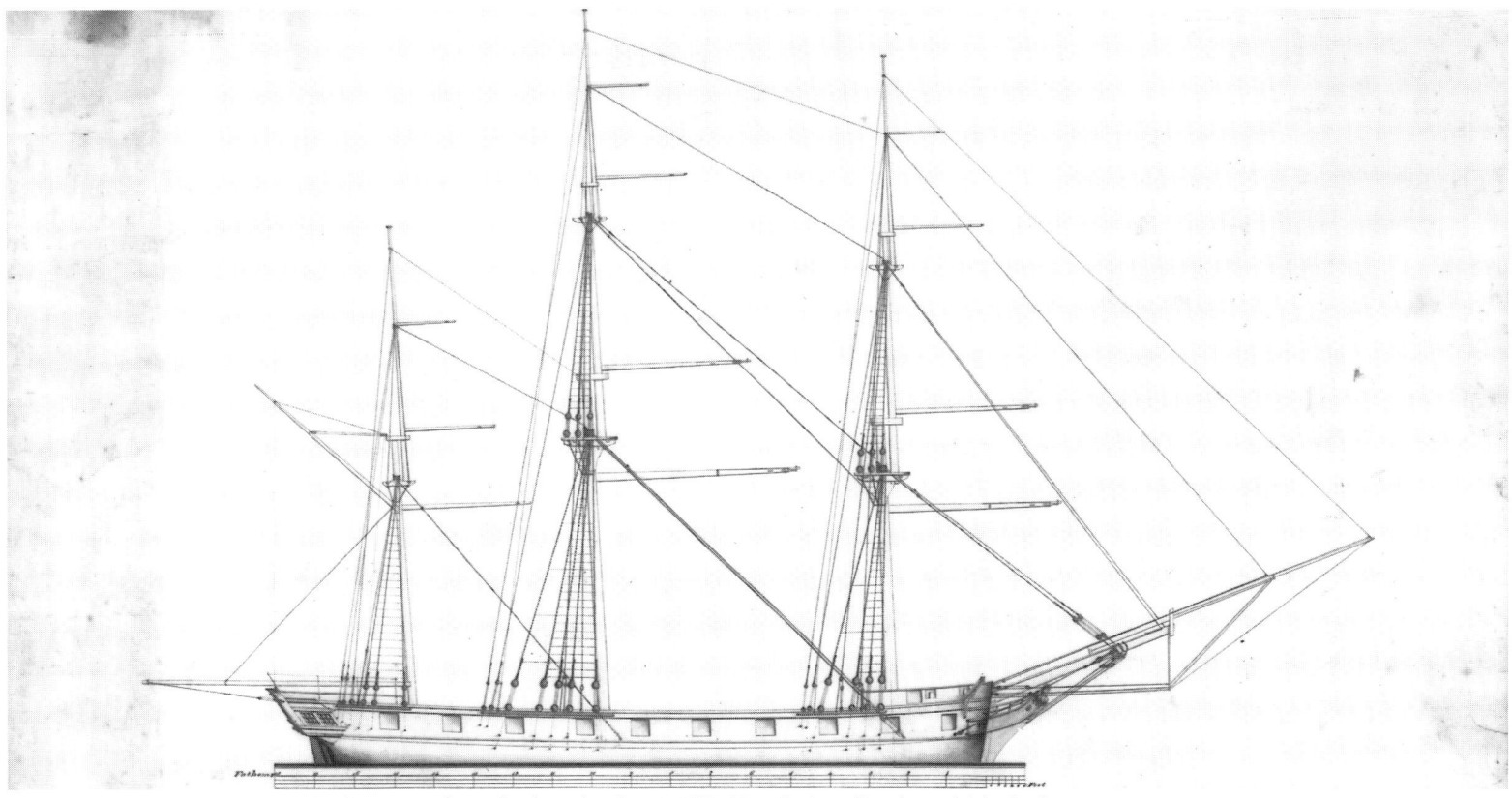

Spar plan (6827) showing the *Cruiser* Class brig *Wasp* of 1812 fitted as a ship sloop in 1828. Note the 'topgallant forecastle' that has been added to the original design

**Rover** Todd, Berwick.
    Ord ____ 1803?; K 06.1804; L 13.2.1808; 1828 sold.
**Avon** Symons, Falmouth.
    Ord ____ 1803?; K 06.1804; L 31.1.1805; 1814 sunk.
**Belette** King, Dover.
    Ord ____ 1805?; K 08.1805; L 21.3.1806; 1812 wrecked.
**Leveret** King, Dover.
    Ord ____ 1805?; K ____ ; L 4.1.1806; 1807 wrecked.
**Mutine** Chapman, Biddeford.
    Ord ____ 1805?; K 09.1805; L 15.8.1806; 1819 sold.
**Emulous** Row, Newcastle.
    Ord ____ 1805?; K 12.1805; L 06.1806; 1812 wrecked.
**Columbine** Adams, Bucklers Hard.
    Ord ____ 1806??; K 01.1806; L 16.7.1806; 1824 wrecked.
**Pandora** Preston, Yarmouth.
    Ord ____ 1806?; K 02.1806; L 11.10.1806; 1811 wrecked.
**Forester** King, Dover.
    Ord ____ 1806?; K 03.1806; L 3.8.1806; 1819 sold.
**Redwing** Warren, Brightlingsea.
    Ord ____ 1806?; K 03.1806; L 30.8.1806; 1824 converted to a ship sloop; 1827 foundered.
**Raleigh** Hurry, Newcastle.
    Ord ____ 1806?; K 03.1806; L 24.12.1806; 1826 converted to a ship sloop; 1839 hulked.
**Frolick** Bools, Bridport.
    Ord ____ 1806?; K 04.1806; L 9.12.1806; 1813 BU.
**Peacock** Bailey, Ipswich.
    Ord ____ 1806?; K 04.1806; L 9.12.1806; 1813 taken.
**Ringdove** Warren, Brightlingsea.
    Ord ____ 1806?; K 04.1806; L 16.10.1806; 1829 sold.
**Sappho** Bailey, Ipswich.
    Ord ____ 1806?; K 04.1806; L 15.12.1806; 1825 laid up as defective; 1830 BU.
**Recruit** Hills, Sandwich.
    Ord ____ 1806?; K 04.1806; L 31.8.1806; 1822 sold.
**Grasshopper** Richards, Hythe.
    Ord ____ 1805?; K 04.1806; L 29.9.1806; 1811 taken.

**Philomel** Bools, Bridport.
    Ord ____ 1806?; K 04.1806; L 11.9.1806; 1817 sold.
**Cephalus** Custance & Shore, Yarmouth.
    Ord ____ 1806?; K 04.1806; L 10.1.1807; 1830 BU.
**Alacrity** Row, Newcastle.
    Ord ____ 1804?; K 05.1806; L 13.11.1806; 1811 taken.
**Clio** Betts, Mistleythorn.
    Ord ____ 1806?; K 05.1806; L 10.1.1807; 1830 converted to a ship sloop; 1833 reconverted to a brig sloop of 16 guns; 1845 BU.
**Procris** Custance & Shore, Yarmouth.
    Ord ____ 1806?; K 05.1806; L 27.12.1806; 1815 sold.
**Royalist** Hills, Sandwich.
    Ord ____ 1806?; K 05.1806; L 10.1.1807; 1819 sold.
**Primrose** Nickells, Fowey.
    Ord ____ 1806?; K 06.1806; L 5.8.1807; 1809 wrecked.
**Foxhound** King, Dover.
    Ord ____ 1806?; K 08.1806; L 30.11.1806; 1809 foundered.
**Carnation** Taylor, Biddeford.
    Ord ____ 1806?; K 08.1806; L 3.10.1807; 1808 taken.
**Derwent** Blackburn, Turnchapel.
    Ord ____ 1806?; K 12.1806; L 23.5.1807; 1817 sold.
**Eclair** (ex *Pelican*) Warren, Brightlingsea.
    Ord ____ 1806?; K 12.1806; L 8.7.1807; 1823 converted to a ship sloop; 1831 BU.
**Eclipse** King, Dover.
    Ord ____ 1806?; K 12.1806; L 4.8.1807; 1815 sold.
**Barracouta** Bailey, Ipswich.
    Ord ____ 1807??; K 01.1807; L 6.7.1807; 1815 sold.
**Nautilus** Betts, Mistleythorn.
    Ord ____ 1807??; K 01.1807; L 5.8.1807; 1823 BU.
**Pilot** Guillaume, Northam.
    Ord ____ 1807??; K 01.1807; L 6.8.1807; 1828 sold.
**Sparrowhawk** Warren, Brightlingsea.
    Ord ____ 1807??; K 01.1807; L 20.8.1807; 1824 converted to a ship sloop; 1834 reconverted to a 16 gun brig sloop; 1841 sold.
**Zenobia** Brindley, Lynn.
    Ord ____ 1807?; K 03.1807; L 7.10.1807; 1835 sold.

*Magnet* Guillaume, Northam.
Ord ____ 1807?; K 04.1807; L 19.10.1807; 1809 wrecked.
*Peruvian* Parsons, Warsash/Bursledon.
Ord ____ 1807?; K 09.1807; L.26.4.1808; 1830 BU.
*Pelorus* Kidwell, Itchenor.
Ord ____ 1808??; K 01.1808; L 25.6.1808; 1826 converted to a ship sloop; 1831 reconverted to a brig sloop; 1841 sold.
*Dotterell* Blake, Bursledon.
Ord ____ 1808?; K 04.1808; L 6.10.1808; 1827 hulked.
*Arachne* Hills, Sandwich.
Ord ____ 1806?; K 09.1808; L 18.2.1809; 1824 converted to a ship sloop; 1837 sold.
*Persian* List, Cowes.
Ord ____ 1808?; K 09.1808; L 2.5.1809; 1813 wrecked.
*Castilian* Hills, Sandwich.
Ord ____ 1808?; K 10.1808; L 29.5.1809; 1829 BU.
*Charybdis* Richards, Hythe.
Ord ____ 1808?; K 10.1808; L 28.8.1809; 1819 sold.
*Trinculo* Tyson, Bursledon.
Ord ____ 1808?; K 10.1808; L 15.7.1809; 1828 converted to a ship sloop; 1832 reconverted to a brig sloop; 1841 BU.
*Crane* Brindley, Frindsbury.
Ord ____ 1808?; K 12.1808; L 29.7.1809; 1814 foundered.
*Echo* Pelham, Frindsbury.
Ord ____ 1808?; K 12.1808; L 1.7.1809; 1817 BU.
*Hecate* King, Upnor.
Ord ____ 1808?; K 12.1808; L 30.5.1809; 1817 sold.
*Scylla* Davy, Topsham.
Ord ____ 1808?; K 12.1808; L 29.6.1809; 1824 converted to a ship sloop; 1846 BU.
*Sophie* Pelham, Frindsbury.
Ord ____ 1808?; K 12.1808; L 8.9.1809; 1825 sold.
*Thracian* Brindley. Frindsbury.
Ord ____ 1808?; K 12.1808; L 19.7.1809; 1822 converted to a ship sloop; 1832 reconverted to a brig sloop; 1841 BU.
*Rifleman* King, Upnor.
Ord ____ 1808?; K 01.1809; L 12.8.1809; 1836 sold.
*Childers* Good, Bridport.
Ord ____ 1811?; K 08.1811; L 9.7.1812; 1822 sold.
*Curlew* Good, Bridport.
Ord ____ 1811?; K 10.1811; L 27.5.1812; 1822 sold.
*Wasp* Davy, Topsham.
Ord ____ 1811?; K 10.1811; L 9.7.1812; 1828 converted to a ship sloop; 1833 reconverted to a 16 gun brig; 1847 BU.
*Arab* Pelham, Frindsbury.
Ord ____ 1811?; K 11.1811; L 22.8.1812; 1823 lost.
*Nimrod* Bailey, Ipswich.
Ord ____ 1811?; K 11.1811; L 25.5.1812; 1827 wrecked.
*Saracen* Bools, Bridport.
Ord ____ 1811?; K 11.1811; L 25.7.1812; 1819 sold.
*Espiegle* Bailey, Ipswich.
Ord ____ 1811?; K 01.1812; L 10.8.1812; 1826 converted to a ship sloop; 1832 sold.
*Bacchus* Chatham Dyd.
Ord ____ 1811?; K 01.1812; L 17.4.1813; 1826 hulked.
*Pelican* Davy, Topsham.
Ord ____ 1811?; K 01.1812; L 08.1812; after 1845 to the Coastguard.
*Fairy* Taylor, Biddeford.
Ord ____ 1811?; K 01.1812; L 11.6.1812; 1821 BU.
*Despatch* King, Upnor.
Ord ____ 1811?; K 02.1812; L 7.12.1812; 1825 converted to a ship sloop; 1832 reconverted to a 16 gun brig; 1836 sold.
*Heron* (ex *Rattlesnake*) King, Upnor.
Ord ____ 1811?; K 02.1812; L 22.10.1812; 1825 converted to a ship sloop; 1831 BU.
*Satellite* List, Fishbourne.
Ord ____ 1811?; K 03.1812; L 9.10.1812; 1824 sold.
*Fly* Bailey, Ipswich.
Ord ____ 1812?; K 06.1812; L 16.2.1813; 1822 converted to a ship sloop; 1828 sold.
*Epervier* Ross, Rochester.
Ord ____ 1811?; K 07.1812; L 21.12.1812; 1814 taken.
*Challenger* Hobbs, Redbridge.
Ord ____ 1812?; K 08.1812; L 15.5.1813; 1813 hulked.
*Grasshopper* Portsmouth Dyd.
Ord ____ 1811?; K 08.1812; L 17.5.1813; 1822 converted to a ship sloop; 1832 sold.
*Jaseur* Bailey, Ipswich.
Ord ____ 1812?; K 08.1812; L 2.2.1813; 1824 converted to a ship sloop; 1834 reconverted to a 16 gun brig; 1845 sold.
*Argus* Hills, Sandwich.
Ord ____ 1812?; K 09.1812; L 11.9.1813; 1828 sold.
*Halcyon* Larking, Lynn.
Ord ____ 1812?; K 09.1812; L 16.5.1813; 1814 wrecked.
*Pandora* Deptford Dyd.
Ord ____ 1812?; K 09.1812; L. 12.9.1813; 1825 converted to a ship sloop; 1831 sold.
*Carnation* Durkin, Southampton.
Ord ____ 1812; K 11.1812; L 29.7.1813; 1826 hulked.
*Penguin* Bottomley, Lynn.
Ord ____ 1812; K 11.1812; L 29.6.1813; 1815 taken.
*Confiance* Ross, Rochester.
Ord ____ 1812; K 12.1812; L 30.8.1813; 1822 wrecked.
*Elk* Hobbs, Redbridge.
Ord ____ 1812?; K 12.1813; L 28.8.1813; 1836 sold.
*Alert* Pitcher, Northfleet.
Ord ____ 1812?; K 01.1813; L 14.7.1813; 1827 converted to a ship sloop; 1832 sold.
*Harlequin* Bailey, Ipswich.
Ord ____ 1812?; K 02.1813; L 15.7.1813; 1825 converted to a ship sloop; 1829 sold.
*Harrier* Bailey, Ipswich.
Ord ____ 1812?; K 02.1813; L 28.7.1813; 1823 converted to a ship sloop; 1829 sold.
*Ontario* (ex *Mohawk*? ) Chapman, Biddeford.
Ord ____ 1812?; K 02.1813; L 26.10.1813; 1832 sold.
*Belette* Larking & Spring, Lynn.
Ord ____ 1813?; K 10.1813; L 18.6.1814; 1828 sold.
*Gannet* Larking & Spring, Lynn.
Ord ____ 1813?; K 12.1813; L 13.11.1814; 1831 converted to a ship sloop; 1834 16 guns; 1838 sold.

## Fir Built Version.
*Beagle* Perry, Blackwall.
Ord ____ 1804?; K 06.1804; L 9.8.1804; 1814 sold.
*Elk* Barnard, Deptford.
Ord —— 1804?; K 06.1804; L 22.8.1804; 1812 BU.
*Harrier* Barnard, Deptford.
Ord ____ 1804?; K 06.1804; L 22.8.1804; 1809 foundered.
*Raven* Perry, Blackwall.
Ord —— 1804?; K 06.1804; L 25.7.1804; 1805 wrecked.
*Reindeer* Brent, Rotherhithe.
Ord ____ 1804?; K 06.1804; L 15.8.1804; 1814 taken.
*Saracen* Perry, Blackwall.
Ord ____ 1804?; K 06.1804; L 25.7.1804; 1812 BU.

## Teak Built Version.
*Victor* Bombay Dyd.
Ord ____ 1812?; K 01.1814; L 29.10.1814; 1823 converted to a ship sloop; 1831 reconverted to a brig; 1842 foundered.
*Zebra* Bombay Dyd.
Ord ____ 1813?; K 12.1814; L 18.12.1815; 1840 wrecked.

## Cancelled 1814?
*Lynx*? (ex *Pandora*?) Woolwich Dyd.
Ord ____ 1812?

FATES: *Ferret* wrecked near Leith 7.1.1813. *Racehorse* wrecked on the Isle of Man 14.12.1822; wreck recently located by divers. *Avon* sunk after action with the American *Wasp* 2.9.1814. *Belette* wrecked in the Kattegat 24.11.1812. *Leveret* wrecked on the Galloper Rock 10.11.1807. *Emulous* wrecked on Sable Island

3.8.1812. *Columbine* wrecked off the Island of Sapienza 25.2.1824. *Pandora* wrecked in the Kattegat 13.2.1811. *Redwing* supposed foundered on the West African Station 06.1827. *Raleigh* became a target at Sheerness; 1841 sold. *Peacock* foundered after being taken by the American *Hornet* 24.2.1813. *Grasshopper* (1806) taken by the Dutch in the Nieuw Diep, Texel; renamed *Irene* in Dutch service. *Alacrity* taken by the French *Abeille* (20) off Corsica; in French service until 1823. *Primrose* wrecked on the Manacle Rocks near Falmouth 22.1.1809. *Foxhound* foundered in the Atlantic 31.8.1809. *Carnation* taken by the French *Palinure* (16) off Martinique 3.10.1808; 1809 burnt at Martinique to prevent recapture. *Magnet* wrecked in the ice off Malmo, Sweden 11.1.1809. *Dotterell* became a receiving and accommodation ship at Bermuda; 1848 BU. *Persian* was wrecked on Silver Keys 16.6.1813. *Crane* foundered in the West Indies 30.9.1814. *Arab* wrecked off Belmullet, West coast of Ireland 12.12.1823. *Nimrod* was bilged in Holyhead Bay 14.1.1827, salvaged and sold. *Bacchus* became a coal depot at Deptford; 1829 became a breakwater. *Pelican* became W.V. No 29 at Rye Harbour; 1865 sold. *Epervier* taken by the USS *Peacock* 29.4.1814; 1815 foundered. *Challenger* hulked at Trincomalee to receive rice; 1824 sold. *Halcyon* wrecked on a reef on the North side of Jamaica 19.5.1814. *Carnation* was hulked at Plymouth for the Breakwater Department; 1836 sold. *Penguin* taken by the USS *Hornet* off Tristan da Cunha and destroyed 23.3.1815. *Confiance* wrecked at Crookhaven, Scotland 21.4.1822. *Harrier* supposed foundered in the Indian Ocean March 1809. *Raven* wrecked in Cadiz Bay whilst attempting to escape from the Spanish fleet 29.1.1805. *Reindeer* taken by the USS *Wasp* in the Channel 28.6.1814, then burnt. *Victor* sailed from Vera Cruz, Mexico on 31.10.1842 for Halifax, and not heard of since. *Zebra* wrecked off Mount Carmel 2.12.1840.

PLANS: Lines/profile/lower deck/upper deck/framing/midships section/spar plan/planking expansion (external)/ (internal)/ specification. *Arachne* as ship sloop 1830: profile & decks. *Grasshopper* 1817: lower deck. 1822: profile/decks. Fir built version: Lines/profile/lower deck/upper deck/framing/midships section. Teak built version: Lines/profile/lower deck/upper deck/framing/midships section.

## *Fly* Class 1804. 16 (later 18) Brigs. Henslow design.
Dimensions & tons: 96', 79'5" x 25'10" x 11'6". 281 85/94.
Men: 94. Guns: 2 x 6 + 14 x 24★ (later extra 2 x 24★ added).
**Fly** Bools & Co, Bridport.
 Ord ____ 1805?; K 03.1805; L 24.10.1805; 1812 wrecked.
**Kite** Warren, Brightlingsea.
 Ord ____ 1805?; K 03.1805; L 13.7.1805; 1815 sold.
**Sparrow** Preston, Yarmouth.
 Ord ____ 1805?; K 03.1805; L 29.7.1805; 1816 sold.
**Raven** Warren, Brightlingsea.
 Ord ____ 1805?; K 04.1805; L 12.8.1805; 1816 sold.
**Wizard** Sutton, Ringmore.
 Ord ____ 1805?; K 04.1805; L 11.1805; 1816 sold.

### 'Fir' Built (actually pitch pine) Version.
**Challenger** Wallis, Blackwall.
 Ord ____ 1805?; K 01.1806; L 30.7.1806; 1811 taken.
**Goshawk** Wallis, Blackwall.
 Ord ____ 1805?; K 04.1805; L 12.8.1805; 1813 lost.
FATES: *Fly* wrecked on Anholt Island 29.2.1812. *Challenger* taken by a French frigate off Mauritius 12.3.1811. *Goshawk* wrecked off Barcelona 21.9.1813.
PLANS: Lines/profile & decks/framing/specification. *Challenger* & *Goshawk*: Lines & profile & decks.

## *Seagull* Class 1805. 16, Brig. Rule design.
Dimensions & tons: 93', 76'0⅝" x 26'5" x 12'. 282 26/94.
Men: 95. Guns: 2 x 6 + 14 x 24★.
**Seagull** King, Dover.
 Ord ____ 1805?; K 02.1805; L 1.7.1805; 1808 taken.
**Oberon** Shepherd, Hull.
 Ord ____ 1805?; K 03.1805; L 13.8.1805; 1816 BU.
**Imogen** Bailey, Ipswich.
 Ord ____ 1805?; K 04.1805; L 11.7.1805; 1817 sold.
**Nightingale** King, Dover.
 Ord ____ 1805?; K 04.1805; L 29.7.1805; 1815 sold.
**Savage** Adams, Southampton.
 Ord ____ 1805?; K 04.1805; L 30.7.1805; 1819 sold.
**Electra** Betts, Mistleythorn.
 Ord ____ 1805?; K 08.1805; L 22.1.1806; 1808 wrecked.
**Orestes** Bailey, Ipswich.
 Ord ____ 1805?; K 08.1805; L 23.10.1805; 1817 sold.
**Paulina** Guillaume, Northam.
 Ord ____ 1805?; K 08.1805; L 7.12.1805; 1816 sold.
**Satellite** Hills, Sandwich.
 Ord ____ 1805?; K 09.1805; L 03.1806; 1810 foundered.
**Julia** Bailey, Ipswich.
 Ord ____ 1805?; K 10.1805; L 4.2.1806; 1817 wrecked.
**Sheldrake** Richards, Hythe.
 Ord ____ 1805?; K 10.1805; L 20.3.1806; 1817 sold.
**Skylark** Row, Newcastle.
 Ord ____ 1805?; K 10.1805; L 02.1806; 1812 burnt.
**Delight** Farrington (Fremington).
 Ord ____ 1805?; K 09.1805; L 06.1806; 1808 taken.
FATES: *Seagull* taken by a Danish flotilla off the Naze 19.6.1808. *Electra* wrecked at Port Augusta, Sicily, and wreck sold 25.3.1808. *Satellite* foundered in the Channel 31.12.1810. *Julia* wrecked off Tristan da Cunha 'in consequence of the sudden rising of the tide' 2.10.1817. *Skylark* grounded and destroyed to prevent capture near Boulogne 3.5.1812. *Delight* stranded, captured by the French, and lost 3.1.1808.
PLANS: Lines/profile & decks.

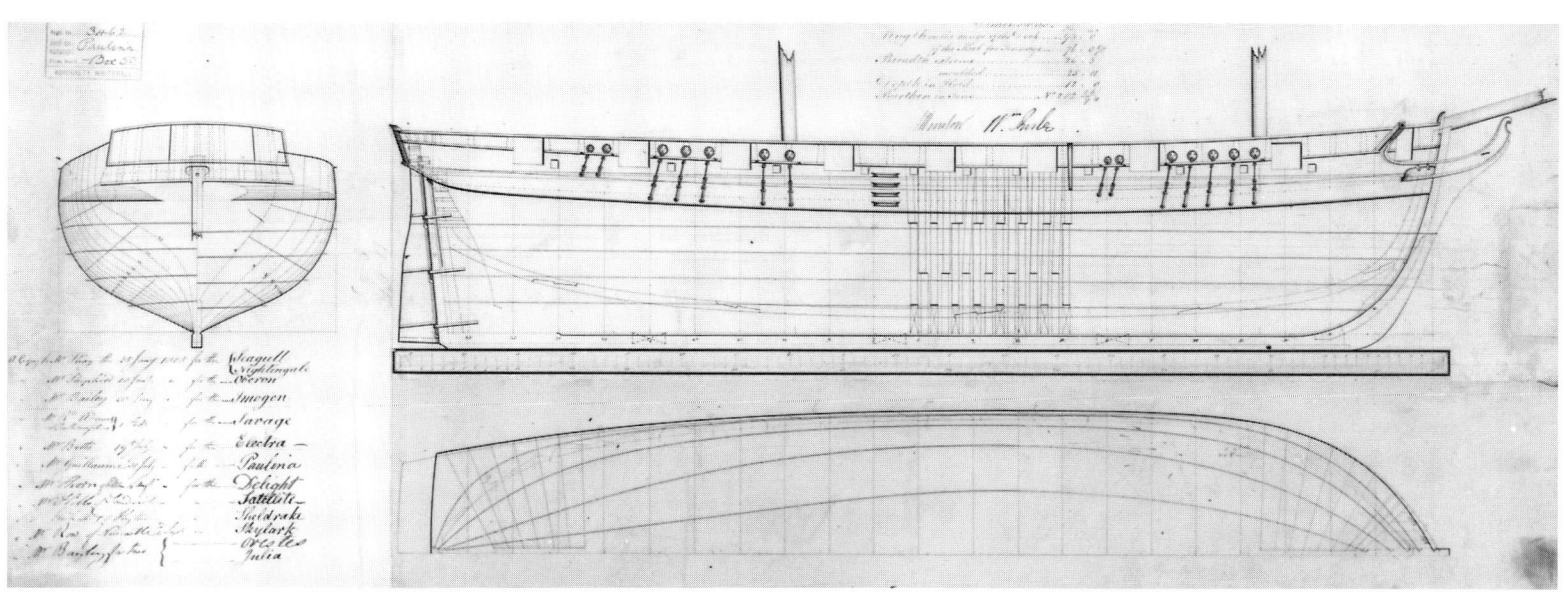

The design lines plan (3462) for building the *Seagull* Class brigs of 1805

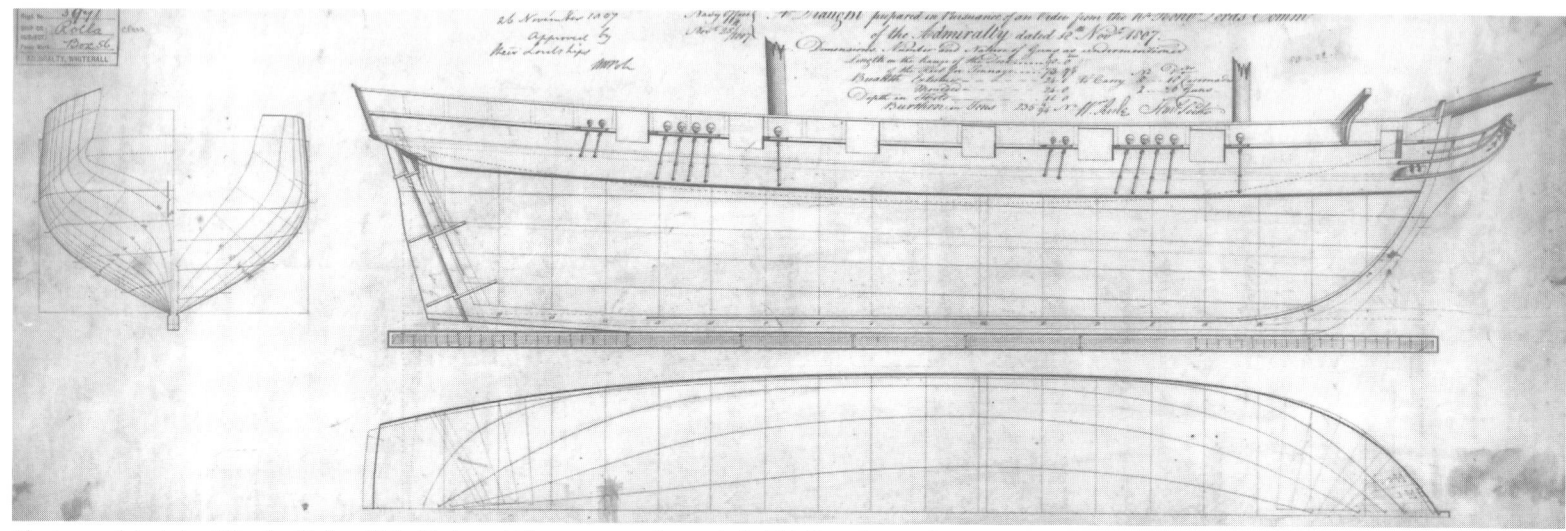

The ten gun brig design (*Cherokee*/*Cadmus*/*Rolla* Class) to which so many vessels were built. This lines plan was the one approved by the Board of Admiralty (3971)

## *Crocus* (*Banterer*) Class 1807. 14 Brigs.

Surveyors of the Navy design.

Dimensions & tons: 92', 72'8¾" x 25'6" x 12'8". 251 41/94.

Men: 85. Guns: 2 x 6 + 12 x 24★.

*Crocus* Plymouth Dyd.
Ord ____ 1807?; K 11.1807; L 10.6.1808; 1815 sold.
*Merope* Chatham Dyd.
Ord ____ 1807?; K 11.1807; L 25.6.1808; 1815 sold.
*Podargus* Portsmouth Dyd.
Ord ____ 1807?; K 11.1807; L 26.5.1808; 1833 sold.
*Apelles* Woolwich Dyd.
Ord ____ 1807?; K 02.1808; L 10.8.1808; 1816 sold.
*Muros* Chatham Dyd.
Ord ____ 1808?; K 06.1808; L 23.10.1809; 1822 sold.
*Prospero* Woolwich Dyd.
Ord ____ 1808?; K 08.1808; L 9.11.1809; 1816 sold.
*Zephyr* Portsmouth Dyd.
Ord ____ 1808?; K 10.1808; L 29.4.1809; 1818 sold.
*Banterer* Woolwich Dyd.
Ord ____ 1808?; K 12.1809; L 2.6.1810; 1817 sold.
*Portia* Deptford Dyd.
Ord ____ 1808?; K 12.1809; L 30.8.1810; 1817 sold.
*Wolf* Woolwich Dyd.
Ord ____ 1810?; K 08.1812; L 16.9.1814; 1825 sold.

PLANS: Lines/profile/decks/lower deck/upper deck/framing/midships section/planking expansion (external)/specification.

## *Cherokee*/*Cadmus*/*Rolla* Class 1807. 10 Brigs/Brig Sloops.

Peake design. There seem to have been a number of slight variants of the basic design amongst these 'coffin brigs'; which, apart from the *Cruiser* Class, formed the largest class of sailing warships built. Some of the vessels fitted as packets or exploring ships (e.g. *Beagle*) were given a small mizzen mast and rigged as barques. Their line of descent was through the earlier gunboats and gunbrigs, though most were classed as brigs.

Dimensions & tons: 90', 73'7⅝" x 24'6" x 11'. 235 9/94.

Men: 75. Guns: 2 x 6 + 8 x 18★ (some of the later vessels fitted as packets only carried 3 guns).

*Cherokee* Perry, Blackwall.
Ord ____ 1807?; K 12.1807; L 24.2.1808; 1828 sold.
*Cadmus* Dudman, Deptford.
Ord ____ 1807?; K 12.1807; L 26.2.1808; 1835 to the Coastguard.
*Achates* Brent, Rotherhithe.
Ord ____ 1807?; K 12.1807; L 1.2.1808; 1810 wrecked.
*Leveret* Perry, Blackwall.
Ord ____ 1807?; K 12.1807; L 24.2.1808; 1822 sold.
*Parthian* Barnard, Deptford.
Ord ____ 1807?; K 12.1807; L 13.2.1808; 1828 wrecked.
*Rolla* Pitcher, Northfleet.
Ord ____ 1807?; K 12.1807; L 13.2.1808; 1822 sold.
*Badger* Brindley, Frindsbury.
Ord ____ 1807?; K 02.1808; L 23.7.1808; 1834 hulked.
*Briseis* King, Upnor.
Ord ____ 1807?; K 02.1808; L 19.5.1808; 1816 wrecked.
*Ephira* King, Upnor.
Ord ____ 1807?; K 02.1808; L 28.5.1808; 1811 wrecked.
*Jasper* Bailey, Ipswich.
Ord ____ 1807?; K 02.1808; L 27.5.1808; 1817 wrecked.
*Onyx* Bailey, Ipswich.
Ord ____ 1807?; K 02.1808; L 8.7.1808; 1819 sold.
*Chanticleer* List, Cowes.
Ord ____ 1807?; K 03.1808; L 26.7.1808; 1828 survey ship; 1833 to the Customs.
*Goldfinch* Warwick, Ealing.
Ord ____ 1807?; K 03.1808; L 8.8.1808; 1824 converted to a packet; 1838 sold.
*Opossum* Muddle, Gillingham.
Ord ____ 1807?; K 03.1808; L 9.7.1808; 1819 sold.
*Rinaldo* Dudman, Deptford.
Ord ____ 1807?; K 03.1808; L 13.7.1808; 1824 converted to a packet; 1835 sold.
*Shearwater* Row, Newcastle.
Ord ____ 1807?; K 03.1808; L 21.11.1808; 1832 sold.
*Wild Boar* Pelham, Frindsbury.
Ord ____ 1807?; K 03.1808; L 9.7.1808; 1810 wrecked.
*Woodlark* Row, Newcastle.
Ord ____ 1807?; K 03.1808; L 17.11.1808; 1818 sold.
*Britomart* Dudman, Deptford.
Ord 03.1807; K 04.1808; L 28.7.1808; 1819 sold.
*Calliope* Dudman, Deptford.
Ord ____ 1807?; K 04.1808; L 8.7.1808; 1822 tender; 1829 BU.
*Hope* Bailey, Ipswich.
Ord ____ 1807?; K 04.1808; L 22.7.1808; 1819 sold.
*Prince Arthur* Dudman, Deptford.
Ord ____ 1807?; K 04.1808; L 28.7.1808; 1808 sold to the Sultan of Morocco.
*Cordelia* King, Upnor.
Ord ____ 1807?; K 05.1808; L 26.7.1808; 1833 sold.
*Helicon* King, Upnor.
Ord ____ 1807?; K 05.1808; L 8.8.1808; 1829 BU.
*Lyra* Dudman, Deptford.
Ord ____ 1807?; K 05.1808; L 2.8.1808; 1818 sold.
*Redpole* Guillaume, Northam.
Ord ____ 1807?; K 05.1808; L 29.7.1808; 1825 converted to a packet; 1828 foundered.
*Bermuda* Pelham, Frindsbury.

Ord ____ 1807?; K 08.1808; L 20.12.1808; 1816 wrecked.
**Drake** Bailey, Ipswich.
Ord ____ 1807?; K 08.1808; L 3.11.1808; 1822 wrecked.
**Renard** King, Upnor.
Ord ____ 1807?; K 08.1808; L 5.12.1808; 1818 sold.
**Rosario** Bailey, Ipswich.
Ord ____ 1807?; K 08.1808; L 7.12.1808; 1832 sold.
**Tyrian** Guillaume, Northam.
Ord ____ 1807?; K 08.1808; L 16.12.1808; 1819 sold.
**Rhodian** Guillaume, Northam.
Ord ____ 1807?; K 08.1808; L 3.1.1809; 1813 wrecked.
**Sarpedon** Warwick, Ealing.
Ord ____ 1807?; K 09.1808; L 1.2.1809; 1812 foundered.
**Beaver** Bailey, Ipswich.
Ord ____ 1807?; K 10.1808; L 16.2.1809; 1829 sold.
**Sphinx** Bombay Dyd. Teak built.
Ord ____ 1812?; K 05.1814; L 25.1.1815; 1825 PACKET; 1835 sold.
**Cameleon** Bombay Dyd. Teak built.
Ord ____ 1812?; K ____ 1815; L 15.1.1816; 1849 BU.
**Eclipse** Plymouth Dyd.
Ord ____ 1817?; K 03.1817; L 23.7.1819; 1823 packet; 1836 to the Coastguard.
**Alacrity** Deptford Dyd.
Ord ____ 1817?; K 10.1817; L 29.12.1818; 1835 sold.
**Bustard** Chatham Dyd.
Ord ____ 1817?; K 11.1817; L 12.12.1818; 1829 sold.
**Brisk** Chatham Dyd.
Ord ____ 1817?; K 11.1817; L 10.2.1819; 1843 sold.
**Cygnet** Portsmouth Dyd.
Ord ____ 1817?; K 11.1817; L 11.5.1819; 1824 packet; 1835 sold.
**Delight** Portsmouth Dyd.
Ord ____ 1817?; K 11.1817; L 10.5.1819; 1824 wrecked.
**Falcon** Pembroke Dyd.
Ord ____ 1817?; K 05.1818; L 10.6.1820; 1827 tender; 1838 sold.
**Emulous** Plymouth Dyd.
Ord ____ 1817?; K 06.1818; L 16.12.1819; completed as a packet; 1834 to the Coastguard.
**Beagle** Woolwich Dyd.
Ord ____ 1817?; K 06.1818; L 11.5.1820; 1823 converted to a barque-rigged survey ship of 6 guns; 1845 to the Coastguard. (This was Fitzroy's and Darwin's *Beagle*.)
**Barracouta** Woolwich Dyd.
Ord ____ 1817?; K 06.1818; L 13.5.1820; 1829 converted to a barque-rigged packet; 1836 sold.
**Frolic/Frolick** Pembroke Dyd.
Ord ____ 1817?; K 08.1818; L 10.6.1820; 1823 converted to a packet; 1838 sold.
**Ariel** Deptford Dyd.
Ord ____ 1817?; K 02.1819; L 28.7.1820; 1828 converted to a packet; 1828 foundered.
**Jasper** Portsmouth Dyd.
Ord ____ 1818?; K 05.1819; L 26.7.1820; 1828 wrecked.
**Lyra** Plymouth Dyd.
Ord ____ 1818?; K 03.1819; L 1.6.1821; 1823 tender; 1829 converted to a packet; 1845 sold.
**Britomart** Portsmouth Dyd.
Ord ____ 1818?; K 06.1819; L 24.5.1820; 1843 sold.
**Opossum** Sheerness Dyd.
Ord ____ 1818?; K 11.1819; L 11.12.1821; 1829 packet; 1841 sold.
**Onyx** Sheerness Dyd.
Ord ____ 1818?; K 11.1819; L 24.1.1822; 1827 tender; 1837 sold.
**Partridge** Plymouth Dyd.
Ord ____ 1818?; K 12.1819; L 22.3.1822; 1824 wrecked.
**Reynard** Pembroke Dyd.
Ord ____ 1818?; K 05.1820; L 26.10.1821; 1825 tender; 1829 packet; 1841 yard craft at Chatham.
**Weazle** Chatham Dyd.
Ord ____ 1818?; K 05.1820; L 26.3.1822; 1844 sold.

**Ferret** Portsmouth Dyd.
Ord ____ 1820?; K 08.1820; L 21.10.1821; 1837 sold.
**Plover** Woolwich Dyd.
Ord ____ 1820?; K 08.1820; L 6.10.1821; 1824 packet; 1836 hulked.
**Kingfisher** Woolwich Dyd.
Ord ____ 1818?; K 12.1820; L 11.3.1823; 1824 packet; 1838 sold.
**Procris** Chatham Dyd.
Ord ____ 1818?; K 03.1821; L 21.6.1822; 1837 sold.
**Algerine** Deptford Dyd.
Ord ____ 1818?; K 04.1821; L 10.6.1823; 1826 foundered.
**Philomel** Portsmouth Dyd.
Ord ____ 1820?; K 06.1821; L 28.4.1823; 1833 sold.
**Magnet** Woolwich Dyd.
Ord ____ 1818?; K 06.1821; L 13.3.1823; completed as a packet; 1847 sold.
**Royalist** Portsmouth Dyd.
Ord ____ 1820?; K 08.1821; L 12.5.1823; 1826 tender; 1838 sold.
**Zephyr** Pembroke Dyd.
Ord ____ 1818?; K 11.1821; L 1.11.1823; completed as a packet; 1836 sold.
**Hope** Plymouth Dyd.
Ord ____ 1822?; K 03.1822; L 8.12.1824; completed as a packet; 1846 hulked.
**Mutine** Plymouth Dyd.
Ord ____ 1822?; K 04.1822; L 19.5.1825; 1826 packet; 1841 sold.
**Tyrian** Woolwich Dyd.
Ord ____ 1823?; K 04.1823; L 16.9.1826; completed as a packet; 1845 hulked.
**Leveret** Portsmouth Dyd.
Ord ____ 1823?; K 05.1823; L 20.2.1825; 1826 tender; 1843 sold.
**Musquito** Portsmouth Dyd.
Ord ____ 1823?; K 05.1823; L 20.2.1825; 1843 sold.
**Hearty** Chatham Dyd.
Ord ____ 1823?; K 07.1823; L 22.10.1824; 1827 packet; 1827 burnt.
**Myrtle** Portsmouth Dyd.
Ord ____ 1823?; K 07.1823; L 14.9.1825; 1827 packet; 1829 wrecked.
**Lapwing** Chatham Dyd.
Ord ____ 1823?; K 09.1823; L 20.2.1825; 1829 packet; 1845 hulked.
**Sheldrake** Pembroke Dyd.
Ord ____ 1823?; K 11.1823; L 17.5.1825; completed as a packet; 1853 sold.
**Rapid** Portsmouth Dyd.
Ord ____ 1823?; K 01.1824; L 17.8.1829; 1838 wrecked.
**Harpy** Chatham Dyd.
Ord ____ 1823?; K 03.1824; L 16.7.1825; 1841 sold.
**Fairy** Chatham Dyd.
Ord ____ 1823?; K 07.1824; L 25.4.1826; 1832 survey ship; 1840 foundered.
**Reindeer** Plymouth Dyd.
Ord ____ 1823?; K 12.1824; L 29.9.1829; completed as a packet; 1841 hulked.
**Espoir** Chatham Dyd.
Ord ____ 1823?; K 01.1825; L 9.5.1826; 1853 sold.
**Recruit** Portsmouth Dyd.
Ord ____ 1823?; K 02.1825; L 17.8.1829; 1831 tender; 1832 foundered.
**Calypso** (ex *Hyena*) Chatham Dyd.
Ord ____ 1823?; K 03.1825; L 19.8.1826; completed as a yacht; 1829 packet; 1833 foundered.
**Skylark** Pembroke Dyd.
Ord ____ 1823?; K 05.1825; L 05.1826; completed as a packet; 1845 wrecked.
**Rolla** Plymouth Dyd.
Ord ____ 1823?; K 06.1825; L 10.12.1829; 1848 training ship; 1868 BU.
**Spey** Pembroke Dyd.
Ord ____ 1823?; K 07.1825; L 6.10.1827; completed as a packet; 1841 wrecked.
**Pigeon** (ex *Variable*) Pembroke Dyd.
Ord ____ 1823?; K 05.1826; L 6.10.1827; 1829 packet; 1847 sold.

*Briseis* Deptford Dyd.
>Ord ____ 1826?; K 08.1827; L 3.7.1829; 1838 foundered.

*Delight* Chatham Dyd.
>Ord ____ 1826?; K 08.1827; L 27.11.1829; completed as a packet; 1844 sold.

*Algerine* Chatham Dyd.
>Ord ____ 1826?; K 10.1827; L 1.8.1829; 1844 sold.

*Thais* Pembroke Dyd.
>Ord ____ 1823?; K 07.1828; L 12.10.1829; 1832 converted to a packet; 1833 foundered.

*Partridge* Pembroke Dyd.
>Ord ____ 1826?; K 08.1828; L 1.10.1829; 1836 tender; 1843 to the Coastguard.

*Curlew* Woolwich Dyd.
>Ord ____ 1825?; K 11.1829; L 25.2.1830; 1847 mortar brig; 1849 BU.

*Nautilus* Woolwich Dyd.
>Ord ____ 1825?; K 04.1829; L 11.3.1830; 1878 BU.

*Savage* Plymouth Dyd.
>Ord ---- 1823?; K 10.1829; L 29.12.1830; 1853 yard craft.

*Wizard* Pembroke Dyd.
>Ord ____ 1826?; K 10.1829; L 24.5.1830; 1859 lost.

*Saracen* Plymouth Dyd.
>Ord ____ 1823?; K 12.1829; L 30.1.1831; 1854 survey ship; 1862 sold.

*Charybdis* Portsmouth Dyd.
>Ord ____ 1826?; K 12.1829; L 28.2.1831; classed as a 3 gun brig; 1843 sold.

*Scorpion* Plymouth Dyd.
>Ord ____ 1823?; K 06.1830; L 28.7.1832; 1849 survey ship; 1858 hulked.

## Completed as Brigantines. *Termagant* Portsmouth Dyd.
>Ord ____ 1825?; K 10.1829; L 26.3.1838; 1845 sold.

*Buzzard* Portsmouth Dyd.
>Ord ____ 1826?; K 12.1829; L 25.3.1834; 1843 sold.

*Griffon* Chatham Dyd (probably begun at Deptford Dyd).
>Ord 23.5.1820 (cancelled 1828 then transferred?)
>Ord ____ 1829?; K 07.1830 L 11.9.1832; 1869 sold.

*Lynx* Portsmouth Dyd.
>Ord ____ 1825?; K 02.1830; L 2.9.1833; 1845 sold.

*Forester* Chatham Dyd.
>Ord ____ 1830?; K 09.1830; L 28.8.1832; 1843 sold.

## Cancelled in 1830? *Forester* Deptford Dyd.
>Ord ____ 1824.

## Cancelled in 1831. *Sealark* Plymouth Dyd.
>Ord ____ 1823?; K 11.1830

*Foxhound* Plymouth Dyd.
>Ord ____ 1824; not begun ?

*Helena* Plymouth Dyd.
>Ord ____ 1826?; not begun?

*Calypso* [renamed *Hyena* 1826] Deptford Dyd.
>Ord ____ 1823; not begun?

*Halcyon* Deptford Dyd.
>Ord 21.11.1818 (cancelled 21.2.1831)

FATES: *Cadmus* to the Coastguard at Whitstable; later WV 24; 1864 sold. *Achates* wrecked at Guadeloupe 7.2.1810. *Parthian* wrecked near Marabout Island, Egypt. *Badger* became a receiving ship at the Cape of Good Hope; 1860 BU. *Briseis* wrecked on Point Pedras, Chile 5.11.1816. *Ephira* wrecked near Cadiz 26.12.1811. *Jasper* wrecked under Mount Batten, Plymouth Sound 20.1.1817. *Chanticleer* to the Coastguard at the River Crouch; later WV 13; 1865 sold. *Wild Boar* wrecked on the Runnelstone, Scilly Isles 15.2.1810. *Redpole* sailed from Rio de Janeiro and sunk in action with the pirate ship *Congress* off Cape Frio 10.8.1828. *Bermuda* wrecked South of Tampico Bar 16.11.1816. *Drake* wrecked on the Eastern end of Saint Shotts 23.6.1822. *Rhodian* wrecked at Jamaica, wreck sold 21.1.1813. *Sarpedon* supposed foundered 12.1812. *Eclipse* to the Coastguard at ____ ; later WV 21; 1865 sold. *Delight* wrecked in a cyclone at Mauritius 23.2.1824. There is a suggestion that *Falcon* may have had an engine fitted 1833-1834 but this is not borne out by the records sighted. After being sold mercantile in 1838 she was renamed *Waterwitch*. *Emulous* to the Coastguard at 'Bugsley Hole' then to 'Haven Hole' ; later WV 13; 1865 sold. *Beagle* to the Coastguard at Paglesham; later WV 7; 1870 sold. *Ariel* foundered in a hurricane near Sable Island 18.12.1828. *Jasper* wrecked off Santa Maura, wreck sold 11.10.1828. *Partridge* stranded on the Island of Vlieland 27.11.1824, wreck later sold. *Reynard* became a mooring vessel at Chatham; 1857 BU. *Plover* became a lazaretto at Sheerness; 1838 lent as an accommodation ship for the Thames Tunnel (Wapping); 1839 to Woolwich as a lazaretto; 1841 sold. *Algerine* capsized in a squall in the Mediterranean 9.1.1826. *Hope* became a lazaretto and guard ship at Pembroke; 1863 fitted to receive the boilers of ships; 1882 BU. *Tyrian* became a lazaretto at the Motherbank, Portsmouth; 1864 receiving ship; 1866 to the Coastguard. *Hearty* was supposed burnt at sea in 09.1827. *Myrtle* wrecked at Halifax 3.4.1829. *Lapwing* became a breakwater at Keyham, Plymouth; 1864 sold. *Rapid* was wrecked at Crete 12.4.1838. *Fairy* sailed from Harwich and not heard of again 12.11.1840. *Reindeer* became a guard ship at Gibraltar; 1857 sold. *Recruit* supposed foundered on passage from Halifax to Bermuda in 1832. *Calypso* foundered on passage between Halifax and England in 02.1833. *Skylark* wrecked on Kimmeridge Ledge near Weymouth 25.4.1845. *Spey* wrecked on Racoon Key 28.11.1841, wreck then sold. *Briseis* supposed foundered in the Atlantic between Falmouth and Halifax 01.1838. *Thais* supposed foundered on passage to Halifax 12.1833. *Partridge* to the Coastguard for Southampton at Netley; 1864 sold. *Savage* became a chain lighter at Malta; 1864 still in service. *Wizard* wrecked at Berehaven 8.2.1859. *Scorpion* became a floating police station for the (Thames) River Police at Blackwall; 1874 BU.

PLANS: Lines/profile & decks/framing/midships section/sails. *Goldfinch* as packet 1823: profile & decks. *Redpole* as packet 1823: Profile & decks. *Delight* as packet 1823: profile & decks. *Barracouta* 1829 as packet: Profile/lower deck/upper deck/tops. *Frolic* as packet 1823: profile & decks. *Procris* 1829: profile/decks. *Algerine* as built as yacht: Profile & decks. *Magnet* as built as packet: profile & decks. *Zephyr* as built as packet: profile & decks. *Hope* as built as packet: profile & decks. *Zephyr* as packet: profile & decks. *Tyrian* as built as packet: profile & decks. As coastguard hulk 1867: lower deck/upper deck. *Curlew* & *Nautilus*: Profile/lower deck/upper deck/mortar bed (side view)/(plan view). As brigantines (*Lynx, Forrester, Buzzard, Termagant, Griffon*): Lines/profile & decks/midships section/sails. *Griffon*: pivot gun mounting.

## *Carron* (*Alban*) Class 1824 Paddle Steamers, Originally Ordered as *Cherokee* Class Brigs [?] and Modified From Their Lines [?]
____ Modified design by Seppings 'for a particular service'.
>Dimensions & tons: Originally: 100', 82'3⅝" x 24'6" x 10'6". 262 72/94.
>Modified to: 109'8", 91'11⅝" x 24'6" x 11'. 293 59/94.
>Men: ____ . Guns: ____ .

*Carron* Deptford Dyd.
>Ord ____ ; K 07.1824; L 9.1.1827; 1830 packet; 1844 hulked.

*African* (ex *Dee*) Woolwich Dyd.
>Ord ____ ; K 09.1824; L 30.8.1825; delivered to the Colonial Department for African service; 1842 tug; 1862 BU.

*Alban* Deptford Dyd.
>Ord ____ ; K 07.1824; L 27.12.1826; later (1840) lengthened to 145'; 1860 BU.

*Echo* Woolwich Dyd.
>Ord ____ ; K 12.1824; L 28.5.1827; 1840 tug; 1885 sold.

*Confiance* Woolwich Dyd.
>Ord ____ ; K 02.1825; L 28.3.1827; 1842 tug; Ca 1872 BU.

*Columbia* Woolwich Dyd.
>Ord ____ ; K 03.1827; L 1.7.1829; lengthened by 19' whilst building; 1834 troopship; 1859 sold.

FATES: *Carron* became a coal hulk on the Thames; 1872 deleted.

PLANS: Lines/profile/lower deck/upper deck. *Alban*: 1835: sails. Lengthened 1840: Lines/profile/lower deck/upper deck/midships section. 1847: sails. *Echo*: 1835: midships section. 1836: Lower deck/upper deck. As tug: Lower deck. *Confiance* 1836: Lower deck/upper deck. *Columbia*: 1834 as troopship: Lines & profile/lower deck/upper deck.

## *Rapid* Class 1808. 16 Brig.
Originally referred to as a 'schooner'. 'Proposed by Mr. Peake .... to fight 12 guns a side.'
>Dimensions & tons: 107', 92'7" x 23'x 12'3". 260 48/94.

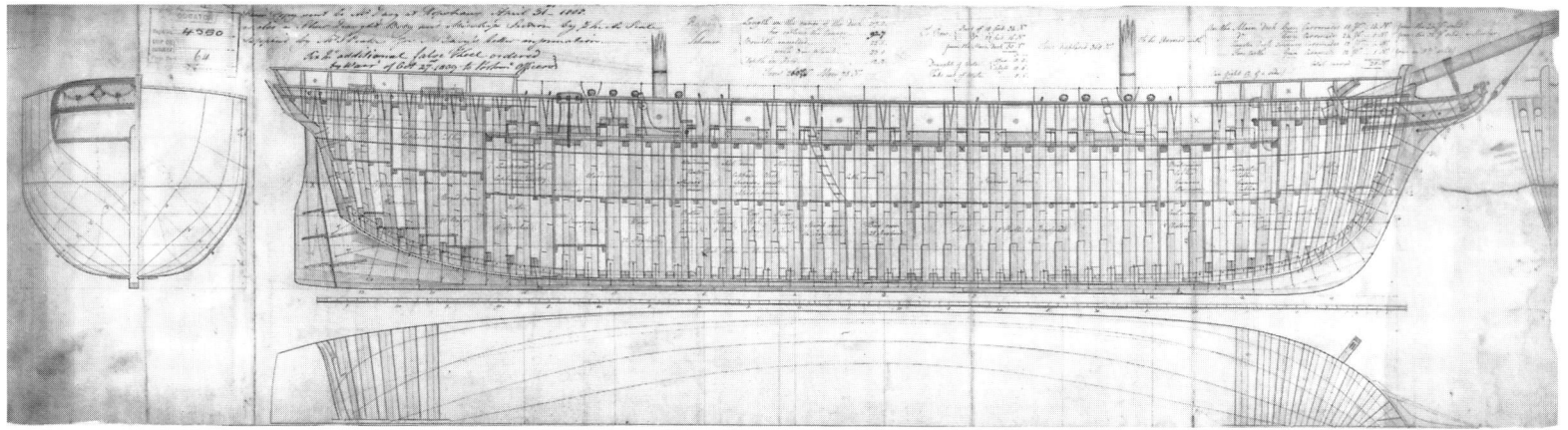

Lines and framing plan (4550) for the brig *Rapid* of 1808

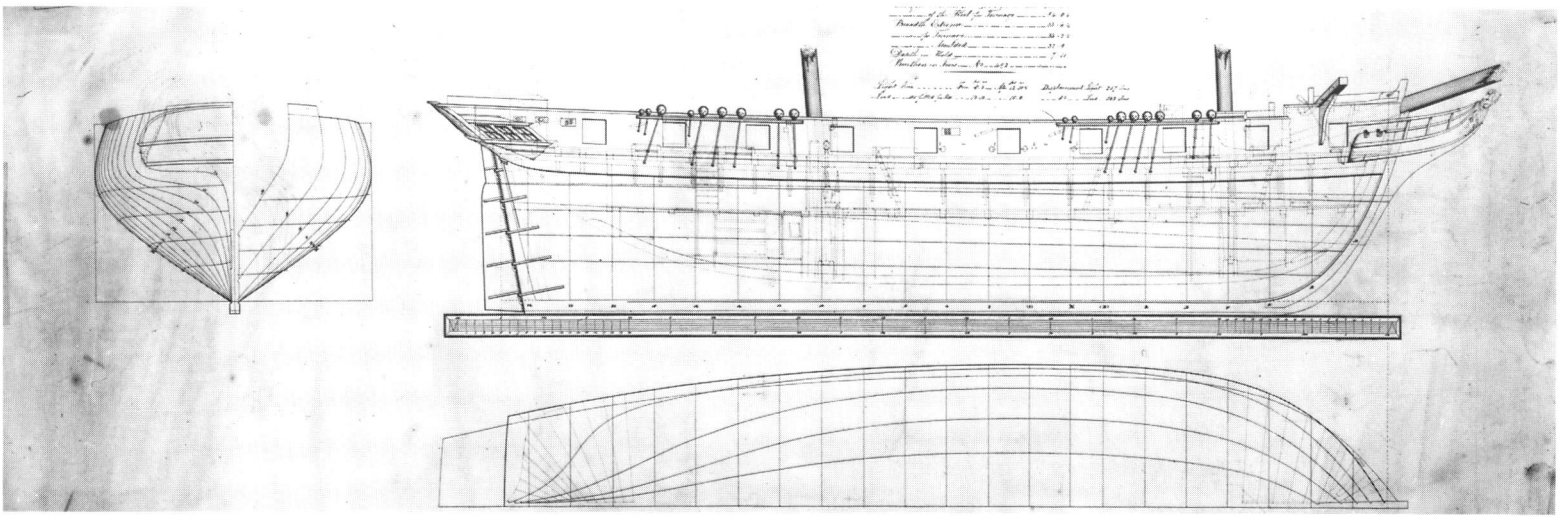

Lines and profile plan (3628) showing the brig *Columbine* of 1826 'as fitted' in 1834

Men: 75. Guns: UD 14 x 18 'gun carronades' + 1 x 24 'as mortar', QD 4 x 12★, Fc 1 x 12 'gun carronade'.
*Rapid* Davy, Topsham.
  Ord ____ 1807; K 05.1808; L 22.11.1808; 1814 sold.
PLANS: Lines & profile/specification.

### *Primrose* Class 1809. 18, Brig Sloop. Peake design.
Dimensions & tons: 108', 87' x 28'9" x 13'6". 382 47/94.
Men: 121. Guns: 2 x 6 + 16 x 32★.
*Primrose* Portsmouth Dyd.
  Ord ____ 1809?; K 05.1809; L 22.1.1810; 1824 converted to a ship sloop; 1832 BU.
PLANS: Lines/profile/lower deck/upper deck.

### *Icarus* Class 1813. 10, Brig. 'The Draught proposed and the ship to be built by the superior class of shipwright apprentices in Portsmouth Yard.'
Dimensions & tons: 90', 71'5¾" x 24'10" x 11'. 234.
Men: 50. Guns: 2 x 6 + 8 x 18★.
*Icarus* Portsmouth Dyd.
  Ord ____ 1812?; K 03.1813; L 18.8.1814; 1838 to the Coastguard.
FATE: To the Coastguard at Lymington; 1861 sold.
PLANS: No plans.

### *Colibri* Class 1813. ____ , Brigs. Designer uncertain. Frames built of fir at Chatham, taken down in 03.1814 and sent to Halifax, Nova Scotia to be sent on to the Great Lakes, to be completed there, or to be assembled at Halifax 'should it not be found practicable and advisable to construct [them] on the Lakes'.
Dimensions & tons: ____ ____ ____ .
Men: ____ . Guns: ____ .
*Colibri, Goshawk* Both Chatham Dyd.
  Ord ____ ; K 01.1814; 03.1814 to Halifax; 1815 sold there.

PLANS: No plans.

### *Columbine* Class 1825. 18, Brig Sloop (later Brigantine).
Design by Lieutenant W.Symonds, RN. On 12.3.1825 the Honourable George Vernon was called upon to give a bond of £20,000 'for the fulfilment of a promised engagement to purchase this vessel [from the Navy] if she did not answer'. She did 'answer', but this gesture of support helps to explain why Symonds named a frigate *Vernon* after his patron (and not, as might have been expected, after Admiral Vernon) after becoming Surveyor of the Navy.
Dimensions & tons: 105'0½", 84'0⅛" x 33'6¼" x 7'11". 492.
Men: 125. Guns: 2 x 6 + 16 x 32★.
*Columbine* Portsmouth Dyd.
  Ord ____ 1826??; K 01.1826; L 1.12.1826; 1854 hulked.
FATE: Became a coal depot at Sheerness; 1892 sold.
PLANS: Lines & profile/profile/lower deck/upper deck & forecastle/sails/detail of chains. As built: Lines & profile & decorations.

## Bomb Vessels

### *Vesuvius* Class 1812. Peake design. Ship rigged vessels, with hulls very much in the frigate/sloop tradition, rather than the hitherto normal pink-sterned pattern of previous bombs. The design was slightly altered an enlarged for the subsequent *Hecla* Class, though *Hecla* herself, *Infernal* and , probably, *Fury* appear to have originally been intended to be built to the *Vesuvius* design, and only altered to the later version whilst building.
Dimensions & tons: 102', 83'10" x 27' x 12'6". 325 7/94.
Men: 67. Guns: 1 x 13" + 1 x 10" mortar, 2 x 6 + 8 x 24★.
*Vesuvius* Davy, Topsham.
  Ord 30.3.1812; K 07.1812; L 1.5.1813; 1819 sold.
*Terror* Davy, Topsham.

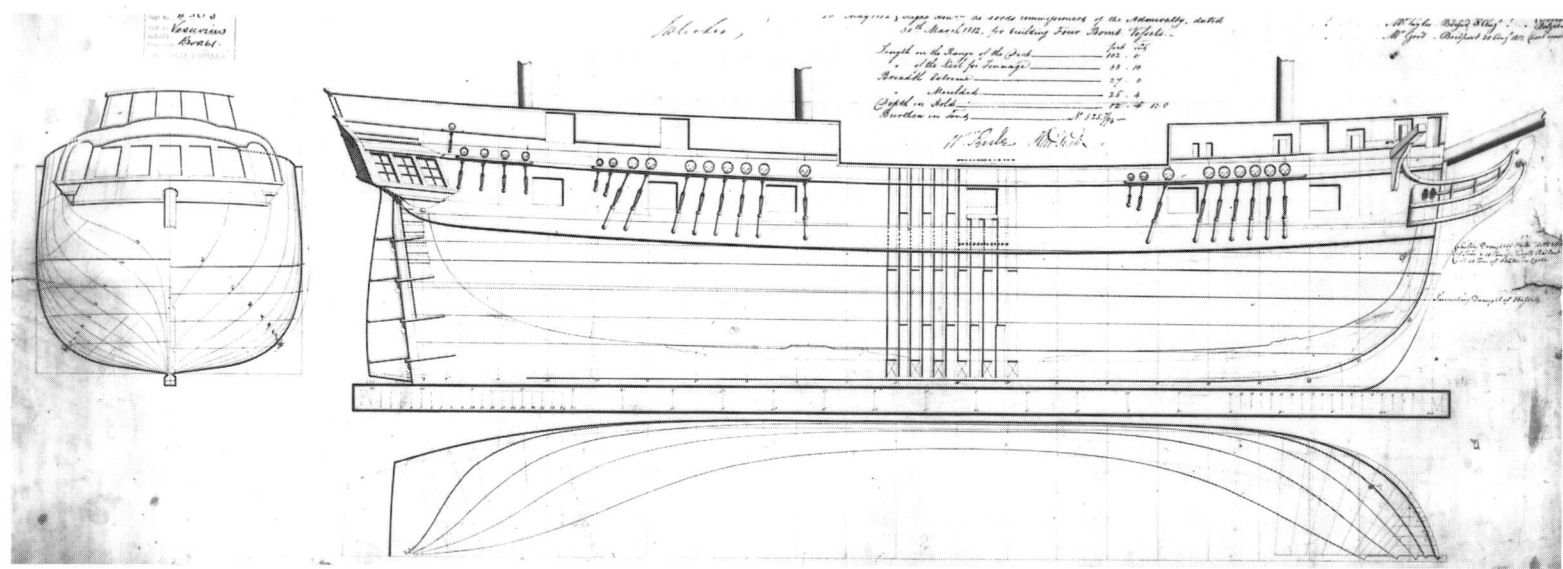

Design lines plan (4305) to which the 1812 *Vesuvius* Class bomb vessels were built

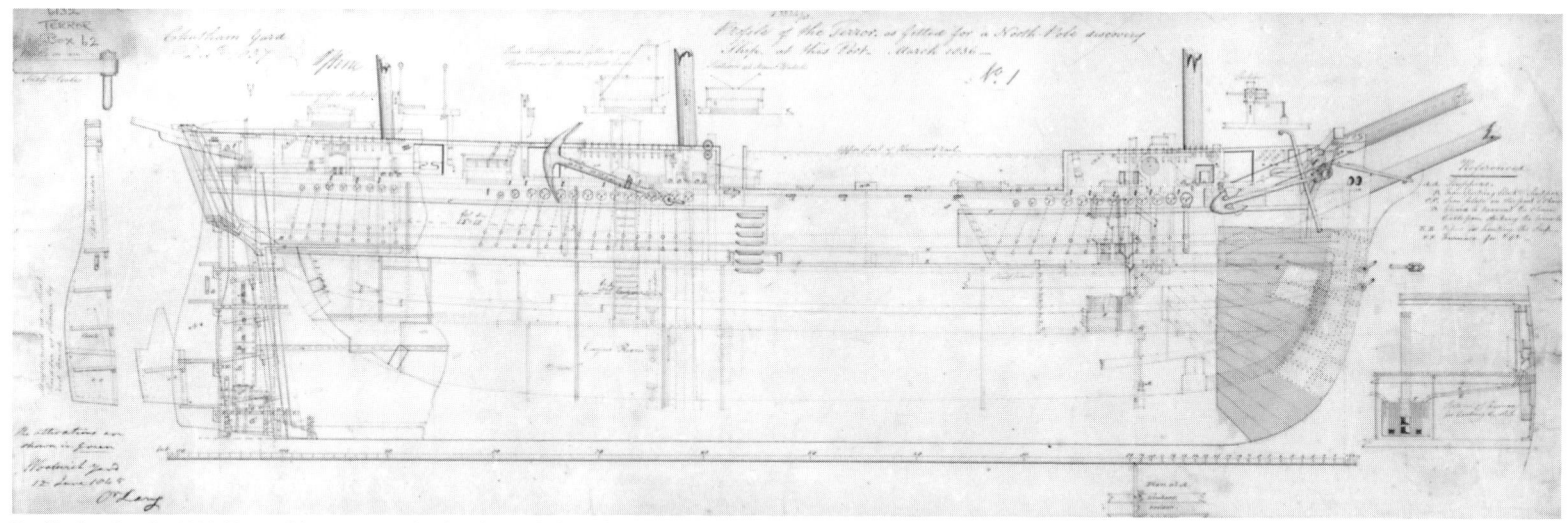

Profile showing the 1813 *Terror* of the *Vesuvius* Class bomb vessels fitted for Arctic explortion in 1837. Confusingly, the plan is altered in green to show alterations made in 1845 for the same purpose (including the addition of an auxiliary steam engine). Note the plated icebreaking bow, the detachable rudder, and the details of the heating system shown in the partial section

Ord 30.3.1812; K 09.1812; L 29.6.1813; 1836 exploration ship; 1845 fitted with auxiliary steam screw engine; 1848 lost.

**Beelzebub/Belzebub** Taylor, Biddeford
Ord 30.3.1812; K 11.1812; L 30.7.1813; 1820 BU.

## Cancelled 1812. **Beelzebub** Good, Bridport.
Ord 30.3.1812; not begun?

NOTE: Dittmar suggests that a vessel named *Thunder* was ordered from Bailey, Ipswich, and cancelled, but this may well be a confusion with the above (which he does not note). Colledge suggests a vessel of the same name was ordered from Brindley, Frindsbury; but this is unlikely.

FATES: *Terror* abandoned in the Arctic ice during the final and fatal Franklin expedition to find the North West Passage 22.4.1848.

PLANS: Lines/profile/orlop/lower deck/upper deck/midships section/specification. *Terror* (plans also apply to *Erebus* of next class) 1837 as Arctic ship: profile/orlop/lower deck/upper deck/midships section. 1839 as Arctic ship: profile/orlop/lower deck/upper deck/propeller aperture. *Beelzebub* 1816: decks.

## Hecla Class 1813. Peake design. Slightly enlarged version of the *Vesuvius* Class (see above).

Dimensions & tons: 105', 86'1¼" x 28'6" x 13'10". 372.
Men: 67. Guns: 1 x 13" + 1 x 10" mortar, 2 x 6 + 8 x 24★.

**Hecla** Barkworth & Hawkes, North Barton, Hull.
Ord ____ 1813?; K 07.1813; L 22.7.1815; 1819 exploration ship; 1831 sold.

**Infernal** Barkworth & Hawkes, North Barton, Hull.
Ord ____ 1813?; K 07.1813; L 26.7.1815; 1831 sold.

**Fury** Mrs. Ross, Rochester.
Ord ____ 1812?; K 09.1813; L 4.4.1814; 1821 exploration ship; 1825 lost.

**Meteor** Pembroke Dyd.
Ord ____ 1819?; K 05.1820; L 25.6.1823; 1832 renamed *Beacon* and converted to survey ship; 1846 sold.

**Etna/Aetna** Chatham Dyd.
Ord ____ 1819?; K 09.1821; L 14.5.1824; 1831 survey ship; 1839 hulked.

**Sulphur** Chatham Dyd.
Ord ____ 1819?; K 05.1824; L 26.1.1826; 1828 emigrant ship 'for New Holland'; 1835 survey ship; 1843 hulked.

**Erebus** Pembroke Dyd.
Ord ____ 1823?; K 10.1824; L 7.6.1826; 1839 exploration ship; 1845 fitted with an auxiliary steam screw engine; 1848 lost.

**Thunder** Deptford Dyd.
Ord ____ 1819?; K 11.1826; L 4.8.1829; 1833 survey ship; 1851 BU.

## Cancelled 1831. **Devastation** Plymouth Dyd.
Ord ____ 1814?; K ____ 1820.

**Volcano** Plymouth Dyd.
Ord ____ 1814?; K ____ 1821.

**Vesuvius** Deptford Dyd.
Ord ____ 1823; transferred 30.8.1828 to: Chatham Dyd.
Ord ____ K 02.1830.

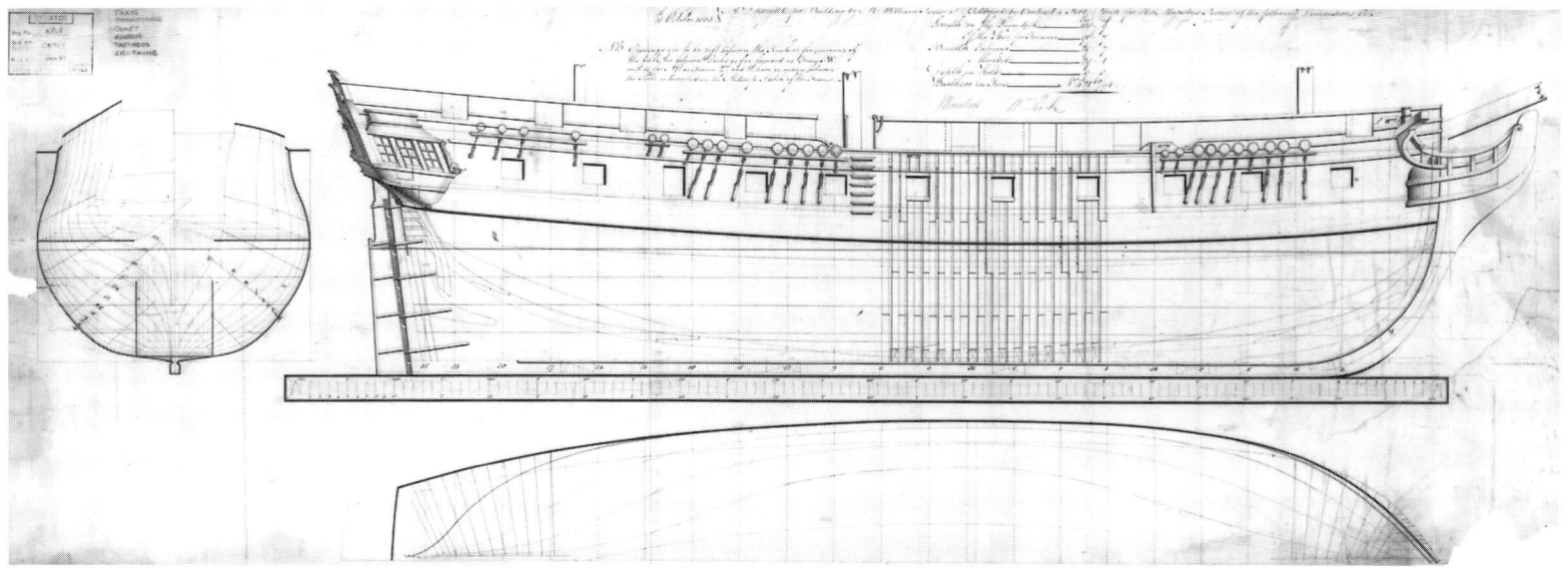

Design lines plan (6868) for fireships of the *Thais* Class of 1805, specifically identified with the *Comet*

**Cancelled 1832?** *Belzebub*? Plymouth Dyd.
   Ord ____ 1821?

FATES: *Fury* bilged in Prince Regent Inlet, Arctic 25.8.1825. *Etna* became a receiving ship at Portsmouth; ? to Liverpool later?; 1846 BU. *Sulphur* became a convict and receiving ship at Woolwich. *Erebus* abandoned in the Arctic ice during the final and fatal Franklin Expedition to find the North West Passage 22.4.1848.

PLANS: Lines/profile/orlop/lower deck/upper deck/midships section. *Fury* 1818: Profile/orlop/lower deck/upper deck. 1821 as discovery vessel: Profile & midships section/orlop/lower deck/upper deck/quarter deck. *Etna* as survey vessel 1822: Profile /orlop/quarter deck & forecastle. *Erebus* as Arctic vessel; see *Terror* of previous class. *Thunderer* as survey vessel 1833: Profile/orlop/lower deck/upper deck.

## Fireships (later reclassed as Sloops)

### *Thais* Class 1805. Adapted by Henslow from the design of the *Tisiphone* Class of 1780.
   Dimensions & tons: 108'9", 90'7" x 29'7" x 9'. 421 63/94.
   Men: 55. Guns: UD 16 x 24★ or 32★, Spar dk. 2 x 9 + 8 x 18★ (later: UD 14 x 18★, Fc 2 x 9).

*Thais* Tanner, Dartmouth.
   Ord ____ 1805?; K 10.1805; L 19.8.1806; 1808 reclassed as a ship sloop (later Sixth Rate); 1814 cut down (spar dk. removed?); 1818 sold.

*Tartarus* Davy, Topsham.
   Ord ____ 1805?; K 11.1805; L 10.1806; 1808 reclassed as a ship sloop (later Sixth Rate); 1816 sold.

*Prometheus* Thompson, Southampton.
   Ord ____ 1805?; K 12.1805; L 27.3.1807; 1808 reclassed as a ship sloop (later Sixth Rate); 1815 cut down (spar dk. removed); 1819 hulked.

*Lightning* Ayles, Topsham.
   Ord ____ 1805?; K 01.1806; L 14.10.1806; 1808 reclassed as a ship sloop (later Sixth Rate); 1816 sold.

*Comet* Taylor, Biddeford.
   Ord ____ 1805?; K 02.1806; L 25.4.1807; 1808 reclassed as a ship sloop (later Sixth Rate); 1815 sold.

*Erebus* Owen, Topsham.
   Ord ____ 1805?; K 01.1806; L 20.8.1807; 1808 reclassed as a ship sloop (later Sixth Rate); 1819 sold.

FATES: *Prometheus* became a lazaretto at Portsmouth; ? later receiving ship?; 1839 renamed *Veteran*; 1852 BU.

PLANS: Lines/profile & midships section/orlop/lower deck/upper deck.

## Gunboats, Gunbrigs, Floating Batteries

### *Conquest* Class 1794. 14, Gunboats/Gun Vessels.
Henslow design 'to row with 18 oars'. Originally schooner or brigantine rig? Some were later fitted with Schanck 'sliding keels' (daggerboards).
   Dimensions & tons: 75', 62'3 1/8" x 21' x 7'. 146 41/94.
   Men: 50. Guns: 2 x 24 bow chase guns + 10 x 18★ broadside guns + 2 x 4 or 18★ stern chase guns.

*Aimwell* Perry, Blackwall.
   Ord ____ 1794?; K 03.1794; L 12.5.1794; 1811 BU.

*Attack* Wilson, Frindsbury.
   Ord ____ 1794?; K 03.1794; L 29.8.1794; 1796 fitted with sliding keels; 1802 sold.

*Borer* Randall, Rotherhithe.
   Ord ____ 1794?; K 03.1794; L 17.5.1794; 1796 fitted with sliding keels; 1808? breakwater.

*Conquest* Brindley, Frindsbury.
   Ord ____ 1794?; K 03.1794; L 07.1794; 1796 fitted with sliding keels; 1817 sold.

*Fearless* Cleverley, Gravesend.
   Ord ____ 1794?; K 03.1794; L 06.1794; 1804 wrecked.

*Force* Pitcher, Northfleet.
   Ord ____ 1794?; K 05.1794; L ____ 1794; 1796 fitted with sliding keels & brig rig; 1802 sold.

*Pelter* Perry, Blackwall.
   Ord ____ 1794?; K 03.1794; L 12.5.1794; 1798 brig rig; 1802 sold.

*Piercer* King, Dover.
   Ord ____ 1794?; K 03.1794; L 2.6.1794; 1797 brig rig; 1802 sold.

*Plumper* Randall, Rotherhithe.
   Ord —— 1794?; K 03.1794; L 17.4.1795; 1798 brig rig; 1802 sold.

*Swinger* Hill, Limehouse.
   Ord ____ 1794?; K 03.1794; L 31.5.1794; 1802 sold.

*Teaser* Dudman, Deptford.
   Ord ____ 1794?; K 03.1794; L 26.5.1794; 1798 `altered' (to brig rig? or to sliding keels? or both?); 1802 sold.

*Tickler* Wells, Deptford.
   Ord ____ 1794?; K 03.1794; L 28.5.1794; 1798 brig rig; 1802 sold.

FATES: *Fearless* wrecked in Cawsand Bay 02.1804.

PLANS: Lines/profile & decks & midships section. Vessels fitted with Shanck sliding keels: profile & decks.

### *Musquito* Class 1794. 4, Rowing Gun Boats. Design by Sir Sidney Smith. Extraordinary experimental vessels with teardrop-shaped plan view (broad forward. tapering aft) flat bottomed, but with a keel. The sides

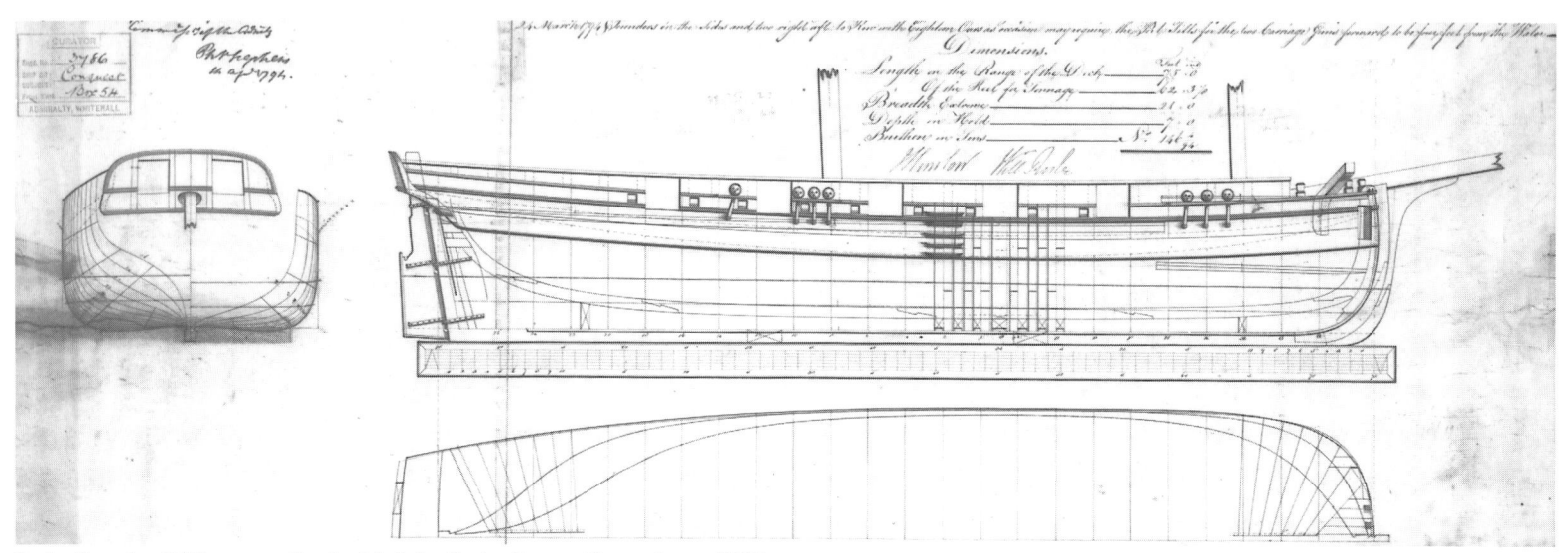

Design lines plan (3786), approved by the Admiralty, for the *Conquest* Class gunboats of 1794

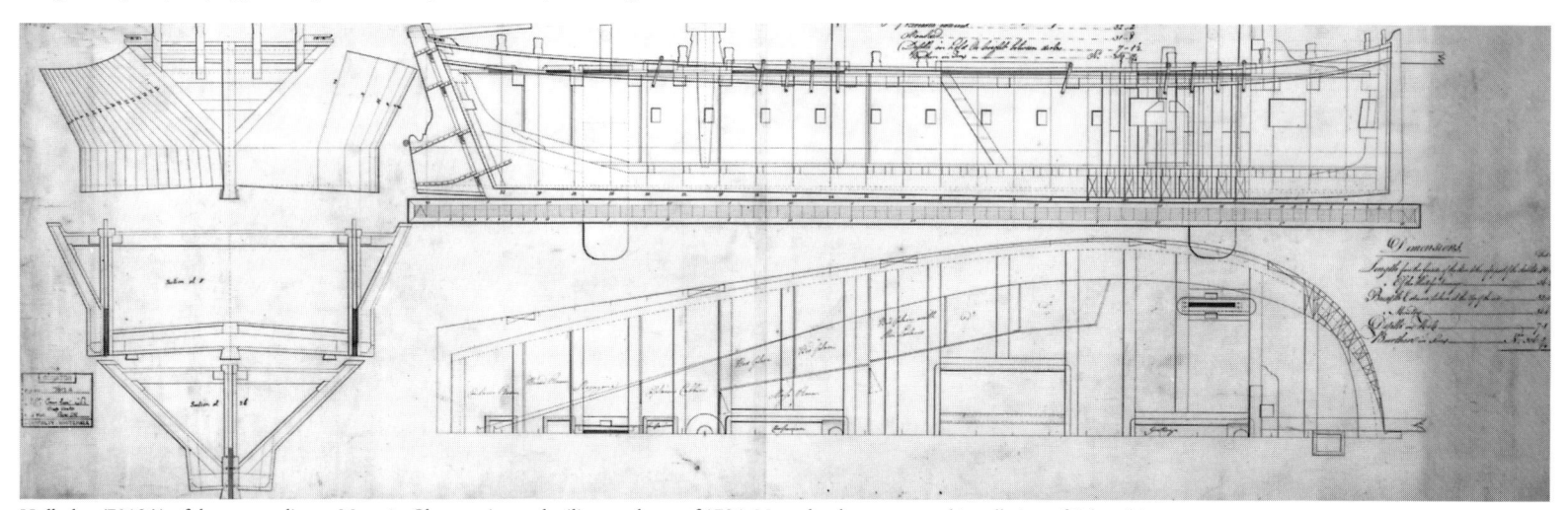

Hull plan (7012A) of the extraordinary *Musquito* Class rowing and sailing gunboats of 1794. Note the three-cornered installation of Schanck's 'sliding keels'

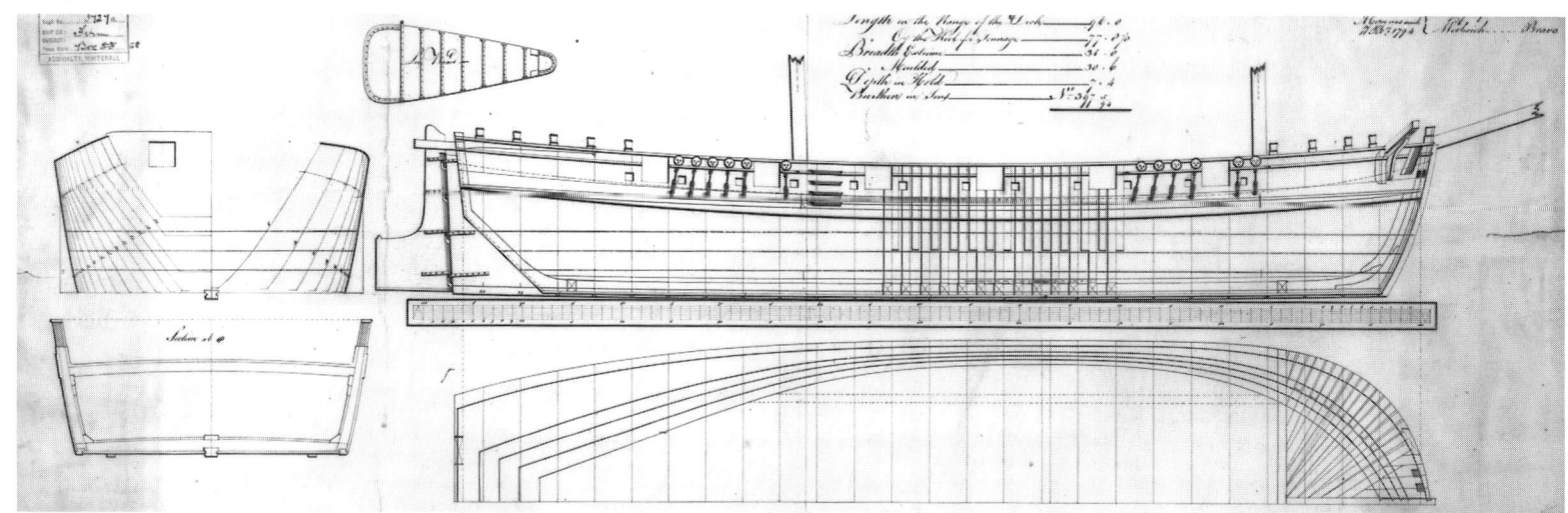

Lines, midsection and leeboard building plan (3727A) for the gunboats or floating batteries of the *Firm* Class of 1794. The similarity to the shape of a Thames sailing barge is emphasised by the leeboards

were straight from the turn of the bilge to half way up, where they flared sharply out. Two side by side Schanck type 'sliding keels' were fitted forward and a single one on the centre line aft. Topsail schooner rig.

    Dimensions & tons: 80', ____ x 32' x 7'10". 306 3/94.
    Men: 149. Guns: UD 2 x 24, QD 2 x 68★.

**Musquito** Wells, Deptford.
    Ord ____ 1794?; K ____ ; L ____ 1794; 1795 wrecked.
**Sandfly** Wells, Deptford.
    Ord ____ 1794?; K ____ ; L ____ 1794; 1803 BU.

FATES: *Musquito* lost on the coast of France 05.1795.

PLANS: Lines & profile & decks & sections/sails.

### *Firm* Class 1794. 16, Floating Batteries/Gunboats.

Henslow design. Flat bottomed, resembled enlarged Thames sailing barges, and probably had a similar rig. *Bravo* was briefly classed as a Sixth Rate in 1794.

    Dimensions & tons: 96', 77'8 1/8" x 31' x 7'4". 397.
    Men: 100. Guns: 16 x 18★.

**Firm** Deptford Dyd.

# HENSLOW TO SEPPINGS 1786-1830

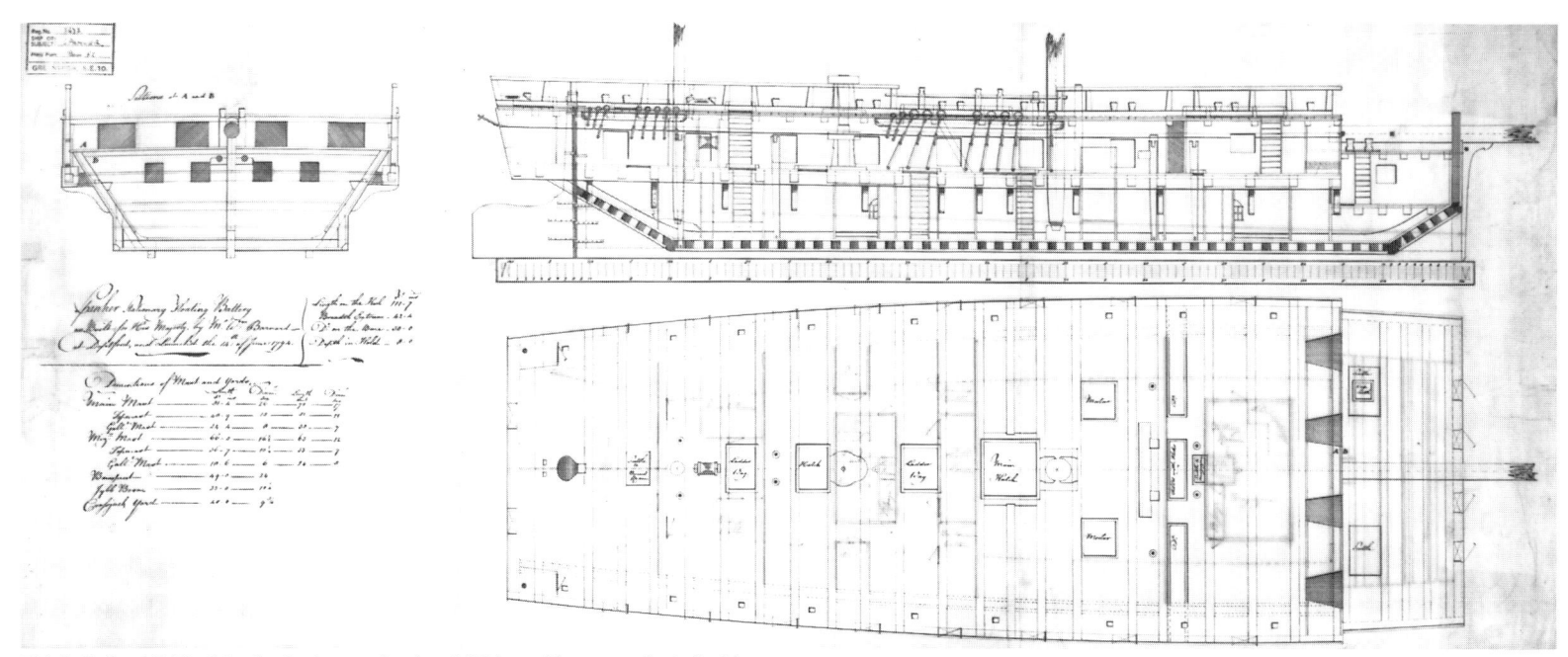

This hull plan (6742) of the floating battery *Spanker* of 1794 would appear to be 'as built'

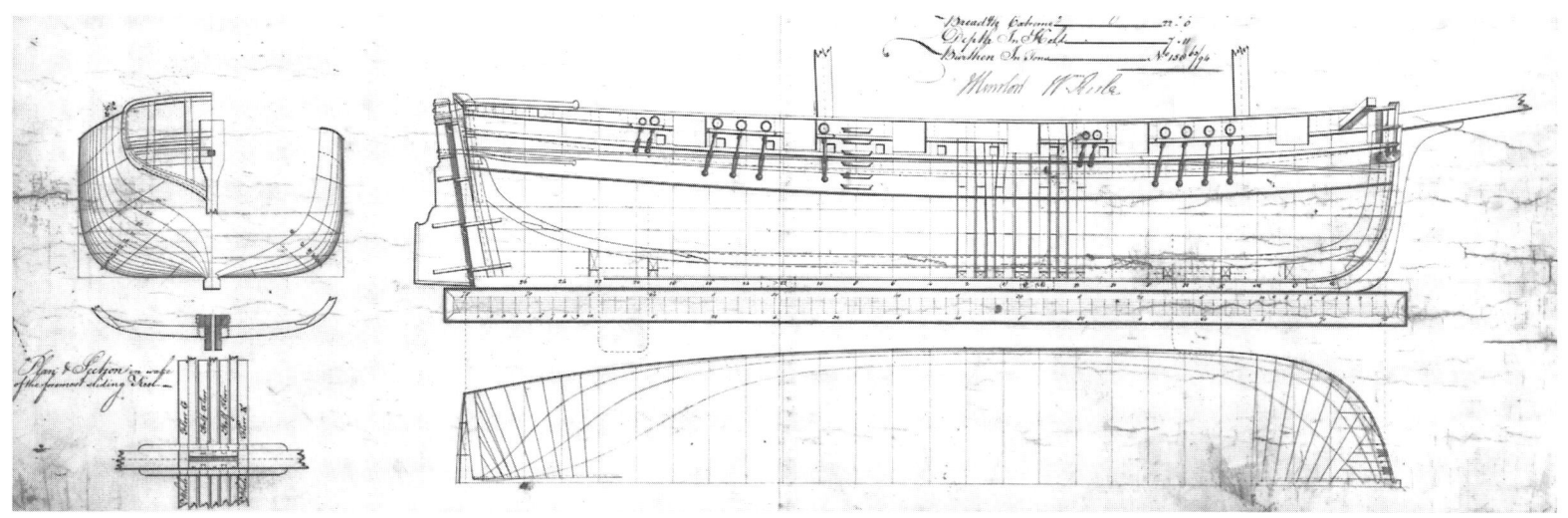

This building lines plan (3782) for gunboats of the *Acute* Class of 1797 also shows the details of their Schanck 'sliding keel' installation. This is for the three vessels built by Dudman - *Blazer, Clinker* and *Cracker*

Ord ____ 1794?; K 3.2.1794; L 1.5.1794; 1802 sold.
**Bravo** Woolwich Dyd.
    Ord ____ 1794?; K 3.2.1794; L 31.5.1794; 1803 deleted.
PLANS: Lines & midships section & leeboard/profile/decks.

## *Spanker* Class 1794. Stationary Floating Battery.

Richard White design. Form basically that of a swim barge, or radeau. Would appear to have been rigged as a ketch or spritsail barge.
    Dimensions & tons: 111'6" (on the keel), ____ x 42'4" x 7'8". ____ .
    Men: 150. Guns: Lower tier forward 4 x 24, Upper tier forward 4 x 42, stern 2 x 24, broadsides 4 x 42★, also 2 x 10" land service mortars carried.
**Spanker** Barnard, Deptford.
    Ord ____ 1794?; K 03.1794; L 14.6.1794; 1795 hospital ship.
FATE: Hospital ship at Sheerness; 1810 struck off the list of the Navy.
PLANS: Profile & lower deck & midships section/upper deck.

## *Acute* Class 1797. 14, Gunboats/Gun Vessels.

Henslow design. Brig rig and fitted with Schanck 'sliding keels'. Initially numbered, but given names shortly after completion.
    Dimensions & tons: 75', 61'7⅝" x 22' x 7'11". 158 63/94.
    Men: 50. Guns: 2 x 24 (in bow) + 12 x 18★.
**Acute** (No 6) Randall, Rotherhithe.
    Ord 7.2.1797; K 02.1797; L 04.1797; 1802 sold
**Adder** (No 17) Barnard, Deptford.

    Ord 7.2.1797; K 02.1797; L 22.4.1797; 1798 lengthened (97' [83'7⅜" keel for tonnage], 215 29/94 tons); 1805 BU.
**Asp** (No 5) Randall, Rotherhithe.
    Ord 7.2.1797; K 02.1797; L 10.4.1797; after 1816 discarded.
**Assault** (No 4) Randall, Rotherhithe.
    Ord 7.2.1797; K 02.1797; L 10.4.1797; 1807 or 1825 yard craft.
**Biter** (No 10) Wells, Deptford.
    Ord 7.2.1797; K 02.1797; L 13.3.1797; 1802 sold.
**Blazer** (No 12) Dudman, Deptford.
    Ord 7.2.1797; K 02.1797; L 14.4.1797; 1803 sold.
**Bouncer** (No 8) Wells, Deptford.
    Ord 7.2.1797; K 02.1797; L 04.1797; 1802 sold.
**Boxer** (No 9) Wells, Deptford.
    Ord 7.2.1797; K 02.1797; L 11.4.1797; 1809 sold.
**Bruiser** (No 11) Wells, Deptford.
    Ord 7.2.1797; K 02.1797; L 11.4.1797; 1802 sold.
**Clinker** (No 14) Dudman, Deptford.
    Ord 7.2.1797; K 02.1797; L 28.4.1797; 1802 sold.
**Contest** (No 16) Barnard, Deptford.
    Ord 7.2.1797; K 02.1797; L 11.4.1797; 1799 wrecked.
**Cracker** (No 13) Dudman, Deptford.
    Ord 7.2.1797; K 02.1797; L 25.4.1797; 1802 sold.
**Crash** (No 15) Barnard, Deptford.
    Ord 7.2.1797; K 02.1797; L 5.4.1797; 1802 sold.
**Sparkler** (No 7) Randall, Deptford.

Ord 7.2.1797; K 02.1797; L 04.1797; 1802 sold.
*Spiteful* (No 18) Barnard, Deptford.
Ord 7.2.1797; K 02.1797; L 24.4.1797; 1804 hulked.

FATES: *Assault* became a lighter at Portsmouth, probably in 1807, though there is a note of her conversion to a lighter in 1825 in the Progress Books; 1827 sold. *Contest* wrecked on the coast of Holland 28.8.1799. *Spiteful* became a convict hospital ship attending *Captivity* at Portsmouth; 1823 BU.

PLANS: Lines/profile & decks/midships section.

### *Courser* Class 1797. 12 Gunboats/Gun Vessels. Rule design. Brig rig and fitted with Schanck 'sliding keels'. Originally numbered but given names shortly after completion.

Dimensions & tons: 76', 62'2⅝" x 22'6" x ___ . 167 50/94.
Men: 50. Guns 2 x 24 bow guns + 10 x 18★.

*Courser* (No 20) Hill & Mellish, Limehouse.
Ord 7.2.1797; K 02.1797; L 25.4.1797; 1803? to the Customs.
*Defender* (No 21) Hill & Mellish, Limehouse.
Ord 7.2.1797; K 02.1797; L 21.5.1797; 1802 sold.
*Eclipse* (No 22) Perry, Blackwall.
Ord 7.2.1797; K 02.1797; L 29.3.1797; 1802 sold.
*Flamer* (No 24) Perry, Blackwall.
Ord 7.2.1797; K 02.1797; L 30.3.1797; 1802 sold.
*Furious* (No 23) Perry, Blackwall.
Ord 7.2.1797; K 02.1797; L 31.3.1797; 1802 sold.
*Furnace* (No 25) Perry, Blackwall.
Ord 7.2.1797; K 02.1797; L 10.4.1797; 1802 sold.
*Gallant* (No 29) Pitcher, Northfleet.
Ord 7.2.1797; K 02.1797; L 04.1797; 1802 sold.
*Grappler* (No 28) Pitcher, Northfleet.
Ord 7.2.1797; K 02.1797; L 04.1797; 1803 blown up.
*Griper* (No 27) Pitcher, Northfleet.
Ord 7.2.1797; K 02.1797; L 10.4.1797; 1802 sold.
*Growler* (No 26) Pitcher, Northfleet.
Ord 7.2.1797; K 02.1797; L 10.4.1797; 1797 taken.
*Hardy* (No 30) Cleverley, Gravesend.
Ord 7.2.1797; K 02.1797; L 10.4.1797; 1802 sold.
*Hasty* (No 33) Wilson, Frindsbury.
Ord 7.2.1797; K 02.1797; L 06.1797; 1802 sold.
*Haughty* (No 31) Cleverley, Gravesend.
Ord 7.2.1797; K 02.1797; L 04.1797; 1802 sold.
*Hecate* (No 32) Wilson, Frindsbury.
Ord 7.2.1797; K 02.1797; L 2.5.1797; 1809? breakwater.

*Steady* (No 19) Hill & Mellish, Limehouse.
Ord 7.2.1797; K 02.1797; L 24.4.1797; 1803 sold.
*Tigress* (No 45) Brindley, Lynn.
Ord ___ ; K 03.1797; L 11.9.1797; 1802 sold.

FATES: *Grappler* grounded, set on fire by the French, and blew up at the Isles de Chosé 30.12.1803. *Growler* taken off Dungeness by two French rowing boats 20.12.1797; 1.8.1809 retaken?; if so not put back in service.

PLANS: Lines/profile & decks/midships section.

### *Bloodhound* Class 1801. 12, Gunboats later Gunbrigs. Henslow design.

Dimensions & tons: 80', 65'6½" x 23' x 8'6". 184 39/94.
Men: 50. Guns: 2 x 18 or 32★ in the bow + 10 x 18★.

*Basilisk* Randall, Rotherhithe.
Ord 10.1.1801; K 01.1801; L 2.4.1801; 1814 tender; 1818 sold.
*Bloodhound* Randall, Rotherhithe.
Ord 10.1.1801; K 01.1801; L 2.4.1801; 1816 sold.
*Censor* Randall, Rotherhithe.
Ord 10.1.1801; K 01.1801; L 2.4.1801; 1816 sold.
*Escort* Perry, Blackwall.
Ord 10.1.1801; K 01.1801; L 1.4.1801; 1815 to the Customs.
*Ferreter* Perry, Blackwall.
Ord 10.1.1801; K 01.1801; L 4.4.1801; 1807 taken.
*Jackall* Perry, Blackwall.
Ord 10.1.1801; K 01.1801; L 1.4.1801; 1807 wrecked.
*Monkey* Nicholson, Rochester.
Ord 10.1.1801; K 01.1801; L 11.5.1801; 1810 wrecked.
*Snipe* Adams, Bucklers Hard.
Ord 10.1.1801; K 01.1801; L 2.5.1801; 1816 yard craft.
*Starling* Adams, Bucklers Hard.
Ord 10.1.1801; K 01.1801; L 4.4.1801; 1804 wrecked.
*Vixen* Adams, Bucklers Hard.
Ord 10.1.1801; K 01.1801; L 9.6.1801; 1815 sold.

FATES: *Ferreter* taken by seven Dutch gunboats in the Ems 3.5.1807. *Jackall* wrecked near Calais 30.5.1807. *Monkey* wrecked on Belle Isle 25.12.1810. *Snipe* became mooring lighter No 13 at Chatham; later renumbered as No 9; 1846 BU. *Starling* wrecked near Calais 18.12.1804.

PLANS: Lines/profile & decks/specification. *Vixen* 1814: decks.

### *Archer* Class 1800. 12, Gunboats, later Brigs. Rule design.

Dimensions & tons: 80', 65'10¼" x 22'6" x 9'5". 177 31/94.

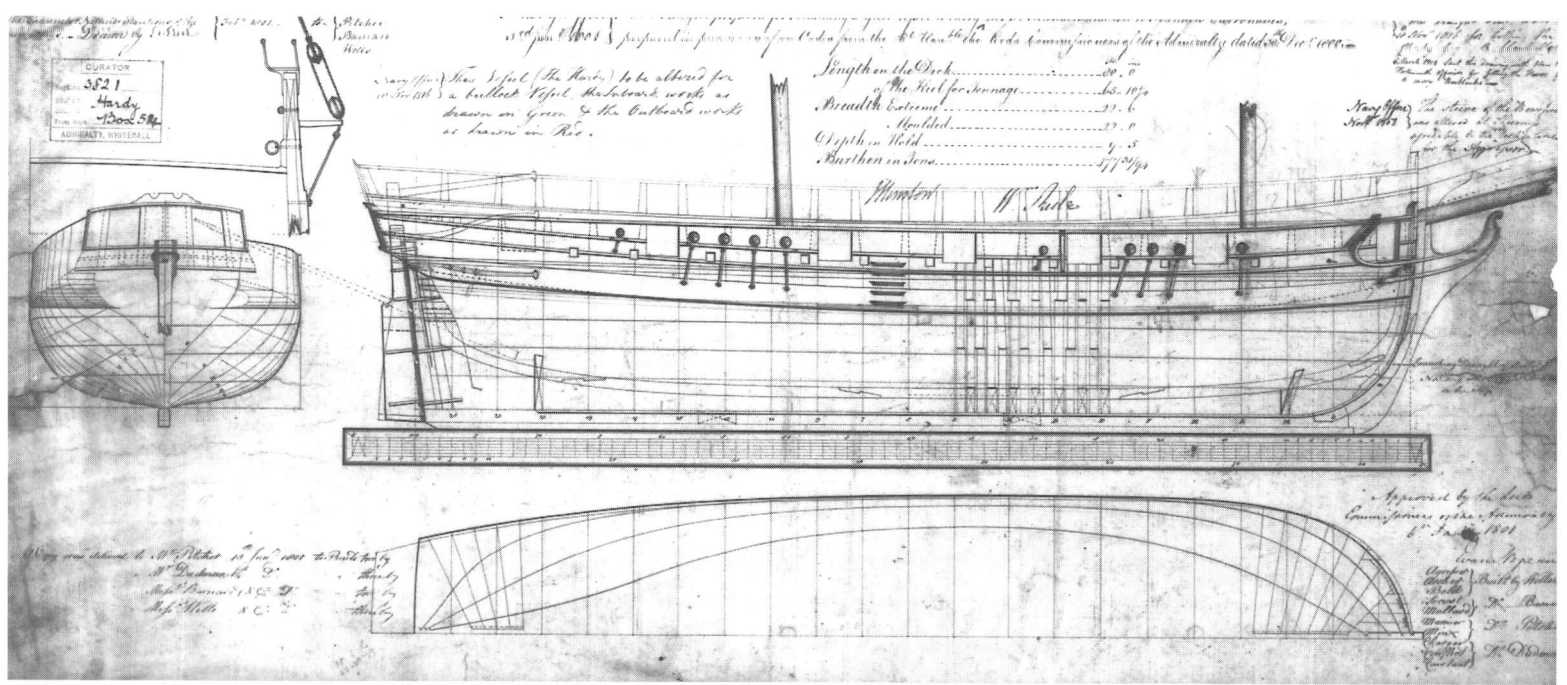

This lines plan (3821) was used for building a number of gunvessels (later classed as gunbrigs) of the *Archer* Class of 1800 and modified in 1816 to show additional upperworks etc. added to the *Hardy* of 1804 as converted in 1816 to a bullock vessel for the Cape of Good Hope

Men: 50. Guns: 2 x 12 or 32★ bow guns + 10 x 18★. Those later fitted as mortar brigs carried 1 x 10" mortar.

***Archer*** Wells, Deptford.
    Ord 30.12.1800; K 01.1801; L 2.4.1801; 1815 sold.
***Aggressor*** Wells, Deptford.
    Ord 30.12.1800; K 01.1801; L 1.4.1801; 1815 sold.
***Bold*** Wells, Deptford.
    Ord 30.12.1800; K 01.1801; L 16.4.1801; 1811 BU.
***Charger*** Dudman, Deptford.
    Ord 30.12.1800; K 01.1801; L 17.4.1801; 1809 mortar brig; 1814 sold.
***Conflict*** Dudman, Deptford.
    Ord 30.12.1800; K 01.1801; L 17.4.1801; 1804 taken.
***Constant*** Dudman, Deptford.
    Ord 30.12.1800; K 01.1801; L 28.4.1801; 1816 sold.
***Locust*** Barnard, Deptford.
    Ord 30.12.1800; K 01.1801; L 2.4.1801; 1814 sold.
***Mallard*** Barnard, Deptford.
    Ord 30.12.1800; K 01.1801; L 11.4.1801; 1804 taken.
***Mariner*** Pitcher, Northfleet.
    Ord 30.12.1800; K 01.1801; L 4.4.1801; 1814 sold.
***Minx*** Pitcher, Northfleet.
    Ord 30.12.1800; K 01.1801; L 14.4.1801; 1809 taken.
***Blazer*** Pitcher, Northfleet.
    Ord ____ 1804?; K 02.1804; L 3.5.1804; 1814 sold.
***Bruizer*** Pitcher, Northfleet.
    Ord ____ 1804?; K 02.1804; L 28.4.1804; 1815 sold.
***Contest*** Courtney, Chester.
    Ord ____ 1804?; K 02.1804; L 06.1804; 1809 foundered.
***Firm*** Brindley, Frindsbury.
    Ord ____ 1804?; K 02.1804; L 2.7.1804; 1811 wrecked.
***Flamer*** Brindley, Frindsbury.
    Ord ____ 1804?; K 02.1804; L 8.5.1804; 1815 hulked.
***Haughty*** Dudman, Deptford.
    Ord ____ 1804?; K 02.1804; L 7.5.1804; 1816 sold.
***Manly*** Dudman, Deptford.
    Ord ____ 1804?; K 02.1804; L 7.5.1804; (1806-1809 in Dutch hands); (1811-1813 in Danish hands); 1813 renamed *Bold*; 1814 sold.
***Pelter*** Dudman, Deptford.
    Ord ____ 1804?; K 02.1804; L 25.7.1804; 1810 foundered.
***Attack*** Adams, Southampton.
    Ord ____ 1804?; K 03.1804; L 9.8.1804; 1812 taken.
***Biter*** Wallis, Blackwall.
    Ord ____ 1804?; K 03.1804; L 27.7.1804; 1805 wrecked.
***Defender*** Courtney, Chester.
    Ord ____ 1804?; K 03.1804; L 28.7.1804; 1809 wrecked.
***Furious*** Brindley, Lynn.
    Ord ____ 1804?; K 03.1804; L 21.7.1804; 1815 sold.
***Safeguard*** Davy, Topsham.
    Ord ____ 1804?; K 03.1804; L 4.8.1804; 1811 taken.
***Steady*** Richards, Southampton.
    Ord ____ 1804?; K 03.1804; L 21.7.1804; 1815 sold.
***Acute*** Row, Newcastle.
    Ord ____ 1804?; K 04.1804; L 21.7.1804; 1813 hulked.
***Bouncer*** Row, Newcastle.
    Ord ____ 1804?; K 04.1804; L 11.8.1804; 1805 taken.
***Cracker*** Pitcher, Northfleet.
    Ord ____ 1804?; K 04.1804; L 31.6.1804; 1815 sold.
***Griper*** Brindley, Lynn.
    Ord ____ 1804?; K 04.1804; L 24.9.1804; 1807 wrecked.
***Growler*** Adams, Bucklers Hard.
    Ord ____ 1804?; K 04.1804; L 10.8.1804; 1815 sold.
***Piercer*** Ayles, Topsham.
    Ord ____ 1804?; K 04.1804; L 29.7.1804; 1814 given to the Government of Hanover.
***Pincher*** Graham, Harwich.
    Ord ____ 1804?; K 04.1804; L 28.8.1804; 1816 sold.
***Plumper*** Dudman, Deptford.
    Ord ____ 1804?; K 04.1804; L 7.9.1804; 1805 taken.
***Staunch*** Tanner, Dartmouth.
    Ord ____ 1804?; K 04.1804; L 21.8.1804; 1811 wrecked.
***Swinger*** Davy, Topsham.
    Ord ____ 1804?; K 04.1804; L 09.1804; 1812 BU.
***Wrangler*** Dudman, Deptford.
    Ord ____ 1804?; K 04.1804; L 28.5.1804; 1815 sold.
***Clinker*** Pitcher, Northfleet.
    Ord ____ 1804?; K 05.1804; L 31.6.1804; 1806 foundered.
***Gallant*** Roxby, Wearmouth.
    Ord ____ 1804?; K 05.1804; L 20.9.1804; 1815 sold.
***Hardy*** Roxby, Wearmouth.
    Ord ____ 1804?; K 05.1804; L 7.8.1804; 1817 storeship; 1822 yard craft.
***Sparkler*** Warren, Brightlingsea.
    Ord ____ 1804?; K 05.1804; L 6.8.1804; 1808 wrecked.
***Teaser*** Dudman, Deptford.
    Ord ____ 1804?; K 05.1804; L 16.7.1804; 1815 sold.
***Tickler*** Warren, Brightlingsea.
    Ord ____ 1804?; K 05.1804; L 8.8.1804; 1808 taken.
***Tigress*** Dudman, Deptford.
    Ord ____ 1804?; K 05.1804; L 1.6.1804; 1808 taken.
***Daring*** Bailey, Ipswich.
    Ord ____ 1804?; K 06.1804; L 10.1804; 1813 wrecked.
***Fearless*** Graham, Harwich.
    Ord ____ 1804?; K 06.1804; L 18.12.1804; 1812 wrecked.
***Attentive*** Bools & Good, Bridport.
    Ord 12.6.1804; K 07.1804; L 18.9.1804; 1812 BU.
***Cheerly*** Bools & Good, Bridport.
    Ord 13.6.1804; K 07.1804; L 10.1804; 1815 sold.
***Rapid*** Davy, Topsham.
    Ord ____ 1804?; K 07.1804; L 20.10.1804; 1808 sunk.
***Urgent*** Bass, Lympston.
    Ord ____ 1804?; K 07.1804; L 2.11.1804; 1816 sold.
***Desperate*** White, Broadstairs.
    Ord ____ 1804?; K 07.1804; L 2.1.1805; 1809 mortar brig; 1814 sold.
***Forward*** Todd, Berwick.
    Ord ____ 1804?; K 07.1804; L 4.1.1805; 1815 sold.
***Dexterous*** Adams, Eling.
    Ord ____ 1804?; K 08.1804; L 2.2.1805; 1816 sold.
***Fervent*** Adams, Bucklers Hard.
    Ord ____ 1804?; K 08.1804; L 15.12.1804; 1816 yard craft.
***Earnest*** Menzies, Leith.
    Ord ____ 1804?; K 08.1804; L 01.1805; 1816 sold.
***Protector*** Warren, Brightlingsea.
    Ord ____ 1804?; K 08.1804; L 1.2.1805; 1817 survey ship; 1833 sold.
***Sharpshooter*** Warren, Brightlingsea.
    Ord ____ 1804?; K 08.1804; L 2.2.1805; 1816 sold.
***Woodlark*** Menzies, Leith.
    Ord ____ 1804?; K 08.1804; L 01.1805; 1805 lost.
***Redbreast*** Preston, Yarmouth.
    Ord ____ 1805??; K 09.1804; L 27.4.1805; 1815 to the Customs.
***Plumper*** Halifax Dyd., Nova Scotia.
    Ord ____ 1805?; K ____ ; L 29.12.1807; 1812 wrecked.

FATES: *Conflict* taken by the French off Brest 24.10.1804; renamed *Le Conflit*; 1814 renamed *Le Lynx*; served the French until 1834. *Mallard* taken by the French after grounding near Calais 24.12.1804; 1815 name changed to *Le Favori*; served the French until 1824. *Minx* taken by six Danish gunboats 2.9.1809. *Contest* supposed foundered in the Atlantic 12.1809. *Firm* wrecked on the French coast 28.6.1811. *Flamer* to the Gravesend Alien service; 1858 sold for BU. *Manly* seized by the Dutch in the Ems; 1809 retaken; 2.9.1811 taken by three Danish brigs; 1813 retaken. *Pelter* foundered in the Atlantic 01.1810. *Attack* taken in the Kattegat by fourteen Danish gunboats 19.8.1812. *Biter* grounded near Calais; got off but then hit by French shell and had to be grounded in sinking condition 10.11.1805. *Defender* wrecked near Folkestone 14.12.1809. *Safeguard* captured by the Danes in the Baltic 29.6.1811. *Acute* became a quarantine hulk at Sheerness; 1864? BU. *Bouncer* run ashore near Dunkirk and taken by the French; 1814 renamed *L'Ecureil*; served the French until 1819. *Griper* wrecked near Ostend 18.2.1807. *Plumper* (1804) taken by five French gunbrigs off Guernsey 10.7.1805; 1814 renamed *L'Argus*; in French

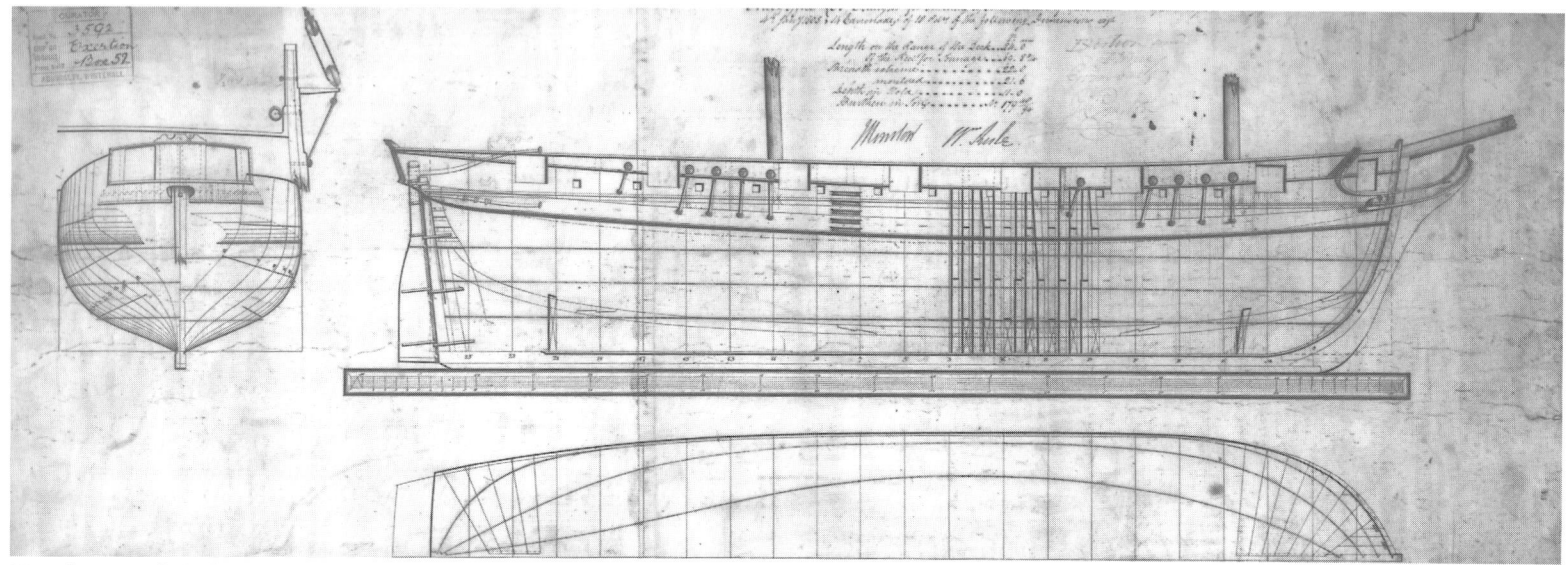

Lines plan (3592) for building *Exertion* and *Fancy*, gunbrigs of the *Confounder* Class of 1804

service until 1820. *Staunch* wrecked off Madagascar 06.1811. *Clinker* foundered off Le Havre 12.1806. *Hardy* became a bullock vessel at the Cape of Good Hope; later convict hospital ship -fate uncertain. *Sparkler* wrecked on the Dutch coast 14.1.1808. *Tickler* taken by four Danish gunboats in the Great Belt 4.6.1808. *Tigress* taken by 16 Danish gunboats in the Great Belt 2.8.1808. *Daring* wrecked and blown up to prevent capture by the French *Rubis* at Tamora, Sierra Leone 27.1.1813. *Fearless* wrecked near Cadiz 8.12.1812. *Rapid* sunk by Spanish shore batteries (in the Tagus?) 18.5.1808. *Fervent* became a lighter at Portsmouth; by 1864 mooring lighter No 3; fate uncertain. *Woodlark* wrecked near Calais 13.11.1805. *Plumper* (1807) wrecked near Saint Johns in the Bay of Fundy 5.12.1812.

PLANS: Lines/profile & decks/specification. *Aggressor* & *Hardy* as bullock transports 1821: Lines. *Charger, Safeguard, Desperate* as mortar brigs 1809: Profile & decks & midships section.

## *Confounder* Class 1804. 12/14 Gunbrigs/Brigs/Brig Sloops.

Rule design. Admiralty Order 20.11.1804: 'small brigs to carry 14 x 18 pounder carronades of nearly the same dimensions and tonnage of the last built gun brigs'. Admiralty Order 16.11.1811: 'gun brigs to carry 10 x 18 pounder carronades and 2 x 6 pounders'.

Dimensions & tons: 84', 69'8¾" x 22 x 11'. 179 48/94.

Men: 50. Guns: 1804 batch: 14 x 18★, later and from the start for the 1811 batch: 2 x 6 + 10 x 18★. As mortar brigs carried 1 x 8" brass mortar as well.

**Confounder** Adams, Southampton.
Ord 20.11.1804; K 01.1805; L 04.1805; 1814 sold.
**Adder** Ayles, Topsham.
Ord 20.11.1804; K 03.1805; L 9.11.1805; 1806 taken.
**Bustler** Ayles, Topsham.
Ord 20.11.1804; K 03.1805; L 12.8.1805; 1808 taken.
**Conflict** Davy, Topsham.
Ord 20.11.1804; K 02.1805; L 14.5.1805; 1810 foundered.
**Dapper** Adams, Southampton.
Ord 20.11.1804; K 01.1805; L 12.1805; 1814 sold.
**Encounter** Guillaume Northam.
Ord 20.11.1804; K 01.1805; L 16.5.1805; 1812 wrecked.
**Exertion** Preston, Yarmouth.
Ord 20.11.1804; K 01.1805; L 2.5.1805; 1812 run ashore.
**Fancy** Preston, Yarmouth.
Ord 20.11.1804; K 01.1805; L 7.1.1806; 1811 foundered.
**Havock** Stone, Yarmouth.
Ord 20.11.1804; K 02.1805; L 25.7.1805; 1818 storeship (bullock vessel); 1821 light vessel.
**Hearty** Bailey, Ipswich.
Ord 20.11.1804; K 01.1805; L 12.4.1805; 1816 sold.
**Indignant** Bools & Good, Bridport.
Ord 20.11.1804; K 01.1805; L 13.5.1805; 1809 mortar brig; 1811 BU.
**Intelligent** Bools & Good, Bridport.
Ord 20.11.1804; K 01.1805; L 26.8.1805; 1816 yard craft.
**Inveterate** Bools & Good, Bridport.
Ord 20.11.1804; K 01.1805; L 30.5.1805; 1807 wrecked.
**Martial** Ross, Rochester.
Ord 20.11.1804; K 01.1805; L 17.4.1805; 1816 fishery protection vessel; 1831 hulked.
**Rebuff** Richards, Hythe.
Ord 20.11.1804; K 01.1805; L 30.5.1805; 1809 mortar brig;; 1814 sold.
**Resolute** King, Dover.
Ord 20.11.1804; K 01.1805; L 15.4.1805; 1814 tender; 1816 hulked.
**Richmond** Greensword, Itchenor.
Ord 20.11.1804?; K 11.1805; L 02.1806; 1814 sold.
**Starling** Row, Newcastle.
Ord 20.11.1804; K 01.1805; L 05.1805; 1814 sold.
**Strenuous** Row, Newcastle.
Ord 20.11.1804; K 02.1805; L 16.5.1805; 1814 sold.
**Turbulent** Tanner, Dartmouth.
Ord 20.11.1804; K 02.1805; L 17.7.1805; 1808 taken.
**Virago** Tanner, Dartmouth.
Ord 20.11.1804; K 02.1805; L 23.9.1805; 1816 sold.
**Bold** Tyson, Bursledon.
Ord 16.11.1811; K 02.1812; L 26.6.1812; 1813 wrecked.
**Borer** Tyson, Bursledon.
Ord 16.11.1811; K 03.1812; L 27.7.1812; 1815 sold.
**Boxer** Hobbs, Redbridge.
Ord 16.11.1811; K 03.1812; L 25.7.1812; 1813 taken.
**Conflict** Good, Bridport.
Ord 16.11.1811; K 02.1812; L 26.9.1812; 1830s hulked.
**Contest** Good, Bridport.
Ord 16.11.1811; K 02.1812; L 24.10.1812; 1828 foundered.
**Hasty** Hills, Sandwich.
Ord 16.11.1811; K 04.1812; L 26.8.1812; 1819 survey ship; 1827 yard craft.
**Manly** Hills, Sandwich.
Ord 16.11.1811; K 02.1812; L 13.7.1812; 1833 sold.
**Plumper** Good, Bridport.
Ord 16.11.1811; K 04.1812; L 9.10.1813; 1833 sold.
**Shamrock** Larking, Lynn.
Ord 16.11.1811; K 03.1812; L 8.8.1812; 1817 survey ship; 1831 hulked.
**Snap** Russell, Lyme Regis.
Ord 16.11.1811; K 02.1812; L 25.7.1812; 1821 survey ship; 1827 hulked.
**Thistle** Ross, Rochester.
Ord 16.11.1811; K 03.1812; L 13.7.1812; 1823 BU.
**Swinger** Good, Bridport.
Ord 16.11.1811; K 04.1812; L 15.5.1813; 1829 yard craft.
**Adder** Davy, Topsham.

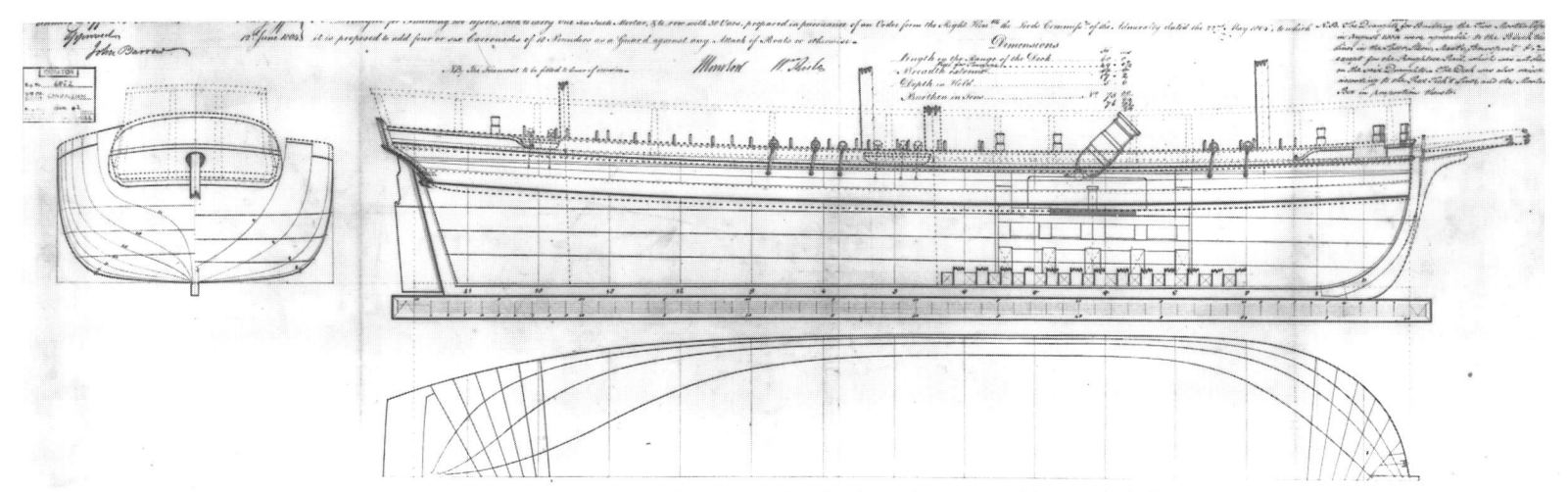

Building lines and part profile plan (6852) for the *Convulsion* Class mortar boats of 1804. Note this plan shows the position of the masts both for the original brig rig and the later ketch rig

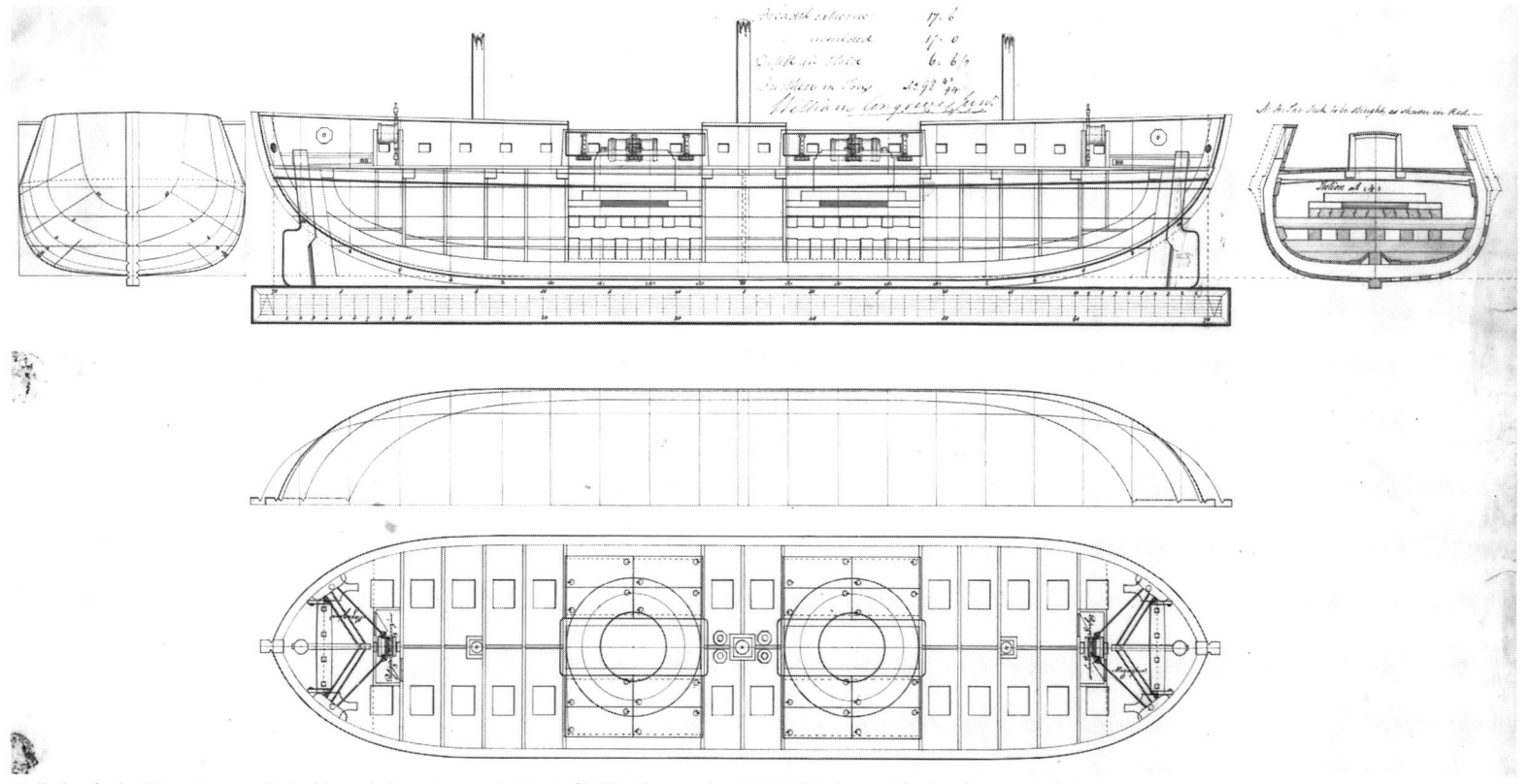

Hull plan for building Congreve's double-ended mortar vessel *Project* of 1806. Plan number 6442. This does not look to be a particularly seaworthy or useful vessel – and her short life of four years would seem to confirm this

Ord ____ 1812?; K 12.1812; L 28.6.1813; 1826 to the Coastguard; 1832 wrecked.

**Clinker** Davy, Topsham.
Ord ____ 1812?; K 01.1813; L 15.7.1813; 1831 to the Coastguard.

**Griper** Richards & Co, Hythe.
Ord ____ 1812?; K 01.1813; L 14.7.1813; 1825 to the Coastguard.

**Mastiff** Taylor, Biddeford.
Ord ____ 1812?; K 01.1813; L 25.9.1813; 1836 survey ship; 1851 BU.

**Pelter** Tucker, Biddeford.
Ord ____ 1812?; K ____ ; L 27.8.1813; 1826 to the Coastguard.

**Snapper** Hobbs, Redbridge.
Ord ____ 1812?; K 04.1813; L 27.9.1813; 1824 to the Coastguard.

FATES: *Adder* driven ashore and taken by the French near Abreval 9.12.1806. *Bustler* stranded near Cape Gris Nez and taken by the French 26.12.1808; 8.12.1813 taken from the French by the Dutch at Zierickzee. *Conflict* (1805) foundered in the Bay of Biscay 9.11.1810. *Encounter* wrecked off San Lucar, Spain 11.7.1812. *Exertion* grounded and destroyed to avoid capture in the Elbe 8.7.1812. *Fancy* foundered in the Baltic 24.12.1811. *Havock* became the light vessel at Bembridge; 1834 Coastguard watch vessel at Hamble; 1859 BU. *Intelligent* became a lighter at Portsmouth; by 1864 Mooring Lighter No 4; fate uncertain. *Inveterate* wrecked off Saint Valéry en Caux 18.2.1807. *Martial* hulked for quarantine service at Sheerness; 1836 sold. *Resolute* became a diving bell ship at Plymouth; 1826 to Bermuda where she combined the function of diving bell ship and receiving ship; 1844 convict hulk; 1852 BU. *Turbulent* was taken by a Danish flotilla in Malmo Bay 18.6.1808. *Bold* was wrecked on Prince Edward Island 27.9.1813. *Boxer* was taken off Portland, Maine, by USS *Enterprise* 5.9.1813; not taken into the USN. *Conflict* (1812) became a receiving ship at Sierra Leone; 1841 sold. *Contest* foundered off Halifax 14.4.1828. *Hasty* became a yard craft at Mauritius; fate uncertain. *Shamrock* became a quarantine service vessel at Woolwich; 1833 to the Coastguard at Rochester; later WV 18?; by 1867 sold. *Snap* became a floating magazine at Woolwich; 1832 sold. *Swinger* became a mooring lighter at Portsmouth; still in service 1864. *Adder* became a watch vessel on the Sussex Coast; wrecked near Newhaven 1832. *Clinker* to the Coastguard at Yantlet Creek; later WV 12; 1867 sold. *Griper* to the Coastguard at Blackwall; later to Sussex; 1860 target for gunnery experiments at Portsmouth; 1862 armour plate experiments; 1863 BU. *Snapper* to the Coast Blockade at Folkestone; 1862 sold. *Snapper* to the Coast Blockade at ____ ; 1861 sold.

PLANS: Lines/profile & decks/planking expansion. *Resolute* 1806: stern with diving bell. *Snap* as survey vessel 1821: decks/midships section.

## Mortar Boats

### *Convulsion* Class 1804.
Henslow & Rule design. 'To row with 30 oars.' Originally seem to have been designed for a brig rig, but this was changed before completion to ketch rig. Basically large rowing gunboats. The design does not seem to have been a success, certainly the two vessels concerned had a very short service life.

Dimensions & tons: 60', 49'0½" x 17'2" x 7'. 76 81/94.
Men ____ . Guns: 1 x 10" mortar. 4 x 18★.

*Convulsion* Brent, Rotherhithe.
  Ord ____ 1804?; K 08.1804; L 31.8.1804; 1806 sold.
*Destruction* Wells (Perry), Blackwall.
  Ord ____ 1804?; K 08.1804; L 3.9.1804; 1806 sold.

PLANS: Lines/profile, deck, sections/midsection/deck beam fastening/contract.

### *Project* Class 1805.
'Proposed by Mr. Congreve' (William Congreve, Jnr). Double-ended, rudder at each end (lightship type hull). 3 masts, 16 oar ports a side.

Dimensions & tons: 70', 60'5¼" x 17'6" x 6'6½". 98 42/94.
Men: ____ . Guns 2 x ____ mortars.

*Project* Woolwich Dyd.
  Ord 19.7.1805; K 08.1805; L 26.3.1806; 1810 BU.

PLANS: Lines, profile, deck, section.

## Small Rowing and Sailing Gunboats

This can only be a selection of those boats that are known to have been built - there being no adequate listing known. The main source of information for the gunboats given here is the Admiralty plans.

### Gunboat, One Gun 1787.
1 x 24 on slide forward. 1 mast, lateen rig. At least one built for Gibraltar 1787. At least one each built by Burr and by Harris. 48'6" x 13'6" x 4'. 39 tons. One ordered on 26.9.1787. It is probably one of these shown in a plan 'As fitted at Portsmouth 1788'. Similar boats, but with a two masted lateen rig were ordered from Burr, Harris, Roberts and also Wallis in 1790, in 1793 from Plymouth and/or Portsmouth and in 1803 two were built at Deptford Dyd and two at Woolwich Dyd.

PLANS: Lines, profile, deck, sections (1787 - three versions)/ditto (1788 as fitted)/rig (1788 as fitted)/lines, profile, deck, sections (1790)/ditto (1793 and 1803).

### Gunboat, Double Ended. 1796.
1796 one built by John Parkin, Cawsand. 'Suggested by Sir Sidney Smith' and with alterations directed by him. 60' x 7'9". 1 x 24 (on field carriage) capable of firing from either end, and usually stowed amidships (or possibly two of these weapons carried - the plan is not altogether clear on this).

PLANS: Lines, profile, deck, section. (There is also another plan, dated 1795, of a very similar boat 61' x 15'8" x 6'8" showing 'gunboats built at Cawsand' under Sidney Smith's supervision.)

### Rowing (Lateen Rig) Gunboats. 1798.
One carronade forward. 'Peg top' midsection. 60' x 13'6" x 7'3". 21.2.1798 order for Pitcher to build two boats.

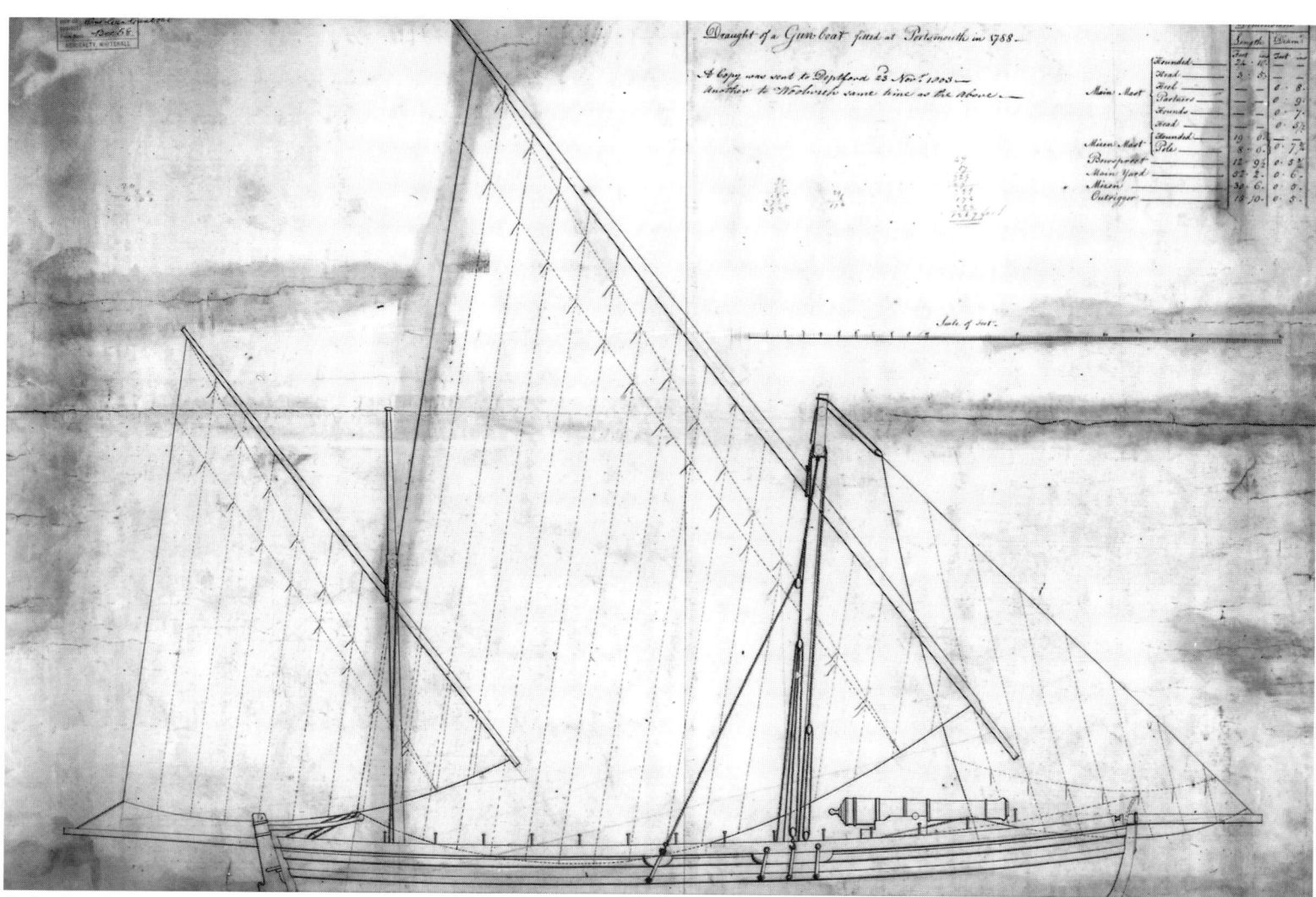

'As fitted' rig plan (4073) for one gun gunboat at Portsmouth 1788 - copies sent to Deptford and Woolwich in 1803

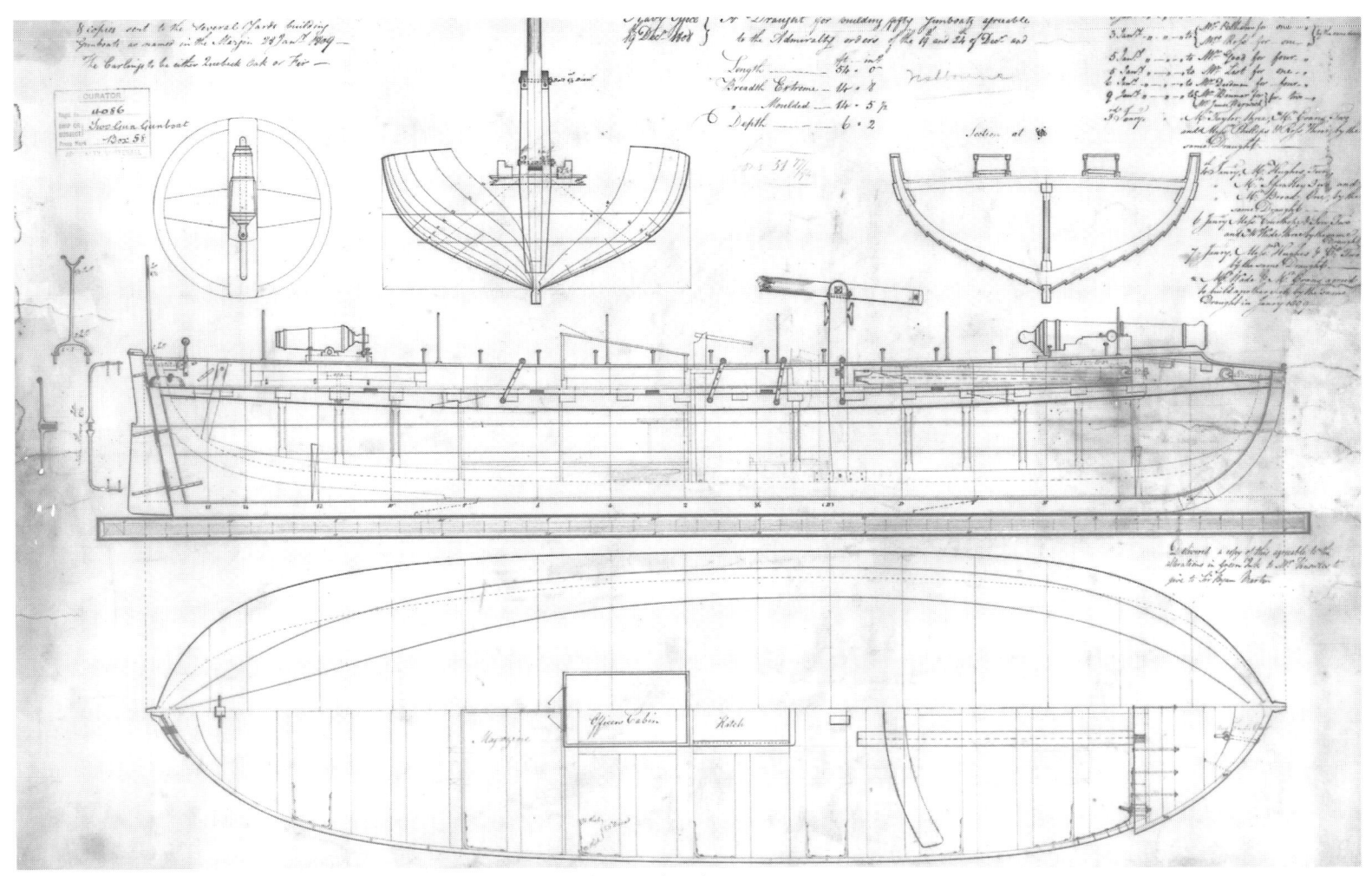

Building hull plan (4086) for the December 1808 orders for Commissioner Hamilton's gunboat design

PLANS: Hull with section.

**Flat Bottom Boats.** Armed with a carronade or howitzer capable of being lowered into the bottom of the boat in bad weather (shipping or unshipping could be done in 5 minutes or less). Henslow design? 31' x 9'9" x 2'11". 12 tons 2 hundredweight. In 1801 flat bottom boats nos. 102, 138, 141, 146, 147, 148 and 149 were fitted as gunboats to this design and appear to have been based on Deal. In 1804 one built by contract by ____ ? Ten built at Plymouth Dyd.

PLANS: (1801): sketch profile. (1804) Profile, decks & midsection.

**Row Galley No 1.** Sheerness Dyd? Pivots forward and aft and amidships (the latter with all-round traverse) and with rails for transferring what was probably a single gun from one to another. 42 rowlocks. 66'4" x 13'1½" x 5'5". Three masts, probably rigged as a lugger. The plan is dated 1801 and shows one long gun.

PLANS: Lines, profile, deck.

**'Launches' – 1 x 12★ in Bow.** 'As suggested by Captain Edward P.Brenton' design by Rule? Double ended and quite unlike ordinary launches. 27' x 7' x 3' (originally 26' x 6'6" x 2'11"). Originally 6 ordered to be built by contract as ordinary launches in 1803 (by Cowthorp [4] and Burn [2]) but the design was revived in 1808 and built by all the home dockyards except Sheerness, and by Malta Dyd, Cowthorp, Wilks & Wade and one other private builder. It may be that another version was built at Sheerness, possibly the one larger boat to the same basic design (28'6" x 7'6" x 4'1") provided 'in lieu of launch' for *Melpomene*. Most if not all seem to have been used as gunboats.

PLANS: Lines & profile/*Melpomene*'s boat: Sheer.

***Experiment* Gunboat 1805.** 1 x 18 + 1 x 18★. Commissioner Hamilton's [original?] design. Built Woolwich Dyd. In service 1805-1807. Possibly the prototype for all the boats of his design?

**'Commissioner Hamilton's' Gunboats.** One mast, lateen rig, clinker built. 50'7", 41'6" x 14' x 5'4½". 43 87/94 tons. 2 x 18 on slides in the bows + 1 x 18★ on pivot aft. Six built in frame 1805 and sent to Gibraltar.

PLANS: Lines, profile, deck.

**'Commissioner Hamilton's' Gunboats. Nos 1 to 85.**

It seems likely that there were three separate batches built. In 02.1808 repeats of the 1805 design with 2½" more breadth were ordered as follows: Wallis (4), Cowthorp or Courthorpe & Son (2), Usmar? (2), Ross & Phillips (3), King & Co (3), Taylor & Co (3), Bailey (2), Brindley (3), Spratley (3), North (2), Evans, Limehouse (2), Rattenbury, Limehouse (1), Hughes, Limehouse (1). Then, in 03.1808, the design was enlarged to 54' x 14'8" x 5'8" equipped with 2 x 24 (short, 36 cwt) and 5 were ordered from Wade & Gledfield (Glanfield?). Later the design was enlarged further by increasing the depth to 6'2" and tonnage to 51 77/94. The armament seems to have reverted to the original form. 50 were ordered to this revised design on 19 & 24.12.1808 to be built as follows: Usmar? (2), Wallis (2), Glanfield (3), Bailey. Ipswich (6), Pelham (1), Mrs Ross, Rochester (1), Good, Bridport (4), List, Fishbourne (1), Dudman (4), Binner (2), Warwick (2), Taylor (3), Evans (2 plus 1), Phillips & Ross, Limehouse (3), Hughes (2), Spratley (2), Broad (1), Courthope & Son, 'Redriff' (2), Wade (3), Hughes & Co (2) [There would appear to be one more boat to account for the numbering of the whole class from 1 to 85.]

FATES of Numbered gunboats (presumably of the Commissioner Hamilton design) include the following: No 2; ('*Camperdown*'); 28.10.1810 wrecked on rocks near Cadiz. No 8; 07.1814 wrecked in the Elbe. No 10; ____ , 1812 foundered at Cadiz. No 11; 19.2.1812 wrecked and abandoned near Cadiz. No 14; ('*Gibraltar*'); 26.11.1809 stranded off San Pedro. No 16; 10.1812 wrecked near Malaga. No 17; 07.1813 swept on to bows of *Devonshire* at the Nore and sank. No 23; 08.1813 destroyed to avoid capture after grounding in the mouth of the Elbe.

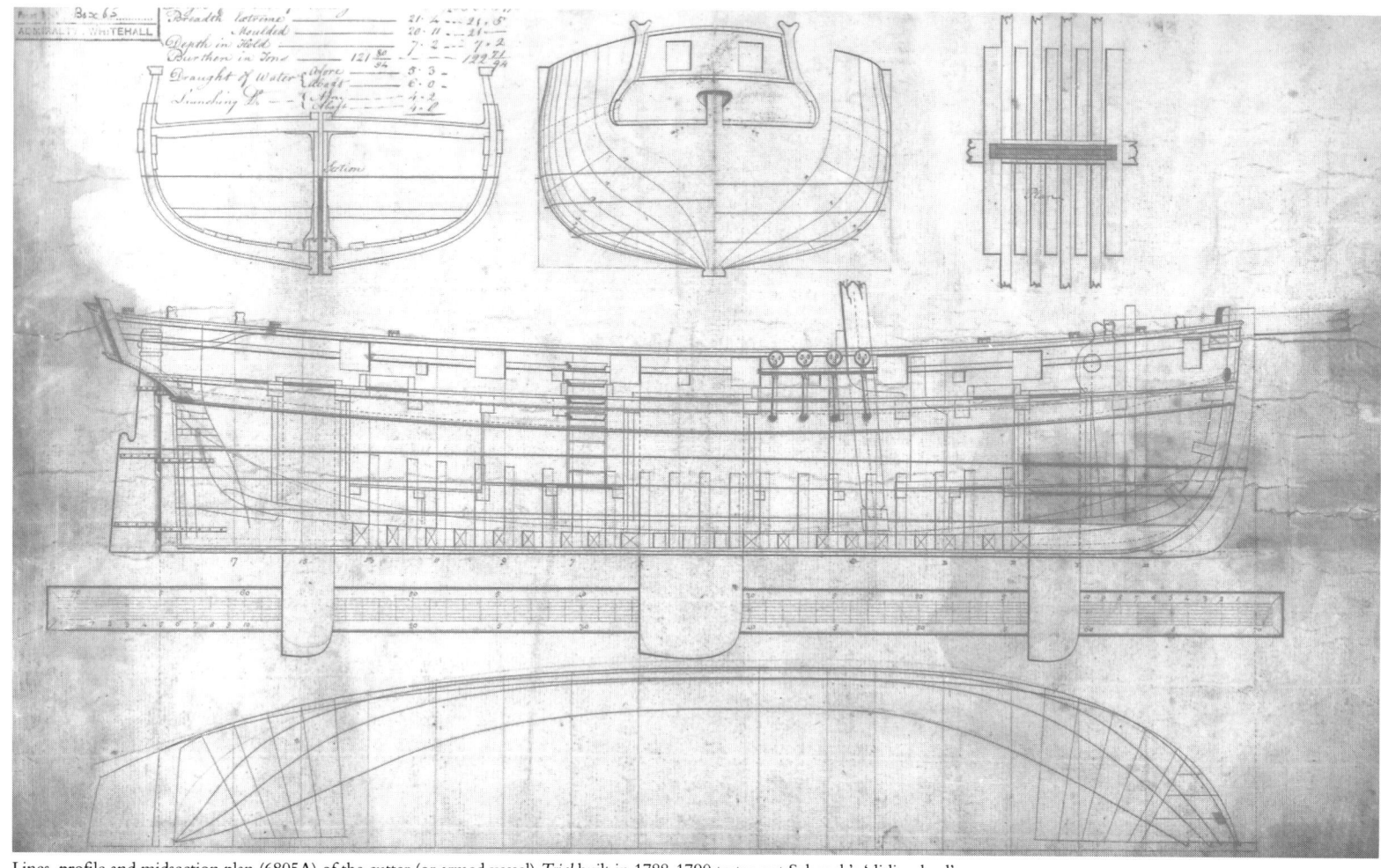

Lines, profile and midsection plan (6805A) of the cutter (or armed vessel) *Trial* built in 1788-1790 to try out Schanck's 'sliding keel'

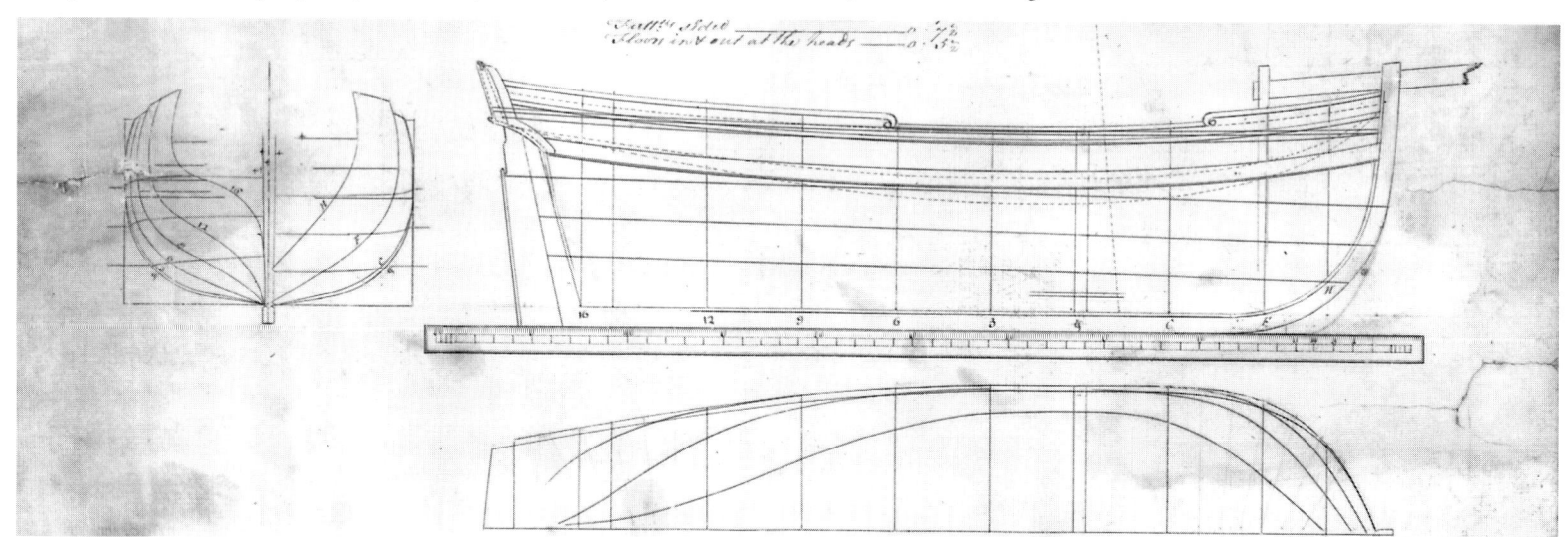

Lines plan (6450) for the *Placentia* Class sloop rigged vessels of 1790 intended for service on the Newfoundland station

PLANS: Rig (54' type)/Lines, profile, deck, section, details (for all three types).

## Cutters, Schooners & Luggers

**Trial Class 1788.** 8, cutter/armed vessel. 'Vessel with three sliding keels agreeable to a plan suggested by Captain Schank.' Originally to be classed as a cutter, but before completion this was changed to an armed vessel of 8 guns. She was cutter-rigged.

Dimensions & tons: 65', 50'4½" x 21'4" x 7'4". 121 80/94.
Men: 45. Guns: 8 x 3 (1793 4 x 12★ added).

**Trial** Dunsterville, Plymouth.
Ord 1.12.1788; K 11.1789; L 9.9.1790; 1810 hulked.

FATE: Became a depot ship at Waterford; by 1840s to Callao in Peru as a coal depot; 1848 sold.

PLANS: as built: Lines, profile & sections/decks/keel details.

**Placentia Class 1789.** Henslow design for vessels rigged as sloops for Newfoundland. These small vessels bore no relationship to naval sloops of this period, their rig was that of the mercantile type of sloop, with one mast and fore and aft sails only.

Dimensions & tons: 44'7", 35'4⅝" x 15' x 8'4". 42.
Men: 30?. Guns: 4 swivels.

**Placentia** Jeffery & Start, Newfoundland.
Ord ____ ; K ____ ; L ____ 1790; 1794 wrecked.

**Trepassey** Lester & Stone, Newfoundland.
Ord ____ ; K ____ ; L ____ 1790; 1803 sold.

FATES: *Placentia* wrecked at Newfoundland 8.5.1794.

HENSLOW TO SEPPINGS 1786-1830

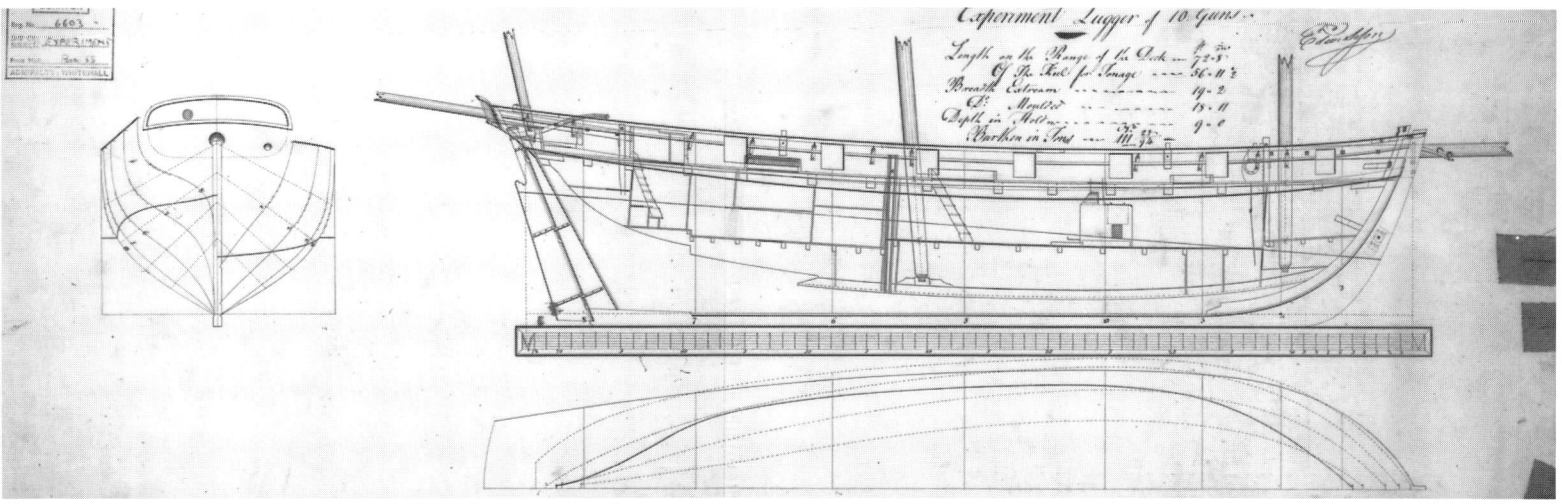

Lines and profile plan (6603) of the only lugger built for the Royal Navy, the appropriately named *Experiment* of 1793

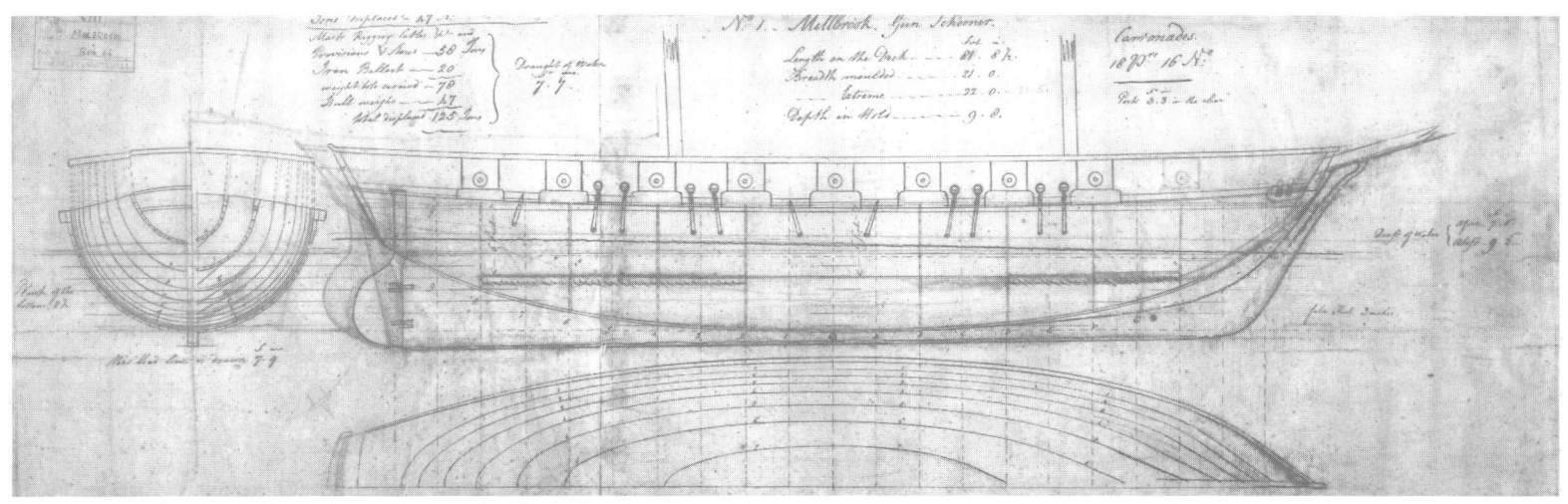

Lines plan (6761) for Bentham's schooner rigged gunvessel *Milbrook* of 1797

PLANS: Lines.

### *Experiment* Class 1792. 10 Lugger. Design origin uncertain - possibly by her builder? She was the only lugger actually designed and built for the Royal Navy.

 Dimensions & tons: 72'6", 56' x 18'6" x 9'. 101.
 Men: 45. Guns: 10 x 12★ + 12 swivels.
**Experiment** Parkins, Plymouth.
 Ord ____ ; K ____ 1792; L 05.1793; 1796 taken.
FATE: Taken by the Spaniards in the Mediterranean 2.10.1796.
PLANS: Lines & profile/decks.

### *Eling* Class 1796. 14, schooner-rigged gunboats. Design by 'General' (Brigadier General in the Russian army) Samuel Bentham. Experimental gunboats (also classed as advice boats) with schooner rig. Rather like smaller versions of the same designer's *Dart* and *Arrow* (see under ship sloops) with the same type of structural bulkheads, but more nearly double-ended, and with a sharper (almost triangular) midsection than the two larger vessels or than the other Bentham designed schooner-rigged vessels (*Milbrook* and *Netley*; see below). Built under Bentham's own supervision and not that of the Navy Board. Originally to be designated Advice Boats Nos 1 & 2; later usually listed with the gunbrigs.

 Dimensions & tons: 80', ____ x 22'2" x 11'6". 158.
 Men: 50. Guns: 2 x 12 + 12 x 18★ (may have been altered later).
**Eling** Hobbs, Redbridge.
 Ord 1796?; K ____ ; L ____ 1796; 1814 BU.
**Redbridge** Hobbs, Redbridge.
 Ord 1796?; K ____ ; L ____ 1796; 1803 taken.
FATES: *Redbridge* taken by the French off Toulon 4.8.1803; in French service until 1814.

PLANS: Lines, profile, framing, midships section.

### *Milbrook* Class 1797. 16, Schooner-rigged Gun Vessel. Bentham design. In most respects can be considered a smaller version of the same designers *Dart* and *Arrow* (see under ship sloops) with the same rounded section, structural bulkheads and almost double-ended form. Construction supervised by the designer.

 Dimensions & tons: 81'8½", ____ x 22' x 9'8". 125.
 Men: 50. Guns: 16 x 18★, may have been altered later.
**Milbrook** Hobbs, Redbridge.
 Ord ____ 1797?; K ____ ; L ____ 1797; 1808 wrecked.
FATE: Wrecked on the Burlings 26.3.1808.
PLANS: Lines.

### *Netley* Class 1797. 16, Schooner Rigged Gun Vessel. Bentham design. Generally similar to *Milbrook* (see above) but with considerable detail differences. The carronades, as with several other Bentham vessels, were mounted on his 'non-recoil' system. Possibly purchased whilst building?

 Dimensions & tons: 86'7", 71'2"x 21'5½" x 11'3". 176.
 Men: 50. Guns: 16 x 24★, may have been altered later.
**Netley** Hobbs, Redbridge.
 Ord ____ ; K ____ ; L ____ 1797; 1806 taken.
FATE: 17.12.1806 Taken by French *Thetis* and consort in the West Indies; 10.7.1808 retaken and wrecked.
PLANS: Lines & profile.

### *Express* Class 1800. 6, Advice Boats. Henslow design, probably schooner rigged.

 Dimensions & tons: 88', 72'7½" x 21'6" x 13'1". 178 48/94.
 Men: 30. Guns: 6 x 12★.

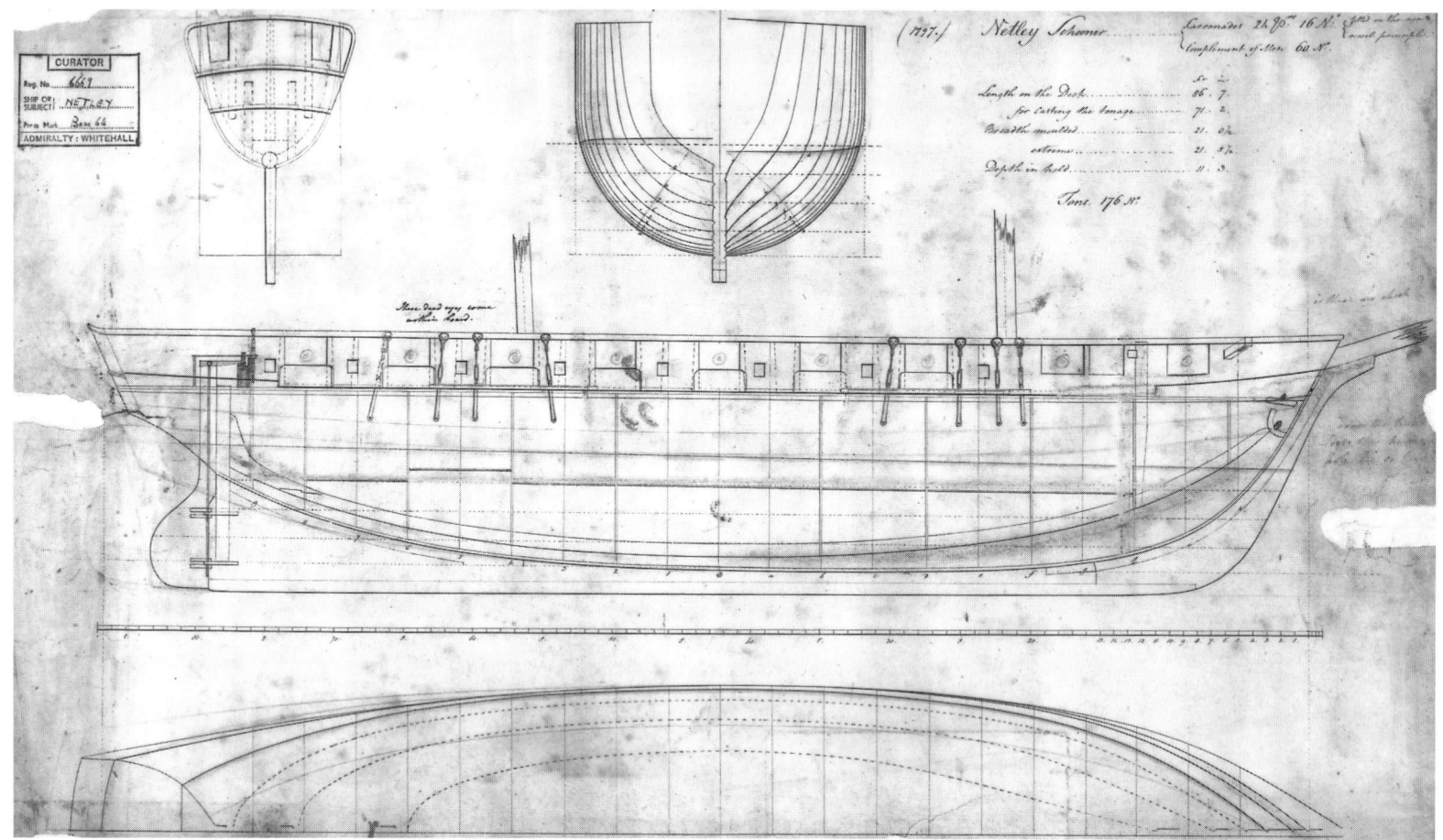

Lines plan (6669) for Bentham's schooner *Netley* of 1797

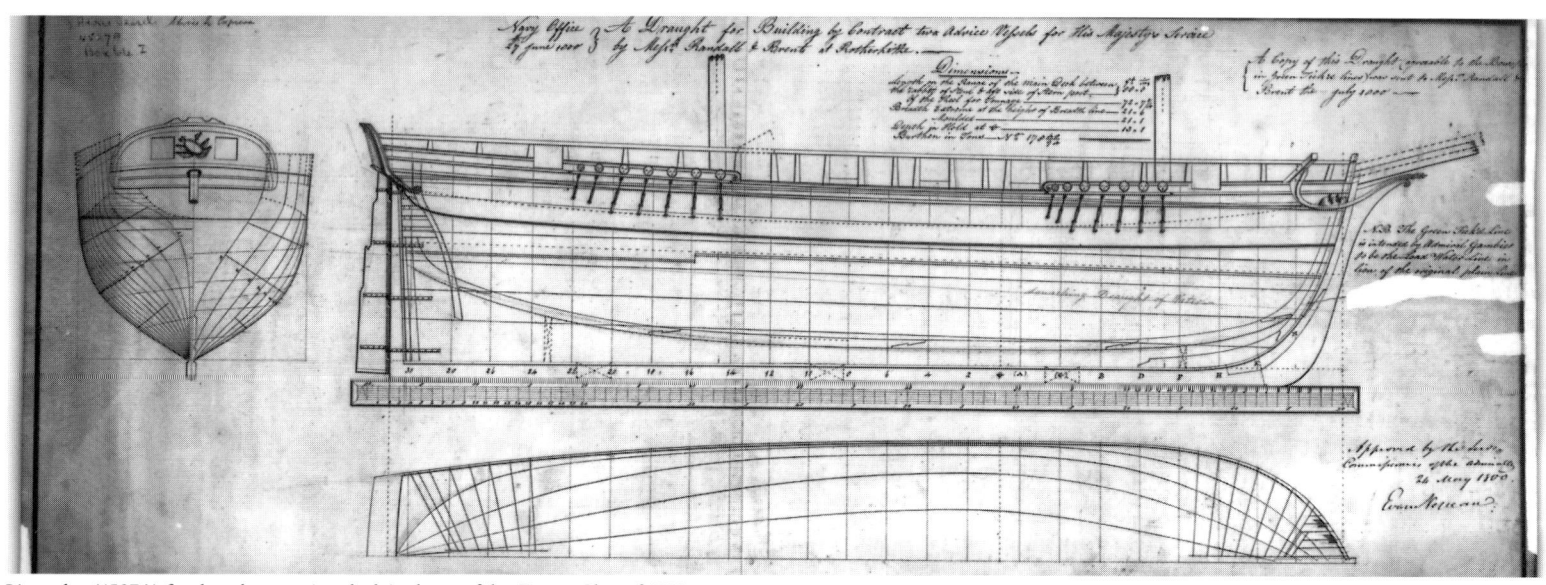

Lines plan (4527A) for the schooner rigged advice boats of the *Express* Class of 1800

**Express** (Advice Boat No 1) Randall & Brent, Rotherhithe.
  Ord ____ K 07.1800; L 30.12.1800; 1813 sold.
**Advice** (Advice Boat No 2) Randall & Brent, Rotherhithe.
  Ord ____ ; K 07.1800; L 30.12.1800; 1805 sold.
PLANS: Lines/profile & decks/framing.

### Ballahoo ('Fish') Class 1803. 4/6, Gun Schooners.

Originally ordered in 06.1803 to be built to an Admiralty design [proposed dimensions & tons: 55'2", 40'10½" x 18' x 9'. 70 41/94] which was to be sent to Bermuda; but then this was changed to a design produced in Bermuda 'similar to a Bermudian dispatch boat'. Contracted for by Goodrich & Co who subcontracted to various Bermudan builders: Brown, McCallin, Outerbridge, Zwill, Hill, Dill, Tynes etc. Built of cedar. The *Cuckoo* ('Bird') class was built to the same Bermudian design, but by British, not Bermudian, builders.
  Dimensions & tons: 55'4", 41'6" x 18' x 9'. 71 47/94.
  Men 20. Guns: 4 x 12★ (in at least one case - *Pilchard* - the armament is given as 2 x 12 + 4 x 12★).

**Ballahoo/Balahou** ____ , Bermuda.
  Ord ____ 1803; K ____ ; L ____ 1804; 1814 taken.
**Capelin** ____ , Bermuda.
  Ord ____ 1803; K ____ ; L ____ 1804; 1808 wrecked.
**Flying Fish** ____ , Bermuda.
  Ord ____ 1803?; K ____ ; L ____ 1804; 1807? wrecked.
**Mackerel** ____ , Bermuda.
  Ord ____ 1803?; K ____ ; L ____ 1804; 1815 sold.
**Snapper** ____ , Bermuda.
  Ord ____ 1803?; K ____ ; L ____ 1804; 1811 taken.

# HENSLOW TO SEPPINGS 1786-1830

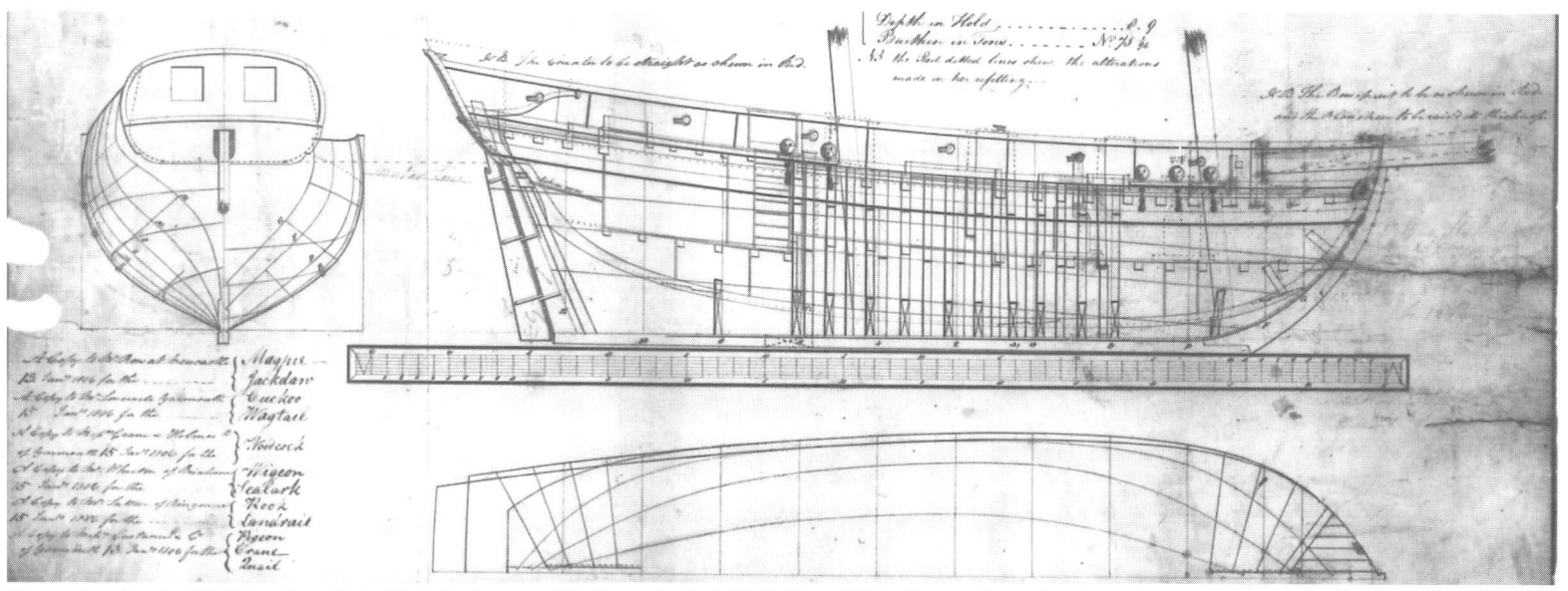

Lines and profile plan (4542) 'as taken off' the *Haddock* schooner, of the Bermudan-built 'Fish' Class of 1803. This was also used as the guidance plan for building the 'Bird' Class schooners in home yards

*Haddock* Skinner, Bermuda.
    Ord \_\_\_\_ 1803?; K \_\_\_\_ ; L \_\_\_\_ 1805; 1809 taken.
*Pilchard* \_\_\_\_ , Bermuda.
    Ord \_\_\_\_ 1803?; K \_\_\_\_ ; L \_\_\_\_ 1805; 1813 sold.
*Whiting* \_\_\_\_ , Bermuda.
    Ord \_\_\_\_ 1803?; K \_\_\_\_ ; L \_\_\_\_ 1805; 1812 taken.
*Pike* \_\_\_\_ , Bermuda.
    Ord \_\_\_\_ 1803?; K \_\_\_\_ ; L \_\_\_\_ 1806?; (1807 briefly in French hands); 1809 foundered.
*Bream* \_\_\_\_ , Bermuda.
    Ord \_\_\_\_ 1805?; K \_\_\_\_ 1806; L \_\_\_\_ 1807; out of service by 1817, fate uncertain.
*Chub* \_\_\_\_ , Bermuda.
    Ord \_\_\_\_ 1805?; K \_\_\_\_ 1806; L 05.1807; 1812 foundered.
*Cuttle* \_\_\_\_ , Bermuda.
    Ord \_\_\_\_ 1805?; K \_\_\_\_ 1806; L 05.1807; 1813 out of commission.
*Grouper* \_\_\_\_ , Bermuda.
    Ord \_\_\_\_ 1805?; K \_\_\_\_ ; L \_\_\_\_ 1807; 1811 lost.
*Porgey* \_\_\_\_ , Bermuda.
    Ord \_\_\_\_ 1805?; K \_\_\_\_ 1806; L 05.1807; 1810 grounded.
*Mullett* \_\_\_\_ , Bermuda.
    Ord \_\_\_\_ 1805?; K \_\_\_\_ 1806; L 05.1807; 1814 sold.
*Tang* \_\_\_\_ , Bermuda.
    Ord \_\_\_\_ 1805?; K \_\_\_\_ 1806; L 09.1807; 1808 foundered.
*Herring* \_\_\_\_ , Bermuda.
    Ord \_\_\_\_ 1805?; K \_\_\_\_ ; L \_\_\_\_ 1807; by 1816 yard craft.

FATES: *Ballahoo* taken by the American privateer *Perry* 29.4.1814. *Capelin* wrecked off Brest 30.6.1808. *Flying Fish* wrecked at Curacoa 17.11.1807? *Snapper* taken by the French lugger *Rapace* off Brest. *Haddock* taken by the French *Génie* in the Channel 30.1.1809. *Whiting* taken by the French privateer *Diligente* off North America 22.8.1812. *Pike* taken by the French privateer *Marat* in the West Indies retaken shortly afterwards by *Moselle*. *Chub* capsized off Halifax 14.8.1812. *Grouper* wrecked at Guadeloupe 21.10.1811. *Porgey* grounded and burnt to avoid capture in the Scheldt 4.6.1810. *Tang* foundered in the Atlantic in 1808. *Herring* became a yard craft at Halifax; fate uncertain.

PLANS: Lines/profile & decks. *Haddock* as built: Lines & profile/decks/specification.

## *Cuckoo* ('Bird') Class 1805. 2, Gun Schooners.
Copies of the Bermudian built *Ballahoo* ('Fish') design; but built in Britain.
    Dimensions & tons: 55'4", 41'6" x 18' x 9'. 71 47/94.
    Men: 20. Guns: 2 x 12★.
*Cuckoo* Lovewell, Yarmouth.
    Ord \_\_\_\_ 1805?; K 01.1806; L 12.4.1806; 1810 wrecked.
*Jackdaw* Row, Newcastle.
    Ord \_\_\_\_ 1805?; K 01.1806; L 19.5.1806; 1810 tender; 1816 sold.
*Landrail* Sutton, Ringmore.
    Ord \_\_\_\_ 1805?; K 01.1806; L 18.6.1806; 1811 tender; 1814 taken.
*Magpie* Row, Newcastle.
    Ord \_\_\_\_ 1805?; K 01.1806; L 17.5.1806; 1807 wrecked.
*Crane* Custance, Yarmouth.
    Ord \_\_\_\_ 1805?; K 02.1806; L 26.4.1806; 1808 wrecked then BU.
*Pigeon* Custance, Yarmouth.
    Ord \_\_\_\_ 1805?; K 02.1806; L 26.4.1806; 1809 wrecked.
*Quail* Custance, Yarmouth.
    Ord \_\_\_\_ 1805?; K 02.1806; L 26.4.1806; 1816 sold.
*Rook* Sutton, Ringmore.
    Ord \_\_\_\_ 1805?; K 02.1806; L 21.5.1806; 1808 taken.
*Wagtail* Lovewell, Yarmouth.
    Ord \_\_\_\_ 1805?; K 02.1806; L 12.4.1806; 1807 wrecked.
*Sealark* Wheaton, Brixham.
    Ord \_\_\_\_ 1805?; K 03.1806; L 1.8.1806; 1809 wrecked?
*Wigeon* Wheaton, Brixham.
    Ord \_\_\_\_ 1805?; K 03.1806; L 19.6.1806; 1808 wrecked.
*Woodcock* Crane & Holmes, Yarmouth.
    Ord \_\_\_\_ 1805?; K 02.1806; L 11.4.1806; 1807 wrecked.

FATES: *Cuckoo* wrecked on the Dutch coast 4.4.1810. *Landrail* taken by the American privateer *Siren* in the Atlantic 12.7.1814. *Magpie* wrecked on the coast of France 18.2.1807. *Crane* wrecked on the rocks at West Hoe 26.10.1808 then BU. *Pigeon* wrecked near Margate 15.1.1809. *Rook* taken by two French privateers off San Domingo 18.8.1808. *Wagtail* wrecked at Saint Michael's, Azores 13.2.1807. *Sealark* wrecked in the North Sea 18.6.1809. *Wigeon* wrecked on the Scottish coast 29.8.1808. *Woodcock* wrecked at St Michael's, Azores 13.2.1807 (with *Wagtail*).

PLANS: Lines & profile/decks/specification.

## *Arrow* Class 1805. 14 or 2, Armed Schooner.
Designed by 'John Peake of the Admiralty'.
    Dimensions & tons: 91', 79'3½" x 19' x 9'. 152.
    Men: \_\_\_\_ . Guns: 14 x \_\_\_\_ or 2 x 6.
Arrow Deptford Dyd.
    Ord \_\_\_\_ 1804?; K 01.1805; L 7.9.1805; 1814 yard craft.

FATE: Became a yard craft at Plymouth for the Breakwater Department; 1828 BU.

PLANS: Framing & decks/midships section.

## *Adonis* Class 1806. 10, Cutters/Schooners.
'Similar to the Bermudian sloop *Lady Hammond*'. Though originally classed as cutters it seems unlikely that they were ever rigged as anything other than their

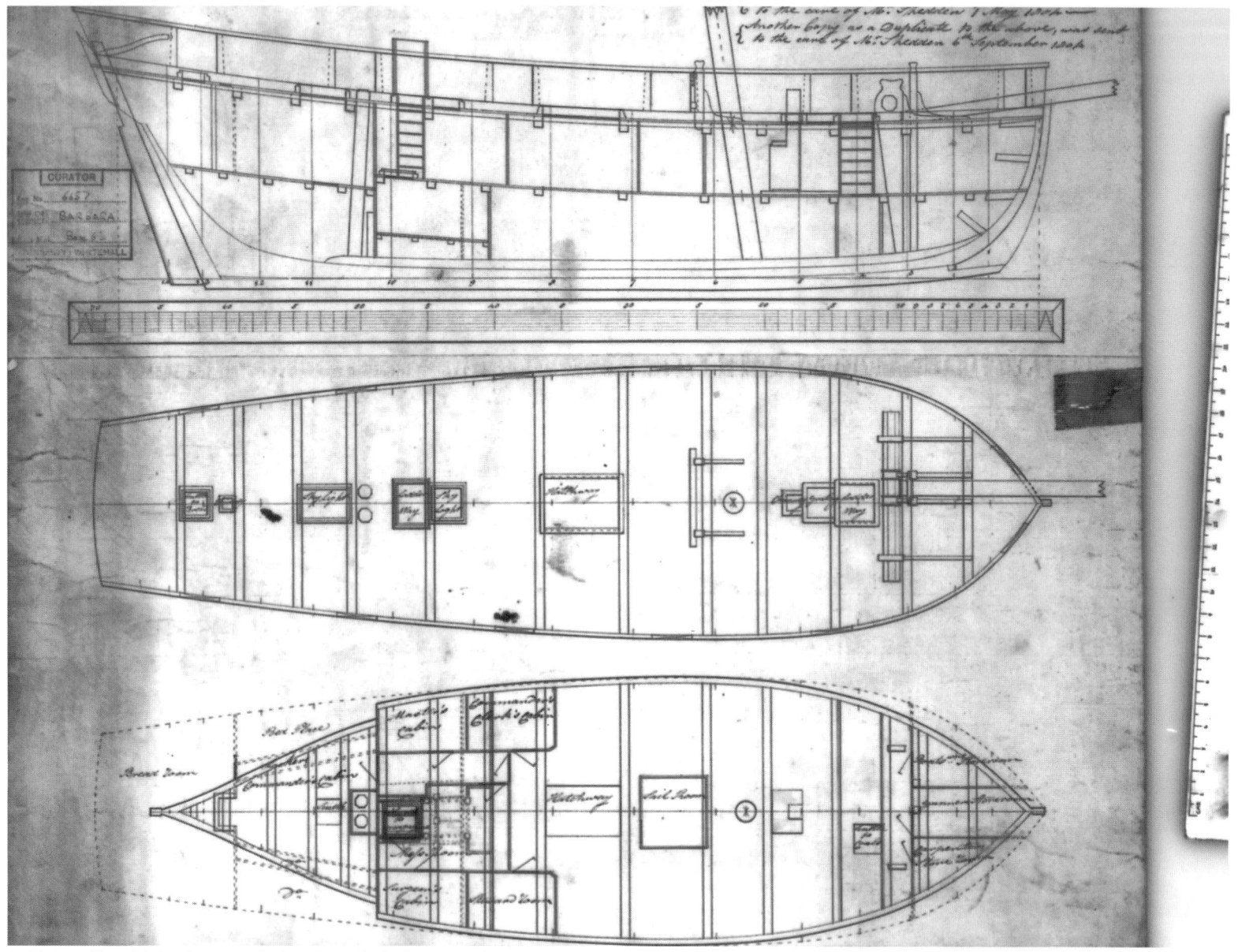

Building profile and decks plan (6657) for the *Adonis* Class built in Bermuda in 1806. Shown here with the original single masted cutter rig, but this seems soon to have been changed to two masted schooner rig

later designation as schooners. Built of cedar. Contracted for by Goodrich & Co, but presumably sub-contracted to various Bermudan builders, like the *Ballahoo* Class.

Dimensions & tons: 72'4", 54' x 22'3" x 9'8". 142 18/94.
Men: 35. Guns: 10 x 18★.

**Adonis** ____, Bermuda.
    Ord ____ 1804?; K ____ ; L ____ 1806; 1814 sold.
**Alban** ____, Bermuda.
    Ord ____ 1804?; K ____ ; L ____ 1806; (1810-1811 in Danish hands); 1812 wrecked.
**Alphea** ____, Bermuda.
    Ord ____ 1804?; K ____ ; L ____ 1806; 1813 blown up.
**Bacchus** ____, Bermuda.
    Ord ____ 1804?; K ____ ; L ____ 1806; 1807 foundered.
**Barbara** ____, Bermuda.
    Ord ____ 1804?; K ____ ; L ____ 1806; (1807-1808 in French hands); 1815 sold.
**Cassandra** ____, Bermuda.
    Ord ____ 1804?; K ____ ; L ____ 1806; 1807 foundered.
**Claudia** ____, Bermuda.
    Ord ____ 1804?; K ____ ; L ____ 1806; 1809 wrecked.
**Laura** ____, Bermuda.
    Ord ____ 1804?; K ____ ; L ____ 1806; 1812 taken.
**Olympia** ____, Bermuda.
    Ord ____ 1804?; K ____ ; L ____ 1806; 1815 sold.
**Sylvia** ____, Bermuda.
    Ord ____ 1804?; K ____ ; L ____ 1806; 1816 sold.
**Vesta** ____, Bermuda.
    Ord ____ 1804?; K ____ ; L ____ 1806; 1816 sold.
**Zenobia** ____, Bermuda
    Ord ____ 1804?; K ____ ; L ____ 1806; 1806 wrecked.

FATES: *Alban* taken by a flotilla of Danish gunboats 24.5.1810; 11.5.1811 retaken. *Alphea* blown up in action with the French privateer *Renard* 9.9.1813. *Bacchus* supposed foundered at sea (in the West Indies?) 08.1807. *Barbara* taken in the West Indies by the French privateer *Général Ernouf*; became French privateer *Peraty*; 17.7.1808 retaken. *Cassandra* foundered off Bordeaux 13.5.1807. *Claudia* wrecked off Norway 20.1.1809. *Laura* taken by the French privateer *Diligente* off North America 10.1812. *Zenobia* ran aground on the Florida coast 29.10.1806.

PLANS: Lines/lines & profile/profile & decks/sails.

### *Cheerful* Class 1806. 12, Cutters. Henslow design.
Dimensions & tons: 63', 46'9⅜" x 23'6" x 10'. 137.
Men: 50. Guns: 12 x 12★.
**Cheerful** Johnson, Dover.
    Ord ____ 1806?; K 06.1806; L 11.1806; 1816 sold.
**Surly** Johnson, Dover.

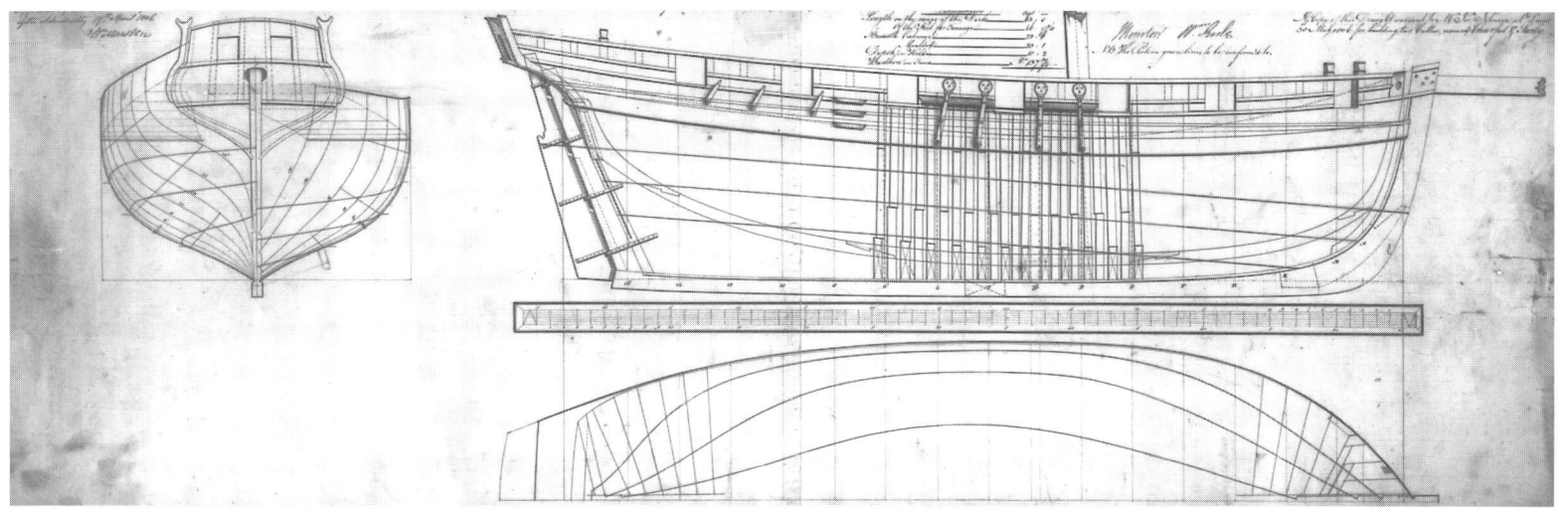

Design lines plan (6370) for the *Cheerful* Class cutters of 1806

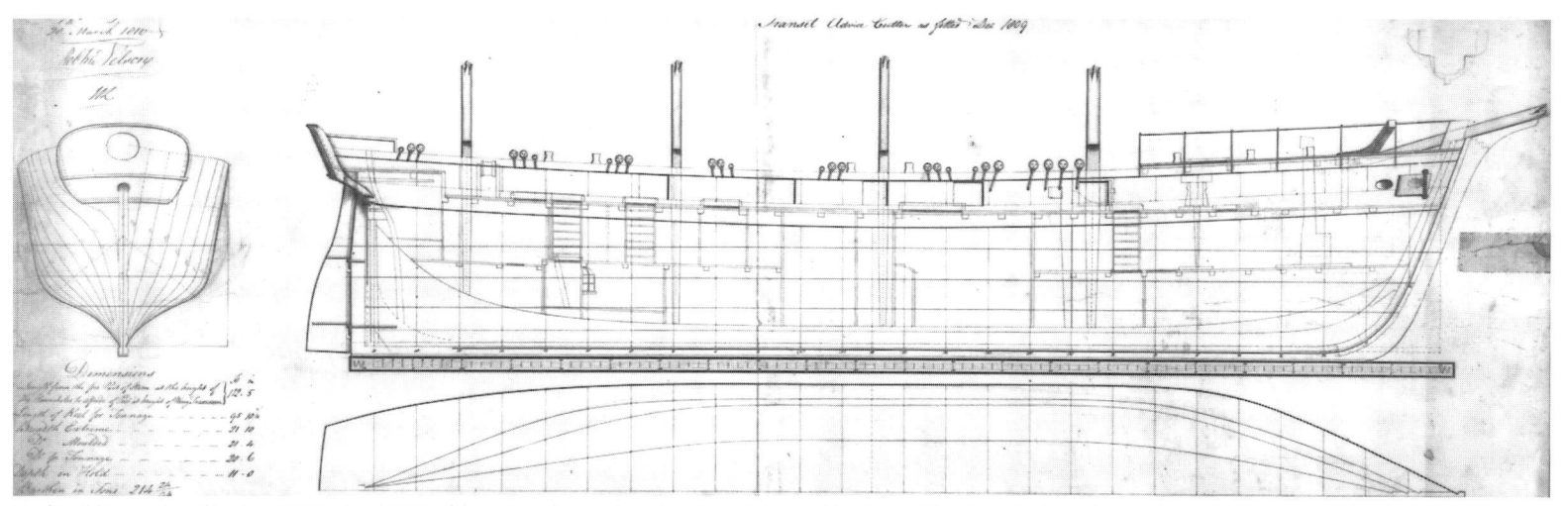

'As fitted' lines and profile plan (6387B) dated 1809 of the extraordinary advice boat *Transit* designed by Gover. This shows both the four masts and the hull shape, but not the totally distinctive rig

Ord ____ 1806?; K 07.1806; L 15.11.1806; 1833 lighter.

FATES: *Surly* became a lighter at Chatham; 1837 sold.

PLANS: Lines/profile/decks/planking expansion (external)/(internal).

### *Shamrock* ('Shrub') Class 1808. 8, Armed Schooners.

'Similar to the [American] *Flying Fish*.' Built of cedar.
Dimensions & tons: 78'8", 60'8" x 21'7" x 7'10". 150 32/94.
Men: 50. Guns: 2 x 6 + 6 x 12★.

*Shamrock* ____ , Bermuda.
Ord ____ 1808?; K ____ ; L 15.9.1808; 1811 wrecked.

*Thistle* ____ , Bermuda.
Ord ____ 1808?; K ____ ; L 27.9.1808; 1811 wrecked.

*Bramble* Dell, Bermuda.
Ord ____ 1808?; K ____ ; L ____ 1809; 1815 sold.

*Holly* Tynes, Bermuda.
Ord ____ 1808?; K ____ ; L ____ 1809; 1814 wrecked.

*Juniper* ____ , Bermuda.
Ord ____ 1808?; K ____ ; L ____ 1809; 1814 sold.

*Mistletoe* Hill, Bermuda.
Ord ____ 1808?; K ____ ; L ____ 1809; 1816 sold.

FATES: *Shamrock* wrecked on Cape Saint Mary 25.2.1811. *Thistle* wrecked near New York 6.3.1811. *Holly* wrecked off San Sebastian 29.1.1814. PLANS: Lines/profile/decks/sails/ballast.

### *Transit* Class 1808. 11, 'Advice Cutter'/Advice Boat.

Designed by Captain Gover of the East India Company. He had already built one *Transit* and this vessel was a modified version of the same design. Several different lines and sets of dimensions were considered before the final ones (the dimensions given below are 'as built') were adopted. Gover claimed that Admiralty interference ruined his conception. The hull was very long in relation to its breadth – and had a very odd cross-section somewhat like a peg-top with straight sides tapering down from the rails and a deep vee-shaped bottom. The rig was even odder, approximating to that of a four-masted barquentine. In 1809 she was shortened by 20', her midships portion being cut out, which explains why later lists give her length as 112'5" and her tonnage as 214 24/94.

Dimensions & tons: 130', 117'2½" x 22'6" x 12'? 262.
Men: 40. Guns: 1 x 12 (brass) + 10 x 12★.

*Transit* Bailey, Ipswich.
Ord ____ 1808?; K 07.1808; L 3.3.1809; 1809 shortened; 1812 impressment tender; 1815 sold.

PLANS: Lines/midships section/specification. As built: Lines & profile.

### *Pigmy* Class 1809. 10, Cutters, Later Schooners. Peake design.

Dimensions & tons: 82'6", 70'6" x 22'8" x 10'6". 192 62/94.
Men: 60 Guns: 10 x 18★.

*Pigmy* King, Upnor.
Ord 10.1809; K 11.1809; L 24.2.1810; 1811 schooner; 1823 sold.

*Algerine* King, Upnor.
Ord 10.1809; K 11.1809; L 3.3.1810; 1813 wrecked.

*Pioneer* King, Upnor.
Ord 10.1809; K 12.1809; L 10.3.1810; 1811 schooner; 1824 to the Coastguard; 1849 sold.

FATE: *Algerine* wrecked 'on Gallipagos Rocks, the North side of Little Bahama Bank' 20.5.1813.

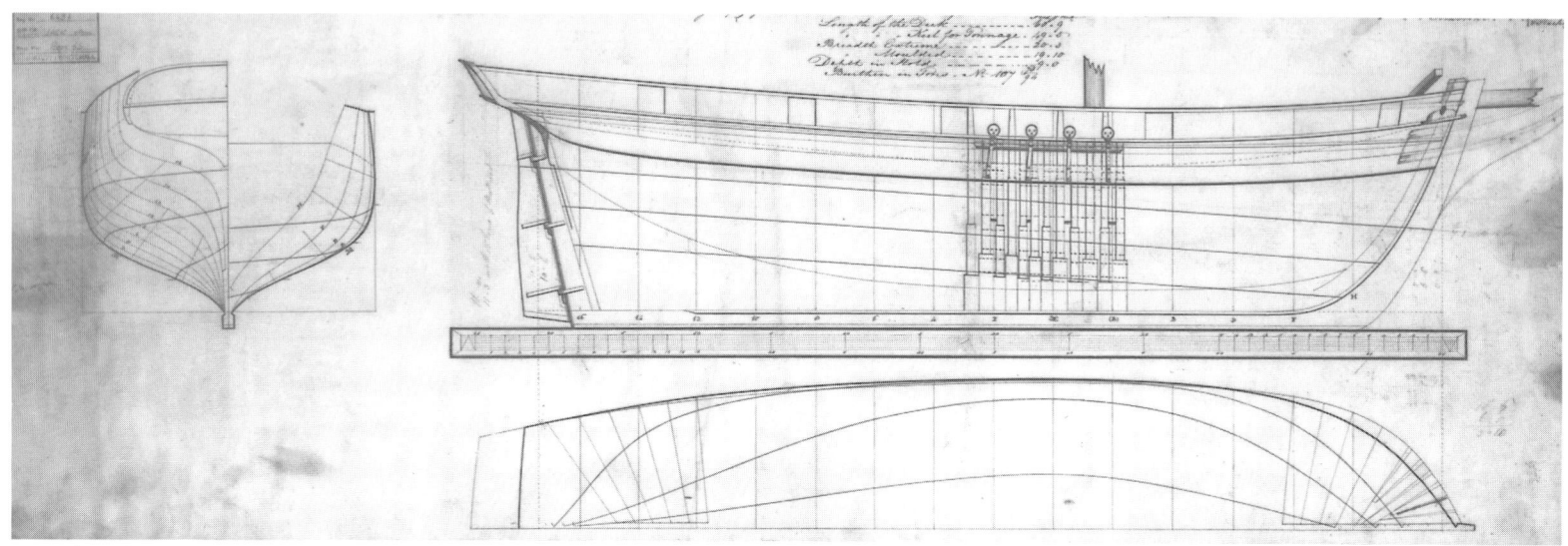

Design lines plan (6436) for the *Quail* Class cutter tenders of 1816

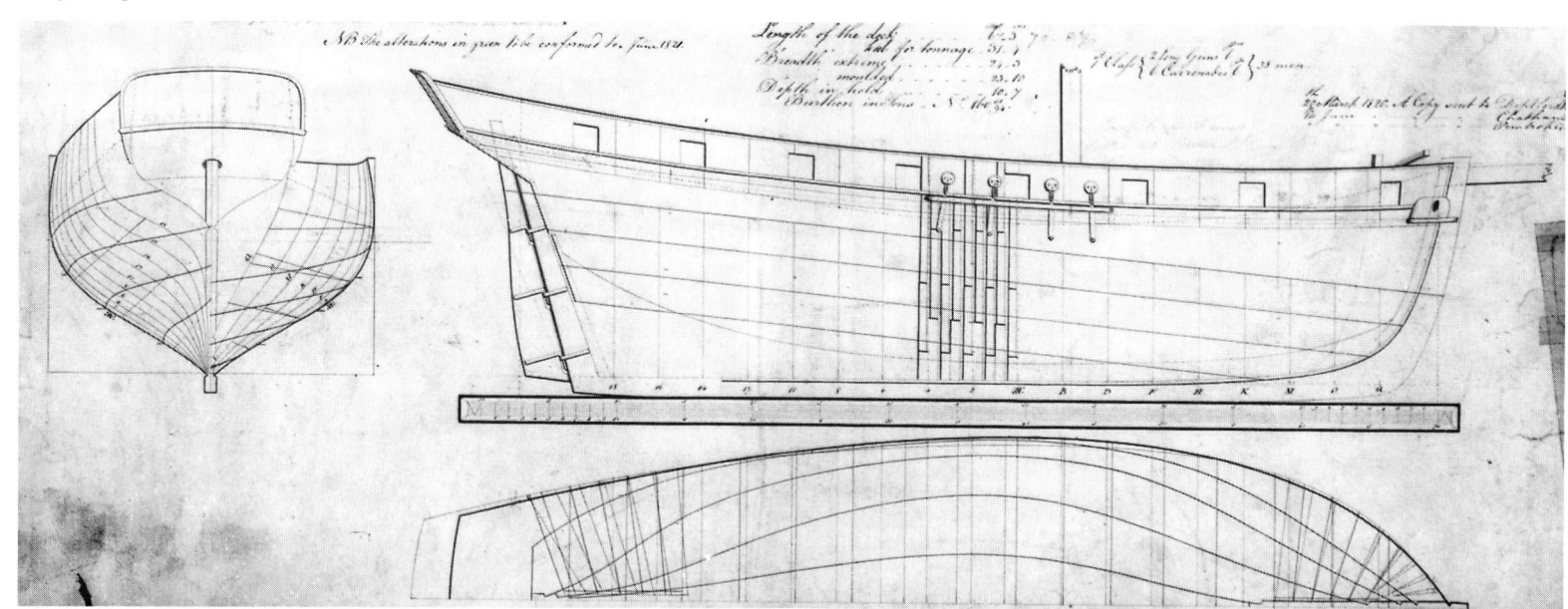

Lines plan (6332) for the *Vigilant* Class cutter *Basilisk* of 1822

PLANS: Lines & profile/decks.

### Decoy Class 1809. 10, Cutters. Rule design, clinker built.
Dimensions & tons: 74'4", 55'10⅝" x 26' x 11'. 200 88/94.
Men: 60. Guns: 10 x 18★.
**Decoy** List, Fishbourne.
  Ord 2.10.1809; K 11.1809; L 22.3.1810; 1814 taken.
**Dwarf** Lawes, Sandgate.
  Ord 2.10.1809; K 11.1809; L 24.4.1810; 1824 wrecked.
**Racer** Baker, Sandgate.
  Ord 2.10.1809; K 11.1809; L 24.4.1810; 1810 taken.
FATES: *Decoy* taken by the French after grounding between Calais and Gravelines on the Waldun Flat 22.3.1814. *Dwarf* wrecked on the pier at Kingstown, Ireland 1.4.1824. *Racer* taken by the French after stranding on the coast of France 24.5.1810.
PLANS: Lines/lines & profile/midships section/specification.

### Nimble Class 1811. 12, Cutters. 'By the draught of an improved revenue cutter.'
Dimensions & tons: 63', 49'1¼" x 22'9" x 10'6". 140 16/94.
Men: 50. Guns: 10 x 12★.
**Nimble** Gely, Cowes.
  Ord ____ 1811?; K 06.1811; L 14.12.1811; 1812 wrecked.
**Swan** Gely, Cowes.
  Ord ____ 1811?; K 06.1811; L 1.11.1811; 1831 fishery protection vessel; 1837 hulked.

FATES: *Nimble* wrecked on a sunken rock near Salo Beacon 6.10.1812. *Swan* to the Seamen's Home Society as a church ship on the Thames; 1874 BU.

PLANS: Lines & profile/decks/specification.

### Quail Class 1816. 2, Cutter Tender. 'By draught of *Watchful* revenue cruiser but of increased dimensions.'
Dimensions & tons: 55'6", 44'5" x 18'7" x 7'6". 80.
Men: 30. Guns: 2 x 6.
**Quail** Deptford Dyd.
  Ord 23.5.1816; K 08.1816; L 3.1.1817; later renamed *Providence*; 1829 BU.
**Redbreast** Woolwich Dyd.
  Ord 23.5.1816; K 11.1816; L 18.2.1817; BY 1830 hulked.
**Linnet** Deptford Dyd.
  Ord 23.5.1816; K 08.1816; L 3.1.1817; 1833 sold.
**Swift** Woolwich Dyd.
  Ord 23.5.1816; K 11.1816; L 15.2.1817; 1821 sold.
**Woodlark** Deptford Dyd.
  Ord ____ ; K 09.1820; L 2.7.1821; 1829 survey ship; 1864 sold.
**Highflyer** Woolwich Dyd.
  Ord ____ 1821??; K 01.1821; L 11.6.1822; 1833 sold.
FATES: *Redbreast* hulked at Liverpool; then to Sheerness as a tender; 1840s to Plymouth as a lazaretto; 1850 sold.

PLANS: Lines/lines & profile/decks/midships section. *Swift* as built: profile & decks.

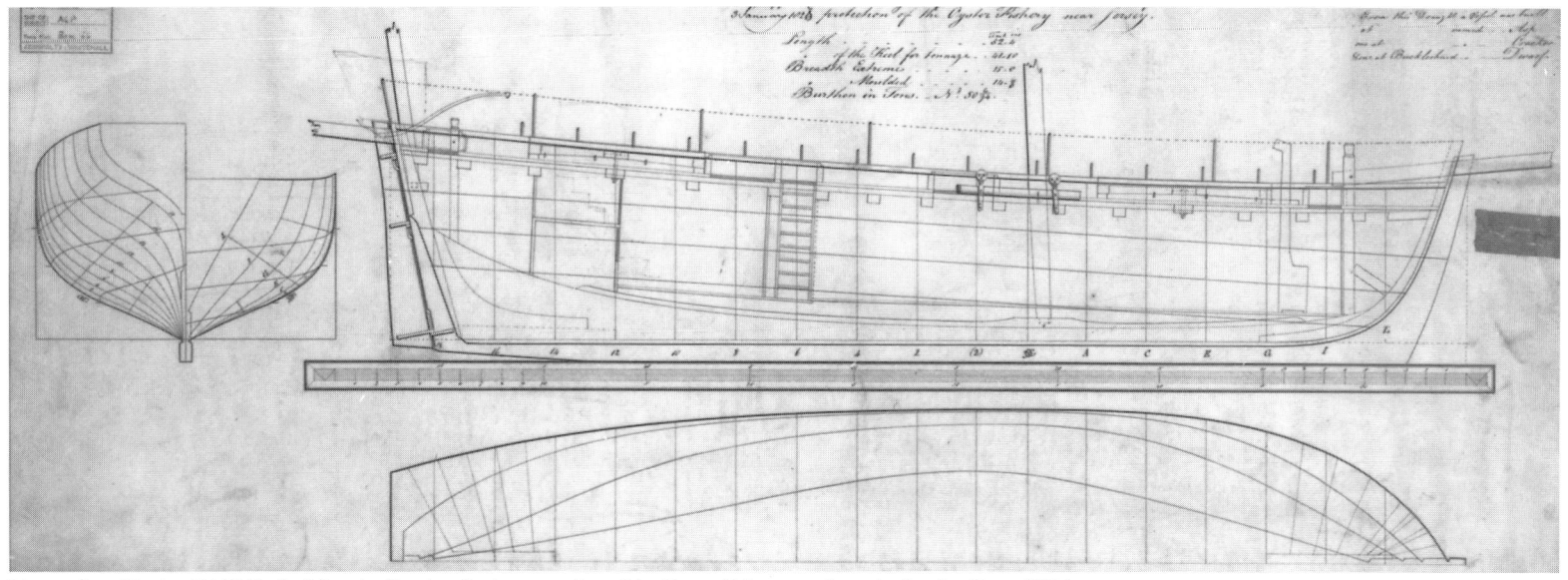

Lines and profile plan (6446) 'for building the Tenders for the protection of the Oyster Fishery near Jersey' – the *Asp* Class of 1826

### *Starling* Class 1816. 10, Cutter. 'Similar to the *Griper* Cutter.'
Dimensions & tons: 71'9", 60'8" x 21'7⅞" x 9'3". 151.
Men: 45. Guns: 2 x 6 (brass) + 8 x 6★.
*Starling* Chatham Dyd.
Ord ____ 1816?; K 09.1816; L 3.5.1817; 1828 BU.
PLANS: Lines/profile & decks/midships section.

### *Vigilant* Class 1819. 12/10, Cutters. 'On the lines of the *Lapwing* revenue cutter enlarged.'
Dimensions & tons: 67'3", 51'4" x 24'3" x 10'7". 160 51/94.
Men: 38. Guns: 2 x 6 (brass) + 10 x 9★ (or 2 x 6 + 8 x 6★?).
*Vigilant* Deptford Dyd.
Ord 3.9.1819; K 08.1820; L 18.4.1821; later ketch rig; 1832 sold.
*Basilisk* Chatham Dyd.
Ord ____ 1819?; K 02.1821; L 7.2.1822; 1835 ketch rig; 1846 sold.
*Skylark* Pembroke Dyd.
Ord ____ ; K 08.1820; L 17.2.1821; fate uncertain.
PLANS: Lines/profile & decks/framing/midships section. *Vigilant* as built: Profile & decks. *Basilisk* as ketch 1835: Profile/decks.

### *Bramble* Class 1820 (or 1819?) 10, Cutters. 'As *Diligence* revenue cutter'.
Dimensions & tons: 70'9", 52'4" x 24'2½" x 11'. 161 (approx).
Men: 50. Guns: 2 x 6 (brass) + 8 x 12★.
*Bramble* Plymouth Dyd.
Ord ____ 1819?; K 08.1820; L 8.4.1822; 1842 survey ship, schooner rig; 1853 to New South Wales Government.
*Sparrow* Pembroke Dyd.
Ord ____ 1819?; K 10.1827; L 28.6.1828; 1837 ketch rig; 1860 BU.
PLANS: Lines & profile/profile/decks. *Bramble* as schooner 1841: sails. *Sparrow* as ketch 1847: profile/decks.

### *Nightingale* Class 1822. 6, Cutter Tenders. 'Similar to the *Dove* revenue cutter.'
Dimensions & tons: 63'9", 46'10" x 22'2" x 9'6". 122.
Men: 34. Guns: 2 x 6 (brass) + 4 x 6★.
*Nightingale* Plymouth Dyd.
Ord ____ 1822?; K 04.1822; L 19.4.1825; later schooner rigged; 1829 wrecked.
*Snipe* Pembroke Dyd.
Ord ____ 1822?; K 10.1827; L 28.6.1828; 1860 BU.
*Speedy* Pembroke Dyd.
Ord ____ 1822?; K 11.1827; L 28.6.1828; 1853 harbour craft.
FATES: *Nightingale* wrecked on the Shingles 17.2.1829. *Speedy* became harbour craft at Plymouth; 1866 BU.
PLANS: Lines & profile/decks/framing/midships section. *Nightingale* as schooner 1828: sails.

### *Arrow* Class 1822 (or 1820?) 10, Cutter. Designed by Captain Hayes.
Dimensions & tons: 64', 48'8⅝" x 24'9" x 9'2½". 157.
Men: 50. Guns: 2 x 6 (brass) + 8 x 6★.
*Arrow* Portsmouth Dyd.
Ord ____ 1820?; K 09.1822; L 14.3.1823; 1831 fitted for the Customs; 1838 ketch rig; 1852 BU.
PLANS: as built: Lines & profile/decks. As ketch 1828: sails. 1851: decks.

### *Magpie* Class 1826 (or 1825?). 3/4, Cutters. 'Similar to the *Assiduous* tender.'
Dimensions & tons: 53'3", 40'8½" x 18' x 7'3". 70.
Men: 35/26. Guns: 2 x 9 + 2 x 18★ (*Magpie*), 2 x 12 + 1 x 5½" howitzer (*Monkey*).
*Magpie* McLean, Jamaica.
Ord ____ 1825?; K ____ ; L 06.1826; 1826 wrecked.
*Monkey* McLean, Jamaica.
Ord ____ 1825?; K ____ ; L 06.1826; 1831 wrecked.
*Nimble*? McLean, Jamaica?
Ord ____ ; K ____ ; L ____ 1826; 05.1826 returned to builder as unfit for service [Presumably this class, dimensions fit]; Dittmar suggests that two members of this class were cancelled in 1825.
FATES: *Magpie* wrecked by a squall in Colorados Road, Cuba 11.1826. *Monkey* wrecked on Tampico Bar 05.1831.
PLANS: No plans.

### *Asp* Class 1826. 2, 'Cutters' (Lug Rig?). Design by Joseph Tucker. 'Not found to answer the purpose intended.' There is some evidence that *Asp* served briefly in the Channel Islands (or was this another, possibly purchased, vessel of the same name?), otherwise the class did not see much, if any, naval service; deleted 1826.
Dimensions & tons: 52'4", 41'10" x 15' x 6'1½". ____ .
Men: 16. Guns: 2 x 12★.
*Asp* Tucker, Gosport.
Ord ____ ; K ____ ; L ____ 1826; 1826 deleted.
*Dwarf* Tucker, Gosport.
Ord ____ ; K ____ ; L ____ 1826; fate uncertain? (possibly sold in 1860?; or is this another vessel of the same name: deleted 1826?).
*Cracker* Tucker, Gosport.
Ord ____ ; K ____ ; L ____ 1826; 1826 deleted.
PLANS: Lines/lines & profile/decks.

### *Pickle* Class 1825? 3, Schooners. 'By a draught prepared at Jamaica.'

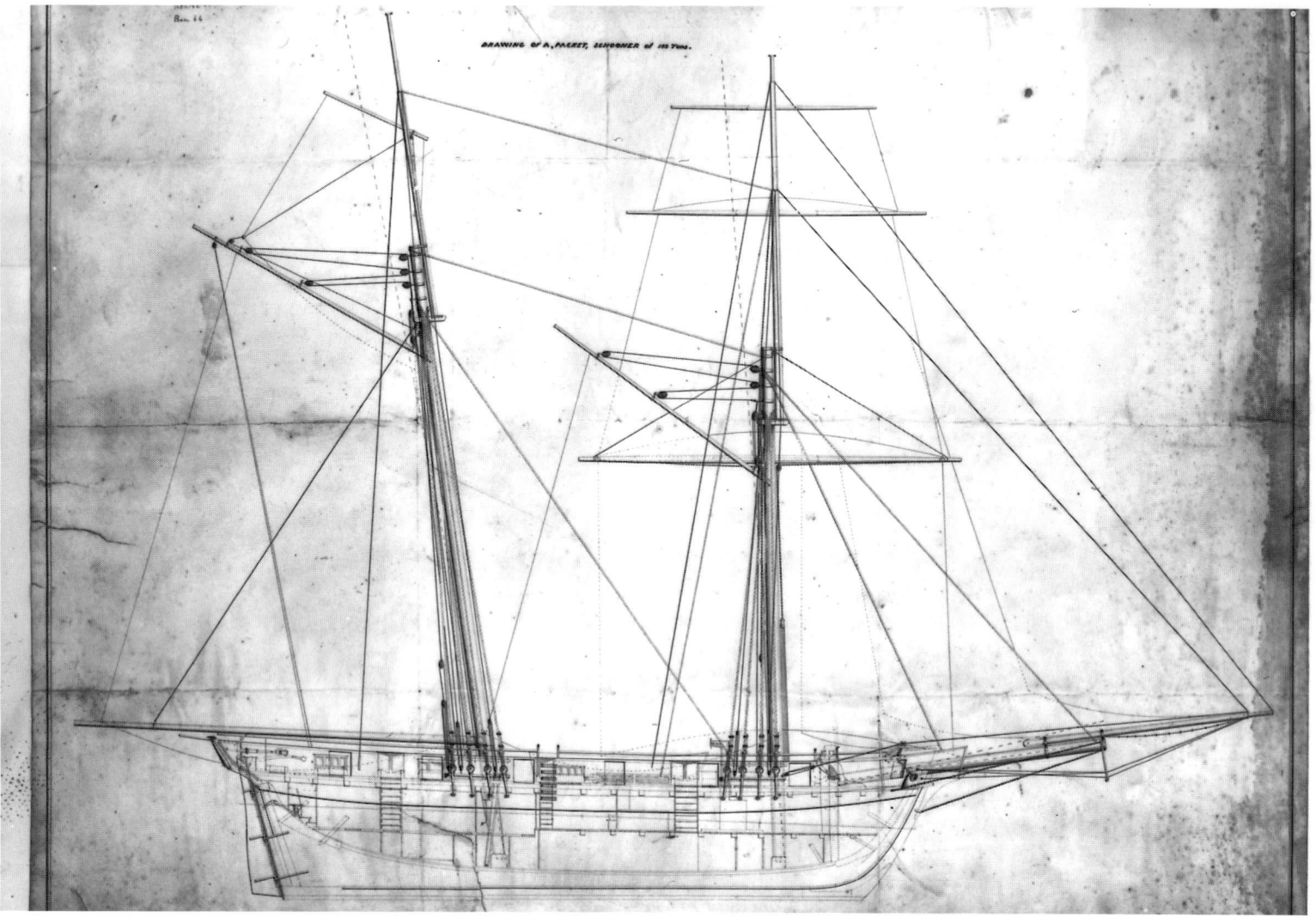

Rig plan (x4593) for the *Hornet* Class schooner/brigantine packets of 1829

Dimensions & tons: 62'8", 54' x 21'2" x 9'9". 118.
Men: 36. Guns: 1 x 18 + 2 x 18★

**Pickle** ____ , Bermuda.
Ord ____ 1825?; K ____ ; L 08.1827; 1847 BU.
**Pincher** ____ , Bermuda.
Ord ____ 1825?; K ____ ; L 08.1827; 1838 wrecked.
**Skipjack** (ex *Skylark*) ____ , Bermuda.
Ord ____ 1825?; K ____ ; L 08.1827; 1841 wrecked.

FATES: *Pincher* capsized off Cowes Light 6.3.1838, raised and sold. *Skipjack* wrecked on the Caymans 06.1841, wreck sold.

PLANS: *Pincher* 1831: Lines & profile/decks/sails.

### Sylvia Class 1820? One Gun, Cutter. Designed by
Commander Symonds, RN, intended for the protection of the oyster fisheries at Jersey.

Dimensions & tons: 52'7", 40'8½" x 18'1" x 8'6". 70.
Men: 25. Guns: 1 x 3 (brass).

**Sylvia** Portsmouth Dyd.
Ord ____ 1820?; K 01.1827; L 24.3.1827; (1833-1836 yard craft, cutter tender, at Portsmouth); 1859 sold.

PLANS: No plans.

### Minx Class 1829. 3, Schooner. 'Similar to the *Union*' [*Union* was a schooner purchased in 1823.]

Dimensions & tons: 59'9", 44'6½" x 18'3½" x 6'10". 84.
Men: 32. Guns: 2 x 12★ + 1 x 5½" howitzer (may, at one time, have carried as many as 6 guns).

**Minx** ____ , Bermuda.
Ord ____ 1828?; K ____ ; L 23.12.1829; 1833 sold.

PLANS: No plans.

### Hornet Class 1829. 6, Schooners. Seppings design. Later mostly brigantine-rigged packets.

Dimensions & tons: 80', 64'5½" x 23' x 9'10". 182 89/94.
Men: ____ . Guns: 2 x 6 + 4 x 12★.

**Hornet** Chatham Dyd.
Ord ____ 1828?; K 12.1829; L 24.8.1831; completed as packet (brigantine); 1845 BU.
**Spider** Chatham Dyd.
Ord ____ 1828?; K 03.1830; L 23.9.1835; 1847 packet; later described as a sloop; 1861 sold.
**Viper** Pembroke Dyd.
Ord ____ 1828?; K 06.1830; L 12.5.1831; later classed as a packet; 185. BU.
**Cockatrice** Pembroke Dyd.
Ord ____ 1828?; K 07.1831; L 14.8.1832; completed as a packet (brigantine); 1846 victualling transport; 1847 store tender for the Pacific; 1858 sold.

PLANS: Lines/profile & decks/lower deck/upper deck/midships section/sails/specification.

### Lark Class 1829. 4, Cutter Tenders. Seppings design.

Dimensions & tons: 60'9", 49'5" x 20'3" x 9'. 107 73/94.
Men: 30 Guns: 2 x 6 ( brass ) + 2 x 6★

**Lark** Chatham Dyd.
Ord ____ 1829?; K 06.1829; L 23.6.1830; 1835 schooner; (1849-1858 loaned to the Liberian Government); 1860 BU.
**Quail** Sheerness Dyd.

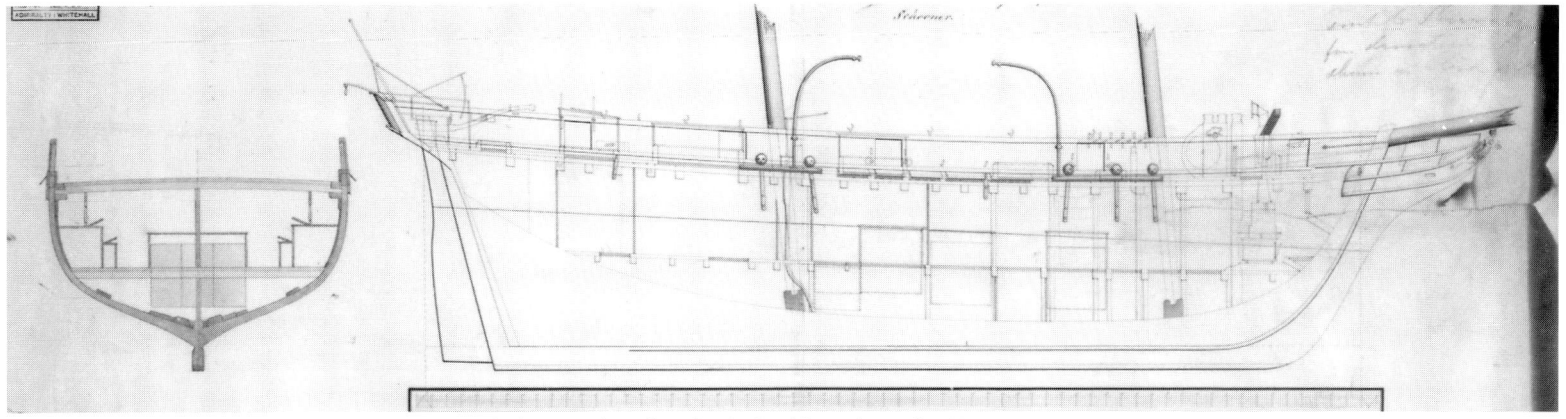

Profile and midsection (6800A) for the cutter/schooner *Lark*, the name ship of her class 'as fitted for a schooner' 1835

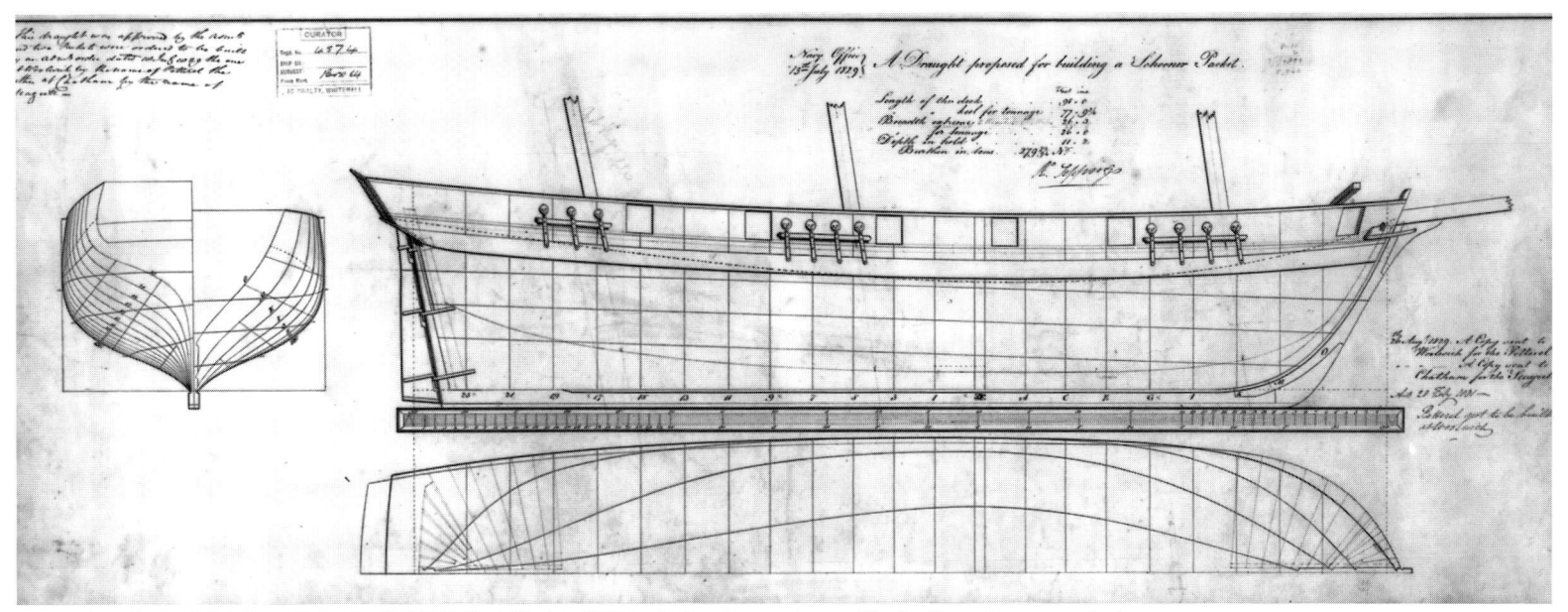

Lines and profile (6453A) 'taken off' the *Quail* of 1830 (*Lark* Class) in 1858 just before she was given to Liberia. Alternative cutter and schooner arrangements are shown

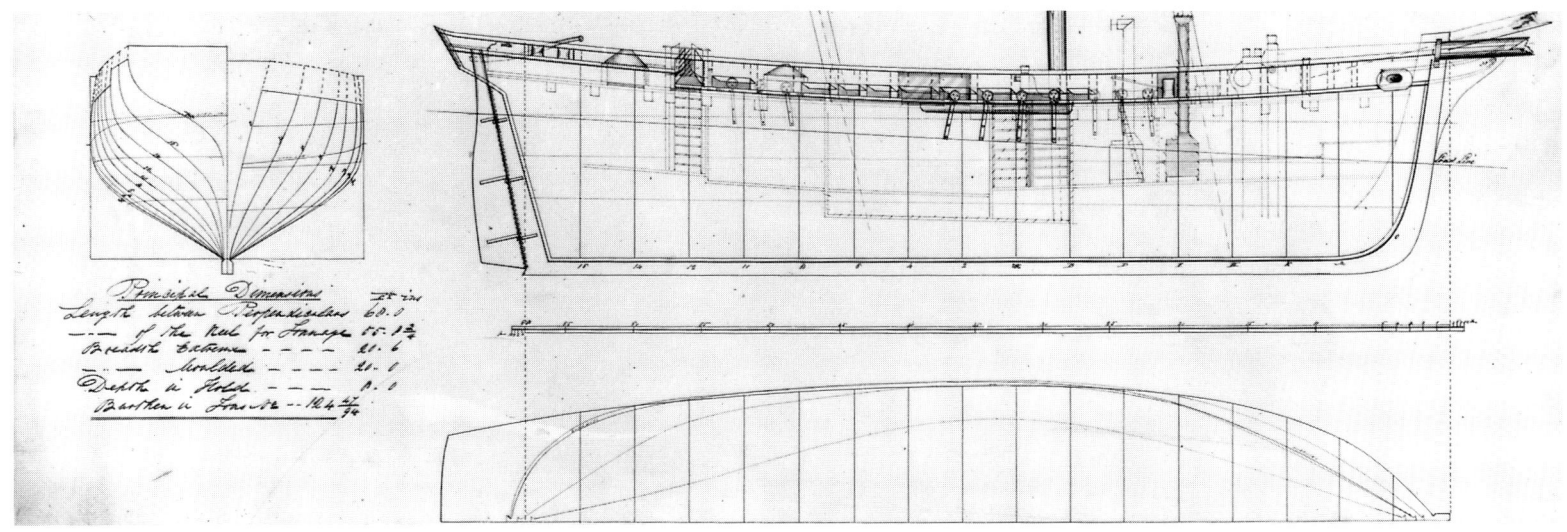

Design lines plan (4574) for building the *Seagull* schooner packet of 1831

Ord ____ 1829?; K 12.1829; L 30.9.1830; 1838 lighter at Sheerness; 1859 lengthened at this stage to 28' (124 tons); as cutter given to Liberian Government.

**Raven** Pembroke Dyd.
Ord ____ 1829?; K 06.1829; L 31.10.1829; 1835 yawl rig; survey ship; 1859 sold.

**Jackdaw** Chatham Dyd.
Ord ____ 1829?; K 08.1829; L 4.8.1830; 1833 schooner; 1835 wrecked.

**Starling** Pembroke Dyd.

Ord ____ 1829?; K 06.1829; L 31.10.1829; ? completed as schooner (certainly one by 1835); survey vessel; 1844 sold.

**Magpie** Sheerness Dyd.
Ord ____ 1829?; K 12.1829; L 30.9.1830; 1836 survey ship; 1846 yard craft.

FATES: *Jackdaw* wrecked off Old Providence Island, Honduras 11.3.1835. *Magpie* became a water tank at Chatham; fate uncertain.

PLANS: Lines/profile & decks/midships section/sails/specification. *Lark* fitted for surveying: profile & decks. As schooner 1835: Profile & decks/profile &

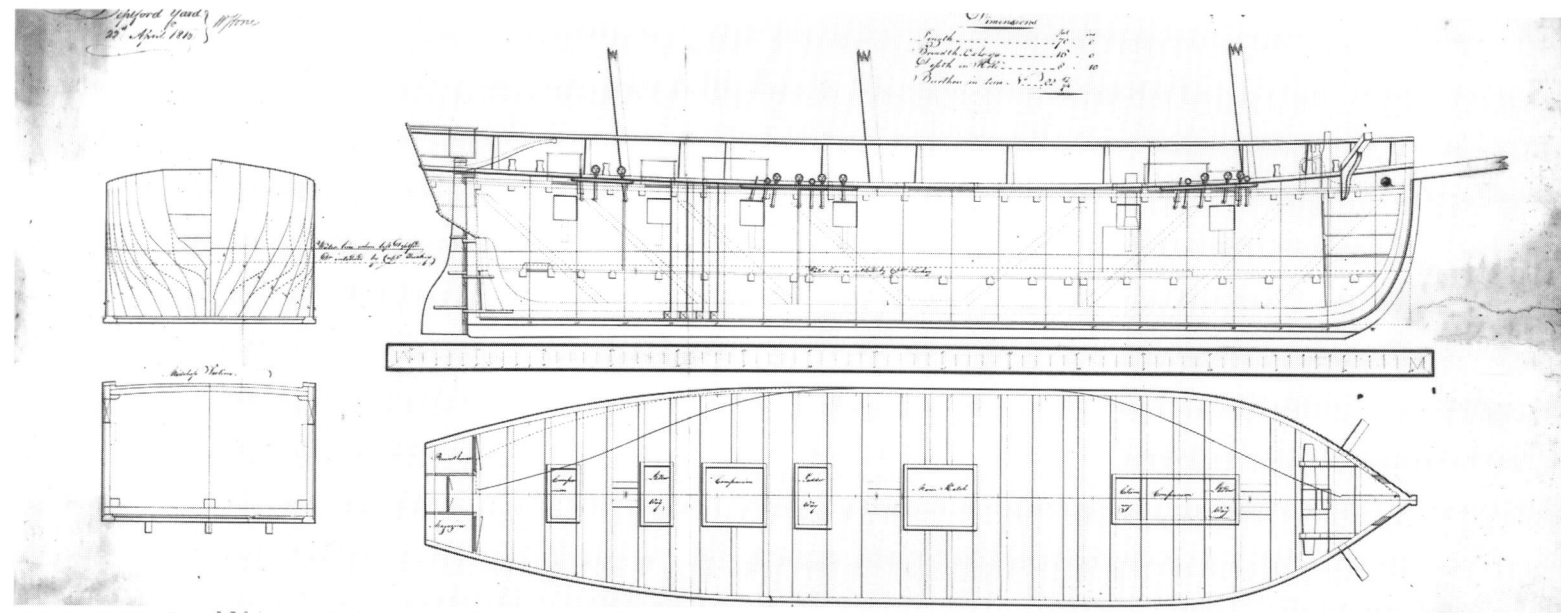

Hull plan (6318) of the steamer *Congo* 'as fitted' at Deptford in 1816 with her engine not fitted and described as a 'sloop'

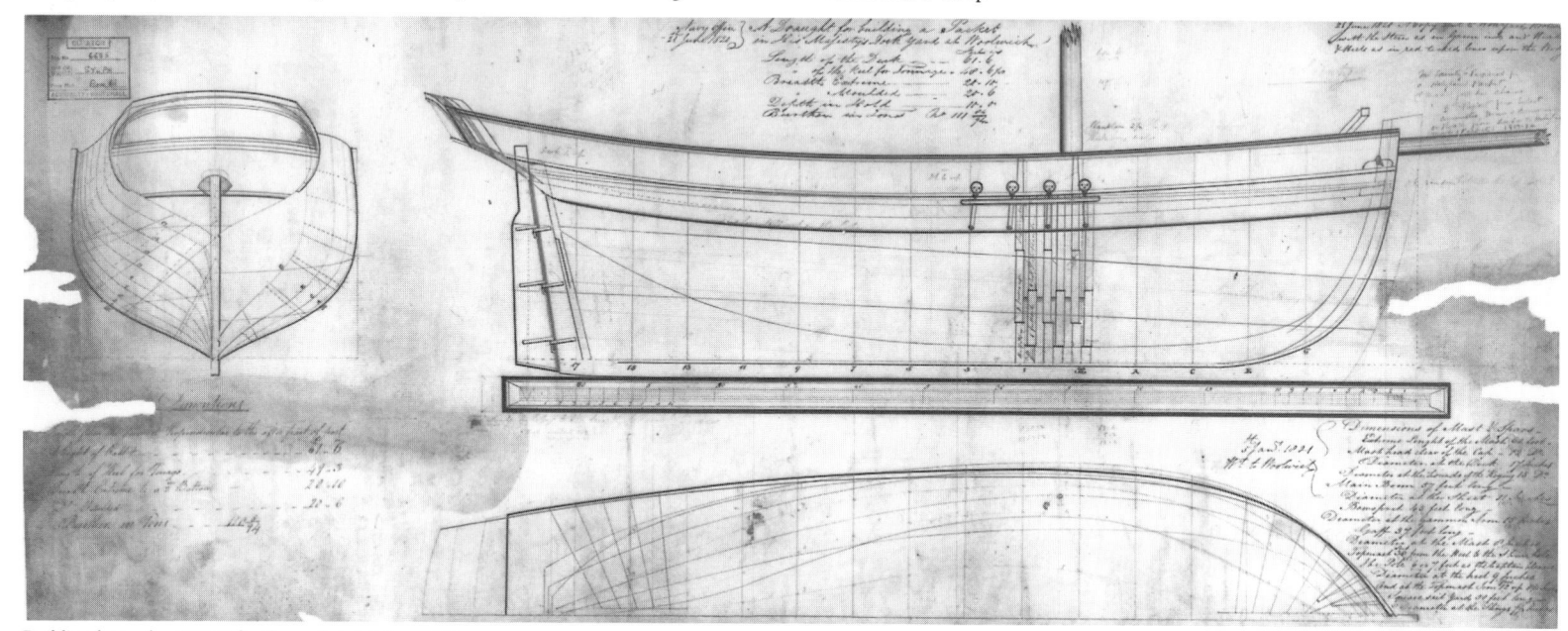

Building lines plan (6635) for the packet *Sylph* of 1821

midships section/sails. *Quail* as schooner 1858: Lines & profile/decks/sails. *Raven* as survey ship: decks. As ketch: sails. As yacht 1838: sails. *Jackdaw* as survey ship: profile & decks. As schooner: sails. *Starling* as schooner: Profile & decks. *Magpie* as survey ship: decks. As schooner: sails. As ketch: sails.

### *Seagull* Class 1829. 6, 3 Masted Schooner; later Brigantine. Seppings design.

Dimensions & tons: 95', 77'9¾" x 26'3" x 11'2". 279 74/94.
Men: 29. Guns: 2 x 6 (brass) + 4 x 12★.

*Seagull* Chatham Dyd.
Ord 18.7.1829; K 08.1830; L 28.11.1831; 1834 packet; 1852 BU.

### Cancelled 1831. *Peterel* Woolwich Dyd.
Ord 18.7.1829; K ____ .

PLANS: Lines/profile & decks/midships section. As brigantine packet 1833: sails. 1835: sails.

## Storeships, etc

### *Porpoise* Class 1797. 10, Armed Vessel. Henslow design 'for the service of New South Wales'.

Dimensions & tons: 96', 76'10⅛" x 28" x 12'. 320 42/94.
Men: 33. Guns: 10 x 6.

*Porpoise* Hill & Mellish, Limehouse.
Ord ____ ; K 12.1797; L 16.5.1798; 1799 (or 1801?) renamed *Diligent*; 1802 sold.

PLANS: Lines/profile/orlop, lower & upper decks/quarter deck & forecastle/framing/stern. 1800: Profile.

### *Aid* Class 1808. Transports. Tucker, Seppings & Peake design. Mostly used as yard craft.

Dimensions & tons: 104', 87'2⅛" x 26' x 17'6". 313 47/94.
Men: 39 (*Aid* as survey ship: 10). Guns: None (*Aid* as survey ship: 4 x 12★ + 2 x 6★)

*Aid* Brindley, Lynn.
Ord ____ ; K 07.1808; L 4.4.1809; 1817 survey ship; 1821 renamed *Adventure*; 1839 transport; 1853 sold.
*Assistance* Dudman, Deptford.
Ord ____ ; K 09.1808; L 7.3.1809; fate uncertain.
*Chatham* Brindley, Frindsbury.
Ord ____ ; K 10.1810; L 2.6.1811; 1825 breakwater.
*Portsmouth* Stone, Milford.
Ord ____ ; K 10.1810; L 28.8.1811; 1828 hulked.
*Diligence* Bailey, Ipswich.
Ord ____ ; K 10.1813; L 30.9.1814; 1862 hulked.

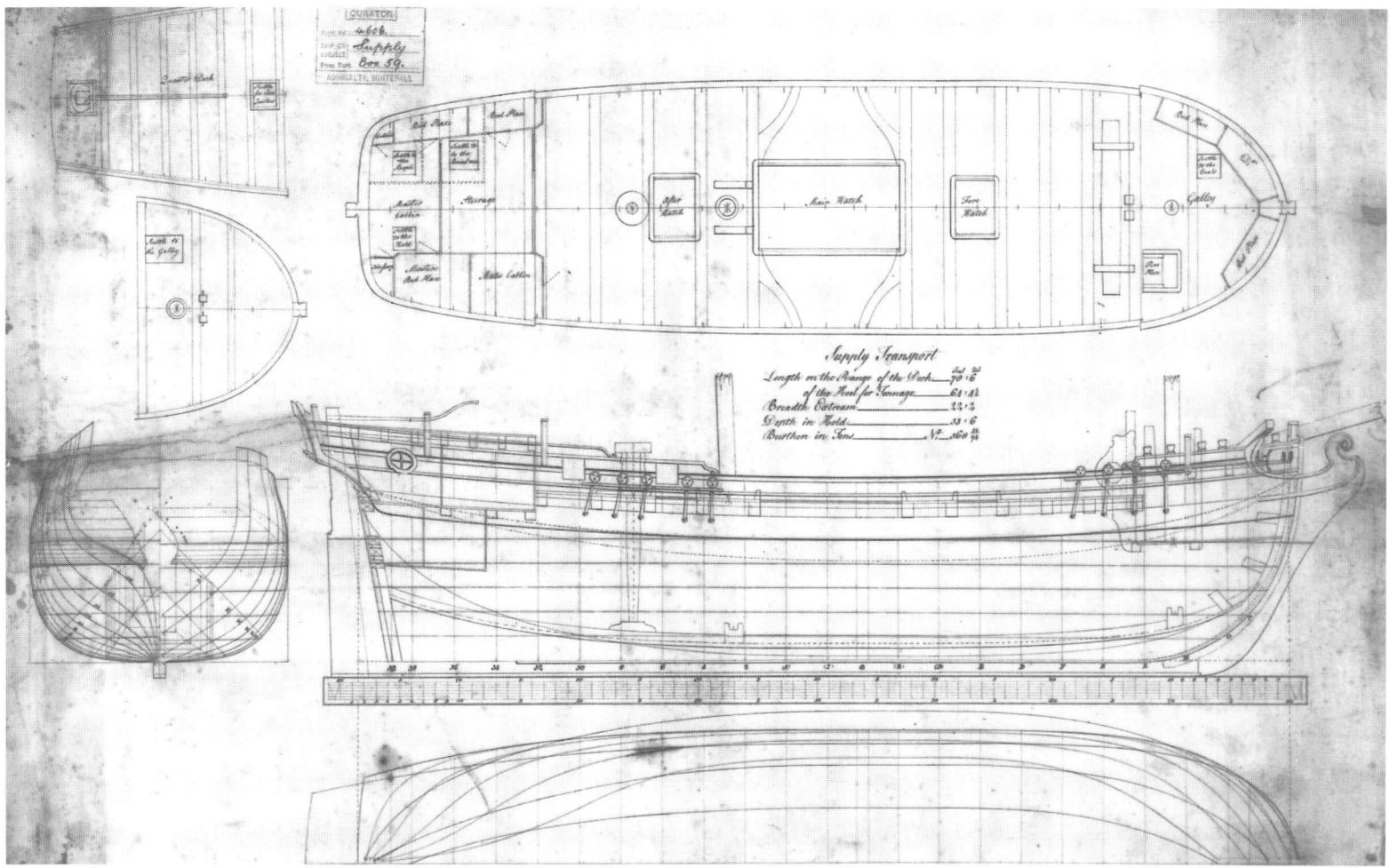

Hull plan (4606) of *Supply* as converted in 1786 from a dockyard craft to an armed tender to accompany the 'first fleet' to Australia

***Industry*** Warwick, Eling.
>Ord ____ ; K 01.1814; L 13.10.1814; 1835 hulked.

FATES: *Portsmouth* became a coal depot at Woolwich; later to Deptford; 1834 BU. *Diligence* became a coal depot at Plymouth; ? later to Portsmouth?; 1904 sold. *Industry* became a church ship at the Isle of Man; 1846 BU.
PLANS: Lines & profile/specification.

### *Investigator* Class 1811. 6, Survey Vessel. Peake design.
>Dimensions & tons: 75'0⅜", 62'10⅞" x 19'2" x 10'10⅝". 121.
>Men: 20. Guns: 6 x 12★.

***Investigator*** Deptford Dyd.
>Ord ____ 1811??; K 01.1811; L 23.4.1811; 1837 hulked.

FATE: To the Thames River Police as a tender off Somerset House; 1867 BU.
PLANS: Lines & profile/decks.

### *Congo* Class 1815. One Gun, Exploration Vessel.
Seppings design 'to be built for the purpose of exploring the interior of Africa by the River Congo'. She was designed as a paddle steamer, the first intended for the Royal Navy (if one excepts the dubious case of the experimental vessel *Kent* built at Blackwall for what are alleged to have been naval purposes, but for which no evidence appears in the Admiralty papers sighted. She [*Kent*] was later supposed to have become a gunboat once her engine was removed, but again no evidence has been found for this in Admiralty lists). *Congo* was flat-bottomed and slab-sided. 'She was found too weak to carry the steam engine which was intended for her', so was converted to a sailing vessel and went to Africa in this guise, on what proved to be a thoroughly disastrous expedition.

>Dimensions & tons: 70', ____ x 16' x 8'10". 82 57/94.
>Men: ____ . Guns: 1 x 12★ + 12 swivels.

***Congo*** Deptford Dyd.
>Ord ____ ; K 10.1815; L 11.1.1816; 1817 fitted for a (sailing) survey vessel; 1819 fitted to lie in the Swale; 1826 sold.

PLANS: Lines.

### *Sylph* Class 1820. 2, Packet/Survey Ship. Designer
uncertain. Intended as a 'model packet for the Post Office'; but never fitted for sea as such.
>Dimensions & tons: 61'7", 48'8" x 21'2" x 10'. 114.
>Men: ____ . Guns: 2 x 3 (brass).

***Sylph*** Woolwich Dyd.
>Ord ____ ; K 10.1820; L 15.6.1821; 1833 ? survey ship ?; 1837 tender , then to Customs [1900 deleted].

PLANS: Lines/profile/midships section.

## Conversion of Yard Craft to Armed Tender.
***Supply*** 1786. 8, armed tender. Deptford yard transport designed by Slade and built by Bird at Rotherhithe in 1759.
>Dimensions & tons: 79'4", 64'11" x 22'6" x 11'6½". 174 76/94.
>Men: 55 (ex: 14) Guns: 4 x 3 + 4 x 12★ + 12 musketoons (ex: 4 x 3 + 6 swivels).

***Supply*** 1786 converted to accompany the 'First Fleet' to Botany Bay; 1792 sold.
PLANS: Lines & profile & decks.

# Chapter 6: Symonds and After 1830-1860

This was the period in which steam advanced from an occasional auxiliary to the main battle propulsion for the fighting fleet. Here we are only concerned with those ships that were ordered or commenced as sailing warships, a diminishing number as the period wore on. Sail retreated to being used only for cruising power for major warships, and then for minor ones. After 1860 very nearly all ships built for the RN and most purchased were steam powered, though it was to be many years before masts and yards disappeared from seagoing warships.

For a full account of the developments with larger ships in this period see Andrew Lambert's two major works, *The Last Sailing Battlefleet* and *Battleships in Transition*; their author very kindly has checked the entries in this chapter, which has greatly benefited from his comments. In his own books he has shown how much design advanced during the period and how political and administrative battles caused some of the delays and alterations which surface in the changes of names and designs seen in the following pages.

The crucial administrative change happened at the beginning of the period covered by this chapter, with the abolition of the Navy Board, the appointment of a naval officer and amateur ship designer (Symonds) to head the ship design section of the Admiralty in place of the last Master Shipwright to be Surveyor of the Navy (Seppings). Symonds had his own formula for building fast ships, broad in the beam and with 'vee' shaped bottoms. This produced generally fast ships but ones which were poor gun platforms. His designs and the various reactions to them dominate the history of British warship design for the next few years until the introduction of screw propulsion and then of iron hull construction (successfully from 1860 after a false start in the 1840s) made the controversy no longer relevant.

Since 1815 the tendency had been towards slow construction of the bigger vessels; sometimes years elapsed between ordering and laying down, and then between that and launching, whilst completion might be indefinitely delayed. This helped to contribute to the great success in building wooden hulls where the problem of decay had been mastered as far as was humanly possible. It also meant that numbers of hulls begun as sailing ships ended up altered to a greater or lesser extent as screw steamers (a far easier conversion than from sail to paddles, as is witnessed by the insignificant number of conversions to the latter system of propulsion).

## First Rates, 120 Guns (three-deckers)

**Royal Albert Class 1843.** Lang design.
   Dimensions & tons (design as sailing ship): 220', 177'2¾" x 60'10" x 25'. 3393 70/94.
   Men: 1000. Guns: GD 4 x 8" + 28 x 32. MD 2 x 8" + 32 x 32. UD 34 x 32. Spar deck: 6 x 32 + 14 x 32 (short).

**Royal Albert** Woolwich Dyd.
   Ord ____ ; K 08.1844; L as screw 13.5.1854; 1883 sold for BU.

PLANS: Lines/Profile/hold & hold sections/orlop/gun deck/main deck/upper deck/quarter deck & forecastle/framing. As built: Lines/profile/hold & hold sections/orlop/gun deck/main deck/upper deck/quarter deck & forecastle/roundhouse.

**Duke of Wellington Class 1848.** Surveyors Department design to the lines of the *Queen* but to carry 120 guns, though possibly originally intended to be of 110 guns (as with most ships of this period the prehistory of the design is complicated and long-drawn out as part of the struggle between Symonds and his opponents - see Lambert *Last Sailing Battlefleet* for elaboration and explanation). The *Prince of Wales* Class were a slight modification of this design, a modification that was probably decided upon at the time that the decision was made to convert them all to screw (see below).
   Dimensions & tons (design as sailing ship) 210', 171'1" x 60' x 24'8". 3185. (as screw ship of the line): 240'6", 201'8" x 60' x 24'8". 3759 4/94.
   Men: 970. Guns: GD 30 x 8", MD 30 x 32, UD 32 x 32. Spar deck 8 x 32 + 14 x 32 (short).

**Duke of Wellington** (ex *Windsor Castle*) Pembroke Dyd.
   Ord ____ 1844?/29.8.1848; K 05.1849; L as screw 14.2.1852; 1863 hulked.

FATE: Hulked for 'harbour service' at Portsmouth; 1909 sold for BU.

PLANS: Lines/profile/hold & hold sections/orlop/gun deck/upper deck/quarter deck & forecastle/roundhouse.

**Prince of Wales Class 1848.** Surveyors Department design (with a complicated prehistory, involving the Committee of Reference), to the lines of the *Queen* but to carry 120 guns, originally part of the *Duke of Wellington* Class. Lengthened bow ordered in 1853, presumably when it was decided to convert them to screw.
   Dimensions & tons (design as sailing ship): 210', 171'1" x 60' x 24'8". 3185. (as screw ships of the line): 252', 213' x 60' x 24'8". 3966 20/94.
   Men: 970. Guns: GD 30 x 8", MD 30 x 32, UD 32 x 32, Spar deck 8 x 32 + 14 x 32 (short).

**Prince of Wales** Portsmouth Dyd.
   Ord ____ 1842?/29.6.1848; K 06.1848; L as screw 25.1.1860; 1869 hulked.

**Royal Sovereign** Portsmouth Dyd.
   Ord ____ 1844?/29.6.1848; K 17.12.1849; L as screw 25.4.1857; 1864 completed as a turret ironclad 1885 sold for BU.

**Marlborough** Portsmouth Dyd.
   Ord ____ 1844?/ ____ 1848?; K 1.9.1850; L as screw 31.7.1855; 1878 hulked.

FATES: *Prince of Wales* became a training ship for cadets renamed *Britannia*; 1909 sold for BU. *Marlborough* became an instruction ship at Portsmouth; 1904 renamed *Vernon II*; 1924 sold and foundered on the way to BU 10.1924.

PLANS: As *Duke of Wellington* (see above). *Prince of Wales*: Lines/profile/hold & hold sections/orlop/gun deck/main deck/upper deck/quarter deck & forecastle/roundhouse. 1869 as cadet ship *Britannia*: Profile/hold & hold sections/orlop/gun deck/main deck/upper deck/quarter deck & forecastle. 1882: Profile/orlop/gun deck/main deck/upper deck. *Marlborough*: as built: Profile/hold & hold sections/orlop/gun deck/main deck/upper deck/quarter deck & forecastle. 1886: Decks. 1888: Decks. *Royal Sovereign* as completed: Profile/hold & hold sections/lower deck/upper deck.

## First Rates, 110 Guns (three-deckers)

**Queen Class 1833 Symonds Design.** *Queen* herself was slightly different in dimensions and tonnage from the rest.
   Dimensions & tons: *Queen*: 204', 166'5¼" x 60' x 23'9". 3099 16/94. Rest of class: 204', 165'11" x 60' x 23'9". 3083.
   Men: 950 Guns: GD 28 x 32 + 2 x 68, MD 28 x 32 + 2 x 68, UD 32 x 32, QD 10 x 32 'gunnades', Fc 2 x 32 + 2 x 68.

**Queen** (ex *Royal Frederick* name changed 1839) Portsmouth Dyd.
   Ord 29.10.1827; K 11.1833; L 15.5.1839; 1858-9 converted to two-decker 86 gun screw line of battle ship; 1871 BU.

**Windsor Castle** (ex *Victoria* name changed 1855; ex *Howe*?) Pembroke Dyd.
   Ord 3.10.1833; K 05.1844; L as screw 26.8.1858; 1869 hulked.

**Frederick William** (ex *Royal Frederick*) Portsmouth Dyd.
   Ord ____ ; K 07.1841; L as screw 24.3.1860; 1876 hulked.

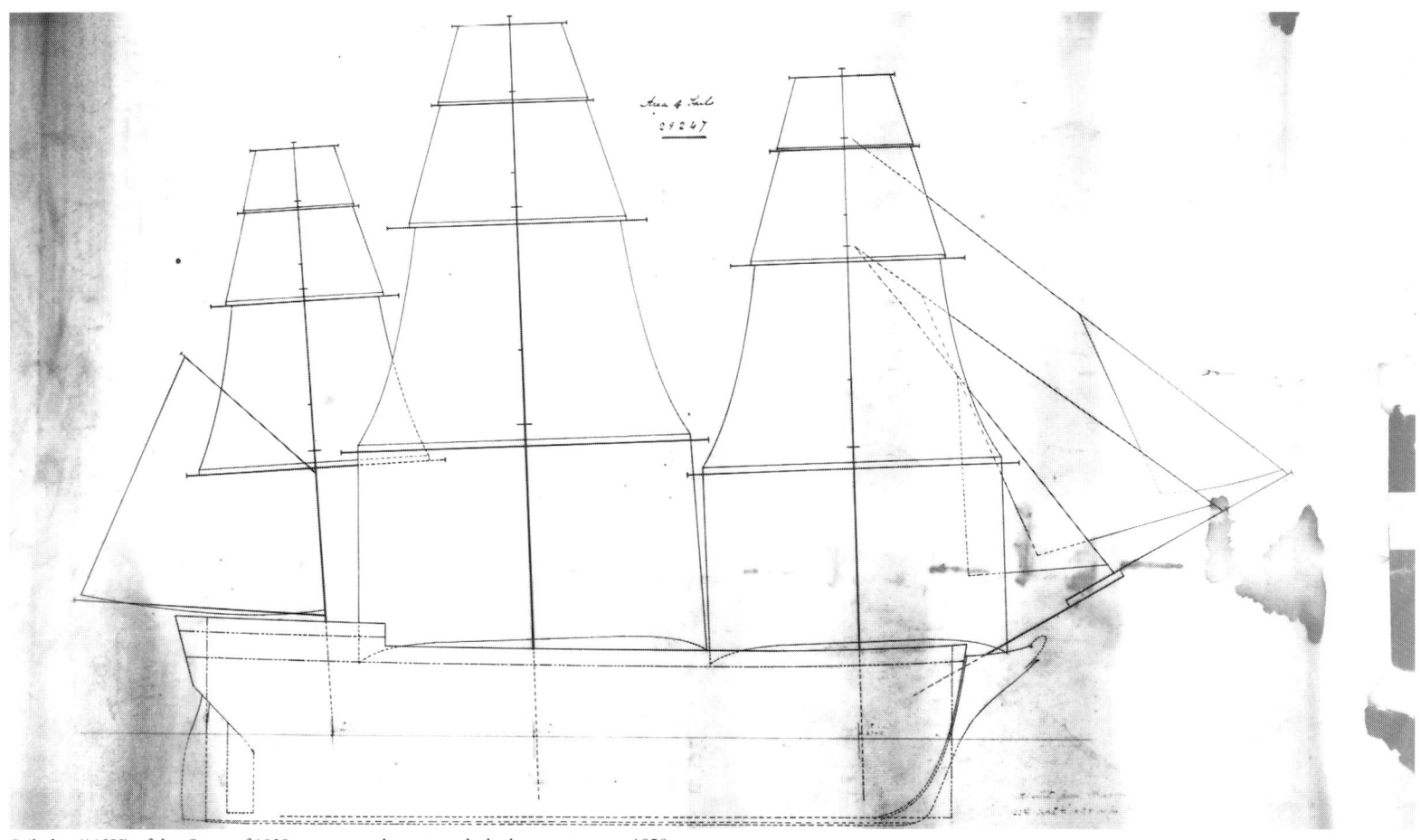

Sail plan (113X) of the *Queen* of 1839 as converted to a two-decked screw steamer, 1858

**Cancelled 11.12.1834.** *Algiers* Pembroke Dyd.
Ord 3.10.1833; not begun.

**Cancelled 1838.** *Royal Sovereign* (ex ____ ) Woolwich Dyd.
Ord ____ 1832; not begun?

FATES: *Windsor Castle* became gunnery training ship at Plymouth, renamed *Cambridge*; 1908 sold. *Frederick William* became the training ship *Worcester*; 1948 sold; foundered in the Thames, raised; 1953 BU at Grays.

PLANS: Lines/profile/upper deck/framing. *Queen* as fitted 1842: cabins on middle deck. 1858: machinery arrangement. 1858 as two decker screw: Lines/hold & hold sections/orlop/lower deck/main deck/spar deck/sails. No date, as two decker screw: Lines. *Windsor Castle* as screw 1858: Profile/hold & hold sections/orlop/lower deck/main deck/middle deck/upper deck/round house/machinery arrangement. *Frederick William* as fitted 1865: Profile/hold & hold sections/lower deck/main deck/upper deck/topsides

## Second Rates, 90 Guns (two-deckers)

***Albion* Class 1839.** Symonds design. *Princess Royal* and *Algiers*, possibly also *Caesar* and the 1854 *Hannibal* originally ordered as members of this class, but designs subsequently modified.
Dimensions & tons: 204', 165'11" x 60' x 23'8". 3083 12/94.
Men: 830. Guns: GD 32 x 8", UD 32 x 32, Spar Dk 26 x 32.

*Albion* Plymouth Dyd.
Ord 18.3.1839; K 13.8.1839; L 6.9.1842; 1861 converted to screw; 1884 BU.

*Aboukir* Plymouth Dyd.
Ord 18.3.1839; K 08.1840; L 4.4.1848; 1858 converted to screw; 1878 sold.

*Exmouth* Plymouth Dyd.
Ord 18.3.1839; K 09.1841; L as screw 12.7.1854; 1877 hulked.

**Cancelled 1845.** *Saint Jean D'Acre* Plymouth Dyd.
Ord 1844; not begun; later reordered (1851) as screw line of battle ship.

**Cancelled 1846.** *Hannibal* Woolwich Dyd.
Ord 18.3.1839; not begun?; transferred to Deptford Dyd; later built to modified design (see below under *Hannibal* Class).

FATES: *Exmouth* became a training ship for the Metropolitan Asylums Board; 1905 sold for BU.

PLANS: Lines/profile/hold & hold sections/orlop/gun deck/upper deck/quarter deck & forecastle/sails. *Albion*: 1860 as screw: profile/hold & hold sections/orlop/gun deck/upper deck/roundhouse. *Exmouth* as screw: profile/hold & hold sections/orlop/gun deck/upper deck/quarter deck & forecastle.

***Princess Royal* Class 1847.** Modification of original *Albion* design by Edye; by Admiralty Order of 25.4.1847 the breadth was reduced. *Hannibal* (see below) is considered by Lambert as a sister; and certainly the two were very closely related.
Dimensions & tons: 204', 166'7$\frac{3}{8}$" x 58' x 24'. 2896 25/94.
Men: 830. Guns: GD 32 x 8", UD 32 x 32, spar deck 26 x 32.

*Princess Royal* (ex *Prince Albert*) Portsmouth Dyd.
Ord 12 & 28.3.1840; K 02.1841; L as screw 23.6.1853; 1872 sold.

PLANS: 1840: Lines/profile/hold & hold sections/orlop/gun deck/upper deck/quarter deck & forecastle/roundhouse/midships section. 1847 as screw: Lines. 1850: Hold & hold sections/orlop. 1853: Profile/hold & hold sections. 1854: Hold & hold sections. No date as screw: Hold & hold sections/gun deck/upper deck.

***Algiers* Class 1847.** Modification of *Albion* design (by Committee of Reference? Andrew Lambert believes that she was Edye's modification of the *Albion* to follow the less controversial [with the Admiralty] lines of the *Queen*, as lengthened).
Dimensions & tons (as designed as sailing vessel): 210', 170' x 60' x 24'5". 3165 42/94. (as screw): 218'7", 179'9" x 60' x 24'5". 3346 93/94.
Men: 830. Guns: GD 32 x 8", UD 32 x 32, spar deck 26 x 32.

*Algiers* Plymouth Dyd (? originally ordered from Pembroke Dyd?)
Ord 21.10.1847; K 07.1848; L as screw 26.1.1854; 1870 sold.

PLANS: Lines/profile/hold & hold sections/orlop/gun deck/upper deck/quarter deck & forecastle/roundhouse.

***Caesar* Class 1847.** Committee of Reference design. Modified from the *Rodney* design (?) as part of the assault by the Committee of Reference upon Symonds and Edye (see Andrew Lambert's *Last Sailing Battlefleet*). 06.1847 ordered to be named *Caesar*. Admiralty order 28.6.1847 to build to this design. Conversion to screw ordered 1852.

    Dimensions & tons: 207'4", 170'7⅜" x 56' x 23'4". 2761 24/94.
Men: 820. Guns: GD 8 x 8" + 24 x 32 (or 18 x 8" + 14 x 32)
UD 6 x 8" + 26 x 32, QD 18 x 32 (or 2 x 8" + 16 x 32), Fc 8 x 32.

***Caesar*** (ex *Algiers*, ex *Caesar*) Pembroke Dyd.
    Ord 11.3.1847; K 08.1848; L as screw 8.8.1853; 1870 sold.

PLANS: Lines/profile/hold & hold sections/orlop/gun deck/upper deck/quarter deck & forecastle/roundhouse.

***Hannibal* Class 1847.** Edye design (apparently modified from the *Albion* design; and very closely related to the *Princess Royal* [see above]). Originally ordered and begun at Woolwich in 1839 as an *Albion* class ship but keel taken up in 1842 to make room for *Prince Albert*; all work suspended 1844 and cancelled 1846. 1852 ordered to be converted to screw.

    Dimensions & tons (design as sailing ship): 208', 170'7⅜" x 58' x 24'. 2965 75/94. (as screw): 217', 179'1¾" x 58' x 24'. 3114 75/94.
Men: 820. Guns: GD 18 x 8" + 14 x 32, UD 6 x 8" + 26 x 32, QD 2 x 8" + 16 x 32, Fc 8 x 32.

***Hannibal*** Deptford Dyd.
    Ord 19.6.1847; K 12.1848; L as screw 31.1.1854; 1874 hulked.

FATE: Became a hulk [?] at Portsmouth; 1904 sold for BU.

PLANS: Lines/specification. As built: profile/orlop/gun deck/upper deck/quarter deck & forecastle/roundhouse. 1857: hold & hold sections.

## Second Rates, 80 Guns (two-deckers)

***Vanguard/Collingwood* Class 1832.** Symonds design. The shape of the bow was altered in later ships of the class. Sometimes classed as 78s.

    Dimensions & tons: 190', 155'3" x 56'9" x 23'4". 2889 61/94.
Men: 630/720. Guns: GD 30 x 32, UD 28 x 32, QD 16 x 32, Fc 6 x 32.

***Vanguard*** Pembroke Dyd.
    Ord 23.6.1832 or 7.7.1832; K 05.1833; L 25.8.1835; 1875 BU.

***Goliath*** Chatham Dyd.
    Ord 7.10.1833; K 03.1834; L 25.7.1842; 1858 converted to screw; 1870 hulked.

***Superb*** Pembroke Dyd.
    Ord ____ 1838?; K 11.1838; L 6.9.1842; 1864 hulked.

***Meeanee*** (ex *Madras*) Bombay Dockyard; Teak built.
    Ord ____ 1842?; K 08.1842; L 11.11.1848; 1857 converted to screw; 1867 hulked.

***Collingwood*** Pembroke Dyd.
    Ord 23.6.1832 or 7.7.1832; K 09.1835; L 17.8.1841; 1861 converted to screw; 1867 sold.

***Centurion*** Pembroke Dyd.
    Ord ____ ; K 07.1839; L 2.5.1844; 1855 converted to screw; 1870 sold.

***Mars*** Chatham Dyd.
    Ord ____ 1839?; K 12.1839; L 1.7.1848; 1856 converted to screw; 1869 hulked.

***Lion*** Pembroke Dyd.
    Ord ____ 1840?; K 07.1840; L 29.7.1847; 1859 converted to screw; 1905 sold.

***Majestic*** Chatham Dyd.
    Ord ____ ; K 02.1841; L as screw 1.12.1853; 1868 BU.

***Colossus*** Pembroke Dyd.
    Ord ____ 1843?; K 10.1843; L 1.6.1848; 1854 converted to screw; 1867 sold.

***Irresistible*** Chatham Dyd.
    Ord ____ ; K 01.1849; L as screw 27.10.1859; 1894 sold.

FATES: *Goliath* lent as a training ship; 22.12.1875 burnt by accident. *Superb* lent as an accommodation ship for Turkish naval crews for ships building on the Thames; by 1866 returned to the Ordinary?; 1869 BU. *Meeanee* passed to the War Department as a hospital ship [?]; 1906 BU. *Mars* became a training ship; 1929 sold.

PLANS: Lines/profile/orlop/gun deck/upper deck/quarter deck & forecastle/roundhouse/midships section. *Vanguard*: as built: Profile/hold & hold sections/orlop/gun deck/upper deck/quarter deck & forecastle/roundhouse. *Goliath*: 1857 as screw: orlop. 1863 as troopship: Profile/hold & hold sections/orlop/gun deck/upper deck/quarter deck & forecastle. *Collingwood*: 1844: orlop. *Centurion*: 1854 as screw machinery arrangement. 1858: Profile/hold/orlop/gun deck/upper deck/quarter deck & forecastle. 1863 as troopship: Profile/hold/orlop/gun deck/upper deck/quarter deck & forecastle. *Mars* as quarantine ship: specification. *Lion*: 1863 as troopship: Profile/hold & hold sections/orlop/gun deck/upper deck/quarter deck & forecastle. *Majestic*: as built: Profile/hold & hold sections /orlop/gun deck/upper deck/quarter deck & forecastle/roundhouse. *Colossus*: 1854 as screw: Profile/hold & hold sections/orlop/gun deck/upper deck/quarter deck & forecastle/roundhouse. . *Irresistible*: 1864: Profile/hold & hold sections/orlop/gun deck/upper deck/quarter deck & forecastle/roundhouse.

***Cressy* Class 1842.** School of Nautical Architecture (Read, Chatfield & Creuze) design.

    Dimensions & tons: 198'5" 162'1⅜" x 55' x 21'8¼". 2537 80/94.
Men: 750. Guns: GD 6 x 8" + 24 x 32, UD 2 x 8" + 30 x 32,
Spar deck 6 x 32 (long) + 16 x 32 (short).

***Cressy*** Chatham Dyd.
    Ord 23.4.1842/23.11.1843; K 04.1846; L as screw 21.7.1853; 1867 sold.

NOTE: There was probably to be one other member of this class which was eventually transmogrified into the screw line of battle ship *James Watt* (originally *Audacious*) (Pembroke). There was also another, similar, 80 ordered on 26.7.1841 to a design by Captain Hayes which was later cancelled and replaced by a screw line of battle ship with the same name; *Agamemnon*; see Lambert's

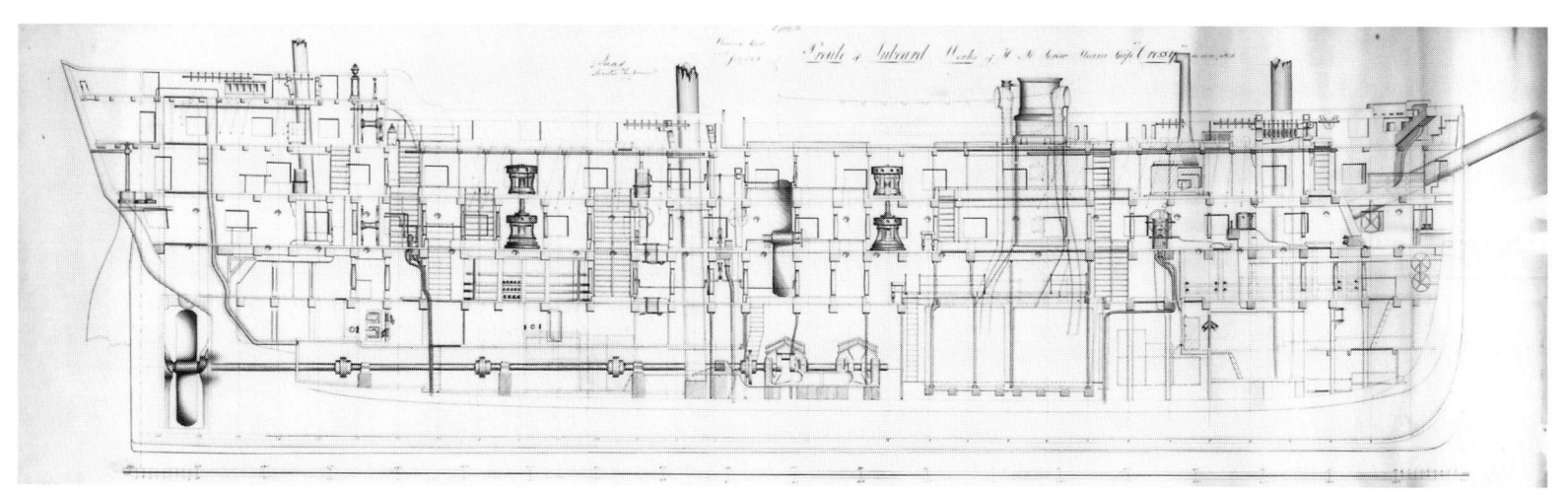

Profile (448) of the *Cressy* 80 gun ship as completed in 1858 as a screw line of battle ship

works.

PLANS: Lines/hold & hold section/orlop/gun deck/upper deck/quarter deck & forecastle/round house/sails/midships section/specification. As built: Profile/hold & hold sections/orlop/gun deck/upper deck/quarter deck & forecastle/roundhouse.

### *Brunswick* Class 1847. *Vanguard* design 'as altered by the Admiralty'.
By Admiralty Order of 27.9.1847 she was to have a fuller midships section but with less breadth.
Dimensions & tons: 190', 154'4¾" x 55'9" x 23'4". 2484 27/94.
Men: 750. Guns: GD 8 x 8" + 20 x 32, UD 4 x 8" + 24 x 32.
Spar deck 24 x 32.

***Brunswick*** Pembroke Dyd.
Ord ____ ; K 08.1847; L as screw 1.6.1855; 1867 sold.

PLANS: 1855 as screw: Lines. 1858: Profile/hold & hold sections/orlop/gun deck/upper deck/quarter deck & forecastle/roundhouse.

### *Sans Pareil* Class 1842. Lines of the old *Sans Pareil* (French, taken 1794).
Dimensions & tons: 193', 158'11½" x 52'1" x 22'8". 2242.
Men: 750. Guns: GD 6 x 8" + 24 x 32, UD 2 x 8" + 30 x 32, Spar deck 6 x 32 (long) + 16 x 32 (short).

***Sans Pareil*** Plymouth Dyd.
Ord 21.9.1842/22.4.1843; K 09.1845; L as screw 18.3.1851; 1867 sold.

PLANS: Lines/profile/hold & hold sections/orlop/gun deck/upper deck/quarter deck & forecastle/roundhouse/midships section/sails. As built: Profile/hold & hold sections/orlop/gun deck/upper deck/quarter deck & forecastle/sails. 1855: Profile/hold & hold sections/orlop/gun deck/upper deck/quarter deck & forecastle.

### *Orion* Class 1848. 'Surveyors Department' (Edye & Watts) design.
Dimensions & tons: (design as sailing ship): 198', 167'7¼" x 55'9" x 23'4". 2600 1/94. (as screw): 234', 200'10¾" x 55'9" x 23'4". 3202 45/94.
Men: 720. Guns: GD 8 x 8" + 20 x 32, UD 4 x 8" + 24 x 32.
Spar deck 24 x 32.

***Orion*** Chatham Dyd.
Ord 30.3.1848; K 02.1850; L as screw 6.11.1854; 1867 BU.

***Hood*** Chatham Dyd.
Ord 30.3.1848; K 08.1849; L as screw 4.5.1859; 1888 sold.

NOTE: Lambert suggests there was a third 'ghost' member of the class named *Edgar*, ordered but not begun.

PLANS: Lines/profile/hold & hold sections/orlop/gun deck/upper deck/quarter deck & forecastle/roundhouse/midships section/specification. *Orion* as screw: Lines/profile/hold & hold sections/orlop/gun deck/upper deck/quarter deck & forecastle/roundhouse. as screw: Profile/hold & hold sections/orlop/gun deck/upper deck/quarter deck & forecastle/roundhouse.

## Third Rates, 70 Guns (two-deckers)

### *Boscawen* Class 1834. Symonds design.
Originally *Boscawen* was ordered as a 74, then an 80 and began building as the latter in 1826. The frame was taken down for conversion to this 70 gun design by Admiralty Order of 3.3.1834. Lavery gives these two ships as two separate classes. The two ships were certainly intended to be built to the same design but the detail form of bow and stern were considerably altered during construction and particularly in the early 1840s by those hostile to Symonds [see Lambert] and the eventual 'as built' dimensions and forms seem to have varied somewhat (the 'taken off' dimensions were 180', 145'5⅜" x 54'3" x 22'4" for *Cumberland* and 187'4½", 153'11¼" x 50'9" x 21'6" for *Boscawen* according to Rif Winfield); which may well reflect the complicated prehistory of the *Boscawen* and the material collected for her as well as subsequent alterations.
Dimensions & tons: 180', 146'8" x 54' x 22'4". 2212 12/94.
Men: 620. Guns: GD 24 x 32 + 2 x 68, UD 28 x 32, QD 8 x 32, Fc 4 x 32, RH 4 x 18 (? carronades?)

***Boscawen*** Woolwich Dyd.
Ord 11.5.1817/3.3.1834; K (original design) 01.1826 L 3.4.1844; 1862 hulked.

***Cumberland*** Chatham Dyd.
Ord ____ 1836?; K 04.1836; L 21.10.1842; 1863 hulked.

FATES: *Boscawen* became a drill ship at Southampton; 1874 to the Tyne as a hulk, renamed *Wellesley*; 1914 burnt & BU. *Cumberland* became a sheer hulk at Sheerness; 1869 training ship In the Clyde for the Clyde Industrial Training Ship Association; 1889 destroyed by fire.

PLANS: Lines/profile/orlop/gun deck/upper deck/quarter deck & forecastle/roundhouse/midships section/specification. *Boscawen*: 1851: profile/hold & hold sections/orlop/gun deck/upper deck/quarter deck & forecastle. *Cumberland* as built: Profile/orlop/gun deck/upper deck/quarter deck & forecastle/roundhouse.

## Large Frigates

### *Vernon* Class 1831. 50. Symonds design.
Dimensions & tons: 176', 144'6¼" x 52'8½" x 17'1". 2082 15/94.
Men: 450/500. Guns: UD 28 x 32, QD 14 x 32, Fc 8 x 32.

***Vernon*** Woolwich Dyd.
Ord ____ 1831?; K 10.1831; L 1.5.1832; 1863 hulked.

FATE: Hulked at Portland; 1873 to Portsmouth as the torpedo and mining school; 1886 renamed *Acteon*; 1923 sold.

PLANS: Lines/hold & hold sections/sails. 1840: hold & hold sections. 1845: lower deck/upper deck/quarter deck & forecastle.

NOTE: A 50 gun 2084 ton frigate to be called *Indefatigable* was ordered from Woolwich Dyd on 29.11.1832 but cancelled in 03.1834. This was the fate in store for the timber of the 80 gun *Boscawen* until Symonds intervened with Hardy to secure her re-ordering as a new 70. The design was to be that of the *Vernon*.

### *Raleigh* Class 1842. 50. Fincham design.
Dimensions & tons: 180'6", 149' x 50' x 16'8". 1943 69/94.
Men: 450/500. Guns: UD 28 x 32, QD 14 x 32, Fc 8 x 32.

***Raleigh*** Chatham Dyd.
Ord ____ 1842?; K 08.1842; L 8.5.1845; 1857 wrecked.

***Severn*** Chatham Dyd.
Ord ____ 1844?; K 13.8.1849; L 26.1.1856; never fitted out as a sailing vessel, instead L as screw frigate 8.2.1860; 1876 BU.

FATES: *Raleigh* lost near Macao 14.4.1857; wreck sold.

PLANS: Lines. *Raleigh* as built: Profile. *Severn* as screw ship: Lines/profile/hold & hold sections/lower deck/upper deck/quarter deck & forecastle/sails/machinery arrangement.

### *Constance* Class 1843. 50. Symonds design.
The *Constance* herself replaced an earlier order for a 36 gun frigate of the same name to be built at Portsmouth (see under the *Pique* Class below).
Dimensions & tons: 180', 146'10¼" x 52'8" x 16'3". 2125 75/94.
Men: 450/500. Guns: UD 28 x 32, QD 14 x 32, Fc 8 x 32.

***Constance*** Pembroke Dyd.
Ord 31.3.1843; K 10.1843; L 12.3.1846; 1860 lengthened as a screw frigate; 1875 BU.

***Arethusa*** Pembroke Dyd.
Ord ____ 1844?; K 01.1846; L 20.6.1849; 1860 lengthened as a screw frigate; 1874 hulked.

***Octavia*** Pembroke Dyd.
Ord ____ 1846?; K 09.1846; L 18.8.1849; 1861 lengthened as a screw frigate; 1876 BU.

FATES: *Arethusa* to the Shaftesbury Homes as a training ship; 1934 BU.

PLANS: Profile/upper deck/specification. *Constance*: Lines/hold & hold sections/upper deck/quarter deck & forecastle. As screw ship 1860: lines/sails. 1861: Profile/hold & hold sections. *Arethusa* as screw ship 1860: Lines/upper deck/sails. 1864: Profile/hold & hold sections/lower deck/upper deck/quarter deck & forecastle. *Octavia* as screw ship 1860: Lines/lower deck/upper deck/quarter deck & forecastle. No date: Profile.

### *Sutlej* Class 1846? 50. 'As altered by Admiralty' (from Symonds design?).

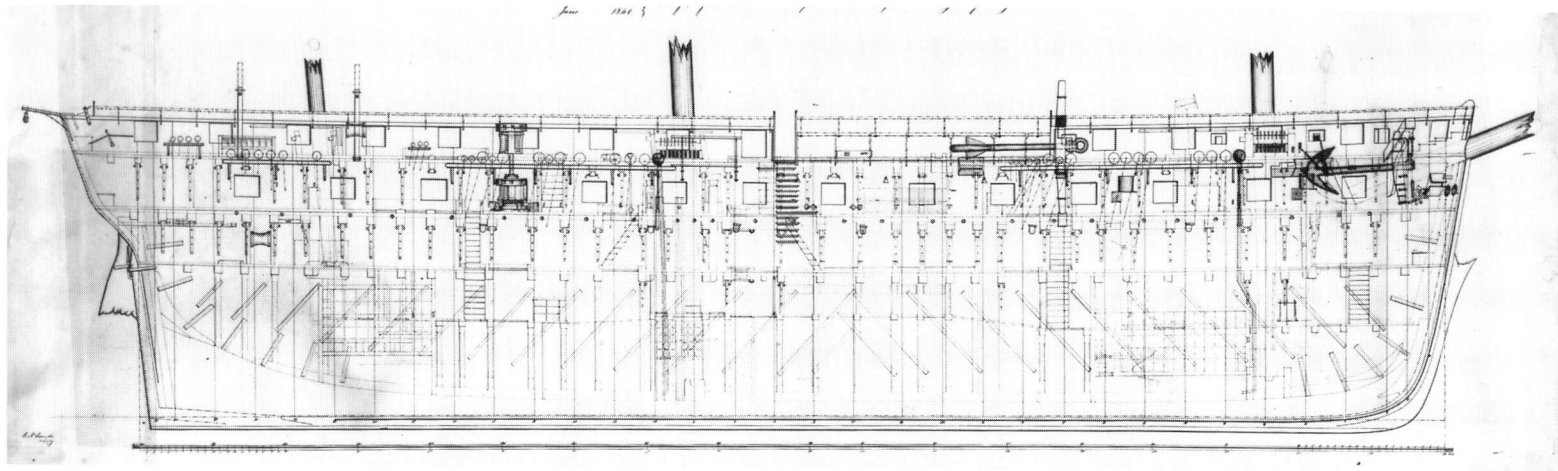

Profile (1424) of the 50 gun frigate *Raleigh* of 1845 as built

Dimensions & tons: ____ , ____ x ____ x ____ . ____ . (as steam frigate: 254'6", 220'1⅛" x 51'8" x 17'1". 3064 81/94).
Men: ____ . Guns: ____ .
**Sutlej** Pembroke Dyd.
Ord: ____ 1846?; K 08.1847; L 17.4.1855; 1859-60 converted to screw; 1869 BU.
PLANS: As screw ship: Lines/profile/hold & hold sections/lower deck/main deck/upper deck/sails/machinery arrangement.

### Leander Class 1845. 50. Blake design.
Dimensions & tons: 181'4", 149'0¾" x 49'10" x 15'8". 1968.
Men: 450/500. Guns: UD 8 x 68 + 22 x 32, QD 2 x 68 + 14 x 32, Fc 2 x 68 + 4 x 32.
**Leander** Portsmouth Dyd.
Ord by 4.7.1843; K 02.1845; L 8.3.1848; 1861 converted to screw frigate; 1867 sold.
PLANS: Lines/profile/hold & hold sections/orlop/lower deck/upper deck/quarter deck & forecastle. As screw ship 1860: Lines/hold & hold sections/lower deck/upper deck/quarter deck & forecastle/machinery arrangement.

### Phaeton Class 1845. 50. White design, three separate stages in design shown in the plans.
Dimensions & tons: (design as a sailing vessel): 184'11", 151'11⅛" x 48'9" x 15'10½". 1921.
Men: 500. Guns: UD 6 x 8" + 22 x 32, spar deck 4 x 8" + 18 x 32.
**Phaeton** (ex *Arrogant*) Deptford Dyd.
Approved: 02.1845; K 1.9.1845; L 25.11.1848; 1859 screw frigate; 1875 BU.
PLANS: Lines/profile/hold & hold sections/lower deck/upper deck/quarter deck & forecastle. 1850: Profile/hold & hold sections/lower deck/upper deck/quarter deck & forecastle. As screw ship 1854: Lines/Profile/hold & hold sections/lower deck/upper deck/quarter deck & forecastle/sails.

### Nankin Class 1846. 50. Lang design ('of the *Barham*'s Class'; following the 74 gun ships cut down to frigates.
Dimensions & tons: 185', 149'10½" x 50'10" x 17'2". 2005 40/94.
Men: 500. Guns: UD 6 x 8" + 22 x 32, Spar deck 4 x 8" + 18 x 32.
**Nankin** Woolwich Dyd.
Approved 01.1846; K 06.1846; L 16.3.1850; 1859 proposed to convert to screw, but not attempted; 1895 sold.
PLANS: Lines/profile/hold & hold sections/lower deck/upper deck/quarter deck & forecastle/midships section/specification. Proposed as screw ship 1859: Lines. 1880: lower deck/upper deck.

### Indefatigable Class 1846. 50. Edye design.
Dimensions & tons: 180', 147'8½" x 51'6" x 16'3". 2043 52/94.
Men: 500. Guns: UD 6 x 8" + 22 x 32, Spar deck 4 x 8" + 18 x 32.
**Indefatigable** Plymouth Dyd.

Ord ____ 1845?; K 22.8.1846; L 27.7.1848; 1864 hulked.
**Phoebe** Plymouth Dyd.
Ord ____ 1847?; K 7.8.1848; L 12.4.1854; 1860 screw frigate; 1875 BU.
FATES: *Indefatigable* to the Committee for the Liverpool Training Ship; 1914 sold.
PLANS: Lines/profile/orlop/lower deck/upper deck/quarter deck & forecastle. As screw ship 1854: Lines/profile/hold & hold sections/lower deck/upper deck/quarter deck & forecastle. 1866-1867: profile/hold & hold sections/orlop/lower deck/upper deck/quarter deck & forecastle/sails.

### Emerald Class 1848. 60. Edye & Watts design.
Dimensions & tons: 185', 152'2⅛" x 52' x 15'8". 2146 81/94.
Men: 500. Guns: UD 30 x 8", QD & Fc 20 x 32 (8½') + 2 x 32 (9½').
Waist 8 x 32 (8').
**Emerald** Deptford Dyd.
Ord ____ ; K 05.1849; L as screw 19.7.1856; 1869 sold.
**Melpomene** Pembroke Dyd.
Ord ____ ; K 09.1849; L as screw 8.8.1857; 1875 sold.
**Impérieuse** Deptford Dyd.
Ord ____ ; K ____ ; L as screw 15.9.1852; 1867 sold.
**Immortalité** Pembroke Dyd.
Ord ____ ; K 11.1849; L as screw 25.10.1859; 1883 sold.
PLANS: Lines/profile/hold & hold sections/lower deck/upper deck/quarter deck & forecastle. *Impérieuse* as screw vessel 1852: Lines/profile. 1853: hold & sections. *Immortalité* as screw vessel: Lines/profile/hold & hold sections/sails, 1874: Profile/hold & hold sections/lower deck/upper deck/quarter deck & forecastle. *Melpomene* 1856: Profile/hold & hold sections/lower deck/upper deck/quarter deck & forecastle. 1857: lines. 1859: Profile/hold & hold sections/lower deck/upper deck/quarter deck & forecastle. *Emerald* as screw 1856: Lines/profile/hold & hold sections/lower deck/upper deck/quarter deck & forecastle/framing/midships section/sails.

### San Fiorenzo Class 1848. 50. School of Naval Architecture (Read, Chatfield & Creuze) design.
Dimensions & tons: 187'4", 155'4" x 50'6" x 15'6½". 2065 56/94.
Men: 500. Guns: UD 28 x 8", Spar deck 22 x 32.
### Cancelled 1851. **Bacchante** Portsmouth Dyd.
Ord 22.6.1848; not begun?
### Cancelled 1856. *San Fiorenzo* Woolwich Dyd.
Ord 22.6.1848; K 06.1850.
PLANS: Lines/midships section.

### Narcissus Class 1849? 50. Surveyor's Department design. There is a possibility that *Liffey* may have belonged to the *Constance* class rather than this one. *Shannon* was cancelled as a sailing frigate in 1851 (?) and *Liffey* also.
Dimensions & tons: 180', 148'1⅞" x 50'10" x 15'10". 1996.
Men: ____ . Guns: ____ .
**Narcissus** Plymouth Dyd.

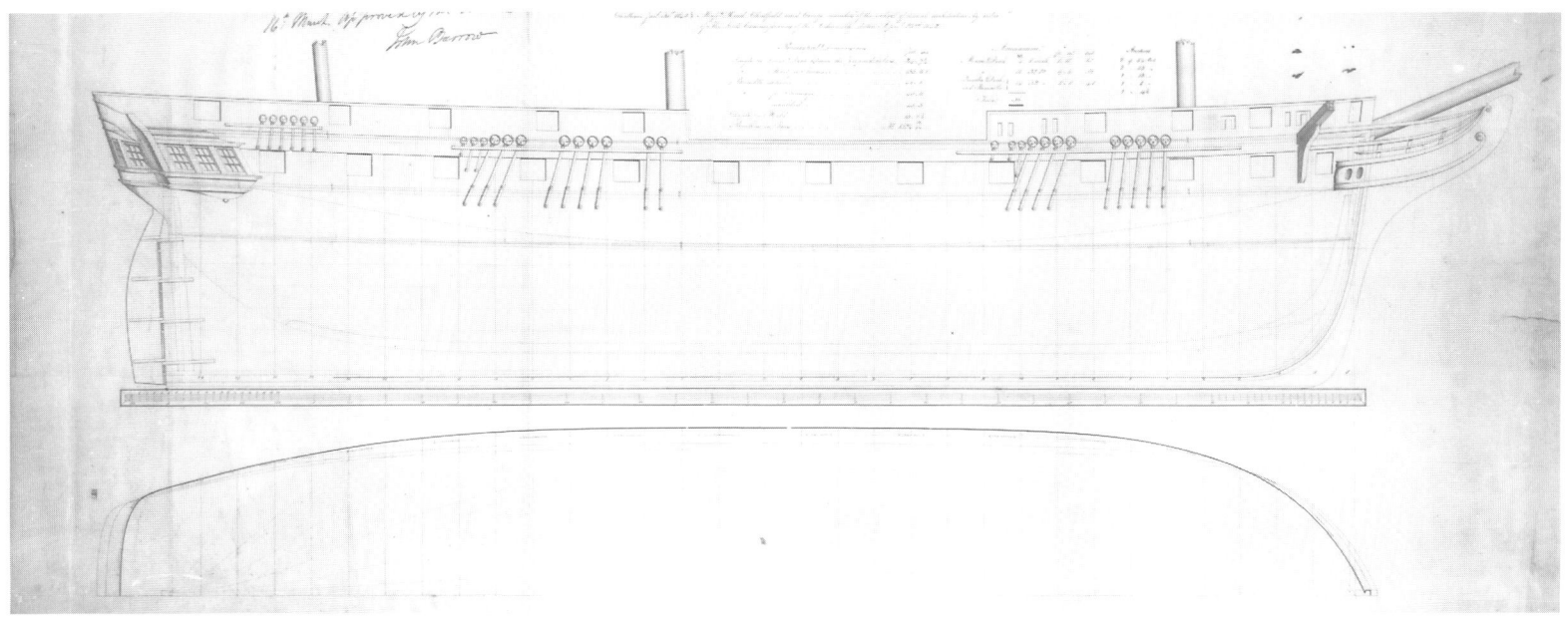

Part (omitting body plan) of the design lines plan (2207) for the 36 gun frigates of the 1842 *Thetis* Class

Ord ____ 1849?; K 11.1849; then cancelled (1857?) broken up and built as a screw frigate (to another design?); L 26.10.1859; 1883 sold.
**Shannon** Portsmouth Dyd.
    Ord 20.3.1844; K ____ ; L as screw 24.11.1855; 1871 sold.
**Liffey** Plymouth Dyd.
    Ord 20.3.1844/29.11.1844; K ____ ; L as screw 6.5.1856; 1877 hulked.
**Topaze** Plymouth Dockyard.
    Ord ____ ; K ____ ; L as screw 12.5.1858; 1884 sold.
FATES: *Liffey* became a store hulk [at Coquimbo?]; sold 04.1903 and became a civilian-owned hulk at Coquimbo.
NOTE: Rif Winfield suggests that the *Euryalus* may be originally been ordered (from Chatham Dyd) as one of this class before being completed as a screw frigate (launched 5.10.1853), also that in some listings the designers of this class are given as Edye and Watts.
PLANS: As screw ships: Profile/Hold & hold sections/lower deck/main deck/upper deck/sails. *Liffey* 1867: Profile/hold & hold sections/lower deck/main deck/upper deck/topsides & part sections.

## Frigates. 32 pounder, 36 Guns (Fifth Rates)

**Pique Class 1832.** Symonds design (original order date for *Pique* probably for a vessel to be built to another design).
    Dimensions & tons: 160', 131' x 48'8" x 14'6". 1622 20/94.
    Men: 275/360. Guns: UD 4 x 8" + 18 x 32. Spar deck 4 x 8" + 14 x 32.
**Pique** Plymouth Dyd.
    Ord ____ 1825?; K 07.1833; L 21.7.1834; 1862 hulked.
**Active** Chatham Dyd.
    Ord ____ 1833?; K 08.1836; L 19.7.1845; 1863 hulked.
**Sybille** Pembroke Dyd.
    Ord ____ 1833?; K 12.1835; L 15.4.1847; 1866 BU.
**Flora** Plymouth Dyd.
    Ord ____ 1833?; K 12.1834; L 11.9.1844; 1865 hulked.
**Cambrian** Pembroke Dyd.
    Ord ____ 1833?; K 08.1837; L 5.7.1841; 1869 hulked.

**Cancelled 1843 [?]**. **Constance?** Portsmouth Dyd.
    Ord ____ 1833?; not begun?; replaced by 50; see above

**Cancelled 1851.** **Chesapeake** Chatham Dyd.
    Ord 10.12.1834; not begun? Eventually replaced by order for screw frigate.

FATES: *Pique* became a hulk at Plymouth; 1871 lent as a cholera hospital ship to the sanitary committee at Plymouth; 1879 lent to the Port Sanitary authority at Plymouth for hospital purposes; 1882 loan to Plymouth town council as a hospital ship; 1910 sold. *Active* became a drill ship for Naval Volunteers at Sunderland (had originally been fitted as a drill ship in 1858?); 1867 renamed *Tyne* but this altered later the same year to *Durham;* 1906 sold. *Flora* hulked as store ship at Ascension Island; 1872 to Simonstown as receiving ship; 1891 sold. *Cambrian* was hulked at Plymouth; 1879 floating factory for the steam reserve at Plymouth; 1892 sold.
PLANS: Lines/profile/lower deck/upper deck/quarter deck & forecastle/framing/midships section. *Flora* as coal depot: profile & decks.

**Inconstant Class 1834?** Captain Hayes' design.
    Dimensions & tons: 160'1", 133'5⅜" x 45'5" x 13'7". 1421 5/94.
    Men: 275/360. Guns: UD 4 x 8" + 18 x 32, Spar deck 4 x 8", 14 x 32.
**Inconstant** Portsmouth Dyd.
    Ord ____ 1826?; K 08.1834; L 16.6.1836; 1854 hulked.
FATE: Temporary hospital ship at Plymouth; 1861 receiving ship and temporary flag ship at Cork; 1862 sold; 1863 handed over; 1866/7 BU.
PLANS: Lines/profile/orlop/lower deck/upper deck/quarter deck & forecastle/framing/midships section.

**Thetis Class 1842.** School of Naval Architecture (Read, Chatfield & Creuze) design.
    Dimensions & tons: 164'7½", 135'10" x 46'3¾" x 13'6½". 1533 17/94.
    Men: 330 Guns: UD 4 x 8" + 18 x 32 (56 cwt),
    QD & Fc 14 x 32 (40 cwt).
**Thetis** Plymouth Dyd.
    Ord 23.4.1842; K 2.12.1844; L 21.8.1846; 1855 to the Prussian Navy (exchange for two steam gunboats).
PLANS: Lines & profile/profile/orlop/lower deck/upper deck/spar deck. As built: Lines & profile/profile/hold, orlop & hold sections/lower deck/upper deck/spar deck/midships section.

## Sixth Rates, 28-24 Guns

('Donkey Frigates'/'Jackass Frigates')

**Vestal Class 1831. 26.** Symonds design.
    Dimensions & tons: 130', 106'1" (or 101'6"?) x 40' x 10'6" (or 11'6"?). 911 75/94 (or 863 78/94?).
    Men: 240. Guns: UD 2 x 8" + 16 x 32, QD & Fc 2 x 32 + 6 x 32★.
**Vestal** Sheerness Dyd.
    Ord ____ 1830?; K 05.1832; L 6.4.1833; 1862 sold.
**Cleopatra** Pembroke Dyd.
    Ord ____ 1832?; K 06.1832; L 28.4.1835; (1857-1858 accommodation ship); 1862 sold.
**Carysfort** Pembroke Dyd.
    Ord ____ 1832?; K 09.1832; L 12.8.1836; 1841 laid up until; 1861 sold.

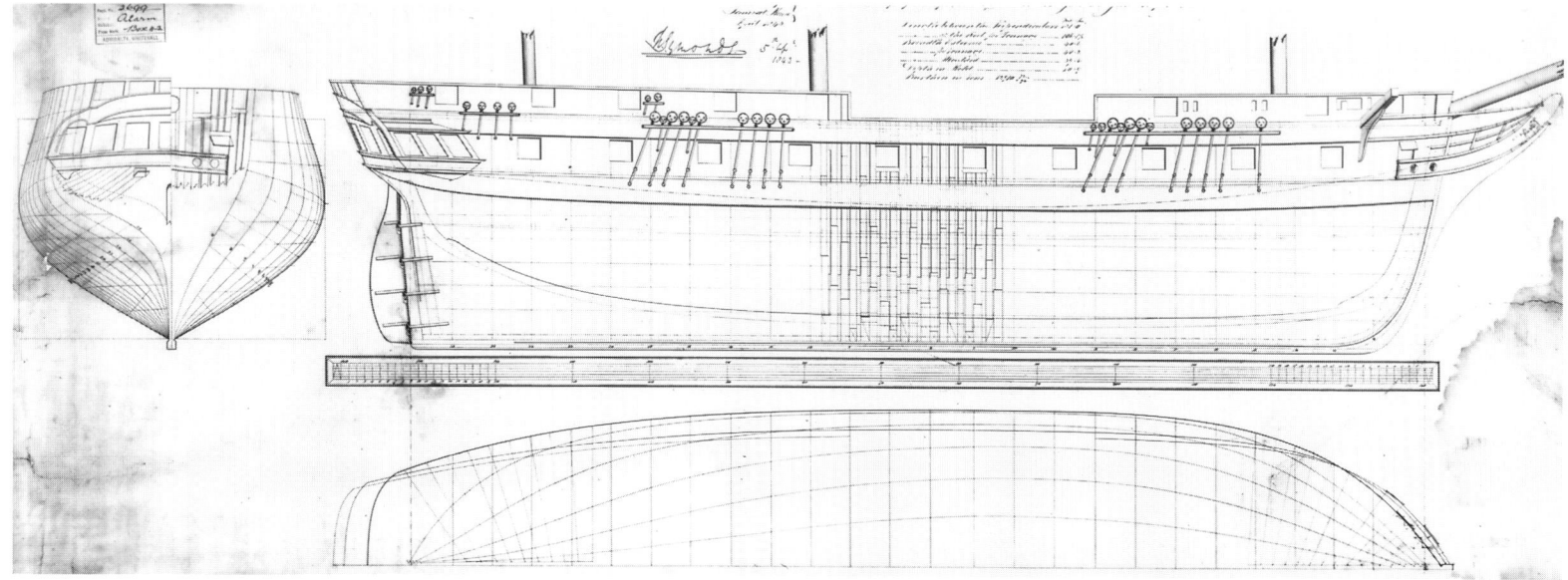

Lines plan (2699) for building the 26 gun Sixth Rate *Alarm* of the 1841 *Spartan* Class

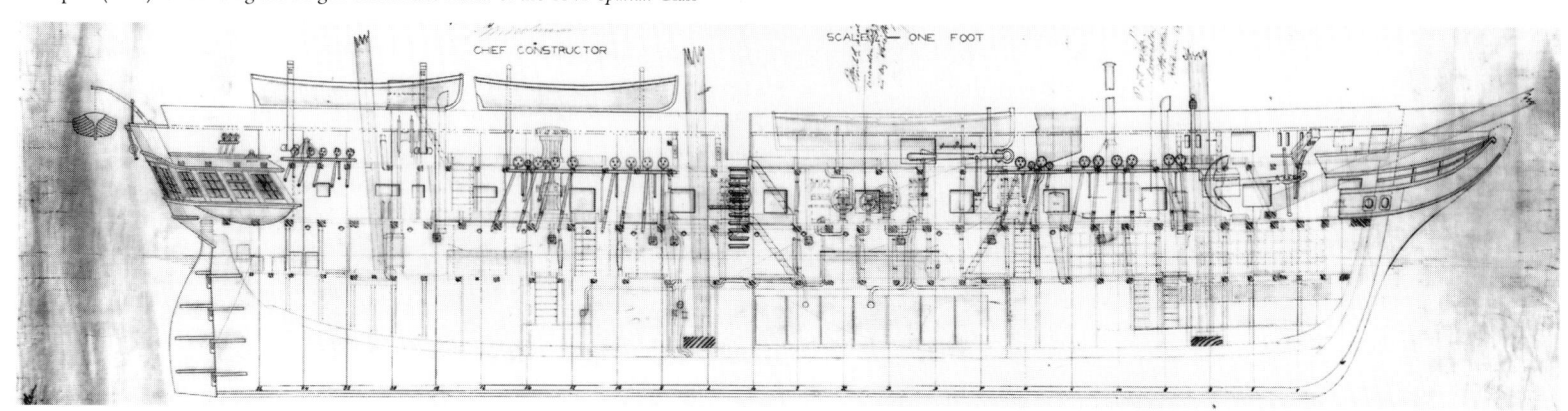

Profile (7510B) of the *Eurydice*, Sixth Rate, 'as fitted' in 1877 as a training ship, just before her tragic loss

**Iris** Pembroke Dyd (? uncertain whether she was of this class - there are some indications that she might belong to the *Spartan* class instead).
Ord ____ 1837?; K 09.1838; L 14.7.1840; 1866 lent.

FATES: *Cleopatra* became an accommodation ship at Blackwall for a Peruvian crew waiting for the completion of their new ship in 1857; 1858. *Iris* lent to the Atlantic Telegraph Co; 1867 lent to Pitcher to assist in the recovery of the *Foyle*; 1867 to the Telegraph Construction & Maintenance Co (and again in 1869 with an intervening period of being laid up?); 1870 purchased by the Telegraph Construction & Maintenance Co.

PLANS: Lines/profile/lower deck/upper deck/spar deck/midships section. *Vestal* as built: Profile/lower deck/upper deck/spar deck.

### Spartan Class 1841. 26. ?Symonds design? *Niobe* (see *Diamond* Class below) was originally to be of this design.
Dimensions & tons: 131', 106'1" x 40'6" x 10'9". 911 33/94.
Men: 240. Guns: UD 2 x 8" + 16 x 32, QD & Fc 2 x 32 + 6 x 32★.
**Spartan** Plymouth Dyd.
Ord 20.2.1837; K 06.1838; L 16.8.1841; 1862 sold.
**Juno** Pembroke Dyd.
Ord ____ 1837?; K 04.1842; L 1.7.1844; (1862-1878 hulked); 1878 seagoing training ship renamed *Atalanta*; 1880 foundered.
**Creole** Plymouth Dyd.
Ord 20.2.1837; K 04.1842; L 1.10.1845; 1870 hulked.
**Amethyst** Plymouth Dyd.
Ord ____ 1840?; K 04.1843; L 7.12.1844; 1864 lent.
**Malacca** Moulmein, Burma; teak built.
Ord ____ ; K ____ ; L 9.4.1853; became screw at Chatham 1854; 1869 sold.
**Alarm** Sheerness Dyd.
Ord ____ 1843?; K 09.1843; L 22.4.1845; 1861 hulked.

FATES: *Juno* became water police ship at Portsmouth (1862-1878); supposed lost at sea 02.1880. *Creole* was lent to the Cork Industrial Training Ship Organisation; 1875 BU. *Amethyst* was lent to the Atlantic Telegraph Co; 1866 lent to Pitcher to assist in the recovery of the *Foyle*; 1869 lent to the Telegraph Construction & Maintenance Co; 1869 sold to that firm. *Malacca* was resold to Japan as *Tsukuba* (*Tsukuba Kan*); in Japanese service until BU 1906. *Alarm* became a coal depot at Plymouth; 1877 to Queenstown; 1900 landing stage at Berehaven; 1904 sold.

PLANS: Lines/profile/lower deck/upper deck/spar deck/midships section/specification. *Alarm* as built: Profile/hold & hold sections/lower deck/upper deck/spar deck. As hulk 1861: Profile/orlop/lower deck/upper deck. *Malacca* as screw sloop: Lines/midships section/specification. As built: Profile/hold & hold sections/orlop/lower deck/upper deck/spar deck.

### Eurydice Class 1842. 24/26. Designed by Admiral the Hon G Elliot.
Dimensions & tons: 141'2", 117'9¾" x 38'10" x 8'9". 911.
Men: 190. Guns: UD 18 x 32, QD 6 x 32 (short ), Fc 2 x 32 (long).
**Eurydice** Portsmouth Dyd.
Ord ____ 1841?; K 04.1842; L 16.5.1843; 1861 harbour training ship; 1877 seagoing training ship; 1878 foundered.

FATE: Foundered in a squall off the Isle of Wight 24.3.1878 with heavy loss of life; wreck salvaged and then BU.

PLANS: Lines/lower deck/quarter deck & forecastle.

### Diamond Class 1846. 28. Symonds design later classed as corvettes.
*Niobe* was originally to be built as one of the *Spartan* Class.
Dimensions & tons: 140', 113'9¾" x 42' x 11'1". 1051 1/94.
Men: 240? Guns: UD 20 x 32 (8'6"), QD 1 x 8" + 4 x 32 (6'), Fc 1 x 8" + 2 x 32 (6').
**Diamond** Sheerness Dyd.
Ord ____ 1845?; K 07.1846; L 29.8.1848; 1866 hulked.

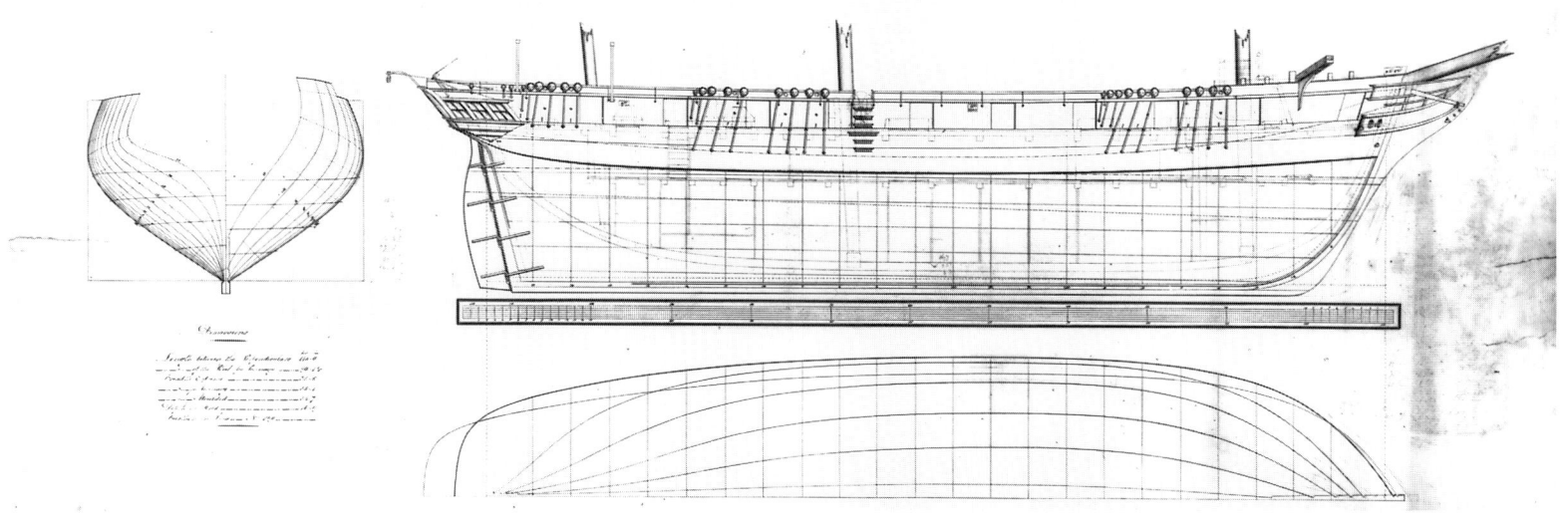

Lines plan of the *Rover* of 1832 'as fitted for sea'. This plan (3235) describes her as a 'corvette'

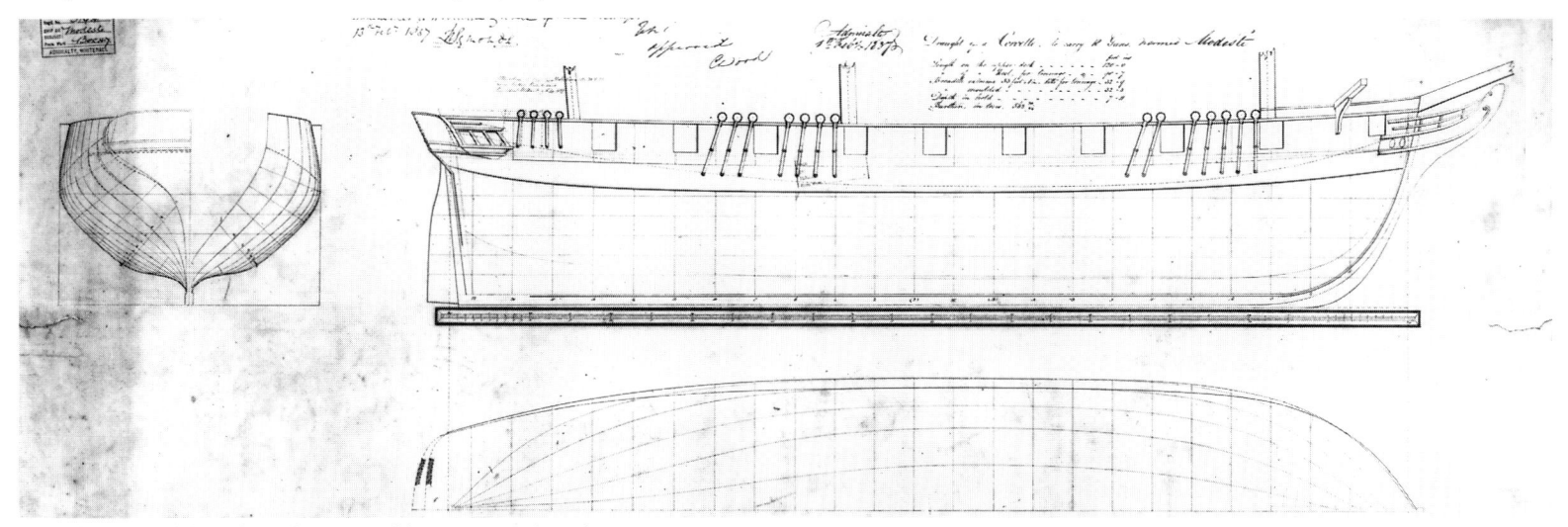

Admiralty approved design lines plan (3194) of the corvette *Modeste* of 1837

*Niobe* Plymouth Dyd.
    Ord ____ 11840?; K 05.1849; 186 sold.

**Cancelled 18 ____ ?** Ship to be built at Sheerness Dyd; probably to be called *Tribune*.

**Cancelled 1848.** *Narcissus* Plymouth Dyd.
    Ord ____ 1845? (Uncertain if she was of this class.)

FATES: *Diamond* lent to the Seamens Mission at 'Shields' and renamed *Joseph Straker*; 1885 sold to BU. *Niobe* sold to the Prussian Navy as a training ship.
PLANS: Lines/profile/lower deck/upper deck/quarter deck & forecastle/midships section.

## Flush Decked Ship Sloops/Corvettes/Sixth Rates

### *Rover* Class 1832. 18. Flush Decked Ship Sloop.
Symonds design. *Rover* was originally ordered as one of the *Scout* Class and *Electra* as one of the *Fly* Class.
    Dimensions & tons: 113', 89'9½" x 35' x 16'9". 585 12/94.
    Men: 120. Guns: 2 x 32 + 16 x 32 (short).
*Rover* Chatham Dyd.
    Ord ____ 1829?; K 02.1832; L 17.7.1832; 1845 BU.
*Electra* Portsmouth Dyd.
    Ord ____ 1828?; K (02.1830; original design); L 28.9.1837; 1862 sold.

PLANS: Lines/profile/midships section/stern. *Rover* as built: Lines & profile/lower deck/upper deck. *Electra* as built: Lines.

### *Dido* Class 1834. 18. Flush Decked Ship Sloop/Corvette/Sixth Rate. Symonds design.
    Dimensions & tons: 120', 99'5½" x 37'6" x 18'. 730 71/94.
    Men: 145/175. Guns: 18 x 32.
*Dido* Pembroke Dyd.
    Ord ____ 1834?; K 09.1834; L 13.6.1836; 1860 hulked.
*Daphne* Pembroke Dyd.
    Ord ____ 1834?; K 12.1835; L 6.8.1838; 1865 BU.
*Calypso* Chatham Dyd.
    Ord ____ 1834?; K 12.1837; L 05.1845; 1866 BU.

**Cancelled in the 1840s (or 1851?).** *Coquette* Chatham Dyd.
    Ord 27.4.1834; not begun?

FATES: *Dido* became a coal hulk at Sheerness; 1903 sold.
PLANS: Lines & profile/profile/lower deck/upper deck/quarter deck & forecastle/midships section.

### *Modeste* Class 1837. 18. Flush Decked Ship Sloop/Corvette. Designed by Admiral the Hon G Elliot.
    Dimensions & tons: 120', 98'7" x 33'1" x 14'2". 562 30/94.
    Men: 120/150. Guns: 2 x 32 + 16 x 32★.
*Modeste* Woolwich Dyd.
    Ord ____ 1837?; K 05.1837; L 31.10.1837; 1866 BU.
PLANS: Lines/profile/lower deck/upper deck/midships section/lines & profile.

### *Arachne* Class 1844. 18. Corvette. Symonds design.
    Dimensions & tons: 115', 90'11" x 35'5" x 16'10". 600 84/94.
    Men: 150. Guns: 16 x 32 (6') + 2 x 32 (7'6").

*Arachne* Plymouth Dyd.
　　Ord 21.12.1844; K 10.1845; L 30.3.1847; 1866 BU
*Terpsichore* Wigram, Blackwall.
　　Ord 15.12.1845; K 01.1846; L 18.3.1847; 1850s? hulked.

FATES: *Terpsichore* became a coal hulk at Chatham; 1865 used for mining experiments, sunk and raised; 1866 BU.

PLANS: lines & profile/lower deck/upper deck/midships section/sails/specification.

### *Challenger* Class 1844. 18. Flush Decked Ship Sloop.

Designed by the Earl of Dundonald (Thomas Cochrane, the great frigate captain).
　　Dimensions & tons: 134', 110'4¾" x 37'6" x 17'6". 810.
　　Men: 175/145. Guns : probably to be 18 x 32.

### Cancelled 22.3.1845? *Challenger* Chatham Dyd.
　　Ord 8.8.1844?; not laid down?
PLANS: Lines.

## Brigs

### *Snake* Class 1831. 16/12. Second Class Brigs. 
Symonds design. The *Racer* class were the Dockyard-built version with slightly modified lines.
　　Dimensions & tons: 100', 79'9½" x 32' x 14'10". 418 24/94.
　　Men: 110. Guns: 2 x 18 (or 9) + 14 x 32★.
*Snake* Fletcher, Limehouse.
　　Ord ____ 1831?; K 12.1831; L 3.5.1832; 1847 foundered.
*Serpent* Fletcher, Limehouse.
　　Ord ____ 1831?; K 02.1832; L 14.7.1832; 1857 gunnery tender; 1861 BU.

FATES: *Snake* wrecked in the Mozambique Channel 29.8.1847.

PLANS: lines/profile/lower deck/upper deck/midships section/specification. *Serpent* 1846: Hold & hold sections.

### *Racer* Class 1832. 16/12. Second Class Brigs.
Symonds design. *Snake* Class with slightly modified lines. See also the slightly modified *Nerbudda*.
　　Dimensions & tons: 100'6", 78'7" x 32'4" x 15'4". 428 2/94.
　　Men: 110. Guns: 2 x 18 (or 9) + 14 x 32★.
*Racer* Portsmouth Dyd.
　　Ord ____ 1832?; K 09.1832; L 18.7.1833; 1852 sold.
*Wanderer* Chatham Dyd.
　　Ord ____ 1832? (20.4.1833?); K 02.1833; L 10.7.1835; 1850 BU.
*Harlequin* Pembroke Dyd.
　　Ord ____ 1832?; K 11.1832; L 18.3.1836; 1859 hulked.
*Ringdove* Plymouth Dyd.
　　Ord ____ 1832?; K 10.1832; L 06.1833; 1850 BU.
*Wolverene* Chatham Dyd.
　　Ord ____ 1832? (20.4.1833?); K 02.1833; L 13.10.1836; 1855 wrecked.
*Sappho* Plymouth Dyd.
　　Ord ____ 1832?; K 12.1834; L 02.1837; 1859 foundered.
*Lily* Pembroke Dyd.
　　Ord ____ 1832?; K 12.1835; L 28.9.1837; 1859 hulked.
*Liberty* Pembroke Dyd.
　　Ord 20.2.1837; K 09.1848; L 11.6.1850; 1861 sail training ship; 1905 sold.
*Squirrel* Pembroke Dyd.
　　Ord 20.2.1837; K 09.1850; L 8.8.1853; 1862 survey ship; 1879 BU.

FATES: *Harlequin* became a coal depot at Plymouth; 1889 sold. *Wolverene* was wrecked at night on the South South East part of the Courtown Bank, 11.8.1855; wreck sold. *Sappho* was supposed foundered off Australia in 1859. *Lily* became a coal depot at Portsmouth Ca 1890 redesignated C.29; later (at Chatham?) redesignated C.15; 1905 BU at Chatham.

Plans: Lines/profile/lower deck/upper deck/midships section/sails/specification.

### *Nerbudda* Class 183. 16/12. Second Class Brig.
Symonds design. Slightly modified from the lines of the *Snake* and *Racer* Classes (see above).
　　Dimensions & tons: 100', 77'0⅞" x 32'4" x 13'6". 419 52/94.
　　Men: 110. Guns: 2 x 18 (or 9) + 14 x 32★.
*Nerbudda* (ex *Goshawk*) Bombay Dyd (teak built).
　　Ord ____ 1843??; K 1.1.1843; L 5.2.1848; 1856 foundered.

FATE: Supposed foundered on the Cape Station 07.1856.

PLANS: Lines/profile/lower deck/upper deck/midships section/sails/specification.

### *Pandora/Bonetta* Class 1832. 10/3. Brigs/Brigantines.
Symonds design. Some differences between them, particularly *Pandora* and the later ships.
　　Dimensions & tons: 90', 71'5" x 29'3" x 13'10". 318 68/94.
　　Men: 50/60. Guns: Various armaments; finally 3 x 32 [except *Rapid* and *Sealark*: Men: 80 Guns: 2 x 18 (long) + 6 x 18 (short).]
*Pandora* Woolwich Dyd.
　　Ord ____ 1831?; K 08.1832; L 4.7.1833; completed as a packet; 1845 survey ship; 1857 to Coastguard.
*Bonetta* Sheerness Dyd.
　　Ord ____ 1834?; K 10.1834; L 5.4.1836; 1861 BU.
*Dolphin* Sheerness Dyd.
　　Ord ____ 1834?; K 10.1834; L 14.6.1836; 1861 to the Coastguard.
*Spy* Sheerness Dyd.
　　Ord 20.2.1837; K 09.1838; L 24.3.1841; 1862 sold.
*Rapid* Portsmouth Dyd.

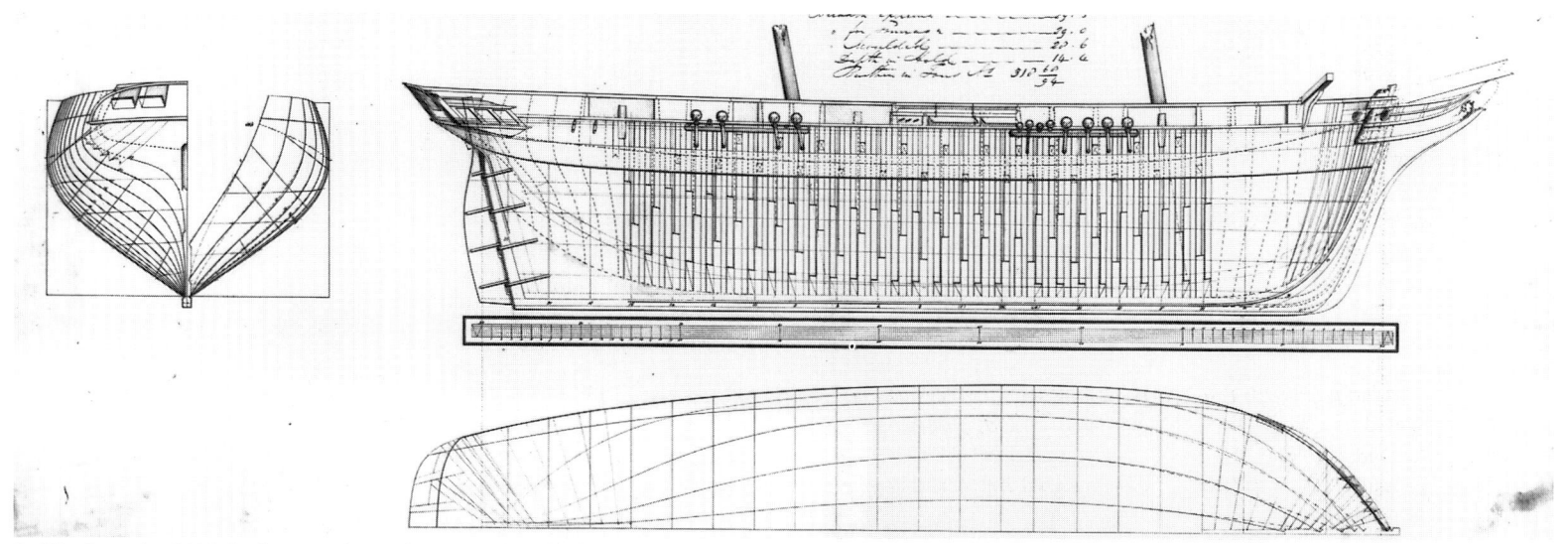

Building lines plan (3920) for the *Spy* and *Dart*, the two *Pandora* Class brigs (described here as brigantines) built at Sheerness Dockyard between 1837 and 1847

SYMONDS AND AFTER 1830-1860

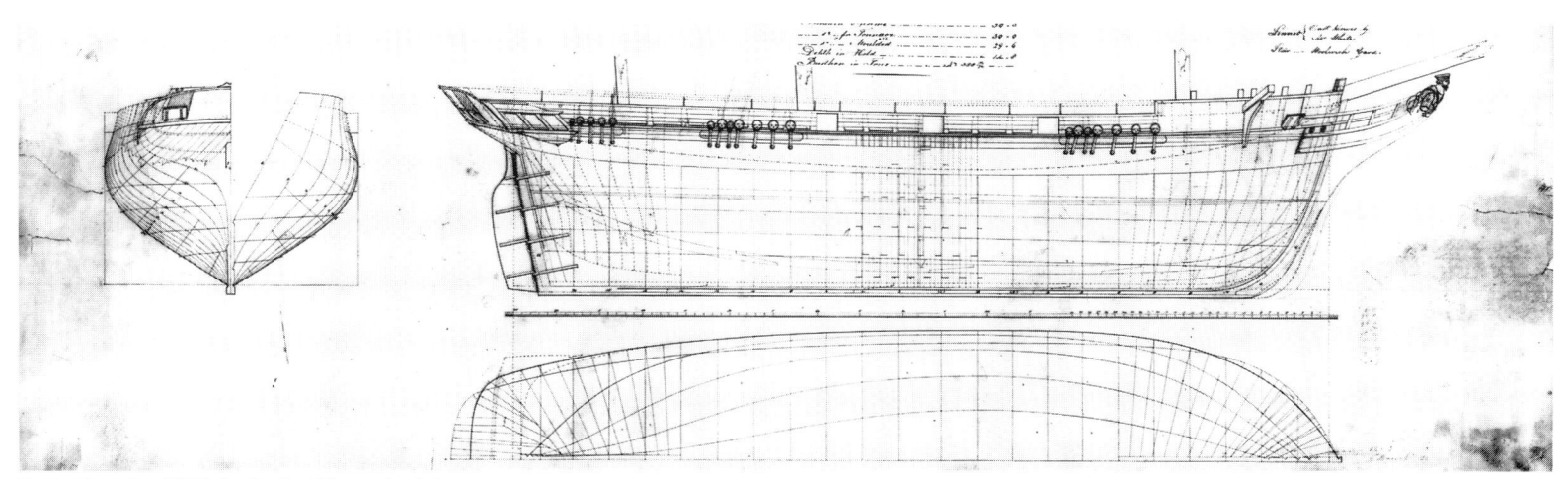

Design lines plan (3995) for the *Star* Class packet brigs

Ord 20.2.1837; K 09.1839; L 3.6.1840; 1856 sold.
**Sealark** Portsmouth Dyd.
Ord 20.2.1837; K 05.1840; L 27.7.1843; 1853 training ship; 1898 sold.
**Dart** Sheerness Dyd.
Ord 20.2.1837; K ____ ; L 17.3.1847; 1857 to the Coastguard.

FATES: *Pandora* to Coastguard at ____ ; 1862 sold at Montevideo. *Dolphin* to the Coastguard at ____ ; WV 3; 1894 BU. *Dart* to the Coastguard at Beresford, Ireland as WV 26; 1875 BU.

PLANS: Lines/lines & profile/profile/lower deck/upper deck/decks/midships section/sails/specification. *Pandora* as built: Lines & profile/profile/lower deck/upper deck. 1837: decks. 1844: lower deck/upper deck. *Bonetta* 1840: hold. *Dolphin* as built: profile/upper deck/lower deck/32 pounder and pivot mounts (separate sheets for heavier and lighter weapons). 1840: lower deck. *Sealark* 1863: sails. *Dart* as built: Profile/lower deck.

**Cancelled 4.9.1843.** Possibly of this class (319 tons). **Daring** Sheerness Dyd.
Ord 14.5.1840.

## *Star* Class 1834. 10/6/3. Brigs/Packets. Symonds design.
*Dispatch, Dove, Mariner, Martin* all originally ordered as members of this class but then ordered to be completed as 16 gun brigs of the *Pilot* Class (see below).
Dimensions & tons: 95', 74'10" x 30'3" x 14'10". 358 18/94.
Men: 44/80. Guns 2 x 6 + 4 x 12★, later 6 x 32. As packet 3 x ?.
**Star** Woolwich Dyd.
Ord ____ 1834?; K 09.1834; L 29.4.1835; completed as a packet; 1843 brig sloop; 1857 to the Coastguard.
**Ranger** Bottomley, Rotherhithe.
Ord ____ 1834?; K 09.1834; L 25.7.1835; completed as a packet; 1845 brig sloop; after 1851 hulked.
**Express** Colson, Deptford.
Ord ____ 1834?; K 09.1834; L 8.10.1835; completed as a three-masted packet; 1862 sold.
**Swift** Colson, Deptford.
Ord ____ 1834?; K 09.1834; L 21.11.1835; completed as a three-masted packet; 1860 yard craft.
**Linnet** (ex *Liberty*) White, Cowes.
Ord ____ 1834?; K 09.1834; L 27.7.1835; completed as a three-masted packet; 1857 to the Coastguard.
**Alert** Bottomley, Rotherhithe.
Ord ____ 1834?; K 09.1834; L 8.10.1835; completed as a packet; 1843 brig sloop; 1851 BU.
**Penguin** Pembroke Dyd.
Ord ____ 1836?; K 11.1836; L 10.4.1838; completed as a packet; 1857 to the Coastguard.
**Peterel** Pembroke Dyd.
Ord ____ 1836?; K 04.1837; L 23.5.1838; completed as a packet; later brig sloop; 1862 sold.
**Crane** Woolwich Dyd.
Ord 20.2.1837; K 01.1838; L 28.3.1839; completed as a packet; 1862 sold.

**Ferret** Plymouth Dyd.
Ord 20.2.1837; K 04.1838; L 1.6.1840; completed as a brig; 1859 sail training ship; 1869 wrecked.
**Philomel** Plymouth Dyd.
Ord 20.2.1837; K 04.1840; L 28.3.1842; completed as a survey vessel; 1847 brig sloop (8 guns); 1857 to the Coastguard.
**Cygnet** Woolwich Dyd.
Ord 20.2.1837; K 05.1838; L 6.4.1840; 1844 sloop (6); 1857 to the Coastguard.
**Heroine** (Packet No 1) Woolwich Dyd.
Ord ____ 1839?; K 05.1840; L 16.8.1841; 1851 laid up; 1878 BU.
**Hound** (Packet No 2) Woolwich Dyd.
Ord 24.5.1839; K. 2.7.1845; L. 23.5.1846; 1852 paid off; 1887 sold for BU.

NOTE: *Hound* (and probably *Heroine*) appears to have been ordered to a different design, cancelled in 06.1844 and re-ordered to this class design in late 1844 or early 1845.

**Cancelled 1843?** *Dove* ? Chatham Dyd?
Ord ____ 1837? (probably of this class).

FATES: *Star* to the Coastguard at Gravesend as WV 11; Ca 1899 BU. *Ranger* became church ship at Kingstown; 1867 sold after sinking. *Swift* became a mooring vessel at the Cape of Good Hope 1869 still in service; eventual fate uncertain. *Linnet* to the Coastguard at ____ as WV 36; 1867 sold. *Penguin* to the Coastguard at the Hamble River as WV 31; 1871 sold. *Ferret* lost alongside the Admiralty pier at Dover 29.3.1869. *Philomel* to the Coastguard at Milton, Southampton as WV 23; 1870 sold. *Cygnet* became Coastguard Watch Vessel No 30 at Chichester in 1863; 1877 BU.

PLANS: Lines/profile/hold & hold sections/lower deck/upper deck/midships section/framing. *Express* as built: decks. *Ranger* 1836: hold & hold sections. *Alert* 1843: profile/lower deck/upper deck. *Star* as built: profile/hold & hold sections/lower deck/upper deck.

## *Pilot/Acorn* Class 1836. 16/12. Second Class Brigs.
Symonds design, *Dispatch, Kangaroo* (ex *Dove*), *Martin* & *Mariner* all ordered as members of the *Star* Class (see above) but order altered in 02.1843. *Nautilus* of 1879 and *Pilot* of 1890 were built to the same design after the end of our period.
Dimensions & tons: 105', 82'2¾" x 33'6" x 14'10", 481 11/94.
Men: 130/110. Guns: 2 x 18 (or 32) + 14 x 32★.
**Pilot** Plymouth Dyd.
Ord ____ 1836?; K 08.1837; L 9.6.1838; 1862 sold.
**Grecian** Pembroke Dyd.
Ord ____ 1836?; K 07.1836; L 24.4.1838; 1865 BU.
**Fantome** Chatham Dyd.
Ord ____ 1837?; K 05.1837; L 30.5.1839; 1865 BU.
**Acorn** Plymouth Dyd.
Ord ____ 1836?; K 11.1837; L 15.11.1838; 1867 hulked.
**Persian** Pembroke Dyd.
Ord ____ 1836?; K 05.1838; L 7.10.1839; 1866 BU.
**Arab** Chatham Dyd.
Ord ____ 1836?; K 02.1838; L 31.3.1847; 1867 to the Coastguard.
**Albatross** Portsmouth Dyd.

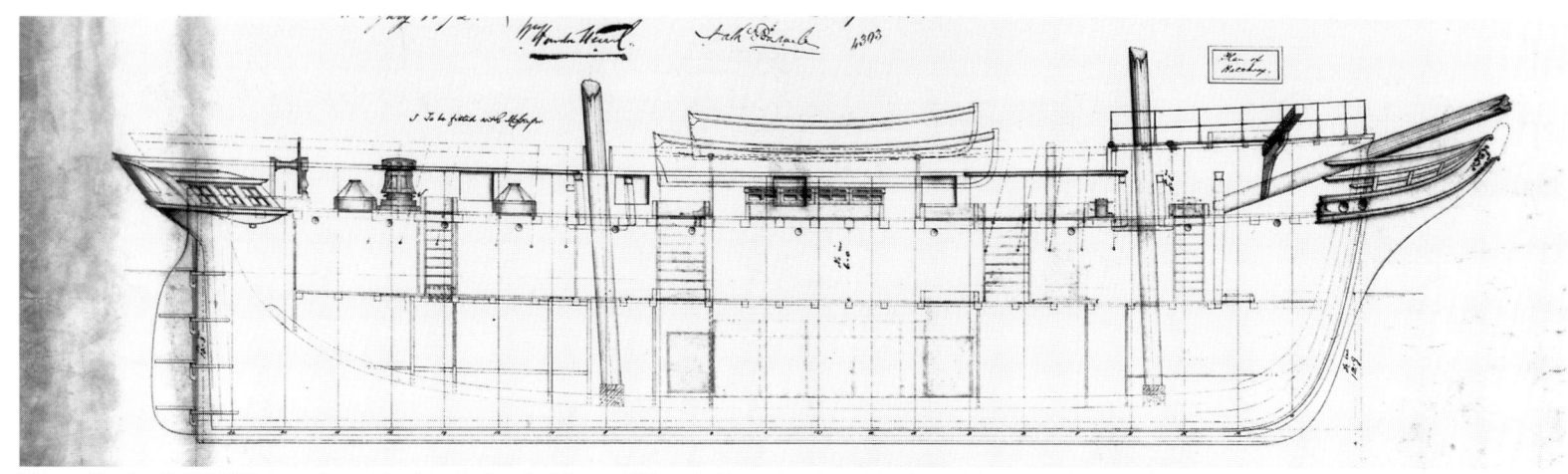

*This profile plan (3000393D) shows the training brig Seaflower building in 1872, probably, as the Martin, Nautilus and the 1880 Pilot certainly were, built to Symonds' design for the Pilot/Acorn Class brigs*

Ord 20.2.1837; K 07.1838; L 28.3.1842; 1860 BU.
**Bittern** Portsmouth Dyd.
Ord 20.2.1837; K 03.1838; L 18.4.1840; 1860 sold.
**Mariner** Pembroke Dyd.
Ord 20.2.1837; K 05.1845; L 19.10.1846; 1865 sold.
**Elk** Chatham Dyd.
Ord 20.2.1837; K 11.1846; L 27.9.1847; 1863 to the Coastguard.
**Despatch** Chatham Dyd.
Ord 20.2.1837; K ____; L 25.11.1851; 1863 to the Coastguard.
**Heron** Chatham Dyd.
Ord 20.2.1837; K 11.1846; L 27.9.1847; 1859 foundered.
**Martin/Marten** Pembroke Dyd.
Ord 20.2.1837; K ____; L 19.9.1850; 1864 sail training ship; 1890 renamed *Kingfisher*, 1907 sold.
**Kangaroo** (ex *Dove*) Chatham Dyd.
Ord 20.2.1837; K 1.4.1850; L 31.8.1852; 1863 to the Coastguard.

FATES: *Acorn* became a hospital ship at Shanghai; 1869 sold. *Arab* to the Coastguard at Queenborough (Swale) as WV 18; 1879 BU. *Elk* to the Coastguard at ____ as WV 13; 1871? WV 28 at Tilbury; 1893 sold. *Despatch* to the Coastguard at the East Swale as WV 24; 1901 sold. *Heron* foundered on the way from Ascension to Sierra Leone 9.5.1859. *Kangaroo* to the Coastguard at Stangate Creek as WV 20; 1870 at the River Crouch; 1897 sold.

PLANS: Lines/profile/lower deck/upper deck/sails/midships section/planking expansion (external)/(internal)/specification. *Martin* 1880: Lines. *Nautilus* (built 1879 to the design): as built; 1880: Profile/hold & hold sections/lower deck/upper deck & forecastle. *Pilot* (built 1880 to the same design): as built: Profile/hold & hold sections/lower deck/upper deck & forecastle.

### *Siren* Class 1839. 16. First Class Brigs. Symonds design.
Dimensions & tons: 110', 86'9" x 34'10" x 15'. 549.
Men: 140. Guns: 16 x 32.
**Siren** Woolwich Dyd.
Ord ____ 1839?; K 06.1839; L 23.4.1841; 1862 hulked.
**Helena** Pembroke Dyd.
Ord ____ 1839?; K 09.1839; L 11.7.1843; 1868 hulked.
**Jumna** (ex *Zebra*) Bombay Dyd (teak built).
Ord ____ 1843? (19.1.1846?); K 12.1843; L 7.3.1848; 1862 sold.
**Camilla** Pembroke Dyd.
Ord ____ 1844?; K 10.1846; L 8.9.1847; 1861 foundered.
**Atalanta** Pembroke Dyd.
Ord ____ 1844?; K 10.1846; L 9.10.1847; 1868 BU.
**Musquito** Pembroke Dyd.
Ord ____ 1847?; K ____; L 29.7.1851; 1862 sold.
**Rover** Pembroke Dyd.
Ord ____ 1837?; K 09.1850; L 21.6.1853; 1862 sold.

FATES: *Siren* became a target ship at Portsmouth; 1868 BU. *Helena* became a chapel ship at Portsmouth; 1879 to Ipswich as a church ship; 1880 to Chatham as a depot ship; 1884 police accommodation ship; 1921 sold. *Camilla* foundered on the China Station 03.1861. *Musquito* and *Rover* sold to the Prussian Navy 1862.

PLANS: Lines/profile/lower deck/upper deck/midships section/sails. *Jumna* 1861: Profile & midships section & roof/lower deck.

### *Frolic* Class 1841. 16. First Class Brig. Designed by Captain W. Hendry.
Dimensions & tons: 108'3", 84'3⅛" x 34' x 15'6". 510 69/94.
Men: 140. Guns: 16 x 32.
**Frolic/Frolick** Portsmouth Dyd.
Ord ____ 1841?; K 02.1841; L 23.8.1842; 1865 BU.

PLANS: Lines/profile/lower deck/upper deck/midships section.

### *Flying Fish* Class 1843. 12. Third Class Brig. Symonds design.
Dimensions & tons: 103', 81'8" x 32'4" x 14'8". 444 76/94.
Men: 130/110. Guns: 2 x 18 + 10 x 32.
**Flying Fish** Pembroke Dyd.
Ord ____ 1843?; K 10.1843; L 3.4.1844; 1852 BU.
**Kingfisher** Pembroke Dyd.
Ord ____ 1843?; K 04.1844; L 8.4.1845; 1890 sold.

PLANS: Lines/profile/lower deck/upper deck/midships section.

### *Daring* Class 1843. 12. Third Class Brig. White design.
Appears originally to have been ordered in 1840 as one of the *Pandora* class at Sheerness but this was cancelled in 1843 and this vessel ordered instead.
Dimensions & tons: 104', 83'0½" x 31'4½" x 15'5½". 426.
Men: 130. Guns: 2 x 18 + 10 x 32.
**Daring** Portsmouth Dyd.
Ord 14.5.1840?; K 10.1843; L 2.4.1844; 1865 BU.

PLANS: Lines/sails. As built: Profile/hold & hold sections/lower deck/upper deck.

### *Mutine* Class 1843. 12. Third Class Brig. Fincham design.
Dimensions & tons: 102', 81'2¼" x 31'10" x 13'7". 428 38/94.
Men: 130/110. Guns: 2 x 18 + 10 x 32.
**Mutine** Chatham Dyd.
Ord ____ 1843?; K 30.10.1843; L 20.4.1844; 1848 wrecked.

FATE: Wrecked near Venice 21.12.1848.

PLANS: Lines. As built: Profile/lower deck/upper deck.

### *Osprey* Class 1843. 12. Third Class Brig. Blake design.
Dimensions & tons: 101'5", 80'6¼" x 31'6" x 13'6". 424 91/94.
Men: 130. Guns: 2 x 18 + 10 x 32.
**Osprey** Portsmouth Dyd.
Ord ____ 1843?; K 11.1843; L 2.4.1844; 1846 wrecked.

FATE: Wrecked off False Hokianga, New Zealand 11.3.1846; there are reports that this wreck may have been found by divers.

PLANS: Lines. As Built: Profile/hold & hold sections/lower deck/upper deck.

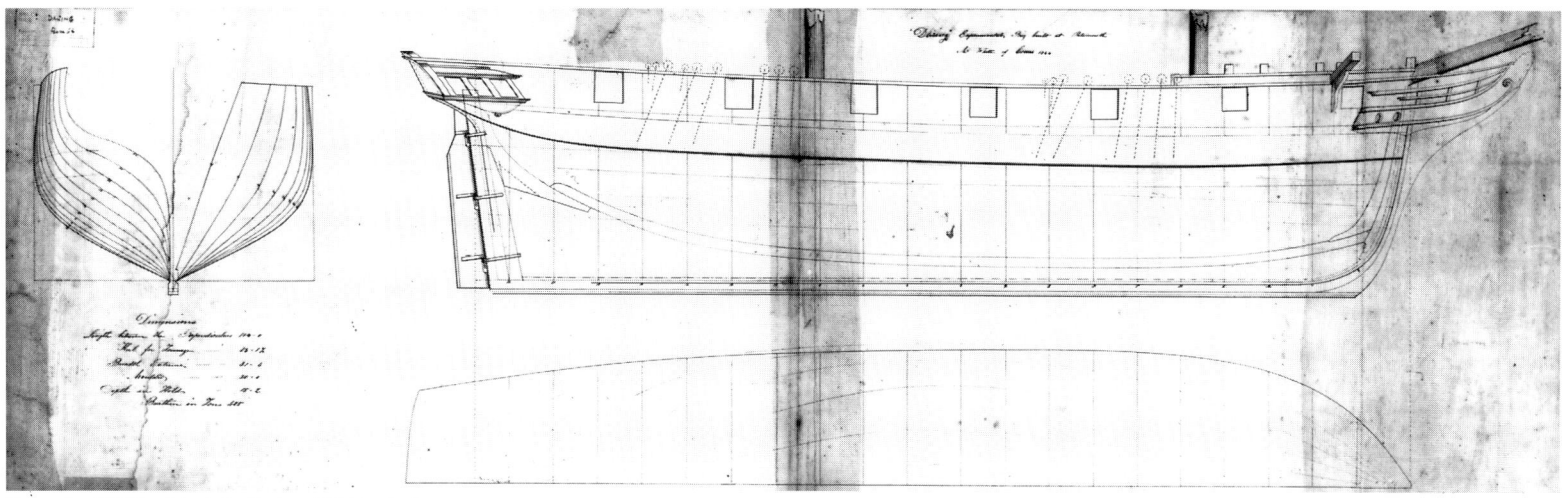

Lines plan (6752) of the brig *Daring* of 1844 as built

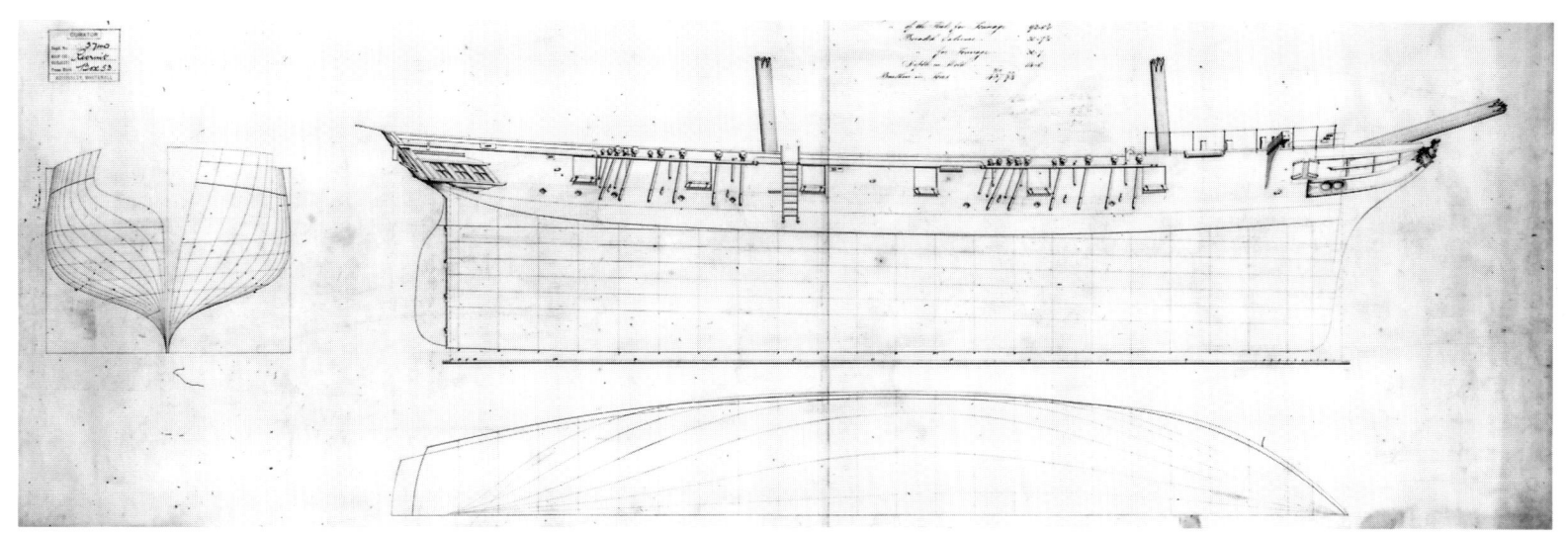

Lines (3740) 'taken off' the iron brig *Recruit* of 1846 after completion

### *Espiegle* Class 1844 (1843?) 12. Brig. 'Naval College' (Reed, Chatfield & Creuze) design.

> Dimensions & tons: 104'8", 82'8" x 31'8" x 13'1¼". 431 86/94.
> Men: 130. Guns: 2 x 18 + 10 x 32.

*Espiegle* Chatham Dyd.
> Ord ____ 1843?; K 01.1844; L 20.4.1844; 1861 sold for BU.

PLANS: Lines/profile/hold/lower deck/upper deck/midships section/specification. As built: Profile/lower deck/upper deck.

### *Contest* Class 1845. 12. Third Class Brig. White (builder's) design.

> Dimensions & tons: 109', 88'8" x 31'4" x 15'2". 446 11/94.
> Men: 130. Guns: 2 x 18 + 10 x 32.

*Contest* White, Cowes.
> Ord ____ 1845?; K 06.1845; L 11.4.1846; 1868 BU.

PLANS: Lines/specification. 1850: Lines & profile/hold & lower deck/upper deck/stowage of hold.

### *Britomart* Class 1846 (1843?). 10/8. Brig. Symonds design.

> Dimensions & tons: 93', 73'5⅛" x 29'3" x 13'5". 328 46/94.
> Men: 80. Guns: 2 x 18 (long) + 6 x 18.

*Britomart* Pembroke Dyd.
> Ord ____ 1843?; K 05.1846; L 12.6.1847; 1863 to the Coastguard.

FATE: To the Coastguard at Beresford, Ireland as WV 25; then at Burnham; 1874 BU.

PLANS: Lines/profile/lower deck/upper deck/midships section.

### *Recruit* Class 1846 (1844?). 12. Third Class Iron Brig.

Ditchburn & Mare (builder's) design. The only iron sailing warship built for the RN (if we exclude some small craft like the Crimean War mortar vessels, which cannot really be counted as seagoing warships). Sold back to her builder, Mare, as a result of the Navy's temporary revulsion against the use of iron for hulls in the late 1840s.

> Dimensions & tons: 113', 94'11" x 30'3" x 14'. 469 67/94.
> Men: 130. Guns: 10 x 32 + 2 x 18.

*Recruit* Ditchburn & Mare, Blackwall.
> Ord 1844?; K ____ ; L 10.6.1846; 1849 sold.

FATE: Became merchantman *Harbinger* in 1850.

PLANS: Lines. As built: Lines, profile & decorations/profile/hold & hold sections/lower deck/upper deck/midships section.

## Cutters and Schooners

### *Fanny* Class 1827. Tender. ? design.

> Dimensions & tons: ____ , ____ x ____ x ____ .
> Men: ____ . Guns: None?

*Fanny* Deptford Dyd.
> Ord ____ ; K ____ ; L 05.1827; 1835 BU.

PLANS: None.

### *Seaflower* Class 1830. 4. Cutter. Designed by Captain Hayes.

> Dimensions & tons: 60', 46'10⅜" x 21'8½" x 10'6". 116.
> Men: 35. Guns: 2 x 6 (brass) + 2 x 6★.

*Seaflower* Portsmouth Dyd.
> Ord ____ 1829?; K 02.1830; L 20.5.1830; 1855 survey ship; 1866 BU.

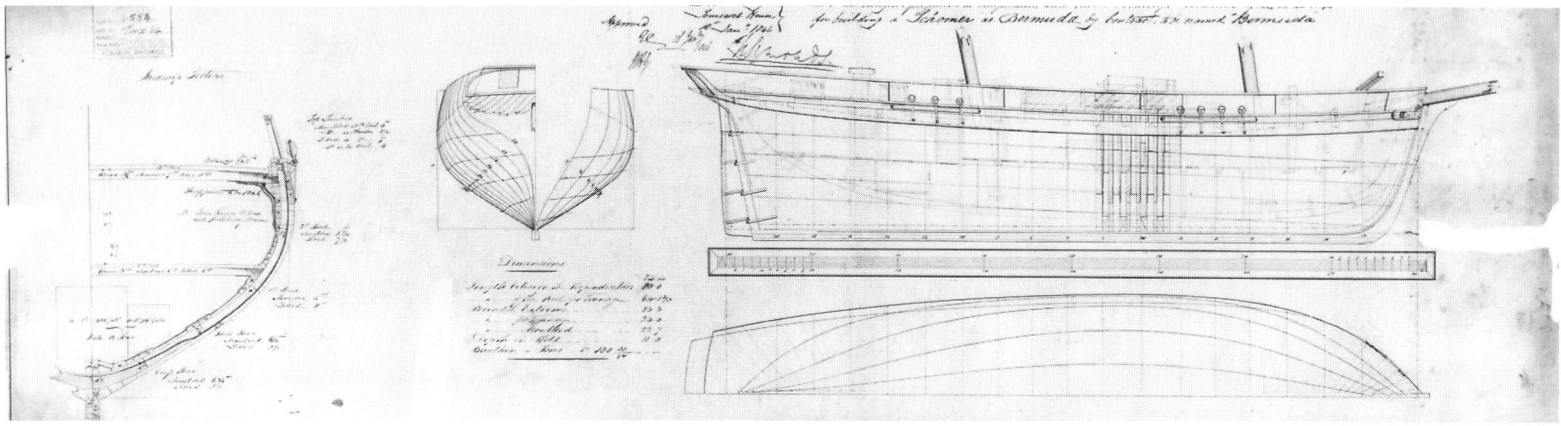

Building lines and midsection plan (4584) for the *Bermuda* schooner of 1848

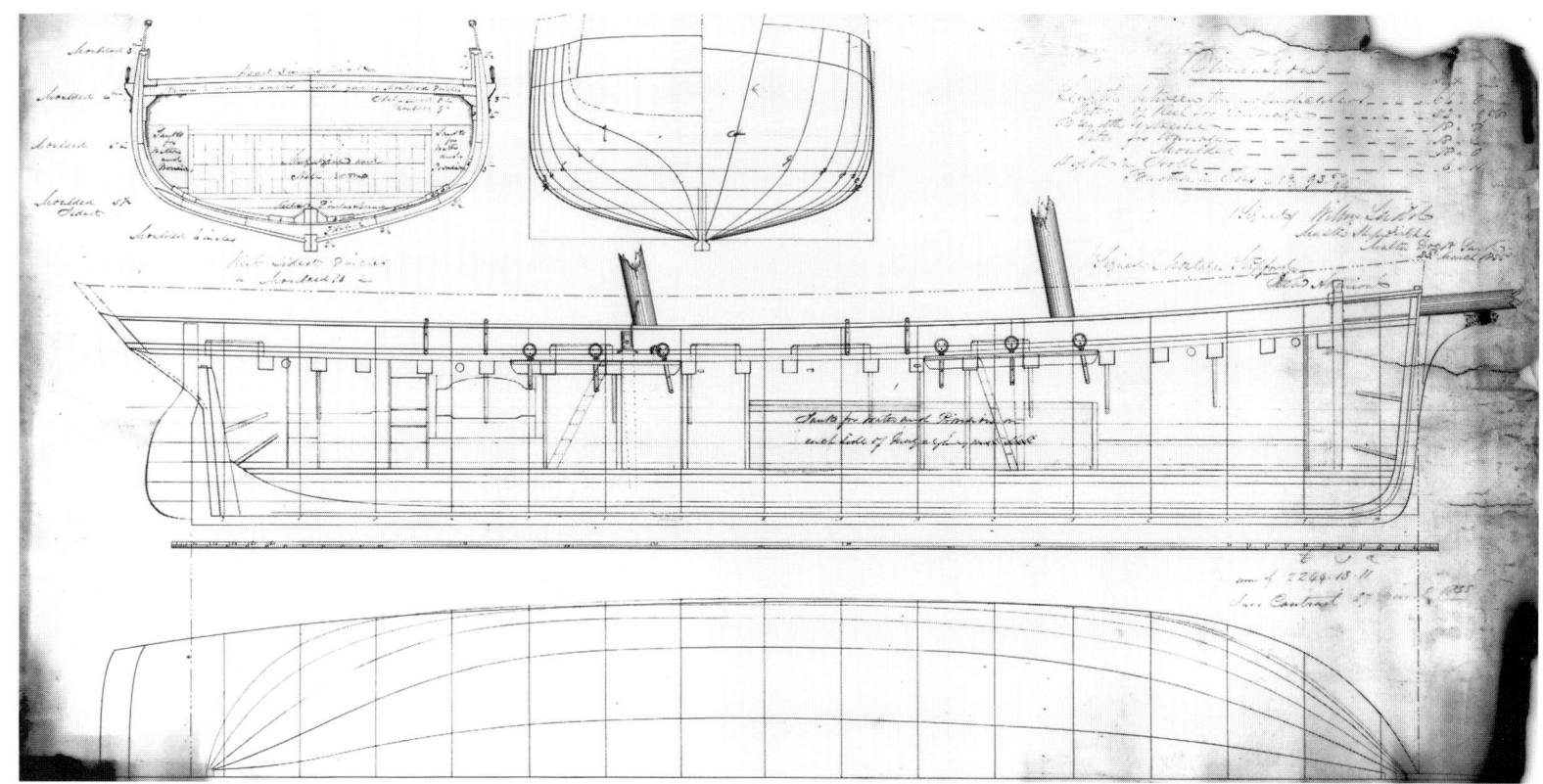

Lines, part profile and midships section (3000346B) for the 1855 schooner gunboats *Azov* and *Kertch* built at Malta

PLANS: As survey ship 1855: Profile & decks.

### Gipsy Class 1836. ? Cutters/tenders. Symonds design.
Dimensions & tons: 54', 41'7⅛" x 18' x 8'7". 70.
Men: 6? Guns: 2 x 3 brass?

**Gipsy** Sheerness Dyd.
Ord ____ ; K 04.1836; L 27.10.1836; 1892 sold.

**Mercury** Chatham Dyd.
Ord ____ ; K 05.1836; L 7.2.1837; by 1860 yard craft.

FATES: *Mercury* became a yard craft at Portsmouth; fate uncertain.

PLANS: Lines/profile/decks/midships section/sails.

### Bermuda Class 1845. 3. Schooner. Symonds design.
Dimensions & tons: 80', 64'1⅜" x 23'3" x 11'. 180 38/94.
Men: 25? Guns: 1 x 32 + 2 x 12★.

**Bermuda** Outerbridge & Hollis, Bermuda.
Ord 24.12.1845; K ____ ; L 03.1848; 1855 wrecked.

FATE: Wrecked near Turks Island on 20.1.1855; wreck sold.

PLANS: Lines & profile/lower deck/upper deck/sails/gun mounting.

### 'Australian' Class 1849. Teak Built schooner. John Edye
design. For the South Australian Government (*Australian* may well be a description rather than a name).
Dimensions & tons: 60', 47'6⅜" x 18'4" x 8'. 82.
Men: ____ . Guns: ____ ?

**Australian** Bombay Dyd.
Ord 1.6.1847; K ____ ; L ____ 1849; may have eventually been transferred to the Royal Indian Marine?

PLANS: Lines, profile, midships section/rig.

### Azov Class 1855. 2. Schooners/Gunboats. Ordered to be built by Admiral Stopford, the Superintendent at Malta (and designed by him?). Originally referred to as 'B' and 'C'. Classed as schooner-rigged gunboats. In 1897 *Azov* was referred to as 'slow and leewardly, steers badly in light winds'.
Dimensions & tons: 64', 52'7⅝" x 18'8" x 6'11". 93 56/94.
Men: ____ . Guns: 2 x 32 (pivot).

**Azov/Azoff** Guman, Malta.
Ord ____ ; K 04.1855; L 14.7.1855; later yard craft.

**Kertch** Guman, Malta.
Ord ____ ; K 04.1855; L 19.7.1855; 1860 hulked.

FATES: *Azov* became a yacht at Malta; 1899 sold. *Kertch* became a water tank vessel at Gibraltar (YC No 1).

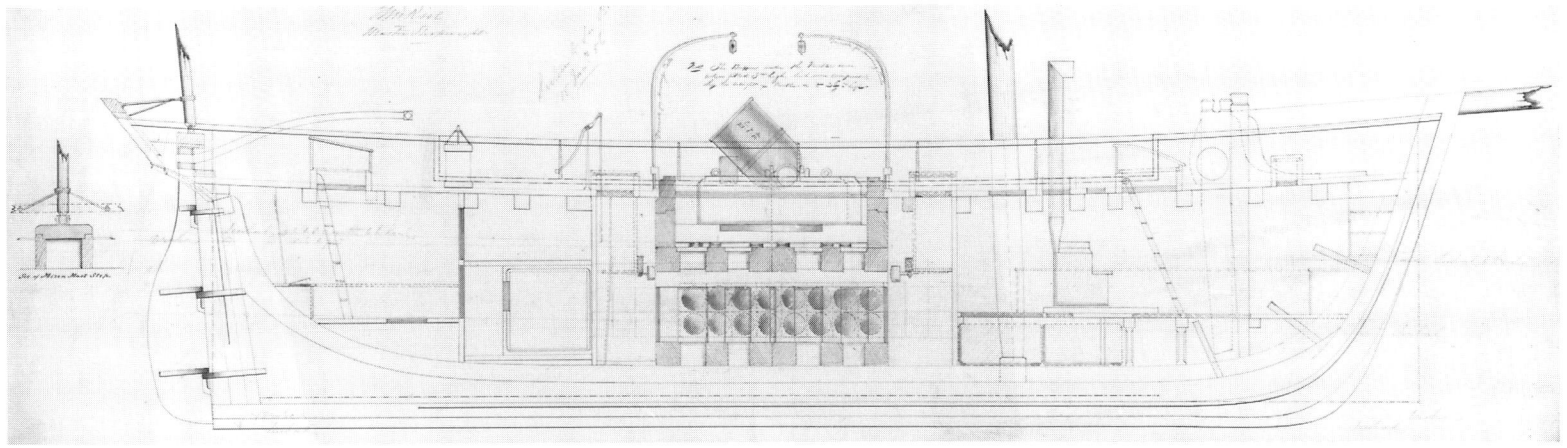

Profile (8038) of the dockyard sailing lighter *Sinbad* of 1834 as converted in 1854 as Mortar Vessel No 2

PLANS: Lines/decks/sails/specification.

**Class Uncertain, Cancelled 1849.** *Hornet* Woolwich Dyd.
Ord 12.8.1847.

**Class Uncertain, Cancelled 1850.** *Cracker* Deptford Dyd.
Ord ____ 1848.

## Mortar Vessels

NOTE: The large programme of mortar vressels and mortar floats built for the Crimean War have not been included here.

**Conversion from Yard Craft.** *Hamoaze* 1824 was originally the sailing lighter of the same name built Oreston 1800.
Dimensions & tons: 69'7", 55'4" x 23'2½" x 9'10". 159.
Men: ____ . Guns: ____ .
Converted 1824 at Deptford; later reconverted; 1843 BU.

**Mortar Vessel No 1 1854** was sailing lighter *Drake* built at Portsmouth Dyd 1834.
Dimensions & tons: 60', 47'5" x 20'8" x 8'10". 105 11/94.
Men: ____ . Guns: ____ .
Converted 1854; 1856 renamed *Sheppey* as sailing lighter again; 1867 BU.
PLANS: As built: Lines, profile, deck/midships section (both also for *Sinbad*).

**Mortar Vessel No 2 1854** was sailing lighter *Sinbad* built Pembroke Dyd 1832-1834.
Dimensions & tons: 60'1", 47'6" x 20'9" x 9'. 105.
Men: ____ . Guns: 1 x 13"? Mortar.
Converted 1854; 1856 reverted to lighter; Ca 1863 Woolwich Yard Craft No 3; 1866 BU.

PLANS: As built: Specification. Otherwise as *Drake*. Conversion (as fitted): Profile/lower deck/upper deck/midships section.

## Rowing Gunboats

**Two Gun Gunboat.** (One gun at each end.) Diagonal planked, two-masted (schooner rig?). Admiralty Order 29.4.1841 to be built at Bermuda; copies of plan sent to Bermuda. 55' x 15'3" x 5'.
PLANS: Lines.

**One Gun Gunboat.** 1 x 8" gun. For Bermuda 1851; also 'Canadian gunboat *Cherokee* or Nos 1, 2, 3 (Canadian lakes?).
PLANS: Lines.

# PART II
# PRIZES AND PURCHASES

The ships designed and built for the Royal Navy were supplemented by other vessels captured from the enemy or purchased from the merchant marine or direct from their builders. These require a somewhat different treatment from the purpose built ships which are listed by class. The prizes and purchases need to be treated individually, so each is made the subject of an individual entry in the following format:

**(1)** Name (s) and dates in Royal Naval service. **(2)** Origins when known (in brackets: nationality or 'mercantile' followed by original name if different from the Royal Naval one, then date(s) of build, where built, and name of designer). If a prize is known to have had a different designation from the British one before capture this is noted here immediately after the nationality. **(3)** Designation/number of guns in RN service. **(4)** Dimensions and tons in the usual form used in this work: length on gun deck; length of keel for tonnage x breadth moulded x depth in hold. Tonnage calculated in 94ths. All these figures are in feet and inches and are those 'taken off' the vessels at the stage when they were being surveyed by the Royal Navy. **(5)** Official complement in British service. **(6)** Armament in British service, listed by deck, working up from the lowest: GD = gun deck, MD = main (middle) deck, UD = upper deck, QD = quarterdeck, Fc = Forecastle, RH = round house (poop). Number of guns are given by (x) their designation in (British) pounds. When this figure is marked by an asterisk (★) it indicates that the weapons in question are carronades. **(7)** Details of acquisition are followed by notes of alterations in status and then ultimate fate [with subsequent history after leaving first-line service or loss by the RN given in square brackets]. **(8)** Finally, plans in the National Maritime Museum's collections are noted when they exist.

Since the vast majority of prizes or purchases were acquired in wartime, the logical way of splitting this part of the list is by making each war a chapter. The comparatively few vessels acquired in the periods of relative tranquility between the wars are listed as a coda to the chapter on the preceding war. Within these chapters the ships are grouped by type in descending order of power and size, with auxiliaries such as storeships and transports coming last. Within each type the order is roughly chronological by date of acquisition by the Royal Navy. In some cases, for example the purchase of 'frigates' (ship sloops) at the outbreak of the Seven Years War and the acquisition of cutters at the end of that war, a number of ships are grouped under the note of the Admiralty Order (AO) authorising their purchase by the Navy Board.

After all the vessels known to have belonged to the Royal Navy in a given period have been listed, there remains a number of vessels whose status (or even existence) is doubtful. These are ships or craft not listed in the main primary sources used (Progress and Dimensions Books, the particular manuscript Navy Lists used or the Admiralty Ship Plans Collection) but whose existence is attested with greater or lesser reliability by some other source. These are listed as 'Unregistered' vessels (this is my own definition for this particular purpose; it does not necessarily mean that the vessels in question are not included on any Admiralty list). Numbers of these vessels (particularly in the later wars from the War of American Independence on) have a well attested existence but were acquired on distant stations, never coming to Britain or therefore to the direct attention of the Admiralty. In other cases involving distant stations, both the Admiralty clerks responsible for listing the Navy and the modern researcher have been in the same dilemma. That is in judging whether one name appearing at different times and in different contexts and often with different designations indicates one or two or even more ships. This uncertainty is particularly noticeable in the War of American Independence but recurs in later conflicts, particularly in the West and East Indies. Another source of uncertainty (particularly in the earlier wars) is whether a particular vessel was purchased or just hired for naval service.

In these circumstances, I have judged it best to exclude from the main part of the listing those vessels whose commissioning in the Navy or whose very existence could not be proven from the primary sources used. I have interpreted this rule very strictly (some may feel too strictly). However, it can be all too easy to include impostors in a work of this kind, as the history of the totally fictitious 28 gun frigate *Ariel* reputedly built in 1785 proves. The existence of the majority of vessels listed in the 'Unregistered' sections, or indeed the fact of their naval service, is not in doubt. I have indicated the more dubious cases where they exist.

Wartime expansion of the Navy did not rely only on purchasing captured and merchant vessels. Other ships were hired (usually with their crews, sometimes with their original captains still in charge), particularly for convoy service and despatch duties. Hiring or requisitioning merchantmen to supplement the King's Ships was a very old practice; the majority of the ships which fought the Spanish Armada were merchantmen with their original crews and captains. There was an echo of this in the first war we consider here, that of the English Succession, when large merchantmen were still being hired to become Fourth or Fifth Rates. By the time the next war broke out in 1702, the Navy was both more securely financed and less threatened. From this time on hired vessels were relegated to lesser positions in the fleet. This was also a reflection of the increasing specialisation of the fighting ship and the decreasing relative fighting value of even the biggest merchantman. From this time on hired ships and vessels were mostly used as 'armed ships' for trade protection, as transports of one kind or another, or were small vessels used for various kinds of communications duties. Vessels were usually hired by the year or some shorter period, and this fact combined with the usual (for this period) duplication of mercantile names makes for considerable confusion (it is quite often unclear whether, for example, one of two vessels called *Truelove* hired in a particular year is the same as one hired a couple of years earlier), particularly if there is a small variation in tonnage given which may or may not be significant. A further source of uncertainty is that although I have comparatively good primary listings of hired vessels for three wars (Spanish Succession, French Revolutionary and Napoleonic) the listings for the other periods have had to be compiled from various secondary sources and are probably both incomplete and unreliable. As with the 'Unregistered' sections, the element of doubt should always be borne in mind when referring to the 'Hired Vessels' parts of these chapters. I would be delighted to hear (c/o the National Maritime Museum) from researchers who have located good *primary* evidence which supplements or alters my notes on vessels in either of these categories (though *not*, please, 'evidence' from printed Navy Lists, twentieth-century listings or similar secondary or tertiary sources).

# Chapter 1: King William's War 1689-1697

This war, sometimes known as the 'War of the League of Augsburg', the 'War of the English Succession', or even the 'Eight Years War', was fought by a coalition of powers bought together by fear of the overweening power of Louis XIV of France and by the negotiations of William of Orange, the Stadthoulder of the United Provinces (The Netherlands), who had just contrived to take over the throne of England in conjunction with his wife, Mary, from his father-in-law James II. The naval war was fought between the alliance of England and Holland against France. At first the war was one of the collision of fleets. France's new and powerful fleet initially had the advantage, winning the battle of Beachy Head, or Béveziers (1690) against the combined Anglo-Dutch fleet. Two years later the verdict of this battle was reversed by the battle of Barfleur, in which a numerically smaller French fleet attempted to engage the Anglo-Dutch fleet. Some of the dispersed French fleet was destroyed in boat actions at Cherbourg and off the island of Tatihou (the 'battle of La Hogue').

After this reverse the French, despite retaining a very powerful fleet of line of battle ships, and actually winning the race to build more of these vessels,[1] gave up the struggle to compete with the English and Dutch in the line of battle and gave themselves over to commerce raiding (the *guerre de course*) with considerable, though not decisive, success. The listing of captures given here, and indeed the losses to the Royal Navy, tells its own story of a war of smaller units: small two-deckers, Sixth Rates and other lesser craft. From the end of 1692 on there were no major actions, but many encounters in home waters and as far afield as Hudson's Bay between ships too small to lie in the line of battle.

As was the case with vessels designed and built for the Royal Navy, prizes and purchases at this time had three different 'Establishments' for men and guns depending on whether they were at war or peace, and, in the former case, in home waters or overseas; this is reflected in the three separate sets of figures given here (when two sets are given the second applies to both war overseas and peace).

## Small Line of Battle Ships

***Content Prize*** 1695-1702 (French *Le Content* 1686 Toulon) 70/62 Third Rate.
Dimensions & tons: ____ , 119'6" x 42'2" x 15'. 1130 1/5. ____ .
Men: 440/320. Guns: GD 24 x 24, UD 26 x 9, QD 12 x 6, Fc 4 x 6, RH 4 x 3. Taken 06.1695 by *Carlisle*; 1703 hulked [at Lisbon; Ca 1715 sold.]

## Small Two-deckers

***Saudadoes/Saudadoes Prize*** 1692-1702 (French ____ ) 40/36 Fifth Rate.
Dimensions & tons: 103'6", 86'x 29'x 12'6". 385 .
Men: 180-150-115. Guns: 40/36. Taken 09.1692 off the 'Saudadoes'; 1702 hulked [at Plymouth as tender to hulk; 1708 hulk; 1712 breakwater.]

***Trident/Trident Prize*** 1695-1702 (French *Le Trident* 1688 Toulon) 58/50 Fourth Rate.
Dimensions & tons: 129'6", 107'1½" x 36'7" x 13'4". 762½.
Men: ____ . Guns: 58/50. Taken 27.6.1695 (?) 'in the Straits' (Mediterranean); 1702 breakwater.

***Lewis Prize/Saint Lewis Prize*** 1696-1701 (French *Le Saint Louis* 1696 Saint Malo) 42/36 Fifth Rate.
Dimensions & tons: 113', 93'x 30'6" x 11'3". 460.
Men: 180-150-115. Guns: 42/36. Taken 1696 [1694?] by *Medway*; 1701 hulked [at Jamaica; 1707 sold.]

***Medway Prize*** 1697; hulked after capture (French ____ ) 50 Fourth Rate.
Dimensions & tons: 116'7", ____ x 34'1" x 13'1". 500 .
Men: 250/200/150. Guns: 50 or 48/42. Taken 20.8.1697 by *Medway*; hulked [at Sheerness; 1709 to be sunk; order then changed to be fitted to receive men; 1712 breakwater.]

## Fifth Rates

***Lively Prize*** 1689-1689 (French ____ ) 30.
Dimensions & tons: ____ , 78'4" x 27'8" x 11'. 309.
Men: 130/110/85. Guns: ____ . Taken 25.7.1689; 4.10.1689 retaken by the French.

***Play Prize*** 1689-1697 (French *Le Jeux* 1689 Dunkirk) 30.
Dimensions & tons: 97'6½", 87' x 28'2" x 10'8". 367 .
Men: ____ . Guns: 30. Taken 1689; 1697 breakwater.

***Conception Prize*** 1690-1694 (French ____ ) 32.
Dimensions & tons: 98', 81' x 29'6" x 11'5". 375 .
Men: ____ . Guns: ____ . Taken 1690; 1694 discarded (or 1694 wrecked?).

***Virgin Prize*** 1690-1698 (French ____ ) 32.
Dimensions & tons: 95', 79' x 27'9" x 11'. 324.
Men: ____ . Guns: ____ . Taken 1690 by *Happy Return*; 1698 sold.

***Dover Prize*** 1693-1698 (French ____ ) 32.
Dimensions & tons: 105', 86' x 26'10" x 12'. 329.
Men: ____ . Guns: ____ . Taken 1693 by *Dover*; 1698 sold.

***Diligent Prize*** 1694-1698 (French *La Diligente* 1691 Le Havre) 32.
Dimensions & tons: 116', 96'x 33'x 12'6". 556.
Men: ____ . Guns: 32. Taken 1694 by *Ruby*; 1698 sold.

***Ruby Prize*** 1694-1698 (French ____ ) 40.
Dimensions & tons: 108', 82'2" x 31'x 13'6". 420 .
Men: ____ . Guns: ____ . Taken 1694 by *Ruby*; 1698 sold.

***Betty*** 1695-1695/1697 ***Betty Prize*** 1702 ( ____ ) 32/28.
Dimensions & tons: 103', 86'x 28'6" x 11'9". 371 65/94.
Men: 135-115-90. Guns: 32/28. Purchased or hired (?) 1695; 09.1695 taken by the French; 1697 (or 1696?) retaken by *Plymouth*; 1702 sold.

***Rainbow Prize*** 1696-1698 (French ____ ) 38/32.
Dimensions & tons: 103'7", 83'4" x 27'7" x 10'10". 345 91/94.
Men: ____ . Guns: ____ . Taken 1696 by *Medway*; 1698 sold.

***Thunderbolt Prize*** 1696-1699 (French bomb vessel *Le Foundroyante*, L'Orient 1695) 32/28.
Dimensions & tons: 119', 99' x 31'9" x 13'3". 530.
Men: 135-115-90. Guns: 32/28. Taken 1696 by *Content*; 1699 hulked [at Plymouth; 1731 BU.]

## 'Corvettes' (large Sixth Rates, 20 Guns or more)

***Swift Prize*** 1689-1695 (French *frégate légère La Railleuse* 1683 Dunkirk) 20.
Dimensions & tons: 87'6", 78' x 26'4" x 9'8". 288.
Men: 80. Guns: 20 x ____ . Taken 1689 by *Nonsuch*; 1695 discarded.

***Henry Prize*** 1690-1698 (French ____ ) 24.
Dimensions & tons: 86', 71'x 25'6" x 12'6" . 246.
Men: 80. Guns: 24 x ____ . Taken 1690 by *Dover*; 1698 sold.

***Crown Prize*** 1691?; 1692 (French ____ ) 26.
Dimensions & tons: 84', 67' x 25'x 11'. 223.
Men: ____ . Guns: 26. Taken 1691?; 9.2.1692 wrecked ['near the Start' off Dartmouth.]

***Sun Prize*** 1692-1701 (French ____ ) 24/22.

---

[1] This point became very clear in the 1992 Anglo-French Historians Conference at Portsmouth, held 300 years after Barfleur/La Hogue was fought, thanks to the contributions of Jan Glete, Martine Acerra and Patrick Villiers (it is intended to publish the proceedings).

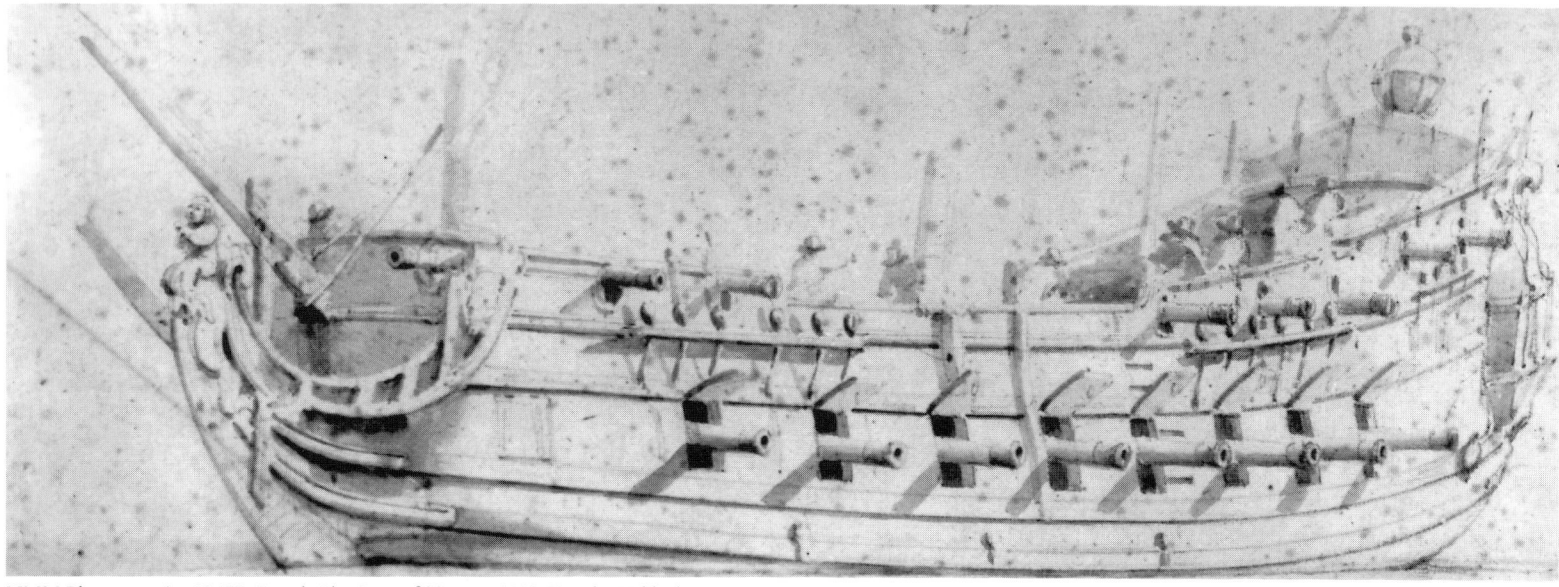

NMM Photo negative V.677. French *Play Prize* of 30 guns, 1689; Van der Velde drawing

   Dimensions & tons: 80'9", 70'x 24'x 9'4". 214.
   Men: 90-60. Guns: 24/22. Taken 08.1692; 1696 briefly in French hands (or 06.1693-1696?); 1701 sold.
*Legore* 1694?- ____ ? (French privateer *La Legore* [?] ____ ) 26.
   Dimensions & tons: 100', 84'x 26'7" x 11'. 315½.
   Men: ____ . Guns: 26 x ____ . Taken 1694 (?) by *York*; uncertain whether she was ever registered in the RN; fate unknown.
*Mary De Ponchartine* 1694?- ____ ? (French privateer *La Marie de Pontchartrain* ____ .) 26?
   Dimensions & tons: 94', 76'x 26'x 10'2". 336.
   Men: ____ . Guns: 26 (?) x ____ . Taken 1694(?) by *Reserve* and *Foresight*; uncertain whether she was ever registered in the RN; fate unknown.
*Queen Prize* 1694?- ____ ? (French privateer ____ ) 26.
   Dimensions & tons: 100', 85'x 25'2" x 10'2". 288.
   Men: ____ . Guns: 26 or 16 x ____ . Taken 1694(?) by *Adventure*; uncertain whether she was ever registered in the RN; fate unknown.

## Small Sixth Rates (less than 20 Guns)

*Blade Of Wheat* 1689-1689 (French ____ ) 10.
   Dimensions & tons: ____ , ____ x ____ x ____ . 150.
   Men: ____ . Guns: 10 x ____ . Taken 1689; 25.12.1689 wrecked [at Mill Bay].
*Dragon Prize* 1689-1690 (French ____ ) ____ .
   Dimensions & tons: ____ , ____ x ____ x ____ . ____ .
   Men: ____ . Guns: ____ . Taken 28.6.1689; 12.1.1690 lost [at Kingsgate, River Thames.]
*Saint Albans Prize* 1691-1698 (French[?] ____ ) 18.
   Dimensions & tons: ____ , 74' x 25'10" x ____ . 263.
   Men: 90. Guns: 18 x ____ . Taken 9.10.1691 by *Saint Albans*; 1698 sold.
*Saint Martin* 1690-1695 (French [?] ____ ) 18.
   Dimensions & tons: 78'6", 63' x 23' x 10'. 177.
   Men: ____ . Guns: 18 x ____ . Taken 1690 by *Pembroke*; 1695 discarded.
*Germoon/Germoon Prize* 1691-1700 (French ____ ) Sixth Rate/advice boat 10.
   Dimensions & tons: 68', 60' x 18' x 7'6". 103.
   Men: 50. Guns: 10 + 4 swivels. Taken 1691 by *Chester*; 4.7.1700 wrecked ['oversett a careening at Porto Bello'].
*Goodwin Prize* 1691-1695 (French ____ ) 6.
   Dimensions & tons: 58'10", 52'10" x 16' x 6'10". 72.
   Men: 35. Guns: 6 x ____ . Taken 11.1691 by *Goodwin*; 03.1695 taken and sunk by French privateer.
*Marianna Prize* 1693-1698 (French *La Marianne* ____ .) 18.
   Dimensions & tons: 80'6", 66' x 24' x 11'. 202.
   Men: 85. Guns: 18 x ____ . Taken 27.1.1693 [02.1692/3?] by *York* and *Dover*; 1698 sold.

*Mermaid* 1692-1693 (French [?] ____ ) 12.
   Dimensions & tons: 78', 62'x 23'x 10'3". 174.
   Men: 40. Guns: 12 x ____ . Taken 8.11.1692 by *Adventure*; 25.2.1693 burnt [by accident at Plymouth].
*Precious Stone/Hyacintha* 1692- ____ ? (French privateer *L'Hyacinthe* ____ ) 12.
   Dimensions & tons: 94', 78'x 27'x 11'. 302.
   Men: 45. Guns: 12. Taken 11.1692 by *Deptford* and *Portsmouth*; uncertain whether she was ever registered in the RN; fate unknown.
*Rupert Prize* 1692-1700 (French ____ ) 18/16.
   Dimensions & tons: ____ , 70' x 22' x 8'8". 180.
   Men: 85-70-50. Guns: 18/16 x ____ . Taken 1692 by *Rupert*; 1700 sold.
*Swallow Prize* 1692-1696 (French [?] ____ ) 18.
   Dimensions & tons: 68', 54'x 20'4" x 13'2½". 119.
   Men: 75. Guns: 18 x ____ . Taken 1692 by *Swallow*; 02.1696 taken by French privateer [off Weymouth].
*Tartan Prize* 1692?-1695 (French ____ ) 10.
   Dimensions & tons: ____ , 36' x 14'6" x 6'. 49.
   Men: 20. Guns: 10 x ____ . Taken 1692 (?); 17.5.1695 retaken by the French.
*Adventure Prize* 1693-1695 (French ____ ) Sixth Rate/hoy.
   Dimensions & tons: 37'4", 30' x 12'6" x 6'3". 24½.
   Men: ____ . Guns: ____ . Taken 1693 by *Germoon*; 1695 pitch boat [at ____ ?; 1698 sold].
*Jolly Prize* 1693-1698 (French *Le Joli* ____ ) 10.
   Dimensions & tons: 67', 56'x 19'6" x 9'. 113.
   Men: 50. Guns: 10 x ____ . French fireship taken 1693 by *Southampton*; 1698 sold.
*Pearl* 1693-1694 (French ____ ) 12.
   Dimensions & tons: ____ , 69'10" x 22'11" x 9'4". 195.
   Men: 65. Guns: 12 + 6 swivels. Taken 1693; 17.5.1694 wrecked [cast away off Goree].
*Essex Prize* 1694-1702 (French [?] frégate légère *La Héroine* [?] 1692 Brest) 16/14.
   Dimensions & tons: 74'10", 57'x 22'3" x 9'9". 152.
   Men: 70-60. Guns: 16/14 x ____ . Taken 16.7.1694; 1702 sold.
*Portsmouth Prize* 1694-1696 (French [?] ____ ) 10.
   Dimensions & tons: 68', 57' x 18'9" x 7'2". 106.
   Men: 55. Guns: 10 x ____ . Taken 21.7.1694 by *Portsmouth*; 28.9.1696 retaken by the French.
*Joyfull Prize* 1694?- ____ ? (French *La Joyeuse* ____ ) 10.
   Dimensions & tons: 68', 62'x 18'x 6'10". 106.
   Men: 45. Guns: 10 x ____ . Taken 1694 (?); out of service before 1699.
*Brilliant Prize* 1695?-1698 (French [?] ____ ) 6.
   Dimensions & tons: ____ , ____ x ____ x ____ .
   Men: 30. Guns: 6 x ____ + 2 swivels. Taken 1695 (?); 1698 sold.
*Concord* 1697- ____ ? (French [?] ____ ) ____ .

# KING WILLIAM'S WAR 1689-1697

Dimensions & tons: 77',64'x 22'6" x 9'6". 172.
Men: ____ . Guns: ____ . Taken 03.1697; out of service before 1699.

***Prince of Wales*** ____ ?- ____? (French [?] ____ ) ____ .
Dimensions & tons: ____ , 63' x 20'4" x 9'1". 163.
Men: ____ . Guns: ____ . Taken during the 1690s by *York*; out of service by 1699.

***Saint Nicholas*** ____ ?- ____? (French ____ ____ Dunkirk) ____ .
Dimensions & tons: 75', 62'x 22'10" x 9'4". 171.
Men: ____ . Guns: ____ . Taken during the 1690s by *Saudadoes*; hulked by 1699 [at ____ ?; fate uncertain].

## Fireships and Machines

(machines are noted as such, fireships not)

***Cadiz Merchant*** 1688-1692 (mercantile *Cadiz Merchant* ____ ) 12.
Dimensions & tons: ____ , 83'6" x 25'4" x ____ . 285 or 320.
Men: 45. Guns: 12 x ____ . Purchased 12.1688; 19.5.1692 expended [at the Battle of Barfleur].

***Charles*** 1688-1695 (mercantile *Charles* ____ ) 6.
Dimensions & tons: ____ , ____ x ____ x ____ . 90.
Men: 20. Guns: 6 x ____ . Purchased 08.1688; 5.7.1695 expended [at Saint Malo].

***Charles and Henry*** 1688-1689 (mercantile *Charles and Henry* —— . ) 6.
Dimensions & tons: ____ , ____ x ____ x ____ . 120.
Men: 25. Guns: 6 x ____ . Purchased 09.1688; 29.11.1689 wrecked [near Plymouth; or burnt by accident?].

***Cygnet*** 1688-1693 (mercantile *Cygnet* ____ ) 8.
Dimensions & tons: ____ , ____ x ____ x ____ . 100.
Men: 25. Guns: 8 x ____ . Purchased 09.1688; 20.9.1693 taken by the French.

***Elizabeth and Sarah*** 1688-1690 (mercantile *Elizabeth and Sarah* ____ ) 6.
Dimensions & tons: ____ , ____ x ____ x ____ . 100.
Men: 25. Guns: 6 x ____ . Purchased 08.1688; 1690 sunk as a foundation at Sheerness (or burnt by accident?).

***Richard and John*** 1688-1692 (mercantile *Richard and John* ____ ) 10.
Dimensions & tons: ____ , ____ x ____ x ____ . 200 or 160.
Men: 40. Guns: 10 x ____ . Purchased 11.1688; 1692 breakwater.

***Roebuck*** 1688-1692 (mercantile *Roebuck* ____ . ) 6.
Dimensions & tons: ____ , 64'x 19'6" x 9'2". 136 or 80.
Men: 16. Guns: 6 x ____ . Purchased 08.1688; 1692 breakwater.

***Speedwell*** 1688-1692 (mercantile *Speedwell* ____ ) 8.
Dimensions & tons: ____ , ____ x 22'6" x ____ . 127 or 120.
Men: 25. Guns: 8 x ____ . Purchased 1688; 1692 breakwater.

***Thomas And Elizabeth*** 1688-1692 (mercantile *Thomas and Elizabeth* ____ ) 10.
Dimensions & tons: ____ , ____ x ____ x ____ . 184 .
Men: 40. Guns: 10 x ____ . Purchased 1688; 24.5.1692 expended [at La Hogue].

***Unity*** 1688-1690 (mercantile *Unity* ____ ) 6.
Dimensions & tons: ____ , 83' x 27'8" x ____ . 120.
Men: 25 men. Guns: 6 x ____ . Purchased 1688; 1690 yard craft [at Portsmouth; 1707 'rebuild'].

***Owners Love*** 1688-1697 (mercantile *Owners Love* ____ ) 10.
Dimensions & tons: ____ , 71'2" x 22'6" x 9'6". 200.
Men: 50. Guns: 10 x ____ . Purchased 1688 from John Watts; 26.8.1697? sunk [by the French in Hudsons Bay].

***Nathaniel*** 1689-1692 (mercantile *Nathaniel* ____ . ) ____ .
Dimensions & tons: ____ , ____ x ____ x ____ .
Men: ____ . Guns: ____ . Purchased 1689 at Liverpool; 1692 breakwater.

***Crescent*** 1690-1698 (French [?] ____ ) 8.
Dimensions & tons: 85', 70'6" x 25' x 10'5". 234 .
Men: 45. Guns: 8 x ____ . Taken 1690 (or 1692?) by *Dover*, 1698 sold.

***Hopewell*** 1690-1692 (mercantile *Hopewell* ____ . N Barrett, Harwich) ____ .
Dimensions & tons: ____ , ____ x ____ x ____ .
Men: ____ . Guns: ____ . Purchased (whilst building?) 11.8.1690; 19.5.1692 expended [at the Battle of Barfleur].

***Pelican*** 1690-1692 (French [?] ____ ) ____ .
Dimensions & tons: ____ , ____ x ____ x ____ . 200.
Men: ____ . Guns: ____ . Taken 7.7.1690 by Sir Cloudesley Shovell; 1692 breakwater.

***Extravigant*** 1691-1692 (French *L'Extravigant* ex French mercantile *La Visitation* purchased at Marseilles 1689) 10.

NMM Photo negative V.681. Van der Velde drawing of the purchased fireship *Thomas and Elizabeth* of 1688

Dimensions & tons: 95', 80'x 25'6" x 11'6". 276½.
Men: 45. Guns: 10 x ____. Taken 1691 by *Saint Albans*; 19.5.1692 expended [at the Battle of Barfleur].

***Fortune/Fortune Prize*** 1692-1698 (French [?] ____ ____ ) 8.
Dimensions & tons: 88'6", 73'x 26'x 11'. 262.
Men: 45. Guns: 8 x ____. Taken 1692 (or early 1693?) by *Deptford* and *Portsmouth*; 1698 sold.

***Machine*** 1692-1696 (French privateer *La Machine* ____ St. Malo) 12.
Dimensions & tons: 109', 92' x 28'3" x 10'6". 391.
Men: 50. Guns: 12 x ____ (on two decks). Taken 15.9.1692 by *Centurion*; 1696 breakwater (?).

***Saint Vincent*** 1692-1698 (French [?] ____ ) 8.
Dimensions & tons: ____, 71'6" x 22'11" x 9'6". 200.
Men: 40. Guns: 8 x ____. Taken 18.6.1692; 1698 sold.

***Joseph*** 1692-1698 (French [?] ____ ) 8.
Dimensions & tons: 99', 74'6" x 26'6" x 11'3". 278.
Men: 45. Guns: 8 x ____. Taken 1692 by *Rupert*; 1698 sold.

***Ruzee*** 1692- ____ ? (French *La Ruzee* [or *Le Soubreau*?] 1689 ex Dutch ____ 1675) ____ .
Dimensions & tons: ____, ____ x ____ x ____ . 135.
Men: ____. Guns: ____. Taken 1692; burnt by accident some time in the 1690s. (Probably the vessel referred to by Pitcairn Jones as *Fuzee*.)

***Owners Goodwill*** 1694- ____ ?, (mercantile [?] smack *Owners Goodwill* ____ ) machine (ex smack).
Dimensions & tons: 38'10", ____ x 12'6" x 6'7". 24¾ .
Men: 4. Guns: No guns ? Purchased 1694; by 1699 hoy at Deptford; 1706 sold.

***Blessing*** 1694-1696 (mercantile smack *Blessing* ____ ) machine.
Dimensions & tons: ____ , ____ x 25' x 11'10". 18⅓.
Men: 4. Guns: No guns? Purchased 1694; 1696 sold.

***Grafton*** 1694-1696 (mercantile smack *Grafton* ____ ) machine.
Dimensions & tons: ____ , 24' x 13'6" x ____ . 18½.
Men: 4. Guns: No guns? Purchased 08.1694; 1696 sold.

***Hopewell*** 1694-1696 (mercantile smack *Hopewell* ____ ) machine.
Dimensions & tons: ____ , 25'x 11'10" x ____ . 18½ .
Men: 4. Guns: No guns ? Purchased 06.1694; 1696 sold.

***John and Martha*** 1694-1698 (mercantile *John and Martha* ____ ) machine.
Dimensions & tons: ____ , 24'6" x 11'8" x ____ . 17⅔.
Men: 4. Guns: No guns? Purchased 1694; 1698 sold.

***William and Mary*** 1694-1696 (mercantile smack *William and Mary* ____ ) machine.
Dimensions & tons: ____ , 24'x 13'6" x ____ . 22⅔.
Men: 4. Guns: No guns? Purchased 1694; 12.9.1694 expended [at Dunkirk].

***Sea Horse*** 1694-1698 (Dutch mercantile ____ ) machine.
Dimensions & tons: 57', 45'6" x 17' x 9'. 69¾ ____ .
Men: ____. Guns: No guns? Purchased from Dutch owners 04.1694; later waterboat [at Sheerness; 1698 sunk as a foundation].

***Saint Nicholas*** 1694-1694 (Dutch mercantile ____ ) machine.
Dimensions & tons: 70'6", 57'x 18'10" x 10'7". 107 50/94.
Men: ____. Guns: No guns? Purchased from Dutch owners 04.1694; 12.7.1694 expended [at Dunkirk].

***Trumpett*** 1694-1695 (Dutch mercantile ____ ) machine.
Dimensions & tons: ____ , ____ x ____ x ____ . 70?
Men: ____. Guns: No guns? Purchased 04.1694 from Dutch owners; 1695 sold.

***Abraham's Offering/Abram's Offering*** 1694-1694 (Dutch mercantile ____ ) machine.
Dimensions & tons: 55', 43'6" x 16'7" x 9'. 62 68/94.
Men: ____. Guns: ____. Purchased from Dutch owners 1694; 12.9.1694 expended [at Dunkirk].

***Castle of Masterland*** 1694-1695? (Dutch mercantile ____ ) machine.
Dimensions & tons: ____ , ____ x ____ x ____ . 100?
Men: ____. Guns: No guns? Purchased from Dutch owners 1694; 1695? sold.

***Crown Herring*** 1694- ____ ? (Dutch mercantile ____ ) machine.
Dimensions & tons: 65', 53'x 15'6" x 9'4". 67 69/94.

Men: ____ . Guns: No guns? Purchased from Dutch owners 1694; fate uncertain.

***Young Lady*** 1694-1695? (Dutch mercantile ____ ) machine.
Dimensions & tons: 63', 52'x 15'1" x 7'7". 62⅞.
Men: ____. Guns: No guns? Purchased from Dutch owners 1694; 1695? sold.

***Vesuvius*** 1694-1705 (building? Barrett, Shoreham) 8.
Dimensions & tons: 92', 77'6" x 25'7"x 10'6". 269.
Men: 45. Guns: 8 x ____. Purchased whilst building (?) 1694; 26/27.11.1703 drove ashore near Southsea Castle during the 'Great Gale', later salved; 1705 BU.

***Endeavour*** 1695-1705 (mercantile *Endeavour* ____ ) machine.
Dimensions & tons: 33', 23'8" x 12'x 5'9". 18.
Men: 4. Guns: No guns ? Purchased 1.3.1695; by 1700 hoy or smack [at Woolwich; 1705 sold].

## Bomb Vessels

***Julian*** 1690-1698 (French [?] ____ ) 8.
Dimensions & tons: 67', 54'2" x 19' x 9'4". 104.
Men: 45. Guns: 1 mortar + 8 x ____. Taken 1690 by *Foresight*; ? may originally have been classed as a 14 gun Sixth rate and converted to a bomb in 1694?; 1698 sold.

***Phoenix*** 1692-1698 (mercantile ____ ) 8.
Dimensions & tons: 55', 44'10" x 19'2½" x 8'7". 86 .
Men: 20. Guns: 8.( + 1 mortar?). Purchased 1692; 1698 sold.

***Endeavour*** 1694-1695 (mercantile ____ ) 4.
Dimensions & tons: ____ , 40' x 16'8" x 9'3". 59 .
Men: 18. Guns: 4 minions (+ 1 mortar?). Purchased 1694; 1695 sold.

***Mary Ann [Mariana]*** 1694-1698 (mercantile *Mary Ann* ____ ) 4.
Dimensions & tons: ____ , 44'6" x 18'8" x 10'4". 82.
Men: 20. Guns: 4 x ____ (+ 1 mortar?). Purchased 04.1694; 1698 sold.

***Owner's Advent*** 1694-1698 (ex mercantile *Owner's Advent* ____ ) 6.
Dimensions & tons: ____ , 52'x 20'5" x 10'2". 115.
Men: 25. Guns: 6 x ____ (+ 1 or 2 mortars). Purchased 1694; 1698 sold.

***Society*** 1694-1698 (mercantile *Society* ____ ) 8.
Dimensions & tons: ____ , 52'6" x 19'1" x 9'3". 102.
Men: 30. Guns: 8 x ____ (+ 1 or 2 mortars). Purchased 04.1694; 1698 sold.

***Angel*** 1694-1696 (mercantile ____ ) 6.
Dimensions & tons: ____ , 35' [probably actually 53'] x 21'3" x 11'. 132.
Men: 25. Guns: 6 minions (+ 1 mortar?). Purchased 1694 from Nicholas Smith; 1696 sold.

***Greyhound*** 1694-1698 (mercantile ____ ) 6.
Dimensions & tons: ____ , 51'x 18'8" x 9'11". 95.
Men: 20. Guns: 6 minions (+ 1 mortar?). Purchased 1694 from Richard Lapthorne; 1698 sold.

***Truelove*** 1694-1698 (mercantile ____ ) 4.
Dimensions & tons: ____ , 42'6" x 17' x 9'4". 65.
Men: 18. Guns: 4 x ____ (+ 1 mortar?). Purchased 1694 from John Gothwaite (?); 1698 sold.

## Miscellaneous Small Vessels

***Providence*** 1691-1707? (French ____ ) ketch ____ .
Dimensions & tons: ____ . 33' x 12'10" x 6'8". 28¾ .
Men: ____. Guns: ____. Taken 19.4.1691; 10.1707 retaken by the French.

***Saint John*** 1695-1696 (French ____ ) advice boat 6.
Dimensions & tons: 59', 50' x 16'4" x 6'6". 70.
Men: ____. Guns: 6 x ____. Taken 09.1695; 5.11.1696 retaken by the French.

## Storeships, etc

***Canterbury*** 1692-1703 (mercantile *Canterbury* 1692 [?] Ipswich) storeship 8.
Dimensions & tons: 96', 82'6" x 28'11" x 11'. 367.

Men: 35. Guns: 8 sakers. Purchased 1692 (two decked ship); 26/27.11.1703 wrecked at Bristol during the 'Great Gale'; remains sold.

*Katherine* 1694-1701 (mercantile ____ foreign built ) storeship 6.
Dimensions & tons: 97', 87'6" x 25'1½" x 11'6". 292.
Men: 30-80. Guns: 6 x ____ . Purchased 1694; 1701 sold.

*Suffolk/Suffolk Hagboat* 1694-1713 (mercantile ____ ) storeship 6/12.
Dimensions & tons: 117', 96' x 30'7" x 12'8". 477.
Men: 50-80. Guns: 6-8-10-12 x ____ . 1708: 16 sakers. Purchased 1694; 1696 hulked at ____ but back in service by 1708; 1712 hospital ship?; 1713 sold.

*Josiah* 1694-1696 (mercantile ____ ) storeship 30.
Dimensions & tons: 130'9", 108'x 34'x 16'10". 664.
Men: 120. Guns: 30 x ____ . Purchased 1694; 1696 hulked [at Woolwich; 1715 sunk as a breakwater at Sheerness].

*Fortune* 1699?-1700 (mercantile ____ ) storeship. ____ .
Dimensions & tons: ____ , ____ x ____ x ____ . ____ .
Men: ____ . Guns: ____ . Purchased 1699 (or 1700?) in America?; 15.12.1700 wrecked on the coast of Cornwall.

## Miscellaneous Unregistered Vessels

[ie. vessels only found in secondary listings - not in the primary Admiralty sources used; existence of some doubtful.]

*Advice Prize* 1693-1695 (French ____ ) sloop. Taken 1693; 1695 sold [given in Colledge only].
*Alexander* 1688?-1689 (French [?] ____ ) fireship 6. 150 tons. Taken 1688 or 1689; 21.6.1689 burnt by accident.
*Basing* 1688-1698 (mercantile [?] ____ ) fireship ____ . Purchased (?) 1688; 1698 sold.
*Basing Galley* 1693-1694 (mercantile [?] ____ ) Sixth Rate 18.
Men: 70. Purchased 1693; ____ 1694 taken by the French.
*Boston* 1694-1695 (American-built in 1692) Fifth Rate, 32. 1694 presented to the Navy by the citizens of Boston, Massachusetts; 4.7.1695 taken by the French in the Atlantic [given in J Colledge only, not found in other listings].
*Dover Prize* 1689-1689 (French ____ ) hulk. Taken 1689; 25.12.1689 wrecked at Plymouth.
*Eaglet* bomb, listed in 1699? [Dubious, given in Pitcairn Jones only.]
*England* 1693-1695 (mercantile [?] ____ ) Fifth Rate 42. 400 tons. [Hired 1692]; Purchased 19.3.1693; 1.2.1695 sunk by the French in action off Cape Clear.
*Ephraim* 1695-1695 (mercantile [?] ____ ) machine. 170 tons. Purchased 1695; 1.8.1695 expended at Dunkirk.
*Fowey* 1692-1693 there is a log for a vessel of this name and dates.
*Gabriel* 1694 (mercantile [?] ____ ) Pink/machine. Purchased 1694; fate uncertain.
*Happy Return* 1695-1695 (mercantile [?] ____ ) fire vessel/machine. 84 tons. Purchased 1695; 1.8.1695 expended at Dunkirk.
*Hopewell* 1689 scout boat.
*Humber* fireship of 254 tons and 8 guns built 1690; no further information; is this a misreading of the *Hunter* of 1690? [Only given in J Colledge].
*John* [*Of Dublin*] 1689-1690 (French [?] ____ ) fireship. Taken from the French 1689; lost 13.1.1690 or sunk as breakwater 1690.
*Joseph* 1694-1695 bought or hired Fifth Rate (same as the 1692 fireship?).
*Joshua* (or *Josiah*?) 1694 storeship (same as the storeship and Woolwich hulk *Josiah*?).
*Magdalene* 1689-1699 (French ____ ) hulk.
Dimensions & tons: ____ , 75' x 26'9" x ____ . 285.
Men: ____ . Guns: ____ . Taken 1689 by *Montague*; 1699 sold.
*Mary* 1694 ( ____ ) ketch, 10. Origins unknown; 19.2.1694 foundered off Gibraltar (or is this a confusion with *William* given as lost on the same date and in the same area?).
*Mayflower* 1694-1696 (mercantile [?]____ ) smack.
Dimensions & tons: ____ , 24'6" x 12' x ____ . 18.
Men: ____ . Guns: ____ . Purchased 1694; 1696 sold.
*Mayflower* 1695-1695 (mercantile [?] ____ ) machine. 109 tons. Purchased 1695; 1.8.1695 expended at Dunkirk.

*Muscovy Merchant* Ca 1694 storeship.
*New Africa* 1690-1695 type and other details uncertain, there is a log for this vessel.
*Olive Branch* 1690-1690 (mercantile [?] ____ ) fireship. Purchased 1690; 1690 taken by the French.
*Pelican* 1697; there is a log for a ship of this name and date; no further information.
*Redbridge* Fifth Rate 32 guns 160 men; listed in 1699? (given by Pitcairn Jones only).
*Saint Malo Prize* 1696 (French ____ ) Fifth Rate 36. Taken 1695 or 1696; in service 1696; fate uncertain.
*Siam/Syam* Ca 1694 storeship 22 (?hired?).
*Speedwell* 1689 scout boat (hired?; given only by Pitcairn Jones).
*Thomas and James* 1691-1697 there is a log for a ship of this name and dates (hired?).
*Wild/Wild Prize* 1692-1694 (French ____ ) Sixth Rate 12.
Dimensions & tons: ____ , 51'3" x 16' or 18'10" x 7'1". 97.
Men: ____ . Guns: 12 x ____ . Taken 1692; 18.6.1694 taken by the French.
*William* 1694 ketch; only note found on her is that she foundered off Gibraltar on 19.2.1694 (is this a confusion with *Mary* lost on same date?).
*William and Elizabeth* 1695-1695 (mercantile [?] ____ ) fire vessel. Purchased 1695; 1.8.1695 expended at Dunkirk.

## Hired Vessels

( [?] not specifically stated to be hired, but probable.)

*Adventure Gally* 1696-1699 ['Kidd the Pirate'].
*Africa* 1692-1695 Fourth Rate, 46 guns.
*Archangel* 1695 Fourth Rate, 48 guns.
*Baltimore* 1691 hospital ship 300/324 tons.
*Betty* [?] 1689 Scout boat.
*Bonaventure* 1689-1697 Fifth/Fourth Rate; 1689: 36 guns. 1692: 50 guns.
*Bristol* 1692-1697 Fifth Rate/Hospital ship; 532 tons; 20 guns.
*Catherine* [?] 1691-1694 storeship.
*Charity* 1689-1692 Sixth Rate, 10 guns; 1692 fitting as a bomb?
*Delaval* 1694 Fifth Rate, 30 guns.
*Edward and Susanna* 1689-1690 hoy, Sixth Rate, 8 guns.
*England* 1692; see the (unregistered) Fifth Rate purchased in 1693.
*Enquiry* 1691 sloop 3 guns. Taken by the French 13.10.1691.
*Europa/Europe* 1689-1690 Fourth Rate, 30 guns.
*Falcon* 1694-1695 Fifth Rate, 38 guns. Taken by the French 3.1.1695; probably a confusion with the 24 gun *Falcon* (see main list).
*Farmers Goodwill* [?] 1699? Fourth Rate, 50 guns, 260 men. [Noted by Pitcairn Jones as in 1699 list?; otherwise unknown.]
*Friendly Society* 1693-1696 Fifth Rate, 36 guns.
*Frog* 1689 dogger, 10 guns.
*George* 1689-1690 Fourth Rate, 48 guns.
*Hannibal* 1689-1691 Fourth Rate, 48 guns.
*Jerusalem* 1689 Sixth Rate.
*Levant* [?] 1689-1690 hoy, Sixth Rate, 8 guns, 40 men; in Irish waters.
*Loyalty* [?] 1694 Fifth Rate, 34 guns.
*Lumley Castle* 1693 Fourth Rate, 56 guns.
*Mary* 1690 Sixth Rate, 12 guns, 105 tons; wrecked 18.3.1690 in Wicklow Bay.
*Modena* 1690-1691 Fourth Rate, 64 guns, 350 men.
*Owners Endeavour* 1689 Sixth Rate.
*Prince George* 1693-1695 Fifth Rate, 36 guns.
*Prince of Orange* [?] 1692-1697 Fourth Rate, 46 guns.
*Princess Ann* 1692-1699 Fourth Rate/hospital ship, 484 tons, 48 guns.
*Prosperous* 1689 ketch/Sixth Rate, 6 guns, 40 men.
*Providence* 1689 hoy, 10 guns.
*Prudence* [?] 1693-1695 42 guns.
*Rebecca* 1691 Fifth Rate, 20 guns.
*Richard and Martha* 1689 ketch, 10 guns.
*Saint Ann* 1693-1703 hospital ship/Fifth Rate.
*Saint Anthony* [?] 1694 Sixth Rate.
*Saint Malo Merchant* 1689-1690 Sixth Rate, 8 guns.
*Sampson* 1689 Fourth Rate, 50 guns.

***Samuel and Henry*** 1689-1695 Fourth Rate, 44 guns.
***Sapphire*** 1689-1690 Fifth Rate, 30 guns.
***Sceptre*** 1689 Fourth Rate, 40 guns.
***Seventh Hoy/Severn Hoy*** 1689-1690 8 guns.
***Smyrna Factor*** 1693-1695 Fifth Rate, 300 tons, 24 guns, 45 men.
***Smyrna Merchant*** 1689-1697 Fifth Rate, 40/34 guns.
***Society*** [?] 1690-1697 Fifth Rate/hospital ship, 357 tons, 30 guns; 23.8.1697 taken by the French [became Société de Grenze; sold 1699?].
***Spencer*** [?] 1691 Fifth Rate/hospital ship, 245 tons.
***Success*** 1689-1692 Fourth Rate, 48 guns; probably the vessel purchased as a hulk at Deptford in 1694 (which see).

***Supply*** 1688-1690 armed ship, 308 tons, 34 guns; wrecked 11.1.1690 at Hoylake.
***Thomas and Daniel*** [?] 1689 scout boat.
***Three Brothers*** [?] 1689 scout boat.
***Tiger*** 1695-1696 Fifth Rate, 32 guns.
***Unity*** 1689 ketch/Sixth Rate, 8 guns.
***Unity*** 1692-1695 Fifth Rate, 240 tons, 36 guns.
***Warrington*** 1692-1693 200 tons, 30 guns; 1693 taken by the French.
***Welcome*** 1689-1690 ketch/Sixth Rate, 10 guns, 45 men.
***William and Mary*** [?] 1689 scout boat.
***William and Rebecca*** [or ***William and Robert***] 1689 dogger/Sixth Rate, 10 guns.
***William and Thomas*** 1689 Sixth Rate, 8 guns.

# Chapter 2: The War of Spanish Succession 1701-1713

The renewed war with Louis XIV's France in effect resumed where the previous struggle had left off: with France continuing, with some success, to concentrate on attacking British and Dutch commerce. The only clash of opposing fleets took place in the Mediterranean with the battle of Velez-Malaga (1704), basically a drawn battle but one which left the British in possession of the recently captured base of Gibraltar. The greatest success of the naval war was at the beginning with the very profitable destruction of the Spanish plate fleet at Vigo (1702). By the end of the war the relative decline of other navies, particularly the Dutch, left the British as unmistakeably the premier naval power in Europe.

The pattern of captures, mostly smaller warships and privateers, is not unlike that of the latter part of the previous war. However, there is a great difference to be noted in the listing of purchased and hired vessels. The former are comparatively few in number, whilst the latter do not include the numbers of larger merchantmen hired in the previous decade to bulk out the numbers of Fourth and Fifth Rates. These are both indications of the increased size and capability of the Royal Navy and of the lesser intensity of the conflict it was involved in. The main reason for the large number of small vessels listed as hired is the availability of annual lists of the tenders which were hired to attend larger warships.

An appendix or coda to the chapter is formed by a list of the small number of vessels purchased or taken in the period between 1713 and the outbreak of the next major European war in which Britain was involved (1739). During this time there were two periods of armed conflict with Spain, the first of which produced a fleet battle at Cape Passaro (1718) in which the Spanish fleet was thoroughly defeated. A number of Spanish ships of the line were taken in this battle, and were apparently kept laid up at Port Mahon for some years, however they were never taken into the Royal Navy and therefore no effort is made to list them here.

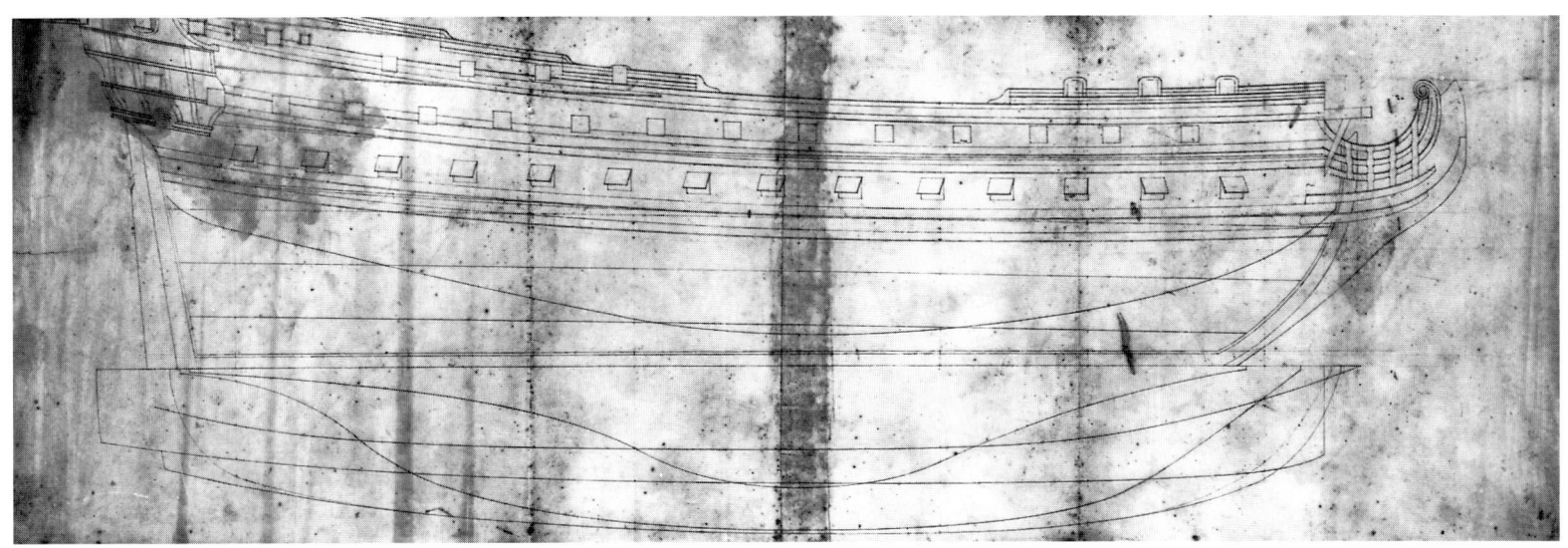

Sheer and half breadth plan (1444) of the 80 gun French prize *Prompt* captured in 1702

## Large two-decker

***Prompt*** 1702-1703 (French *Le Prompt* 1693 Dunkirk) Third Rate 80/76.
Dimensions & tons: 157'9", 129'2" x 45' x 15'4½". 1391.
Men: ____ . Guns: 80/76 x ____ . Taken 12.10.1702 at Vigo; 1703 BU.
PLANS: Sheer & half breadth/body plan.

## Small Line of Battle Ships (two-deckers)

***Assurance*** 1702-1712 (French *L'Assuré* 1696 Toulon) Third Rate 70/64.
Dimensions & Tons: 145'10", 118' x 41'11" x 16'11". 1102 71/94.
Men: 440/400/365. Guns: 70/66/64. Taken at Vigo 1702; 1712 BU.

***Firme*** 1702-1713 (French *Le Ferme* 1699 Rochefort) Third Rate 70.
Dimensions & tons: 156'2", 129' x 43'4" x 17'11". 1288.
Men: 440/380. Guns: 72/70 (Dimensions Book B gives 80, but this would appear to be an error). Taken at Vigo 1702; 1713 sold.

***Moderate*** 1702-1712? (French *Le Modéré* 1685 Dunkirk) Fourth Rate 56/64.
Dimensions & tons: 132'6", 107' x 39'6" x 14'8". 887 85/94.
Men: 365/240. Guns: 56 x ____ (from 1703 64 x ____ ; but seems usually to have been classed as a Fourth and not a Third Rate ship). Taken at Vigo 1702; by 1712 hulked [at Portsmouth; 1713 sold].

***Hazardous*** 1703-1706 (French *Le Hasard/Hazardeux/Hazardeus* 1699 Lorient) Fourth Rate 54.
Dimensions & tons: 137', 114' x 38' x 15'. 875.
Men: 320/240. Guns: 1703 & 1705: GD 22 x 18, UD 22 x 12, QD 8 x 6, Fc 2 x 6. Taken 1703 by *Warspight, Orford* & others; 19.11.1706 run ashore near Selsey Bill [this wreck is being excavated, having recently been located].

***Falkland Prize*** 1704-1705 (Possibly French *La Seine* 1688 Rochefort) Fourth Rate 54.
Dimensions & tons: 132'6", 112'4" x 35' x 15'1". 732.
Men: 280. Guns: GD 22 x 12, UD 22 x 6, QD 8 x 6, Fc 2 x 6. Taken 1704 by *Falkland* and others; 19.12.1705 run ashore in a storm in Sandwich Bay; wreck sold for BU.

***August*** 1705-1716 (French *L'Auguste* 1704-1705 Brest) Fourth Rate 60/56.
Dimensions & tons: 141'6", 115'2½" x 39' x 16'. 932.
Men: 365/240. Guns: GD 24 x 24, UD 26 x 9, QD 8 minions, Fc 2 minions. Taken 19.8.1705 by *Chatham*; 10.11.1716 wrecked on the Island of Anholt returning from the Baltic.
PLANS: Lines.

***Arrogant*** 1705-1709 (French *L'Arrogant* 1681 Le Havre) Fourth Rate 60.
Dimensions & tons: 137'6" or 137'8", 115'4" x 29'6" x 19'8". 928.
Men: 365/240. Guns: GD 24 x 24, UD 26 x 9, QD 8 minions, Fc 2 minions. Taken 10.3.1705 at Leake's relief of Gibraltar; 5/6.1.1709 lost in a violent storm going from Lisbon to Port Mahon.

***Sweepstakes*** 1709-1716 (French *Gloire* [38] 1707 L'Orient) Fifth Rate 40.
Dimensions & tons: 122', 102'4" x 34'9" x 13'. 657 24/94.

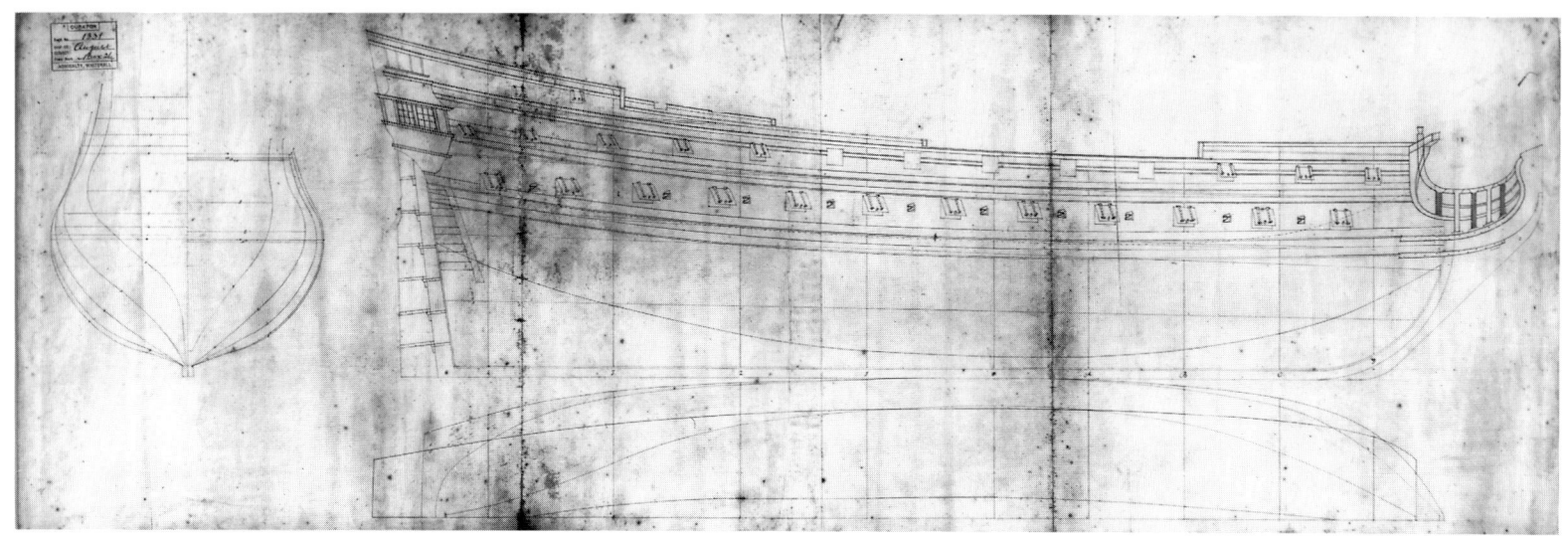

Lines plan (1331) of the French 60 gun ship *August* captured in 1705

Men: ____ . Guns: 40/42. Taken 6.5.1709 by *Chester* of Lord Dursley's squadron; 1716 sold.

***Moor*** 1710-1716 (French *Le Maure* 1688 Toulon) Fourth Rate 54.
Dimensions & tons: 135'4", 116'2" x 36'2¾" x ____ . 811 or ?1020?
Men: 320. Guns: 54. Taken 13.12.1710 by *Berwick, Bredah, Warspight*; 1716 breakwater.

***Superb/Superbe*** 1710-1733 (French *Le Superbe* 1708 Lorient) Third Rate 64.
Dimensions & tons: 143'6", 119'3" x 40'2" x 15'6½". 1020 33/94.
Men: 365. Guns: 64. Taken 29.7.1710 by *Kent*; 1733 BU.
PLANS: Lines & profile. 1720: Lines.

## Small Two-deckers

***Triton*** 1702-1709 (French *Le Triton* 1696 Brest) Fourth Rate 42.
Dimensions & tons: 128', 105'7" x 34'4" x 13'4". 661 73/94.
Men: ____ . Guns: 42. Taken at Vigo 1702; 1709 sold.

## Small Fifth Rates

***Litchfield Prize*** 1703-1706 (French ____ ) 36.
Dimensions & tons: 100', 80'6" x 30'5½" x 10'2". 397.
Men: 155/110. Guns: 18 x 8, 18 x 4. Taken 29.7.1703 by *Litchfield*; 1706 sold.

***Orford Prize*** 1703-hulked (French ____ ) ____ .
Dimensions & tons: ____ , ____ x ____ x ____ . ____ .
Men: ____ . Guns: ____ . Taken 07.1703 by *Orford*; 'appears not fit for Naval service'; 09.1703 hulked [at ____ ?; fate uncertain]. Note that there is an '*Oxford Prize*' a Fifth Rate, listed by Pitcairn Jones; 330 tons; captured in 1701 [?] and sold in 1703 which may be a confusion with this vessel, or, less likely, a separate one.

***Triton Prize*** 1705-1713 (French privateer *Le Royal* ____ St. Malo) 30.
Dimensions & tons: ____ , 75' x 26'3" x 12'6". 274.
Men: 115/100/65. Guns: 26 x 6 + 4 x 3. 1705: 6 sakers, 20 minions, 4 x 3. Taken 3.3.1705 by *Triton*; 1713 sold.

***Swallow Prize*** 1705-1711 (French *l'Etoile?* 1703 ____ ) 32/28.
Dimensions & tons: ____ , ____ x ____ x ____ .
Men: 145/110. Guns: 4 x 9, 22 x 6, 6 x 4. Taken early 10.11.1704(?) by *Swallow*; entered into the RN 14.3.1705; 29.7.1711 lost going into the port of Alaza, Corsica.
PLANS: Lines.

***Edinburgh*** 1707-1709 (Scots *Royal William* 1696 Thames) 32/28.
Dimensions & tons: 99', 83'4½" x 28'8" x 11'2". 364½.
Men: 145/100. Guns: GD 10 x 9, UD 18 x 6, QD 4 x 4.
Transferred under the Act of Union 1707; 1708 to be sold; 1709 sunk as a breakwater at Harwich.

***Winchelsea*** 1708-1735 (mercantile ____ ____ Bier, ____ ) 36/32/30.
Dimensions & tons: 108'2½", 87'7½" x 29'10" x 12'9½". 414 78/94.
Men: 155/110. Guns: GD 8 x 12, UD 22 x 6, QD 6 x 4. Purchased 1708; 3.2.1709 taken 'off the Northward Cape' by the French then retaken by *Chichester*; 1716 reduced to Sixth Rate 20; 1735 BU.

## 'Corvettes', Sixth Rates of 20 Guns or more

***Sun Prize*** 1704-1713 (French ____ ) 22.
Dimensions & tons: 82'8", 69'3" x 24'2" x 9'7½". 215.
Men: 110. Guns: 1705: 18 sakers + 4 minions. Taken 4.7.1704 by *Litchfield*; 18.1.1708 retaken by French 36 gun privateer off the Needles.

***Medway Prize*** 1704-1713 (French ____ ) 28.
Dimensions & tons: 92'8", 72'6" x 25' x 10'6". 241.
Men: 110. Guns: LD 6 x 6. UD 22 x 6. Taken 6.9.1704 by *Medway*; 1713 sold.

***Dunkirk Prize*** 1705-1708 (French privateer *Le Hocquart* 1704 St. Malo) 24.
Dimensions & tons: ____ , 70'8" x 27'10" x 12'6". 299.
Men: 115/85. Guns: 1703 establishment: UD 20 x 6, QD 6 x 3. Taken 1705 by *Dunkirk*; 18.10.1708 wrecked on the rocks near Cape François (Haiti).

***Cruiser/Cruizer Prize*** 1705-1708 (French *Le Meric* ____ ) 24.
Dimensions & tons: ____ , 75' x 26'6" x 11'3½". 280.
Men: 115. Guns: 24. Taken 5.5.1705 by *Triton*; 15.12.1708 wrecked on the Island of Terceira.

***Valeur/Valure*** 1705-1718 (French *Valeur* 1704? Brest) 24.
Dimensions & tons: 110'9", 81' x 27'4" x 11'8". 321 64/94.
Men: 115-110. Guns: 24. Taken 5.5.1705 by *Worcester*, 6.9.1710 retaken by French *Surprize* in Newfoundland; later the same year recaptured by *Essex*; 1717 briefly converted as a fireship and then reconverted to a 24; 1718 BU.

***Fox/Fox Prize*** 1705-1706 (French ____ ) 24.
Dimensions & tons: 93', 76' x 26' x ____ . 273.
Men: 115. Guns: 24. Taken 05.1705 by *Triton*; 28.8.1706 wrecked off 'Hollyhead'.

***Enterprize*** 1705-1707 (French *L'Enterprise* 1705 ____ ) 24.
Dimensions & tons: ____ , 79'9" x 27'6" x 11'5". 320.
Men: 115. Guns: 24. Taken 05.1705 by *Triton*; 2.10.1707 foundered off Tholon 'in the Streights' (Mediterranean).

***Childs Play*** 1706-1707 (French *le Jeux* ex *l'Opiniâtre* 1689 Dunkirk) 24.
Dimensions & tons: 103' x 80'7" x 29'6" x 10'7". 373.
Men: 115/85. Guns: 20 x 6, 4 x 4. Taken 1706 by *Tartar*, 29.8. (or 2.9.) 1707 foundered in a hurricane off Saint Kitts.

***Dumbarton Castle*** 1707-1708 (Scots *Dumbarton Castle* 1696 Thames) 24.
Dimensions & tons: ____ ____ ____ .
Men: 115. Guns: 24 Transferred
under the Act of Union 1707; 26.4.1708 taken by a French 44 off

# THE WAR OF SPANISH SUCCESSION 1701-1713

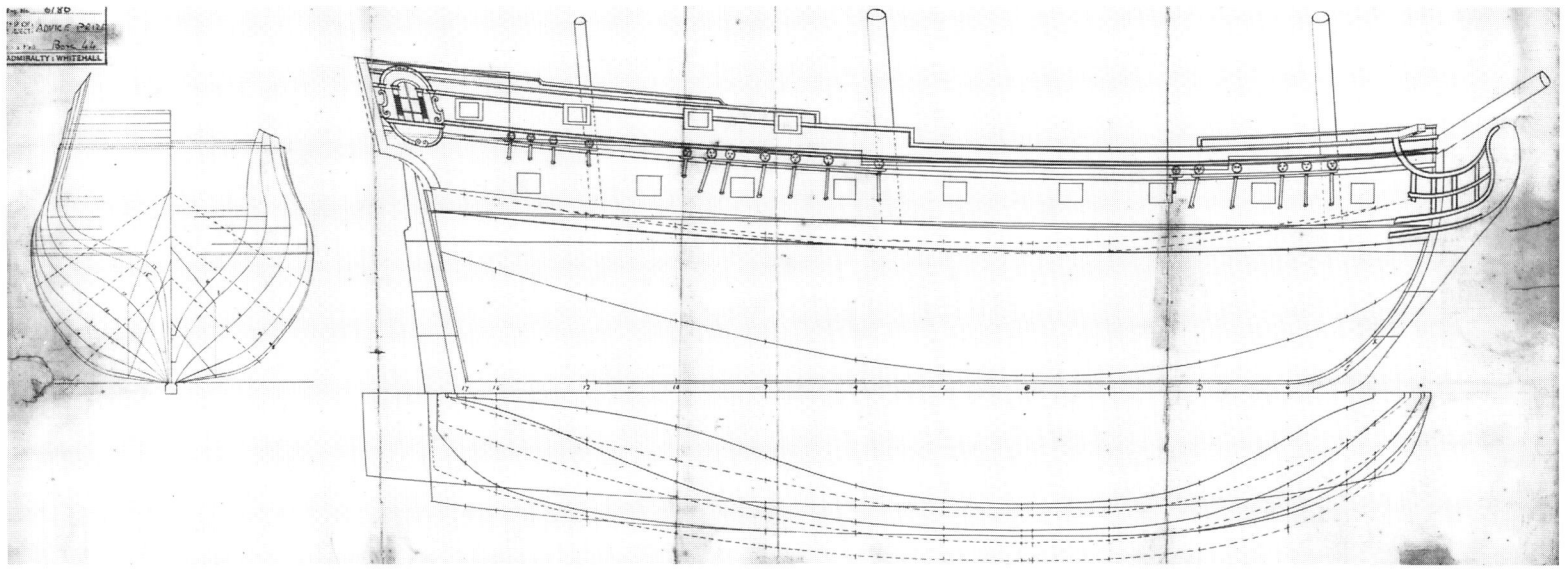

Lines plan (6186) of the 18 gun French built *Advice Prize* of 1704

Waterford.

**Glasgow** 1707-1719 (Scots *Royal Mary* 1696 Thames) 24.
Dimensions & tons: 92'6", 77' x 26'4" x 10'10". 284.
Men: 105/100. Guns: 24. Transferred under the Act of Union 1707; 1719 sold.

**Orford Prize** 1708-1709 (French [?] ____ ) 24.
Dimensions & tons: ____ , 74'11" x 26'8½" x 11'6". 283 34/94.
Men: 115. Guns: 24. Taken 21.10.1708; 27.5.1709 taken by two French 30 gun privateers.

**Fame** 1709-1710 (French ____ ) 24.
Dimensions & tons: 106', 88' x 26' x 11'. 316 24/94 [?].
Men: 100/115. Guns: 24 Taken 07.1709; 21.9.1710 retaken by French off Port Mahon.

## Small Sixth Rates, under 20 Guns

**Postillion Prize** 1701-1709 (French *Le Postillion* 1691 ____ ) 10.
Dimensions & tons: 65'4", 53'4" x 19'4" x 7'10". 105 47/94.
Men: 50/40/30. Guns: 1703 establishment: 10 sakers. Taken 1701 (or 09.1702) by *Worcester*; 7.5.1709 lost near Ostend.

NOTE: She was a French 'barque longue'.

**Rochester Prize** 1702-1712 (French *La Gracieuse* ____ ) 18.
Dimensions & tons: 87', 73' x 22'9" x 9'11". ____ .
Men: 85/70/50. Guns: 1703 establishment: 16 x 6, 2 minions. Taken 18.5.1702 by *Rochester*; 1712 sold.

**Chatham Prize** 1704-1708 (French *Le Mince* ____ ) Sixth Rate/advice boat 8.
Dimensions & tons: 53'4", 44' x 16'8" x 6'7". 65.
Men: 55. Guns: 8. Taken 03.1704 by *Chatham*; 1708 sold.

**Nottingham Prize** 1704-1706 (French [?] ____ ) 8.
Dimensions & tons: 46', 37'6" x 14'3" x 5'. 40 44/94.
Men: 34. Guns: 1703 establishment: 2 falcons, 2 sakers, 4 patteroes. Taken early 1704 by *Nottingham*; 1706 breakwater.

**Advice Prize** 1704-1712 (French ____ ) 18.
Dimensions & tons: 82', 63' x 24'6" x 10'7". 200.
Men: 100/90. Guns: 1703 establishment: 18 minions. Taken 19.6.1704 by *Advice*; 1712 sold.
PLANS: Lines.

**Worcester Prize** 1705-1708 (French ____ ) 14.
Dimensions & tons: 72'10", 60'6" x 20'10½" x 8'10". 140¼.
Men: 75/60/40. Guns: 12 x 4 + 2 minions. Taken 01.1705 by *Worcester*; 27.5.1708 taken by the French; 14.6.1708 retaken by a Dutch privateer and bought back; 6.10.1708 retaken by the French.

**Seaford Prize** 1708-1712 (French [?] ____ ) 12/10. Dimensions & tons: 62'7", 50'11¼" x 17'10½" x 7'4". 86 53/94.
Men: 70/60/50. Guns: 12/10.

Taken 04.1708 by *Seaford*; 1712 sold.

**Speedwell Prize** 1708-1712 (French [?] ____ ) 20?
Dimensions & tons: 75'4", 60' x 22'1" x 9'10½". 155 60/94.
Men: 100. Guns: ____ . Taken 4.10.1708 by *Speedwell*; 1712 sold.

**Jolly** 1709-1714 (building [?] Johnson, Blackwall [?]) 8 [or 14?].
Dimensions & tons: ____ , 61'10½" x 22'7¼" x 10'1". 168 15/94.
Men: 70. Guns: 8. Purchased from Johnson whilst building 1709 (?); 1714 sold.

**Shoreham Prize** 1709-1712 (French [?] privateer ____ , ) 12.
Dimensions & tons: 56'10½", 46'7" x 17'1¾" x 9'. 72 79/94.
Men: 70. Guns: 12. Taken 26.8.1709 by *Shoreham*; 1712 sold.

**Hare** 1709-1712 (French privateer ____ ) 10?
Dimensions & tons: 53'5", 43'11" x 15'4¼" x 7'1". 55 34/94.
Men: 50. Guns: ____ . Taken by *Speedwell* 8.9.1709; 1712 sold.

**Monk Prize** 1709-1712 (French privateer ____ ) 16?
Dimensions & tons: ____ , 59'4,' x 20'9" x 9'11". 135 80/94.
Men: 90. Guns: ____ . Taken 9.9.1709 by *Monck*; 1712 sold.

**Hind** 1709-1711 (French privateer ____ ) 16?
Dimensions & tons: ____ , 66'3" x 23'3" x 10'7". 190 40/94.
Men: 90. Guns: ____ . Taken 21.9.1709 by *Medway*; 29.11.1711 sunk 'coming into Poolbigg Harbour near Dublin'.

**Rose** 1709-1711 (mercantile pink *Anthelope* ____ ) 14.
Dimensions & tons: ____ , 62'2" x 21'5" x 9'7½". 151 63/94.
Men: 70. Guns: 14. Purchased 05.1709; 1711 sold.

**Diligence/Diligence Gally** 1709-1712 (mercantile *Queen Anne Gally* ____ ) 12.
Dimensions & tons: ____ , 57'1'" x 22'4½" x 10'2". 152.
Men: 50. Guns: 12. Purchased 23.5.1709 from Johnson (her builder?); 1712 sold.

**Success** 1709-1709/1710-1710 (mercantile *Swift Gally* ____ ) 10.
Dimensions & tons: ____ , 56'2½" x 19'3" x 8'10". 110 70/94.
Men: 50. Guns: ____ . Purchased 1.6.1709; 13.9.1709 taken about 14 leagues South South East from the Lands End by a French privateer of 30 guns; 02.1710 (new style) retaken by Dutch privateer and repurchased by order dated (?) 22.9.1710; 11.4.1710 (?) retaken by the French 70 leagues North North West from the Rock of Lisbon.

**Discovery Dogger** 1711-1712 (French [?] ____ ) 10.
Dimensions & tons: ____ , ____ x ____ x ____ , ____ .
Men: 80/50. Guns: 10 x ____ + 6 swivels. Taken by *Canterbury* 1711; 1722 sold.

**Whiteing** 1711-1712 (French [?] ____ ) 4.
Dimensions & tons: ____ , ____ x ____ x ____ . 45.
Men: 30. Guns: 4 x ____ + 2 swivels. Taken 06.1711 by *Winchester*; probably never registered in the RN; 1712 sold.

## Sloop

**Saint Antonio** 1700-1707 (pirate ____) 4. Sloop (probably North American type of mercantile sloop - single masted).
Dimensions & tons: 55', 42'6" x 17'2" x 7'4½". 67 5/94.
Men: 15/20/32. Guns: 4 x ____. Taken 04.1700 from 'Kidd ye pyrat'; before 24.8.1707 sunk at Jamaica.

## Fireships

**Earl Gally/Earl** 1702-1707? (____ ____) ____.
Dimensions & tons: ____, ____ x ____ x ____. ____.
Men: ____. Guns: ____. Presented to the Crown by the Government of Jamaica; 1704 entered into the Navy; last listed 1707 (?sold 25.2.1705/6?).
**Harman** 1702-1705 (____ ____) ____.
Dimensions & tons: ____, ____ x ____ x ____. ____.
Men: ____. Guns: ____. Presented to the Crown by the Government of Jamaica 02.1702; 02.1705 sunk as unserviceable at Jamaica.

## Storeships

**Elephant/Elephant Flyboat** 1705-1709 (French *L'Eléphant* 1701 Bayonne) storeship/Fifth Rate 12.
Dimensions & tons: 102'6", 86'7½" x 26'5¼" x 11'10". 322 5/94.
Men: 40/45. Guns: 12 x 6. French 'flûte' taken 25.2.1705; 1709 hulked [at Port Mahon; fate uncertain].

## Miscellaneous Unregistered Vessels

[ie. vessels found only in secondary listings, not in the primary Admiralty sources consulted; existence of some of these craft is doubtful.]

**Content** 1708-1715 (mercantile hoy *Content*? ____) storeship. 100 tons. Purchased 05.1708; 15.12.1715 sold.
**Cruizer** 1709-1712 (mercantile *Unity Gally* ____) Sixth Rate 12.
Dimensions & tons: ____, 56'7" x 20'3" x 10'1". 123.
Men: 60. Guns: 12. Purchased 1709 from Winder; 1712 sold.
**Fame** 1702-1703 there is a log for a vessel of this name and dates; no further information.
**Fortune** 1700-1702 Storeship noted as captured [?] 1700?; in service 1702; no further information; possibly a confusion with the 1699 storeship (see earlier).
**Grampus** 1707-1708 there is a log for a vessel of this name and dates.
**Griffin/Griffon** 1713 1714 (French *Le Griffon* 1704 Port Louis) Fifth rate 44. Taken off Finisterre 1713 (or 19.8.1712 new style); 1714 restored to the French at the conclusion of peace.
**Hare** bomb listed 1703-1704; no further information.
**Julian** fireship noted in 1704 (only by Pitcairn Jones); no further information.
**Jupiter** there is a log for an unidentified vessel of this name dated 1707-1708.
**Oxford Prize★** 1701?-1703 Fifth Rate, 330 tons; captured [?] 1701?; 1703 sold; but is this just a confusion with *Orford Prize*? see above.
**Plymouth Prize** 1709-1709 (French 44/flute/20 ?*La Dryade*? 1703 Le Havre) Sixth Rate 16.
Dimensions & tons: 56'6", ____ x 21'2" x 8'8". 134.
Men: 90. Guns: 16. Taken 19.7.1709; 21.12.1709 retaken by the French.
**Saint Joseph** 1704-1710 (mercantile [?] ____) hoy/Sixth Rate. 70 tons. Purchased Lisbon 24.11.1704; 04.1710 sold.
**Snake** 1711-1714 there is a log for a vessel of this name and dates.
**Thoulouse** 1711-1712 (French *Le Comte de Toulouse* 1703 Toulon) 50. Taken 2.12.1711 by *Stirling Castle* and *Hampton Court*; 1712 discarded.
**Trident Prize** 1705-1713 there is a log for a vessel of this name and dates but no further information.

## Hired Vessels

NOTE: This list is taken from copies held at the NMM of the *Admiralty Progress and Dimensions Book 1699-1715* at the PRO (the PRO ADM reference is not given), which gives annual listings of hired vessels. These indicate which individual warships each tender was attached to. With differing tonnages often given for different years there is always the possibility that one entry may cover more than one ship of the same name, equally well separate entries for vessels with different tonnages and/or type names may represent the same ship - and this is not taking the possibility of changes of name into account! Vessels marked by an asterisk [★] are not from the above lists, but noted in Pitcairn Jones or similar listings.

**Adventure** 1704-1705 Ketch (tender) 4. 60 tons. 8 men.
**Adventure** 1704-1705 Ketch (tender) 4. 80/90 tons. 8 men.
**Adventure** 1704-1706, 1708, 1711 Smack (tender) 6. 45/46 tons. 6 men.
**Amity** 1703-1705 Timber transport 8. 18 men.
**Amity** 1706-1707 Hoy (tender) 4. 60/55 tons. 8 men.
**Ann** 1703-1704 Timber transport 12. 30 men.
**Ann/Anne** 1704-1708 Yacht (tender) 4. 70/65 tons. 8 men.
**Ann** 1705 Hoy (tender) 4. 60 tons. 8 men.
**Ann** 1705 Ketch (tender) 4. 70 tons. 8 men.
**Ann and Francis** 1705-1707, 1711 Ketch (tender) 4. 60/70/51 tons. 8 men.
**Ann of Sandwich** 1711 Smack (tender) 2. 40 tons. 6 men.
**Anthelope** 1702-1708 Hospital ship 24. 550 tons. 70 men
**Benjamin** 1701-1703 Storeship 20. 50 men.
**Benjamin** 1704 Ketch (tender) 4. 55 tons. 8 men.
**Benjamin** 1705 Ketch (tender) 4. 80 tons. 8 men.
**Blackwell Hall** 1700-1701 Sloop 4 + 4 swivels. 12 men.
**Bonadventure** 1705 Hoy (tender) 4. 80 tons. 6 men.
**Bristol** 1703-1704 Timber transport 8. 35 men.
**Burchett** 1708, 1711. Hoy (tender) 2. 48/40 tons. 6 men.
**Burchett** 1708-1709. Sloop 4 + 8 swivels. 50 tons; 6.2.1709 taken by a French privateer.
**Canada** 1711 Hospital ship 26. 65 men.
**Canterbury** 1711 Hoy (tender) 4. 54 tons. 8 men.
**Cholmondy** 1704 Pink (tender) 4. 90 tons. 8 men.
**Churchill** 1705-1708 Smack (tender) 2. 40/39/38 tons. 6 men.
**Content** 1705-1706 Smack (tender) 2. 40 tons. 6 men.
**Coronation** (*Coronacon*) 1705-1707 Pink (tender) 4. 100/88 tons. 8 men.
**Daniel and Christopher** 1706 Smack (tender) 2. 40 tons. 6 men.
**Delight Frigat** 1701-1702 Storeship 16. 32 men.
**Dispatch** 1711 Transport (hired in America) 24, 254 tons.
**Dorothy** 1707-1708 Pink (tender). 8 men.
**Elizabeth** 1705 Hoy (tender) 2. 45 tons. 6 men.
**Elizabeth** 1708 Pink (tender) 4. 100 tons. 8 men.
**Endeavour** 1704 Ketch (tender) 4. 103 tons. 8 men.
**Endeavour** 1704-1706 Smack (tender) 2. 35 tons. 6 men.
**Endeavour** 1705-1707 Pink (tender) 4. 65/56/ 70 tons. 8 men.
**Endeavour** 1711 Ketch (tender) 4. 60 tons. 8 men.
**Fordwich** 1704 Ketch (tender) 4. 90 tons. 4 men.
**Fortune** 1706 Pink (tender) 4. 59 tons. 8 men.
**Fortune** 1707 Pink (tender), 110 tons.
**Francis and Lewis** (or **Francis and Lucy**) 1704-1706 Brigantine (tender) 4. 80 tons. 8 men.
**Friends Adventure** 1705-1706 Hoy (tender) 4. 80 tons. 8 men.
**Friends Adventure** 1711 Ketch (tender) 2. 47 tons. 6 men.
**Friends Goodwill** 1705 Pink (tender) 4. 60 tons. 8 men.
**Friends Goodwill** 1707 Hoy (tender) 4. 60 tons. 8 men.
**Friendship** 1702-1704 Timber transport 6. 20 men.
**Friendship** 1704-1705 Ketch (tender) 4. 80 tons. 8 men.
**Friendship** 1706-1708, 1711. Hoy (tender) 2. 50/60/40 tons. 6 men.
**George** 1705-1708, 1711 Hoy (tender) 2. 48/46/40 tons. 6 men.
**George** 1706 Smack (tender) 2. 40 tons. 6 men.
**George and Ann** 1704-1705 Smack (tender) 2. 40 tons. 6 men.
**George and Mary** 1704-1706 Hoy (tender) 2. 40/70 tons. 8 men.
**Gloucester** 1704-1707 Yacht (tender) 4. 70 tons. 8 men.
**Goodwill** 1705, 1707-1708, 1711 Smack (tender) 2. 40 tons. 6 men.
**Gravesend** 1704-1705 Yacht (tender) 4. 70 tons. 8 men.
**Henry and Jane** 1704 Smack (tender) 2. 40 tons. 6 men.
**Henry and Richard** 1711 Hoy (tender) 2. 48 tons. 6 men.
**Henry and Susann** 1704 Smack (tender) 2. 36 tons. 6 men.
**Holygoland** 1704-1706 Hoy (tender) 4. 100 tons. 8 men.

*Hope* 1704 Storeship 6, 17 men.
*Hopewell* 1704 Ketch (tender) 4. 80 tons. 8 men.
*Hopewell* 1704, 1706 Smack (tender) 2. 40/48 tons. 6 men.
*Hurley/Hurly* 1704-1706 Smack (tender) 2. 40/46 tons. 6 men.
*Jacob and John* 1704 Hoy (tender) 4. 55 tons. 8 men.
*James and Mary* 1703-1704 Timber transport 6. 17 men.
*Jane* 1706-1708, 1711 Ketch (tender) 4. 85/92/82 tons. 86 men.
*Jefferies* 1701-1706 Hospital ship 20. 510 tons. 60 men.
*Johanna* 1711 Pink (tender) 4, 96 tons. 8 men.
*John* 1705 Ketch 4, 70 tons. 8 men.
*John and Ann* 1705-1706, 1708, 1711 Smack (tender) 2. 35 tons, 6 men.
*John and Ann* 1711 Hoy (tender) 2. 46 tons. 6 men.
*John and Bridget* 1704-1708, 1711 Yacht (tender) 2. 40/41/39 tons. 6 men.
*John and Elizabeth* 1705-1706 Smack (tender) 2. 40/36 tons. 6 men.
*John and Elizabeth* 1711 Ketch (tender) 2. 52 tons. 8 men.
*John and Francis* 1711 Hoy (tender) 4. 56 tons. 8 men.
*John and Jane* 1706-1708 Ship (tender) 4. 100 tons. 8 men.
*John and Joseph* 1705 Hoy (tender) 4. 67 tons. 8 men.
*John and Mary* 1705 Ketch (tender) 2. 36 tons. 6 men.
*John and Mary* 1705-1706 Pink (tender) 4. 100 tons. 8 men.
*John and Mary* 1706, 1708 Smack (tender) 2. 40 tons. 6 men.
*John and Mary* 1707 Smack (tender) 2. 38 tons. 6 men.
*John and Mary* 1711 Hoy (tender) 4. 64 tons. 8 men.
*John and Robert* 1707 Hoy (tender) 2. 20 tons. 6 men.
*John and Sarah* 1706 Hoy (tender) 4. 70 tons. 6 men.
*John and Sarah* 1707-1708 Smack (tender).
*John and Susann* 1703-1704 Timber transport 6. 20 men.
*John and Susan* 1705-1706 Smack (tender) 2. 40 tons. 6 men.
*Jumper* 1705 Hoy (tender) 4. 75 tons. 8 men.
*Lawrell* 1704-1705 Ketch (tender) 4. 70 tons. 8 men.
*Leak/Leake* 1705-1707 Pink (tender) 4. 80/60/56 tons. 8 men.
*Leak*★ 1709-1712 Hospital ship.
*Lemon Flower* 1705 Pink (tender) 4. 100 tons. 8 men.
*Little Sarah* 1708 Ship (tender) 4. 80 tons. 8 men.
*London* 1704-1706, 1711 Yacht (tender) 4. 70/75 tons. 8 men.
*Loyal Merchant* 1701-1703 Storeship 24. 50 men.
*Lydall* 1705 Pink (tender) 4. 80 tons. 8 men.
*Maidstone* 1705 Galley (tender) 4. 80 tons. 8 men.
*Margate* 1706 Pink (tender) 4. 76 tons. 8 men.
*Martha*★ 1707-1710 hired hospital ship, 70 men + 14 attendants?
*Mary* 1704 Sloop (tender) 4. 80 tons. 8 men; 1704 burnt.
*Mary* 1704-1705 Yacht (tender) 4. 70 tons. 8 men.
*Mary* 1705 Pink (tender) 4. 100 tons. 8 men.
*Mary* 1706 Dogger (tender) 4. 8 men.
*Mary* 1706 Hoy (tender) 2. 19 tons. 6 men.
*Mary* 1706, 1708 Yacht (tender) 4. 60 tons. 8 men.
*Mary* 1707-1708 Ketch (tender) 4. 65/60 tons. 8 men.
*Mary* 1711 Pink (tender) 4. 61 tons. 8 men.
*Mary and Sarah* 1703-1704 Storeship 4. 13 men.
*Mathews* 1706-1708 Hospital ship 24. 446 tons. 60 men.
*Mayflower* 1705, 1709-1711 Pink (tender) 4. 80 tons. 8 men.
*Mudd* 1704-1705 Pink (tender) 4. 80 tons. 8 men.
*Muscovia Merchant* 1701-1703 Storeship. 45 men; 1703 taken by the French.
*Owner's Adventure* 1704-1705 Smack (tender) 2. 35/40 tons. 6 men.
*Owners Goodwill* 1706-1708 Hoy (tender) 4. 60 tons. 8 men.
*Owners Goodwill* 1706 Ship (tender) 4. 70 tons. 8 men.
*Peace Flyboat* 1703-1704 Timber transport 6. 16 men; Commanded by Captain Woods Rogers – presumably the famous privateer captain.
*Pelican* 1706 Ketch (tender) 4. 100 tons. 8 men.
*Pembroke* 1709, 1713 Hospital ship. 74 men.
*Peter* 1706 Hoy (tender) 2. 28 tons. 6 men.
*Phenix* 1704 Storeship? 6. 28 men.
*Plow* 1706 Hoy (tender) 2. 48 tons. 6 men.
*Postillion* 1709-1711 Sloop (tender) 2. 37 tons. 6 men.
*Princess Ann* 1701-1706 Hospital ship 30/22. 484 tons. 70/80 men.
*Prosperous* 1704-1706 Smack (tender) 2. 30 tons. 6 men.
*Prosperous* 1711 Hoy (tender) 2. 49 tons. 6 men.
*Prosperous* 1711 Hoy (tender ) 4. 62 tons. 8 men.

*Providence* 1704-1706, 1711. Pink (tender) 4. 80/85 tons. 8 men.
*Richard and John* 1704 Pink (tender) 4. 90 tons. 8 men.
*Richard and Margate* 1705 Ketch (tender) 4. 100 tons. 8 men.
*Rochester* 1705 Ketch (tender) 2. 40 tons. 6 men.
*Rochester* 1706 Ketch (tender) 2. 50 tons. 6 men.
*Rochester* 1707 Ketch (tender). 60 tons.
*Samuell* 1704 Pink (tender) 4. 51 tons. 8 men.
*Sandwich* 1711 Pink (tender) 4. 80 tons. 8 men.
*Sarah* 1704-1705 Ketch (tender) 4. 90/100 tons. 8 men.
*Sarah* 1704-1711 Pink (tender) 4. 96/100 tons. 8 men.
*Sarah and Betty* 1701-1703 Hospital ship 20. 45 men.
*Seaflower* 1704-1706 Smack (tender) 2. 36/35 tons. 6 men.
*Smirna Factor* 1701-1705 Hospital ship 24. 380 tons. 45 men.
*Society* 1709-1711 Pink (tender) 4. 59 tons. 8 men.
*Southampton* 1704 Sloop (tender) 2. 47 tons. 6 men.
*Southampton* 1708 Galley (tender ) 4. 70 tons. 8 men.
*Speedwell* 1706-1707 Hoy (tender) 2. 43 tons. 6 men.
*Success* 1704-1708, 1711 Smack (tender) 2. 36 tons. 6 men.
*Success* 1711 Pink (tender) 4. 100 tons. 8 men.
*Supply* 1705 Bark (tender) 2. 46/40 tons. 6 men.
*Susanna* 1704-1707 Smack (tender) 2. 40 tons. 6 men.
*Syam* 1701-1703 Hospital ship 24. 340 tons. 50 men.
*Theophilus* 1705 Pink (tender) 4. 70 tons. 8 men.
*Thomas* 1704-1705 Ketch (tender) 2. 40 tons. 6 men.
*Thomas and Elizabeth* 1711 Pink (tender) 4. 103 tons. 8 men.
*Thomas and Hester* 1704-1706, 1711 Smack (tender) 2. 36/38/35 tons. 6 men.
*Thomas and Katherine* 1705, 1707 Smack (tender) 2. 40/38 tons. 6 men.
*Thomas and Mary* 1705-1708 Hoy (tender) 2. 50 tons; 6 men.
*Thomas and Mary* 1708 Smack (tender).
*Three Brothers*★ 1704-1706 Smack (tender) 2. 35/38 tons. 6 men.
*Truelove* 1704-1708 Smack (tender) 2. 30/28 tons. 6 men.
*Two Brothers* 1706 Dogger (tender) 4. 90 tons. 8 men.
*Unity* 1704-1706 Ketch (tender) 4. 70/73/80 tons. 8 men.
*Unity* 1704-1706 Smack (tender) 2. 38 tons. 6 men.
*Welcome* 1704 Smack (tender) 2. 37 tons. 6 men.
*Welcome* 1704, 1708 Smack (tender) 2. 35 tons. 6 men.
*Welcome* 1705-1706 Smack (tender) 2. 37 tons. 6 men.
*William and Ann* 1704-1706 Pink (tender) 4. 80 tons. 8 men.
*William and Elizabeth* 1707, 1711 Ketch (tender) 4. 100/102 tons. 8 men.
*William and James* 1705 Hoy (tender) 2. 55 tons. 6 men.
*William and Margate* 1706 Smack (tender) 2. 35 tons. 6 men.
*William and Mary* 1704, 1707-1708, 1711 Smack (tender) 2. 38/45 tons. 6 men.
*William and Mary* 1705-1706, 1708, 1711 Hoy (tender) 2. 50/46 tons. 6 men.
*William and Mary* 1706 Smack (tender) 4. 100 tons. 8 men.
*Woolwich* 1704 Smack (tender) 2. 38 tons. 6 men.
*Woolwich* 1705-1708, 1711 Smack (tender) 2. 40/38/35 tons. 6 men.

# VESSELS TAKEN OR PURCHASED 1714-1738

## Bomb Vessel

***Thunder*** 1718-1734 (Spanish _____ ) 6.
Dimensions & tons: 82', 63'5½" x 27'5" x 10'7½". 253 68/94.
Men: 40. Guns: 6 + 6 swivels (+ one mortar?). Taken off Sicily by Byng, 10.1718; 1734 BU.

## Miscellaneous Unregistered Vessels 1714-1738

(not listed in the primary Admiralty sources consulted.)

***Bonetta*** in service 1718; 28.8.1719 sold; type uncertain.
***Brilliant*** 1729 Sloop in service in this year (only noted by Colledge?)
***Cruizer*** 1721-1724 (Spanish _____ ) sloop 8. Taken 1721; _____ 1724 foundered.
***Discovery*** 17____-1719 (mercantile _____ ) discovery vessel. Purchased (?);

\_\_\_\_ 1719 lost in the Arctic.

**Elizabeth** Smack; 1721-1722 quarantine guard at Plymouth (only noted Pitcairn Jones).

**Hester** Smack; 1720-1721 quarantine guard at Yarmouth (only noted by Pitcairn Jones).

**Swift's Prize** (Spanish?) Fireship taken 1717 (only noted by Pitcairn Jones).

**Thetis** Storeship 22 guns 720 tons, Plymouth 1717? (only noted by Colledge).

**Thomas and Mary** Smack; 1721 quarantine guard at Dover (only noted by Pitcairn Jones).

**Unity** Smack; 1721-1722 quarantine guard at Sandgate (only noted by Pitcairn Jones).

# Chapter 3: Wars of Jenkins' Ear & of Austrian Succession 1739-1748

Greed for the supposed riches of Spain's empire caused Britain to use colonial and trading friction as an excuse to go to war with Spain in 1739. By the beginning of 1744 the maritime struggle had spread to include France. In the preceding period both France and Spain had not only rebuilt their navies but also produced new and more powerful types of warship. The large Spanish 70 gun ship *Princesa* was captured in 1740, but it took three British ships of nominally the same power though considerably smaller size to subdue her. The shock of meeting such powerful vessels was repeated when the British met the new French 74 gun ships in battle. The British had no equivalent at all to the other new type of vessel the French had produced, the 'true' frigate. Examples of both types of ship were captured during this war and would eventually be used as models for the design of British equivalents. The same would happen with French corvettes, one of which, the *Amazon* (ex *Panthère*) would still be interesting British designers half a century later. The beginnings of the war with Spain, and to a lesser extent that with France, are marked by the purchase of merchantmen to be converted to sloops and fireships and achieve a rapid enhancement of numbers of such smaller vessels before the emergency orders for purpose built ships could be completed.

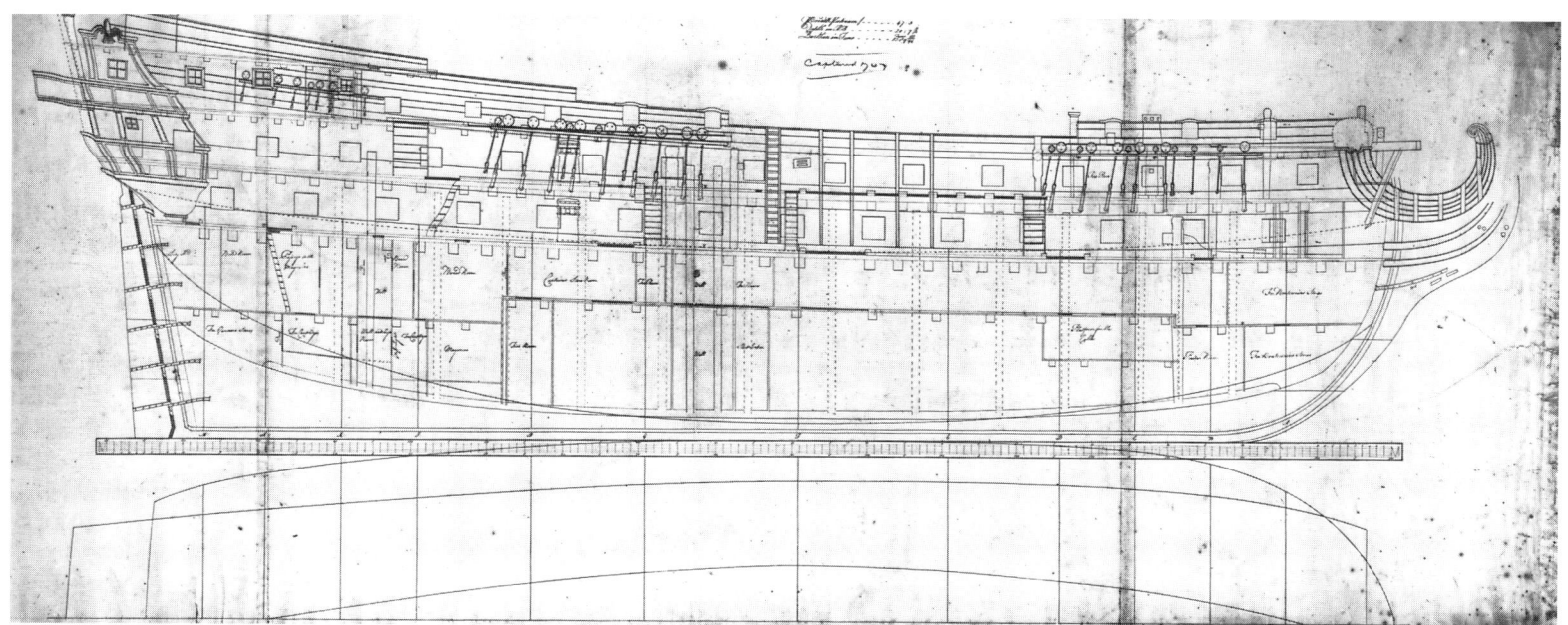

Part of lines and profile plan (omitting body plan of plan number 933) of the French *Terrible,* 74 guns, captured in 1747

## 74 Gun Ships (Third Rates)

***Invincible*** 1747-1758 (French *L'Invincible* 1744 Rochefort; Morineau design).
Dimensions & tons: 171'3", 139'0½" x 49'3" x 21'3". 1793.
Men: 700. Guns: GD 28 x 32, UD 30 x 18 (replaced by 24s in 1755), QD & Fc 16 x 9. Taken 3.5.1747 by Anson's fleet off Finisterre; 19.2.1758 wrecked on the Dean Sand near Portsmouth; the wreck was located by Arthur Mack and has been excavated by a team under Commander John Bingeman. [See Brian Lavery's *The Royal Navy's First Invincible*, Portsmouth 1988.]

***Monarch*** 1747-1760 (French *Le Monarque* 1747 Brest).
Dimensions & tons: 175'3", 149'10" x 47'2½" x 20'1½". 1775 48/94.
Men: 650. Guns: GD 28 x 32, UD 30 x 18, QD 10 x 9, Fc 6 x 9. Taken 14.10.1747 off Finisterre by Hawke; 1760 sold.
PLANS: Lines, profile & decorations/head & stern decorations.

***Terrible*** 1747-1763 (French *Le Terrible* 1739 Toulon).
Dimensions & Tons: 164'3", 133'11" x 47'3" x 20'7½". 1589 83/94.
Men: 650. Guns: GD 28 x 32, UD 30 x 18, QD 10 x 9, Fc 6 x 9. Taken 14.10.1747 off Finisterre by Hawke; 1763 BU.
PLANS: Lines & profile/decks.

***Magnanime*** 1748-1775 (French *Le Magnanime* 1744 Rochefort; Geslain design) 70/74.
Dimensions & Tons: 173'7", 146'2" x 50' x 21'7". 1943 66/94.
Men: 700 (1755: 650). Guns: GD 28 x 32 (1755: x 24), UD 30 x 18, QD 10 x 9, Fc 2 x 9. Taken 31.1.1748 by *Nottingham, Portland*; 1775 BU.
PLANS: Lines, profile & decorations.

## Small Line of Battle Ships

***Princesa*** 1740-1761 (Spanish *Princesa* 1730 Santander [Guarnizo]) Third Rate 70.
Dimensions & Tons: 165'4", 130'3" x 49'9" x 22'1". 1714.
Men: 650. Guns: GD 28 x 32, UD 28 x 18, QD 10 x 9, Fc 4 x 9. Taken 8.4.1740 by *Kent, Lenox, Orford*; 1761 hulked [sheer hulk at Portsmouth; 1784 sold].
NOTE: A large and powerful ship, as was shown by the fact that it took three British two-deckers to capture her. She was well thought of in the RN, and played a major part in the evolution of the British 74 gun ship.
PLANS: Lines & profile.

***Vigilant*** 1745-1759 (French *Le Vigilant* 1745 Brest) Fourth Rate 58.
Dimensions & Tons: 153'9", 130' x 43'8" x 18'2½". 1318 48/94.
Men: ____ . Guns: GD 24 x 24, UD 24 x 12, QD 8 x 6, Fc 2 x 6. French 64 taken by Warren's squadron off Cape Breton in 05.1745; 1759 sold.

***Mars*** 1746-1755 (French *Le Mars* 1740 Brest) Third Rate 64.
Dimensions & tons: 159'3", 128'10⅜" x 44'9¾" x 18'4". 1374.
Men: 470. Guns: GD 26 x 24, UD 28 x 12, QD 8 x 6, Fc 2 x 6. Taken 10.1746 by *Nottingham*; 06.1755 wrecked going into Halifax harbour.

***Intrepid*** 1747-1765 (French *Le Sérieux* 1740 Toulon) Third Rate 64.
Dimensions & tons: 151'6", 124'11½" x 44' x 19'4½". 1300.

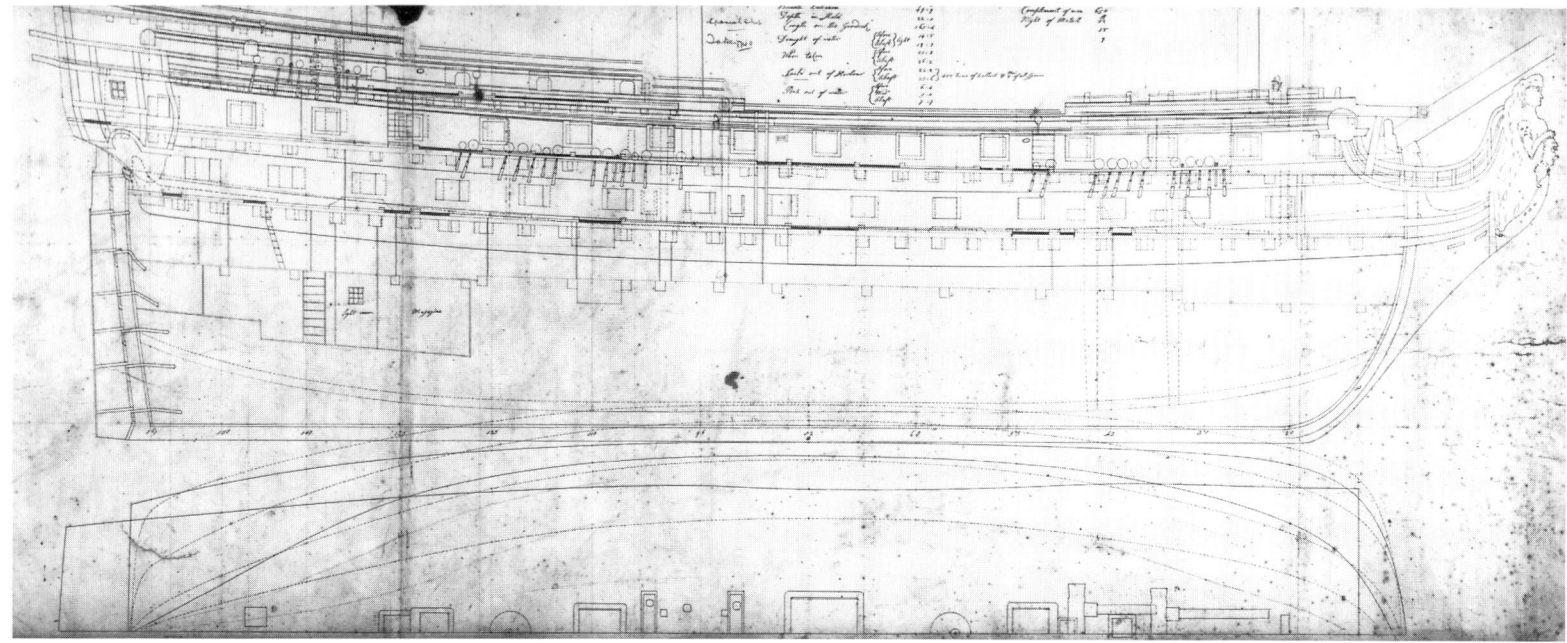

Part of lines, profile and main deck plan (1061, omitting body plan) of the large Spanish 70 gun ship *Princesa* taken in 1740

Men: 470 (1757: 420). Guns: 64 (1757: 60). Taken 3.5.1747 by Anson off Finisterre; 1757 reduced to Fourth Rate 60; 1765 BU.

***Fougeaux*** 1747-1759 (French *Le Fougeux* 1747 Brest) Third Rate 64.
 Dimensions & tons: 159'11", 135' x 44'2½" x 18'6¼". 1402 82/94.
 Men: 520. Guns: GD 26 x 32, UD 28 x 18, QD 6 x 12, Fc 4 x 12. Taken by Hawke off Finisterre 14.10.1747; 1759 BU.

PLANS: [incomplete lines plan in the Danish Rigsarkivet].

***Trident*** 1747-1763 (French *Le Trident* 1742 Toulon) Third Rate 64.
 Dimensions & tons: 151'6", 123'7" x 43'9" x 19'8". 1258.
 Men: 520. Guns: GD 26 x 32, UD 28 x 18, QD 6 x 9, Fc 4 x 9. (1755: GD 26 x 24, UD 28 x 12, QD 6 x 6, Fc 4 x 6). Taken by Hawke off Finisterre 14.10.1747; 1763 sold.

## Small Two-deckers

***Portlands Prize*** 1746-1749 (French *L'Auguste* 1740 Brest) Fourth Rate 50.
 Dimensions & tons: 134'2", 109'5" x 38'7" x 15'3½". 866 50/94.
 Men: 300. Guns: GD 22 x 18, UD 24 x 9, QD 4 x 6. Taken 9.2.1746 by *Portland*; 1749 sold.

***Glory*** 1747-1763 (French *La Gloire* 1727 Le Havre, 1740-1742 'Refondu' ['Rebuilt'] ____ ) Fifth Rate 44.
 Dimensions & tons: 130'7", 108'10" x 35'11½" x 15'7½". 748 48/94.
 Men: 280. Guns: GD 22 x 18, UD 22 x 9 (1757: GD 20 x 18, UD 10 x 9?). Taken by Anson off Finisterre 3.5.1747; 1766 sold.

***Isis*** 1747-1766 (French *Le Diamant* 1730-1733 Toulon) Fourth Rate 50.
 Dimensions & tons: 142'3", 117'3" x 40'3¾" x 17'2". 1013.
 Men: ____ . Guns: GD 24 x 24, UD 24 x 9, QD 2 x 6. Taken by Anson off Finisterre 3.5.1747; 1766 sold.

***Jason*** (French 52 *Le Jason* 1724 Le Havre, 1744-1745 'Refondu' ['Rebuilt'] ____ ) Fourth Rate 44/40.
 Dimensions & tons: 131'3", 108'4" x 37'6" x 16'0½". 810.
 Men: 280 (1757:200). Guns: GD 20 x 18, UD 20 x 9, QD 4 x 6 (1757: GD 20 x 18, UD 10 x 9). Taken by Anson off Finisterre 3.5.1747; 1763 Sold.

## 12 pounder Frigate (Fifth Rate)

***Ambuscade*** 1746-1762 (French *L'Embuscade* 1745 Le Havre) 40.
 Dimensions & tons: 132'6", 107'5½" x 36' x 10'6". 740 72/94.
 Men: 250. Guns: UD 26 x 12, QD 10 x 6, Fc 4 x 6. Taken 21.4.1746 by *Defiance*; 1762 sold.

PLANS: Lines, profile, lower & upper decks/lines. As fitted: Lines, profile & decorations.

## 9 pounder Frigates (Sixth Rates)

***Medway's Prize*** 1745-1750 (French *Faonette* or *Fauvette/Fauvrette* 1743 [?] ____ ) 30.
 Dimensions & tons: 127'7", 109' x 35'1" x 17'. 744 10/94.
 Men: 200. Guns: 30. Taken by *Medway* in the East Indies 01.1745; 1750 sold.

NOTE: Though described by her captors as 'a fine ship ..... capable of mounting 44 guns, is a strong well built ship, and with little expense may be made a compleat 44 gun ship' she was found on survey in Britain to be a 'weak ship' and was sold shortly after.

***Bellona*** 1747-1748 (French privateer *La Bellone* 1745 Nantes) 30.
 Dimensions & tons: 112'3½", 92'3" x 33'2½" x 10'9". 541.
 Men: 220. Guns: UD 24 x 9, QD 4 x 4, Fc 2 x 4. Taken 2.2.1747 by *Nottingham* and others; 1748 sold.

***Ranger*** 1747-1749 (French privateer *'Two Crowns'* = ? *Deux Couronnes*?) 30.
 Dimensions & tons: 122'5½", 98'5" x 34'11½" x 14'0¼". 639.
 Men: 220. Guns: UD 24 x 9, QD 4 x 6, Fc 2 x 6 + 20 swivels. Taken 5.5.1747 by *Gloucester*; 1749 sold.

***Renown*** [ex *Fame*] 1747-1771 (French *La Renommée* 1744 Brest) 30.
 Dimensions & tons: 126'2", 103'7" x 34'10" x 11'8". 694.
 Men: 220. Guns: UD 24 x 9, QD 6 x 4. Taken 27.9.1747 by *Dover*; [reclassed as a Fifth Rate on 22.4.1757 according to Rif Winfield]; 1771 BU.

PLANS: Lines/lines.

## 'Corvettes' (Sixth Rates)

***Rippon's Prize*** 1744-1747 (French or Spanish *Conde de Chincan* [?] ____ ) 20.
 Dimensions & tons: 105'6", 83'8" x 29'10" x 8'9". 396.
 Men: 150. Guns: 20. Taken 1744 by *Rippon* and purchased at Jamaica; 1747 condemned.

***Grand Turk*** 1745-1749 (French privateer *Le Grand Turque* 1745 ____ ) 22.
 Dimensions & tons: 100'9", 81'10¼" x 29' x 9'5". 366 15/94.
 Men: 160. Guns: 20 x 9, 2 x 3. Taken 26.5.1745 by *Captain*; 1749 sold.

***Amazon*** 1745-1763 (French *La Panthère* 1744 Brest) 20/22.
 Dimensions & tons: 115'6", 92' x 31' x 10'2". 481 15/94.
 Men: 160. Guns: 20/22. Taken 07.1745 by *Monmouth*; 1763 sold.

NOTE: Half a century after her capture her design was revived for large classes of sloops and fireships.

PLANS: Lines, profile & decorations/lines & profile.

***Lys*** 1745-1749 (French privateer *Lys* ____ ) 24.
 Dimensions & tons: 105'9", 82' x 29' x 10'9½". 366.

NMM Photo Negative A.3608 of a contemporary print showing three prizes taken in 1747, the Spanish *Glorioso*, 74 (see the note on her under 'unregistered vessels' towards the end of this chapter) and the French small two-deckers *Jason* 44 and *Gloire* (*Glory* in RN service) 44

Men: 160. Guns: as French privateer: 2 x 9, 22 x 6, 8 x 3, 5 swivels (British armament uncertain). Taken 12.1745 by *Hampton Court*; 1749 sold.

PLANS: [Lines in the Swedish Chapman Collection].

***Richmond*** 1746-1749 (French *Le Dauphin* 1743 ____ ) 24.
Dimensions & tons: 102', 80' x 30'6" x 12'6". 396.
Men: 120. Guns: 24. Taken 01.1746 in the Leeward Islands; 1749 sold.

***Inverness*** 1746-1750 (French privateer *Le Duc de Chatres* ____ , ) 22.
Dimensions & tons: 104'7", 82' x 28'6" x 10'5". 354.
Men: 160. Guns: UD 20 x 9, QD 2 x 3. Taken 18.1.1746 by *Edinburgh*; 1750 BU.

***Nightingale*** 1746-1770 (building by Bird, Rotherhithe) 24.
Dimensions & tons: 113'3", 94'5" x 32'2¾" x 11'1". 522.
Men: 160. Guns: 24. Purchased 18.6.1746 from Bird whilst building;; launched 6.10.1746; 1770 hulked [at Woolwich as receiving ship for impressed men; 1776 to the Tower as a receiving and hospital ship; 1783 BU].

***Margate*** 1746-1749 (French privateer *Le Léopard* ____ ) 24.
Dimensions & tons: 107'8", 86'7½" x 30'10" x 13'9". 438.
Men: 160. Guns: 22 x short 9, 2 x 4. Taken 27.10.1746 by *Windsor*, 1749 sold.

## Ship Sloops

***Weazle*** 1745-1779 (building Taylor & Randall, Rotherhithe) 16.
Dimensions & tons: 94'6¾", 76'4⅜" x 27'6½" x 12'. 308.
Men: 125/120. Guns: 16 x 6 + 16/14 swivels. Purchased whilst building as a two masted sloop 04.1745; launched 22.5.1745; completed with three masts as a ship sloop; 13.1.1779 taken by the French *Boudeuse* in the West Indies [taken into Guadeloupe].

PLANS: As fitted: Lines & profile/lines, profile & upper deck. 1749: decks. 1770: decks.

***Porcupine*** 1746-1763 (building Taylor, Rotherhithe) 16.
Dimensions & tons: 94'4¼", 76'3⅞" x 27'9⅝" x 12'0¾". 314.
Men: 110. Guns: 16 x 6 + 14 swivels. Purchased whilst building 07.1746; launched 20.9.1746 (probably converted from two to three masts whilst fitting out); 1763 sold.

## Two Masted Sloops

***Saltash*** 1741-1742 (building Bird, Deptford) 8.
Dimensions & tons: 89', 71'3⅝" x 24'1⅝" x 7'11½". 220 86/94.
Men: 90. Guns: 8 x 4 + 12 swivels. Purchased on the stocks 28.8.1741; launched 3.9.1741; 18.4.1742 wrecked on the coast of Portugal whilst chasing a privateer.

***Ferret*** 1743-1757 (building Bird, Rotherhithe) 14.
Dimensions & tons: 88'4½", 75'2¾" x 25'3¼" x 6'9½". 255 51/94.
Men: 125. Guns: 14 x 4 + 14 swivels. Purchased whilst on the stocks 6.4.1743; launched 10.5.1743; 1755 fitted as a ship sloop; 24.9.1757 foundered in a hurricane off Halifax.

***Fame*** 1744-1745 (mercantile [?] ____ ) 12.
Dimensions & tons: ____ , ____ x ____ x ____ . 272.
Men: 80. Guns: 12 x 3 + 20 swivels. Purchased at Antigua 1744; 22.7.1745 foundered in the Atlantic on passage from the West Indies to England.

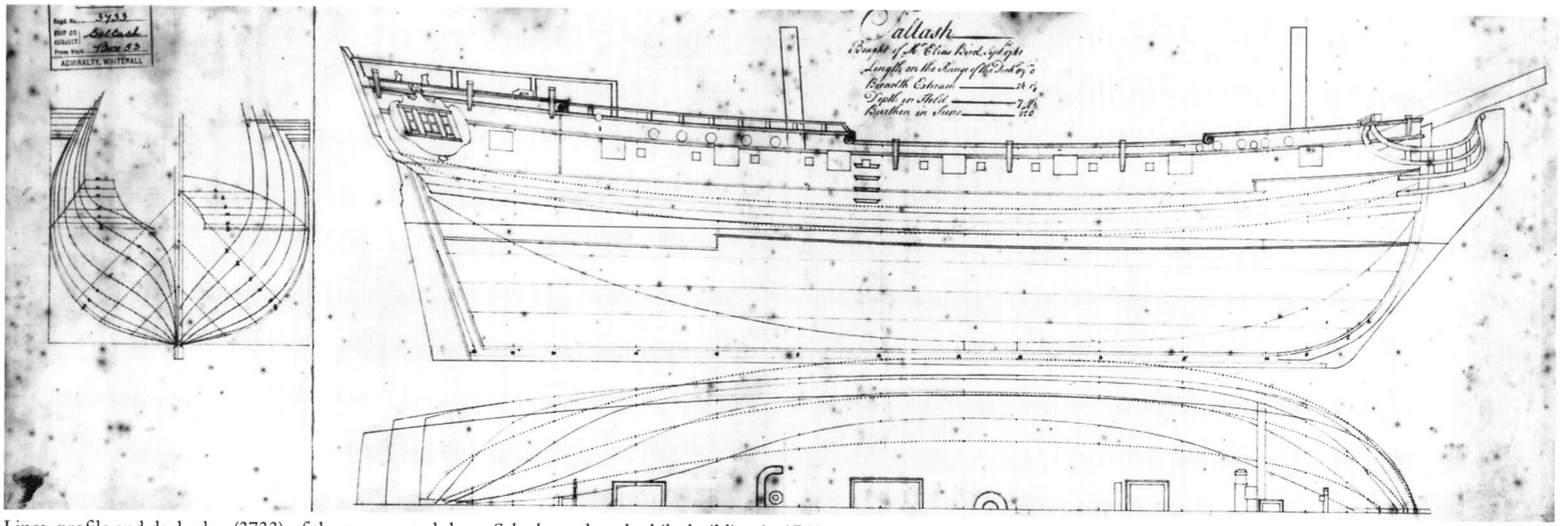

Lines, profile and deck plan (3733) of the two masted sloop *Saltash* purchased whilst building in 1741

***Shoreham's Prize*** 1746-1747 (Spanish privateer . ____ Bilbao) 10 snow.
Dimensions & tons: ____ , ____ x ____ x ____ . ____ .
Men: 100. Guns: 10 x ____ + 18 swivels. Taken 24.4.1746 by small vessel fitted out as a tender by *Shoreham;* 7.6.1747 sunk in Oporto harbour.

## Sloops, Rig Uncertain

***Deptford's Prize*** 1740-1744 (Spanish ____ ) 8 probably two masted.
Dimensions & tons: 73'7", 59'8" x 21'6" x ____ . 146 69/94.
Men: 60/70. Guns: 8 x 4 + 10 swivels. Taken 23.5.1740 by *Deptford;* 1744 sold.

***Pembroke's Prize*** 1740-1744 (Spanish ____ ) 8 probably two masted.
Dimensions & tons: 79'10", 65'10" x 23'8" x 10'. 196 13/94.
Men: 80. Guns: 8 x 4 + 12 swivels. Taken 09.1740 by *Pembroke*; 1744 sold.

***Rupert's Prize*** 1741-1743 (French or Spanish ____ ) 14.
Dimensions & tons: 71', 52'11¼" x 21'5½" x 9'8". 142 36/94.
Men: 100/80 (later 55). Guns: 14 x 4 + 8 swivels (later 8 x 6 + 10 swivels?). Taken 05.1741 by *Rupert;* 1743 sold.

***Galgo/Galgo Prize*** 1742-1743 (Spanish Galgo [?] ____ ) 12.
Dimensions & tons: 78'4", 62'3¼" x 22'3" x ____ . 164.
Men: 80. Guns: 12 x 4 + 12 swivels. Taken 04.1742 by *Hampshire*; 1743 sold.

***Peregrine/Peregrine Prize*** 1742-1743 (Spanish *Peregrina*? ____ ) 8.
Dimensions & tons: 76'7", 61'11" x 22'2¾" x 10'10". 162 74/94.
Men: 80. Guns: 8 x 4 + 12 swivels. Taken 08.1742 by *Launceston;* 1743 sold.

***Saphire's Prize*** 1745-1745 (Spanish privateer ____ ) 10.
Dimensions & tons: 78'6", 63'4" x 22'1" x 11'1½". 164 21/94.
Men: 100. Guns: 10 x 6 + 10 swivels. Taken 05.1745 by *Saphire;* 15.9.1745 wrecked on the coast of Ireland.

***Dreadnought's Prize*** 1748-1748 (French ?*Palme*? 1745 Brest) ____ .
Dimensions & Tons: 61'4", 47'4" x 20'9½" x 9'. 109.
Men: ____ . Guns: ____ . Taken 1748 in the West Indies by *Dreadnought;* 1748 sold.

## Fireships

***Anne Galley*** 1739-1744 (mercantile ____ ) 8.
Dimensions & Tons: 97'9", 80' x 26'7¾" x 12'3". 302 16/94.
Men: 55. Guns: 8 x 6. Purchased 22.6.1739; 11.2.1744 expended off Toulon.

***Duke*** 1739-1742 (mercantile ____ ) 8.
Dimensions & tons: 83'1", 66'2⅛" x 23'9½" x 11'1". 199 25/94.
Men: 45. Guns: 8 x 6. Purchased 22.6.1739; 14.6.1742 expended at Saint Tropez Bay in destroying five Spanish galleys.

***Eleanor*** 1739-1742 (mercantile ____ ) 8.
Dimensions & tons: 85'6", 63'8½" x 23'9¾" x 11'6". 192 65/94.
Men: 45. Guns: 8 x 6. Purchased 22.6.1739; 1742 breakwater.

***Mercury*** 1739-1744 (mercantile ____ ) 8.
Dimensions & tons: 89'11", 71'10¼" x 23'10¼" x 11'1". 217 45/94.
Men: 45. Guns: 8 x 6. Purchased 22.6.1739; 12.12.1744 'Came to anchor in Margate Roads the wind blew a storm and she has not been heard of since therefore it's supposed she foundered at anchor'.

***Cumberland*** 1739-1742 (mercantile ____ ) 8.
Dimensions & tons: 79', 62'9¼" x 23'3½" x 11'4". 181 13/94.
Men: 45. Guns: 8 x ____ . Purchased 29.6.1739; 1742 BU.

***Blaze*** 1739-1742 (mercantile *America* ____ ) 8.
Dimensions & tons: 79'9½" , 61'8" x 23'6" x 11'2". 181 11/94.
Men: 45. Guns: 8 x 6. Purchased 7.9.1739; 1742 BU.

***Strombolo*** 1739-1744 (mercantile *Mollineux* ____ ) 8.
Dimensions & tons: 88', 70'6" x 24'0½" x 11'6". 216 68/94.
Men: 45. Guns: 8 x 6. Purchased 7.9.1739; 1744 sold.

***Vulcan*** 1739-1743 (mercantile *Hunter* ____ ) 8.
Dimensions & tons: 88'10", 70'9¾" x 25'11⅛" x 11'6". 253 21/94.
Men: 45. Guns: 8 x 6. Purchased 7.9.1739; 1743 hulked [at Jamaica; fate uncertain].

***Vesuvius*** 1739-1742 (mercantile *Worcester* ____ ) 8.
Dimensions & tons: 83', 65'8¾" x 23'11" x 11'8". 199 93/94.
Men: 45. Guns: 8 x 6. Purchased 12.9.1739; 1742 BU.

***Etna*** 1739-1746 (mercantile *Mermaid* ____ ) 8.
Dimensions & tons: 81'8", 66'5½" x 22'9" x 11'. 183 13/94.
Men: 45. Guns: 8 x 6. Purchased 14. 9. 1739; 1746 sold.

***Firebrand*** 1739-1743 (mercantile *Charming Jenny* ____ ) 8?
Dimensions & tons: 87'3", 69'7" x 24'3½" x 11'7". 221 14/94.
Men: 45. Guns: 8 x 6. Purchased 14.9.1739; 1743 sold.

***Phaeton*** 1739-1743 (mercantile *Poole* ____ ) 8.
Dimensions & tons: 84'1", 64'4½" x 24'6" x 11'6". 214 11/94.
Men: 45. Guns: 8 x 6. Purchased 14.9.1739; 1743 sold.

***Scipio*** 1739-1747 (mercantile *Scipio* ____ ) 8.
Dimensions & tons: 71'7", 61'7¾" x 22'10" x 10'2". 170 81/94.
Men: 45. Guns: 8 x 6. Purchased 28.9.1739; 1747 sold.

***Pluto*** 1745-1747 (mercantile *Roman Emperor* ____ ) 8.
Dimensions & tons: 85'6¾", 69'5⅛" x 26'11½" x 11'0½". 268 36/92.
Men: 45. Guns: 8 x 6. Purchased 21.8.1745 from Grevill & Whetstone (her builders?) and fitted as a sloop (100 men, UD 14 x 4, QD 2 x 2, 12 swivels); 1746 fitted as a fireship; 1747 sold.

***Vulcan*** 1745-1749 (mercantile *Mary* ____ ) 8.
Dimensions & tons: 87'0¾", 71'3⅜" x 24'4¼" x 10'. 224 83/94.
Men: 45. Guns: 8 x 6. Purchased 21.8.1745 and fitted as a sloop (100 men, UD 10 x 4, QD 6 x 2 , 12 swivels); 1746 fitted as fireship; 1749 sold.

***Salamander*** 1745-1748 (mercantile *Pelham* 1744 [?] America) 8.
Dimensions & tons: 95'7½", 74'6¾" x 27'8½" x 11'9¼". 304.

Men: 45/55. Guns: 8 x 6. Purchased 6.9.1745 and fitted as a sloop (120 men, UD 16 x 6 , 12 swivels); 1746 fitted as a fireship; 1748 sold.

NOTE: Referred to as 'Plantation built' - presumably meaning that she was built in the American colonies.

***Conqueror*** 1746-1749 (French or Spanish ____ ) 8.
Dimensions & tons: 94'2", 75'8" x 27'9" x 11'9". 308.
Men: 45. Guns: 8 x 6. Taken in the Mediterranean 1746 (? or 1745?); 1749 (? or 1748?) sold.

***Duke*** 1746-1749 (French ____ ) 8.
Dimensions & tons: 106'10", 83'8" x 32'6" x 15'4". 469.
Men: 55. Guns: 8 x 6. Taken 1746 by *Faversham* in the Mediterranean; 1749 sold.

## Miscellaneous Small Vessels

***Margaretta*** 1744-1744 (ex Spanish or French? ____ ) tender (sloop rig?) 4.
Dimensions & tons: ____ , ____ x ____ x ____ . ____ .
Men: ____ . Guns: 4 swivels? Taken 1744?; Purchased at Jamaica by order of Sir Challoner Ogle 1744?; 1744 sold.

***Mediator*** 1745-1745 (mercantile *Mediator* [?] 1742 Virginia) sloop rig 10.
Dimensions & tons: 61'4", 44' x 21'2" x 9'9". 104 74/94.
Men: 80/60. Guns: 10 x short 4, 18 swivels. Purchased 1745; 29.7.1745 sunk in Ostend harbour.

NOTE: She was built in the Chesapeake Bay area and had the typical American single masted sloop rig (in other words a mercantile sloop and *not* a sloop of war).
PLANS: Lines, profile & decks.

## Storeships, etc

***Princess Royal*** 1739-1750 (mercantile *Princess Royal* [?] ____ ) 24 storeship/hospital ship.
Dimensions & tons: 123', 108'11½" [or 100'11½"] x 31'9" x 13'10". 541 28/94.
Men: 120. Guns: 24 x ____ . Purchased 21.11.1739 for use as a storeship but fitted out for use as a hospital ship; 1743 storeship; 1750 sold.

***Scarborough*** 1739-1744 (mercantile *Scarborough* [?] ____ ) 18 storeship/hospital ship.
Dimensions & tons: 116'8", 96'6" x 31'3" x 13'8". 501 25/94.
Men: 60. Guns: 18 x ____ . Purchased 21.11.1739 for use as a storeship but fitted out as a hospital ship; 1744 sold.

***Wager*** 1739-1741 (mercantile ____ ) 22 storeship (Sixth Rate).
Dimensions & tons: 123', 101'4⅛" x 32'2⅜" x 14'4". 558 82/94.
Men: 110. Guns: 22 x ——. Purchased 21.1.1739; 14.5.1741 wrecked on the coast of Southern Chile.

***Astrea Prize*** 1740-1744 (Spanish ____ , French built [?]) 20 storeship.
Dimensions & tons: ____ , 100'x 31'4" x 13'. 522.
Men: 110. Guns: 20 x ____ . Taken by Vernon 03.1740; 16.1.1744 burned by accident at Piscatagua, New England.

NOTE: 'French ship taken from Spaniards'.

***Discovery*** 1741-1750 (mercantile ____ 1739 ____ ) 12/10 transport (pink).
Dimensions & tons: 74'6", 57'1⅛" x 22'6½" x 8'0¾", 154 31/94.
Men: 30/20/14. Guns: 12 (later 10) x ____ , 12 swivels. Purchased 20.4.1741; employed at Gibraltar; 1750 sold.

***Trelawny*** 1744-1747 (mercantile ____ ) 14 storeship.
Dimensions & tons: 105'10", 79'2" x 28'10" x 11'6". 350.
Men: 40. Guns: 10 x 9 + 4 x 3 + 6 swivels. Purchased 27.2.1744; (based at Deptford); 1747 sold.

***Bien Aime*** 1743-1746 (French *Bien Aimé* [?] ____ ) 24 storeship.
Dimensions & tons: 112', ____ x 34'7" x 16' ____ .
Men: 120. Guns: 24 x ____ . Purchased at Antigua in 1743; 1746 condemned at Boston, Massachusetts; ? sold 1748, 1749 or 1750?

***Apollo*** 1747-1749 (French East Indiaman *L'Apollon* [30] 1716 Hamburg) 20 hospital ship.
Dimensions & tons: 127'2½", 105'6" x 36'5" x 15'2". 744.

Men: 120. Guns: UD 16 x 9, QD 4 x 6. Taken 3.5.1747 by Anson and Warren; 12/14.4.1749 lost in a hurricane of wind in the Road of Fort Saint David, India.

***Brave*** 1747-1748 (Spanish ____ ) 3 transport (xebec).
Dimensions & tons: 75'9", 56' x 22'9" x ____ . 154 .
Men: 70. Guns: 3 x ____ + 15 swivels. Taken 1747 by *Blandford*; 1748 sold.

## Miscellaneous Unregistered Vessels

(ie. not listed in the primary Admiralty sources consulted)

***Achille*** 1744-1744 (French ____ ) sloop. Taken 1744. Taken by Spaniards 1744.

***Achilles*** 1747-1748 (French or Spanish [?] ____ ) schooner 8. Taken on the Jamaica Station 1745; 1747 purchased; ____ 1748 taken by Spain. (There may have also been another *Achilles*, also a schooner, taken from the French in 1744 and recaptured by them in 1748.)

***Anna*** or ***Anne*** 1739?; 1741 purchased pink/storeship 400 tons, 8 guns; 1741 BU at Juan Fernandez as unserviceable.

***Barrington*** 1747/8-1748 sloop bought or captured at Jamaica, paid off 1748.

***Blaze*** 1746-1750? purchased fireship; in service 1750? (J Colledge only).

***Carmela*** 1741-1742 (Spanish *Nuestra Senora del Monte Carmen* ____ ) Ca 400 tons, 75 men, 4 guns. Taken off Juan Fernandez by *Centurion* 12.10.1741; 29.4.1742 destroyed at Chequeten?

***Carmin*** 1741-1742 (Spanish *Nuestra Senora del Carmin* ____ ) 300 tons. Taken 6.11.1741 by *Centurion*; 29.4.1742 destroyed at Chequeten.

***Centurion Prize*** 1743-1743 (Spanish *Nuestra Senora del Covadonga* 17 ____ . Cavite, Philippines). Ca 700 tons. 50 guns. Taken by *Centurion* 20.6.1743; 1743 sold to the Portuguese at Macao.

***Cruizer*** 1746-1746 (?French or Spanish? ____ ) 12 guns, 80 men. Taken 1746 in the Mediterranean; 1746 sold.

***Cumberland*** ?Fireship in service 1745? (J Colledge only).

***Dragon's Prize*** 1742?-1745? (Spanish or French? ____ ) 60 men. 8 guns. Taken by boats of *Dragon* in the Mediterranean 1739-48.

***Elizabeth*** 1741 storeship in the West Indies (Pitcairn Jones only).

***Enterprize*** 1743-1748 (Spanish or French? ____ ) Barcalonga/sloop 60 men 8 guns. Taken 1743?; served in Mediterranean; Summer 1748 for sale at Port Mahon.

***Fogo*** 1747-1747 (mercantile? ____ ) Fireship; bought in East Indies?; 2.12.1747 wrecked.

***Galicia*** 1741-1741 (Spanish *Galicia* 70 guns ____ ). Taken at Carthagena 25.3.1741; 16.4.1741 expended as a floating battery.

***Garland's Prize*** 1741-1744 (French or Spanish? ____ ) Xebec. 40 men. Taken 1741, served in the Mediterranean.

***Glorioso*** 1747-1749 (Spanish *Glorioso* ____ ) Third Rate 74. Taken off Lagos by *Russell* 5.10.1747; 1749 sold.

***Goodly*** 1740-1741? Purchased bomb tender.

***Guernsey's Prize*** 1746-1747 (French or Spanish? ____ ) Barcalonga. Taken 1746; served in Mediterranean in 1747.

***Hind*** sloop in service 1741; 6.10.1743 sold (J.Colledge only).

***Jane*** 1747-17____ ( ____ ). Transport acquired 1747.

***Louisbourg*** 1746-1747 (French ____ ) fireship. Taken at the fall of Louisbourg 1746; 4.1.1747 retaken by French privateers in the Channel.

***Madelena*** 1748-1748 polacca 16 guns 80 men; bought or captured in the Mediterranean 1748; 1748 for sale.

***Maidstone's Prize*** taken 04.1746 by *Maidstone*; nothing further (Pitcairn Jones only).

***Maria Teresa*** 1748 purchased or captured polacca 14 guns 80 men, serving in the Mediterranean.

***Medea*** 1744-1745 (French *La Médée* 1741 Brest) 26 Sixth Rate.
125'5", 101'6" x 23'4" x 16'8". 600 tons. Taken 4.4.1744 by *Dreadnought*; ? never commissioned into the RN?; sold 1745 and became the privateer *Boscawen*.

***Mercury*** 1744-1745 (mercantile [?] ____ ) 16 brigantine/sloop. Guns 16 x 6. Purchased 1744; 15.4.1745 taken by the French coming from Jamaica [in French service as *Le Mercure* until 1757].

NOTE: At this period 'brigantine' rig , it would seem, was what would later be called 'brig' rig. In any case this vessel would not have the same rig as a nineteenth-century brigantine.

*Merton* Pitcairn Jones notes a (?) sloop of this name in service in 1744; no other record.

*Neptune* Pitcairn Jones notes a (?) sloop of this name in service 1742; 8 guns + 10 swivels

*Pansy/Pancey* 1747 captured or purchased settee, 80 men; 1748 for sale.

*Pompey* 1740-1741? Purchased bomb tender.

*Postillion* 1745-1748 bought or captured xebec; 12 guns 60 men.

*Rochester* Pitcairn Jones notes a cutter in service in 1746 (hired?).

*Royal George* Pitcairn Jones notes a possible cutter of this name in service in 1746.

*Rubis* 1747-1748 (French *Rubis* [52] 1728 ____ ) Sixth Rate [transport?] 26. Taken 1747; out of service and BU after 1748.

*Savage* 1748-1748 sloop 14 guns purchased 1748; 1748 wrecked on the Lizard.

*Saxon* 1747-1748 sloop 16 guns + 14 swivels, 110 men; acquired in the Leeward Islands 1747 [?]; 1748 sold.

*Scorpion* 1742 there is a log for a vessel of this name and date.

*Shipley* or *Shirley* [probably the latter] 1747-1748 (Mercantile *Shipley* or *Shirley* ____ ) Galley 14. Purchased 1747; at Annapolis Royal; out of service after 1748.

*Spry* 1742 fireship, 6 guns (Pitcairn Jones only).

*Triton* sloop in commission 1741? (Noted only in Colledge; doubtful).

*Triumph* 1739-1740 (Spanish *Triumfo* ____ ) Sloop ____ . Taken 1739; 1740 deleted.

*Tryal Prize* 1741-1742 ( ____ ) Sixth Rate 20. 600 tons. Taken 1741; 1742 burnt.

*Vainquer* or *Vanquer* 1748 (French? ____ ) sloop [?] taken; based at Jamaica.

*Viper* 1746-1746 (French *Volage* 1741 Rochefort) Fifth Rate 32. Taken 13.4.1746 by *Stirling Castle*; 1746 retaken [and resumed her original name; in French service until 1750].

*Whitehaven* 1745-1747 ( ____ ) Sloop/armed ship ____ . Acquired 1745; 1747 burnt.

## Hired Vessels

*Baltimore* 1745 Armed vessel? 16.

*Boscawen* 1748-? Armed ship, 16 (also hired 1759-1760?).

*Carlisle* [?] 1745 10 guns + 16 swivels.

*Charles* 1745-1747 Sloop 14 or 6 guns, 100 men.

*Duke William* 1746 Armed ship 50 guns.

*Eagle* 1745 Fireship 10 guns + 20 swivels (also hired 1757?) or sunk 1745 as a breakwater.

*Fraternity* 1739 Sloop hired by Admiral Vernon (in West Indies, presumably).

*Happy Jennet* [?] 1745-1747 Armed ship? 116 men. 20 guns + 12 swivels. Hired at Leith?

*Hardwick* 1748 Hired privateer? 44 guns.

*Industry* 1740 Pink; with Anson's squadron & discharged at Santa Catharina (Brazil).

*Intrepid* 1747-1748 Armed ship 18 guns + 6 swivels. 116 men.

*James and Mary* [?] 1744-1748 50 men, 8 guns + 6 swivels.

*John and Anne* 1744-1746 106/100 men, 16 guns + 6 swivels.

*Kingston* 1746-1748 Armed ship 14 guns + 10 swivels, 86 men.

*Kingston* 1748 Privateer (?hired?) 36 guns.

*Martin* [?] 1744 10 guns.

*Mayflower* [?] 1745 14 guns + 4 swivels.

*Revenge* 1740 Brigantine.

*Saint Michael* [?] 1748 Hired privateer. 12 guns.

*Saint Quintin* 1744-1748 100 men. 16 guns + 6 swivels.

*Stork* 1746-1747; in Mediterranean.

*Swift* 1748 Privateer (?hired?) 12 guns (Pitcairn Jones only).

*Tavistock* 1746-1747 Cutter.

*Ursula* 1745 Armed ship 16 guns + 6 swivels, 100 men.

*York* 1744-1745 16 guns + 6 swivels. 100 men.

*York* 1748? Hired privateer? 12 guns.

# VESSELS ACQUIRED 1750-1755

## Hired Vessels

*Merlin* Sloop (?hired?). In service 1753 (according to Colledge who is the only one to note this vessel; perhaps a clerical error?).

# Chapter 4: The Seven Years' War 1756-1763

Despite its beginning in the failure and humiliation of the fall of Minorca and defeat in North America, this war rapidly became one of triumph following on triumph, as can be seen by the captures listed here. France's navy was shattered at Lagos, Quiberon Bay (both 1759) and elsewhere. Spain made the bad mistake of entering the war alongside France (1762) only to provide another series of targets for successful British aggression.

As in the previous war, merchant vessels were purchased in some numbers to produce an instant increase in the numbers of sloops and fireships. These included the first ship sloops (confusingly referred to at first as 'frigates') at the outbreak of war. During the war the first cutters were acquired by the Navy, and in 1763 there was a massive purchase of vessels of this type, presumably because hired cutters had performed well.

The overwhelming success of Britain in the maritime and colonial struggle produced its own nemesis very rapidly. The problems of policing the American colonies show up in the final section of this chapter. This records the comparatively large number of vessels acquired in the period of growing colonial tension which culminated in the American Revolution and the rapid intervention of Britain's defeated rivals against her in the subsequent conflict. As most of these vessels were not only intended for use in the colonies but also purchased there, it was natural that the most popular local type of vessel would be strongly represented. So it was the schooner entered Royal Naval service. A less ominous note is sounded by the inclusion of two of the most famous ships purchased for the Royal Navy at this or any other time, Captain Cook's *Endeavour* and *Resolution*. These, though sometimes classed as sloops, were in effect storeships and this is where they are listed rather than amongst the 'sloops of war'. This is a case of 'sloop' being used as an indicator of the rank of the commanding officer, just as the larger Royal Yachts commanded by a Captain counted as rated ships.

## Large Two-deckers (Third Rates)

***Foudroyant*** 1758-1787 (French *Le Foudroyant* 1750 Toulon) 80.
Dimensions & tons: 180'8½", 147'3" x 50'3" x 23'. 1977 69/94.
Men: 700. Guns: GD 30 x 32, UD 32 x 24, QD 10 x 9, Fc 8 x 9. Taken 28.2.1758 in the Mediterranean; 1787 BU.
PLANS: Lines & profile.

***Formidable*** 1759-1768 (French *Le Formidable* 1751 Brest) 80.
Dimensions & tons: 188', 155'4½" x 49'2¾" x 21'4½". 2002 88/90.
Men: 650. Guns: GD 30 x 32, UD 32 x 24, QD 12 x 9, Fc 6 x 9. Taken 20.11.1759 at Quiberon Bay; never fitted for sea by the RN; 1768 BU.

## 74 Gun Ships (Third Rates)

***Centaur*** 1759-1782 (French *Le Centaure* 1757 Toulon).
Dimensions & tons: 175'8", 148'3¼" x 47'5" x 20'. 1773 19/94.
Men: 600. Guns: GD 28 x 32, UD 30 x 18, QD 12 x 9, Fc 4 x 9. Taken 18.8.1759 at Lagos; 24.9.1782 foundered in a gale of wind on the Banks of Newfoundland.
PLANS: Lines, profile & decorations/orlop, gun & upper decks/quarter deck, forecastle & roundhouse. As altered 1760: profile /decks.

***Temeraire*** 1759-1784 (French *Le Téméraire* 1749 Toulon).
Dimensions & tons: 169'2", 137'2⅜" x 48'0¾" x 20'9½". 1685 80/94.
Men: 600. Guns: GD 28 x 32, UD 30 x 18, QD 10 x 9, Fc 6 x 9. Taken 18.8.1759 at Lagos; 1784 sold.
PLANS: Lines, profile & decorations/decks.

***Courageux*** 1761-1796 (French *Le Courageux* 1753 Brest).
Dimensions & tons: 172'3", 140'1⅛" x 48'0¾" x 20'10½". 1721 30/94.
Men: 650. Guns: GD 28 x 32, UD 28 x 18, QD 14 x 9, Fc 4 x 9. Taken 13.8.1761 by *Bellona*; 10.12.1796 wrecked beneath Ape's Hill, Barbary Coast (North Africa).
PLANS: Lines, profile & decorations/decks.

***Infanta*** 1762-1775 (Spanish *Infante/San Luis Gonzaga* 1750 Havanna).
Dimensions & tons: 171'6", 137'4" x 51'4" x 22'5". 1918 64/94.
Men: 650. Guns: ____ . Taken at the surrender of Havanna 1762; 1775 sold.
PLANS: Lines & profile (incomplete).

***Moro*** 1762-1770 (Spanish *Aquilon/San Damaso* [68] 1750 Havanna).
Dimensions & tons: 175'3", 143'7" x 49'7½" x 20'5½". 1880½.

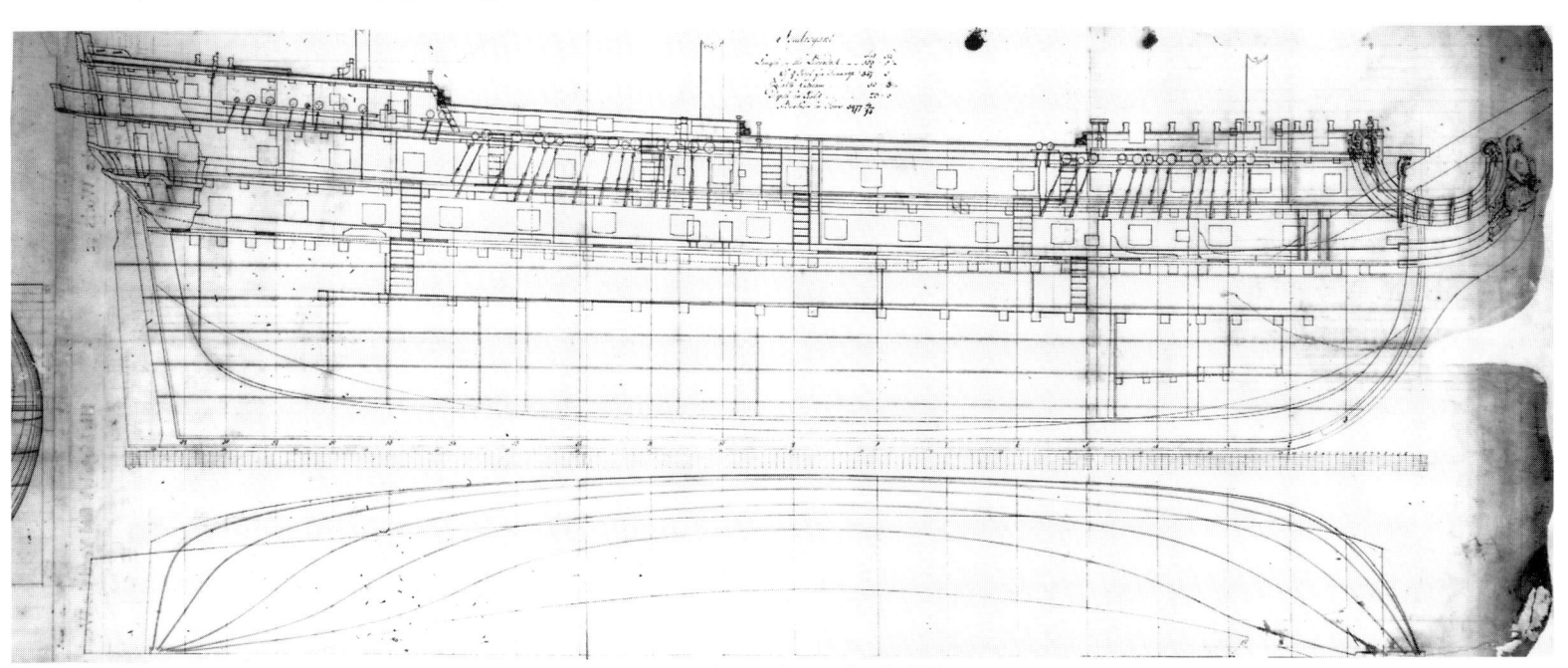

Part (omitting the body plan) of the lines and profile plan (478) of the French 80 gun ship *Foudroyant* taken in 1758

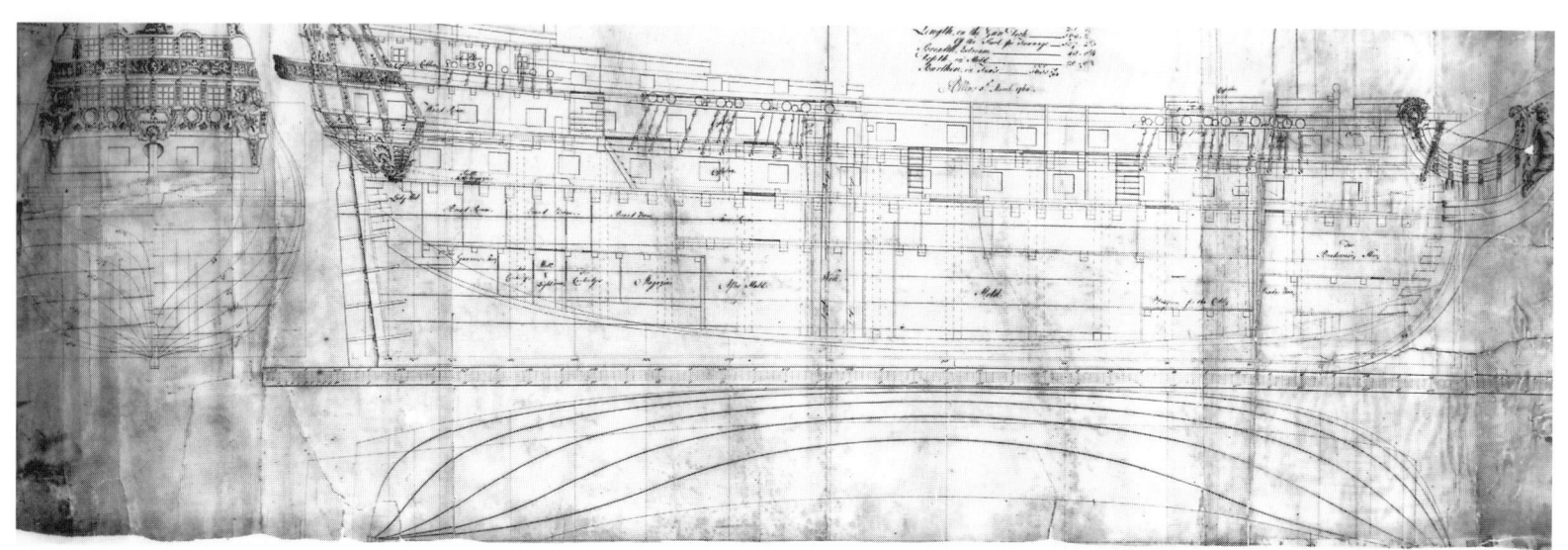

Lines and profile (260) of the 74 gun French prize *Temeraire* of 1759

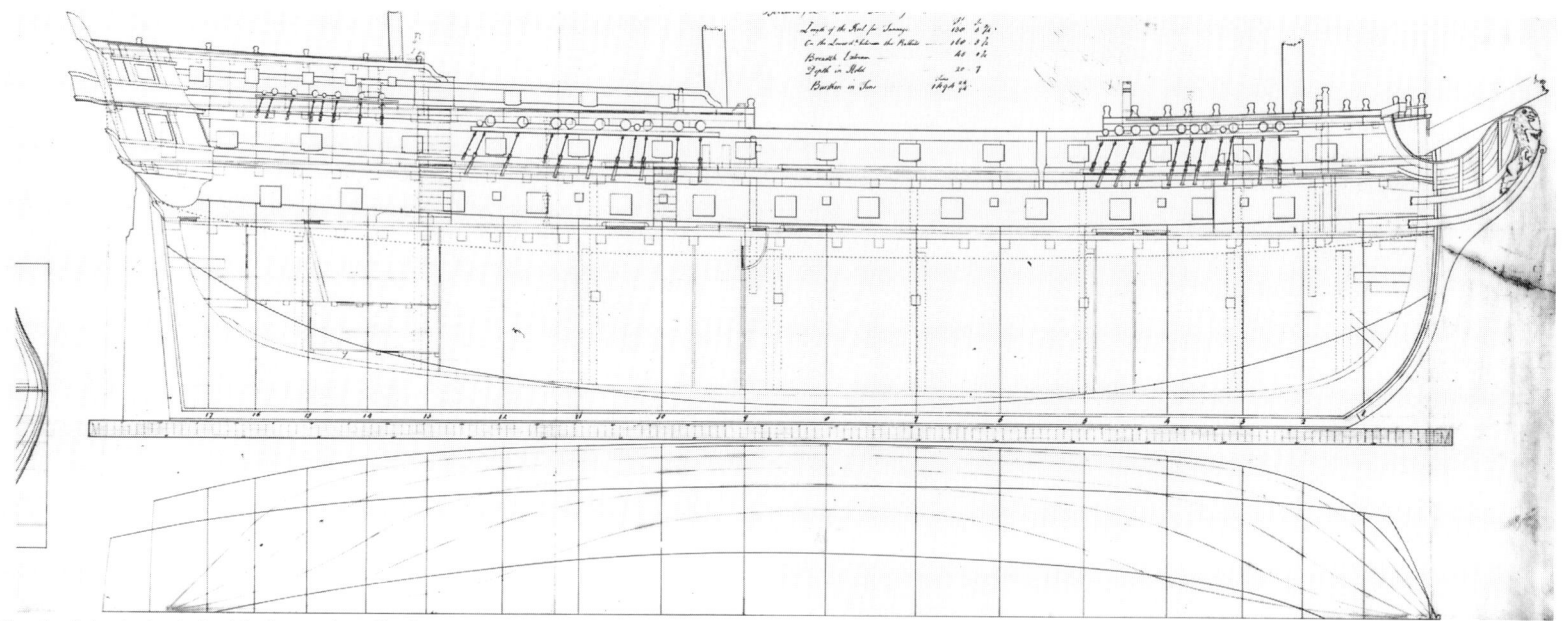

Part (omitting body plan) of the lines and profile plan (1110) of the 64 gun French *Bertin*, renamed *Belleisle*, captured in 1761

Men: 650. Guns: ____ . Taken at the surrender of Havanna 1762; 1770 BU.

***Reyna*** 1762-1772 (Spanish *Reina* [70] 1743 Havanna).
Dimensions & tons: 173'2", 131'8½" x 48'7" x 22'1". 1653½ [Winfield suggests 1779 14/94 with a keel length of 141'8½".]
Men: 650. Guns: ____ . Taken at the surrender of Havanna 1762; 1772 sold.

***Soverano*** 1762-1770 (Spanish *Soberano/San Gregorio* 1755 Ferrol).
Dimensions & tons: 175', 143'5" x 49'7" x 21'. 1875½.
Men: 650. Guns: ____ . Taken at the surrender of Havanna 1762; 1770 BU.

***Tiger*** 1762-1780 (Spanish *Tigre/San Lorenzo* 1747 Havanna).
Dimensions & tons: 169'3", 136'4" x 51' x 21'8". 1886 17/94.
Men: 650. Guns: ____ . Taken at the Surrender of Havanna 1762; 1780 hulked [at Portsmouth as receiving ship for sick men; 1783 sold].
PLANS: 1778 as hospital ship at Plymouth: decks.

## Small Line of Battle Ships

***Alcide*** 1755-1772 (French *L'Alcide* 1741 Brest) Third Rate 64.
Dimensions & tons: 159', 128'4½" x 44'10¾" x 18'2⅜". 1375 75/94.
Men: 500. Guns: GD 26 x 24, UD 28 x 12, QD 6 x 6, Fc 4 x 6. Taken by Boscawen in America (before the official declaration of war) 6.6.1755; 1772 sold.

PLANS: Lines, profile & decorations/decks.

***Duke of Aquitaine*** 1757-1761 (French *Le Duc de Aquitaine* 1754 L'Orient for the French East India Co) Third Rate 64.
Dimensions & tons: 159'5", 129'11¼" x 44'4" x 19'5". 1358.
Men: 580. Guns: GD 24 x 24, UD 26 x 12, QD 12 x 6, Fc 2 x 9. Taken 06.1757 by *Eagle* & *Medway*; 1.1.1761 foundered at anchor about two leagues south of Pondicherry during the Pondicherry hurricane.

***Raisonable*** 1758-1762 (French *Le Raisonable* 1755 Rochefort) Third Rate 64.
Dimensions & tons: 159'2", 129'7" x 43'10½" x 19'6¾". 1326 80/94.
Men: 500. Guns: GD 26 x 24, UD 28 x 12, QD 6 x 6, Fc 4 x 6. Taken 29.5.1758 by *Dorsetshire*; 7.3.1762 lost off Martinique.
PLANS: Lines, profile & decorations.

***Bienfaisant*** 1758-1783 (French *Le Bienfaisant* 1754 Brest) Third Rate 64.
Dimensions & tons: 158'9", 129'1½" x 44'6" x 19'4¼". 1360 7/94.
Men: 500. Guns: GD 26 x 24, UD 28 x 12, QD 6 x 6, Fc 2 x 6. Cut out of Louisbourg by the boats of Boscawen's fleet (in the only RN cutting out action ever to capture a capital ship) on 26.7.1758; 1783 laid up and subsequently hulked [at Plymouth as a prison ship; 1814 BU].
PLANS: 1794 as prison ship at Plymouth: decks.

***Belliquex*** 1758-1772 (French *Le Belliquex* 1756 Brest) Third Rate 64.
Dimensions & tons: 157'10½", 128'0⅞" x 44'10½" x 19'10½". 1372.
Men: 511. Guns: GD 26 x 24, UD 28 x 12, QD 6 x 6, Fc 4 x 6. Taken by *Antelope* 3.11.1758; 1772 BU.

***Saint Florentine*** 1759-1771 (French *Le Comte de Saint Florentine* [? privateer?,

THE SEVEN YEARS' WAR 1756-1763

not naval anyway] ____ ) Fourth Rate 60.
Dimensions & tons: 147'9½", 120'6½" x 41'7" x 17'10". 1108 70/94.
Men: ____ . Guns: GD 24 x 24, UD 26 x 12, QD 8 x 6, Fc 2 x 6. Taken by *Achilles* 05.1759; 1771 breakwater.
PLANS: Lines & profile.

*Modeste* 1759-1778 (French *Le Modeste* 1759 Toulon) Third Rate 64.
Dimensions & tons: 158'6", 129' x 44'5¾" x 19'8". 1357 47/94.
Men: 500. Guns: GD 26 x 24, UD 28 x 12, QD 6 x 6, Fc 4 x 6. Taken at Lagos 1759; 1778 hulked [at Portsmouth as a receiving ship; 1800 BU].
PLANS: Lines & profile/lines, profile & decorations/decks.

*Belleisle* 1761-1784 (French East Indiaman *Le Bertin* 1760 L'Orient) Third Rate 64.
Dimensions & tons: 168'5½", 138'5¾" x 45'0½" x 20'7". 1494 26/94.
Men: 500 (1782: 420). Guns: GD 26 x 24, UD 28 x 18, QD 8 x 9, Fc 2 x 6 (1782: GD 24 x 68★, UD 26 x 42★, QD & Fc 8 x 24★). Taken by *Hero* and *Venus* 3.4.1761; 1784 hulked [at Sheerness as a lazaretto; 1819 sold].
PLANS: Lines & profile/lines & profile.

*Saint Ann* 1761-1775 (French *La Sainte Anne* 1760 [built 1756 ____ ]) Third Rate 64.
Dimensions & tons: 165', 136'2" x 44'1" x 19'7½". 1407 50/94.
Men: 500. Guns: GD 26 x 24, UD 28 x 18, QD 8 x 9, Fc 2 x 9. Taken 5.6.1761 by Holmes in the West Indies (had been purchased for the French Navy at Lisbon from an unknown owner in 1760, builder also unknown); 1775 hulked [at Plymouth as a receiving ship; 1784 sold].
PLANS: Lines, profile, decorations/orlop/gun deck/upper deck/quarter deck & forecastle (all incomplete).

*Saint Antonio* 1762-1775 (Spanish *San Antonio* [60] 1761 Havanna) Third Rate 64.
Dimensions & tons: 159'6", 132'2" x 44'6" x 20'. 1392 12/94.
Men: ____ . Guns: ____ . Taken at the surrender of Havanna 1762; 1775 sold.
PLANS: Lines & profile/gun deck/quarter deck & forecastle.

*Conquestadore* 1762-1775 (Spanish *Conquistador* 1758 Carraca, Cadiz) Fourth Rate 60.
Dimensions & tons: 155'9", 128'6" x 43'3" x 19'3". 1278 51/94.
Men: ____ . Guns: ____ . Taken at the fall of Havanna 1762; 1775 hulked [at Sheerness as a receiving ship for the Nore; 1782 BU].
PLANS: Lines & profile.

## Small Two-deckers

*Arc-en-Ciel* 1756-1759 (French *L'Arc en Ciel* 1745 Bayonne) Fourth Rate 50.
Dimensions & tonnage: 146', 119'3" x 41'2½" x 18'4". 1077 12/94.

Men: 350. Guns: GD 22 x 24, UD 22 x 12, QD 4 x 6, Fc 2 x 6. Taken 12.6.1756 by a squadron in North American waters 1756; 1759 sold.

## 18 pounder Frigates (Fifth Rates)

*Aurora* 1757-1763 (French *L'Abénakise* 1756 Quebec) 38.
Dimensions & tons: 144', 118'9" x 38'8½" x 15'2". 946 37/94.
Men: 250. Guns: LD 8 x 18, UD 28 x 12, Fc 2 x 6. Taken by *Chichester* 12.1757; 1763 BU.
NOTE: Though sometimes given in British sources as 'La Bien Aquise' her real French name was taken from the Abenaki tribe of Canadian Indians who were staunch allies of the French. Her 18 pounder guns were in a typically French 'demie batterie' installation on the lower deck aft. Though neither solidly or well built, her lines much impressed both Thomas Slade and the Admiralty, and were used as the basis for subsequent British designs.
PLANS: Body plan & stern & bow lines.

## 12 pounder Frigates (Fifth Rates)

*Melampe* 1757-1764 (French privateer *Le Mélampe* 1757 Bayonne) 36.
Dimensions & tons: 134'6", 111'6½" x 35'6" x 11'3½". 747 66/94.
Men: 240. Guns: UD 26 x 12, QD 8 x 6, Fc 2 x 6. Taken by *Tartar* 2.11.1757; 1764 laid up; no further information.

*Crescent* 1758-1777 (French privateer *Rostan* ____ ____ ) 32.
Dimensions & tons: 130'5" , 107'6½" x 35' x 11'2". 731 8/94.
Men: 220. Guns: UD 24 x 12, QD 6 x 6, Fc 2 x 6. Taken by *Torbay* and *Chichester*, 1777 sold.
PLANS: 1771: decks.

*Repulse* 1759-1776 (French *La Bellone* 1755 Rochefort) 32.
Dimensions & tons: 122'7", 104'0½" x 34'11½" x 10'10½". 676 67/94.
Men: 220. Guns: UD 26 x 12, QD 4 x 6, Fc 2 x 6. Taken by *Vestal* 21.2.1759; 12.1776 foundered off Bermuda.
PLANS: Lines, profile & decorations/decks. 1770: Profile/orlop/gun deck/upper deck/quarter deck/forecastle. 1776: decks.

*Danae* 1759-1771 (French *La Danaé* 1756 Le Havre) 38.
Dimensions & tons: 147'6", 123'11" x 37'9½" x 11'4½". 941 35/94.
Men: 250. Guns: UD 30 x 12, QD 6 x 6, Fc 2 x 6. Taken 27 3.1759 by *Southampton* and *Melampus*; 1771 BU.
PLANS: Lines, profile & decorations/lines, profile & decorations/decks.

*Arethusa* 1759-1779 (French *L'Aréthuse* 1757 Le Havre) 32.
Dimensions & tons: 132'2", 110'10⅜" x 34'5½" x 10'8". 700 31/94.
Men: 220. Guns: UD 26 x 12, QD 4 x 6, Fc 2 x 6. Taken by *Thames* 18.5.1759; 19.3.1779 wrecked off Ushant.
NOTE: Building as a privateer named *La Pélérine*, purchased for the French navy

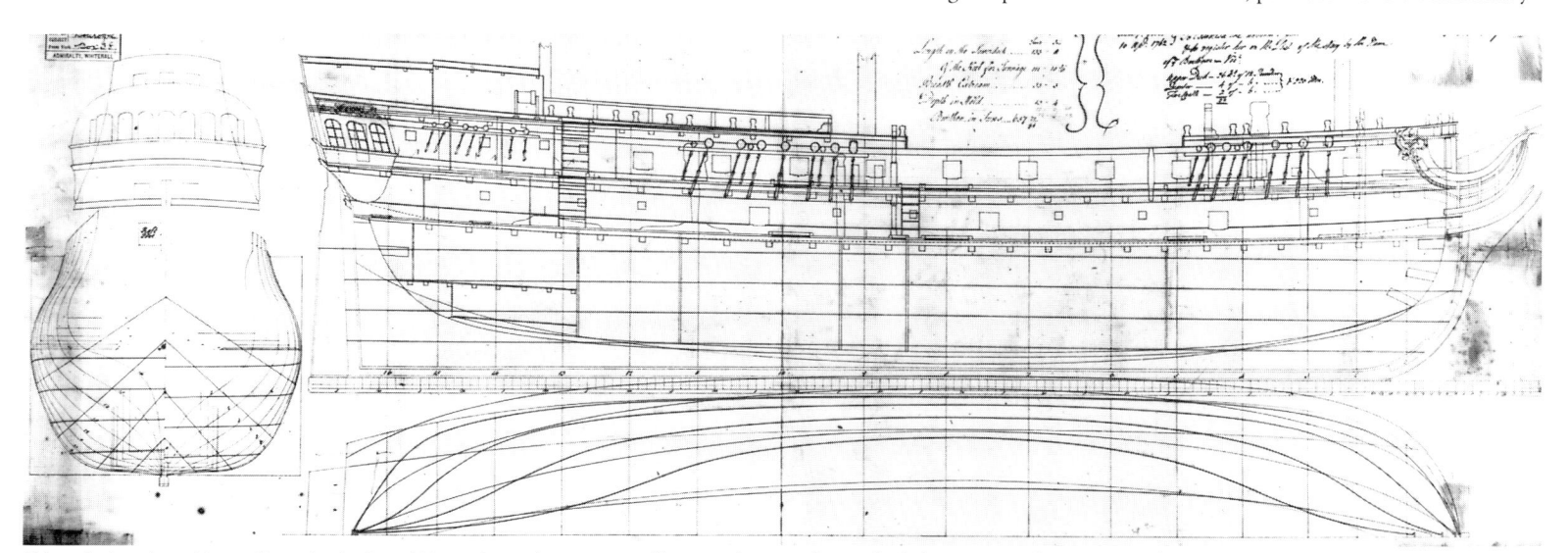

Although classed as a 32 gun frigate by the Royal Navy, the *Boulogne* was actually a French East Indiaman, built the same year she was captured (1762), this lines and profile plan (2226) being 'taken off' her soon after capture whilst completing and name changed.

PLANS: Lines & profile/cabins.

**Blonde** 1760-1782 (French *Blonde* 1753-1755 Le Havre) 32.
   Dimensions & tons: 133', 109'0¼" x 34'10" x 10'7". 703 57/94.
   Men: 220. Guns: UD 26 x 12, QD 4 x 6, Fc 2 x 6. Taken by *Eolus* 28.2.1760; 10.5.1782 wrecked 'off Boston ..... near Cape Sable'.

**Baleine** 1760-1767 (French *La Baleine* 1758 _____ ) Fifth Rate 32.
   Dimensions & tons: 149'10⅞" , 129'0¼" x 32' x 12'9". 702 ½.
   Men: 220. Guns: UD 26 x 12, QD 4 x 6, Fc 2 x 6. French 'flûte' (storeship) taken at Pondicherry 11.1760; 1767 sold.

**Brune** 1761-1792 (French *La Brune* 1755 Le Havre) 32.
   Dimensions & tons: 130'6", 108'11¾" x 34'7½" x 10'7½". 694 86/94.
   Men: 220. Guns: UD 26 x 12, QD 4 x 6, Fc 2 x 6. Taken by *Venus* 30.1.1761; 1792 sold.

PLANS: Lines, profile & decorations. 1776: decks.

**Flora** 1761-1778 (French *La Vestale* 1757 Le Havre) 32.
   Dimensions & tons: 131'7", 110'1¼" x 34'6½" x 10'9". 698 67/94.
   Men: 220. Guns: UD 26 x 12, QD 4 x 6, Fc 2 x 6. Taken by *Unicorn* 8.1.1761; 5.8.1778 scuttled at Rhode Island to prevent capture [later raised by the Americans and sold by them to the French in 1784 who renamed her *La Flore* (or *La Flore Américaine*); 1793 fitted as a privateer, continued to serve as such until taken by *Phaeton* on 7.9.1799 and then sold].

**Boulogne** 1762-1776 (French East Indiaman *Le Boullogne* 1762 L'Orient) 32.
   Dimensions & tons: 133'8", 111'10¾" x 33'3" x 13'4". 657 71/94.
   Men: 220. Guns: UD 26 x 12, QD 4 x 6, Fc 2 x 6. Taken by *Venus* 03.1762; 1776 hulked [as storeship at Halifax, Nova Scotia; 1784 used for lengthening the wharf there].

NOTE: The French name came from a nobleman concerned with the East India Company, and not from the town of Boulogne, hence the different spelling.

PLANS: Lines & profile.

## 9 pounder Frigates (Sixth Rates)

**Emerald** 1757-1761 (French *L'Emeraude* 1744 Le Havre) 28.
   Dimensions & tons: 115'4", 93'1⅝" x 33'11½" x 9'3¾". 571.
   Men: 180-200. Guns: UD 28 x 9. Fc 4 x 4. Taken by *Southampton* 21.9.1757; 1761 BU.

**Vengeance** 1758-1766 (French privateer *La Vengeance* 1757 St. Malo) 28.
   Dimensions & tons: 116'11", 95'10⅜" x 32'4" x 11'3½". 533.
   Men: 200. Guns: UD 24 x 9, QD 4 x 4. Taken by *Hussar* 02.1758; 1766 breakwater.

**Valeur** 1759-1764 (French *La Valeur* 1753 Rochefort ) 28.
   Dimensions & tons: 115'6", 93'4" x 32'6" x 10'10". 524½.
   Men: 200. Guns: UD 18 x 9 + 6 x 6, QD 4 x 3. Taken 18.10.1759 by *Lively* of Boscawen's squadron in the Mediterranean; 1764 sold.

## 'Corvettes' (Sixth Rates)

**Tartar's Prize** 1757-1760 (French privateer *La Marie Victoire* 1757 _____ ) 24.
   Dimensions & tons: 117'3", 99'5½" x 28'4" x 13'3". 424 65/94.
   Men: 160. Guns: 24 x 9. Taken by *Tartar* 03.1757; 2.3.1760 foundered, sprang a plank and sank off Sardinia.

**Eurus** 1758-1760 (French privateer ?*Le Dragon*? or ?*L'Eurus*? 1758) 24.
   Dimensions & tons: 121'9½", 101'6" x 31'10" x 10'6". 546 72/94.
   Men: 160. Guns: 24 x 9. Taken by *Coventry* 1758; 26.6.1760 wrecked in the Saint Lawrence.

NOTE: Her original French name is uncertain as is whether she was a naval vessel or privateer when still in French hands.

**Echo** 1758-1770 (French *L'Echo* 1757 ex privateer *Le Maréchal de Richelieu* 1756 Nantes?) 24.
   Dimensions & tons: 118'2", 96'9⅜" x 32'6" x 9'11½". 539 54/94.
   Men: 160. Guns: 24 x 9. Taken by Boscawen at Louisbourg 29.5.1758; 1770 sold.

**Terpsichore** 1760-1766 (French *La Terpsichore* 1757 _____ ) 24.
   Dimensions & tons: 113'9", 93'2½" x 30'9" x 14'3". 468 74/94.
   Men: 160. Guns: 24 x 6. Taken by *Eolus* and others on 28.2.1760; 1766 sold.

## Ship Sloops

NOTE: There was an Admiralty Order of 15.4.1757 to purchase ships to serve as 'frigates' (ie as ship sloops) which was the authority for buying all the 1757 purchases listed below, including the prizes bought in.

**Escorte** 1757-1768 (French privateer *L'Escorte* 1756 St.Malo) 14.
   Dimensions & tons: 84'6", 69'3⅞" x 24'8½" x 9'. 225 11/94.
   Men: 100-110. Guns: 14 x 6 + 10 swivels. Taken by *Badger* 24.2.1757; 1768 sold.

NOTE: In the survey made before her sale she is referred to as 'French built and very slight'.

PLANS: Lines & profile/orlop & lower deck.

**Postillion** 1757-1763 (French privateer *Le Duc d'Aiguillon* _____ . St. Malo) 18.
   Dimensions & tons: 97'1½", 78'6¼" x 29'6½" x 13'0½". 365.
   Men: 120. Guns: 18 x 6 + 14 swivels. Taken by *Tartar* 15.4.1757; 1763 sold.

**Beaver** 1757-1761 (French privateer *Trudaine*? 1754? _____ ) 18.
   Dimensions & tons: 94'11½", 76'5¼" x 28'7¾" x 11'1". 333.
   Men: 120. Guns: 18 x 6 + 12 swivels. Taken _____ ?; purchased 28.4.1757; 1761 sold.

**Cormorant** 1757-1762 (French privateer? *Le Machault*? 1755 Nantes) 18.
   Dimensions & tons: 100'8", 81'3⅝" x 30'7" x 13'6". 408.
   Men: 120. Guns: 18 x 6 + 14 swivels. Taken _____ ; purchased 28.4.1757; 1758 fireship; 1762 sold.

**Merlin** 1757-1777 (building Randall, Rotherhithe) 18.

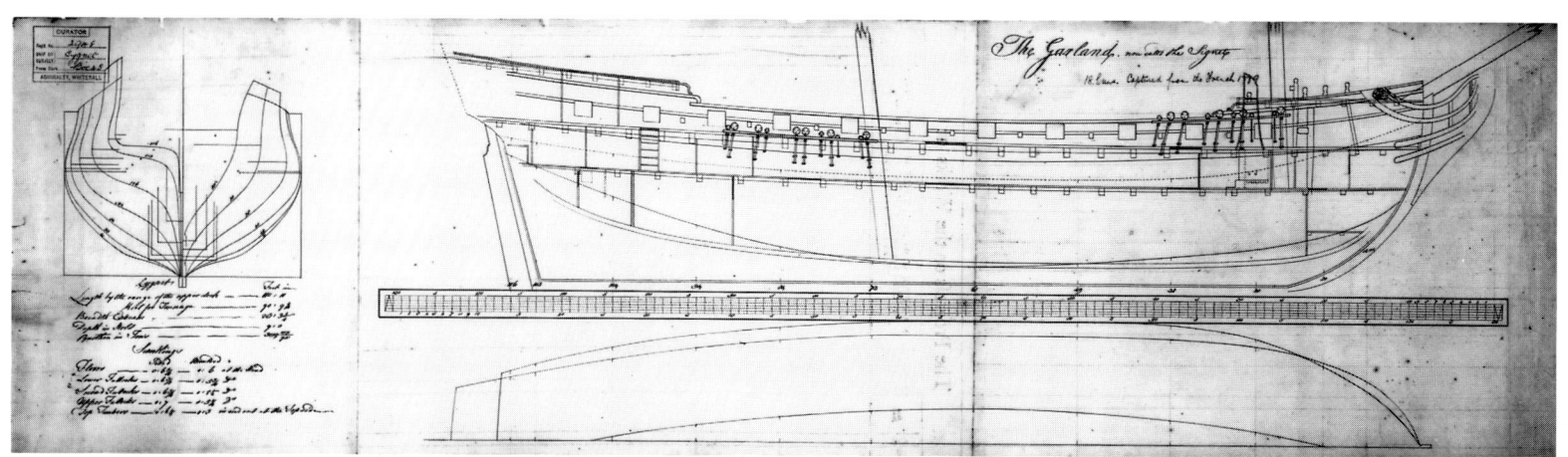

This incomplete lines and profile plan (2948) of the French sloop *Guirlande* ('*Garland*') shows her with a two masted rig as captured in 1758. She would serve the RN rerigged as a ship sloop under the name *Cygnet*.

Dimensions & tons: 100'2¼", 83'7⅞" x 26'1" x 12'. 304.
Men: 130. Guns: 18 x 6 + 14 swivels. Purchased on the stocks 20.4.1757; launched 1.7.1757; 23.10.1777 grounded on Mud Island in the Delaware under enemy fire and burnt (by red hot shot).

**Pelican** 1757-1763 (mercantile *Saint George* 1754 Shoreham) 16.
Dimensions & tons: 86'10½", 71'7½" x 24'5" x 11'1". 234.
Men: 80 (as bomb: 60). Guns: 16 x 4 + 10 swivels (as bomb: 1 x 13" mortar + 8 x 4 + 12 swivels). Purchased 28.4.1757; 1758 bomb vessel; 1760 sloop; 1762 bomb vessel; 1763 sold.
PLANS: 1758 as bomb: profile, orlop, upper deck/midships section.

**Roman Emperor** 1757-1763 (mercantile *Roman Emperor* 1749 River Thames) 16.
Dimensions & tons: 91'10", 75'7¾" x 26' x 11'6". 272.
Men: 100 (as fireship: 45). Guns: 16 x 6 + 12 swivels (as fireship: 8 x 4 + 8 swivels). Purchased 15.5.1757; 1758 fireship; 1758 sloop; 1762 fireship; 1763 sold.

**Gramont** 1757-1762 (French privateer *Gramont* ____ ) 18.
Dimensions & tons: 98'1", 80'5" x 27'6½" x 8'8". 324 41/94.
Men: 125. Guns: UD 18 x 6. Taken by *Tartar* 12.1757; 27.6.1762 retaken by French at Saint Johns harbour, Newfoundland.

**Cygnet/Signet** 1758-1768 (French *La Guirlande* 1757 Brest) 18.
Dimensions & tons: 110'11", 90'9¼" x 28'3¼" x 9'. 393 93/94.
Men: 130. Guns 18 x 6 + 12 swivels. Taken 11.1758 [?]; 1768 sold.
PLANS: Lines & profile.

**Epreuve/Epreuve** 1760-1764 (French *L'Eprevue* [?] 1759 or *L'Epreuve* ex privateer *L'Observateur* 1759 ____ ) 14.
Dimensions & tons: 92'9½", 74'2⅛" x 25'9" x 13'7½". 261 47/94.
Men: 125. Guns: 14 x 6 + 10 swivels. Taken by *Niger* late 1760; 03.1764 supposed foundered coming home from Georgia.
NOTE: Originally the privateer *L'Observateur* before being purchased by the French Navy.
PLANS: Taken off & as fitted: Lines & profile.

**Pomona** 1761-1776 (French privateer *Le Chevert* ____ ____ ) 18.
Dimensions & tons: 108', 91'8" x 27'4" x 12'6¼". 364 26/94.
Men: 130. Guns: 18 x 6 + 12 swivels. Taken 30.1.1761; 7.9.1776 supposed to be lost in the West Indies.
PLANS: Lines, profile & decorations. 1773: Lines, profile & decorations/decks.

**Pheasant** 1761-1761 (French *Le Faisan* 1761 ____ ) 14.
Dimensions & tons: 106'2½", 90'6⅝" x 24'7½" x 10'7". 291 54/94.
Men: 125. Guns: 14 x 6 + 10 swivels. Taken by *Albany* 04.1761; 10.10.1761 supposed to have foundered in the Channel.

**Sardoine** 1761-1768 (French *La Sardoine* 1757 ____ ) 14.
Dimensions & tons: 94'4½", 78'9¼" x 24'8½" x 10'1½". 255 74/94.
Men: 110. Guns: 14 x 4 + 10 swivels. Taken 04.1761; 1768 sold.
PLANS: Lines, profile & decorations/lines, profile & decorations/decks.

## Brigs, etc

**Dolphins Prize** 1757-1760 (French ____ ____ ____ ) 12 snow rig.
Dimensions & tons: 67', 54'8½" x 21'4" x 10'2". 132½.
Men: 80 Guns: 12 x 4. Taken by *Dolphin* 06.1757; 1760 sold.

**Goree Brig** 1759-1763 (French ____ ____ ____ ) ____ brig sloop.
Dimensions & tons: ____ , ____ x ____ x ____ . ____ .
Men: ____ . Guns: ____ . Taken at the surrender of Goree 1759; 1763 condemned.

## Sloops, Rig Uncertain

**Etna** 1756 See under Fireships.

**Hazard Prize** 1756-1759 (French privateer ____ ____ Dunkirk) 8.
Dimensions & tons: 61'9", 48'7" x 19'9" x 8'11½". 100 72/94.
Men: 60. Guns: 8 x 4 + 12 swivels. Taken by *Hazard*; 1759 sold.

**Gibraltar's Prize** 1757-1761 (French privateer *Le Glaneur* ____ ____ ) 14.
Dimensions & tons: 63'10", 59'1" [or 49'1"?] x 18'7⅛" x 8'2". 117 22/94.

Men: 50-40. Guns: 14 x 6 + 12 swivels. Taken by *Gibraltar* 02.1757; 1761 sold.

**Flamborough Prize** 1757-1763 (French privateer *Le Général Lally* 1757 Boulogne) 10.
Dimensions & tons: 66'0½", 52'8⅝" x 20'3" x 9'3". 114 82/94.
Men: 70. Guns: 10 x 4 + 12 swivels. Taken by *Flamborough* early 1757; 1763 sold.

**Antigua** 1757-1763 (privateer ____ ____ ____ ) 14.
Dimensions & tons: 71'6", 53' x 23'7½" x 10'6". 157.
Men: 110. Guns: 14 x 4. Purchased at Antigua 11.6.1757; 1763 sold.

**Barbadoes** 1757-1763 (Mercantile ____ ____ ____ ) 14.
Dimensions & tons: 80', 52'10" x 21'6" x ____ . 130.
Men: 100-80. Guns: 14 (both 6 and 4 pounders). Purchased at Antigua 25.11.1757; 1763 sold.
NOTE: Alleged to be schooner-rigged.

**Laurel** 1759-1763 (privateer *Beckford* ____ , ____ ) 12.
Dimensions & tons: 66', 53' [or 53'9"?] x 19'1" x 8'10½". 104 9/94.
Men: 80. Guns: UD 12 x 4 + 10 swivels. Purchased mid-1759; 1763 sold.

## Fireships

**Proserpine** 1756-1756 ( ____ ____ ____ ____ ) ____ .
Dimensions & tons: ____ , ____ x ____ x ____ . ____ .
Men: ____ . Guns: ____ . Fitted as a fireship at Port Mahon 1756; 11.6.1756 lost when the French took Port Mahon.

**Blaze** 1756-1757 (mercantile ____ ____ ____ ) ____ .
Dimensions & tons: ____ , ____ x ____ x ____ . ____ .
Men: ____ . Guns: ____ . Purchased in the East Indies 22.7.1756; 1757 taken out of service; 1759 sold.

**Etna** 1756-1763 (mercantile *Charlotte* ____ ____ ) 8.
Dimensions & tons: 98'5", 79'9¼" x 27'4" x 11'8½". 317.
Men: 45 (as sloop: 80). Guns: 8 x 4 + 6 swivels (as sloop: 14 x 6 + 12 swivels). Purchased when 'about 6½ years old' on 10.11.1756 and fitted as a sloop; 1758 fireship; 1763 sold.

**Vesuvius** 1756-1763 (mercantile *King of Portugal* ____ ____ ) 8.
Dimensions & tons: 91'6", 72'3¼" x 27'10½" x 12'. 298 65/94.
Men: 45 (as sloop: 80). Guns: 8 x 4 + 6 swivels (as sloop: 14 x 6 + 12 swivels). Purchased 17.11.1756 and fitted as a sloop; 1758 fireship; 1763 sold.

**Strombolo** 1756-1768 (mercantile *Owner's Goodwill* ____ ____ ) 8.
Dimensions & tons: 92'10", 74'2¾" x 26'1" x 11'4". 268½.
Men: 45 (as sloop: 100). Guns: 8 x 6 + 8 swivels (as sloop: 14 x 6 + 12 swivels). Purchased 31.12.1756 from Randall (her builder?); 1759 began refitting as a sloop, but this was cancelled; 1768 sold.

**Pluto** 1756-1762 (mercantile *New Concord*, ____ ____ ) 8.
Dimensions & tons: 86'3", 67'10⅜" x 27'4½" x 12'4". 270½.
Men: 45 (as sloop: 100). Guns: 8 x 6 + 8 swivels (as sloop: 16 x 6 + 8 swivels). Purchased 31.12.1756; 1762 sold.

**Proserpine** 1757-1763 (mercantile *Maryland Planter* ____ ____ ) 8.
Dimensions & tons: 90'9", 75'6½" x 25'4½" x 11'6". 253.
Men: 45 (as sloop: 90). Guns: 8 x 4 + 8 swivels (as sloop: 16 x 4 + 12 swivels). Purchased 3.1.1757 (was she originally hired 1755-1756 as an armed ship under her mercantile name?; it seems likely that this is the same ship); 1763 sold.

**Salamander** 1757-1762 (Mercantile *Applewhite & Frere* ____ ____ ) 8.
Dimensions & tons: 88'7½", 72'2¼" x 25'7¾" x 11'8½". 252½.
Men: 45 (as sloop: 90). Guns: 8 x 4 + 8 swivels (as sloop: 16 x 4 + 12 swivels). Purchased 5.1.1757 and fitted as a fireship; 1758 fireship; 1761 sloop; 1762 sold.

## Bomb Vessels

**Racehorse** 1757-1775 **Thunder** -1778 (French *Le Marquis de Vaudreuil* 1754 [?] Nantes) 8.
Dimensions & tons: 96'7", 77'1¼" x 30'8" x 13'4". 385.
Men: 70 (as sloop: 120, as fireship: 45).

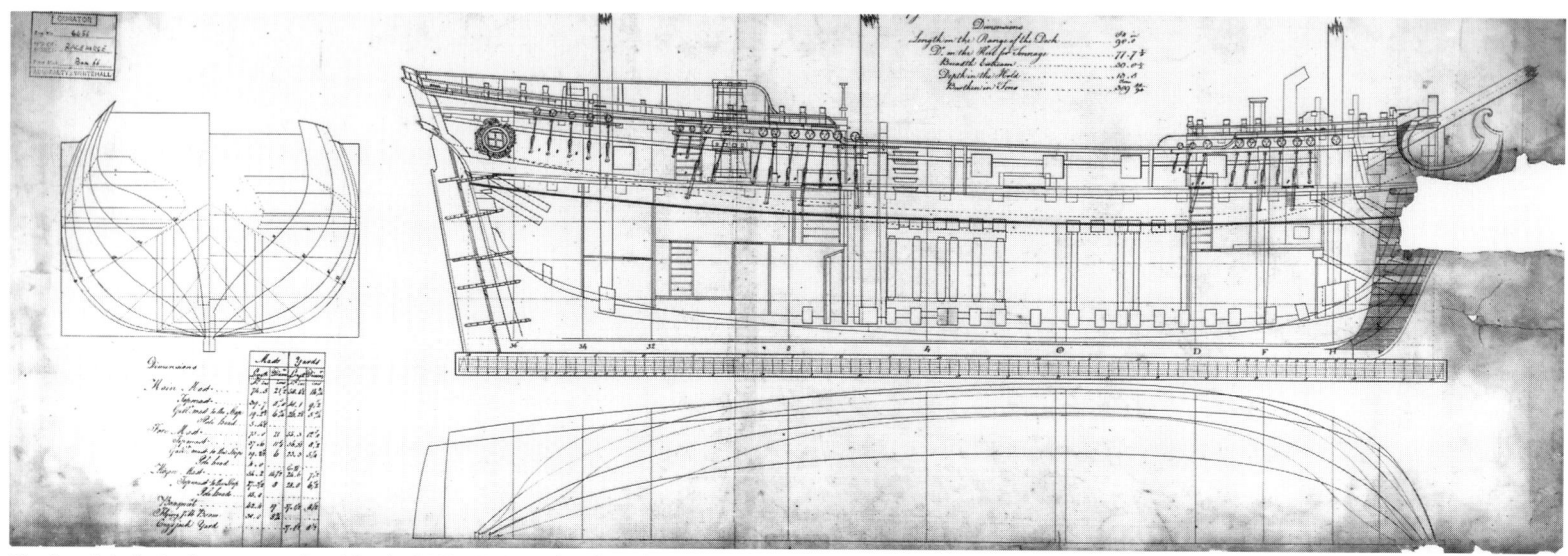

The French built *Racehorse* was purchased for the RN in 1757 as a ship sloop, but rapidly reclassed, first as a fireship and then as a bomb. This lines and profile plan (6456) shows her 'as fitted' for Arctic exploration in 1773. Notice the reinforcing of the bow to deal with ice, the reinforced mortar 'beds' which were retained, also the list of spar dimensions

Guns: 1 x 13" & 1 x 10" mortars + 8 x 6 + 12 swivels (as sloop: 18 x 6 + 14 swivels, as fireship: 8 x 6 + 8 swivels). Purchased 28.4.1757 as ship sloop; 1758 fireship; 1758 bomb; 1773 exploration vessel; 1775 bomb & renamed *Thunder*, 14.8.1778 taken by French *Le Vaillant*.

PLANS: As fitted as a fireship: Profile, lower & upper decks/quarter deck & forecastle. As bomb: Profile & upper deck/ midships section. 1773 fitted for Arctic exploration: Lines & profile/orlop/upper deck/quarter deck & forecastle.

## Cutters

An Admiralty Order of 16.1.1761 stated that 'two sailing cutters' were to be purchased. It is ironical that the first pair of this quintessentially English type of vessel to be commissioned into the Royal Navy were French prizes.

**Lurcher** 1761-1765 (French *La Comtesse D'Ayen* ____ ) 6 carvel-built.
   Dimensions & tons: 54', 40'10½" x 19'4½" x 8'5". 81 57/94.
   Men: 30-35. Guns: 6 x 3 + ____ swivels. Taken ____ ; 1761 purchased; 1763 or 1765 sold.

**Swift** 1761-1762 (French privateer *Le Comte de Vallance* ____ ____ ) 10.
   Dimensions & tons: 53'10", 40'4⅜" x 19'7½" x 8'4½". 82 69/94.
   Men: 30. Guns: 10 x 3 + ____ swivels. Taken 1760; 1761 purchased; 30.6.1762 taken by 22 gun French privateer off Ushant.

In 1763 there was a massive purchase of cutters. It is likely that most, if not all, of these craft had already served as hired vessels, and no doubt most retained their original names. The majority will have been of mercantile origin.

**Adventure** 1763-1768 (____) 6.
   Dimensions & tons: 47'11", 35'1⅞" x 18'0¾" x ____ . 61 1/94.
   Men: 26. Guns: 6 x 3 + 8 swivels. Purchased 02.1763; 1768 sold.
PLANS: Lines, profile & decks.

**Alarm** 1763-1780 (____) 6.
   Dimensions & tons: 53'3", 39'11" x 19'6½" x 8'3". 81¾.
   Men: 30. Guns: 6 x 3 + 10 swivels. Purchased 02.1763; 1780 sold.

**Anson** 1763-1774 (____) 6.
   Dimensions & tons: 51'3", 36'9" x 22'1" x 8'2". 95.
   Men: 30. Guns: 6 x 3 + 12 swivels. Purchased 02.1763; 1774 sold.

**Boscawen** 1763-1773 (____) 4.
   Dimensions & tons: 42'5", 30'9" x 17'2" x 7'6". 48.
   Men: 24. Guns: 4 x 3 + 8 swivels. Purchased 02.1763; 1773 sold.

**Charlotte** 1763-1770 (____) 4.
   Dimensions & tons: 46'7", 32'2½" x 20'4" x 7'10" . 70.
   Men: 26. Guns: 4 x 3 + 8 swivels. Purchased 02.1763; 1770 sold.

**Cholmondley** 1763-1771 (____) 4.
   Dimensions & tons: 57'4", 44'0⅞" x 18'5" x ____ . 79¾.
   Men: 28. Guns: 4 x 3 + 10 swivels. Purchased 01/02 1763; 1771 sold.

**Duke of York** 1763-1766 (____) 4.
   Dimensions & tons: 40', 31' x 18' x 7'4". 53 40/94.
   Men: 24. Guns: 4 x 3[?] + 8 swivels. Purchased in early 1763; 1766 sold.
PLANS: Lines, profile & decks.

**Duke William** 1763-1768 (____) 4.
   Dimensions & tons: 43'2, 32'1½" x 19'6¼" x 7'6". 65 10/94.
   Men: 26. Guns: 4 x 3[?] + 8 swivels. Purchased in early 1763; 5.10.1768 lost.
PLANS: Lines, profile & decks.

**Endeavour** 1763-1771 (____) 4.
   Dimensions & tons: 44'5", 28'2½" x 18'10" x 7'6". 53 10/94.
   Men: 24. Guns: 4 x 3 + 8 swivels. Purchased 02.1763; 1771 sold.
PLANS: Taken off: Lines, profile & decks. As fitted: Lines, profile & decks.

**Esther/Hester** 1763-1779 (____) 6.
   Dimensions & tons: 50'1½", 33'6¾" x 23'10" x 8'3¼". 101¼.
   Men: 30. Guns: 6 x 3 + 12 swivels. Purchased 01.1763; 1779 sold.

**Fly** 1763-1771 (____) 4.
   Dimensions & tons: 46', 33'10½" x 20'10¼" x 8'1¼". 78 34/94.
   Men: 28. Guns: 4 x 3 + 12 swivels. Purchased 01.1763; 1771 sold.
PLANS: Lines, profile & decks.

**Friendship** 1763-1771 (____) 4.
   Dimensions & tons: 40', 28'11" x 19'10" x 7'6". 60 47/94.
   Men: 24. Guns: 4 x 3 + 6 swivels Purchased in early 1763; 1771 sold.

**Goodwill** 1763-1786 (____) 4.
   Dimensions & tons: 44'4", 30'3" x 17'7" x 7'6". 49 70/94.
   Men: 24. Guns: 4 x 3 + 8 swivels. Purchased 01.1763; 1786 sold.
PLANS: Lines, profile & decks/lines, profile & decks.

**Grace** 1763-1772 (____) 6.
   Dimensions & tons: 56'1½", 40'3½" x 22'5½" x 7'9". 108.
   Men: 30. Guns: 6 x 3 + 12 swivels. Purchased 01.1763; 1772 BU.
PLANS: Lines & profile.

**Greyhound** 1763-1780 (____) 4.
   Dimensions & tons: 52'2", 38'7½" x 18'10" x 7'2". 72¾.
   Men: 28. Guns: 4 x 3 + 8 swivels. Purchased in early 1763; 1780 sold.

**Hector** 1763-1773 (____) 4.
   Dimensions & tons: 48'10", 32'7" x 19'9½" x 7'10". 67¾.
   Men: 26. Guns: 4 x 3 + 8 swivels. Purchased in early 1763; 1773 sold.

**Hornet** 1763-1772 (____) 6.
   Dimensions & tons: 50'1", 36'7⅜" x 22'3¾" x ____ . 97.
   Men: 30. Guns: 6 x 3 + 12 swivels. Purchased 01.1763; 1772 sold.

**Hunter** 1763-1771 (____) 4.

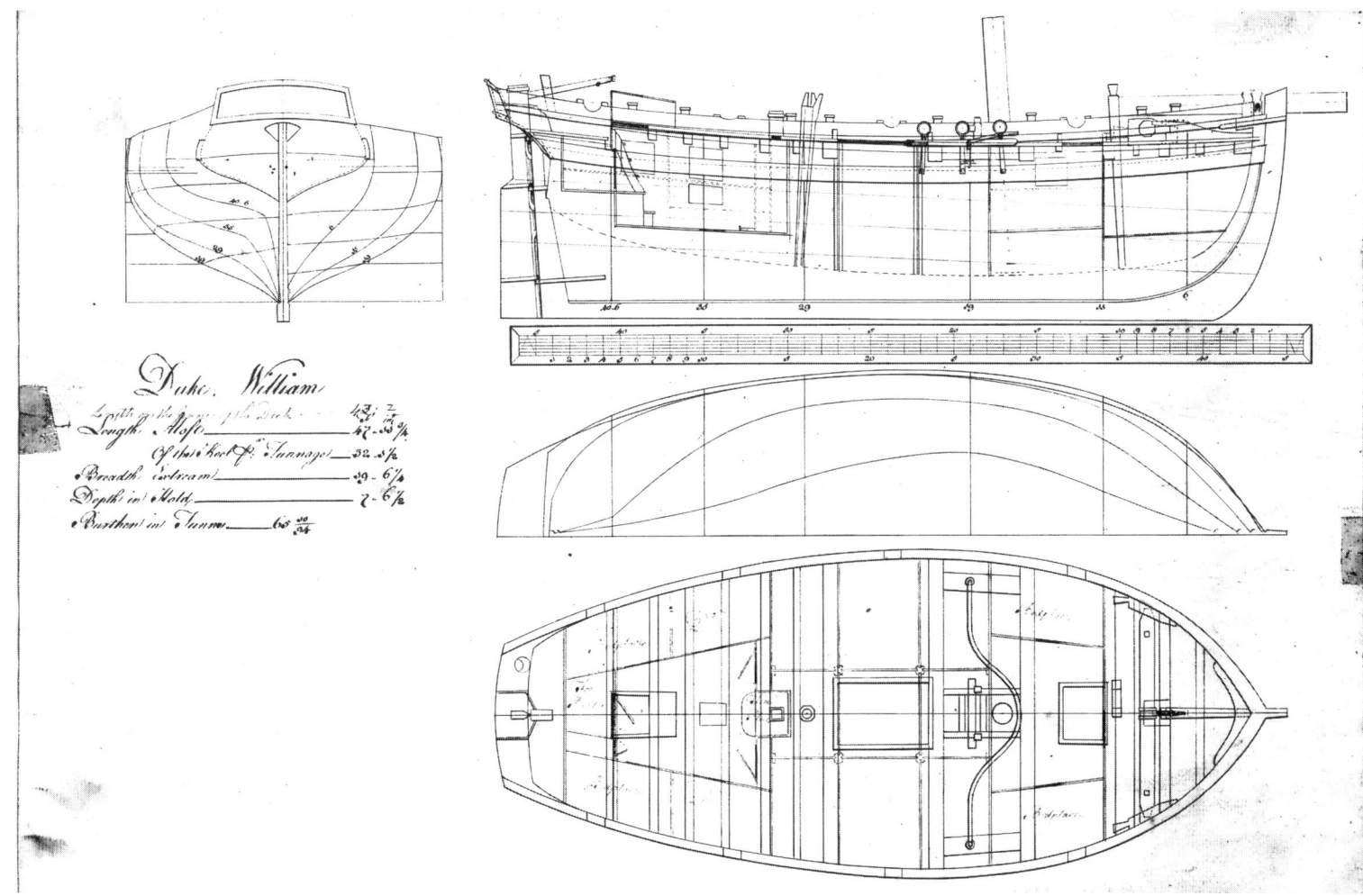

Hull plan (6376) of the cutter *Duke William* purchased in 1763

Dimensions & tons: 50', 35'10½" x 19'6½" x 8'. 72¾.
Men: 28. Guns: 4 x 3 + 10 swivels. Purchased 02.1763; 1771 sold.

**King of Prussia** 1763-1765 ( ___ ) 6.
Dimensions & tons: 50'10", 37'4½" x 21'3½" x 7'10". 90.
Men: 30. Guns: 6 x 3 + 10 swivels. Purchased in early 1763; 6.2.1765 wrecked on the Cross Ledge near Ramsgate.
PLANS: Lines, profile & decks.

**Laurel** 1763-1771 ( ___ ) ___ .
Dimensions & tons: 45'8½", 33'11" x 17'3½" x ___ . 53 37/94.
Men: 24. Guns: 8 swivels. Purchased 12.1763; 1771 sold.

**Lion** 1763-1771 ( ___ ) 4.
Dimensions & tons: 49', 35'11¾" x 17'10¾" x ___ . 61¼.
Men: 26. Guns: 4 x 3 + 8 swivels. Purchased in early 1763; 1771 sold.

**Lord Howe** 1763-1771 ( ___ ) 6.
Dimensions & tons: 52'9", 38'2½" x 20'2¼" x 7'11". 82¾.
Men: 30. Guns: 6 x 3 + 10 swivels. Purchased in early 1763; 1771 sold.

**Mecklenburgh** 1763-1773 (French ___ ) 6.
Dimensions & tons: 54', 48'8" x 19'7" x 8'4½". 82 89/94.
Men: 30. Guns: 6 x 3 + 10 swivels. Purchased 02.1763 (presumably a French prize, certainly French-built); 1773 Breakwater.
PLANS: Lines, profile & decks.

**Meredith** 1763-1784 ( ___ ) 6.
Dimensions & tons: 54'6", 40'1½" x 19'9½" x 8'5½". 83¼.
Men: 30. Guns: 6 x 3 + 10 swivels. Purchased in early 1763; 1784 sold.

**Morning Star** 1763-1773 ( ___ ) 4.
Dimensions & tons: 47'6", 38'3¾" x 18'6" x ___ . 69 14/94.
Men: 26. Guns: 4 x 3 + 8 swivels. Purchased 01/02.1763; 1773 sold.
PLANS: Lines & profile.

**Pitt** 1763-1766 ( ___ ) 6.
Dimensions & tons: 58'8", 44' x 20'8" x 9'1½". 99 90/94.
Men: 30. Guns: 6 x 3 + 12 swivels. Purchased 01.1763; 5.8.1766 supposed to be lost coming from Africa.
PLANS: Lines, profile & decks.

**Prince George** 1763-1771 ( ___ ) 4.
Dimensions & tons: 47', 32'3½" x 19' x 8'. 62.
Men: 26. Guns: 4 x 3 + 8 swivels. Purchased in early 1763; 1771 sold.

**Spy** 1763-1773 ( ___ ) 4.
Dimensions & tons: 47'6½", 34'10" x 17'9¾" x 7'1". 58 ¾.
Men: 24. Guns: 4 x 3 + 8 swivels. Purchased 01/02.1763; 1773 sold.

**Swift** 1763-1773 ( ___ ) 4.
Dimensions & tons: 49'9", 35'8½" x 16'10½" x 7'11". 54.
Men: 24. Guns: 4 x 3 [?] + 8 swivels. Purchased 01/02.1763; 1773 sold.

**Tartuffe** 1763-1770 ( ___ ) 6.
Dimensions & tons: 52', 36'5¾" x 22' x 7'3". 93¼.
Men: 30. Guns: 6 x 3 [?] + ___ swivels. Purchased 01/02.1763; 1770 sold.

**Winchelsea** 1763-1774 ( ___ ) 4.
Dimensions & tons: 44', 32'7¼" x 19'11¼" x 7'10½". 68 87/94.
Men: 26. Guns: 4 x 3 + 8 swivels. Purchased 01/02.1763; 1774 breakwater.
PLANS: Lines, profile & decks.

## Schooners

**Grenville** 1763-1775 (American mercantile *Sally* 1754 Massachussetts Bay) 12.
Dimensions & tons: 54'11", 43'0⅝" x 17'2½" x 7'4". 67 76/94.
Men: 20. Guns: 12 x ___ . Purchased 7.8.1763 in Newfoundland; 1775 BU.

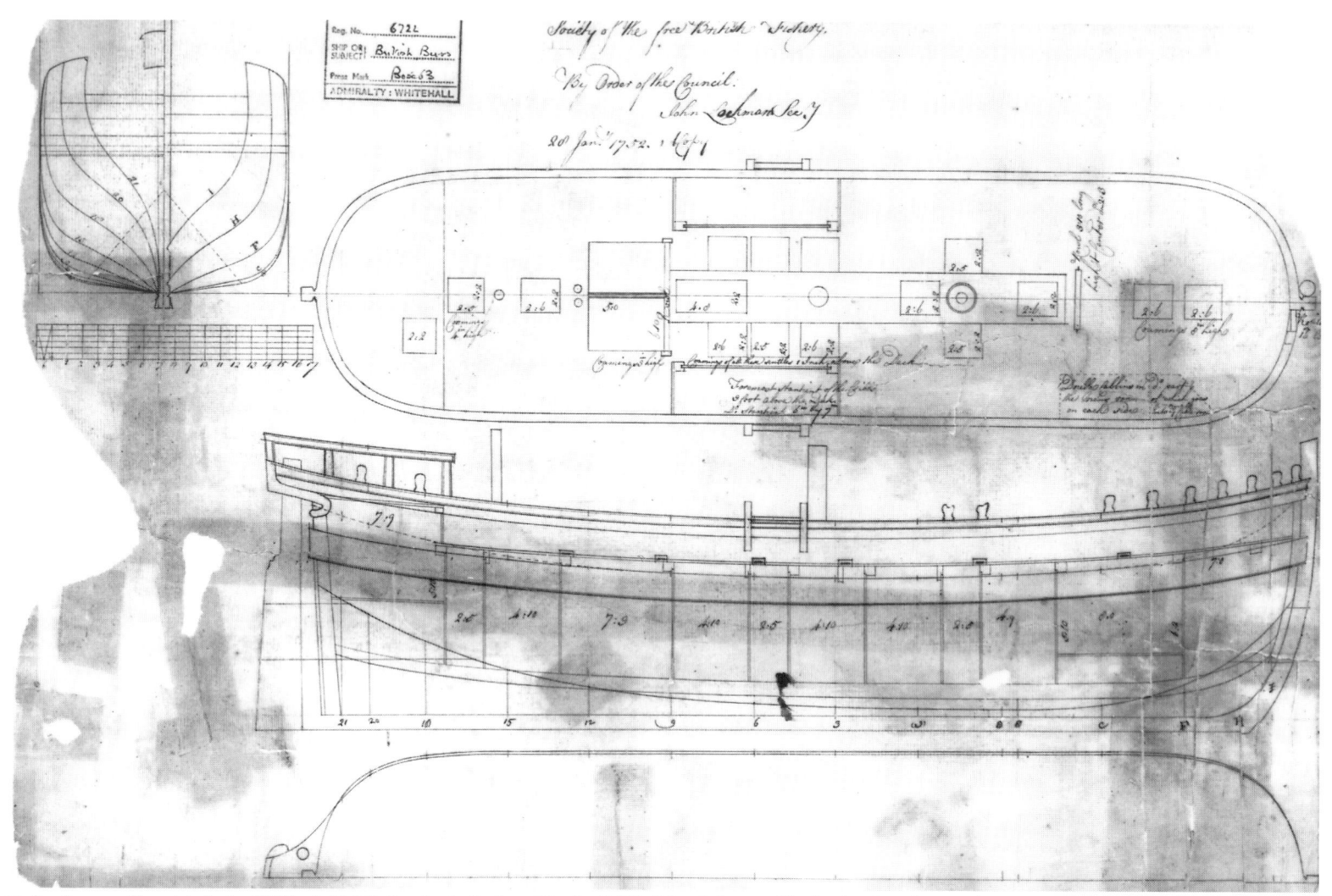

Outline lines, profile and deck plan (6722) of a 'British Buss for the Society of the free British Fishery'; the storeships *London, Canterbury, Medway, Portsmouth* purchased in 1756 were all probably built to this design

## Miscellaneous Small Vessels

**Polacca** 1756-1761 (French [?] ____ Ca 1752 ____ ) Polacca 4.
Dimensions & tons: 67'6", ____ x 22'2" x 8'7½". 146.
Men: 40. Guns: 4 x 4 + 16 swivels. Taken by *Dolphin* 1756; 1769 sold.

**Diligence** 1759-1769 ( ____ ) sloop-rigged vessel.
Dimensions & tons: ____ , 29'7" x 14'6" x 7'8". 33.
Men: ____ . Guns: ____ . Purchased 29.1.1759; 1769 sold.

**Mediterranean/Mediterranean Xebec** 1757 - ____ ? (French [?] ____ ) xebec 12.
Dimensions & tons: 92'6", ____ x 20'6" x 7'9". ____ .
Men: ____ . Guns: 2 x 6 + 10 x 3 + 19 swivels. Oars: 20. Taken (from the French?) 1757; no further information.

## Storeships, etc

**London** 1756-1758 (fishing buss *Helden* [*Holden*?] [? Dutch-built?]) storeship 6.
Dimensions & tons: 64', 54'8⅝" x 16'7" x 8'6". 80 3/94.
Men: 45 Guns: 6 x 4 + 10 swivels. Purchased from the Society for the Free British Fishery 12.11.1756; 02 or 03.1758 lost at Senegal.

PLANS: (only identified as 'British buss') Lines, profile & decks. There is a detailed and attractive sail plan for what may well be the same; or at least a very similar; vessel in the Chapman Collection at the Sjohistoriskas Museum at Stockholm, with an engraved version in Chapman's 'Architectura Navalis Mercatoria'.

**Canterbury** 1756-1760 (fishing buss ____ ) storeship 6.
Dimensions & tons: 64', 54'9⅛" x 16'6" x 8'9". 79 79/94.
Men: 45. Guns: 6 x 4 + 10 swivels. Purchased from the Society for the Free British Fishery 12.11.1756; 1760 yard craft [at Sheerness; 1764 sold].

PLANS: See *London*.

**Medway** 1756-1764 (fishing buss ____ ) storeship 6.
Dimensions & tons: 64'10", 55'5" x 16'9½" x 9'. 83 10/94.
Men: 45. Guns: 6 x 4 + 10 swivels. Purchased from the Society for the Free British Fishery 10.11.1756; 1760 yard craft [ at Sheerness; 1764 sold].

PLANS: See *London*.

**Portsmouth** 1756-1759 (fishing buss *Beckford*) storeship 6.
Dimensions & tons: 63', 54'6" x 16'6" x ____ . 78 or 79.
Men: 45. Guns: 6 x 4 + 10 swivels. Purchased 10.11.1756 from the Society for the Free British Fishery; 1759 lost on the Bar at the capture of Senegal.

**Port Antonio** 1757-17____ (Bermuda sloop ____ ____ ) storeship ____ .
Dimensions & tons: ____ , ____ x ____ x ____ . 80?
Men: ____ . Guns: ____ . Purchased at Jamaica 1757; 1763 registered; fate uncertain.

**Port Royal** 1757-1763 (Bermuda sloop ____ ____ ) storeship ____ .
Dimensions & tons: ____ , ____ x ____ x ____ . 80?
Men: ____ . Guns: ____ . Purchased at Jamaica 1757; 1763 registered, sold.

**Virgin** Ca 1760-1764 ( ____ ) sloop-rigged storeship ____ .
Dimensions & tons: ____ , ____ x ____ x ____ . ____ .
Men: ____ . Guns: ____ . Probably purchased or taken in the West Indies, date uncertain; 1760 briefly in French hands as *La Vierge* but recaptured; 1764 sold.

**Manilla** Ca 1760-1763 (French or Spanish [?] ____ ) storeship (?sloop rig?).
Dimensions & tons: ____ , ____ x ____ x ____ . ____ .
Men: ____ . Guns: ____ . Taken before 1763, probably in the East Indies, and presented to the RN by her captors; 1763 sold.

## Unregistered Vessels

(ie. not mentioned in the primary Admiralty sources consulted.)

*Comet* Galley in service 1756? [dubious, only in Colledge].

*Comet* 1761-1763 (French *Comète* 1752 Brest) Frigate, Fifth Rate 32. 647 tons. Taken 16.3.1761 by *Bedford*; 1763 returned to the French [as Comete and in their service until 1767].

*Duke* 1760-1761 Storeship 10. Purchased on the East India Station 1.1.1761 lost off Pondicherry.

*Erynnis* 1758-1760 (French privateer [?] ____ ) Sixth Rate 24. Dimensions & tons: 121'9", 101'6" x 31'10" x 10'6". 547. Men: 160 Guns: 24. Taken 1758; listed until 1760.

*Garland* 1762 - ____ ? (French *La Guirlande* 1759 Nantes) Sixth Rate 22. Taken 18.8.1762; no evidence of active service in the RN; fate unknown.

NOTE: The suggestion that she was sold in 1783 is probably due to confusion with the 1748 *Garland*. It is unlikely that this vessel lasted that long.

*Guadeloupe* Ca 1760-1762 Sloop in the West Indies. 80 men.

*Hermione* 1760-1763 (French ____ ) 28 guns. 812 tons. Taken outside Pondicherry 10.1760; in service in the East Indies in 1763.

*Industry* - there is a log for a vessel in service 1760-1762 also there was a sloop in service in 1765; which may be the same as the brig *Industry* whose lines were taken off at Deptford in 1765 (83'9", 68'11" x 24'7" x 12'8". 229 50/94 tons).

PLANS: [Brig *Industry*] Lines & profile/sections (2).

*London* buss 1759; served until 1764?

*Lys* 1755 (French *Lys* 1744 Brest) 64 gun ship. Taken 10.6.1755 (before the official outbreak of war) by Boscawen's squadron in the Atlantic when armed 'en flute' and commissioned for the voyage home, but not otherwise employed by the RN.

*New Augusta* 1763-1765 there is a log for a vessel of this name and dates.

*Oiseau* (French *Oiseau* 1757 Toulon) 26 gun Sixth Rate taken by *Brune* in the Mediterranean 23.10.1762; no further information (not taken into the Navy?).

*Oriflamme* (French *Oriflamme* 1744 Toulon) 50 gun Fourth Rate taken by *Isis* 1.4.1761; no further information (not taken into the Navy?).

*Port Royal* 18 guns, dating from 1762; 8.5.1781 taken by the Spaniards at Pensacola [?].

*Protector* 1756-1761 ( ____ ) Sloop? ____ . Acquired 1756; 1761 lost.

*Robust/Robuste* 1758 (French ____ ) taken en flute in 1758 by *Alcide* and *Actaeon*; not taken into service.

*Rodney* 1759-17____ ( ____ ) Cutter 4. No information.

*San Genaro* 1762-1763 Spanish 60 gun Fourth Rate taken at Havanna 13.8.1762; 01.1763 wrecked in the Downs.

*Speaker* 1756-1758 (mercantile? ____ Bermuda built?) brig rigged sloop, 90 men, 10/14 guns; presented by the Barbados merchants 10.1756; in service until 1758.

## Hired Vessels

*Adventure* 1756-1756 Brig, 6 x 3 pdrs. Taken 1756 by French privateer *l'Infernale* (12) off Bamborough.

*Adventure* [?] 1757-1758 Armed ship, 18 guns

*Boscawen* 1759-1760 Armed ship, 16 guns, 100 men; same as that hired in previous war [?].

*Brilliant* 1755-1756 Armed ship, 16 guns + 6 swivels.

*Broderick* [?] 1759-1760 ____ ?

*Cruizer* [?] 1759 Cutter.

*Cypress* 1760 Cutter.

*Drake* 1759 Storeship; 1.1.1760 foundered at Pondicherry (only noted in Colledge).

*Esther* 1760-1762 Armed ship.

*George* [?] 1759-1762 Armed vessel 60 men 24 guns.

*Halifax* [?] 1756-1757 Armed ship 80 men 12 guns.

*Julines and Elizabeth* [?] 1758-1762 transport.

*London* 1757-1758 Armed ship 18 guns.

*Macclesfield* 1755-1758 Armed ship 80 men 16 guns + 6 swivels.

*Maryland/Maryland Planter* [?] 1755-1756 Armed ship 80 men 16 guns + 6 swivels [it seems likely that this was the ship then purchased as a fireship in 1757 and renamed *Proserpine*; see above].

*Prince of Wales* [?] 1756-1758 Armed ship 120 men 18 guns.

*Roast Beef* 1760-1761 Hired privateer.

*Roehampton* 1755-1756 Armed ship 16 guns + 6 swivels.

*Royal Exchange* 1759 ____ ?

*Saint Anne?* 1756-1758 Armed ship 18 guns (built 1756).

*Sarah* 1756-1758 Armed ship 110 men 16 guns.

*Speedwell?* Cutter taken 4.4.1761 by French *Achille* at Vigo (only noted in Colledge).

*Taylor* 1758 Armed ship.

*Union* [?] 1759-1760 Galliott; on W Coast of Africa?

*William and Anne* 1757 armed ship 80 men 8 guns.

# VESSELS ACQUIRED DURING THE PERIOD OF GATHERING TENSION IN AMERICA 1764-1775

## Ship Sloops

*Raven* (ex *Vesuvius*) 1771-1780 (Building Randall, Rotherhithe) 14. Dimensions & tons: 93'5", 76'9" x 27'1⅜" x 11'9½", 300. Men: 125. Guns: 14 x 6? + 14 x ½pdr swivels. Order 25.12.1770 to purchase three ships for use as fireships. Order 4.3.1771 not to purchase the third (the other was *Scorpion* - see below). Purchased whilst building for £2500 and to be named *Vesuvius*; L 15.5.1771; 24.7.1771 ordered to be renamed *Raven* and to be fitted as a sloop; 1780 sold.

*Scorpion* (ex *Etna*) 1771; 1780 (Mercantile *Borryan* [?] ____ ____ ) 14. Dimensions & tons: 94'4", 76'0⅜" x 27'5" x 12'3". 294 84/94. Men: 120. Guns: 14 x 6 + 14 swivels Purchased 01.1771 for use as a fireship and named *Etna*; 07.1771 to be a sloop and named *Scorpion*; 1780 hulked [at New York; 1780 sold].

## Brig

*Dispatch* Ca 1770-1771 (Mercantile ____ ) 8 [?]. Dimensions & tons: ____ , ____ x ____ x ____ . ____ . Men: ____ . Guns: 8? x ____ . Purchased at Antigua Ca 1770; 1771 lost.

## Fireships

*Etna, Vesuvius* 1771 See ship sloops *Scorpion, Raven* above.

## Schooners

(all American built, and purchased there.)

*Egmont* 1764-1776 ( ____ ) schooner or snow 10? Dimensions & tons: 62', 48'9⅝" x 19'7" x 10'4". 99 57/94. Men: ____ . Guns ____ . Purchased by Commodore Palliser 1764; 12.7.1776 lost in Trepassey Bay.

The following 1764 purchases were by Lord Colvill acting under an Admiralty Order dated 7.1.1764:

*Chaleur* 1764-1768 ( ____ ) 12? Dimensions & tons: 70'8". 55' x 20'4" x 7'9½". 120 85/94. Men: 30. Guns: ____ . Purchased 1764; 1768 sold.

PLANS: 1768: Lines, profile & decks (3 copies).

*Gaspee* 1764-1772 ( ____ ) ____ . Dimensions & tons: ____ , 49' x 19'10" x 7'10", 102 44/94. Men: 30. Guns: ____ . Purchased 1764; 9.6.1772 burnt by a mob at Rhode Island.

*Hope* 1764-1776 ( ____ ) ____ . Dimensions & tons: ____ , 45'5" x 20'2" x 8'4". 105 40/94.

Men: 30. Guns: ____ . Purchased at Halifax 1764; 1776 put out of commission.

*Saint John* 1764-1777 ( ____ ) ____ .
Dimensions & tons: ____ , 50'6" x 20'8" x 8'. 114 65/94.
Men: 30. Guns: ____ . Purchased at Boston 1764; 1777 condemned.

*Saint Lawrence* 1764-1766 ( ____ ) ____ .
Dimensions & tons: ____ , 50'6" x 20'8" x 8'. 114 65/94.
Men: 30. Guns: ____ . Purchased 1764; 16.7.1766 struck by lightning & blew up off Cape Breton.

*Magdalen* 1764-1777 ( ____ ) ____ .
Dimensions & tons: ____ , 48'8" x 18'8" x 8'4". 90 3/94.
Men: 30. Guns: ____ . Purchased 1764 (or 1769 in Canada); 1777 sold at Quebec.

*Sultana* 1768-1773 ( ____ , America).
Dimensions & tons: 50'6", 38'5⅛" x 16'0¾" x 8'4". 52 68/94.
Men: 25. Guns: 8 x ½pdr swivels. Purchased 18.3.1768; 1773 sold.
PLANS: Taken off & as fitted: Lines, profile & upper deck/lower deck.

*Halifax* 1768-1775 ( ____ , America).
Dimensions & tons: 58'3", 46'10½" x 18'3" x 8'10". 83 4/94.
Men: ____ . Guns: 6 swivels [?]. Purchased 10.1768; 1775 lost near Machias Harbour, New England.
PLANS: Taken off & as fitted: Lines, profile & decks.

*Sir Edward Hawke* 1768-1773 ( ____ America [Marblehead?]) 8.
Dimensions & tons: ____ , ____ x ____ x ____ . ____ .
Men: ____ . Guns: 8 x ____ . Purchased 1768; 1773 sold.
PLANS: Lines (?).

*Earl of Northampton* 1769-1774 ( ____ America) 'sloop on survey'; schooner rig 6.
Dimensions & tons: ____ , ____ x ____ x ____ . 48.
Men: 16. Guns: ____ . Purchased 1769; used on the coast of Florida; 1774 sold.

*Egmont/Earl of Egmont* 1770-1773 ( ____ America) 6.
Dimensions & tons: ____ , ____ x ____ x ____ . ____ .
Men: 30. Guns: 6 x ____ . Purchased 1770 (or 03.1767?); 1773 sold (or 1777?).

*Endeavour* 1775-1782 ( ____ America) 14.
Dimensions & tons: ____ , ____ x ____ x ____ . ____ .
Men: 45. Guns: 14 x ____ . Purchased 1775; 1782 sold (also stated to have foundered at Jamaica during the hurricane of 11.10.1780; possibly salved after this?).

*Halifax* 1775-1779 ( ____ America) 6.
Dimensions & tons: 58'3", 46'10½" x 18'3" x 8'10". 83.
Men: 30. Guns: 6 x ____ + 12 x ½ pdr swivels. Purchased 1775; 1779 condemned; 1780 sold.

*Saint Laurence* 1775-1776 ( ____ America).
Dimensions & tons: 61'10", 49'11" x 20'0½" x 7'6½". 106½.
Men: 30. Guns: ____ . Purchased 1775; 1776 sold *or* 1777 put out of service at New York? (possibly two separate schooners of the same name involved?).

## Storeships, etc

*Canceaux* 1764-1782 (Mercantile *William* ____ ) 'sloop on survey'; two masted rig [brig?] 6.
Dimensions & tons: 80'6", 64'10¼" x 23'1" x 10'9". 183 77/94.
Men: 55. Guns: 6 x ____ + 8 swivels. Purchased for surveying the coast of North America in early 1764; 1770s classed as an armed ship; 1782 sold.
PLANS: Lines, profile & decks.

*Bird* 1764-1775 ( ____ ) 'sloop on survey'; rig uncertain.
Dimensions & tons: 58'6", 45'5" x 17'7¾" x 8'1" . 75 20/94.
Men: 30. Guns: 8 x ____ . Purchased from Henry Bird (her builder?) 05.1764; 1775 BU.

*Florida* 1764-1772 (Mercantile *Gloucester* ____ ) storeship.
Dimensions & tons: ____ , 72' x 27'11½" x ____ . 299 40/94.
Men: 24-26-36 Guns: 10 swivels. Purchased 08.1764; carried stores to 'the Gulf of Mexico and the Coast of Florida'; 1772 BU.

*Experiment* 1765-1768 (Mercantile *Experiment* 1758 Wells, Deptford) storeship.
Dimensions & tons: 96', 80'5⅜" x 28'3½" x 13'5". 342 47/94.
Men: 30. Guns: 10 x ____ . Purchased 08.1765; 1768 sold.
NOTE: Her lines resemble those of contemporary French vessels, with sharp turn of the bilge and much tumble-home, despite her British build.
Plans: Lines & profile.

*Endeavour* 1768-1775 (collier *Earl of Pembroke* 1768 Whitby) 'bark'/sloop.
Dimensions & tons: 97'8", 81' x 29'2" x 11'4". 366½.
Men: 85. Guns: 6 x 6 + 8 (later 4) swivels. Purchased 03.1768 at James Cook's urging for his first voyage; 1773 storeship; 1775 sold.
NOTE: Sometimes classed as a 'bark' (or 'cat bark') which was a description of the type of hull - full form, no head, etc - rather than of the rig which was that of a 3 masted ship.
PLANS: Lines & profile/lines & profile/lines, profile & decks/lower deck/main deck/cabins/quarter deck/decks/decks.

*Earl of Northampton* 'sloop on survey' See under schooners above.

*Adventure* (ex *Raleigh*) 1771-1783 (collier *Marquis of Rockingham* 1770? Whitby) sloop [ship-rigged exploration vessel].
Dimensions & tons: 99'3", 76'9½" x 28'4" x 13'. 336 41/94.
Men: 80. Guns: 10 x 4 + 8 swivels. Purchased for Cook's voyage; 1775 storeship; 1783 sold.
PLANS: Lines & profile. 1771: decks. 1772: decks.

*Resolution* (ex *Drake*) 1771-1782 (collier *Marquis of Granby* ____ ) sloop

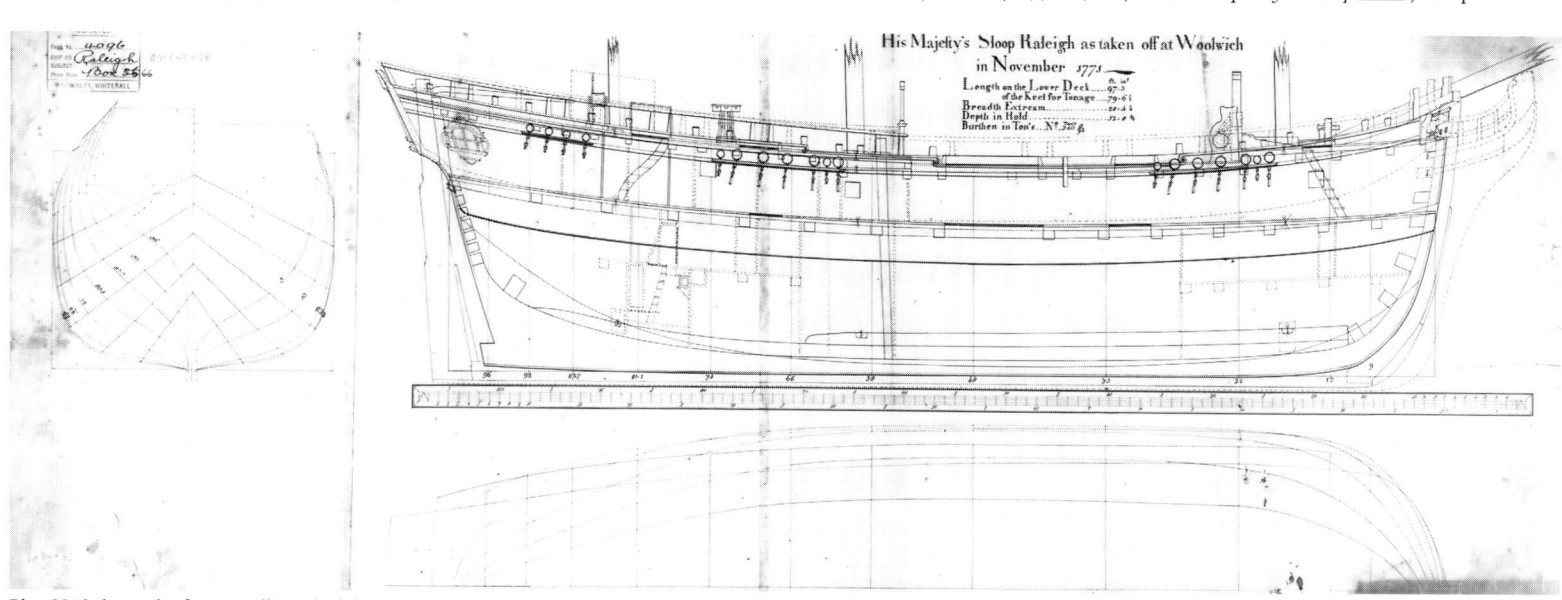

Plan 3962 shows the former collier which became Cook's *Adventure* 'as taken off' in 1771 whilst still bearing her intermediate name of *Raleigh*. Notice the dotted addition of a full head to the plain rounded 'cat bark' bow

[ship-rigged exploration vessel].
Dimensions & tons: 110'8", 93'6" x 30'5⅛" x 13'1½". 461 24/94.
Men: 110 Guns: 12 x 6 + 12 swivels. Purchased 1771 for Cook's voyage; later armed transport; 9.7.1782 taken by the French *Sphinx* in the East Indies; [in French service until 1783].
PLANS: Lines & profile. As fitted: Lines & profile/orlop & lower deck/upper deck.

*Florida* 1774-1778 ( ___ ) 'sloop on survey' [? schooner ?].
Dimensions & tons: 56', 41' x 17'6" x 6'. 68.
Men: 16. Guns: 6 x ___ + 8 swivels. Purchased in the West Indies 1774; 1778 sold at Pensacola (or 1776 at Jamaica?).
NOTE: Is this the same as the *West Florida* allegedly captured at Pensacola in 1779 by the Spaniards (which seems likely)?

## Unregistered Vessels

(ie. not recorded in primary Admiralty sources consulted.)

*Diana* schooner 6. About 120 tons. 30 men. On 8.1.1775 Admiral Graves, the commander on the American station, wrote: '....I have taken upon me to purchase the *Diana* schooner of 120 tons, about 8 months old, so exceedingly well built that she is allowed to be the best Vessel of the Kind that has been yet in the King's service, her first cost is £750 Sterling, and as I have thought it best for His Majesty's service that she should be an established armed cruiser. I have directed the necessary alterations to be made in her Hull, and for her to be fitted in all respects like other Vessels of her class; She will have the Saint Lawrence's guns.'; 28.5.1775 abandoned and burnt at Boston, Massachussetts.

*Diligent* 1770-1775 schooner 8 guns, 45 tons, 30 men; purchased 1770; 15.6.1775 captured at Machias.

*Endeavour* Ca 1770 [?] possibly a schooner of this name in service on the American Station?

*Fortune* 1770-1772? ( ___ ) brig sloop 10.
Dimensions & tons: 70', ___ x 24' x ___ . 120.
Men: ___ . Guns: 10 x ___ . Purchased 1770; fate unknown, though still in service in 1772.

*Germain* 1763 schooner or brig 54'11", 43'1" x 17'2" x 7'4". 67½. Built 1763? Purchased 1764?; no further information.

*Hawk* schooner in service 1775; 4.4.1776 taken by the Americans.

*Leopard* 1770-1772 there is a log for a vessel of this name and dates.

*Margaretta* 1775?-1775 cutter. Purchased 1775?; 12.6.1775 taken by the American 'rebels'.

*Prince Edward* 1772-1775 there is a log for a vessel of this name and dates.

*Saint Lawrence* 1767-1775 (Mercantile *Sally* ___ ) schooner 6.
Dimensions & tons: ___ , 50'6" x 26'8" x 8'. 114 [?]
Men: ___ . Guns: 6 x ___ + 8 swivels. Purchased from Mr Gerrish at Halifax in 1767; served in North American waters; 1775 lost?; or 1776 sold.

*Swift* 1773-1784 sloop, 8. Purchased 03.1773 at Antigua; 13.5.1784 sold? (Colledge only).

## Hired Vessels

*Dispatch* 1764 Sloop 14 guns; suggested that this was the vessel which capsized in the Saint Lawrence in 1780 which seems unlikely but not impossible.

*Liberty* 1768-1769 Sloop; lost by fire 1769.

# Chapter 5: The American War of Independence 1776-1783

This war began as a colonial conflict with insurgents who from the start could commission numbers of privateers and who rapidly proved capable of building warships up to the size of frigates.[1] By the time the war had reached its half-way point, Britain stood alone against the three next biggest naval powers, France, Spain and the Netherlands. In the end there was no irretrievable naval disaster, though a drawn battle at the Capes of the Chesapeake ensured that Yorktown surrendered and the survival of the United States was made certain. As will be seen from the listing below, the Royal Navy did none so ill in the exchange of prizes. The largest French warship was captured, though soon lost through stress of weather. For once the Royal Navy had taken the lead in building the first larger frigates armed with 18 pounder guns in the main battery, but French vessels of this type were captured before the end of the war. Before then the Dutch had joined the anti-British alliance, a decision which the evidence of the prizes noted here shows was not entirely advantageous to them.

This was the one war of the period in which there were more fleet actions overseas than in European waters, where the strategic importance of the West Indies and the struggle for independence of the colonies meant that the focus of naval interest was mostly on the Western side of the Atlantic, shifting North from the West Indies to North American waters as the hurricane season approached, then back again to the Caribbean once the danger from tropical storms was past. Though on a lesser scale, there were more major actions in the Indian ocean - the series of ding-dong battles between Hughes and Suffren - than in any other war of the sailing era. The result of all this extra-European activity is that it is particularly difficult to document all the prizes and purchases for this war. More were acquired in the East and West Indies and in North American waters than at any other time; thus the Admiralty listings, at the mercy of slow and uncertain communications, of distant Admirals double-booking names and sending inadequate information. The sheer volume of business falling on the shoulders of a small number of clerks did not help either. The usual sources for this work are particularly fragmentary, confusing and inadequate for this period, and the record presented here for the smaller and less important vessels are correspondingly incomplete and lacking in certainty.

Another consequence of the widespread nature of this war, and particularly an indication of the enormous effort involved in supporting a major military and naval effort on the other side of the Atlantic, can be seen in the large number of storeships and transports purchased for the Navy and listed here. These were only a small proportion of a much greater number of merchantmen purchased or hired for the government in this conflict.[2]

As usual the beginning of the war was marked by the purchase of merchantmen to be converted to sloops, fireships and the like, whilst the number of schooners and cutters was also increased in the same way. The menace of the Combined fleets of France and Spain cruising in the Channel and outnumbering the defending fleet (1778) was marked by further purchases of ships to be used as fireships. However, mercantile conversions proved too slow to keep up with the defending fleet, which helps to explain the adoption of a very fine-lined design for the *Tisiphone* Class of purpose built fireships ordered in that year.

Other trends which surface in this chapter are the increasing numbers of brigs coming into service from various sources. Also a general increase in the numerical strength of navies in general and the Royal Navy in particular is reflected in the length of this chapter when compared with its predecessors. One type name that appears at this time but then disappears again is 'galley', used, apparently, to describe coast defence gunboats, probably of a variety of types, different rigs, some purpose built and some probably conversions of merchant craft. Some appear to combine oars and sails, others to have a variety of sailing rigs. Though a number of vessels called 'galleys' are listed as serving in the Royal Navy, all of them acquired on the North American coast, there are, alas, no surviving plans of any of these; though there is a plan for a captured American galley serving on Lake Champlain, the one thing one can be sure of is that it is not representative of the seagoing vessels given the same name. The common factors that are identifiable are (comparatively) small size and an armament consisting of one or a small number of heavy guns. Probably some were fairly similar to the gunboats of the next war. It is likely that several were the galleys purpose built for American States navies and captured from their original owners. Unfortunately, I have not, so far, been able to make a direct identification or connection here.

The short list of vessels obtained between the American and French Revolutionary wars is of particular interest because it includes one of the only two frigates that the Royal Navy purchased from the builder, who had constructed her 'on spec'. As Robert Gardiner has observed, it is probably significant that both were the products of the same builder, Adams of Bucklers Hard. Also the developing non-military role of the Royal Navy is seen in the listing of Vancouver's *Discovery*, Bligh's *Bounty* and *Providence*, the surveying ship *Sorlings*, not to mention the conversion of the storeship *Berwick* to become the flagship of Australia's 'First Fleet' under the new name of *Sirius*.

## Large Three-deckers

**Ville De Paris** 1782-1782 (French *La Ville de Paris* 1764 Rochefort, Clairin Deslauriers design) First Rate 100.
  Dimensions & tons: 185'7½", 153' x 53'8½" x 22'2". 2347.
  Men: 850. Guns: ____ . Taken 12.4.1782 at the Saintes; Ca 30.9.1782 foundered on the Banks of Newfoundland in a gale.
NOTE: Began building as the 90 gun ship *L'Impetueux* but name changed before launching. Converted to 100/104 gun ship during her 1779 repair.

## Large Two-deckers

**Gibraltar** 1780-1813 (Spanish *Fenix/San Alejandro* 1749 Havanna) Third Rate 80.
  Dimensions & tons: 178'10¾", 144'5¾" x 52'11¾" x 22'1¾" 2157.
  Men: ____ . Guns: GD 30 x 24, UD 32 x 18, QD 12 x 9, Fc 6 x 9.
  Taken by Rodney off Cape St. Vincent 16.1.1780; Ca 1805 reclassed as a Second Rate; 1813 hulked [at Plymouth as a powder hulk; 1824 to Pembroke as a lazaretto; 1836 BU].
PLANS: Taken off: decks. Taken off & as fitted: Lines, profile & decorations. As fitted: decks. 1815 as powder magazine: Profile/hold/orlop/lower deck/upper deck/quarter deck & forecastle/roundhouse.

---

1  By the end of the conflict the Americans were completing a 74 gun ship, but this they never commissioned. Instead she was given to the French.
2  See David Syrett, *Shipping and the American War 1775-83* (London 1970) for further details of this massive effort.

## 74 Gun Ships (Third Rates)

**Hector** 1782-1782 (French *L'Hector* 1755 Toulon, P B Coulomb design).
  Dimensions & tons: 170'2½" , 144' x 48'2½" x 18'11". 1783.

# THE AMERICAN WAR OF INDEPENDENCE 1776-1783

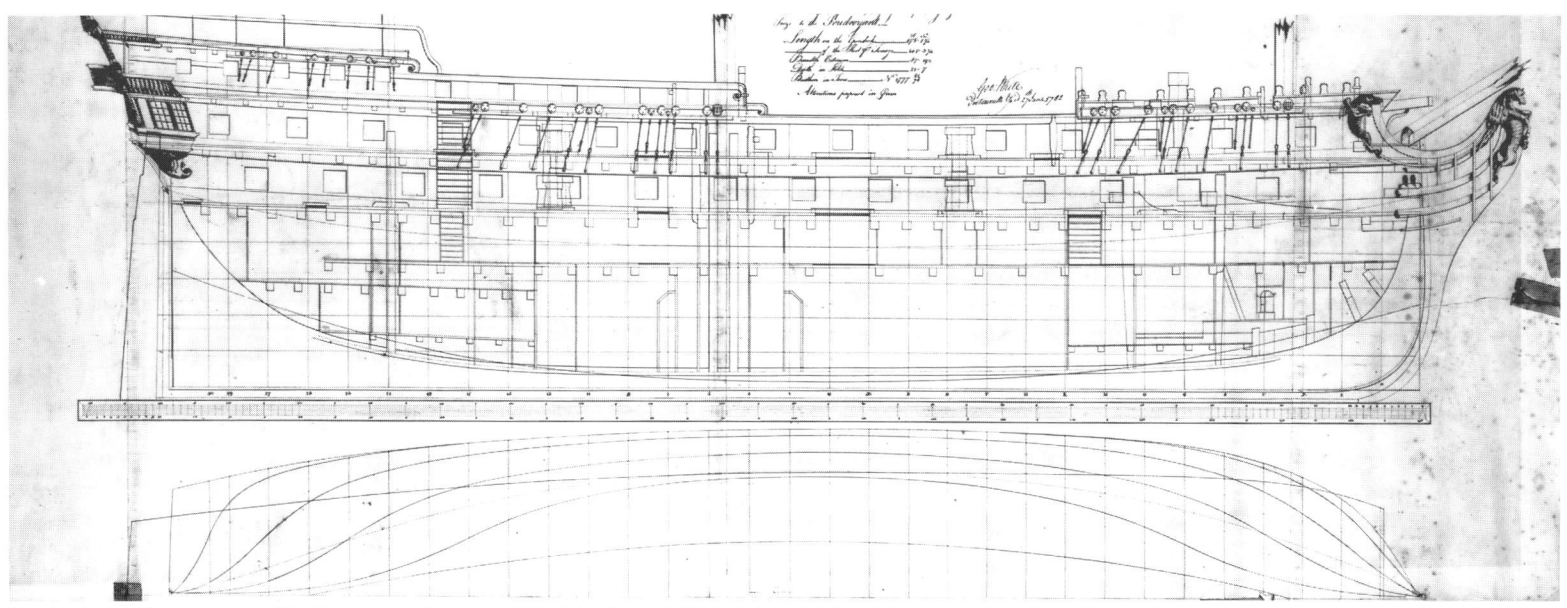

Part (excluding the body plan) of the lines and profile plan (x685) of the French 74 *Pegase* taken 1782

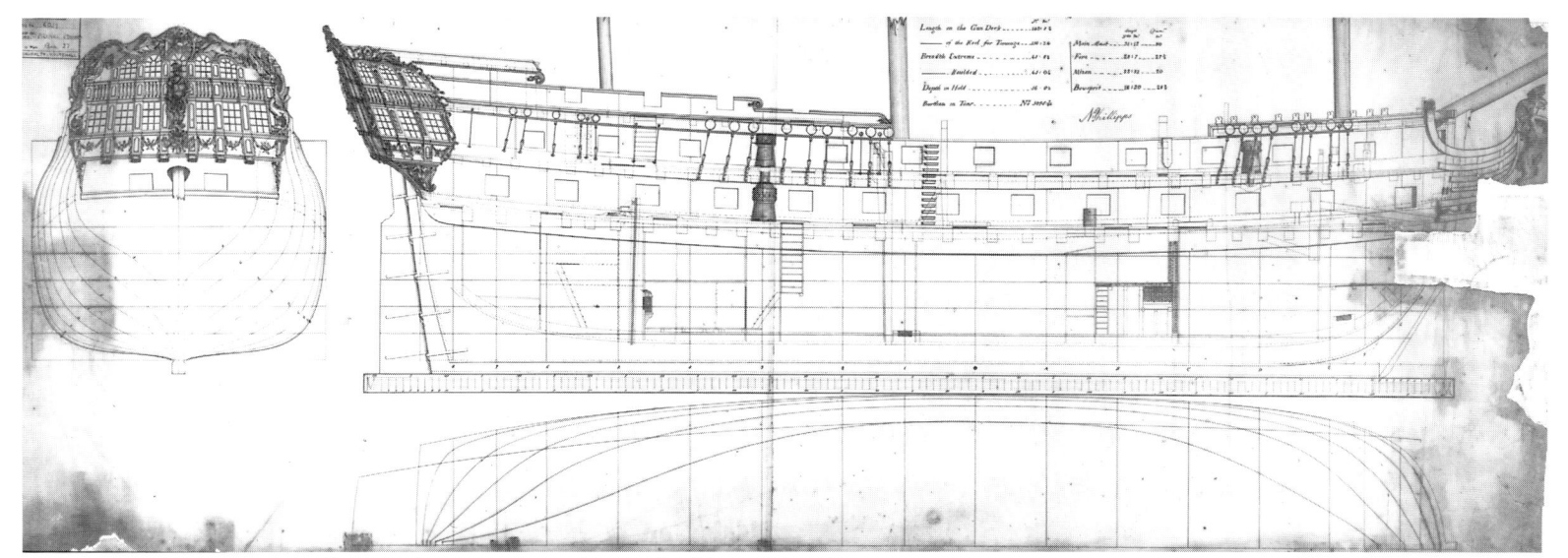

Lines and profile plan (6029) 'taken off' the Dutch 62 or 56 gun vessel *Prince Edward* (ex *Mars*) in 1781

Men: 600. Guns: GD 28 x 32, UD 28 x 18, QD 14 x 9, Fc 4 x 9. Taken at the Saintes 12.4.1782; 5.9.1782 retaken by the French *Aigle* and *Gloire*; [3.10.1782 sank in a gale].

**Glorieux** 1782-1782 (French *Le Glorieux* 1756 Rochefort, Clairin Deslauriers design).
Dimensions & tons: 175', 144'3" x 47'4½" x 21'3". 1718.
Men: 600. Guns: ____ . Taken by Rodney's fleet near the Saintes 15.4.1782; 16.9.1782 foundered in a gale on the Banks of Newfoundland.

**Pegase** 1782-1794 (French *Le Pégase* 1781 Brest, Groignard design).
Dimensions & tons: 178'1¾", 145'3¾" x 47'11½" x 21'7". 1777 68/94.
Men: 600. Guns: GD 28 x 32, UD 28 x 18, QD 14 x 9, Fc 4 x 9. Taken off Ushant by *Foudroyant* 21.4.1782; 1794 hulked [at Portsmouth as a prison ship; 1801 prison hospital ship; 1815 BU].
PLANS: Taken off & as fitted: Lines, profile & decorations. 1799 as hospital/prison hulk: decks. 1804: decks. No date as hospital ship: decks.

**San Miguel** 1782-1791 (Spanish *San Miguel* 1773 Havanna).
Dimensions & tons: 182'1½", 149'9⅞" x 48'11⅛" x 20'8". 1908 31/94.
Men: 650. Guns: GD 28 x 32, UD 30 x 18, QD 12 x 9, Fc 4 x 9. Taken after grounding at Gibraltar 10.9.1782; 1791 sold.
PLANS: Lines, profile & decorations/orlop/gun deck/upper deck/quarter deck & forecastle/roundhouse.

## Small Line of Battle Ships

**Prince William** 1780-1788 (Spanish *Guipuscano* ____ ) Third Rate 64.
Dimensions & tons: 153'2¼", 130'3⅜" x 44'1" x 19'9¼". 1346 61/94.
Men: 500. Guns: GD 26 x 24, UD 28 x 12, QD & Fc 10 x 9. Taken by Rodney with the Caracas Convoy 8.1.1780; 1788/1791 hulked [sheer hulk at Portsmouth; later fitted for the reception of guns; 1817 BU].
PLANS: Lines & profile.

**Monarca** 1780-1791 (Spanish *Monarca* [70/74] 1756 Ferrol) Third Rate 68.
Dimensions & tons: 174'10¼", 145'9½" x 49'11½" x 20'8". 1934 46/94.
Men: 520. Guns: GD 28 x 24, UD 30 x 18, QD & Fc 10 x 9. Taken by Rodney in the Moonlight Battle 16.1.1780; 1791 sold.
PLANS: Lines, profile & decorations.

**Princessa** 1780-1784 (Spanish *Princessa/Santa Barbara* [74] 1750 Havanna) Third Rate 70.
Dimensions & tons: 170'2¼", 138'3" x 51'2¼" x 22'1¼". 1926 67/94.
Men: 560. Guns: GD 28 x 24, UD 30 x 18' QD & Fc 12 x 9. Taken by Rodney in the Moonlight Battle 16.1.1780; 1784 hulked [sheer hulk at Plymouth; 1809 BU].
PLANS: Lines/main mast. 1809 as sheer hulk: profile above water line/main deck/quarter deck/forecastle. No date: profile/main deck.

**Diligent** 1780-1784 (Spanish *Diligente* ____ ) Third Rate 70.
Dimensions & tons: 175'1", 147'7" x 49'10½" x 20'6½", 1953.

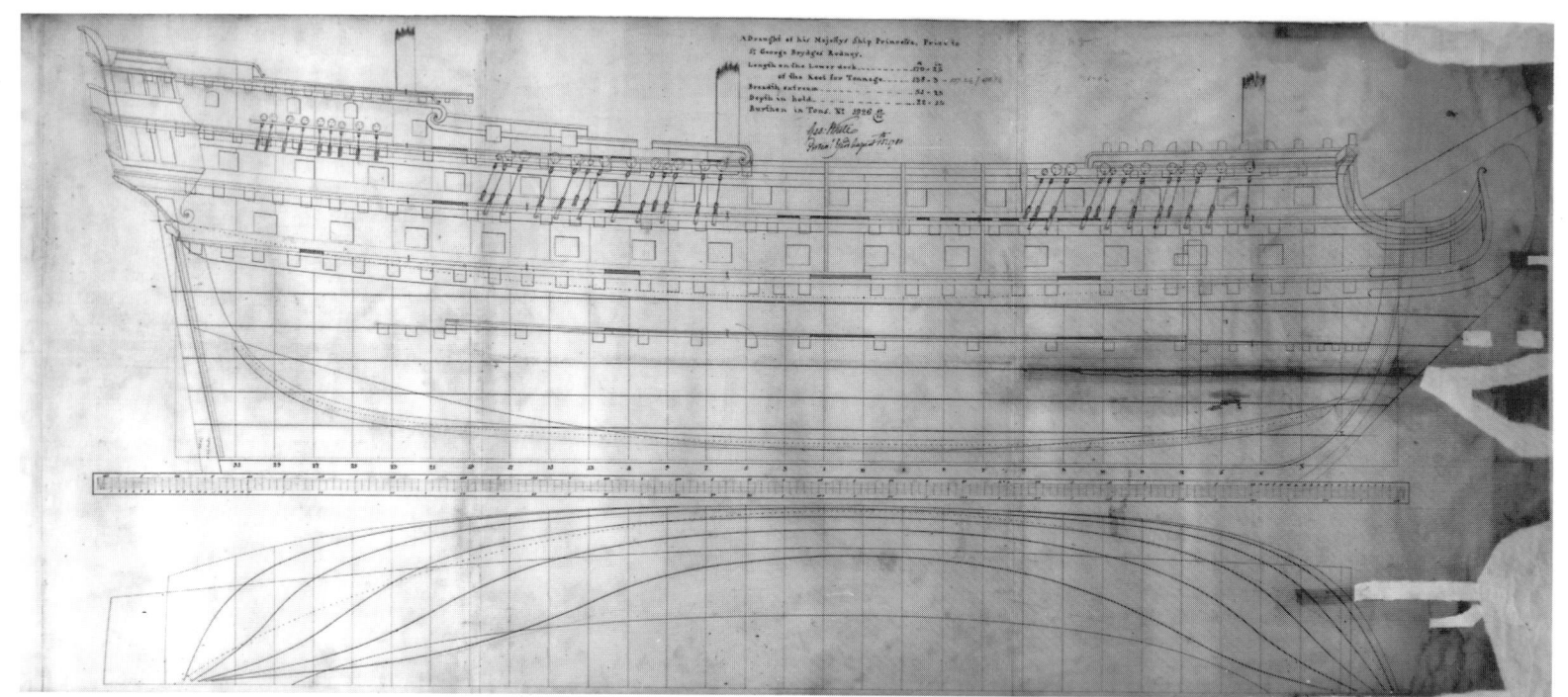

Part (excluding the body plan) of the lines and profile plan (x855) of the Spanish prize 70 gun ship *Princessa* taken 1780

Men. 520/560. Guns: LD 28 x 24, UD 30 x 12, QD & Fc 10 x 9. Taken by Rodney in the Moonlight Battle 16.1.1780; 1784 sold.

***Prothee*** 1780-1795 (French *Le Protée* 1772 Brest, J Ollivier design) Third Rate 64.
Dimensions & tons: 164'8½", 135'1½" x 44'6" x 19'2½", 1423 38/94.
Men: 500. Guns: GD 26 x 24, UD 26 x 18, QD 10 x 9, Fc 2 x 9. Taken by Digby's Squadron 23.2.1780; 1795 hulked [at Portsmouth as a prison ship; 1815 BU].

PLANS: Lines, profile & decorations.

***Prince Edward*** 1781-1782 (Dutch *Mars* 1763 Rotterdam Admiralty) Fourth Rate 62/56.
Dimensions & tons: 142'8¾", 113'3⅝"x 41'8¾" x 16'0¼", 1075 7/94.
Men: 420. Guns: GD 24 x 24, UD 26 x 12, QD 8 x 6, Fc 2 x 6. Taken by Rodney near Saint Eustatius 4.4.1781; 1782 hulked [at Chatham as a receiving ship; 1802 sold].

PLANS: Lines & profile/gun deck/upper deck/quarter deck & forecastle/midships section.

***Argonaut*** 1782-1797 (French *Le Jason* 1779 Toulon, J Coulomb design) Third Rate 64.
Dimensions & tons: 166'7". 136'6⅛" x 44'8⅛" x 19'1". 1451 75/94.
Men: 500. Guns: GD 26 x 24, UD 26 x 18, QD 10 x 9, Fc 2 x 9. Taken in the Mona Passage 19.4.1782; 1797 hulked [at Chatham as a hospital ship; 1831 sold].

PLANS: Lines, profile & decorations/decks. 1797 as hospital ship: profile (incomplete)/decks.

***Solitaire*** 1782-1786 (French *Le Solitaire* 1774 Brest, Groignard design) Third Rate 64.
Dimensions & tons: ____ , ____ x ____ x ____ . 1397.
Men: 500. Guns: ____ . Taken 6.12.1782 by *Ruby* and *Polyphemus*; 1786 sold.

***Caton*** 1782-1794 (French *Le Caton* 1770-1777 Toulon, J Coulomb design) Third Rate 64.
Dimensions & tons: 166', 136'4¾" x 44'0½" x 19'4". 1407.
Men: ____ . Guns: ____ . Taken in the Mona Passage 19.4.1782; 1794 hulked [at Plymouth as a hospital ship; 1815 sold].

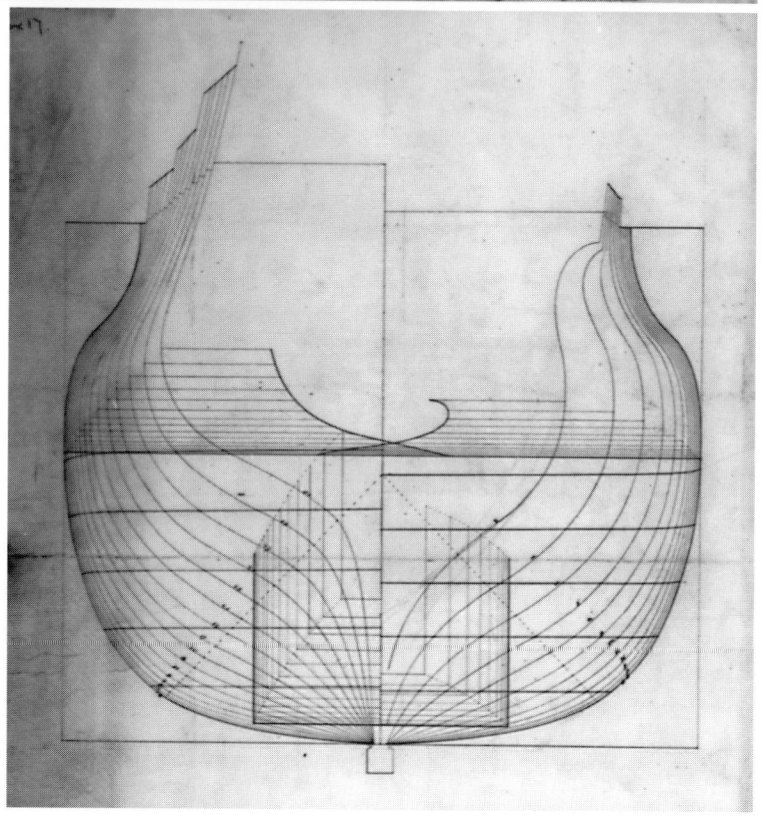

## Small Two-deckers

***Princess Caroline/Princess Carolina*** 1780-1783 (Dutch *Princessen Carolina* [54] 1748 Rotterdam Admiralty) Fifth Rate 44.
Dimensions & tons: 129'1", 107'5" x 38'10" x 15'6". 862.
Men: 280. Guns: ____ . Taken by *Bellona* 31.12.1780; 1783 hulked [at Sheerness as a receiving ship; 1799 BU].

PLANS: Lines & profile/gun deck/upper deck/quarter deck & profile.

***Rotterdam*** 1781-1785 (Dutch *Rotterdam* 1761 Rotterdam Admiralty) Fourth Rate 50.
Dimensions & tons: 135'8", 114'4" x 39'5" x 15'3". 877 73/94.
Men: 350. Guns: ____ . Taken 5.1.1781 by *Warwick*; 1785 hulked [sheer hulk and receiving ship at Chatham; 1806 sold].

PLANS: Lines & profile/Lines, profile & decorations. 1785 as sheer hulk: Profile (incomplete) & upper deck/quarter deck & forecastle.

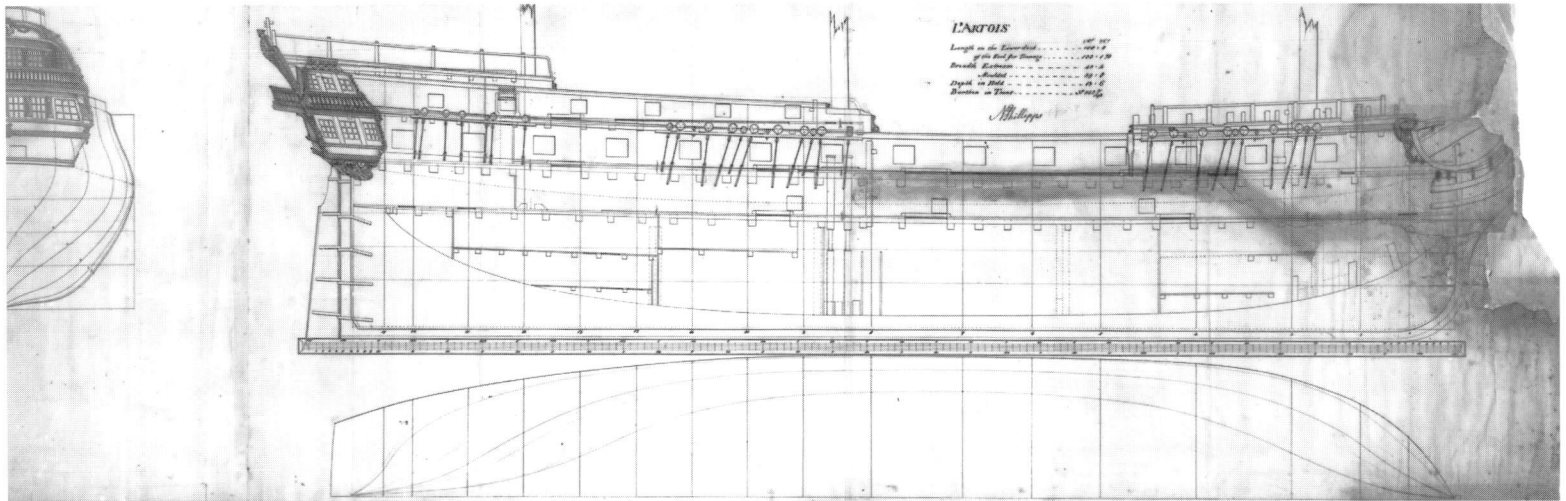

# THE AMERICAN WAR OF INDEPENDENCE 1776-1783

Lines and profile plan (6025) of the French large frigate *Artois* taken in 1780. She has a double-level galleried stern like a ship of the line and what appear to be three small loading ports a side on the lower deck, both of which seem to confirm that she was not built for the French navy

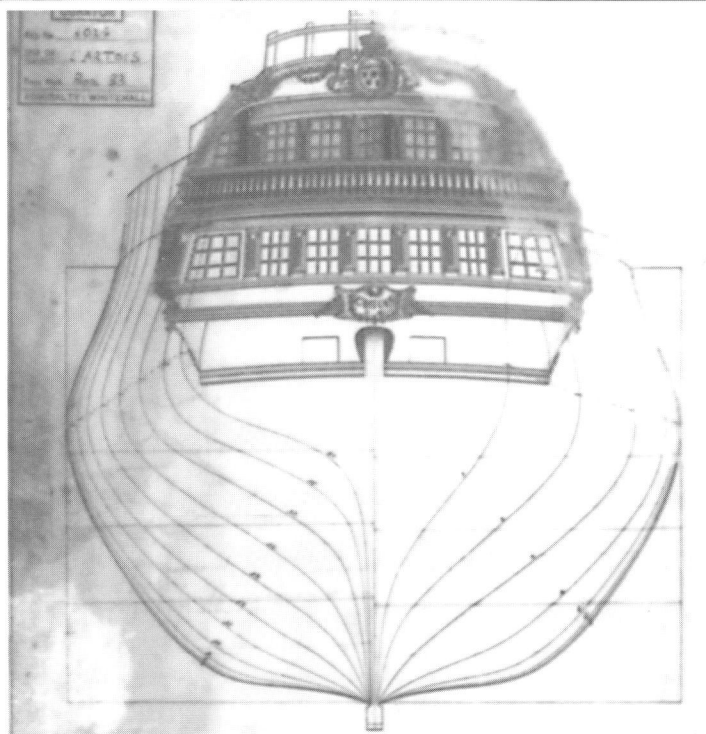

## 18 pounder Frigates (Fifth Rates)

***Artois*** 1780-1786 (French privateer *L'Artois* 1780 Arnous, Lorient) 40.
  Dimensions & tons: 158'8", 133'1⅜" x 40'4" x 13'6". 1152.
  Men: 280. Guns: UD 28 x 18, QD 10 x 9, Fx 2 x 12 (10. 1781 2 x 68★
  Taken by *Romney* whilst acting as a privateer fitted out by Prince Charles [the future Charles X], the Count of Artois, though manned by the French Navy (460 men); 1786 sold.
  PLANS: Lines, profile & decorations.

***Hebe*** 1782-1805 ***Blonde?*** -1811 (French *L'Hébé* 1782 Saint Malo. Sané design) 36/38.
  Dimensions & tons: 150'1½", 125'4½" x 39'11" x 12'9". 1062 51/94.
  Men: 250/300. Guns: UD 28 x 18, QD 8 x 9 + 6 x 32★,
  Fc 4 x 9 + 2 x 32★. Taken by *Rainbow* (which had an all-carronade armament, the Captain of *Hébé* was so impressed by the size of the shot from *Rainbow*'s bow chasers that he promptly surrendered rather than risk her broadside) on 4.9.1782; 1798-1806 troopship; 1805 renamed *Blonde* [?]; 1811 BU.
  NOTE: A very fine design, not only were several other French frigates built to her lines, but the design was also adapted as the basis for the large and successful *Leda* class in the Royal Navy, two of which; *Unicorn* and *Trincomalee/Foudroyant*; are still afloat today. The design, therefore, can count as that for the largest class of sailing frigates ever built.
  PLANS: As fitted: Lines, profile & decorations/decks.

## 12 pounder Frigates (Fifth Rates)

***Iris*** 1777-1781 (American *Hancock* 1776 Newburyport, Mass.) 28/32.
  Dimensions & tons: 136'7", 115'10⅞" x 35'2" x 11'0½". 762 62/94.
  Men: 200 (1779:220). Guns: UD 28 x 9, (1779 26 x 12, QD 4 x 6, Fc 2 x 6). Taken 1777 by *Rainbow* and classed as a Sixth Rate 28; 1779 reclassed as a Fifth Rate 32; 11.9.1781 taken by the French under De Grasse in the Chesapeake; (1785 sold by the French Navy; 1792 repurchased; 1794 burnt].
  PLANS: Lines, profile & decorations/decks.

***Raleigh*** 1778-1783 (American *Raleigh* 1776 Portsmouth, New Hampshire) 32.
  Dimensions & tons: 131'5", 110'7¼" x 34'5" x 11'. 696 80/94.
  Men: 220. Guns: UD 26 x 12, QD 6 x 6. Taken 29.8.1778 by *Experiment* and *Unicorn*; 1783 sold.
  PLANS: Lines, profile & decorations/decks.

***Convert*** 1778-1782 (French *La Pallas* 1777 Saint Malo, Guignace design) 32.
  Dimensions & tons: 127'9", 106'2⅛" x 35'4¼" x 13'9¼". 706.
  Men: 220. Guns: UD 26 x 9, QD 4 x 4, Fc 2 x 4 (later only 24 x 9) - though designed for 12 pounders by the French and therefore listed in that category here. Taken 19.6.1778 by *Victory*; 1782 sold.

***Licorne*** 1778-1783 (French *La Licorne* 1755 Brest, Geffrey Snr design) 32/36.
  Dimensions & tons: 127'1", 106'7½" x 34'7⅞" x 10'6¼". 679 82/94.
  Men: 220. Guns: UD 26 x 12, QD(& Fc?) 6 x 6 or QD 8 x 6, Fc 2 x 6 (2.11.1778 12 pounders replaced by 9s and 6 pounders by 4s). Seized by *Keppel* in the Channel 18.6.1778; 1783 sold.
  PLANS: Lines, profile & decorations/profile/lower deck/upper deck/quarter deck & forecastle/midships section.

***Santa Monica/Santa Ammonica*** 1779-1782 (Spanish *Santa Monica* [34] 1777 Cartagena) 36.
  Dimensions & tons: 145'0½", 120'2⅞" x 38'8" x 11'10". 956 18/94.
  Men: 240. Guns: UD 26 x 12, QD 8 x 9, Fc 2 x 6 (by 01.1780 QD 8 x 6 + 6 x 18★, Fc 2 x 6 + 2 x 18★ [later 24★]). Taken off the Azores by *Pearl* 1779; 1.4.1782 wrecked 'near Newman's Isle one of the Keys of the Virgin Islands'.
  PLANS: Lines, profile & decorations (incomplete).

***Prudente*** 1779-1796 (French *La Prudente* 1777 Saint Malo, Guignace design) 36/38.
  Dimensions & tons: 140'6½", 118'8⅛" x 37'11½". x 11'8". 909 52/94.
  Men: 240 Guns: UD 26 x 12, QD 10 x 6 + 8 x 18★
  (later 12 x 6 + 6 x 18★, later still 10 x 6, Fc 2 x 12). Taken near Jamaica 2.6.1779; 1796 hulked [at Portsmouth as a receiving ship; 1802/3 sold].

NMM negative number 1051. A French prize taken in one war takes another at the beginning of the next conflict. This engraving by Dodd shows the *Nymphe* (36) taken in 1780 capturing the *Cléopâtre* (see under her RN name of *Oiseau* at the beginning of the 12 pounder frigates section in the next chapter) in 1793 – the first major single-ship action of the French Revolutionary war, and the portent of many such successful actions to come

PLANS: Taken off & as fitted: Lines & profile.

**Santa Margarita** 1779-1814 (Spanish *Santa Margarita* [34] 1774 Ferrol) 36.
Dimensions & tons: 145'6", 123'6⅛" x 38'10½" x 11'8½". 992 78/94.
Men: 240. Guns: UD 26 x 12, QD & Fc 10 x 6 + 8 x 18★ (by 1807: QD 8 x 24★, Fc 2 x 24★). Taken off Portugal 11.11.1779; 1814 hulked [at Pembroke as a lazaretto; 1836 sold].

PLANS: Lines. 1791: Profile/orlop/lower deck/upper deck/quarter deck & forecastle/sheer. 1815 as lazaretto: Lower deck/upper deck/quarter deck & forecastle. No date as hulk: decks.

**Fortunee** 1779-1785 (French *La Fortunée* 1777 Brest, Forfait design) 36/38/40.
Dimensions & tons: 141'3", 124'5" x 38'1¼" x 12'1". 948.
Men: 300 .Guns: UD 28 x 12, QD 8 (later 10) x 6, Fc 2 x 6. Taken by Parker's fleet in the West Indies 22.12.1779; 1785 hulked [place and ultimate fate unknown].

**Monsieur** 1780-1783 (French privateer *Le Monsieur* 1779 _____ ) 36.
Dimensions & tons: 139'2¼" , 115'3⅜" x 36'6" x 17'9½". 819.
Men: 240 (later 255). Guns: UD 26 x 12, QD 8 x 6, Fc 2 x 6. Taken early 1780; 1783 sold.

**Providence** 1780-1784 (American *Providence* 1776 Providence) 32.
Dimensions & tons: 126'6½", 104'10¾" x 33'8" x 10'5". 632.
Men: _____ . Guns: _____ . Taken at Charleston 12.5.1780; never fitted for sea by the R.N. after her arrival in Britain. On her survey at Plymouth in 12.1780 the Dockyard Officers wrote; '... her construction being very full and burdensome (we) ... are of opinion she is not fit for a ship of war'.

**Proselyte** 1780-1785 (French privateer *Le Stanislaus* 1780 [?] Le Havre [?]) 32.
Dimensions & tons: 134'4", 113'11⅞" x 33'8" x 10'10". 687 33/94.
Men: 220. Guns: UD 26 x 12, QD 4 x 6, Fc 2 x 6 (later the 12 pounders were replaced by 32 pounder carronades, and the 6 pounders by 18 pounder carronades). Taken 1780; purchased 12.1780; 1785 sold.

PLANS: Lines & profile. No date: Lines & profile.

**Belle Poule** 1780-1796 (French *La Belle Poule* 1766 Bordeaux, Guignace design) 36.
Dimensions & tons: 140'0½", 118'7⅛" x 37'10" x 11'11". 902 82/94.
Men: 284. Guns: UD 26 x 12, QD 8 x 6, Fc 2 x 6. Taken by *Nonsuch* 14.7.1780; 1796 hulked [at Sheerness as a temporary receiving ship; 1818 sold].

PLANS: Lines, profile & decorations.

**Nymphe** 1780-1810 (French *La Nymphe* 1777 Brest) 36.
Dimensions & tons: 141'5½", 120'4½" x 38'3¼" x 11'9". 937 72/94.
Men: _____ . Guns: UD 26 x 12, QD 8 x 6, Fc 2 x 6 (6 pounders later replaced, first by 12 pounder carronades, then by 32 pounder carronades). Taken by *Flora*; 18.12.1810 wrecked in the Firth of Forth.

PLANS: Lines, profile & decorations.

**Clinton** 1780-1784 (French *L'Esperance* 1779 Bordeaux) 32.
Dimensions & tons: 134', 113' x 35' x 13'9". 736 28/94.
Men: _____ . Guns: UD 26 x 12, QD &/or Fc 2 x 6 (as transport 12 x 32★). Taken by *Pearl* 30.9.1780; 1783 armed transport; 1784 sold.

**Mars** 1781-1784 (Dutch *Mars* [38] 1769 Amsterdam Admiralty) 32.
Dimensions & Tons: 130'9", 108'10" x 34'10" x 11'10". 702 86/94.
Men: 220. Guns: _____ . Taken 3.2.1781 at the fall of Saint Eustatius; 1784 sold.

PLANS: Lines & profile/lower deck/upper deck/quarter deck & forecastle.

**Confederate** 1781-1782 (American *Confederacy* 1778 Norwich, Connecticut) 36/32.
Dimensions & Tons: 154'9", 133'5" x 37' x 12'3". 970 86/94.
Men: _____ . Guns: _____ . Taken 14.4.1781 by *Roebuck* and *Orpheus*; 1782 BU.

PLANS: Lines, profile & decorations/orlop/lower deck/upper deck/quarter deck & forecastle.

**Leocadia** 1781-1794 (Spanish *Santa Leocadia* [34] 1777 Ferrol) 36.
Dimensions & Tons: 144'10" , 119'8⅛" x 38'8" x 11'7¼". 951 67/94.
Men: 240-255. Guns: UD 26 x 12, QD & Fc 10 x 6. Taken by the *Canada* in the North Atlantic 1.5.1781; 1794 sold.

NOTE: In Spanish service carried 34 guns, though pierced for 40. Interestingly she was coppered when taken – an indication that the Spaniards were following the British example in this respect.

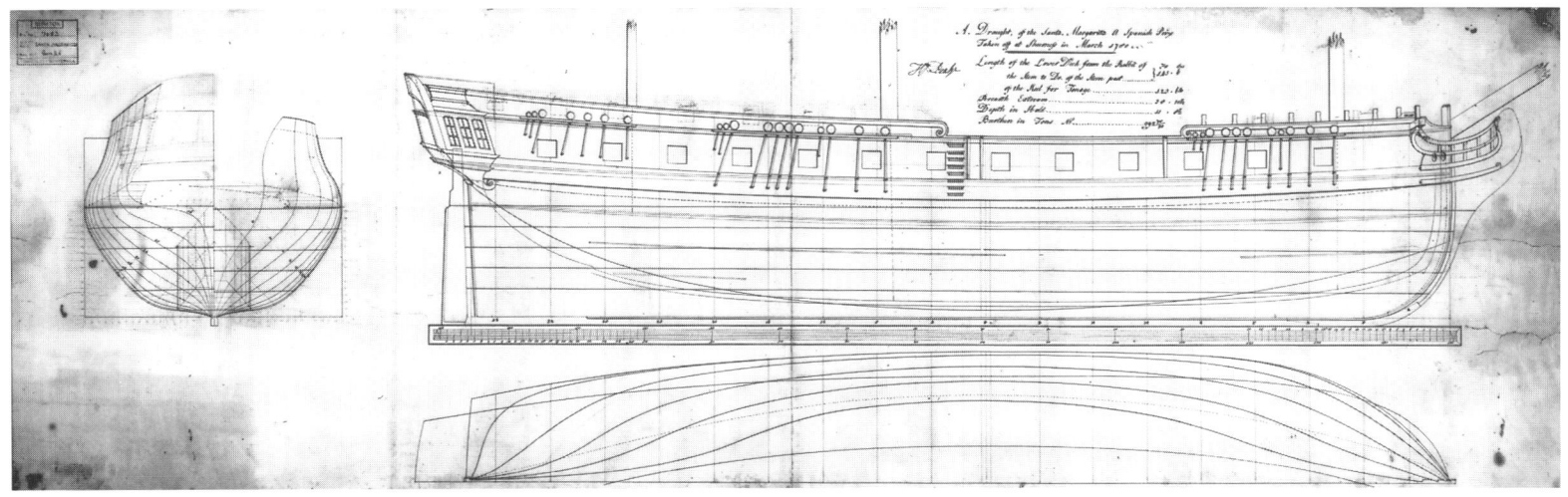

Lines (7083) 'taken off' the Spanish frigate *Santa Margarita* in 1780

PLANS: Lines, profile & decorations.

***Magicienne*** 1781-1810 (French *La Magicienne* 1777 Toulon J M B Coulomb design) 32.
  Dimensions & Tons: 143'9", 118'4¼" x 39'2½" x 12'4½". 967 74/94.
  Men: 220. Guns: UD 26 x 12, QD 4 x 6, Fc 2 x 6. Taken by *Chatham* 2.9.1781; 27.8.1810 grounded and burnt to avoid capture in Grand Port, Mauritius; [her wreck has since been found and salvaged].

PLANS: Lines, profile & decorations/orlop/lower deck/upper deck/quarter deck & forecastle. 1791: profile/orlop/lower deck/upper deck/quarter deck & forecastle/sheer.

***Aimable*** 1782-1814 (French 8 pounder frigate *L'Aimable* 1776 Toulon, Groignard design) 32.
  Dimensions & Tons: 133'5", 109'5" x 36'8" x 11'. 782.
  Men: 220-215. Guns: UD 26 x 12, QD 4 x 6, Fc 2 x 6. Taken by Rodney's fleet 19.4.1782; 1814 BU.

PLANS: 1791: profile/sheer/orlop/lower deck/upper deck/quarter deck & forecastle.

***Aigle*** 1782-1798 (French *L'Aigle* 1779 Saint Malo) 36/38.
  Dimensions & Tons: 147'5", 122'3" x 39'3" x 12'2". 1001 72/94.
  Men: ____. (1793:280). Guns: UD 26 x 12, QD 8 x 6, Fc 2 x 6 (1793:UD 28 x 18, QD 8 x 9, Fc 2 x 9). Taken after going aground 14.9.1782; 1793 re-armed as an 18 pounder frigate; 18.7.1798 wrecked on Cape Farina 'Barbary'.

PLANS: 1790: Lines & profile/planking expansion (internal)/(external).

***Concorde*** 1783-1811 (French *La Concorde* 1777 Rochefort, Chevillard Snr. design) 32.
  Dimensions & tons: 142'11", 118'10" x 37'6" x 11'7". 888 83/94.
  Men: 220. Guns: UD 26 x 12, QD 4 x 6, Fc 2 x 6. Taken 15.2.1783 by *Magnificent*; 1811 sold.

PLANS: Lines & profile/lower deck/upper deck/quarter deck & forecastle. 1791: profile/orlop/lower deck/upper deck.

***Heroine*** 1783-1803 (building Adams, Bucklers Hard) 32.
  Dimensions & tons: 130'11½", 107'7⅜" x 36'10¾" x 13'. 779.
  Men: 220. Guns: UD 26 x 12, QD 4 x 6 (1783 6 x 18* added), Fc 2 x 6. Purchased 05.1782; launched 08.1783; 1800 troopship; 1803 hulked [location and ultimate fate uncertain].

PLANS: building: Lines & profile/orlop/lower deck/upper deck/quarter deck & forecastle.

## 9 pounder Frigates (Sixth Rates, except where stated)

***Iris*** See under 12 pounder frigates.

***Delaware*** 1777-1783 (American *Delaware* 1777 Philadelphia) 28.
  Dimensions & tons: 117'0½", 98'0½" x 32'10½" x 9'8½". 563 49/94.
  Men: 200. Guns: 28 x 9. Taken by the Army in the Delaware 27.9.1777; 1783 sold.

***Virginia*** 1778-1782 (American *Virginia* 1776 North Point, near Baltimore) 28.
  Dimensions & tons: 126'3½", 105'7¼" x 34'10" x 10'5½". 681 53/94.
  Men: 200. Guns: 28 x 9 (later 24 x 9 [?]). Captured after grounding in the Chesapeake 30.3.1778; 1782 BU.

PLANS: Lines & profile/lower deck/upper deck/quarter deck & forecastle.

***Sartine*** 1778-1780 (French *Le Sartine* 1776 ____ ) 28.
  Dimensions & tons: 132'6", 118' x 35'9" x 15'3". 802 [?].
  Men: ____. Guns: ____. Taken 25.8.1778 by *Seahorse* and *Coventry*; 26.11.1780 wrecked off Mangalore, India.

NOTE: Believed to have been an English-built vessel.

***Oiseau*** 1779-1783 (French *L'Oiseau* 1769 Toulon) Fifth Rate 32.
  Dimensions & tons: 146'3", 126'9⅝" x 34'1" x 9'10¾". 783 49/94.
  Men: 220. Guns: UD 26 x 9, QD & Fc 6 x 6. Taken by *Apollo* 31.1.1779; 1783 sold.

PLANS: Lines, profile & decorations. As fitted: Profile & lower deck/orlop/upper deck & quarter deck & forecastle/decks.

***Danae*** 1779-1797 (French *La Danaé* 1763 Nantes, Groignard design) Fifth Rate 32.
  Dimensions & tons: 129'3", 107'2⅞" x 34'9" x 10'6½". 688 77/94.
  Men: 220. Guns: UD 26 x 9, QD 4 x 6, Fc 2 x 6. Taken by *Experiment* 13.5.1779; 1797 sold.

PLANS: Lines, profile & decorations/decks. As fitted: decks.

***Alcmene*** 1779-1784 (French *L'Alcmène* 1774 Toulon, Groignard design) Fifth Rate 32.
  Dimensions & tons: 131', 111'1¾" x 35'8" x 11'6". 731 9/94.
  Men: 220. Guns: UD 26 x 9, QD 2 x 6, Fc 4 x 6. Taken in the West Indies 21.10.1779; 1784 sold.

PLANS: As fitted?: Profile/lower deck/upper deck/quarter deck & forecastle. 1781: Lines, profile & decorations/orlop & lower deck/upper deck, quarter deck & forecastle.

***Charlestown*** 1780-1783 (American *Boston* 1777 Newburyport) 28.
  Dimensions & tons: ____, ____ x ____ x ____. 562.
  Men: ____. Guns: ____. Taken at Charleston 12.3.1780; 1783 sold.

***Albermarle/Albermarle*** 1781-1784 (French *La Menagère*? 1776 [?] ____ ) 28.
  Dimensions & tons: 125'3½", 102'2⅝" x 31'7¼" x 13'7½". 543 6/94.
  Men: ____. Guns: ____. Taken by *Mediator* 1781; 1784 sold.

PLANS: Lines, profile & decorations/orlop/lower deck/upper deck/quarter deck & forecastle.

***Abondance*** 1781-1784 (French 'gabare' *L'Abondance* 1780 ____ ) 28.
  Dimensions & tons: 132'6", 113'11⅞" x 29'5⅝" x 11'9". 526 50/94.
  Men: 200 (as storeship: 52). Guns: UD 24 x 9, QD 4 x 4 (As storeship: 14 x ____ ). Taken by Kempenfelt's squadron 12.12.1781; 1783 storeship; 1784 sold.

PLANS: 1781 as storeship: Lines & profile.

***Sophie*** 1782-1784 (French merchantman *La Sophie* ____ ) 28.
  Dimensions & tons: ____, ____ x ____ x ____. 422.

Men: ____ . Guns: ____ . Taken by *Warwick* 12.9.1782; 1784 sold.

## Frigates, Calibre of Guns Uncertain

*Shark* 1780[?]-1780 ( ____ ) Sixth Rate 28.
Dimensions & tons: ____ , ____ x ____ x ____ . ____ .
Men: ____ . Guns ____ . Purchased by Rodney in the West Indies 1780 [?]; ____ 1780 foundered in a gale off North America.

## 'Corvettes' (all Sixth Rates)

*Camel* 1776-1784 (Mercantile *Yorkshire* ____ ) 24/26.
Dimensions & tons: 113', 96'11½" x 31'7¾" x 13'3½". 516.
Men: ____ (as transport: 52). Guns: UD 20 x 9, QD 4 x 3 (1780: UD 22 x 9, QD 4 x 6. As transport: UD 14 x 6). Purchased 1776; 1783 armed transport; 1784 sold.

*Proteus* 1777-1779 (East Indiaman *Talbot* ____ ) 26.
Dimensions & tons: ____ , 108'10" x 34'2" x 13'11". 676.
Men: 160. Guns: UD 20 x 9, QD 6 x 4. Purchased 02.1777; 1779 hulked [at Newfoundland as a prison ship; 1783 sold].

*Hydra* 1778-1783 (building Adams & Barnard, Deptford) 22/24.
Dimensions & tons: 109'4½", 90'7¾" x 30'8¾" x 10'6½". 454.
Men: 160. Guns: ____ (mostly 9 pounders). Purchased whilst on the stocks in 04.1778; launched 8.8.1778; 1783 sold.
PLANS: Lines/profile/orlop/lower deck/upper deck/quarter deck & forecastle.

*Termagant* 1780-1795 (building Hillhouse, Bristol) 22.
Dimensions & tons: 110'5", 90'6 5/8" x 28' x 8'7". 377 58/94.
Men: 160. Guns: 22 x 6 + 4 x 12★ (as sloop: 18 x 6). Purchased whilst on the stocks 1780; launched 3.6.1780; 1782 reduced to an 18 gun ship sloop; 1795 sold.
PLANS: Lines, profile & decorations/lines, profile & decorations/decks.

*Hussar* 1780-1783 (Massachussetts State Navy *Protector* 1776 ____ ) 22.
Dimensions & tons: ____ , ____ x ____ x ____ . ____ .
Men: ____ . Guns: ____ . Taken by *Roebuck* 05.1780; 1783 sold.

*Saint Eustatius* 1781-1783 (Dutch ____ ) 20.
Dimensions & tons: 104', ____ x 27'6" x ____ . ____ .
Men: ____ . Guns: 20 x 9. Taken 1781; 1783 sold.

*Grana* 1781-1793 (Spanish *Grana*/*Nuestra Senora de la Paz* 1778 Ferrol) 28.
Dimensions & tons: 117'10", 97', x 31'11¾" x 9'4". 527 60/94.
Men: ____ . Guns: UD 22 x 6 , QD 4 x 4, Fc 2 x 4. Taken off Finisterre by *Cerberus* 25.2.1781; 1793 hulked [at Sheerness as a convalescent ship; 1800 sold].

NOTE: Though a 28 she is listed as a corvette and not a frigate because of her 6 pounder main armament.

PLANS: Lines, profile & decorations/decks. As fitted: Lines, profile & decorations/decks.

*Bellisarius* 1781-1783 (American privateer [?] *Bellisarius* ____ ) 20.
Dimensions & tons: 118'9", 100'8" x 30'2" x 15'. 487.
Men: 140. Guns: ____ . Taken 7.8.1781 by *Medea*; 1783 sold.

*Naiade* 1783-1784 (French *La Naiade* [20] 1779 Toulon, Joseph-Coulomb design) 26/30/22.
Dimensions & tons: 126'8", 105'10⅞" x 33'8½" x 10'2". 640 7/94.
Men: ____ . Guns: UD 22 x 12, QD 6 x 12★, Fc 2 x 18★. Taken in the East Indies by *Sceptre*; never registered in the RN; 1784 sold.
PLANS: Lines, profile & decorations.

## Ship Sloops

*Shark* 1776-1778 *Salamander* -1783 (building Randall, Rotherhithe) 16.
Dimensions & tons: 96'3", 78'4½" x 29'5" x 9'. 313.
Men: 125 (as fireship: 45 [?]) Guns: 16 x ____ (as fireship: 8 x 4 + 8 swivels). Purchased whilst on the stocks 30.10.1775; launched 9.3.1776; 1779 fireship, renamed *Salamander*; 1783 sold.
PLANS: As fitted: Lines & profile/orlop/upper deck/quarter deck & forecastle.

*Druid* 1776-1779 *Blast* -1783 (mercantile ____ American-built?) 16.
Dimensions & tons: 89', 74'2 5/8" x 27' x 11'10". 287 75/94.
Men: 100 (as fireship: 45). Guns: 16 x ____ . (as fireship: 8 x 4 + 6 swivels). Purchased mid-1776; 1779 fireship, renamed *Blast*; 1783 sold.
PLANS: Taken off & as fitted: Lines & profile/orlop & lower deck/upper deck, quarter deck & forecastle.

*Sylph* 1776-1779 *Lightning* -1783 (Mercantile *Lovely Lass* ____ ) 14.
Dimensions & tons: 85', 68' x 27'6" x 11'8½", 273 80/94.
Men: 90 (as Fireship: 45). Guns: 14 x ____ (as fireship: 8 x 4 or x 3 + 8 swivels). Purchased mid-1776; 1779 fireship, renamed *Lightning*; 1783 sold.
PLANS: Lines & profile/lower deck/upper deck.

*Grasshopper* 1776-1779 *Basilisk* -1781 (Mercantile ____ ) 14.
Dimensions & tons: 92'6", 74'3" x 26'9" x 10'10". 282 44/94.
Men: ____ (as fireship: 45). Guns: 14 x ____ (as fireship: 8 x ____ ). Purchased 1776; 1779 hulked; 1779 fireship , renamed *Basilisk*; 1781 hulked [at Plymouth as a slop ship; 1783 sold].
PLANS: Taken off & as fitted: Lines & profile/orlop.

*Porpoise* 1777-1778 *Firebrand* -1781 (Mercantile *Annapolis* ____ ) 16.
Dimensions & tons: ____ , 86' x 29'7⅝" x 11'4¾" . 402.
Men: ____ (as fireship: 50/55). Guns: 16 x ____ . (as fireship: 8 x ____ + 8 swivels). Purchased early 1777; 1778 fireship, renamed *Firebrand*; 11.10.1781 accidentally burnt getting under way in Falmouth harbour.

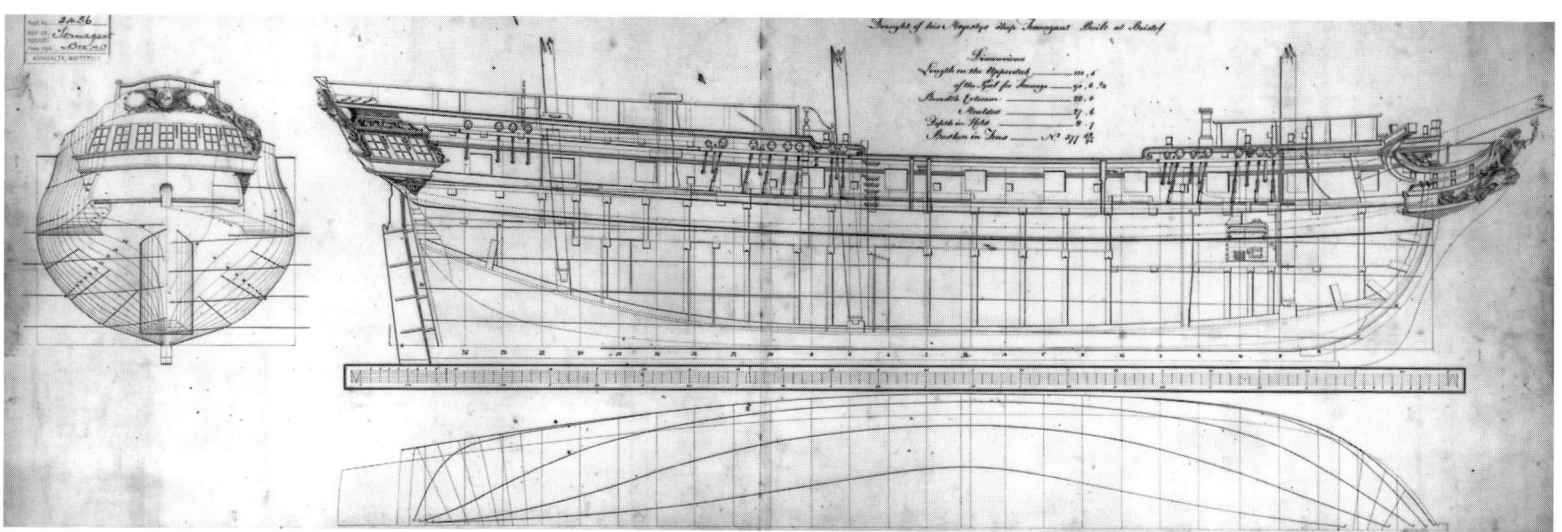

Lines and profile plan (2456) of the purchased 22 gun ship *Termagant* of 1780. Her very fine lines would seem to make it likely she was building as a privateer, and also serve to provide a likely explanation for her being reduced in armament and status to an 18 gun sloop within two years of launching. Notice the position of the galley stove forward on the lower deck instead of the (from the beginning of the eighteenth century) more usual location one deck up

# THE AMERICAN WAR OF INDEPENDENCE 1776-1783

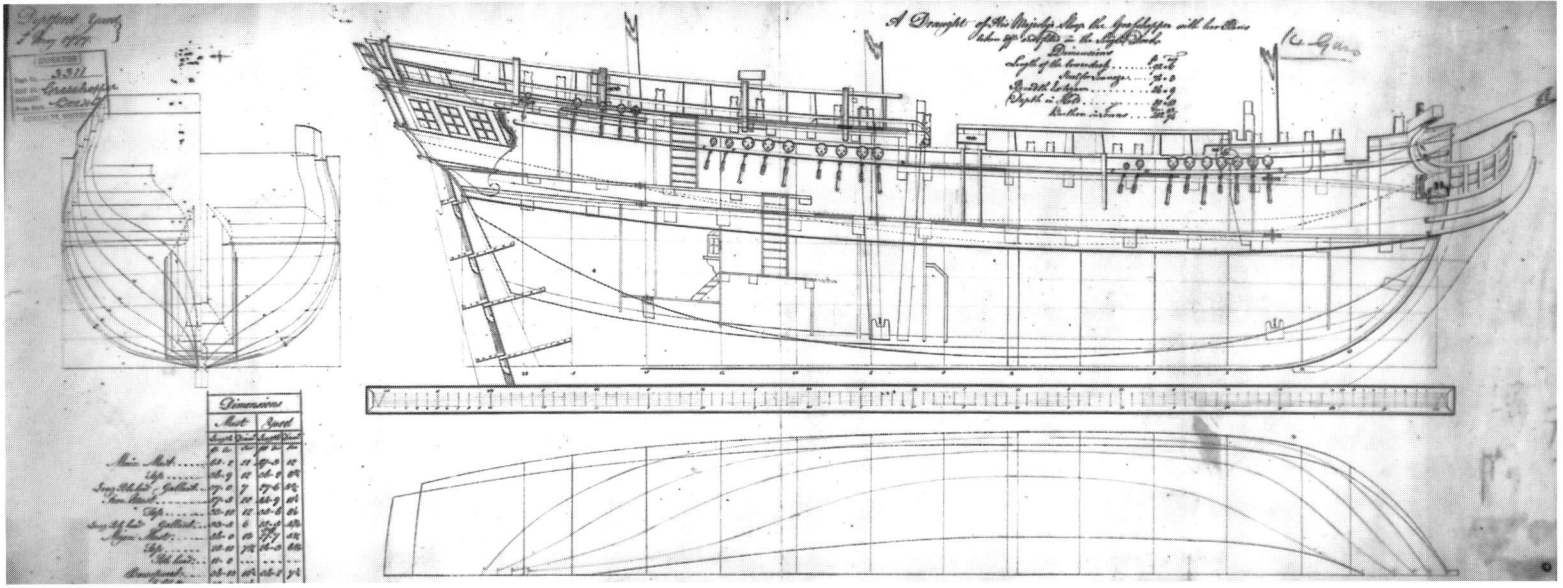

A lines and profile plan with spar dimensions (3311) 'taken off' the ship sloop *Grasshopper* purchased in 1776

**Lynx** 1777-1783 (building Randall, Rotherhithe) 16.
  Dimensions & tons: 95', 77'7½" x 28' x 11'11". 324.
  Men: ____ (as hospital ship: 50/55). Guns: 16 x ____ (as hospital ship: 10 x 18★, later additional 4 x 12★ on QD and 2 x 12★ on Fc). Purchased whilst on stocks in early 1777; 10.3.1777 launched; 1780 hospital ship; 1783 sold.

**Harpy** 1777-1783 (building Fisher, Liverpool) 18.
  Dimensions & tons: 102'9", 85' x 28'6" x 11'. 367 22/94.
  Men: ____ (as fireship: 50). Guns: 18 x ____ (as fireship: 10 x ____ ). Purchased whilst on stocks in early 1777; launched 8.5.1777; 1779 temporary fireship; 1783 sold.

PLANS: Building & as fitted: Profile.

**Convert** 1777-1778 **Beaver's Prize** -1780 (American privateer *Oliver Cromwell* ____ Philadelphia) 24/16.
  Dimensions & tons: 85'9", 69' x 26' x 12'5". 248 10/94.
  Men: ____ (1778: 110). Guns: UD 12 x 6, QD & Fc 6 x 6 + 6 x 4 (1778: 16 x 6). Taken by Beaver 19.5.1777 and initially classed as a Sixth Rate; 1778 arrived in Britain, reclassed as a sloop and renamed *Beaver's Prize*; 11/12.10.1780 lost at Saint Lucia.

PLANS: 1778: Lines, profile & decoration/decks.

**Cupid** 1777-1778 ( ____ ) 14.
  Dimensions & tons: 92'1", 74'9¾" x 27' x 12'2". 290 5/94.
  Men: ____ . Guns: 14 x ____ . Purchased 1777; 28.12.1778 foundered off Newfoundland.

PLANS: Taken off & as fitted: Lines & profile/orlop & lower deck/upper deck & quarter deck & forecastle.

**Hinchinbrook** 1777-1780? (American privateer *American Tartar* ____ ) 16.
  Dimensions & tons: ____ , ____ x ____ x ____ . 318.
  Men: 125. Guns: 16 x 6. Taken 1777 by *Bienfaisant*; 1780 [?] hulked [at Sheerness as a slop ship; 1783 sold].

**Lucifer** 1778-1779 **Avenger** 1783 (building Randall & Co, Rotherhith) 14 (ex fireship).
  Dimensions & tons: 97'10", 79'4¼" x 28'4¼" x 12'11". 339.
  Men: As sloop: 120. Guns: As sloop: 14 x 6 + 14 x ½ swivels. Purchased in frame from Randall & Co for use as a fireship, 10.1778; L 9.10.1778; 1779 made a sloop and renamed *Avenger* by order of Admiral Arbuthnot on the North American Station; 12.6.1783 condemned and sold at New York.

**Barbadoes** 1778-1784 (American [?] *Barbadoes* ____ ) 12.
  Dimensions & tons: 97'8", 81'9" x 25'2⅛" x 10'4". 275 59/94.
  Men: ____ . Guns: 12 x ____ . Purchased 1778; never registered upon the list of the Navy [or purchased in 1780?]; 1784 sold.

PLANS: Lines & profile/decks/lines & profile.

**Guay Trouin** 1780-1783 (French privateer *Le Du Guay Trouin* ____ ) 18.
  Dimensions & tons: 83'11", 68'10½" x 26'2⅝" x 12'8½". 251 66/94.
  Men: 110. Guns: 14 x 6 + 4 x 18★ + 10 swivels. Taken by *Surprise* 1780; 1783 sold.

PLANS: Lines, profile & decorations.

**Halifax** 1780-1781 (American *Ranger* 1777 Hackett, Portsmouth, New Hampshire) 14.
  Dimensions & tons: ____ , ____ x ____ x ____ . ____ .
  Men: ____ . Guns: 14 x ____ . Taken at Charleston 12.5.1780; 1781 sold.

**Merlin** 1780-1789 (building King, Dover) 20.
  Dimensions & tons: 101', 81'5¼" x 28' x 12'6". 360.
  Men: 125. Guns: 18 x 6, 2 x 3 chase guns. Purchased whilst building late 1780; 1789 hulked [at Sheerness as an ordnance hulk; 1795 sold].

**Marquis De Seignelly** 1780-1786 (French privateer *Le Marquis De Seignelay* 1780 ____ ) 14.
  Dimensions & tons: 97'2", 76'10¾" x 26'3" x 14'1". 281 74/94.
  Men: 110. Guns: 14 [or 16?] x 6. Taken by *Portland* and *Solebay* 10.12.1780; 1786 sold.

PLANS: Lines & profile.

**Porto** 1780-1782 (American *Harlequin*, French *Arlequin* ____ ) 12.
  Dimensions & tons: 80'1", 78'6" x 20'5" x 9'3". 141 47/94.
  Men: ____ . Guns: 12 x ____ . Taken 1780 by Johnson's Squadron; 1782 sold.

NOTE: Although a number of secondary sources suggest that she was renamed *Harlequin* in 1782, I can find no evidence for this having been done.

PLANS: Lines, profile & decorations.

**Duc De Chartres** 1781-1784 (French privateer *Le Duc De Chatres* ____ ) 18.
  Dimensions & tons: 109'2", 86'5⅝" x 30'5¼" x 11'11½". 426 7/94.
  Men: ____ . Guns: 18 x 6. Taken by Darby's Squadron 1781; 1784 sold.

PLANS: Lines, profile & decoration.

**Duc D'Estissac** 1781-1783 (French privateer *Le Duc D'Estissac* ____ ) 16.
  Dimensions & tons: 94'7½", 79'0⅜" x 24'0½" x 11'1½". 242 91/94.
  Men: 125. Guns: UD 14 x 6, QD? 2 x 4, also 10 swivels. Taken 6.6.1781 by *Cerberus*; 1783 sold.

PLANS: Lines, profile & decoration/decks.

**Racehorse** 1781-1799 (Building by Fisher, Liverpool) 16.
  Dimensions & tons: 101'7", 83'3½" x 28'3½" x 13'9". 354.
  Men: 125. Guns: 16 x 6. Purchased whilst building in 1781; L 20.10.1781; 1799 BU.

**Cormorant** 1781-1783 **Rattlesnake** 1786 (American privateer *Rattlesnake* ____ ) 12.
  Dimensions & tons: 89'3", 74'11" x 22'4" x 8'10½". 198 70/94.

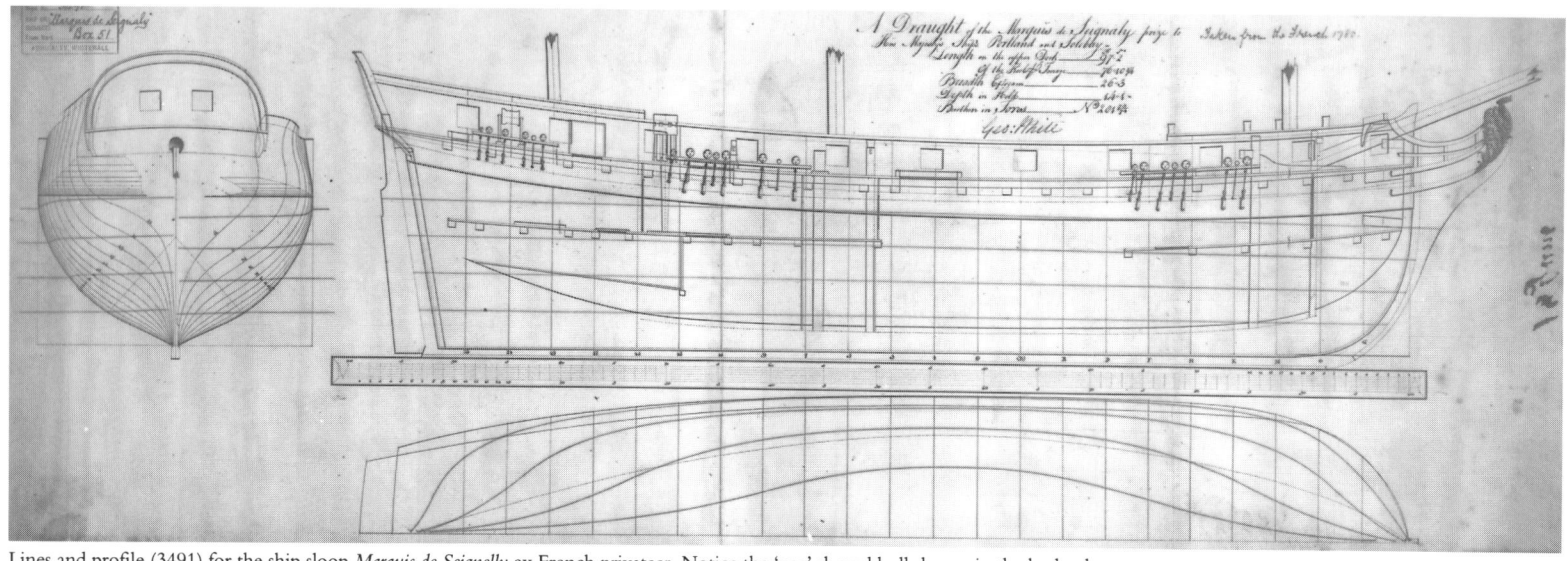

Lines and profile (3491) for the ship sloop *Marquis de Seignelly* ex French privateer. Notice the 'vee' shaped hull shown in the body plan

Men: 90. Guns: 12 x 4 (short) + 10 swivels. Taken by *Assurance* in American waters; 1783 renamed *Rattlesnake;* 1786 sold.

PLANS: Lines, profile & decorations/decks.

*Rover* 1781-1781 (American ____ ) 20.
  Dimensions & tons: 98', 90' x 24' x 8'3". 194.
  Men: 125. Guns: 2 x 9 + 6 x 6 + 12 x 4. Taken 1781; ____ 1781 lost in America.

*Albecore* 1781-1784 (American privateer *Royal Louis* ____ ) 16.
  Dimensions & tons: 96'9", 80' x 27'4" x 12'7". 319 22/94.
  Men: ____ . Guns: 16 x ____ . Purchased 1781; 1784 sold.

*Pelican* 1781-1783 (French privateer *Le Fredrick* ____ ) 16.
  Dimensions & tons: 85'4¾", 69'7" x 23'6" x 9'4". 204 37/94.
  Men: ____ . Guns: ____ . Taken 1781; 1783 sold.

*Mentor* 1781-1783 (American *Aurora* ____ ) 18?
  Dimensions & tons: ____ ____ ____ ____ .
  Men: 125. Guns: 18 ? Taken 1781; 04.1783 supposed to have foundered near Bermuda.

*Termagant* See under corvettes above.

*Aglaia* 1782-1783 (French privateer *L'Aglaia?* ____ ) 18.
  Dimensions & tons: 94'11", 75'9¼" x 27'6½" x 14'4¼". 306.
  Men: 125. Guns: 18 x 6? 18.4.1782 taken by *Aeolus;* 1783 sold.

*Robecque* 1782-1783 (French *Le Robecq* ex *La Comptesse de Provence* 1779 ____ ) 18.
  Dimensions & tons: 111'5½", 92'1⅞" x 28'1⅜" x 12'6½". 388.
  Men: ____ . Guns: 18 x 6. Taken 1782; 1783 sold.

*Mohawk* 1782-1783 (American [?] ____ ) 14.
  Dimensions & tons: 95'3", ____ x 26'4" x ____ . ____ .
  Men: ____ . Guns: 14 x ____ . Taken 1782; purchased 10.1782; never registered on the list of the Navy; 1783 sold.

PLANS: Lines.

# Brigs, etc

*Badger* 1776-1777 (American [?] *Pitt?* ____ ) brig sloop 12.
  Dimensions & tons: 68'6", 54'3" x 21'10" x 9'4". 137 52/94.
  Men: 90. Guns: 12 x 4 + 12 swivels. Purchased 1776 (probably an American prize); 1777 condemned.

PLANS: Lines & profile/decks.

*Badger* 1777-1783 (American [?] *Pitt?* ____ ) brig sloop ____ .
  Dimensions & tons: ____ , ____ x ____ x ____ . ____ .
  Men: ____ . Guns: ____ . Purchased 1777 (probably an American prize); 1783 sold.

NOTE: Nelson's first commissioned command.

*Cabot* 1777-1783 (American [?] *Cabot* ____ ) brig/brig sloop 14.
  Dimensions & tons: 74'9¾", 58'7" x 24'8" x 11'4". 190.

Men: 75 (1779:80). Guns: 14 x 6 (1779: 12 x 4). Purchased by Lord Howe 1777; 1783 sold.

*Lowestoffe's Prize* 1777-1779 (American [?] ____ ) brig 8.
  Dimensions & tons: 49'2", 39'9" x 16'4" x 7'1". 61.
  Men: 30. Guns: 8 x ____ + 12 swivels. Taken by *Lowestoffe* 1777 and purchased at Jamaica; 1779 condemned.

*Porcupine* 1777-1788 (Mercantile: *Union* ____ ) brig sloop 16.
  Dimensions & tons: ____ , 54' x ____ x 11'. 160.
  Men: 100. Guns: 16 x 4 + 16 swivels. Purchased at Jamaica 1777; 1788 sold.

*Hinchinbrook* 1777?-1778 (American privateer ____ ) armed brig/brig 14.
  Dimensions & tons: ____ , ____ x ____ x ____ . ____ .
  Men: 65-75. Guns: 14 x 4. Taken 1777 (?) and purchased by Lord Howe; 04.1778 taken by American privateers.

*Helena* 1778-1778/1779-1798 (French? *Hélène* ____ ) brig sloop/brig 14.
  Dimensions & tons: 76'1½", 56'5½" x 26'9" x 10'8". 214 84/94 or 220.
  Men: 80/70. Guns: 14 x 6. Purchased early 1778; 16.9.1778 taken by the French; 1779 retaken, still as Helena; 7.10.1798 'sailed from England ....... supposed to have foundered in the North Seas'.

*Keppel* 1778-1783 (American *Keppel?* ____ ) brig 14.
  Dimensions & tons: ____ , ____ x ____ x ____ .
  Men: 75. Guns: 14 x 4. Taken 1778; 1783 sold.

*Swift* 1778-1781 (American ____ ) brig sloop 14.
  Dimensions & tons: ____ , ____ x ____ x ____ . 140.
  Men: ____ . Guns: 14 x ____ . Taken 1778; 1781 sold.

*Swallow* 1779-1781 (building as a cutter by Ladd, Dover) brig sloop 14.
  Dimensions & tons: 79'5", 60'2" x 26'7" x 10'2". 226.
  Men: 80/90. Guns: 10 x 4 + 12 swivels. (1780 4 x 18★ to be added). Purchased whilst on the stocks (?) early 1779; L 2.4.1779; 26.8.1781 driven ashore by 4 American privateers on Fire Island near Long Island.

PLANS: Lines & profile (*Drake* noted as similar).

*Drake* 1779-1800 (building as a cutter by Ladd, Dover) brig sloop 14.
  Dimensions & tons: 78'10½", 59'8" x 26'4½" x 10'9½". 221.
  Men: 80. Guns: 14 x 4 (later 14 x 6 + 12 swivels). Purchased whilst on the stocks 1779; 05.1779 launched; by 1800 out of commission.

PLANS: See under *Swallow* above.

*Zephyr* 1779-1783 *Dispatch* -1798 (building by Barnard, Deptford) brig sloop 14.
  Dimensions & tons: 75'5¾" , 60'6" x 24'1" x 9'3". 187.
  Men: 70 (as transport: 16). Guns: 14 x 4 (as transport: no guns?). Purchased whilst on the stocks in early 1779; 31.5.1779 launched; 1783 fitted as naval transport and renamed *Dispatch;* 1798 sold.

PLANS: As built: profile & decks. 1783: Lines, profile & decks.

*Lively* 1779-1782 (building King, Dover) brig/brig sloop 12.
  Dimensions & tons: 78'3", 55'8" x 25'8¾" x 10'10". 206.

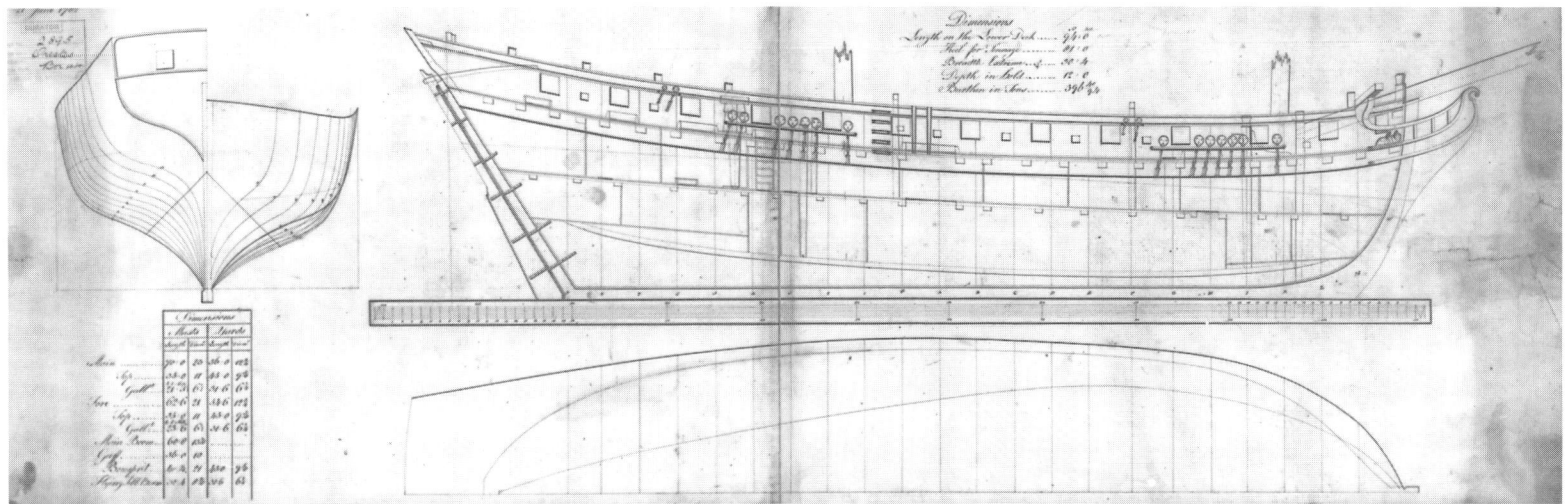

Lines and profile plan with spar dimensions (2895) 'taken off and as fitting' for the Dutch privateer brig *Orestes* captured in 1781. Notice the 'wine glass' shape of the sections shown in the body plan

Men: ____ . Guns: 12 x ____ (later 12 x 18★). Purchased (or possibly built for the Navy from the start?) 1779; 9.12.1782 taken by American prisoners who took her into Havanna.

**Scourge** 1779-1795 (building as a cutter by Allin, Dover) brig sloop 16.
Dimensions & tons: 80'6", 60'8¾" x 26'11" x 11'1". 234 1/94.
Men: 80. Guns: 16 x 6 + 12 swivels (originally meant to have been 14 x 6, by 1794: 16 x 24★). Purchased whilst on the stocks 1779; 26.10.1779 launched; 7.11.1795 foundered on the coast of Friesland.
PLANS: Lines & profile.

**Gibraltar** 1779-1781 (American? *Virginia*? ____ ) brig 10.
Dimensions & tons: 63', 54' x 21' x 7'6". 85.
Men: 45. Guns: 10 x ____ . Purchased 1779; 17.4.1781 taken by Spanish xebec off Gibraltar [served as Spanish *Salvador* until 1800].

**Active** 1779-1780? ( ____ ) brig 14.
Dimensions & tons: ____ , ____ x ____ x ____ . ____ .
Men: 60. Guns: 14 x ____ . Purchased (in North America?) 1779; ____ 1780? taken by the Americans off New York?

**San Firmin** 1780-1780 (Spanish *San Firmin* ____ ) brig sloop? 14.
Dimensions & tons: 90', ____ x 25' x 12'. 250.
Men: 100. Guns: 14 x ____ . Taken 8.1.1780; 04.1780 retaken by the Spaniards off Gibraltar.

**Scout** 1780-1794 (building Jacobs, Folkestone) brig sloop 14.
Dimensions & tons: 82'. 59'7½" x 29'6" x 11'2". 276.
Men: 80. Guns: 14 x 4 (she was to have had an extra 2 guns, but these were never fitted). Purchased whilst on the stocks 28.3.1780; 30.7.1780 launched; 24.8.1794 taken by two French frigates off Cape Bon; [in French service until wrecked at Cadiz 12.12.1795].
PLANS: Lines & profile/decks.

**Speedwell** 1780-1807? ( ____ ) cutter (brig rig)/brig/brig sloop 14.
Dimensions & tons: 75'3", 54'6" x 25'10" x 10'2". 193.
Men: 70. Guns: 14 x 4 + 10 swivels. Purchased 1780 and classed as a cutter, though rigged as a brig; 1783 classed as a brig sloop; 1783 classed as a cutter (and probably rigged as such); 1797 altered to a brig; 18.2.1807 foundered off Dunkirk.

**Fortune** 1780-1797 (building Stewart, Sandgate) brig sloop 16.
Dimensions & tons: 85'2", 61'6" x 28'10½" x 11'10". 273.
Men: 90. Guns: 16 x 6 (later 2 x 12 added?). Purchased whilst on the stocks 1780; 08.1780 launched; 14.6.1797 wrecked near Oporto.

**Hope** 1780-1781 ( ____ ) brig 14.
Dimensions & tons: 91'6", 75'6" x 26'6" x 10'4½". 282.
Men: 125. Guns: 14 x 4. Purchased 1780 (in North America?); ____ 1781 run ashore to prevent sinking off Savannah.

**Prince Edward** 1780-1782 ( ____ ) brig 14.
Dimensions & tons: ____ , ____ x ____ x ____ . ____ .
Men: ____ . Guns: 14 x 4. Taken 1780 (and purchased in North America?); ____ 1782 taken by American prisoners.

**Placentia** 1780-1782 ( ____ ) brig ____ .
Dimensions & tons: ____ , ____ x ____ x ____ . ____ .
Men: ____ . Guns: ____ . Purchased 1780 (at Newfoundland?); 30.9.1782 supposed lost off Newfoundland.

**Cameleon** 1781-1783 (Mercantile *Hawke* ____ ) brig 16/14 clinker-built.
Dimensions & tons: 78', 58'7⅛" x 29'3½" x 12'. 297 37/94 [or 267 37/94].
Men: 90. Guns: 16 or 14 x ____ . Purchased 01 or 02.1781; 1783 sold.
PLANS: Lines & profile.

**Pelican** 1781 See under ship sloops.

**Swallow** 1781-1793 (building as a 'brigantine cutter' by Fabian, Cowes) brig sloop 16.
Dimensions & tons: 80', 58'7¼" x 29' x 10'5". 262 15/94.
Men: 90-86. Guns: 16 x 4. Purchased whilst on the stocks 1781; 10.1781 [?] launched; 1793 sold.
PLANS: Lines & profile.

**Orestes** 1781-1799 (Dutch privateer *Mars* ____ ) brig sloop 18.
Dimensions & tons: 100'11" , 73'2⅝" x 30'4" x 14'. 396 40/94.
Men: 120. Guns: 18 x 9 (short) + 12 swivels. (1792: 9 pounders replaced by short 6 pounders). Taken 3.12.1781 by *Artois*; 5.11.1799 foundered in a hurricane in the Indian Ocean.
NOTE: Captured in company with the *Pylades* which was apparently a near sister. Reported by their captor: 'to be quite new and the completest privateers he ever saw'.
PLANS: Lines & profile/lower deck/upper deck. 1789: Lines & profile.

**Pylades** 1781-1790 (Dutch privateer *Hercules* ____ ) brig sloop 18.
Dimensions & tons: 90'2", 81'6⅝" x 30'4" x 12'. 399.
Men: 125. Guns: 18 x 9 (short) + 12 swivels. Taken 3.12.1781 by *Artois* (see *Orestes* above); 1790 BU.

**Swan** (ex *Bonetta*) 1781-1782 (Mercantile *Roebuck* ____ ) brig sloop 14.
Dimensions & tons: 84', 64'6¼" x 28'2" x 11'3". 272 26/94.
Men: 110. Guns: 14 x 6 (1782: 18 x 18★). Purchased 12.1781; 08.1782 capsized off Waterford 'in a sudden gust of wind'.
PLANS: Lines & profile/decks.

**Bustler** 1782-1788 (cutter ____ ) brig sloop 14.
Dimensions & tons: 76'7½", 60'11⅝" x 25'4½" x 10'2". 208 76/94.
Men: 90. Guns: 14 x 4. Purchased 01.1782 and fitted out as a cutter sloop but immediately afterwards converted instead to a brig sloop; 1788 sold.
PLANS: Lines & profile.

**Wasp** 1782-1800 (building ____ , Folkestone) brig sloop 16.
Dimensions & tons: 78'8", 56'11⅝" x 26'2" x 10'1". 207 43/94.
Men: 90. Guns: 16 x 4 (by 1794: 8 x 18★). Purchased whilst building 1782; 1798 converted to a fireship; 7.7.1800 expended in Dunkirk Road.

**Kingsfisher** 1782-1798 (building ____ , Rochester) brig sloop 18.
Dimensions & tons: 95'1", 73'3½" x 30'0½" x 7'6¼", 369 57/94.

Men: ____ . Guns: 18 x 6 (short). Purchased whilst building 1782; 3.12.1798 wrecked on Lisbon Bar.

***Espion*** 1782-1784 (French *L'Espion* ____ Dunkirk) brig (or cutter) sloop 14.
Dimensions & tons: ____ , ____ x ____ x ____ . 269.
Men: ____ . Guns: 14 x ____ . Taken by *Lizard* off Saint Kitts 24.1.1782; 1784 sold.

***Swift*** 1783?-1784 (American or French *Dauphin?* ____ ) brig sloop ____ .
Dimensions & tons: 75'6", 62'4" x 20'10" x 7'9". 143 88/94.
Men: ____ . Guns: ____ . Taken ____ ?; 1784 sold.
PLANS: Lines & profile.

## Sloops, Rig Uncertain

***Albany*** 1776-1782 (American *Rittenhouse* ____ ) 14.
Dimensions & tons: ____ , ____ x ____ x ____ . ____ .
Men: 125. Guns: 14 x ____ . Purchased (? taken) 1776; 28.12.1782 run aground and lost in Penobscot Bay.

***Hope*** 1776-1779/1781 ***Recovery*** -1783 ( ____ ) 8.
Dimensions & tons: ____ , 45'5" x 20'2" x 8'4". 105.
Men: 50. Guns: 6 x ?4 + 12 swivels. Purchased (probably in North America) 1776; 1779 taken by the Americans; 1781 retaken and named *Recovery*; 1783 sold. Probably schooner rigged.

***Racehorse*** 1776-1776 ( ____ ) 10.
Dimensions & tons: 59', 43' x 20'9" x 9'. 98.
Men: ____ . Guns: 10 x ____ . Purchased (in North America?) 1776; 12.1776 taken by the Americans.

***Drake*** 1777-1778 (Mercantile *Resolution?* ____ ) 16.
Dimensions & tons: 91'5", 75'10¾" x 26'1" x 18'3½". 275.
Men: 100. Guns: 16 or 14 x 4. Purchased 1777; 24.4.1778 taken by American 'privateer' (J Paul Jones in the USS *Ranger*) off Belfast.

***York*** 1777-1779 (American sloop *Betsey* ____ ) sloop (1 mast) rigged? 12.
Dimensions & tons: 65', ____ x 22'x 10'. ____ .
Men: 65 (later 75, later 50). Guns: 12 x 4 (later 10 x ____ , later 10 x 3). Purchased by Earl Howe in North America 29.3.1777; 10.7.1778 taken by the French; 23.8.1778 retaken by Lord Howe; 07.1779 taken by the French fleet under D'Estaing. [May have at times been classed as a cutter. Rif Winfield has a reference to her as a brig.]

***Ostrich*** 1777-1782 (Mercantile *Hector?* ____ ) 14 possibly ship-rigged.
Dimensions & tons: 94', 77'10½" x 26' x 11'7". 280.
Men: 110. Guns: 14 x ____ (in 1779 2 x 4 added). Purchased 1777 (whilst building?); 1782 sold.

***Comet*** 1777 1778 (*Mercantile* [?] ____ ) 10.
Dimensions & tons: ____ , ____ x ____ x ____ . ____ .
Men: 100. Guns: 10 x ____ . Purchased 1777; 1778 sold. (May have served for some of the time as a fireship; certainly her name is one traditionally appropriated for fireships.)

***Snake*** 1777-1780 (Mercantile [?] ____ or French *Seine?*) 14.
Dimensions & tons: ____ , ____ x ____ x ____ . 287.
Men: 125. Guns: 14 x ____ . Purchased 1777; 1779 temporary fireship; 1780 hulked [at Portsmouth as slop ship; 1783 BU].

***Pelican*** 1777; later *Earl Denbigh*; 1787 See under schooners.

***Racehorse*** 1777-1778 (American? *Polly?* ____ ) 12.
Dimensions & tons: 68', 49'6" x 20'6" x 6'9". 111.
Men: ____ . Guns: 12 x ____ . Purchased (in the West Indies ?) 1777; 1778 sold.

***Surprise*** 1778-1783 (American privateer *Bunkers Hill* ____ ) 18.
Dimensions & tons: 96'9", 80'3⅝" x 22'10" x ____ . 222 63/94.
Men: 125. Guns: 18 x 6 + 8 swivels. Taken 4.12.1778 off Saint Lucia by Barrington; 1783 sold.

***Star*** 1778-1783 (French ____ ) 10.
Dimensions & tons: ____ , ____ x ____ x ____ . ____ .
Men: 70. Guns 10 x 4. Taken 1778; 1783 sold after apparently spending her naval career in the West Indies.

***Greenwich*** 1778-1779 (American ____ ) 12.
Dimensions & tons: ____ , ____ x ____ x ____ . ____ .

Men: 50. Guns: 12 x ____ . Taken 1778 in North America; 22.5.1779 lost on the Carolina coast.

***North*** 1778-by 1783 (American ____ ) 18.
Dimensions & tons: ____ , ____ x ____ x ____ . ____ .
Men: ____ . Guns: 18 x ____ . Taken 1778; still in service 1780 but out of service by 1783. (Or lost, wrecked at Halifax 12.12.1779?; Possibly two separate vessels?).

***Otter*** 1778-1783 (American *Gleneur?* ____ ) 14.
Dimensions & tons: ____ , ____ x ____ x ____ . ____ .
Men: ____ . Guns: 14 x ____ . Taken 1778; 1783 sold.

***Allegiance*** 1779-1782 (American *King George* ____ ) 14.
Dimensions & tons: ____ , ____ x ____ x ____ . ____ .
Men: ____ . Guns: 14 x 6. Taken 1779; 6.8.1782 taken by the French (or Americans?).

***Fortune*** 1779-1780 (American ____ ) 10.
Dimensions & tons: 70', ____ x 24' x 8'6½". 120.
Men: 90/45. Guns: 10 x 4. Taken 1779; ____ 1780 lost.

***Rover*** 1779-1780 (American *Cumberland* [?] 1777 ) 16.
Dimensions & tons: 90', 76'5" x 22'8" x 9'. 106.
Men: 125. Guns: 16 x 6. Taken 1779; 13.9.1780 taken by the French *Junon* in the West Indies; 02.1781 retaken by an English privateer and later wrecked in American waters (as the ship sloop *Rover* 20?; but the dimensions are rather too different for this to be likely).

***Trepassey*** 1779-1781/1782-1784 (American ____ ) 14 probably brig-rigged.
Dimensions & tons: ____ , ____ x ____ x ____ . ____ .
Men: 80. Guns: 14 x 6. Purchased 1779; 28.5.1781 taken by the American *Alliance*; 1782 retaken; 1784 sold.

***Victor*** 1779-1780 (American or French *Victor?* 1779 ____ ) 18?
Dimensions & tons: ____ , ____ x ____ x ____ . ____ .
Men: ____ . Guns: 18 (?) x ____ . Taken 1779; 5.10.1780 foundered in the Great West Indian Hurricane.

***Loyalist*** 1779-1781 (ex *Restoration?* ____ ) 14.
Dimensions & tons: 99', 82'5" x 27'8" x 7'8". 320.
Men: 125. Guns: 14 x ____ . Purchased (in North America?) 1779; 30.7.1781 taken by the French *Glorieux* in the Chesapeake.

***Port Antonio*** 1779-1783 (French? *Theresa?* ____ ) perhaps a brig? 12.
Dimensions & tons: 59'6", 44'10" x 21'x 8'6". 69.
Men: 110. Guns: 12 x ____ . Purchased (possibly at Jamaica?) 1779; 1783 sold.

***Echo*** 1780-1781 (French *Le Hussard* 1779 Lorient, Segondat design) 18.
Dimensions & tons: 84'7", 64'2" x 28'2¼" x 11'7". 271.
Men: 90. Guns: 18 x 6. Taken 5.7.1780 by *Nonsuch* rigged as a brig though built as a cutter (converted before completion); 11.2.1781 wrecked 'in a gale of wind' in Deadman's Bay, Plymouth Sound.

***Hannibal*** 1780-1780s ( ____ ) 14.
Dimensions & tons: 94'6", 79'8" x 24'8½" x 10'6". 219 76/94.
Men: 110. Guns: 14 x 6 (+ 14 swivels?). Purchased(in North America?) 1780 or 1782?; out of service well before 1788.

***Pacahunta*** 1780-1782 ( ____ ) 14.
Dimensions & tons: 98', 82'10½" x 25'4" x 10'10". 242.
Men: ____ . Guns: 14 x ____ . Purchased (in North America?) 1780; 1782 sold. (May have served for a while as a fireship.)

***Beaumont*** 1780-1783 (French *Zephir?* ____ ) 14.
Dimensions & tons: ____ , ____ x ____ x ____ . ____ .
Men: ____ . Guns: 14 x ____ . Taken 05.1780 at Charleston; 1783 sold at New York.

***San Vincente*** 1780-1783 (Spanish *San Vincente* [?] ____ ) 14.
Dimensions & tons: 83', ____ x 24' x 12'. ____ .
Men: ____ . Guns: 14 x 6. Taken in the West Indies 1780; 1783 sold.

***Germain*** 1780/1-1784 (Mercantile? *Germain?* ____ ) 14.
Dimensions & tons: 88', ____ x 26'4" x ____ . 261.
Men: 80/110. Guns: 14 x ____ . Purchased 1780 or 1781; 1784 sold.

***Chaser*** 1781-1782/1783-1784 (French? *Chasseur?* ____ ) ____ .
Dimensions & tons: ____ , ____ x ____ x ____ . ____ .
Men: ____ . Guns: ____ . Purchased in the East Indies 1.1.1781; ____ 1782 taken by the French in the Bay of Bengal renamed *Chasseur?*; 1783 retaken and renamed *Chaser*; 1784 sold.

***Expedition*** 1781-1783 ( ____ ) ____ .

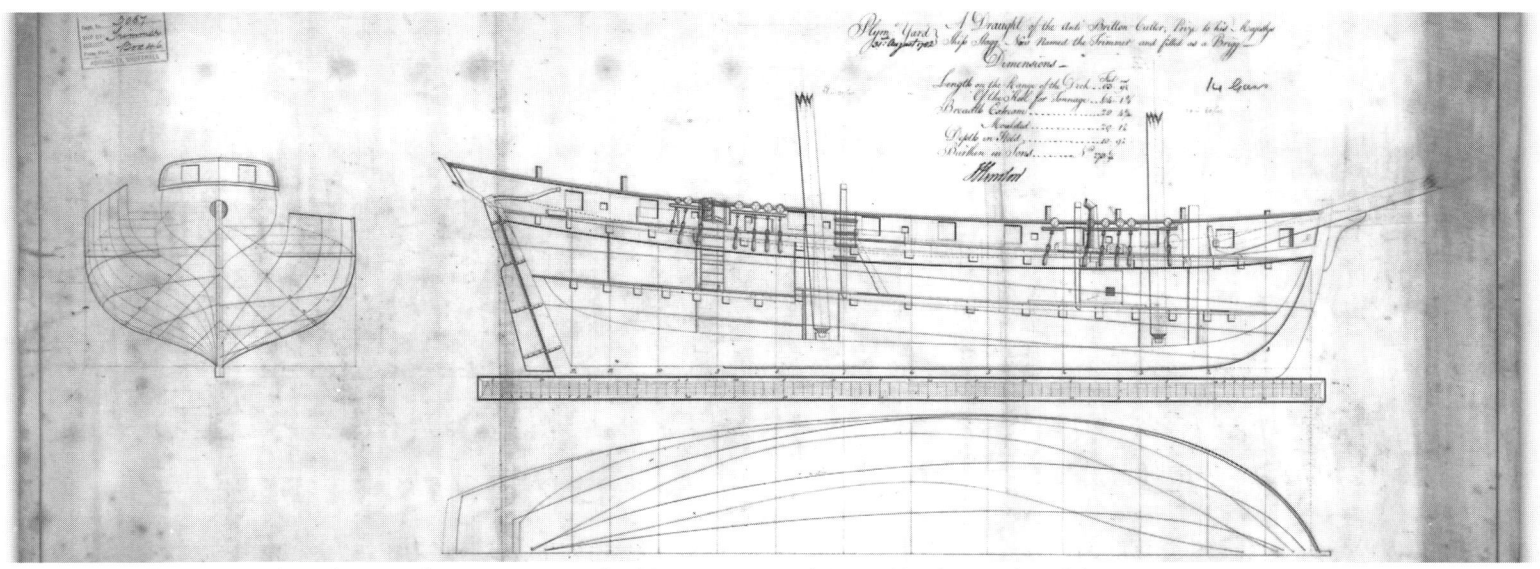

This lines and profile plan (3087) shows the captured cutter *Anti-Briton* as fitted for RN service as a brig in 1782 and renamed (surely by someone with a sense of humour) *Trimmer*

Dimensions & tons: ____ , ____ x ____ x ____ . ____ .
Men: ____ . Guns: ____ . Purchased (in North America?) 1781; ____ 1783 lost.

***Experiment*** 1781-1785 ( ____ ) 14.
Dimensions & tons: 90', 78' x 26' x 9'. 176.
Men: 110. Guns: 14 x ____ . Purchased (in North America? ) 1781; 1785 sold at Antigua.

***Jane*** 1781-1782 ( ____ ) ____ .
Dimensions & tons: ____ , ____ x ____ x ____ . ____ .
Men: ____ . Guns: ____ . Purchased (in North America?) 1781; ____ 1782 taken by the French.

***Tickler*** 1781-1783 ( ____ ) 12.
Dimensions & tons: 77', ____ x 25'3" x ____ . ____ .
Men: 100. Guns 12 x ____ . Purchased (in North America?) 1781; ____ 1783 taken by a French frigate in the West Indies.

***Carolina*** 1781-1784 (Mercantile? *Springfield*? ____ ) sloop/armed vessel ____ .
Dimensions & tons: ____ , ____ x ____ x ____ . 317.
Men: ____ . Guns: ____ . Purchased in North America 1781; 1784 sold.

***Lisborne*** 1781-1783 ( ____ ) 12.
Dimensions & tons: ____ , ____ x ____ x ____ . 187.
Men: 90. Guns: 12 x 3 (originally 10 x ____ ?). Purchased 1781; 1783 sold.

***Morning Star*** 1781-1782 (American privateer? ____ ) 16.
Dimensions & tons: ____ , 72'3¼" x 22'3¼" x 8'3½". 292.
Men: 100. Guns: 16 x ____ . Purchased 1782 (? had been taken by *Medea* on 14.1.1781?); 1782 sold.

***Saint Philips Castle*** 1782-1783 ( ____ ) ____ .
Dimensions & tons: ____ , ____ x ____ x ____ . ____ .
Men: ____ . Guns: ____ . Purchased at Minorca by General Murray ____ ?; 1783 sold.

***Stormont*** 1781-1784 ( ____ ) 16?
Dimensions & tons: 80', ____ x 23' 7" x ____ . ____ .
Men: 100. Guns. 16? x ____ . Purchased 1781; 1784 sold. Dittmar suggests taken by the French in 1782 and served as their *Stormon* until 1787; (there could be two sloops of the name but it seems more likely there was only one).

***Vaughan*** 1781-1783 ( ____ ) 16? Probably brig-rigged.
Dimensions & tons: 80', 59'2" x 23' 7" x ____ . 175.
Men: ____ . Guns: 20 x ____ . Purchased 1781; 1783 sold.

***Tobago*** 1782-1783 (Mercantile? *Minerve*? ____ ) 16?
Dimensions & tons: ____ , ____ x ____ x ____ . ____ .
Men: ____ . Guns: 26? x ____ . Purchased (in the West Indies?) 1782; 1783 sold.

***Trimmer*** 1782-1801 (American or French cutter *Anti-Briton* ____ ) brig sloop? 14.
Dimensions & tons: 83'9½", 64'1⅜" x 28'4⅞" x 10'9¼". 275.

Men: 110. Guns: 14 x 6 + 10 swivels (later 8 x 18★). Taken by *Stag*; fitted as a brig;; 1798 temporary fireship; 1801 sold.

***Bloodhound*** 1780s-1783 ( ____ ) ____ .
Dimensions & tons: ____ , ____ x ____ x ____ . ____ .
Men: ____ . Guns: ____ . Purchased or taken ____ ?; 1783 sold.

***Observer*** 1783-1784 ( ____ ) ____ .
Dimensions & tons: ____ ____ x ____ x ____ . ____ .
Men: ____ . Guns: ____ . Purchased or taken ____ ?; never registered on the list of the Navy; 1784 sold.

## Fireships

***Vulcan*** (ex *Vesuvius*) 1777-1781 (American ____ ) 8.
Dimensions & tons: 91'6", 72'0" x 27'9" x 12'3½". 296 26/94.
Men: 45. Guns: 8 x 6 + 6 swivels. Taken 1777; 22.9.1782 destroyed at Yorktown to prevent capture.
PLANS: Lines & profile & midsection/decks.

***Firebrand*** 1778 See ship sloop *Porpoise*.
***Salamander*** 1778 See ship sloop *Shark*.
***Lucifer*** 1778 See ship sloop *Avenger*.
***Sulphur*** 1778-1783 ( ____ ) 8.
Dimensions & tons: ____ , ____ x ____ x ____ . ____ .
Men: 45. Gun: 8 x 4 + 8 swivels. Purchased 07.1778 by Lord Howe in North America; 1783 sold.

***Volcano*** 1778-1781 ( ____ ) ____ .
Dimensions & tons ____ ____ ____ x ____ x ____ . ____ .
Men: 45. Guns: ____ . Purchased by Lord Howe in North America; 1781 sold.

***Infernal*** 1778-1783 (building Perry, Blackwall) 8.
Dimensions & tons: 97'6", 79'6¾" x 26'11" x 13'9". ____ .
Men: 45. Guns: 8 x 4 + 8 swivels. Purchased whilst on stocks 11.1778; 6.11.1778 launched; 1783 sold.
PLANS: Profile/decks.

***Furnace*** 1778-1783 (building Fisher, Liverpool) 8.
Dimensions & tons: 100'7", 82'6" x 28'10" x 13'2". 365.
Men: 45. Guns: 8 x 4 + 8 swivels. Purchased whilst on stocks 1778; 6.12.1778 launched; 1783 sold.
PLANS: Sheer & profile/decks.

***Harpy*** 1779 See ship sloops.
***Blast*** 1779 See ship sloop *Druid*.
***Lightning*** 1779 See ship sloop *Sylph*.
***Basilisk*** 1779 See ship sloop *Grasshopper*.
***Snake*** 1779 See under sloops, rig uncertain.
***Antigua*** 1779-1783 ( ____ ) temporary fireship 8; brig rig?
Dimensions & tons: ____ , ____ x ____ x ____ . ____ .

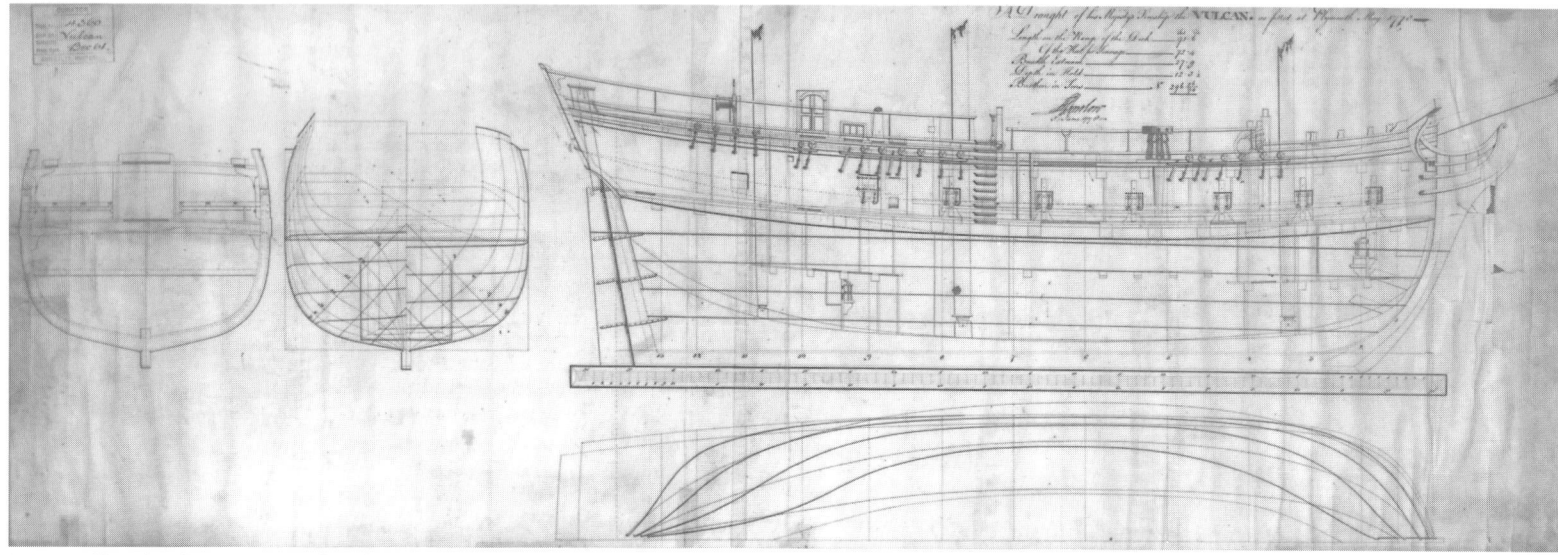

Lines, profile and midsection (4360) of the fireship *Vulcan* as fitted in 1778

Men: 50. Guns: 8 x ____ . Purchased (? taken) 1779; 1780 brig; 1783 sold.

**Combustion** 1782-1784 ( ____ ) ____ .
Dimensions & tons: ____ , ____ x ____ x ____ . ____ .
Men: ____ . Guns: ____ . Purchased 1782; 1784 sold.

**Volcano** 1780-1784 (Mercantile? *Empress of Russia*? ____ ) ____ .
Dimensions & tons: ____ , ____ x ____ x ____ . ____ .
Men ____ . Guns ____ . Purchased 1780; never registered in the RN?; 1784 sold.

**Lucifer** 1780?-1784 (American? *Elizabeth*? ____ ) ____ .
Dimensions & tons: ____ , ____ x ____ x ____ . ____ .
Men: ____ . Guns: ____ . Purchased 1780?; never registered in the RN; 1784 sold.

## Cutters

**Griffin** 1778-1786 (Mercantile? *Charming Molly*? ____ ) 10 clinker-built.
Dimensions & tons: 73'4", 53'2½" x 25'7" x 10'6". 186.
Men: 60 or 55. Guns: 10 x 4 + 12 swivels (1779: 12 x 4). Purchased early 1778; 1786 sold.

**Jackall** 1778-1779/1781-1785 (Mercantile? *Active*? ____ ) 10 clinker-built.
Dimensions & tons: 72'10", 54'3" x 25'4" x 10'3". 187 21/94.
Men: 55 or 60. Guns: 10 x 4 + 12 swivels. Purchased early 1778; 27.11.1779 crew mutinied and handed her to the French, who renamed her *Le Boulogne*; 1781 retaken by *Prudente* and named *Jackal* 6/1785 sold.

PLANS: Lines & profile/decks.

**Kite** 1778-1793 (Mercantile? *Cruizer*? ____ ) 10 clinker-built.
Dimensions & tons: 77'5½", 55'11" x 27'1" x 10'6". 218.
Men: 60. Guns: 10 x 4 + 10 swivels. Purchased early 1778; 1779 to 1783 rated as a sloop; 1793 BU.

**Nimble** 1778-1781 (Mercantile? *Mary*? ——— ) 10 clinker-built.
Dimensions & tons: 72'7", 53'11⅞" x 22'2¾" x 10'5". 182 73/94.
Men: ____ . Guns: 10 x 4 + ____ swivels (1779: 12 x 4). Purchased early 1778; 11.2.1781 lost in Mounts Bay.

PLANS: Lines & profile/decks.

**Pheasant** 1778-1781 (Mercantile? *Prince of Orange*? ____ ) 10 clinker-built.
Dimensions & tons: 65'8½", 47'8½" x 24'3" x 9'6". 149 24/94.
Men: 50. Guns: 10 x ____ + 12 swivels (1780: 12 x ____ ). Purchased early 1778; 20.6.1781 capsized 'in a sudden gale of wind' in the Channel.

PLANS: Lines, profile, decks.

**Rambler** 1778-1787 (ex *Good Intent* ——— ) 10.
Dimensions & tons: 65'7", 49' x 22'6" x 9'6". 139.
Men: 50/45. Guns: 10 x 3 + 12 swivels. Purchased early 1778; 1787 sold.

**True Briton** 1778-1780/1782-1785 (Mercantile? *Tartar*? ——— ) 10 clinker-built.
Dimensions & tons: 74'6¾" , 54'4" x 25'6⅛" x 9'11". 188 9/94.

Men: 60. Guns: 10 x 4 + 12 swivels (1782?: 2 x 12★ added). Purchased early 1778; 5.12.1780 taken by French privateer and renamed *Le Tartare*; 02.1782 retaken by *Arethusa* and renamed *True Briton*; 1785 sold.

PLANS: Lines & profile.

**Busy** 1778-1792 (building Farley, Folkestone) 10 clinker-built.
Dimensions & tons: 73'6½", 53'6¼" x 25'8" x 9'7". 188.
Men: 55. Guns: 10 x 4 + 12 swivels. (1780: 2 x 4 and 2 x 12★ added. By 1783: 14 x 4). Purchased whilst on the stocks 1778; 06.1778 launched; 1792 sold.

PLANS: Lines & profile/decks.

**Advice** 1779-1793 (French? ____ ) 10 clinker-built.
Dimensions & tons: 56', 38'2" x 21'8" x ____ . 93 or 95 29/94.
Men: 50/45. Guns: 10 x 3 + 10 swivels. Purchased 1779; 1793 sold.

**Liberty** 1779-1816 ( ____ ) 14 clinker-built.
Dimensions & tons: 74'3", 53'9" x 25'6¼" x 10'6½". 187.
Men: 70 (later 55). Guns: 14 x 4 + 10 swivels (1780: 12 x 4 + 2 x 12★. Later: 16 x 12★, 1789: reverted to old armament of 4 pounders and swivels). Purchased 1779; 1790 rigged as brig, though still classed as cutter; later re-rigged as cutter; 1816 sold.

**Repulse** 1779-1782 ( ____ ) 12 clinker-built.
Dimensions & tons: 64'5", 49'3⅛" x 22'9" x ____ . 136.
Men: 60. Guns: 12 x 4 + 10 swivels. Purchased 1779; Ca 17.3.1782 wrecked off Yarmouth.

**Tapageur** 1779-1780 (French fireship? or privateer? *Le Tapageur* ____ ) 14.
Dimensions & tons: 73'6", 54'10¼" x 27'9" x 10'10". 224 64/94.
Men: 70. Guns: 14 x 4 + 10 swivels. Taken by *Milford* 04.1779; ____ 1780 lost in warping in to the careenage at Saint Lucia.

PLANS: Lines & profile/profile & decks.

**Duc de la Vaugignon [Duke de la Vaugignon]** 1779-1779 (French privateer *Le Duc de la Vauguyon* 1779 Honfleur) 10.
Dimensions & tons: 68'1", 51'9¾" x 24'5" x 9'7". 164 30/94.
Men: 60. Guns: 10 x 4 + 10 swivels (later 12 x 4). Taken 1779; 12.1779 foundered.

PLANS: Lines & profile/decks.

**Mutin** 1779-1798 **Pigmy** 1805 (French *Le Mutin* 1778 ____ ) 14 clinker-built.
Dimensions & tons: 79'11¼" x 26'1" x 10'1", 215.
Men: 70 Guns: 14 x 4 + 10 swivels. Taken 2.10.1779 by *Jupiter*; 20.1.1798 renamed *Pigmy*; 9.8.1805 wrecked in Saint Aubin's Bay, Jersey.

**Resolution** 1779-1792 ( ____ ) 14 (brig?).
Dimensions & tons: ____ , 200 or 198.
Men: 70. Guns: 14 x 4 + 10 swivels. Purchased 1779; 01.1797 foundered in the North Sea.

**Surprize** 1779/1780-1786 ( ____ ) 10 clinker-built.
Dimensions & tons: 67'3", 48'3" x 22'11" x 9'. 135.

THE AMERICAN WAR OF INDEPENDENCE 1776-1783

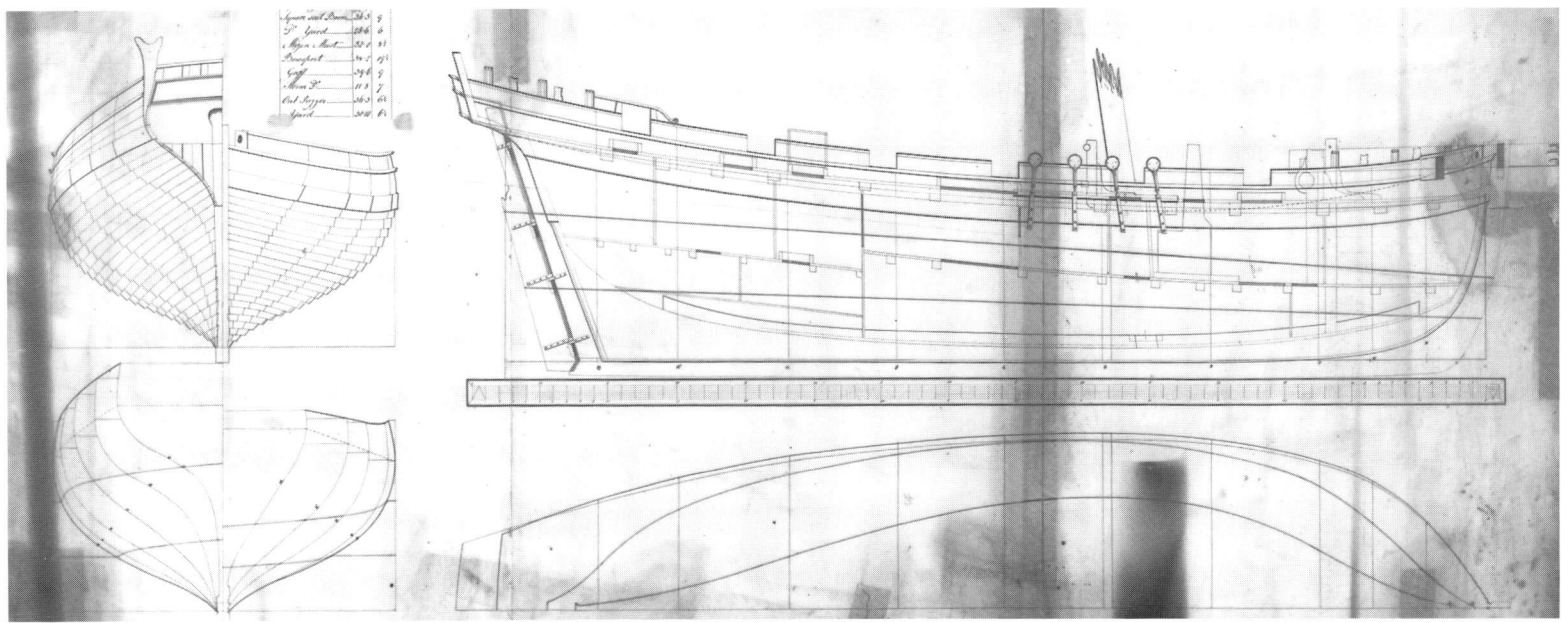

This 'taken off and fitted' lines and profile plan (6607) of the purchased cutter *Nimble* of 1778 shows very clearly in the stern and bow views that this was a clinker built vessel

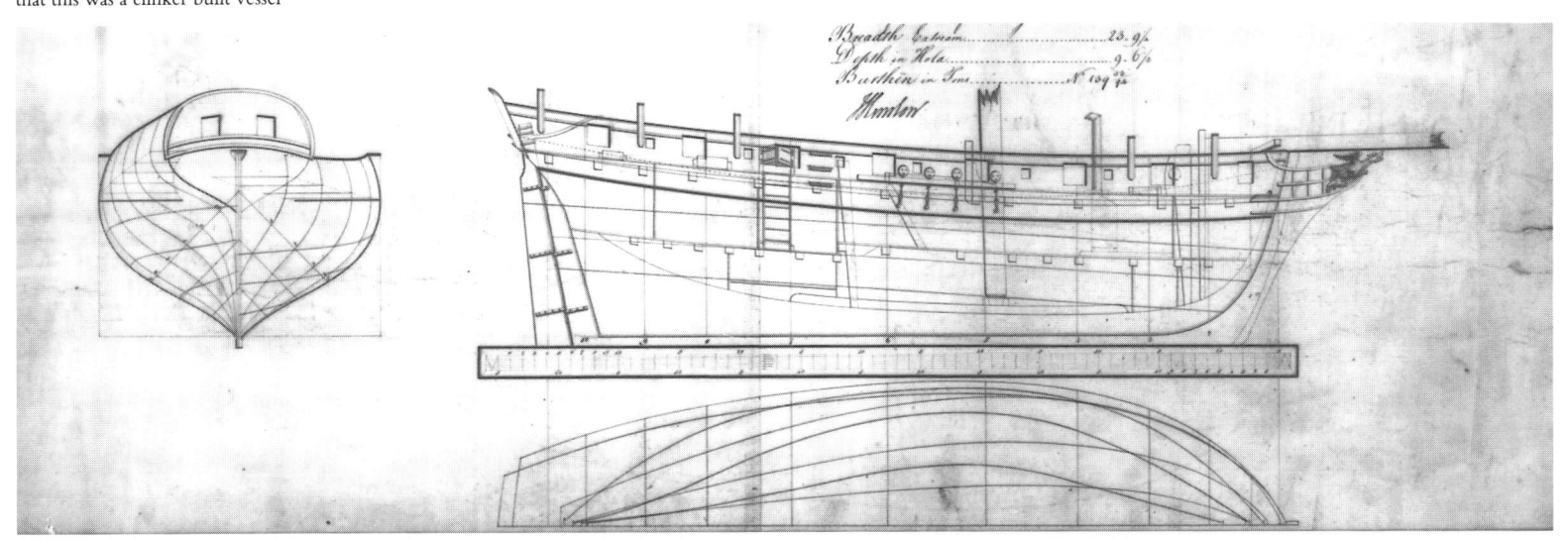

The purchased cutter *Dragon* of 1782, as shown in this lines and profile plan (6811), was unusual amongst cutters in having a full head with rails, beak and figurehead

Men: 50. Guns: 10 x 3 + 8 swivels (later two carronades added). Purchased late 1779 or early 1780; 1786 sold.

**Pilote** 1779-1799 (French *Le Pilote* 1779 Dunkirk) 14.
Dimensions & tons: 78'6¾", 69'5⅜" x 26'0¾" x 10'2½". 218.
Men: 70. Guns: 14 x 4 + 10 swivels (in the 1790s the 4 pounders were replaced by 12 pounder carronades). Taken 2.10.1779 by *Jupiter*; 1794 fitted as a brig; 1799 sold.

**Sylph** 1780-1782 (Mercantile *Active* ____ ) 18 classed as a sloop.
Dimensions & tons: 80'4", 60'5" x 26'5" x 10'6". 224.
Men: 80. Guns: 18 x 4 + 12 swivels. Purchased 05.1780; 3.2.1782 taken by the French at the surrender of Demara [in French service until 1786].

**Cruizer** 1780-1791 (Mercantile *Cruizer* ____ ) 14.
Dimensions & tons: 73'7", 54'3" x 26'3" x 10'. 199.
Men: 70. Guns: 14 x 4 + 10 swivels. Purchased early 1780; 4.5.1791 sailed from Plymouth, lost without trace.

**Hope** 1780-1785 (Mercantile *Hope* ____ ) 12 clinker-built.
Dimensions & tons: 67'9", 48'2½" x 24'8" x 10'4", 156.
Men: 60/55. Guns: 12 x 4 + 8 swivels. Purchased early 1780; 1781 taken by French privateer; 6 days later retaken by *Stag*; 1785 sold. (Dittmar suggests she was the American *Lady Washington* taken in 1779.)

**Sultana** 1780-1799 (Mercantile *Sprightly* ____ ) 12 clinker-built.
Dimensions & tons: 65'7", 48'9" x 24'3" x 9'7". 152 40/94.
Men: 60. Guns: 12 x 4 + 8 swivels (later 10 x 12★ + 12 swivels). Purchased 06.1780; 1799 sold.

**Lark** 1780-1784 (Mercantile *Resolution* ____ ) 16 classed as a sloop, clinker-built.
Dimensions & tons: 74'4", 54'5½" x 26'2" x 10'4". 198.
Men: 70. Guns: 16 x 4 + 2 x 12★ + 10 swivels. Purchased early 1780; 1784 sold.

**Monkey** 1780-1786 (Mercantile *Lark* ____ ) 12.
Dimensions & tons: 72'10", 54'3" x 25'4" x 10'3". 187.
Men: 60. Guns: 12 x 4 + 8 swivels. Purchased mid-1780; 1786 sold. (Dittmar suggests that her mercantile name was *Jackal*.)

**Brazen** 1781-1799 ( ____ ) 10 clinker-built.
Dimensions & tons: 58', 44'4" x 22'10" x 9'6". 112.
Men: 60. Guns: 10 x ____ + 12 swivels ( 1790: 12 x 12★). Purchased 06.1781; 1799 sold.

**Nimble** 1781-1808 (building Jacobs, Folkestone) 12 clinker-built.
Dimensions & tons: ____ , 58'7" x 27'4" x ____ . 168.
Men: ____ . Guns: 10 x 4 + 12 swivels (later 12 x + 12 swivels (later: 10 x 18★ + 12 swivels 1790: 2 x 3 added). Purchased whilst on the stocks 1781; launched 6.7.1781; 1805 converted to a brig; ____ 1808 run aground near Stangate Creek.

**Cupid** 1781 (181 tons) 'never purchased tho' agreed for ..... supposed to have been some disagreement between the [Navy] Board and the owner.'

**Dragon** 1782-1785 (Mercantile *Dragon* ____ ) 16 clinker-built.
Dimensions & tons: 61'2", 46'10" x 23'9¼" x 9'6½". 139 33/94.
Men: 50. Guns: 10 x 4 + 10 swivels. Purchased Spring 1782; 1785 sold.

***Seaflower*** 1782?-1814 (Mercantile *Swiftsure?* ____ ) 16 clinker-built.
Dimensions & tons: 72'5", 56'9¾" x 25'11" x 10'7½". 203.
Men: 70. Guns: 16 x 4 (later 16 x 12★). Purchased 04.1782?; 1791 repaired and 'her bottom planked the carvel way'; 1814 sold.

***Barracouta*** 1782-1792 (Mercantile ____ ? Stewart, Sandgate? ) 14.
Dimensions & tons: 75'2", 54'11" x 25'11¾" x 10'6". 197.
Men: 90 (1783: 60). Guns: 14 x 4. Purchased 06.1782 and registered as a cutter-rigged sloop; 1783 reclassed as a cutter; 1792 sold.

***Spider*** 1782-1806 (French privateer *L'Araignée* ____ ) 12.
Dimensions & tons: 69', 54'1¾" x 24'3" x 10'7". 169 32/94.
Men: 50. Guns: 12 x 4. Taken 09.1782; 1797 fitted as a schooner; 1806 sold. (Dittmar suggests that her French name was *Victoire*.)

***Substitute*** 1782-1783 ( ____ ) ____ .
Dimensions & tons: ____ , ____ x ____ x ____ . ____ .
Men: ____ . Guns: ____ . Purchased at Rio de Janeiro by Bickerton 1782; 1783 sold.

## Schooners

***Diligent*** 1776-1777 ( ____ ) ____ .
Dimensions & tons: ____ , ____ x ____ x ____ . ____ .
Men: ____ . Guns: ____ . Purchased (in North America?) 1776; 1777 wrecked.

***Dispatch*** 1776-1776 ( ____ ) ____ .
Dimensions & tons: ____ , ____ x ____ x ____ . ____ .
Men: ____ . Guns: ____ . Purchased (in North America?) 1776; ____ 1776 taken by the Americans.

***Pelican*** 1777; later ***Earl Denbigh***; 1787 (Mercantile *Earl of Denbigh?* ____ ) ____ .
Dimensions & tons: ____ , ____ x ____ x ____ . 150.
Men: 45. Guns: ____ . Purchased 1777; at some time appears to have been classed as a sloop 'she has been called *Earl Denbigh* almost from the time when bought' (1783 Dimensions Book); 25.6.1787 employed in clearing Antigua Harbour when sunk.

***Viper*** 1777-1779 (American? *Viper?* ____ ) ____ .
Dimensions & tons: ____ , ____ x ____ x ____ . 149?
Men: ____ . Guns: ____ . Purchased in North America 1777; 1779 condemned; 1780 sold.

***Coureur*** 1778-1780 (French 'lugger' *Coureur* 1776 ____ ) 8.
Dimensions & tons: 68'10", 52'5¾" x 22'3¼" x 9'10½". 138 42/94.
Men: 45. Guns: 8 x 4. Taken by Keppel's fleet 17.6.1778; ____ 1780 taken by two American privateers off Newfoundland. She was clinker planked on the lower hull.
PLANS: Lines & profile.

***Florida*** 1778-1778 ( ____ ) schooner/'sloop on survey' 12.
Dimensions & tons: 63', 47'6" x 19'2" x 7'8". 92.
Men: ____ . Guns: 12 x ____ . Purchased in North America 1778; 1778 put out of commission.

***Racehorse*** 1779-1781 (Mercantile *Racehorse* ____ ) 10.
Dimensions & tons: ____ , ____ x ____ x ____ . ____ .
Men: 45. Guns: 10 x 3 + 12 swivels. Purchased 1779 (or 1778?); 01.1781 wrecked near Seaford (off Beachy Head).

***Sprightly*** 1779-1782 ( ____ ) ____ .
Dimensions & tons: ____ , ____ x ____ x ____ . ____ .
Men: ____ . Guns: ____ . Purchased (in North America?) 1779; 1782 sold.

***Kitty*** 1780- ____ ? (Mercantile? *Kitty?* ____ ) ____ .
Dimensions & tons: ____ , ____ x ____ x ____ . ____ .
Men: ____ . Guns: ____ . Purchased (in North America?) 1780; out of service at some time before 1788.

***Independence*** 1780- ____ ? (Mercantile? *Independence* ____ ) ____ .
Dimensions & tons: ____ , ____ x ____ x ____ . ____ .
Men: ____ . Guns: ____ . Purchased (in North America?) 1780; out of service at some time before 1788.

***Berbice*** 1780-1796 ( ____ ) 8.
Dimensions & tons: 72'9", 54' x 20'6" x 8'. 120 66/94.

Men: 42. Guns: 6 x 3 + 2 x 12★ + 8 swivels + 6 musketoons. Purchased (in America or the West Indies?) 1780; 1793 registered; 3.11.1796 wrecked at San Domingo.

***Tryal*** 1780s-1783? ( ____ ) schooner/advice boat ____ .
Dimensions & tons: ____ , ____ x ____ x ____ . ____ .
Men: ____ . Guns: ____ . Purchased ____ ?; out of service by 1783.

***Gros Isle*** 1781?-1783? ( ____ ) ____ .
Dimensions & tons: ____ , ____ x ____ x ____ . ____ .
Men: ____ . Guns: ____ . Origins and fate uncertain, but probably acquired in 1781 and deleted in 1783. Never on the list of the Navy.

***Lion*** 1782-1785 (Mercantile? *Neptune?* ____ ) 10.
Dimensions & tons: 71'6", 59'2½" x 20'6" x 8'. 132 50/94.
Men: 40/55. Guns: 10 x 3. Purchased 1782; 1785 sold.

## Galleys, etc

These, except for the last, all appear to be captured American vessels. The individual states built numbers of these vessels for coastal defence.

***Dependance*** 1776-1786 (American ____ ) galley ____ .
Dimensions & tons: ____ , ____ x ____ x ____ . ____ .
Men: 40. Guns: ____ . Taken 1776; 1786 sold.

***Alarm*** 1777-1778 (ex East Indiaman *Mountfield* ____ ) galley ____ .
Dimensions & tons ____ ____ x ____ x ____ ____ .
Men: 40. Guns: ____ . Purchased 4.6.1777 in North America by Lord Howe and cut down; 5.8.1778 abandoned and burnt at Rhode Island when blockaded there by the French.

***Cornwallis*** 1777-1782 (American ____ ) galley 5.
Dimensions & tons ____ , ____ x ____ x ____ . ____ .
Men: 40 Guns: 1 x 24 + 4 x 4. Taken 1777; 1782 put out of commission.

***Crane*** 1777-1783 (American ____ ) galley 1.
Dimensions & tons: ____ , ____ x ____ x ____ . ____ .
Men: 30. Guns: 1 x 24. Taken 1777; 1783 sold.

***Spitfire*** 1778-1778 (American? ____ ) galley ____ .
Dimensions & tons: ____ , ____ x ____ x ____ . ____ .
Men: 40. Guns: ____ . Purchased in North America by Lord Howe 25.1.1778; 5.8.1778 abandoned and burnt at Rhode Island when blockaded there by the French.

***Pigot*** 1778-1778 (American? ____ ) galley ____ .
Dimensions & tons: ____ , ____ x ____ x ____ . ____ .
Men: 40. Guns: ____ . Purchased by Lord Howe in North America 19.7.1778; 5.8.1778 abandoned and burnt at Rhode Island when blockaded there by the French.

***Hussar*** 1780-1786 (American *Hussar?* ____ .) galley 1.
Dimensions & tons: ____ , ____ x ____ x ____ . ____ .
Men: 30. Guns: 1 x 24. Taken 1780; 1786 sold.

***Lee*** 1780-1784 (American? *Adder* ____ ) galley ____ .
Dimensions & tons: ____ , ____ x ____ x ____ . ____ .
Men: ____ . Guns: ____ . Purchased (in North America?) 06.1780; 1784 sold.

***Marquis De Bretigny*** 1780-1780s ( ____ ) galley ____ .
Dimensions & tons: ____ , ____ x ____ x ____ . ____ .
Men: ____ . Guns: ____ . Purchased (in North America?) 1780; out of service before 1788.

***Revenge*** 1781-1782 ( ____ *Firefly?* ____ ) galley 8.
Dimensions & tons: ____ , ____ x ____ x ____ .
Men: 30. Guns: 8 x ____ . Origin and fate uncertain, in service 1781-1782.

***Gloire*** 1781-1783 (French *La Gloire?* ____ ) lugger 8.
Dimensions & tons: 70'3", 60'7½" x 18'10" x 7'10". 114.
Men: ____ . Guns: 8 x ____ . Taken 1781; 1783 BU.

## Storeships, Transports, etc

***Elephant*** 1776-1779 (Mercantile *Union* ____ ) storeship 10.
Dimensions & tons: 103'3", 85'0⅛" x 29'1" x 11'9". 382 42/94.
Men: 40 Guns: 10 x 4 + 8 swivels. Purchased 07.1776; 1779 taken by an American privateer, retaken and sold.

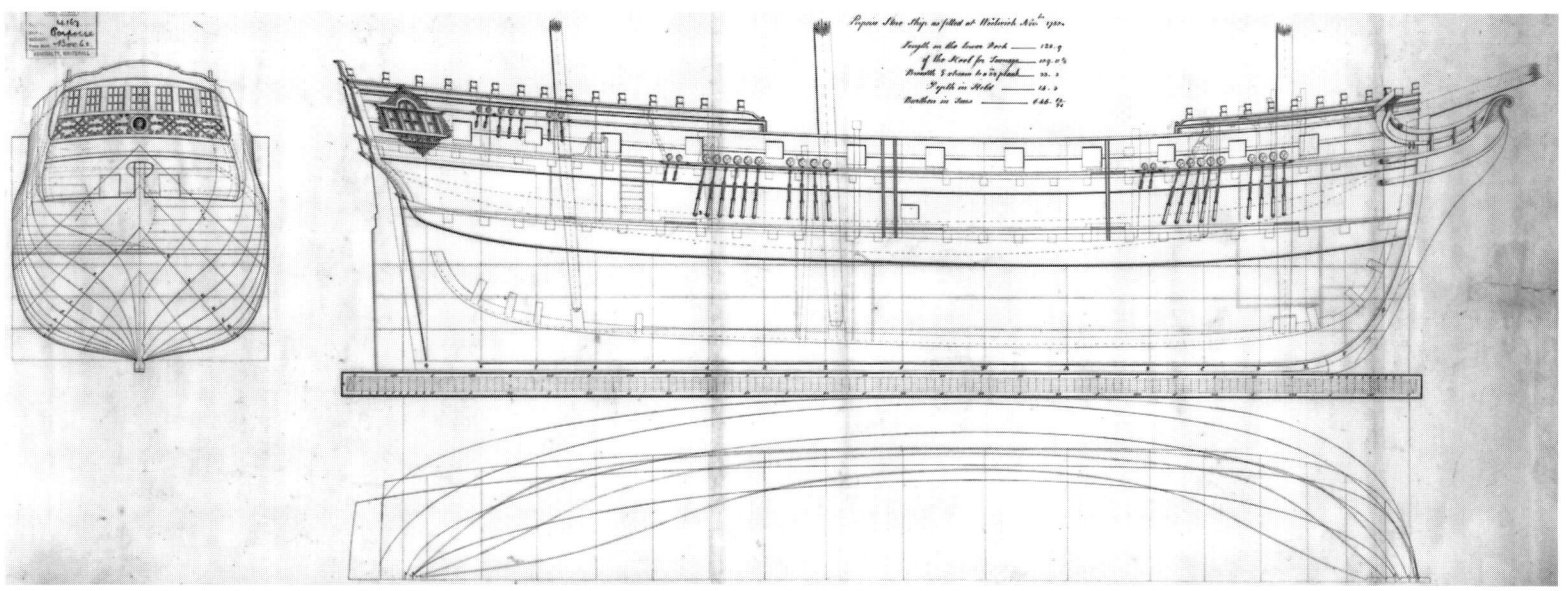

# THE AMERICAN WAR OF INDEPENDENCE 1776-1783

This lines and profile plan (4167) shows the storeship *Porpoise* 'as fitted' in 1780 with her upperworks far more substantial and utterly changed from her original appearance as a Swedish merchantman

PLANS: Taken off & as fitted: Lines & profile/orlop/lower deck/upper deck.

***Discovery*** 1776-1797 (Mercantile *Diligence*? or *Bloodhound*? 1774 Whitby) ship sloop/exploration ship 8.
Dimensions & tons: 91'5", 74'9" x 27'5" x 11'5". 299.
Men: 70 (as transport: 15/18/24). Guns: 8 x ____ + 8 swivels. Purchased 01.1776; 1780 naval transport; 1797 BU.

***Heart of Oak*** 1777-1782? ( ____ ) armed ship 20.
Dimensions & tons: ____ , ____ x ____ x ____ . ____ .
Men: 120. Guns: 20 x ____ . Purchased (or hired?) 1777; still in service 1782.

***Peterel*** 1777-1788 (Mercantile *Duchess of Manchester*? ____ ) 'sloop on survey' 4.
Dimensions & tons: 58'9", 56'0½" x 21'6" x 8'6". 138.
Men: 18/27. Guns: 4 x 3 + 6 swivels. Purchased 04.1777 and fitted for surveying the coast of Great Britain; 1784 sold.

PLANS: As *Duchess of Manchester*: Lines & profile/decks.

***Diligent*** 1777-1779 ( ____ ) armed vessel [brig rig] 12.
Dimensions & tons: ____ , ____ x ____ x ____ . ____ .
Men: 50. Guns: 12 x 3. Purchased by Lord Howe in North America 1777; 7.5.1779 taken by the American *Providence* off Newfoundland; [destroyed by the British at the Penobscot? 15.8.1779].

***Tortoise*** 1777-1780 (Mercantile *Grenville*? ____ ) storeship ____ .
Dimensions & tons: ____ , 110'10" x 34'7" x 14'2½". 705.
Men: ____ . Guns: ____ . Purchased mid-1777; 11.1779 foundered off Newfoundland.

***Nabob*** 1777-1780 (East Indiaman *Triton* ____ ) storeship 26.
Dimensions & tons: 136'5", 110' x 35' x 14'. 637.
Men: 72. Guns: UD 22 x 9, QD 4 x 6. Purchased 1777; 1780 hulked [at Chatham as a convalescent ship; ?later to Sheerness?; 1783 sold].

***Lioness*** 1777-1779 (Mercantile *Lioness*? ____ ) storeship 26.
Dimensions & tons: 140'9", 116' x 33'8" x 13'. 712.
Men: 72. Guns: UD 22 x 9, QD 4 x 6. Purchased 1777; 1780 hulked [at Portsmouth as a convalescent ship; 1783 sold].

***Greenwich*** 1777-1779 (East Indiaman *Greenwich* ____ ) storeship 26?
Dimensions & tons: ____ , ____ x ____ x ____ . 687.
Men: 72? Guns: UD 22 x 9, QD 4 x 6. Purchased 1777; 1779 hulked [as a receiving ship classed as a Fifth Rate at Sheerness; 1783 sold].

***Supply*** 1777-1779 (Mercantile *Prince of Wales* ____ ) storeship 26.
Dimensions & tons: ____ , ____ x ____ x ____ . ____ .
Men: 72? Guns: UD 22 x 9, QD 4 x 6 [?]. Purchased 10.1777; 14.6.1779 burnt by accident at Saint Kitts.

***Pacific*** 1777-1781 (French East Indiaman *Pacifique*? ____ ) storeship 26.
Dimensions & tons: 137'7", 115'3" x 33'1¾" x 13'8". 674.
Men: 160. Guns: 26 x ____ . Purchased (Taken?) 1777; 1781 breakwater.

***Dromedary*** 1777-1783 (Mercantile *Duke of Cumberland* ____ ) storeship 26.
Dimensions & tons: 140'0½", 115'10" x 37'10½" x 16'4". 884.
Men: 120 (as guard ship: 200). Guns: 26 x ____ (as guard ship 22 x 9 + 8 x 6). Purchased 1777; 1779 fitted as a guard ship [30] for the Downs and classed as a Fifth Rate; 1783 BU.

***Vigilant*** 1777-1780 (Mercantile [hired transport] *Empress of Russia* ____ ) armed ship/Sixth Rate 22/20.
Dimensions & tons: 122'6", ____ 34'10½" x ____ . 684.
Men: 150/180. Guns: UD 14 x 24, QD 6 x 6, Fc 2 x 9 (Fc guns later removed). Purchased at New York by Lord Howe 1777; ____ 1780 condemned as unserviceable, later deliberately burnt at Beaufort, South Carolina.

NOTE: She appears to have been fitted as a fire support ship for amphibious operations on the coast of North America. However the guns proved to be too heavy for her hull, and she became excessively leaky. (Information from Professor David Syrett.)

***Haerlem*** 1778-1779 (possibly American? —— ) armed ship/sloop 10.
Dimensions & tons: ____ , ____ x ____ x ____ .
Men ____ . Guns: 10 x ____ . Taken 1778; 07.1779 taken by an American privateer.

***York*** 1779-1781 (Mercantile *York*? ____ ) storeship 14.
Dimensions & tons: ____ , 112'2" x 33'4½" x 17'1½". 664.
Men: 60. Guns: UD 14 x 6 + 12 swivels. Purchased 03.1779?; 1781 sold at Bombay.

***San Carlos*** 1779-1784 (Spanish privateer *San Carlos* ____ ) armed ship 22.
Dimensions & tons: 125'9", 104'3⅜" x 34'11" x 14'. 676 36/94.
Men: 60. Guns: UD 14 x 6, QD 6 x 18★, Fc 2 x 18★. Taken in the West Indies by *Salisbury* 12.12.1779 [then carried 397 men and 50 guns]; 1780 purchased; 1784 sold.

***Manilla*** 1780-1782 (Mercantile? ____ ) armed transport 10.
Dimensions & tons: ____ , 87'0⅛" x 29'7½" x 12'11½". 406 40/94.
Men: 40. Guns: UD 8 x 18★, QD 2 x 18★. Purchased 11.1780; ____ 1782 lost in the East Indies.

***Porpoise*** 1780-1783 (Swedish? mercantile *Hertigennan*?/*Hertigennan af Sodermanland* ____ ) storeship/armed transport 18.
Dimensions & tons: 128'9", 109'11⅝" x 33'3" x 14'3". 646 64/94.
Men: 65. Guns: UD 18 x 6. Purchased 08.1780; 1783 sold 'being a very heavy sailer and much out of repair'.

PLANS: Proposed: Profile/decks. As fitted: Lines & profile/decks.

***Raikes*** 1780-1782 (Mercantile *Raikes*? ____ ) armed transport 10.
Dimensions & tons: ____ , 91'8" x 29'9" x 13'. 384 36/94.
Men: 40. Guns: UD 10 x 18★. Purchased 09.1780; 6.6.1782 taken by the French *Sphinx* and *Artésien* of Suffren's squadron in the East Indies; [in French service until 1784].

***Pondicherry*** 1780-1784 (Mercantile? ____ ) armed transport 18.

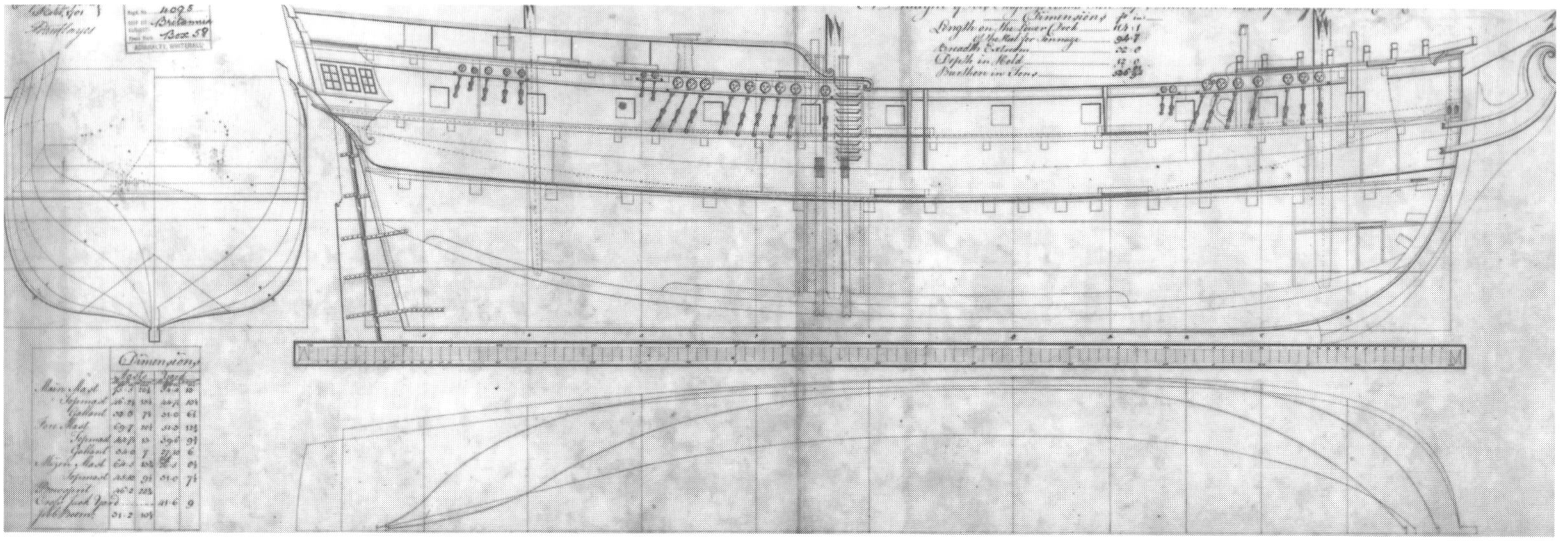

Lines and profile plan with spar dimensions (4095) for the purchased armed storeship *Britannia* of 1781

Dimensions & tons: ____ , 109'10¼" x 34'6½" x 11'10". 697 19/94.
Men: 65. Guns: 18 x 6 + 10 swivels (later: 10 x 18★ instead). Purchased 11.1780; 1784 sold.

***Royal Charlotte*** 1780-1783 (French privateer *Charlotte* ____ ) armed transport 10.
Dimensions & tons: ____ , 92'6" x 32'6" x 13'1". 520.
Men: 50. Guns: 10 x 6. Taken 1780; 1783 sold.

***Dauphine*** 1780?-1784 (French *Dauphine* ? ____ ) storeship ____ .
Dimensions & tons: ____ , ____ x ____ x ____ . ____ .
Men: ____ . Guns: ____ . Taken 1780; 1783 sold.

***Port Morant*** 1780?-1784 (Spanish? *Hermosa Mariana*? ____ ) armed ship 10.
Dimensions & tons: 54'7½", 31'1" x 20'x 9'. 58.
Men: 45. Guns: 10 x ____ . Purchased 1780 (or 1779?) possibly at Jamaica; 1784 sold at Jamaica. Possibly schooner rig.

***Sandwich*** 1780-1781 (Mercantile *Marjory* ____ ) armed ship 24.
Dimensions & tons: ____ , ____ x ____ x ____ . 150.
Men: ____ . Guns: 24 x ____ . Purchased (in North America?) 1780; 24.8.1781 taken by the French (De Grasse's fleet).

***Whitby*** 1781-1785 (Mercantile *Whitby* ____ ) armed storeship 14.
Dimensions & tons: 100'6", 90'1⅜" x 30'1" x 11'10". 433 74/94.
Men: 40. Guns: UD 14 x 6 (later: 6 x ____ ). Purchased at the beginning of 1781; 1785 sold.

PLANS: Lines & profile/decks.

***Cornwallis*** 1781-1782 (Mercantile ____ ) armed storeship 14.
Dimensions & tons: 107', 87'10⅝" x 30'9" x 12'9". 442 23/94.
Men: 40. Guns: 14 x ____ . Purchased 04.1781?; Ca 30.9.1782 foundered in the Atlantic.

PLANS: Taken off & as fitted: Lines & profile.

***Achilles*** 1781-1784 (Mercantile ____ ) armed storeship 14.
Dimensions & tons: 101'3", 82'0⅛" x 31'1" x 12'6½". 421 43/94.
Men: 40. Guns: UD 14 x 6. Purchased 05.1781?; 1784 sold.

PLANS: Taken off & as fitted: Lines & profile.

***Sally*** 1781-1783 (Mercantile *Sally* ____ ) armed storeship 14.
Dimensions & tons: ____ , 85'0⅜" x 29'8" x 12'11". 398 7/94.
Men: 40. Guns: UD 14 x 6. Purchased 05.1781?; 1783 sold.

***Minerva*** 1781-1783 (Mercantile ____ ) armed storeship 34.
Dimensions & tons: 130', 111'2⅝" x 34'1" x 12'7". 687 22/94.
Men: 60. Guns: UD 26 x 6, QD 6 x 18★, Fc 2 x 18★. Purchased 06.1781; 1783 sold.

PLANS: Taken off & as fitted: Lines & profile.

***Britannia*** 1781-1782 (Mercantile *Britannia*? ____ ) armed storeship 20.
Dimensions & tons: 114'1", 94'7" x 32'x 12'. 535.
Men: 40. Guns: 20 x ____ + 6 swivels. Purchased mid-1781 for service in the East Indies; 11.4.1782 wrecked on the Kentish Knock.

PLANS: Taken off & as fitted: Lines & profile.

***Harriott*** 1781-1784 (Mercantile *Harriott* ____ ) armed storeship 22.
Dimensions & tons: 104'4", 86'5⅜" x 29'11" x 12'1". 411 60/94.
Men: ____ . Guns: 20 x ____ + 2 x 18★. Purchased mid-1781 for service in the East Indies; 1784 sold.

***Berwick*** 1781-1786 ***Sirius*** -1790 (Mercantile *Berwick*, Watson, Rotherhithe) armed storeship ____ .
Dimensions & tons: 110'5", 89'8¾" x 32'9" x 12'11". 511 83/94.
Men: 50 (1786:160). Guns: ____ (1786: 4 x 6 + 6 x 18★). Purchased 1781; 1786 classed as a Sixth Rate and renamed *Sirius* as the flagship of the 'First Fleet' for Botany Bay; 19.3.1790 swept onto a reef of rocks off Norfolk Island, New South Wales [This wreck has recently been located and dived on. See G Henderson & M Stanbury's book on the ship 'The *Sirius* past and present' (Sydney, NSW 1988); which also shows that the usual description of her as an East Indiaman results from a misunderstanding; she was built for the 'Eastern' (ie. Baltic) trade. The story that she suffered a fire just before completion also seems to be a legend.

PLANS: As fitted: Lines & profile.

***Rhinoceros*** 1781-1784 (Mercantile *Rhinoceros*? or *Birch*? ____ ) armed vessel 10.
Dimensions & tons: ____ , ____ , ____ x ____ x ____ . 711.
Men: ____ . Guns: UD 10 x ____ ★ Purchased 1781; 1784 sold.

***Supply*** 1782-1784 (Mercantile *Supply*? or *Admiral Parker*? ____ ) armed storeship 16.
Dimensions & tons: 114'3", 95'1⅝" x 31'2" x 12'6". 491 33/94.
Men: ____ . Guns: UD 16 x ____ . Purchased 01.1782; 1784 sold.

PLANS: Taken off & as fitted: lines & profile.

***Tortoise*** 1782?-1785 (Mercantile *Russian Eagle* or *Prussian Eagle* armed storeship 16.
Dimensions & tons: 123'8", 102'6⅛" x 33'9¼" x 12'6". 621 92/94.
Men: 75. Guns: UD 16 x 24★. Purchased 01.1782?; 1785 sold.

PLANS: Taken off & as fitted: Lines & profile.

***Bountiful*** 1782-1783 (Mercantile *Minerva* ____ ) armed storeship 1782.
Dimensions & tons: ____ x 18'6½". 777 68/94.
Men: 75. Guns: 24 or 22 x 42★? Purchased 5.2.1782; 1783 or 1784 sold at Bombay.

PLANS: Taken off & as fitted: Lines & profile.

***Prosperity*** 1782-1784 (Mercantile *Prosperity* or *Fame*? ____ ) armed storeship 22.
Dimensions & tons: 130'10", 109'7⅞" x 34'4" x 12'5". 687 57/94.
Men: 75 [60 as receiving ship?]. Guns: UD 22 x 24★. Purchased 03.1782; 1784 hulked [at Sheerness as receiving ship; 1796 BU].

PLANS: Taken off & as fitted: Lines & profile.

***Steady*** 1782-1784 (Mercantile? *Steady*? or Russian? *Waldimar Melgounoff*? ____ ) armed vessel/transport 10.
Dimensions & tons: 107'3", 88'10¾" x 30'9" x 11'2½". 447 10/94.

THE AMERICAN WAR OF INDEPENDENCE 1776-1783

Men: 40. Guns: 10 x ____ *. Purchased early 1782; 1784 sold.
PLANS: Taken off & as fitted: lines & profile.

*Providence* 1782-1784 (Mercantile *Keppel*? ____ ) armed storeship 16.
Dimensions & tons: 111'8½", 93'1⅛" x 30'6½" x 12'1". 461 84/94.
Men: 50. Guns: UD 16 x 6. Purchased early 1782; 1784 sold.
PLANS: Taken off & as fitted: lines & profile.

*Cyrus* 1782-1786 (Mercantile *Cyrus*? ____ ) armed transport 16.
Dimensions & tons: 118'5", 99'9⅜" x 34'8" x 12'7". 625 6/94.
Men: 62, later 72. Guns: UD 16 x 6. Purchased 1782; 22.4.1786 wrecked on the Barbadoes.
PLANS: Taken off & as fitted: Lines & profile.

## Unregistered Vessels

(ie. not recorded in primary Admiralty sources consulted.)

*Active*: There appear to have been at least two, probably three, vessels of this name in service at this time:
(1) 12 gun brig of 55 men (possibly purchased 1776?).
(2) 16/14 gun cutter of 40 men (purchased or hired 1779).
One of these had the following dimensions: 71', 59'1" x 20'2" x 8'1", 109 tons, though it is not clear which. A cutter called *Active* was captured in the Channel on 18.8.1779 by the French *Mutine* [14]; though this could have been either a merchant vessel or a privateer and not a naval vessel. It is possible that the brig mentioned above was captured by the Americans in 1780. However a brig called *Active* was in service on the North American Station in 1781 under the command of Lt Delance. This was possibly the ex-*Rosebud* which was acquired in 1780 and deleted in 1782.

*Adder* Galley 8. 1782-1787. In service on the North American Station in 1782. Origins and dimensions obscure; 1787 sold.

*Amphitrite* (French? privateer *Solitaire* ____ ) Sloop, rig uncertain, 18. Taken by *Ruby* and others on 6.12.1782, but does not appear to have been commissioned in the RN.

*Antelope* Sloop, rig uncertain, 14. In service 1783; 30.7.1784 lost in a hurricane off Jamaica.

*Antigua* 1777-1781? Brig or schooner in the West Indies; 05.1781 captured or 1781; 1782 sold.

*Arbuthnot* Ca 1782 galley in North America.

*Argo* 1780-1783 Schooner, 10 guns? Purchased 1780; 1783 sold at New York.

*Arrow* Sloop, rig uncertain, 18. Purchased 1782; served at Jamaica 1782-1783; ultimate fate unknown.

*Barbadoes* 1782-1784 (American *Rhodes* ____ ) Brig sloop? Taken in the West Indies 15.2.1782; 2.12.1784 sold.

*Barbuda* Purchased sloop 14. Taken 1780? (American? *Charming Sally*? ____ ). Serving in the West Indies during 1781; 3.2.1782 taken by the French at the surrender of Demarara [served as French *Barboude* until 1786].

*Berbice* Schooner 8. Purchased in America in 1782; still in service in 1785?

*Blanche* 1779-1780 (French *La Blanche* 1766 Le Havre, Giroux design) 12 pounder Frigate 32/36 guns. Taken 21.10.1779 by *Magnificent* and others in the West Indies; 11/12.10.1780 foundered during the Great West Indian Hurricane.

*Bolton* 1775-1776 Brig 12. 30 men. Purchased whilst building (for John Hancock) and nearly complete, at Boston, Mass. in 06 or 07.1775. Purchase authorised by Admiral Graves; 5.4. 1776 taken by an American squadron.

*Bonavista* Brig 10. In service 1778-1779 at Newfoundland.

*Canada* 1775-1775/1776 Schooner in service in Newfoundland; wrecked 1775 or 1776.

*Canada* ? 1777-1779? Schooner in service in Newfoundland.

*Castor* 1781-1781 (Dutch *Castor* ____ ) 36. Taken 23.5.1781; 20.6.1781 retaken by the French.

*Clinton* Armed galley 8. 45 men. In service on the North American station in 1782.

*Comet* Galley 8. 40 men. In service 1778, also recorded in 1782 on the North American station.

*Comte D'Estaing* Fireship ____ (?). Purchased (French prize?) in the West Indies in 1782?; fate uncertain.

*Conflagration* 1781-1781 (mercantile *Loyal Oak* ____ ) Fireship 8 guns. Purchased in North America 1781; 1781 lost on the North American coast.

*Congress* Galley 8. 30 men. Presumably an American prize. In service on the North American station in 1781.

*Coquette* 1783-1785? (French 20 1779 Toulon, Coulomb design) taken by *Resistance* 2.3.1783; not taken into service and sold mercantile in 1783 (noted as being listed in 1785; presumably a clerical error); [I owe this note to Iain Mackenzie].

*Delight* 1782 Sloop 14. Purchased in North America 1782; no further information.

*Diligent* Schooner in service 1781-1790? (listed in Colledge only).

*Dispatch* Schooner or sloop (sloop-rigged vessel?) 6 guns; captured 12.7.1776 by American *Tyrannicide* [may not have been RN at all].

*Dispatch* Schooner ____ . Purchased 1780 or 1781; tender?; fate uncertain (out of service 1795?).

*Dispatch* Brig 10 guns, 50 men. Purchased 1781; 1782 lost in hurricane.

*Espoir*? Sloop sold 25.3.1784? (listed in Colledge only).

*Etna* Bomb 8 guns, purchased 1781? (listed in Pitcairn Jones only).

*Europa* Gun boat at Gibraltar; commissioned 1782.

*Fair Rhodian* Brig 8. In service in North American waters 1781-1782.

*Firefly* Armed ship or galley 8/10 . Serving in North American waters 1781-1782.

*Flirt* 1778-1782 there is a log for a vessel of this name and dates but no further information.

*Fly* Cutter 8. 50 men. Purchased 1780 (or 1779?) commissioned 05.1780; 05.1781 taken by the French in America. (Dittmar gives acquired 1778 and 14 guns).

*Flying Fish* 1778-1782 Cutter (sloop) 10/12/14 . 75', 51'5" x 25'8" x 10'8". 190 tons. 60 men. 12 carronades + 12 swivels. A.O. 17.10.1780 to add 2 x 4pdrs. Purchased 04.1778; 3.12.1782 wrecked near Calais

*Flying Fish* 1782-1783 Armed vessel, 10 guns, in North American waters [Is this the same as the hired vessel of the same dates?].

?*Fracour*? 14 guns; 1782 in service In North America?; name uncertain, probably a misreading or distortion of the real one.

*Fury* Gunboat 1. Commissioned at Gibraltar 1782; fate unknown.

*Gaspee* 1774-1775/1776-1777 Brig or schooner, 6 guns. Taken 11.1775 by Americans; 04.1776 recaptured.

*General Monck* 1781-1782 Purchased sloop/armed ship 20/18?; ex American privateer *General Washington* ex *Congress*? taken 09.1781 by *Chatham*; taken 8.4.1782 by American 'privateer' (the State of Pennsylvania's ship) *Hyder Ali* [1784 sold].

*George* Tender. Origins obscure (in service by 1770?); 26.12.1776 wrecked at Piscatagua?

*Germaine* 1779-1781 Armed ship ( ____ *Germaine*? ____ ) ____ . Purchased 1779; ____ 1781 taken by the French.

*Grampus* [?] 1777-1778 Transport; foundered off Newfoundland 1777 or 1778.

*Greenwich* [?] 1778-1782? Brig 14 guns. 50 men.

*Greyhound* 1780-1781 *Viper* -1803? Cutter 20/12. 148 tons, 20 x 4 later 12 guns only. Purchased 06.1780; 1781 renamed *Viper;* listed until 1803; 1809 sold?

*Hammond* 1782-1783 Galley 8 guns, 45 men; in North American waters.

*Harlequin* 1781- ____ ? (Spanish? ____ ) Sloop 16.
Dimensions & tons: 80', ____ x 20'6" x ____ . 141.
Men: ____ . Guns: 16 x ____ . Taken by *Surprize* 1781; ? sold 19.6.1782.

*Hinchinbrook* Schooner 8? 30 men. In service at Boston 1775-1777.

*Hinchinbrook* (*Hinchingbroke*) 1779-1782 (French *L'Astrée*, ____ ) Sixth Rate 28 guns. 115', 94'9" x 33'3" x ____ 557 18/94; taken 1778 or 1779; 18.1.1782 foundered off Jamaica or 19.1.1783 wrecked in Saint Anne's Bay.

*Holderness* Cutter 8. Origins uncertain; ____ 1779 taken by the Franco-Spanish fleet in the Channel.

*Hornet* 1779 Galley.

*Hussar* 1778 Galley in ? North American waters.

*Jack* 1780-1781 Sloop 14 guns purchased (or hired?) 1780; 21.6.1781 taken by the French off Cape Breton (listed in Colledge only).

*Jamaica* Sloop 16. (Mercantile *Joseph*? ____ ). In service on the Jamaica station 1780-1783 (local purchase?).

*Justitia* 1777-1798 Prison ship 260 tons. Purchased (as prison ship) 1777; 1798 still in service (listed in Colledge only).

*Keppel* Galley 8. 75 men. 1778 taken from the Americans (?); 1779 serving in North America; fate uncertain.

*Labrador* Schooner 10. Serving at Newfoundland 1777-1779.

*Lawrence* (*Saint Lawrence*?) Brig 14. 1782 at Newfoundland; still in service 1785?

*Little Lucy* 1777-1778 Schooner ____ (tender). 1777 taken by *Lowestoffe* in the West Indies; served till 1778.

NOTE: Lt Horatio Nelson's first command.

*Little Sally* Advice boat 1778. Commissioned by Admiral Barrington in the Leeward Islands; fate unknown.

*Lizard* Schooner 10. Acquired 1781? In service in the West Indies 1782-1783.

*Lizard* 1782?-1785? (French *Le Lezard* 1777 ____ cutter 10. 110 men. 10.1782 taken by *Sultan* in the East Indies; 1786? sold in the East Indies.

*Lord Howe* Armed ship ____ (or cutter or brig). In service from 1777 to 1781 (?) on the North American station. Still listed in 1789 but may well have been disposed of or lost long before; or there may have been more than one vessel of this name.

*Mackworth* See *Lady Mackworth* under hired ships.

*Magdalen* 90 tons, purchased in North America 1780; fate unknown (listed in Colledge only).

*Marechal De Bretagne*? purchased 1780? (listed in Pitcairn Jones only).

*Mentor* 1780-1781 Sixth Rate/armed ship 20. 160 men. (American? ____) Taken 1778? Probably purchased in the Western Hemisphere; commissioned 03.1780; 8.5.1781 burnt at Pensacola to prevent capture by the Spaniards.

NOTE: A coppered vessel.

*Mersey* Storeship; 22.6.1782 foundered off Port Royal?

*Musquito*? Vessel in service 1777? (listed in Colledge only).

*Necker* (French *Le Necker* ____) Armed ship 22. 128 tons. Taken by *Hannibal* off the Cape of Good Hope 26.10.1781; 1781-1782 serving in the East Indies; fate unknown (deleted 1789?).

*Nestor* 1781-1783 (American privateer *Franklin* ____) Sixth or Fifth Rate 24/28/32. Taken by *Ramilles* and *Ulysses* in the West Indies 1781; 1783 paid off and sold.

*North American* 18. Purchased in 1778? No other mention and possibly only a confusion with the sloop *North*; or just a clerical error.

*Oronoque* (French privateer? ____) Sloop 18. Taken 1782?; served in the Leeward Islands; 3.2.1782 taken by the French at Demerara.

*Pearl* (French? *La Perle* ____) Sloop 16. Taken by Rodney off the coast of Portugal 5.7.1780; served during the rest of 1780 and perhaps until 1782.

*Penguin* Schooner 10. Purchased 1777; served at Newfoundland; out of service by 1782.

*Philadelphia* 1778?-1782? Galley 8 guns 30 men; in North America.

*Polecat* Brig sloop 14. Purchased or taken in 1782? Served on the North American station and taken there later in the same year by the French.

*Port Royal* 1779-1783 (French privateer *La Comtesse de Maurepas* ____) Sloop 14. Taken by the *Countess of Scarborough* 06.1779; 1781 possibly temporarily in Spanish hands (though this could be another vessel of the same name); 1783 out of service.

*Providence* (American ____) sloop 12. Taken from the Americans at the Penobscot 14.8.1779; no evidence that she actually served in the RN.

*Quebec* Schooner 10/14. Purchased 1775? in North America (?). Was at Newfoundland in 1777 and probably continued to serve in that area until 1782.

*Racoon* Brig sloop 14. Serving in North America in 1782 and captured by the French *Aigle* and *Gloire* off the Delaware on 12.9.1782.

*Railleur* French sloop taken by *Cyclops* 11.1.1783; no further information (returned to the French?), (listed in Colledge only).

*Ranger* 1779-1784 Purchased cutter. 80'4", 55'3" x 26'1½" x ____ 200 79/94 tons. Men: 70. Rerated as a sloop on 13.2.1782.

*Rattlesnake* 1779-1781 Sloop 84'(?), 74'9" x 22'4" x 8'10" (?) 198 tons. 14 or 12 guns, 70 men. In the East Indies? Prize?

*Recovery* American sloop 14 guns taken 1782; in the West Indies; 1785 sold.

*Renard/Reynard* Sloop, rig unknown 18. Taken (French? *Renard*? ____) by *Brune* in the West Indies, 07.1781; 1782/1783 convalescent ship at Antigua; 1784? BU?

*Repulse* 1781-1783 Gunboat 5. (5 x 26 pdrs.) ? Spanish prize? purchased at Gibraltar 1781; 1783 sold.

*Resolution* 1780-1782 Captured (?) transport ____, 97'9" x 34'8" x 12'7", 625 tons, 8 guns, 45 men.

*Revenge* Brig sloop 14. Purchased ____ ? and in service by 1778; ____ 1779 taken by the Americans.

*Rodney* 1781-1782 Brig 16. Purchased in the West Indies (?) 1781; 3.2.1782 taken by the French at Demerara.

*Saint John* Sloop 14. 90 tons. 1782-1783 serving at Newfoundland (local purchase in 1780?).

*Saint Lawrence* Schooner 10. 1782 serving in North America (possibly this is a confusion with the brig *Lawrence*?).

*San Bruno* (Spanish *San Bruno* ____) 26. Taken by Rodney's fleet 8.1.1780; not taken into the RN?

*San Rafael* (Spanish *San Rafael* ____) 30. Taken by Rodney's fleet 8.1.1780; not taken into the RN?

*Santa Lucia* Brig 14. Serving in the West Indies 1780-1783.

*Santa Teresa* (Spanish *Santa Teresa* ____) 28. Taken by Rodney's fleet 8.1.1780; not taken into the RN?

*Savannah* Brig/armed ship? 14. Serving in North America 1779; 09.1779 sunk as a blockship at Savannah.

*Scourge* Galley 8. 30 men. 1779-1783 served in North America.

*Sea Nymph*? Cutter 8 guns in service 1782?

*Selby* (*Shelby* - which see - is probably a clerical error for this) 1781; 1783 purchased storeship. Purchased 04.1781; 2.12.1783 sold.

*Shelanagig* Tender (schooner?) 16? 1781 serving in the West Indies and taken there ____ 1781.

*Shelby* 1781-1783 (probably a clerical error for *Selby* - which see) storeship ____. Purchased 1781; 1783 sold.

*Snake* 1776-1781 Sloop rig unknown 12. 125 men. Purchased in the West Indies 13.6.1776; 1781 taken by two American privateers *Pilgrim* and *Wanderer* in the West Indies.

NOTE: Possibly a confusion with the other sloop *Snake* of the same period.

*South Carolina* (American *South Carolina* 1778 Holland) Frigate. Taken by *Diomede*, *Astrea*, *Quebec* 20.12.1782; not taken into the RN.

*Stanley* Brig 10. In service 1778; 1778 taken by French.

*Stork* Sloop rig uncertain 10. In service at Jamaica 1778-1780.

*Sultana* 1776-1783 Vessel in service between these dates (there is a log but no further information).

*Tapigeur/Tapageur*? Cutter ____. 1781 serving in the West Indies?

*Tarleton* 1782/1793-1794? Brig sloop 14 guns? in service 1782; 1782 taken by the French; 1793 handed over at Toulon; 1794 still in service; fate unknown.

*Terrier* Sloop, rig uncertain 16. 1782 at Jamaica, 1783 at New York.

*Tobago* 1778?-1781 Sloop, rig uncertain 14/16. Taken in 1779 (American? *Governor Trumbull* ____) ____ ?; served in the Leeward Islands; 1781 taken (or 1783 sold?).

*Tryal* Sloop 16. Served at Newfoundland at some time during this war.

*Tryal* Advice boat (schooner) 2. 20 men. Serving in North America 1776; 1777 lost (burnt)?

*Vanguard* Gunboat 4 (2 x 26 + 2 x 12). Probably a Spanish prize. Purchased in 1781, probably at Gibraltar; 1782 still in service; 1783 sold?

*Vaughan* Galley 8. In service 1782; 1783 paid off and sold.

*Vindictive* Galley 8. Taken from the Americans (*Vindictive* ____) 1779; served in North America until 1783.

*Viper* Purchased galley 8 guns 1779-1783?

*Vulcan* 1780 Armed ship 20 guns, at Newfoundland.

*Waterwitch* 1781 ____ in the Leeward Islands. (listed in Pitcairn Jones only).

*Wolf* Armed ship? 8 guns. Origins obscure; 07.1780 wrecked off Newfoundland. (listed in Colledge only).

# Hired vessels

(and ones that were probably hired.)

*Admiral Barrington* Hired brig 14. Hired at Jamaica 1782-1783.

*Adventure* 1780 Armed ship, 20 guns

# THE AMERICAN WAR OF INDEPENDENCE 1776-1783

*Alert* (?) 1778-1780 probably Hired cutter 78', ____ x 25', 220 tons. Built at Dover 1778; 10.1780 taken by the French in the Bay of Biscay. [Only noted in Colledge.]

*Alfred* Armed ship 20. Purchased, or more probably hired, 1778. Sold or otherwise disposed of in 1782. Probably the first ship of the American Navy which was captured by the RN in 1778 (American *Alfred* 1775 ex mercantile *Black Prince* Philadelphia 1774?).

*Amelia* 1782 Cutter.

*Augustus* Hired cutter 8. Served in the Downs 1782.

*Bellona* Armed ship 20/18. 120 men. Built 1778 by Pye at Stockton-on-Tees and purchased or hired the same year; wrecked in the Texel in 1779 or in the mouth of the River Elbe in 1780.

*Betsy* Hired Cutter ____. Served in the Downs in 1782.

*Champion* Hired cutter ____. In service 1783?

*Content* Armed ship 20. 120 men. Purchased or hired 1777 and commissioned in November of that year; still in service (In the North Sea) in 1780.

*Countess of Scarborough* Armed ship 20. Hired (?) and commissioned in 11.1777; 23.9.1779 taken by Jones' Franco-American squadron off Flamborough Head.

*Dart* Hired cutter 8. Serving in the Channel 1782.

*Defiance* Armed ship 20 or sloop 18. 120 men. Hired 1781; 1782 serving in the Channel.

*Defiance* Brig 14. Hired vessel serving in the Channel in 1782 (possibly the same vessel as the one noted above?).

*Dorset* Hired cutter; 8. 1782 in the Channel.

*Drake* 1779 Brig; 6 gun. 130 tons.

*Duchess of Cumberland* 1783? Cutter 8 guns.

*Earl of Bute/Bute* Hired armed ship 10 (Mercantile *Earl of Bute*? ____). 160 men. Hired 1776 and commissioned 01.1777; 1777 foundered.

*Flora* 1783 Cutter (also in service 1796?)

*Fly* 1778 Hired lugger 16 guns.

*Flying Fish* 1782-1783 Cutter [is this the same as the unregistered armed ship in service at the same dates?].

*Fortune* 1782 Cutter 10 guns.

*Fox* Hired cutter 12/10. Serving 1783. (May have been three separate cutters, two of 12 and one of 10 guns.)

*Friendship* [?] 1776-1777 Armed ship 22 guns.

*George* 1782 Cutter 8 guns.

*Grace* [?] 1781 Cutter.

*Greyhound* 1783 Cutter, 12 guns.

*Hawk* 1782-1796 Cutter, 8 guns (listed in Colledge only).

*Jackal* 1781-1782 Hired armed ship 8 guns. 35 men. Commissioned 01.1781; 11.4.1782 taken by the American *Deane* in the West Indies.

*Kent* Armed ship 20. 160 men. Commissioned 09.1776; 1777 in the West Indies; fate uncertain.

*Lady Mackworth/Mackworth* Hired armed ship 20/22. 90 men. Commissioned 06.1779; still serving in the Irish Sea 1782. Possibly two separate ships; *Lady Mackworth* and *Mackworth*, but the references are more likely to be to two variant forms of the name of the one ship.

*Leith* 1777-1782 Hired armed ship 20. 120 men. Commissioned 11.1777; 07.1782 paid off.

*Lily* Hired cutter ____. In service 1782.

*Loudon* Hired armed ship 26. 140 men. Commissioned 08.1779; 1781 still serving in the North Sea.

*Lucy* 1777 Tender.

*Merchant* 1777-1782 Hired armed ship 20. 120 men. Commissioned 12.1777; 1782 still serving in the North Sea.

*Molly* Armed ship 20. Purchased or hired, date unknown (1781?); 05.1781 accidentally burnt; or 18 guns taken 10.1782 by the French *Semillante* [or are these two separate ships of the same name?].

*Nimrod* Armed ship 14. Probably hired in the West Indies 1777 and served during that year in the Leeward Islands; fate unknown.

*Peggy* [?] 1779 Cutter.

*Princess of Wales* Armed ship 20. Commissioned 11.1777; served in the North Sea during 1778; fate unknown.

*Queen* 1778-1782 Hired armed ship 20 guns, 120 men (also hired 1793-4?).

*Rambler* [?] Cutter 1778-1779 10 guns; blown up in action 6.10.1779.

*Ranger* Armed ship 30. 150 men. Purchased or hired 1777; 1781 still in service; fate unknown.

*Sandwich* Cutter 8 guns, hired 1782-1783.

*Satisfaction* Armed ship 20. 120 men. Hired or purchased? 1777-1782 served in the Clyde area.

*Sea Nymph* Armed ship 20. 1780 serving at Jamaica.

*Symetry* Transport 18. In service 1776; fate unknown. Probably hired.

*Three Brothers* 1777-1782 Hired armed ship 20. 120 men. 11.1777 commissioned; served in the Irish Sea until 1782.

*Three Sisters* 1777-1780 Hired armed ship 20. 120 men. Commissioned 11.1777; served in the North Sea until 1780.

*Union* 1782 Cutter.

*Vernon* 1781-1782 Hired armed ship.

*William* [?] 1778-1783 hired armed ship 20 guns 120 men.

*Young Hazard* 1779 Cutter (listed in Pitcairn Jones only).

# VESSELS ACQUIRED 1784-1792

## 18 pounder Frigate (Fifth Rate)

*Beaulieu* 1791-1806 (building Adams, Bucklers Hard) 40.
Dimensions & tons: 147'3", 122'10⅝" x 39'6" x 15'2 5/8". 1019 79/94.
Men: 280 (later 274). Guns: UD 28 x 18, QD 8 x 9, Fc 4 x 9 (1793 to add 2 x 32★ on UD and 6 x 18★ on QD. Later these carronades were removed). Purchased whilst on the stocks 06.1790; launched 4.5.1791; 1806 BU.

PLANS: Lines/profile/orlop/lower deck/upper deck/quarter deck & foc'sle.

## Sloops, Rig Uncertain

*Eclipse* 1787-1787 ( ____ ) ____ .
Dimensions & tons: ____ , ____ x ____ x ____ . ____ .
Men: ____ . Guns: ____ . Purchased 1787; not registered; 1787 sold.

## Cutters

*Ranger* 1787-1794/1798 *Venture* -1802 (Revenue cutter *Rose* ____ ) 12.
Dimensions & tons: 74'7¼", 56'0⅝" x 25'7" x 9'5". 195 5/94.
Men: 55. Guns: 12 x 4. Purchased 02.1787; 1794 taken by the French, became *Le Ranger*, 1798 retaken by *Galatea* and renamed *Venture*; Ca 1802 out of service (1803 sold?).

## Schooners

*Chatham* 1790-1794 (Mercantile *Earl of Chatham* ____ ) armed schooner 4.
Dimensions & tons: ____ , ____ x ____ x ____ . 93.
Men: 20. Guns: 4 x 3. Purchased at Halifax, Nova Scotia 1790 'to prevent a contraband trade being carried out in Nova Scotia'; 1794 sold.

*Diligent* 1790-1794 ( ____ ) schooner 8.
Dimensions & tons: ____ , ____ x ____ x ____ . 89.
Men: 20. Guns: 4 x 3. Purchased at Halifax, Nova Scotia 1790; 20.11.1794 sold.

## Storeships, etc

*Bounty* 1787-1789 (Mercantile *Bethia* 1784 Hull) armed vessel 4.
Dimensions & tons: ____ , 69'11⅜" x 24'4" x 11'4". 220.
Men: 45. Guns: 4 x 4 (short) + 10 swivels. Purchased 23.5.1787 'to go to the Society Islands to carry bread fruit trees to the West Indies ____ '; 1789 mutinied [later burnt and sank at Pitcairn Island; the wreck has been found and dived on].

PLANS: As fitted: Lines & profile/deck/arrangement for breadfruit/lines of launch.

*Chatham* 1788-1830? (Mercantile brig, just built by King, Dover) tender brig-rig 4.

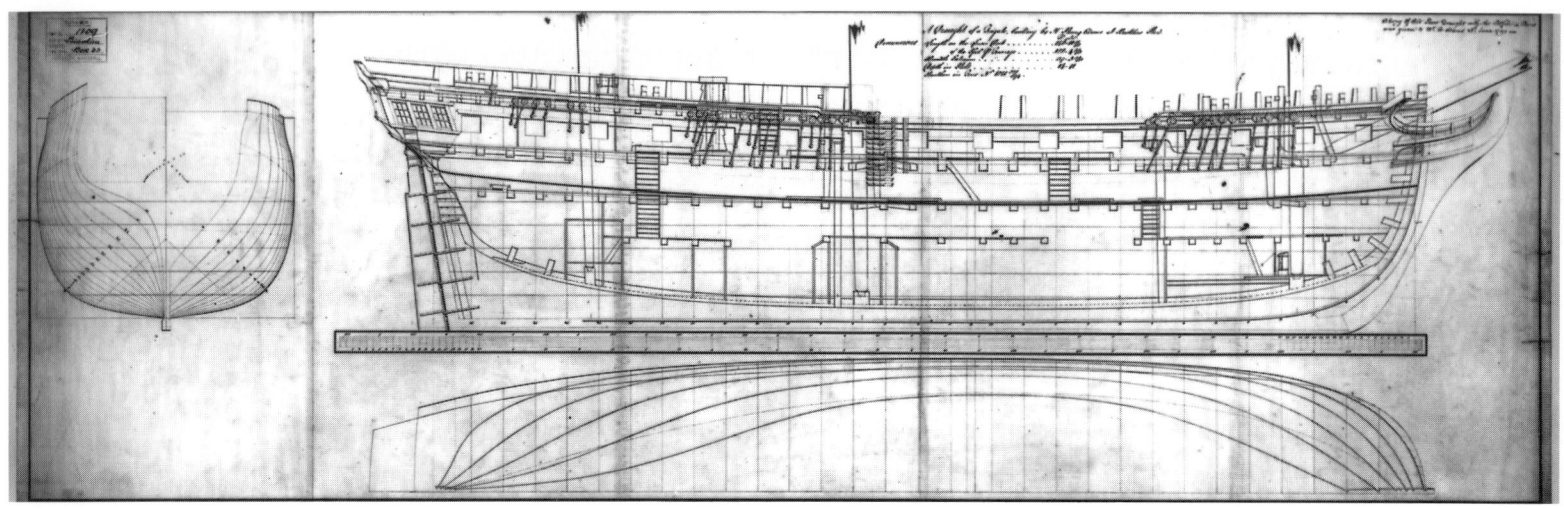

Lines (1809) of the 40 gun frigate *Beaulieu* built 'on spec' and purchased for the Navy in 1791

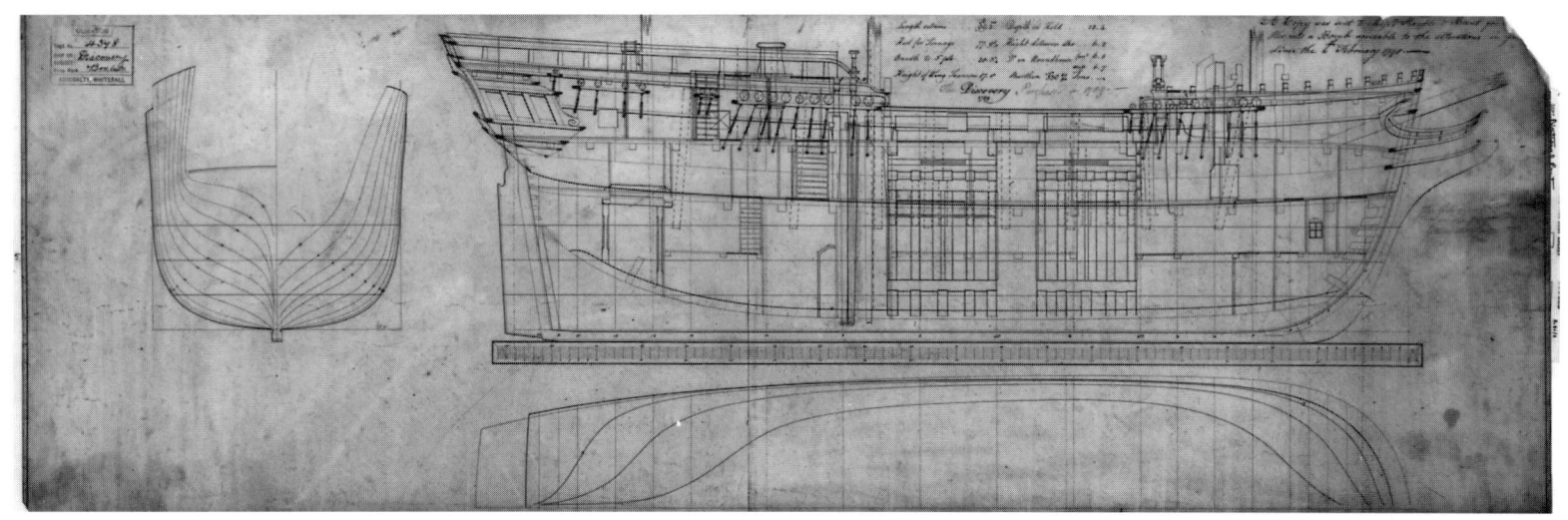

Lines and profile plan (4378) for Vancouver's exploration ship *Discovery* of 1789 showing her with the alterations (in green on the original plan) for converting her to a bomb vessel in 1798

Dimensions & tons: ____, 53'1¾" x 21'6¾" x 10'1". 131.
Men: 12/45. Guns: 4 x 3 + 10/6 swivels. Purchased 12.2.1788 from King for use as a tender; 1790 fitted for 'remote service' (Vancouver's expedition); 1830 sold at Jamaica (not certain this is still the same vessel).
PLANS: Decks.

**Woolwich** 1788-1800? (Mercantile *Marianne* ____ Thames) tender ____.
Dimensions & tons: ____, 59'6" x 23'1" x 10'8". 169.
Men: ____. Guns: ____. Purchased 1788; 1800 still in service; fate uncertain.

**Sorlings** 1789-1809 (Mercantile *Elizabeth* 1788 St Davids?) survey vessel.
Dimensions & tons: 52'1½", 41'2⅞" x 17'0½" x 7'11". 63 66/94.
Men: 14. Guns: 4 swivels. Purchased 05.1789 for surveying the Scilly Islands; 1809 dockyard lighter [at Deptford by 1816; fate uncertain].

**Discovery** 1789-1808 (building Randall, Rotherhithe) ship sloop (exploration ship) 10.
Dimensions & tons: 99'2", 77'9⅝" x 28'3¼" x 12'4". 330 65/94.
Men: 100. Guns: 10 x 4 (short) + 10 swivels. Purchased whilst building 11.1789; 1798 bomb vessel; 1808 hulked [at Sheerness as a convict ship; 1818 to Woolwich as a hulk; 1834 BU].
PLANS: Building & alterations for naval service: Profile/decks (incomplete). As fitted & 1798 conversion to bomb: Lines & profile/decks. 1798 conversion to bomb: decks/midships section.

**Providence** 1791-1797 (building Perry, Blackwall) exploration vessel (Sixth Rate) 10/12.
Dimensions & tons: 107'10¼", 89'6" x 29'2½" x 12'3¼". 406 12/94.
Men: 100. Guns: 10/12 x ____ (1797: 18 x 4). Purchased 02.1791 whilst on the stocks; launched 23.4.1791; 1793 classed as a sloop and fitted for 'remote parts' (Bligh's second breadfruit voyage); 16.5.1797 wrecked on Formosa.

PLANS: Lines & profile/decks.

**Assistant** 1791?-1802 (Mercantile? *Gosport Packet* ____) armed tender ____.
Dimensions & tons: ____, ____ x ____ x ____. ____.
Men: ____. Guns: ____. Purchased 1791?; 1794 naval transport at Deptford; 1802 sold.

## Unregistered Vessels
(ie. not listed in Admiralty sources used.)

**Advice** 1792-1793 Schooner (or cutter?) 4 x 3 pdrs., 30 men; 1.3.1793 lost on Key Bostell, Bay of Honduras.

**Alert** 1790-1799 Armed schooner 4. (4 x 3). 88 tons. 20 men. Purchased 1790; 1799 BU

**Bombay** 1790 Storeship (listed in Colledge only).

**Diligence** Thames-built Sloop purchased 1789. 96', 79' x 29'4" x 12'4". 337 tons. 10 guns. No further information.

**Hydra** 1791-1792 there is a log for a vessel, type unknown, of this name and these dates.

**Indus** Storeship, ex-East Indiaman, purchased 1790; no further information (noted in Colledge only).

**Jackal** 1792-1794 Purchased brig, tender to *Lion* on her voyage to China with Lord Macartney's embassy. 101 tons, 10 guns.

**Salamander** 1787?-1799 Brig sloop 93'6", 79' x 24'8" x 10'. 240 tons. Taken by ____ ? 1799 (existence doubtful).

**Surprise** 1786-1792 (Mercantile? *Surprise* ____) Cutter 10.

Dimensions & tons: 68'9", 49'7" x 22'10" x 9'. 134.
Men: ____ . Guns: 10 x ____ + 8 swivels. Purchased 10.1786; 1792 sold.

**Swan** 1788 purchased for Revenue service; to Admiralty 1790; 26.5.1792 wrecked off Shoreham. 10 guns 90 tons. [Noted in Colledge only.]

## Hired Vessels

*Daedalus* 1790-1795 Storeship.

*Duke of Rutland* 1784 Armed ship, may have been in service earlier, during the American War; 30.7.1784 foundered off Jamaica in a hurricane.

# Chapter 6: The French Revolutionary War 1793-1801

The immense advantage the Royal Navy had over its French rival in this conflict showed from the start. The effect of the Revolution on the French Army was to produce the most effective land force in Europe. Its effect on the French Navy was disastrous. The well organised and effective fighting force of the previous war had become a shadow of its former self by the time war broke out between France and Britain in 1793. The officer corps was almost ruined, the dockyard organisation badly damaged. The results can be seen in the list of prizes that follows. The first large windfall occurred when the French dockyard town of Toulon was taken over by counter-revolutionaries who surrendered it and all the ships based there to the British fleet. Not all of these vessels were eventually lost to France as some were neither taken away nor destroyed in the chaotic fall of Toulon to the young Napoleon's artillery. However, the loss was significant and was added to at the Glorious First of June (1794) and subsequent actions small and large culminating in the crushing loss of Aboukir Bay ('The Battle of the Nile' 1798). Meanwhile, Spain and Holland came in on France's side only to add to Britain's prize list at Saint Vincent and Camperdown (1797). The process had already started which after a barely interrupted twenty years of war would leave Britain as no longer *primus inter pares* as the leading navy, but instead incomparably the largest and most powerful navy, a giant which had grown whilst all its competitors had shrunk.[1]

The increasing domination of the Continent by the French Army is equally apparent in the numbers of purchases of merchantmen for coastal defence gunboats and coastal offence fireships and bombs, assembled against the French invasion flotilla. The extent to which the Navy was stretched by the struggle can be seen in the continuing purchases of mercantile hulls throughout this period, including vessels which were converted to small ships of the line (intended for North Sea service where their chief antagonists were likely to be the small Dutch vessels). One of the more interesting trends at this stage, observable both amongst the captured and purchased vessels and the purpose built ships of this period, is the increasing proportion of smaller vessels with a hull possessing much deadrise, ie. with 'vee' shaped sections, usually of a 'peg top' shape. It rather seems that this was originally a transatlantic development (though whether it came from Bermuda, from what had become the United States, or from elsewhere is unclear). Contrasting with the more rounded or flat bottomed sections usual for European warships up to this stage (smaller ships might often have finer lines, but these would usually be reduced and more elegant versions of lines used for the larger vessels), this type of hull form seems to have evolved in small craft, but at this stage was being considerably enlarged for brigs, sloops, corvettes and vessels of this type. It was usually associated with flush decks and the absence of poop and forecastle (even if these were fitted they would usually be the restricted platforms at the extreme ends commonly associated with the expression 'topgallant forecastle' or poop). In the post-Napoleonic period this hull form would be pushed to extremes by Symonds who would use it for frigates and then ships of the line. However, though it made for speed it also produced an uneasy motion and extremes of heel, neither making for a good gun platform, which probably explains why it was restricted to the smaller types of warship in the period covered by this chapter and the next.

## Large Three-deckers (First Rates)

***Commerce de Marseilles*** 1793-1796 (French *Le Commerce de Marseilles* 1788 Toulon, Sané design) 120.
Dimensions & tons: 208'4", 172'0⅛" x 54'9½" x 25'0½". 2746 73/94.
Men: 875 Guns: GD 34 x 32, MD 34 x 24, UD 34 x 12, QD 14 x 12, Fc 4 x 12 + 2 x 32*, RH 8 x 24*. Surrendered at Toulon 27.8.1793; 1796 hulked [at Plymouth as prison ship?; 1802 BU].

PLANS: 1796: Lines & profile & decorations/orlop & gun deck/middle, upper & quarter decks & forecastle.

***Salvador Del Mundo*** 1797-1815 (Spanish *Salvador del Mundo* 1786 Havanna) 112.
Dimensions & tons: 190', 152'11" x 54'3½" x 23'3". 2397 47/94.
Men: 839. Guns: GD 30 x 32, MD 32 x 24, UD 32 x 12, QD 12 x 9, Fc 6 x 9. Taken 14.2.1797 at St Vincent; never fitted for sea by the RN; 1815 BU.

PLANS: 1806: Lines & profile.

***San Josef*** 1797-1849 (Spanish *San Josef* 1783 Ferrol) 114.
Dimensions & tons: 194'3", 156'11¼" x 54'3" x 24'3½". 2456 24/94.
Men: 839. Guns: GD 32 x 32, MD 32 x 24, UD 32 x 12, QD 12 x 9, Fc 6 x 9. Taken at St Vincent 14.2.1797; 1849 BU.

PLANS: 1797: Lines & profile & decorations. 1801: Lines & profile & decorations/orlop, gun deck & middle deck/upper deck, quarter deck & forecastle.

## Large Two-deckers (Third Rates)

***Pompee*** 1793-1811 (French *Le Pompée* 1791 Toulon, Sané design) 80.
Dimensions & tons: 182'2", 148'7¼" x 49'0½" x 21'10½". 1901.
Men: 640 Guns: GD 30 x 32, UD 30 x 18, QD 12 x 32*, Fc 4 x 32*, RH 8 x 18*. Surrendered at the taking of Toulon 27.8.1793; 1811 hulked [at Portsmouth as a prison ship; 1817 BU; her name appears to have been the origin of the sailors' name for Portsmouth: 'Pompey'].

PLANS: Lines, profile & decorations/upper deck/quarter deck & forecastle/roundhouse.

***Achille*** 1794-1796 (French *L'Achille* ex *L'Annibal* [74] 1778 Brest, Sané design) 78.
Dimensions & tons: 177'8", 147'1⅜" x 48'2½" x 21'1½". 1818 60/94.
Men: ____ . Guns: ____ . Taken at the Glorious First of 06.1794; never fitted for sea by the RN; 1796 BU.

PLANS: Lines, profile & decorations/orlop & gun deck/upper deck/quarter deck & forecastle.

***Impetueux*** (ex *Impetuous*, ex *America*) 1794-1813 (French *L'Amérique* 1788 Brest) 78/80.
Dimensions & tons: 182', 149'7⅞" x 48'7½" x 21'7".1879 69/94.
Men: 670. Guns: GD 30 x 32, UD 30 x 18, QD 2 x 18 + 14 x 32*, Fc 2 x 18 + 4 x 32*. Taken at the Glorious First of 06.1794; 1813 BU.

PLANS: Lines & profile/orlop/gun deck/upper deck/quarter deck & forecastle.

***Juste*** 1794-1811 (French *Le Juste* ex *Les Deux Frères* 1784 Brest, Groignard design) 84.
Dimensions & tons: 193'4", 159'3¾" x 50'3½" x 22'5½". 2143 18/94.

---

1 On this see Jean Meyer and Martine Acerra's excellent *Marines et Révolution* (Rennes 1988) which makes this point of the decline of virtually all other navies, whilst the Royal Navy reached its apogee. The authors also believe that the battle of the Nile destroyed any final, faint, chance that France could successfully challenge Britain's command of the sea. Trafalgar merely and inevitably confirmed this verdict. I would agree with this verdict; too many recent historians writing in English have been over-impressed by the danger posed by Napoleon's widespread and numerically impressive shipbuilding activities. It should be understood that the difficulties of adequate manning, training and co-ordination would make it extremely unlikely that the potential numerical superiority would prove of use. Towards the end of the struggle the 74 gun *Rivoli* was completed and fitted out at vast expense of time and effort at Venice. Shortly after she emerged into the Adriatic, she met the British 74 *Victorious* and after a smart little action was added to the RN (see the next chapter). This is a far better indicator of what is likely to have happened than the nightmares of contemporary British officers and the arithmetical computations of modern historians.

# THE FRENCH REVOLUTIONARY WAR 1793-1801

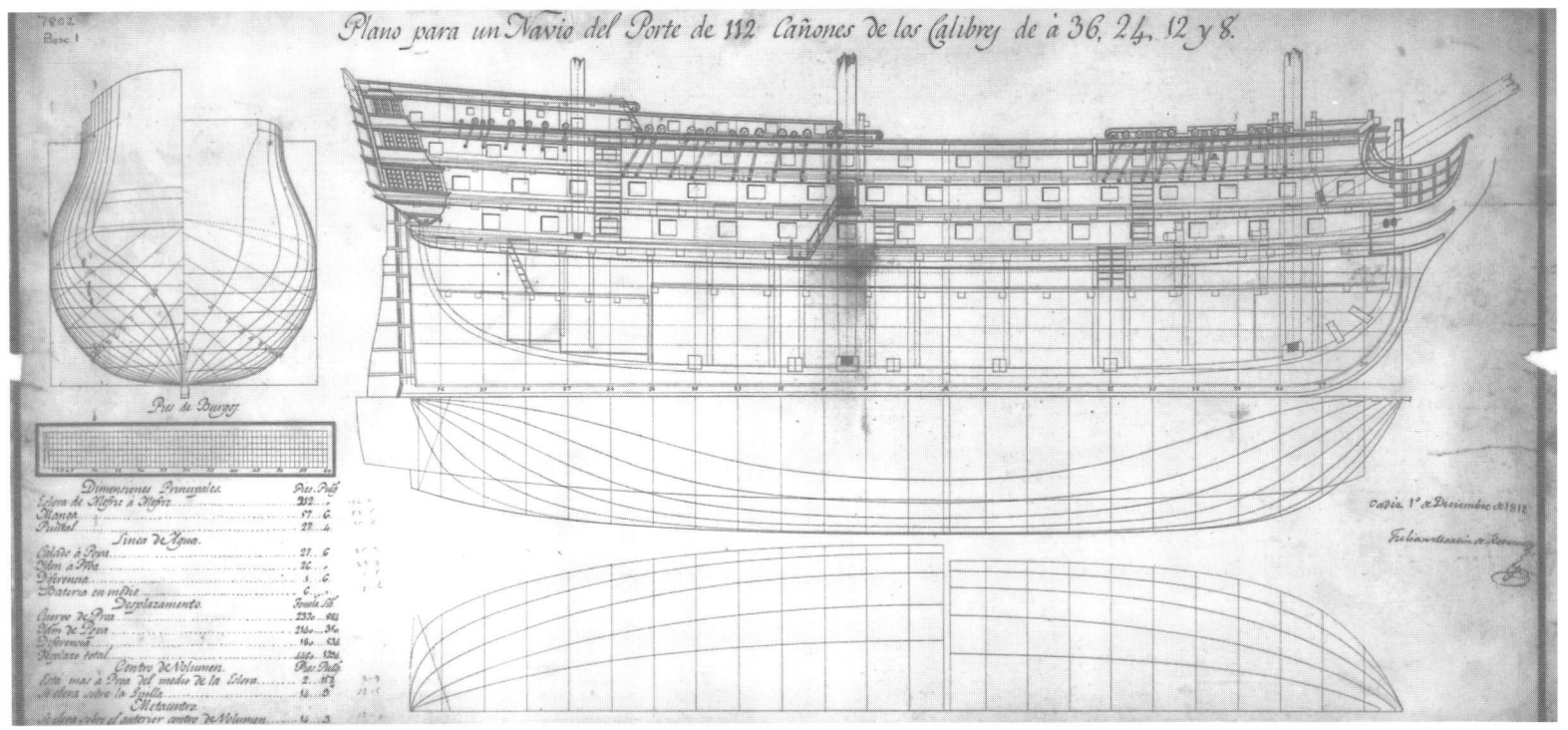

Spanish lines plan of a 112 gun ship dated 1812 (Admiralty number 7802). Though of a later date this will serve to illustrate the Spanish three-deckers captured at the battle of Saint Vincent (1797)

Men: 738. Guns: GD 30 x 32, UD 32 x 18, QD 2 x 24 + 14 x 32★, Fc 6 x 12 + 2 x 32★. Taken at the Glorious First of 06.1794; laid up from 1802; 1811 BU.

PLANS: Lines & profile/orlop/gun deck/upper deck/quarter deck & forecastle.

**Northumberland** 1794-1795 (French *Le Northumberland* [74] 1779 Brest, Sané design) 78.
Dimensions & tons: 178'8", 140'6" x 48'2½" x 21'1". 1827 87/94.
Men: ____ . Guns: ____ . Taken at the Glorious First of 06.1794; never fitted for sea by the RN; 1795 BU.

PLANS: Lines, profile & decorations/decks.

**Sans Pareil** 1794-1810 (French *Le Sans Pareil* 1793 Brest, Groignard design) 84.
Dimensions & tons: 193', 158'11¼" x 51'6" x 23'4". 2342.
Men: 738. Guns: GD 30 x 24, UD 30 x 24, QD 2 x 24 + 12 x 24★, Fc 2 x 24 + 4 x 24★. Taken at the Glorious First of 06.1794; 1810 sheer hulk [at Plymouth; 1842 BU].

NOTE: Her design was used in the 1840s to build one of the RN's last sailing ships of the line. This ship was completed as a screw vessel but retained the name of her French predecessor.

PLANS: Lines. 1842 as sheer hulk: Lines/profile & details of stern structure/quarter deck/forecastle/lower deck/orlop/midships section/scantlings.

**San Nicholas/Saint Nicholas** 1797-1800 (Spanish *San Nicolas* [80] 1769 Cartagena) 82.
Dimensions & tons: 179'9½", 148'4¾" x 49'7¼" x 20'1¼". 1942.
Men: 719. Guns: GD 30 x 32, UD 32 x 18, QD 14 x 9, Fc 6 x 9. Taken 14.2.1797 at St Vincent; never fitted for sea by the RN; ?1800 hulked [at Plymouth as a prison ship; 1814 sold].

**Canopus** 1798-1862 (French *Le Franklin* 1797 Toulon) 80.
Dimensions & tons: 193'4½", 159'10" x 52'4" x 23'. 2269 46/94.
Men: 700. Guns: GD 32 x 32, UD 32 x 18, QD 2 x 18 + 12 x 32★, Fc 2 x 9 + 4 x 32★. Taken 1.8.1798; later reclassed as an 84 gun Second Rate; 1862 hulked [at Plymouth as a receiving ship; 1869 mooring hulk; 1887 sold].

NOTE: Considered an excellent ship and as such worth several expensive repairs. She still outperformed all comers in sailing trials when 40 years old, and was used as the model for the large class of 84s (*Formidable* Class) built for the RN in the years after the end of the Napoleonic War.

PLANS: 1802: decks. 1814: Profile/profile. 1815: sheer/orlop/gun deck/upper deck/quarter deck & forecastle/detail of laying decks. 1840: Lines & profile/lines & profile (incomplete)/body/orlop/gun deck/upper deck/quarter deck & forecastle/roundhouse. 1880: position of capstans. No date (Ca 1810?): Lines (incomplete).

**Spartiate** 1798-1842 (French *La Spartiate* 1797 Toulon) 80 later 76.
Dimensions & tons: 182'7", 150'4" x 49'4½" x 21'7". 1949 41/94.
Men: 640. Guns: GD 28 x 32. UD 30 x 18, QD 2 x 18 + 14 x 32★, Fc 2 x 18 + 6 x 32★. Taken at Aboukir Bay 1.8.1798; 1842 hulked [at Plymouth as a temporary sheer hulk; 1857 BU].

PLANS: As fitted: Lines, profile & decorations/orlop & gun deck/upper deck & quarterdeck & forecastle. 1835: Orlop (incomplete).

**Tonnant** 1798-1821 (French *Le Tonnant* 1789 Toulon) 80.
Dimensions & tons: 194'2", 160' x 51'9¼" x 23'3". 2281 3/94.
Men: 700. Guns: GD 32 x 32, UD 32 x 18, QD 2 x 18 + 14 x 32★, Fc 4 x 32★. Taken at Aboukir Bay 1.8.1798; 1821 BU.

PLANS: 1805: decks. 1806: lines & profile. 1813: upper deck & quarter deck & forecastle.

**Donegal** 1798-1845 (French *Le Hoche* ex *Le Pégase* ex *Le Barras* 1794 Toulon) 76.
Dimensions & tons: 182', 150'5" x 48'9" x 21'10". 1901 43/94.
Men: 640. Guns: GD 30 x 32, UD 30 x 18, QD 10 x 9, Fc 6 x 9 + 2 x 32★, RH 6 x 18★. Taken off Ireland 12.10.1798; 1845 BU.

PLANS: Lines & profile/orlop & gun deck/upper deck/quarter deck & forecastle.

**Genereux** 1800-1816 (French *Le Généreux* 1785 Rochefort, Sané design) 80.
Dimensions & tons: 185'6", 149'10" x 49'2"x 20'9". 1926.
Men: ____ . Guns: ____ . Taken in the Mediterranean 18.2.1800; 1802 laid up until; 1816 BU.

**Malta** 1800-1831 (French *Le Guillaume Tell* 1795 Toulon) 84.
Dimensions & tons: 194'4", 159'9⅜" x 51'7½" x 23'4". 2265 9/94.
Men: 780. Guns: GD 30 x 32 + 2 x 68★, UD 30 x 24, QD 18 x 24 + 6 x 24★, Fc 2 x 12, RH 2 x 68★ + 2 x 24★. Taken 30.3.1800 in the Mediterranean; 1831 hulked [as an ordinary (= reserve) depot ship at Plymouth; 1840 BU].

PLANS: Lines & profile.

## 74 Gun Ships (Third Rates)

**Puissant** 1793-1796 (French *Le Puissant* 1782 Lorient, Groignard design).
Dimensions & tons: 178'9", 146'5½" x 48'0¾" x 21'4". 1799 49/94.

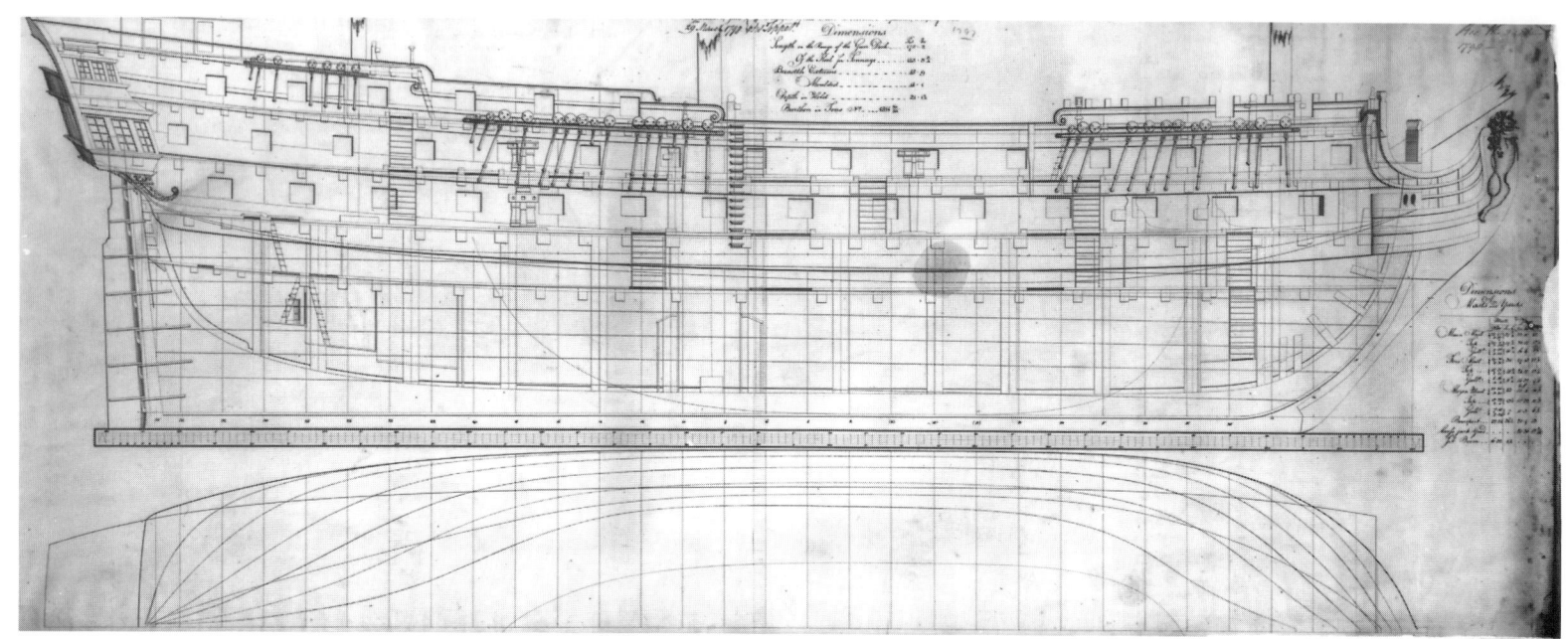

Part lines and profile plan (plan 928 omitting body plan) of the Spanish 74 *San Damaso* taken in 1797

Men: ____ . Guns: ____ . Taken at the surrender of Toulon 27.8.1793; never fitted for sea by the RN; 1796 hulked [at Portsmouth as a receiving ship and hulk for Spithead; 1816 sold].

PLANS: 1810: Lines & profile.

***Belleisle*** 1795-1814 (French *Le Formidable* ex *Le Marat* ex *Le Lion* 1793 Rochefort).
Dimensions & tons: 184'5", 149'5¼" x 48'9" x 21'7½". 1889.
Men: 700. Guns: GD 30 x 32, UD 30 x 24, QD 2 x 9 + 14 x 32★, Fc 2 x 9 + 2 x 24★, RH 6 x 24★. Taken off Lorient 25.6.1795; 1814 BU.

***Tigre*** 1795-1817 (French *Le Tigre* 1793 Brest) 74/76.
Dimensions & tons: 182', 149' x 48'9½" x 21'7⅜". 1886 67/94.
Men: 640. Guns: GD 28 x 32 + 2 x 68★, UD 28 x 18 + 2 x 68★, QD 4 x 18 + 10 x 32★, Fc 2 x 18 + 2 x 32★. RH 6 x 18★. Taken off Lorient 25.6.1795; 1817 BU.

PLANS: As fitted: Lines & profile/orlop/gun deck/upper deck/quarter deck & forecastle.

***San Damaso*** 1797-1800 (Spanish *San Damaso* 1775 Cartagena).
Dimensions & tons: 175'2", 143'3¾" x 48'9" x 21'1½". 1811 60/94.
Men: 590. Guns: ? as *San Isidro*?. Taken at the capture of Trinidad, she had been scuttled to block the harbour entrance, but was subsequently raised; never fitted for sea by the RN; ?1800 hulked (at Portsmouth as a prison ship?; 1814 sold].

PLANS: Lines & profile/orlop/gun deck/upper deck/quarter deck & forecastle.

***Hercule*** 1798-1810 (French *L'Hercule* 1797 Lorient).
Dimensions & tons: 181'3", 149'8½" x 48'6½" x 21'9". 1876 33/94.
Men: 640. Guns: GD 28 x 24, UD 28 x 24 (Gover's), QD 2 x 24 (Gover's) + 12 x 24★, Fc 2 x 24 (Gover's) + 2 x 24★. Taken by *Mars* off Brest 21.4.1798; 1810 BU.

PLANS: As fitted: Lines & profile & decorations/orlop & gun deck/upper deck & quarter deck & forecastle.

***Conquérant*** 1798-1803 (French *Le Conquérant* 1765 Brest, J Olivier design).
Dimensions & tons: 181'4", 147'9" x 46'3" x 21'10". 1681.
Men: 590. Guns: ____ . Taken at Aboukir Bay 1.8.1798; never fitted for sea by the RN; 1803 BU.

***Guerier*** 1798 & hulked after capture (French *Le Peuple Souverain* ex *Le Souverain* 1757 Toulon, Pomet design).
Dimensions & tons: 174'9", 140'5" x 47'6" x 21'10". 1685.
Men: ____ . (as hulk: 7) Guns: ____ . Taken at Aboukir Bay 1.8.1798; hulked on arrival at Gibraltar; [1810 BU].

***Aboukir*** 1798-1802 (French *L'Aquilon* 1789 Rochefort).
Dimensions & tons: 185'5", 150'5" x 48'4" x 21'. 1869. 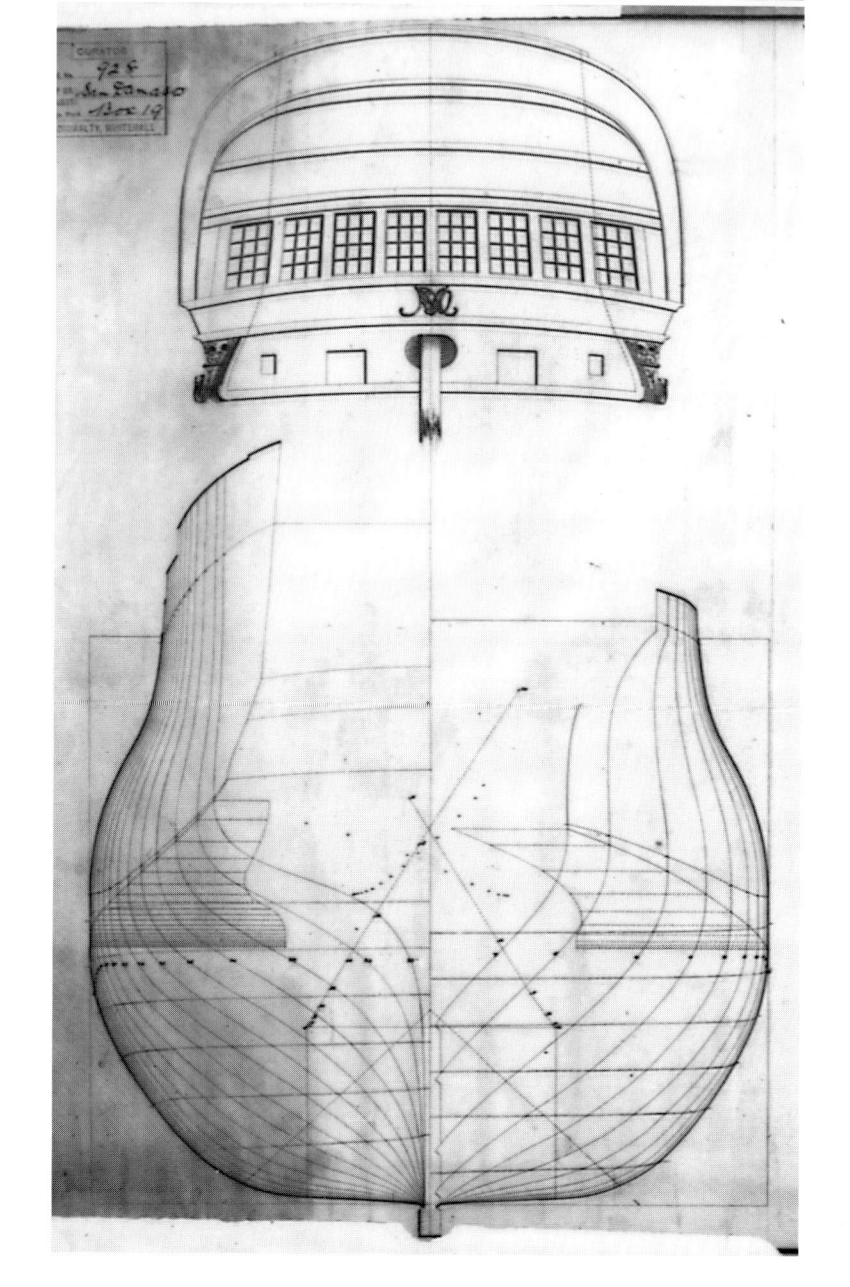
Men: 590. Guns: ____ . Taken at Aboukir Bay 1.8.1798; never fitted for sea by the RN; 1802 BU.

***Prince of Orange/Princess of Orange*** 1799-1806 (Dutch *Washington* 1795

# THE FRENCH REVOLUTIONARY WAR 1793-1801

Amsterdam Dyd).
Dimensions & tons: 168'5", 138'0¾" x 46'2" x 18'5½". 1565.
Men: 590. Guns: GD 28 x 32, UD 28 x 18, QD 2 x 18 + 10 x 32★,
Fc 2 x 18 + 4 x 32★. Surrendered at the Texel 1799; never actually registered in the RN as a 74; 1806 hulked [at Chatham as a 'stationary ship' (though possibly initially at Sheerness?); 1811 powder magazine; 1822 sold].

NOTE: There is some confusion over the name; some sources give *Prince* and some *Princess*; it is possible that originally she was to be *Prince* _____ but the name used in practice (for example in her logs) was *Princess* _____ .

**San Antonio** (**Saint Antonio**) 1801 & hulked after capture (French *Saint Antoine* 1800 or 1801 transferred Spanish *San Antonio* 1785 Cartagena).
Dimensions & tons: 174'10", 139'8" x 47'10" x _____ . 1700.
Men: 590. Guns: GD 28 x 32, UD 28 x 18, QD 4 x 9 + 10 x 32★,
Fc 2 x 9 + 2 x 32★. Taken 12.7.1801 soon after transfer from Spain to France; never fitted for sea by the RN after her arrival in Britain and soon hulked [at Portsmouth as a prison ship; 1814 powder magazine; 1827 sold; 1828 resold].

## Small Line of Battle Ships (Third Rates 72s, 70s, 64s & Fourth Rate 60s)

**Overyssel** 1795-1805 (Dutch *Overijssel* 1794 purchased Spanish *San Felipe Apostol* 1781 Ferrol) 64.
Dimensions & tons: 153'6⅛", 124'8½" x 43' x 18'1½". 1226.
Men: 491. Guns: GD 26 x 24, UD 26 x 18, QD 8 x 9, Fc 4 x 9. Taken 22.10.1795 by *Polyphemus*; 1805 hulked [at Sheerness as a provision receiving ship; 1809 became a breakwater].

**Zealand** 1796-1803 (Dutch *Zeeland* 1784 Zeeland Admiralty) 64.
Dimensions & tons: 156'8¼", 127'3⅝" x 44'4" x 18'4". 1330.
Men: 491. Guns: GD 26 x 24, UD 26 x 18, QD 10 x 9, Fc 2 x 9. Taken in harbour at Plymouth 19.1.1796; 1803 hulked [at Sheerness as a receiving ship; 1809 convict ship; 1812 renamed *Justitia*; 1830 sold].

**York** 1796-1804 (East Indiaman *Royal Admiral* building Barnard, Deptford) 64.
Dimensions & tons: 174'3", 144'4" x 43'2½" x 19'7½". 1433.
Men: 491. Guns: GD 26 x 24, UD 26 x 18, QD 10 x 9,
Fc 2 x 9 (+ 2 x 18★?). Purchased whilst on the stocks 1795; launched 24.3.1796; 01.1804 foundered in the North Sea.

PLANS: Building: Lines/profile/orlop/gun deck/upper deck/quarter deck & forecastle.

**Ardent** 1796-1813 (East Indiaman *Princess Royal* building Pitcher, Northfleet) 64.
Dimensions & tons: 173'3", 144' x 43' x 19'10". 1416 24/94.
Men: 491. Guns: GD 26 x 24, UD 26 x 18, QD 10 x 9, Fc 2 x 9. Purchased whilst on the stocks 1795; launched 9.4.1796; 1812 troopship; 1813 hulked [at Bermuda as a prison ship; 1824 BU].

PLANS: Building: Lines & profile/orlop/gun deck/upper deck/quarter deck & forecastle. 1804: quarter deck & forecastle.

**Agincourt** 1796-1812 **Bristol** -1814 (East Indiaman *Earl Talbot* building Perry, Blackwall) 64.
Dimensions & tons: 172'8", 143'9½" x 43'4½" x 19'8¾". 1439.
Men: 491. Guns: GD 28 x 24, UD 28 x 18, QD 6 x 9, Fc 2 x 9.
Purchased whilst on the stocks 1795; launched 23.7.1796; 1808 victualler; 1810 troopship & renamed *Bristol*; 1814 sold.

PLANS: Building: Lines & profile/orlop/gun deck/upper deck/quarter deck & forecastle. 1808 as victualler: profile & decks.

**Monmouth** 1796-1834 (East Indiaman *Belmont* building Randall, Rotherhithe) 64.
Dimensions & tons: 173'1", 144'1½" x 43'4" x 19'8". 1439.
Men: 491. Guns: GD 26 x 24, UD 26 x 18, QD 10 x 9, Fc 2 x 9.
Purchased whilst on the stocks 1795; launched 23.4.1796; 1815 hulked [at Deptford as a sheer hulk; 1828-1833 at Woolwich; 1834 BU].

PLANS: Lines & Profile/orlop/gun deck/upper deck/quarter deck & forecastle (also for *Lancaster*). 1809 as victualler: Profile & decks,

**Dordrecht** 1796-1800 (Dutch *Dordrecht* 1782 Dort) 64.
Dimensions & tons: 159', 130' x 45' x _____ . 1437.
Men: _____ (as troopship: 215). Guns: _____ (as troopship: UD 18 x 12, QD 6 x 6, Fc 6 x 6). Taken at Saldanha Bay 17.8.1796; 1799 troopship; 1800 hulked [at Chatham as a receiving ship; 1814 to Sheerness; 1823 sold].

**Prince Frederick** 1796-1800 (Dutch *Revolutie* ex *Prins Frederick* 1779? _____ ) 64.
Dimensions & tons: 156'9'; 129'4" x 42'11" x 16'3½". 1267 7/94.
Men: 491 (as storeship: 230). Guns: GD 24 x 24, UD 26 x 18, QD 10 x 9, Fc 2 x 9 + 2 x 24★, RH 6 x 18★ (as storeship. UD 24 x 12, QD 6 x 6, Fc 6 x 6). Taken at Saldanha Bay 17.8.1796; 1797 storeship; 1800 hulked [at Plymouth as a convalescent ship; 1804 to Berehaven as a hospital ship; 1817 sold].

PLANS: Lines & profile & decorations/orlop/gun deck/upper deck/quarter deck & forecastle. 1806 as hospital ship: orlop & gun & upper decks.

**Tromp** 1796-1800 (Dutch *Marten Harpentzoon Tromp* 1779 _____ ) 60/50.
Dimensions & tons: 143'10½", 117'10" x 40'8¾" x 15'3". 1039 65/94.
Men: _____ (as troopship: 215). Guns: _____ . Taken at Saldanha Bay 17.8.1796; 1798 troopship; 1800 hulked [at Chatham as a prison ship; 1803 to Portsmouth; 1811 receiving ship; 1815 sold].

PLANS: Lines & profile/orlop/gun deck/upper deck/quarter deck & forecastle.

**Lancaster** 1797-1807 (East Indiaman *Pigot* building Randall, Rotherhithe) 64.
Dimensions & tons: 173'6", 144'3" x 43'3" x 19'9". 1430.
Men: 491. Guns: GD 26 x 24, UD 26 x 18, QD 10 x 9, Fc 2 x 9.
Purchased whilst on the stocks 1797; launched 29.1.1797; 1807 hulked [as a receiving ship for Malta but instead fitted as a victualler; 1815 given to the West India Dock Co. for the reception of boys belonging to that trade; 1832 returned to the Admiralty and sold].

PLANS: Lines & profile/orlop/gun deck/upper deck/quarter deck & forecastle (also for *Monmouth*).

**San Ysidro** 1797 & hulked after capture (Spanish *San Isidro* [74] 1768 Ferrol) 72.
Dimensions & tons: 176'0½", 144'1⅜" x 48'11¼" x 20'1½". 1836.
Men: 590. Guns: GD 28 x 32, UD 30 x 18, QD 8 x 9, Fc 6 x 9. Taken at Saint Vincent 1797; hulked at Plymouth on arrival there [as a prison ship; 1814 sold].

**Alkmaar** 1797-1815 (Dutch *Alkmaar* 1782 Enkhuizen) 60/50.

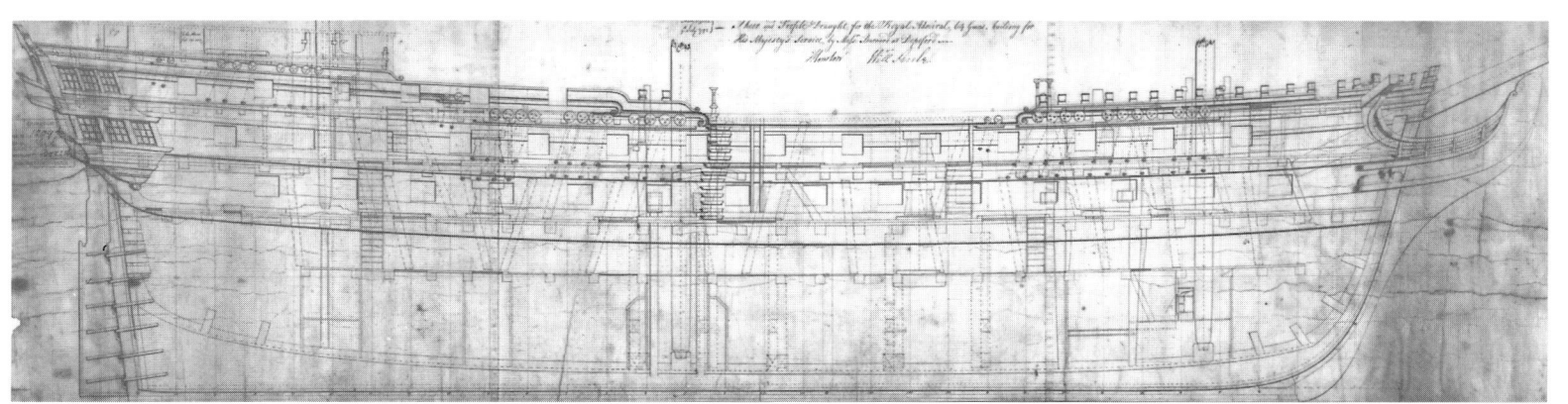

Profile plan (1209) showing the 64 gun ship *York* purchased whilst still on the stocks. Here she is still given her original East Indiaman name as *Royal Admiral*

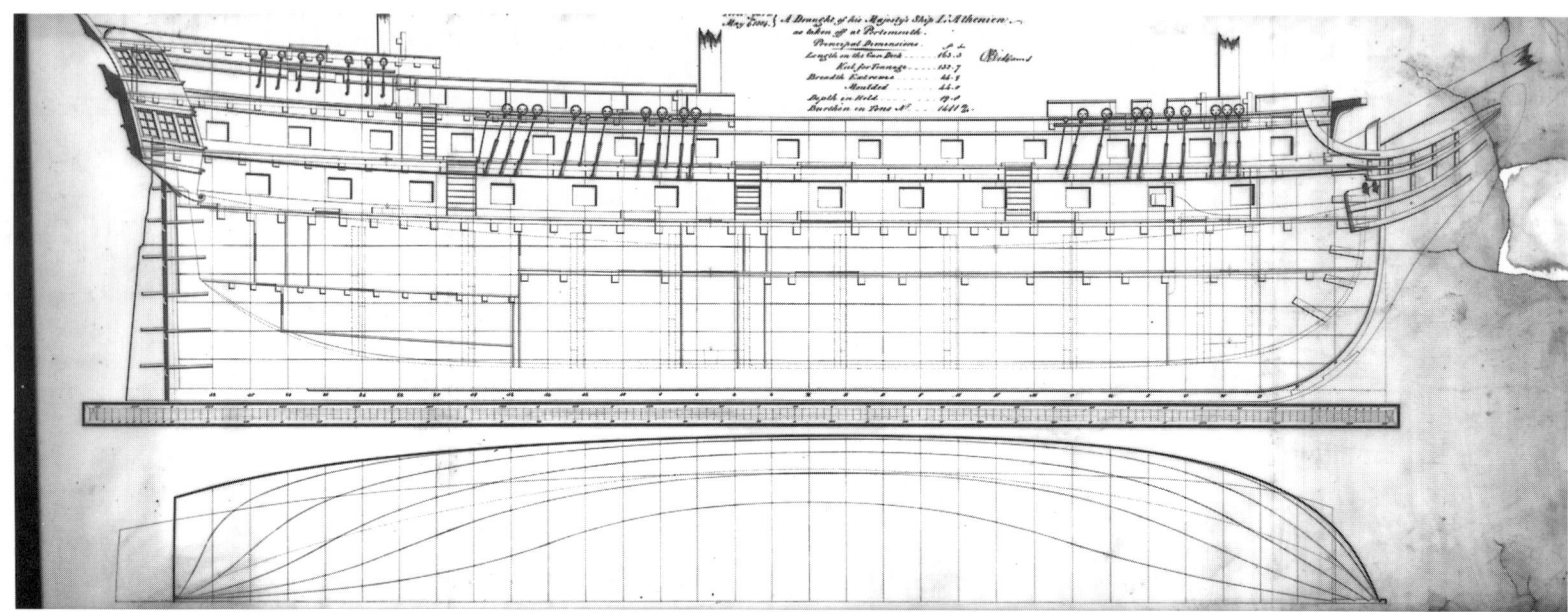

Lines and profile plan (7260) 'taken off' the Maltese built 64 captured from the French at the fall of that island and renamed *Athenian*

Dimensions & tons: 142'4", 117'10" x 40'9" x 15'6½". 1041.
Men: 324. (as troopship: 215). Guns: ____ . (as hospital ship 18 x 9 + 6 x 6). Taken at Camperdown 1797; 1798 troopship; 1801 hospital ship; 1805 storeship for the Downs; 1807 laid up until; 1815 sold.

**Admiral De Vries** 1797-by 1806 (Dutch *Admiraal Tjerk Hiddes de Vries* 1783 ____ ) 64.
Dimensions & tons: 157'5", 127'11⅝" x 44'8½" x 16'2½". 1360.
Men: 491 (as troopship: 250). Guns: GD 26 x 24, UD 24 x 18, QD 8 x 9 (as troopship: UD 28/24 x 18, QD 8 x 9, Fc 8 x 9). Taken at Camperdown 1797; 1798 troopship; before 1806 hulked [at Jamaica as a receiving ship; 1806 sold].

**Camperdown** 1797-1802 (Dutch *Jupiter* [74] 1782 Arie Staats, Amsterdam) 64.
Dimensions & tons: 167'4", 136' x 46'5" x 18'2½". 1559.
Men: 491 (as hulk: 73). Guns: GD 28 x 32, UD 28 x 18, QD & Fc 10 x 12 + 4 x 8. Taken at Camperdown 1797; never fitted for sea by the RN; 1802 hulked [at Chatham as a powder hulk and prison ship; 1817 sold]. May have originally been classed as a 74 in the RN?

**Delft** 1797-1802 (Dutch *Hercules* 1781 Spaan, Dordrecht) 64.
Dimensions & tons: 157'2", 129'6¼" x 42'10½" x 16'11½". 1266.
Men: 491 (as troopship: 250/215). Guns: GD 26 x 24, UD 26 x 18, QD 4 x 9 + 8 x 24★, Fc 4 x 9 + 2 x 24★. Taken at Camperdown 1797; 1799 troopship; 1802 hulked [at Chatham as a prison ship; 1822 became a breakwater at Harwich].

PLANS: Lines & profile /profile & decks.

**Gelykheid** 1797-1798 (Dutch *Gelykheid* 1795 ex *Prins Frederick Willem* 1788 Amsterdam Admiralty) 64.
Dimensions & tons: 155'8", 126'8⅛" x 44' x 17'1". 1305.
Men: 491. Guns: GD 26 x 24, UD 26 x 18, QD 10 x 9, Fc 4 x 9. Taken at Camperdown 1797; 1798 hulked [at Chatham as a prison ship; by 1814 at Portsmouth; 1814 sold].

**Haerlem** 1797- 1805 (Dutch *Haarlem* 1785 Amsterdam Admiralty) 64.
Dimensions & tons: 156'7", 127'1¾" x 44'3" x 17'3". 1322.
Men: 491. (as troopship: 250) Guns: GD 26 x 24, UD 28 x 18, QD 8 x 9, Fc 8 x 9. Taken at Camperdown 1797; 1798 troopship; 1805 hulked [at Chatham; lay at Stangate Creek?; 1811 receiving ship; 1816 sold].

**Vryheid** 1797-1799 (Dutch *Vrijheid* 1782 Amsterdam Admiralty) 72.
Dimensions & tons: 167'7½", 138'5" x 40'3" x 18'9". 1562.
Men: ____ . Guns: GD 28 x 32, UD 28 x 18, QD & Fc 16 x 12. Taken at Camperdown 1797; laid up on arrival in Britain; 1799 hulked [at Chatham as a prison ship; 1811 sold].

**Wasanaar/Wassenaar** 1797-1802 (Dutch *Wassenaar* 1781 ____ ) 64.
Dimensions & tons: 159'2", 131'1¼" x 42'8" x 20'2½". 1269. 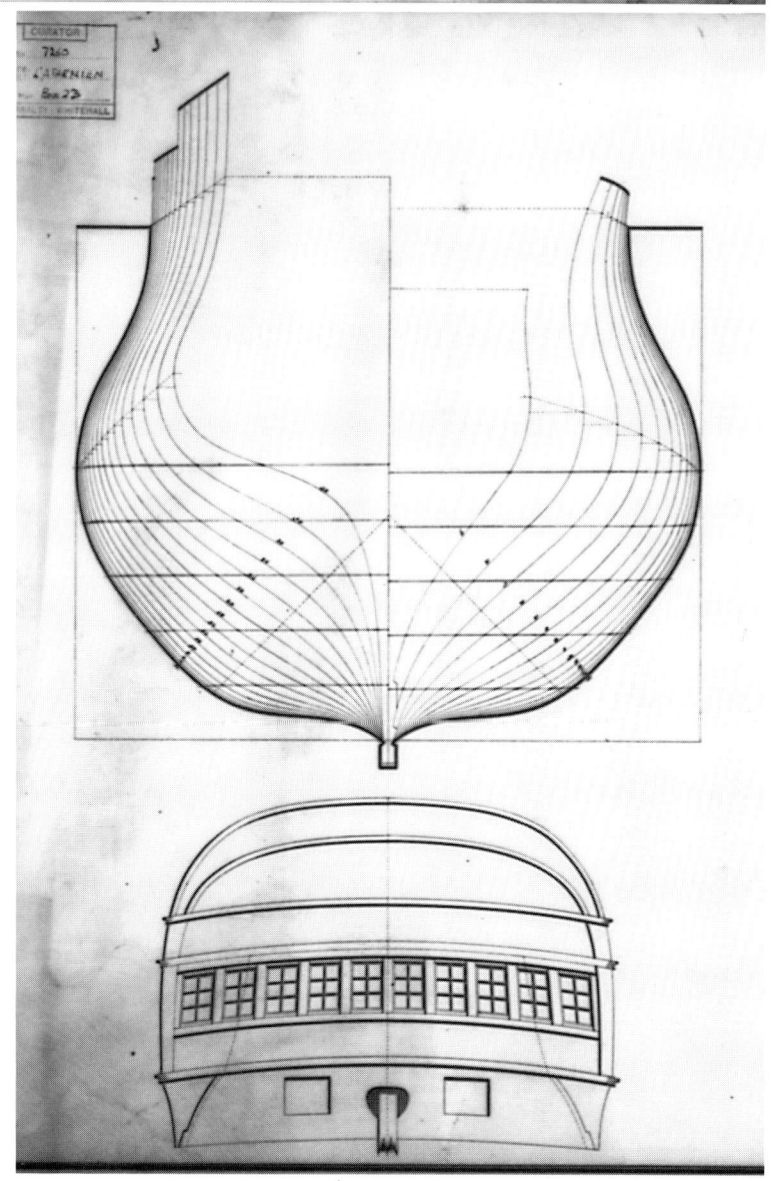
Men: 491 (as troopship: 250) Guns: GD 26 x 24, UD 24 x 18, QD 8 x 9, Fc 8 x 9 (as troopship: UD 28 x 18, QD 8 x 9, Fc 2 x 9). Taken at Camperdown 1797; 1798 troopship; 1802 hulked [at Chatham as a powder hulk; 1818 sold].

**Guelderland** 1799-1811 (Dutch *Gelderland* 1781 Amsterdam) 64.
Dimensions & tons: 157'2", 128'3" x 44'11¼" x 16'7". 1342.
Men: ____ . Guns: ____ . Taken in the Texel 1799; never fitted for sea

THE FRENCH REVOLUTIONARY WAR 1793-1801

by the RN; 1811 hulked [at Chatham as a receiving ship; 1817 sold].

**De Ruyter** 1799-1801 (Dutch *Admiraal De Ruijter* 1776 Amsterdam Admiralty) 64.
Dimensions & tons: 151'2", 122'6⅛" x 44'0½" x 17'6". 1264.
Men: ____ . Guns: ____ . Taken in the Texel 1799; 1801 hulked [at Antigua as a prison ship (and storeship?); 3.9.1804 wrecked in a hurricane].

**Leyden** 1799-1815 (Dutch *Leyden* 1786 Amsterdam Admiralty or 1790 Rotterdam) 64.
Dimensions & tons: 155'11½", 127'5⅛" x 43'11" x 17'2". 1307.
Men: 250 (Ca.1807: 350). Guns: GD 26 x 24, UD 26 x 24 (Gover's), QD 2 x 24 (Gover's) + 2 x 24★, Fc 2 x 24 (Gover's) (Ca.1807: GD 26 x 24, UD 26 x 24★, QD 2 x 24, Fc 2 x 24★). Taken in the Texel 1799; 1800 floating battery; 1810 troopship; 1815 sold.

**Texel** 1799-1818 (Dutch *Cerberus* 1784 Amsterdam Admiralty) 64.
Dimensions & tons: 156', 126'11¼" x 44'2" x 17'1. 1317.
Men: 250. Guns: GD 26 x 24, UD 26 x 18, QD 2 x 9. Taken in the Texel 1799; 1803 Floating battery (?); 1818 sold.

**Utrecht** 1799-1815 (Dutch *Utrecht* 1781 Amsterdam) 64.
Dimensions & tons: 156'8", 127'9½" x 44'3" x 17'1". 1331.
Men: 491. Guns: GD 26 x 24, UD 26 x 18, QD 10 x 9, Fc 2 x 9. Taken in the Texel 1799; 1815 sold.

**Athenian** 1800-1806 (French *L'Athénienne* taken 1798 from the Knights of St.John at Malta as *Saint Jean/San Giovanni* [70] building 1796, Malta) 64.
Dimensions & tons: 163'3", 132' x 44'9" x 19'8". 1411 89/94.
Men: 491. Guns: GD 26 x 20, UD 26 x 18, QD 2 x 9 + 8 x 24★, Fc 2 x 9 + 4 x 24★. Taken at the surrender of Malta 1800; 27.10.1806 wrecked off Sicily.

NOTE: Site recently investigated by Robert Sténuit, who reports little surviving of the wreck.
PLANS: Lines & profile/orlop/gun deck/upper deck/quarter deck & forecastle.

**Dego** 1800 & hulked after capture (French *Dego* taken 1798 from the Knights of Saint John at Malta as *Zachari* ____ ) 60.
Dimensions & tons: ____ , ____ x ____ x ____ . ____ .
Men: ____ . Guns: ____ . Taken at the surrender of Malta 1800; hulked at Malta [as a prison ship; 1802 sold].

**Holstein** 1801-1805 *Nassau* -1814 (Danish *Holstein* 1772 Copenhagen, Krabbe design) 64.
Dimensions & tons: 161'3", 131'8⅛" x 44'7½" x 19'2". 1395.
Men: 491. Guns: GD 26 x 24, UD 24 x 18, QD 12 x 24★, Fc 2 x 9 + 2 x 24★. Taken at Copenhagen 1801; 1805 renamed *Nassau*; 1814 sold.

## Small Two-deckers (Fourth Rate 56s & 54s, Fifth Rate 44s)

**Abergavenny** 1795-1807 (East Indiaman *Abergavenny* fitted – and built? – by Pitcher, Northfleet) 54.
Dimensions & tons: ____ , 131'6" x 41'1½" x 17'. 1182.
Men: 324. Guns: GD 28 x 18, UD 26 x 32★, Fc 2 x ____ 'bow chacers'. Purchased 1795; 1807 sold.

**Calcutta** 1795-1805 (East Indiaman *Warley* Perry, Blackwall?) 54.
Dimensions & tons: 150'11", 129'7¾" x 41'3½" x 17'2". 1176.
Men: 324. Guns: GD 28 x 18, UD 26 x 32★, Fc 2 x 32★ 'bow chacers'. Purchased 1795; 1796 transport; 1805 taken by the French (Allemand's squadron) off Sicily; [12.4.1809 destroyed in action by British ships].

**Glatton** 1795-1830 (East Indiaman *Glatton* Wells, Blackwall) 56.
Dimensions & tons: 163'11¼", 133'4¼" x 42'1" x 17'. 1256.
Men 344. Guns: GD 28 x 68★, UD 28 x 42★ (later: GD 28 x 18, UD 28 x 32★). Purchased 1795; 1809 laid up until; 1830 breakwater.

**Grampus** 1795-1799 (East Indiaman *Ceres* Thames?) 54.
Dimensions & tons: 157'1", 130'4" x 41'3" x 15'6". 1165.
Men: 324. Guns: GD 28 x 18, UD 26 x 32★. Purchased (whilst building?) 1795; 1797 storeship; 12.4.1799 wrecked on Barking Shelf 'afterwards set on fire by persons unknown and burnt to the water's edge'; then BU (?).

**Hindostan** 1795-1804 (East Indiaman *Hindostan* Barnard, Deptford) 54.
Dimensions & tons: 160'3", 132' x 42'1" x 17'1". 1249.
Men: 324. Guns: GD 28 x 18, UD 26 x 32★, Fc 2 x 18 'bow chacers' (as storeship: UD 2 x 18 + 26 x 32★). Purchased 1795; 1802 storeship; 2.4.1804 burnt by accident in Rosas Bay.

**Malabar** 1795-1796 (East Indiaman *Royal Charlotte* ____ ) 54.
Dimensions & tons: 161', 132'3¼" x 42'2¼" x 17'6". 1252.
Men: ____ . Guns: ____ . Purchased 1795; 10.10.1796 foundered coming from the West Indies.

**Weymouth** 1795-1800 (East Indiaman *Earl of Mansfield* building Wells, Deptford) 56.
Dimensions & tons: 175'5", 144'1" x 43'3" x 17'6". 1416.
Men: 344. Guns: GD 28 x 18, UD 28 x 32★. Purchased whilst on the stocks 1795; 13.9.1795 launched; 1796 transport; 21.1.1800 wrecked on Lisbon Bar.

**Coromandel** 1795-1807 (East Indiaman *Winterton* building Perry, Blackwall) 56.
Dimensions & tons: 169', 140' x 42'5" x 17'2". 1340.
Men: 324. Guns: GD 28 x 18, UD 28 x 32★. Purchased whilst on the stocks 1795; launched 9.5.1795; 1796 transport; 1802 troopship; 1807 hulked [at Jamaica as an army convalescent ship; 1813 sold].

**Madras** 1795-1807 (East Indiaman *Lascelles* building Wells, Deptford) 56.
Dimensions & tons: 175'1½", 144' x 43'1¾" x 17'6". 1426.
Men: 344. Guns: GD 28 x 18, UD 28 x 32★. Purchased whilst on the stocks 1795; launched 4.7.1795; 1803 storeship; 1807 BU.

**Brakel** 1795-1807 (Dutch *Brakel* 1784 Rotterdam Admiralty) 54.
Dimensions & tons: 148'3½", 117'7" x 42'1½" x 17'2¾". 1110.
Men: 355. (later: 240). Guns: GD 22 x 24, UD 24 x 12, QD 6 x 6, Fc 2 x 6. Seized at Plymouth 4.3.1795; 1799 troopship; 1807 hulked [at Chatham as a receiving ship; 1814 sold].

**Batavier** 1799-1817 (Dutch *Batavier* 1778 Amsterdam) 56.

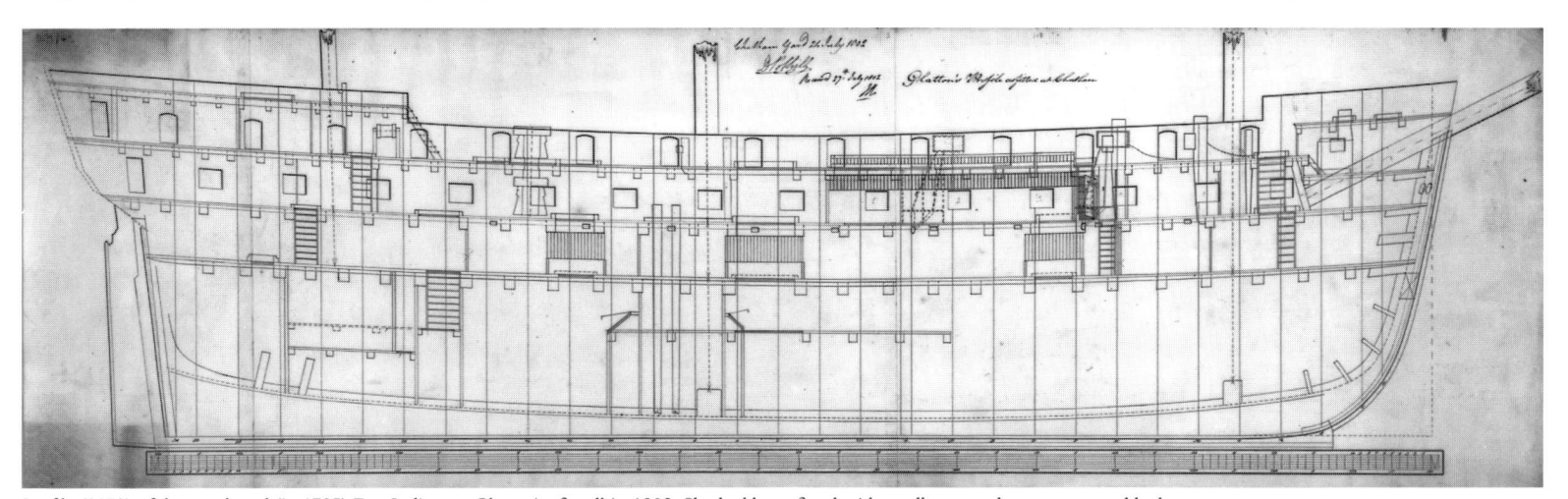

Profile (8073) of the purchased (in 1795) East Indiaman *Glatton* 'as fitted' in 1802. She had been fitted with an all-carronade armament and had fired incendiary projectiles ('carcasses') to good effect at the 1801 battle of Copenhagen. Note the high arched gunports which allowed the upper deck carronades extra elevation

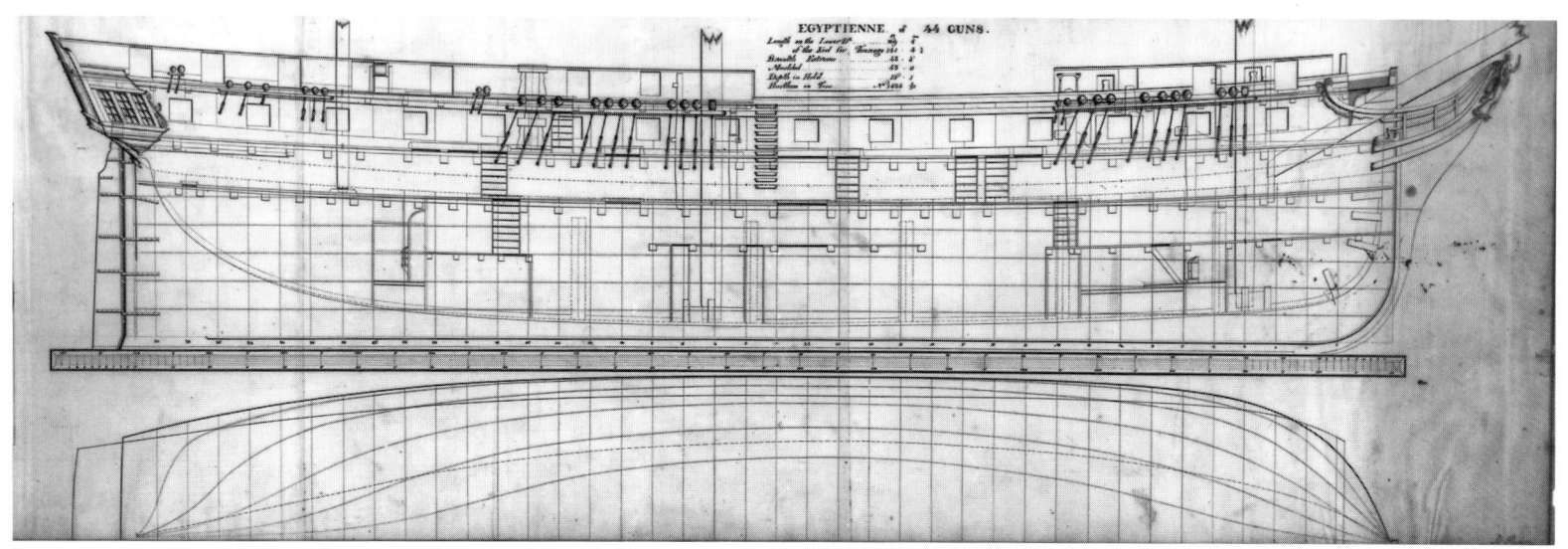

Lines and profile (6787) of the French 24 pounder frigate *Egyptienne* taken 1801

Dimensions & tons: 144'7", 118'1⅞" x 40'10" x 16'5". 1048.
Men: 350. Guns: GD 20 x 24, UD 28 x 18, QD 2 x 9. Taken in the Texel 1799; 1801 floating battery; 1809 laid up until; 1817 hulked [at Sheerness as a convict ship; 1823 BU].

*Beschermer* 1799-1806 (Dutch *Beschermer* 1781 North Holland or 1784 Enkhuizen?) 54.
Dimensions & tons: 145'1", 118'7" x 40'10" x 16'4". 1052.
Men: 350. Guns: GD 20 x 24, UD 20 x 18, QD 2 x 9. Taken in the Texel 1799; 1801 floating battery; 1806 hulked [lent to the East India Company at Blackwall; 1838 returned and sold].

*Broaderscarp* 1799-1805 (Dutch *Broderschap* 1795 ex *Prinses Louisa* 1760 Amsterdam Admiralty) 54/50.
Dimensions & tons: 140'1", 114'11" x 41'8½" x 8'8½". 1063.
Men: ____ . Guns: ____ . Taken in the Texel 1799; hulked on arrival at Sheerness (for use at the Cape of Good Hope?); 1803 floating battery; 1805 BU.

*Drochterland* 1799 & hulked after capture (Dutch *Unie* 1795 ____ ) 44.
Dimensions & tons: 135', 109'10" x 38'7½" x 14'6". 871.
Men: ____ . Guns: ____ . Taken in the Texel 1799; hulked on arrival at Sheerness [as a receiving ship; 1815 BU].

*Pandour* 1799-1805 (Dutch *Hector* 1784 Amsterdam) 44.
Dimensions & tons: 134'3", 108'11" x 39'3½" x 15'2". 894.
Men: 294. Guns: GD 20 x 12, UD 22 x 9, QD 2 x 6. Taken in the Texel 1799; 1801 troopship; 1803 floating battery; 1805 to Customs (as hulk).

*Vlieter* 1799-by 1808 (Dutch *Mars* 1780 Haarlem) 44.
Dimensions & tons: 155'8", 127'1½" x 44'9½" x 16'9". 1357.
Men: 185. Guns: UD 26 x 24, QD 10 x 18★. Taken in the Texel 1799 (as a razee); converted to a floating battery on arrival in Britain; by 1808 hulked [at Chatham as a powder magazine; 1809 to Sheerness as a sheer hulk; 1817 BU].

## 24 pounder Frigates (Fifth Rates)

*Pomone* 1794-1802 (French *La Pomone* 1785 Rochefort, Bombelle design) 44.
Dimensions & tons: 159'2⅜", 132'4¼" x 41'11⅜" x 12'9½". 1238 67/94.
Men: 300. Guns: UD 26 x 24, QD 14 x 32★, Fc 2 x 9 + 4 x 32★ (originally intended to arm her with 18 pdrs. on the UD in the RN). Taken 23.4.1794 by *Flora* and others (Warren's squadron); 1802 BU.
PLANS: Taken off/as fitted: Lines, profile & decorations.

*Amphitrite* 1799-1801 *Imperieuse* 1805 (Dutch *Amfitrite* 1797 Amsterdam) 40.
Dimensions & tons: 150'9½", 123'7⅝" x 42'5" x 12'9". 1, 183 19/94.
Men: 320. Guns: UD 28 x 24, QD 10 x 24★?, Fc 2 x 9 + 4 x 24★? (In 1804 the 24 pounders were replaced by 18 pounders.) Taken in the Texel 30.8.1799; 1801 renamed *Imperieuse*; 1805 BU.

*Egyptienne* 1801-1806 (French *L'Egyptienne* 1799 Lorient, Caro design) 38/44.
Dimensions & tons: 169'8", 141'4¾" x 43'8" x 15'1". 1434 4/94. 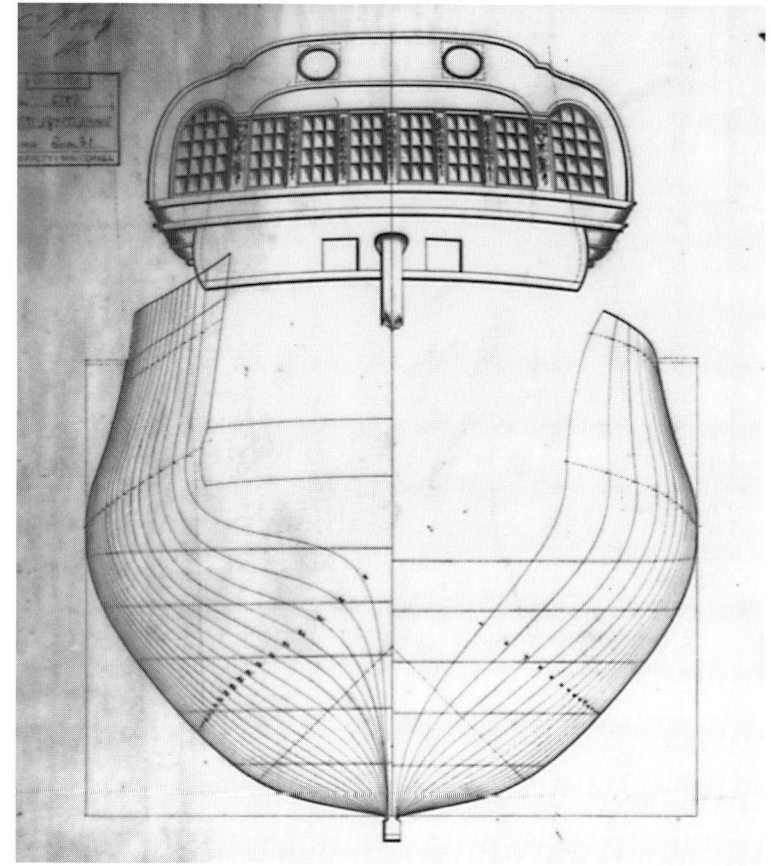
Men: 330 Guns: UD 28 x 24, QD 2 x 9 + 12 x 32★, Fc 2 x 9 + 4 x 32★. Taken 2.9.1801 at the fall of Alexandria; 1806 hulked [at Plymouth as a receiving ship; 1817 sold].
NOTE: Converted whilst still on the stocks from a 74 gun ship.
PLANS: 1802: profile/lower deck/upper deck. 1807?: Lines, profile & decorations.

## 18 pounder Frigates (Fifth Rates)

*Amethyst* 1793-1795 (French *La Perle* 1790 Toulon, Coulomb design) 40/38.
Dimensions & tons: 150'4", 124'6" x 39'5" x 12'10½". 1028 80/94.
Men: 300. Guns UD 28 x 18, QD 10 x 6 + 6 x 32★ (later just 8 x 6), Fc 2 x 6 + 2 x 32★ (later just 2 x 6). Surrendered at the capitulation of Toulon 1793; 29.12.1795 wrecked at Alderney.
PLANS: Lines & profile/orlop/lower deck/upper deck/quarter deck & forecastle.

*Arethuse* 1793-1795 *Undaunted* -1796 (French *L'Aréthuse* 1791 Brest, Ozanne design) 38.
Dimensions & tons: 152', 126'10" x 39'8½" x 12'4". 1064.

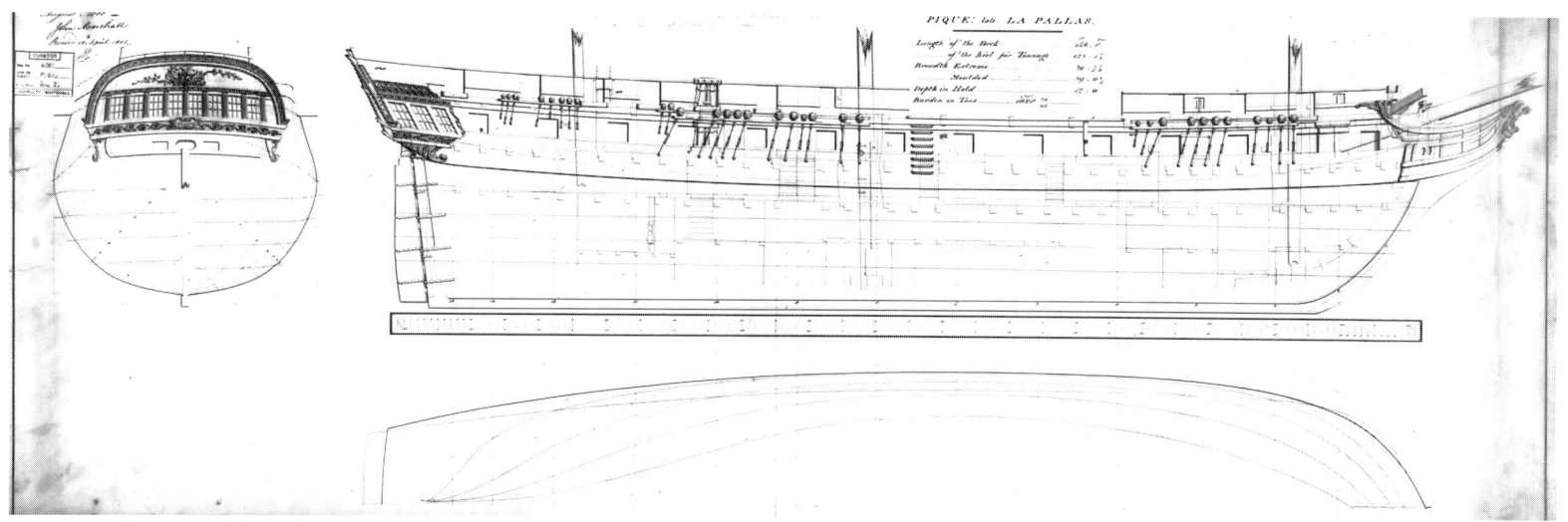

Lines and profile (6080) of the French 18 pounder frigate *Pique* of 1800

Men: 286. Guns: UD 28 x 18, QD 6 x 6 + 4 x 24★, Fc 4 x 6. Taken at the surrender of Toulon 1793; 1795 renamed *Undaunted*; 27.8.1796 wrecked on Morant Keys, West Indies.

PLANS: Lines, profile & decorations/orlop/lower deck/upper deck/quarter deck & forecastle.

***Imperieuse*** 1793-1803 ***Unite*** -1836 (French *L'Impérieuse* 1787 Toulon, Coulomb design) 38.
  Dimensions & tons: 148'6", 124'10" x 39'7" x 12'8". 1040 32/94.
  Men: 315 (by 1807: 284). Guns: UD 28 x 18, QD 8 x 9 + 6 x 32★, Fc 2 x 9 + 2 x 32★ (by 1807: UD 26 x 24, QD 12 x 32★, Fc 2 x 9 + 2 x 32★). Taken off La Spezia 10.10.1793; 1803 renamed *Unite*; hulked for Trinity House; 1805 refitted as frigate; 1836 hulked [as a convict hospital ship at Woolwich; 1858 BU].

PLANS: Lines, profile & decorations/lower deck/upper deck/quarter deck & forecastle.

***Modeste*** 1793-1814 (French *La Modeste* 1786 Toulon ___ ) 36.
  Dimensions & tons: 143'6⅝", 118'3" x 38'8" x 12'1½". 940 35/94.
  Men: 270. Guns: UD 26 x 18, QD 14 x 32★, Fc 2 x 9 + 2 x 32★. Cut out of Genoa 17.10.1793; 1803 hulked for Trinity House; 1805 refitted as frigate; 1814 BU.

PLANS: Lines, profile & decorations.

***San Fiorenzo/Saint Fiorenzo*** 1794-1812 (French *La Minerve* 1782 Toulon, Coulomb design) 34.
  Dimensions & tons: 148'8", 124'4⅛" x 39'6" x 13'3". 1031 86/94.
  Men: 274. Guns: UD 26 x 18, QD 6 x 6 + 6 x 32★, Fc 2 x 9 + 2 x 32★. Damaged by British shore batteries at San Fiorenzo, Corsica, on 18.2.1794; then scuttled; raised by the British; 1810 troopship; 1812 hulked [at Woolwich as a receiving ship; 1818 to Sheerness as a lazaretto; 1837 BU].

PLANS: Lines, profile & decorations/upper deck/quarter deck & forecastle.

***Sybille*** 1794-1831 (French *La Sybille* 1791 Toulon, Sané design) 40/44/48.
  Dimensions & tons: 154'3", 127'4¾" x 40'1½" x 12'4". 1090 91/94.
  Men: 300. Guns: UD 28 x 18, QD 12 x 9 (later 8 x 9 + 6 x 32★), Fc 4 x 9 (later: 4 x 9 + 2 x 32★). Taken by *Romney* in the Mediterranean 17.6.1794; 1831 hulked [as a lazaretto at Dundee; 1833 sold].

NOTE: Major repairs in 1804-1805 and 1815-1816 help to explain her longevity.

PLANS: Lines & profile/lower deck/upper deck/quarter deck & forecastle. 1815: profile.

***Melpomene*** 1794-1815 (French *La Melpomène* 1789 Toulon, Coulomb design) 38.
  Dimensions & tons: 148'2", 123'8¼" x 39'3" x 13'6". 1014.
  Men: 284. Guns: UD 28 x 18, QD 8 x 9 + 6 x 32★, Fc 2 x 9 + 2 x 32★. Taken in the harbour of Calvi, Corsica 10.8.1794; 1810 troopship; 1815 sold.

PLANS: Lines, profile & decorations/lower deck/upper deck/quarter deck & forecastle.

***Revolutionaire*** 1794-1822 (French *La Révolutionnaire* 1794 Le Havre, Forfait design) 38/46.
  Dimensions & tons: 157'2", 131'9⅞" x 40'5½" x 12'6". 1147 68/94.
  Men: 280. Guns: UD 28 x 18, QD 8 x 9 + 6 x 32★, Fc 2 x 9 + 2 x 32★ (later UD 18 pdrs. replaced by 32 pdr. carronades). Taken 21.10.1794 by *Artois* and others; 1822 BU.

PLANS: as fitted: Lines, profile & decorations/lower deck/upper deck/quarter deck & forecastle. 1812: Profile/orlop.

***Minerve*** 1795-1803 (French *La Minerve* 1794 Toulon, Sané design) 38.
  Dimensions & tons: 154'4½", 130'0⅛" x 39'11" x 13'. 1101 79/94.
  Men: 300. Guns: UD 28 x 18, QD 8 x 9 + 6 x 18★, Fc 2 x 9 + 2 x 18★. Taken by *Lowestoft* and *Dido* in the Mediterranean 24.6.1795; 2.7.1803 stranded near Cherbourg and taken by the French *Chiffonne* and *Terrible*; [renamed *La Cannonière* in French service; 1809 sold to French commercial interests, renamed *La Confiance*; 1810 taken by *Valiant* but not taken back into the RN].

PLANS: Lines, profile & decorations/orlop/lower deck/upper deck/quarter deck & forecastle.

***Virginie*** 1796-1811 (French *La Virginie* 1794 Brest, Sané design) 38.
  Dimensions & tons: 151'3¾", 126'3¼" x 39'10" x 12'8". 1065 62/94.
  Men: 315. Guns: UD 28 x 18, QD 12 x 32★, Fc 2 x 9 + 2 x 6 + 2 x 32★. Taken by *Indefatigable*; 1811 hulked [at Plymouth as a receiving ship; 1827 sold; 1828 resold].

PLANS: Lines, profile & decorations/decks. As fitted: profile/decks.

***Tholin*** 1796 & hulked after capture (Dutch *Thulen/Tholen* 1794 or 1783 Zeeland Admiralty) 36.
  Dimensions & tons: 143'9½", 118'11⅛" x 39'11¾" x 13'1½". 1011.
  Men: 265. Guns: UD 24 x 18, QD 10 x 6, Fc 2 x 6. Seized at Plymouth 8.6.1796; never fitted for sea by the RN; soon hulked [at Plymouth as temporary receiving ship; 1811 BU].

***Amelia*** 1796-1816 (French *La Proserpine* 1785 Brest, Sané design) 38.
  Dimensions & tons: 151'4", 126'1⅜" x 39'8⅞" x 12'6½". 1059 35/94.
  Men: 284. Guns: UD 28 x 18, QD 14 x 32★, Fc 2 x 9 + 4 x 32★. Taken by *Dryad* 13.6.1796; 1816 BU.

PLANS: Lines, profile & decorations/decks. As fitted: profile/decks.

***Saldanha*** 1796-1798 (Dutch *Castor* 1780 Rotterdam) 40.
  Dimensions & tons: 147'3", 123'1" x 40'4" x ___ . 1065.
  Men: 284. Guns: UD 26 x 18, QD 10 x 9, Fc 4 x 9. Surrendered at Saldanha Bay 14.8.1796; 1798 hulked [at Plymouth as a receiving ship; 1811 BU].

***Urania/Uranie*** 1797-1807 (French *Le Tartu* ex *L'Uranie* 1788 Lorient, Segondat design) 38.
  Dimensions & tons: 154'5", 128'3½" x 40'1¾" x 13'. 1099 76/94.
  Men: 280. Guns: UD 28 x 18, QD 8 x 9 + 4 x 24★, Fc 2 x 9 + 4 x 24★. Taken by *Polyphemus* 5.1.1797; 1807 sold.

PLANS: Lines, profile & decorations/decks. As fitted: Lines, profile &

decorations/decks.

**Fishguard/Fisgard** 1797-1814 (French *La Résistance* 1793-1794 Paimbeuf, Degay design) 38.
Dimensions & tons: 160'6", 134'2¼" x 40'8½" x 13'3½". 1182 10/94.
Men: 280. Guns: UD 28 x 18 (later all 32★), QD 4 x 9 + 10 x 32★, Fc 4 x 32★. Taken by *San Fiorenzo* and others on 9.3.1797; 1814 sold.

PLANS: Lines, profile & decorations. As fitted: Lines, profile & decorations.

**Seine** 1798-1803 (French *La Seine* 1793 Le Havre, Forfait design) 40.
Dimensions & tons: 156'9", 131'4⅛" x 40'6" x 12'4½". 1145 87/94.
Men: 284. Guns: UD 28 x 18, QD 8 x 9, Fc 4 x 9. Taken by *Jason* and *Pique* 30.6.1798; 21.7.1803 wrecked off the Texel and burnt to prevent capture.

PLANS: Lines, profile & decorations/orlop/lower deck/upper deck/quarter deck & forecastle.

**Loire** 1798-1818 (French *La Loire* 1795 Nantes, Degay design) 40.
Dimensions & tons: 153'6", 128'2⅝" x 40'2" x 12'11¾". 1100 31/94.
Men: 284. Guns: UD 28 x 18, QD 8 x 9 + 4 x 24★, Fc 4 x 9 + 2 x 24★. Taken by *Anson* of Warren's squadron; 1818 BU.

PLANS: Decks. As fitted: Lines, profile & decorations/decks.

**Immortalite** 1798-1806 (French *L'Immortalité* 1795 Lorient) 36.
Dimensions & tons: 145'2", 123'9¾" x 39'2" x 11'5". 1010 25/94.
Men: 264. Guns: UD 26 x 18, QD 8 x 9 + 4 x 24★, Fc 2 x 9 + 2 x 24★. Taken by *Fishguard* 20.10.1798; 1806 BU.

PLANS: As fitted: Lines, profile & decorations/decks.

**Princess Charlotte** 1799-1812 **Andromache** -1828 (French *La Junon* 1782 Toulon, Coulomb design) 38/46.
Dimensions & tons: 148'10", 124'9" x 39'4½" x 12'10". 1028 73/94.
Men: 264. Guns: UD 26 x 18, QD 2 x 9 + 12 x 32★, Fc 2 x 9 + 4 x 32★. Taken by Lord Keith in the Mediterranean 18.6.1799; 1828 BU.

PLANS: taken off/as fitted: Lines, profile & decorations/orlop/lower deck/upper deck/quarter deck & forecastle.

**Pique** 1800-1819 (French *La Pallas* 1779 Saint Malo ____ ) 32/36.
Dimensions & tons: 146'8", 123'1½" x 39'7½" x 12'. 1028 29/94.
Men: 274. Guns: UD 26 x 18, QD 2 x 9 + 10 x 32★, Fc 4 x 9 + 2 x 32★. Taken off the French Coast 6.2.1800; 1819 sold.

PLANS: As fitted: Lines, profile & decorations/decks/profile (incomplete)/ decks (incomplete).

**Vengeance** 1800-1801 (French *La Vengeance* 1793 Paimboeuf, Degay design [40])
____ .
Dimensions & tons: 160', 147'10" x 41'9" x 15'9". 1390.
Men: 284. Guns: UD 28 x 18, QD 8 x 9, Fc 2 x 9. Taken by *Seine* in the Mona Passage 23.4.1800; ?became a prison ship at ____ ?; 1801 stranded and then BU.

**Desiree** 1800-1823 (French *La Désirée* [40, 24 pdr. frigate?] 1796 Dunkirk) 36.
Dimensions & tons: 147'3", 124'2¼" x 39'2¾" x 11'9". 1016 50/94.
Men: 264. Guns: UD 24 x 18, QD 2 x 9 + 12 x 32★, Fc 2 x 9 + 2 x 32★. Cut out of Dunkirk Roads by *Dart* 8.7.1800; 1823 hulked [at Sheerness as a slop ship; 1832 sold].

PLANS: Lines & profile/lower deck/upper deck/quarter deck & forecastle.

**Niobe** 1800-1816 (French *La Diane* 1796 Toulon, Sané design) 38.
Dimensions & tons: 155'10", 129'6⅞" x 40'8½" x 12'7". 1142 15/94.
Men: 284. Guns: UD 28 x 18, QD 2 x 9 + 10 x 32★, Fc 2 x 9 + 2 x 32★. Taken off Malta 25.8.1800; 1814 troopship; 1816 BU.

PLANS: 1803: stern decorations. 1810: Lines & profile/orlop/lower, upper, quarter decks & forecastle. 1814 as troopship: Profile (incomplete)/lower deck/upper deck/quarter deck & forecastle.

**Africaine** 1801-1816 (French *L'Africaine* 1798 Rochefort) 38.
Dimensions & tons: 153'10", 128'0⅜" x 39'11" x 12'6". 1085.
Men: 284. Guns: UD 28 x 18, QD 2 x 9 + 10 x 32★, Fc 2 x 9 + 2 x 32★. Taken by *Phoebe* 19.2.1801; 1809 retaken by French *L'Iphigénie* but immediately recaptured by *Boadicea*); 1816 BU.

PLANS: Lines & profile/orlop/lower deck/upper deck/quarter deck & forecastle.

**Carrere** 1801-1814 (French *Le Carrière* [40, 12 pdr frigate] ex Venetian ____ 1797 Venice) 44/38.
Dimensions & tons: 150'10", 122'3½" x 39'5½" x 12'9". 1013.
Men: 284. Guns: UD 28 x 18, QD 2 x 9 + 10 x 32★, Fc 2 x 9 + 2 x 32★. Taken off Elba by *Pomone* 12.8.1801; never fitted for sea by the RN; 1814 sold.

## 12 pounder Frigates (Fifth Rates)

**Oiseau** 1793-1816 (French *La Cléopatre* 1781 Saint Malo, Sané design) 36/38.
Dimensions & tons: 144'7¾", 120'8⅞" x 37'8½" x 12'3". 913 16/94.
Men: 255. Guns: UD 26 x 12, QD 8 x 24★, Fc 2 x 6 + 2 x 24★. Taken by *Nymphe;* 1816 sold.

PLANS: Lines, profile & decorations/orlop/lower deck/upper deck/quarter deck & forecastle. 1800: Lines.

**Lutine** 1793-1799 (French *La Lutine* 1779 Toulon, Joseph-Coulomb design) 36/38.
Dimensions & tons: 143'3", 118'6" x 38'10" x 12'1½". 950 50/94.
Men: 240. Guns: UD 26 x 12, QD 4 x 6 + 4 x 24★, Fc 2 x 6 + 2 x 24★. Taken at the surrender of Toulon 1793; 9.10.1799 wrecked off Vlieland [the attempts to salvage her cargo of bullion have been many and ultimately successful].

NOTE: Her salvaged bell hangs at Lloyds of London.

PLANS: Lines, profile & decorations/orlop/lower deck/upper deck/quarter deck & forecastle.

**Topaze** 1793-1814 (French *La Topaze* 1789 Toulon, J M B Coulomb design) 38.

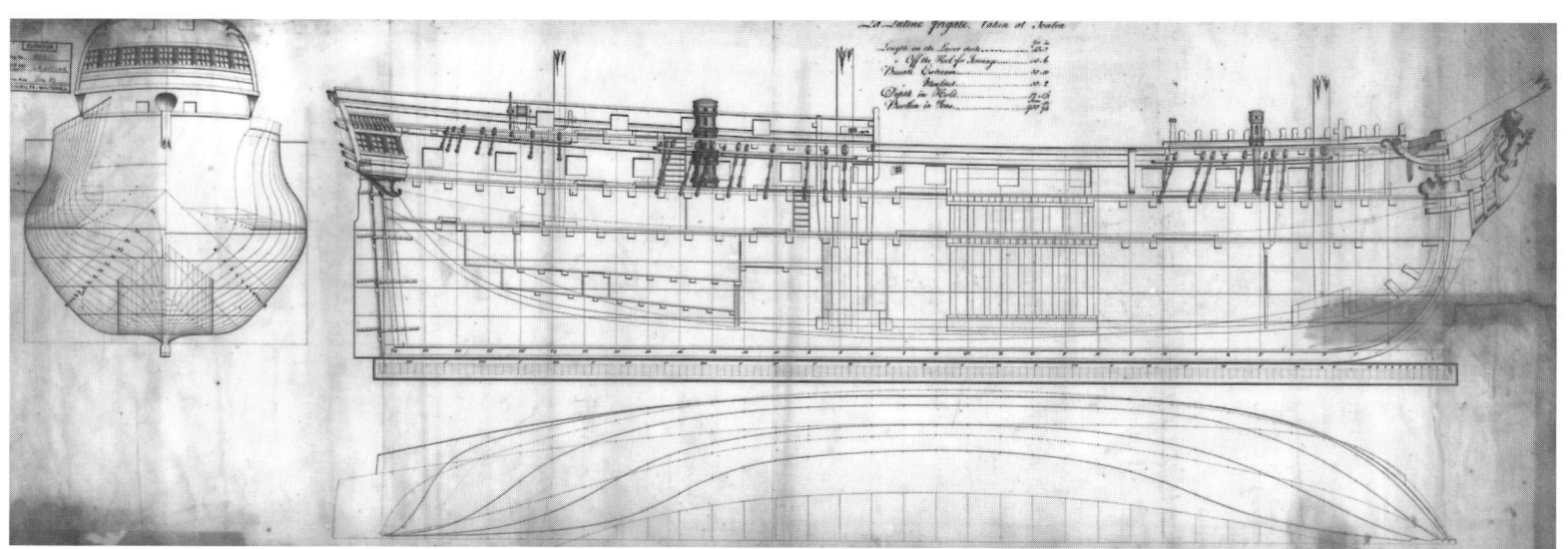

Lines and profile (6081) of the French 12 pounder frigate *Lutine* of 1793

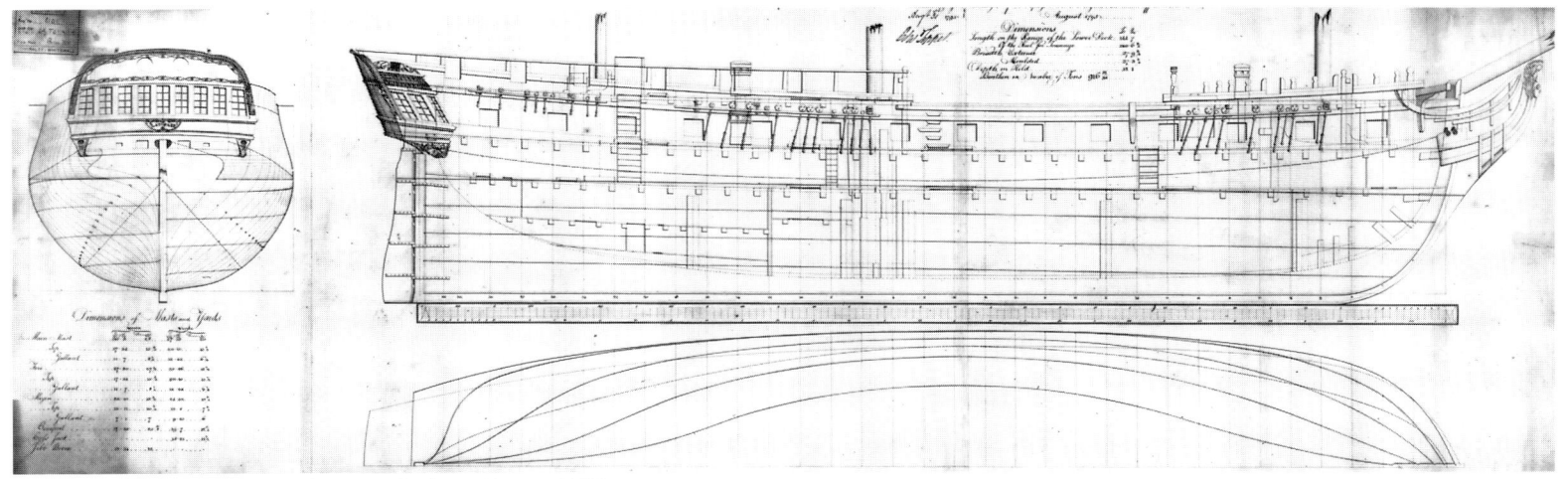

# THE FRENCH REVOLUTIONARY WAR 1793-1801

Lines and profile (6101) of the French 12 pounder frigate *Topaze* of 1793

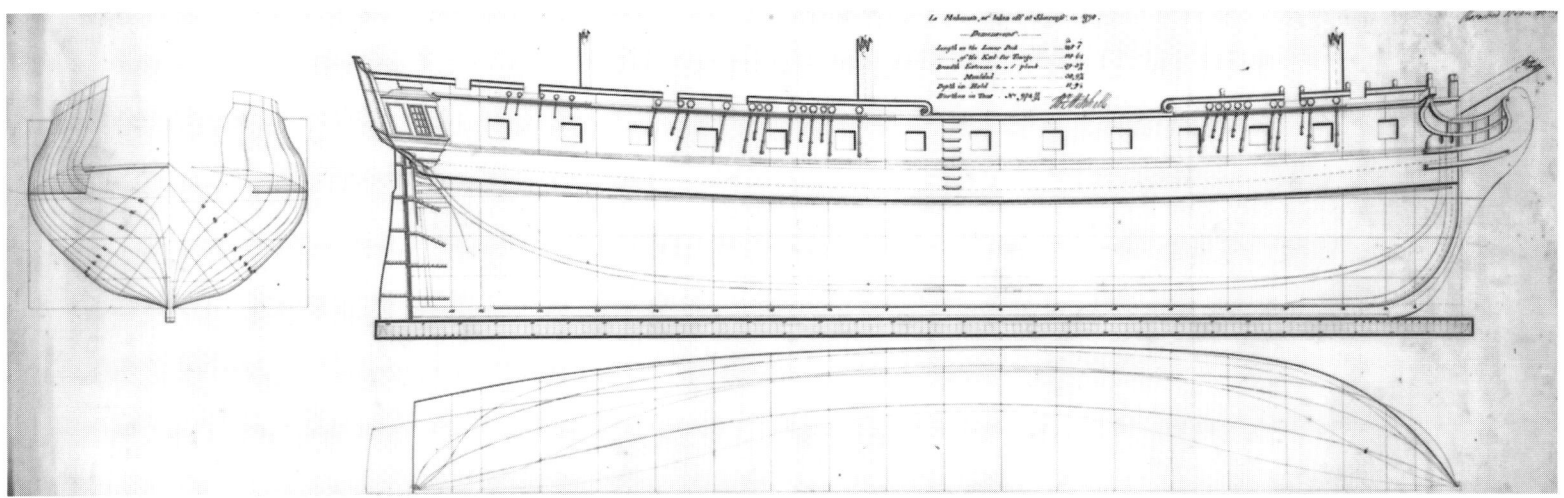

Lines (6207) of the Spanish 12 pounder frigate *Mahonesa* captured in 1796 and 'taken off' in 1798

Dimensions & tons: 144'7", 120'6¾" x 37'9¾" x 12'1". 916 82/94.
Men: 280/274. Guns: UD 28 x 12, QD 12 x 32★, Fc 2 x 12 + 4 x 32★. Taken at the surrender of Toulon 1793; 1814 sold.
PLANS: Lines, profile & decorations/lower deck/upper deck/quarter deck & forecastle. 1800: figurehead.

***Reunion*** 1793-1796 (French *La Réunion* 1785-6 Toulon, J M B Coulomb design) 36.
Dimensions & tons: 144', 118'4⅜" x 38'10½" x 12'1". 951 43/94.
Men: 255. Guns: UD 26 x 12, QD 8 x 6, Fc 2 x 6. Taken 20.10.1793 by *Crescent*; 7.12.1796 wrecked on the South Sand.
PLANS: Lines, profile & decorations/orlop/lower deck/upper deck/quarter deck & forecastle.

***Engageante*** 1794 & hulked after capture (French *L'Engageante* 1766 Toulon, Estienne design) 38.
Dimensions & tons: 139'7½", 116'6½" x 38'9" x 12'3". 931.
Men: ____ (as hospital ship: 70). Guns: ____. (as hospital ship: 8 x 4) Taken by *Concorde* 23.4.1794 and fitted as a hospital ship [at Portsmouth; 1811 BU].

***Espion*** 1794-1799 (French *L'Atalante* 1784 Rochefort, Chevillard cadet design) 32.
Dimensions & tons: 148'9", 120'6" x 39'2" x 11'10½". 983 20/94.
Men: 215 later 121. Guns: UD 26 x 12, QD 4 x 6, Fc 2 x 6. Taken by *Swiftsure* 7.5.1794; 1798 floating battery; 1799 troopship; 16.11.1799 wrecked on the Goodwins.
PLANS: Lines.

***Pique*** 1795-1798 (French *La Pique* ex *La Fleur de Lys* 1785 Rochefort, Haran design) 34.
Dimensions & tons: 144'1½", 119'5¼" x 37'9¼" x 11'8". 906 21/94.

Men: 274. Guns: UD 26 x 12, QD 6 x 6 + 4 x 24★, Fc 2 x 6 + 2 x 24★. Taken in the Windward Islands by *Blanche* 6.1.1795; 2.7.1798 grounded on the French coast whilst engaging *La Seine*, could not be got off and was destroyed.
PLANS: Lines & profile/lower deck/upper deck/quarter deck & forecastle.

***Gloire*** 1795-1802 (French *La Gloire* 1778 Saint Malo, Guignace design) 36.
Dimensions & tons: 141'2¼", 116'8¼" x 37'7¼" x 11'11½". 877.
Men: ____. Guns: UD 26 x 12, 'between decks' 10 x 24★, QD 8 x 6 + 4 x 24★, Fc 2 x 6 + 2 x 24★. Taken by *Astrea* 10.4.1795; never fitted for sea by the RN; 1802 sold.

***Gentille*** 1795 & hulked after capture (French *La Gentille* 1778 Saint Malo, Guignace design) 40.
Dimensions & tons: ____, ____ x ____ x ____. ____.
Men: ____. Guns: 40 x ____. Taken by *Hannibal* 11.4.1795; hulked on arrival [at Portsmouth as a receiving ship; 1802 sold].

***Prevoyante*** 1795-1819 (French flute *La Prévoyante* 1793 Bayonne) 36.
Dimensions & tons: 143', 121'11½" x 35'2½" x 13'4". 804.
Men: 284 (as storeship: 90). Guns: UD 30 x 12, QD 4 x 9 + 8 x 18★, Fc 2 x 9 (short) + 2 x 18★ (as storeship: UD 14 x 6, QD 4 x 4). Taken by *Thetis* and *Hussar* 17.5.1795; 1801 storeship; 1819 sold.

***Alliance*** 1795-1802 (Dutch *Alliante* ____) 36.
Dimensions & tons: 130'9", 107'11¼" x 34'10" x 12'1". 696 81/94.
Men: ____. (as storeship: 121). Guns: ____ (as storeship: UD 14 x 6). Taken by *Stag* and others 22.8.1795; fitted as a storeship; 1802 sold.
PLANS: As storeship: Lines, profile & decorations/orlop/lower deck/upper deck/quarter deck & forecastle.

***Eurus*** 1796-1806 (Dutch *Zefir* 1786 Amsterdam Admiralty) 32.
Dimensions & tons: 126'7½, 107'8" x 35' x 12'6½". 702.
Men: 215 (as storeship: 58). Guns: UD 26 x 12, QD 4 x 6,

Fc 2 x 6 (as storeship: UD 20 x 24★, QD 2 x 24★). Taken in port in the Forth 6.3.1796; 1799 troopship; 1803 storeship; 1806 hulked [at Haulbowline as a storeship; 1834 BU].

**Unite** 1796-1802 (French *L'Unité* ex *La Gracieuse* 1785-7 Rochefort, Chevillard cadet design) 32.
Dimensions & tons: 142'5½", 118'5⅛" x 37'8" x 11'. 873 71/94.
Men: 254. Guns: UD 26 x 12, QD 4 x 6, Fc 2 x 6 + 4 x 32★. Taken by *Revolutionaire* 13.4.1796; 1802 sold.

PLANS: Lines, profile & decorations. As fitted: profile/decks.

**Janus** 1796-1798 (Dutch *Argo* 1791 or 1789 Amsterdam Admiralty) 32.
Dimensions & tons: 131', 108' x 35' x 11'10¼". 704.
Men: 215. Guns: UD 26 x 12, QD 4 x 6, Fc 2 x 6. Taken 12.5.1796 by *Phoenix*; 1798 hulked [at Deptford as a receiving ship 1811 sold].

**Tribune** 1796-1797 (French *La Tribune* ex *La Charente-Inférieure* 1794 Rochefort) 34.
Dimensions & tons: 143'7½", 119'0⅝" x 38'0½" x 11'6½" 916 34/94.
Men: 244. Guns: UD 26 x 12, QD 6 x 6, Fc 2 x 6. Taken 7.6.1796 by *Unicorn*; 23.11.1797 wrecked at the entrance to Halifax Harbour; [the wreck has been found and dived on].

PLANS: Lines & profile/orlop/lower deck/upper deck/quarter deck & forecastle.

**Proselyte** 1796-1801 (Dutch *Jason* 1770 Rotterdam Admiralty) 32.
Dimensions & tons: 133'1", 110'8" x 35'8" x 12'. 748 35/94.
Men: 244. Guns: UD 26 x 12, QD 4 x 6, Fc 2 x 6. Brought into Greenock by her crew 8.6.1796; 4.9.1801 wrecked on a sunken rock off Saint Martin, West Indies; [reported located by divers].

PLANS: Lines.

**Renommee** 1796-1810 (French *La Renommée* ex *La Républicaine Française* 1793 ____ ) 44/38.
Dimensions & tons: 140'6½", 119'4⅝" x 38'1½" x 11'7½" . 823.
Men: 264. Guns: UD 26 x 12, QD 12 x 32★, Fc 2 x 9 + 2 x 32★. Taken in the West Indies by *Alfred* 20.7.1796; 1800 troopship; 1810 BU. [Winfield suggests the tonnage is a clerical error/miscalculation and should be 923 given the dimensions.]

**Braave** 1796-1811 (Dutch *Braave* 1795 ex *Prinses Frederika Louise Wilhelmina* 1789 Rotterdam Admiralty) 38.
Dimensions & tons: ____ , ____ x ____ x ____ . ____ .
Men: ____ . Guns: UD 26 x 12, QD & Fc ____ . Taken at Saldanha Bay 17.8.1796; 1803 arrived in Britain and laid up until; 1811 hulked [at Portsmouth as a receiving ship; 1825 sold].

**Mahonesa** 1796-1798 (Spanish *Mahonesa* [34] 1789 Mahon) 36.
Dimensions & tons: 145'1", 118'6¼" x 39'3¾" x 11'9½". 974 29/94.
Men: 264. Guns: UD 26 x 12, QD 8 x 6, Fc 2 x 6. Taken by *Terpsichore* 13.10.1796; 1798 BU.

PLANS: Lines.

**Hamadryad** 1797-1798 (Spanish *Ninfa* 1794-5 Port Mahon) 36.
Dimensions & tons: ____ , ____ x ____ x ____ . 890.
Men: 264. Guns: UD 26 x 12, QD 8 x 6, Fc 2 x 6 + 2 x 32★. Taken by *Irresistible* 26.4.1797; 24.12.1797 lost off the Portuguese coast.

**Nereide** 1797-1810 (French *La Nereide* 1779 Saint Malo, Sané design?) 36.
Dimensions & tons: 142'5", 118'10⅝" x 37'6¾" x 11'9½". 892 22/94.
Men: 254. Guns: UD 26 x 12, QD 8 x 6 + 4 x 24 (Later just 12 x 24★), Fc 2 x 6 + 2 x 24★ (later just 4 x 24★). Taken by *Phoebe* 22.12.1797; 28.8.1810 run aground and taken by the French at Mauritius; 6.12.1810 retaken at the capture of Mauritius but not put back into service; 1816 sold.

PLANS: Lines, profile & decorations/decks. As fitted: Lines, profile & decorations/decks.

**Sensible** 1798-1802 (French *Le Sensible* 1786 Toulon, J M B Coulomb design) 36.
Dimensions & tons: 146'3", 118'11" x 38'8" x 11'10". 946.
Men: originally 264 proposed? later 115. Guns: 'no establishment of guns'. Taken by *Seahorse* in the Mediterranean 27.6.1798; 1799 troopship; 2.3.1802 wrecked off Ceylon near Trincomallee.

**Santa Dorothea** 1798-1814 (Spanish *Santa Dorotea* 1775 Ferrol) 34.
Dimensions & tons: 146'9", 118'4½" x 39' x 12'. 958.
Men: 240. Guns: UD 26 x 12, QD 6 x 6 + 6 x 24★, Fc 2 x 6 + 2 x 24★. Taken by *Lion* off Alicante 13.7.1798; 1802 laid up until; 1814 BU.

**Decade** 1798-1811 (French *La Décade Française* ex *La Macreuse* 1794 Bordeaux) 36.
Dimensions & tons: 143'8½", 119'1¼" x 38' x 11'8". 914 77/94.
Men: 255. Guns: UD 26 x 12, QD 2 x 6 + 10 x 24★, Fc 4 x 6 + 2 x 24★. Taken by *Magnanime* and *Naiad* 24.8.1798; 1811 sold.

PLANS: As fitted: lines & profile/decks.

**Ambuscade** 1798-1802 **Seine** -1813 (French *L'Embuscade* 1789 Rochefort, Vial du Clairbois design) 36.
Dimensions & tons: 142'4", 120'0⅜" x 37'8" x 11'6½". 905 78/94.
Men: 254. Guns: UD 26 x 12, QD 8 x 6 + 6 x 32★, Fc 2 x 6 + 2 x 32★. Taken by Warren's Squadron 12.10.1798; 1813 BU.

PLANS: As fitted: Lines, profile & decorations.

**Resolue** 1798 & hulked after capture (French *La Résolue* 1777 Saint Malo, Guignace design) 40.
Dimensions & tons: 140'2", 116'3" x 37'8" x 11'9½". 877.
Men: ____ . Guns: ____ . Taken by *Melampus* 13.10.1798 and hulked soon after her arrival in Britain [at Plymouth; 1811 BU].

**Wilhelmina** 1798-1813 (Dutch *Die Furie* 1795 ex *Wilhelmina* 1787 Vlissingen) 32.
Dimensions & tons: 133', 109'1" x 37'9" x 12'4". 827.
Men: 244 (121 as troopship). Guns: UD 26 x 12, QD 4 x 6 + 4 x 24★, Fc 2 x 6 + 2 x 24★ (18 guns as troopship). Taken by *Sirius* 24.10.1798; 1800 troopship; 1813 sold.

**Helder** (ex *Ambuscade*) 1799-1807 (Dutch *Helder* or *Embuscade* 1795 Rotterdam) 32.
Dimensions & tons: 127', 103'4¾" x 37'5" x 13'0½". 770.
Men: 244 (155 as floating battery). Guns: UD 26 x 12, QD 4 x 6 + 4 x 24★, Fc 2 x 6 + 2 x 24★ (1803: UD 24 x 24★, QD 8 x 18★). Taken 1799; 1803 floating battery (?); 1807 sold.

**Santa Teresa** 1799-1802 (Spanish *Santa Teresa* 1787 Ferrol) 30.
Dimensions & tons: 143'10", 120'7" x 38'6" x ____ . 952.
Men: 250. Guns: UD 26 x 12, QD 2 x 6 + 10 x 32★, Fc 2 x 6 + 2 x 32★. Taken near Majorca by *Argo* 6.2.1799; 1802 sold.

**Alceste** 1799-1802 (French *L'Alceste* L 28.10.1780 Toulon, Joseph Coulomb design) 32.
Dimensions & tons: 145', 121'8" x 39' x 11'6". 932.
Men: 215 (as floating battery: 96). Guns: ____ . Taken by Keith's squadron in the Mediterranean 18.6.1799; 1801 floating battery, registered as sloop; 1802 sold.

NOTE: Originally surrendered to the British at the handing over of Toulon in 1793, then handed to the Sardinians; 1794 retaken by the French.

**Heldin** (ex *Helder*, ex? *Alarm*?) 1799-1802 (Dutch *Heldin* 1796 Amsterdam Dyd) 28.
Dimensions & tons: 122'1", 102'3⅝" x 36'2¼" x 11'4¾". 635 91/94.
Men: 207. Guns: UD 24 x 12, QD 4 x 6. Taken in the Helder 1799; 1802 sold.

PLANS: Lines & profile & decoration/lower deck/upper deck/quarter deck & forecastle.

**Florentina** 1800-1803 (Spanish *Santa Florentina* 1786 Carthagena) 36.
Dimensions & tons: 146'8", 119'6" x 37'8" x 10'6". 902.
Men: ____ . Guns: 'by the old establishment' UD 26 x 12, QD 6 x 6, Fc 2 x 6. Taken 7.4.1800 by *Leviathan* and *Emerald* off Cadiz; 1803 sold.

**Carmen** 1800-1802 (Spanish *Carmen/Nuestra Senora del Carmen* 1770 Ferrol) 36.
Dimensions & tons: 147'2", 119'9" x 37'9" x 11'. 908.
Men: ____ . Guns: 26 x 12, QD 6 x 6, Fc 2 x 6. Taken 7.4.1800 by *Leviathan* and *Emerald* off Cadiz; 1802 sold.

**Dedaigneuse** 1801-1810 (French *La Dédaigneuse* 1797 Bayonne, Haran design) 32.
Dimensions & tons: 143'10½", 119'9½" x 37'6¼" x 11'9". 897 3/94.
Men: 215. Guns: UD 26 x 12, QD 2 x 12 + 10 x 24★, Fc 2 x 6 + 2 x 24★. Taken by *Oiseau* and others 29.1.1801 (?); 1810 hulked [at Deptford as a receiving ship; 1823 sold].

PLANS: Lines & profile & decoration/orlop & lower deck/upper deck/quarter deck & forecastle.

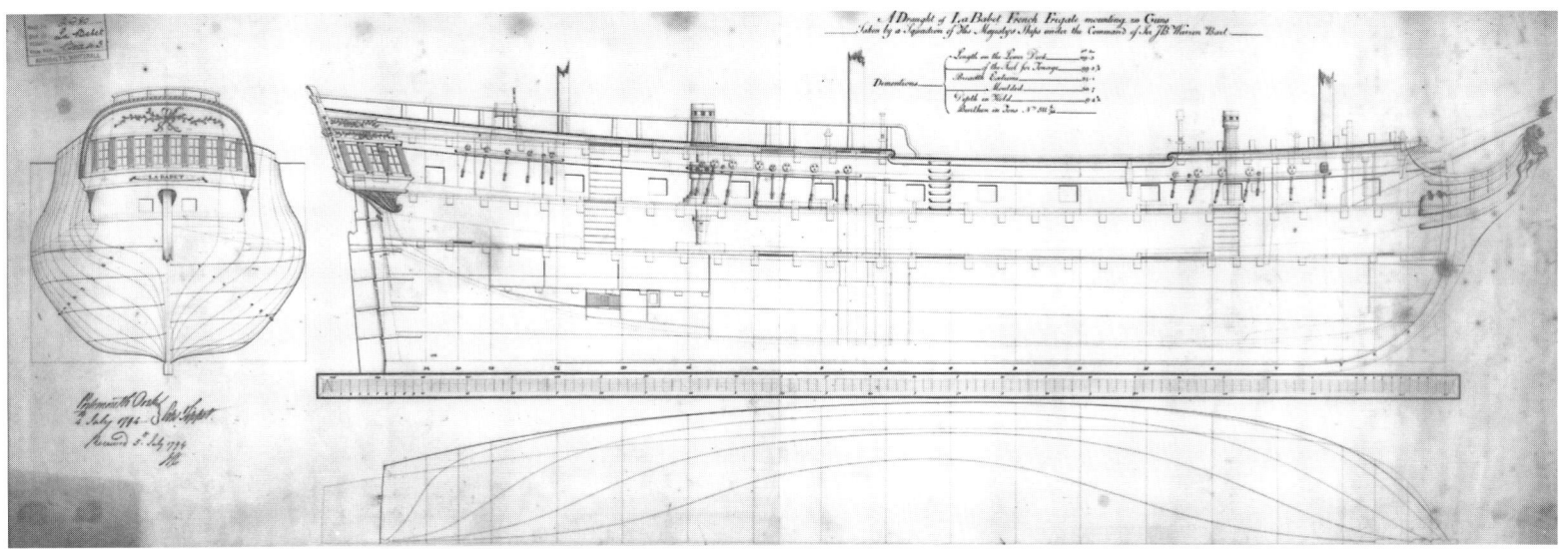

Lines and profile (2980) of the French 9 pounder frigate *Babet* of 1794

***Chiffone*** 1801-1814 (French *La Chiffone* 1795 Nantes) 36.
  Dimensions & tons: 144'1", 120'6¼" x 37'11" x 12'. 921 60/94.
  Men: 264. Guns: UD 26 x 12, QD 2 x 9 + 10 x 32★, Fc 2 x 9 + 2 x 32★.
  Taken by *Sybille* in the Seychelles 20.8.1801; 1814 sold.
PLANS: Lines & profile & decoration/orlop/lower deck/upper deck/quarter deck & forecastle.

***Alexandria*** 1801-1804 (French *La Régenerée* 1793 \_\_\_\_ ) 36/38.
  Dimensions & tons: 144'3", 119'8½" x 37'6" x 11'8". 895 20/94.
  Men: 200. Guns: UD 28 x 12, QD 8 x 6, Fc 2 x 6. Taken at the surrender of Alexandria 1801; not fitted for sea after arrival in Britain; 1804 BU.
PLANS: 1804: Lines & profile.

## 9 pounder Frigates

(Sixth Rates except where stated otherwise)

***Prompte*** 1793-1813 (French *La Prompte* corvette [20] 1792 Le Havre) 28.
  Dimensions & tons: 118'9", 99'3¾" x 31'3¾" x 9'5½". 509 1/94.
  Men: 165. Guns: UD 20 x 9 (later changed to 6s), QD 6 x 12★, Fc 2 x 12★. Taken by *Phaeton* 18.5.1793; 1801 laid up (?) until; 1813 BU.
PLANS: Lines & profile/orlop/lower deck/upper deck/quarter deck & forecastle.

***Undaunted*** 1794-1795 (French 'flûte' *La Bien Venue* ex *La Royaliste* 1788 \_\_\_\_ ) 28.
  Dimensions & tons: \_\_\_\_ , \_\_\_\_ x \_\_\_\_ x \_\_\_\_ . \_\_\_\_ .
  Men: 195. Guns: \_\_\_\_ . Taken by Jervis' squadron at Martinique 17.3.1794; 1795 sold.

***Babet*** 1794-1801 (French *La Babet* 1793 \_\_\_\_ ) 20.
  Dimensions & tons: 119'3", 99'5⅜" x 31'1" x 9'4½". 511 1/94.
  Men: 165. Guns: UD 20 x 9, QD 6 x 12★, Fc 2 x 12★. Taken by *Flora* and others 23.4.1794; just after 25.10.1801 supposed foundered on passage from Martinique to Jamaica.
PLANS: Lines & profile & decoration/lower deck/upper deck/quarter deck & forecastle.

***Matilda*** 1794-\_\_\_\_ ? (French *Le Jacobin* 1794 \_\_\_\_ ) Sixth Rate 28.
  Dimensions & tons: 129'2", 105'7" x 32'10" x 9'10". 573.
  Men: \_\_\_\_ . Guns: \_\_\_\_ . Taken by *Ganges* and *Montague* in the West Indies 30.10.1794; never fitted for sea after her arrival in Britain; ?1799 ? hulked [at Woolwich as hospital ship; 1810 BU].

***Tourterelle*** 1795-1816? (French *La Fidèle*/*La Tourterelle* 1794 \_\_\_\_ ) 28.
  Dimensions & tons: 125'11", 108'3¼" x 31'9⅛" x 9'10". 580 90/94.
  Men: \_\_\_\_ . Guns: UD 24 x 9, QD 4 x 6 + 4 x 24★, Fc 2 x 24★. Taken by *Lively* 13.3.1795; by 1816 breakwater.
PLANS: Lines & profile & decoration/decks. As fitted: Profile/decks.

***Surprize*** 1796-1802 (French *L'Unité* 1794 Le Havre) Fifth Rate 34.
  Dimensions & tons: 126', 108'6⅛" x 31'8" x 10'0½" 578 73/94.
  Men: 200. Guns: UD 24 x 9, QD 8 x 4 + 4 x 12★, Fc 2 x 4 + 2 x 12★ (actually appears to have been armed with: UD 24 x 32★, QD 8 x 18★, Fc 2 x 6). Taken by *Inconstant* in the Mediterranean 20.4.1796; 1802 sold.
PLANS: Lines & profile & decoration/quarter deck & forecastle.

***Vindictive*** 1796-1798 (Dutch *Bellona* 1794 Rotterdam) 28.
  Dimensions & tons: 112', 92'5" x 32'1" x 10'6". 502.
  Men: 121. Guns: UD 20 x 9, QD 2 x 6, Fc 2 x 6. Taken in Saldanha Bay 17.8.1796; 1798 hulked [at Sheerness as Port Admiral's ship; 1816 sold].

***Amaranthe*** 1799-1804 (Dutch *Venus* 1778 Amsterdam) 28.
  Dimensions & tons: 112'6", 92'10½" x 31'9" x 10'6". 498.
  Men: 105. Guns: UD 24 x 9, QD 2 x 6, Fc 2 x 6. Taken in the Texel 1799; 1804 BU.

## Corvettes (Sixth Rates)

***Laurel*** 1795-1797 (French *Le Jean Bart* 1795 \_\_\_\_ ) 22.
  Dimensions & tons: 107', 82'6½" x 30' x 8'7". 423 81/94.
  Men: 160. Guns: 22 x 9. Taken by Warren's squadron 15.4.1795; 1797 sold.
PLANS: Lines & profile & decoration/decks. As fitted: profile/decks.

***Raison*** 1795-1801 (French flute *La Raison* ex *Le Necker* 1794? \_\_\_\_ ) 26.
  Dimensions & tons: \_\_\_\_ , \_\_\_\_ x \_\_\_\_ x \_\_\_\_ . 472.
  Men: 195. Guns: UD 20 x 9 + 2 x 18★ (in bow ports), QD 4 x 6 + 2 x 12★, Fc 2 x 6. Taken by *Thetis* and *Hussar* off the Chesapeake 17.5.1795; 1801 hulked [at Sheerness as receiving ship; 1802 sold].

***Perdrix*** 1795-1799 (French gabare *La Perdrix* 1784 \_\_\_\_ ) 24/28.
  Dimensions & tons: 118'5½", 98'7⅜" x 31'4½" x 9'. 516 31/94.
  Men: 155. Guns: \_\_\_\_ . Taken by *Vanguard* off Antigua 06.1795; 1799 BU.
PLANS: Lines & profile & decoration/lower deck/upper deck.

***Superbe*** 1795 & hulked after capture (French *Le Superbe* \_\_\_\_ ) 22.
  Dimensions & tons: 121', 95' x 35' x 12'6". 619.
  Men: \_\_\_\_ . Guns: \_\_\_\_ . Taken by *Vanguard* 10.10.1795 in the Windward Islands; 1796 hulked [as a prison ship at Martinique; 1798 sold].

***Jamaica*** 1796-1814 (French *La Perçante* 1795 Bayonne) 26/24.
  Dimensions & tons: 119'8½", 100'6⅞" x 31' x 8'5½". 514 8/94.
  Men: \_\_\_\_ . Guns: 20 x 9 + 6 x 2 (brass). Taken by *Intrepid* off Port Plata 02.1796; 1814 sold.
PLANS: Lines & profile/decks. 1800: profile/decks.

***Daphne*** 1796-1798 ***Laurel*** -1798 (Dutch *Sirene* \_\_\_\_ ) 24.
  Dimensions & tons: 117'10½", 97'1" x 33'4½" x 11'10". 574 45/94.

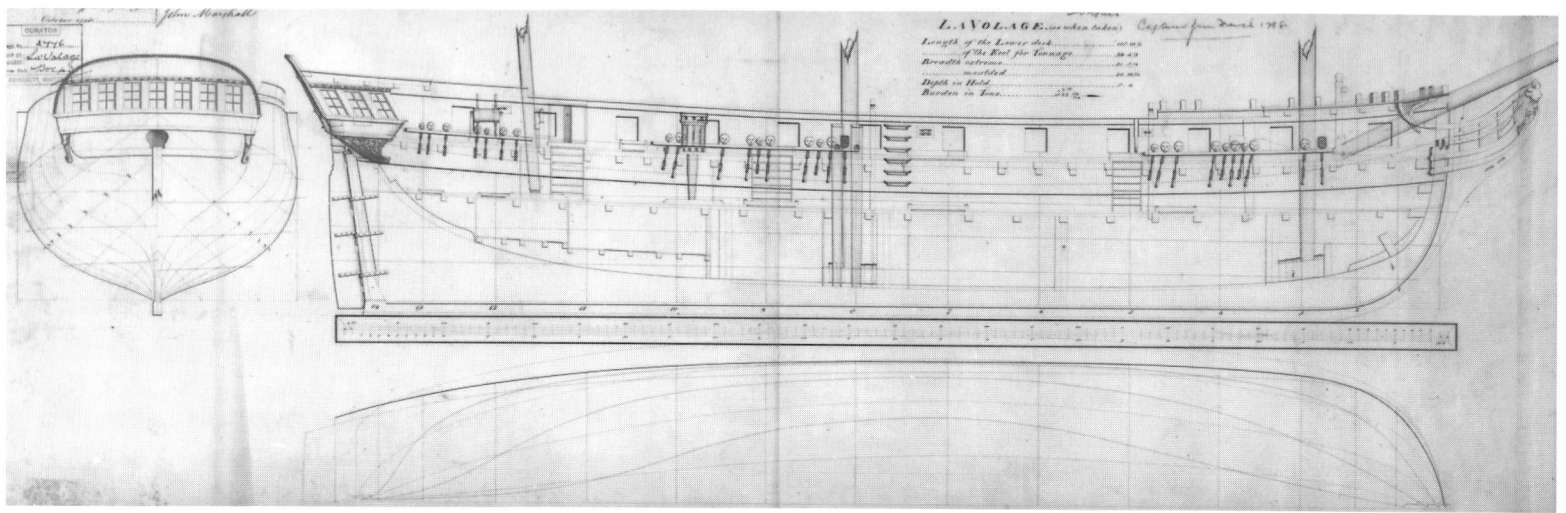

Lines and profile 'as taken' (2776) of the French privateer corvette *Volage* of 1798

Men: 155. Guns: UD 22 x 9, QD 2 x 6 + 6 x 18★, Fc 2 x 18★. Surrendered at Saldanha Bay 1796; 1798 renamed *Laurel;* 1798 hulked [as a convict ship at Portsmouth; 1821 sold].

PLANS: 1798: Lines & profile & decoration/lower deck/upper deck/quarter deck & forecastle.

***Utile*** 1796-1798 (French 'gabare' *L'Utile* 1785 ____ ) 24.
  Dimensions & tons: ____ , ____ x ____ x ____ . ____ .
  Men: ____ . Guns: ____ . Taken by *Southampton* near Toulon 10.6.1796; never fitted for sea after arrival in Britain; 1798 sold.

***Cormorant*** 1796-1800 (French *L'Etna* 1794 Honfleur) 20.
  Dimensions & tons: 119'4", 98'2½" x 32'10½" x 14'9½". 564 40/94.
  Men: 155. Guns: 18 x 9 + 2 x 32★ (later 2 x 9 + 18 x 32★). Taken by *Melampus* 13.11.1796; 20.5.1800 wrecked on the coast of Egypt.

PLANS: Lines & profile & decoration/lower deck/upper deck.

***Constance*** 1797-1806 (French *La Constance* 1794 Le Havre) 22.
  Dimensions & tons: 121'8", 103'4¾" x 31'1¼" x 9'8". 532 9/94.
  Men: 170. Guns: 22 x 9. Taken by *San Fiorenzo* and *Nymphe* 9.3.1797; 14.8.(or 12.10.)1806 grounded, taken by the French and wrecked near Cape Fréhel, France.

PLANS: Lines & profile & decoration/decks. As fitted: Lines & profile & decoration/decks.

***Volage*** 1798-1804 (French privateer *Le Volage* 1797 ____ ) 22/20.
  Dimensions & tons: 118'10½", 99'5½" x 31'5¼" x 8'4". 522 79/94.
  Men: 155. Guns: 22 x 32★. Taken by *Melampus* 23.1.1798; 1804 BU.

PLANS: Lines & profile & decoration/decks. As fitted: decks.

***Coureur*** 1798-1801 (French privateer *Le Coureur* ____ ____ ) 20.
  Dimensions & tons: 110'11½", 89'3⅝" x 27'4½" x 13'6". 355.
  Men: 135. Guns: 20 x 18★. Taken by *Jason* 23.2.1798; never fitted for sea by the RN; 1801 sold.

***Arab*** 1798-1810 (French privateer *Le Brave* ____ ) 22.
  Dimensions & tons: 109'11", 88'10" x 32'8½" x 14'3". 505 48/94.
  Men: 155. Guns: 20 x 9 + 2 x 32★. Taken by *Phoenix* 24.4.1798; 1810 sold.

PLANS: Lines & profile & decoration/decks. As fitted: Lines & profile & decoration/decks.

***Danae*** 1798-1800 (French *La Vaillante* 1795 Bayonne) 20.
  Dimensions & tons: 119'2", 99'7¼" x 30'11¼" x 8'11". 507 8/94.
  Men: 155. Guns: UD 20 x 32★, QD 6 x 12, Fc 6 x 12 (QD & Fc guns probably, but not certainly, carronades). Taken by *Indefatigable* 7.8.1798; 17.3.1800 mutinied and taken into Brest by the mutineers; [handed to the French; who sold her out of service in 1801].

PLANS: Lines & profile & decoration/decks. As fitted: Lines & profile & decoration/decks.

***Waaksamheidt*** 1798-1802 (Dutch *Waaksamheid* 1769 ____ ) 24.
  Dimensions & tons: 114'6", 94'8⅝" x 31'7½" x 10'6". 504.
  Men: 155. Guns: UD 20 x 9, QD 2 x 6 + 4 x 18★, Fc 2 x 6 + 2 x 18★. Taken by *Sirius* 24.10.1798; 1802 sold.

***Poulette*** 1799-1814 (French privateer *Le Foudroyant* ____ ) 20/16.
  Dimensions & tons: 120'8", 100'4½" x 31' x 13'4". 513.
  Men: 155. Guns: 2 x 9 + 16 x 18★ (later just 16 x 18★). Taken by *Phoenix* 23.1.1799; 1805 laid up until; 1814 sold.

***Determinee*** 1799-1803 (French privateer *La Determinée* ____ ) 20.
  Dimensions & tons: 124'5", 103'5" x 31'5⅝" x 15'4". 544 67/94.
  Men: 145. Guns: UD 18 x 32★, Fc 2 x 9. Taken by *Revolutionaire* 29.6.1799; 26.3.1803 wrecked on Jersey.

PLANS: Lines & profile & decoration/orlop/upper deck.

***Braak*** 1799-1802 (Dutch *Minerva* ____ Zeeland) 24.
  Dimensions & tons: 116'6½", 95'8⅛" x 34'8½" x 10'6". 613 6/94.
  Men: 155. Guns: UD 22 x 32★, QD 2 x 6. Taken in the Texel 1799; 1802 sold.

PLANS: Lines & profile & decoration/lower deck/upper deck/quarter deck & forecastle.

***Bourdelois*** 1799-1804 (French privateer *Le Bordelais* 1799 Nantes) 24.
  Dimensions & tons: 138'6", 116'6" x 31'9" x 15'1". 624 63/94.
  Men: 195. Guns: 22 x 32★ + 2 x 9. Taken by *Revolutionaire* 11.10.1799; 1804 BU.

PLANS: As fitted: Lines & profile & decorations (2 copies)/decks.

***Heureux*** 1799 1806 (French privateer *L'Heureuse* ____ ) 22.
  Dimensions & tons: 127'8½", 102'9" x 33'1" x 16'2". 598 18/94.
  Men: 155. Guns: UD 20 x 32★, Fc 2 x 9. Taken by *Stag* 19.10.1799; ____ 1806 foundered in the Atlantic.

PLANS: As fitted: Lines & profile & decoration/orlop & lower deck/upper deck.

***Garland*** 1800-1803 (French privateer *Le Mars* 1799 Bordeaux) sloop/Sixth Rate 24.
  Dimensions & tons: 124'4", 100'4⅝" x 31'5¾" x 14'1". 529 11/94.
  Men: 135. Guns: UD 2 x 9 + 22 x 32★. Taken by *Amethyst* 1.4.1800; 16.11.1803 wrecked off Cape Francois.

PLANS: As fitted: Lines, profile & decoration/lower deck/upper deck.

# Ship Sloops

***Eclair*** 1793-1797 (French 'barque'/corvette *L'Edair* 1770 Toulon) 18.
  Dimensions & tons: 105'1", 85'3⅜" x 29'8" x 12'5". 399 5/94.
  Men: ____ . Guns: UD 18 x 6, QD 6 x 12★. Taken in the Mediterranean by *Leda* 9.6.1793; 1797 hulked [as powder hulk (?) at Sheerness; 1806 sold].

PLANS: 1797: Lines/lines.

***Espion*** 1793-1794/1795 ***Spy*** -1801 (French privateer *Le Robert* ____ Nantes) 16.

Dimensions & tons: 86'5½", 69'6⅜" x 27'3¾" x 13'0½". 275 83/94.
Men: 120. Guns: 16 x 6. Taken by *Siren* 13.6.1793; 22.7.1794 retaken by the French *Tamise* and two other frigates; 3.3.1795 taken by *Lively* and renamed *Spy*; 1801 sold.

PLANS: Lines & profile & decoration/decks. As fitted: profile/decks.

***Avenger*** 1794-1802 (French *La Marseillaise?* ____ ) 16?
Dimensions & tons: ____ , ____ x ____ x ____ . 330.
Men: ____ . Guns 16? x ____ . Taken at Martinique by Jervis' squadron; never fitted for sea by the RN after her arrival in Britain; 1802 sold.

***Moselle*** 1793-1794/1794-1802 (French 'gabare' *La Moselle* 1788 ____ ) 18?
Dimensions & tons: 120', ____ x 31' x ____ . 520.
Men: 126. Guns: 18 x ____ ? Taken at the surrender of Toulon 1793; 7.1.1794 taken when entering Toulon by mistake after its capture by the Republicans; taken by *Aimable* off the Hyères Islands 23.5.1794; 1797 laid up until; 1802 sold.

***Hobart*** 1794-1803 (French privateer *La Revanche* ____ ____ ) 18.
Dimensions & tons: ____ , ____ x ____ x ____ . ____ .
Men: 121. Guns: 18 x ____ . Taken off Sunda by *Resistance* 21.10.1794; 1803 sold.

***Arab*** 1795-1796 (French *Le Jean Bart* 1793 ____ ____ ) 16.
Dimensions & tons: 90'8½", 69'11⅜" x 29'1⅜" x 11'4". 315 34/94.
Men: 100. Guns: 16 x 6. Taken by *Cerberus* and others 19.3.1795; 10.6.1796 wrecked on the Glénan Isles.

PLANS: Lines & profile/orlop/upper deck.

***Beaver*** 1795-1808 (building by Graham, Harwich) 14.
Dimensions & tons: 94'3", 77'6" x 25'6½" x 13'4". 269.
Men: 116. Guns: UD 14 x 6, QD 4 x 12★, Fc 2 x 12★. Purchased whilst on the stocks (possibly even before she was laid down) in early 1795; Launched 29.9.1795; 1808 sold.

PLANS: Building: Profile, forecastle & upper deck/lower deck & orlop.

***Spencer*** 1795-1804 ***Lilly***-1804 (Mercantile *Sir Charles Grey* ____ ) 16?
Dimensions & tons: 92'6", 72' x 22'11" x 12'. 200.
Men: 121. Guns: 16 x ? Purchased at Bermuda 10.1795; 15.7.1804 taken by French privateer *Dame Ambert* off Georgia; became French privateer *General Ernouf*; 20.3.1805 retaken by *Renard* and blew up.

NOTE: Also listed as a brig sloop, but probably ship rigged in RN service.

***Bonne Citoyenne*** 1796-1819 (French *La Bonne Citoyenne* 1794 ____ ) 20.
Dimensions & tons: 120'1", 100'6¼" x 30'11" x 8'7". 511 4/94.
Men: 125 (later: 121). Guns: UD 18 x 6, QD 2 x 32★ (later: UD 2 x 9 + 18 x 32★). Taken by *Phaeton* 10.3.1796; 1819 sold.

NOTE: The basis for the design of two separate classes of ship sloops. She was later classed (as were the designs based on her) as a Sixth Rate.

PLANS: Lines & profile. 1810: Lines & profile.

***Scourge*** 1796-1802 (French *La Robuste* 1793 ____ ) 18.
Dimensions & tons: 102'9", 83'5" x 28'11⅝" x 12'10½". 372 34/94.
Men: 125. Guns: 18 x 6 (later 18 x 24★). Taken by *Pomone* 15.4.1796; 1802 sold.

PLANS: Lines & profile & decoration. Taken off & as fitted: decks. As fitted: decks.

***Legere*** 1796-1801 (French *La Légère* ex *L'Inabordable* ex *La Légère* 1784 ____ ) 18.
Dimensions & tons: 116'2½", 94'8" x 30' x 9'7½". 453 17/94.
Men: 121. Guns: UD 18 x 6, QD 6 x 12★, Fc 2 x 12★ (later UD 16 x 18★, QD 6 x 18★, Fc 2 x 6 + 2 x 18★). Taken by *Apollo* and *Doris* 22.6.1796; 2.2.1801 wrecked near Cartagena, South America.

PLANS: Lines & profile & decoration/decks. As fitted: Profile/orlop & lower deck/upper deck & quarter deck & forecastle.

***Havick*** 1796-1800 (Dutch *Havik* ____ ) 16.
Dimensions & tons: 101'10", 83'5" x 28'8" x 12'9". 364 58/94.
Men: 121. Guns: 16 x 6. Surrendered at Saldanha Bay 1796; 9.11.1800 wrecked in Saint Aubin's Bay, Jersey.

PLANS: Decks. As fitted: Lines & profile & decoration/decks.

***Hunter*** 1796-1797 (building by Pender, Bermuda) 16.
Dimensions & tons: 102'9", 80'11" x 26' x 16'1½". 336.
Men: 80. Guns: 16 x 24★. Purchased whilst building (probably as a brig; design dimensions of this ship and her sister *Rover* seem to have been: 95', 80' x 26' x 14'. 288 tons) 10.1795; Goodrich & Co acting for the Admiralty as contractors; 1796 launched after lengthening; 27.12.1797 wrecked on Hog Island, Virginia.

***Rover*** 1796-1798 (building by Tynes, Bermuda) 16.
Dimensions & tons: 104', 80'1½" x 26'1" x 16'0¼". 356.
Men: 80. Guns: 16 x 24★. Purchased whilst building (as a brig - like her sister *Hunter*, see above) 10.1795; Goodrich & Co acting as contractors; 1796 launched after lengthening; 23.6.1798 wrecked in the Gulf of Saint Lawrence.

***Musette*** 1796-1801 (French privateer *La Musette* 1793 Nantes) 20.
Dimensions & tons: 102'0½", 80'9⅜" x 26'11¼" x 13'3¾". 312.
Men: 126. Guns: UD 20 x 24★. Taken by *Hazard* 21.12.1796; never fitted for sea by the RN; 1801 hulked [at Plymouth as a receiving ship; 1803 floating battery for the River Yare; 1806 sold].

***Sardine*** 1796-1806 (French *La Sardine* [22 or 14] 1771 Toulon, Broquier design) 16?
Dimensions & tons: ____ , ____ x ____ x ____ . 300.
Men: ____ . Guns: ____ . Taken by *Egmont*, *Barfleur* and others; 1806 sold.

***Trompeuse*** 1797-1800 (French privateer *Le Mercure* 1797 Nantes) 18.
Dimensions & tons: 103'5", 83'8¼" x 27'6½" x 12'10¾". 337 77/94.
Men: 96. Guns: 2 x 6 + 16 x 24★. Taken by *Melampus* 05.1797; 17.5.1800 supposed foundered in the Channel after sailing from Plymouth.

PLANS: As fitted: Lines & profile & decoration/lower deck/upper deck.

***Gayette*** 1797-1808 (French *La Gaiete* 1795 ____ ) 20.
Dimensions & tons: 120'3½", 100'0¾" x 31'1" x 8'8". 514 20/94.
Men: 125. Guns: UD 2 x 9 + 18 x 32★. Taken in the Atlantic by *Arethusa* 20.8.1797; 1808 sold.

PLANS: Lines & profile/lower deck/upper deck/forecastle.

***Epervoir*** 1797-1801 (French privateer *L'Epervier* ____ ) 18.
Dimensions & tons: 94', 76'3" x 25'0¼" x 10'7". 254.
Men: 125. Guns: 18 x 18★. Taken 14.11.1797 (?) by *Cerberus*; never fitted for sea by the RN; 1801 sold.

***Renard*** 1797-1805 (French privateer *Le Renard* ____ ) 16.
Dimensions & tons: 103', 81'5" x 28'3" x 11'6½". 345 57/94.
Men: 90. Guns: UD 16 x 12★. Taken 14.11.1797 by *Cerberus*; 1805 sold.

PLANS: Lines & profile & decoration/upper deck. As fitted: Lines & profile & decoration/lower deck/upper deck.

***Railleur*** 1797-1800 (French privateer *Le Railleur* ____ ) 14.
Dimensions & tons: 89'6", 71'7¼" x 26'2¼" x 10'1½". 261 18/94.
Men: 76. Guns: 14 x 12★. Taken 17.11.1797 by *Boadicea*; 05.1800 foundered, sailed from Plymouth and not heard of since.

PLANS: Lines & profile/decks. As fitted: Lines & profile/decks.

***Voltigeur*** 1798-1802 (French privateer *L'Audacieux* ____ ) 20.
Dimensions & tons: 116'6", 95'3⅛" x 28'4⅜" x 12'2". 407 62/94.
Men: 121. Guns 20 x 18★. Taken by *Magnanime* 2.4.1798; 1802 sold.

PLANS: Lines & profile & decoration/orlop/upper deck. As fitted: Lines & profile & decoration/lower deck/upper deck.

***Albion*** 1798-1802? (Mercantile? *Albion*? ____ ) 20?
Dimensions & tons: 93'10½", 76'11⅛" x 29'8" x 13'0½". 360 10/94.
Men: ____ . Guns: ____ . Purchased (originally as armed vessel) 04.1798 (had been a hired ship 1793-1794?); 1802? sold.

PLANS: Lines & profile/decks.

***Ann*** 1798-1802? (Mercantile? *Ann*? ____ ) ____ .
Dimensions & tons: 104'2", 101' x 28'6" x 12'9½". 345.
Men: ____ . Guns: ____ . Purchased 04.1798? (originally as an armed vessel); 1802? sold.

***Hermes*** 1798-1802 (Mercantile? *Hermes*? ____ ) ____ .
Dimensions & tons: 100', ____ x 28'5" x ____ . 331.
Men: 76. Guns: ____ . Purchased 1798 (originally as an armed vessel); 1802 sold.

***William*** 1798-1810 (Mercantile? *William*? ____ Whitby?) ____ .

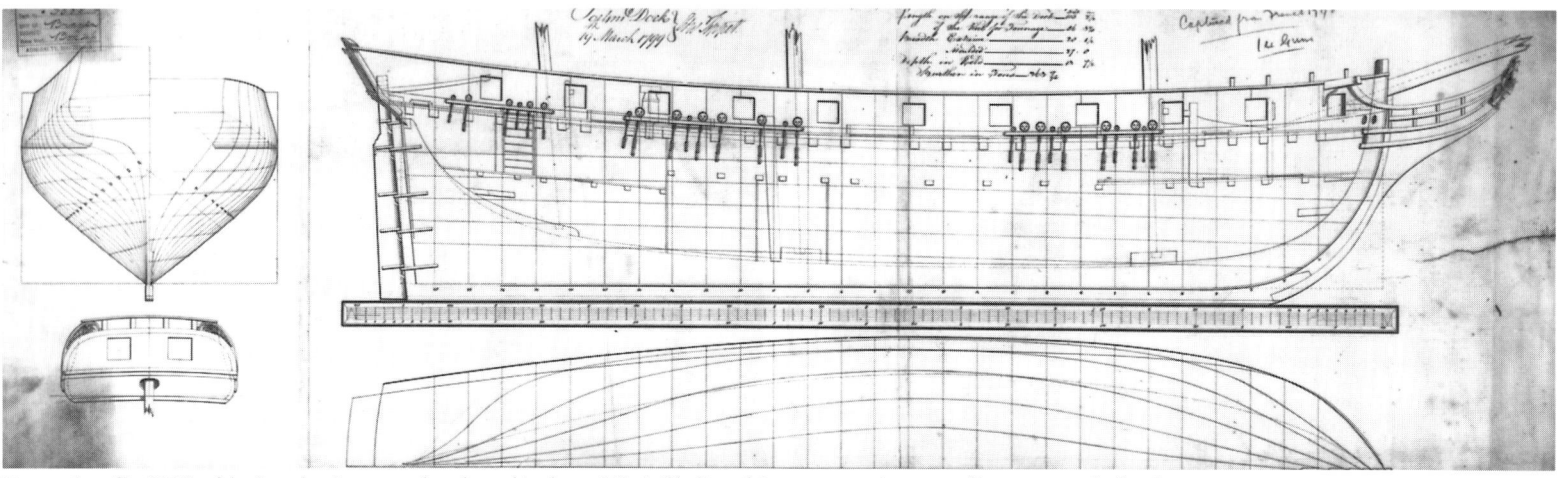

Lines and profile (3088) of the French privateer, reclassed as a ship sloop, *L'Invincible General Buonaparte* - a nice sense of humour was displayed in her renaming as *Brazen* (1798)

Dimensions & tons: 99'8", 82'9½" x 29'1" x 12'10⅛". 374.
Men: 32? (as storeship: 46). Guns: ____ (as storeship UD 4 x 32★).
Purchased 04.1798? (originally as armed vessel); 1800 storeship; 1810 BU.

***Selby*** 1798-1801 (Mercantile *Selby* 1786 Whitby) 20.
Dimensions & tons: 100', 83'1⅛" x 28'3½" x 12'7½". 354.
Men: 90. Guns: UD 20 x 32★. Purchased 1798 (originally as armed vessel); 1801 sold.

***Xenophon*** 1798- by 1801 ***Investigator*** 1810 (Mercantile *Xenophon* 1795 Sunderland) armed vessel/sloop 8.
Dimensions & tons: ____ , ____ x ____ x ____ . ____ .
Men: ____ . Guns: ____ . Purchased 1798 as an armed vessel then registered as a ship sloop; 1801 fitted for a voyage of discovery (Flinders'); 1810 sold.
PLANS: Upper deck/quarter deck.

***Bonetta*** 1798-1801 (French privateer *Les Huit Amis* 1798 ____ ) 18.
Dimensions & tons: 103'1", 85'5" x 27'8" x 13'0½". 347 70/94.
Men: 121. Guns: 18 x 6. Taken by *Endymion* 10.5.1798?; 25.10.1801 wrecked on the Jardines, Cuba.
PLANS: Lines & profile & decoration/decks/profile (incomplete)/decks.

***Brazen*** 1798-1800 (French privateer *L'Invincible General Bonaparte* ____ ____ ) 16.
Dimensions & tons: 105'2½", 86'3⅞" x 28'1½" x 13'7½". 363 17/94.
Men: ____ . Guns: 16 x 24★. Taken by *Boadicea* 1798?; 26.1.1800 wrecked near Brighton.
PLANS: Lines & profile & decoration/orlop/upper deck.

***Fleche*** 1798-1810 (French privateer *La Caroline* ____ ) 18.
Dimensions & tons: 92'0¼", 74'4¼" x 26'6¼" x 10'11¾". 279.
Men: 75. Guns: UD 16 x 18★, Fc 2 x 18★. Taken by *Phoenix* 31.5.1798; 24.5.1810 wrecked off the mouth of the Elbe.

***Sophie*** 1798-1809 (French privateer *Le Premier Consul* 1788 ____ ) 18.
Dimensions & tons: 108'8", 87'3⅞" x 28'11" x 13'11½". 388 38/94.
Men: 120. Guns: UD 2 x 6 + 16 x 24★. Taken by *Endymion* 09.1798; 1809 BU.
PLANS: Lines, profile & decoration.

***Dispatch/Despatch*** 1799-1801 (French privateer *L'Indefatigable* ____ ) 18.
Dimensions & tons: 90'4", 71'0⅞" x 25'0⅞" x 11'3". 238.
Men: 90. Guns: 2 x 6 + 16 x 18★. Taken by *Ethalion* 6.3.1799 (or 6.5.1799?); not fitted for sea in the RN; 1801 sold.

***Argus*** 1799-1811 (French privateer *L'Argus* 1798 Bordeaux) 18.
Dimensions & tons: 102'9", 82'3½" x 27'3½" x 13'. 326.
Men: 86. Guns: UD 2 x 6 + 16 x 18★. Taken by *Pomone* 3.4.1799; 1811 BU.

***Lutine*** 1799-1802 (French privateer *Le Courageux* 1779? ____ ) ____ .
Dimensions & tons: 145', 121'8" x 39' x 11'6" (note that these dimensions are most unlikely - they are appropriate to a frigate, not a 332 ton sloop. It would seem quite possible that there is a confusion here with the 32 gun prize frigate *Courageuse* [see Unregistered Vessels]). 332.

Men: 121. Guns: ____ . Taken 18.6.1799 in the Mediterranean by Lord Keith's squadron; 1802 sold.

***Surinam*** 1799-1803 (French *Le Hussard* 1797 ____ ) 18.
Dimensions & tons: 105', 86'3" x 30'2" x 8'5". 413.
Men: 121. Guns: 18 x 6. Taken at Surinam by Seymour's squadron 20.8.1799; 07.1803 seized in harbour by the Dutch at Curacoa; 1.1.1807 retaken by the RN but subsequent fate unknown.

***Nimrod*** 1799-1811 (French *L'Eole* 1799 ____ ) 18.
Dimensions & tons: 99', 89' x 29' x ____ . 395.
Men: 121. Guns: UD 2 x 6 + 16 x 24★. Taken by *Solebay* 22.11.1799 in the West Indies; 1811 sold.

***Trompeuse*** 1800-1811 (French privateer *La Trompeuse* 1799 Nantes) 18.
Dimensions & tons: 100'10", 80'4½" x 29'10" x 13'1¾". 380.
Men: 96. Guns: UD 2 x 6 + 16 x 24★. Taken by *Revolutionaire* 4.3.1800; 1811 BU.

***Rosario*** 1800-1809 (French privateer *Le Hardi* 1800 Bordeaux) 18.
Dimensions & tons: 111'2", 89' x 29'11¾" x 13'6½". 425 44/94.
Men: 121. Guns: 2 x 6 + 16 x 24★. Taken by *Anson* 29.4.1800; 1809 BU.
PLANS: As fitted: Lines & profile.

***Wasp*** 1800-1811 (French privateer *La Guêpe* 1798 Bordeaux) 18.
Dimensions & tons: 101'9", 83'5" x 25'11" x 12'2". 298 2/94.
Men: 105. Guns: 2 x 9 + 16 x 24★. Taken by boats of Warren's squadron 29.8.1800; 1811 sold.
PLANS: As fitted: Lines, profile & decoration/decks.

***Scout*** 1800-1801 (French *La Vénus* ex *La Vengeance* 1793 Bordeaux or *La Vengeance* 1799 Nantes) 20.
Dimensions & tons: 111', 89'10⅝" x 29'1½" x 13'4". 405 53/94.
Men: 121. Guns: UD 18 x 24★, Fc 2 x 6. Taken 22.10.1800 (or, possibly, early 1800?); 25.3.1801 wrecked on the Shingles, Isle of Wight.
PLANS: Lines & profile/lower deck/upper deck.

***Imogen*** 1800-1805 (French privateer *Le Diable à Quatre* 1792 Bordeaux) 20.
Dimensions & tons: 108'2", 87'3" x 29'4" x 15'. 399.
Men: 121. Guns: UD 18 x 24★, Fc 2 x 6. Taken by *Thames* 26.10.1800; 1.3.1805 foundered in the Atlantic.

***Charwell*** 1801-1813 (French *L'Aurore* 1799 Le Havre) 16.
Dimensions & tons: 102'1", 78'8" x 28'9" x 13'1¼". 346.
Men: 96. Guns: UD 14 x 18★, Fc 2 x 6. Taken by *Thames* 18.1.1801; 1810 laid up until; 1813 sold.

***Delight*** 1801-1803 (French *Le Sans Pareil* 1798 La Ciotat) 18.
Dimensions & tons: 97'5", 77'3" x 28'7" x 8'2". 336.
Men: 121. Guns: 18 x 24★. Taken off Sardinia by *Mercury* 20.1.1801; 1802 laid up until; 1805 sold.

***Alonzo*** 1801-1815 (Mercantile *Alonzo* ____ Portsmouth?) 16.
Dimensions & tons: 102'3", ____ x 29'10" x ____ . 384.
Men: 100. Guns: UD 14 x 24★ + 2 x 18★. Purchased from Dudman (? the builder?) 8.2.1801; 1813 hulked [at Woolwich as convict hospital ship; 1828 to Portsmouth as convict ship; 1835 to Leith Seamens Society as chapel ship; 1842 deliberately sunk].

PLANS: 1814: Part profile & decks.

***Autumn*** 1801-1810 ***Strombolo*** 1815 (Mercantile *Autumn* 1800 South Shields) 14.
Dimensions & tons: 98'11", 78'6" x 28'4" x ____ . 335.
Men: 70. Guns: UD 12 x 24★, QD 2 x 18★. Purchased from Perry? (? the builder?) 28.2.1801; 1805 hulked [at Portsmouth as receiving ship]; 1810 converted to bomb vessel and renamed *Strombolo*; 1815 sold.

***Falcon*** 1801-1810 (Mercantile *Diadem* 1801? ____ ) 16.
Dimensions & tons: 102'10", ____ x 29'3½" x ____ . 368.
Men: 75. Guns: UD 14 x 24★, QD 2 x 18★. Purchased from Hill 28.2.1801; 1810 hulked [at Sheerness as a military depot and hospital ship; 1816 sold].

***Diligence*** 1801-1812 (Mercantile *Union* ____ ____ ) 16.
Dimensions & tons: 99', ____ x 29'1½" x ____ . 361.
Men: 75. Guns: UD 14 x 24★, QD 2 x 18★. Purchased from Randall (?the builder?) 5.2.1801; 1812 sold.

***Hound*** 1801-1812 (Mercantile *Hound*? ex *Monarch* ____ ) 14.
Dimensions & tons: 103'3", 81' x 27'9½" x ____ . 333.
Men: 100 (as bomb: 67). Guns: UD 12 x 24★, QD 2 x 18★ (as bomb: 1 x 13" + 1 x 10" mortars + 10 x 24★). Purchased 02.1801; 1807 converted to bomb vessel; 1812 BU.

***Galgo*** 1801-1814 (Mercantile *Garland* ____ ____ ) 16.
Dimensions & tons: 102', ____ x 28'7" x ____ . 353.
Men: 75. Guns: UD 14 x 24★, QD 2 x 18★. Purchased 2.3.1801 from Randall (? the builder?); 1814 hulked [as a receiving ship for Gravesend; 1814 sold].

***Scout*** 1801-1802 (French *Le Premier Consul* 1800 Nantes) 20.
Dimensions & tons: 113'8", 91'9" x 30'3½" x 11'3". 447 73/94.
Men: 120. Guns: UD 18 x 24★, Fc 2 x 6. Taken 5.3.1801 by *Dryad*; 1802 supposed to have foundered off Newfoundland.

PLANS: Lines & profile/decks.

## Brigs, etc

***Goelan*** 1793-1794 (French 'Aviso' *Le Goëland* 1787 Bayonne, Haran design) brig sloop ____ .
Dimensions & tons: ____ , ____ x ____ x ____ . 248.
Men: ____ . Guns: ____ . Taken by *Penelope* 26.4.1793; 1794 sold.

***Lutin*** 1793-1796 (French *Le Lutin* 1788 Bayonne, Haran design) brig sloop ____ .
Dimensions & tons: ____ , ____ x ____ x ____ . ____ .
Men: ____ . Guns: ____ . Taken off Newfoundland by *Pluto* 25.7.1793; 1796 sold.

***Espiegle*** 1793-1802 (French *L'Espiègle* 1788 Bayonne, Haran design) brig sloop 16.
Dimensions & tons: 91'7", 74'5½" x 26'2" x 11'3". 271 18/94.
Men: 96. Guns: 16 x 6. Taken by *Nymphe* and *Circe* 30.11.1793; 1802 sold.

PLANS: Lines, profile & decoration.

***Trompeuse*** 1794-1796 (French *Le Trompeuse* 1793 Le Havre) brig sloop 16.
Dimensions & tons: 91'9", 73'0¾" x 29'7½" x 11'7". 342.
Men: 110. Guns: 16 x 6? (originally an armament of 18 pdr carronades was proposed). Taken by *Sphinx* 12.1.1794; 15.7.1796 wrecked near Kinsale, wreck sold.

PLANS: Lines & profile/decks.

***Viper*** 1794-1797 (French *La Vipère* 1793 Le Havre) brig sloop 16.
Dimensions & tons: 95'7", 78' x 26'5½" x 12'3". 290 41/94.
Men: 100. Guns: 16 x 6 + 2 x 12★. Taken 23.1.1794 by *Flora*; 21.1.1797 foundered off the mouth of the Shannon.

PLANS: Taken off & as fitted: Lines & profile/lower deck/upper deck.

***Actif*** 1794-1794 (French *L'Actif* ____ ____ ) brig sloop 10.
Dimensions & tons: ____ , ____ x ____ x ____ . ____ .
Men: 60. Guns: 10 x 4. Taken by Ford's squadron in the West Indies 16.3.1794; 26.11.1794 foundered off Bermuda.

***Jack Tar*** 1794-1794 (French ____ ____ ____ ) brig sloop 14.
Dimensions & tons: ____ , ____ x ____ x ____ . 193.
Men: 90. Guns: 14 x 4. Taken by Ford's squadron in the West Indies 16.3.1794?; 1794 sold.

***Requien*** 1795-1801 (French *Le Requin* 1793 Rochefort) armed brig 10.
Dimensions & tons: 71'2", 53'11½" x 24' x 10'3". 165 29/94.
Men: 59. Guns: 10 x 4. Taken by *Thalia* 20.2.1795; 11.1.1801 wrecked near Quiberon.

NOTE: A listing of a brig called *Requin* in 1802 may just be a clerical error - an accidental carrying over of this vessel. However the number of guns is given as 16, so this may be a reference to a totally different vessel.

PLANS: Lines & profile/orlop/upper deck.

***Zephyr*** 1795-1808 (building by King, Dover) brig sloop 14.
Dimensions & tons: 82'6", 62'7" x 27'1" x 10'8". 244 9/94.
Men: 76. Guns: 14 x 6. Purchased 02.1795 whilst building; 1801 fireship; 1808 sold.

PLANS: Building: Lines & profile.

***Hope*** 1795-1807 (Dutch East India Company brig *Ster* ____ ____ ) brig sloop 14.
Dimensions & tons: 77', ____ x 24'6" x 14'1". 230.
Men: 60. Guns: 14 x ____ . Taken at Simons Bay 18.8.1795; 1807 sold.

***Braak/De Braak*** 1795-1798 (Dutch cutter *De Braak*? ____ ____ ) brig sloop 18.
Dimensions & tons: 84', 57'4¾" x 28'11" x 11'2". 255 28/94.
Men: 86. Guns: 16 x 24★ + 2 x 6 chase guns (later two of the carronades were removed). Taken 20.8.1795 and refitted as a brig; 23.5.1798 capsized in the Delaware [wreck recently located and salvaged; an attempt is now being made to record what has been brought up in proper archaeological terms - somewhat too late to be satisfactory].

PLANS: Lines & profile/decks. As fitted: profile/decks.

***Suffisante*** 1795-1803 (French *La Suffisante* 1793 Le Havre) brig sloop 14.
Dimensions & tons: 86'1", 67'4⅝" x 28'3¼" x 12'7¾". 286 4/94.
Men: 90/86. Guns: 14 x 6. Taken by Admiral Duncan 25.8.1795; 15.12.1803 wrecked off Queenstown.

PLANS: Lines & profile/decks.

***Victorieuse*** 1795-1805 (French *La Victorieuse* 1794 Honfleur) brig sloop 14.
Dimensions & tons: 103'1", 86'6⅞" x 27'8½" x 12'10". 349.
Men: 130. Guns: 12 x 12 + 2 x 32★. Taken by Admiral Duncan 25.8.1795; 1805 BU.

PLANS: Lines & profile/lower deck/upper deck.

***Comeet*** 1795-1798 ***Penguin*** -1809 (Dutch *Komeet* ____ ____ ) brig sloop 16.
Dimensions & tons: 92'9⅛", 73'3⅞" x 29'4⅜" x 16'4½". 336 26/94.
Men: 125. Guns: 14 x 9 + 2 x 18★. Taken by *Unicorn* 28.8.1795; 1798 renamed *Penguin*; 1809 sold.

PLANS: Lines, profile & decoration/lower deck/upper deck.

***Pandour*** (***Pandora***) 1795-1797 (French *Le Pandour* 1780 ____ ____ ) armed brig 14.
Dimensions & tons: 78', 60'0¼" x 26'11" x 10'6". 231 27/94.
Men: 75. Guns: 14 x 4. Taken 31.8.1795 by *Caroline*; 06.1797 foundered 'in the North Seas'.

PLANS: Lines & profile/decks.

***Miermen*** 1796-1801 (Dutch *Meermin* ____ ____ ) brig sloop 16.
Dimensions & tons: ____ , ____ x ____ x ____ . 203.
Men: 86. Guns: 16 x 6. Seized at Plymouth 4.3.1796; not fitted for sea by the RN; 1801 sold.

***Pyl*** 1796-1801 (Dutch *Pijl* 1784 ____ ) brig sloop 12.
Dimensions & tons: ____ , ____ x ____ x ____ . 200.
Men: 70 (as fireship: 45). Guns: 12 x 6 (as fireship: 8 x 18★). Seized at Plymouth 4.3.1796; 1798 fireship; 1801 sold.

***Hermes*** 1796-1797 (Dutch *Mercurius* ____ ____ ) brig sloop 16.
Dimensions & tons: ____ , ____ x ____ x ____ . 210.
Men: 80. Guns: 14 x 6 (or 14 x 24★?) + 2 x 6 'for chace'. Taken by *Sylph* 12.5.1796; 31.1.1797 sailed, not heard of since, presumed foundered.

***Rambler*** 1796-1810 ( ____ King, Dover) brig-rigged cutter 14.
Dimensions & tons: 74'5¼", 56'5½" x 25'4" x 10'1½". 193.
Men: 86. Guns: 14 x 6. Purchased from King (whilst building?) in 02.1796; 1798 classed as brig sloop; 1810 sold.

***Amaranthe*** 1796-1800 (French privateer *L'Amarante* 1793 Le Havre, Forfait

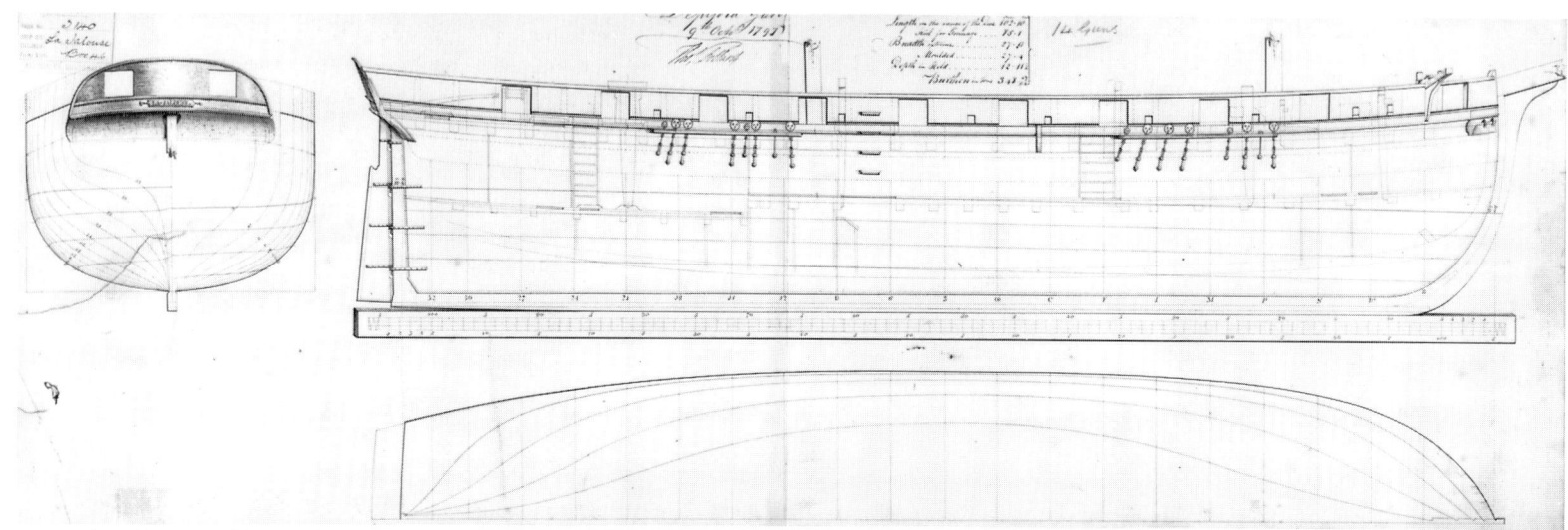

Lines and profile (3140) of the French brig sloop *Jalouse* taken in 1797

NMM photo negative B2290. The NMM's model of the purchased brig *Wolverene* of 1798 with her centre-line armament of powerful carronades

design) brig sloop 12.
Dimensions & tons: 86'1¼", 68'8⅜" x 28'2½" x 13'2¾". 290 30/94.
Men: 86. Guns: 12 x 24★. Taken 31.12.1796 by *Diamond*; 25.10.1799 'lost in the Gulph of Florida'.
PLANS: Lines & profile/orlop/upper deck.

**Atalanta** 1797-1807 (French *L'Atalante* 1794 Bayonne, Haran design) brig sloop 16.
Dimensions & tons: 99', 78'0⅜" x 12'2¼". 309 80/94.
Men: 90. Guns: 16 x 6. Taken by *Phoebe*? 10.1.1797; 12.2.1807 wrecked off Rochefort.
PLANS: Lines, profile & decoration/decks. As fitted: Profile & decoration/decks.

**Transfer** 1797-1802 (French privateer *Les Quatres Frères* ____ ) armed brig 12.
Dimensions & tons: 80', 63' x 23'3" x 14'6". 181.
Men: 70. Guns: 12 x 6. Taken 1797; 1802 sold.

**Eugenia** 1797-1802? (French privateer *La Nouvelle Eugénie* ____ ) ____ brig sloop 16.
Dimensions & tons: 84'9", 66'3" x 26'2" x 13'6". 241 26/94.
Men: 90. Guns: 16 x 6. Taken by *Indefatigable* and others 11.5.1797; 1802? sold.
PLANS: Lines, profile & decoration/decks. As fitted: Lines, profile & decoration/decks.

**Jalouse** 1797-1807 (French *La Jalouse* 1794 Dunkirk or Honfleur?) brig sloop 18.
Dimensions & tons: 102'10", 85'4" x 27'9" x 12'11½". 348 47/94.
Men: 121. Guns: 18 x 6. Taken 13.5.1797 by *Vestal*; 1807 BU.
PLANS: Lines & profile/decks.

**Mutine** 1797-1807 (French *La Mutine* 1794 ____ ) brig sloop/armed brig 18.
Dimensions & tons: 104'6", 84'10" x 27'10" x ____ . 349.
Men: 121. Guns: 18 x 6. Cut out of Teneriffe by the boats of *Lively* 12.6.1797; 1807 sold.

**Espoir** 1797-1804 (French *L'Espoir* ex *Le Lazovski* ex *L'Espoir* 1789 Bayonne, Haran design) brig sloop 14.
Dimensions & tons: 92'10", 75'1" x 25'1" x ____ . 251.
Men: 80. Guns: 14 x 6. Taken by *Thalia* in the Mediterranean 11.9.1797; 1804 sold.

# THE FRENCH REVOLUTIONARY WAR 1793-1801

***Mary*** 1797-1800 ***Halifax*** -1801 (French privateer *La Marie* ____ ) armed brig 12.
    Dimensions & tons: 71', 55'1½" x 21'7" x 9'8". 136 55/94.
    Men: 55. Guns: 12 x 12★. Taken by *Jason* 21.11.1797; 1800 renamed *Halifax*; ____ 1801 lost.
PLANS: Profile/lower & upper decks. As fitted: Lines & profile/decks.

***Belette*** 1798-1801 (French privateer *Le Belliquex* ____ ____ ) brig sloop 20.
    Dimensions & tons: 104'4½", 85'7½" x 27'7" x 7'5". 346.
    Men: 125. Guns: 20 x 24★. Taken by *Seahorse* 10.1.1798; not fitted for sea by the RN?; 1801 sold.

***San Leon*** 1798-1800 (Spanish *San Leon?* ____ ____ ) brig sloop 12.
    Dimensions & tons: ____ , ____ x ____ x ____ . ____ .
    Men: 78. Guns: 16 x 6. Taken by *Santa Dorothea* 24.1.1798; (employed in the Mediterranean?); 1800 sold.

***Pandour*** 1798-1800 ***Wolf*** -1802 (French *L'Eugénie* 1798? ____ ) brig sloop 16.
    Dimensions & tons: 85'10", 67'9¼" x 25'11¾" x 11'11". 243.
    Men: 86. Guns: 16 x 6. Taken by *Magnanime* 16.3.1798; never fitted for sea by the RN?; 1800 renamed *Wolf*; 1802 BU.

***Wolverene*** 1798-1804 (Mercantile *Rattler* of London 1798 ____ ). Gunboat/brig/brig sloop.
    Dimensions & tons: ____ , ____ x ____ x ____ . ____ .
    Men: 70. Guns: 14 (UD 6 x 24 pounder carronades on pivot mounts on the centre line + 2 sided 18★, QD 4 x 12★, Fc 2 x 12★). Purchased by Captain Schank 03.1798 and armed in the above experimental manner according to his instructions; 24.3.1804 taken and sunk by French privateer *Blonde* in the Western Atlantic.

***Arrogant*** (?) 1798-1801 ***Insolent*** -1818 (French *L'Arrogante* ex *L'Insolent* ex *La Brave* 1794 ____ ) gunboat/gunbrig/brig sloop 12.
    Dimensions & tons: 91'9", 72'1" x 25'11⅜" x 11'2". 258 14/94.
    Men: 55 (later, as brig sloop: 85). Guns: 2 x 18 + 10 x 32★ (later as brig sloop 2 x 6 + 12 x 24★). Taken by *Jason* 19.4.1798; not renamed *Insolent* until fitted for sea in 1801?; 1818 sold.

***Mendovi*** 1798-1811 (French *Le Mendovi* 1797 taken from Venetians as ____ ____ Venice) brig sloop 14.
    Dimensions & tons: 81'6", 63'8" x 25' x ____ 211.
    Men: 76. Guns: 14 x 6. Cut out of Cerigo by the boats of *Flora* 13.5.1798; 1802 laid up until ; 1811 BU.

***Port Mahon*** 1798-1815 (Spanish ____ building at Port Mahon) brig sloop 18.
    Dimensions & tons: 91'5½", 82'1" x 25'2" x 12'8". 276½.
    Men: 121. Guns: 18 x 6. Taken at the capture of Port Mahon whilst still on the stocks there 15.11.1798; 1798 launched; 1815 hulked [as a receiving ship at Gravesend; 1817 to the Thames Police; 1837 sold].
PLANS: Lines & profile (also for *Vensejo*).

***Weazel*** 1799-1804 (built by King, Dover) brig/brig sloop 12.
    Dimensions & tons: 77', 59'1" x 26'1" x 10'8½". 213 82/94.
    Men: 70. Guns: 12 x 18★ (probably + 4 x 6). Purchased 'when built' in 03.1799; 1.3.17804 wrecked near Gibraltar.
PLANS: Lines & profile/decks.

***Vincejo/Vensejo*** 1799-1804 (Spanish *El Vensejo* 1799 ____ ____ ) brig sloop 18.
    Dimensions & tons: 91'5½", 82'1" x 25'2" x 12'8". 276½.
    Men: 100. Guns: UD 16 x 18★, QD 2 x 6. Taken by *Cormorant* 19.3.1799 in the Mediterranean; 1804 taken by French gunboats in Quiberon Bay; [sold by the French].
PLANS: Lines & profile (also for *Port Mahon*).

***Utile*** 1799-1801 (French privateer *L'Utile* ____ ____ ) brig sloop 18.
    Dimensions & tons: 89'6", 74'0½" x 26'7" x 12'9½". 278 88/94.
    Men: 76. Guns: UD 14 x 24★, Fc 2 x 6. Taken by *Boadicea* 1.4.1799; 11.1801 capsized in the Mediterranean.
PLANS: Lines & profile/orlop/upper deck.

***Minorca*** 1799-1802 (French *L'Alerte* 1789 ____ ____ ) brig sloop.
    Dimensions & tons: 85', 70'1" x 26'7" x 10'. 248.
    Men: ____ . Guns: ____ . Taken by Lord Keith's Squadron in the Mediterranean 18.6.1799; 1802 sold.

***Salamene*** 1799-1802 (French *La Salamene* taken 1798 from Spain as *Infanta* ?1787? ____ ) brig sloop ____ .
    Dimensions & tons: 93'6", 79' x 24'8" x 10'. 240.
    Men: 86. Guns: ____ . Taken by Lord Keith's squadron in the Mediterranean 18.6.1799; 1802 sold.

***Anacreon*** 1799-1802 (French privateer *L'Anacréon* 1797 Dunkirk) brig 14.
    Dimensions & tons: 76'5½", 60'5½" x 21'8" x 9'. 150 80/94.
    Men: 60. Guns: 14 x 4.
PLANS: Lines & profile.

***Camphaan/Kemphaan*** 1799-1801 (Dutch *Kemphaan?* ____ ) brig sloop ____ .
    Dimensions & tons: ____ , ____ x ____ x ____ . ____ .
    Men: ____ . Guns: ____ . Taken at Surinam 08.1799; 1801 sold.

***Gier*** 1799-1803 (Dutch *De Gier* ____ ) brig sloop 14.
    Dimensions & tons: 91'7", 71'8⅛" x 29'2" x 12'3". 324 30/94.
    Men: 96/86. Guns: UD 12 x 32★, Fc 2 x 6. Taken by *Wolverene* 12.9.1799; 1803 BU.
PLANS: Lines, profile & decoration/decks.

***Raven*** 1799-1804 (French *L'Aréthuse* 1798 Nantes or L'Orient) brig sloop 18.
    Dimensions & tons: 107'7", 85'0¾" x 29'4¼" x 13'7½". 389 81/94.
    Men: 96. Guns: 16 x 18★ + 2 x 6 'for chase guns'. Taken 11.10.1799 by *Crescent*; 6.7.1804 wrecked near Mazari, Sicily.
PLANS: As fitted: Lines, profile & decoration/lower deck/upper deck.

***Gannet*** 1800-1814 ( ____ ?King, Dover) brig sloop 16.
    Dimensions & tons: 88'7", 69' x 28'1½" x 12'6". 290 29/94.
    Men: 86. Guns: UD: 14 x 18★, Fc 2 x 6. Purchased from King (whilst building?) 03.1800; 1814 sold.
PLANS: Lines & profile/decks.

***Gironde*** 1800-1801 (French privateer ?*La Gironde* 1793 Bordeaux) brig sloop 16.
    Dimensions & tons: 91'7", 72'6" x 25'11" x 12'1¾". 259.
    Men: 67. Guns: UD 14 x 18★, Fc 2 x 6. Taken by *Boadicea* 08(?).1800; not fitted for sea by the RN; 1801 sold.

***Vivo*** 1800-1801 (Spanish *El Vivo* 1794 Cadiz) brig sloop 14.
    Dimensions & tons: 80'10¼", 60'6" x 25'11" x 10'4". 218 or 216.
    Men: 67. Guns: 14 x 18★. Taken by *Fisguard* 30.9.1800; not fitted for sea by the RN; 1801 sold.

***Calpe*** 1800-1802 (Spanish polacca *San Josef* 1796 Greece) brig sloop 14.
    Dimensions & tons: 75'4", 57' x 26'3" x 12'6". 209.
    Men: ____ . Guns: 14 x ____ . Cut out of Malaga by the boats of *Phaeton* 27.10.1800; 1802 sold.

***Morgiana*** 1800-1801 (French privateer *L'Actif* 1800 Bordeaux) brig sloop 16.
    Dimensions & tons: 95'8", 76' x 26'5½" x 11'11½". 282 93/94.
    Men: 90. Guns: UD 14 x 18★, QD 2 x 6. Taken by *Thames* 30.11.1800; 1807 laid up until; 1811 BU.
PLANS: Lines, profile & decoration/decks.

***Goree*** 1800-1806 (French ____ ) brig sloop 12.
    Dimensions & tons: 88'7¾", 68'7½" x 24'3½" x 11'8". 245.
    Men: 70. Guns: UD 12 x 12★. Taken ____ ?; 1800 purchased; 1801 laid up until; 1806 disposed of (Dittmar gives 'delivered to claimant').

***Moucheron*** 1801-1807 (French privateer *Le Moucheron* 1799 ____ ) brig sloop 18.
    Dimensions & tons: 93', 76'1⅛" x 26'7" x 12'. 286.
    Men: 96/90. Guns: UD 14 x 18★ + 2 x 6. Taken 16.2.1801 by *Revolutionaire*; 04.1807 wrecked in the Mediterranean.

***Hunter*** 1801-1809 ( ____ ?King, Dover) brig/brig sloop 16.
    Dimensions & tons: 90'8", 71'8¼" x 28'6" x ____ . ____ .
    Men: ____ . Guns: 14 x 6. 2 x 24★. Purchased 05 or 06.1801 from King (whilst building?); 1809 BU.
PLANS: Lines & profile.

***Guachapin*** 1801-1811 (Spanish *Guachapin* 1800 Bayonne, France) brig sloop 4.
    Dimensions & tons: 80'5", 63'4" x 23'1" x 11'. 176.
    Men: 26. Guns: 4 x 12 or 18. Taken 1801; 29.7.1811 wrecked by running ashore in a hurricane at Jamaica, remains sold.

## Sloops, Rig Uncertain

***Reprisal*** 1794-1794 ( ____ ) ____ .
    Dimensions & tons: ____ , ____ x ____ x ____ . ____ .

Men: ____ . Guns: ____ . Origins uncertain; 1794 sold.

***Serin/Sirene*** 1794-1796 (French *Le Serin* 1787 ____ ) brig sloop? 14 or 16?.
Dimensions & tons: 92'5", 88'11" x 26' x 11'. 320.
Men: 90. Guns: 14 x ____ . Taken off San Domingo 08.1794; 26.7.1796 sailed 'to convoy the Jamaica convoy ... and to have joined off Cape Antonio, not having done so is supposed to have foundered'.

***Esperance*** 1795-1803 (French *L'Esperance* [22] ex Spanish ____ 1793 ____ ) 16? quarter decked ship sloop?
Dimensions & tons: ____ , ____ x ____ x ____ . ____ .
Men: ____ . Guns: ____ . Taken by *Argonaut* off the Chesapeake; 1798 sold.

***Corso*** 1796-1803 (Spanish *El Corso* ____ ____ ) brig sloop? 18.
Dimensions & tons: ____ , 72'8" x 24'7½" x 11'7". 234.
Men: 100. Guns: 18 x 6. Taken 8.12.1796 in the Mediterranean by *Southampton*; 1803 hulked [at Gravesend as a receiving ship; 1814 sold].

***Rosario*** 1797-1800 (Spanish *Nuestra Senora del Rosario* ____ ) ____ .
Dimensions & tons: 89', 71'4" x 23'6" x ____ . 209.
Men: ____ . (as fire vessel: 45). Guns: ____ (as fire vessel: 6 x 18★).
Taken by *Romulus* and others 24.5.1797; 1799 temporary fire vessel; 7.7.1800 expended in Dunkirk Roads.

***Galathee*** 1799-1807 (Dutch *Galathea/Galathee?* ____ ) brig sloop? 16?
Dimensions & tons: ____ , ____ x ____ x ____ . 358?
Men: ____ . Guns: ____ . Taken in the Texel 1799; probably laid up on arrival in Britain; 1806 still laid up?; 1807 sold to BU at Rotherhithe [or Chatham?].

## Fire Vessels

Twelve vessels ordered to be purchased as such on 6.3.1794. (Fire vessels were smaller and less well armed than fireships.)

***Amity*** 1794-1802 (Mercantile *Amity?* ____ ____ ) fire vessel ____ .
Dimensions & tons: 67'6", 54'6" x 19'2" x 9'. 106.
Men: 9. Guns: None? Purchased 04.1794; 1802 BU.

***Catherine*** 1794-1801 (Mercantile *Catherine?* ____ ____ ) fire vessel ____ .
Dimensions & tons: 58'6", 43'1½" x 20'4" x 10'5". 95.
Men: 9. Guns: None? Purchased 03.1794; 1801 sold.

***Experiment*** 1794-1801 (Mercantile *Experiment?* ____ ____ ) fire vessel ____ .
Dimensions & tons: 62'9", 50'7½" x 17'9" x 9'7". 85.
Men: 9. Guns: None? Purchased 03.1794; 1801 sold.

***Firebrand*** 1794-1796 (Mercantile ____ ____ ____ ) fire vessel ____ .
Dimensions & tons: 60'1", 46'5¼" x 19' x 12'7½". 89.
Men: 9. Guns: None. Purchased 05.1794; 1796 lighter/slop ship [at Plymouth; 1800 BU].

***Friendship*** 1794-1801 (Mercantile *Friendship?* ____ ____ ) fire vessel ____ .
Dimensions & tons: 50'9", 36'9" x 16'11" x 9'5½". 56.
Men: 9. Guns: None? Purchased 03.1794; 1801 BU.

***Heart of Oak*** 1794-1796 (Mercantile *Heart of Oak?* ____ ____ ) fire vessel ____ .
Dimensions & tons: 51', 37'5¼" x 16'5" x 9'2". 54.
Men: 9. Guns: None? Purchased 03.1794; 1796 sold.

***Industry*** 1794-1795 (Mercantile *Industry?* ____ ____ ) fire vessel ____ .
Dimensions & tons: 59'6", 45'2½" x 17' x 8'5". 70.
Men: 9. Guns: None? Purchased 03.1794; 1795 BU.

***Lively*** 1794-1795 (Mercantile *Lively?* ____ ____ ) fire vessel ____ .
Dimensions & tons: 59'5¾", 48'1½" x 17' x 8'9". 74.
Men: 9. Guns: None? Purchased 03.1794; 1798 sold.

***Olive Branch*** 1794-1802 (Mercantile *Olive Branch?* ____ ____ ) fire vessel ____ .
Dimensions & tons: 62'5½", 48'5" x 20' x 9'. 103.
Men: 9. Guns: None. Purchased 03.1794; 1802 sold.

***Satisfaction*** 1794-1802 (Mercantile *Satisfaction?* ____ ____ ) fire vessel ____ .
Dimensions & tons: 64'3", 49' x 18' x 10'4". 84.
Men: 9. Guns: None? Purchased 03/04.1794; 1802 sold.

***Nancy*** 1794-1801 (Mercantile *Nancy?* ____ ____ ) fire vessel ____ .
Dimensions & tons: 55'6½", 43'0½" x 17'9" x 10'1½". 72.
Men: 9. Guns: None. Purchased 03/04.1794; 1801 sold.

***Akers*** 1794-1801 (Mercantile *Akers?* ____ ____ ) fire vessel ____ .
Dimensions & tons: 57'6", 43'11" x 19'0½" x 9'9½". 85.
Men: 9. Guns: None? Purchased 04.1794; 1801 sold.

***Victoire*** 1797-1801 (French privateer *La Victoire* 1793 ____ ____ ) temporary fire vessel, schooner rig? 2.
Dimensions & tons: 62'10", 50'8" x 16'5" x 7'6". 73.
Men: 20. Guns: 2 x 4. Taken by *Termagant* 26.12.1797; 1801 sold.

## Bomb Vessels

***Explosion*** 1797-1807 (Mercantile *Gloucester* ____ ____ ) ship rig 10.
Dimensions & tons: 96'4", 80'4½" x 27'6" x 12'9". 323 29/94.
Men: 67. Guns: GD 2 x 10" mortars, UD 4 x 68★, QD 4 x 18★, Fc 2 x 18★. Purchased 04.1797; 10.9.1807 wrecked on a reef near Heligoland.

PLANS: Lines & profile/decks & midships section.

***Hecla*** 1797-1813 (Mercantile *Scipio* ____ ) ship rig 10.
Dimensions & tons: (?) 92'9" (given as 62'9" - but this must be an error as it is much too short for a vessel with this tonnage, breadth and armament), 76'5" x 27'2" x 12'5". 300.
Men: 67. Guns: GD 2 x 10" mortars, UD 4 x 68★, QD 4 x 18★, Fc 2 x 18★. Purchased 04.1797; 1813 BU.

***Strombolo*** 1797-1809 (Mercantile *Leander* 1795 North Shields) ship rig 10.
Dimensions & tons: 92'5", 75'2" x 29'0¼" x 12'6½". 323 64/94.
Men: 67. Guns: GD 2 x 10" mortars, UD 4 x 68★, QD 4 x 18★, Fc 2 x 18★. Purchased 04.1797; 1802 laid up until; 1809 BU.

PLANS: As fitted: Profile/midships section.

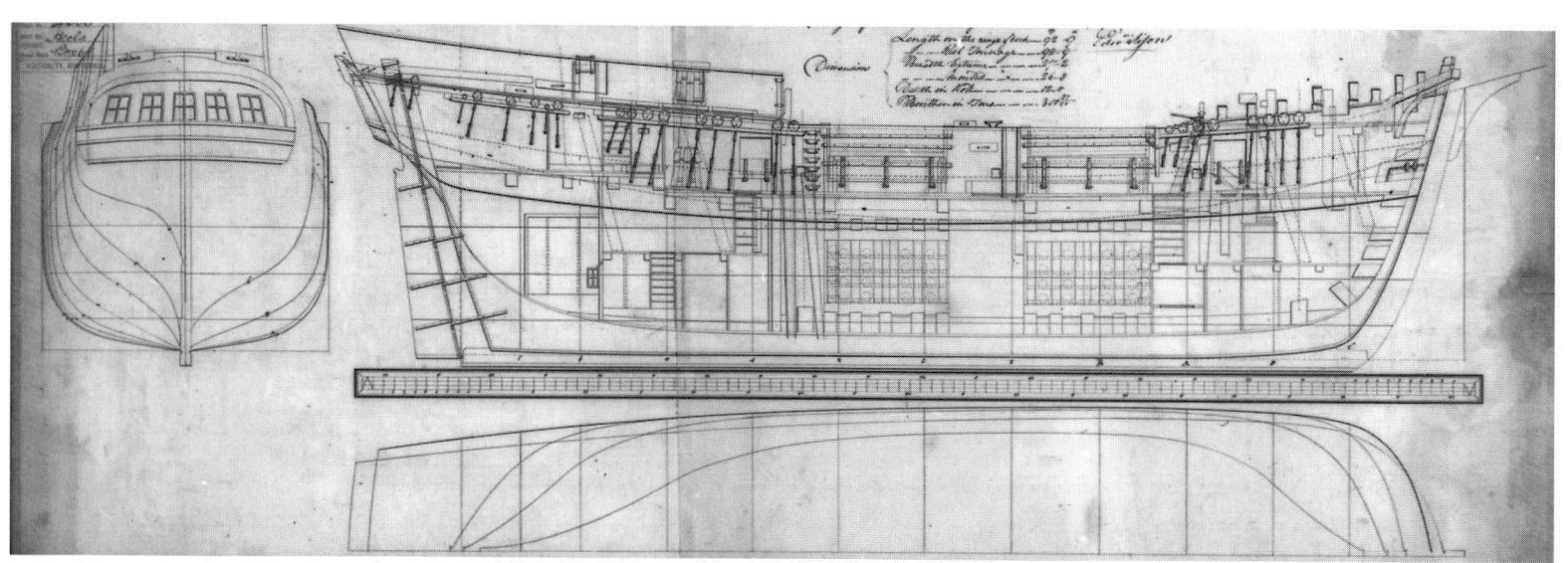

'As fitted' lines and profile (4303) of the purchased bomb vessel *Hecla* of 1797. Note the hinged bulwarks in the way of the mortar mountings

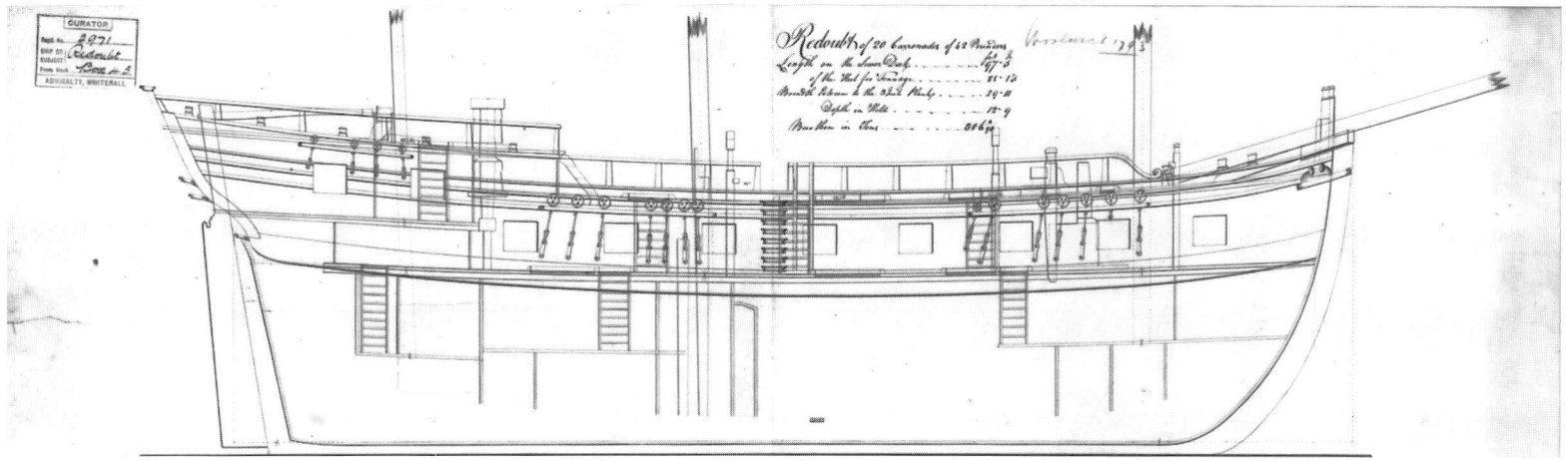

Profile (2971) of the purchased (in 1793) floating battery *Redoubt*

***Sulphur*** 1797-1805 (Mercantile *Severn* ____ ____ ) ship rig 10.
Dimensions & tons: 96'10¼", 79'3½" x 29' x 13'. 354 61/94.
Men: 67. Guns: GD 2 x 10" mortars, UD 4 x 68★, QD 4 x 18★,
Fc 2 x 18★. Purchased 04.1797; 1805 hulked [at Deptford as receiving ship; 1816 sold].
PLANS: Profile. As fitted: Lines & profile/decks and midships section.

***Tartarus*** 1797-1804 (Mercantile *Charles Jackson* ____ ____ ) ship rig 10.
Dimensions & tons: 94'6", 79'10" x 28'6" x 12'2", 344 26/94.
Men: 67. Guns: GD 2 x 10" mortars, UD 4 x 68★, QD 4 x 18★,
Fc 2 x 18★. Purchased 04.1797; 20.12.1804 wrecked on Margate Sands.
PLANS: As fitted: Lines & profile/decks & midships section.

***Volcano*** 1797-1810 (Mercantile *Cornwall* 1796 North Shields) ship rig 10.
Dimensions & tons: 99'9", 83' x 28'10½" x 12'11". 367 52/94.
Men: 67. Guns: GD 2 x 10" mortars, UD 4 x 68★, QD 4 x 18★,
Fc 2 x 18★. Purchased 04.1797; 1810 sold.
PLANS: As fitted: Profile/quarter deck & forecastle/midships section.

***Thunder*** 1797-1802 (Dutch *Die Guter Envatung* ____ ____ ) 8.
Dimensions & tons: ____ , ____ x ____ x ____ . 230.
Men: ____ . Guns: ____ . Taken 1797; 1802 sold.

## Cutters

***Advice*** 1796-1807 (Mercantile *Brilliant* ____ Itchen Ferry?) Advice cutter [schooner rig?] 4.
Dimensions & tons: 45'4", 34'9¾" x 15'10¾" x 6'9½". 47.
Men: 10. Guns: 4 x 3. Purchased in frame (?) late 1796; 1807 sold.

***Fulminante*** 1798-1801 (French *Fulminante?* ____ ____ ) 8.
Dimensions & tons: 44', 33'0½" x 15'2" x ____ . 40.
Men: 40. Guns: 8 x 4. Taken by *Espoir* in the Mediterranean 29.10.1798; 16.3.1801 wrecked on the coast of Egypt.

***Entreprenante*** 1798-1812 (French *Entreprenante* ____ ____ ) 10.
Dimensions & tons: 67', 51'6" x 21'6" x ____ . 123.
Men: 40. Guns: 10 x 12★. Taken 1798; 1812 BU.

***Dolphin*** 1801-1802 (Hired cutter *Dolphin* first hired 1793 ____ ) 12.
Dimensions & tons: 58', 43'4¼" x 20'0½" x 8'6". 93.
Men: 36. Guns: 12 or 4 x 12★ or 6 x 3. Purchased (hired cutter from 1793) in 1801; 1802 sold.

## Schooners

***Spitfire*** 1793-1794 (French *Poulette* ____ ) 4.
Dimensions & tons: 59'4", 53' x 14' x 5'4". 60 or 61.
Men: 40 or 35. Guns: 4 x 3. Taken 1793; 02.1794 capsized off Santo Domingo.

***Flying Fish*** 1793-1795/1796 or 1797-1799 (French *L'Esperanza* ____ ____ ) 4.
Dimensions & tons: 62'7", 52' x 17' x 6'. 80.
Men: 30. Guns: 4 x 3. Taken early 1793 in the West Indies; purchased 06.1793 at Jamaica; 16.6.1795 taken by two French privateers off the Island of Gonaive; renamed *Poisson Volante*?; retaken 6.5.1796 (or 1797?) by *Magicienne* (or *Esperance*?); 1799 out of service.

***Musquito*** 1794-1799 (French privateer *La Venus* ____ ____ ) 4.
Dimensions & tons: 55'4", 46'6" x 16'11" x 6'4". 71.
Men: 30. Guns 4 x 3. Taken 1793; 1794 purchased; 1799 taken by two Spanish frigates off Cuba.

***Swift*** 1794-1802 ( ____ built as a Virginia pilot boat) ____ schooner tender.
Dimensions & tons: 44'4½", 35'10½" x 15'7¾" x 5'3½". 46 67/94.
Men: ____ . Guns: ____ . Engaged in America then taken into service at Cork [?] 1794; 1802 BU.

***Marie Antoinette*** 1795-1797 (French *La Convention Nationale* ex *La Marie Antoinette* ____ ____ ) 10.
Dimensions & tons: ____ , ____ x ____ x ____ . ____ .
Men: 50. Guns: 10 x 4. Taken by Commodore Ford at Cape Saint Nicholas, 09.1795; 09.1797 mutinied and the mutineers handed her to the French in the West Indies.

***Coureuse*** 1795-1799 (French *La Coureuse* 1799 New York) armed schooner/despatch boat 2.
Dimensions & tons: 55'10", 41'11⅝" x 15'9" x 6'5". 55 85/94.
Men: 35. Guns: 2 x 24★ + 12 swivels. Taken 26.2.1795 by *Pomone*; 1799 sold.
PLANS: Taken off & as fitted: Lines, profile & decks.

***Deux Amis*** 1796-1799 (French privateer *Les Deux Amis* ____ ____ ) armed schooner 14?
Dimensions & tons: ____ , ____ x ____ x ____ . 220.
Men: 27. Guns: ____ . Taken by *Polyphemus* 12.1796; 23.5.1799 wrecked on the Isle of Wight.

***Gozo*** 1800-1804 (Spanish *Malta* 1797 American-built) 10.
Dimensions & tons: 80'5", 64'11" x 21'8" x 12'. 162.
Men: 50. Guns: 10 x 4. Taken 1800; sold 1804.

***Sting*** 1800-1802 *Pickle* -1808 (Mercantile *Sting* ____ Bermuda) schooner/armed tender 6.
Dimensions & tons: 73', 56'3¾" x 20'7½" x 9'6". 127.
Men: 30 or 35. Guns: 6 x 18★ or 12★. Purchased 1800; 1802 name changed to *Pickle*; 27.7.1808 wrecked off Cadiz.
NOTE: She was the vessel that carried the news of the victory of Trafalgar and of Nelson's death home to England.

## Floating Batteries

***Redoubt*** 1793-1802 (Mercantile *Rover* ____ ____ ) Sixth Rate 20.
Dimensions & tons: 97'5", 81'1⅛" x 29'11" x 12'9". 386 4/94.
Men: 150 or 146. Guns: 20 x 42★ (alternative 20 x 24 armament may have been fitted at some time in her career). Purchased 1793; 1802 sold.
PLANS: Profile/platforms/lower deck & cabins/upper & quarter decks.

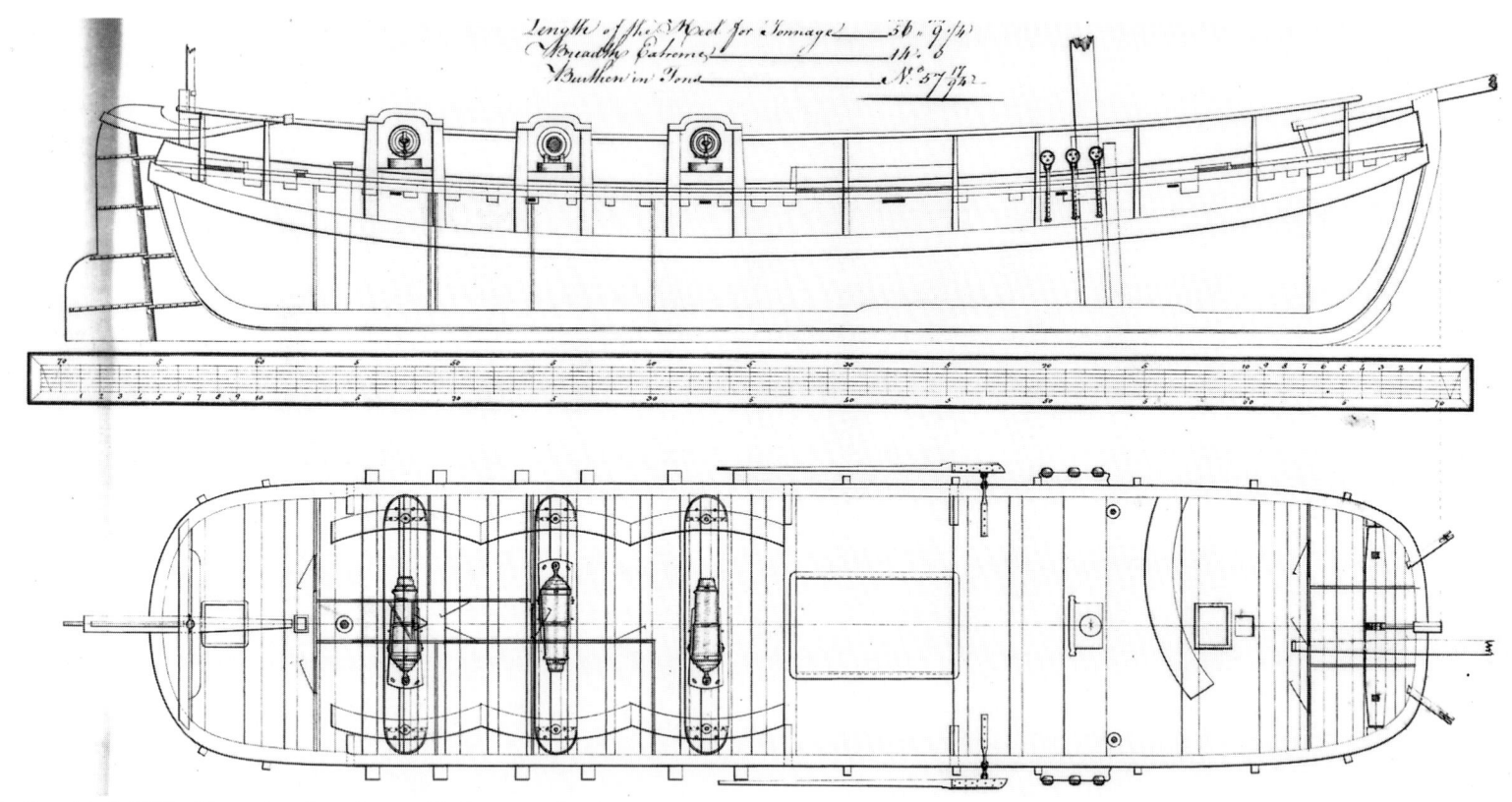

Profile and deck (6420) of the gunboat *Dover* (presumably the one purchased in 1794)

## Gunboats, etc

NOTE: Usual armament for converted hoys [A] = 1 x 24 + 3 x 32★ and for converted barges [B] = 2 x 18 + 1 x 32★.

**Vengeance** 1793-1804 (Mercantile *Lady Augusta?* ____ ____ ) armed galleot or cutter/schooner. ____ .
   Dimensions & tons: ____ , ____ x ____ x ____ . ____ .
   Men: ____ . Guns: ____ . Purchased 21.11.1793; 11.1804 (or 1803?) sold.

**Badger** 1794-1802 (Dutch *Badger?* ____ ____ ) gunboat 3.
   Dimensions & tons: 61', 54'1" x 14'4" x 6'2". 59.
   Men: 30. Guns: [A]. Purchased 02.1794; 1802 sold.

**Bellona** 1794-1799 (river barge ____ ) gunboat 3.
   Dimensions & tons: 61'9", 49'3" x 18'2" x 5'5½". 86.
   Men: 25. Guns: [B]. Purchased 02.1794; 1799 mud barge [at Woolwich; 1805 BU].

**Benjamin and Ann** 1794-1800 (river barge *Benjamin and Ann?* ____ ____ ) gunboat 3.
   Dimensions & tons: 59'3¼", 47'11⅞" x 16'9" x 5'4½". 72.
   Men: 25. Guns: [B]. Purchased 02.1794; 1800 sold.

**Bulldog** 1794-1794 (Dutch hoy ____ ) gunboat 4.
   Dimensions & tons: 64'4½", 56'4" x 13'11" x 6'. 58.
   Men: 30. Guns: [A]. Purchased 10.1794; 1794 sold.

**Caroline** 1794-1802 (river barge *Caroline* ____ ____ ) gunboat 3.
   Dimensions & tons: 64', 50'8⅛" x 19'4" x 5'9". 101.
   Men: 25. Guns: [B]. Purchased 1794; 1802 sold.

**Crown** 1794-1800 (river barge *Crown* ____ ____ ) gunboat 3.
   Dimensions & tons: 57'11", 47'6⅜" x 17'0½" x 5'5½". 73.
   Men: 25. Guns: [B]. Purchased 1794; 1800 sold.

**Dover** 1794; ____ ( ____ ____ ____ ) gunboat 3?
   Dimensions & tons: ____ , 56'9¼" x 14'6" x ____ . 57 17/94.
   Men: ____ . Guns: 3 x ____ . Purchased 1794; not registered; fate unknown. This is possibly the vessel for which a plan exists, taken off at Woolwich Dockyard and dated 1798. The dimensions shown above are taken from this plan which shows a hull built like a Dutch Schuyt or similar barge, with three carronades mounted on slides on the centre line abaft the midships hatch and capable of firing on either broadside.
   PLANS: See above: Profile & deck.

**Defiance** 1794-1797 (Dutch hoy ____ ____ ) gunboat 4.
   Dimensions & tons: 67', 58'0¾" x 15'2¼" x 7'0¾". 71.
   Men: 30. Guns: [A]. Purchased 02.1794; 1797 sold.

**Eagle** 1794-1804 (Dutch *Eagle?* ____ ____ ) gunboat 4.
   Dimensions & tons: 67'9", 60'2⅝" x 14'10½" x 7'3". 71.
   Men: 30. Guns: [A]. Purchased 02.1794; 1796 transport; 1804 sold.

**Ferret** 1794-1802 (Dutch hoy ____ ) gunboat 4.
   Dimensions & tons: 64'1", 57'5⅛" x 14'7¼" x 6'5". 66.
   Men: 30. Guns: [A]. Purchased 02.1794; 1796 transport; 1804 sold.

**Forrester** 1794-1800? (Mercantile *Forrester* ____ ____ ) gunboat 4.
   Dimensions & tons: 71'10", 57'7⅜" x 23'8" x 8'. 71 66/94. [?].
   Men: 31. Guns: 2 x 24 + 2 x 42★. Purchased 03.1794; out of service after 1800. Thames barge type hull ('stumpy' type with swim ends). Fitted under the direction of sir Sidney Smith. The plan shows the two long guns side by side on slides forward.
   PLANS: Profile & deck.

**Four Brothers** 1794-1801 (Mercantile *Four Brothers* ____ ____ ) gunboat 3.
   Dimensions & tons: 60', 46'7½" x 19'6" x 6'. 94.
   Men: 28. Guns: [B]. Purchased 02.1794?; 1801 sold.

**Fury** 1794-1802 (Dutch hoy *Furie?* ____ ____ ) gunboat 4.
   Dimensions & tons: 57'10", 50'2½" x 14'6" x 5'4". 56.
   Men: 30. Guns: [A]. Purchased 04.1794; 1802 sold.

**George** 1794-1798 (barge *George* ____ ____ ) gunboat 3.
   Dimensions & tons: 55', 43'8⅜" x 16'3" x 4'11½". 61.
   Men: 25. Guns: [B]. Purchased 02.1794; 1798 sold.

**Grace** 1794-1798 (river barge *Grace* ____ ____ ) gunboat 3.
   Dimensions & tons: 57'1½", 46'7¼" x 16'9¼" x 4'11". 69.
   Men: 25. Guns: [B]. Purchased 02.1794; 1798 sold.

**Hawke** 1794-1796 (Dutch ____ ____ ) gunboat/galley 4.
   Dimensions & tons: 66'10", 57'9⅝" x 14'0½" x 7'3½". 68.
   Men: 30. Guns: [A]. Purchased 02.1794; 1796 (as storeship?) wrecked at Hayling (or sold?)

**Hope** 1794-1799? (river barge *Hope* ____ ____ ) gunboat 3.
   Dimensions & tons: 57'7", 47'3¼" x 16'6" x 5'3". 68.
   Men: 25. Guns: [B]. Purchased 02.1794; out of service after 1799.

**Hornet** 1794-1795 (Dutch hoy ____ ____ ) gunboat 4.
   Dimensions & tons: 63'6", 53'11½" x 14'3" x 5'11". 60.
   Men: 30. Guns: [A]. Purchased 02.1794; 1795 BU.

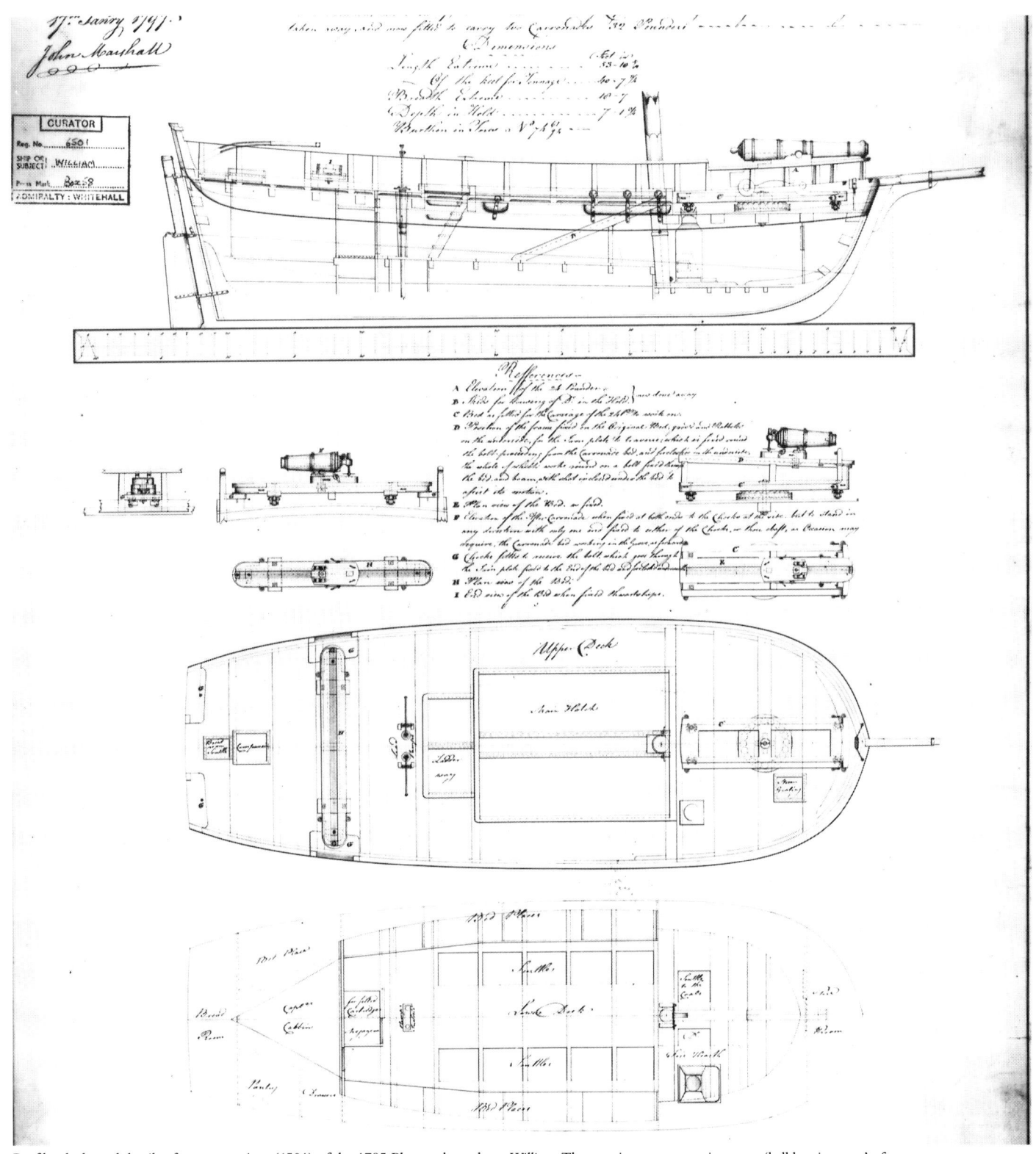

# THE FRENCH REVOLUTIONARY WAR 1793-1801

Profile, decks and details of gun mountings (6501) of the 1795 Plymouth gunboat *William*. The rotating gun mountings on a 'ball bearing race' of cannon balls are of considerable interest as an early and ingenious form of turret mounting

**James and William** 1794-1799 (river barge *James and William* ____ ____ )
 gunboat 3.
 Dimensions & tons: 56'4", 45' x 16'3½" x 5'1½". 63.
 Men: 25. Guns: [B]. Purchased 1794; 1799 sold.

**Leopard** 1794-1796? (Dutch ____ ____ ____ ) gunboat 4.
 Dimensions & tons: 66'4", 57'11½" x 14'6" x 6'3½". 65.
 Men: 30. Guns: [A]. Purchased 02.1794; 1796 BU [?]

**Lion** 1794-1795 (Dutch hoy ____ ____ ) gunboat 4.
 Dimensions & tons: 67', 60'9" x 15'1 ¼" x 6'7". 74.
 Men: 30. Guns: [A]. Purchased 02.1794; 1795 sold.

**Louisa** 1794-1798 (river barge *Louisa* ____ ) gunboat 3.
 Dimensions & tons: 66', 54'3" x 18'1½" x 4'10". 95.
 Men: 25. Guns: [B]. Purchased 02.1794; 1798 sold.

**Mary** 1794-1798 (river barge *Mary* ____ ) gunboat 3.
 Dimensions & tons: 54'7⅞", 44'2¼" x 16'1½" x 4'7". 61.
 Men: 25. Guns: [B]. Purchased 02.1794; 1798 sold.

**New Betsy** 1794-1798 (river barge *New Betsy* ____ ) gunboat 3.
 Dimensions & tons: 59', 46'7⅜" x 18' x 5'4½". 80.
 Men: 25. Guns: [B]. Purchased 02.1794; 1798 sold.

**Nottingham** 1794-1800 (river barge *Nottingham* ____ ) gunboat 3.
 Dimensions & tons: 57', 46'4½" x 16'7¼" x 5'5½". 67.
 Men: 25. Guns: [B]. Purchased 02.1794; 1800 sold.

**Repulse** 1794-1795 (Dutch hoy ____ ____ ) gunboat 4.
 Dimensions & tons: 63', 55'2" x 13'7½" x 6'1". 54.

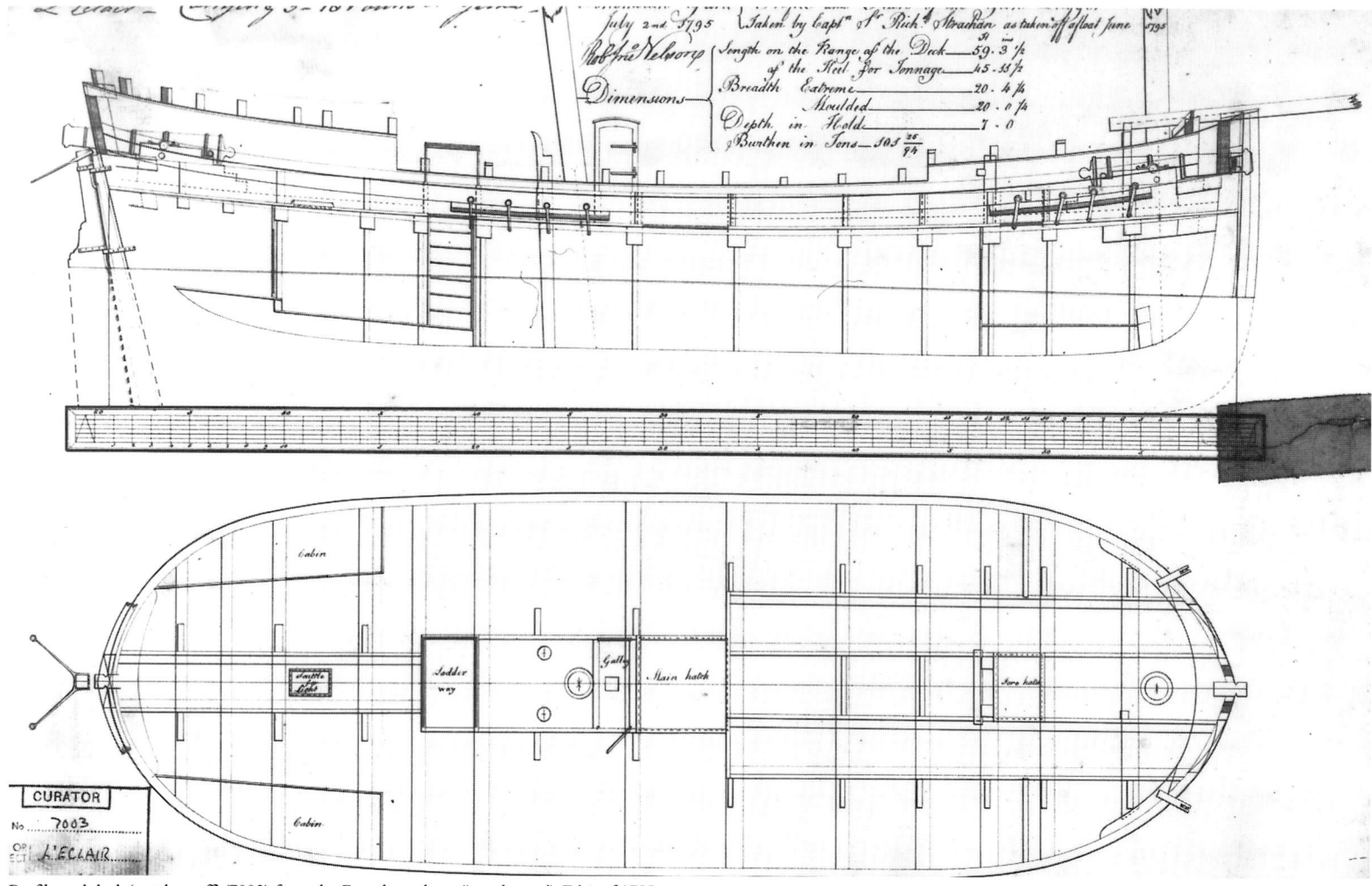

Profile and deck 'as taken off' (7003) from the French gunboat ('gun lugger') *Eclair* of 1795

    Men: 30. Guns: [A]. Purchased 02.1794; 1795 BU.
*Robert* 1794-1799 (river barge *Robert* ____ ) gunboat 3.
    Dimensions & tons: 65'1", 52'10⅛" x 16'6⅝" x 4'4½". 87.
    Men: 25. Guns: [B]. Purchased 02.1794; 1799 sold.
*Scorpion* 1794-1804 (Dutch hoy *Scorpion* ____ ) gunboat 4.
    Dimensions & tons: 66'5", 58'9⅝" x 14'11" x 6'8½". 70.
    Men: 30. Guns: [A]. Purchased 02.1794; 1804 sold.
*Scourge* 1794-1803 (Dutch ____ ) gunboat 4.
    Dimensions & tons: 66'2", 58'7⅜" x 14'8" x 6'7". 67.
    Men: 30. Guns: [A]. Purchased 02.1794; 1803 BU.
*Serpent* 1794-1802? (Dutch ____ ) gunboat 4.
    Dimensions & tons: 64'10", 55'1" x 13'11" x 6'6½". 57.
    Men: 30. Guns: [A]. Purchased 02.1794; Ca 1802 sold?
*Seven Sisters* 1794-1800 (river barge *Seven Sisters* ____ ) gunboat 3.
    Dimensions & tons: 58'1", 47'2½" x 16'8½" x 5'6". 71.
    Men: 25. Guns: [B]. Purchased 02.1794; 1800 sold.
*Shark* 1794-1795 (Dutch ____ ) gunboat 4.
    Dimensions & tons: 64'8", 57'4½" x 14'3¼" x 6'4". 63.
    Men: 30. Guns: [A]. Purchased 02.1794; 11.12.1795 crew mutinied and the mutineers handed her to the French at La Hogue.
*Tiger* 1794-1798 (Dutch hoy ____ ) gunboat 4.
    Dimensions & tons: 67'8", 59'0⅞" x 16' x 6'8". 80.
    Men: 30. Guns: UD: [A]. Purchased 02.1794; 1798 sold.
*Union* 1794-1800 (river barge *Union* ____ ) gunboat 3.
    Dimensions & tons: 59'3", 46'10⅜" x 18' x 5'6". 81.
    Men: 25. Guns: [B]. Purchased 02.1794; 1800 sold.
*Viper* 1794-1801 (Dutch hoy *Viper* ____ ) gunboat 4.
    Dimensions & tons: 66'5", 57'9" x 15' x 6'6". 69.
    Men: 30. Guns: [A]. Purchased 02.1794; 1796 transport; 1801 BU.
*Wasp* 1794-1801 (Dutch hoy *Wasp* ____ ) gunboat 4.
    Dimensions & tons: 64'3", 55'10½" x 14'6¼" x 6'5". 63.

    Men: 30. Guns: [A]. Purchased 02.1794; 1801 sold.
*William* 1794-1801 (river barge *William* ____ ) gunboat 3.
    Dimensions & tons: 59', 47'7½" x 17'9" x 5'6". 80.
    Men: 25. Guns: [B]. Purchased 02.1794; 1801 sold.
*Wilson* 1794-1797? (river barge *Wilson* ____ ) gunboat 3.
    Dimensions & tons: 56'10", 45'2" x 16'10½" x 5'3". 68.
    Men: 25. Guns: [B]. Purchased 02.1794; 1797? transport; fate uncertain.
*Wolf* 1794-1803 (Dutch hoy ____ ) gunboat 4.
    Dimensions & tons: 61'4", 53' x 15'6" x 6'2½". 68.
    Men: 30. Guns: [A]. Purchased 02.1794; 1803 BU.

The following gunboats were all barges purchased in 1795 and both converted and based at Plymouth. No dimensions or other data are available for them - only names and some dates:

*Anna Teresa* No record after 1801.
*Augusta* No record after 1801.
*Brothers* 1802 sold.
*Caroline* 1802 sold.
*Elizabeth* No record after 1801.
*Fowey* No record after 1801.
*Friendship* 2 guns; 9.11.1801 driven ashore near St Malo.
*Martha and Mary/Mary and Martha* No record after 1801.
*Mary and Betsy* No record after 1801.
*Morwelham* No record after 1801.
*Success* 1802 sold.
*William* 1802 sold.
*William and Lucy* 4.11.1801 wrecked in Guernsey Roads.
PLANS: *William*: Profile & decks (2 versions). These show a hoy or cutter-like vessel. One plan shows 1 x 24 forward and 2 carronades aft. A modification to the plan notes the removal of the 24 pounder and 2 x 32★ instead.

*Dixmunde* 1795?-1803? (French *Dixmunde*? ____ ) gunboat ____.

# THE FRENCH REVOLUTIONARY WAR 1793-1801

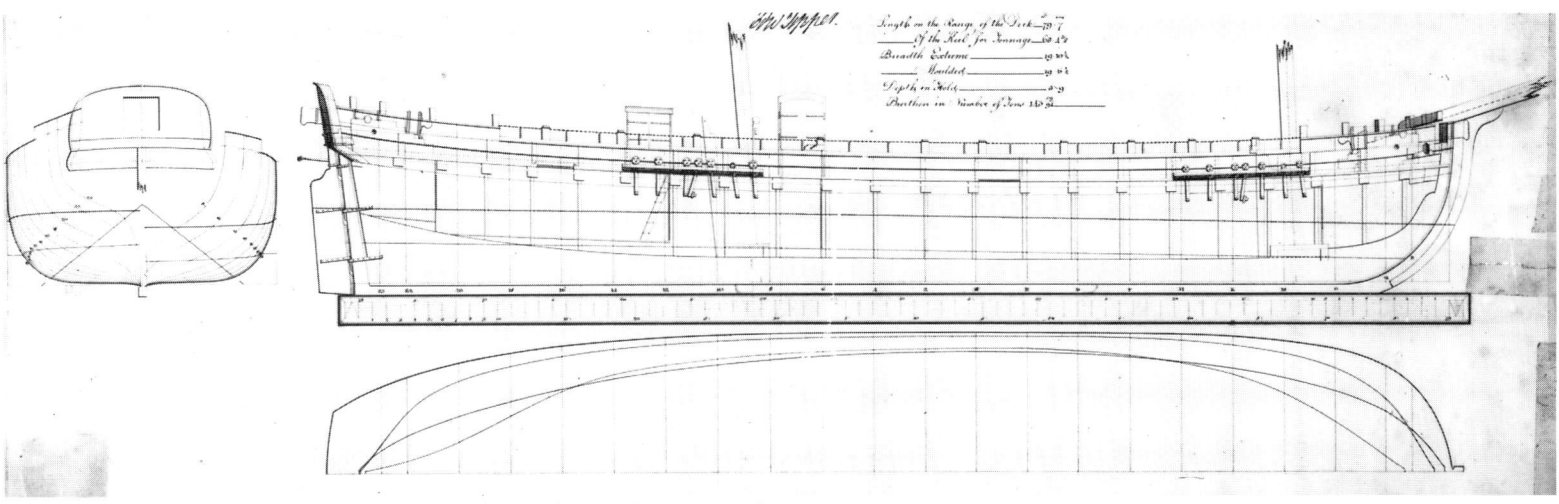

Lines and profile (6361) of the French invasion vessel (gunboat/gunbrig) *Crache Feu* captured in 1795

Dimensions & tons: ____ , ____ x ____ x ____ . ____ .
Men: ____ . Guns: ____ . Purchased or taken 1795?; 1803 still in service; fate uncertain.

***Nieuport*** 1795?-1810? (French? *Nieuport*? ____ ) gunboat 14?
Dimensions & tons: ____ , ____ x ____ x ____ . ____ .
Men: ____ . Guns: 14 x ____ . Purchased or taken 1795?; 1810 still listed; fate uncertain.

***Ostend*** 1795?-1809? (French? ____ ) gunboat 1.
Dimensions & tons: 48'6", ____ x 13'6" x 4'. ____ .
Men: 12 . Guns: 1 x 24. Taken or purchased 1795?; still in service 1809; fate uncertain.

***Eclair*** 1795-1802 ***Safety*** -1802 (French *L'Eclair* ____ ) gunboat (schooner or lugger rig) 3.
Dimensions & tons: 59'3½", 45'11½" x 20'4½" x 6'11". 101.
Men: 46. Guns: 3 x 18*. Taken by Strachan's squadron 9.5.1795; 1796 classed as a schooner; 1802 hulked [at Tortola; renamed *Safety*; 1875 or 1879 BU].
PLANS: 1797: Lines/lines.

***Crache Feu*** 1795-1797 (French *Le Crache Feu* 1795 ____ ) gunboat 3.
Dimensions & tons: 79'7", 68'4⅜" x 19'10½" x 6'3". 144.
Men: 56. Guns: 3 x 18. Taken by Strachan's squadron 9.5.1795; 1797 BU
PLANS: Lines, profile & deck.

***Vesuve*** 1795-1802 (French *Le Vesuve* 1795 Saint Malo) gunboat 3.
Dimensions & tons: 73'6¾", 60'6" x 22'4" x 7'8". 160.
Men: 68. Guns: 3 x 18. Taken by Strachan's squadron 3.7.1795; 1802 sold.

***Manly*** (ex No 37) 1797-1802 (Mercantile *Experiment* ____ ) gunboat 12.
Dimensions & tons: 78', 65'3" x 21'3" x 8'11½". 157.
Men: 50. Guns: UD 10 x 18★, Fc 2 x 18. Purchased 04.1797 at Leith; 1802 sold.

***Mastiff*** (ex No 35) 1797-1800? (Mercantile *Herald* ____ ) gunboat 12.
Dimensions & tons: 71'7", 57'10¼" x 23' x 10'1¼". 163.
Men: 50. Guns: UD 10 x 18★, Fc 2 x 18. Purchased 03.1797 at Leith; 5.1.1800 wrecked on Yarmouth Sands.

***Meteor*** (ex No 34) 1797-1802 (Mercantile *Lady Cathcart* ____ ) gunboat 12.
Dimensions & tons: 74'6", 61'7¼" x 21'8" x 9'. 154.
Men: 50. Guns: UD 10 x 18★, Fc 2 x 18. Purchased 03.1797 at Leith; 1802 sold.

***Minx*** (ex No 36) 1797-1802 (Mercantile *Tom* ____ ) gunboat 12.
Dimensions & tons: 64'4", 51' x 20'10½" x 7'10½". 118.
Men: 50. Guns: UD 10 x 18★, Fc 2 x 18. Purchased 03.1797 at Leith; 1802 sold.

***Pincher*** (ex No 39) 1797-1802 (Mercantile *Two Sisters* ____ ) gunboat 12.
Dimensions & tons: 73'8", 58'10½" x 22'7⅛" x 9'. 160.
Men: 50. Guns: UD 10 x 18★, Fc 2 x 18. Purchased 03.1797 at Leith; 1802 sold.

***Pouncer*** (ex No 38) 1797-1802 (Mercantile *David* ____ ) gunboat 12.
Dimensions & tons: 76'10", 63'8" x 22'1¾" x 8'9½". 165.
Men: 50. Guns: UD 10 x 18★, Fc 2 x 18. Purchased 03.1797 at Leith; 1802 sold.

***Rattler*** (ex No 41) 1797-1802? (Mercantile *Hope* ____ ) gunboat 12.
Dimensions & tons: 72'7", 59'3" x 22'4" x 9'4¾". 158.
Men: 50. Guns: UD 10 x 18★, Fc 2 x 18. Purchased 03.1797 at Leith; 1802? sold.

***Ready*** (ex No 42) 1797-1802 (Mercantile *Minerva* ____ ) gunboat 12.
Dimensions & tons: 70'2", 56'5" x 22'6" x 9'6". 152.
Men: 50. Guns: UD 10 x 18★, Fc 2 x 18. Purchased 03.1797 at Leith; 1802 sold.

***Safeguard*** (ex No 43) 1797-1802 (Mercantile ____ ) gunboat 12.
Dimensions & tons: 78'9", 66'9" x 22' x 8'10½". 172.
Men: 50. Guns: UD 10 x 18★, Fc 2 x 18. Purchased 03.1797 at Leith; 1802 sold. [Winfield has sighted evidence to suggest she carried 24 pounders, not 18s, on the Fc.]

***Wrangler*** (ex No 40) 1797-1802 (Mercantile *Fortune* ____ ) gunboat 12.
Dimensions & tons: 72'3", 60' x 20'9½" x 8'4". 138.
Men: 50. Guns: UD 10 x 18★, Fc 2 x 18. Purchased 03.1797 at Leith; 1802 sold.

***Staunch*** (ex No 44) 1797-1803 (building by Nicholson, Rochester) gunboat 12.
Dimensions & tons: 68'4", 53'11" x 23'1" x 9'10". 153.
Men: 50. Guns: UD 10 x 18★ + 2 x 24 as chase guns. Purchased in frame 03/04.1797; launched 1.5.1797; 1803 sold.

***Royalty*** 1797; ____ ? ( ____ ____ ) gunboat ____ .
Dimensions & tons: ____ , ____ x ____ x ____ . ____ .
Men: ____ . Guns: ____ . Purchased 1797 at Portsmouth; no further record.

***Judith*** 1798; ____ ? ( ____ ____ ) gunboat ____ .
Dimensions & tons: ____ , ____ x ____ x ____ . ____ .
Men: ____ . Guns: ____ . Purchased 1798?; fitted at Portsmouth; no further record.

***Liberty*** 1798?; ____ ? ( ____ ____ ) gunboat ____ .
Dimensions & tons: ____ , ____ x ____ x ____ . ____ .
Men: ____ . Guns: ____ . Purchased 1798?; fitted at Portsmouth; no further record.

***Polly*** 1798?; ____ ? ( ____ ____ ) gunboat?
Dimensions & tons: ____ , ____ x ____ x ____ . ____ .
Men: ____ . Guns: ____ . Purchased 1798?; fitted at Portsmouth; no further record.

***Two Brothers*** 1798?; ____ ? (French? *Deux Frères*? ____ ____ ) gunboat ____ .
Dimensions & tons: ____ , ____ x ____ x ____ . ____ .
Men: ____ . Guns: ____ . Purchased 1798? (or taken 1799?); fitted at Portsmouth; 1802 recruiting tender; no further record.

***Netley*** 1798-1806; it is uncertain whether there was a separate gunboat of this name at the time or whether there is just a confusion with Bentham's schooner gunboat of this name.

***Contest*** 1799-1803 (Dutch *Hell-Hound* ____ ) gunboat 5.
Dimensions & tons: ____ , ____ x ____ x ____ . ____ .

Men: 35. Guns: UD 3 x 24★, FC 2 x 32★. Taken 1799; 1803 BU.
***Cruelle*** 1800-1801 (French 'chaloupe cannonière' ____ 1793 ____ ) gunboat, cutter rig 8.
Dimensions & tons: ____ , ____ x ____ x ____ . ____ .
Men: 30. Guns: UD 8 x 4. Taken by *Mermaid* 1.6.1800; 1801 sold.
***Cerbere*** 1800-1804 (French *Le Cerbère* 1793 Dunkirk or *Le Chalier* 1794 Cherbourg?) gunboat 10.
Dimensions & tons: 79'6½", 66'7" x 19' x 6'1". 138.
Men: 50. Guns: UD 10 x 18★. Taken by *Viper* off Mauritius 29.7.1800; 19.2.1804 wrecked on Beachy Head.
***Dangereuse*** 1800-1802 (French *Dangereuse*? 1796 ____ ) gunboat ____ .
Dimensions & tons: ____ , ____ x ____ x ____ . ____ .
Men: ____ . Guns: ____ . Taken 1800; 1802 (or 1800?) sold.
***Havre de Grace*** 1800- ____ ? (French *Havre de Grace*? ____ ) gunboat ____ .
Dimensions & tons: ____ , ____ x ____ x ____ . ____ .
Men: ____ . Guns: ____ . Taken 1800; no further information.
***Prince of Wales*** Ca 1801 ( ____ ) 'Row Galley' (gunboat) ____ .
Dimensions & tons: 67'9", ____ x 16'4" x 6'5½". ____ .
Men: ____ . Guns: ____ . At Sheerness in 1801 when the plan was made.
PLANS: Lines & profile.

## Storeships, etc

***Supply*** 1793-1802? (Mercantile *New Brunswick* ____ American-built) armed vessel 10.
Dimensions & tons: 97'4", 84' x 29'5½" x 16'5½" 388.
Men: 50. Guns: UD 10 x 4. Purchased 10.1793 ('... almost a new ship when purchased and is said to have been built in America of Black Birch'); 1801 still in service; fate unknown but probably discarded in 1802.
***Reliance*** 1793-1800 (Mercantile *Prince of Wales* ____ South Shields) armed vessel 10.
Dimensions & tons: 96'1", 80'7¼" x 30'4" x 13'. 394.
Men: 87. Guns: 10 x 4. Purchased 12.1793; 1800 hulked [at Sheerness as a receiving ship; sold 1815].
PLANS: 1802 as receiving ship: Profile/orlop & lower deck/upper decks.
***Etrusco*** 1795?-1798 (Spanish or French or Mercantile *Etrusco*? ____ ) storeship 16.
Dimensions & tons: 137'8", 116'5⅞" x 38'6" x 13'6". 919.
Men: 125. Guns: 16 x 6. Taken 1795?; 1795 registered; 25.7.(or 15.8)1798 foundered coming from the West Indies.
***Princess*** 1795-1797 (Dutch *Willemstadt* or *Willemstadten Boetzelaer* ____ ) storeship 26/18.
Dimensions & tons: 140'6", ____ x 32'2" x ____ . 677.
Men: ____ . Guns: 26/18 x 6. Taken at Simons Bay 14.9.1795; 1797 hulked [in Ireland as a flagship and receiving ship at ____ ; ?1814? out of service?].
***Euphrosyne*** 1796-1802 ( ____ ) armed vessel 14? (brig rig?).
Dimensions & tons: ____ , ____ x ____ x ____ . ____ .
Men: 125. Guns: 14 x ____ . Purchased 1796; may have served as a fireship for some time; 1802 sold.
***Buffalo*** 1797-1809 (Mercantile *Fremantle* building Dudman, Deptford) armed transport/storeship 10.
Dimensions & tons: 109'2", 90'6¼" x 31' x 13'2½". 462 66/94.
Men: 33. Guns: 10 x 6. Purchased whilst on the stocks 1797; 3.11.1797 launched; 1809 hulked [as an Army prison ship at Cowes, Isle of Wight; 1817 sold].
PLANS: As fitted: Lines & profile/deck.
***Investigator*** (ex ***Xenophon***), ***Selby*** and ***William*** etc. See under ship sloops above.
***Empress Mary*** 1799-1804 (Mercantile *Empress Mary* 1797 ____ ) storeship 12?
Dimensions & tons: 123', 105'3¼" x 34'1½" x 13'7". 652 2/94.
Men: 62. Guns: 12 x 24★? Purchased 17.4.1799; 1804 breakwater.
***Abundance*** 1799-1816 (building Adams, Bucklers Hard) storeship 16.
Dimensions & tons: 142'4", 119'2¼" x 32'7" x 15'. 673.
Men: ____ . Guns: UD 16 x 24★. Purchased whilst completing 06.1799; 1816 hulked [for the Committee for Distressed Seamen as an accommodation ship; 1819 storeship for Saint Helena (never sent?); 1821 laid up at Deptford; 1823 sold].
PLANS: Building: Lines/orlop/main deck/upper deck/stern/after part of spar deck.
***Porpoise*** 1799-1803 (Spanish sloop *Infanta Amelia* ____ ) transport 10.
Dimensions & tons: 93', 73'10" x 28' x ____ . 308.
Men: ____ . Guns: 10 x 6. Taken in the West Indies 1800; 1803 wrecked on a reef off the coast of New South Wales.

## Unregistered Vessels

(ie. those not listed in the main Admiralty sources consulted.)
***Abeille*** 1796- ____ ? (French *Abeille* ex *Bonnet Rouge* 1795 ____ ) cutter 14. Taken off the Lizard by *Dryad* 2.5.1796; no further record.
***Active*** 1800 taken in the West Indies; no other information.
***Admiral Rainier*** 1800 (Dutch ____ ) brig taken at Batavia 08.1800 by Rainier's squadron.
***Advice*** (?) 1793-1793 (French? ____ ) schooner 4. Purchased Jamaica 05.1793; 12.1793 wrecked on the Motherbank.
***Albanaise*** 1800-1800 (French *L'Albanaise* 1799 ____ ) brig 14. 88'7¾", 68'7½" x 24'3½" x 11'8". 238. Taken off Cape Fano by *Phoenix* 3.6.1800; 23.11.1800 seized by mutineers and handed to the Spaniards at Malaga, who probably returned her to the French.
***Alerte*** 1793-1793 (French *Alerte*? 1787 ____ ) Brig 14. Taken 08.1793; retaken by the French 18.12.1793.
***Alexander*** or ***Alexandria***? 1796-1802 tender 6. Purchased in the West Indies 1796; 1802 BU.
***Amboyna*** 1796-1802 (Dutch *Harlingen/Haarlem* ____ ) brig 10, 180 tons, taken at Amboyna 02.1796; 1802 BU.
***Ant*** 1798-1815 (French ____ or Mercantile *Lively Lass*? ____ ) schooner/armed tender 4. 60'7", 45'5" x 18'10¼" x 8'7". 86 37/94 tons. 4 x 12? Taken 06.1797; 1798 purchased; 1815 sold.
PLANS: Taken off 1803: Lines and profile/deck.
***Argus*** 1794-1798 purchased lugger, 8 guns; 1798 taken by French privateer *Vendemaire*.
***Athenienne*** 1796-1802 (French *Athénienne*? 1796 ____ ) Gun brig 14. 200? tons. Taken 1796; 1802 BU.
***Augustus*** 1796-1801 gunvessel 1 or 2 guns; 7.7.1801 wrecked in Plymouth Sound.
***Aurore*** 1793-1795 (French 12 pounder frigate *L'Aurore* ex *L'Envieuse*? 1769 Rochefort, Chevillard Jnr design) frigate 36/32 . 130', ____ x 33'6" x 17'. 700. Taken at the fall of Toulon 1793, 1795 prison ship [at Gibraltar fate uncertain (1802 sold? or 1803 BU?)].
***Aventurier***? 1798-1799 (French *Aventurier*? 1793 ____ ) Gunbrig ____ . 192 tons? Taken 1798; 1799 deleted.
***Barbara***? 1796- ____ (French? ____ ) Schooner ____ . Acquired 1796; out of service by 1800.
***Belette*** 1793-1796 (French corvette *La Belotte* 1781 Toulon, Coulomb design) armed ship/sloop 16/Sixth Rate 28 (?). 120', ____ x 32 'x 16'. 600. Taken 1793; 20.10.1796 burnt as unserviceable at Ajaccio, Corsica.
***Berbice***? 1793-1796 (French *Berbice* ____ ) Schooner? 8? 121 tons. Taken 1793; 1796 lost?
***Bermuda*** 1795-1796 (Building Bermuda) Brig sloop 14. 1795 purchased whilst building; 1796 lost.
***Bien Aime*** 1793-1794 (French *Bien Aimé*? ____ Indiaman) sloop 20 guns. French Indiaman taken at Calcutta 1793; 1794 BU.
***Bishop*** 1797-1802 Gun vessel 1, on the River Shannon. 6 men? Acquired 1797; 1802 sold.
***Blonde*** 1793-1794 (French *Blonde* 1781 Toulon, Coulomb design) Sixth Rate 28. Taken off Ushant by *Latona* and *Phaeton* 28.11.1793; 1794 sold.
***Bloom*** 1795-1797 Tender 14. Purchased 1795; 24.2.1797 taken by the French off Holyhead. (Probably same ship as, or confused with, *Brighton*; see below.)
***Brighton*** 1795-1797 tender 14. purchased 1795; 24.2.1797 taken by the French off Holyhead. (See *Bloom* above.)

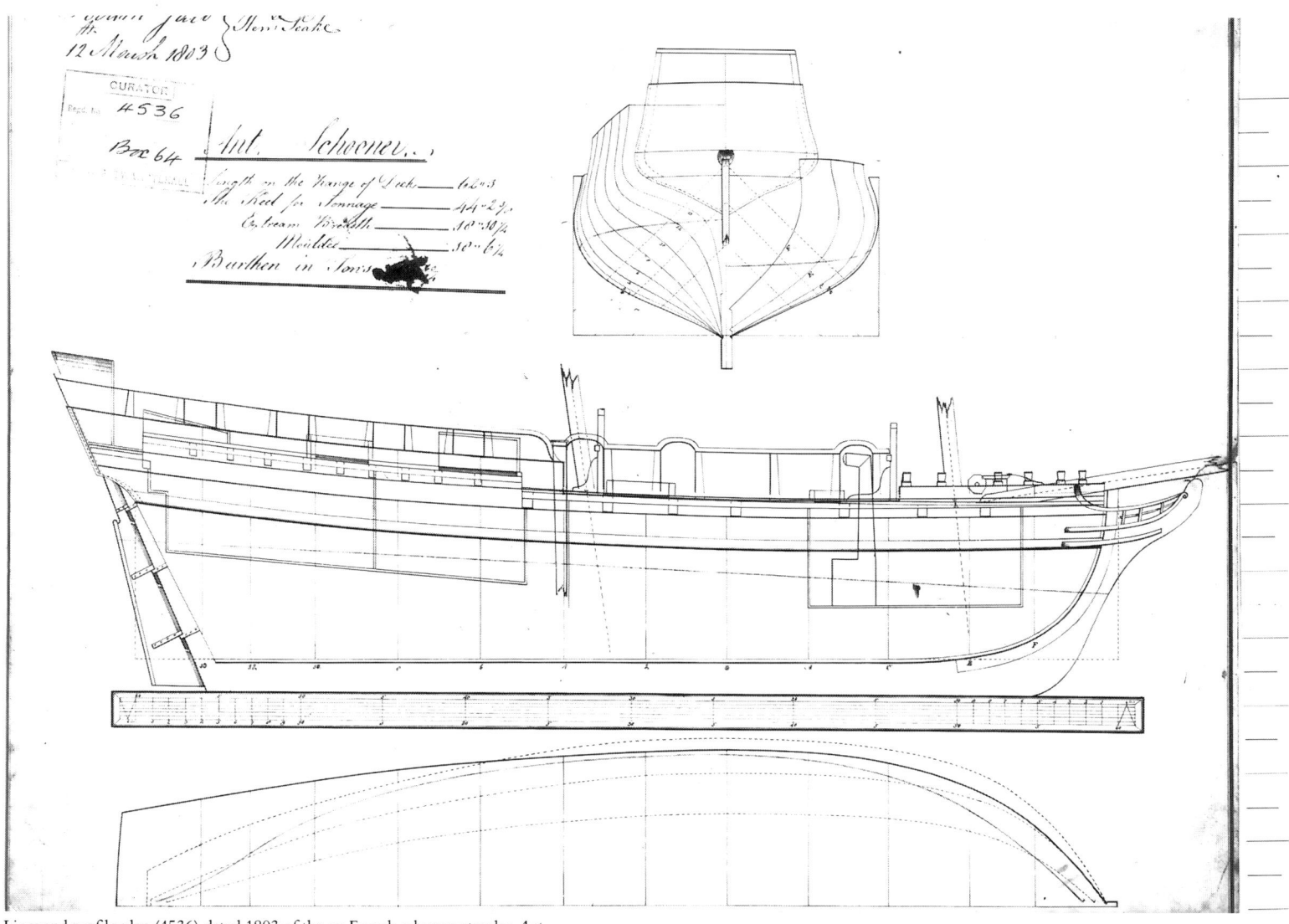

Lines and profile plan (4536) dated 1803 of the ex French schooner tender *Ant*

***Cacafuego*** 1797-1804? captured (Spanish?) mortar vessel; served as a gunboat in the Mediterranean in 1799; still in service 1804?

***Ca Ira*** 1795-1796 (French *Le Ça Ira* ex *La Couronne* 1766 Brest, Groignard design) 80, Third Rate. 2210 tons. Guns: GD 30 x 32, UD 32 x 24, QD 14 x 12 + 8 x 32★ + 2 x 42★, Fc 4 x 12. Taken in Hotham's Action 14.3.1795; 11.4.1796 burnt, by accident in San Fiorenzo Bay.

***Camden*** 1797-1801 Gunvessel 1 on the River Shannon. Acquired 1797; 1801 deleted.

***Campbell*** 1796-1803 Schooner or cutter 4? Purchased in the West Indies 1796; 1803 sold.

***Censeur*** 1795-1795 (French *Le Censeur* 1782 ____ ) Third Rate 74. 1820 tons. Taken in Hotham's Action 14.3.1795; 7.10.1795 retaken by French squadron off Cape Saint Vincent; renamed *La Révolution*, 1799 transferred by France to Spain. Found to be rotten and never served in the Spanish Navy.

***Chance*** 1799-1800 (Spanish *Galgo* ____ ) brig/sloop 18/16/14. 99', 89' x 29' x 13'. 395 tons. Taken by *Crescent* on passage to the West Indies 15.11.1799; 9.10.1800 capsized in a squall in the West Indies.

***Charlotte*** 1798-1799 schooner 10/8, 60 men. Purchased 1798; 02.1799 taken by the French off Cape Francois and renamed *Le Vengeur*, 1799 retaken by *Solebay* off Cape Tiberon and then BU?

***Charlotte*** 1800-1801 purchased schooner 6 x 3. 60 men. Purchased 1800; 28.3.1801 wrecked on the Isles des Vaches/Isle of Ash, West Indies.

***Contest*** 1799-1799 Gunboat? 14 x 24★ 109 tons. Purchased 1799; 1799 BU (Colledge only?).

***Convert*** 1793-1794 (French *L'Inconstante* 1790 ____ ) frigate 36. 930 tons. Taken off San Domingo by *Penelope* and *Iphigenia* 23.11.1793; 18.3.1794 wrecked on a reef of rocks off Grand Cayman [site now being investigated by archaeologists].

***Coquille*** 1798-1798 (French *Coquille* ____ ) 36 guns 916 tons. 12.10.1798 Taken off the coast of Donegal; 14.12.1793 burnt by accident at Portsmouth (before actually being taken into the RN?).

***Cornwallis*** brig, purchased Ca 1798 in the East Indies or at the Cape; 07.1798 paid off.

***Courageuse/Courageux*** French 32 guns taken by Markham's squadron 18 or 19.6.1799 in the Mediterranean; receiving ship in the Mediterranean; 1803 still in service.

***Cruelle*** 1798-1801 (French *Cruelle* ____ ) Schooner/lugger? 10? 110 tons. Taken 1798; 1801 sold.

***Dame de Grace*** 1799-1799 (French *La Dame de Grace*? ____ ) gunbrig 4. 87 tons. Taken at Mount Carmel by the boats of *Tigre*; 8.5.1799 retaken and sunk by the French *Salamine*.

***Dauphin Royal*** 1796-1801 schooner 8. Purchased in the West Indies 1796; 1801 out of service.

***Desperate*** 1799-1811 schooner 8. In service 1799-1811, no further information.

***Diligent*** 1800-1814 French storeship. Taken 1800; 11.8.1814 sold.

***Dispatch*** 1797-1800 tender 6. Allocated to the Transport Office.

***Dolphin*** 1799-1802 (Dutch *Dolfijn* 1774 ____ ) Sixth Rate/troopship 20. 505 tons. Taken by *Wolverine* and *Arrow* on 15.9.1799; 1800 troopship; 1802 storeship; no further record.

***Drake*** 1798-1804 (French privateer *Tigre* 1798 or 1793 ____ ) brig sloop 16/14 x 6 + 12 swivels. 79', 76'3" x 23'9" x 10'6". 212 tons. 86 men. Taken 1798 by *Melpomene*; 12.7.1804 or 3.9.1804 wrecked on a shoal off Nevis.

***Earl of Chatham*** 1792?-1799? gun vessel noted as purchased 1792, which seems not particularly likely; 1793 being much more probable (which is why the vessel is noted in this chapter); out of service by 1800.

***Eclair*** 1801-1807 ***Safety*** -1807 (French *Edair* 1799 Nantes) Schooner. 10 x 18★ + 2 x 24 (or 12 x 12★?) 145 tons. 1801 taken; 1807 renamed *Safety* and then sold (or name not changed until 1809 when she was renamed *Pickle*, not be sold till 1818).

*Enterprise* 1801 schooner in the West Indies.

*Espiegle* 1794-1802 (French *Espiegle* ____ ). Gunbrig 10. 135 tons. Taken 1794; 1802 deleted.

*Eveille*? 1795-1796 (French *Eveille* 1778) Brig sloop 16? Taken 1795; 1796 deleted.

*Ferret* 1799-1799 schooner 6. Origins obscure (purchased 1799?) , in service in the West Indies when taken ____ 1799 by the Spaniards [incorrect report?].

*Fleche* 1794-1795 (French *La Flèche* 1768 Toulon, Chapelle Jnr design) ship sloop 18. 92'9", 74'4" x 26'6" x 11'. 379 tons. 75 men. 16 x 18★ + 2 x 18. Taken at Bastia 21.5.1794; 12.11.1795 wrecked in San Fiorenzo Bay.

*Forte* 1799-1801 (French *La Forte* 1794 Rochefort or Lorient, Caro design?) 24 pounder, 50. 170', 139'2" x 43'6" x ____ 1401 tons. 343 men. UD 30 x 24, QD 16 x 32★, Fc 2 x 12 + 4 x 32★. Taken 18.2-1.3.1799 by *Sybille* off the Sand Heads; 28.1.1801 wrecked on a rock at the entrance to the harbour at Jeddah.

*Fortune* 1798-1799 (French 'demi-chebec' *La Fortune* 1798 Egypt) sloop 18. 150 tons. Taken off the Nile by *Swiftsure* 11.8.1798; 8.5.1799 retaken by the French *Salamine* and sunk off the Syrian coast at Jaffa.

*Foudre* 1800-1801 (French *Foudre* ____ ). Gunbrig 10. 192 tons? Taken 1800; sold 1801.

*Fox* 1799-1799 schooner 18/14. 150 tons. Origins uncertain (taken 1799? or 1797?); 28.9.1799 lost in the Gulf of Mexico on a sandbank near the Island of Providence.

*Frederick*? 1795-1799 (Mercantile *Frederick*? ____ ). Armed ship/schooner 10 guns, 60 men. Acquired (or hired?) 1795, Leeward Islands; out of service 1799?

*Garland* 1798-1803 Cutter/schooner? 6? Acquired 1798; 1803 sold.

*General Duff* 1797-1802 Gunboat 1 on the River Shannon. Acquired 1797; 1802 sold.

*General Lake* 1797-1802 Gunboat 1 on the River Shannon. Acquired 1797; 1802 sold.

*George* 1796-1798 cutter/sloop (sloop-rigged vessel) 6. 105 tons. 44 men. Origins uncertain (taken from the French 1796?); 3.1.1798 taken by two Spanish privateers en route from Martinique to Demerara.

*Gipsy* schooner. Two vessels listed, which may possibly be references to the same ship: (1) 10 x 4, 40 men. In service 1799 as a tender to the flagship at Jamaica and still in service in 1801. (2) 69' length. 121 tons. Purchased 1803 or 1804?; 1808 sold.

*Good Intent* gunboat/gunvessel. Origins obscure, possibly purchased in 1792; ____ 1793 taken by the French.

?*Goonong Assi*? 1797-1800 Fire vessel. Acquired 1797; 1800 deleted.

*Harlequin* 1796-1802 schooner 8. Purchased (in the West Indies?) 1796; in service until 1802.

*Hart* 1793-1817 cutter (built at Deptford?); 1817 sold.

*Hawke* 1795-1796 cutter in service 1795; 1796 sold to Mr Power (is this the same as the 1794-1796 gunboat?).

*Hector* 1797 bomb vessel. 92'9", 76' x 27'2" x 12'5". 300 tons. 60 men. Purchased 1797; no further information.

*Hind* 1793 schooner; no further information (listed in Pitcairn Jones only).

*Hosier* 1801-1802 brig?; no further information.

*Hussar* 1798-1800 (French privateer *Le Hussard* ____ ) sloop 14. 105'2", 86'3" x 30'2" x 8'5". 413. (Note that these figures are virtually identical with those given for the ship sloop *Surinam*; there is probably confusion here as both are given as having the same French name; it seems likely that they are correct for *Surinam* but not for this vessel.) Taken 10.1798 by *America*; 1800 sold.

*Impetueux* 1794-1794 (French *Impetueux* 1787 Rochefort) Third Rate 74. 182', 149'8¼" x 48'7¾" x 21'6". 1884 16/94 tons. Taken 1794; 24.8.1794 accidentally burnt in Portsmouth Harbour.

*James and Elizabeth* 1796-1802 (Mercantile *James and Elizabeth* ____ ) Gunboat ____ . Acquired 1796; 1802 deleted.

*Janissary* 1801-1802 Gunboat ____ . Acquired 1801; 1802 sold.

*Jean Bart* 1793 (French) gunboat 2 x 8 + 2 x 6. Taken at Toulon 1793; no further information.

*Kent* 1798-1802 Gunboat ____ . Acquired 1798; 1802 deleted. (Presumably this is the pioneer - and unsuccessful - steamboat *Kent* in whose trials for the Earl of Stanhope the Admiralty had taken an interest. Built Marmaduke Stalkartt, Rotherhithe. K 10.1792; L 03.1793. 111' x 21' x 10'. 200 tons. The steam machinery - which drove a number of 'duck paddles' and consisted of two cylinders providing about 12 HP - was removed before or during her conversion to a gunboat. See HP Spratt, *The Birth of the Steamboat* London, 1958).

*Kingsmill* 1797-1802 Gunboat 1 on the River Shannon. Acquired 1797; 1802 sold.

*Lacedemonian* 1796- 1797 (French privateer *La Lacédemonienne* 1794/5 ____ ) brig/sloop 16. 195 tons. 120 men. Taken in the West Indies by *Charon* and *Pique* 05.1796; 17?.5.1797 retaken by the French in the West Indies.

*Lady Nelson* 1800-1825 brig with sliding keels 10. 52'6", ____ x 17'6" x ____ 60 tons. Designed by Captain Schank and built at Deptford in 1799; Purchased 1800; 1825 destroyed by Natives at Babber Island, near Timor; [this wreck has been dived on recently.]

*Legere* 1798-1801 (French *La Légère* ____ ) gunvessel 6. Taken by *Alcmene* 22.8.1798; in service until 1801.

*Leighton* 1798-1803? Sixth Rate 22 (all-carronade armament). Purchased 1798; out of service by 1804 (deleted 1800?).

*Lively* 1799-1800? purchased storeship on American station; in service 1800.

*Lord Nelson* 1801?-1807? storeship. Purchased 1800 or 1801 in the East Indies; sold about 1807.

*Louisa* 1798-1802 (Mercantile *Louisa* ____ ) Gunboat. Acquired 1798; 1802 deleted.

*Love and Friendship* 1796-1803 (Mercantile *Love and Friendship* ____ ) gunboat acquired 1796; 1803 deleted.

*Margaret* 1795-1798 tender, 25 men. Acquired 1795; 11.1798 lost on the Irish Coast, driven ashore near Innishboffin Island.

*Marianne* 1799-1801 (French) gunbrig, 12. Taken 18.3.1799; 09.1801 sold (Impress Service tender?).

*Marie Rose/Mary Rose* 1799-1801 (French *Marie Rose* ____ ) Gunboat/gunbrig 4 guns. Taken 18.7.1799 off Acre; 1801 sold.

*Marsouin* 1795-1798 (French *Marsouin* 1787 Bayonne) Sloop 16? Taken 1795-1798 deleted.

*Mary* 1797-1807? (Mercantile *Mary*? ____ ) tender 6. Purchased 1797; in service until 1807? (or deleted 1800?).

*Mary and Lucy* 1796-1799 gunboat. 19 men.

*Mary Ann* 1800-18 ____ (French ____ ) Gunboat ____ . Taken 1800; no further information.

*Mary Anne* armed cutter, origins obscure; 25.7.1795 wrecked on the North Sand Head.

*Medee* 1800-1802 (French 12 pounder frigate *La Médée* 1778 Saint Malo, Guignace design) frigate 36. Taken by two East Indiamen off Rio 5.8.1800; 1802 prison ship; 1805 sold.

*Mermaid* (No 3) 1799-1800 gunboat. Acquired 1799 for the defence of Belize; 1800 sold.

*Mignonne* 1794-1797 (French corvette *La Mignonne* 1783 ____ ) frigate 32 or 28 (9 pounder). 684 tons. Taken at Calvi 10.8.1794; 31.7.1797 burnt as unserviceable at Porto Ferrajo.

*Minorca* sloop 18 laid down at Port Mahon 1799; handed over to Spain incomplete at the cession of Minorca 1802 (had been laid down at the instigation of Duckworth to keep the men busy; information from Captain A B Sainsbury, RNR).

*Mohawk* 1798?-1799 sloop in service 1798?; 1799 taken (listed in Colledge only).

*Mongoose* 1800-1803 (Dutch ____ ) Gunboat/armed ship/sloop 12 guns, 35 men, 140 tons? Taken 1799; Acquired 1800; served in the East Indies; 1803 sold.

*Montego Bay* 1796-1800 schooner 10. Purchased 1796 on the Jamaica station; in service until 1800.

*Mosquito*? 1799-1803? ( ____ *Hunter* ____ ) Schooner 12. Taken 1799; 1803 sold?

*Mulette* 1793-1796 (French 'gabare' *Le Mulet* 1784 ____ ) armed ship. 112', ____ x 27' x 13'6". 400 tons. Handed over at Toulon 1793; in service until 1796.

*Mustico/Manstice* 1800-1816 (Spanish *Manstice/Mustico?* 1796 Catalonia) lateen settee/cutter. 49'6", 38'3" x 17' x ____ . 59 tons. Taken 1800; in harbour service until 1816.

*National Cockade*? 1795-1796? (French *Cocarde Nationale*? ____ ) Schooner 14. Taken 1795; out of service by 1796.

*Negresse* 1799-1803 (French *Négresse* 1798 ____ ) Gunboat ____ . Taken 1798; 1803 deleted.

*New Adventure* 1799-1801? tender/transport; purchased 1799; still listed in 1801.

*Pakenham* 1797-1802 gunboat 1 at Waterford. Purchased 1797; still in service 1800; 1802 sold?

*Pedro* 1796?-1801 schooner 8. Purchased (in the West Indies?) 1796?; in service until 1801.

*Petit Boston* 1793-1793 schooner 2 x 4 + 4 swivels, 10 men. Taken at Toulon 1793; 1793 abandoned at Toulon.

*Petit Victoire* 1793-1793 brig 2 x 6. Taken at Toulon 1793; 1793 lost off Cape Corse (loss not recorded in Gossett).

*Pickle* 1800-1801 tender? 6 guns? 35 men. In the West Indies.

*Plym* 1795?-1802 (hired packet brig *Plym* or *Plymouth Packet* ____ ) gunvessel 181 tons? Purchased 1795? (or 1796?); 1802 sold.

*Portland* 1796-1802 (barge *Portland* ____ ) Gunboat 3? Acquired 1796; 1802 sold.

*Port Royal* 1796-1797/1797 *Recovery* -1801 schooner 10. Purchased 1796; 30.3.1797 taken by the French after being driven ashore on the coast of Haiti, became privateer *Peuple*; 18.10.1797 retaken by *Pelican* and renamed *Recovery*; 1801 sold.

*Poulette* 1793-1796 (French corvette *La Poulette* 1781 Toulon) Sixth Rate 26. 480 tons. Taken at the surrender of Toulon 1793; 20.10.1796 burnt as unserviceable at Ajaccio.

*Proselyte* 1793-1794 (French *La Proselyte* 1781-1785 Le Havre) 12 pdr. 36 gun frigate/floating battery. 950 tons. 1793 taken at the surrender of Toulon; 11.4.1794 set on fire by red-hot shot whilst bombarding Bastia, Corsica, abandoned and sank.

*Proserpine* 1798-1799 (French 12 pounder frigate *La Bellone* 1778 Saint Malo, Guignace design) frigate 32. 141', 116'2¼" x 37'10" x 11'6½", 888. Taken by *Ethalion* off the Irish Coast 12.10.1798; 1799 hulked [at Plymouth; 1811 BU].

*Providence* (ex *Prince William Henry* ____ ) 1798-1804 schooner 14. 85 tons. Purchased 1798; 2.10.1804 expended as a fireship at Boulogne.

*Prudence* 1800-18 ____ . Gunboat ____ . Acquired 1800; no further information.

*Republican* 1795-1803? (French *Républicain* 1793 Le Havre) Schooner 18? Taken 13.10.1795 by *Mermaid* and *Zebra* off Grenada; 1803 still in service?; no further information.

*Resolue* 1795-1802 (French corvette *Résolue*/*L'Hydra* ____ ) brig? 10? Taken 16.8.1795 in the Bay of Alassio by Nelson's squadron; in service until 1802.

*Resolution* 1798 purchased cutter. 200 tons. 14 x 4 + 10 swivels. 70 men. No further information (in Pitcairn Jones only).

*Revenge* 1796-1798 Cutter/schooner 8? Acquired 1796; 1798 deleted.

*Revolutionnaire* 1794-1795 *Reprise* -1797 (French *Révolutionnaire* 1794 ____ ) Sloop 16. Taken 1794; 1795 renamed *Reprise*?; 1797 deleted.

*Rhinoceros* 1793 purchased floating battery/Sixth Rate. 97'5", 81'1" x 27'11" x 12'9". 386 tons. 20 x 42★. Fate unknown.

*Rosa* schooner 12 guns listed 1802-1803 (Colledge only).

*Royalist* 1797-1801 Schooner ____ . Purchased 1797; in service until 1801. Is this the same as a 4 gun schooner (or gun vessel?) purchased in 08.1798 and out of service by 1800, or a separate vessel?

*St Croix* 1793-1794? schooner. 8 men. 4 swivels. Taken at Toulon 1793; 1794 still in service.

*St Pierre* 1796-1796 sloop obtained 1796; 12.2.1796 wrecked off Port Negro? (Colledge only).

*St Thome* 1800-18 ____ . Gunboat ____ . Acquired 1800; no further information.

*Scipion* 1793-1793 (French *Le Scipion* 1790 ____ ) Third Rate 74. 1810 tons. Handed over at Toulon 1793; 20.11.1793 burnt by accident off Leghorn (Livorno).

*Self* 1801 brig in service in the Mediterranean? (In Pitcairn Jones only and very doubtful.)

*Serpent* 1793-1793 gunboat 1 x 18. 15 men. 1793 taken at Toulon; 1793 abandoned at Toulon.

*Shannon* 1797-1802 Gunboat 1, on the River Shannon. Acquired 1797; 1802 sold.

*Sincere* 1793-1797 (French 'gabare' *Sincère* 1784 Toulon) Sloop 14? 112', ____ x 27' x 13'6". 400 tons. Taken 11.1793; 1797 paid off and sold.

*Sparrow* 1796-1805 (mercantile *Rattler* 1780 ____ ) cutter 12. 66'3", 46'6" x 22'3" x 8'9¾". 123 10/94. Men: 40. Guns: 10 x 12★ + 2 x 6. Purchased 1796; 1805 BU.

*Speaker* 1798-1802 gunboat ____ at Waterford. Acquired 1798; 1802 sold.

*Spiteful* 1794-1800? gunvessel 12 guns ex Dutch hoy, purchased 1794 and listed in 1800. (Colledge only, doubtful.)

*Spitfire* 1798-1801 captured schooner 64 tons. 7.7.1801 wrecked near Admiralty Islands. (Colledge only.)

*Strombolo* gunboat taken and sunk by Spanish gunboat 19.1.1799.

*Surprise* 1799-1800 (French merchant ____ ) schooner, 10 guns. 01.1799 taken by *Braave* in the East Indies; 1800 sold. (Colledge only.)

*Swan* 1795-1795 (built in 1792 for the Revenue) transferred 1795; 1795 taken by the French.

*Swinger* (No 4) 1798-1799 gunboat ____ in the West Indies. Acquired 1798; 1799 deleted.

*Tarleton* 1793-1794? See American War of Independence chapter.

*Teazer* (No 5) 1798-1799 gunboat/gunvessel/schooner ____ in the West Indies. Acquired 1798 at Honduras for local defence; 1799 deleted.

*Terror* 1794-1804 (Dutch hoy ____ ) gunboat 4. 69 tons? Acquired 1794; 1804 sold.

*Thetis* 1796-1802 cutter/schooner 8? 110 tons? Acquired 1796; 1802 BU.

*Thomas* 1796-1798 cutter/schooner 8? Acquired 1796; 1798 deleted.

*Tickler* (No 2) 1798-1800 gunboat ____ . In the West Indies. Acquired 1798; 1800 deleted.

*Tormentor* 1795 gunvessel in the West Indies (listed in Colledge only).

*Torridge* 1799-1801? (French 'chaloupe cannonière' *Torride*? ____ 1795 ____ ) ketch 7. Taken off Aboukir Castle by the boats of *Goliath* 1799; 1801? converted to fireship; fate unknown (deleted 1802?).

*Towzer* (No 1) 1798-1799 gunboat ____ . In the West Indies. Acquired 1798; 1799 deleted.

*Trincomalee* 1799-1799 (French or Dutch? ____ ) sloop 18/16 (carronades?). 315 tons. 100 men. Taken 1799; 12.10.1799 blown up in action with the French privateer *Iphigénie* (also destroyed) in the Straits of Bab el Mandeb.

*Trincomalee* 1801-1803/1806-1808? (French privateer *La Gloire* 1800 Bayonne) ship sloop 16. 98'3", ____ x 27'9" x 14'. 320 tons. 121 men. 16 x 6. Taken in the Indian Ocean by *Albatross* 23.3.1801; ____1803 retaken by the French and named *Emilien*; 25.9.1806 taken in the East Indies by *Culloden*; sold about 1808?

*Tyrone* 1798-1802 gunboat ____ at Waterford. Acquired 1798; 1802 sold.

*Undaunted* 1799-1800 (Dutch) gunvessel. Taken by the boats of *Pylades* on the Dutch coast 13.8.1799; 1800 sold.

*Urchin* 1797-1800 gunboat/gunvessel ____ . 155 tons. 121 men. Acquired 1797; 12.10.1800 under tow by *Hector* in Tetuan Bay she took a sheer, filled with water and sank.

*Vanneau* 1793-1796 (French 'brick' *Le Vanneau* 1781 ____ ) cutter 6/gun brig 8. 120 tons. Taken by Colossus in the Bay of Biscay 6.6.1793; 21.10.1796 wrecked at Porto Ferrajo.

*Venom* 1794-1800? (French *Génie* ____ ) gunbrig 4. 65'1", 47' x 22'8" x' 9'. 128 tons. 45 men. 4 x 18 (carronades?). Taken in the West Indies 1794; no further information. [Dittmar gives: taken 1796; 1800 deleted.]

*Vesuvius* 1797 bomb vessel, 8 guns. 293 tons. Purchased 04.1797 [a note gives her as sold in 1812; but this seems likely to be a confusion with the 1775 bomb of this name].

*Victoire* 1795-1800? lugger, 14 guns. Purchased 1795; listed until 1800 (Colledge only).

*Vidette* 1800-1802 (French *Vedette* ex *Pensée* of 1793 renamed 1796 ____ ) brig 14. 234 tons? Taken 1800; 1802 deleted.

*Vigilant* 1798-1800 schooner ____ . 55'9", 44'8" x 16'x 5'4". In service 1798; 1800.

*Vigilante* 1793-1793 (French *Alerte*? ____ ) brig sloop 14. 248 tons. Taken 1793; 18.12.1793 burnt in Toulon Dockyard during evacuation; but salved and repaired by French.

*Vipere* 1793-1793 (French *Vipère*? ____ ) cutter/schooner 4? 60 tons? Taken 1793; 12.1793 wrecked in Hières Bay.

*Virago* 1793-1800 (French ____ ) gunboat ____ . Taken 1793; 1800 BU.

*Virginia* 1796-1800 Schooner? Acquired 1796; 1800 deleted.

*Vulcan* 1796-1801 (American *Hector*? ____ ) bomb vessel. 320 tons. Purchased in the East Indies 1796; 1802 sold in Madagascar.
*Wasp* 1793 gunboat at Toulon.
*William and Ann* 1796-____ ( ____ ) gunboat. Acquired 1795; later dirt boat [BU 1819]; (possibly a confusion with *William and Lucy* of 1795; see earlier entry under Gunboats).
*Wolf* 1801-1829 cutter tender [also Revenue?] 55', ____ x 19' x ____ 81 tons. 6 x 6 + 2 x 3. Built by White, Cowes 1801; 1829 BU at Portsmouth.
**Gunboats Nos 1, 2, 3.** All acquired 1793; 1799 all out of service. ?2 x 18 + 10 x 18*. Ca 140 tons.

## Hired Vessels

NOTE: This is taken from the list in ADM/BP/24A 'An Account of Hired Armed Vessels' held in the National Maritime Museum's Admiralty Collection (I would like to thank my friend and colleague Dr Roger Morriss for drawing my attention to this document), supplemented from the list by F Dittmar in *The Belgian Shiplover* No 129 which was extracted from 'letters, annual reports and payment record books [Unspecified] at the Public Record Office ....'. He also adds vessels recorded in 'other reference sources' (presumably published lists: Steele etc) which are asterisked [*] in front of the name here, and whose existence can be regarded as 'not proven' - though also not impossible. Those craft noted by Dittmar which do not come into this category but are not in the NMM list are marked with a percentage [%] mark after the name thus: ____ % here. Dates given after the name are of first hiring and last discharge. It is quite possible that there was a large overlap with the vessels hired after the ending of the Peace of Amiens but no effort to link them has been made here as there is too much guesswork involved for certainty. Numbers of the vessels noted as armed cutters in the NMM list are given by Dittmar as brigs, schooners or luggers and are marked here as Cutter [type?]. Vessels neither in the NMM list nor Dittmar but found in Pitcairn Jones' list are marked by a plus sign [+] here, whilst vessels noted only in Colledge's *Ships of the Royal Navy* are listed with a dollar sign [$].

*Active* 1794-1800 Cutter 10 guns, 35 men. 71 tons. Taken 9.10.1800.
*Admiral Mitchell* 1800-1801 Cutter 12 guns, 43 men. 134 tons.
+*Admiral Pasley* 1801-1802 brig.
★*Advice* 1801-1802 Cutter 12.
★*Agenoria* 1801-1801 Bomb tender 6.
*Albion* 1793-1794 Armed sloop 20, men found by His Majesty. 393 tons. Purchased 1798 (see Main List).
*Alert* 1798-1801 Cutter 12 guns, 40 men. 120 tons.
*Alfred* 1793-1801 Armed vessel [brig?] (built 1787) 8 guns, 17 men. 135 tons.
*Alligator* 1793-1801 Cutter 6 guns, 14 men. 42 tons.
*Amphitrite* 1793-1794 Armed ship 20, 'men found by His Majesty' 328 tons.
*Ann* 1796-1801 Cutter 12 guns, 50 men.
*Argus* 1794-1799 Cutter [lugger?] 14/16 guns, 53 men, 148 tons; taken 6.2.1799 by French 'Vendemaire' [?Privateer *Vendémiaire*?] in the Atlantic and brought in to Corunna.
*Aristocrat* 1794-1798 Cutter [lugger?] 22 guns, 90 men, 172 tons.
*Aristocrat* 1799-1801 Cutter [brig?] 20 (14?) guns, 50 men, 162 tons.
*Atlantic* 1793-1799 Armed vessel (built 1792) 10 guns, 25 men, 195 tons.
★*Bab El Mandeb*? 1801-1802 Armed ship 26.
+*Beaver* 1801-1801 Gun vessel.
%*Betsey* 17____ - 1____ Smack.
★*Beulah* 1798-1802 Troopship 6 guns.
$*Black Joke* 1793-1800 Cutter.
$*Brave* 1798-1799 Lugger 12 guns; sunk 12.4.1799 in collision with transport *Eclipse* in the Channel.
$*Britannia* 1793-1796 Cutter 6 guns.
$*British Fair* (*Faire*) 1798-1800 Cutter.
+*Caroline* 1798-1798 Tender 6; 1798 lost with all hands in the East or West Indies.
$*Ceres* 1793-1794 Sloop 10 guns.
$*Ceres* 1801 Gun vessel.
$*Champion* 1793-1800 Cutter.
+*Chapman*? 1793-1798 Armed ship (purchased or hired?).
★*Charlotte* Cutter; lost 1801 (is this a confusion with the purchased schooner of 1800-1801?).

*Charming Molly* 1796-1801 Cutter 8 guns; 02.1801 foundered on passage from Jersey.
$*Chatham* 1793-1795 Brig 10 guns 184 tons.
*Cockchafer* 1793-1801 Armed vessel 8 guns; 1.11.1801 lost off Guernsey.
★*Constance* 1798-1801 lugger 12 guns.
★*Constance* 1799-1804 cutter/tender 6 guns [wrecked 17.1.1805 on Galway coast].
$*Constitution* 1796-1801 Cutter 14 guns; 9.1.1801 taken by two French cutters off Portland; 10.1.1801 recaptured.
*Content* [?] 1799-1799 Brig wrecked ____ 1799 on the Dutch coast.
$*Courier* 1798-1801 Cutter 12 guns.
+*Cracker* 1797 Lugger.
*Crown* 1801-1801 Gunbarge, 4 men.
*Cygnet* 1796-1799 Cutter 12 guns, 40 men. 114 tons.
*Cygnet* 1800-1801 Cutter 12 guns, 40 men. 120 tons.
*Daphne* 1794-1796 Cutter [lugger?] 22 guns, 85 men. 150 tons.
*Diamond* 1793-1794 Armed vessel [brig?] (built 1786) 10 guns, 28 men. 191 tons.
★*Diligence* 1800-1800 Cutter.
*Diligent* 1793-1801 Cutter 6 guns, 20 men.
★*Dispatch* 1797-1801 Cutter 6.
*Dolly* 1796-1801 Cutter 8 guns, 33 men.
*Dolphin* 1793-1796 Cutter [lugger?] 2 guns, 15 men. 150 tons.
*Dolphin* 1793-1801 Cutter 12 guns, 36 men. 92 tons. Purchased 1801 (see under cutters earlier in this chapter).
*Dorset* 1793-1802 Cutter 8 guns, 25 men.
*Dover* 1793-1794 Armed vessel [brig?] (built 1787?) 10 guns, 25 men. 177 tons.
*Dover* 1794-1801 Cutter 14 guns, 50 men. 115 tons.
*Drake* 1799-1801 Cutter 12 guns, 42 men. 130 tons.
*Duchess of Cumberland* 1793-1801 Cutter 8 guns, 22 men. 66 tons.
*Duchess of York* 1796-1800 Cutter 6 guns, 26 men.
*Duke of Clarence* 1794-1801. Cutter 8 guns, 24 men. 65 tons.
*Duke of York* 1794-1799 Cutter [lugger?] 6 guns, 24 men. 57 tons. Lost 2.1.1799.
*Earl St Vincent* 1799-1801 Cutter [schooner?] 16 guns 56 men 185 tons.
*Earl Spencer* 1799-1801 Cutter 14 guns, 50 men. 164 tons.
*Earl Spencer* 1799-1801 Cutter 14 guns, 48 men. 142 tons [Are these two cutters actually the same listed twice through clerical error?]
★*Edwinstone* 1801-1802 Ship (tender) 10.
*Fanny* 1799-1801 Cutter [lugger?] 16 guns. 50 men. 177 tons.
*Farmer's Increase* 1801-1801 Gunbarge, 4 men.
★*Findon* 1801-1801 Transport 6.
*Flirt/Royalist* 1794-1798 Cutter 16 guns, 65 men 128 tons.
*Flirt* 1798-1800 Cutter 12 guns, 40 men.
*Flirt* 1801-1801 Cutter 12 guns, 40 men.
*Flora* 1793-1795 Armed vessel [brig?] (built 1792) 12 guns, 22 men. 202 tons.
*Flora* 1794-1798 Cutter 14 guns, 61 men. 158 tons. Lost 1.12.1798, taken by the French.
*Flora* 1800-1801 Cutter [brig?] 14 guns, 48 men. 147 tons.
*Fly* 1795-1801 Cutter 8 guns, 26 men. 84 tons.
★*Fortune* 1797-1798 Cutter 8 guns.
★*Fortunee* 1801 Armed ship 18 in Mediterranean?
*Fowey* 1798-1800 Cutter 14 guns, 40 men.
*Fox* 1793/4-1797 Cutter 10/12 guns, 56 men. 124 or 184 tons. Sunk 24.7.1797 engaging Spanish shore batteries at Santa Cruz, Teneriffe.
*Fox* 1794-1801 Cutter 10 guns, 45 men. 94 tons.
*Fox* 1796-1801 Cutter 12 guns, 47 men.
*Fox* 1796-1802 Cutter 10 guns, 40 men. 104 tons.
%*Fox* ____ Smack.
*Friendship* 1793-1794 Armed sloop 20, 'men found by His Majesty'. 323 tons.
*Friendship* 1801-1801 Gunbarge, 4 men.
★*Galatea* 1795-1805? Lugger.
*Gannet* 1796-1798 Cutter 10 guns, 40 men. 128 tons. Lost 7.6.1798.
*General Elliot* 1801-1801 Gunbarge, 4 men.
*General Small* 1797-1801 Cutter 6 guns, 15 men. 60 tons.
*George* 1798-1801 Cutter 12 guns, 40 men. 125 tons.
*Good Design* 1797-1802 Armed ship (built 1797) 14 guns 44 men 320 tons.
*Good Intent* 1801-1801 Gunbarge, 4 men.

THE FRENCH REVOLUTIONARY WAR 1793-1801

*Grace* 1793-1801 Cutter 10 guns, 34 men.
*Grand Falconer* 1796-1801 Cutter 10 guns, 36 men.
*Greyhound* 1798-1799 Cutter 12 guns, 40 men, 115 tons.
*Griffin* 1794-1801 Cutter 10 guns, 36 men. 71 tons.
★*Hannah* 1798-1802 Transport 6.
+*Hannibal* 1793 46 guns 450 tons.
*Harman* 1801-1801 Gunbarge, 4 men.
*Hart* 1796-1801 Cutter [brig?] 14 guns, 60 men. 278 tons.
*Hawke* 1798-1802 Cutter 12 guns, 40 men.
*Hazard* 1793-1801 Cutter 6 guns, 20 men.
*Heart of Oak* 1801-1801 Gunbarge, 4 men.
★*Hebe* 1798-1802 Transport 6.
★*Hexham* 1798-1802 Bomb tender 6.
*Hind* 1794-1801 Cutter 14 guns, 61 men. 134 tons.
*Hirondelle* 1801-1801 Cutter 14 guns, 50 men.
*Hope* 1795-1797 Cutter [lugger?] 12 guns, 56 men. 130 tons. Lost 25.11.1797.
*Hope* 1798-1800 Cutter [schooner?] 12 guns, 46 men.
★*Hope* 1799-1802 Cutter 6.
*Hudson* 1793-1794 Armed vessel 10 guns, 22 men. 173 tons.
*Hurley* 1798-1801 Cutter 14 guns, 50 men. 162 tons.
*Industrious Ann* 1801-1801 Gunbarge, 4 men.
*Industry* 1801-1801 Gunbarge, 4 men.
*Jane* 1798-1801 Cutter [lugger?] 14 guns, 43 men.
*Joseph* 1796-1801 Cutter 12 guns, 40 men. 102 tons.
*Kent* 1798-1801 Cutter 12 guns, 41 men. 121 tons.
*King George* 1796-1801 Cutter 12 guns, 40 men. 128 [ 133?] tons.
*Lady Ann* 1799-1801 Cutter [lugger?] 12 guns, 46 men. 141 tons.
*Lady Charlotte* 1799-1801 Cutter [schooner?] 12 guns, 40 men. 121 tons.
*Lady Duncan* 1798-1801 Cutter [lugger?] 12 guns, 40 men. 104 tons.
*Lady Jane* 1795-1800 Cutter 6 or 8 guns, 28 men. 53 tons; 17.5.1800 parted company in a gale in the channel & presumed to have foundered
★*Lady Nelson* 1801-1802 Cutter. (This is probably the brig with Schank keels; see under unregistered vessels above).
*Lady Taylor* 1793-1794 Armed sloop 20, men found by His Majesty. 379 tons.
*Lark* 1799-1801 Cutter [lugger?] 14 (2 x 4 + 12 x 12★), 50 men. 170 tons.
*Lark* 1801-1801 Gunbarge, 4 men.
★*Lascelles* 1800-1801 ____ (built 1799).
★*Laurel* 1802-1802 Brig 18.
★*Leith* 1798-1802 Cutter (tender) 14.
*Liberty* 1793-1801 Cutter 6 guns, 21 men. 46 tons. (?given as 10 guns from 1799; or is this another vessel?).
*Linnet* 1801-1801 Gunbarge, 4 men.
*Lion* 1793-1794 Armed vessel (built 1789) 10 guns, 21 men. 152 tons.
*Lion* 1793-1801 Cutter 10 guns, 36 men. 86 tons.
*London/London Packet* 1793-1801 Armed vessel/sloop (built 1792) 10 guns, 25 men. 191 tons.
*Lord Duncan* 1798-1801 Cutter 8 guns, 40 men.
*Lord Hood* 1797-1798 Armed ship (built 1797) 14 guns 45 men 361 tons.
*Lord Mulgrave* 1793-1799 Armed sloop 20 guns; 'men found by His Majesty'. 429 tons. Lost 10.4.1799.
*Lord Nelson* 1798-1801 Cutter 12 guns, 52 men.
★*Lord Saint Vincent* 1801-1802 Schooner 16.
*Louisa* 1799-1801 Cutter [brig?] (built 1799) 16 guns, 56 men, 185 tons.
*Lurcher* 1796-1801 Cutter 12 guns, 40 men. Taken 15.1.1801.
*Marechal de Coburg* 1794-1801 Cutter 12 guns. 66 men. 205 tons.
★*Maria* 1802-1802 Bomb tender 4.
*Mary* 1793-1794 Armed vessel (built 1788) 8 guns, 66 men, 205 tons.
*Mary* 1800-1801 Cutter, 12 guns, 34 men, 106 tons.
*Mary* % ____ Smack.
★*Mary Ann* 1802-1802 Brig.
*Mary Ann*% ____ Smack.
*Mediterranean Packet* 1793-1794 Armed vessel (built 1792) 10 guns, 22 men, 159 tons.
*Mentor* 1793-1801 Armed vessel (built 1792) 10 guns, 22 men, 192 tons.
*Minerva* 1793-1794 Armed vessel (built 1787) 10 guns, 28 men. 200 tons.
*Minerva* 1793-1801 Cutter 8 guns, 22 men. 68 tons.
*Minerva* % ____ Smack.
*Minion* % 1793-1794 Armed vessel 8. 207 tons.

*Nancy* 1793-1794 Cutter 6 guns, 20 men. Lost 10.3.1794.
$*Narcissus* 1796-1801 Cutter.
$*Nelson* 1799-1801 ? ?
★*Nemesis* 1798-1804? Armed ship 22 guns.
+*Neptune* 1798 Lugger hired 06.1798 (or 1796?); 1798 run down and sunk off Beachy Head.
$*Neptune* 1798-1801 Cutter?
*New Union* ???
$*Nile* Cutter.
$*Nile* 1799-1800 Brig 16 guns.
$*Nimrod* 1794-1802 Cutter.
$*Nox* 1798-1801 Cutter 14 guns.
$*Nymphe* 1793-1794 Sloop 10 guns.
*Pasley* ???
$*Peggy* 1796-1799 Lugger 8 guns; 1799 taken by the French.
$*Penelope* 1795-1799 Cutter; 7.7.1799 taken by Spanish schooner.
$*Phoenix* 1794-1801 Cutter 10 guns.
$*Phoenix* 1799-1801 Lugger 14 guns.
*Pleasant Hill* ???
$*Plymouth* 1797-1801 Lugger 16 guns.
*Polly* 1801-1801 Gunbarge, 4 men.
*Pomona* 1797-1801 Armed ship (built 1797) 14 guns, 45 men. 351 tons.
*Prestwood* 1793-1801 Cutter 6 guns, 21 men. 51 tons.
★*Prince Cobourg* 1795-1796 Cutter 16. (Could this be a confusion with the *Marechal de Cobourg*?; see above.)
*Prince D'Auvergne* 1794-1801 Cutter 8 guns, 40 men. 50 tons.
*Prince Edward* 1793-1794 Armed sloop 14, 'men found by His Majesty'. 290 tons.
+*Prince of Wales* 1795-1801 Transport 38 guns?
*Princess of Wales* 1797-1802 Cutter 10 guns, 40 men. 105 tons.
*Princess Royal* 1793-1797 Cutter 8 guns, 25 men? 07.1798 taken by French privateer in the North Sea.
*Prince William* 1797-1801 Armed ship (built 1797) 14 guns, 42 men. 306 tons.
%*Prosperous* ____ Smack.
*Providence* 1800-1801 Cutter 14 guns, 50 men. 152 tons.
★*Providence* 1801-1802 Transport.
*Queen* 1793-1794 Armed sloop (built 1792) 20 guns, 'men found by His Majesty'. 416 tons.
*Queen* 1793-1794 Armed vessel 10 guns, 24 men. 175 tons.
*Queen* 1794-1801 Cutter 14 guns, 60 men. 135 tons.
★*Queen* 1801-1802 Bomb tender 6.
*Queenborough* 1800-1801 Cutter 12 guns, 50 men. 182 tons.
★*Queen Charlotte* 1800-1801 Cutter 12 guns. 140 tons.
★*Racehorse* 1801-1802 Bomb tender 4.
*Rattler* 1793-1796 Cutter 10/16 guns, 41 men. 124 tons.
★*Regard* 1798-1802 Transport 6.
*Resolution* 1795-1801 Cutter [lugger?] 12 guns, 41 men.
*Robert* 1801-1801 Gunbarge, 4 men.
*Rose* 1793-1801 Cutter 8 x 4, 21 men. 55 tons.
*Rose* 1794-1800 Cutter 10 guns, 41 men. 97 tons. Taken 13.10.1800.
*Rover* 1798-1801 Cutter [lugger?] 12 guns, 40 men.
*Rowcliffe* 1799-1801 Armed vessel 18 guns. 28 men.
★*Royalist* 1798-1800 Lugger 8.
★*Saint George* 1802-1802 Cutter.
*Saint Vincent* 1798-1802 Cutter 14 guns, 60 men. 194 tons.
*Sally* 1797-1801 Armed ship 14 guns, 45 men. 340 tons.
★*Sally* 1800-1800 Lugger 14.
★*Sally* 1801-1801 Bomb tender 6.
*Sandwich* 1798-1801 Cutter [lugger?] 14 guns, 50 men.
*Sandwich* 1798-1799 Cutter 12 guns, 40 men. 111 tons. Taken 14.6.1799.
★*Sea Nymph* 1800-1800 Armed ship.
★*Severn* 1801-1802 Bomb tender 6.
*Sheerness* 1800-1801 Cutter 12 guns, 34 men. 101 tons.
*Sidney Smith* 1799-1801 Cutter [schooner?] 12 guns, 51 men. 151 tons.
*Sincerity* 1793-1794 Cutter 6 guns, 20 men. 47 tons.
★*Sir Thomas Pasley* 1800-1800 Brig 16. 40 men; taken 9.12.1800 by two Spanish gunboats in the Mediterranean.
*Sir Thomas Pasley* 1801-1802 Cutter 14 guns, 54 men.

*Sly* 1798-1800 Cutter [lugger?] 12 guns, 40 men. 119 tons.
*Speculator* 1794-1801 Cutter [lugger?] 12 guns, 41 men. 93 tons.
*Speedwell* 1797-1801 Cutter 14 guns, 50 men. 152 tons.
*★Speedwell* 1801-1802 Bomb tender 6.
*Spider* 1794-1796 Cutter [lugger? ex French *Victoire*?] 18 guns, 76 men. Lost 4.4.1796 in collision with the *Ramillies*.
*Stag* 1794-1801 Cutter 14 guns, 60 men. 134 tons.
*Supply* 1801-1801 Gunbarge, 4 men.
*Sussex* ???
*Swan* 1799-1801 Cutter 14 guns, 40 men. 130 tons.
*Swift* 1793-1794 Armed vessel 10 guns, 24 men. 167 tons.
*Swift* 1796-1801 Cutter 12 guns, 40 men. 104 tons.
*Swift* 1799-1801 Cutter 14 guns, 50 men. 161 tons.
*Tartar* 1794-1801 Cutter 12 guns, 48 men. 91 tons.
*Teaser* 1801-1801 Gunbarge, 4 men.
*Telegraph* 1798-1801 Cutter [brig?] (built 1798) 14 guns, 61 men. Lost 14.2.1801.
*Telemachus* 1796-1801 Cutter 14 guns, 60 men.
*Terrier* 1798-1800 Cutter [brig?] (built 1798) 12 guns, 52 men, 173 tons.

*Thetis* 1793-1794 Armed vessel (built 1792) 10 guns, 25 men, 179 tons.
*Thetis* 1793-1794 Armed vessel (built 1790) 10 guns, 28 men. 201 tons.
*Thetis* 1798-1800 Cutter [lugger?] 12 guns, 40 men, 104 tons.
*★Union* 1793-1793 Brigantine (built 1787) 12; Burnt/blown up at the evacuation of Toulon 8.12.1793.
*Union* 1793-1794 Armed vessel 8 guns, 29 men.
*Union* 1799-1801 Cutter 12 guns, 45 men. 137 tons.
*Union* 1801-1801 Gunbarge, 4 men.
*Valette* [or *Vadette*?] 1800-1801 Cutter 14 guns, 58 men, 198 tons.
*Valiant* 1794-1801 Cutter [lugger?] 11 guns, 45 men, 110 tons.
*Venus* 1793-1797 Cutter 8 guns, 25 men.
*Venus* 1794-1801 Cutter 6 guns, 25 men. 56 tons.
*Venus* 1796-1801 Cutter 8 guns, 31 men.
*★Venus* 1798-1802 Transport 6.
*Vigilant* 1793-1801 Cutter 6 guns, 20 men.
*William Pitt* 1796-1799 Lugger 12 guns, 40 men 108 ton; taken 5.6.1799.
*★Willington* 1801-1802 Bomb tender 6.
*Wright* 1797-1801 Armed ship (built 1797) 14 guns, 45 men, 341 tons.

# Chapter 7: The Napoleonic War 1802-1815

The recommencement of war with Imperial France after the breakdown of the Peace of Amiens saw the need for large purchases of small craft, thanks to the Earl of Saint Vincent's 'bull-in-a-china-shop' approach to reform and retrenchment in the period of truce. His inflexible and unfortunate attempts to root out corruption and instigate economies in the Royal Dockyards and in the Navy's relationships with private shipbuilders and other contractors resulted in much disruption, ill-will and consequent delays and shortages when it became evident that rearmament was a vital and urgent necessity. Fortunately, even this self-inflicted wound was not to prove fatal. The number of ex-enemy vessels listed here is evidence of the continuing success of the Royal Navy in maintaining its predominance. Even the excellent United States Navy eventually contributed ships to the total, though it gave the Royal Navy its most formidable challenge in individual ship actions, and administered a major, and much needed shock to what had been in danger of becoming an automatic assumption of superiority. It is arguable that the influence of the extremely powerful American frigates, with their line of battle ship scantlings and their 24 pounder main battery, distorted subsequent frigate development in the direction of over-large 'battle cruiser' equivalents not particularly suited to the Royal Navy's 'sea control' requirements, however desirable they were for the 'hit and run' tactics of a less powerful fleet.

An increasing proportion of the lesser captures were of the flush decked, 'peg top' section (with much dead-rise) type that we have already noted as appearing in numbers in the previous chapter. The increasing standardisation of French warship designs in this period is particularly noticeable in the plans taken off the larger frigates, the majority of which were built to the designs of Sané. It is noticeable that, whilst adapted French designs continued to be used by the Royal Navy at this period, the majority were of an earlier generation. Another influence at this period was an excellent Danish design, that of the *Christian VII*, the only non-French line of battle ship to be the direct basis of designs built for the Royal Navy. Earlier Spanish designs (eg. *Princesa*) had been influential but had not been specifically copied. This Danish ship of the line was, with most of the rest of the navy of that unfortunate country, seized after the brutally effective attack on Copenhagen launched in 1807 to pre-empt Napoleon's attempt to take over the Danish fleet. Whether necessary or not (and it is difficult to see what other alternatives were left to the British government if it wished to continue to fight effectively) this act drove the Danes into being Napoleon's most loyal ally and simultaneously forced them into becoming very effective proponents of gunboat tactics, since they had virtually no warships left. They made great use of vessels like the *Steece* (see below) and, as can be seen by looking at the list of Royal Naval losses, did much damage to British small craft.

## Large Two-deckers (Third Rates)

***San Rafael*** 1805-1810 (Spanish *San Rafael* 1772 Havanna) 80.
Dimensions & tons: 188'6", 155'3⅜" x 50'10" x 20'9". 2130.
Men: 640. Guns: ____ . Taken in Calder's Action 22.7.1805; never fitted for sea in the RN; probably hulked in 1806; 1810 sold.

***Implacable*** 1805-1855 (French *Le Duguay-Trouin* 1800 Rochefort) 76/74.
Dimensions & tons: 181'6", 148'11⅝" x 48'11" x 22'. 1896 22/94.
Men: 670. Guns: GD 30 x 32, UD 30 x 18, QD 2 x 12 + 12 x 32★, Fc 2 x 12 + 2 x 32★. Taken in Strachan's Action 3.11.1805; 1855 hulked [at Plymouth as a training ship; 1911 lent to Wheatley Cobb at Falmouth; 1949 scuttled in the Channel; a plan to preserve her at Greenwich in the dock now occupied by the *Cutty Sark* having failed; the rotten state of her timbers was supposed to have prevented her from being preserved].
PLANS: 1810: Lines, profile & decorations. 1815: Gun deck/upper deck/cabins. 1949: Plans of carvings (3 sheets).

***Brave*** 1805-1808 (French *Le Formidable* ex *Le Figuières* 1795 Toulon) 80/84.
Dimensions & tons: 194'6", 159'7¾" x 51'5½" x 26'6". 2249.
Men: 690. Guns: GD 32 x 32, UD 30 x 18, QD 2 x 12 + 14 x 32★, Fc 2 x 12 + 4 x 32★. Taken in Strachan's Action 3.11.1805; never fitted for sea in the RN; 1808 hulked [at Plymouth as a prison ship; 1814 powder ship; 1816 BU].

***Maida*** 1806-1814 (French *Le Jupiter* ex *Le Constitution* ex *Le Viala* ex *Le Voltaire* 1794-5 Lorient) 76/74.
Dimensions & tons: 181'9⅝", 148'11½" x 48'11½" x 21'5½". 1899.
Men: ____ . Guns: GD 30 x 24, UD 30 x 24, QD 2 x 24 + 12 x 24★, Fc 2 x 24 + 2 x 24★. Taken at Santo Domingo 6.2.1806; 1814 sold.

***Marengo*** 1806 & hulked after capture (French *Marengo* ex *Le Jean Jacques Rousseau* 1795 Rochefort) 80.
Dimensions & tons: 182'6", 150'3⅛" x 49'1⅝" x 21'3". 1929 56/94.
Men: 640. Guns: GD 30 x 32, UD 30 x 18, QD 8 x 12 + 8 x 32★, Fc 2 x 12 + 2 x 32★. Taken by *London* and *Amazon* 13.3.1806; hulked on arrival [at Portsmouth, as a temporary prison ship; 1816 BU].
PLANS: 1817: Lines.

***Alexandre*** 1806-1808 (French 80 *L'Alexandre* ex *L'Indivisible* 1799 Brest) 80.
Dimensions & tons: 195'2", 158'11¾" x 51'4½" x 23'2". 2231.
Men: 590. Guns: GD 28 x 32, UD 28 x 18, QD 4 x 12 + 10 x 32★, Fc 2 x 12 + 2 x 32★, RH 6 x 18★. Taken at Santo Domingo 6.2.1806; never fitted for sea in the RN; 1808 hulked [at Plymouth as a powder ship; 1822 sold].

***Christian VII*** (ex *Blenheim*) 1807-1814 (Danish 90 *Christian VII* 1803 Nyholm, Hohlenberg design) 80.
Dimensions & tons: 187'2", 154'7⅜" x 50'10½" x 20'10". 2128 58/94.
Men: 670. Guns: GD 30 x 32, UD 32 x 18, QD 2 x 18 + 12 x 32★, Fc 4 x 12★, RH 3 x 18★. Taken at the surrender of Copenhagen 1807; 1814 hulked [at Sheerness as a lazaretto; 1838 BU].
NOTE: The only non-French two-decker prize to be copied directly by the RN. One 80 (*Cambridge*) was built to her lines, whilst the *Black Prince* Class 74s were slightly smaller versions, and the *Jupiter* Class of 50s much smaller ones. She had what was virtually a 'pink' stern in an attempt to produce a stronger above-water stern structure than was traditional in line of battle ships. This provides an interesting contrast to Seppings' round stern which was developed for much the same reason; as an attempt to make 'raking' less easy.
PLANS: Lines & profile.

***Waldemar*** (ex *Yarmouth*) 1807-1812 (Danish *Waldemar* 1797 ____ , Stibolt design) 80.
Dimensions & tons: 185'6¼", 152'11⅜" x 50'10¼" x 20'6". 2104.
Men: 670. Guns: GD 30 x 32, UD 32 x 18, QD 4 x 12 + 10 x 32★, Fc 2 x 12 + 4 x 32★, RH 6 x 18★. Taken at the surrender of Copenhagen 1807; not fitted for sea after arrival in Britain; 1812 hulked [as a prison ship at Portsmouth; 1816 BU].

***Genoa*** 1814-1838 (Franco-Italian *Le Brilliant* building at Genoa) 78/74.
Dimensions & tons: 182', 150'4¾" x 48'6¾" x 21'7". 1886 56/94.
Men: 600. Guns: GD 28 x 32, UD 30 x 18, QD 4 x 12 + 10 x 32★, Fc 2 x 12 + 4 x 32★. Taken on the stocks at Genoa when that city surrendered in 1814; launched [1815?] and completed for the RN; 1838 BU.
PLANS: Lines & decoration/orlop/gun deck/upper deck/quarter deck & forecastle. 1821: Upper deck.

## 74 Gun Third Rates

[For largest 74s see previous section.]
***Duquesne*** 1803-1805 (French *Le Duquesne* 1788 Toulon).

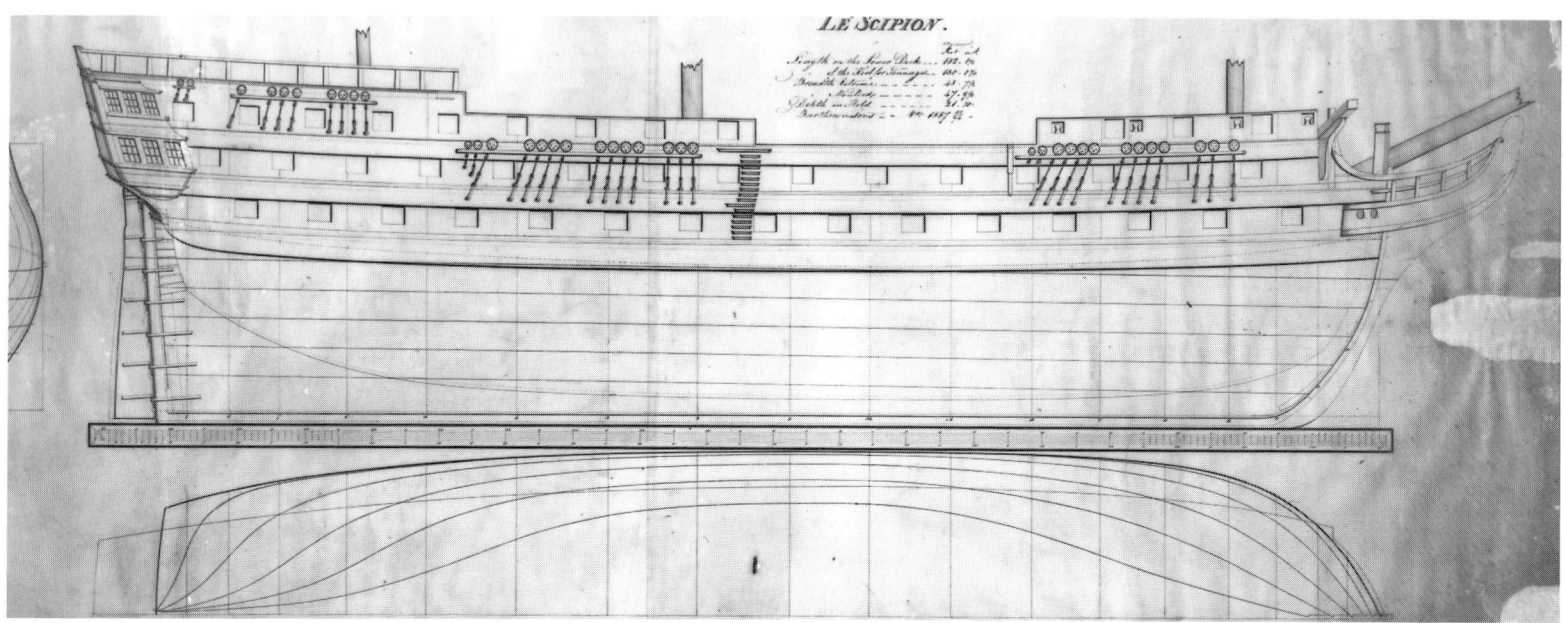

Lines (7462) of the French 74 *Scipion* taken in 1805

Dimensions & tons: 182'3", 150'4½" x 48'9½" x 15'1". 1903 17/94.
Men: ____ . Guns: ____ . Taken off Santo Domingo by *Bellerophon* and *Vanguard* 24.7.1803; 1804 stranded but got off; 1805 BU.

PLANS: Lines & profile.

***Firme*** 1805-1814 (Spanish *Firme* 1765 Carraca, Cadiz).
Dimensions & tons: 174'5½", 138'7⅛" x 49'5½" x 19'9". 1803.
Men: ____ . Guns: ____ . Taken in Calder's Action 22.7.1805; never fitted for sea in the RN; 1814 sold.

***Ildefonso*** 1805-1808 (Spanish *San Ildefonso* 1785 Cartagena).
Dimensions & tons: 175'3", 140'2⅞" x 48'5½" x 20'10". 1753.
Men: 640. Guns: GD 28 x 32, UD 30 x 18, QD 6 x 12 + 8 x 32★, Fc 2 x 12 + 2 x 32★. Taken at Trafalgar 21.10.1805; never fitted for sea in the RN; 1808 hulked [at Portsmouth as victualler at Spithead; 1816 BU].

PLANS: 1809 as victualling depot: Profile & orlop/decks.

***Bahama*** 1805-1807 (Spanish *Bahama/San Cristobal* 1784 Havanna).
Dimensions & tons: 175'1", 144'9½" x 48'2" x 20'4½". 1786 75/94.
Men: 640. Guns: GD 28 x 32, UD 30 x 18, QD 6 x 12 + 2 x 32★, Fc 2 x 12 + 2 x 32★. Taken at Trafalgar 21.10.1805; never fitted for sea in the RN; by 1807 hulked [at Chatham as a prison ship; 1814 BU].

PLANS: 1810. Lines/orlop/gun deck/upper deck/quarter deck & forecastle.

***San Juan*** [***Berwick***] 1805 & hulked after capture (Spanish *San Juan Nepomuceno* 1766 Guarnizo, Santander).
Dimensions & tons: ____, ____ x ____ x ____ . ____ .
Men: ____ . Guns: ____ . Taken at Trafalgar 21.10.1805; hulked [at Gibraltar, classed as a sloop ( = a commander's command); 1818 sold].

***Mont Blanc*** 1805-1811 (French *Le Mont Blanc* ex *Le Republicain* ex *Le Trente-et-Un Mai* ex *Le Pyrrhus* 1791 Rochefort).
Dimensions & tons: 183'2", 149'7⅜" x 48'8¼" x 20'6". 1886.
Men: 640. Guns: GD 30 x 32, UD 30 x 18, QD 2 x 12 + 12 x 32★, Fc 2 x 12 + 2 x 32★. Taken in Strachan's Action 4.11.1805; never fitted for sea in the RN; 1811 hulked [at Plymouth as a powder hulk; 1819 sold].

***Scipion*** 1805-1819 (French *Le Scipion* 1801 L'Orient) 74.
Dimensions & tons: 182'1½", 150'0⅞" x 48'7½" x 21'10". 1887 39/94.
Men: 640. Guns: GD 30 x 32, UD 30 x 18, QD 2 x 12 + 12 x 32★, Fc 2 x 12 + 2 x 32★, RH 2 x 18★. Taken in Strachan's action 4.11.1805; 1819 BU.

PLANS: Lines.

***Denmark/Dannemark*** (ex *Marathon*) 1807-1815 (Danish *Danemark* 1795 Copenhagen, Stibolt design).
Dimensions & tons: 179'3½", 148' x 48'3½" x 19'8". 1836 14/94.
Men: 590. Guns: GD 28 x 32, UD 30 x 18, QD 4 x 12 + 10 x 32★, Fc 2 x 12 + 2 x 32★, RH 6 x 18★. Taken at the fall of Copenhagen 1807; 1815 sold.

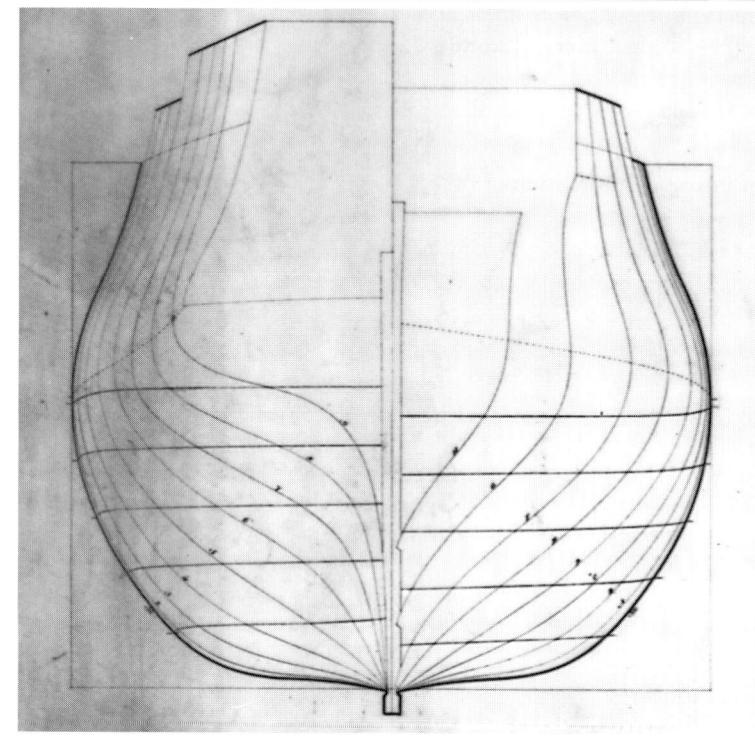

PLANS: Lines & profile.

***Fyen*** (ex *Bengal*) 1807-1809 (Danish *Fyen* 1787 Copenhagen, Gerner design).
Dimensions & tons: 174'6", 143'1⅞" x 48'6⅛" x 20'4½". 1791 87/94.
Men: 590. Guns: GD 28 x 32, UD 28 x 18, QD 4 x 12 + 10 x 32★, Fc 2 x 12 + 2 x 32★. Taken at the fall of Copenhagen 1807; never fitted for sea in the RN; 1809 hulked [at Chatham as a prison ship; 1814 sold].

PLANS: Lines & profile (also applies to *Kronprincessen*).

***Heir Apparent Frederick*** (ex *Cornwall*) 1807-1811 ***Arve Princen*** -1815 (Danish *Arve Princen Frederick* 1782 Copenhagen, Gerner design).
Dimensions & tons: 174'6¾", 143'6½" x 47'10" x 19'6". 1747.
Men: 590. Guns: GD 28 x 32, UD 30 x 18, QD 4 x 12 + 10 x 32★, Fc 2 x 12 + 2 x 32★, RH 6 x 18★. Taken at the fall of Copenhagen 1807; never fitted for sea in the RN; 1815 hulked [at Portsmouth as a victualling depot ship; 1817 sold].

***Justitia*** (ex *Oxford*) 1807-1817 (Danish *Justitia* 1777 Copenhagen, Gerner design).
Dimensions & tons: 174'3", 144'2⅛" x 47'10½" x 19'9". 1758.
Men: 590. Guns: GD 28 x 32, UD 28 x 18, QD 4 x 12 + 10 x 32★,

# THE NAPOLEONIC WAR 1802-1815

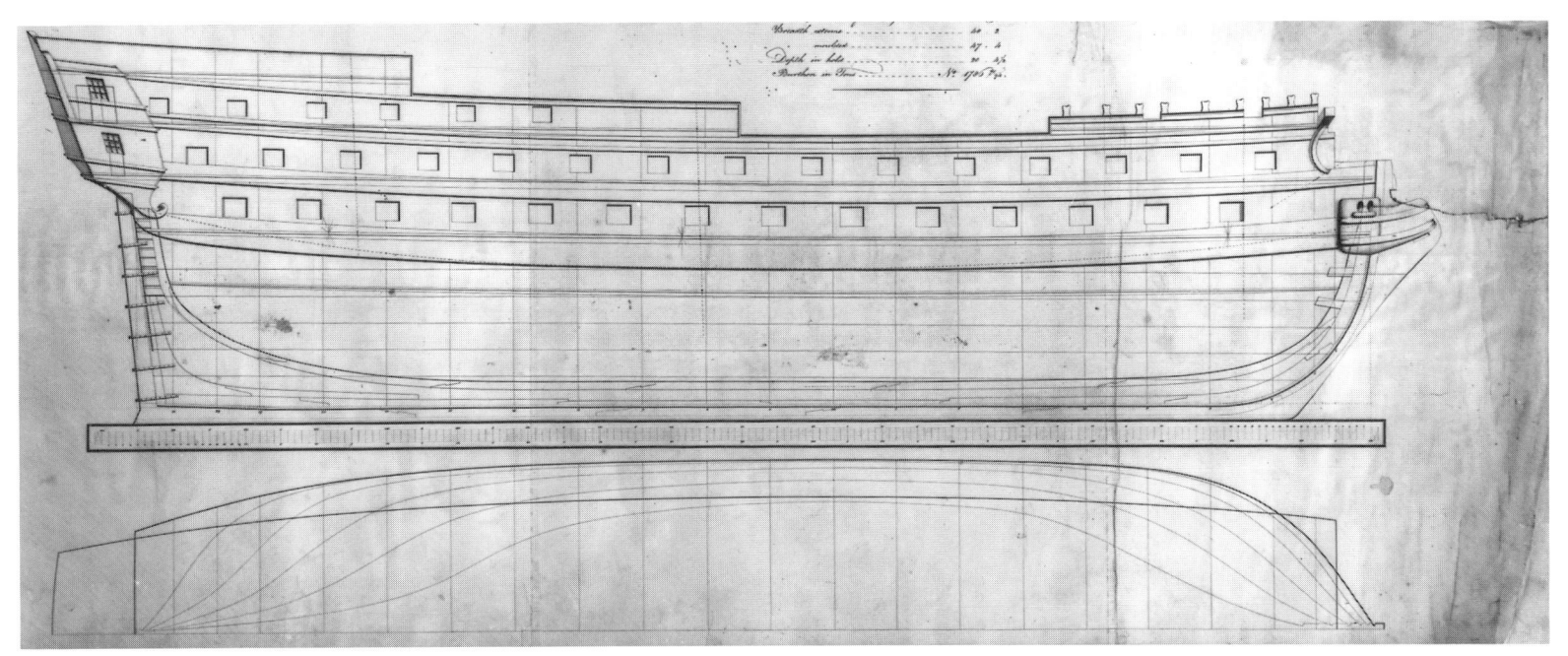

Lines (611) of the Spanish 74 *Bahama* of 1805

Fc 2 x 12 + 2 x 32★, RH 6 x 18★. Taken at the fall of Copenhagen 1807; never fitted for sea in the RN; 1817 used for an experiment with Seppings' diagonal braces; 1817 BU.

PLANS: 1817: Trusses.

**Kron Princen** (ex *Norfolk*) 1807-1809 (Danish *Kronprinds Frederick* 1783 Copenhagen, Gerner design).
Dimensions & tons: 175'1", 143'5⅜" x 47'10½" x 19'5½". 1749.
Men: 590. Guns: GD 28 x 32, UD 28 x 18, QD 4 x 12 + 10 x 32★, Fc 2 x 12 + 2 x 32★, RH 6 x 18★. Taken at the fall of Copenhagen 1807; never fitted for sea in the RN; 1809 hulked [at Chatham as a prison ship; 1814 sold].

PLANS: Lines.

**Kron Princessen** (ex *Torbay*) 1807-1814 (Danish *Kronprincessen Marie* 1791 Copenhagen, Gerner design).
Dimensions & tons: 175'3¼", 143'5⅛" x 48'0¼" x 19'8". 1759.
Men: 590. Guns: GD 28 x 32, UD 28 x 18, QD 4 x 12 + 10 x 32★, Fc 2 x 12 + 2 x 32★, RH 6 x 18★. Taken at the fall of Copenhagen 1807; never fitted for sea in the RN; 1814 sold.

PLANS: See *Fyen*.

**Norge** (ex *Nonsuch*) 1807-1816 (Danish *Norge* 1799 Copenhagen, Hohlenberg design).
Dimensions & tons: 183'7", 151'5¼" x 49'4" x 20'5". 1960 39/94.
Men: 640. Guns: GD 28 x 32, UD 32 x 18, QD 4 x 12 + 10 x 32★, Fc 2 x 12 + 2 x 32★ (Fx carronades later replaced by 3 x 18 'elevating guns'). Taken at the fall of Copenhagen 1807; 1816 sold.

PLANS: Lines & profile.

**Odin** 1807-1811 (Danish *Odin* 1787 Copenhagen, Gerner design).
Dimensions & tons: 174'7¾", 143'2¼" x 47'10¾" x 19'7". 1747 12/94.
Men: 590. Guns: GD 28 x 32, UD 28 x 18, QD 4 x 12 + 10 x 32★, Fc 2 x 12 + 2 x 32★, RH 6 x 18★. Taken at the fall of Copenhagen 1807; never fitted for sea in the RN; 1811 hulked [at Portsmouth as a receiving ship; 1825 sold].

PLANS: Lines & profile/upper deck/quarter deck & forecastle.

**Princess Carolina** (ex *Braganza*) 1807-1813 (Danish *Prindsesse Carolina* 1805 Copenhagen, Hohlenberg design).
Dimensions & tons: 173'1½", 143'4¾" x 46'3⅛" x 19'2". 1636 59/94.
Men: 534. Guns: GD 28 x 24, UD 28 x 24, QD 2 x 24 + 10 x 24★, Fc 2 x 24 + 4 x 24★. Taken at the fall of Copenhagen 1807; 1815 sold.

PLANS: Lines & profile.

**Princess Sophia Frederica** (ex *Cambridge*) 1807-1816 (Danish *Prindsesse Sophie Frederica* 1773 or 1775 Copenhagen, Gerner design).

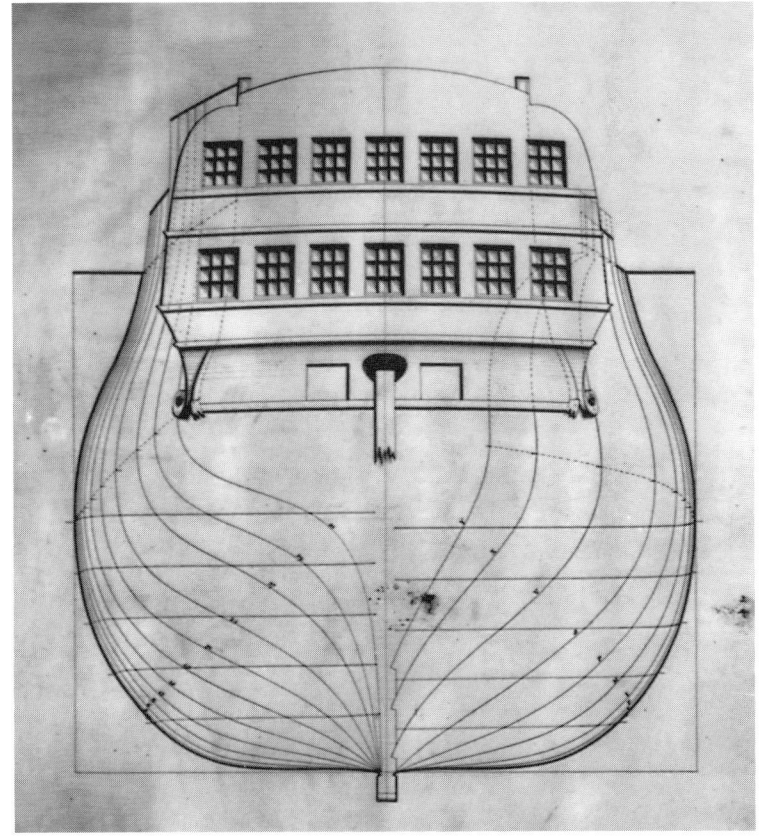

Dimensions & tons: 175'1", 144'1⅛" x 47'11½" x 19'8". 1763.
Men: 590. Guns: GD 28 x 32, UD 28 x 18, QD 4 x 12 + 10 x 32★, Fc 2 x 12 + 2 x 32★, RH 6 x 18★. Taken at the fall of Copenhagen 1807; never fitted for sea in the RN; 1816 BU.

**Skiold** (ex *Somerset*) 1807-1811 (Danish *Skiold* 1791 or 1792 Copenhagen, Gerner design).
Dimensions & tons: 174'5½", 143'3⅜" x 47'10½" x 19'7". 1747.
Men: 590. Guns: GD 28 x 32, UD 28 x 18, QD 4 x 12 + 10 x 32★, Fc 2 x 12 + 2 x 32★, RH 6 x 18★. Taken at the fall of Copenhagen 1807; never fitted for sea in the RN; 1811 hulked [at Portsmouth as a receiving ship; 1825 sold].

**Tre Kronen** (ex *Medway*) 1807 & hulked after capture (Danish *Trekroner* 1789 or 1790 Copenhagen, Gerner design).
Dimensions & tons: 175'5⅛", 143'4" x 47'10½" x 19'9". 1746.
Men: 590. Guns: GD 28 x 32, UD 28 x 18, QD 4 x 12 + 10 x 32★, Fc 2 x 12 + 2 x 32★, RH 6 x 18★. Taken at the fall of Copenhagen 1807;

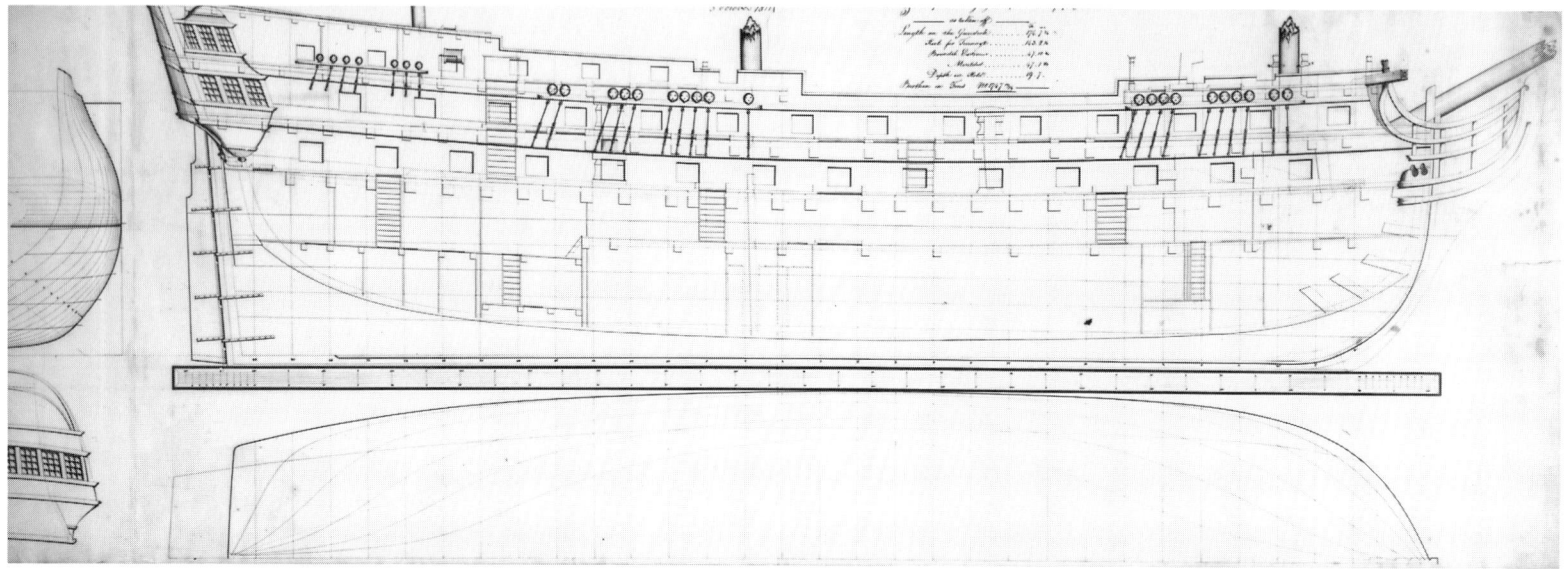

Part lines and profile (missing most of the body plan and stern of plan 858) of the Danish 74 *Odin* taken in 1807

never fitted for sea in the RN; and soon hulked [at Portsmouth as receiving ship; 1825 sold].

***Abercrombie*** 1809-1817 (French *D'Hautpoult* 1807 L'Orient).
Dimensions & tons: 181'6", 149'4" x 48'6¾" x 21'9". 1870 78/94.
Men: 640. Guns: GD 30 x 32, UD 30 x 18, QD 4 x 12 + 10 x 32★, Fc 2 x 12 + 2 x 32★. Taken in the North Atlantic 17.4.1809; 1817 sold.

PLANS: 1813: Lines & profile/orlop/gun deck/upper deck/quarter deck & forecastle.

***Rivoli*** 1812-1819 (Franco-Italian *Le Rivoli* 1810 Venice).
Dimensions & tons: 176'5½", 144'8⅝" x 48'5" x 21'3¼". 1804 41/94.
Men: 590. Guns: GD 28 x 32, UD 28 x 18, QD 4 x 18 + 10 x 32★, Fc 2 x 12 + 2 x 32★, RH 6 x 18★. Taken by *Victorious* in the Adriatic 22.2.1812; 1819 BU.

PLANS: Lines & profile/orlop/gun deck/upper deck/quarter deck & forecastle.

## Small Line of Battle Ships (64 Gun Third Rate)

***Syeren*** (ex *Behemoth*) 1807-1815 (Danish *Seijeren* 1795 Copenhagen, Stibolt design).
Dimensions & tons: 165'0⅛", 134'7¾" x 45'7½" x 19'4". 1491.
Men: 491. Guns: GD 26 x 24, UD 26 x 18, QD 4 x 9 + 6 x 24★, Fc 2 x 9 + 2 x 24★, RH 6 x 18★. Taken at the fall of Copenhagen 1807; never fitted for sea in the RN; 1815 sold.

## Small Two-deckers

***Malabar*** 1804-1815 *Coromandel* -1827 (East Indiaman *Cuvera* 1798 Calcutta) Fourth Rate 56.
Dimensions & tons: 168'6", 127'4" x 37'2" x ___. 936.
Men: ___ (as storeship: 150). Guns: GD 28 x 18, UD 28 x 24★ (as storeship: 2 x 9 + 8 x 32★). Purchased 05.1804; 1808 storeship (10); 1815 renamed *Coromandel*; 1819 seagoing convict ship; 1853 BU].

PLANS: As convict ship 1828: profile/profile & decks/orlop/main deck/upper deck.

***Mediator*** 1804-1809 (East Indiaman *Ann and Amelia* 1799 Calcutta) Fifth Rate 44.
Dimensions & tons: 134'8", ___ x 34' x ___. 683
Men: 254 Guns: GD 26 x 18, UD 18 x 24★. Purchased 1804; 1809 fireship; 11.4.1809 expended at Basque Roads.

***Weymouth*** 1804-1828 (East Indiaman *Wellesley* 1797 Calcutta) Fifth Rate 44; teak built.
Dimensions & tons: 136', 121' x 37' x 12'4". 826.
Men: ___ (as storeship: 121) Guns: ___ (as storeship: GD 10 x 24★, QD 6 x 24★, Fc 2 x 9). Purchased 1804; 1807 storeship (18;) 1828 hulked [at Bermuda as a receiving and convict ship; 1865 sold].

PLANS: 1811 as storeship: profile & decks. 1813 as storeship: profile & decks.

***Hindostan*** 1804-1815 *Dolphin* - 1830 (East Indiaman *Admiral Rainier* 1799 *Calcutta*) Fourth Rate 54.
Dimensions & tons: 158'6", 121'9" x 37' x ___ . 887.
Men: 294 (as storeship). Guns: GD 26 x 18, UD 26 x 24★. Purchased in 1804; 1811 storeship 20; 1819 renamed *Dolphin*; 1830 hulked [at Woolwich as a convict ship, renamed *Justitia*; 1855 sold].

## 24 pounder Frigates

(Fifth Rates unless otherwise noted)

***Surveillante*** 1803-1814 (French *La Surveillante* 1802 Nantes, Sane design) 38.
Dimensions & tons: 151'6", 126'11¼" x 40'3" x 12'8". 1093 80/94.
Men: 300. Guns: UD 28 x 24, QD 12 x 32★, Fc 2 x 9 + 2 x 32★. Taken at the capitulation of San Domingo 1803; 1814 BU.

PLANS: 1807: Lines & profile/orlop/gun deck/upper deck/quarter deck & forecastle.

***Cornwallis*** 1805-1812 *Akbar* -1824 (Bombay Marine frigate [Hon East India Co] *Marquis of Cornwallis* 1801 Bombay) 50/38.
Dimensions & tons: 166'4½", 142'6¼" x 43'4½" x 15'3". 1388.
Men: 430. Guns: UD 30 x 24, QD 26 x 42★, Fc 1 x 18 (as troopship: UD 22 x 32★ + 2 x 9, QD 6 x 32★, Fc 2 x 9). Purchased from the East India Co in India 1805, 18___ troopship, 1824 hulked [at Pembroke as a lazaretto; 1827 to Liverpool; 1864? sold].

***Daedalus/Dedalus*** 1811-1813 (Franco-Italian *Corona* 1809 Venice) 38.
Dimensions & tons: 152'9", 126'11¼" x 40'3" x 12'0½". 1094.
Men: 315/274. Guns: UD 28 x 24, QD 14 x 24★, Fc 2 x 6 + 2 x 24★. Taken at the Battle of Lissa 13.3.1811; 2.7.1813 wrecked off Ceylon.

***President*** (American 44 *President* 1800 Doughty & Bergh, New York) Fourth Rate 60.
Dimensions & tons: 173'3", 146'4¾" x 44'4" x 13'11". 1533 7/94.
Men: 450. Guns: GD 30 x 24, UD 2 x 24 + 28 x 42★. Taken by a squadron off New York 15.1.1815 but strained by grounding just before capture; never fitted for sea by the RN after arrival in Britain; 1818 BU. [Design copied for the next RN *President*.]

PLANS: Lines & profile.

## 18 pounder Frigates (Fifth Rates)

***Clorinde*** 1803-1807 (French *La Clorinde* ex *La Havraise* 1799 ___ ) 38.
Dimensions & tons: 158'6", 133'2" x 40'10" x 12'2". 1181 54/94.
Men: 330. Guns: UD 28 x 18, QD 16 x 32★, Fc 2 x 12 + 2 x 32★. Taken at the surrender of Santo Domingo 30.11.1803; 1817 sold.

PLANS: 1808: Figurehead/stern decorations. 1810: Lines & profile & decorations/decks.

THE NAPOLEONIC WAR 1802-1815

*Virtu* 1803-1810 (French *La Vertu* 1794 L'Orient, Segondat design) 40/38.
Dimensions & tons: 151', 127'11⅝" x 37'8½" x 11'10". 1073.
Men: 330. Guns: UD 28 x 18, QD 16 x 32★, Fc 2 x 12 + 2 x 32★. Taken at the fall of Santo Domingo 1803; never fitted for sea in the RN after her arrival in Britain; 1810 BU.

*Fama* 1804-1812 (Spanish *Fama* 1795 Carthagena) 38.
Dimensions & tons: 145'2¼", 119'5¾" x 39'3" x 11'9". 979.
Men: 284. Guns: UD 28 x 18, QD 2 x 9 + 8 x 32★, Fc 2 x 9 + 2 x 32★. Taken before the declaration of war on Spain by *Medea* and *Lively* on 5.1.1804 whilst she was part of the Plate fleet from South America; never fitted for sea in the RN; 1812 sold.

*Imperieuse* (ex *Iphigenia*?) 1804-1818 (Spanish *Medea* 1797 Ferrol) 38.
Dimensions & tons: 148'1", 121'11⅞" x 40'1" x 12'4". 1042 1/94.
Men: 284. Guns: UD 28 x 18, QD 12 x 32★, Fc 2 x 9 + 2 x 32★. Taken whilst part of the Plate fleet on 5.1.1804, before the declaration of war on Spain; 1818 hulked [at Sheerness as a lazaretto; 1838 sold].
NOTE: This was Cochrane's *Imperieuse*.
PLANS: 1806: Figurehead/stern & quarter decorations/decks. 1810: Lines & profile.

*Amphitrite* 1804-1805 *Blanche* -1807 (Spanish *Anfitrite/Santa Ursula* 40/42 1797 Havanna) 38.
Dimensions & tons: 150'1", 122'3⅝" x 39'11" x 12'6". 1036.
Men: 284. Guns: UD 28 x 18, QD 8 x 9 + 4 x 32★, Fc 2 x 9 + 2 x 32★. Taken by *Donegal* off Cadiz 25.11.1804; 1805 renamed *Blanche*; 4.3.1807 wrecked off Ushant.

*Hamadryad* 1804-1815 (Spanish *Santa Matilde* 1778 Havanna) 38.
Dimensions & tons: 145', 120'2⅜" x 38'10½" x 11'10½". 966 20/94.
Men: 284. Guns: UD 28 x 18, QD 2 x 9 + 10 x 32★, Fc 2 x 9 + 2 x 32★. Taken by *Donegal* and *Medusa* off Cadiz 25.11.1804; 1815 sold.
PLANS: 1810: Lines & profile/quarter deck & forecastle.

*Milan* 1805-1815 (French *La Ville de Milan* ex *L'Hermione* 1803 L'Orient, Geoffroy design) 38.
Dimensions & tons: 153'1", 128'4⅞" x 39'10½" x 12'10". 1085 91/94.
Men: 300. Guns: UD 28 x 18, QD 12 x 32★, Fc 2 x 32★. Taken by *Leander* and *Cambridge* 23.2.1805; 1815 BU.
PLANS: Lines & profile/quarter deck & forecastle.

*Didon* 1805-1811 (French *La Didon* 1799 Saint Malo) 38.
Dimensions & tons: 153', 127'7⅛" x 40'1" x 12'10". 1090 42/94.
Men: 300. Guns: UD 28 x 18, QD 6 x 9 + 6 x 32★, Fc 2 x 9 + 2 x 32★. Taken by *Phoenix* 10.8.1805; 1811 BU.
PLANS: Lines & profile/decks.

*Volontaire* 1806-1816 (French *La Volontaire* 1794-1796 Bordeaux) 38.
Dimensions & tons: 151'9", 127'1⅜" x 40'0½" x 19'5". 1084 3/94.
Men: 284. Guns: UD 28 x 18, QD 14 x 32★, Fc 2 x 9 + 2 x 32★. Taken 4.3.1806 entering the roadstead at the Cape of Good Hope, unaware of its capture by the British; 1816 BU.
PLANS: Lines & profile/quarter deck & forecastle.

*Belle Poule* 1806-1816 (French *La Belle Poule* 1802 Nantes, Sané design) 38.
Dimensions & tons: 151'6", 127'7⅝" x 39'11" x 13'4". 1076 64/94.
Men: 284. Guns: UD 28 x 18, QD 14 x 32★, Fc 2 x 9 + 2 x 32★. Taken by a squadron in the central Atlantic 13.3.1806; 1814 troopship; 1816 sold.
PLANS: Lines & profile/lower deck/upper deck/quarter deck & forecastle.

*Guerriere/Guerrier* 1806-1812 (French *La Guerrière* 1798-1799 Cherbourg, Lafosse design) 38.
Dimensions & tons: 155'9", 129'11½" x 39'9" x 12'10". 1092.
Men: 284. Guns: UD 28 x 18, QD 14 x 32★, Fc 2 x 9 + 2 x 32★. Taken by *Blanche* 19.7.1806; 19.8.1812 taken by USS *Constitution* and burnt.

*Rhin* 1806-1838 (French *Le Rhin* 1802 Toulon, Sané design) 38/46.
Dimensions & tons: 152'1", 127'1½" x 39'11½" x 13'. 1079 62/94.
Men: 284. Guns: UD 28 x 18, QD 14 x 32★, Fc 2 x 9 + 2 x 32★. Taken by *Mars* off Rochefort 28.7.1806; 1838 hulked [at Sheerness as a lazaretto; 1871 to the Thames as a cholera ship; 1884 sold].
PLANS: 1817: Lines & profile. No date: Lines & profile.

*Immortalite* 1806-1811 (French *L'Infatigable* 1799 Le Havre, Forfait design) 38.
Dimensions & tons: 156'6", 132'0⅝" x 40'7" x 12'. 1157.
Men: 284. Guns: UD 28 x 18, QD 14 x 32★, Fc 2 x 9 + 2 x 32★. Taken 24.9.1806 by a squadron off Rochefort; never fitted for sea in the RN; 1811 BU.

*Armide* 1806-1815 (French *L'Armide* 1804 Rochefort, Rolland design) 38.
Dimensions & tons: 152'11½", 128'1" x 40'4" x 12'3½". 1104 30/94.
Men: 284. Guns: UD 28 x 18, QD 14 x 32★, Fc 2 x 9 + 2 x 32★. Taken 25.9.1806 by a squadron off Rochefort; 1815 BU.
PLANS: As taken off/as fitted: Lines & profile/decks.

*Alceste* 1806-1817 (French *La Minerve* 1804-5 Rochefort, Rolland design) 38.
Dimensions & tons: 152'5", 128'8⅝" x 40'0½" x 12'8". 1097 71/94.
Men: 284. Guns: UD 28 x 18, QD 14 x 32★, Fc 2 x 9 + 2 x 32★. Taken 25.9.1806 by a squadron off Rochefort; 1814 troopship; 18.2.1817 wrecked by striking a rock in the Straits of Gaspar.
PLANS: As fitted: Lines, profile & decoration/decks. 1814 as troopship: Profile/orlop & lower deck/upper deck & quarter deck & forecastle. 1816 as fitted for an ambassador: decks.

*Gloire* 1806-1812 (French *La Gloire* 1803 Nantes, Forfait design) 38.
Dimensions & tons: 158', 131'8⅝" x 49'7½" x 12'10". 1156 29/94.
Men: 284. Guns: UD 28 x 18, QD 14 x 32★, Fc 2 x 9 + 2 x 32★. Taken by *Mars* and *Centaur* off Rochefort 25.9.1806; 1812 BU.
PLANS: 1810: Lines & profile/orlop/lower deck/upper deck/quarter deck & forecastle.

*President* 1806-1815 *Piedmontaise* -1815 (French *Le Président* 1804 L'Orient, Forfait design) 36.
Dimensions & tons: 158', 132'4¾" x 40'6¾" x 12'2". 1147 61/94.
Men: 284. Guns: UD 28 x 18, QD 14 x 32★, Fc 2 x 9 + 2 x 32★. Taken by *Canopus* off Morbihan 28.9.1806; 1815 BU.
NOTE: Her lines were used as the basis for the design of a large class of British 38s; the *Seringapatam* Class.
PLANS: As fitted: Lines, profile & decoration/decks/lines & profile.

*Freya/Freija* (ex *Hypollitus*) 1807-1816 (Danish *Freya* 1793 Copenhagen, Stibolt design) 36.
Dimensions & tons: 148'9", 124' x 39'4½" x 10'9". 1022 56/94.
Men: 264. Guns: UD 26 x 18, QD 12 x 32★, Fc 2 x 9 + 2 x 32★. Taken at the fall of Copenhagen 1807; 1811 troopship; 1816 sold.
PLANS: 1810: Lines & profile (incomplete).

*Nymphen/Nymphin* (ex *Determinee*) 1807-1816 (Danish *Nymphen* 1807 Copenhagen, Hohlenberg design) 36/32.
Dimensions & tons: 140'4", 117'10⅛" x 38'0½" x 10'3". 907 11/94.
Men: 264. Guns: UD 26 x 18, QD 12 x 32★, Fc 2 x 9 + 2 x 32★. Taken at the fall of Copenhagen 1807; 1816 sold.
PLANS: Lines & profile/orlop.

*Pearlen/Perlen/Perlin* (ex *Theban*) 1807-1813 (Danish *Perlen* 1804 Copenhagen, Hohlenberg design) 38.
Dimensions & tons: 156', 133' x 41'3" x 12'2". 1203 71/94.
Men: 284. Guns: UD 28 x 18, QD 14 x 32★, Fc 2 x 9 + 2 x 32★. Taken at the fall of Copenhagen 1807; 1813 hulked [at Pembroke as a lazaretto; 1821 to Liverpool; 1846 sold].
PLANS: Lines & profile.

*Rota* (ex *Sensible*) 1807-1816 (Danish *Rota* 1802 Copenhagen, Hohlenberg design) 38.
Dimensions & tons: 153'6", 128'11½" x 40'1" x 10'10". 1102.
Men: 284. Guns: UD 28 x 18, QD 14 x 32★, Fc 2 x 9 + 2 x 32★. Taken at the fall of Copenhagen 1807; 1816 sold.

*Hasfruen* (ex *Boreas*) 1807-1814 (Danish *Havfruen* 1789 Copenhagen, Stibolt design) 36.
Dimensions & tons: 148'7", 124'5½" x 39'4½" x 11'3". 1028.
Men: 264. Guns: ____. Taken at the fall of Copenhagen 1807; never fitted for sea in the RN; 1814 sold.

*Iris* (ex *Marie*, ex *Alaric*) 1807-1816 (Danish *Iris* 1795 Copenhagen, Stibolt design) 38.
Dimensions & tons: 148'6", 125'0¾" x 39'5" x 11'1". 1033 47/94.

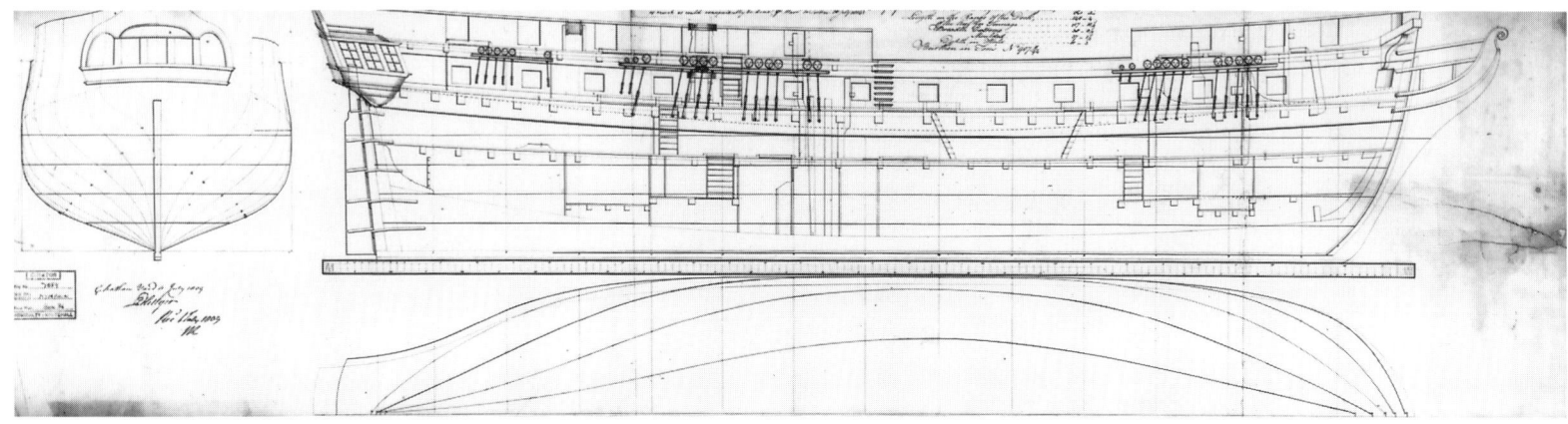

Lines and profile (7059) of the Danish 18 pounder frigate *Nymphen* of 1807. In the form of the hull and the shape of the stern this ship shows considerable originality

Men: ____ . Guns: ____ . Taken at the fall of Copenhagen 1807; 1816 sold.

PLANS: Lines & profile/orlop/lower deck/upper deck/quarter deck & forecastle.

**Nyaden/Nayaden** (ex *Hephaestion*) 1807-1812 (Danish *Najaden* 1795 Copenhagen, Hohlenberg design) 38.
  Dimensions & tons: 140'7", 118'7⅝" x 38'2½" x 10'6". 827 38/94.
  Men: ____ . Guns: ____ . Taken at the fall of Copenhagen 1807; 1812 BU.

PLANS: Profile. As fitted: Lines & profile/orlop/lower deck/upper deck/quarter deck & forecastle.

**Venus** (ex *Levant*) 1807-1815 (Danish *Venus* 1805 Copenhagen, Hohlenberg design) 36.
  Dimensions & tons: 143'4¾", 120'10½" x 38'3¼" x 10'8¾". 942.
  Men: 254. Guns: UD 26 x 18, QD 12 x 32★, Fc 2 x 9 + 2 x 32★. Taken at the fall of Copenhagen 1807; 1815 sold.

**Piedmontaise** 1808-1813 (French *La Piémontaise* 1804 Saint Malo, Pestel design) 38.
  Dimensions & tons: 157'5", 128'10" x 39'11" x 12'8". 1092.
  Men: 300. Guns: UD 28 x 18, QD 14 x 32★, Fc 2 x 9 + 2 x 32★. Taken off Ceylon by *San Fiorenzo* 8.3.1808; 1813 BU.

**Brune** 1808-1816 (French *La Thetis* 1787 Brest, Lamoth Jnr design) 38.
  Dimensions & tons: 153'9½", 126'8½" x 40'2⅝" x 12'8". 1090.
  Men: 284. Guns: UD 28 x 18, QD 14 x 32★, Fc 2 x 9 + 2 x 32★. Taken by *Amethyst* 10.11.1808; 1810 troopship; 1816 hulked [at Sheerness as a victualling depot; 1829 to Chatham; 1838 sold].

PLANS: 1820 as depot ship: Lower deck/upper deck/quarter deck & forecastle.

**Doris** 1808 See Main List under *Phoebe* Class.

**Alcmene** (ex *Jewel*?) 1809-1816 (French *La Topaze* 1804 Nantes, Forfait design) 38.
  Dimensions & tons: 156'1", 129'5¼" x 40'6½" x 13'. 1131 90/94.
  Men: 300. Guns: UD 28 x 18, QD 14 x 32★, Fc 2 x 9 + 2 x 32★. Taken by two frigates off Guadeloupe 22.1.1809; 1816 BU.

PLANS: 1816: Lines & profile/orlop/lower deck/upper deck/quarter deck & forecastle.

**Niemen** 1809-1815 (French *Le Niémen* 1808 Bordeaux or L'Orient, Rolland design) 38.
  Dimensions & tons: 154'2½", 129'1½" x 39'10¾" x 12'5⅞". 1093 37/94.
  Men: 300. Guns: UD 28 x 18, QD 14 x 32★, Fc 2 x 9 + 2 x 32★. Taken by *Amethyst* and *Emerald* 6.4.1809; 1815 BU.

PLANS: As fitted: Lines & profile/orlop/lower, upper and quarter decks and forecastle.

**Furieuse** 1809-1816 (French *La Furieuse* 1797 Cherbourg, Forfait design) 38.
  Dimensions & tons: 157'2½", 133'2" x 39'1¼" x 12'6". 1083 14/94.
  Men: 284. Guns: UD 28 x 18, QD 12 x 32★, Fc 2 x 12 + 2 x 32★ + 1 x 18★ + 1 x 12★. Taken by *Bonne Citoyenne* whilst sailing *en flûte* in the Atlantic 6.7.1809; 1816 BU.

PLANS: 1811: Lines & profile.

**Bourbonnaise/Bourbonnoise** 1809-1817 (French *La Caroline* 1802 or 1806 ____ ) 38.
  Dimensions & tons: 151'6", 127'4⅞" x 39'10⅝" x 12'2". 1078 10/94.
  Men: 300. Guns: UD 28 x 18, QD 14 x 32★, Fc 2 x 9 + 2 x 32★. Taken at Reunion 21.9.1809; 1810 arrived in Britain and never again fitted for sea; 1817 BU.

**Laurel** 1811-1812 (Franco-Dutch *La Fidèle* building at Flushing 1809) 36/38.
  Dimensions & tons: 152'1", 126'8⅛" x 40'5" x 12'4". 1104 19/94.
  Men: 300. Guns: UD 26 x 18, QD 14 x 32★, Fc 2 x 9 + 2 x 32★. Taken 1809 on the stocks at Flushing during the Walcheren Expedition; launched there and taken to Deptford where she was fitted out, completing in 1811; 31.1.1812 wrecked on the Govivas Rock, Teigneux Passage.

PLANS: As fitted: Lines & profile/orlop/lower deck/upper deck/quarter deck & forecastle.

**Nereide** 1810-1816 (French *La Vénus* 1806 Le Havre, Forfait design) 38.
  Dimensions & tons: 157', 132'2⅛" x 40'8½" x 12'2½". 1165.
  Men: 300. Guns: UD 28 x 18, QD 14 x 32★, Fc 2 x 9 + 2 x 32★. Taken off Reunion by *Boadicea* 18.9.1810; never fitted for sea after arrival in Britain; 1816 BU.

PLANS: Lines & profile.

**Pomone** 1810-1816 (Franco-Italian *L'Astrée* 1808 Genoa, Sané design) 38.
  Dimensions & tons: 152', 127'6" x 40'2" x 12'9". 1093 42/94.
  Men: 300. Guns: UD 28 x 18, QD 14 x 32★, Fc 2 x 9 + 2 x 32★. Taken at the fall of Mauritius 4.12.1810; 1816 BU.

PLANS: Lines & decorations.

**Junon** 1810-1817 (French *La Bellone* 1803 Saint Malo, Pestel design) 38.
  Dimensions & tons: 154', 129'5¼" x 40'3" x 12'5". 1116 44/94.
  Men: 295. Guns: UD 28 x 18, QD 14 x 32★, Fc 2 x 9 + 2 x 32★. Taken at the fall of Mauritius 4.12.1810; 1817 BU.

PLANS: Lines.

**Java** 1811-1812 (French *La Renomée* 1805 or 1807 Nantes) 38.
  Dimensions & tons: 152'5½", 126'5½" x 39'11⅜" x 12'9". 1073.
  Men: 300. Guns: UD 28 x 18, QD 14 x 32★, Fc 2 x 9 + 2 x 32★. Taken off Madagascar 20.5.1811; 29.12.1812 taken by USS *Constitution* then foundered near Bahia.

**Madagascar** 1811-1819 (French *La Néréide* 1809 Saint Malo, Pestel design) 38.
  Dimensions & tons: ____, ____ x ____ x ____. ____.
  Men: ____ . Guns: UD 28 x 18, QD 14 x 32★, Fc 2 x 9 + 2 x 32★. Taken at the capture of Tamatave, Madagascar, 1811; 1819 BU.

PLANS: Lines & profile.

**Ambuscade** 1811-1812 (Franco-Italian *La Pomone* 1804 Genoa, Sané design) 38.
  Dimensions & tons: 152'10½", 126'7" x 40'1¾" x 12'10½". 1085.
  Men: 300. Guns: UD 28 x 18, QD 14 x 32★, Fc 2 x 9 + 2 x 32★. Taken near Corfu by *Active* and *Alceste* 29.11.1811; 1812 arrived in Britain and broken up soon after arrival.

**Chesapeake** 1813-1819 (American *Chesapeake* 1799 Norfolk, Virginia) 38.
  Dimensions & tons: 151', 127'5" x 40'11" x 13'9". 1135.

Men: 315. Guns: UD 28 x 18, QD 14 x 32★, Fc 2 x 9 + 2 x 32★. Taken off Boston by *Shannon* 1.6.1813; 1819 sold.

PLANS: Lines.

***Weser*** 1813-1817 (Franco-Dutch *La Weser* 1812 Amsterdam, Sané design) 38/46.
Dimensions & tons: 152'6", 127'2" x 39'11¾" x 12'6". 1081.
Men: 300. Guns: UD 28 x 18, QD 14 x 32★, Fc 2 x 9 + 2 x 32★. Taken off the Isle Bas 21.10.1813; 1814 troopship; 1817 sold.

***Trave*** 1813-1821 (Franco-Dutch *La Trave* 1810 Amsterdam, Sané design) 38.
Dimensions & tons: 151'5¼", 126'5⅛" x 39'10½" x 12'4". 1069 26/94.
Men: 300. Guns: UD 28 x 18, QD 14 x 32★, Fc 2 x 9 + 2 x 32★. Taken by *Andromache* 23.10.1813; 1814 troopship; 1821 sold.

PLANS: As fitted as troopship: Lines & profile/orlop & lower deck/upper deck, quarter deck & forecastle.

***Seine*** 1814-1823 (French *La Cérès* 1810 Brest. Sané design) 38.
Dimensions & tons: 152', 126'11½" x 39'10½" x 12'8". 1073 70/94.
Men: 315. Guns: UD 28 x 18, QD 14 x 32★, Fc 2 x 9 + 2 x 32★. Taken by *Niger* and *Tagus* 6.1.1814; 1823 BU.

PLANS: Lines & profile/lower deck/upper deck/quarter deck & forecastle.

***Gloire*** (ex *Palma*?) 1814-1817 (French *L'Iphigénie* 1810 Cherbourg) 38.
Dimensions & tons: 154'5", 126'10¼" x 39'9" x 12'7½". 1066.
Men: 315. Guns: UD 28 x 18, QD 14 x 32★, Fc 2 x 9 + 2 x 32★. Taken by *Venerable* and *Cyane* 20.1.1814; 1817 sold.

***Modeste*** 1814-1816 (French *Terpsichore* 1812 Antwerp, Sané design) 38.
Dimensions & tons: 152'2", 127'7½" x 39'10⅞" x 12'4". 1081 6/94.
Men: 315. Guns: UD 28 x 18, QD 14 x 32★, Fc 2 x 9 + 2 x 32★. Taken by *Majestic* 3.2.1814; never fitted for sea after arrival in Britain; 1816 BU.

PLANS: Lines & profile.

***Immortalite*** (ex *Dunira*) 1814-1822 (French *L'Alcmène* 1811 Cherbourg, Rolland design) 38.
Dimensions & tons: 152'8", 127'11⅜" x 39'10" x 12'7½". 1079 78/94.
Men: 315. Guns: UD 28 x 18, QD 14 x 32★, Fc 2 x 9 + 2 x 32★. Taken by *Venerable* 16.2.1814; never fitted for sea after arrival in Britain; 1822 hulked [at Portsmouth as a receiving ship; 1837 sold].

PLANS: Lines & profile/lower deck/upper deck/quarter deck & forecastle.

***Sultane*** 1814-1819 (French *La Sultane* 1813 Nantes, Sané design) 38.
Dimensions & tons: 151'6", 127'0¼" x 39'9" x 12'6". 1067 51/94.
Men: 315. Guns: UD 28 x 18, QD 14 x 32★, Fc 2 x 9 + 2 x 32★. Taken 26.3.1814 by *Hannibal*; 1819 BU.

PLANS: Lines & profile/lower deck/upper deck/quarter deck & forecastle.

***Topaze*** 1814-1823 (French *L'Etoile* 1814 Nantes, Sané design) 38.
Dimensions & tons: 151'5⅜", 126'8⅛" x 39'8" x 12'5¼". 1060 23/94.
Men: 315. Guns: UD 28 x 18, QD 14 x 32★, Fc 2 x 9 + 2 x 32★. Taken by *Hebrus* 26.3.1814; 1823 hulked [at Portsmouth as a receiving ship; 1851 target then BU].

PLANS: Lines & profile/lower deck/upper deck & quarter deck & forecastle.

***Essex*** 1814-1824 (American *Essex* 1799 Salem) 36/42.
Dimensions & tons: 138'7", 117'2⅞" x 37'3½" x 11'9". 867.
Men: 280. Guns: UD 26 x 18, QD 12 x 32★, Fc 2 x 9 + 2 x 32★. Taken 28.3.1814 off Valparaiso by *Phoebe* and *Cherub*; never fitted for sea after arrival in Britain; 1824 hulked [at Cork as a convict ship; 1837 sold].

***Aurora*** 1814-1832 (French *La Clorinde* 1808 Paimboeuf, Gauthier design) 38/46.
Dimensions & tons: 152'1", 127'0⅛" x 40'0½" x 12'7½". 1083 14/94.
Men: 300. Guns: UD 28 x 18, QD 14 x 32★, Fc 2 x 9 + 2 x 32★. Taken by *Dryad* and *Eurotas* 20.2.1814; 1832 hulked [as a coal depot at Falmouth; 1851 BU].

PLANS: Lines & profile (incomplete)/lower deck/upper deck/quarter deck & forecastle.

***Melpomene*** 1815-1821 (French *La Melpomène* 1810 Toulon) 38.
Dimensions & tons: 152'10½", 127'0⅞" x 40'1¼" x 12'6¼". 1087.
Men: 300. Guns: UD 28 x 18, QD 14 x 32★, Fc 2 x 9 + 2 x 32★. Taken off Ischia by *Rivoli* 30.4.1815; never fitted for sea after arrival in Britain; 1821 sold.

## 12 pounder Frigates (Fifth Rates)

***Franchise*** 1803-1815 (French *La Franchise* 1798 Rochefort) 36.
Dimensions & tons: 143', 119'8¼" x 37'6¾" x 11'8". 898 16/94.
Men: 264. Guns: UD 26 x 12, QD 2 x 12 + 10 x 24★, Fc 2 x 12 + 4 x 18★. Taken by a squadron 28.5.1803; 1815 BU.

PLANS: 1810: Lines & profile/orlop/lower deck, upper deck, quarter deck & forecastle.

***Barbadoes*** 1804-1812 (French privateer *Le Brave* 1801 ____ ) 32/28.
Dimensions & tons: 139'8", 117'6½" x 35'2⅝" x 10'3". 775 43/94.
Men: ____ . Guns: (Later?) UD 24 x 9, QD 8 x 24★, Fc 2 x 6 + 2 x 24★. Taken 05.1803?; 1804 presented to the RN by the inhabitants of Barbados; 28.9.1812 wrecked on Sable Island. Sometimes referred to in later lists as a 28 gun Sixth Rate. It seems quite likely that this reflected an actual reduction in number of guns and rate; possibly occurring when she arrived in Britain from the West Indies.

PLANS: 1810: Lines & profile /orlop/lower deck/upper deck/quarter deck & forecastle.

***Amsterdam*** 1804-1806 (Dutch *Proserpine* 1801 Amsterdam Dyd) 32.
Dimensions & tons: 140'8", 113'6" x 37'6" x 11'11". 849.
Men: ____ . Guns: ____ . Taken at Surinam 4.5.1804; 1806 hulked [at Cork as a storeship; 1811 to Plymouth; 1815 sold].

***Clara*** 1804-1811 (Spanish *Santa Clara* 1784 Ferrol) 38/36.
Dimensions & tons: 144'6", 120'2⅜" x 38'8¼" x 11'8". 958.
Men: 284. Guns: UD 26 x 12, QD 10 x 32★, Fc 2 x 9 + 2 x 32★. Taken in the Atlantic by *Indefatigable, Lively, Amphion, Medusa* 5.10.1804; 1811 hulked [at Plymouth as a receiving ship; 1815 sold].

***Santa Gertruyda*** 1804-1807 (Spanish *Santa Gertruyda* 1768 Guarnizo, Santander) 32/36.

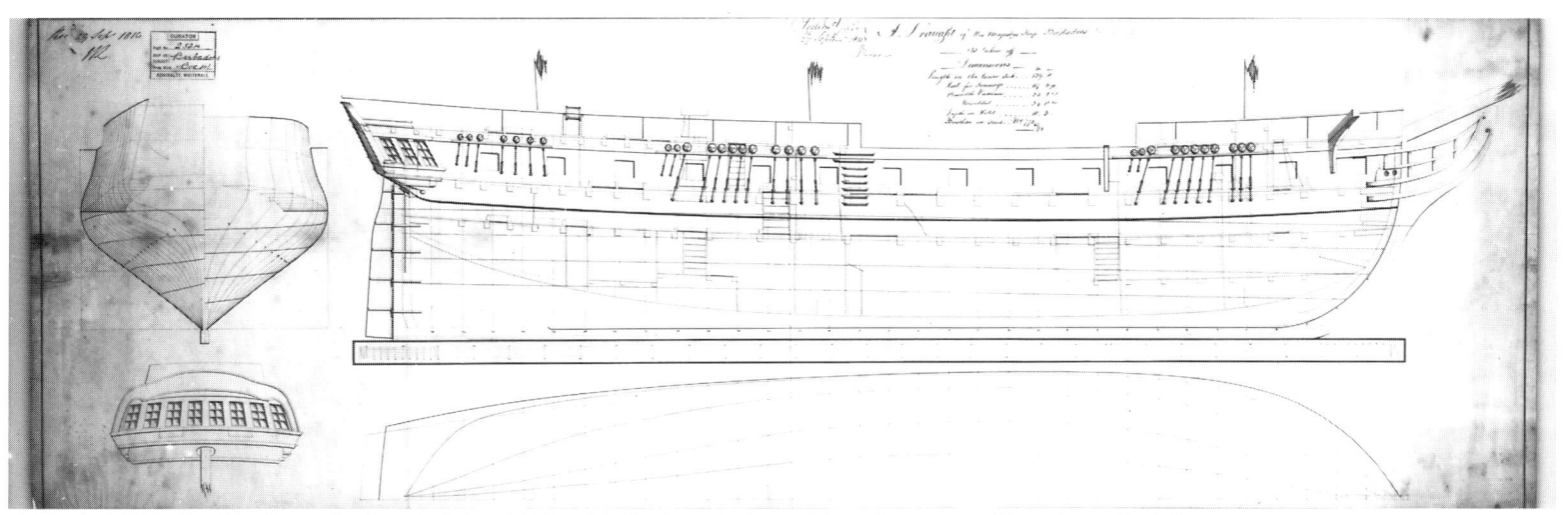

Lines and profile (2524) of the 12 pounder frigate *Barbadoes* of 1804, originally a French privateer. Notice the pronounced 'vee' shape of the hull

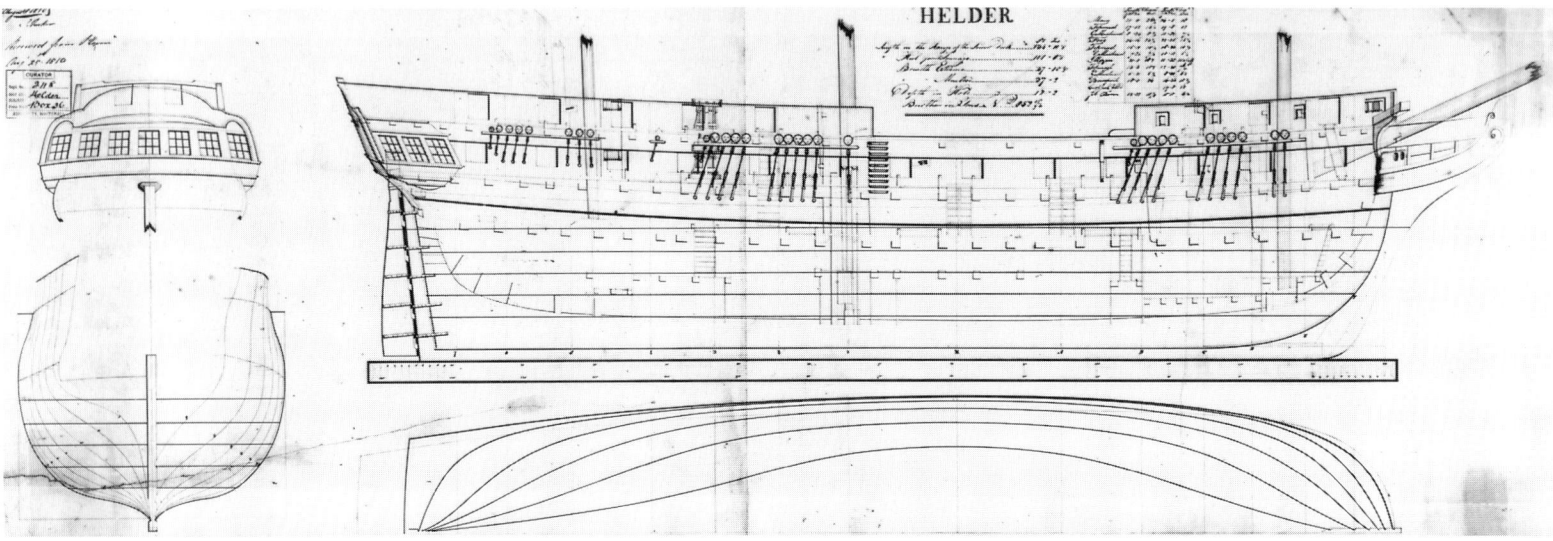

Lines, profile and spar dimensions (2118) for the Dutch 12 pounder frigate *Helder* taken in 1808

Dimensions & tons: ____ , 106'11⅝" x 36'8¼" x 10'6". 766.
Men: 200. Guns: UD 26 x 24★, QD 2 x 9 + 4 x 24★, Fc 2 x 9. Taken 7.12.1804 by *Polyphemus* and *Lively*; 1807 hulked [at Plymouth as a receiving ship; 1811 BU].

*Psyche* 1805-1812 (French *La Psyché* 1798 Nantes) 32.
Dimensions & tons: 138'6", 117' x 36'10⅛" x 10'5". 846 22/94.
Men: ____ . Guns: UD 24 x 12, QD 8 x 18★, Fc 2 x 6 + 2 x 18★. Taken off India by *San Fiorenzo* 14.2.1805; 1812 sold.
PLANS: As fitted: Lines & profile/decks.

*Bombay* 1805-1808 *Ceylon* -1832 (East India Company's *Bombay* 1793 Bombay) 32.
Dimensions & tons: 130', 104'10¼" x 34'8½" x 11'8". 672.
Men: 215 (1830: 115). Guns: UD 26 x 12, QD 12 x 24★, Fc 2 x 6 + 2 x 24★ (1830: UD 18 x 32★ + 2 x 9, QD 6 x 32★, Fc 2 x 9). Purchased 1805 in India from the East India Co; 1808 renamed *Ceylon*; 1810 taken by the French (whilst serving as a troopship) then retaken; 1814 troopship; 1816 laid up until; 1832 hulked [at Malta as a receiving ship; 1857 sold].

*Cuba* 1806-1811 (Spanish *Pomona* 1794 Ferrol) 36.
Dimensions & tons: 142'4", 116'2" x 37'7" x 11'. 873.
Men: 254. Guns: UD 26 x 12, QD 6 x 24★, Fc 2 x 6 + 2 x 24★. Taken by *Arethusa* off Havanna; 1811 hulked [at Portsmouth; 1817 sold].

*Frederickstein* (ex *Teresa*) 1807-1813 (Danish *Frederickstein* 1800 Copenhagen, Hohlenberg design) 28/32.
Dimensions & tons: 128'8", 108'10" x 34'3" x 9'4". 679 17/94.
Men: 215. Guns: UD 26 x 12, QD 6 x 24★, Fc 2 x 6 + 2 x 24★. Taken at the fall of Copenhagen 1807; 1813 BU.
PLANS: Lines & profile.

*Frederickswaern* 1807- 1811 (Danish *Frederickswaern* 1778 or 1783 Copenhagen, Gerner design) 32.
Dimensions & tons: 130'2", 107'9¾" x 36'9¾" x 9'10". 776.
Men: 240. Guns: As *Frederickstein*. Taken 16.8.1807 by *Comus*; not fitted for sea by the RN; 1811 hulked [at Chatham as a receiving ship; 1814 sold].

*Helder* 1808-1813 (Dutch *Gelderland* 1806, ex *Orpheus* 1803 Amsterdam Dyd) 32.
Dimensions & tons: 134'11½", 111'8⅛" x 37'10½" x 12'2". 852 16/94.
Men: 215. Guns: UD 26 x 12, QD 10 x 32★, Fc 2 x 6 + 2 x 32★. Taken by *Virginie* 19.5.1808; 1813 hulked [at Woolwich as receiving ship; 1816 receiving ship for distressed seamen; 1817 BU].
PLANS: Figurehead/stern & quarter decorations. 1810: Lines & profile/orlop/lower deck, upper deck, quarter deck & forecastle.

## 9 pounder Frigates

*Hyena* 1804-1822 (Mercantile *Hope* 1800 India) Sixth Rate 20 teak built.
Dimensions & tons: 122'4", 102'1⅛" x 30'1" x ____ . 519.
Men: 165. Guns: GD 20 x 9, QD 12 x 18★, Fc 2 x 9 (as storeship: 2 x 9 + 12 x 18★). Purchased 06.1804; 1809 storeship; 1822 sold.

*Proselyte* 1804-1808 (Mercantile *Ramillies* 1804 North Shields) Sixth Rate 28.
Dimensions & tons: 107'6", 87'3¼" x 29'6" x ____ 404.
Men: 155 (as bomb vessel: 70). Guns: UD 24 x 9, QD 2 x 6, Fc 2 x 6 (as bomb vessel: UD 1 x 13" & 1 x 10" mortar + 6 x 24★, QD 2 x 18★, Fc 2 x 18★). Purchased 06.1804; 1808 converted to bomb vessel; 5.12.1808 caught in the ice and then wrecked on Anholt Reef.

*Dover* 1811-1825 (Franco-Italian *Bellona* 1806 Venice) Fifth Rate 32.
Dimensions & tons: 131'6", 110'4⅛" x 34'4" x 8'11". 692.
Men: ____ (as troopship: 115). Guns: ____ (as troopship: UD 12 x 9, QD 4 x 18★, Fc 2 x 6). Taken by Hoste at Lissa 13.3.1811; 1812 troopship; 1819 guard ship; 1825 hulked [at Deptford as a receiving ship; 1831 for quarantine service; 1836 sold].

NOTE: Quite probably this vessel carried 12 pounder guns when serving as a frigate; also Winfield suggests that her original allocated complement was 215 men which would indicate that this was so.

## Frigates, Guns Uncertain

NOTE: Though the three converted Indiamen listed here appear to have been rated with a frigate armament, the appearance of the one of which plans survive (*Sir Edward Hughes* in her later role as the storeship *Tortoise*) clearly shows her to have had the appearance of a small two-decker, which was probably the case with all three.

*Sir Francis Drake* 1805-1825 (East Indiaman *Asia* ____ ) Fifth Rate 38.
Dimensions & tons: 132'3½", 112'4⅛" x 35'5½" x 14'0½". 751.
Men: ____ (as storeship: 88) Guns: ____ (as storeship: UD 2 x 9 + 10 x 24★, QD 4 x 24★, Fc 2 x 9). Purchased in India 1805; 1813 storeship; 1825 sold.

*Sir Edward Hughes* 1806?-1809 *Tortoise* -1824 (East Indiaman *Sir Edward Hughes*? 1787 Bombay) Fifth Rate 38 teak built.
Dimensions & tons: 147'2" , 118'4⅞" x 39'1" x 19'4½". 962.
Men: ____ (as storeship: 90). Guns: ____ (as storeship: UD 18 x 9, Fc 2 x 9). Purchased 1806 or 1807 in India?; 1808 storeship; 1809 renamed *Tortoise*; 1824 hulked [at Pembroke as a coal depot; 1841 to Sheerness; ?1844 to Ascension as a receiving ship?; 1863 BU].
PLANS: As fitted 1842: Lines and profile/decks.

*Dover* (ex *Duncan/Lord Duncan*) 1807?; 1811 (East India Company's *Carron* 1804 Bombay?) Fifth Rate 38 teak built.
Dimensions & tons: ____ , ____ x ____ x ____ . ____ .
Men: ____ . Guns: ____ . Purchased 1807 (or 1804?); 2.5.1811 wrecked in Madras Road.

THE NAPOLEONIC WAR 1802-1815

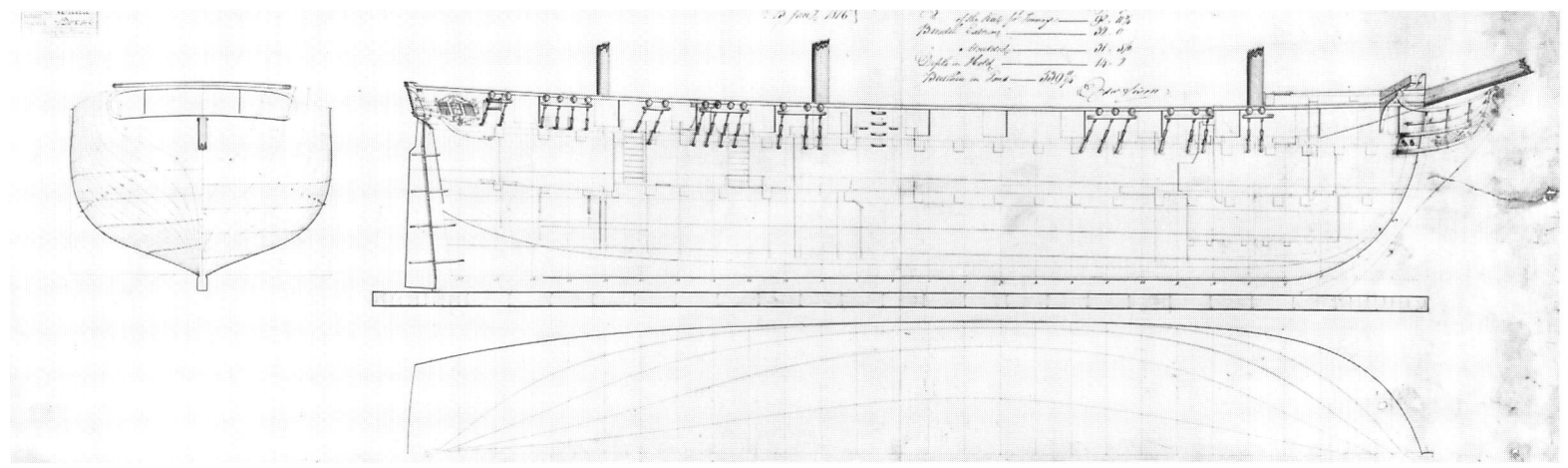

Lines and profile (3085) of the 20 gun Sixth Rate *Florida*, taken from the Americans (as the *Frolic*) in 1814

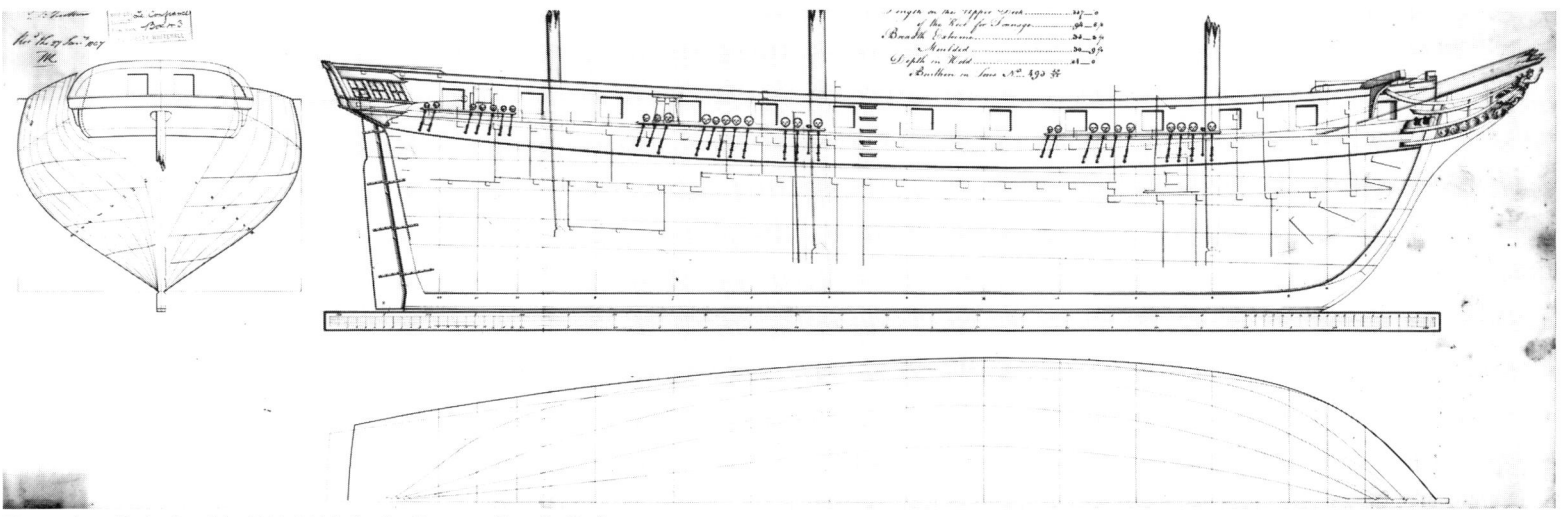

Lines and profile 'as fitted' in 1806 (2803) for the 22 gun ex French *Confiance*

# Corvettes (Sixth Rates)

**Bacchante** 1803-1809 (French *La Bacchante* 1795 ___ ) 20.
Dimensions & tons: 131'6", 111'8⅛" x 32'10½" x 14'8¾". 642.
Men: 175. Guns: UD 2 x 12 + 18 x 32★. Taken by *Endymion* 25.6.1803; 1809 sold.

**Sagesse** 1803-1805 (French *La Sagesse* 1794 Bayonne) 20.
Dimensions & tons: 112'4", 91'0⅜" x 31'9" x 9'. 480 61/94.
Men: ___. Guns: ___. Taken by *Theseus* in the West Indies 8.9.1803; 1805 hulked [at Portsmouth as a convict hospital ship; 1821 sold].
PLANS: Lines & profile & decorations/lower deck/upper deck/quarter deck & forecastle.

**Ligera** 1804-1814 (Spanish *Diligentia* ___ ) 22.
Dimensions & tons: 106', 100' x 28'9" x 7'7". 440.
Men: 155. Guns: UD 20 x 9, QD 10 x 12★, Fc 2 x 6 + 6 x 12★. Taken by *Diana* and *Pique* in the West Indies 12.1804; 1805 laid up until 1814 sold.

**Confiance** 1805-1810 (French privateer *La Confiance* 1796 Bordeaux) ship sloop/Sixth Rate 22.
Dimensions & tons: 117', 94'8½" x 31'2⅞" x 14'. 491 59/94.
Men: 140. Guns: UD 2 x 6 + 22 x 18★ (actually 24 guns, though classed as a 22). Taken by *Loire* at Mudros 1805; 1810 sold.
PLANS: Lines & profile & decorations/decks.

**Barbette** 1805-1811 (French privateer *La Vaillante* 1801 Bordeaux) 22.
Dimensions & tons: 124', 101'5¼" x 33'6¼" x 16'6". 605.
Men: 170. Guns: UD 22 x 9, QD 8 x 18★. Taken 1805; never fitted for sea by the RN; 1811 BU.

**Muros** 1806-1808 (French privateer *L'Alcide* 1802 ___ ) 22.
Dimensions & tons: 107'11", 89'8⅞" x 30'6" x 14'5¼". 444 4/94.
Men: 140. Guns: UD 2 x 12 + 20 x 32★, QD 4 x 12★, Fc 2 x 12★. Taken by *Egyptienne* 8.3.1806; 24.3.1808 wrecked in Honda Bay, Cuba.
PLANS: As fitted: Lines & profile & decorations/decks.

**Fuerte** 1807-1812 (Spanish *Fuerte* ___ ) 20 1807.
Dimensions & tons: ___ , ___ x ___ x ___ . 490.
Men: ___ . Guns: 20 x ___ . Taken at the fall of Montevideo 3.2.1807; 1808 arrived in Britain; never fitted for sea by the RN afterwards; 1812 BU.

**Little Belt** (ex *Espion*) 1807-1811 (Danish *Lille Belt* 1801 Copenhagen, Hohlenberg design) 24.
Dimensions & tons: 116'4", 94' x 30'4" x 12'5½". 460.
Men: 121. Guns: UD 2 x 9 + 18 x 32★. Taken at the fall of Copenhagen 1807; 16.5.1811 taken by the USS *President* in time of peace (reprisal for the *Leopard/Chesapeake* affair); returned to the RN and sold.

**Fylla/Fiila** 1807-1814 (Danish *Fylla* 1802 Copenhagen, Hohlenberg design) Sixth Rate/sloop 22/26.
Dimensions & tons: 115'11", 94'4⅞" x 30'3½" x 7'2". 460 72/94.
Men: 140. Guns: UD 2 x 6 + 18 x 32★. Taken at the fall of Copenhagen 1807; 1814 sold.
PLANS: 1809: Lines & profile.

**Rainbow** 1809-1815 (French *L'Iris* 1806 Dunkirk) 28.
Dimensions & tons: 123'9", 105'0⅝" x 32'5" x 10'1½". 587 17/94.
Men: 155. Guns: UD 20 x 9, QD 6 x 18★, Fc 2 x 6. Taken 3.2.1809 by *Aimable*; 1815 sold.
PLANS: As fitted: Lines & profile/orlop/lower deck/upper deck/quarter deck & forecastle.

**Ganymede** 1809-1819 (French *L'Hébé* 1808 ___ ___ ) 24/20.
Dimensions & tons: 126'7", 105'5⅛" x 32'8⅞" x 10'4". 601 17/94.
Men: 175. Guns: UD 20 x 9, QD 2 x 6 + 2 x 18★, Fc 2 x 6 Taken by *Loire* 5.2.1809; 1819 hulked [at Portsmouth as a convict ship (?); 1838 capsized at Woolwich, raised &; 1840 BU].
PLANS: Lines & profile/lower deck/upper deck/quarter deck & forecastle.

**Andromeda** 1812-1816 (American *Hannibal* 1810 Maryland) 24.

Dimensions & tons: 135'6", 108'6¼" x 37'5½" x 10'11". 809 4/94.
Men: 195. Guns: UD 22 x 32★ + 2 x 12. Taken 1812; 1816 sold.

PLANS: As fitted: Lines & profile.

**Florida** 1814-1819 (American *Frolic* 1813 Charlestown, Massachussetts) 20.
Dimensions & tons: 119'5½", 98'11¾" x 32' x 14'2". 539 11/94.
Men: 135. Guns: UD 2 x 9 + 18 x 32★. Taken 20.4.1814 by *Orpheus*; 1819 BU.

PLANS: Lines & profile & decorations/tiller & wheel.

## Ship Sloops

**Curlew** 1803-1810 (Mercantile *Leander* 1801 ____ 'Shields' ) 16.
Dimensions & tons: 98'7", 75'7½" x 29'6" x 12'9". 350.
Men: 100. Guns: UD 6 x 24 (Gover's) + 8 x 24★, Fc 2 x 6. Purchased 06.1803; 1810 sold.

**Vulture** 1803-1814 (Mercantile *Warrior* 1801 ____ 'Shields' ) 18.
Dimensions & tons: 105" 81'11" x 29'11½" x 12'9½". 391.
Men: 100. Guns: UD 14 x 24★ + 2 x 12★, Fc 2 x 6. Purchased 06.1803; 1808 floating battery; 1814 sold.

**Hermes** 1803-1810 (Mercantile *Majestic* 1801 ____ Whitby) 16.
Dimensions & tons: 107" 84'3" x 27'6" x 12'9". 339.
Men: 100. Guns: UD 14 x 24★, Fc 2 x 6. Purchased 07.1803; 1809 storeship; 1810 sold.

**Merlin** 1803-1810 (Mercantile *Hercules* 1801 ____ 'Shields') 20.
Dimensions & tons: 104' 79'10"(?) x 30'6" x 12'10". 395.
Men: 100. Guns: MD 14 x 32★, Spar dk. 2 x 9 + 4 x 24★. Purchased 07.1803; 1810 hulked [at Portsmouth as a receiving ship; 1817 sold].

**Scourge** 1803-1816 (Mercantile *Herald* 1801 ____ Whitby) 16.
Dimensions & tons: 107" 84'3" x 27'6" x 12'9". 339.
Men: 100. Guns: UD 14 x 32★, Fc 2 x 6. Purchased 06.1803; 1804 laid up until; 1816 sold.

**Speedy** 1803-1811 (Mercantile *George Hibbert* 1803 ____ Newcastle) 16.
Dimensions & tons: 101'5", 82'6" x 29'4½" x 12'10". 379.
Men: 100. Guns: UD 14 x 24★, Fc 2 x 6. Purchased 07.1803; 1811 hulked [at Portsmouth as a receiving ship; 1817 sold].

**Bonetta** 1803-1810 (Mercantile *Adamant* 1798 ____ Bridlington) 14.
Dimensions & tons: 86'3", ____ x 24'3" x 10'10". 208.
Men: 80. Guns: UD 14 x 24★. Purchased 08.1803; 1807 laid up until; 1810 sold.

**Hawk** 1803-1804 (French privateer *L'Atalante* ____ ) 18.
Dimensions & tons: 90'11", 76'11½" x 27'11¾" x 8'2". 320.
Men: 125. Guns: UD 16 x 24★, QD 2 x 9. Taken by *Plantagenet* 08.1803; 12.1804 foundered in the Channel.

**Inspector** 1803-1810 (Mercantile *Amity* 1801 ____ near Colchester) 14.
Dimensions & tons: 96'5", ____ x 25'1" x 10'6". 249.
Men: 80. Guns: UD 14 x 24★. Purchased 08.1803; 1810 sold.

**Avenger** 1803-1803 (Mercantile *Elizabeth* 1801 Bridlington) 14/16.
Dimensions & tons: ____ , ____ x ____ x ____ . 264 (or 330?).
Men: 80. Guns: 14 x 24★. Purchased 08.1803; 5.12.1803 wrecked at the mouth of the Elbe.

**Orestes** 1803-1805 (Mercantile *Ann*? 1803 Harwich) 14/16.
Dimensions & tons: ____ , ____ x ____ x ____ . 280.
Men: 80. Guns: 14 x 24★. Purchased 08.1803; 12.7.1805 wrecked off Dunkirk and then burnt.

**Hippomenes** 1803-1813 (Dutch *Hippomenes*? 1796 Amsterdam) 18.
Dimensions & tons: 95'10½", 85' x 30'1" x 7'5½". 407.
Men: 121. Guns: UD 2 x 9 + 16 x 32★. Taken 20.9.1803 at Demerara; 1813 sold.

**Renard** 1803-1809 (French lugger or schooner? *Le Renard* 1793 Dieppe) 18.
Dimensions & tons: 101" 81'10¼" x 28'2⅞" x 11'2¼". 347.
Men: 90 Guns: UD 16 x 12, Fc 2 x 6. Taken in the Mediterranean 11.11.1803; 1809 BU.

**Drake** 1804-1808 (Mercantile, East India Company packet *Earl of Mornington* 1799 ____ Thames ) 16.
Dimensions & tons: 104', ____ x 24'6" x ____ . 253.
Men: 75. Guns: spar deck 16 x 18★. Purchased 02.1804?; 1808 sold.

**Alert** 1804-1812 (collier *Oxford* 1803 ____ Howden Pans, Tyne) 18.
Dimensions & tons: 105', ____ x 29'4" x ____ 393.
Men: 80. Guns: UD 2 x 9 + 16 x 18★. Purchased 05.1804; 13.8.1812 taken by USS *Essex*.

**Avenger** 1804-1812 (collier *Thames* 1803 ____ Thames ) 20.
Dimensions & tons: ____ , ____ x ____ x ____ . 390.
Men: 80. Guns: UD 2 x 9 + 18 x 24★. Purchased 05.1804; 8.10.1812 wrecked in the Narrows of Saint John's Harbour, Newfoundland.

**Cormorant** 1804-1817 (Mercantile *Blenheim* 1803 ____ Howden Pans, Tyne) 16.
Dimensions & tons: 100'9", ____ x 27'7" x ____ . 328.
Men: 70. Guns: UD 2 x 9 + 14 x 18★. Purchased 06.1804; 1809 storeship; 1817 sold.

**Espiegle** 1804-1811 (Mercantile *Wimbury* 1803 ____ Barnstaple) 18.
Dimensions & tons: 98', ____ x 27' x ____ . 305.
Men: 65. Guns: UD 2 x 9 + 14 x 18★ Purchased 06.1804; 1811 BU.

**Eugenie** 1804-1810 (Mercantile *Friends* 1800 ____ Ipswich) 16.
Dimensions & tons: 90'6", 73'1" x 26'6" x 17'. 273.
Men: 65. Guns: UD 2 x 9 + 14 x 18★. Purchased 06.1804; 1810 sold.

**Heron** 1804-1811? *Volcano*? -1816 (Mercantile *Jason* 1803 ____ Newcastle) 16.
Dimensions & tons: 97'6", ____ x 29' x ____ . 339.
Men: 70. Guns: UD 2 x 9 + 14 x 18★. Purchased 06.1804; 1811 bomb vessel and renamed *Volcano*?; 1816 sold.

**Railleur** 1804-1810 (Mercantile *Henry* 1804 ____ Gainsborough) 16.
Dimensions & tons: 96'1", ____ , x 26'1" x ____ . 271.
Men: 65. Guns: UD 2 x 9 + 14 x 18★. Purchased 06.1804; 1810 sold.

**Spy** 1804-1813 (Mercantile *Comet* 1800 ____ . New Shoreham) 16.
Dimensions & tons: 95'7", ____ x 26'2" x 10'7". 274
Men: 65. Guns: UD 2 x 9 + 14 x 18★. Purchased 06.1804; 1810 storeship; 1813 sold.

**Swift** 1804-1814 (Mercantile *Pacific* 1802 ____ Rotherhithe) 18. Sloop 1804.
Dimensions & tons: 101'10", 84'10¼" x 26'11" x ____ . 327.
Men: 70. Guns: UD 16 x 12★, QD 2 x 4, Fc 2 x 4. Purchased 06.1804; 1810 storeship; 1814 sold.

PLANS: 1810 as storeship: profile (incomplete)/lower deck/upper deck.

**Utile** 1804-1814 (Mercantile *Volunteer* 1803 ____ South Shields) 16.
Dimensions & tons: ____ , ____ x ____ x ____ . 340.
Men: 70. Guns: UD 2 x 9 + 14 x 18★. Purchased 06.1804; 1814 sold.

**Porpoise** 1804-1814 (Mercantile Lord Melville 1804 ____ South Shields) 'sloop registered as a Sixth Rate' 10.
Dimensions & tons: 100'1", ____ , x 30'10" x 13'. 399.
Men: 70. Guns: UD 2 x 6 + 8 x 18★ (as storeship: 2 x 9 + 10 x 18★). Purchased before 09.1804; 1805 storeship for Australia; 1814 hulked [at Woolwich as a hulk; 1816 sold].

PLANS: Profile/midships section/stern. As fitted; profile/orlop & lower deck/upper deck & quarter deck & forecastle. 1811: profile & decks.

**Lilly** 1804-1811 (East India Company packet *Swallow* 1779 Bombay) 24.
Dimensions & tons: 99', ____ x 29'5" x ____ . 331.
Men: 121. Guns: MD 16 x 24★, spar dk 2 x 6 + 6 x 18★. Purchased before 08.1804; 1811 sold.

**Bergere** 1806-1811 (French *La Bergère* 1793 Rochefort) 18.
Dimensions & tons: 112'2", 94'5⅜" x 29'8" x 18'. 442.
Men: 121. Guns: UD 18 x 9. Taken 14.4.1806? by *Sirius* in the Mediterranean; 1807 arrived in Britain and laid up until; 1811 BU.

**Thrush** 1806-1809 (Revenue brig *Prince of Wales* 1794 Scotts, Greenock) 18.
Dimensions & tons: 96'4", 78'10½" x 27'0⅝" x 13'3½". 307.
Men: 121. Guns: UD 18 x 6. Purchased before 06.1806; 1809 hulked (as a depot for gunpowder at Port Royal, Jamaica; 1819 wrecked and sold).

NOTE: This is the vessel on which Scotts of Greenock have based their claim to be the oldest established shipbuilder to the RN; the claim seems a little dubious as, though she was built for a government agency she was not built for the Navy.

**Eijderen** (ex *Utile*) 1807-1813 (Danish *Eijderen* 1802 Gammeholm, Hohlenburg design) 18.
Dimensions & tons: 99'9", 82'7" x 27'7½" x 10'10". 335.
Men: 100. Guns: UD 2 x 6 + 16 x 24★. Taken at the fall of Copenhagen 1807; 1810 laid up until; 1813 BU.

NOTE: Same class as *Elven* (see below) and *Glückstadt* (see brigs).

*Elven* (ex *Harlequin*) 1807-1814 (Danish *Elven* 1800 Copenhagen, Hohlenburg design) 18.
Dimensions & tons: 101'1", 83'7½" x 27'4½" 11'3". 333.
Men: 100 Guns: UD 2 x 6 + 16 x 24★. Taken 1807; 1810 laid up until; 1814 sold.

NOTE: Sister of *Eijderen*.

*Scipio* 1807-1808? *Samarang* -1814 (Dutch *Psyche* ____ ) 18.
Dimensions & tons: ____ , ____ x ____ x ____ . 408.
Men: 121. Guns: 16 x 32★ + 2 x 9? Taken 1807; 1808? renamed *Samaran* ; 1814 sold.

*Saint Pierre* 1809-1814 (French *La Diligente* 1809 ____ ) ____ .
Dimensions & tons: ____ , ____ x ____ x ____ . 371.
Men: ____ . Guns: ____ . Taken at the surrender of Saint Pierre in 02.1809; 1814 sold.

*Peacock* 1812-1814 (American *Wasp* 1806 Washington Navy Yard) 16.
Dimensions & tons: 105'10½", 83'10½" x 30'10" x 14'. 434.
Men: 121. Guns: UD 2 x 6 + 14 x 32★. Taken by *Poictiers* 18.10.1812; 08.1814. Foundered off South Carolina. (Dittmar suggests that she was initially named *Loup Cervier* in RN service.)

# Brigs, etc

*Colombe* 1803-1811 (French *La Colombe* 1795 Cherbourg) brig sloop 16.
Dimensions & tons: 108'2", 89'10" x 29'0½" x 8'1". 403.
Men: 96. Guns: 2 x 6 + 14 x 32★. Taken by *Dragon* and *Endymion* 18.6.1803; not fitted for sea by the RN?; 1811 BU.

*Eclipse* (ex *Eagle*) 1803-1807 (French *Le Ventueux/Le Venteux* ex? *Le Volage*? 1795 ____ ) gunbrig/gunboat 12.
Dimensions & tons: 74'2½", 59'8¾" x 22'3⅝" x 7'5". 158.
Men: 55 Guns: UD 10 x 32★, Fc 2 x 18★. Taken by the boats of the *Loire* 27.6.1803; 1807 sold.

*Suffisante* 1803-1807 (French *La Vigilante* 1802 ____ ) brig sloop 16.
Dimensions & tons: 90'3", 72'4¾" x 28'8" x 12'10". 316 40/94.
Men: 121. Guns: 16 x ____ . Taken at the capture of San Domingo 30.6.1803; Not fitted for sea by the RN after her arrival in Britain; 1807 sold.

PLANS: Lines & profile.

*Superieure* 1803-1814 (French schooner *La Supérieure* 1801 Maryland) brig sloop (though still schooner rigged?) 10.
Dimensions & tons: 87', 67' x 23'6" x 9'5". 197.
Men: 70. Guns: UD 2 x 12 + 8 x 18★ (by 1810 12 x 4). Taken 30.6.1803 off San Domingo; 1814 sold.

PLANS: Lines & profile.

*Halcyon/Alcion* 1803-1812 (French *L'Alcion* 1802 Nantes) brig sloop 16.
Dimensions & tons: 91'6", 71'4¾" x 28'x ____ . 298.
Men: 95. Guns: UD 14 x 24★ + 2 x 6. Taken by *Narcissus* in the Mediterranean 8.7.1803; 1812 BU.

*Epervier* 1803-1811 (French *L'Epervier* 1801 Nantes, Gréan design) brig sloop 18.
Dimensions & tons: 89'10", 73'2¼" x 28'6" x 8'9½". 315 25/94.
Men: ____ . Guns: 2 x 6 + 16 x 32★. Taken in the West Indies by *Egyptienne* 27.7.1803; 1811 BU.

PLANS: Lines, profile & decorations/lower deck/upper deck.

*Morne Fortunee* 1804?-1804 (French privateer *Morne Fortunée*? ____ Bermuda?) armed brig 6.
Dimensions & tons: 65'6", 45' x 21'0½" x ____ 105 90/94.
Men: 35. Guns: UD 6 x 12★. Taken 1803 in the West Indies: 'reported to be Bermuda built with frames of cedar'; 1804? purchased in Barbados; 6.12.1804 wrecked on Attwood Key, West Indies.

*Goelan* 1803-1810 (French *Le Goëland* 1801 ____ ) brig sloop 16?
Dimensions & tons: ____ , ____ x ____ x ____ . ____ .
Men: ____ . Guns: ____ . Taken 13.10.1803 by *Pique* and *Pelican* at Santo Domingo; 1810 BU.

*Cerf* 1803-1806 (French *Le Cerf* 1801-1803 ____ ) gun brig/brig sloop 14.
Dimensions & tons: 74', 66'9" x 22'x 9'. 172
Men: ____ . Guns: ____ . Taken at the capitulation of Santo Domingo 1803; 1806 sold.

*Curieux* 1804-1809 (French *Le Curieux* 1800 Saint Malo, Pestel design) brig sloop 18.
Dimensions & tons: 97', 77'3" x 28'6" x 13'. 329 8/94.
Men. 67. Guns: UD 8 x 6 + 10 x 24★. Cut out of Martinique by the boats of *Centaur* 4.2.1804; 3.11.1809 wrecked in the West Indies.

PLANS: Taken off/as fitted: Lines, profile & decorations/decks.

*Sentinel/Centinel* 1804-1812 (Mercantile *Friendship* 1800 ____ 'Little Yarmouth') brig 12.
Dimensions & tons: 80'10", 63'4" x 24' x 14'5½". 194.
Men: 45. Guns: UD 2 x 9 + 10 x 18★. Purchased 06.1804; 10.10.1812 wrecked off Rügen Island.

*Thrasher* 1804-1814 (Mercantile *Adamant* 1804 Warren, Brightlingsea) brig 12.
Dimensions & tons: ____ , ____ x ____ x ____ . 154.
Men: 45. Guns: UD 2 x 9 + 10 x 18★. Purchased 06.1804; 1814 sold.

*Volunteer* 1804-1812 (Mercantile *Harmony* 1804 ____ Whitby) brig 12.
Dimensions & tons: ____ , ____ x ____ x ____ . 135.
Men 50. Guns: 2 x 18 + 10 x 18★. Purchased 06.1804; 1812 sold.

*Watchful* 1804-1814 (Mercantile *Jane* 1795 ____ Norfolk) brig 12.
Dimensions & tons: 76'3", 59'2" x 23'2" x ____ . 169.
Men: 45. Guns: 2 x 9 + 10 x 18★. Purchased 06.1804; 1811 tender; 1814 sold.

*Enchantress* 1804-1813 (Mercantile ____ 1802 ____ Ringmore) armed vessel/brig sloop 4.
Dimensions & tons: 79'8", 61'4¼" x 23'2¼" x 16'4". 176.
Men: 30. Guns: 4 x 6. Purchased as an armed vessel 1804 but then classed as a brig sloop despite her small armament; 1813 hulked [at Bristol (?) as a receiving ship; 1816 quarantine service at Milford; 1817 for revenue service at the Blackwater; ____ ].

*Leocadia* 1805-1814 (Spanish *Leocadia*? 1802 ____ ) brig sloop 16.
Dimensions & tons: 85'6", 67'3¼" x 24'6¼'" x 12'. 215.
Men: 70. Guns: UD 2 x 4 + 14 x 12★. Taken by *Helena* 5.6.1805; not fitted for sea by the RN; 1814 sold.

*Acteon* 1805-1816 (French *L'Actéon* 1804 Bayonne, Rolland design) brig 14.
Dimensions & tons: 94', 73'6⅝" x 29'3" x 8'1". 335.
Men: 121 Guns: 14 x 24★. Taken by *Egyptienne* 27.9.1805; 1816 BU.

*Melville* 1805-1808 (French *La Naiade/La Nayade* 1793 ____ ) brig sloop 18?
Dimensions & tons: ____ , ____ x ____ x ____ . 353.
Men ____ . Guns: ____ . Taken by *Jason* in the West Indies 13.10.1805; 1808 sold.

*Rolla* 1806-1810 (French ____ ) brig ____ .
Dimensions & tons: 80'4", 64'10" x 21' x 11'1". 152.
Men: ____ . Guns: ____ .Taken at the Cape 21.2.1806; 1807 arrived in Britain and laid up until; 1810 sold.

*Hawk* 1806-by 1813 *Buzzard* -1814 (French *Le Lutin* 1804 Saint Malo. Pestel design) brig sloop 16?
Dimensions & tons: 91'5", 72'11⅞" x 28'3½" x 7'5". 311.
Men: 120. Guns: 14 x 24★ + 2 x 6 (had been 16 x 6 in French service). Taken near Mauritius by *Carysfort* 24.3.1806; by 1813 renamed *Buzzard*; 1814 sold.

*Musette* (ex *Mignonne*?) 1805-1814 (French *Le Phaeton* 1804 Antwerp, Pestel design) brig sloop 16.
Dimensions & tons: 97', 77' x 28'4" x 7'. 329.
Men: 121. Guns: 2 x 6 + 14 x 32★. Taken 26.1.1805 near Santo Domingo by *Pique*; possibly briefly known as *Mignonne* after her capture; 1810 laid up until; 1814 sold.

*Pelican* 1806-1812 (French *Le Voltigeur* 1805 Antwerp, Pestel design) brig sloop 16.
Dimensions & tons: 95'9", 76'4½" x 28'5" x 13'1½". 328.
Men: 121. Guns: 2 x 6 + 16 x 32★. Taken off Santo Domingo by *Pique* 26.3.1806; 1812 sold.

PLANS: Taken off/as fitted: Lines & profile/decks.

*Nearque* 1806-1814 (French *Le Néarque* 1804 L'Orient, Pestel design) brig sloop 16.

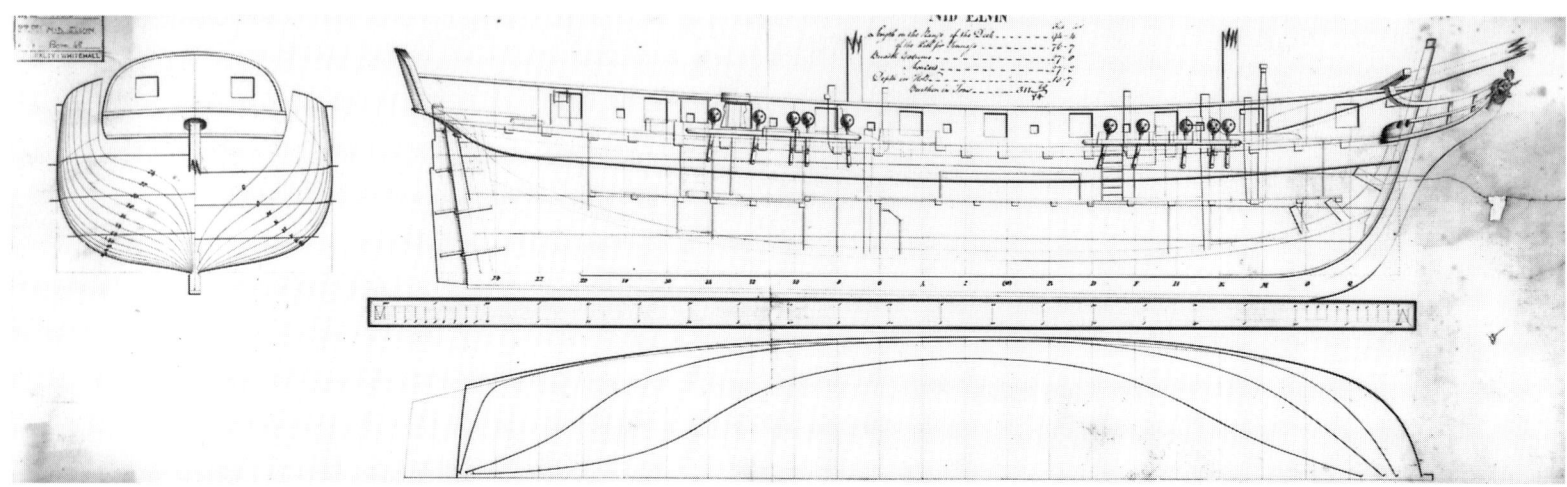

Lines and profile (7153) of the Danish brig *Nid Elvin* of 1807

Dimensions & tons: 94'4", 72'7⅛" x 28'4½" x 7'6". 309.
  Men: 110. Guns: 2 x 9 + 14 x 24★. Taken by *Niobe* 28.3.1806; not fitted for sea by the RN; 1814 sold.

*Spider* 1806-1813 (Spanish *Vigilante* ____ ) brig sloop 16.
  Dimensions & tons: ____ , ____ x ____ x ____ . ____ .
  Men: 85. Guns: UD 2 x 6 + 14 x 18★. Taken 4.4.1806 by *Renommee*; 1813 hulked [at Antigua as station ship at Saint John's Harbour; 1815 BU].

*Wolf* (ex *Prudente*- ex *Diligent*) 1806-1811 (French *Le Diligent* 1799 ____ ) brig sloop 16.
  Dimensions & tons: 93'6", 75'4" x 28'2" x 7'3". 317.
  Men: 95. Guns: 14 x 24★ + 2 x 6. Taken near Guadeloupe by *Renard* 28.5.1806. Briefly bore her French name and then renamed *Prudente* (10.1806) before being renamed *Wolf* early in 1807. 1811 BU.

*Bustard* 1806-1815 (Revenue brig *Royal George* 1803 ____ Cowes) brig sloop 16.
  Dimensions & tons: 82'11½", 68'11½" x 27'1⅜" x 11'6". 270.
  Men: 90. Guns: UD 2 x 6 + 14 x 18★. Purchased early 1806; 1815 sold.

*Observateur* 1806-1814 (French *L'Observateur* 1799 or 1800 Le Havre, Forfait design) brig sloop 16.
  Dimensions & tons: 90'6", 72'8" x 28' x 7'6". 303.
  Men: 95. Guns: 2 x 6 + 14 x 24★. Taken near Bermuda by *Tartar* 9.6.1806; 1810 laid up until; 1814 sold.

*Heureux* 1807-1814 (French *Le Lynx* 1804 Bayonne, Rolland design) brig sloop 16.
  Dimensions & tons: 93'10", 72'8⅜" x 29'6" x 8'6". 337.
  Men: 100. Guns: 2 x 6 + 14 x 24★. Taken by the boats of *Galatea* off Caracas 21.1.1807; 1810 laid up until; 1814 sold.

*Gluckstadt/Glykstat* (ex *Raison*) 1807-1814 (Danish *Glückstadt* 1805 Copenhagen, Hohlenberg design) brig sloop 18.
  Dimensions & tons: 102'2", 82'5⅝" x 27'9" x 11'2". 337 75/94.
  Men: 100. Guns: UD 16 x 18★ + 2 x 6. Taken at the fall of Copenhagen 1807; 1811 laid up until; 1814 sold.

NOTE: Sister of ship sloops *Eijderen*, *Elven*.
PLANS: Lines, profile & decorations.

*Mercurius* (ex *Transfer*) 1807-1815 (Danish *Mercurius* 1807 Copenhagen, Stibolt design) brig sloop 18.
  Dimensions & tons: 94'3", 77'4¾" x 27'4" x 10'3". 308.
  Men: 100. Guns: UD 2 x 6 + 16 x 24★. Taken at the fall of Copenhagen 1807; 1815 sold.

*Nid Elvin* (ex *Legere*) 1807-1814 (Danish *Nid Elvin* 1792 Copenhagen, Stibolt design) brig sloop 18.
  Dimensions & tons: 94'4", 76'7" x 27'8" x 10'7". 311 68/94.
  Men: 95. Guns: UD 2 x 6 + 16 x 24★. Taken at the fall of Copenhagen 1807; 1809 laid up until; 1814 sold.

PLANS: Lines & profile/lower deck/upper deck.

*Sarpen* (ex *Voltigeur*) 1807-1811 (Danish *Sarpen* 1791 Copenhagen, Stibolt design) brig sloop 18.
  Dimensions & tons: 94'5", 77'6⅛" x 27'4" x 10'3". 309.
  Men: 100. Guns: UD 2 x 6 + 16 x 24★. Taken at the fall of Copenhagen 1807; 1811 BU.

*Warning* 1807-1814 (Danish *Steece* 1798 ____ ) gun brig ____ .
  Dimensions & tons: 70'4½", 56'9¼" x 17'11" x 4'8". 97.
  Men: ____ . Guns: ____ . Taken at the fall of Copenhagen 1807; 1811 signal vessel; 1814 sold. Plans: Lines & profile/decks.

*Brev Drageren* (ex *Cockatrice*) 1807-1815 (Danish *Brev Drageren* 1801 Copenhagen, Hohlenberg design) brig sloop/gunbrig/armed vessel 12.
  Dimensions & tons: 82'8½", 66'0⅛" x 22'9" x 10'3½". 181 68/94.
  Men: 50/60. Guns: UD 2 x 6 + 10 x 18★. Taken at the fall of Copenhagen 1807; 1815 hulked [as a tender for Hull (but not sent there?); 1818 army prison ship in the Thames; 1825 sold].

PLANS: 1809: Lines & profile. 1819: decks.

*Flying Fish* (ex *Venture*) 1807-1811 (Danish *Flyvende Fisk* 1788 Copenhagen, Stibolt design) brig sloop ____ .
  Dimensions & tons: 77'3", 58'5½" x 25'9½" x 10'. 213.
  Men: ____ . Guns: ____ . Taken at the fall of Copenhagen 1807; not fitted for sea by the RN; 1811 sold.

*Delphinen* (ex *Mondovi*) 1807-1808 (Danish *Delphinen* 1805 Copenhagen, Stibolt design) brig sloop 16.
  Dimensions & tons: 96'7⅞", 76'10⅞" x 27'4¼" x 9'6". 306.
  Men: 100. Guns: UD 2 x 6 + 14 x 24★. Taken at the fall of Copenhagen 1807; 4.8.1808 wrecked on the Dutch coast.

*Glommen* (ex *Britomart*) 1807-1809 (Danish *Glommen* 1790 Copenhagen, Stibolt design) brig sloop 18.
  Dimensions & tons: 94'2½", 77'5⅝" x 27'1½" x 10'6". 303.
  Men: 100. Guns: UD 2 x 6 + 16 x 24★. Taken at the fall of Copenhagen 1807; 11.1809 wrecked in Carlisle Bay, Barbadoes.

*Alert* (ex *Cassandra*) 1807-1809 (Danish *Alaart* Copenhagen , Stibolt design) brig sloop 18.
  Dimensions & tons: 94'7½", 77'4¾" x 27'3½" x 10'8". 306 47/94.
  Men: 100. Guns: UD 2 x 6 + 16 x 24★. Taken 1807; 10.8.1809 retaken by Danes off the Danish coast.

PLANS: 1809: Lines & profile.

*Putulsk* 1807-1810 (French *Austerlitz* 1795 American-built) brig sloop 18.
  Dimensions & tons: 83'7", 65'9½" x 23'10" x 9'9½". 199.
  Men: 86. Guns: UD 2 x 6 + 16 x 24★. Taken 1807; 1810 BU.

*Virginie* 1808-1812 (French transport *La Virginie* 1803 ____ ) brig ____ .
  Dimensions & tons: ____ , ____ x ____ x ____ . ____ .
  Men: ____ . Guns: ____ . Taken 14.3.1808; 1809 arrived in Britain and laid up until; 1812 BU [or sold 1811?]

*Delight* 1808-1814 (Franco-Italian lugger *Le Friedland* ex *Vendecare* ex *L'Illyric/L'Illyrien* 1808 Venice) brig sloop 14.
  Dimensions & tons: 97'6", 76'1" x 29' x 19'. 340.
  Men: 100. Guns: UD 2 x 6 + 12 x 32★. Taken by *Standard* off Cape Blanco 26.3.1808; 1810 laid up until; 1814 sold.

*Tuscan* 1808-1818 (Franco-Italian *Le Ronco* 1807 Venice) brig sloop 16.

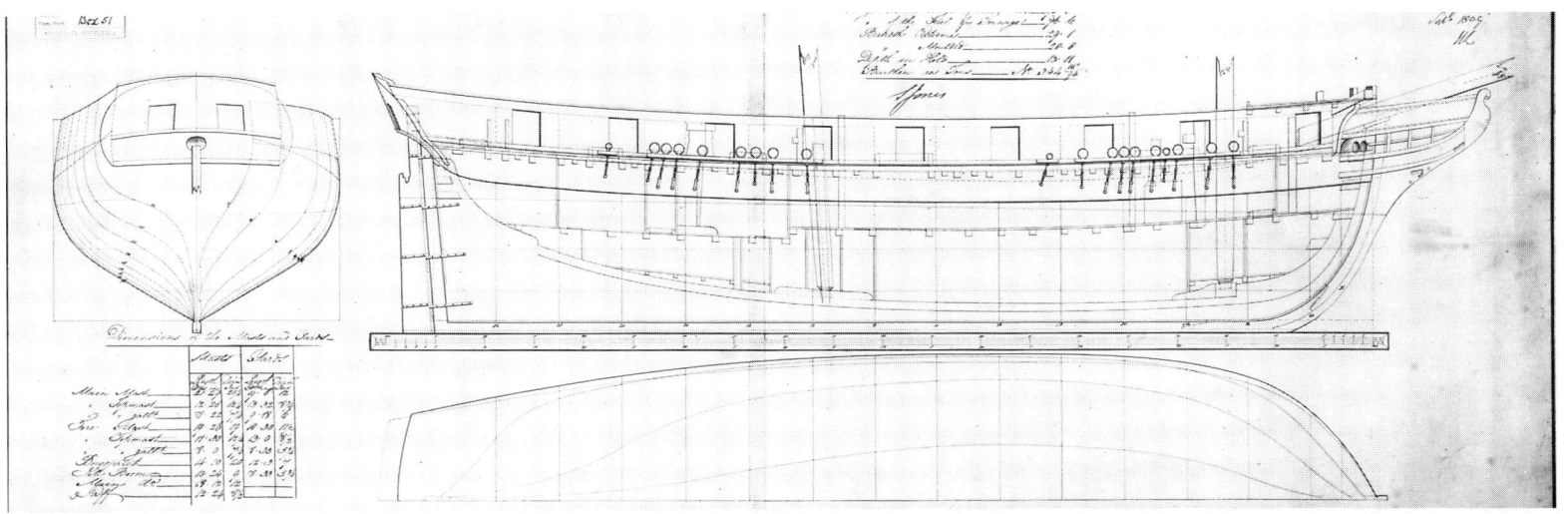

Lines, profile and spar dimensions (3492) taken off in 1809 for the Venetian built brig *Cretan*

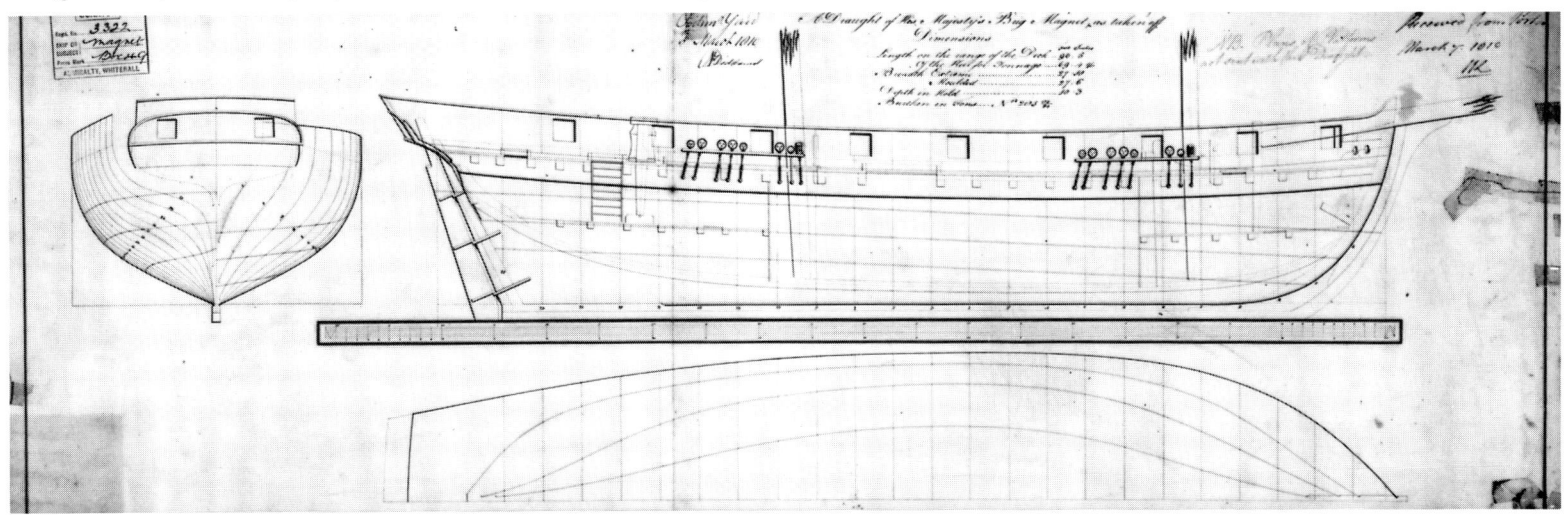

Lines (3322) of the French privateer brig *Magnet* taken in 1809

Dimensions & tons: 97'6", 76'5" x 28'8" x 18'3". 334.
Men: ____ . Guns: 2 x 9 . 14 x 24★. Taken in the Gulf of Venice by *Unite* 2.5.1808; 1818 sold.

**Griffon** 1808-1819 (French *Le Griffon* 1806 Rochefort, Pestel design) brig sloop 16.
Dimensions & tons: 92'6", 80'10" x 29'4" x 8'2". 368.
Men: 100. Guns: UD 2 x 6 + 14 x 24★. Taken by *Bacchante* near Cape Saint Antony 12.5.1808; 1819 sold.

**Cretan** 1808-1814 (Franco-Italian *Nettuno* 1807 Venice) brig sloop 16.
Dimensions & tons: 95'3", 76'6" x 29'1" x 13'11". 344 15/94.
Men: 100. Guns: UD 2 x 6 + 14 x 24★. Taken by *Unite* off Zara 1.6.1808; 1814 sold.
PLANS: Lines & profile/decks.

**Roman** 1808-1814 (Franco-Italian *Le Teulié* 1807 Venice) brig sloop 16.
Dimensions & tons: 97'3", 76'6" x 28'7" x 18'6". 333.
Men: 100. Guns: 14 x 24★ + 2 x 9. Taken by *Unite* in the Adriatic 1.6.1808; 1810 laid up until; 1814 sold.

**Asp** 1808-1814 (French *Le Serpent* ex *Le Rivoli* 1807 Nantes, Pestel design) brig sloop 16.
Dimensions & tons: 98'8½", 78'7½" x 28'3⅛" x 18'. 333.
Men: 100. Guns: 14 x 24★ + 2 x 6. Taken by *Acasta* and others 17.7.1808; 1810 laid up until; 1814 sold.

**Sabine** 1808-1818 (French *Le Requin* 1806 Rochefort, Rolland design) brig sloop 18.
Dimensions & tons: 96', 77'0⅞" x 28'5½" x 13'1". 332 1/94.
Men: 100. Guns: UD 2 x 6 + 16 x 32★. Taken by *Volage* in the Mediterranean 28.7.1808; 1818 sold.
PLANS: Lines & profile/decks.

**Electra** 1808-1816 (French *L'Espiègle* 1804 Saint Malo) brig sloop 16.

Dimensions & tons: 93'3", 74'6⅝" x 28'2⅛" x 12'9". 315.
Men: 95. Guns: UD 2 x 6 + 14 x 24★. Taken by *Sybille* 16.8.1808; 1816 sold.

**Seagull** 1808-1814 (French *Le Sylphe* 1804 Dunkirk, Sané design) brig sloop 16.
Dimensions & tons: 98'5", 79'11⅝" x 28'4⅝" x 12'10". 343.
Men: 100. Guns: UD 2 x 9 + 14 x 24★. Taken off Martinique by *Comet* 18.8.1808; not fitted for sea by the RN after her arrival in Britain; 1814 sold.

**Morne Fortunee** 1808-1813 (French privateer *Le Morne Fortunée* or *La Joséphine*? ____ ) gunbrig/armed brig 10.
Dimensions & tons: ____ , ____ x ____ x ____ . 184.
Men: 55. Guns: UD 2 x 6 + 8 x 18★. Taken by *Belette* 08.1808; 1813 BU.

**Vimiera** 1808-1814 (French *Le Pylade* 1804 Le Havre, Chaumont design) brig 16.
Dimensions & tons: 91', 73' x 28' x 6'10". 304.
Men: 100. Guns: UD 2 x 6 + 14 x 24★. Taken in the West Indies by *Pompee* 20.10.1808; 1810 laid up until; 1814 sold.

**Snap** 1808-1811 (French *Le Palinure* 1804 L'Orient, Pestel design) brig sloop 16.
Dimensions & tons: 91'1", 73'1¼" x 28'1½" x 12'9". 319 56/94.
Men: 100. Guns: 2 x 6 + 14 x 24★. Taken by *Circe* 1.11.1808; 1811 BU.
PLANS: Lines & profile/decks.

**Netley** 1808-1810 (American Merchantman *Nimrod* 1804 ____ ) brig 12.
Dimensions & tons: 76', 57' x 21'6" x 7'10". 140.
Men: 65. Guns: UD 2 x 6 + 10 x 12★. Purchased 1808; 1810 BU.

**Magnet** 1809-1812 (French privateer *Le Saint Joseph*/*San Josepho* 1808? ____ ) brig sloop 14.
Dimensions & tons: 90'5", 69'4⅝" x 27'10" x 10'3". 285 85/94.
Men: 90. Guns: UD 2 x 6 + 12 x 18★. Taken by *Undaunted* 12.2.1809; 09.1812 supposed foundered in the Atlantic on passage to America.

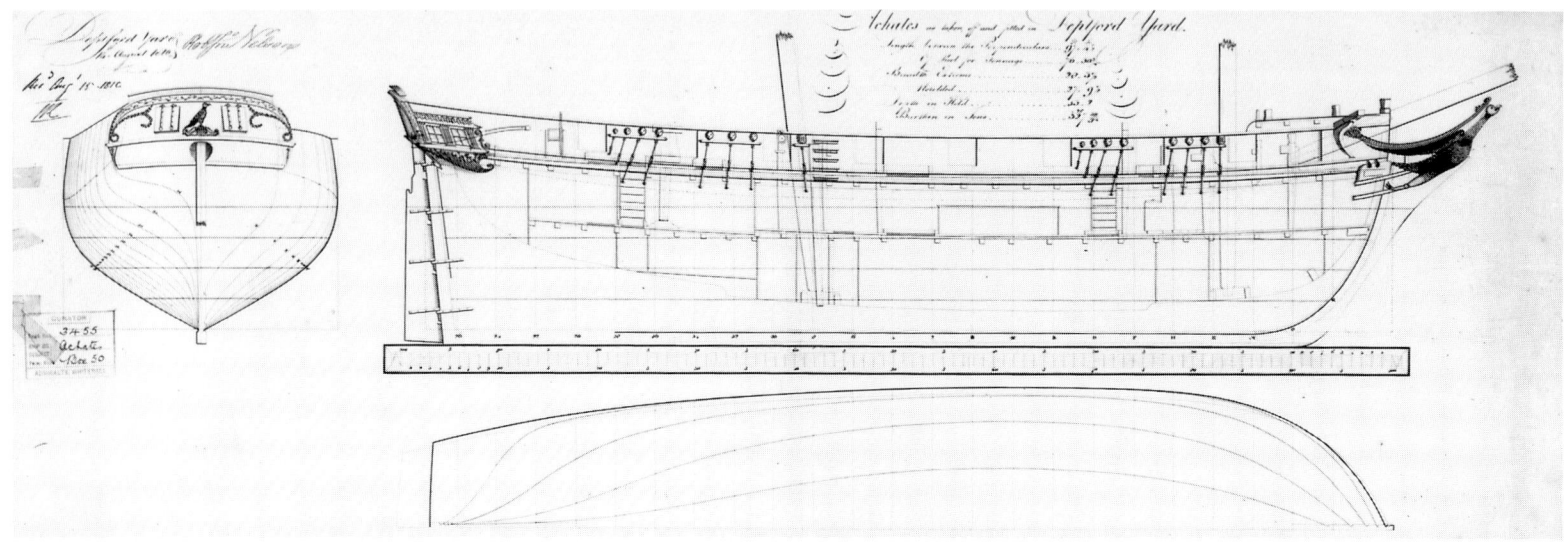

Lines and profile (3455) of the French brig *Achates* 'as taken off and fitted' in 1810

PLANS: Lines & profile/upper deck.

***Vautour*** 1809/1810-1813 (Franco-Dutch *Vautour* building at Flushing [Vlissingen]) brig 16.
Dimensions & tons: 95', 75'11¼" x 28'10" x 12'7¼". 335 73/94.
Men: 100. Guns: 2 x 6 + 14 x 24★. Taken on the stocks at the capture of Flushing 1809; 9.10.1809 launched there, incomplete; to Chatham, completed on the slip there and launched 15.9.1810; 08.1813 supposed foundered.

PLANS: As fitted: Lines & profile/lower deck/upper deck.

***Achates*** 1809-1818 (French *Le Milan* 1807 Saint Malo, Pestel design) brig 16.
Dimensions & tons: 97'4½", 76'10¾" x 28'3½" x 13'2". 327 39/94.
Men: 95. Guns: UD 2 x 6 + 14 x 24★. Taken by *Surveillante* and *Seine* 30.10.1809; 1818 sold.

PLANS: Taken off/as fitted: Lines, profile & decorations/decks.

***Foxhound*** 1809-1816 (French *Le Basque* 1808 Bayonne, Pestel design) brig sloop 16.
Dimensions & tons: 95'6", 78'0¾" x 28'11¼" x 8'1". 348.
Men: 106. Guns: 2 x 9 + 14 x 24★. Taken by *Druid* 13.11.1809; 1816 sold.

***Papillon*** 1809-1815 (French *Le Papillon* 1807 Le Havre, Sané design) brig sloop 18.
Dimensions & tons: 95'7⅝" , 76'9¼" x 28'11⅞" x 13'6¾". 343.
Men: 121. Guns: UD 2 x 6 + 16 x 32★. Taken by *Rosamond* 19.12.1809; 1815 sold.

***Guadeloupe*** 1809-1814 (French *La Nisus* 1806 St Servan or Granville, Pestel design) brig 16.
Dimensions & tons: 98'8⅝", 78'3½" x 28'4¾" x 13'10". 335 14/94.
Men: 100. Guns: 2 x 6 + 14 x 24★. Cut out of Guadeloupe by the boats of *Thetis* 12.12.1809; 1814 sold.

PLANS: Lines, profile & decorations/lower deck/upper deck.

***Curieux*** 1809-1814 (French *La Bearnais* 1808 Bayonne, Sané design) brig sloop 16.
Dimensions & tons: 94'6½", 73'9½" x 24'8" x 11'. 336.
Men: 100. Guns: UD 2 x 6 + 14 x 24★. Taken by *Melampus* off Guadeloupe 14.12.1809; 1810 arrived in Britain and laid up until; 1814 sold.

***Wellington*** (ex *Orestes*) 1810-1812 (French 'dogre' *L'Oreste* 1800 ___ ) brig sloop 16.
Dimensions & tons: 94'10", 73'11" x 28'2¼" x 7'5½". 312.
Men: 100. Guns: 2 x 6 + 14 x 24★. Taken near Guadeloupe by *Scorpion* 12.1.1810; not fitted for sea by the RN after her arrival in Britain; 1812 BU.

***Fantome*** 1810-1814 (French privateer *Le Fantôme* 1809 ___ ) brig sloop 18.
Dimensions & tons: 94'1", 75'6½" x 30'11" x 13'. 384 6/94.
Men: 121. Guns: 2 x 6 + 16 x 32★. Taken by *Melampus* 05.1810; 24.11.1814 wrecked on the coast of Nova Scotia.

PLANS: Lines & profile/decks.

***Carlotta*** 1810-1812 [or 1815] (Franco-Italian *Carlotta*? or *Pylade*? 1807 Venice) gun brig/brig sloop ___ .
Dimensions & tons: 90'6", 74'8" x 22'8" x ___ . 204.
Men: ___ . Guns: ___ . Taken 1810; 26.1.1812 wrecked near Cape Passaro, Sicily (or 1815 BU?).

***Emulous*** 1812-1817 (American *Nautilus* purchased as a schooner [12] 1803) brig 14.
Dimensions & tons: [American measurements] 87'6", ___ x 23'8" x 9'10". 213.
Men: ___ . Guns: 12 x 12★ + 2 x 6. Taken 6.7.1812 by *Shannon*; 1817 sold.

***Nova Scotia*** 1812-1813 ***Ferret*** -1820 (American privateer *Rapid* ___ ) brig sloop 14.
Dimensions & tons: 84', 66'3" x 24'8" x 10'4". 214.
Men: 75. Guns: UD 2 x 6 + 12 x 12★. Taken 17.10.1812 by *Maidstone* off North America; 1820 sold.

***Columbia*** 1812?-1820 (American privateer *Curlew* 1812 ___ ) brig sloop 18.
Dimensions & tons: 94'4", 80'3" x 26'3" x 13'. 294.
Men: ___ . Guns: UD 2 x 6 + 16 x 18★. Taken 1812 or 1813 by *Acasta* off Cape Sable; 1820 sold.

***Anaconda*** 1813-1815 (American privateer *Anaconda* 1812 New York) brig sloop 18.
Dimensions & tons: 102', 85'10" x 29' x 10'2" 384.
Men: ___ . Guns: 18 x 9. Taken 1813 in Chesapeake Bay; 1815 sold.

***Barbadoes*** 1813 & hulked after capture (American *Herald* 1811 ___ ) brig sloop ___ .
Dimensions & tons: 102', 84'6⅝" x 26'2" x 13'. 308.
Men: ___ . Guns: ___ . Taken 1813; hulked at Jamaica soon after capture [as a powder ship; 1815 wrecked].

PLANS: Lines/lines & profile.

***Variable*** 1814?-1817 (American *Edward* 1812 ___ ) brig sloop [or schooner?] 14.
Dimensions & tons: 104'6", 82'2" x 27'3" x 13'. 324.
Men: 95. Guns: UD 2 x 6 + 12 x 24★. Taken ___ ; 1814 purchased; 1817 BU.Sloops, Rig Unknown

***Caledon*** 1808-1811 (French privateer *L'Henri* ___ ) ___ .
Dimensions & tons: ___ , ___ x ___ x ___ . ___ .
Men: ___ . Guns: ___ . Taken 1808; 1811 sold.

## Fire Vessels

***Eruption*** 1804-1807 (Mercantile *Unity* 1770 ___ Yarmouth) ___ .
Dimensions & tons: 65'4", ___ x 16'6¾" x 8'6¾" . 74 .
Men: ___ . Guns: none? Purchased 05.1804; 1807 sold.

***Incendiary*** 1804-1812 (French [probably] *Diligence* ___ ) ___ .
Dimensions & tons: 53'6", ___ x 17'4" x ___ . 62.

//THE NAPOLEONIC WAR 1802-1815

Men: ____ . Guns: none? Purchased 05.1804; 1812 sold.

***Phosphorus*** 1804-1810 (Dutch *Haasje* ____ ) 4.
    Dimensions & tons: ____ , 55'7" x 19'9" x 12'6". 115.
    Men: 18. Guns: 4 x 4. Purchased 06.1804; 1810 sold.

***Sophia*** 1804-1807 (Mercantile *Sophia* ____ ) ____ .
    Dimensions & tons: ____ , ____ x ____ x ____ . ____ .
    Men: ____ . Guns: none? Purchased abroad(?) 06.1804; never went to sea; 1807 sold.

***Ignition*** 1804-1807 (Mercantile *Jeany* 1803 Quebec) 4.
    Dimensions & tons: 69'3", ____ x 21'1" x 11'7". 130.
    Men: 18. Guns: 4 x 4 (or 4 x 12★?). Purchased 06 or 07.1804; 19.2.1807 wrecked off Dieppe.

***Beacon*** 1804-1808 (Mercantile *Duff* ____ French-built) 4.
    Dimensions & tons: ____ , 56'0½" x 22'4" x 8'10". 149.
    Men: 18. guns: 4 x 4. Purchased 07.1804?; never went to sea; 1808 sold.

***Firebrand*** 1804-1804 (Mercantile *Waller/Walter* ____ French-built) 14?
    Dimensions & tons: 80'2", ____ x 20'1½" x ____ . 140.
    Men: ____ . Guns: ____ . Purchased 07 or 08.1804; 13.10.1804 wrecked near Dover.

***Flambeau*** 1804-1807 (Mercantile *Good Intent* 1776 Folkestone) ____ .
    Dimensions & tons: 50'3½", ____ x 16'1½" x ____ . 51.
    Men: ____ . Guns: ____ . Purchased 10.1804; 1807 sold.

***Flash*** 1804-1807 (Mercantile *James* 1774 ____ Whitby) ____ .
    Dimensions & tons: 53'5", ____ x 15'7½" x ____ . 53.
    Men: ____ . Guns: ____ . Purchased 10.1804; 1807 sold.

***Fuze*** 1804-1807 (Mercantile *William Adventure?* 1763 ____ Lowestoft) ____ .
    Dimensions & tons: ____ , 24'6"[?] x 15'10½" x ____ . 55.
    Men: ____ . Guns: ____ . Purchased 10.1804; 1807 sold.

***Rocket*** 1804-1807 (Mercantile *Busy* ____ ) ____ .
    Dimensions & tons: 52'4½", ____ x 17'0½" x ____ . 62.
    Men: ____ . Guns: ____ . Purchased 10.1804?; 1807 sold.

***Salamander*** 1804-1807 (Mercantile *United* 1777 ____ Yarmouth) ____ .
    Dimensions & tons: 59'5", ____ x 18' x 8'6". 78.
    Men: ____ . Guns: ____ . Purchased 10.1804; 1807 sold.

***Squib*** 1804-1805 (Mercantile *Diligent* or *Defence* ____ ) ____ .
    Dimensions & tons: ____ , ____ x ____ x ____ . ____ .
    Men: ____ . Guns: ____ . Purchased 10.1804; 2.10.1805 driven ashore off Deal and bilged.

***Torch*** 1804-1811 (Mercantile? *Fortune* ____ ) ____ .
    Dimensions & tons: 62'6", ____ x 19'3" x ____ . 91.
    Men: ____ . Guns: ____ . Purchased 10.1804?; 1811 BU.

***Firedrake*** 1804-1807 (Mercantile *Ann* ____ ) ____ .
    Dimensions & tons: ____ , ____ x ____ x ____ . ____ .
    Men: ____ . Guns: ____ . Purchased 11.1804?; 1807 sold.

***Wildfire*** 1804-1807 (Mercantile *John* 1766 ____ Yarmouth) ____ .
    Dimensions & tons: 61', ____ x 16'3½" x 8'. 64.
    Men: ____ . Guns: ____ . Purchased 11.1804?; 1807 sold.

***Firebrand*** 1804-1807 (Mercantile *Lord Lenox* ____ ) ____ .
    Dimensions & tons: 55', ____ x 18'5" x ____ . 73.
    Men: ____ . Guns: ____ . Purchased 1804; 1807 sold.

***Rifleman*** 1804-1809 (Mercantile *Telegraph* 1799 [or 1803?] Perry, Blackwall?) 4.
    Dimensions & tons: 80'2", ____ x 22'3" x ____ . 160.
    Men: 20. Guns: 4 x 12★. Purchased (initially as a gunbrig?) 1804; 27.7.1809 sold.

## Bomb Vessels

***Meteor*** 1803-1811 (Mercantile *Sarah Ann* 1800 ____ Newcastle) 8.
    Dimensions & tons: 102'6", 80' x 29'3" x 12'11". 364.
    Men: 67. Guns: GD 1 x 13" & 1 x 10" mortar, UD 8 x 24★. Purchased 10.1803; 1811 sold.

***Acheron*** 1803-1805 (Mercantile *New Grove* 1799 ____ Whitby) 8.
    Dimensions & tons: 108'3", 85'9" x 29'2" x 12'9". 388.
    Men: 67. Guns: GD 1 x 13" & 1 x 10" mortar, UD 8 x 24★. Purchased 10.1803; 3.2.1805 taken by the French *Hortense* and *Incorruptible* in the Mediterranean then burnt.

***Devastation*** 1803-1816 (Mercantile *Intrepid* ____ ) 8.
    Dimensions & tons: 108'5", 85'9" x 29'2" x 12'9". 388.
    Men: 67. Guns: GD 1 x 13" & 1 x 10" mortar, UD 8 x 24★. Purchased 10.1803; 1816 sold.

***Etna*** 1803-1816 (Mercantile *Success* ____ ) 8.
    Dimensions & tons: 102', 81' x 29'2½" x 12'6". 368.
    Men: 67. Guns: GD 1 x 13" & 1 x 10" mortar, UD 8 x 24★. Purchased 10.1803; 1816 sold.

***Lucifer*** 1803-1811 (Mercantile *Spring* ____ ) 8.
    Dimensions & tons: 110', 87'9" x 29'2" x 13'. 397.
    Men: 67. Guns: GD 1 x 13" & 1 x 10" mortar, UD 8 x 24★. Purchased 10.1803; 1811 sold.

***Prospero*** 1803-1807 (Mercantile *Albion* 1800 ____ South Shields) 8.
    Dimensions & tons: 107', 87' x 30'5" x 13'6". 400.
    Men: 67. Guns: GD 1 x 13" & 1 x 10" mortar, UD 8 x 24★. Purchased 10.1803; 18.2.1807 foundered off Dieppe.

***Thunder*** 1803-1814 (Mercantile *Dasher* ____ ) 8.
    Dimensions & tons: 95', 75'10" x 28'1" x ____ . 318.
    Men: 67. Guns: GD 1 x 13" & 1 x 10" mortar, UD 8 x 24★. Purchased 10.1803; 1814 sold.

## Cutters

***Betsy*** ____ ?; 1810 (Mercantile? ____ ) ____ .
    Dimensions & tons: ____ , ____ x ____ x ____ . ____ .
    Men: ____ . Guns: ____ . Purchased ____ ?; 1810 BU.

***Union*** ____ ?; 1810 (Mercantile? ____ ) ____ .
    Dimensions & tons: ____ , ____ x ____ x ____ . ____ .
    Men: ____ . Guns: ____ . Purchased ____ ?; 1810 BU.

***Carrier*** 1805-1808 (Mercantile *Frisk* 1803 Bools & Good?, Bridport) despatch

Profile and sections 'as fitted' in 1804 (4335) for the purchased bomb vessel *Acheron*

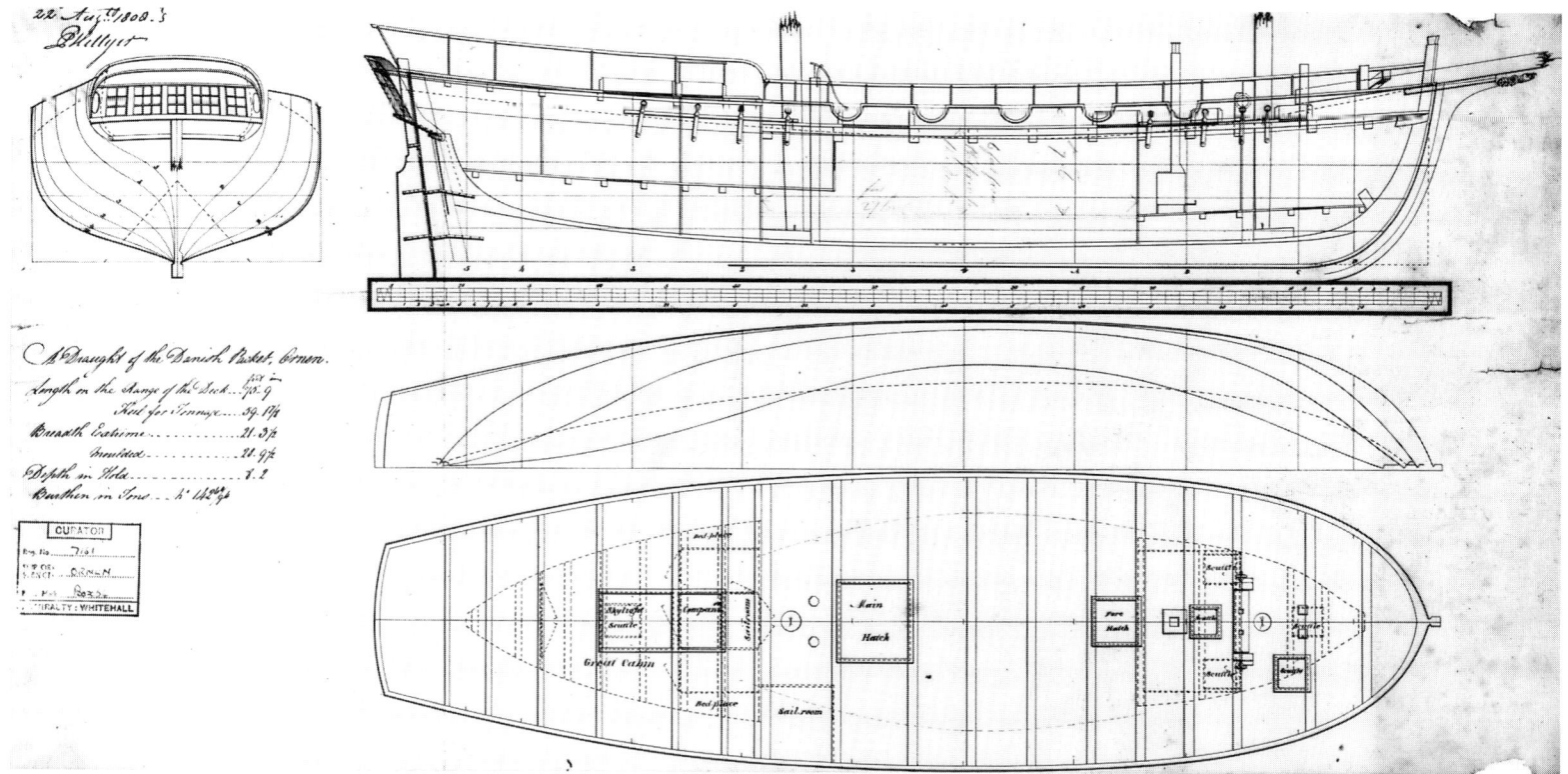

cutter 4.
   Dimensions & tons: ____ , 47'7½" x 17'2" x ____ . 54.
   Men: 17. Guns: 4 x 12★. Purchased early 1805; 24.1.1808 lost on the coast of France.
*Dove* 1805-1805 (Mercantile? *Ariadne* 1803 ____ Cowes?; or is this the later *Dove*?) despatch cutter 4.
   Dimensions & tons: ____ , 68' x 19'4½" x ____ . 103.
   Men: 17. Guns: 4 x 12★. Purchased 05.1805; 5.8.1805 taken by the French Rochefort squadron off that port.
*Pigeon* 1805-1805 (Mercantile *Fanny* 1801 ____ ) despatch cutter 4.
   Dimensions & tons: ____ , 57'6" x 17'6" x ____ . 72.
   Men: 17. Guns: 4 x 12★. Purchased early 1805; 13.11.1805 wrecked off the Texel.
*Dove* (or *Flight*?) 1805-1806 (Mercantile? *Ariadne* 1803 ____ , Cowes?; or is this the earlier *Dove*?) despatch cutter/advice boat 4.
   Dimensions & tons: ____ given as the earlier *Dove*; is this a confusion between the two?
   Men: 17. Guns: 4 x 12★. Purchased 07.1805; 09.1806 supposed foundered.
*Sprightly* 1805-1815 (Mercantile *Lively* 1805 ____ ) 12.
   Dimensions & tons: 63'9", 45'6¼" x 22'3⅛" x 10'10". 120.
   Men: 50. Guns: 12 x 12★. Purchased 09? 1805 (hired during 08.1805 before purchase?); 1815 BU.
*Ranger* 1806-1807 *Pigmy* -1814? (building Avery, Dartmouth) 16.
   Dimensions & tons: 79', 60'4⅞" x 26' x 11'. 217.
   Men: 60. Guns: 14 x 12★ + 2 x 6. Purchased whilst building in 05.1806; 1807 renamed *Pigmy*; 06.1814 lost.
*Linnet* 1806-1813 (revenue cutter *Speedwell* 1797 ____ Cowes) possibly brig rigged but more likely to be cutter rigged 14.
   Dimensions & tons: 77'9½", 57'7" x 25'4" x 10'9". 197.
   Men: 60. Guns: UD 12 x 18★ + 2 x 6. Purchased late 1806; 25.2.1813 taken by the French *Gloire* off Madeira.
*Nile* 1806?-1811 (Mercantile: *Nile*? ____ ) 12.
   Dimensions & tons: ____ , ____ x ____ x ____ . 166.
   Men: 55. Guns: UD 12 x 12★. Purchased late 1806?; 1810 sold, but the purchaser refused to take her away, so; 1811 BU.
*Tickler* 1808-1816 (Mercantile? *Lord Duncan* 1798 ____ Dover) 8.
   Dimensions & tons: 62'6", 44'5" x 21'11½" x ____ . 114.
   Men: 30. Guns: 8 x 12★. Purchased mid 1808; 1816 sold.
*Tigress* 1808-1814? *Algerine* -1818 (French *Pierre Czar* 1805 Baltimore, USA)
14.
   Dimensions & tons: 92'9", 72'9¾" x 24'4" x 10'9". 229.
   Men: 50 Guns: UD 14 x 12★. Taken 1808; 1814 renamed *Algerine*; 1818 sold.
*Baltic* 1808-1810 (Russian *Apith* ____ ) ____ .
   Dimensions & tons: ____ , ____ x ____ x ____ .
   Men: ____ . Guns: ____ . Taken 24.6.1808 by *Salsette*; 1810 sold.
*Viper* 1809-1814 (Mercantile *Niger* 1809 White, Cowes) 8.
   Dimensions & tons: 63'3", 47'5⅛" x 20'4" x 9'9". 104.
   Men: 45. Guns: 8 x 12★. Purchased late 1809; 1814 sold.
*Trial* 1810?-1818? ( ____ ) ____ .
   Dimensions & tons: ____ , ____ x ____ x ____ . ____ .
   Men: ____ . Guns: ____ . Purchased 1810?; ? hulked; 1818 sold.
*Nimble* 1812?-1816 ( ____ ) 12.
   Dimensions & tons: 68'5", 50'11⅝" x 23'3½" x 10'2". 147.
   Men: 50. Guns: 12 x 12★. Purchased late 1812? (or 1813?); 1816 sold.

## Schooners

*Felix* 1803-1807 (French privateer *Le Felix* ____ ) armed schooner 14.
   Dimensions & tons: 90'4½", 60'7½" x 22'2" x 8'5". 158.
   Men: 60. Guns: 14 x 12★. Taken 26.7.1803; 23.1.1807 wrecked near Santander.
*Pitt* 1805-1807 *Sandwich* -1809 (Mercantile *William and Mary* ____ ) 12.
   Dimensions & tons: ____ , ____ x ____ x ____ . ____ .
   Men: 45. Guns: 12 x 12★. Purchased 1805; 1807 renamed *Sandwich*; 1809 BU.
*Favorite* 1806-1807 ( ____ ) ____ .
   Dimensions & tons: ____ .
   Men: ____ . Guns: ____ . In service by 1806; 1807 sold.
*Flying Fish* 1805-1808 (American *Revenge* 1805 Baltimore) 12.
   Dimensions & tons: 78'8", 60'8" x 21'7" x 7'10". 150 32/94.
   Men: ____ . Guns: 10 x 9 or 12★ (?). Acquired 1805; 15.12.1808 grounded and bilged on a reef off San Domingo. Note that this was a three masted vessel, and she was used as the model for building the *Shamrock* class schooners (see Part I Chapter 5).
   PLAN: Lines, profile/decks/ballast/sails.
*Viper* 1807-1809 (Mercantile *Princess Charlotte* ____ ) 4.
   Dimensions & tons: ____ .

Hull plan (7161) of the Danish packet schooner *Omen* of 1807

THE NAPOLEONIC WAR 1802-1815

Men: 25. Guns: 4 x 12★. Purchased 1807; ____ 1809 foundered off Gibraltar.

***Nonpareil*** 1807-1813 (American blockade runner ____ ) 12.
  Dimensions & tons: ____ . 210.
  Men: 40. Guns: 12 x 12★. Taken at the fall of Montevideo 1807; condemned by prize court and purchased 1808; 1813 sold after storm damage.

***Paz*** 1807-1814? (Spanish ____ ) 12.
  Dimensions & tons: ____ .
  Men: 40. Guns: 2 x 6 + 10 x 12★. Taken at the fall of Montevideo 1807; 1814? sold.

***Ornen*** 1807-1814 (Danish *Ornen* ____ ) schooner/armed vessel/packet ____ .
  Dimensions & tons: 75'9", 59'1⅞" x 21'3½" x 8'2". 142 60/94.
  Men: ____ . Guns: ____ . Taken at the fall of Copenhagen 1807; 1814 hulked [at Greenock as a receiving ship; 1815 converted for the use of the Clyde Marine Society ____ ].

PLANS: Lines, profile & decorations/deck.

***Subtle*** 1808-1812 (Danish ____ , American built [Baltimore?]) ____ .
  Dimensions & tons: ____ .
  Men: ____ . Guns: ____ . Taken 1808; 30.11.1812 'every reason to suppose that she upset in a very heavy squall near the island of Bartholomew'.

PLANS: As fitted: Lines & profile/decks.

***Theodosia*** 1808-1814 (Franco-Italian *La Nouvelle Entreprise* 1808 Leghorn) ____ .
  Dimensions & tons: 72', 57' x 20'5" x ____ . 126.
  Men: ____ . Guns: ____ . Taken 1808; 1814 sold.

***Dominica*** 1809-1813/1814-1814 (French privateer *Le Duc de Wagram* 1805 ____ ) 14.
  Dimensions & tons: 89'6½", 71'8⅝" x 23'1" x 9'3¾". 203.
  Men: 45 Guns: UD 2 x 6 + 12 x 12★. Taken 1809; 8.8.1813 taken by the American privateer *Decatur* off Charlestown; 22.5.1814 retaken; 18.8.1814 foundered off Dominica.

***Phipps*** 1808-1812 (Dutch? ____ ) schooner/'sloop' (?rig?) ____ .
  Dimensions & tons: ____ , ____ x ____ x ____ . ____ .
  Men: ____ . Guns: ____ . Taken 1808 as a 'sloop' (? sloop-rigged vessel?); 1809 converted to a schooner; 1812 BU.

***Sealark*** 1811-1820 (American *Fly* 1811 ____ ) 10.
  Dimensions & tons: 79'6", 68'0⅞" x 22'8" x 9'10". 178.
  Men: 50. Guns: 10 x 12★. Taken 29.12.1811 by *Scylla*; 1820 sold.

PLANS: Lines & profile/decks.

***Racer*** 1812-1814 (American privateer *Independence* 1811 New York) 14.
  Dimensions & tons: 93'4", 75'11½" x 24'10½" x 9'8". 250.
  Men: 60. Guns: 2 x 6 + 12 x 12★. Taken 9.11.1812; 10.10.1814 wrecked in the Gulf of Florida.

***Whiting*** 1812-1816 (American *Arrow* 1812 Baltimore) 12.
  Dimensions & tons: 98', 75'8⅞" x 23'7⅝" x 9'10". 225.
  Men: 50. Guns: 2 x 6 + 10 x 12★. Taken 1812; 15.9.1816 wrecked on the Dunbar Sand; wreck sold.

***Telegraph*** 1812?-1817 (American privateer *Vengeance* 1812 New York) schooner/gun brig 12.
  Dimensions & tons: 83'7", 66'5¾" x 22'6½" x 10'6". 180.
  Men: 60. Guns: 12 x 12★. Taken at the end of 1812 or 01.1813; 20.2.1817 wrecked under Mount Batten, Plymouth Sound.

***Canso*** 1813-1816 (American *Lottery* 1811 ____ ) ____ .
  Dimensions & tons: 93', 75'9" x 23'8" x 10'2". 225.
  Men: ____ . Guns: ____ . Taken in the Chesapeake 8.2.1813; 1816 sold.

***Alban*** 1813-1822 (American *William Bayard* 1812 New York) schooner [or brig sloop?] 14.
  Dimensions & tons: 94'4½", 78'6½" x 24'7⅛" x 10'6". 252 75/94.
  Men: 60. Guns: 2 x 6 + 12 x 12★. Taken by the *Warspite* 16.3.1813; 1822 BU.

PLANS: 1817: Lines & profile/decks.

***Musquidobit*** 1813-1820 (American privateer *Lynx* 1812 ____ ) 10.
  Dimensions & tons: 94'7", 73'1¼" x 24' x 10'3". 224.
  Men: 50. Guns: 2 x 6 + 8 x 12★. Taken in the Chesapeake 16.3.1813; 1820 sold.

PLANS: Lines & profile/decks.

***Shelburne*** 1813-1817 (American privateer *Racer* 1811 ____ ) ____ .
  Dimensions & tons: 93'8", 72'8" x 23'11" x 10'8". 221.
  Men: ____ . Guns: ____ . Taken 16.3.1813; 1817 sold.

***Cockchafer*** 1813-1815 (American *Spencer* 1811 ____ ) ____ .
  Dimensions & tons: 69'4", 55'6" x 18'10" x 8'. 104.
  Men: ____ . Guns: ____ . Taken 1813; 1815 sold.

***Highflyer*** 1813-1813 (American privateer *Highflyer* ____ ) 8.
  Dimensions & tons: 80', 67'10⅝" x 20' x ____ 144.
  Men: ____ . Guns: 8 x ____ . Taken 1813; 9.9.1813 retaken by USS *President*.

***Pike*** 1813-1836 (American *Dart* 1813 New Orleans) 14.
  Dimensions & tons: 93', 78'7" x 24'8" x 10'6". 251.
  Men: 60. Guns: 2 x 6 + 12 x 12★. Taken 1813; 5.2.1836 wrecked on Bare Bust Key, Jamaica, and wreck sold.

***Saint Lawrence*** 1813-1815 (American *Atlas* ____ ) 12/14.
  Dimensions & tons: ____ , ____ x ____ x ____ . ____ 240.
  Men: ____ . Guns: 12 x 12★ + 2 x 6. Taken 20.4.1813; 26.2.1815 retaken by American privateer *Chasseur*.

***Pictou/Picton*** 1813-1814/1814-1818 (American *Leyson* 1812. ____ ) ____ .
  Dimensions & tons: 83', 65'2" x 24'8" x 10'6". 211 or 215.
  Men: ____ . Guns: ____ . Taken 1813; 14.2.1814 retaken by USS *Constitution*; 1814 taken again (there is a possibility that this was a different vessel - the 243 tone American *Zebra*); 1818 sold.

***Grecian*** 1814-1822 (American privateer *Grecian* 1813 Baltimore) 10.
  Dimensions & tons: 95'1", 74'3" x 23'10" x 10'5". 224
  Men: 50. Guns: 2 x 6 + 8 x 18★. Taken by *Jaseur* 2.5.1814; 1822 sold.

PLANS: Lines & profile/decks.

***Express*** 1815?-1827 (American *Anna Maria* ____ ) schooner advice boat 4.
  Dimensions & tons: 64'6½", 53'6⅝" x 18' x 7'1". 92.
  Men: 26. Guns: 4 x 12★. Taken ____ ?; 1815 purchased; 1816 tender; 1827 sold [Dittmar suggests she was acquired in 1809?] .

***Speedwell*** 1815-1834 (Mercantile? *Royal George* ____ ) tender ____ .
  Dimensions & tons: 80', 61'2" x 25' x 11'3½". 203 32/94
  Men: ____ . Guns: ____ . Purchased 1815; 1834 sold.

PLANS: Lines.

## Gunboats, luggers, etc

***Dart*** 1803-1808 (French *Le Dart* 1802 ex British privateer *Dart*? ____ ) lugger ____ .
  Dimensions & tons: ____ , ____ x ____ x ____ . ____ .
  Men: ____ . Guns: ____ . Taken by *Apollo* 29.6.1803; 1805 laid up until; 1808 sold.

***Violet*** 1806-1812 (customs vessel *Violet* ____ ) lugger 10.
  Dimensions & tons: 60'2¾", 44'3⅝" x 18'2½" x 8'8" . 82.
  Men: 30. Guns: 10 x 12★. Transferred by the Commissioners of the Customs 1806; 1812 BU.

***Gibraltar gunboat*** 1807; ____ ? ( ____ ) lateen-rigged on one mast, Mediterranean-built Felucca type. One long slide gun in bow. Purchased from Mr Levy at Gibraltar.

PLANS: Outline profile & deck.

***General Concleux*** 1808?-1811 (French *Le General Concleux*? ____ ) lugger ____ .
  Dimensions & tons: ____ , ____ x ____ x ____ . ____ .
  Men: ____ . Guns: ____ . Taken 1808; 1808 tender; 1811 BU.

***Defender*** 1809-1814 (French privateer *Bonne Marseilles*/*Beau Marseilles* ____ ) lugger 8.
  Dimensions & tons: 65'0½", 54'2⅛" x 17'2½" x 7'4". 139 23/94.
  Men: 40. Guns: 8 x 12★. Taken by *Royalist* 12.1809; 1814 sold.

PLANS: Lines & profile/upper deck. As fitted: Profile, decks, midships section (incomplete).

***Alarm*** 1810-1811 (revenue vessel *Alarm* ? ____ ) lugger 8.

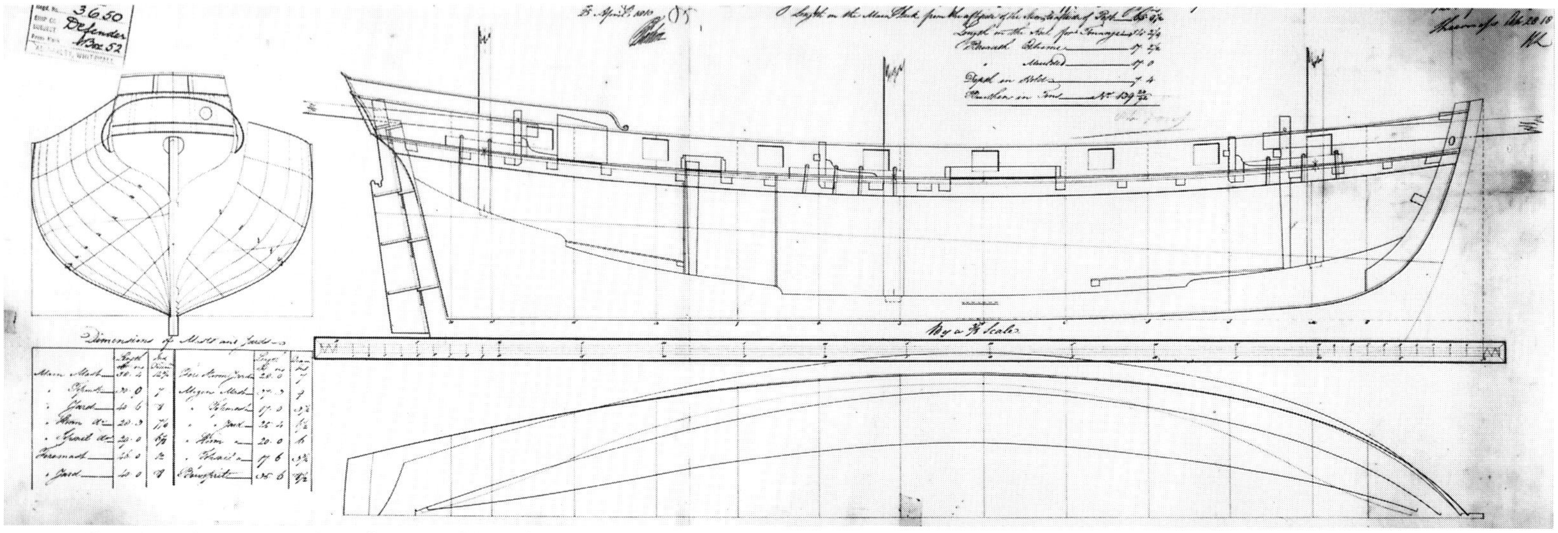

Lines, profile and spar dimensions as taken off in 1810 of the *Defender* French lugger/gunboat

Dimensions & tons: 71'9", 60'8" x 21'8" x 9'3". 151.

Men: 40. Guns: 8 x 12★ 'Given up by Customs' 1810; 1811 'given up to Customs'.

PLANS: Lines, profile, decks, midsection (illustrated in the chapter on Coastguard and customs vessels, which see).

***Staverens*** ____ -1811 ( ____ ) gunboat or gunbrig ____ .

Dimensions & tons: ____ .

Men: ____ . Guns: ____ . Origins unknown; 1811 sold.

***Diligent*** 1813-1814 (French *La Diligente* 1810 ____ possibly originally British *Black Joke* or *Thistle* ____ ) lugger ____ .

Dimensions & tons: ____ 100.

Men: ____ . Guns: ____ . Taken 6.1.1813; 1814 sold.

***Scrub*** 1815-1827 (Mercantile ____ ) tender/survey schooner 2.

Dimensions & tons: ____ 80.

Men: ____ . Guns: ____ . Purchased in Newfoundland 1815; 1827 sold in Newfoundland.

## Storeships, etc

***Porpoise*** 1804-1814 see under ship sloops.

***Ranger*** 1805?- ____ ? ( ____ ) ____ .

Dimensions & tons: ____ .

Men: ____ . Guns: ____ . Purchased 1805?; fate uncertain.

***Howe*** 1805-1808 ***Dromedary*** -1819 (East Indiaman *Kaikusroo* 1799 Bombay) storeship 24.

Dimensions & tons: 150', 124'3⅞" x 39'9¾" x 16'11¾". 1048.

Men: 100 Guns: UD 20 x 9, QD 4 x 6. Purchased (for use as a frigate) in India 1805; 1808 renamed *Dromedary* 1819 hulked [as a convict ship at Deptford; 1825 convict ship for Bermuda; 1864 sold].

***Gleaner*** 1809-1814 (mercantile? ____ 1802? ____ ) survey vessel 2.

Dimensions & tons: ____ 755.

Men: 19/30. Guns: 2 or 4 x 12★. Hired 12.7.1808; 1809 purchased; 1811 fitted, first as a yard lighter and then as a lightship for the Galloper Sand; 24.4.1814 foundered at anchor in a sudden gale at St Jean de Luz.

***Chichester*** 1809-1811 (French 'gabare' *Le Var* 1806 ____ ) storeship 18.

Dimensions & tons: 140'10", 118'10⅜" x 35'0⅝" x 11'6½". 777.

Men: 88. Guns: UD 14 x 6, QD 4 x 4. Taken by *Belle Poule* 15.2.1809; 2.5.1811 wrecked in Madras Road.

***Sydney*** 1813-1825 (mercantile *Sydney* ____ Yarmouth) survey vessel 6.

Dimensions & tons: 71'10", 56'5⅛" x 21'1½" x 12'1¾". 139.

Men: 26. Guns: UD 6 x 12★? Purchased 1813; 1823 hulked [at Woolwich as convict ship; 1825 sold].

***Camel*** 1813-1819 (East Indiaman *Severn* 1812 Calcutta) storeship 18 [teak built].

Dimensions & tons: 115'6", 94'5¼" x 33'8" x 15'1", 369 44/94.

Men: 55. Guns: UD 2 x 9 short + 16 x 24★. Purchased 1813; 1819 hulked [at St Helena; by 1831 returned home & sold]

PLANS: Profile (incomplete)/lower deck/upper deck.

***Buffalo*** 1813-1840 (East Indiaman *Hindostan* 1813 Calcutta) storeship 18 [teak built].

Dimensions & tons: 120', 98'8⅞" x 33'6" x 15'8". 589.

Men: 55. Guns: 2 x 9 short + 16 x 24★. Purchased 1.11.1813; 1833 fitted to bring spars from New Zealand 28.7.1840 wrecked off New Zealand.

PLANS: Profile & decks. 1835 as emigrant ship: bed places, etc 1836 as emigrant ship: decks.

## Unregistered Vessels

(ie. not listed in the official Admiralty sources referred to.)

***Abigail*** 1812-1814 ( ____ ) cutter 3 (only noted in Colledge).

***Acertif*** 1808-1809 (Danish ____ ) brig 18. Taken by *Daphne* in the Baltic 08.1808; 1809 sold.

***Affiance*** 1806-1810? ( ____ ) schooner. Commissioned in the Leeward Islands.

***Ambush*** 1815-1815 ( ____ ) gunboat ____ . Acquired 1815; 1815 sold.

***Anacreon*** Colledge notes a sloop, 18 guns, in service 1804-1805 and a schooner built in 1815 and transferred to Customs in 1816; neither noted elsewhere.

***Anholt*** 1811 schooner; tender to Anholt Island (off Denmark; occupied by a British force).

***Anna***? 1805-1809 schooner tender? 72'8", ____ x 18'8" x 5'7". 106 tons. BU 1809.

***Antelope***? 1805?- 1832 ( ____ built 1800) schooner or cutter. In service 1805; sold 1832.

***Antigua*** 1804-1816 (French 36 gun privateer *L'Egyptienne* ____ ( prison ship. Taken 27.3.1804 by *Hippomenes*; 1816 BU.

***Atalanta*** 1814-1814 (American *Siro* ? ____ Baltimore?) schooner ____ 225 tons. Taken 1814; 1814 retaken.

***Attentive*** 1812-1817 (American *Magnet* ____ ) prison ship.
96'6", 84' x 28'4" x 13'6". 359 tons. Taken 1812 (or acquired as a schooner in the Leeward Islands in 1810); 1817 sold.

***Bacchus*** 1807-1811 (Danish ____ ) schooner? 141 tons? Taken 1807; 1811 deleted.

***Barracouta*** 1804-1807 or 1804-1805 schooner 10 or 4. Taken into service at Jamaica 1804; in service until 1807 (10 guns). Either this or another vessel of the same name is noted as being built at Halifax in 1804. Certainly a 4 gun 78 ton schooner called *Barracouta*; taken into service in 1804; was wrecked off the southern coast of Cuba on 2.10.1805 (Gossett). On balance it looks as if there were probably two vessels of this name in service at the same time.

***Belem*** 1806-1806 (Spanish *Belem*? ____ ) schooner 4. Taken at Montevideo 27.6.1806; 12.8.1806 retaken by the Spaniards at Buenos Aires.

***Bellone*** 1806-1809 ***Blanche*** -1814 (French *Bellone* 1797 Bordeaux) Sixth Rate 28. 132', 107'2" x 33'7" or 32' x 11'4". 648 or 643 tons. Taken by *Powerful* off Ceylon 9.7.1806 renamed *Blanche* 1809; 1814 BU.

***Berbice*** 1805-1806 (Dutch *Berbice* ____ ) schooner 4. 78 tons. Acquired 1805;

# THE NAPOLEONIC WAR 1802-1815

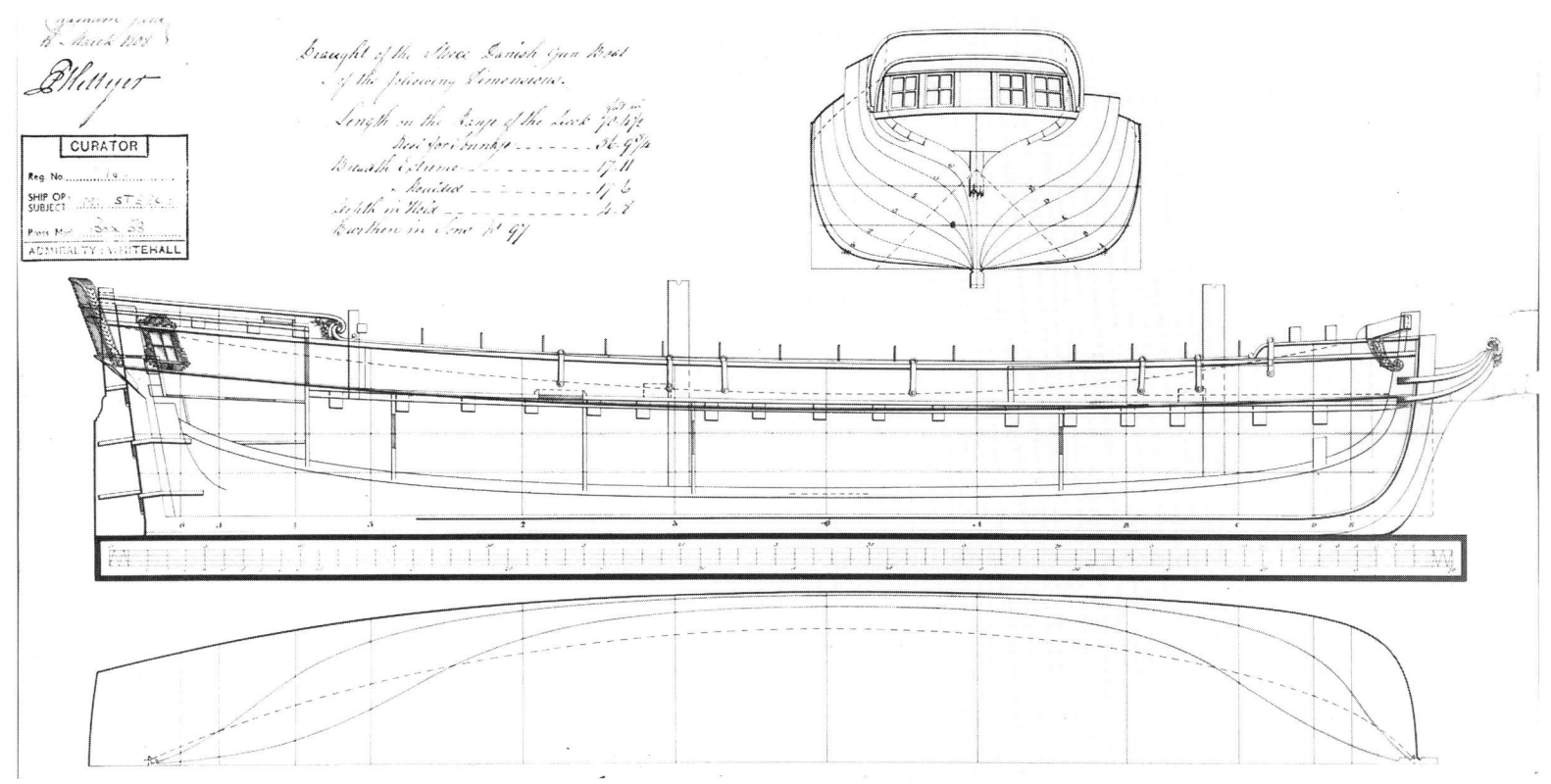

This lines and profile plan (7165) of the Danish gunboat *Steece* was 'taken off' the captured vessel at Chatham. Though this vessel was actually taken into service with the RN as the brig *Warning* (see the entry on her in the section on brigs earlier in this chapter) her plan can represent the large number of captured vessels whose fate is less certain

1806 foundered in the Demerara River.

***Bermuda*** 1813-1841 (American *Delaware* pilot boat ____ ) schooner/yacht 46'2", ____ x 15' x 5'6". 43 tons. Presented by her captors 1813; 1841 BU.

***Brave*** 1806-1806 (French *Le Brave* ex *Le Dix Août* ex *Le Cassard* 1795 ____ ) Third Rate, 74. 1890 tons. Taken 6.2.1806 at the battle of Santo Domingo 12.4.1805 foundered off the Azores on passage to England.

***Brooke*** 1803-1809 type unknown, 16 guns. Purchased 1803; 1809 BU.

***Caroline*** 1809-1814? (French? ____ ) schooner 10 guns. Taken 1809; 1814 still in service?

***Celebes*** 1806-1807 (Dutch *Pallas*? ____ ) frigate 36. Taken by *Greyhound* and *Harrier* in the East Indies 26.7.1806; unsuccessful survey: not considered suitable for the Service; 1807 deleted.

***Cesar/Caesar*** 1806-1807 (French *César*? 1802 ____ ) Brig 16. 320 or 330 tons? Taken 1806; 03.1807 driven ashore and lost in the Gironde.

***Clyde*** 1805-1826 (mercantile *Atalanta*? ____ ) tender to the *Clyde* (frigate). Purchased 1805; 1826 sold.

***Colibri*** 1809-1813 (French 'brick' *La Colibri* 1808 Le Havre) brig sloop 16. 98'9", 79'4" x 29'5" x 13'5". 365 tons. Taken by *Melampus* on the Halifax station 16.1.1809; 22.8.1813 wrecked at Port Royal, Jamaica.

***Cornelius*** 1804-1805? captured? schooner?; commissioned 1804-1805 at Jamaica; no further information.

***Crafty*** 1804-1807 (French *Le Renard* ____ ) schooner 12. 48 men. 2 x 4 + 10 x 12★. Taken in the Mediterranean 25.11.1803; 1804 sold to the RN 9.3.1807 taken by Spanish gun vessels near Tetuan.

***Creole*** 1803-1804 (French 18 pounder frigate *La Créole* 1797 Nantes. Lamothe Jnr. design) frigate 40 guns. Taken by Bayntun's Squadron off San Domingo 30.6/1.7.1803; 2.1.1804 foundered on passage from the West Indies to England.

***Crib*** 1804-1805 there is a log for a vessel of this name and dates.

***Cumberland*** 1803-1804/1809-1810 schooner purchased at Port Jackson 1803; 1804 to 1809 in French hands; 1810 sold.

***Curacoa*** sloop, 18 guns; at Jamaica 1802?

***Cynthia*** 1810-1815 sloop, 16 guns (is this merely a failure to strike the 1796-1809 *Cynthia* off the lists?).

***Dart*** 1810-1813 (mercantile *Belerina*? 1809 ____ Mevagissey) cutter 10. 62'7", 47'7¾" x 22'4¾" x 10' 1". 127. Men: 40. Guns: 10 x 12★ (?). Acquired 1810; 27.10.1813 sailed and not heard from again, presumed to have foundered on the Halifax station.

***Decouverte*** 1803-1808? (French ____ schooner 8. 81'6", 70'2" x 21' x 7'4". 164 56/94 tons. Guns: 8 x 12★. Taken at Santo Domingo 30.11.1803; 1808 sold (or is this a confusion with following vessel?).

***Decouverte*** 1806-1816 (French *Eclipse* or *Encounter*? ____ ) brig or schooner 12. 80'3", 67'3" x 22'10" x 10'2". 181 tons. Men: 55 Guns: 2 x 6 + 8 or 10 x 12★. Taken 1806 (at Santo Domingo?); served in the West Indies; 1816 (or 1808?) sold.

***Defence*** 1804 (or built 1802?) cutter; no further note.

***Demerara*** 1804-1804 ( ____ *Anna*? ____ ) schooner 6 guns. 106 tons. Acquired 1804; 14.7.1804 taken by French privateer *Grand Decide* [*Decade*?] in the West Indies.

***Demerara*** 1806-1813 (French *Cosmopolite*? ____ ) brig 14 guns. 220 tons? Taken or otherwise acquired 1806; 1813 sold.

***Destruction*** 1815-1815 ( ____ ) gunboat ____ . Acquired 1815; 1815 sold.

***Devonshire*** 1804-1804 fireship. Purchased (after being hired?) 1804 2/3.10.1804 expended at Boulogne.

***Diamond Rock/Fort Diamond*** 1804-1805 shore battery on Diamond Rock off Martinique; 1805 taken by the French fleet.

***Diana*** 1807-1810 brig or cutter 10 guns (10 x 6), 150 tons. 50 men. Purchased in the East Indies; 1810 condemned at Rodriguez or 05.1810 wrecked there.

***Diligence*** 1812-1814 (mercantile? *Thebe* ____ ) lugger. Purchased 1812; 1814 sold.

***Dixmund*** 1803-1807 gunbrig 10 guns. In service at Jersey.

***Dolores*** 1807-1807? (Spanish *Dolores* ____ ) schooner? ____ . Taken 1807; no further information.

***Dolphin*** 1813 tender, North American waters.

***Dominica*** 1805-1809 (French privateer *Tape de l'Oeuil*? ____ ) schooner 10/14. 45/55 men. Taken 1805; 08.1809 capsized in a hurricane off Tortola.

***Dominica*** 1805-1806/1806-1808 (French ____ ) sloop (sloop-rigged vessel?) 75 tons. Acquired 1805; 11.5.1806 mutinous crew took her into Guadeloupe and handed her to the French who renamed her *Napoléon*; 24.5.1806 retaken by *Wasp* and renamed *Dominica*; 1808 BU.

***Don Carlos*** ____ -1811 sloop?; 1811 BU; probably identical with the *Infante Don Carlos*; which see (Pitcairn Jones only).

***Duke of Kent*** design for 170 gun 4 decker line of battle ship proposed by Tucker in 1809; not proceeded with. There is a model of this ambitious

(NMM negative A324) The NMM's model of the proposed 170 gun ship *Duke of Kent* of 1809

design in the National Maritime Museum

**Eagle** 1815-1815/1816 (_____) gunboat _____. Acquired 1815; 1815 or 1816 sold.

**Earl Moira** 1812 tender to *Guerrière* (Pitcairn Jones only).

**Egremont** 1810- _____ (American? _____) schooner. Captured; in East Indies no further note.

**Elizabeth** 1806-1807 (French or Spanish? *Elizabeth* _____) schooner/cutter 10/12. 72'8", 57'2" x 21'6" x 9'. 141/110 tons. 35/55 men. 10 x 12★. Taken in the West Indies 1806; 1807 foundered in the West Indies (without trace? and with all hands).

**Elizabeth** 1808-1814 (French? _____) schooner 10/12 guns, 141 tons. Commissioned (& purchased?) 1808 at Antigua; 31.10.1814 capsized whilst chasing an American privateer in the West Indies.

**Ems** 1809-1815 (Dutch schuyt *Ems* _____) gunboat? _____. Taken 1809; based at Yarmouth; 1815 sold.

**Falcon** 1807-1808? (Danish _____) sloop 16. Abandoned by the Danes at Danzig and taken by *Sally* 14.4.1807; in service 1808 but no further mention.

**Fama** 1808-1808 (Danish *Fama*? _____) brig sloop 16/18, 320 tons. Taken by the boats of *Edgar* at Nyborg 18.8.1808; 23.12.1808 wrecked on the island of Bornholm in a snowstorm.

**Favourite** 1805-1813? cutter. Purchased 1805; 1813? sold.

**Fawn** 1805-1806? (French *Le Faune* 1804 Nantes) brig sloop 16. Taken 15.8.1805 by *Camilla* and *Goliath* off Rochefort; in service 1806 but no further record.

**Felicite** 1809-1809 (French *La Félicité* 1785 Brest, Forfait design) 12 pdr frigate 40. Armed 'en flûte' when taken by *Latona* in the West Indies 6.7.1809; sold to Christophe (ruler of Haiti) and named *Amethyste* by him; seized by rebels against Christophe and renamed *Heureuse Révolution* 3.2.1812 taken by *Southampton* and handed back to Christophe.

**Fierce** 1806?-1813 (mercantile? *Desperate* 1797 Bermuda?) schooner _____. 58'8", 45' x 19'3" x 9'6". 82 tons. Purchased 1806 (or 1804?); 1813 BU.

**Firebrand** 1815- _____ ? (_____) gunboat _____. Acquired 1815; then became a tender at Bermuda?

**Firefly** 1803-1806? (mercantile *John Gordon*? 1801 Bermuda) armed vessel 6 guns. _____, 46' x 20' x 10'4". 98 tons, 35 men. 6 x 12★. Purchased 1803 out of service by 1807 (lost?).

**Firefly** 1807?-1812 **Antelope** -1814 (French or Spanish? *Antelope* _____) schooner 10 guns. Taken 1807 or 1808; served at Jamaica; 1812 renamed *Antelope*; 1814 BU.

**Fleur de la Mer** 1808 - later **Pike**? - 1811 (French *Fleur de la Mer* ex *Gipsy*? 1805/1806 American-built) schooner/armed ship 10 guns. 72'3", 58'10" x 19'4" x 9'6". 117 tons. 2 x 6 + 9 x 12★. Taken 1808; 8.1.1811 foundered in the Atlantic.

**Flirt** 1812- _____ ? (Danish _____) gunboat _____. Taken 1812; no further information.

**Flying Fish** 1803-1807 **Firefly**-1807 (French *Poisson Volante* _____) schooner 10/12. Taken by Bayntun's squadron off Santo Domingo 30.6.1803; renamed *Firefly* 1807 (or is this a confusion with the 1804 *Flying Fish*?) 17.11.1807 foundered in the West Indies.

**Fort Diamond** 1804 captured privateer; armed sloop (sloop-rigged vessel) in service 1804; presumably tender to *Diamond Rock*.

**Fowey** 1806 there is a log for a vessel of this name and date.

**Furnace** 1807-1811/1813? bomb vessel 8 guns; at Gibraltar; 1811 or 1813 BU.

**Gipsy** 1805-1808 (Dutch *Antilope*? _____) schooner or cutter? 121 tons. Acquired 1805; 1808 sold.

**Gracieuse** 1804-1807? (French privateer *la Gracieuse*? _____ Bermuda?) schooner 8 guns. 72', 50' x 21'2" x 9'. 119 tons. 40 men. 8 x 18★. Taken 21.10.1804 off Curaçoa by *Blanche*; 1807 still in service; 1808 deleted?

**Grenada** 1804-1809? (French *Jena* 1800) brig or schooner 16/10/4. 71'6", 55' x 21'11" x 8'8". 141 tons. Men: 40. Guns: _____. Taken 1803; 1804 presented to the RN by the inhabitants of Grenada; sold about 1809 (or BU 1811?).

*Grenada* 1807-1814? (French privateer brig *Jena*? ____ ) sloop 16 guns. 327 tons? Taken by *Cruiser* in the North Sea 6.1.1807; still in service 1814? fate uncertain.

*Halstarr* 1807-1809 (Dutch *Hasselaar/Kenau Hasselaar* ____ ) 12 pounder frigate 32 guns. 700 tons? Taken at Curaçoa 1.1.1807; in service until 1809.

*Harlequin* 1815-1816 gunboat/tender. Purchased in the West Indies 1815; at Bermuda 1816; deleted the same year?

*Harriet* 1805 armed ship? (Spanish?).

*Hart* 1805-1810 (French privateer *l'Empereur* 1789 ____ ) brig sloop 16 guns. 68' or 78'6", 63'9" x 21'2" x 10'9". 152 tons. Taken by *Eagle* in the West Indies 06.1805; 1810 sold.

*Hirondelle* 1804-1808 (French *l'Hirondelle*? ____ ) gun brig 14 guns. 216/210 tons. Taken in the Mediterranean by *Bittern* 28.4.1804; 23.2.1808 wrecked near Tunis (near Cape Bon).

*Humber* 1806-1808 French sloop 16 guns; taken 1806; listed until 1808.

*Hurd* ____ ; 1808 cutter 8 guns; origins obscure; 5 or 10.8.1809 wrecked near Flushing.

*Hussar* 1812- ____ ? (Danish ____ ) gunboat ____ . Taken 1812; no further information.

*Indefatigable* 1804-1805 armed ship. Purchased 1804; 1805 sold.

*Industry* 1807 gun vessel in the River Shannon.

*Infante Don Carlos* 1804-1810/1811 (Spanish *Infante don Carlos* ____ ) sloop 16 guns. Taken by *Diamond* 7.12.1804; in service until 1810; 1811 BU?

*Integrity* 1805?-1809? cutter? Purchased in Australia 1805; 1806 at New South Wales; still in service 1809?

*Jahde* 1809-1815 (Dutch *Jahde* ____ ) gunboat? ____ . Taken 1809; 1815 sold.

*Jaseur* 1807-1808 (French *Jaseur* 1806 ____ ) cutter or brig sloop 10/12 guns. 50 men. Taken by *Bombay* off the Andamans 10.7.1807; 08.1808 supposed to have foundered on passage from Bengal to Prince of Wales Island (Penang or Cochin?).

*Jason* 1813 (French privateer ____ ) Gunbrig 12 guns. Taken 31.12.1813; in service 1814.

*Java* 1806-1807 (Dutch *Maria Reijersbergen* ____ ) frigate 36/32 guns. 850 tons. 300 men. UD 26 x 12, QD 10 x 32★, Fc 2 x 9 + 2 x 32★. Taken by *Caroline* in the Indian Ocean 18.10.1806; 28.2.1807 supposed to have foundered in a storm off Rodriguez in the Indian Ocean.

*Junon* 1809-1809 (French *le Junon* 1806 Le Havre. Forfait design) frigate 38. 1100 tons. Men: 295. Guns: UD 28 x 18, QD 14 x 32★, Fc 2 x 9 + 2 x 32★. Taken 10.2.1809 on the Halifax station by *Horatio*, *Latona* etc; 13.12.1809 retaken by the French *Renommée* and *Clorinde* and destroyed by her captors as being too heavily damaged to retain.

*Kingfish* 1806-1812 ( ____ ) schooner 6. 75 tons. Acquired 1806 (or 1804) 1808 taken by a French privateer and taken to Guadeloupe; later retaken by *Pheasant*; 1812 taken [?].

*Lemon* 1812-1813 schooner at Trinidad.

*Loberon* 1812- ____ ? (Danish ____ ) gunboat ____ . Taken 1812; no further information.

*Macassar* 1807-1807 (Dutch ____ ) frigate 32. In service 01-02.1807 in the East Indies.

*Mandarin* 1810-1810 (Dutch *Mandurese* ____ ) brig 12. Scuttled at Amboyna and then raised in 02.1810; 9.11.1810 wrecked near Singapore (on Red Island); 1812 BU.

*Maria/Marie* 1805-1807 (French *La Constance* 1801 ____ ) schooner 12/10. 130 tons. Men: 46. 12 x 12★. Taken 21.6.1805 by *Circe*; 16.10.1807 lost with all hands in the West Indies (foundered).

*Maria* 1808?-1808 (French? ____ ) brig 12. Taken 1808?; 29.9.1808 taken by the French brig *Département des Landes* (22) and grounded at Guadeloupe to prevent sinking.

*Maria* 1809?-1814? schooner 14 or brig 16. Purchased 1809 or 1810?; still in service 1814; fate unknown.

*Mariana/Marianne* 1805-1806? (Spanish ____ ) schooner 10. Taken 1805 by *Swift*; 1806 listed; no further information.

*Mary* cutter 16 guns presumed foundered in the Channel in 02.1805 (is this one of the hired *Mary's*?)

*Matilda* 1805- ____ ? (French privateer *Mathilde* ____ ) schooner 10. Taken by *Cambrian* 3.7.1805; no further information.

*Mercury* 1807-1835 (mercantile ____ 1807 ____ ; Rotherhithe) tender. 44'3", 35'10" x 16'1" x 7'10". 50 tons. Purchased 1807; 1835 BU.

*Mignonne* 1803-1804 (French corvette *La Mignonne* 1795 Cherbourg) sloop 16/18. 462 tons. Men: 121. Guns: 18 x 9. Taken 28.6.1803 by *Goliath* at San Domingo; 12.1804 grounded at Port Royal, Jamaica, then condemned.

*Minerve* 1803-1811 (French ____ ) prison hospital ship. 246 tons? Purchased 1803; 1811 BU.

*Mohawk* 1813-1814 (American *Viper* originally *Ferret* renamed 1810. 1804 Norfolk Navy Yard) gunbrig 10. 148 tons. Taken 1813; 1814 sold.

*Morne Fortunee* 1804-1809 (French privateer *Josephine* ex *Regulus* ____ ) schooner 10/12. 184 tons. 60/65 men. 2 x 6 + 8 x 18★. Taken in the West Indies by *Princess Charlotte* 13.12.1804; 9.1.1809 foundered in a squall off Martinique.

*Mors Aut Gloria* 1810- ____ ? (Spanish *Mors aut Gloria*? ____ Cadiz?) gunboat 2. 35 men. 1 x 36 + 1 x 6" howitzer. Taken 1810?; fate uncertain.

*Mosambique/Mozambique* 1804-1810 (French privateer *Mozambique*? 1798 ____ ) schooner 10. 67'6", 52' x 20'2" x 8'3". 111 tons. 50 men. 10 x 18★. Cut out by crew of *Emerald* near Saint Pierre 13.3.1804; 1810 sold.

*Mulgrave* 1807- ____ ( ____ *Alert* ____ ) sloop 18. 1807 acquired on the Leeward Islands station; no further information.

*Musette* 1803-1808 cutter; no further information.

*Nancy* 1809-1813 (mercantile *Nancy*? ____ ) gunbrig 14. 210 tons? Purchased 1809; 1813 sold.

*Netley* 1807-1808 (French *Determinée* ____ ) gunbrig 10/14. 173 tons. 65 men. Taken 1807; 10.7.1808 wrecked on the Leeward Islands station.

*Orinoco* 1805-1811 ( ____ ) gunvessel 12. 250 tons? Purchased in the West Indies 1805; 1811 deleted.

*Ortenza* 1808-1812 (French *Ortenzia* ____ ) schooner 14. 172 tons? 1808 taken; 1812 sold.

*Papillon* 1803-1806 (French aviso *Le Papillon* 1786 ____ ) sloop 14 (schooner rig?) or gunbrig 10. 145 tons. 121 men. Taken by *Vanguard* off Santo Domingo 4.9.1803; 01.1806 presumed to have foundered in the Atlantic on passage to the Jamaica station.

*Patriot* 1808-1813 (Dutch ____ ) gunbrig? 10. Taken 1808; 1813 sold.

*Peacock* 1815? sloop? 14 guns. No further information.

*Peggy* 1804-1804 fireship/fire vessel. Purchased 1804; 2.10.1804 expended at Boulogne.

*Pert* 1804-1807 (French privateer *Buonaparte* ____ ) brig sloop 14/16. 206 tons. 70 men. Taken by *Cyane* in the West Indies 12.11.1804; 16.10.1807 wrecked on Santa Margarita.

*Pert* 1808-1813 (French *Le Serpent* ex *Le Rivoli* 1807 American-built) brig sloop 10? 97', 79'6" x 23'9" x 10'3". 239 tons. Taken by *Acasta* off La Guaira 1808; 1813 BU.

*Pickle* 1808-1818 (French *Eclair*? 1808 ____ ) schooner? 136 tons. Taken 1808; 1818 sold.

*Pigmy* 1806-1807 (1806 Dartmouth) gunbrig 12/14/16 guns. 79', 60'3" x 26'6" x 11'. 216 16/94 tons. 2 x 6 + 14 x 12★. 60 men. Purchased 1806; 2.3.1807 wrecked near Rochefort.

*Pioneer* 1804-1807 gunbrig; listed in these years at Plymouth; no further information.

*Port d'Espagne* 1806-1811 brig sloop or schooner? Presented to the RN by the inhabitants of Trinidad 1806; 1811 sold.

*Prevost* 1804-1806 ( ____ ) schooner. 145 tons. Acquired 1804; taken 31.8.1806 by French brig *Austerlitz* (18) to the South of Martinique [is this the same as the *Provo* purchased in Bermuda and listed for 1805-1807?].

*Prince de Neufchatel* 1814-1815 (American privateer *Prince de Neufchâtel* 1813 Brown Brothers, New York) schooner 18. 110'8", 93'8¼" x 25'8" x 11'6". 328 22/94. Guns: 16 x 12★ + 2 x 6. Taken in the Atlantic 28.12.1814; not taken into service (reported to have been damaged whilst docked for survey?); 1815 BU.

PLANS: Lines/lines & profile/sails.

*Princess Charlotte* schooner, 8 guns. Noted by Pitcairn Jones in 1807 (possibly 1805-1807?); the same as the hired schooner of 1804-5?

*Princess Mary* Unknown vessel in service 1808-1809?

*Proselyte* 1804-1809 (mercantile *Ramillies* 1804 South Shields) bomb vessel 4. 404/405 tons. 78 men. Purchased 1804; 5.12.1808 caught in ice and wrecked on Anholt Reef.

*Queen Mab* (ex *Courier*?) 1807-1809? (Dutch *Coureer*? ____ ) armed ship 20. 490 tons? Taken 1807; by 1809 deleted [or sold 1812?].

*Rainbow* 1806-1807 (French ____ ) brig sloop 16. Taken 1806; 1807 sold.

*Rapide* 1808-1814 (French *Le Rapide* ex *Vilaret* ex English prize 1803?) schooner/tender 12. 90 tons. 130 men. Taken 1808; 03.1814 wrecked on the Saintes (West Indies).

*Raposa/Raposo* 1806-1808 (Spanish? ____ ) gunbrig 10. 173 tons. Taken at Campeachy by the boats of *Franchise* on 7.1.1806; 15.2.1808 grounded and destroyed to prevent capture at Carthagena.

*Redbridge* 1804-1805 schooner 10. 131 tons. 60 men. Built 1804; 1804 commissioned (purchased?) at Jamaica; 1.3.1805 sprang a leak and foundered near Jamaica.

*Redbridge* 1804-1806 (French *Oiseau* 1801 Nantes) schooner 12. 131 or 170 tons? Taken 1804; 4.11.1806 wrecked near New Providence, West Indies.

*Redbridge* 1807-1808: *Variable* -1814 (American mercantile *Aristotle* 1806 ____ USA) schooner 10. 80'5", 67' x 21'10" x 8'. 172 tons. 50 men. 2 x 6 + 8 x 12★. Purchased at Jamaica (?) in 1807; in 1808 or about 1811 renamed *Variable*; in service until 1814.

*Regulus* 1804-1806 (French privateer ____ ) brig sloop 14. Taken by *Princess Charlotte* in the West Indies 13.12.1804; in service until 1806.

*Renara*? schooner 14 - prize - in service 1806?

*Revenge* 1806 purchased vessel; type uncertain 81'1", 65'11" x 21'8" x 7'11". 165 tons. Fate unknown.

*Rose* 1805-1806 cutter listed for these years (only noted in Colledge possibly one of the cutters of this name which had been hired earlier?)

*Saint Christopher* 1807-1811 (French privateer *Mohawk* ____ ) sloop 18. Presented 1807 by the inhabitants of Saint Kitts; 1811 BU at Antigua.

*Saint Lucia* 1803-1807 (French *Enfant Prodigue* 1794 Bordeaux) gunbrig 14. 183 tons. Taken 1803; 29.3.1807 retaken by the French schooners *Vengeance* (12) and *Friponne* (5) off the Western end of Guadeloupe.

*Saloman/Salorman* 1808-1808 (Danish *Saloman*? ____ ) cutter (schooner?) 10. 120 tons. 30 men. Taken 10.8.1808; 23.12.1808 wrecked near Ysted in the Baltic.

*Sandwich* 1803-1805 (mercantile *William and Mary*? ____ ) cutter? 113 tons. Acquired 1803; 1805 sold.

*Scourge* 1804-1806 gunbrig 10; no further information; but would appear to be a separate vessel from the 1803-1816 purchased sloop.

*Seaforth* 1805-1806 (probably French privateer *Dame Ernouf* ____ ) brig 16/gunbrig 14. 215 tons. 86 men. Taken by *Curieux* 8.2.1805; 02.1806 capsized in a squall in the West Indies (Leeward Islands station).

*Sir Andrew Mitchell* 1806-1807 (mercantile *Sir Andrew Mitchell* ____ ) schooner or cutter? 134 tons. Acquired 1806; 1807 deleted.

*Siren* 1814-1815 (American *Syren* 1803 Hutton, Philadelphia) hospital hulk. 298 tons. Taken 1814; 1815 deleted.

*Skylark* 1812 sloop or schooner? Pitcairn Jones notes a vessel of this name as commissioned in 1812; no further information and doubtful.

*Skipjack* 1807-1812 (French *Confiance* 1800 USA) schooner 12. 71'4", 60'5" x 18'11" x 7'8". 115 ton. Purchased (taken?) in the West Indies 1807; 1812 BU.

*Sombrero* 1811 sloop? (probably sloop-rigged vessel) noted in the West Indies (?) in this year by Pitcairn Jones.

*Somers* 1807-1812 schooner 10; noted by Colledge but not elsewhere - dubious.

*Subtle* (ex *Vigilant*) 1806-1807 (French? *Impériale*? ____ ) schooner 10/8. 102 tons. 50 men. Taken 24.5.1806 by *Cygnet* off Dominica and initially named *Vigilant*; 20.11.1806 renamed *Subtle*; 26.10.1807 lost on a reef off Bermuda.

*Sultan* Ca 1802-1804 gunboat? in service in the Mediterranean? (Pitcairn Jones only).

*Surinam* 1804-1808 (Dutch *Pylades* 1804? ____ ) sloop 18. Taken 4.5.1804 at Surinam; in service until 1808 (a *Surinam* also an 18 gun sloop is noted as being taken at Curaçoa in 1807; but is this just a confusion with this vessel?).

*Surveyor* American schooner 6 guns; taken by the boats of the *Narcissus* 12.8.1813; out of service (if ever in it) by 1814 (Colledge only).

*Swaggerer* 1809-1815 (French privateer *Bonaparte* ____ ) gunbrig/brig 10. 300 tons? 2 x 6 + 8 x 18★. 60 men. Taken 1809; 1815 BU.

*Tapageuse* 1806 (French *Tapageuse*? ____ ) brig sloop. Taken 6.4.1806 by the boats of *Pallas* off Bordeaux; fate unknown (probably never taken into the Navy).

*Thomas* 1808-1809 fireship purchased 1808; 11.4.1809 expended in Basque Roads (Colledge only).

*Tiger* 1808?-1812 gunbrig; vessel of this name in service 1808-1812 (according to Colledge); possibly the revenue cutter of this name built at Bridport in 1805 and transferred. Alternatively there is an indication that a *Tiger* was in service between 1804 and 1806.

*Tobago* 1805-1806/1807 *Vengeur* -1807 ( ____ ) schooner 10. 127/120 tons. Acquired 1805; 24.10.1806 taken by French brig *General Ernouf* off Guadeloupe; became French *Vengeur*, 1807 retaken and kept French name; 1807 deleted.

*Tobago* 1807 schooner, 10, in service in the Leeward Islands (captured 1806?) according to Pitcairn Jones.

*Torch* 1805-1811 (French corvette-gabare 1804 Honfleur) 20 guns. 557 tons? Taken 16.8.1805 by *Goliath* off Cape Finisterre; 1811 BU.

*Trial* 1807-1815 gunvessel 6. No further information.

*Trinidad* 1805-1806 (French? *Elizabeth* ____ ) schooner? Taken 1805; 1806 deleted?

*Trinidad* 1808-1809 ( ____ ) schooner. 127 tons. Acquired 1808; 1809 deleted.

*Tritona* 1811 hospital ship at Lisbon noted by Pitcairn Jones.

*Unique* 1803?-1806 (French? ____ ) schooner 8/10. 120 tons ──. Taken in the West Indies 1803 or 1805?; 23.2.1806 taken by French privateer in the West Indies.

*Unique* 1808-1809 (French? *Harmonie* [or *Duquesne*?] ____ ) brig or schooner 12. 74', 60'3" x 20'8" x 7'4". 121/183 tons. 1808 taken; commissioned at Antigua; later converted to a fireship; 31.5.1809 expended at Basse Terre.

*Urquijo/Orquixo*? 1805-1805 (Spanish *Urquijo* ____ ) sloop 18, 384 tons. Taken 1805; 7.11.1805 foundered in a squall off Jamaica.

*Ventura* 1806 schooner at Malta noted by Pitcairn Jones.

*Venturer* 1807-1814? (French *La Nouvelle Entreprise* 1807 ____ ) schooner 10. 72', 56'4" x 20'6" x ____. 126 tons. Taken 27.12.1807 by *Nimrod*; out of service 1812? or (Colledge) renamed *Theodocia* in 1812 and sold in 1814.

*Victor* 1808-1809/1810-1810? (French *Iéna* ex *Revenant* 1805 Saint Malo) brig sloop 18. 400 tons. Taken in the Bay of Bengal 8.10.1808 by *Modeste*; 2.11.1809 retaken by the French *Bellone* in the Bay of Bengal and renamed *Le Revenant*; 1810 captured at the fall of Mauritius and (?) condemned.

*Vigilant* 1803-1808? schooner. Purchased (in the West Indies?) 1803; 1807 still in service; 1808 sold?

*Violet* 1813-1814 lugger 4. Built 1812; in service 1813-1814.

*Volador* 1807-1808 brig sloop 16, 273 tons, 121 men. Taken into service in the West Indies 1807; 24.10.1808 wrecked in the Gulf of Coro.

*Volcano* 1804-1810 purchased bomb?; 22.12.1810 sold. Confusion with 1797-1810 vessel?

*Wasp* 1809 schooner with Rowley's Squadron at the Isle de Bourbon noted by Pitcairn Jones.

*Weazle* 1808-1812 ( ____ ) schooner? 141 tons. Acquired 1808; 1812 deleted.

*William* 1808-1810 storeship 12. 08.1810 BU at Woolwich, noted by Colledge.

*Zephyr* Fireship expended 11.4.1809 in Basque Roads.

## Hired Ships

NOTE: See same section in previous chapter for sources and other information. Additional information from papers in NMM document class ADM/BP/23A on ships hired in 1803. It is interesting to see from one of the lists in this group that many of the cutters hired in 1803 were from a small group of owners (King - 6 vessels, Iggulden - 3, Collett & Thomsett 8, a firm whose name appears to be Minet & Feetor - 4, H Latham - 6).

Name% = from Dittmar's articles in *The Belgian Shiplover* (stated to be as listed in primary Admiralty sources). ★Name = from unofficial sources - existence not certain. It is likely that quite a few of the vessels noted here are the same as those of the same name listed for 1790-1801; however, with the paucity of information available it is safer to make no attempts at connecting them.

***Active*** 1803-1803/1804 **Lord Keith** %; 1808 Cutter 6 x 4 (or 10 guns). 24 men. 71 73/94 tons. Owned by J Iggulden; driven into Cuxhaven during a gale and taken by the French 11.1.1808.

***Active*** 1803-1814 Cutter 8 x 4. 27 men. 77 59/94 tons. Carvel-built. Owned by Collett & Thomsett.

***Adelphi*** % 1804-1805 Armed ship 10 x 12★. 165 tons.

# THE NAPOLEONIC WAR 1802-1815

★*Adelphi* 1810-1814 Tender 8.
★*Admiral Gardner* 1809-1810 [Irish Gunboat].
*Admiral Mitchell* 1803-1805 Cutter 12 x 12★. 40 men. 132 86/94 tons.
★*Admiral Whitshed* 1807-1808 _____ .
*Adventure* % 1813-1814 Transport.
★*Advice* 1804-1804 _____ 10. Lost 1804 (not recorded in Gosset).
*Africa* 1803-1810 [Irish Gun cutter] 5 men. 70 80/94 tons.
★*Alarm* 1812-1814 Lugger 8.
*Albion* 1803-1808 Cutter 6 x 4. 27 men. 79 17/94 tons. Clinker-built. Owned by J King.
*Albion* % 1808-1812 Cutter 6 x 4. 78 24/94 tons.
*Alert* % 1804-1804 *Lucy* -1805 Lugger 16 (6 x 12 + 8 x 6 + 2 x 12★), 119 80/94 tons.
*Alert* % 1804-1804 Cutter 6 x 12★. 44 49/94 tons.
*Alert* % 1808-1812 Cutter. 44 89/94 tons.
*Alert* % 1809-1809 Smack. Sunk 1809 (not recorded in Gosset).
*Alert* % 1812-1814 Cutter. 117 70/94 tons.
*Alnwick Packet* % 1809-1809 Smack.
*Althorp/Althorpe* % (ex *Earl Spencer*?) 1804-1804 Cutter 14 x 12★ or 16. 30 men. 163 73/94 tons. 21.12.1804 lost (Gossett gives believed to have foundered in the Channel 9.8.1805).
*Amity* % 1804-1804 Schooner fire vessel 14. Expended at Boulogne 3.10.1804.
*Amy* 1803-1804 Tender 4. 156 70/94 tons. At Bristol.
*Ann* 1803-1809 [Irish Gun vessel] 5 men. 70 37/94 tons.
*Ann* % 1804-1809 Brig 10 x 12★. 120 56/94 tons.
*Ann* % 1807-1812 Tender 154 35/94 tons.
*Ann* % 1809-1809 Smack.
*Ant* 1803-1806 Cutter 4 guns. 16 men. 27 43/94 tons. Captured 16.3.1806 by the French.
*Ant* % 1808-1810 Cutter. 40 tons.
★*Ariadne* 1804-1814 Armed ship 16.
★*Ariadne* 1806-1811 Cutter 8.
★*Athens* 1804-1805 Cutter?
★*Atlas* 1804-1804 Armed ship 16.
★*Aurora* 1804-1804 Armed ship 16.
*Auxiliary* % 1809-1809 Smack.
*Badger* % 1812-1814 Cutter 107 35/94 tons.
*Barbara* 1803; _____ ? Tender at Greenock.
*Bardsea* 1802-1810 [Irish Gunboat] 3 men. 50 47/94 tons.
*Bee* % 1809-1809 Smack.
★*Bellona* 1804-1804 Armed ship 18.
*Berwick* Packet★ 1809-1809 Smack.
*Betsey* 1803-1804 Cutter 6 x 3. 20 men. 50 66/94 tons. Owned by Minet & Feetor (?).
*Betsey* % 1804-1804 *Phoebe* -1805 Cutter 10 (2 x 4 + 8 x 12★), 84 32/94 tons.
*Betsey* % 1804-1804 *Jennet* -1804 Cutter 6 x 12★. 60 72/94 tons.
*Betsey* % 1804-1810 [Irish Gunboat] 51 tons.
*Betsey* 1806-1810 Cutter 6, 50 62/94 tons.
*Betsey Gaines* % [or *Cains*] 1808-1810 Cutter (tender) 16, 173 71/94 tons.
*Black Joke* % 1808-1810 Lugger 4. 100 92/94 or 109 tons. Taken 1.7.1810 (Gossett gives just the year 1811) by the French in the Channel.
*Blessing* % 1804-1806 Armed ship 14 x 18★. 310 tons.
*Bolina* % 1803-1807 Tender 6 x 3. 181 12/94 tons. Lost 20.11.1807 by being run aground near Padstow.
*Britannia* 1803-1811 Cutter 6 x 3, 24 men. 68 77/94 tons. Owned by H Latham.
*Britannia* % 1804-1809 [Irish Gunboat] 77 tons.
*British Fair* 1803-1805 Cutter 6 x 3. 24 men. 70 53/94 tons. Owned by H Latham.
*British Fair* % 1807-1814 Cutter 6 x 3. 70 70/94 tons.
★*Broke* 1814 cutter 14.
*Camperdown* 1804-1804 Cutter 14 x 12★. 158 40/94 tons.
★*Canada*? 1805? _____ ?
*Caroline* % 1804-1807 (French *Affronteur* _____ ) Lugger 12 x 12★. 158 tons.
*Caroline* % 1809-1809 Smack.
*Carteret* 1803-1805 Cutter 6 x 3. 22 men. 59 71/94 tons. Owned by Messrs Le Mesurier.
*Cecilia* % 1806-1811? Schooner (tender) 6. 186 69/94 tons. 1810-1811 tender at Cork.
*Champion* 1803-1804 *Sabina* -1804 Cutter 6 x 3. 20 men. 50 31/94 tons. Clinker-built. Owned by J Iggulden.
*Chance* % 1815-1815 Schooner 131 69/94 tons.
★*Chapman* 1804-1806 Armed ship 16.
*Charles* % 1804-1814 Armed ship 14 x 18★. 309 3/94 tons.
*Charles* % 1811-1814 Schooner 118 53/94 tons.
*Charlotte* 1803-1806 'Armed cutter' (schooner) 8/10 (8 x 12★ or 2 x 3 + 8 x 12★). 26 men. 77 14/94 tons. Owned by Heather.
*Charlotte* % 1805-1806 cutter 21 tons.
*Cleveland* 1803-1806 Tender 4 x 2. 68 69/94 tons. At Bristol.
★*Cleveland* 1809-1810 Storeship.
*Cockatrice* % 1804-1808 Brig 12 x 18★. 183 92/94 tons.
*Colpoys* % 1804-1807 Armed ship/brig 16 (2 x 4 + 14 x 12★), 158 20/94 tons.
*Commerce* % 1803-1807 Tender 4 x 3. 116 51/94 tons.
*Concord* 1803-1810 Tender 6 x 3. 152 70/94 tons. At Dublin.
*Constance* See entry in previous chapter on French Revolutionary War.
*Constitution* % 1804-1804 Cutter 10. 120 24/94 tons. 26.8.1804 sunk by a mortar shell off Ambleteuse.
*Conway Castle* % 1804-1809 [Irish Gun vessel] 53 66/94 tons.
*Countess of Elgin* 1803-1814 Cutter 8 x 3. 25 men. 77 88/94 tons. Carvel built. Owned by H Latham.
*Courier* % 1804-1806 Cutter 12 x 4. 114 57/94 tons.
*Cricket* % 1808-1808 Ketch. 96 73/94 tons. Lost 31.10.1808.
*Cruizer* % 1804-1808 _____ .
*Dart* 1803-1805 Cutter 6 x 3. 21 men. 55 83/94 tons. Owned by Minet & Feetor (?)
*Dart* % 1805-1814 Smack. 56 tons.
★*Diadem* 1804-1806 Armed ship 16.
★*Diana* 1804-1805 Armed ship 10. 179 tons.
*Dolly* 1803-1803 Cutter 6 x 3. 22 men. 60 72/94 tons. Owned by Minet & Feetor (?).
*Dover* % 1808-1810 Cutter 4 (2 x 3 + 2 x 12★). 47 tons.
*Drake* % 1804-1805 Cutter 12 (2 x 6 + 10 x 12★). 129 87/94 tons.
*Dublin* 1803-1810 [Irish Gun vessel] 5 men. 73 tons.
★*Duchess of Bedford* 1804-1808 Armed ship 16; stationed at Hoseley Bay [?]
*Duchess of Cumberland* 1803-1805 Cutter 6 x 3. 23 men. 65 82/94 tons. Carvel-built. Owned by H Latham.
★*Duckworth* 1803-1803 _____ .
★*Duke Bronte* 1803-1803 _____ .
*Duke of Clarence* 1803-1804 Cutter 6 x 3. 23 men. 65 tons. Owned by Messrs Le Mesurier. Lost 24.11.1804 by hitting a rock whilst in chase off the coast of Normandy.
★*Duke of Cumberland* 1803-1805 Cutter 6. 60 tons.
★*Duke of Cumberland* 1803-1805 Brig hired as a packet.
*Duke of Kent* % 1804-1805 Armed ship 14 x 18★. 272 tons.
*Duke of York* 1803-1810 Cutter 8 x 4. 27/30 men. 62 27/94 tons. Owned by Ward.
*Earl Saint Vincent* % 1804-1806 Cutter 14 x 12★. 194 16/94 tons.
*Earl Spencer* 1803-1814 Cutter 12 x 12★. 42 men. 141 30/94 tons.
*Ebrington* % 1805-1808 Tender 10. 105 tons.
*Eliza* 1803-1812 Cutter (tender) 6 x 3. 185 tons. Tender at 'Shields'.
*Eliza* % 1806-1811 Tender. 140 61/94 tons.
*Eliza* % 1804-1814 Smack 8. 45 tons.
*Eliza and Jane* % 1803-1812 Tender 4 x 3. 110 11/94 tons.
*Elizabeth* % 1805-1812 Tender 10. 161 22/94 tons.
*Elizabeth* % 1808-1809 Tender 10. 204 58/94 tons.
*Elizabeth* % 1809-1809 Smack.
*Elizabeth and Mary* % 1804-1814 Smack. 45 tons.
*Ellen* % 1804-1809 Smack. 45 tons.
*Ellens* 1803-1809 [Irish Gun vessel] 5 men. 76 tons.
★*Emma* 1809-1811 Armed ship.
*Endeavour* % 1804-1812 Brig 12 x 18★. 168 66/94 tons.
*Esdaile* % 1808; 1809 Transport 14. 210 tons.
*Expedition* 1803-1804 Tender. 70 tons. At Waterford.
★*Fame* 1806-1814 Schooner 8.
*Fancy* % 1808-1809 Cutter 43 tons.
*Fancy* % 1809-1811 Cutter 10 x 12★. 111 80/94 tons.

*Fancy* % 1811-1814 Cutter 10. 117 59/94 tons.
*Fanny* % 1804-1804 Brig 12. 181 tons.
★*Fanny* 1805-1807 Lugger 12.
*Favourite* % 1803-1806 Tender 4 x 2. 66 tons.
*Favourite* 1803-1804 *Florence* -1806 Cutter 8 x 3. 25 men. 76 18/94 tons. Carvel-built. Owned by H Latham
*Favourite*★ 1807-1811 Cutter 6 x 3. 72 38/94 tons.
*Ferret* % 1808-1808 Lugger. 68 52/94 tons. Taken 22.11.1808.
*Findon* % 1804-1810 Cutter/transport 10. 127 tons.
*Flirt* 1803-1806 Cutter 10 x 12★. 40 men. 118 86/94 tons.
*Flora* % 1809-1809 Smack.
*Fly* % 1804-1808 Cutter 10 x 12★. 83 18/94 tons.
*Fly By Night* 1804-1804 Cutter (lugger?) 6 x 12★. 24 men. 71 62/94 tons.
*Flying Fish* % 1804-1804 *Gertrude* -1804 Schooner 12 x 12★. 147 44/94 tons. 16.12.1804 sank in Channel collision with the frigate *Aigle*.
*Flying Fish* % 1809-1814 Schooner 6 x 3. 74 tons.
*Flying Fish*★ 1813-1814 Schooner, 78 tons, purchased 1817 [see next chapter].
*Folkestone*★ 1804-1805/1807-1814 Lugger 12 x 12★. 131 36/94 tons.
*Fox* 1803-1804 *Frisk* -1806 Cutter 6 x 4. 30 men. 99 73/94 tons. Clinker built. Owned by J King.
*Fox* 1803-1804 Cutter 8. 30 men. 95 2/94 tons. Clinker-built. Owned by Iggulden.
*Fox* 1803-1805 Cutter 8 x 4. 30 men. 98 38/94 tons. Clinker-built. Owned by Collett & Thomsett.
*Fox* % 1812-1812 Cutter.
*Friendship* 1803-1805/1807-1807 Tender 6 x 3. 174 64/94 tons. At Liverpool.
*Gambier* % 1808-1814 Cutter 12/10 x 12★. 109 87/94 tons.
*General Coote* % 1804-1804 Lugger 6. 49 41/94 tons.
*Gleaner* % 1808-1814 Ketch tender. 14 x 12★. 153 79/94.
*Glory* 1804-1809 [Irish Gun vessel] 4 men. 72 tons.
*Green Linnet* 1803-1809 [Irish Gun vessel] 5 men. 76 tons.
*Griffin* 1803-1805 Cutter 6 x 3. 24 men. 70 6/94 tons. Owned by J King.
*Hannah* 1803-1809 [Irish Gun vessel] 4 men. 59 tons.
★*Hannibal* 1804-1804 Armed ship 16. 11.1804 wrecked near Sandown Castle, Isle of Wight.
*Happy Return* % 1804-1805 Cutter 10 x 12★, 93 74/94 tons.
*Harlequin* % 1804-1809 Armed ship 18 (10 x 6 + 8 x 12★). 185 27/94 tons; 7.12.1809 went aground in fog and wrecked near Seaford.
*Hawke* 1803-1805 Cutter 10 x 12★. 40 men, 124 29/94 tons.
*Hebe* % 1804-1812 Armed ship 16 x 18★. 266 87/94 tons.
*Hero* % 1804-1805 Cutter 8 x 12★. 72 90/94 tons.
*Hero* % 1809-1811 Cutter 12. 109 58/94 tons, ?23.4.1811 captured and sunk by Danish gunboats.
*Hind* 1803-1804 Cutter 8 x 4. 30 men. 120 tons. Clinker-built. Owned by Collett & Thomsett.
*Hind* % 1804-1805 Armed ship 14 x 12★. 265 tons.
*Hope* % 1803-1807 Tender 4 x 3. 124 74/24 tons.
*Hope* % 1803-1811 Tender 4 x 2. 80 13/94 tons.
*Hope* 1803-1805 Cutter 8 x 12★. 30 men. 84 3/94 tons.
*Hope* % 1809-1809 Smack.
*Humber* % 1804-1811 Armed ship 16 (2 x 6 + 14 x 12★). 257 52/94 tons.
*Idas* % 1808-1812 Cutter 10/12 x 6. 142 tons.
*Idas* % 1809-1810 Cutter 10. 102 18/94 tons. Taken 4.6.1810 after grounding at the entrance to the Scheldt.
*Industry* 1803-1809 [Irish Gun vessel] 5 men, 78 tons.
*Industry* % 1804-1804 *Rhoda* -1804 Cutter 6 x 12★. 45 88/94 tons.
*Industry* % 1804-1804 *Adrian* -1810 Cutter 8 x 12★. 84 15/94 tons.
*James* % 1803-1807 Tender 4 x 2. 86 tons.
★*Jason* 1803 Cutter.
*John* % 1804-1805 Armed ship 14 x 18★. 244 tons.
*John and Mary* 1803-1803 [Irish Gun vessel] 5 men.
*John Bull* % 1804-1806 Cutter 16 x 12★. 119 36/94 tons.
*Joseph* 1803-1805 Cutter 8 x 4. 30 men. 102 tons. Clinker-built. Owned by Gilbee.
*Joseph* % 1807-1809 Cutter. 100 tons.
*Kate Karney* % 1808-1808 Brig. 149 68/94 tons.
*Kendal* 1809-1810 [Irish Gun vessel] 5 men. 77 tons.
*Kent* 1803-1804 Cutter. 121 tons.

*King George* 1803-1804 *Georgiana* -1804 Cutter 8. 40 men. 129 27/94 tons; 25.9.1804 grounded and set afire to avoid capture near Harfleur.
★*Kingston* 1804-1805 Armed ship 16.
*Kitty* % 1804-1805 Armed ship 16 x 18★. 310 tons.
*Lady Melville* % 1804-1805 Armed ship 14 x 18★. 237 17/94 tons.
*Lady Warren* % 1804-1807 Armed ship 32 x 18★. 315 tons.
*Leith* % 1804-1805 Armed ship 16 (6 x 6 + 10 x 18★). 294 85/94 tons.
*Lion* % 1804-1805 Cutter 8 x 12★. 86 66/94 tons.
*Lion* % 1804-1805 Armed ship 14 x 18★. 280 33/94 tons.
*Lively* % 1805-1805 Cutter 12. 120 tons. Purchased later as *Sprightly* (see cutters earlier in this chapter).
*Lord Cochrane* % 1808-1814 Brig 14 (2 x 6 + 12 x 12★) 103 15/94 tons.
★*Lord Eldon* 1804-1807 Armed ship 16. Captured by Spanish gunboats 12.11.1804 and then recaptured shortly afterwards.
*Lord Melville* % 1804-1811 Tender 4 x 2. 103 77/94 tons.
*Lord Nelson* 1803-1804 *Frederick* -1804 Cutter 6 x 4. 24 men. 67 74/94 tons. Clinker-built. Owned by J King.
*Lord Nelson* % 1804-1804 *Julia* -1804/5 Schooner 12 x 12★. 156 46/94 tons. 24.12.1804 (or 14.1.1805) wrecked on the Castle Rocks leaving Portsmouth.
*Lord Nelson* % 1804-1808 Armed ship 16. Taken 11.1804 and retaken later the same year (not noted in Gosset).
*Lord Nelson* % 1807-1809 Cutter 8 x 12★. 68 54/94 tons. 5 or 15.8.1809 wrecked near Flushing (Vlissingen).
★*Lord Sidmouth* 1812-1812 ____ .
*Louisa* 1803-1811 Tender 4 x 3. 120 1/94 tons. At Cork.
*Lyre* or *Lyra* 1803-1812 Tender 4 x 2. 111 68/94 tons. At Liverpool, then Dublin?
*Magdalen* % 1804-1805 Armed ship 14 x 18★. 215 5/94 tons.
★*Majestic* 1804-1805 Armed ship 16.
★*Manchester* 1814-1815 Storeship.
*Margaret* 1803-1809 [Irish Gun vessel] 4 men. 60 40/94 tons.
*Margaret and Ann* 1802-1807 [Irish Gun vessel] 4 men. 60 40/94 tons.
*Maria* 1803-1805 Tender 4 x 3. 153 57/94 tons. At Cork or Belfast (see below).
*Maria* 1803-1812 Tender 6 x 3. 176 23/94 tons. At Belfast or Cork (see above).
*Maria* % 1813-1814 Smack.
*Maria* % 1813-1814 Schooner 104 12/94 tons.
★*Mars* 1804-1805 Armed ship 16.
*Mary* 1803-1808 Tender 6 x 3. 211 tons. At Leith.
*Mary* % 1804-1807 Tender 6 x 12★. 100 tons.
*Mary* 1803-1804 Cutter 8 x 12★. 30 men. 105 tons. This is the only cutter hired in 1803 which is noted as being coppered. Owned by Bentham.
*Mary* % 1809-1812 Cutter 8 x 4. 79 23/94 tons.
*Mary* 1804-1804 *Marcia* -1804 Schooner 6 x 12★. 73 25/94 tons.
*Mary and Ellen* % 1804-1810 [Irish Gun vessel] 75 tons.
*Marys* 1803-1810 [Irish Gun vessel] 4 men. 59 tons.
*Mercator* % 1804-1805 Armed ship 14 x 18★. 257 tons.
*Minerva* 1803-1804 Cutter 6 x 3. 24 men. 67 92/94 tons Owned by H Latham.
*Minerva* % 1804-1805 Tender 4 x 2. 87 21/94 tons.
*Morriston* % 1804-1812 Brig 12 x 18★. 164 tons.
*Mother Ramsay* 1803- ____ ? tender at Dublin.
*Nancy* % 1804-1804 *Frances* -1805 Cutter 6 x 4. 46 27/94 tons.
*Nancy* % 1804-1804 *Chance* -1806 Cutter 10 (2 x 3 + 8 x 12★). 121 15/94 tons.
*Nelson* % 1804-1805 Cutter 10 (2 x 6 + 8 x 12★). 124 44/94 tons.
*Nepean* 1809-1809 Smack.
*Neptune* % 1803-1812 Tender (sloop-rigged?) 4 x 3. 113 39/94 tons.
*Neptune* % 1803-1812 Tender (sloop-rigged?) 4 x 3. 107 44/94 tons.
★*Neptune* 1807-1814 Tender 10.
*Nile* % 1804-1806 Lugger 14 x 12★. 170 22/94 tons.
*Nile* % 1804-1805/1806-1806 Cutter 12 (2 x 6 + 10 x 12★). 166 20/94 tons.
★*Nimble* 1803-1814 Cutter 6 x 4. 70 tons.
*Nimrod* 1803-1808 Cutter 6 x 3. 24 men. 69 78/94 tons. Clinker-built; owned by Jos Sladen.
*Nimrod* % 1808-1814 Cutter 6 x 3. 75 62/94 tons.
★*Norfolk* 1804-1808 Armed ship 14.
★*Norfolk* 1809-1814 Cutter 8.
*Nymph* (or *Nymphe*) 1803-1804/1807-1814 Cutter 8 x 4. 23/29 men. 63 38/94 tons. Carvel-built. Owned by Collett & Thomsett.

***Osborne** 1804-1805 Armed ship 16.
**Ox** % 1807-1807 Lugger 6. 55 73/94 tons.
***Paragon** 1804-1805 Armed ship 16.
**Peggy** % 1804-1804 _____ .
***Perseus** 1804-1805 Armed ship 16.
**Perseverance** % 1806-1808 Tender. 160 tons.
***Phoebe** 1810-1814 Cutter 12.
**Phoenix** 1803-1804 Cutter 8 x 3. 25/26 men. 79 24/94 tons. Clinker-built. Owned by Collett & Thomsett.
**Phoenix** 1804-1804 _____ ??
**Phoenix** 1810-1814 Tender 10.
***Pitt** 1809-1812 Brig 12.
**Pleasant Hill** % 1804-1810 Transport. 75 tons.
**Pocock** % 1808-1809 Cutter 4. 43 tons.
**Poll** % 1808-1809 Cutter 4. 43 tons.
***Porcupine** 1805-1807 Schooner.
***Porpoise** 1804-1805 Armed ship 16.
**Pretty Lass** % 1804-1805 Armed ship 14 x 18★. 259 2/94 tons.
**Prince of Orange** % 1804-1805 Tender.
**Prince William** % 1804-1812 Armed ship 14 x 18★. 307 38/94 tons.
**Princess Augusta** 1803-1814 Cutter 8 x 3 (4?). 25/26 men. 73 (70 55/94?) tons. Carvel built. Owned by J King.
**Princess Charlotte** % 1804-1805 Schooner 8 x 12★. 95 64/94 tons.
**Princess of Wales** 1803-1814 Cutter 10 x 12★. 36 men. 105 63/94 tons.
***Princess Royal** 1804-1805 Cutter.
**Providence** % 1804-1804 _____ .
**Providence** % 1804-1812 Armed ship 14 x 18★. 291 5/94 tons.
**Queen** % 1804-1805 Cutter. 130 65/94 tons.
**Queen Charlotte** 1803-1805/1807-1814 Cutter 8 x 4 (or 2 x 12★ + 8 x 4). 24/25 men. 75 14/94 tons. Carvel-built. Owned by Collett & Thomsett.
***Queen Charlotte** 1803-1812 Cutter 6 x 12★. 60 tons.
**Queen Charlotte** 1804-1805 Cutter 12. 140 tons.
**Queenborough** % 1804-1805 Cutter 12 (2 x 4 + 10 x 12★). 181 69/94 tons.
**Ranger** % 1810-1810 Armed ship 16.
**Rebecca** 1804-1810 [Irish Gun vessel] 4 men. 50 tons.
***Redesdale** 1809-1810 Cutter 8.
***Reliance** 1812-1814 Tender 12.
**Resolution** % 1807-1814 Cutter 8 x 4, 86 5/94 tons.
**Reward** % 1804-1806 _____ .
**Rosa** 1804-1812 Tender 4 x 3. 154 57/94 tons. At Liverpool.
**Rose** 1803-1804 Cutter 6 x 3. 20 men. 52 5/94 tons. Owned by J King.
**Rose** % 1804-1804 **Harriet** -1804 Cutter 4 x 12★, 44 71/94 tons.
**Rose** % 1804-1804 **Beaumont** -1805 Cutter 10 (2 x 4 + 8 x 18★). 104 75/94 tons.
**Rosina** % 1804-1805 Armed ship 14 x 18★. 315 tons.
**Rover** % 1804-1804 Tender. 100 tons.
**Rowena** % 1808-1809 Brig 14. 150 83/94 tons.
***Royalist** 1804-1805 Armed ship 16.
***Saint Vincent** 1803-1806 Cutter 14 (first hired 1800).
**Sally** % 1804-1807 Armed ship 14 x 18★. 310 tons.
**Sally** % 1809-1809 Smack.

**Samuel Braddick** % 1804-1804? **Braddick** -1805 Armed ship 14 x 18★. 241 tons.
**Sandwich** % 1804-1804/1808-1814 Lugger 12 x 12★. 166 59/94 tons.
**Saumerez** % 1804-1805 Cutter 10 x 12★. 106 12/94 tons.
***Sceptre** 1805-1806 Armed ship 16.
**Sheerness** 1803-1805 Cutter 8 x 4. 30 men. 103 tons. Owned by Bentham.
***Sir Andrew Mitchell** 1804-1805 Armed ship 16.
**Sir Thomas Troubridge** % 1804-1806 Armed ship 26 (18 x 6 + 8 x 18★). 315 tons.
**Speculator** 1803-1813 Cutter (lugger?) 10 x 4. 33 men. 93 53/94 tons.
**Speedwell** % 1804-1804 _____ .
**Spider** 1803-1804 Cutter 10 x 12★. 38 men, 114 1/94 tons.
**Sprightly** 1803-1809 [Irish Gun vessel] 4 men. 60 tons.
**Stag** % 1804-1804 Cutter 6 x 3. 57 25/94 tons.
***Suffolk** 1804-1814 Armed ship 16.
**Susannah** % 1804-1804 Fire vessel; 8.12.1804 expended at Calais.
**Sussex Oak** % 1804-1808 Transport/tender. 124 24/94 tons. 02.1808 taken by the French.
**Swan** 1803-1805/1807-1811 Cutter 10 x 12★. 40 men. 119 27/94 tons; 25.4.1811 taken by three Danish gunboats in 'the Sleeve' then sank.
**Swift** 1803-1806 cutter 8 x 12★, 30 men. 100 30/94 tons. Clinker built. Owned by Minet & Feetor (?).
**Swift** 1803-1804 Cutter 8 x 4. 25/26 men, 76 82/94 tons. Owned by Blake.
**Sylph** % 1803-1809 Cutter 19 24/94 tons.
**Telemachus** 1803-1804 + Cutter 10. 40 tons.
**Thames** % 1804-1805 Armed ship 10 x 18★. 118 50/94 tons.
**Thomas and Eleanor** % 1804-1805 _____ .
**Tigre** % 1809-1809 Cutter 46 23/94 tons.
***Traveller** 1804-1805 Armed ship 16.
***Troubridge** 1804-1810 Armed ship 16.
**True Blue** % 1807-1809 Tender 8. 114 67/94 tons.
***True Briton** 1812-1814 Schooner 12. 183 tons.
**Tryall** 1803-1807 [Irish Gun vessel] 5 men. 79 tons.
**Two Friends** 1803-1807 Tender 4 x 3. 145 76/94 tons. At Liverpool.
**Union** 1803-1809 [Irish Gun vessel] 7 men. 106 tons.
**Union** % 1809-1811 Transport. 130 23/94 tons.
**United Brothers** 1803-1806/7 Tender 4 x 3. 143 11/94 tons. At Swansea. Taken 18.12.1806 or 6.1.1807 by a French privateer off the Lizard.
**Unity** 1804-1810 [Irish Gun vessel] 5 men. 67 tons.
**Venus** 1803-1806 Cutter 10 x 4. 20 men, 50 68/94 tons. Taken 31.1.1806 by French *L'Ami-Nationale*.
**Venus** 1803-1804 **Arthur** -1805 Cutter 6 x 4 (3?). 24 men. 70 89/94 tons. Carvel-built. Owned by Collett and Thomsett, Taken 19.1.1805 by the French in the Mediterranean.
**Venus** 1804-1804 *Agnes* -1806 cutter (lugger?) 6 x 12★. 23 men. 66 87/94 tons. 25.3.1806 lost off the Texel.
**Venus Packet** % 1815-1815 _____ 77 84/94 tons.
***Walker** 1804-1805 Armed ship 16.
**Weazle** % 1808-1811 Schooner 6 x 6★, 69 87/94 tons.
***Wellington** 1804-1805 Armed ship 16.
**Ythan** % 1809-1809 Smack.

# Chapter 8: The Period After 1815

The activities of the Royal Navy during the post-Napoleonic era show up in the origins of the vessels listed here: line of battle ships presented by Eastern potentates, captured slavers, vessels purchased for survey work or for polar exploration. The fact that such a small number of vessels had to be purchased into the Navy is in itself a commentary on the way that the RN maintained its predominance as the most powerful maritime force the world had known with comparatively little effort during this period, despite occasional scares and panics, mostly occasioned by French rivalry, though some anxiety was experienced at American growth. By the time war did come with a European enemy it was in alliance with France, at a time when sail was taking a second place to steam in ships intended to fight. It also happened when British industrial development was, for the time being, making her naval predominance even more secure.[1] The requirements of the Russian ('Crimean') War for a large and specialised shore bombardment fleet were largely met by purpose building during that war rather than by mercantile purchase; the Russian fleet kept itself in harbour, thus preventing the Allies taking any substantial number of prizes.

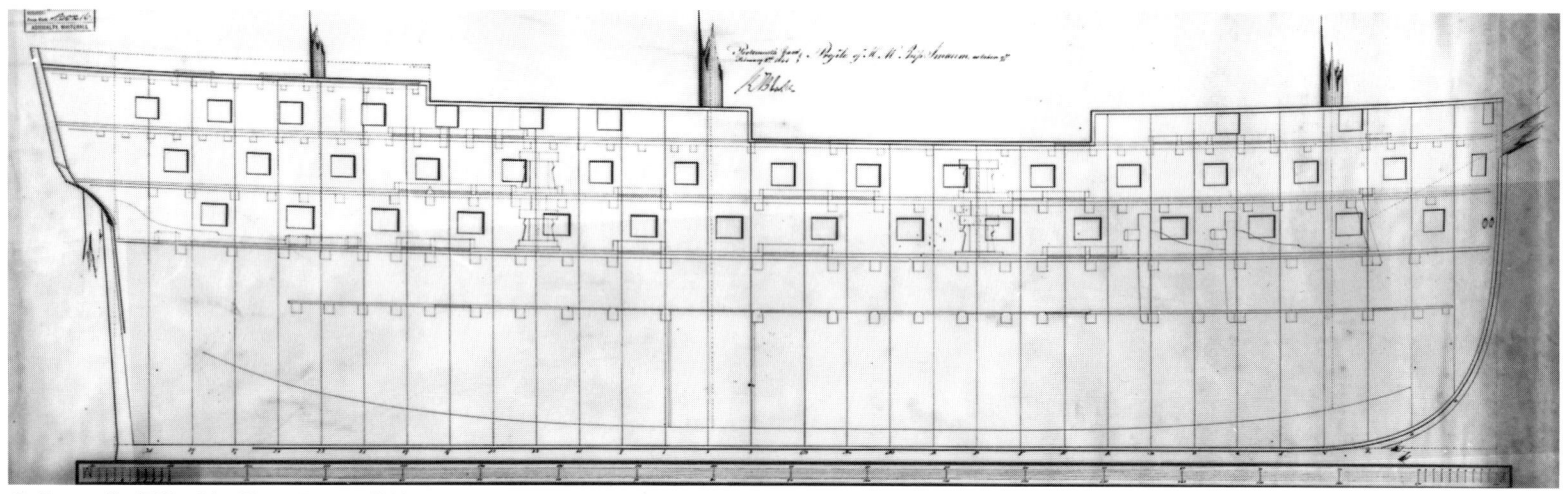

Outline profile (799) of the 70 gun *Imaum* of 1836

## 74 Gun Ship (Third Rate)

**Hastings** 1819-1870 (East India Company ship *Hastings* 1818 Calcutta).
Dimensions & tons: 176'10½", 145'4" x 48'6" x 21'. 1763.
Men: 600. Guns: GD 28 x 32, UD 28 x 18, QD 4 x 12 + 10 x 32★, Fc 2 x 12 + 2 x 32★. Purchased from the East India Co on her arrival in Britain 1819; 1855 converted to screw line of battle ship; 1857 to the Coastguard; 1870 hulked [ at ____ ? as a coal hulk; 1885 sold].

## 70 Gun Ship (Third Rate)

**Imaum** 1836-1842 (Imam of Muscat's *Liverpool* 1826 East India Co, Bombay).
Dimensions & tons: 177', 145'5¾" x 48'4" x 21'. 1776.
Men: ____ . Guns: ____ . Presented to the RN by the Imam of Muscat 1836; 1842 hulked [at Jamaica as a Receiving Ship; 1862/1866 BU].
PLANS: Lines & profile/profile/orlop/gun deck/upper deck/quarterdeck & forecastle.

## Brigs

**Bathurst** 1821-1824 ( ____ ____ India?) brig sloop/brig (teak built) ____ .
Dimensions & tons: ____ . 170.
Men: ____ . Guns: ____ . Purchased at Port Jackson, Australia, in 1821; 1824 to the Coast Blockade [ to lie in the Swale; 11.4.1858 sold to Castle, Charlton, for BU].

**Pantaloon** 1831-1852 (Duke of Portland's yacht *Pantaloon* 1831 Troon, Symonds design) brig 10.
Dimensions & tons: 90', 71'4¼" x 29'4¼" x 12'8". 323.
Men: 80/68/60. Guns: 2 x 6 + 8 x 18★. Purchased 5.12.1831; 1852 BU.

**Waterwitch** 1834-1861 (Earl of Belfast's yacht *Waterwitch* ____ . White, Cowes) brig 10.
Dimensions & tons: 90'6", 71'7" x 29'4" x 12'8". 324.
Men: 55/60. Guns: 2 x 6 + 8 x 18★. Purchased 10.1834; 1861 sold.
PLANS: Lines & profile/profile/hold/profile & decks.

**Royalist** 1841-1857 (mercantile *Mary Gordon* 1839 Bombay) brig/survey vessel (teak-built) ____ .
Dimensions & tons: 87'4", ____ x 25'6" x 6'8". 250.
Men: 60. Guns: UD 4 x 12, QD 2 x 6, Fc 2 x 6. Purchased in China 9.7.1841; 1857 hulked [for the River Police; lay off Somerset House; 1895 sold].

**Pickle** 1852-1854? (slaver *Eolo* ____ ) brig ____ .
Dimensions & tons: ____ . ____ .
Men: ____ . Guns: ____ . Taken by *Orestes* 1852; 1854? sold.

## Schooners

**Nimble** 1826-1834 ( ____ *Bolivar* 1822 Jamaica) schooner 5.
Dimensions & tons: 83'7", 64'7⅝" x 22'2" x 9'5". 168.
Men: 41. Guns: 4 x 18★ + 1 x 18. Purchased 1826; 4.11.1834 wrecked on a reef off Cuba.

**Adelaide** 1827-1833 (slaver *Bella Josephine* 1823 Sardinia) ____ .
Dimensions & tons: 66'8", ____ x 18'7" x ____ . 95.
Men: ____ . Guns: ____ . Purchased 1827; tender to flagship at Rio de Janeiro; 1833 sold at Rio.

**Black Joke** 1827-1832 (slaver *Henriquetta* ____ Baltimore) ____ .
Dimensions & tons: ____ ____ .
Men: ____ . Guns: ____ . Acquired 1827; 1832 scuttled.

---

[1] It was at the end of the 1850s, and of our period, that the classic example of industrial leadership enhancing naval predominance occurred. The French built the wooden hulled *Gloire* as the first seagoing ironclad. The British answer was the *Warrior*, much larger, faster and more powerfully armed. With her sister *Black Prince* she is perhaps the only clear example of a warship design making all other ones instantly obsolete and being potentially able to deal with all ships built before her.

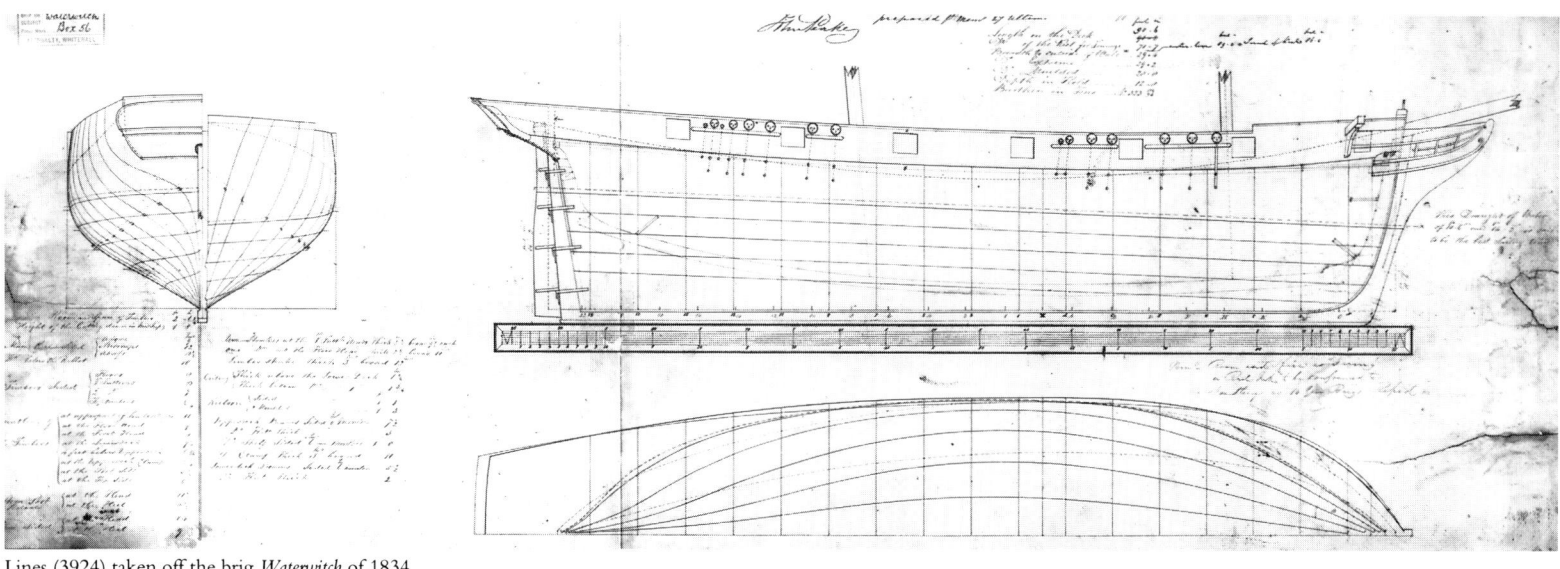

Lines (3924) taken off the brig *Waterwitch* of 1834

*Fair Rosamond* 1831-1845 (slaver *Dos Amigos* ____ Baltimore) 2.
  Dimensions & tons: 74'11½", 64'1½" x 23'2½" x 10'4½". 172.
  Men: 40. Guns: 1 x 18 (bored up) + 1 x 18★. Purchased on the coast of Africa after capture 1831; 1845 BU.
PLANS: Lines/profile & decks.

*Prompt* 1840/1842-1845 (Portuguese *Josephine* ____ ____ ) ____ .
  Dimensions & tons: ____ ____ .
  Men: ____ . Guns: ____ . Taken? 1840?; 22.1.1842 purchased; 1845 sold.

*Fawn* 1839-1842 (Slaver *Caroline* 1835 ____ ) 3?
  Dimensions & tons: 75', 60' x 22'10" x 7'9". 169.
  Men: 60/40 Guns: ?2 x 24 + ?1 x 32. Purchased at Rio de Janeiro in 1839; 1842 (or 1840) became a tank vessel at the Cape of Good Hope; sold to the Natal Colonial Government.

*Young Hebe* 1843-1850 (mercantile ____ ) survey schooner ____ .
  Dimensions & tons: ____ 45.
  Men: 16 Guns: ____ . Purchased 1843; on surveying duties in China; 1850 sold.

*Puck* 1855-1856 schooner gunboat 1 x 32 purchased at Constantinople 08.1856 foundered on passage from Balaklava to be handed over to the Turks.

*Kingston* 1858-1867 (slaver *Orestes* ____ ) schooner/tender.
  Dimensions & tons: ____ ____ .
  Men: ____ . Guns: ____ . Detained by *Forward* 1858; tender to *Imaum* at Jamaica; 1861 appropriated as a bathing place for the crews of ships at Jamaica, being unfit for service; 1865/1867 BU at Jamaica.

*Cuba* 1857-1867 (slaver ____ ____ ) schooner/tender.
  Dimensions & tons: ____ . 67.
  Men: ____ . Guns: ____ . Detained by *Arab* and condemned at Jamaica 1857; tender to *Imaum*; 1862 BU.

*Rose* 1857-1862 (yacht *Lord Star*? 1846 ____ Ardrossan) yacht.
  Dimensions & tons: ____ . 37.
  Men: ____ . Guns: ____ . Purchased 1857; 1862 sold.

## Cutters

*Quail* 1817-1822 *Providence* -1829 (transferred revenue cruiser *Providence* 1804 ____ Dover) tender.
  Dimensions & tons: 54'2", 38'6½" x 20'2" x 8'2½". 82.
  Men: ____ . Guns: ____ . Transferred from the Revenue 1817; 1822 renamed *Providence* (but then renamed *Quail* again?); 1829 BU.

*Emerald* 1820-1847 (built Portsmouth [by Sainty?] 1820) tender.
  Dimensions & tons: 57'3", 45'8¾" x 18'9½" x 9'1". 86.
  Men: ____ . Guns: ____ . Purchased 1820 as tender to the flagship [at Portsmouth?]; 1847 BU.

*Gossamer* 1823-1861 (purchased whilst building at Gosport) tender.
  Dimensions & tons: 46'4", 35'4⅞" x 16'0⅞" x 7'11". 48.
  Men: ____ . Guns: ____ . Purchased 1823 as tender to the flagship at Sheerness; 1861 out of service.

*Netley* 1826-1859 (built at Plymouth as a revenue cruiser in 1823) tender. 1.
  Dimensions & tons: 63'9", 46'11½" x 22'3¼" x 9'6". 122.
  Men: 65. Guns: 1 x 6. Begun 06.1821; launched 13.3.1823; 1826 transferred as tender at Plymouth; 10.1848 foundered at Spithead; raised; 1850 buoy boat for Halifax, Nova Scotia; 1850-1851 at Bermuda; 1851-1852 at Jamaica; 1859 sold at St Johns, Newfoundland.

*Hind* 1827?-1844 (transferred revenue cruiser *Hind* 1790 Cowes) tender.
  Dimensions & tons: 71'8", 52'3¾" x 24'3¾" x 9'7¼". 191.
  Men: ____ . Guns: ____ . Transferred 1827?; became tender to the Plymouth Ordinary; 1844 sold.

*Cracker* 1826-1842 (purchased whilst building at Gosport) tender.
  Dimensions & tons: 50'2", 37'3" x 16'5" x 7'11½". 54.
  Men: ____ . Guns: ____ . Purchased 1826; tender to *Seaflower* at Portsmouth; 1842 sold.

*Cerus* 1830-1860 *Ceres* -1877 (ex buoy boat *Ceres* at Devonport?) tender.
  Dimensions & tons: 38'5½", ____ x 12'8½" x 7'8". 25.
  Men: ____ . Guns: ____ . Purchased 1830 from Mr Little of Anderton as tender for the Sheerness Ordinary; 1845 to Woolwich; 1853 to Portsmouth; 1858 to Plymouth; 1860 renamed *Ceres*; 1862 to Harbour Master at Falmouth; later Devonport Yard Craft No 3; 1877 sold.

*Monkey* 1831-1833 (mercantile *Courier* 1827 Bermuda) tender.
  Dimensions & tons: 56', ____ x 17'10" x 9'4". 68.
  Men: 35. Guns: 2 x 12 + 1 x 5½" howitzer. Purchased (at Bermuda?) 1831; 08.1833 [?] sold.

*Louisa* 1835-1841 (mercantile ____ ) tender.
  Dimensions & tons: ____ ____ .
  Men: ____ . Guns: ____ . Purchased at Canton as tender to the British Factory there; 21.7.1841 wrecked on an island between Macao and Hong Kong during a typhoon. The crew had to promise a ransom to the local pirates before being allowed to escape.

*Adelaide* 1848-1850 (slaver ____ 3 or 4 years old) cutter/tender.
  Dimensions & tons: ____ ____ . 140.
  Men: ____ . Guns: ____ . Taken ____ ; purchased 1848; 9.10.1850 wrecked on Banana Island, West Africa; 9.11.1850 remains sold.

## Storeships, etc

*Dispatch* 1816?-1826 (ex *Cornwallis* 1798? ____ ) brig transport.
  Dimensions & tons: ____ ____ .
  Men: ____ . Guns: ____ . Transferred by the Transport Office Ca 1816; 1826 hulked [at Bermuda as a receiving and slop ship (also sheer hulk?); ?later converted to a diving bell ship?; 1846 wrecked at the Naval Wells then BU.

*Marchioness of Queensbury* 1829-1830 (Post Office packet *Marchioness of Queensbury* ____ ) packet.

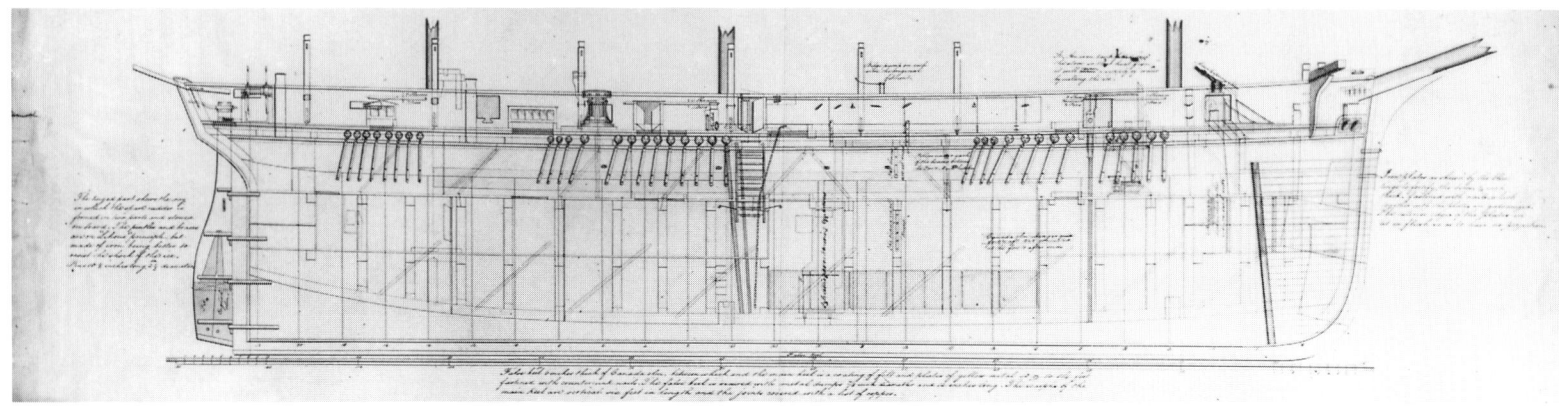

Profile (6382) as fitted for the Arctic Expedition in 1848 for the storeship *Enterprize*

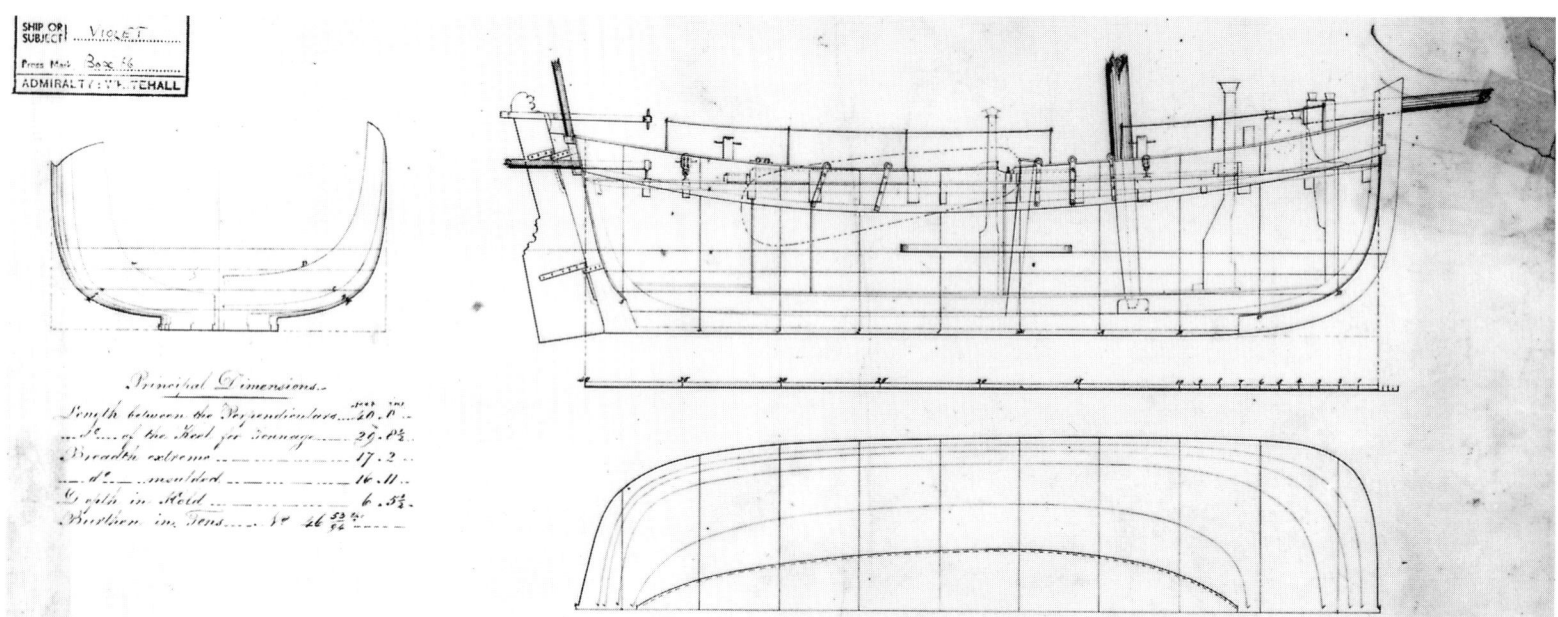

Lines and profile (6817) of the *Violet* acquired as a surveying tender whilst building in Holland

Dimensions & tons: ____ ____ .
Men: ____ . Guns: ____ . Purchased 1829; 1830 police service, to lie at Gibraltar; fate uncertain.

**Nightingale** 1829-1839 (Post Office packet *Marchioness of Salisbury* ____ ) packet brig/storeship 6.
Dimensions & tons: 82'1½", 66'5⅛" x 23'10" x 14'11". 198 2/94.
Men: ____ . Guns: 6 x ____ . Purchased 1829; 1839 hulked [as a depot at Plymouth; 1842 sold].
PLANS: Lines & profile/lower deck/upper deck.

**Kestrel** 1846-1852 (mercantile *Kestrel* 1837 White, Cowes) packet.
Dimensions & tons: ____ , 62'6" x 23'7" x 11'3". 202.
Men: ____ . Guns: ____ . Purchased 1846; 1852 BU.

**Renira** 1847-1850 (mercantile ____ ) ____ ?
Dimensions & tons: ____ ____ . 86.
Men: ____ . Guns: ____ . Purchased 1847; 1850 sold.

**Enterprise** 1848-1860 (mercantile ____ 1848 Wigram, Blackwall) Arctic service.
Dimensions & tons: 125'7½", 108'6" x 28'8½" x 20'. 471.
Men: ____ . Purchased 1848 and fitted for Arctic expedition; 1860 coal depot [1903 sold].

**Investigator** 1848-1853 (mercantile ____ 1848 Scott, Greenock) Arctic service.
Dimensions & tons: 118', 101'2½" x 28'3" x 18'11". 422.
Men: ____ . Purchased 1848 & fitted for Arctic expedition at Green's yard, Blackwall 3.6.1853 abandoned in the ice.
PLANS: Profile/orlop/lower deck/upper deck/midships section.

**Resolute** 1850-1879 (mercantile *Ptarmigan* 1849 Smith [?] 'Shields') Arctic service, barque. Dimensions & tons: 115', 111' x 28'4" x 11'6", 422.

Men: ____ . Purchased from Mr Smith 1850; fitted at Blackwall for Arctic 15.5.1854 abandoned in the ice; 1856 discovered free of the ice by an American whaler; 1857 back in service having been purchased by the USGovernment and presented by the President to Queen Victoria; 1879 BU. [A desk was made from her timbers and presented by Queen Victoria to the President of the USA; it is still in use in the White House.]
PLANS: Profile/hold /lower deck/upper deck/desk [see above].

**Assistance** 1850-1854 (mercantile *Baboo* teak built 1835 at Howrah, Calcutta) Arctic service, barque 2.
Dimensions & tons: 117'4". 115'7" x 28'5" x 13'7". 423.
Men: 58 Guns: 2 x ____ . Purchased from Mr Kincade 1850; 25.8.1854 abandoned in the ice.
PLANS: (As *Baboo*) profile/decks.

**Sophia** 1850-1853 (mercantile *Sophia* 1850? ____ Dundee) Arctic service.
Dimensions & tons: 83', 70'6" x 20' x 11'9". 150.
Men: ____ . Purchased 1850; 1853 sold.

# Surveying Vessels, etc

**Kangaroo** 1818-1828 ( ____ ) survey vessel/brig.
Dimensions & tons: 87', 75'5⅞" x 22'7" x 7'10". 204 74/94. 'Doubled' in 1823: breadth 22'11"; 210 30/94 tons.
Men: ____ . Acquired 1818?; appears to have been 'doubled' and rigged as a ship at Deptford Dockyard in 1823; 18.12.1828 wrecked (as a brig?) in the Bahamas.
PLANS: Lines & profile/decks.

**Violet** 1835-1848 (Dutch fishing boat building at Scheveninge) survey vessel.

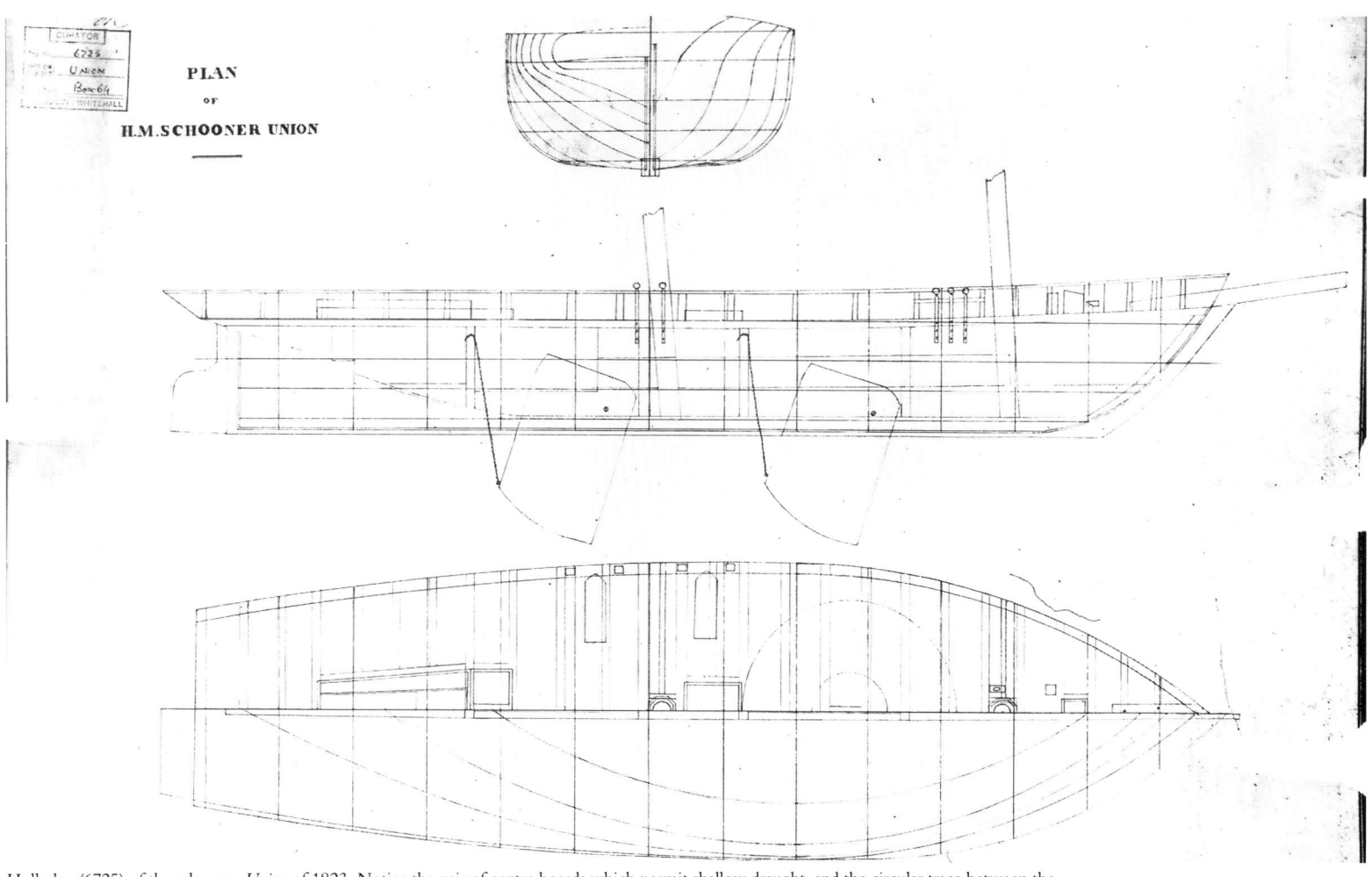

Hull plan (6725) of the schooner *Union* of 1823. Notice the pair of centre boards which permit shallow draught, and the circular trace between the masts on the deck plan which indicates the mounting of a large pivot gun there

Dimensions & tons: 40'2", 29'8½" x 17'2" x 6'5⅛". 46.
Men: ____ . Guns: None. Purchased as tender to HMS *Fairy* whilst building at Scheveningen 1835; 1848 sold.
PLANS: Lines/profile & deck.

**Speedwell** 1841-1855 (mercantile ____ Hull) survey vessel/cutter.
Dimensions & tons: 55'11½", 45'6⅛" x 17'7½" x 8'3". 73.
Men: ____ . Purchased 1841; 05.1855 sold in Canada.

**Plover** 1841-1854 (mercantile *Bentinck* ____ ) survey/Arctic vessel [cutter] teak-built.
Dimensions & tons: 82'2", 64'0½" x 25' x 14'7½". 213.
Men: ____ . Purchased 1841; fitted for Arctic expedition; 1854 sold.

**Castlereagh** 1846-1848 (mercantile ____ ) survey vessel. 4.
Dimensions & tons: ____ ____ .
Men: 50. Guns: 4 x 18★. Purchased at Sydney 1846; 1848 sold at Sydney.

**Research** 1846-1859 ( ____ ) tender.
Dimensions & tons: ____ ____ .
Men: ____ . Guns: ____ . Purchased at Malta as tender to *Beacon* 12.3.1846; 06.1859 BU at Malta.

**Wee Pet** 18____ ; 1849 ( ____ ) tender?
Dimensions & tons: ____ ____ .
Men: ____ . Guns: ____ . Purchased ____ ?; 1849 sold at Malta.

## Unregistered Vessels

(ie. those vessels mentioned in secondary sources only, and not in the Admiralty sources consulted.)

**Albatross** 1828-1833 (mercantile ____ 1826 ____ Scotland) tender. 64 tons. Purchased 1828; 1833 sold.

**Alert** 1848-1850 (slaver ____ ) brig. Taken by *Bonita* 1848; 1850 sold (noted in Colledge only).

**Assiduous** 1823-1827 (pirate *Jackal* ____ ) schooner ____ . Taken 1823; 1825 sold.

**Atalanta** 1816-1817 tender. Deptford Dockyard 1816; 1817 sold to Customs.

**Augusta** 1819-1823 (mercantile *Policy* ____ ) tender Acquired 1819; 1823 sold.

**Bermuda** 1819-1821 ( ____ 1818 ____ ) tender. 50 men. Acquired 1819; 03.1821 foundered between Halifax and Bermuda.

**Caroline** 1859-1863 sailing gunboat in New Zealand. Purchased 1859; 1863 sold.

**Cockburn** 1822-1823 (steam vessel *Braganza* ____ ) schooner 05.1822 purchased at Rio de Janeiro; tender to *Leven*; 1.4.1823 wrecked near Simonstown.

**Constitution** schooner 3 guns. Purchased 24.8.1835; not listed

**Cove** ('**Coventry**') 1835 purchased storeship?

**Cynthia** 1826-1827 (Post Office packet *Prince Regent* 1821 Falmouth) packet brig 6. 232 tons. Purchased 1826; 6.6.1827 wrecked on Barbados.

**Dove** 1823-1829 (Post Office packet *Manchester* 1805 Falmouth) packet brig 6. 187 tons. Acquired 1823; 1829 sold.

**Duchess of York** 1835-1836 (mercantile *Duchess of York* ____ ) tender. 50 tons. Acquired 1835; 1836 sold [?as Army depot at Portsmouth?].

**Emilia** 1840 (Brazilian slaver ____ ) brig. Purchased 29.12.1840; fate uncertain.

**Fanny** 1845 1 gun tender; taken from Argentines; noted by Pitcairn Jones.

**Felicidade** 1845-1845 (Brazilian slaver *Felicidade* ____ ) schooner. Taken 27.2.1845 off Lagos by *Wasp*; 5.4.1845 foundered.

**Firefly** 1828-1835 schooner 6 or 3 (2 x 12 + 1 x 5½" howitzer). 32 men. Acquired (built?) Bermuda 1828; 27.2.1835 wrecked on the coast of British Honduras.

**Fly** explosion vessel (ordnance sloop) expended at Algiers 25.8.1816.

**Flying Fish** schooner purchased 1817 (after being hired 1813-1814) no further information.

**Forte** 1832 governor's yacht at Newfoundland.

**Grecian** 1821-1827 (ex revenue cutter *Dolphin* ____ ) cutter. 1821 acquired 1827 sold.

**Grinder** tender sold 22.8.1832; origins unknown (Colledge only?).

**Gulnare** survey tender in service 1827 to 1842?

**Gulnare** 1855-1863 purchased survey cutter; 1863 sold.

***Happy Ladd*** 1855-1856 gunboat purchased at Constantinople; 1856 sold to the Turks.
***Hope*** 1826 schooner; tender to *Maidstone* 5 guns. 26 men.
***Hope*** Ca 1830-1860 tender. Ca 1830 tender to Hastings. 1834-1836 tender at Sheerness. 1855-1860 tender to convict hulks at Chatham.
***Indian*** 1855-1856 purchased schooner.
***Inspector*** survey cutter 60 tons in service 1822 (Colledge only).
***Jane*** 1818-1821 tender. 1821 sold.
***Kangaroo*** 1829-1834 ( ____ *Las Damas Argentinas* ____ ) schooner tender 3 (2 x 12 + 1 howitzer) 84 tons. Purchased at Jamaica 1829; 1834 sold.
***Lady Louisa*** brig at Plymouth 1829.
***Landrail*** 1817-by 1822 tender 4. At Jamaica.
***Lion*** 1823-1826 (pirate ____ 1821 USA) schooner. 88 tons. Taken 1823; 1826 sold.
***Lysander*** listed (Colledge only) 1842-1844 brig; no further information.
***Martin*** cutter listed 1817 and sold 22.10.1817?
***Mermaid*** 1817-1820 (mercantile *Mermaid* 1816 ____ ) survey ship. 84 tons. Acquired 1817; 1820 deleted.
***Narcissus*** brig 10; noted at Portsmouth 1845 (dubious; Pitcairn Jones only).
***Parthian*** 1829-1830 noted at Cork (dubious; is this a confusion with the brig of 1808-1828?).
***Prince Augustus Frederick*** listed 1816-1821 cutter 8 x 4. 69', 56' x 18' x ____ ; no further information (Colledge only).
***Renegade*** 1823-1826 (pirate/slaver *Zaragozana* 1820 USA) schooner. 115 tons. Taken 1823; 1826 sold.
***Rose*** 1838-1864 cutter, surveying vessel built Cowes 70'2", ____ x 20'10". 37 tons. 20.7.1864 stranded; wreck sold 9.9.1864.
***Sabrina*** schooner noted as being in service 1838 by Colledge but nothing else; dubious.
***Saint Lawrence*** an 18 gun ship of this name noted as being at Halifax in 1817; nothing further.
***Satellite*** 1825 (ex *Larne & Sophie*?) transport armed with 10 x 32* for the Burmese war.
***Scorpion*** 1827-18____ (mercantile *Scorpion* ____ ) tender. Acquired 1824 or 1827; no further information; is this the same as the revenue cutter of this name and period?
***Scrub*** 1823-1828 (building ____ Cowes) tender. Purchased whilst building 1823; 1828 sold.
***Snap*** 1845-1847 (slaver *Cacique* ____ ) troopship. Acquired 1845; 1847 deleted.
***Swallow*** cutter (built? 1811) 46 tons, tender to *Eden*; 30 11 1825 lost.
***Swallow*** 1824-1836 (Post Office packet *Marquis of Salisbury* 1820 Falmouth) packet brig 10. 236 tons. Purchased 1824; 1836 sold.
***Union*** 1823-1828 (mercantile *City of Kingston* 1821 Kingston, Jamaica) schooner. 84 tons. Acquired 1823; 17.5.1828? wrecked on a reef in the West Indies.
PLANS: Lines, profile and deck.
***Wasp*** 1822-1829 (mercantile *Wasp* 1801 White, Cowes) survey ship 91 tons Acquired 1822; 1829 BU. '
***Wizard*** 1824-1826 (slaver ____ ) tender. Acquired 1824; 1826 sold.
***Young Hebe*** 1839-1842 tender. 45 tons. Acquired 1839; 1842 sold.

## Hired Vessels

***Alexander*** 1817-1819 (Aberdeen 1813) brig 253 tons. Arctic exploration.
***Bredalbane*** 184__ -1843 ( ____ 1843) barque, transport. 428 tons. 21.8.1853 crushed in the ice in Lancaster Sound [wreck recently located by sidescan sonar and investigated].
***Diamond*** 1832-1833 schooner 4.
***Dorothea*** 1817-1819 (Jarrow 1812) transport 370 tons. Arctic exploration. 1818 too damaged to repair and return and so purchased; 1819 sold.
***Douro*** 1831-1833 survey vessel.
***Eliza*** 1821-1822 tender (sloop-rigged vessel) 1 x 18* or 1 x 12* tender to Tyne in the West Indies.
***Gulnare*** ? 1844-1850 survey vessel hired? 270 tons.
***Isabella*** 1817-1819 (Hull 1813) transport 383 tons. Arctic exploration.
***Trent*** 1817-1818 (Gainsborough 1811) brig 250 tons. Arctic exploration.
***Vansittart*** 1821-1827 cutter 10.

# PART III
# AUXILIARIES & ANCILLARIES

## Chapter 1: Vessels on the Canadian Great Lakes

*It is quite impossible ...... to fully detail the make, rig, armament and complement of all the vessels employed [on the Lakes] for some of the regularly built warships and many of the sloops and schooners purchased and used as such, changed from time to time, not only in their rig, their armament and their complement, but even in their names.*

President Theodore Roosevelt (in Laird Clowes, *Royal Navy* Vol VI, p110).

This is very much a provisional and tentative list, and should be treated with extreme caution. The listings from which it is taken very often seem to have been compiled by Admiralty clerks who were none too sure of what they were listing themselves. The usual sources (plans, dimensions books, etc) have been supplemented by the use of a typescript list compiled by C H J Snider and a *List of vessels employed on British Naval Service on the Great Lakes* compiled by K R Macpherson and published by the Ontario Historical Society. The names of those vessels taken from these lists for which I have no confirmation from an original source are asterisked [*]. Up to 1765 the Provincial Marine which controlled these ships was an offshoot of the Admiralty. In 1765 it become a department under the Quartermaster-General of the Army. In 1813 the Admiralty took over completely from the Provincial Marine.

It will be noted that there is considerable confusion between lists of different dates. Minor prizes and mercantile vessels temporarily taken over are omitted. It appears that the word 'sloop' was mostly (but not always) used in the mercantile sense of a small, single masted, fore and aft rigged ship. Ships designed for the lakes would be shallower, lighter built and carry more guns and men but less stores than their sea-going equivalents. Use was made of local types such as 'gundalows' (many of the 'sloops' were probably of this type). Those names of vessels whose existence seems to be certain as opposed to possible or likely are given in bold type whereas asterisked [*] names are the more dubious (see above - from modern lists not confirmed in contemporary lists that I have seen). I would like to acknowledge the generous help of Dr Alec Douglas in providing additional material for this list.

The vessels are listed in chronological order under the lake upon which they served (the upper lakes and Lake Erie were interconnecting and for our purposes can be considered as one body of water as the ships concerned could have appeared on any of these lakes). At the end is a listing of vessels whose lake is uncertain, followed by a number of gunboats which served on the River Saint Lawrence.

## Lake Champlain

**Duke of Cumberland*** 1759-17____ . Sloop 20. Built Ticonderoga 1759; 1778 list: 'laid up and decay'd'.

**Boscawen*** 1759-17____ . Sloop 16. Built Ticonderoga 1759; 1778 list: 'laid up and decay'd'.

**Brochette*** 1759-17____ . Sloop 6, [French, built St John's]. Taken 1759; 1778 list: 'In service till decay'd'.

**Lochegeon*** 1759-17____ . Sloop 6, [French, built St John's]. Taken 1759; 1778 list: 'In service till decay'd'.

**Musquenonge*** 1759-17____ . Sloop 6, [French, built St John's]. Taken 1759; 1778 list: 'In service till decay'd'.

**Vigilant*** 1760-17____ . Schooner 8, [French, built St John's]. Taken 1760; 1778 list: 'Laid up till decay'd'.

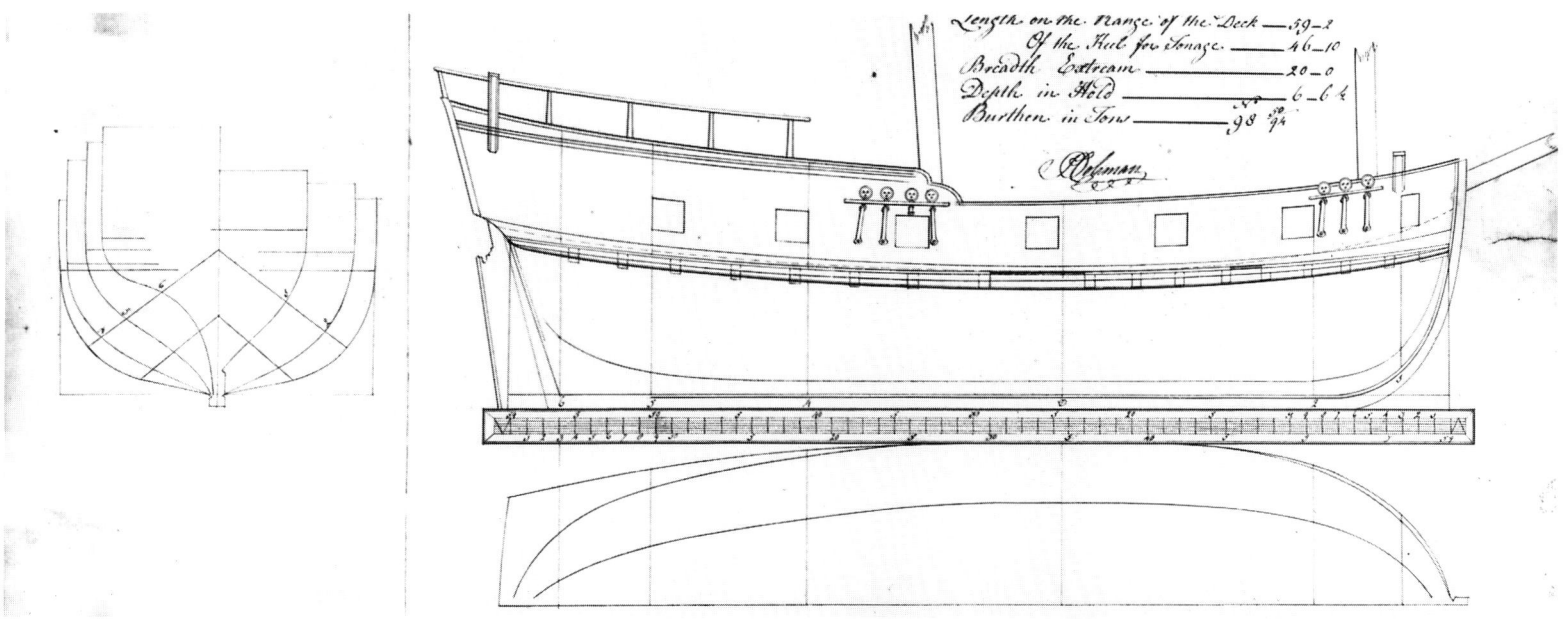

**Waggon*** 1760-17____ . Sloop 6, [French, built St John's]. Taken 1760; 1778 list: 'Laid up till decay'd'.

Lines plan (6369) for the *Carleton* brig 'rebuilt' in 1776

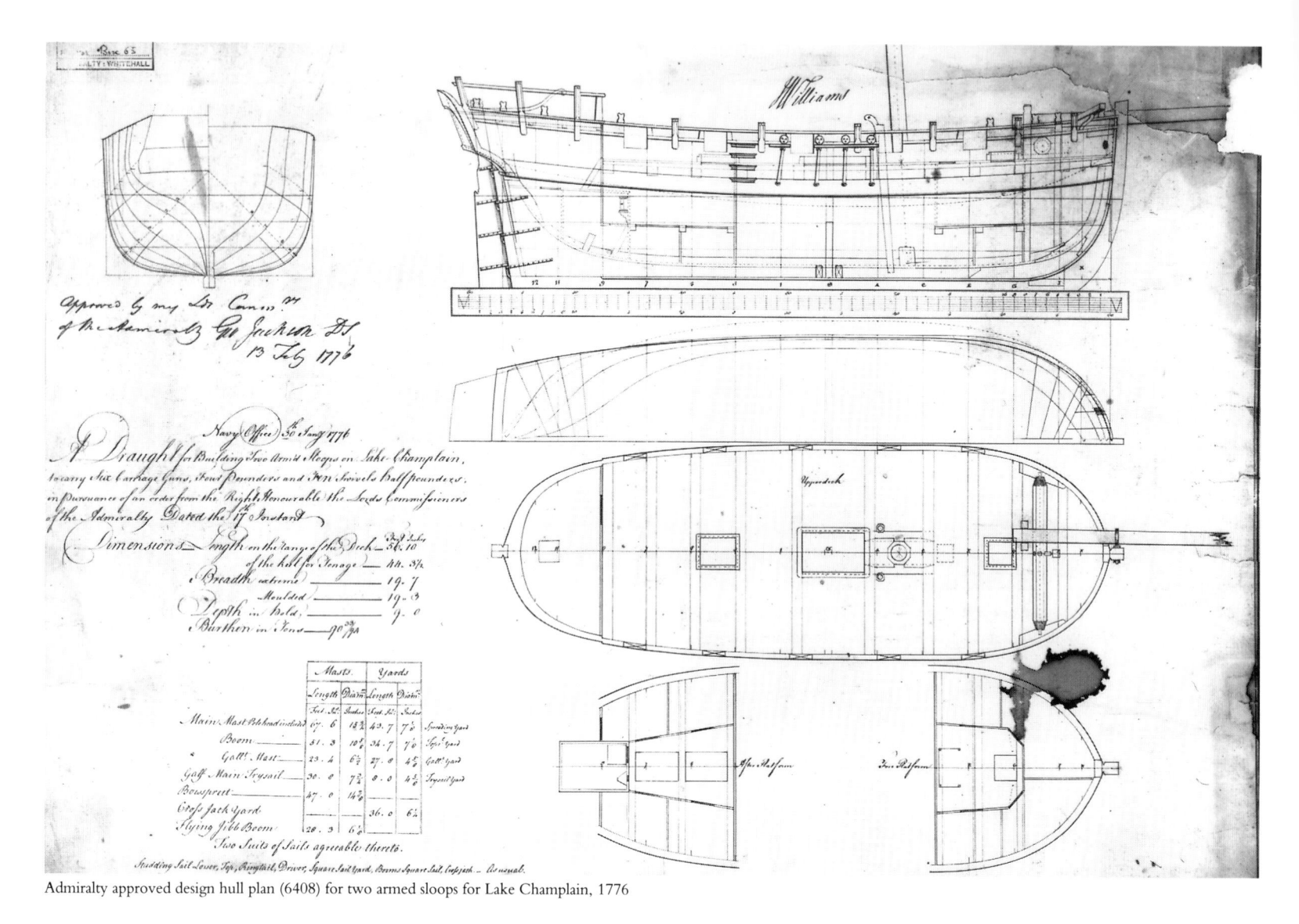

Admiralty approved design hull plan (6408) for two armed sloops for Lake Champlain, 1776

*Betsy** 17____ -1775? Sloop 6. Built St John's 17; 1775? taken by the 'Rebels'.

*Royal Savage* 17____ -1775 (ex *Brave Savage*). Sloop/brig . Built at Saint John's (Saint Jean); 10.1775 sunk by American batteries in the Richelieu river; later salved by the Americans and named *Yankee*.

*Inflexible* 1776-17____ . Ship. Guns 18 x 12 Built 1776 (by Schanck); in service 1779; fate uncertain.

*Carleton* 1776-17____ . Brig. ____ [Mercantile? ____ ]. 59'2", 46'10" x 20' x 6'6½", 99 tons. Acquired and rebuilt at St John's 1776; in service 1779; fate uncertain.
PLANS: Lines.

*Lee* 1776-17____ . Cutter 8 [American Sloop: *Lee*] 43'9", 34' x 16'3 1½" x 4'8", 48 tons. Taken 1776; in service 1779; fate uncertain.
PLANS: Lines.

*Washington* 1776-17____ . Galley [American galley *Washington* 1776 Skenesboro] 72'4", 60'6" x 19'7" x 6'2". 123 20/94 tons. Taken 1776; no further note.
PLANS: Lines.

*Maria* 1776-17____ . Schooner 14 [Mercantile? ____ ] 66', 52'2¼" x 21'6" x 8'2½", 128 tons. Acquired 1776 and rebuilt at St John's; in service 1779; fate uncertain.
PLANS: Lines.

*Thunderer* 1776-17____ . Ketch-rigged 'radeau' (flat bottomed, swim bows and stern, floating battery). Guns: 12 x ____ + 2 howitzers. Built at St John's 1776; fate uncertain, but not listed in 1779.
PLANS: Profile, section, deck.

*Loyal Convert* (*Convert*) 1776-____ . Hoy [American Gundalow ____ ] 62'10", 50'8" x 20'3" x 3'7½", 109 tons. Taken on the River St Lawrence 1776; in service 1779; fate uncertain.

PLANS: Outline profile & section & deck.

*Royal George* 1777- ____ . Ship . 96'6", 77'11½" x 30'6" x 10'. 386 tons. Built St John's 1777; in service 1779; fate uncertain.
PLANS: Lines.

NOTE: In 02.1776 the Admiralty approved the building of two 'armed sloops' to be built on the lake. 56'10", 44'3½" x 19'7" x 9'. 90 tons 6 x 4 + 10 x ½ pounder swivels. The original design appears to have been for two masted (schooner?) rig, but this seems to have been changed to a single masted rig. It is not known if these plans were ever implemented.

Plans: Lines, profile, decks, spar dimensions (preliminary design?)/Lines, profile, decks (2 copies - final design?).

*Camel* Hoy; in service 1779; fate uncertain.

*Ration* Lugger; in service 1779; fate uncertain.

*Chub* (ex *Shannon*) 1813-1814? Cutter 13. Tons: 110. [American Sloop *Growler*]. Taken 3.6.1813; 11.9.1814 retaken by the Americans at Plattsburg? However in 1830 a schooner *Chub* [ex *Eagle*] 12 x 18____ , 50 men is noted as still in service, but this is most likely a different vessel.

*Finch* (ex *Broke*?) 1813-1814? Cutter 11. Tons: 111. [American Sloop *Eagle*.] 1813 taken in the Sorrel River; 11.9.1814 retaken by the Americans at Plattsburg. However in 1830 a schooner *Finch* [ex Government Schooner No 14] 4 x 18____ , 40 men is noted as still in service, but this is most likely a different vessel.

*Linnet* ____ -1814? Brig 16. Built Isle aux Noix ; 1814 taken by the Americans at Plattsburg, at which stage she was armed with 30 x 24. However in 1830 a brig *Linnet* [ex *Growler*] 12 x 18____ , 80 men is noted as still in service, but this is most likely a different vessel.

*Icicle** [American Sloop *President*] 2 guns. Taken 1814; fate uncertain.

*Canada* 1814- ____ . Gunboat. Guns: 3 x 6 + 10 swivels. In service in 1814; fate uncertain.

**Confiance** 1814-1814. Ship 37. Tons: 1200. Men: 270. Built at Isle aux Noix; L 08.1814; 11.9.1814 taken by the Americans at Plattsburg at which stage she was armed with 16 x 12.

**Sir George Prevost** 1814- ____ . Gunboat. Guns: 1 x 24 + 1 x 32* + 46[?] swivels. In service 1814; fate uncertain.

**Sir Home Popham** 1814- ____ . Gunboat. Guns: 1 x 32*. In service 1814; fate uncertain.

**Sir James Yeo** 1814- ____ . Gunboat. Guns: 1 x 24 + 1 x 32*. In service 1814; fate uncertain.

**Tecumseth** 1814- ____ . Gunboat. Guns: 1 x 6 + 26 swivels. In service 1814; fate uncertain.

**Blucher** [*Blutcher*] 1814- ____ . Gunboat. Guns: 1 x 18 + 1 x 18* + 34 swivels. In service 1814; fate uncertain.

**Axeman** 1815- ____ . Gunboat, 2 long guns, one at each end, carronade amidships. 63'10", 54'1⅝" x 16'2" x ____ . Tons: 75 31/94. Built at Isle Aux Noix 1815; fate uncertain.

**Caustic** 1815- ____ . Gunboat as *Axeman* except long gun amidships. 62'3", 52'4⅝" x 16'2" x 4'1". 73 54/94. Built at Isle Aux Noix 1815; fate uncertain.

**General Brock**★ 1814-1815 Gunboat 2.
**Sir Sidney Beckwith**★ Ca 1815 Gunboat 2.
**Marshal Beresford**★ Ca 1815 Gunboat 1.
**Lord Cochrane**★ Ca 1815 Gunboat 1.
**General Drummond**★ Ca 1815 Gunboat 1.
**Colonel Murray**★ Ca 1815 Gunboat 2.
**General Simcoe**★ 1814-1815 Gunboat 1.
**Lord Wellington**★ Ca 1815 Gunboat 1.

NOTE: In early 1815 a design was produced at Kingston Navy Yard for a brig of 10 guns to be built at Isle aux Noix. 85', 68'9½" x 28' x 7'3", 287 tons. 10 x 24pdr. Later three brigs in all were to be built to this design, but it is not known whether this was put into effect.

PLANS: Lines (preliminary design)/lines.

**Champlain** 1816?-Ca 1832. Ship 32 (?). 60'9", ____ x 20' x ____ . 111 tons. In service 1816? or acquired 1819?; about 1832 sold.

Ca 1830 Sloop **Linnet**, Schooners **Chub**, **Finch** in service; see under these names above (Ca 1813-4).

# Lake George

**Sloop 5** 1755 Built at Fort George; 1779 'laid up and decay'd'.

# Lake Ontario

**Oswego**★ 1755-1756 Sloop 5. Built Oswego 1755; taken by the French 14.8.1756.

**Ontario** 1755-1756 Sloop ____ . Built Oswego 1755 (possibly purchased); 14.8.1756 taken by the French at Fort Oswego.

**Lively** 1755-1756/1759-17____ *Farquer*. Schooner ____ . Built Oswego 1755; 1756 taken by the French; 1759 retaken at Niagara and renamed *Farquer*; fate unknown. Noted by Snider but not Macpherson.

**George**★ 1755-1756 Schooner 8. Built Oswego 1755; taken by the French 1756.

**Vigilant**★ 1755-1756? Schooner 8. Built Oswego 1755; taken by the French 1756.

**London** 1756-1756 Brigantine 16. Built Oswego 1756; 14.8.1756 taken by the French.

**Halifax**★ 1756-1756 Snow 22. Built Oswego 1756; 14.8.1756 taken by the French. Despatch vessel★ ____ . Unfinished 1756; never completed?

**Mohawk** 1756-1756 Sloop-rigged vessel 6. Built Oswego 1756 (ex *Mercantile*?). Purchased? 1756; 14.8.1756 taken by the French.

**Mohawk** 1759-1764 Sloop (snow) 18/16. Built Oswego 1759 (possibly an uncompleted French prize); 1764 lost ('cast away').

**Onondaga** 1759-1764 Sloop 22. Built Oswego 1759; 1764 lost.

**Missasago** or **Missisaga**★ 1759-1765 Sloop 8 or Snow 16. Built Oswego 1759/60; 1765 'cast away'.

**Murray**★ 1759?-17____ . Schooner ____ . Built Oswego Ca 1759; fate unknown.

**Johnston**★ 1760-1764 Snow 12/10. [French *L'Ouataouaise* 1759 Pointe au Baril] Taken 1760; 11.1764 lost.

**Williamson**★ 1760-1761 Barque 10. [French *L'Iroquoise* 1759 Pointe au Baril]; 1761 lost.

**Row galleys**★ (five, un-named) 1760 Built at Oswego for Amherst's Expedition.

**Mercury**★ 1763-17____ . Schooner 6. Built Oswego 1763; 1779 'laid up and decay'd'.

**Brunswick**★ 1765-1778 Schooner. Built Oswego 1765; 1778 condemned.

**Charity**★ 1770-1777 Sloop 6 swivels. Built Niagara 1770; 1777 'cast away'.

**Caldwell**★ 1770-17____ . Sloop 2. Built Niagara 1770; still in service 1779; fate unknown.

**Haldimand**★ 1771-17____ . Snow 18. Built Oswegatchie 1771; still in service 1779; fate unknown but out of service by 1787.

**Seneca** 1771-17____ . Sloop (Snow) 18. Built Oswegatchie 1771; still in service 1779; ditto 1788; fate unknown.

**Duke of Kent**★ 1776-17____ . Schooner ____ . Built Carleton Island 1776; fate unknown.

**Ontario** 1780-1780. Brig sloop. 16/22 80', 64'3⅝" x 25'4" x 9'. 386 tons. 16 x 6 + 6 x 4. Built Carleton Island; L 10.5.1780; 1.11.1780 foundered with all hands.

PLANS: Lines, profile.

**Limnade**★ 1781-17____ . Snow or Ship 16. Built Carleton Island 1781; fate unknown.

**Mohawk**★ 1781-17____ . Sloop ____ . Built Carleton Island 1781; fate unknown but out of service by 1787.

**Onondaga** 1790-1793 Schooner 6. Built at Raven Creek 1790; 18.12.1793 lost at York (Toronto) but salved in 1794.

**Mississauga** 1793- ____ . Schooner ____ . Built at Kingston/Raven Creek; fate uncertain.

**Onondaga**★ 1793- ____ . Schooner ____ . Built at Kingston; fate uncertain.

**Sophia**★ 1792 or 1795- ____ . Sloop or gunboat ____ . Built at Kingston 1792 or 1795; fate unknown.

**Mohawk** 1795-1803 Schooner ____ . Built at Kingston; L 14.5.1795; 1803 condemned.

**Speedy** 1796?-1804 Schooner. Built 1796? (or 1804) at Kingston; 8.10.1804 foundered with all hands.

**Duke of Kent**★ 1801-18____ . Snow ____ . Built at Kingston; receiving hulk by 1812; fate uncertain but still in existence in 1816.

**Moira/Earl of Moira** 1805-1814 *Charwell* -1837. Schooner 14. Dimensions: 70', 61' x 23'8" x 2'3". 169 tons? 1830: 2 x 9 + 12 x 24*. Men: 1830: 86. Built at Kingston 1805 (purchased?); 1805 converted to schooner?; 1813 converted to brig 1814 renamed *Charwell*; by 1816 powder hulk; by 1827 accommodation vessel; 1830 referred to as a sloop; 1837 sold. Presumably an entirely different vessel to the *Charwell* of 1815. Just possibly the vessel depicted in the unidentified brig plan of 1815 (see below).

**Gloucester/Duke of Gloucester** 1807-1813 Schooner or Brig 6. Built Kingston; launched 05.1807; 27.4.1813 taken by the Americans; 29.5.1813 burnt at Sacketts Harbour.

**Royal George** (ex *Wolfe*?) 1809-1814 *Niagara* -1837 Sixth Rate/Sloop 20/22. Dimensions: 97', ____ x 28' x ____ . 340 tons. UD 8 x 24 (Congreve) + 2 x 12 + 10 x 24*, 1816: 6 x 18 (Congreve) + 2 x 9 + 12 x 24*. 1830: 1 x 24 + 2 x 18 + 18 x 32* or 4 x 32* only.
Men: 135, 1816: 95, 1830: 175. Built at Kingston, L 07.1809; 1814 renamed *Niagara* (but there is considerable confusion between this vessel and the *Wolfe* of 1813); 1837 sold.

**Prince Regent** 1812-1813 *Beresford/General Beresford* - 1814 *Netley* - 1837/1838 *Niagara* -1843. Schooner ('as a sloop' 1830) 10. 72' (or 71'9"), 60' x 21' x 2'. 187 tons. 1830: 2 x 6 + 8 x 18*.
Men: 1830: 70. Built at York (Toronto); launched 06.1812; 1813 renamed *Beresford* or *General Beresford*; 1814 renamed *Netley* and altered to brig (ditto); 1837 sold; 1838 repurchased and renamed *Niagara* as accommodation vessel; 1843 BU. Is it possible that this vessel is the one depicted in the Lines Plan of a 'brig to carry 10 long 24 pounders' dated 1815 'as built at Kingston' noted below?

**Royal George** (ex *Sir George Prevost*?) 1813-1814 *Montreal* 1832 Ship 22. 101'9", 86'1" x 30'6" x 11'. 426 tons. 22 x 32 ____ . Built at Kingston 1813; 1814

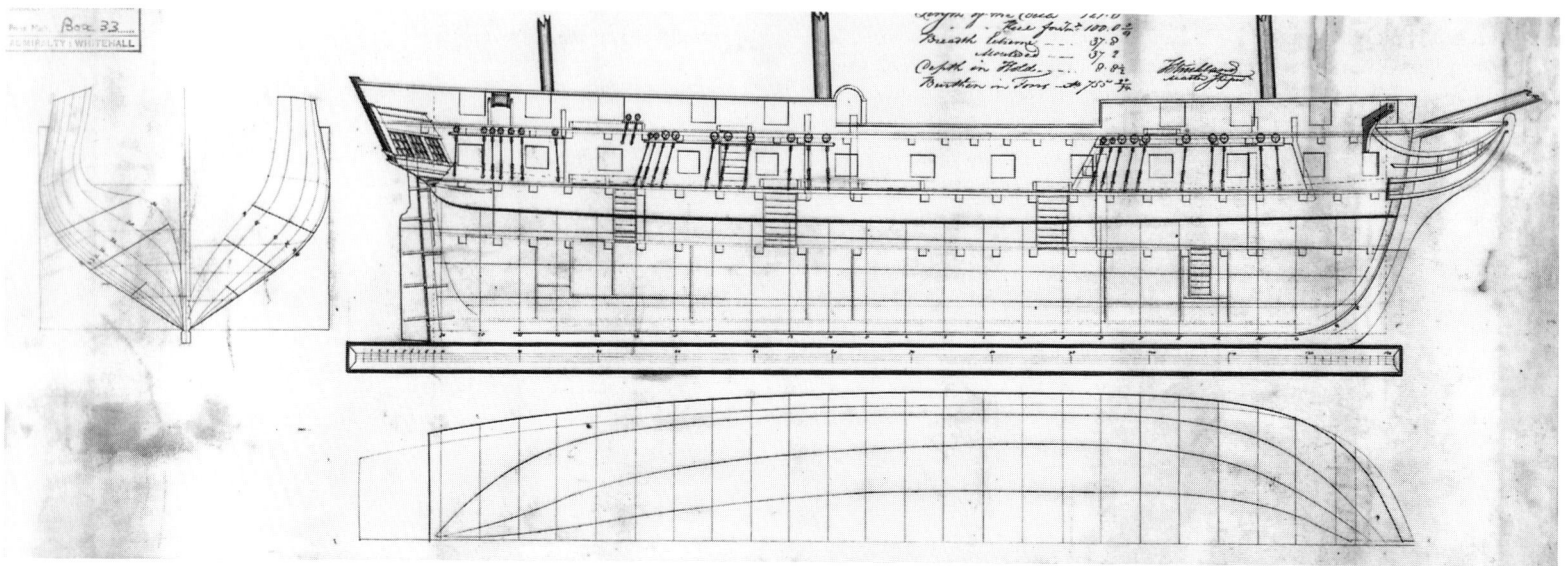

Lines and profile (6013) of the frigate *Princess Charlotte* built at Kingston in 1814

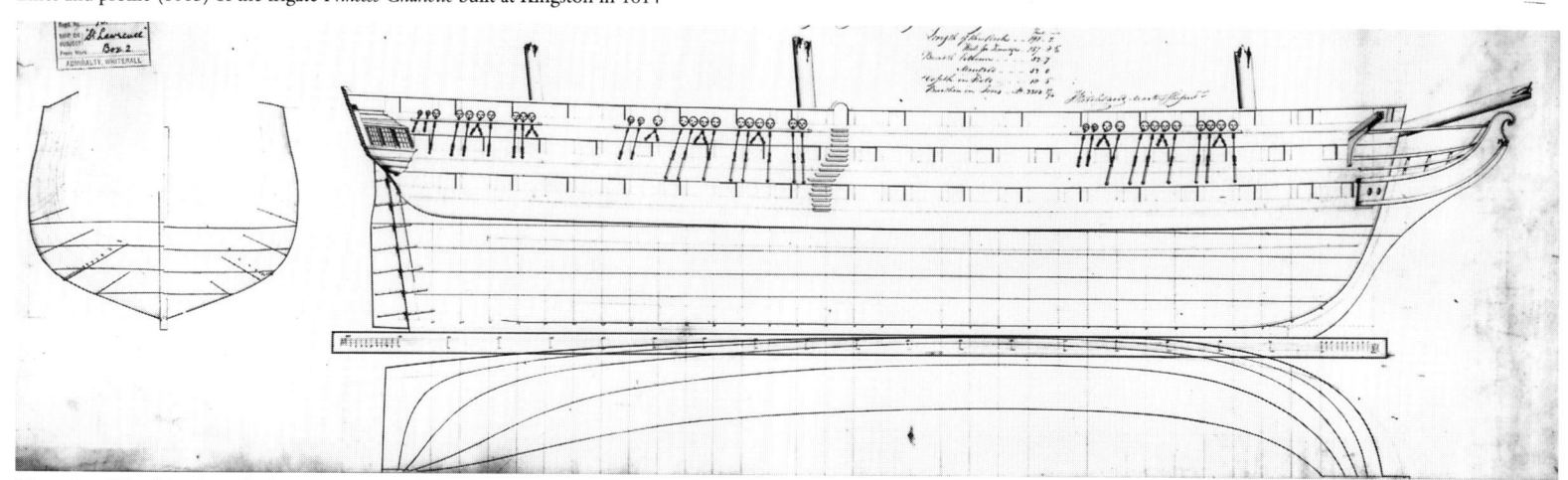

Lines plan (73) of the three-decker *Saint Lawrence* built in 1814. Note that she is considerably smaller and much shallower draughted than the seagoing vessels which carried a comparable number of guns. In a fresh water lake the need for supplies, and particularly of drinking water, was considerably less

renamed *Montreal*; 1832 sold. As the dimensions and other data tie in with the *Montreal* ex *Wolfe* (see below) whose existence is demonstrated by a plan, either this vessel did not exist at all or she was an exact sister whose later history (and renaming) was different from that given here. Another possibility is that the data itself is wrong.]

**Confiance** 1813-1814 Brig or schooner 2. (American *Julia* 1812 Oswego) Taken 10.8.1813; 5.10.1813 retaken by the Americans.

**Sir Isaac Brock** building 1813. Ship 28. Building at York (Toronto) and destroyed on the stocks by the Americans 27.4.1813.

**Montreal** (ex *Wolfe*) 1813-1832. Sloop 22/20. 101'9", 86'1⅛" x 30'6" x 4'6". 426 tons. 8 x 24 Congreves + 2 x 12 + 10 x 24★. 1830: 4 x 32★. 135 men. Built Kingston; L? 04.1813; 1814? renamed *Montreal* from *Wolfe*? (see under *Royal George* above); 1815 altered to have a spar deck added, ventilation ports only on the original upper deck (originally pierced for 22 guns plus 2 chase ports) and pierced for 8 guns on the new spar deck; 1832 sold.

PLANS: Lines, profile (as built)/Lines, profile (as altered 1815).

**Melville/Lord Melville** 1813-1814 **Star** -18____ . Sloop ____ . 71'7", 56'9½" x 24'8" x 8'. 186 tons. 4 x 18 (Congreve) + 2 x 9 + 8 x 24★, 1830: 14 x 32★. Men: 95, 1830: 80. Built Kingston 1813; still in service 1830; fate unknown.

**Wolfe** 1813-1814 **Niagara**? -1842 Schooner ____ . 96'9", 81'11" x 27'7" x 11'. 330 tons. Built at Kingston 1813; 1814 renamed *Niagara*; 1842 out of service.

**Prince Regent** 1814-1815? **Kingston** -1832 Frigate 60. 155'10", 131'1" x 43'1" x 9'2". 1294 tons. Actually noted as having 56 guns: 28 x 24 2 x 68★, 26 x 32★. 1830: GD 30 x 24, UD 2 x 24 + 6 x 68★ + 22 x 32★. 280 men. Built at Kingston; L 14.4.1814; 1815? (lists give 1814, but the plan is dated 1815 and shows the name *Prince Regent*) renamed *Kingston*; 1832 sold.

PLANS: Lines & profile.

**Princess Charlotte** (ex *Vittoria*?) 1814-1815 **Burlington** -1833. Frigate 42/40. 121', 100'0⅜" x 37'8" x 8'8½", 756 tons, 40; GD 28 x 24, UD 2 x 68★ + 16 x 32★. 1830: GD 24 x 24, UD 2 x 68★ + 16 x 32★. 280 men. Built at Kingston L 15.4.1814; 1815 or later renamed *Burlington*; 1833 sold & BU.

PLANS: Lines & profile.

**Saint Lawrence** 1814-1832 Three-decker 112/100. 191'2", 157'8⅝" x52'7" x 18'6". 2305 tons. GD 28 x 32 + 4 x 24 + 4 x 32★, MD 36 x 24, UD 32 x 32★ + 3 x 68★. 700 men. Built Kingston, L 10.9.1814; 1832 sold.

PLANS: Lines/profile/orlop/gun deck/middle deck/upper deck/midships section.

**Psyche** 1814-18____ . Frigate 54/52. 130', 108'0½" x 36'7" x 10'3", 769 tons. 28 x 24, 28 x 32★. 280 men. Built Kingston 1814; still in service 1830; fate uncertain. It would appear that the designs used to prefabricate the never-completed *Psyche* and *Prompte* at Chatham Dockyard (see main list), which were sent to Strickland, the Master Shipwright at Kingston, were used and modified by him to produce this frigate.

PLANS: Lines & profile ('showing the alterations').

**Julia** 1814-18____ . Schooner ____ . 63'10½", 52'3¼" x 16'8½" x 8'7". 77 tons. Built at Kingston 1814 (or seized at Sacketts Harbour; in which case she might be the earlier *Confiance*?); fate uncertain.

**Magnet** (ex *Sir Sidney Smith*) 1814?-1846? Gunboat/schooner? 12. ____ 2 x 18 + 10 x 32★. Men: 75. In service by 1830 and until at least 1846. There was a schooner called *Sir Sidney Smith* first listed in 1813 which may have

VESSELS ON THE CANADIAN GREAT LAKES

been a rebuild (in 1806-7) of the merchantman *Governor Simcoe*★ of 1794, taken into Government service in 1812. However, since this vessel is also reported as run ashore and burnt to avoid capture on 5.8.1814 this seems unlikely - though a successful salvage is a possibility.

**Prefabricated** at Chatham Dockyard and shipped 1814 - but never assembled [uncertain which of the Great Lakes they were intended for] Frigates 32 *Prompte*, *Psyche*. Brigs *Colibri*, *Goshawk* (see Main List).

*Wolfe* building 1815. Three decker, 112/104. 191'3", 157'8" x 50'8" x 18'4". 2152 tons. Building at Kingston in 1815. Originally referred to as 'Ship No 1'; never completed.

PLANS: Lines & profile/frame (also for *Canada*).

*Canada* building 1815 112/104 as *Wolfe* (above), referred to as 'Ship No 2'; never completed.

*Charwell* 1816-18 ____ . Sloop 14. 107'5", 85'8" x 30' x 12'6", 437 tons. 14 x 32★. 80 men. Built at Kingston 1816; still in service 1830 (as transport?); fate uncertain. Presumably an entirely different vessel from *Charwell* ex *Moira* (see above).

Also **Brig 10** 'As built at Kingston' by 1815. 77'3", 61'10½" x 22'5" x 7'3". 166 tons. Does not seem to fit any of the listed vessels, but the *Prince Regent* and *Niagara* of 1812 appear to be the closest, or just possibly the *Moira*/*Charwell* of 1805.

PLANS: Lines.

Also following **gunboats** built at Kingston by 1815: *Crysler*★ (1 gun), *Buffalo*★ (2 guns), *Niagara*★ (2 guns), *Queenston*★ (1 gun lugger)

*Beckwith* 1816-1822. Schooner (transport?). Built at Kingston 1816; out of service by 1822.

*Brock* 1817-1837. Schooner 2. 69'6", 55'9⅞" x 22'3" x 10'. 141 tons. Built Kingston 1817; 1837 sold.

*Toronto* 1817-18 ____ . Schooner transport. 57'11", 47 x 18'6" x 18'(?) 81 tons. Built Kingston 1817; fate uncertain.

PLANS: Lines.

## Lake Erie and Upper Lakes

(Huron, Michigan, Superior)

NOTE: Those vessels known to have served on the upper lakes, which had direct communication with Lake Erie via Lake St Clair for smaller vessels, are noted as such.

**Sloop**★ 8 1763-1764. Built Navy Island, Erie 1763; 1764 'cast away'.

*Victory*★ 1763-17____ . Schooner 6. Built Navy Island 1763; 17____ 'lay'd up, burn'd by accident' (before 1779).

*Huron*★ [Upper Lakes] 1763-17____ . Schooner built Navy Island; no further information.

*Michigan*★ [Upper Lakes] 1763-17____ . Schooner built Navy Island; no further information.

*Boston*★ 1764-1768 Schooner 8. Built Navy Island 1764; 1768 'lay'd up burn'd by accident'.

*Gladwin*★ 1764-17____ . Schooner 8. Built Navy Island 1764; 'In service until decay'd' (before 1779).

*Royal Charlotte/Charlotte* 1764-17____ . Sloop 10. Built Navy Island; in service (still in 1770) 'until decay'd' (by 1779).

*Chippawa*★ 1769-1775 Sloop 4 swivels. Built Detroit 1769; 1775 'cast away'.

*Hope*★ 1771-17____ . Schooner 6. Built Detroit 1771; still in service 1779; fate uncertain.

*Angelica*★ 1771-1783 Sloop 4 swivels. Built Detroit 1771; still in service 1779; 12.1783? wrecked.

*Dunmore* 1772-1797? Schooner 4/12. Built at Detroit 1772; 1796 still in service, no further information.

*Faith*★ 1774-17____ . Schooner 4 swivels. Built at Detroit 1774; still in service 1779; fate uncertain.

*Archangel*★ [Michigan] 1774-17____ . Sloop ____ . Built Detroit 1774; still in service 1779; fate uncertain.

*Felicity*★ 1775-17____ . Sloop 4 swivels. Built at Detroit 1775; still in service 1779; fate uncertain.

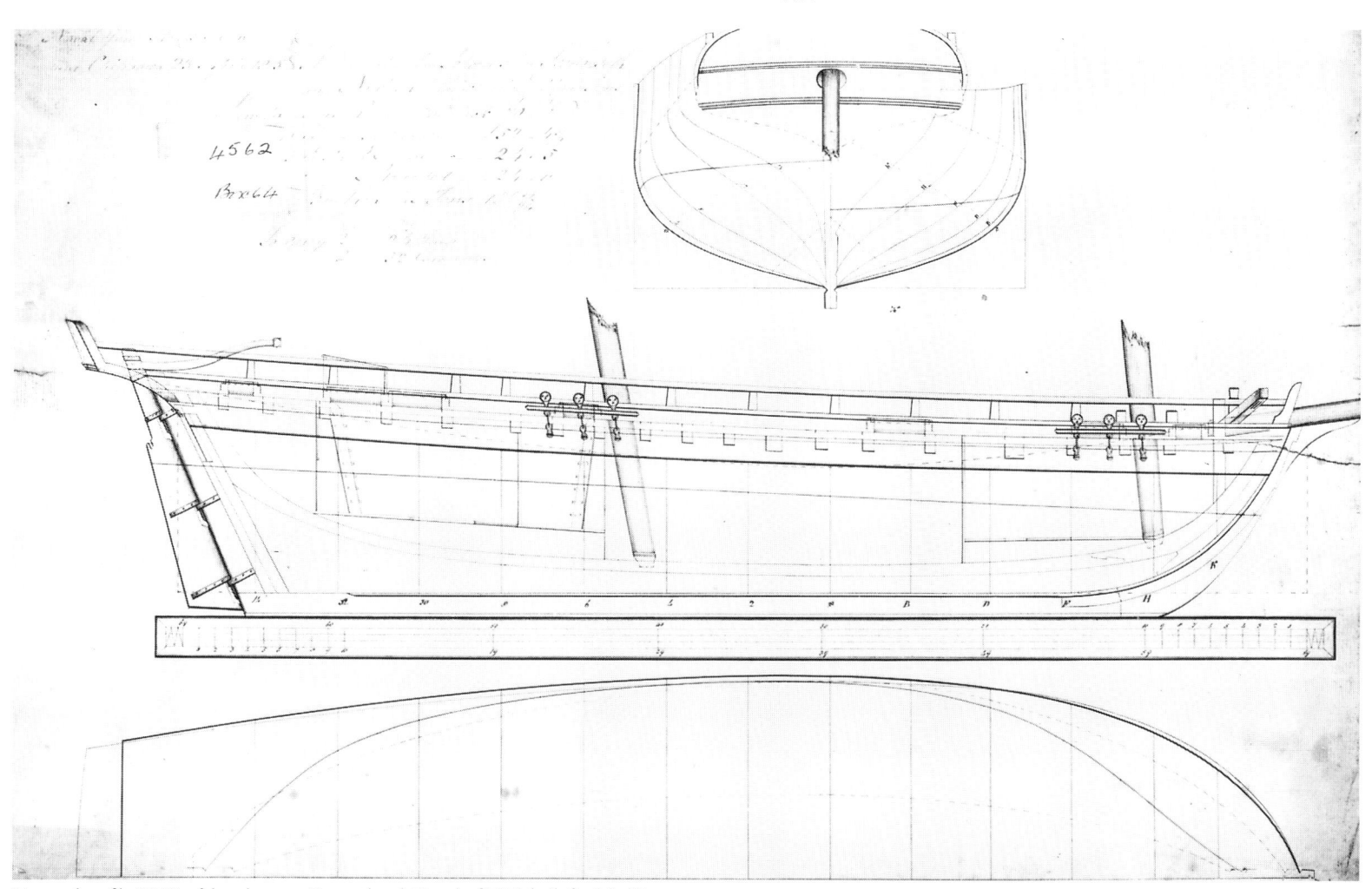

Lines and profile (4562) of the schooners *Tecumseth* and *Newash* of 1815, built for Lake Huron

*Welcome*★ [Huron] 1775-17____ . Sloop 4 swivels. Built Michilimackinac 1775; still in service 1779; fate uncertain.

*Adventure*★ 1776-17____ . Sloop ____ . Built at Detroit 1776; still in service 1779; fate uncertain.

*Gage*★ 1776-17____ . Schooner/brig 16. Built Detroit 1776; still in service 1779; fate uncertain.

*Wyndatt*★ Building 1779 Sloop packet; at Detroit; no further information.

*Ottawa*★ Building 1779 Sloop; at Detroit?; no further information.

*Nancy*★ 1789-1814 Schooner ____ . Launched Detroit 24.11.1789; 14.8.1814 burnt by the Americans.

*Ottawa*★ 1794- ____ . Snow ____ . Built Detroit 1794; no further information.

*Chippewa*★ 1795- ____ . Snow (yacht?) 16 swivels? do tons? Built Detroit 1795; no further information (but see under *Chippeway* of 1812).

*Francis* 1796- ____ . sloop. Built at Detroit 1796 (or 1794); no further information.

*Queen Charlotte* 1809-18____ . Ship 16. Built Amherstburg 1809; no further information.

*Hunter* 1812-1813 Brig 10. 2 x 6 + 4 x 4 + 2 x 12 + 2 x 12★ Built 1812 (is this the same as the *General Hunter*★ said to be built at Amherstburg in 1805?); 10.9.1813 taken by the Americans at Put-in Bay.

*Chippeway* 1812-1813 Schooner 2 x ____ (or 1 x 9?) Built 1812 (or is this the same as the *Chippewa* of 1795?); 10.9 1813 taken by the Americans at Put-in Bay.

*Detroit* 1812-1812 Brig 6. (American *Adams*) Taken at Detroit 1812; 9.10.1812 retaken and burnt by the Americans at Fort Erie.

*Little Belt* 1812-1813 sloop 2 or 3. Built 1812 (or converted from mercantile *Friends Goodwill*?); 10.9.1813 taken by the Americans at Put-in Bay.

*Caledonia* 1812? -1815? Schooner/brig 2/6. Ex Northwest Fur Co. May have been taken and retaken in 1812. 1813 at Put-in Bay; 1815 sold mercantile.

*Lady Prevost* 1812-1813 Schooner 12. Tons: Ca .230. Built Amherstburg? Launched 13.7.1812; 10.9.1813 taken by the Americans at Put-in Bay.

*Detroit* 1813-1813 Ship 19. Tons: 490. Built 1813 at Amherstburg; 10.9.1813 taken by the Americans at Put-in Bay.

*Confiance* [Upper Lakes] 1814-18____ (American *Scorpion*?) Schooner 2. 67'2½", 56'6¾" x 17'10" x 5'2½". 95 tons. 1 x 24 + 1 x 24★. Taken at Mackinack 1814; in service 1830 (or is this a confusion with the 1816 ship of this name?); no further information.

*Surprise* [Upper Lakes] 1814-18. (American *Tigress*) Schooner 2 56'11", 46'2⅜" x 17'5" x 5'3". 74 tons. Taken at Mackinack 1814; still in service 1830; Sunk in the Penetang and hull recovered in 1935?

*Sauk* [Huron] 1810-18____ . (American *Somers* or *Ohio*) Schooner 2. 58'8", 45'10⅝" x 18'11" x 7'2", 87 tons. 1 x 24 + 1 x 32★. 30 men. Taken at Fort Erie 1814; still in service 1830; no further information.

*Huron* [Huron] 1814-18____ . (American *Somers* or *Ohio*) Schooner 2. 53'11", 41'3⅝" x 17'3½" x 8'. 66 tons. 1 x 24 + 1 x 24★. 30 men. Taken at Fort Erie 1814; still in service 1830; no further information.

*Tecumseth* [Huron] 1815-18____ . Schooner 4. 70'6", 52'4¾" x 24'5" x 9', 166 tons. 2 x 24 + 2 x 32★. Men: 1830: 48. Built Streets Farm near Chippewa; by 1830 classed as a brig sloop; fate uncertain.

PLANS: Lines, profile/upper deck; (both also for *Newash*).

*Newash* [Huron] 1815-18____ as *Tecumseth* except in 1830 still classed as a schooner with 37 men.

*Confiance* 1816-1831 Ship 32. Built 1816; in service up to 1831.

*Montreal* 1836-1848 Schooner 1. 82', 76' x 20'6" x 9'. 145. Men: 35 Guns: 1 x 18 pivot. Built 1836; 1848 sold.

## Lake Oswego

*Halifax* 1756-1756 Brigantine. Dimensions & tons: 57', 46'6" x 16' x 7'8". 63. Built on Lake Oswego 1756; 08.1756 taken by the French at Fort Oswego.

## Uncertain Location

*Loudon*★ 1756 brigantine.

*Royal George* Ca 1776 Sloop, 20. Listed in 1776 (probably the vessel completed 1777 on Lake Champlain).

*Jersey* Ca 1776 Sloop-rigged vessel, 7? In service 1776.

*Mary* 1812-1813 Schooner. In service 1812; 1813 taken by the Americans.

**Gunboats** at Coteau-de-Lac Ca 1815: *Brock*★ sloop (1), *Blacksnake*★ lugger (1), *Cornwall*★ sloop (1), *Glengarry*★ sloop (1), *Kingston*★ lugger (1), *Nelson*★ schooner (2), *Quebec*★ lugger (1), *Thunder*★ lugger (1), *York*★ sloop (1).

*Cockburn* 1827-1837 Schooner 1. Tons: 70. 40 men. Built 1827; in service until 1837.

*Bullfrog*★ 1838-1841 Schooner 4. 96 tons. Purchased 17.8.1838; 1841 sold.

'Lateen Barge' 42' x 11' x ____ 1 x 18 medium pivot gun; in service 1805.

**Gunboat Nos 1, 2, 3**. 49' x 13'4" x ____ ; in service 1845.

## St Lawrence River

**Gunboats** in service 1815 *Belabourer*★ (1), *Bloodletter*★ (1), *Cleopatra*★ (2), *Nelly*★ (2), *Retaliation*★? (1), *Spitfire*★? lugger (1), *Ernestown*★? (1).

# Chapter 2: Royal And Other Yachts

This appendix is organised in a similar manner to the main list, giving those vessels in service in 1688 as a simple listing of basic historical data, then those vessels built after that date, with brief technical details. In both cases they are given in chronological order of building.

The name and concept of the yacht came into England from the Netherlands with the Restoration of Charles II in 1660. Specifically, Charles was given a yacht by the Dutch in 1660; she was named *Mary*. The Dutch word *jacht* (like the German *jagd*) came from hunting, and came to be used for small fast craft used for raiding and, increasingly, for communications duties and for carrying important passengers in some style. The Dutch were already using them for pleasure and for ceremonial duties before the words 'yacht' and 'yachting' appeared in English.[1] The King and his brother James, Duke of York (later James II), became fond of the sport of racing such craft, and so had several in service. Most of these are noted here, though the *Mary* herself had been wrecked before 1688. During the Anglo-Dutch wars, yachts were often drafted into service for employment as despatch boats and the like. By the end of the seventeenth century, however, this use of yachts as fighting vessels had become increasingly rare, and yachts were henceforward to be confined to VIP transport and the like, mostly for royalty. Some of the smaller yachts were not royal ones, the vessels being used for the Viceroy of Ireland, the Governor of the Isle of Wight, and for the Commissioners of the Navy and the various dockyards. In either case, they were not fully part of the fighting navy and so are confined to this appendix rather than being dealt with in the main list. The royal yachts were normally based in the Thames, moored off Greenwich, and maintained by Deptford Dockyard. One peculiarity remains to be noted: because of the prestige of royal yachts they were commanded by full captains, and therefore counted as rated vessels. The larger ones were usually Fifth Rates, the smaller Sixth Rates.

## In service 1688

*Jemmy* 1662 Pett, Lambeth; 1722 BU.
*Merlin* 1666 Shish, Rotherhithe; 1698 sold.
*Monmouth* 1666 Castle, Rotherhithe; 25.11.1698 sold.
*Kitchen/Kitchin* 1670 Castle, Rotherhithe; 1692 converted to bomb (which see); 1698 sold.
*Cleveland/Cleaveland* 1671 Portsmouth Dyd (Surveyor: Deane); 1715 sold.
*Queenborough/Quinborough* 1671 Chatham Dyd (Surveyor: Pett); 1718 BU.
*Katherine/Catherine* 1671 Chatham Dyd (Surveyor: ____ ); 1720 BU.
*Isle of Wight/Ile of Wight* 1673 Portsmouth Dyd (Surveyor: Furzer); 1701 BU.
*Navy* 1673 [6 guns, 74 tons] Portsmouth Dyd (Surveyor: Deane); 14.4.1698 sold.
*Portsmouth* 1674 Woolwich Dyd (Surveyor: ____ ); 1679 RB at Woolwich Dyd (Surveyor: ____ ); 1694 converted to a bomb of 143 tons; 27.11.1703 wrecked at the Nore.
*Mary* 1677 Chatham Dyd (Surveyor: Pett); 1726 BU.
*Charlotte/Charlot* 1677 Woolwich Dyd (Surveyor: Shish); 1710 RB at Debtford.
*Fubbs* 1682 Pett, Greenwich; possible RB 1701 at Woolwich Dyd (Surveyor: Lee); dimensions not altered, however; 1723 BU.
*Henrietta* 1682 Woolwich Dyd (Surveyor: Shish); 1721 sold.
*Isabella* 1683 Pett, Greenwich; 1702 BU.

## From 1688

*Soesdyke* purchased 1692-1713 (built 1692 by Freame, Wapping).
    Dimensions & tons: 63'2", 58'9" x 19'10" x 9'1½". 116.
    Men: 35/30/12. Guns: 8 (+ 6 swivels). Purchased from the Earl of Monmouth 1692; 1713 sold.
*Squirrell/Squirill* 1694-1714 Chatham Dyd (Surveyor: Lee).
    Dimensions & tons: 47', 36' x 14' x 6'. 37½.
    Men: 4. Guns: 4/2 Chatham Yard yacht; 1714 sold.
*William and Mary* 1694-1801 Chatham Dyd (Surveyor: Lee).
    Dimensions & tons: 76'6", 61'5 ¼" x 21'7" x 9'6". 152 18/94.
    Men: 40/30 Guns: 10/12. Used as a warship during the War of Spanish Succession later Royal Yacht - ketch rig (by 1764); 1764-5 Great Repair at Deptford; by 1782 ship rig; 1801 sold.
PLANS: 1764: Lines/lines & profile/upper deck. 1782: profile & upper deck.

*Scout* 1695-1703 Portsmouth Dyd (Surveyor: Stigant).
    Dimensions & tons: ____ , 38'6" x 13'8' x 6'4'. 38 23/94.
    Men: 6/2. Guns: 4. Portsmouth Yard yacht; 1703 sold.
*Isle of Wight* 1701[RB]-1712 Portsmouth Dyd (Surveyor: Waffe).
    Dimensions & tons: 46'8", 36'9" x 14' x 6'10". 38 31/94.
    Men: 6/5 (1709:30). Guns: 4 (1709: 8 x 3 - 4½' long). For the use of the Governor of the Isle of Wight; in 1709 used for Channel service; 1712 sold.
*Saint Loe* 1701-1716 Plymouth Dyd (Surveyor: Podd).
    Dimensions & tons: ____ , 42' x 14'6" x 6'. 47.
    Men: 2/4. Guns: 4. Based at Plymouth; 1709 to Chatham; 1716 sold.
*Isabella* 1703 [RB]-1715 Deptford Dyd (Surveyor: Harding).
    Dimensions & tons: 63'4", 52'3" x 19'5" x 9'1¾". 104 73/94.
    Men: 45/30/20. Guns: 8/6. Channel service (as warship); 1715 sold.
*Portsmouth* 1703-1741 **Old Portsmouth** - 1772 **Medina** - 1832 Portsmouth (Surveyor: Podd) Classed as 'small yacht'.
    Dimensions & tons: 52'10", 42'10" x 17' x 8'6½". 66 (earlier 50?)
    Men: 15/6 Guns: 4/6 x 2. Launched 01.1703; Portsmouth Superintendent's Yacht; 1741 ordered to replace her & she was renamed *Old Portsmouth*; 1752 fitted for the use of 'the young gentlemen of the (naval ) Academy (at Portsmouth)'; 1772 Great Repair, renamed *Medina* and fitted for the Governor of the Isle of Wight 1832 BU.
*Drake* 1705-1727 Plymouth Dyd (Surveyor: Lock).
    Dimensions & tons: ____ , 34'9" x 16'6" x 8'11½". 50 27/94.
    Men: 4/17 Guns: 6? Plymouth Superintendent's yacht; 1727 BU.
*Bolton* 1709-1817 Portsmouth Dyd (Surveyor: Podd). Later classed as a 'small yacht'.
    Dimensions & tons: 53'2", 38' x 14'6' x 7'6". 42 46/94.
    Men: 12/5 Guns: 6 x 2. Portsmouth Yard yacht; 1763 found to be 'entirely decayed'; 1773 fitted for the use of the 'young gentlemen of the Academy' in lieu of the *Old Portsmouth/Medina* (see above); 1817 BU.
*Dublin* 1709-1752 Deptford Dyd (Surveyor: Allin) ketch-rigged.
    Dimensions & tons: 73'2", 59'8" x 21'7½" x 9'6". 148 39/94.
    Men: 50/45. Guns: 12/10. Launched 13.8.1709; based at Dublin for the use of the Irish Viceroy and Government; 1732 converted to ship rig; 1752 BU.
PLANS: Lines/lines/profile & deck.
*Irish* 1700s [ RB]-1712 Plymouth Dyd (Surveyor: Lock) Classed as a small yacht.
    Dimensions & tons: ____ ____ .
    Men: ____ . Guns: ____ . Rebuild (exact date uncertain) of the 'Irish lighter'; 1712 sold.
*Charlotte/Charlot* `1711-1761 **Augusta** - 1773 **Princess Augusta** - 1818 Deptford Dyd. (Surveyor: Allin ) Royal yacht.

---

[1] There are some indications that the word 'yacht' might have been used in English a little while before 1660 in the same sense as it would be from that date, but there is not enough information to be certain of overthrowing the commonly accepted version of the appearance of the word given above (see A P McGowan, *Royal Yachts* (HMSO for the National Maritime Museum 1975?)

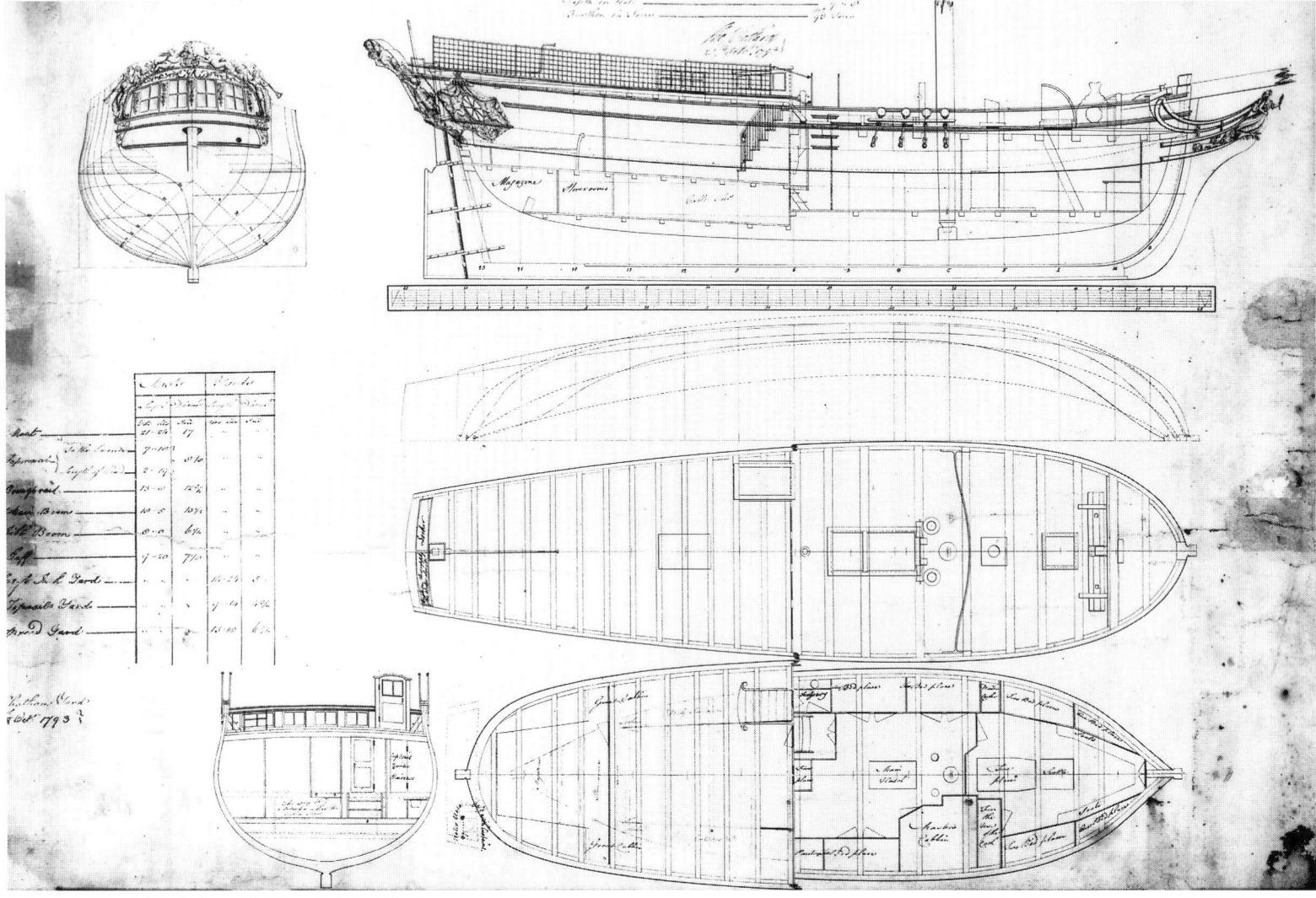

Hull plan (Y70) of the *Chatham* yacht as repaired in 1793

Dimensions & tons: 73'8", 57'7½" x 22'6¼" x 9'6". 155 30/94. Lengthened in 1747 to 79'6"; 184 tons. As taken off in 1800: 80'9", 65'2½" x 23'1¼" x 10'10". 185 10/94.

Men: 25/40. Guns: 8/6. Launched 10.3.1711; 1737 rigged as a ketch; 1761 renamed *Augusta*; 1773 renamed *Princess Augusta* and reclassed as a Sixth Rate; 1818 sold.

PLANS: Lines & profile/decks.

***Carolina/Royal Caroline*** 1716-1732 ex Sixth Rate [20] *Peregrine Galley* Royal Yacht.
Dimensions & tons: 86'6", 71' x 22'10" x 10'. 196 84/94.
Men: ____ . Guns: ____ . Converted during a Great Repair at Woolwich and Deptford Yards; 1732 BU.

PLANS: 1730: Lines, profile & sections/profile, upper deck, breaks in deck & sections/lines, profile & decks/decks/profile, upper decks, sections & breaks in deck.

***Chatham*** 1716-1743 Chatham Dyd (Surveyor: Rosewell) classed as a small yacht.
Dimensions & tons: 56', 44'6" x 16' x 7'6". 60.
Men: 6 Guns: 4. Launched 27.6.1716; Chatham Yard yacht; 1743 sold.

***Queenborough*** 1718[RB]-1777 Sheerness Dyd (Surveyor: Ward) small yacht.
Dimensions & tons: 51'6", 37'3" x 15'2½" x 6'7". 46 4/94.
Men: 7 (13 as survey ship). Guns: 6 x 2. Based at Sheerness; 1775 fitted as a survey ship (classed as a sloop); 1777 sold.

PLANS: Lines, profile & decks.

***Katherine/Catherine*** 1721 [RB]-1801 Deptford Dyd (Surveyor: Stacey) Royal Yacht.
Dimensions & tons: 76'6", 61'6" x 22'4½" x 9'6". 160 68/94.
Men: 40 Guns: 8 x 3 (later 6 x 3). Launched 16.1.1721; 1737 rigged as a ketch; 1801 sold.

***Fubbs*** 1724 [RB]-1781 Deptford Dyd (Surveyor: Stacey) Royal Yacht.
Dimensions & tons: 76'9", 61' x 22' x 9'8". 157.
Men: 40. Guns: 6 x 3. Launched 22.10.1724; 1781 BU.

PLANS: As in 1734? Lines, profile & decks/lower deck & breaks in deck.

***Drake*** 1727 [RB]-1749 Plymouth Dyd. (Surveyor: Lock) small yacht.
Dimensions & tons: 58'6", 45' x 16'10" x 8'5". 67 75/94.
Men: 4. Guns: 6. Launched 6.9.1727; 1749 found 'intirely decay'd' and sold.

PLANS: Lines & profile/lines, profile & upper deck/profile & breaks in deck.

***Mary*** 1728 [RB]-1816 Deptford Dyd (Surveyor: Stacey) Royal Yacht.
Dimensions & tons: 76'6", 61'6" x 22'4" x 9'8". 163 72/94.
Men: 40. Guns: 8 x 3 + 10 swivels. Launched 16.3.1728; 1736 rigged as a ketch; 1783 Rebuild? re-rigged as a ship; 1816 BU.

NOTE: Though her construction was supervised by Stacey, she was probably an Acworth design.

PLANS: 1727?: Lines. 1764: profile. 1772: Lines & profile. 1783 Profile & upper deck.

***Royal Caroline/Carolina*** 1734 [RB]-1749 *Peregrine* -1762 Deptford Dyd (Surveyor: Stacey) Royal Yacht.
Dimensions & tons: 86'6", 70'6" x 24' x 10'6". 216.
Men: 70 (100 as sloop). Guns: 10. Launched 5.2.1734; 1736 registered as a Sixth Rate; 1749/50 renamed *Peregrine* and converted to a ship sloop; 1762 foundered.

PLANS: Lines & profile/profile, upper deck & breaks in deck.

***Chatham*** 1741-1793 Chatham Dyd (Surveyor: Ward) small yacht.
Dimensions & tons: 59'6", 47' x 17'3" x 7'6". 74 36/94.

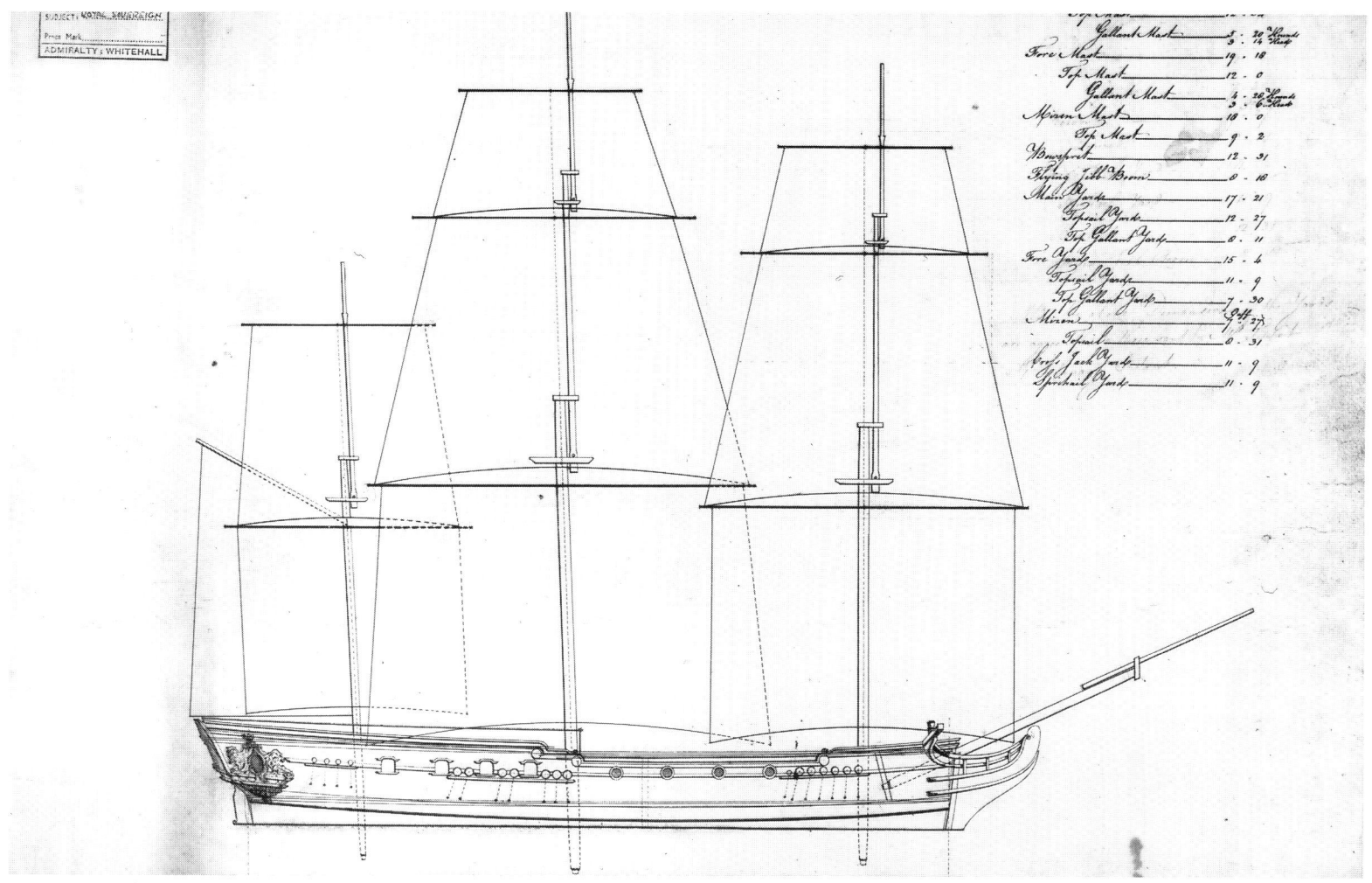

Sail plan (Y31) of the royal yacht *Royal Sovereign* of 1801

Men: 10/9. Guns: 6 x 2. Launched 1.10.1741; Chatham Yard yacht; 1793 began rebuild.

PLANS: Lines & profile/sheer showing decoration (of model)/lines & profile.

**Portsmouth** 1742-1793 Portsmouth Dyd (Surveyor: Lock) small yacht.
Dimensions & tons: 59'6", 48'5" x 18' x 8'6". 83.
Men: 10. Guns: 6 x 2. Launched 30.9.1742; Portsmouth Yard yach ; 1793 began rebuild.

**Royal Caroline** 1750-1761 **Royal Charlotte** -1820 Deptford Dyd (Allin design) ship-rigged Royal Yacht.
Dimensions & tons: 90'1", 72'2½" x 24'7" x 11'. 232.
Men: ____ . Guns: 10. Launched 29.1.1750; classed as a Sixth Rate; 1761 renamed *Royal Charlotte*; 1820 BU.

NOTE: The design of this vessel was the result of a design competition between the Master Shipwrights of the various yards, the Surveyor of the Navy and his assistant. The Surveyor won.

PLANS: Lines, profile & upper deck/profile deck & sections [As fitted: Lines & profile & decorations - original plan - presumably stolen from the Admiralty by Chapman; in the Chapman Collection in the Sjohistoriskas Museum at Stockholm; this is the plan copied in Chapman's 'Architectura Navalis Mercatoria']. 1801: King's apartments.

**Dorset** (ex *Dublin*) 1753-1815 Deptford Dyd (Allin design) Large yacht.
Dimensions & tons: 78', 64'10½" x 21'11" x 10'10". 164.
Men: 50. Guns: 14 swivels (later also 4 x 12*). Launched 17.7.1753 as a replacement for the *Dublin* as the yacht for the Viceroy of Ireland; 1815 sold.

PLANS: Lines & profile/profile & upper deck/lines & profile/lines.

**Plymouth** 1755-1793 Plymouth Dyd (Admiralty design) small yacht, ketch rig.
Dimensions & tons: 64'6", 52'6" x 17'10" x 10'. 88 68/94.
Men: 10 Guns: 6 x 2. Launched 3.12.1755 as a replacement for *Drake* as the Plymouth Yard yacht; 1793 BU.

PLANS: Lines, profile & breaks in deck (two versions).

**Denmark** 1785-1785/1807 **Prince Frederick/Frederick** -1816 **Princess Amelia** 1818 Deptford Dyd Royal Yacht for Denmark.
Dimensions & tons: 89'5", 74'10¾" x 23'6" x 10'. 218.
Men: ____ . Guns: ____ . Laid down: 03.1785; launched 20.8.1785; presented to the Prince Royal of Denmark (for whom she was built) 10.1785; 1807 returned by the Danes as a demonstration of anger and disapproval at the bombardment of Copenhagen and the taking of the Danish fleet; put into service as the *Prince Frederick*; 1816 designated as a Fifth Rate and renamed *Princess Amelia*; 1818 sold.

PLANS: Lines/framing/profile/hold/lower & upper decks/quarter deck & forecastle.

**Chatham** 1793 [RB]-1867 Chatham Dyd Small yacht.
Dimensions & tons: 59'6½", 48'5¼" x 19'/18'7" x 9'8". 93.
Men & guns: as before? Rebuild or repair? 1793-1867 BU.

PLANS: Lines, Profile & decks & midships section. 1826 as fitted: Lines & profile/decks. 1826 proposed for refit: Lines/decks

**Portsmouth** 1794 [RB]-1869 Portsmouth Dyd Small yacht.
Dimensions & tons: 70'4", 53'9" x 18'11" x 11'8". 102.
Men: ____ . Guns: 6 x 2. Launched 1794; Portsmouth Yard yacht; 1869 BU.

PLANS: Lines & profile/decks.

**Plymouth** 1796-1830 Plymouth Dyd (Henslow design) small yacht.
Dimensions & tons: 64'0½", 52'7¼" x 18'6" x 11'8". 102.
Men: ____ . Guns: ____ . Launched 2.11.1796; Plymouth Yard yacht; 1830 BU.

PLANS: Lines (2 versions)/profile & midships section (2 versions)/decks (2 versions). As fitted: Lines, profile & decoration/decks.

**Royal Sovereign** 1804-1850 Deptford Dyd (Henslow design) Royal Yacht, ship rigged.
Dimensions & tons: 96', 80'5" x 25'8" x 10'6". 278.

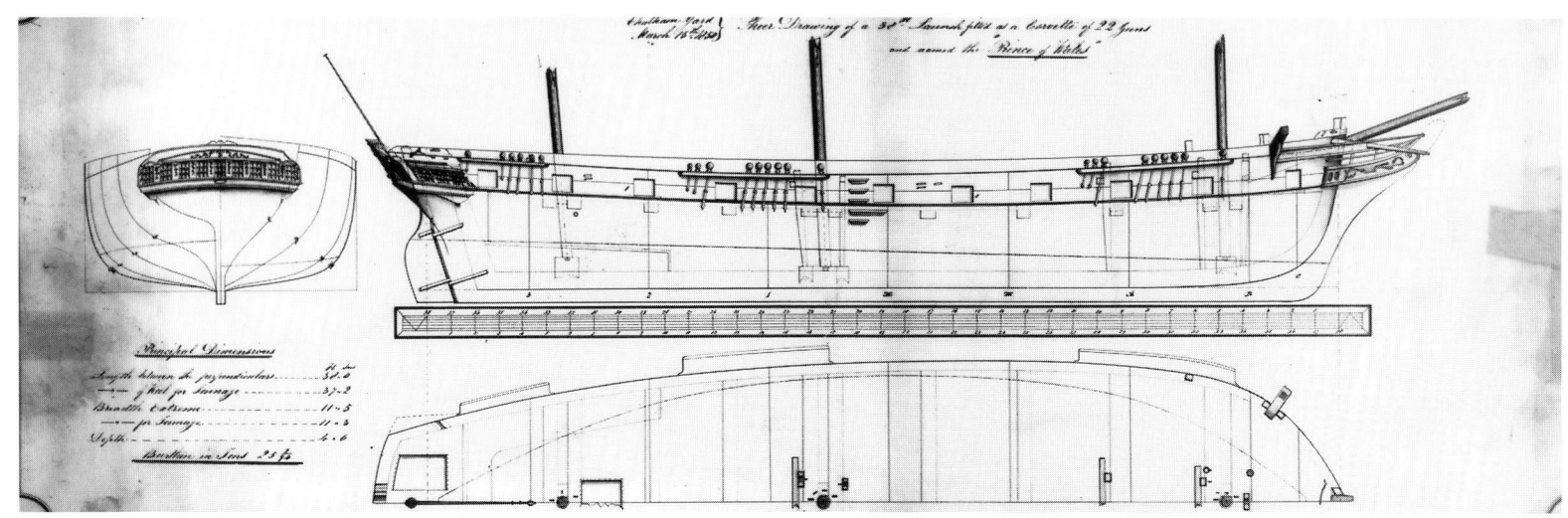

Hull plan (Y86) of a 30 foot launch fitted out to resemble a miniature 22 gun corvette and named *Prince of Wales* at Chatham in 1850

Men: ____ . Guns: ____ . Laid down 11.1801; launched 12.5.1804; 1850 BU.

PLANS: Lines/profile & decks/orlop/lower deck/profile & upper deck/framing/windlass. As fitted: Lines, profile & decorations. 1805 main deck. 1818: lower deck. No date: sails/midships section.

*William and Mary* 1807-1849 Deptford Dyd (Henslow design) Royal Yacht, ship rigged.
Dimensions & tons: 85'0½", 70'3½" x 23'2½" x 11'1". 199
Men: ____ . Guns: ____ . Launched 14.11.1807; 1849 BU.

PLANS: Lines/framing/ decks/profile.

*Admiralty Yacht* 1814-1830 *Plymouth* -1870 Woolwich Dyd (Peake design) small yacht.
Dimensions & tons: 68'7", 54'5⅝" x 20'1" x 12'8". 115.
Men: 6. Guns: None? Launched 21.5.1814; 1830 renamed *Plymouth* as that yard's yacht; by 1860s Devonport Yard Craft No 1; 1870 sold.

*Royal George* 1817-1905 Deptford Dyd (Peake design) Royal Yacht, ship rigged.
Dimensions & tons: 103', 88'4¼" x 26'8" x 11'6". 330.
Men: ____ . Guns: ____ . Laid down: 05.1814; launched 17.7.1817; 1821 registered as a Third Rate; 1905 BU.

PLANS: Lines/profile /orlop/lower deck/upper deck. As Fitted: ballast. 1818: apartments. 1821: lower deck/main deck/upper deck. 1822: method of fitting decks. 1822: lower deck/main deck/upper deck/ventilation plan. No date: decorations.

*Prince Regent* 1820-1847 Portsmouth Dyd (Portsmouth Academy Apprentices' design) Royal Yacht.
Dimensions & tons: 96'0½", 81'3" x 25'8¼" x 10'0¾". 282.
Men: ____ . Guns: ____ . Laid down: 09.1815 (temporarily housed over on the stocks); launched 12.6.1820; 1836 to be a present for the Imam of Muscat (but not sent?); 1847 BU?

PLANS: Lines/profile & upper deck/lower & main decks /broadside decoration.

*Navy Board/Hart* (ex revenue cutter *Hart*?) 1822-1825 [1875] Woolwich Dyd ( ____ ? design) Navy Board yacht, cutter rig.
Dimensions & tons: 53'6", 44'5" x 18'8" x 7'6½".
Men: ____ . Guns: ____ . Launched 16.2.1822; 1825 yard craft at Sheerness [which see].

*Royal Charlotte* 1824-1832 Woolwich Dyd (Seppings design) Royal Yacht ship rigged.
Dimensions & tons: 85'8", 72'8⅜" x 23'0½" x 8'2". 202.
Men: 15 Guns: 6 x 1 pdr swivels. Laid down: 04.1820; launched 22.11.1824; 1832 BU.

PLANS: Lines/framing/midships section/frame sections/profile & upper deck/orlop & lower deck/stern/panelling. 1825 proposal to fit with paddle engine: profile /midships section.

Punt, hand-operated paddles, for Virginia Water 1824. Deptford Dyd.
Dimensions: 40' x 12'9" x 3'4". Double-ended, hand-cranked paddles on the quarters.

PLANS: Profile, section & outline deck.

*Fanny* 1831-1878 Portsmouth Dyd (Captain Hayes' design) Admiralty yacht.
Dimensions & tons: 68'6", 55'3" x 21'8½"x 10'4" ____ (1857 given as 75'6", 62'1⅜" x 21'8¼" x 10'4". 153).
Men: ____ . Guns: ____ . Launched: 28.2.1831; 1834 tender at Portsmouth; 1857 lengthened; 1862 to Coastguard service at Kingstown; 31.10.1878 sunk off the Tuskar Rock in collision with SS *Helvetia*.

*Royal Adelaide* 1834-1877 Sheerness Dyd (Symonds design) miniature frigate.
Dimensions & tons: 50', 42'8½" x 15' x 8'. 50.
Men: ____ . Guns: 22 x 1 pdr brass guns. Laid down: 09.1833; Completed in 12.1833, taken to pieces and transported to Virginia Water where she was assembled and launched on the lake 13.5.1834. She was a complete miniature frigate with a coppered bottom and was sailed for the amusement of the Royal Family upon the lake; 1877 BU.

PLANS: Lines/lines & sails/sails.

'*Prince of Wales*' 1850- ____ ? 38 foot launch fitted at Chatham Dyd as a miniature version of a 22 gun corvette (purpose unknown; but possibly for Virginia Water?).
Dimensions & tons: 38', 37'2" x 11'5" x 4'. 25 2/94.
Men: ____ . Guns: ____ ; fate unknown.

PLANS: Lines, profile, deck.

# Chapter 3: Conversions and Reclassifications

See the original entries for these ships for full details. Some minor reclassifications are not counted here (such as that of *Victory* to a 98 gun Second Rate in 1808), nor are the 1816 reclassifications (38 to 46 gun ships, etc) which are noted in the first section of this work. Also ship sloops rerigged as brigs or vice versa, and similar cases with cutters and schooners, as well as brigs rigged as brigantines or cutters as ketches, do not feature here. Nor is notice taken of fireships or bomb vessels serving as sloops (which they did for much of the time).

## 92 Gun Two-deckers, Second Rate

1847   *Prince Regent* from 120 (1823); 1861 screw

## 84 Gun Three-deckers, Second Rate

1756   *Royal William* from 100 (1719); 1790 hulked

## 80 Gun Three-deckers, Third Rate

1753   *Barfleur* from 90 (1716); 1764 hulked
1755   *Prince George* from 90 (1723); 1758 lost
1756   *Marlborough* from 90 (1732); 1762 lost

## 80 Gun Two-deckers, Third Rate

1821   *Ocean* from 110 (1805); 1831 hulked
1830   [80/76] *Union* from 110 (1811); 1833 BU

## 74 Gun Two-deckers, Third Rate

1746   *Namure* from 90 (1729); 1749 lost
1749   *Torbay* from *Neptune* 90 (1730); 1784 hulked
1781   [for local service] *Yarmouth* from 70 (1745); 1783 hulked
1801   *Blenheim* from 90/98 (1761); 1807 lost
1804   *Atlas* from 90/98 (1782); 1814 hulked
1805   *Namure* from [72] from 90/98 (1756); 1807 hulked
1811   *Queen* from 90/98 (1769); 1821 BU
1814   *Windsor Castle* from 90/98 (1790); 1839 BU

## 66 Gun Two-deckers, Third Rate

1747   *Devonshire* from 80 (1745); 1772 BU
1748   *Cumberland* from 80 (1739); 1760 lost

## 64 Gun Two-deckers, Third Rate

1755   *Ipswich* from 70 (1730); 1757 hulked

## 60 Gun Two-deckers, Fourth Rate

1757   *Intrepid* from French 64; 1765 BU

## Screw Line of Battle Ships [guns]

1847   [Block ship] *Ajax* from 74 (1809); 1864 BU
       [Block ship] *Blenheim* from 74 (1813); 1858 hulked
1848   [Block ship] *Hogue* from 74 (1811); 1865 BU
1851   *Sans Pareil* from uncompleted 80; 1867 sold
1852   [Block ship] *Edinburgh* from 74 (1811); 1866 sold
       *Duke of Wellington* from uncompleted 120; 1863 hulked
1853   *Princess Royal* from uncompleted 90; 1872 sold
       *Caesar* from uncompleted 90; 1870 sold
       *Majestic* from uncompleted 80; 1868 BU
       *Cressy* from uncompleted 80; 1867 sold 1854
       *Royal George* from 120 (1827); 1875 sold
       *Nile* from 90/92 (1839); 1876 hulked
       *Royal Albert* from uncompleted 120; 1883 sold for BU
       *Exmouth* from uncompleted 90; 1877 hulked
       *Algiers* from uncompleted 99; 1870 sold
       *Hannibal* from uncompleted 90; 1874 hulked
       *Colossus* from 80 (1848); 1867 sold
       *Orion* from uncompleted 80; 1867 BU
1855   *Pembroke* from 74 (1812); 1873 hulked
       *Russell* from 74 (1822); 1858 Coastguard guard ship; 1865 BU
       *Hawke* from 74 (1820); 1865 BU
       *Cornwallis* from 74 (1813); 1865 hulked
       *Marlborough* from uncompleted 120; 1878 hulked
       *Centurion* from 80 (1844); 1870 sold
       *Brunswick* from uncompleted 80; 1867 sold
       *Hastings* from purchased 74 (1819); 1857 to the Coastguard; 1870 hulked
1856   *Mars* From 80 (1848); 1869 hulked
1857   *Royal Sovereign* from uncompleted 120; 1864 finally fitted out as a turret ironclad
       *Meeanee* from 80 (1848); 1867 hulked
1858   *Nelson* from 120 (1814); 1867 to New South Wales
       *London* from 90/92 (1840); 1874 hulked
       *Windsor Castle* from uncompleted 110; 1869 hulked
       *Aboukir* from 90 (1848); 1878 sold
       *Goliath* from 80 (1842); 1870 hulked
1859   *Saint George* from 120 (1840); 1883 sold
       *Neptune* [72] from 120 (1832); 1875 BU
       *Waterloo* from 120 (1833); 1862 *Conqueror*; 1876 hulked
       *Trafalgar* from 120 (1841); 1873 *Boscawen*; 1906 sold
       *Queen* [86] from 110 (1839); 1871 BU
       *Lion* from 80 (1847); 1905 sold
       *Irresistible* from uncompleted 80; 1894 sold
       *Hood* from uncompleted 80; 1888 sold
1860   *Royal William* [72] from 120 (1833); 1885 hulked
       *Rodney* from 90/92 (1833); 1882 BU
       *Prince of Wales* from uncompleted 120; 1869 hulked
       *Frederick William* from uncompleted 110; 1876 hulked
1861   *Prince Regent* from 92 ex 120 (1823); 1873 BU
       *Bombay* from 84 (1828); 1864 burnt
       *Albion* from 90 (1842); 1884 BU
       *Collingwood* from 80 (1841); 1867 sold

## 58 Gun Frigates, Fourth Rate

1813   *Goliath* from 74 (1781); 1815 BU
       *Saturn* from 74 (1786); 1825 hulked
1813   *Majestic* from 74 (1785); 1816 BU
1818   *Elephant* from 74 (1786); 1830 BU
1825   *Excellent* from 74 (never completed conversion?); 1830 hulked

## 54 Gun Two-decker (all carronade armament)

1782   *Rippon* from 60 (1758); 1788 hulked

## 50 Gun Frigates, Fourth Rate

1826   *Dublin* from 74 (1812); 1845 laid up

1827 ***Barham*** from 74 (1811); 1840 BU
1827 ***Centurion*** from ***Clarence*** 74 (1812); Not carried through and 1828 BU
***Greenwich*** from ***Rodney*** 74 (1809); conversion never completed? 1836 sold
1828 ***Alfred*** from 74 ex ***Asia*** (1811); 1858 hulked
1830 ***Cornwall*** from 74 (1812); 1859 hulked
1831 ***Eagle*** from 74 (1804); 1857 hulked
***Conquestadore*** from 74 (1810); 1856 hulked
1832 ***Gloucester*** from 74 (1812); 1861 hulked
1833 ***Vindictive*** from 74 (1813); 1862 hulked
1836 ***America*** from 74 (1810); 1864 hulked
1840 ***Warspite*** from 74 (1807); 1862 hulked
1845 ***Grampus*** from ***Tremendous*** 74 (1784); 1866 hulked

## 50 Gun Two-deckers, Fourth Rate

1720 ***Monck*** from 60 (1702); 1720 lost
1744 ***Centurion*** (temporarily renamed ***Eagle***) from 60 (1733); 1769 BU
1752 ***Deptford*** from 60 (1732); 1767 sold
1778 ***Buffalo*** from storeship ex ***Captain*** 70 (1743); 1783 BU
1779 ***Leviathan*** from storeship ex ***Northumberland*** 70 (1750); 1780 lost

## 44 Gun Two-deckers, Fifth Rates

1742 ***Chester*** from 50 (1708); 1743 hospital ship
1744 ***Ruby*** from 50 (1708); 1748 sold
***Milford*** from 50 ***Advice*** (1712); 1749 sold
***Enterprise*** from 50 ***Norwich*** (1718); 1771 BU
1747 ***Romney*** from 50 (1726); 1757 sold

## 40 Gun Two-deckers, Fifth Rates

1716 ***Dover*** from 48/42 Fourth Rate (1695); 1730 BU
1726 ***Southampton*** from 48/42 Fourth Rate (1699) [Fifth Rate 1716]; 1728 hulked

## 38 Gun frigates, Fifth Rates

1794 ***Anson*** from 64 (1781); 1807 lost
1795 ***Indefatigable*** from 64 (1784); 1816 BU
***Magnanime*** from 64 (1780); 1813 BU

## 32 Gun One-decker (proto-frigate), Fifth Rate

1757 ***Saphire*** from 44 (1741); 1784 sold
***Adventure*** from 44 (1741); 1770 sold

## 32 Gun Frigates

1792 ***Venus*** from 36 (1756); 1809 ***Heroine***; 1824 hulked

## 28 Gun Frigates

1777– ***Boston*** 32 (1762)
1779 ***Jason*** 32 (1763)
***Stag*** 32 (1758)
***Quebec*** 32 (1760)
1804 ***Negro***? from troopship ex ***Niger*** 32 (1759); 1814 sold

## Paddle Frigates

1843 ***Penelope*** from 38/46 (1829); 1864 sold

## Screw Frigates

1846 ***Amphion*** from 36/42 (1830); 1862 sold
1847 ***Seahorse*** from 38/46 (1830); 1856 steam mortar frigate; 1870 hulked
***Forth*** from 38/46 (1833); 1856 steam mortar frigate; 1869 hulked
1848 ***Eurotas*** from 38/46 (1829); 1856 steam mortar frigate; 1865 sold
1851 ***Horatio*** from 38 (1807); 1855 steam mortar frigate; 1865 sold
1852 ***Imperieuse*** from uncompleted 60; 1867 sold
1855 ***Shannon*** from uncompleted 50; 1871 sold
1856 ***Fox*** from 38/46 (1829); (also used as a transport); 1882 BU

NMM photo negative X.1286. Sketch by R S Thomas of the *Penelope* of 1829 as a paddle frigate

CONVERSIONS AND RECLASSIFICATIONS

|  | *Emerald* from uncompleted 60; 1869 sold |
|  | *Liffey* from uncompleted 50; 1877 hulked |
| 1857 | *Melpomene* from uncompleted 60; 1875 sold |
| 1859 | *Phaeton* from 50 (1848); 1875 BU |
|  | *Immortalite* from uncompleted 60; 1883 sold |
|  | *Narcissus* ? from uncompleted 50?; 1883 sold |
| 1860 | *Severn* from uncompleted 50; 1876 BU |
|  | *Constance* from 50 (1846); 1875 BU |
|  | *Arethusa* from 50 (1849); 1874 hulked |
|  | *Sutlej* from 50 (1855); 1869 BU |
|  | *Phoebe* from 50 (1854); 1875 BU |
| 1861 | *Octavia* from 50 (1849); 1876 BU |
|  | *Leander* from 50 (1848); 1867 sold |

## 28/24 Gun, Fifth Rate

| 1704 | *Strombolo* from Fireship |

## 24 Gun, Sixth Rate

| 1710 | *Hunter* from fireship (1690); 1710 taken |
| 1717 | *Lyme* from 32 (1694); 1720 BU |
| 1744 | *Fowey* from 40 (1709); 1746 sold |
| 1762 | *Pearl* from 44 (1744); 1771 BU |
| 1777 | *Convert* American; 1778 *Beaver's Prize* sloop |
| 1793 | *Daphne* from 20 (1776); 1802 sold |
|  | *Ariadne* from 20 (1776); 1814 sold |

## 20 Gun Sixth Rates

| 1716/20 | *Tartar* from 32 (1702); 1732 BU |
| Ca 1716? | *Dolphin* from 32 (1711); 1730 BU |
| 1717 | *Experiment* from 32 (1689); 1724 BU |
|  | *Sheerness* from 32 (1690); 1730 BU |
| 1718 | *Rye* from 32/28 (1696); 1727 BU |
| 1798 | *Hyena* from 24 (1778); 1802 sold |

## Corvettes, Sixth Rate [Guns]

| 1820 | *Ariadne* [28/26] from sloop; 1837 hulked |
| 1821 | *Semiramis* [24] from 36; 1844 BU |
|  | *Valorous* [26] from sloop; 1829 BU |
| 1831 | *Aigle* [24] from 36 (1801); 1852 hulked |
|  | *Curacoa* [26/24] from 36 (1809); 1849 BU |
|  | *Magicienne* [24] from 36 (1812); 1845 BU |
|  | *Tweed* [20] from 28 (1823); 1852 BU |
| 1833 | *Tribune* [24] from 36 (1803); 1839 lost |
| 1844 | *Daedalus* [19] from 38/46 (1826); 1861 hulked |
| 1845 | *Amazon* [26] from 38/46 (1821); (1848-1852 lent to Liberia); 1863 sold |
|  | *Havannah* [19] from 36 (1811); 1860 hulked |
| 1846 | *Amphitrite* [26] from 38/46 (1816); 1857 hulked |
|  | *Brilliant* [22] from 36/42 (1814); 1860 hulked |
| 1847 | *Trincomalee* [26] from 38/46 (1817); 1861 hulked |

## Screw Corvettes

| 1854 | *Malacca* from 24 (1853); 1869 sold |

## Registered as Sloops

See also Fireships and Bombs

| 1777 | *Alert* cutter (1777); 1778 taken |
| 1779-1783 | *Kite* purchased cutter (1778) |
| 1779 | *Rattlesnake* cutter (1777); 1781 lost |
| 1782 | *Termagant* from purchased 22 (1780); 1795 sold |
| 1801-1806 | *Flora* 36 (1780) |

## Fireships

| 1688 | *Pearl* from 32 (1651); 1689 32 |
|  | *Garland* from 38 (1654); 1689 38 |
|  | *Guernsey* from 30 (1654); 1689 30 |
|  | *Dartmouth* from 32 (1655); 1689 32 |
|  | *Richmond* from 28 (1656); 1689 28 |
|  | *Swan* from purchased 32; 1689 32 (or remained a fireship?) |
|  | *Saint Paul* from Algerine 32; 1698 sold |
| 1689 | *Rose* from 28 (1674); 1698 sold |
| 1690 | *Roebuck* and *Speedwell* began building as fireships but completed as Fifth Rates (26 and 28/24 guns respectively) |
| 1701 | *Terrible* from 28/24 (1694); 1710 Fifth Rate again |
| 1717 | *Bedford Galley* from 32 (1709); 1725 breakwater |
| 1717-1717 | *Valeur* French 24 |
| 1719 | *Poole* from 32/28 (1696); 1737 BU |
| 1727 | *Bridgewater* from 36/32 (1698); 1733 BU |
| 1734 | *Solebay* from bomb ex 24 (1711); 1735 24 again |
| 1737 | *Alborough* from 20 (1727); Ca 1740 20 again |
| 1739 | *Success* from 20 (1712); 1743 BU |
| 1747 | *Dolphin* from 20 (1732); 1755 renamed *Firebrand*; 1757 20 again as *Penguin* |
| 1755 | *Lightning* from *Viper* sloop (1746); 1762 sold |
| 1758 | *Cormorant* from French sloop (1757); 1762 sold |
|  | *Roman Emperor* from purchased (1757); 1758 sloop; 1762 fireship; 1763 sold |
| 1762 | *Raven* from sloop (1745); 1763 sold |
|  | *Grampus* from sloop (1746); 1772 *Strombolo*; 1780 hulked |
| 1778 | *Pluto* from *Tamar* sloop (1758); 1780 taken |
|  | *Firebrand* from *Porpoise* purchased sloop (1777); 1781 lost |
| 1779-1783 | *Explosion* from *Swan* sloop (1767) |
| 1779 | *Spitfire* from *Speedwell* sloop (1752); 1780 sold |
|  | *Comet* from *Diligence* sloop (1756); 1780 sold |
|  | *Salamander* from *Shark* purchased sloop (1776); 1783 sold |
|  | *Blast* from *Druid* purchased sloop (1776); 1783 sold |
|  | *Lightning* from *Sylph* purchased sloop (1776); 1783 sold |
|  | *Basilisk* from *Grasshopper* purchased sloop (1776); 1781 hulked |
|  | *Harpy* [temporary] from purchased sloop (1777); 1783 sold |
|  | *Snake* [temporary] from purchased sloop (1777); 1780 hulked |
| 1798 | *Wasp* from purchased brig (1782); 1800 expended |
|  | *Trimmer* [temporary] from purchased cutter (1782); 1801 sold |
|  | *Pyl* from Dutch brig (1796); 1801 sold |
| 1799 | *Otter* from brig (1782); 1801 sold |
|  | *Rosario* [temporary fire vessel] from Spanish brig; 1800 expended |
| 1800 | *Falcon* from brig (1782); 1800 expended |
| 1801 | *Zephyr* from purchased brig (1795); 1808 sold |
| 1809 | *Mediator* from Purchased 44 (1804); 1809 expended |

## Bomb Vessels

| 1692 | *Kitchen* from yacht (1670); 1698 sold. |
| 1694? | *Julian* from a French 14 (1690); 1698 sold |
| 1694 | *Portsmouth* from yacht (1674); 1703 wrecked |
| 1719 | *Speedwell* from 20 (1716); 1720 wrecked |
| 1726 | *Solebay* from 24 (1711); 1734 fireship |
| 1727 | *Seaford* from 20 (1724); 1740 BU |
| 1737 | *Shoreham* from 20 (1720); 1744 sold |
| 1758 | *Baltimore* from sloop (1742); 1762 sold |
|  | *Falcon* from sloop (1745); 1759 lost |
|  | *Kingfisher* from sloop (1745); 1760 sloop again |
|  | *Pelican* from purchased sloop (1757); 1760 sloop; 1762 bomb; 1763 sold |

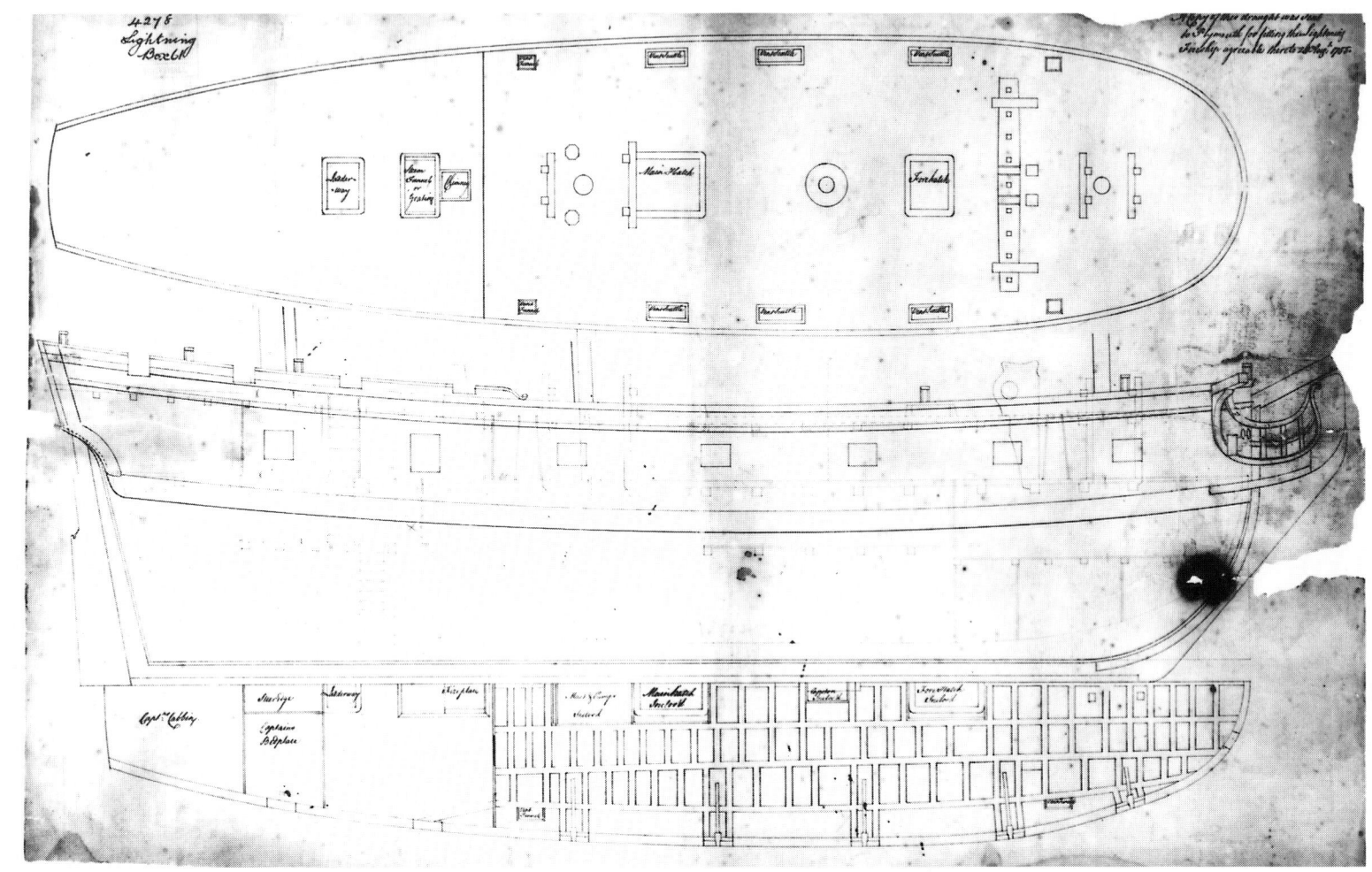

Profile and decks plan (4278) of the fireship *Lightning* as fitted in 1756. She was originally built as the sloop *Viper* in 1746. The closely subdivided nature of the 'fire deck' is shown clearly here

| | | | |
|---|---|---|---|
| 1779–<br>1779? | (?)***Sphinx*** 20 (1775) | 1768–<br>1770 | ***Superb*** 74 (1760) |
| 1798 | ***Perseus*** from 20 (1776); 1805 BU | 1768 | ***Dorsetshire*** from 70 (1757); 1775 BU |
| | ***Zebra*** from sloop (1780); 1812 sold | 1769 | ***Dragon*** from 74 (1760); 1780 hulked |
| | ***Bulldog*** from sloop (1782); 1801 hulked | 1787 | ***Chichester*** from 44 (1785); 1794 storeship |
| | ***Fury*** from sloop (1790); 1811 BU | | ***Sheerness*** from 44 (1787) on completion; 1805 lost |
| 1807 | ***Hound*** from purchased sloop (1801); 1812 BU | 1787–<br>1789 | ***Gorgon*** 44 (1785) |
| 1808 | ***Proselyte*** from purchased 28 (1804); 1808 lost | | |
| 1810 | ***Strombolo*** from ***Autumn*** hulk ex purchased sloop (1801); 1815 sold | 1789 | ***Janus*** from 44 (1778); 1800 wrecked |
| 1811 | ***Heron*** from purchased sloop (1803); 1816 sold | 1790–<br>1802 | ***Ulysses*** 44 (1779) |
| 1812 | ***Meteor*** from ***Star*** sloop (1805); 1816 sold | 1791 | ***Assurance*** from 44 (1780); 1796 transport |
| | | | ***Argo*** from 44 (1781); 1816 sold |

## Mortar Brigs

| | | | |
|---|---|---|---|
| 1809 | ***Charger*** from gunbrig (1801); 1814 sold | 1793 | ***Regulus*** 44 (1785); 1816 BU |
| | ***Desperate*** from gunbrig (1805); 1814 sold | | ***Experiment*** from 44 (1784); 1805 hulked |
| | ***Indignant*** from gunbrig (1805); 1811 BU | | ***Woolwich*** from 44 (1785); 1798 storeship |
| | ***Rebuff*** from gunbrig (1805); 1814 sold | 1798–<br>1803 | ***Dictator*** 64 (1783) |
| 1847 | ***Curlew*** from brig (1830); 1849 BU | 1798–<br>1805 | ***Druid*** 32 (1783) |

## Steam Mortar Frigates

See under Steam Frigates above

| | |
|---|---|
| 1798–<br>1806 | ***Inconstant*** 36 (1783) |
| | ***Hebe*** French 36/38 (1805 renamed ***Blonde***?) |
| 1798 | ***Europa*** from 50 (1783); 1814 sold |

## Troopships

NOTE: These were mostly temporary conversions of serving vessels which would then revert to their original function.

| | | | |
|---|---|---|---|
| 1764–<br>1769 | ***Burford*** 70 (1757) | | ***Expedition*** from 44 (1784); 1810 hulked |
| | | | ***Blonde*** from 32 (1787); 1803 hulked |
| 1764–<br>1770 | ***Fame*** 74 (1759) | | ***Tromp*** from Dutch 60/50 (1796); 1800 hulked |
| | | | ***Alkmaar*** from Dutch 60/50 (1797); 1801 hospital ship |
| | | | ***Admiral De Vries*** from Dutch 64 (1797); before 1806 hulked |
| | | | ***Haerlem*** from Dutch 64 (1797); 1805 hulked |
| | | | ***Wasanaar/Wassenaar*** from Dutch 64 (1797); 1802 hulked |
| | | 1799–<br>1805 | ***Stately*** 64 (1784) |

# CONVERSIONS AND RECLASSIFICATIONS

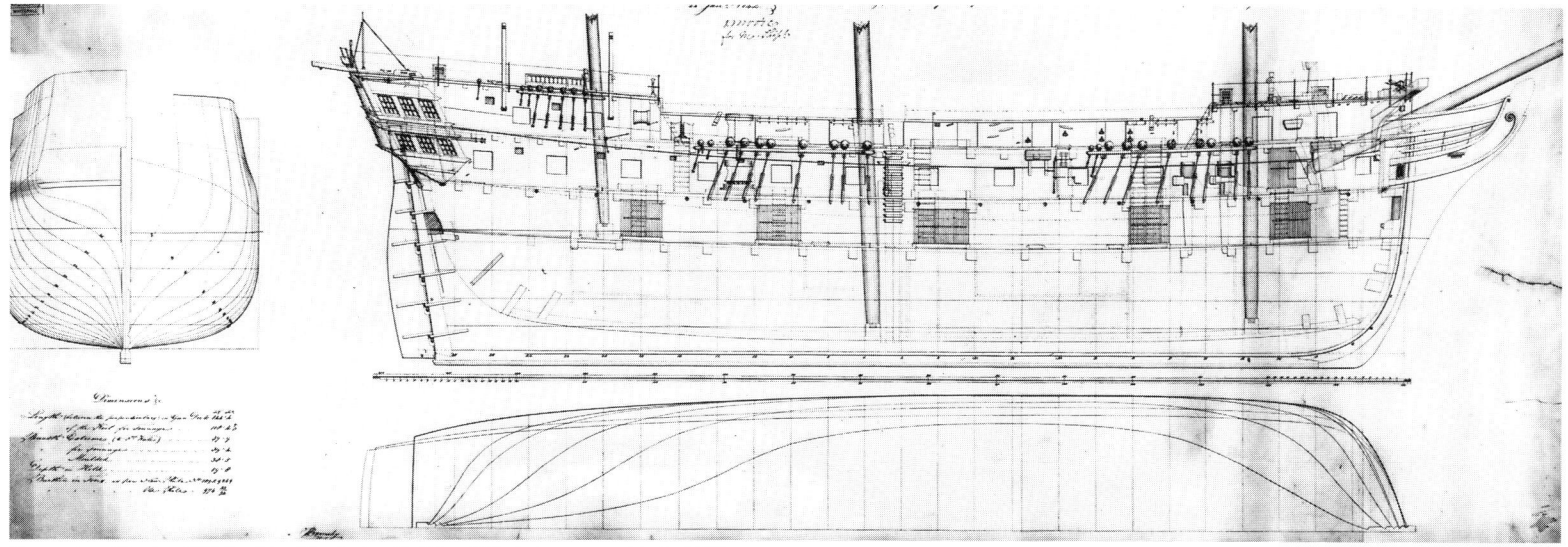

Line and profile plan (1514) of the *Tortoise* (ex *Sir Edward Hughes*) as a storeship in 1842

| | |
|---|---|
| 1799 | *Nassau* from 64 (1785); 1799 lost |
| | *Trusty* from 50 (1782); 1809 hulked |
| | *Adventure* from 44 (1784); 1801 hulked |
| | *Romulus* from 36 (1785); 1803 floating battery |
| | *Niger* from 32 (1759); 1804 28 (*Negro?*) |
| | *Blanche* from 32 (1786); 1799 lost |
| | *Resource* from 28 (1778); 1803 hulked |
| | *Delft* from Dutch 64 (1797); 1802 hulked |
| | *Brakel* from Dutch 54 (1795); 1807 hulked |
| | *Espion* from floating battery ex French 32 (1794); 1800 wrecked |
| | *Eurus* from Dutch 32 (1796); 1803 storeship |
| | *Sensible* from French 36 (1798); 1802 wrecked 1800-1807 |
| 1800-1807 | *Inflexible* 64 (1780) |
| 1800-1805 | *Thetis* 38 (1782) |
| | *Astrea* 32 (1781) |
| 1800 | *Dolphin* from hospital ship ex 44 (1781); 1804 storeship |
| | *Charon* from hospital ship ex 44 (1783); 1805 BU |
| | *Winchelsea* from 32 (1764); 1803 hulked |
| | *Cyclops* from 28 (1779); 1807 hulked |
| | *Vestal* from 28 (1779); (1803-1810 hulked); 1814 hulked |
| | *Pegasus* from 28 (1779); 1814 hulked |
| | *Dido* from 28 (1784) 1804 hulked |
| | *Thisbe* from 28 (1783); 1815 sold |
| | *Heroine* from purchased 32 (1783); 1803 hulked |
| | *Renommee* from French 44/38 (1796); 1810 BU |
| | *Wilhelmina* from Dutch 32 (1798); 1813 sold |
| 1801-1804 | *Thalia* 36 (1782) |
| 1801 | *Iphigenia* from hulk ex 32 (1780); 1801 burnt |
| | *Pandour* from Dutch 44 (1799); 1803 floating battery |
| 1802 | *Coromandel* from transport ex purchased 56 (1795); 1807 hulked |
| By 1810 | *Ceylon* from *Bombay* purchased 32 (1805); 1810 taken and retaken; 1816 laid up; 1832 hulked |
| 1810 | *Latona* from 38 (1781); 1813 hulked |
| | *Romulus* from floating battery ex 36 (1785); 1813 hulked |
| | *Mercury* from floating battery ex 28 (1779); 1814 BU |
| | *Leyden* from floating battery ex Dutch 64 (1799); 1815 sold |
| | *San Fiorenzo* from French 34 (1794); 1812 hulked |
| | *Melpomene* from French 38 (1794); 1815 sold |
| | *Brune* from French 38 (1808); 1816 hulked |
| 1811 | *Leopard* from 50 (1790); 1814 lost |
| | *Mermaid* from 32 (1784); 1815 BU |
| | *Freya/Freija* from Danish 36 (1807); 1816 sold |
| 1812 | *Diomede* from 50 (1798); 1816 BU |
| | *Success* from 32 (1781); 1813 hulked |
| | *Fox* from 32 (1780); 1816 BU |

| | |
|---|---|
| | *Nemesis* from 28 (1780); 1814 sold |
| | *Ardent* from purchased 64 (1796); 1813 hulked |
| | *Bristol* from *Agincourt* victualler ex purchased 64 (1796); 1814 sold |
| | *Dover* from Franco-Italian 32 (1811); 1819 guardship; 1825 hulked |
| 1813 | *Dictator* from 64 (1783); 1817 BU |
| | *Woolwich* from storeship ex 44 (1785); 1813 wrecked |
| | *Hydra* from 38 (1797); 1820 sold |
| 1814 | *Penelope* from 36 (1798); 1815 wrecked |
| | *Bucephalus* from 36 (1809); 1822 hulked |
| | *Thames* from 32 (1805); 1816 BU |
| | *Niobe* from French 38 (1800); 1816 BU |
| | *Belle Poule* From French 38 (1806); 1816 sold |
| | *Alceste* from French 38 (1806); 1817 lost |
| | *Weser* from French 38 (1813); 1817 sold |
| | *Trave* from French 38 (1813); 1821 sold |
| 1822 | *Romney* from 50 (1815); 1837 hulked |
| 1832 | *Jupiter* from hulk ex 50 (1813); 1846 hulked |
| | *Atholl* from 28 (1820); 1851 hulked |
| 1838 | *Hercules* from 74 (1815); 1852 emigrant ship |
| | *Apollo* from 38 (1805); 1853 storeship |
| 1839 | *Rattlesnake* from 28 (1822); 1846 survey ship |
| | *Sapphire* from 28 (1827); 1847 hulked |
| 1841 | *Belleisle* from 74 (1819); 1854 hulked |
| 1842 | *Resistance* from 38 (1805); 1858 BU |
| | *Crocodile* from 28 (1825); 1850 hulked |

## Storeships

| | |
|---|---|
| 1757 | *Crown* (24) from 44 (1747); 1770 sold |
| 1760 | *Southsea Castle* (24) from 44 (1745); 1762 lost |
| 1777 | *Buffalo* (30) from *Captain* 70 (1743); 1778 50 |
| | *Leviathan* (30) from *Northumberland* 70 (1750); 1779 50 |
| | *Grampus* (30) from *Buckingham* 70 (1751); 1779 lost |
| 1783 | *Abondance* from French 28 (1781); 1784 sold |
| 1788 | *Camel* from *Mediator* 44 (1782); 1810 BU |
| 1793 | *Gorgon* from 44 (1785); 1805 floating battery |
| 1793-1795 | *Inflexible* 64 (1780) |
| 1794 | *Chichester* from troopship ex 44 (1785); 1810 hulked |
| 1795 | *Serapis* from 44 (1782); (1801-1803 floating battery); 1819 hulked |
| | *Alliance* from Dutch 36 (1795); 1802 sold |
| 1797 | *Prince Frederick* from Dutch 64 (1796); 1800 hulked |
| | *Grampus* from purchased 54 (1795); 1799 lost |
| 1798 | *Woolwich* from troopship ex 44 (1785); 1813 troopship |
| 1799 | *Roebuck* from hospital ship ex 44 (1774); 1803 floating battery |
| 1800 | *William* from purchased sloop (1798); 1810 BU |
| 1801 | *Portland* from 50 (1770); 1802 hulked |

|  |  |
|---|---|
|  | *Prevoyante* from French 36 (1795); 1819 sold |
| 1802 | *Hindostan* from purchased 54 (1795); 1804 lost |
| 1803 | *Madras* from purchased 56 (1795); 1807 BU |
|  | *Eurus* from troopship ex Dutch 32 (1796); 1806 hulked |
| 1804 | *Dolphin* from troopship ex 44 (1781); 1817 BU |
| 1805 | *Alkmaar* [for the Downs] from hospital ship ex Dutch 60/50 (1797); 1807 laid up; 1815 sold |
|  | *Porpoise* [for Australia, registered as a Sixth Rate] from purchased sloop; 1814 hulked |
| 1807 | *Weymouth* [18] from purchased 44 (1804); 1828 hulked |
| 1808 | *Malabar* [10] from purchased 56 (1804); 1815 *Coromandel*; 1819 convict ship |
|  | *Sir Edward Hughes* [20] from purchased 38 (1806); 1809 *Tortoise*; 1824 hulked |
| 1809 | *Yena* [14] from purchased 20 (1804); 1822 sold |
|  | *Hermes* from purchased sloop (1803); 1810 sold |
|  | *Cormorant* from purchased sloop (1803); 1817 sold |
| 1810 | *Spy* from purchased sloop (1804); 1813 sold |
|  | *Swift* from purchased sloop (1804); 1814 sold |
| 1811 | *Hindostan* [20] from purchased 54 (1804); 1815 *Dolphin*; 1830 hulked |
| 1813 | *Sir Francis Drake* [18] from purchased 38 (1805); 1825 sold |
| 1817 | *Hardy* from gunbrig (1804); 1822 yard craft |
| 1818 | *Havock* [bullock vessel] from gunbrig (1805); 1821 light vessel |
| 1847 | *Cockatrice* [store tender for the Pacific] from victualling transport ex schooner/brigantine (1832); 1858 sold |
| 1849 | *North Star* from 28 (1824); for the Arctic; 1860 BU |
| 1853 | *Apollo* from troopship ex 38 (1805); 1856 BU |
| 1854 | *Talbot* from 28 (1824); 1855 hulked |

## Transports

|  |  |
|---|---|
| 1782 | *Hind* from 24 (1749); 1784 sold |
| 1783 | *Clinton* [armed transport] from French 32 (1780); 1784 sold |
|  | *Camel* [armed transport] from purchased 24/26 (1776); 1784 sold |
|  | *Despatch* from *Zephyr* purchased brig (1779); 1798 sold |
| 1795? | *Dover* from 44 (1786); 1806 burnt |
| 1796 | *Assurance* from troopship ex 44 (1780); 1799 hulked |
|  | *Calcutta* from purchased 54 (1795); 1805 taken |
|  | *Weymouth* from purchased 54 (1795); 1800 lost |
|  | *Coromandel* from purchased 56 (1795); 1802 troopship |
|  | *Eagle* from purchased gunboat (1794); 1804 sold |
|  | *Ferret* from purchased gunboat (1794); 1804 sold |
|  | *Viper* from purchased gunboat (1794); 1801 BU 1797? |
|  | *Wilson* from purchased gunboat (1794); fate uncertain |
| 1807 | *Lancaster* [victualler (hulk?)] from purchased 64 (1797); 1815 hulked |
| 1808 | *Agincourt* [victualler] from purchased 64 (1796); 1812 troopship |
| 1846 | *Cockatrice* [victualling transport] from packet ex schooner/brigantine (1832); 1847 store tender |

## Exploration Ships

|  |  |
|---|---|
| 1740 | *Furnace* from bomb (1740); 1756 bomb again |
| 1773–1775 | *Carcass* bomb (1759) |
| 1775 | *Lion/Lyon* (ex Portsmouth hoy) from brig tender at the Tower; 1777-83 tender; 1783 Surrey slip |
| 1817 | *Congo* from exploration vessel originally intended for steam power (1816); 1819 fitted to lie in the Swale |
| 1825 | *Blossom* from sloop/24 (1806); 1829 survey ship |

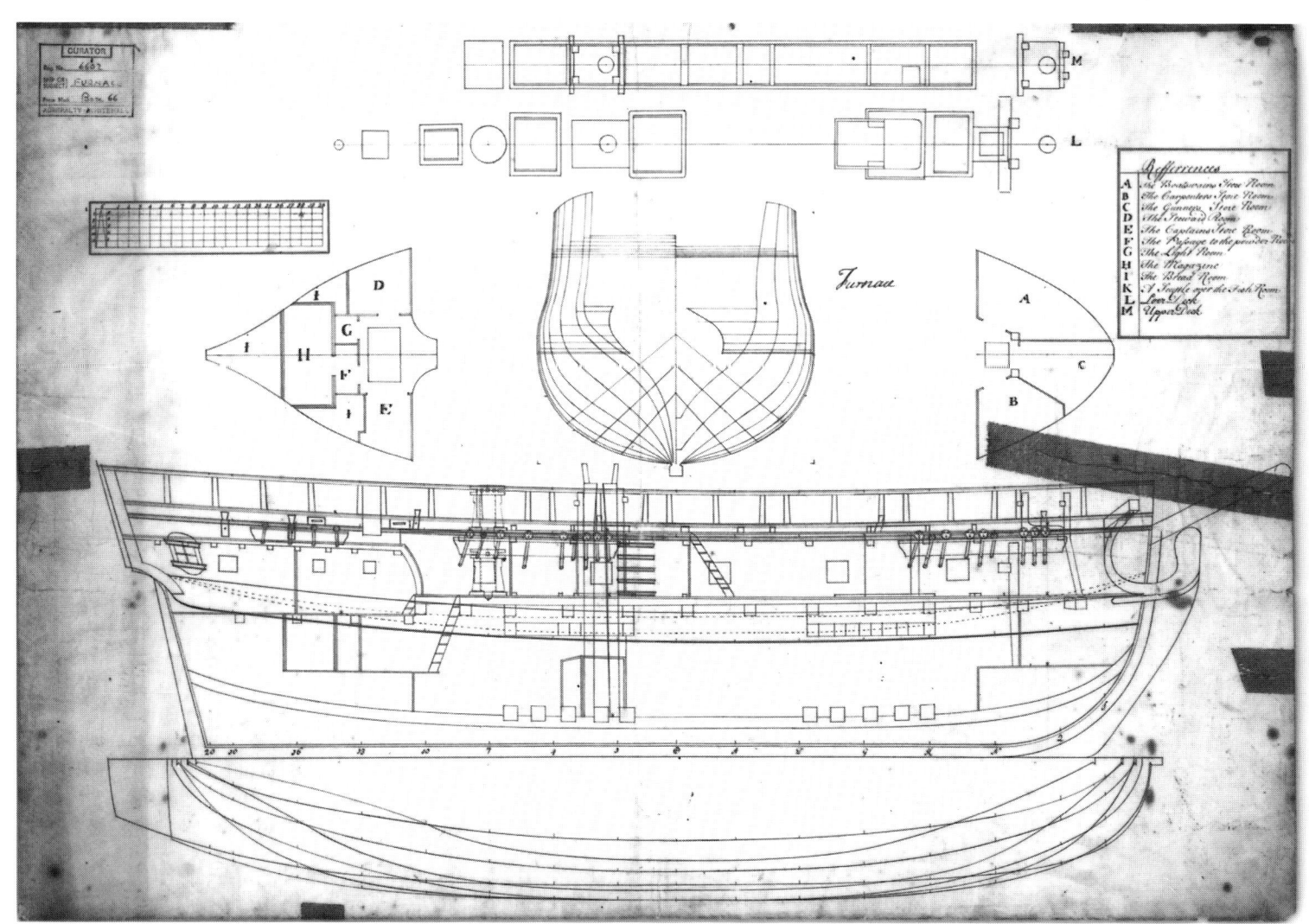

Hull plan (6602) of the bomb vessel *Furnace* of 1740 as converted in 1741 for exploration work. See the section on bomb vessels in Part I Chapter 3 for an illustration of the same ship fitted as a bomb

CONVERSIONS AND RECLASSIFICATIONS

1819 *Hecla* from bomb (1815); 1831 sold
1821 *Fury* from bomb (1814); 1825 lost
1836 *Terror* from bomb (1813); 1848 lost
1839 *Erebus* from bomb (1826); 1848 lost

## Survey Ships

(also see Exploration Ships above)

1783 *Lion/Lyon* ex tender; 1786 sold
1817 *Protector* from gunbrig (1805); 1833 sold
*Shamrock* from gunbrig (1812); 1831 hulked
*Aid* from transport (1809); 1821 *Adventure;* 1839 transport again
1819 *Hasty* from gunbrig (1812); 1827 yard craft
1821 *Snap* from gunbrig (1812-); 1827 hulked
1823 *Beagle* from brig (1820); 1845 to the Coastguard
1828 *Chanticleer* from brig (1808); 1833 to the Customs
1829 *Blossom* from exploration ship ex sloop/24 (1806); 1833 hulked
*Woodlark* from cutter tender (1821); 1864 sold
*Starling* [schooner rig] from cutter tender on completion?; 1844 sold
1831 *Acteon* [18] from 26 on completion; 1866 hulked
*Etna* from bomb (1824); 1839 hulked
1832 *Fairy* from brig (1826); 1840 lost
*Beacon* from *Meteor* bomb (1823); 1846 sold
1833 *Thunder* from bomb (1829); 1851 BU?
*Sylph*? from model packet (1821); 1837 tender
1835 *Sulphur* from emigrant ship ex bomb (1826); 1843 hulked
*Raven* from cutter tender (1829); 1859 sold
1836 *Mastiff* from gunbrig (1813); 1851 BU
*Magpie* from cutter tender (1830); 1846 yard craft
1842 *Bramble* [schooner rig] from cutter (1822); 1853 to New South Wales;
*Philomel* from uncompleted brig; 1847 brig
1845 *Herald* from 28 (1822); 1861 hulked
*Pandora* from packet ex brig (1833); 1857 to the Coastguard
1846 *Rattlesnake* from troopship ex 28 (1822); 1860 BU
1847 *Volage* from 28 (1825); 1855 hulked
1849 *Scorpion* from brig (1832); 1858 hulked
1854 *Saracen* from brig (1831); 1862 sold
1855 *Seaflower* ex cutter (1830); 1866 BU
1862 *Squirrel* from brig (1853); 1879 BU

## Packets

1819 *Emulous* brig on completion; 1834 to the Coastguard
1823 *Eclipse* from brig (1819); 1836 to the Coastguard
*Frolic/Frolick* from brig (1820); 1838 sold
*Magnet* from brig on completion; 1847 sold
*Zephyr* from brig on completion; 1836 sold
1824 *Goldfinch* from brig (1808); 1838 sold
*Rinaldo* from brig (1808); 1835 sold
*Cygnet* from brig (1819); 1835 sold
*Plover* from brig (1821); 1836 hulked
*Kingfisher* from brig (1823); 1838 sold
*Hope* from brig on completion; 1806 hulked
1825 *Redpole* from brig (1808); 1828 lost
*Sphinx* from brig (1815); 1835 sold
*Sheldrake* from brig on completion; 1853 sold
1826 *Mutine* from brig (1825); 1841 sold
*Tyrian* from brig on completion; 1845 hulked
*Skylark* from brig on completion; 1845 lost
1827 *Hearty* from brig (1824); 1827 burnt
*Myrtle* from brig (1825); 1829 wrecked
*Spey* from brig on completion; 1841 lost
1828 *Ariel* from brig (1820); 1828 lost
1829 *Barracouta* from brig (1820); 1836 sold
*Lyra* from tender ex brig (1821); 1845 sold
*Opossum* from brig (1821); 1841 sold
*Reynard* from tender ex brig (1821); 1841 yard craft at Chatham

*Lapwing* from brig (1825); 1845 hulked
*Reindeer* from brig on completion; 1841 hulked
*Calypso* from yacht ex brig (1826); 1833 lost
*Pigeon* from brig (1827); 1847 sold
*Delight* from brig on completion; 1844 sold
1831 *Hornet* from schooner/brigantine on completion; 1845 BU
1832 *Thais* from brig (1829); 1833 lost
*Cockatrice* from schooner/brigantine on completion; 1846 victualling transport
1833 *Andora* from uncompleted brig; 1845 survey ship
1834 *Seagull* from schooner/brigantine (1831); 1852 BU
1835 *Star* from uncompleted brig; 1843 brig
*Ranger* from uncompleted brig; 1845 brig
*Express* from uncompleted brig; 1862 sold
*Swift* from uncompleted brig;; 1860 yard craft
*Linnet* from uncompleted brig; 1857 to the Coastguard
*Alert* from uncompleted brig; 1843 brig
18..? *Viper* from schooner/brigantine (1831); 1851 BU
1838 *Penguin* from uncompleted brig; 1857 to the Coastguard
*Peterel* from uncompleted brig; later brig
1839 *Crane* from uncompleted brig; 1862 sold
1847 *Spider* from schooner/brigantine (1835) 1861 sold

## Fishery Protection Vessels

1816 *Martial* from gunbrig (1805); 1831 hulked
1827 see *Sylvia* cutter
1831 *Swan* from cutter (1811); 1837 hulked

## Training Ships

1848 *Rolla* from brig (1829); 1868 BU
1853 *Sealark* from brig (1843); 1898 sold
1859 *Ferret* from brig (1840); 1869 lost
1861 *Liberty* from brig (1850); 1905 sold
1864 *Martin/Marten* from brig (1850); 1890 *Kingfisher*; 1907 sold
1877 *Eurydice* from harbour training ship ex 24/26 (1843); 1878 lost
1878 *Atalanta* from *Juno* 26 (1844); 1880 lost

## Hospital Ships

1688 *Helderenberg* from Duke of Monmouth's 30; 1688 lost.
1717 *Looe* from 40 (1707); 1724 40 again
1721 *Portsmouth* from 40 (1717); 1728 BU
1735 *Looe* (as above); 1737 breakwater
1743 *Chester* from 44 ex 50 (1708); 1750 BU
1757 *Thetis* (22) from 44 (1747); 1767 sold
1771 *Jersey* from 60 (1736); Ca 1780 hulked
1781 *Dolphin* ex 44 (1781); 1800 troopship
1790- (only partially fitted) *Conflagration* fireship (1783)
1790
1790 *Roebuck* from 44 (1774); 1799 troopship
1794 *Charon* from 44 (1783); 1800 troopship
1795- *Africa* 64 (1781)
1805
1797 *Medusa* 50 (1785); 1798 wrecked
1801 *Alkmaar* from troopship ex Dutch 60/50 (1797); 1805 storeship
1803 *Trent* from 36 (1796); 1815 hulked
1805- *Jupiter* 50 (1778)
1807

## Floating Batteries

1757- *Chester* [at Milford] 50 (1744)
1762
1779 *Dromedary* [Guard ship for the Downs, Fifth Rate 20] from purchased storeship; 1783 BU

| | | |
|---|---|---|
| 1794 | ***Albion*** [60] from 74 (1763); 1797 lost | |
| | ***Nonsuch*** from 64 (1774); 1802 BU | |
| 1798 | ***Espion*** from French 32 (1794); 1799 troopship | |
| 1799 | ***Vlieter*** from Dutch 44 (1799); by 1808 hulked | |
| 1800 | ***Leyden*** from Dutch 64 (1797); 1810 troopship | |
| 1801–1803 | ***Serapis*** storeship ex 44 (1782) | |
| 1801 | ***Batavier*** from Dutch 56 (1799); 1809 laid up; 1817 hulked | |
| | ***Beschermer*** from Dutch 54 (1799); 1806 hulked | |
| | ***Alceste*** [registered as a sloop] from French 32 (1799); 1802 sold | |
| 1803–1805 | ***Dictator*** 64 (1783) | |
| 1803–1807 | ***Saint Albans*** 64 (1764) | |
| 1803 | ***Roebuck*** from troopship ex 44 (1774); 1811 BU | |
| (?) | ***Romulus*** from troopship ex 36 (1785); 1810 troopship | |
| | ***Mercury*** from 28 (1779); 1810 troopship | |
| | ***Tisiphone*** from fireship (1781); 1816 sold | |
| ? | ***Texel*** ? from Dutch 64 (1799); 1818 sold | |
| | ***Broaderscarp*** from hulk ex Dutch 54/50 (1799); 1805 BU | |
| | ***Pandour*** from troopship ex Dutch 44 (1799), 1805 hulked | |
| ? | ***Helder*** ? from Dutch 32 (1799); 1807 sold | |
| | ***Musette*** [for the River Yare] from hulk ex French sloop; 1806 sold | |
| 1805 | ***Gorgon*** from storeship ex 44 (1785); 1817 BU | |
| 1808 | ***Vulture*** from purchased sloop (1803); 1814 sold | |

## Emigrant Ships

| | | |
|---|---|---|
| 1828 | ***Sulphur*** from bomb (1826); 1835 survey ship |
| By 1836 | ***Buffalo*** from convict ship ex purchased storeship (1813); 1840 lost |
| 1852 | ***Hercules*** from troopship ex 74 (1815); after 1853 hulked |

## Convict Ships (seagoing)

| | |
|---|---|
| 1819 | ***Coromandel*** from ***Malabar*** troopship ex purchased 56 (1804); 1827 hulked |
| 1833 | ***Buffalo*** from purchased storeship (1813); by 1836 emigrant ship |

## Signal Vessel

| | |
|---|---|
| 1811 | ***Warning*** from Danish gunbrig (1807); 1814 sold |

## Yachts

| | |
|---|---|
| 1822 | ***Apollo*** from 38 (1805); conversion never completed |
| 1826 | ***Calypso*** from brig on completion; 1829 packet |

## Tenders

| | |
|---|---|
| 1810 | ***Jackdaw*** from schooner (1806); 1816 sold |
| 1811 | ***Landrail*** from schooner (1806); 1814 taken |
| | ***Watchful*** from purchased brig (1804); 1814 sold |
| 1812 | ***Bittern*** from sloop (1796); 1833 sold |
| | ***Transit*** [impressment tender] from advice boat (1809); 1815 sold 1814 |
| | ***Basilisk*** from gunbrig (1801); 1818 sold |
| | ***Resolute*** from gunbrig (1805); 1816 hulked |
| 1816 | ***Express*** from purchased schooner (1815); 1827 sold |
| 1822 | ***Calliope*** from brig (1808); 1829 BU. |
| 1823 | ***Ariel*** from brig (1821); 1829 packet |
| 1825 | ***Reynard*** from brig (1821); 1829 packet |
| 1826 | ***Royalist*** from brig (1823); 1838 sold |
| | ***Leveret*** from brig (1825); 1843 sold |
| 1827 | ***Alcon*** from brig (1820); 1838 sold |
| | ***Onyx*** from brig (1822); 1837 sold |
| 1831 | ***Recruit*** from brig (1829); 1832 lost |
| 1836 | ***Partridge*** from brig (1829); 1843 to the Coastguard |
| 1837 | ***Sylph*** from ?survey ship? ex model packet (1821); 1837 to Customs |
| 1857 | ***Serpent*** [gunnery tender] from brig (1832); 1861 BU |

## Light Vessel

| | |
|---|---|
| 1811 | ***Gleaner*** [for the Galloper Sand] from purchased survey vessel (1809); 1814 lost |

## Experimental Propelling Machinery Fitted

| | |
|---|---|
| 1829 | ***Galatea*** 36 (1810) |

## Loan as Privateer

| | |
|---|---|
| 1763–1764 | ***Roebuck*** 44 (1743) to the 'Antigallican Private Ship of War Association'. |

# Chapter 4: Yard Craft – Listed under their Yards

The Royal Navy was a considerable shipowner in its own right, even ignoring its fighting ships, as its dockyards required large numbers of small cargo vessels to transport naval stores to, from and between the various yards. There were also numbers of boats for laying buoys, dredging, towing ships, ferrying personnel and all the other multifarious tasks required in a large organisation working in a maritime environment. The performance of the Royal Navy in our period owed much to its infrastructure, of which these craft were a vital, though somewhat un-noticed, part.

The vessels are listed under the yards to which they were attached, and then in chronological order of build or of transfer to the yard concerned. When known, outline technical data are given, and (when they exist) plans in the Admiralty Collection at Greenwich are also noted. The plans themselves are an important source for this appendix, sometimes there being no note found elsewhere for craft shown in such plans. It would seem that the sources used did not generally note either victualling craft (which came under the Victualling Board rather than the Navy Board, though both were subordinate to the Admiralty) or ordnance vessels (the Ordnance Board being a separate organisation from either Navy or Army for much of our period, then becoming part of the War Office until the 1890s) until after the end of the Napoleonic War (1815). So this list does not claim to be a complete list of all of these, but it at least includes all non-mechanically propelled (ie. not steam, though early non-self propelled steam dredgers are recorded, these craft having the distinction of being the first Royal Naval steam vessels!) dockyard craft of which a note has been seen from 1660 to the end of sail in the RN. The list of identified vessels for each yard is followed by notes on plans of vessels associated with that yard which it has not been possible to identify with a noted vessel.

In the following list the **first** column gives the **dates** at which service in the yard being listed commenced and ended. The **second** gives (when there is one) the **name** or other designation of the vessel. The **third** column gives the **type** of craft (sometimes, in the case of numbered vessels, the second and third columns are combined in one entry), whilst the **fourth** details the arrival and departure of the vessel from the yard in question, starting with the **builder** (when known) if the vessel was built for that yard (usually at one of the royal dockyards, though often not by the one in question). This section also includes other '**historical**' details such as change of name, designation or function, plus details, when appropriate, of fate. When known, **technical data** and **plans** are listed in the normal way used in this work, on separate lines beneath the rest of the entry.

## Deptford Dockyard

| | | | |
|---|---|---|---|
| 1660-1714 | *Royal Escape* | smack | Purchased (was the smack in which Charles II escaped after the Battle of Worcester); 1714 rebuilt |
| By 1694 | *Sophia* | hoy, 135 tons | In service 1694; later to Portsmouth (BU 1712/1713); is this the fireship captured from the Duke of Argyle in 1685? |
| By 1699 -1706 | *Owners Goodwill* | hoy, 24 tons | Ex purchased machine [see Main List]; 1706 sold |
| 1702-1707? | *Deptford* | small yacht (transport); | Purchased mercantile *Expedition* by 1708 at Lisbon |

52'5", 42'10½" x 16' x 6'7½". 58 36/94 tons

| 1705-1730 | *Navy [Transport]* transport | | Deptford Dyd (Surveyor: Harding); 1730 rebuilding |

63'11", 51'5½" x 19'9½" x 9'. 107 20/94 tons. 9 men. 2 guns + 6 swivels

| 1706-1708 | *Runner [Transport]* | hoy | Deptford Dyd (Surveyor: Allin), replacement for *Owners Goodwill*; 1708 at Lisbon |

49'6", 40' x 15'2" x 7'. 48 88/94 tons. 6/9 men

| 1709-1752 | *Lion/Lyon* | hoy, | 108 tons, 4 guns. Deptford Dyd; 1752 lost |

PLANS: (presumably of this *Lyon*): Lines, profile, decks

| 1713-1714 | *Harwich* | hoy/lighter | From Harwich; replacing No 1 Sailing Lighter; 1714 sold |
| 1714-1736 | *Royal Escape* | smack | Deptford Dyd; [RB]; 1736 rebuilt |
| 1716-1811? | *Deptford* | lighter | Sheerness Dyd (Surveyor: Ward); (gone by 1740? unlikely) or BU 1811? (likely) |

52', 41'2½" x 18' x 6'6". 71 tons. 2 guns + 6 swivel

| 1730-1742 | *Navy Transport* transport | | Deptford Dyd (Surveyor: Stacey); [RB] 1742 BU |

63'11", 50'9½" x 20'1½" x 10'. 109 39/94 tons. 9 men. 2 guns + 6 swivels

| 1736-1750 | *Royal Escape* | store vessel | Deptford Dyd (Surveyor: Stacey); laid up from 1743; 1750 sold |

52', 39'0½" x 15'5" x 7'6". 49 1/4 ton

PLANS: Lines.

| 1740s-1757 | *Supply* | hoy | From Chatham; 1757 taken by the French |
| 1744-1791 | (*Royal Escape*) | lighter | Carter, Limehouse; 1749 named *Royal Escape* (previously un-named); 1791 BU. |

63'2", 50'3" x 20' x 8'4". 104 74/94 tons. 7 men

PLANS: Lines, profile & decks (also for Minorca lighter)

| 1752-1791 | *Navy Transport* | hoy | Purchased when 4½ years old; 1791 sold |
| 1757-1768 | Close lighter | | Purchased; 1768 sold. |

39', 26'4⅝" x 18'3¾" x 6'11". 47 tons

| 1759-1786 | *Supply* | transport | Bird, Deptford (Slade design); 1786 armed tender |

(see Main List for details & plans in this latter role)

PLANS: Profile & decks

| 1770-1822 | Close lighter | | Deptford Dyd; 1822 BU. |

40', ___ x 18'2" x 7'6". 48 tons. 2 men

| 1774-1794 | Storehouse longboat | | Deptford Dyd; 1794 rebuild |

39'2", ___ x 13'8" x 6'. 30 tons. 4 men

| 1781-1830s | *Deptford* | sailing lighter, 105 tons | Muddle?, Gillingham; 1830s to Sheerness. |
| 1784 | *Deptford* | transport | To Plymouth |
| 1787-1820? | Chain Boat No 1 chain boat | | Wallis, Thames; by 1821 out of service. |

44'3", 35'1¾" x 15'3" x 6'2". 43 44/94 tons

PLANS: Lines, profile, deck/specification

| 1788- Ca 1840 | Chain Boat No 2 chain boat | | Batson, Thames; Ca 1840 out of service. |

___ , 31'4½" x 15'6" x ___ 40 8/94 tons

PLANS: Lines, profile, deck

| 1792-by 1830 | *Royal Escape* | sailing lighter/ transport | Nowlan, Northam; by 1830 to Chatham. Built to the plan of the previous *Royal Escape* |

63'5¼", 50'10⅛" x 20'1⅜" x 8'4", 109 82/94 tons

PLANS: Lines, profile & deck

| 1794-1802 | *Assistant* | naval transport | ex armed tender (1791); 1802 sold. One mast, pierced for 8 guns. |

___ , 51'4¼" x 29'1" x 19'8", 110 37/94 tons. 25 men, 4 x 3 + 8 swivels as an armed tender

PLANS: 1794: Lines & profile/decks

| 1794-1811 | Storehouse longboat | | Deptford Dyd [RB]; 1811 BU [As before RB] |
| 1794-___ ? | Launch | | James Wade, Rotherhithe |

37', ___ x 10' x 3'9½". ___

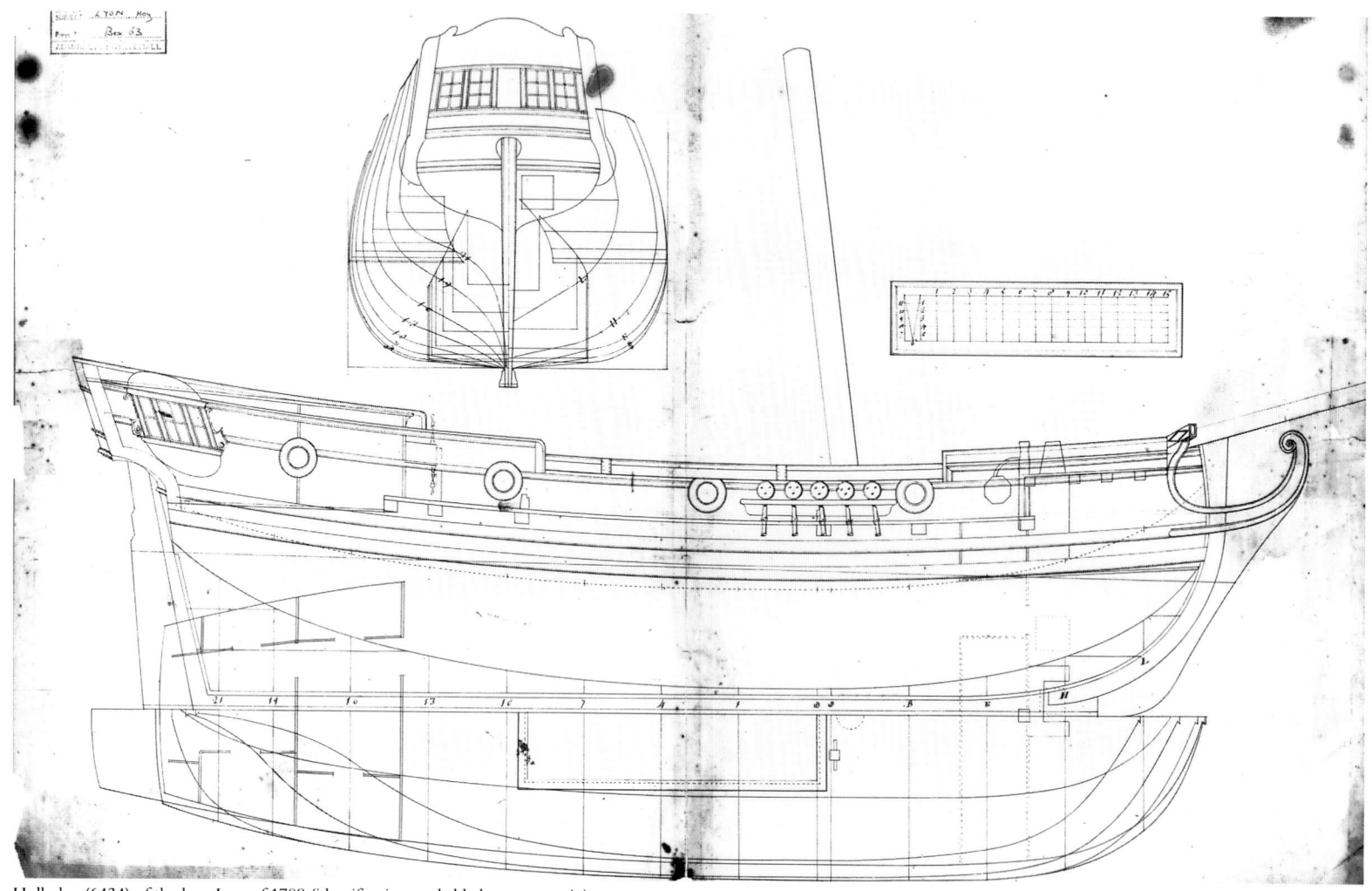

Hull plan (6434) of the hoy *Lyon* of 1709 (identification probable but not certain)

|  |  | PLANS: Lines, profile, deck |  |
|---|---|---|---|
| 1799-1812 | *Grasshopper* | longboat | ____, Gillingham; 1812 BU |
| 1800-____ ? | *Yarmouth* | lighter | Preston, Great Yarmouth (?) later to Woolwich (may have served at Chatham for some time) |
|  |  | 60'3", 49' x 20' x 9'5". 104 24/94 tons |  |
|  |  | PLANS: As 1798 Chatham lighter; also: Bow horse |  |
| 1800-____ ? | Pinnace for the Commissioners of the Navy | | Deptford Dyd, but not necessarily based there. |
|  |  | 30', ____ x 6'4" x 2'4". ____ |  |
|  |  | PLANS: Lines, profile, deck |  |
| Ca 1800 | Pilot boat for the Master Attendant | | Deptford Dyd |
|  |  | 21', ____ x 6'8" x 2'5". ____ |  |
|  |  | PLANS: Lines |  |
| 1801-1839 | *Mary* | victualling lighter | Muddle, Gillingham 1801; 9.2.1839 sold to Castle to BU |
|  |  | 37', ____ x 16'6" x ____ 54 tons |  |
| 1806-____ | steam dredger ['for raising mud/digging soil'] | | Deptford Dyd [Launched 19.4.06, cost £2,986; transferred to Woolwich; later to Portsmouth. |
|  |  | 96', ____ x 26' x 7'6". 9 men |  |
| 1808-by 1830 | *Bee* | longboat 43 tons | Deptford Dyd; by 1830 to Woolwich |
| 1809-1821 | *Assistance* | transport | see Main List for details |
| 1811-by 1830 | *Swallow* | longboat | Deptford Dyd; by 1830 tender (decked, single masted, cutter-like) in the West Indies (or 1825 lost?) |
|  |  | 45'1", 35'2" x 15'10" x 8'2". 46 83/94 tons |  |
|  |  | PLANS: Lines & profile/decks |  |
| 1812-by 1830 | *Deal* | lugger | ____, Sandwich; by 1830 to the Preventive Service; 1832? sold. |
| 1814-by 1830 | *Dart* | sloop? | Deptford Dyd; by 1830 to Portsmouth (though is there a confusion here with the longboat of the same name noted under Sheerness and at the end of this list?) |
| 1815-by 1845 | *Ant* | sailing barge swim ended type | Deptford Dyd; by 1845 to Woolwich (or is this a confusion with the victualling lighter of 1834?) |
|  |  | 72', ____ x 20'6" x 7'4". ____ |  |
|  |  | PLANS: Profile, deck, sections. |  |
| 1815-____ ? | Open sailing barge, swim ended | | Deptford Dyd; fate uncertain |
|  |  | (?) 66', ____ x 18'9" x 5'7" ____ . |  |
|  |  | PLANS: Profile, plan sections |  |
| By 1816-1821 | *Ranger* | brig, transport [225 tons] | Purchased [built 1797 on the Thames]; 1821 sold |
| By 1816-____ ? | *Sorlings* | lighter | Purchased (built at Saint Davids 1788; is this the purchased survey ship of 1789?); fate uncertain |
| 1821-____ ? | Chain Boat No 1 | | Deptford Dyd; fate uncertain |
| 1823-____ ? | (Old) *Truelove* | hoy for Victualling Department (1720) | From Portsmouth; later to Woolwich |
| 1823-____ ? | ex Chatham | longboat | From Chatham; fate uncertain |
| 1823?-1864 | *Wight* | victualling lighter | White, Cowes; 1864 BU. |
|  |  | 51'2", 40'2" x 18'3⅛" x ____ 71 49/94 ton |  |
|  |  | PLANS: Lines & profile/decks |  |
| ____?-1831 | *Thames* | longboat | From Woolwich; 1831 to Sheerness |
| 1825-by 1845 | *Fly* | sailing barge [132 tons] | Chatham Dyd; to Woolwich by 1845 |
| 1827 | *Louisa* | victualling lighter | ____, Northfleet; still in service 1854 |
| By 1830-Ca 1830 | *Ann* | barge | From Plymouth; Ca 1830 to Woolwich |
| By 1830-____ ? | *Dart* | longboat | From Chatham; fate uncertain |

# YARD CRAFT

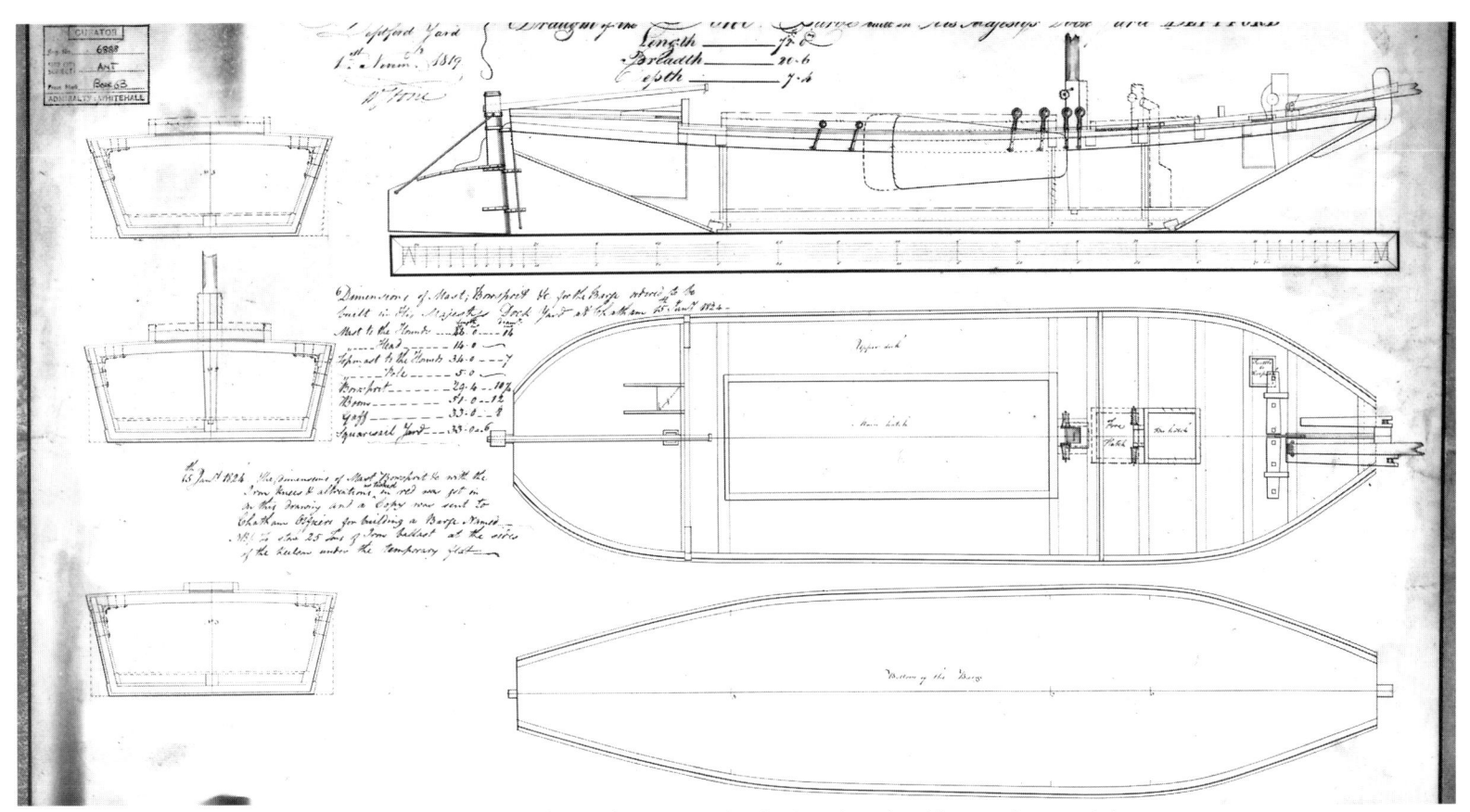

Profile, decks and sections (6888) of the sailing barge *Ant* built at and for Deptford in 1815. This plan shows her to be of the original swim-ended form of Thames sailing barges. She was presumably sprit-rigged, but unlike later Thames barges, does not have a small mizzen mast

| | | | |
|---|---|---|---|
| Ca 1830-1853 | **Sheppy/Sheppey** | sailing barge | From Woolwich; BU 04.1853. |
| 1834-by 1845 | **Ant** | victualling lighter | Portsmouth Dyd (Surveyor Symonds); by 1845 to Woolwich (or is this a confusion with the earlier *Ant*?) |
| 1838-____ | **Mary** | victualling hoy | Woolwich Dyd, (Surveyor Symonds); later to Woolwich. |
| By 1845-1866 | **Van** | barge/lighter | From Woolwich; 1866 BU. |
| 1854-1870 | **Staunch** | barge | Sheerness Dyd; 1865 mooring lighter; Deptford Yard Craft No 2; 1870 sold |
| 1855?-1864 | **Ann and Eliza** | hoy | Purchased; 1864 BU. |
| 1855?-1869 | **Lion** | hoy ('billy boy') | Purchased for victualling yard; by 1860 Royal Victoria Victualling Yard Craft No 5; 1869 sold. |
| 18__-1887 | No 4 Victualling Yard Lighter | | Origins unknown; A.O.17.11.1870 renamed *Bonita*; 1882 laid up at Chatham; 11.1887 sold to Perry & Sons, Bristol, for £115. |

**Also the following plans show vessels apparently associated with the Yard:**

'Clinch work ballast lighter taken off at Deptford' 1786.
42'7½", 28'8⅝" x 20'1" x 5'10". 61 52/94 tons
PLANS: Lines, profile, deck, dredging equipment

*Harmony* Barge 'not approved' 1797.
66'3", ____ x 18'8" x 5'0½". 65 87/94 tons; flat bottomed with mast well forward.
PLANS: Outline profile, deck, midsection

## Woolwich Dockyard

| | | | |
|---|---|---|---|
| 1694-1705 | **Endeavour** | purchased smack | ex mercantile? purchased from John Brown [?]; 1705 sold. |
| | | | 33', ____ x 12' x 5'9". 18 tons |
| 1705-1726 | **Woolwich** | transport | 165 tons, 2 guns Woolwich Dyd; 1706 rebuilt |
| 1726-1767 | **Woolwich** | transport | (2 guns) Woolwich Dyd [RB]; 1767 sold |
| 1730-1813 | Lighter No 1 | sailing lighter | Woolwich Dyd (Hayward); 1813 BU. |
| | | | 60'9", 49'11⅝" x 19'0½" x ____. 96 tons. 5 men |
| 1730-1737? | Lighter No 2 | sailing lighter | Woolwich Dyd (Hayward); fate uncertain, still in service 1737 |
| | | | ____, 32'10" x 18' x ____ 56 55/94 tons |
| 1739-1813 | **Woolwich** | [?] 87 tons | Woolwich Dyd; 10.1813 BU |
| Ca 1749-____? | mooring lighter | | Bennett, Faversham?; existence and fate uncertain. |
| 1758-____? | mooring lighter | | Bennett, Faversham, fate obscure, may be a confusion with the chain boat noted below |
| | | | 53', 41'4" x 19'6" x 6'3". 83½ tons |
| 1758-1786 | Chain boat | | Bennett, Faversham; 1786 too rotten to be repaired |
| | | | 45', 34'6" x 15' x 6'. 41 27/94 tons |
| | | | PLANS: Lines, profile, deck (another boat was built to this plan by Bennett, probably not for this yard) |
| 1777-1783 | **Dispatch** | transport | ex sloop *Cherokee* built 1774; 1783 sold (see Prizes & Purchases list) |
| 1780-1797 | **Discovery** | naval transport | ex sloop of 1776; 1797 BU |
| 1783-1798 | **Dispatch** | naval transport | ex brig *Zephyr* of 1779; 1798 sold |
| 1786-1826 | chain boat | | Cleverley, Gravesend; 1826 BU. (Copy in dimensions and scantlings of the previous boat of this type.) |
| | | | 46'8", 35'1¾" x 15'3" x 6'2". 43 44/94 tons |
| | | | PLANS: Lines, profile, deck |
| 1788-Ca 1805 | Sailing lighter No 2 | | Woolcombe, Rotherhithe; Ca 1805 BU |
| | | | 50'2", 39' x 18'4" x 8'. 69 67/94 tons |
| | | | PLANS: Lines, profile/decks |
| 1794-1808 | **Variation** | purchased close lighter 47 tons | Ex mercantile?; 1808 sold |
| 1798-1834 | **Supply** | transport 223 tons | Pitcher, Northfleet (Henslow design); 1834 BU |
| 1798-1820 | **Dispatch/Despatch** | transport | Purchased sloop ex mercantile [?] *Cornwallis* 1798?; 1820 hulked at Bermuda as diving bell vessel. |

317

| | | | |
|---|---|---|---|
| 1799-1805 | **Bellona** | mud boat | 238 tons, 6 guns ex gunboat; 1805 BU |
| 1799-____ ? | **Mayflower** | mud boat | ex gunboat; fate uncertain |
| 1804-1824 | Lug boat No 1 | (luggage boat) | Rawlinson, Westminster; 1824 sold. |
| | 35', ____ x 13' x ____ 23 53/94 tons | | |
| | PLANS: Lines | | |
| 1804-1824 | Lug boat/Lighter No 2 | | Rawlinson, Westminster; 1824 sold [as No 1 above] |
| After 1805 -by 1830 | **Thames** | longboat | ____, Rochester 1805; from Chatham; by 1830 to Deptford |
| 1806-1815 | **Plymouth** | purchased. transport (lighter?) 195 tons | Ex mercantile (built ____ Whitby 1805); 1815 sold |
| 1806- by 1816 | 'Steam engine' | (dredger) | from?) Deptford Dyd; by 1816 to Portsmouth |
| 1807-1824/ ____?- after 1854 | **Falmouth** | sailing lighter | ____, Topsham; 1824 temporary mortar vessel; later Woolwich Yard Craft No 1; after 1854 to Portsmouth |
| | 70'3", 55'10¼" x 23'1" x 11'. 158 27/94 | | |
| | PLANS: (as taken off 1829) Lines/profile & decks | | |
| 1807-1838 | Chain boat No 2 | | ex Danish gunboat; 1838 BU |
| 1808-1828 | **Star** | longboat 114 tons | Woolwich Dyd; 1828 sold to Levy, Rochester, to BU |
| 1809-1815 | **Aid** | transport | see Main list for details |
| 1811-1818 | **Kent** | lighter 114 tons | Woolwich Dyd; 1818 BU |
| 1814-1862 | **Diligence** | transport | see Main List for details |
| 1814-1835 | **Industry** | transport | see Main list for details |
| 1815- by 1830 | **Woolwich** | lighter 114 tons | Woolwich Dyd; by 1830 transferred to Jamaica as *Port Royal* |
| After 1816 | steam dredger | | From Portsmouth; then to Chatham |
| Ca 1820?- by 1830 | **Sheppy/Sheppey** | sailing barge | From Sheerness; by 1830 to Deptford |
| 1820s-1828 | **Yarmouth** | lighter | From Deptford; 1828 to Customs |
| 1824- by 1845 | **Van** | barge/lighter | Chatham Dyd; by 1845 to Deptford 132 tons |
| 1827- after 1854 | **Beaver** | 'lump' | ____, Harwich; still in service 1854 |
| By 1830- after 1855 | **Ann** | barge | From Deptford; still in service 1855 |
| By 1830- 1847 | **Bee** | longboat | From Deptford; 1847 to Chatham |
| By 1833- 1835 | **Rennie** | lighter | From Plymouth; 1835 to Sheerness |
| 1834- by 1845 | Chain lighter No 1 | | From Chatham; deleted by 1845 |
| 1834-1854 /1856-1866 | **Sinbad** (ex *Ant*) | sailing lighter | Pembroke Dyd [Launched 27.2.34]; 1854 Mortar Vessel No 2; 1856 Woolwich Yard Craft No 3; 1866 BU |
| | 60', 47'5" x 20'5" x 8'10". 105 11/94 | | |
| | PLANS: Specification/lines, profile & deck/midships section (also for *Duck, Drake, Ant* and *Fountain* tank vessel). For plans as mortar vessel see main list | | |
| 1837- after 1854 | Mooring lighter No 1 | | Woolwich Dyd (Symonds design); still in service 1854 |
| 1837-1886 | Mooring lighter No 2 | | Woolwich Dyd (Symonds design); 1886 to be burnt |
| | 65', 53' x 20'4" x 6'6". 112 72/94 tons | | |
| | PLANS: Lines & profile | | |
| 1838-1840 | **Nelson** | longboat | Purchased 07.1838; 1840 sold |
| 1830s- after 1854 | dumb barge | | Origin uncertain; still in service 1854 |
| By 1845 -1857 | **Ant** | sailing barge | From Deptford; 1857 BU |
| By 1845- 1863 | **Fly** | sailing barge | From Deptford; 1863 BU |
| 1845- by 1867 | **Manoeverer** | barge | Woolwich Dyd; later Woolwich Yard Craft No 4; by 1867 to Sheerness |
| 1847- after 1854 | **Hope** | lighter | Received from the Convict Establishment to take the place of *Bee*; still in service in 1854 |
| 1864-1871 | **Julia** (Woolwich sailing barge YC No 2) | | 122 tons Purchased by A.O.5.4.1864 from Dodds of Gravesend to replace *Fly* [built 1863 by Joshua Wright, Milton, Kent]; A.O.6.3.1871 to be broken up at Sheerness |

NOTE: There is a plan [lines & profile] dated 1806 for building a sailing lighter for Woolwich by Bools & Good of Bridport [63', 50'10" x 20'2" x 9'3", 112 4/94 tons] it is not clear if this is what became *Fal*, was a cancelled design, or what

# Chatham Dockyard

| | | | |
|---|---|---|---|
| 1673-1713 | **Flemish** | longboat | Purchased?; 1713 sold. |
| | 33', 23' x 9'10" x 4'4". 11 tons. 2 or 3 men | | |
| 1665?-1712 | **Unity** | horseboat | Built Chatham?; 1712? BU or sold. 80 or 68 tons?; used for stores |
| 1677-1703 | **Transporter** | hoy 92 tons | Sheerness Dyd (Surveyor: James Shish); 1703 rebuilding |
| 1691- by 1701 | **Supply** | hoy 94 tons, 4 guns | Chatham Dyd (Surveyor: Lee); by 1701 to Sheerness |
| 1693-1713 | **Unity** | hoy 96 tons | Chatham Dyd; 1707 lighter [96 tons] 1713 sold |
| 1693-1704 | Tow Boat No 1 ('Tow engine') | | Chatham Dyd (Surveyor: Lee); 1704 deleted. |
| | 80', 66'10" x 16' x 3'4". 91 tons | | |
| 1693-1704 | Tow Boat No 2 | | Chatham Dyd (Surveyor: Lee) 1704 deleted [As No 1 above] |
| 1693 | ?Savary's 4-horse powered boat | | tried at Chatham 1693; is this Tow Boat No 1 or conversion of *Unity* 'horseboat'? |
| 1694-1813 | **Chatham** | sheer hulk | Chatham Dyd (Surveyor: Lee) 1813 BU. With her successor of 1813 and one other built for Plymouth the only Royal Navy sheer hulks built as such. |
| | PLANS: Lines, profile & deck | | |
| 1703-1708 | **Lyon** | hoy 100 tons | Purchased; 1708 taken by the French |
| 1704-1713 | **Transporter** | hoy | Deptford Dyd (Surveyor: Harding); [RB]; 1713 sold. |
| 1725-1740s | **Supply** | hoy | Chatham Dyd (Surveyor: Rosewell); 1740s to Deptford |
| | 72', 56'5" x 20'2" x 10'. 122 4/94 tons. 11 men. 4 guns + 4 swivels | | |
| 1734-1770 | Chatham Lighter No 1 | | Chatham Dyd (Surveyor: Ward); 1770 BU |
| | 51'6", ____ x 19' x 8'10". 73 tons. 8 men | | |
| 1739-1798 | Chatham longboat | | Chatham Dyd (Surveyor: Ward); 1798 sold |
| | 39'5¾" 31'6⅛" x 13'3" x 5'11". 29 19/94 tons. 4 men | | |
| 1748-1828 | Sailing lighter No 2 | | Chitty & Vernon, Chichester; 1828 to the Coast Blockade |
| | 65'11¾", 53'1½" x 20'2¾" x 8'10". 116 tons. 6/8 men | | |
| 1760-1764 | **Canterbury** | storeship | (or at Sheerness?) see main list (purchased 1756); 1764 sold |
| By 1765?- 1797 | Mooring lighter No 2 | | Origins unknown; 1797 BU |
| | 49'7½", 38'2⅜" x 19'7¼" x 5'4". 78 tons | | |
| By 1765?- 1825 | Mooring lighter No 3 | | Origins unknown; 1825 BU |
| | 52'4¾", 40'5½" x 18'4½" x 4'7½". 72 tons | | |
| By 1765?- 1820 | Mooring lighter No 4 | | Origins unknown; 1820 BU. |
| | 49'6", 37'3¾" x 19'8½" x 5'6⅝". 77 tons | | |
| By 1765?- by 1830 | Mooring lighter No 5 | | Origins unknown; deleted by 1830 |
| | 54'7", 43'11⅜" x 18'6¾" x 6'9". 81 tons | | |
| By 1765?- after 1854 | Mooring lighter No 6 | | Origins unknown; still in service 1854 |
| | 53'4¾", 41'10¼" x 19'5½" x 4'9¼". 84 tons | | |
| By 1765?- after 1864 | Mooring lighter No 7 | | Origins unknown; still in service 1864 |
| | 56'2", 46'2⅝" x 18'1¼" x 5'9¼". 81 tons | | |
| 1771-1824 | Sailing lighter No 1 | | Chatham Dyd; 1824 BU |
| | 66'3¾", 52'8⅛" x 21'7¾" x 9'5". 130 tons. 8 men | | |
| 1781-1785 | **Medway** (ex Dutch) | lighter | Taken 1781; 1785 sold. 191 tons. 12 men |
| 1782-1820 | Mooring lighter No 8 | | Muddle, Gillingham; 1820 BU |
| | 56'7½", 45'2⅜" x 19'8¼" x 6'6½". 93 tons | | |
| 1785-1820 | Mooring lighter No 9 | | Chatham Dyd; 1820 BU |
| | 64'11", 51'10½" x 20'10½" x 7'2". 120 tons | | |
| 1785-1841 | Mooring lighter No 10 | | Chatham Dyd; 1841 sold |

## YARD CRAFT

| Date | Name | Type | Notes |
|---|---|---|---|
| | | | 64'11", 51'10½" x 20'10½" x 7'2". 120 tons |
| 1785-___? | | Sailing lighter | ___, Gillingham; fate uncertain |
| | | | 60'3", 49' x 20' x 9'5". 104 24/94 tons |
| | | | PLANS: Lines & profile |
| 1789-1816 | | Mooring lighter No 1 | Bennet, Faversham (Henslow design); 1816 BU. |
| | | | 57'1½", 44'6⅜" x 20'11" x 6'8½". 96 tons |
| | | | PLANS: Lines & profile. |
| 1798-after 1864 | | Mooring lighter No 2 | Cleverley, Gravesend (Henslow design); 1845 lighter; still in service 1864 |
| | | | 56'4", 44'5" x 19'10" x ___ 92 89/94 tons |
| | | | PLANS: Lines & profile. (repeat of No 1) |
| 1798-___? | | Sailing lighter [No 2?] | Brindley, Lynn (Henslow design); fate unknown |
| | | | 60'3", 49' x 20' x 9'5". 104 24/94 tons |
| | | | PLANS: Lines, profile, deck/specification |
| 1798-1823 | | Chatham longboat | Ross, Rochester; 1823 to Deptford |
| | | | 45', 34'8¼" x 15'8" x 7'9". 45 22/94 tons |
| | | | PLANS: Lines & profile |
| 1800-___? | *Yarmouth* | sailing lighter | Preston, Yarmouth; from Deptford (date of service - if any - at Chatham unknown); fate unknown [As 1785 and 1798 sailing lighters above] |
| 1805-by 1830 | *Thames* | longboat | ___, Rochester; later to Woolwich; by 1830 to Deptford; 1831 to Sheerness |
| 1807-after 1864 | | Mooring lighter No 11 | Preston, Yarmouth; still in service 1864 |
| | | | 62'1", 48'6" x 22' x ___ 124 81/94 tons. Seppings design |
| | | | PLANS: Lines & profile/deck |
| 1807-after 1864 | | Mooring lighter No 12 | Preston, Yarmouth; still in service 1864. [As No 11 above] |
| 1809-___ | *Adventure* | transport | see Main List for details |
| 1810-1824 | *Goodwill* | sailing lighter | Mrs Ross, Rochester; 1824 converted to mortar vessel (see Main List) later lighter at Sheerness. 127 tons. See *Saltash* under Plymouth for details |
| 1813-1879 | *Chatham* | sheer hulk | Chatham Dyd; 1879 BU. With her predecessor of 1694 and one similar vessel built for Plymouth the only sheer hulks built as such |
| | | | 145'4", 146' x 46' x 14'3". 1691 23/94 tons |
| | | | PLANS: Lines & profile |
| 1814-after 1854 | | Mud boat No 1 | Chatham Dyd; still in service 1854 |
| | | | PLANS: Lines (for one of these boats) |
| 1814-after 1854 | | Mud boat No 2 | Chatham Dyd; still in service 1854 |
| 1816-1846 | | Mooring lighter No 13 | ex gunbrig *Snipe* (1801); later No 9; 1846 BU. |
| After 1816-by 1830? | | steam dredger | From Woolwich; out of service before 1830 |
| 1818-after 1854 | | Mud boat No 3 | Chatham Dyd; still in service 1854 |
| 1819-after 1854 | | Mud boat No 4 | Chatham Dyd; still in service 1854 |
| 1820-after 1854 | | Open barge No 1 | Chatham Dyd; still in service 1854; Built to the design of the Deptford Open barge of 1815 - which see |
| 1821-after 1854 | | Open barge No 2 | Chatham Dyd; still in service 1854. As No 1 above |
| 1824-after 1854 | | Mud boat No 5 | Sheerness Dyd; still in service 1854 |
| 1824-after 1854 | | Mud boat No 6 | Sheerness Dyd; still in service 1854 |
| 1828-___? | *Hope* | tender | From Leith; later convict ship; later to Portsmouth |
| 1828-1872 | *Dove* (ex *Van*) | sailing barge | Chatham Dyd; 05.1872 [or 1823?] completed BU |
| 1828-after 1854 | | Pitch boat | Chatham Dyd; still in service in 1854 |
| By 1830-by 1845 | *Royal Escape* | sailing lighter | From Deptford; by 1845 to Sheerness |
| By 1830 | *Sheerness* | sailing lighter | From Sheerness; fate uncertain |
| 1833- | *Duck* | sailing lighter | Portsmouth Dyd (Symonds design); after 1901 also known as Chatham Yard Craft No 4; still in service 1901. (As *Sinbad*, Woolwich) |
| 1833-1837 | *Surly* | lighter | ex cutter (see Main list); 1837 sold |
| 1833-1877 | *Rochester* | sailing lighter | Chatham Dyd (Symonds design); also known as Chatham Yard Craft No 2; 04.1877 BU |
| | | | 68'6", 54'10" x 23'3" x 11'. 154 26/94 tons |
| | | | PLANS: Lines & profile/sails/decks/midsection/Specification. *Lively* at Portsmouth and *Devon* at Plymouth similar, also *Aid* |
| 1833-after 1854 | | Mud barge No 8 | Chatham Dyd; still in service in 1854 |
| 1833-after 1854 | | Mud barge No 9 | Chatham Dyd; still in service in 1854 |
| 1830s-1841 | | Mooring lighter No 1 | ex Sheerness lighter from Sheerness; 1841 sold |
| 1830s-after 1854 | | Mud boat No 7 | Origins uncertain; still in service 1854 |
| 1830s-1854 | | Mud punt | Origins uncertain; still in service in 1854 |
| 1838-1878 | *Aid* | sailing lighter | Chatham Dyd (Symonds design); also known as Chatham Yard Craft No 3; 1878 BU. [Also see *Rochester* above.] |
| 1830s | *Hope* | buoy boat | ___; 1830s to Portsmouth |
| 1841-1857 | *Renard* | mooring vessel | ex brig (1821); 1857 BU. [see Main List] |
| 1843-___? | *Fly* | river swim barge | employed in the conveyance of stores from Chatham, Sheerness etc' |
| | | | 71'10", 57'8¾" x 20'7" x 7'. 131 93/94 tons |
| | | | PLANS: Lines, profile, deck. Also refers to *Staunch* & to *Ant*. |
| By 1845-1863 | *Rennie* | lighter | from Sheerness; 1863 BU |
| By 1845-1865 | *Deptford* | lighter | from Sheerness; 1865 sold |
| 1846-1908? | *Magpie* | water tank | ex cutter; sold 1908? |
| 1847-1849 | *Bee* | longboat | From Woolwich; 1849 BU |
| By 1851-1877 | *Camel* | lighter (Y.C.1) | in service 1851; 5.3.1877 BU completed |

**Other vessels for which there are plans apparently connected to this yard include:**

| Date | Name | Type | Notes |
|---|---|---|---|
| Ca 1800? | | Chatham sailing lighter | |
| | | | ___ 66'. 53'2" x 20'6" x 9'6". 118 79/94 tons |
| | | | PLANS: Lines & profile |
| No date | *Toliapis* | Chatham wherry | Mercantile? |
| | | | 42'10½", 26'2⅜" x 15'1" x ___. 31 67/94 tons |
| | | | PLANS: (Read Collection) lines, profile, deck |
| No date | 'A Chatham wherry' | | Mercantile? |
| | | | 36'6", 24'3" x 13'6" x ___. 23 55/94 tons |
| | | | PLANS: Lines & profile & deck. This would appear to be the same as a plan in the Danish (Rigsarkivet) collection which is described as 'built at Chatham and taken off November 1758'. |

# Sheerness Dockyard

| Date | Name | Type | Notes |
|---|---|---|---|
| 1691-___? | *Sheerness* | smack | Purchased; deleted by the early eighteenth century |
| Ca 1695-1698 | *Sea Horse* | waterboat | ex purchased Dutch machine; 1698 sunk as foundation |
| 1698-1724 | | Sheerness waterboat | Sheerness Dyd (Surveyor: Shortiss); 1724 BU |
| | | | 54', ___ x 16' x 7'6". 69 26/94 tons. 5 men |
| By 1701-1747 | *Supply* | hoy, later water boat | Chatham (Lee); 1747 BU |
| | | | 63', 54'6" x 18' x 7'6". 94 tons. 5 men |

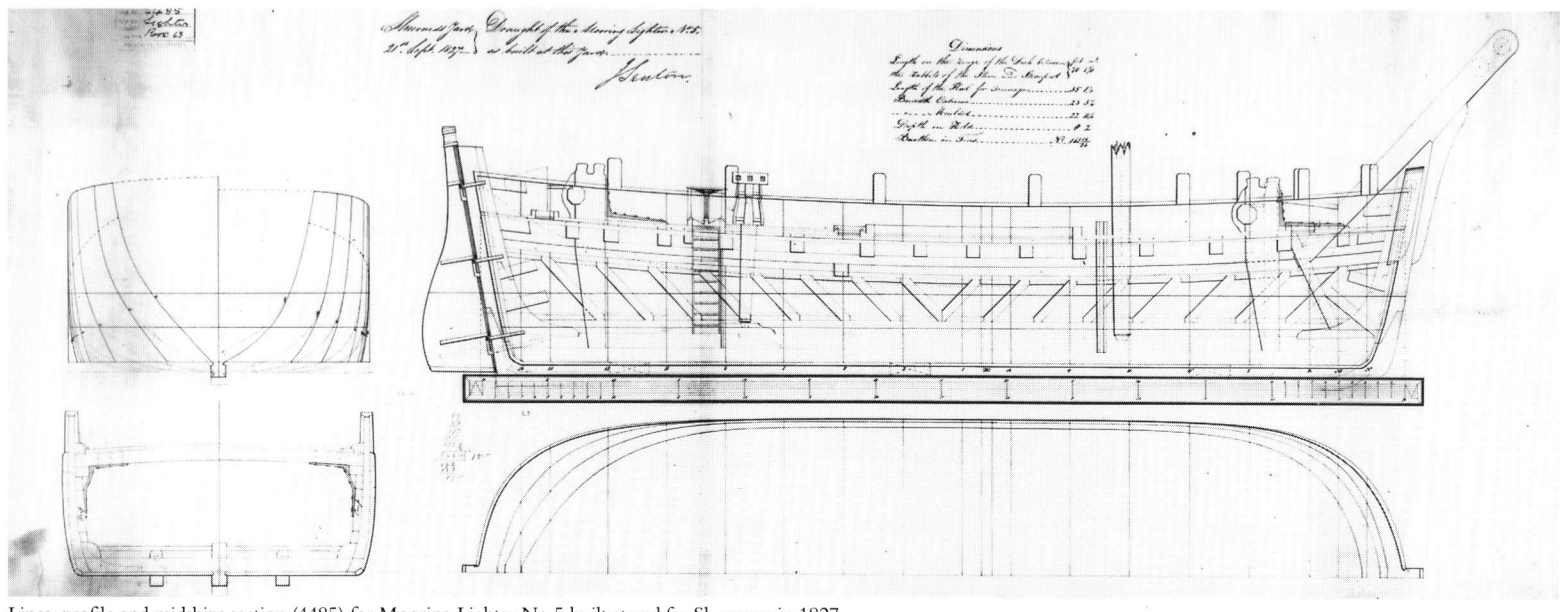

Lines, profile and midships section (4485) for Mooring Lighter No 5 built at and for Sheerness in 1827

| | | | |
|---|---|---|---|
| 1706-1791 | *Goodwill* lighter | Sheerness Dyd (Surveyor Acworth); 1791 breakwater | |

51'6", 39' x 18'10" x 6'6". 73 54/94 tons

| | | | |
|---|---|---|---|
| 1724-1796 | Sheerness longboat No 1 | Sheerness Dyd (Surveyor: Ward); by 1796 pitch boat; 1796? sold | |

41'10", 32'6¼" x 12'6" x 6'. 27 2/94 tons. 4 men

| | | |
|---|---|---|
| 1724-1727? | Sheerness Longboat No 2 | |
| Ca 1725-1747 | Sheerness waterboat | ex hoy *Supply* [see above]; 1747 BU |
| 1742-1796 | Mooring lighter No 2 Sheerness Dyd; 1796 lost | |

50'10", 38'10¾" x 19'5" x 7'. 78 tons

| | | |
|---|---|---|
| 1747-1772 | Sheerness waterboat | Purchased mercantile timber hoy *Duke of Cumberland* built 1747 on the Thames; 1772 sold |

55', 41'9⅞" x 19'8" x 9'9". 86 tons. 5 men

| | | |
|---|---|---|
| 1758-____? | Two longboats | Sheerness Dyd; ? possibly one of these was No 2 longboat (see below) |

One 34', ____ x 10'4" x 4'11"
PLANS: Lines

| | | | |
|---|---|---|---|
| 1758-by 1830 | *Sheerness* sailing lighters | Bennett, Faversham; by 1830 to Chatham | |

109 tons

| | | | |
|---|---|---|---|
| 1758-____? | Mooring lighter | Bennett, Faversham; to Woolwich soon after completion? | |

53', 41'4" x 19'6" x 7'5". 83 56/94 tons
PLANS: Lines, profile

| | | | |
|---|---|---|---|
| 1760-1764 | *Canterbury* storeship (buss) | See storeships under Main List; possibly at Chatham?; 1764 sold | |
| 1760-1764 | *Medway* storeship (buss) | See storeships under main list; 1764 sold | |
| 1761-____ | Rigger's launch | Burr, ____; no further information. | |

34' x 9' x 3'7½"
PLANS: Lines, profile, midsection, deck/scantlings

| | | |
|---|---|---|
| By 1760s?-1797 | Sheerness longboat No 2 | Origins unknown (but see under 1758 above); 1797 sold |

40', 29'7½" x 13'11½" x 6'4". 30 tons. 4 men

| | | |
|---|---|---|
| By 1760s?-1790 | Mooring lighter No 3 | Origins unknown; 1790 breakwater |

49'2", 38'5¾" x 18'0½" x 6'0½". 70 1/94 tons

| | | |
|---|---|---|
| 1765-1820 | Mooring lighter No 1 | Sheerness Dyd; 1820 BU |

56'6", 44'7½" x 20'4½" x 7'2½". 96 87/94 tons
PLANS: Lines, profile, deck, midsection

| | | |
|---|---|---|
| 1770-until | Sheerness waterboat | Sheerness Dyd; in service until 1830 1830 [or sold 1802?] |

56'1", 43'2" x 21' x 9'2". 101 24/94 tons. 9 men
PLANS: Lines, profile, deck

| | | |
|---|---|---|
| 1791-after 1816 | *Goodwill* sailing lighter | Nicholson, Jnr, Chatham; still in service 1816 |

63'5", 51'11" x 20'5" x 9'5½". 115 9/94 tons. 8 men

PLANS: 1810 alterations: Profile/profile & forward platform

| | | |
|---|---|---|
| 1791-____? | Sheerness Long Boat No 2 | Sheerness Dyd? |

42', 41'8" x 15'8" x 7'11½". 42 1/94 tons
PLANS: Lines/profile & deck.

| | | |
|---|---|---|
| 1792-1829 | Mooring lighter No 3 | Woolcombe, Thames; 1818 No 2; 1829 BU |

57'9", 43'7¼" x 20'0½" x 8'3". 93 tons

| | | | |
|---|---|---|---|
| 1796-by 1830 | *Medway* sailing lighter | Nicholson, Chatham; deleted by 1830 | |

63', 51'7" x 20'7" x 9'3". 116 tons

| | | |
|---|---|---|
| 1796-1836 | Longboat No 1 | Ross, Rochester; BU 1836 |

41'9½", 31'0¼" x 15'9½" x 7'6". 41 tons
PLANS: Profile & deck

| | | |
|---|---|---|
| 1797-after 1854 | Longboat No 2 | Ross, Rochester; still in service in 1854 |

42', 41'8" x 15'8" x 7'11½". 42 1/94 tons
PLANS: Lines/profile & deck

| | | |
|---|---|---|
| 1798-____? | Launch for the Master Attendant | Sheerness Dyd |

35'6", ____ x 9'7" x 3'9". ____
PLANS: Lines, profile, deck

| | | |
|---|---|---|
| 1798-____? | Launch for the Master Attendant | Sheerness Dyd |

31'6", ____ x 9'7" x 3'9". ____
PLANS: Lines, profile, deck

| | | |
|---|---|---|
| 1800s?-after 1830 | Mooring lighter No 3 | Origins uncertain, still in service 1830 |
| 1811-after 1854 | Mooring lighter No 4 | Pelham, Frindsbury; by 1845 No 3; still in service 1854 |

70', 54'11¾" x 23'5" x 8'5". 160 34/94 tons (similar to 1796 Plymouth mooring lighter of 1796)
PLANS: Lines & profile/decks/specification

| | | |
|---|---|---|
| 1813-____? | *Sheppy/Sheppey* Barge | Purchased mercantile barge *Citizen*; later to Woolwich |
| 1814-by 1830 | *Dart* longboat | Deptford Dyd; by 1830 to Deptford |
| 1814-by 1830 | Boat No 1 | Chatham Dyd; by 1830 to Chatham |
| ____?-1821 | Mooring lighter No 5 | Origins uncertain; 1821 BU |
| ____?-1821 | Tank vessel | Origins uncertain; 1821 BU |
| ____?-by 1830 | Schooner | Origins uncertain; deleted by 1830 |
| ____?-by 1830? | Pitch boat | Origins uncertain; deleted by ____? |
| 1820-after 1854 | Tank schooner | Chatham Dyd; by 1830 transferred to the Victualling Department; still in service 1854 |
| 1825-1875 | Longboat | ex Navy Board (ex *Hart*) yacht; 1833 Admiralty tender at Sheerness; Yard Craft No 1; 1870 renamed *Drake*; 1872 superintendent's yacht; 1875 BU |
| 1827-after 1854 | Mooring lighter No 5 | Sheerness Dyd; later No 4?; still in service in 1854 |

# YARD CRAFT

| | | | |
|---|---|---|---|
| | 70'1⅜", 55'1⅛" x 23'5¼" x 8'2". 161 16/94 tons | | |
| | PLANS: Lines, profile & midships section/upper deck | | |
| 1827-after 1854 | Mooring lighter No 6 | | Sheerness Dyd; later No 5? still in service in 1854 |
| After 1827-1876 | *Louisa* ____ | | ____, Northfleet 1827; ex Deptford; 09.1876 BU at Sheerness |
| By 1830-1858 | *Goodwill* | sailing lighter | ex mortar vessel (before that lighter at Chatham); 1858 BU |
| 1830-1832 | *Medway* | sailing lighter | Sheerness Dyd,[Launched 30.9.30]; 1832 to Jamaica |
| | 71'11", 58'8" x 23'4½" x 12'6". 167 tons | | |
| 1830s?-by 1845 | *Deptford* | lighter | From Deptford; to Chatham by 1845 |
| 1831-1872 | *Thames* | longboat fitted as a tender | Origins uncertain; later Sheerness Yard Craft No 2; sold 1872 |
| 1835-by 1845 | *Rennie* | lighter | From Woolwich; by 1845 to Chatham |
| 1836-after 1854 | Mooring lighter No 1 | | Chatham Dyd (Symonds design); still in service in 1854 |
| | 70'3", 56' x 23'8" x 8'10". 166 78/94 tons | | |
| | PLANS: Lines & profile | | |
| 1836-after 1854 | Mooring lighter No 2 | | Chatham Dyd (Symonds design); still in service in 1854.[As No 1 above] |
| 1838-1859 | *Quail* | sailing lighter | ex cutter; 1859 fitted for the Liberian Government |
| By 1845-after 1854 | Pitch boat | | From Falmouth; still in service in 1854 |
| By 1845-1877 | *Royal Escape* | sailing lighter | From Chatham; 02.1877 BU at Sheerness |
| 1853-after 1860 | Sheerness Yard Craft No 11 Temporary mooring lighter | | ex cutter tender *Speedy*; still in service 1860 |
| By 1867-1894 | *Manoeverer* | barge | From Woolwich; 05.1894 sold |

**Other vessels for which there are plans apparently connected with this yard include:**

| | | | |
|---|---|---|---|
| Late C18th? | Sailing lighter | | Proposed to be built for Sheerness |
| | 60'3", 49' x 20' x 9'5". 104 24/94 tons | | |
| | PLANS: Lines, profile, deck | | |
| 1790 | Sheerness lighter | | Proposed |
| | 57'6", 43'5¾" x 19'11" x ____ 91 68/94 tons | | |
| | PLANS: Lines, profile, deck | | |
| 1796 | Vessel for yard service | | Proposed by Mitchell, not approved |
| | 43'8", 32'4⅜" x 15'6" x 7'4". 41 33/94 tons | | |
| | PLANS: Lines & profile | | |
| 1831 | Sailing Lighter | | ____ (Seppings design) |
| | 71'9", 58'0⅝" x 24' x 12'6". 177 80/94 tons | | |
| | PLANS: Lines/profile & decks | | |

# Portsmouth Dockyard

| | | | |
|---|---|---|---|
| 1655-1712 | *Marigold/Marygold* | hoy | Portsmouth Dyd; 1712 BU |
| | ____, 37' x 14'8" x 8'3". 42 31/94 tons. 5 men | | |
| 1672-1713 | lighter | | Portsmouth Dyd (Surveyor: Tippetts); 1713 sold |
| | ____, 40' x 20'3" x 8'1". 87 23/94 tons. 5 men | | |
| 1686-1714 | *Nonsuch* | hoy | Portsmouth Dyd. (Surveyor: Lucas); 1714 sold |
| | ____, 52' x 18'10½" x 8'4½". 95 92/94 tons. 5 men | | |
| 1686?-1714 | *Delight* | hoy | Purchased 1686?; 1714 sold |
| | ____, 47'6" x 18'5" x 8'6". 84 49/94 tons. 4 men | | |
| 1690-1707 | *Unity* | hoy | ex fireship; 1707 rebuilding |
| After 1694 | *Sophia* | hoy | From Deptford after 1694; 1712 BU -1712 |
| 1694/1698-1714 | *Truelove* | hoy | Purchased ex bomb vessel?; 1707 lighter; 1714 BU? |
| | ____, 48' x 17'6" x 9'. 76 tons. 4 men | | |
| 1693-1752 | *Forrester* | hoy | Portsmouth Dyd (Surveyor: Stigant); 1752 wrecked. 125 tons. 2 guns |
| By 1699-____? | Buoy boat | | Origins and fate unknown |
| | ____, 31' x 9'6" x 4'3". 12 tons | | |
| 1705-1729 | *Hayling/Heyling* | hoy | Purchased; 1729 rebuilding. 117 tons. 4 guns |
| 1700s?-1713 | *Dolphin* | wellboat | Origin uncertain; 1713 sold |
| 1707-1773 | *Unity* | hoy | Portsmouth Dyd (Surveyor: Podd) [RB]; sold 1773? There is a plan of a vessel called *Unity* dated 1759 which may be of this vessel: |
| | 66'9", ____ x 21'4" x ____ 128 tons | | |
| | PLANS: Lines, profile, decks? | | |
| 1707-Ca 1770 | *Truelove* | hoy | Portsmouth Dyd (Surveyor: Podd); deleted Ca 1770 |
| | PLANS?: Lines | | |
| 1720-Ca 1830 | [Old] *Truelove* | hoy | Portsmouth Dyd (Surveyor: Naish) [RB of 1690s *Truelove*]; 1823 to the Victualling Department; Ca 1830 to Deptford |
| | 54'2⅜", 41'2" x 16'5" x 7'4". 59 tons. 3 men | | |
| 1729-1759 | *Hayling* (*Heylin*) | hoy | Portsmouth Dyd (Surveyor: Allin) [RB]; 1759 converted to a sloop/floating battery for the Coast of Africa (see Main List) |
| 1736 or 1756? | Mooring lighter | | Proposed |
| | PLANS: Lines & profile | | |
| 1748-after 1771 | Sailing lighter No 1 | | Ewer, Bursledon; deleted at some time after 1771 [or sold 1767] |
| | 62'11¾", 51'9" x 20'2" x 8'8". 111 89/94 tons | | |
| | PLANS: Lines, profile, deck (two built to this design by Ewer and two by Chitty & Vernon) | | |
| 1748-1828 | *Forrester* | hoy | Ewer, Bursledon; 1828 given up to the Preventive Service |
| 1753-1775 | *Lyon* | transport, two masted | Portsmouth Dyd (Surveyor: Lock); 1775 tender to the Tower of London |
| | PLANS: (Charnock Collection) Lines & profile (presumably of this *Lyon*) | | |
| 1757-by 1830 | Sailing lighter No 2 | | Ewer, Bursledon; probably this one which had a Great Repair at Portsmouth 1785/6; by 1830 dismasted and laid up |
| | 60'6¾", 49'2¼" x 29'7¾" x 7'4". 111 49/94 tons | | |
| 1757-after 1771 | Sailing lighter No 3 | | Adams, Southampton; sunk at some time after 1771 at Portsmouth |
| | 42'2", 33'9" x 17'7¾" x 4'10½". 55 84/94 tons | | |
| 1757-1778 | Sailing longboat No 1 | | Ewer, Bursledon; 12.1778 lost |
| | 43', 34'2½" x 13' x 6'9". 30 3/4 tons | | |
| 1757-1787 | Sailing longboat No 2 | | Ewer, Bursledon; 1787 BU |
| | 47', 38'6" x 12'2" x 6'6". 30½ tons | | |
| 1757-____? | Sailing longboat No 3 | | Adams, Bursledon (Southampton); fate uncertain |
| | 43', 34'2½" x 13' x 6'9". 30¾ tons | | |
| 1757-1787 | Sailing longboat No 4 | | Adams, Southampton; 1787 BU |
| | 47', 38'6" x 12'2" x 6'6". 30½ tons | | |
| 1760-1782 | *Hayling* (*Heyling*) | hoy or transport) | Adams, Bucklers Hard; 1782 foundered in the Channel |
| | 66'8", 52'11⅝" x 21'6" x 10'1". 130 22/94 tons | | |
| | PLANS: Lines, profile, deck | | |
| 1772-1817 | Mooring lighter No 4 | | Portsmouth Dyd 1790 No 2; 1817 BU |
| | 60'9½", 48'2⅝" x 20'9" x 7'5". 115 66/94 tons | | |
| 1777-by 1830 | Well boat | | Portsmouth Dyd; by 1830 tender at Newfoundland |
| | 51'7", 40'7½" x 16'11½" x 8'4". 62 4/94 tons | | |
| | PLANS: Lines & deck | | |
| 1778-1817 | Mooring lighter No 1 | | ____, Northam; 1790 renumbered as No 3; 1817 BU |
| | 64'4", 52' x 20'9½" x 7'4½". 119 51/94 tons | | |
| Ca 1780-1795 | Buoy boat | | Taken from the French; 1795 sold |
| By 1785-____? | Mooring lighter No 3 | | Origins uncertain |
| | ____ 60'9½", 48'2⅝" x 20'9" x 7'5". 115 66/94 tons | | |
| 1785-after 1864 | Mooring lighter No 2 | | Portsmouth Dyd; 1790 renumbered to No 4; by 1830 No 9; by 1845 No 3; 1854 open lighter No 3 still in |

| Date | Name | Type | Builder / Notes |
|---|---|---|---|
| | | | service 1864. |
| | | | 64'4", 52'x 20'9½" x 7'4½". 119 51/94 tons |
| 1789-by 1854 | Chain boat No 1 | | Adams, Bucklers Hard; out of service |
| | | | 39'4", 31'1⅜" x 13'8" x 5'4½". 30 70/94 tons |
| | | | PLANS: Lines & profile |
| 1789-after 1864 | Mooring lighter No 1 | | Adams, Bucklers Hard; still in service in 1864, by then known as 'Old No 1' |
| | | | 64'3½", 52'0¼" x 20'9" x 7'4½". 119 10/94 tons |
| | | | PLANS: Lines & profile/deck |
| 1794-after 1864 | Lighter No 7 | | Edwards, Shoreham; still in service, classed as an open lighter, in 1864 |
| | | | 51'8", 40'8½" x 21'3½" x 6'. 98 tons |
| | | | PLANS: Lines & profile/deck |
| 1794-after 1864 | Lighter No 8 | | Edwards, Shoreham; by 1854 Open lighter No 1; still in service 1864 |
| | | | 52'7½", 39'11⅛" x 21'2" x 5'10½". 95 tons |
| | | | PLANS: As No 7 above |
| 1795-by 1814 | Buoy boat | | Portsmouth Dyd; between 1807 & 1814 taken by escaping French prisoners |
| | | | 50'3", 37'8" x 17'10½" x 9'2". 43 61/94 tons |
| | | | PLANS: Lines & profile/deck |
| 1795-after 1864 | Lighter No 11 | | Parsons, Bursledon; by 1850 Open lighter No 5; 1864 still in service |
| | | | 51'7¼", 39'11½" x 21'1½" x 7'7¼". 95 tons |
| 1795-after 1864 | Lighter No 12 | | Adams, Bucklers Hard; by 1850 Open lighter No 2; 1864 still in service |
| | | | 52', 39'1⅝" x 21'1" x 7'5". 92 tons |
| 1795-1800 | *Firebrand* | lighter/slop ship | Ex purchased fire vessel; 1800 sold? |
| 1797-after 1830 | *Lively* | lighter/storeship | Purchased whilst building by Parsons, Bursledon; still in service 1830 |
| | | | 63'6", 50'11½" x 20' x 9'5" or 63'10¾", 51'7½" x 20'1" x 9'8". 108 32/94 tons |
| | | | PLANS: Lines, profile, deck |
| 1797-after 1864 | Lighter No 6 | | Parsons, Bursledon; still in service 1864 |
| | | | 43'4", 34'6½" x 14'8" x 5'10". 40 18/94 tons |
| | | | PLANS: Lines, profile |
| 1798 | Six launches, 6 oared. | | Portsmouth Dyd |
| | | | 33' x 8'8" x 3'1". ____ |
| | | | PLANS: Lines, profile, deck, midsection |
| 1801-after 1854 | Mooring lighter No 5 | | ____, Bursledon; by 1854 No 6 |
| 1800 | Two flat bottom mud barges to serve the steam dredger (1802), Bentham design. | | |
| | | | 100' x 25' x 7'5½". 200 tons |
| | | | PLANS: Profile, deck, midsection. (2 alternative ways of dumping the spoil are shown, hopper doors in the bottom, and an inclined tipping plane) |
| 1800s?-after 1830 | Chain boat No 2 | | Origin uncertain; 1830 still in service |
| 1800s?-by 1830 | Chain boat No 3 | | Origin uncertain; out of service by 1830 |
| 1800s?-after 1830 | Chain boat No 4 | | Origin uncertain; still in service in 1830 |
| 1800s?-after 1830 | Chain boat No 5 | | Origin uncertain; still in service in 1830 |
| 1800s?-after 1864 | Lighter No 10 | | Origin uncertain; by 1854 Open lighter No 4; still in service 1864 |
| 1800s?-1808 | *Leopard* | hoy | Ex pitch boat; origins uncertain; 1808 sold |
| 1802-Ca 1816? | 'Steam engine' | (dredger) | Portsmouth Dyd [Keel laid 07.01, launched 11.02 cost £1,685]; about 1816 transferred to Woolwich. 'Vessel for digging soil' |
| | | | 99'6", 83'6½" x 26'6" x 7'6". 315 tons |

NOTE: This was the Royal Navy's first steam engine afloat (if one ignores the dubious case of the 1790s *Kent* which was certainly never registered in the Navy as a steamer) and was part of the major series of developments sponsored by Samuel Bentham when he was at Portsmouth. The steam engine was used to power the continuous chain of buckets and was not, of course, for propulsion.

PLANS: Profile, deck & section (original design; not built but showing dredging and steam machinery)/Lines, profile & deck (hull, probably as built, not showing machinery).

| Date | Name | Type | Builder / Notes |
|---|---|---|---|
| 1803-___? | Steam dredger | | Portsmouth Dyd [Launched 07.1803. Cost £2,180; fate uncertain [as 1802 dredger?] |
| 1806-after 1864 | Pitch boat No 2 | | ____, Cowes; by 1845 Lighter No 9; by 1854 Open lighter No 7; still in service 1864 |
| 1807-after 1830 | Pitch boat No 1? | | Purchased in frame whilst building by ____; later No 1 (Master Attendant's); still in service in 1830 |
| 1807-after 1864 | Long boat No 2 | | Purchased in frame whilst building by ____; by 1845 decked sailing lighter; by 1854 Lighter No 8; still in service 1864 |
| 1808-after 1864 | Buoy boat | | Portsmouth Dyd; still in service in 1864 |
| | | | 51'3", 38'2⅜" x 18'1" x 9'4". 66 41/94 tons |
| | | | PLANS: Lines, profile/decks (dated 1815; presumably this boat) |
| 1809-after 1864 | *Bounty* | victualling lighter | ____, Fishbourne, Isle of Wight; still in service 1864 |
| By 1810-___? | *Providence* | victualling hoy | |
| | | | ____ 50'6½", 39'5¼" x 17'9" x 7'3⅛". 66 3/94 tons |
| | | | PLANS: (As taken off) Lines, profile, midsection/deck/scantlings (plans also sent to List in 1810 and Richards & Co in 1812 for building repeats) |
| 1812-1863 | *Sisters* | victualling lighter | ____, Hythe; 1863 sold |
| 1814-___ | *Industry* | transport | see Main List for details |
| 1816-1843 | *Supply* | victualling lighter | ____, Hythe; 1843 BU |
| 1816-after 1864 | *Fervent* | lighter | Ex gunbrig; by 1864 Mooring lighter No 3 |
| 1816-after 1864 | *Intelligent* | lighter | Ex gunbrig; by 1864 Mooring lighter No 4 |
| After 1816 | Steam dredger | | From Woolwich, ex Deptford; not listed after 1830? |
| 1819/1823 | Ordinary boats | | Portsmouth Dyd |
| | | | 19' x 6'10" x 2'7" and 21' x 7' x 2'10½" |
| | | | PLANS: Lines |
| 1826-___? | Tank vessel No 2 | | ____, Gosport; fate uncertain |
| 1826-1838 | *Nimble* | victualling lighter | ____, Gosport; 1838 BU |
| 1827-___? | Commissioner's barge | | Portsmouth Dyd |
| | | | 37' x ____ x ____ |
| | | | PLANS: Sails |
| 1827-___? | Commissioner's cutter | | Portsmouth Dyd '12 oared cutter or pay boat' |
| | | | 35' x ____ x ____ |
| | | | PLANS: Sails [(3 masted lugger)] |
| 1829-after 1854 | *Swinger* | mooring lighter | Ex gunbrig; still in service 1854 (and 1864?) |
| 1830 | Sailing lighter | | Building Portsmouth Dyd (Seppings & Tucker design) |
| | | | 71'9", 58'8" x 23'1" x 12'6". 166 26/94 tons |
| | | | PLANS: Lines/profile & decks/scantlings |
| 1832-1842 | *Cracker* | | ____ Ex cutter; 1842 sold |
| 1833-1859 | *Sylvia* | cutter tender | Ex cutter; 1836 cutter again?; 1859 sold at Londonderry |
| 1834-1857 | *Ant* | lighter/barge | Portsmouth Dyd, for the Victualling Yard; 1857 BU at Woolwich. [As *Sinbad*; see under Woolwich] |
| 1834-1854 | *Drake* | sailing lighter | Portsmouth Dyd (Symonds design); 1854 converted to mortar vessel No 1 (see Main List) [As *Sinbad*; see under Woolwich] |
| 1834-___ | *Sinbad* | sailing lighter | Pembroke Dyd (Symonds design); soon to Woolwich (which see for details) |
| 1836-1890 | *Bat* | victualling lighter | Chatham Dyd (Symonds design); |

# YARD CRAFT

| | | | |
|---|---|---|---|
| 1838-1894 | *Lively* (ex *Portsea*) sailing lighter | | 1890 sold<br>Portsmouth Dyd (Symonds design);<br>1863 Yard Craft No 3; 1870 renamed *Dromedary*; 1886 handed over to the Royal Engineers; 1894 sold [As *Rochester*, see under Chatham]. |

PLANS: Lines, profile/decks/midsection/specification

| | | | |
|---|---|---|---|
| 1839-after 1864 | *Alpheus* | victualling lighter | Portsmouth Dyd (Symonds design); by 1854 converted to tank vessel; 1864 still in service |
| 1839-after 1864 | Mud punt No 1 | | Portsmouth Dyd; still in service in 1864 |

?51'2", ____ x 17'x 2'8". ____
PLANS: Profile, section, deck?

| | | | |
|---|---|---|---|
| 1839-after 1864 | Mud punt No 2 | | Portsmouth Dyd; by 1854 No 17; still in service in 1864 |
| 1839-after 1864 | Mud punt No 3 | | Portsmouth Dyd; by 1854 No 18; still in service in 1864 |
| 1839-after 1864 | Mud punt No 4 | | Portsmouth Dyd; by 1854 No 19; still in service in 1864 |
| 1839-after 1864 | Mud Punt No 5 | | ____, Emsworth; still in service in 1864 |
| 1839-after 1864 | Mud punt No 6 | | ____, Emsworth; still in service in 1864 |
| 1839-after 1864 | Mud punt No 7 | | ____, Emsworth; still in service in 1864 |
| 1839-after 1864 | Mud punt No 8 | | ____, Emsworth; still in service in 1864 |
| 1839-after 1864 | Mud punt No 9 | | ____, Emsworth; still in service in 1864 |
| 1841-after 1864 | Mud punt No 10 | | ____, Littlehampton; still in service in 1864 |
| 1841-after 1864 | Mud punt No 11 | | ____, Littlehampton; still in service in 1864 |
| 1841-after 1864 | Mud punt No 12 | | ____, Littlehampton; still in service in 1864 |
| 1841-after 1864 | Mud punt No 13 | | ____, Littlehampton; still in service in 1864 |
| 1841-after 1864 | Mud punt No 14 | | ____, Littlehampton; still in service in 1864 |
| 1841-after 1864 | Mud punt No 15 | | ____, Littlehampton; still in service in 1864 |
| 1841-after 1864 | Mud punt No 16 | | ____, Littlehampton; still in service in 1864 |
| 1841-after 1864 | Mud punt No 17 | | ____, Littlehampton; still in service in 1864 |
| 1841-after 1864 | Mud punt No 18 | | ____, Littlehampton; still in service in 1864 |
| 1841-after 1864 | Mud punt No 19 | | ____, Littlehampton; still in service in 1864 |
| 1850s-1883 | *Falmouth* | lighter | From Woolwich; Portsmouth Yard Craft No 46; 1883 sold |
| 1850s-after 1879 | *Hope* | tank vessel at Portland | From Chatham; tank vessel in place of *Rennie* and used at Portland; still in service in 1879 as Portsmouth Yard Vessel No 42 (Portland tank vessel) |
| 1856-____? | *Sheppey* | lighter | Ex Mortar Vessel No 1 (see Main List); previously lighter *Drake* (see above); Portsmouth Yard Craft No 1; fate uncertain |
| 1861-1874 | Mooring lighter A (Y.C.No 11) | | Portsmouth Dyd [Ordered 11.9.60, Keel laid 10.9.60. Launched 9.7.61]; 1874 BU at Portsmouth |
| 1861-____? | Mooring lighter B (Y.C.No 12) | | Portsmouth Dyd [Ordered 11.9.60, Keel laid 10.9.60, Launched 21.8.61]; ____ |
| 1861-____? | *Bessy/Bessie* | hoy (Royal Clarence Y.C. No.2) | J & R White. Cowes [Keel laid 4.2.61, Launched 07.61]; still in service 1871 |
| ____-1894 | Yard Craft No 4 | | Origins uncertain; A.O.17.11.1870 renamed *Chub*; 23.11.1894 sold |

**The following vessels also apparently have some connection with this Yard:**

| | | | |
|---|---|---|---|
| No date (Mid C18th?) | Buoy boat | | French prize |

41'10", 33'4" x 13'10" x 7'8". 33 tons

| | | | |
|---|---|---|---|
| 1795-1827 | *Eling* | [? 50 tons] | Purchased 1795; 1827 sold |
| Ca 1840? | 30 ton mud punt | | (White?) |

37'6" x 13'6" x ____
PLANS: Outline profile, deck & section

| | | | |
|---|---|---|---|
| Ca 1840? | 20 ton mud punt | | (White?) |

35'5" x 11'6" x ____
PLANS: Profile, deck, section/scantlings

# Plymouth Dockyard

| | | | |
|---|---|---|---|
| 1702-1713 | *Hamoaze* | transport/hoy | Plymouth Dyd (Surveyor: Podd); 1713 sold. 111 tons. 4 guns |
| 1703-1713 | *Tryall* | sloop (= sloop rigged vessel?) | Origin uncertain [possibly ex naval sloop?]; 'to employ her at Plymouth till ye square stern hoy be built'; 1713 sold |
| 1704-1742 | *Plymouth* | transport | Plymouth Dyd (Surveyor: Rosewell); 1742 BU. 110, later 166 tons? 4 guns. Fisher suggests there was a RB at Chatham in 1730 |
| 1708-1713 | *Endeavour* | transport | Plymouth Dyd (Surveyor: Lock); 1713 sold. 211 tons |
| 1712-1728 | *Mary* | hoy | Plymouth Dyd (Surveyor: Hayward); 1728 rebuilding |
| 1720-1788 | *Unity* | lighter | Plymouth Dyd (Surveyor: Pownell); 1788 sold |

60', 44'3" x 18'6" x 9'6". 80 tons

| | | | |
|---|---|---|---|
| 1728-1779 | *Mary* | hoy | Plymouth Dyd (Surveyor: Lock) [RB]; later called a smack; 1779 lost in Plymouth Sound? |

____, 39' x 16' x ____. 51 70/94 tons
PLANS: Lines, profile, deck

| | | | |
|---|---|---|---|
| 1728-1813 | *Betty* | smack | Plymouth Dyd (Surveyor: Lock); later cook boat; 1813 BU |

47'10¾", 37'10¾" x 18'6⅞" x 6'2". 69 50/94 tons

| | | | |
|---|---|---|---|
| 1742-1806 | *Plymouth* | transport | Blaker, Shoreham; 1806 sold |

77', 63'4" x 22'x 11'. 163 tons. Pierced for 4 guns
PLANS: Lines & profile/(1759) profile & decks

| | | | |
|---|---|---|---|
| 1743-1833 | *Dennis* | lighter | Fellows, Plymouth; sold 30.8.1833 |

58'6", 43'x 20'x 8'. 91 tons. 3 men

| | | | |
|---|---|---|---|
| 1746-1765 | *Culloden* | smack/sloop (sloop-rigged vessel) | Plymouth Dyd; 1748 in use on the Coast of Scotland (surveying); 1765 sold 36 tons. 2 guns |
| 1747-1767 | *Portsmouth* | transport | Purchased ex French timber vessel *William and Jean* (?*Guillaume et Jean*?); 1767 sold |

63'7", 47'9¾" x 21'2¾" x 11'3⅞". 114 tons. 11 men, 4 guns. 4 swivels

| | | | |
|---|---|---|---|
| 1748-1752 | *Saltash* | lighter | Chitty & Vernon, Chichester; 26.8.1752 wrecked [on the coast of Cornwall] |

64'1", 50'10" x 20'2" x 8'1¼". 109 tons

| | | | |
|---|---|---|---|
| 1748-1767 | *Tavistock* | lighter (No 1) | Chitty & Vernon, Chichester; 1767 sold |

63'2½", 52'3" x 20'1½" x 8'8". 112½ tons. 12 men

| | | | |
|---|---|---|---|
| 1756-1775 | *Saltash* | hoy/lighter | Purchased from John Richardson at Plymouth; 1775 BU |

54'9", 43'6" x 19'5" x 7'10". 85 tons

| | | | |
|---|---|---|---|
| 1759-1767 | *Diligence* | sloop (sloop-rigged vessel) | Purchased; 1767 sold [33 tons] |
| 1760-after 1816 | | Buoy boat | Origin uncertain; fitted for this purpose in 1760; still in service 1816 |

36'9½", 29'2" x 11'6⅜" x 5'3¼". 20 tons

| | | | |
|---|---|---|---|
| 1771-after 1864 | *Assistance* | lighter | Plymouth Dyd; by 1854 probably reclassed as 'Lump' No 4; still in service 1864; 94 tons |
| 1778-1830s? | *Plympton* | lighter | Plymouth Dyd; used in building the breakwater; by 1820s described as 'light sloop';1830s out of |

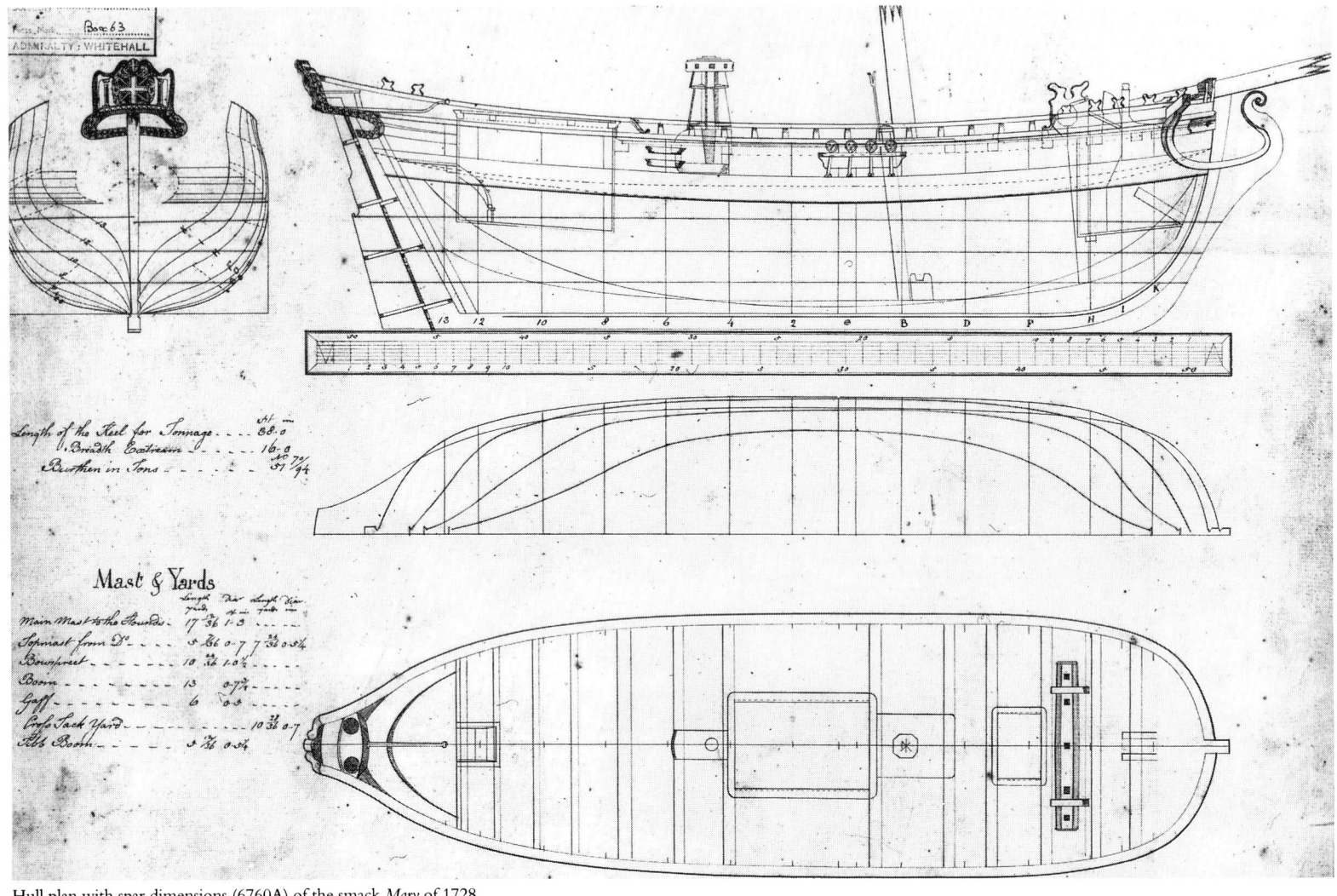

Hull plan with spar dimensions (6760A) of the smack *Mary* of 1728

| | | | |
|---|---|---|---|
| | | | service? 64 tons. At one time had 6 guns |
| 1779-1798 | **Plymouth** | lighter | ____, Plymouth; 1798 BU |
| | 59'2¼", 47'6⅛" x 20'2" x 7'. 102 73/94 tons. 3 men | | |
| 1780-1787 | **Tortoise** | lighter | Barnard, Deptford; 1787 lost |
| | 60'4½", 49'8¾" x 20'4" x 9'5". 109 tons. 11 men | | |
| 1781-by 1830 | **Ann** | barge | Purchased; later armed as a gunboat?; by 1830 to Deptford |
| | 47'6", 35'10⅛" x 17'10¼" x 6'5". 61 5/94 tons | | |
| 1784-1816 | **Deptford** | transport | Batson, Thames; 1816 BU. 198 tons |
| 1785-by 1816 | **Security** | lighter | ____, Franks Quarry, Plymouth; by 1816 to Falmouth. 141 tons |
| 1785-1816 | **Morwelham** | dirt boat | ____, Plymouth; 1816 sold |
| 1785-____? | Harbour boat | | Plymouth Dyd |
| | 21' x 7'6" x 3' | | |
| | PLANS: Lines, profile, deck, section/scantlings | | |
| 1786-1815 | **Plymouth** | transport 8 | ____; 14.12.1815 sold |
| 1788-1878 | **Unity** | lighter | Begun Hooper, Torpoint; completed by Leigh, Torpoint; by 1845 Mooring lighter No 4; by 1854 Lump No 4? then YC 13?; BU 1878. 140 tons |
| 1788-____? | **Sheerness** | tender (brig rig) | Purchased (built by Wilson, Sandgate in 1786); fate uncertain |
| | ____ 58'4⅛" x 21'10¼" x 10'. 148. 14 men. 6 x 3 + 6 x ½ pdr swivels | | |
| 1789-1863 | **Tortoise** | sailing lighter | Sibrell. Plymouth; 1863 BU. 140 tons. [As the 1786 Chatham sailing lighter.] Later armed with 10 guns |
| 1790-____? | Riggers launch | | Plymouth Dyd; no further information |
| | 36' x 10' x 3'4" | | |
| | PLANS: Lines, profile, midsection | | |
| 1793-____? | Riggers launch (two?) | | Plymouth Dyd; No further information |
| | 36' x 10'3" x 4'. 20 oars | | |

| | | | |
|---|---|---|---|
| | | PLANS: Lines, profile, section, deck. (also for 1814 Halifax boat) | |
| 1795-1798 | **Tamar** | lighter? | Parkin, Cawsand; 1798 BU |
| | 64'10", 51'6" x 21' x 9'1". 120 75/94 tons | | |
| | PLANS: Lines & profile | | |
| 1795-1819 | **William and Ann** | dirt boat | Purchased sailing barge; 1819 BU |
| 1795-by 1830 | **Elisabeth** | sailing barge | Purchased; deleted by 1830 |
| 1795-1798 | **Cawsand** | hoy | ____, Cawsand 1795; 1798 BU. 126 tons |
| 1796-____? | Water boat | | ____ 'The Point near Plymouth' (Samuel Bentham design); ____; Semicircular body plan almost double-ended. presumably schooner-rigged |
| | 80'2", 59'3" x 21' x 10'3". 138 92/94 tons | | |
| | PLANS: Lines | | |
| 1797- after 1864 | **Tavy** | lighter | ____ Franks Quarry. Plymouth; by 1830 mooring lighter No 2; by 1854 Lump No 2; still in service 1864. 171 tons |
| 1798-by 1830 | **Camel** | 'stump' sailing lighter | Purchased from Mr Graham; built ____, Thames; By 1830 to Cork |
| | 61'1½", 46'8" x 19'10" x ____ 105 29/94 tons | | |
| | PLANS: Lines, profile, deck | | |
| 1799- after 1864 | **Plymouth** | 'stump' sailing lighter | ____, Franks Quarry, Plymouth; by 1845 Mooring lighter No 1; by 1854 Lump No 1; still in service 1864 |
| 1800-1878 | **Wasp** | victualling vessel | Portsmouth Dyd; 11.1878 BU |
| 1800-1843 | **Hamoaze** | sailing lighter | Soley, Oreston; 1824 mortar vessel; later reconverted; 1843 BU |
| | 69'7". 55'4" x 23'2½" x 9'10". 158 tons | | |
| | PLANS: As fitted 1828: Lines & profile/decks | | |

# YARD CRAFT

| Dates | Name | Type | Notes |
|---|---|---|---|
| 1804-1852? | **William** | pitch boat | Purchased; 04.1852? BU |
| 1806-___? | | Water vessel | |
| | | 64'6", 52'4" × 19' × 10'. 100 46/94 tons | |
| | | PLANS: Specification/lines, profile, midsection/rig/decks | |
| 1806-1872 | **Hythe** | victualling vessel | ____, Hythe; 05.1872 BU? |
| 1806-___? | | Pitch boat No 2 | Built Plymouth?; No further information |
| 1807-after 1830 | | Lighter No 3 | ____, Topsham; still in service 1830 |
| 1807-after 1830 | | Lighter No 7 | Plymouth Dyd?; still in service 1830 |
| 1807-after 1864 | | Mooring lighter No 9 | ____, Dartmouth; by 1854 Lump No 7; still in service 1864 |
| 1807-after 1864 | | Buoy boat for Master Attendant | ____, Franks Quarry, Plymouth; still in service 1864 (it was probably this vessel that became the cutter tender *Cerus/Ceres* in 1830; which by the 1870s was back at Plymouth as Devonport Yard Craft No 3 and sold 1877). |
| | | ?49'9", 35'2½" × 16'4" × 7'5". 49 89/94 tons | |
| | | PLANS: Scantlings/lines, profile (dated 1804) | |
| 1808-1831 | **Saltash** | sailing vessel | Plymouth Dyd; 12.7.1831 sold |
| | | 63'9½", 46'8" × 22'5½" × 10'7¼". 125 18/94 tons. 4 × 18★. | |
| | | PLANS: Lines & profile. (also for vessel for Jamaica, and *Goodwill* for Sheerness).] | |
| 1808-by 1830 | | Water vessel | ____, Topsham; by 1830 to the Victualling Department |
| 1809-after 1845 | | Mooring lighter No 8 | ____, Dartmouth; by 1845 renumbered No 6 |
| Ca 1810?-after 1864 | **Harbour** | pitch boat | Ex Spanish launch; still in service 1864 |
| Ca 1810?-1813 | | Dirt boat | Origin uncertain; 1813 BU |
| 1811-1864 | **Chatham** | transport | see Main List for details |
| 1811-1828 | **Portsmouth** | transport | see Main List for details |
| 1812-____ | | Stone Vessels, Woolwich Nos 1 & 2. | Woolwich Dyd. Like all the following Stone Vessels had stern dropping gear for the blocks of stone from which the breakwater at the entrance of Plymouth Sound would be built |
| | | 63', 51'10" × 20'2" × 9'4". 112 4/94 tons | |
| | | PLANS: Lines, profile/upper deck/bed places (also for all the similar vessels noted below). | |
| 1812-____ | | Stone Vessel, Portsmouth No 1 | Portsmouth Dyd. Stern dropping gear. As above |
| | | PLANS: Specification/upper deck | |
| 1812-____ | | Stone Vessels, Davy Nos 1 & 2 | Davy, Topsham. Stern dropping gear; No 2 by 1845 Plymouth No 8 mooring lighter. As above |
| | | PLANS: Lines & profile/upper deck | |
| 1812-____ | | Stone Vessel, Bailey No 1 | Bailey, Ipswich. Stern dropping gear. As above |
| 1812-____ | | Stone Vessels, Deptford Nos 1,2,3,4 | Deptford Dyd. Stern dropping gear. As above |
| | | PLANS: Lines & profile/lower deck/upper deck | |
| Ca 1813?-1827 | | Dirt boat | Origin uncertain; 1827 sold. [? is this the vessel shown in a plan of 1819 (Profile, deck, sections) showing the boat 'built on the principle suggested by R. Pering Esq. for conveying and discharging mud'? This is shown as having a series of individually tipping buckets along the side. Another plan dated 1830 (Profile, deck, sections) shows a vessel fitted with a similar spoil dumping system |
| | | 90', ____ × 25' × 7' | |
| 1813-after 1864 | | Light vessel for the breakwater | Purchased (built 1813); later No. 9; still in service 1864 |
| 1813-____ | | Sailing vessel for the accommodation of the Officers inspecting the Breakwater | Plymouth Dyd; ____ [It seems likely this is the same vessel as the *Rennie* 1813-1833 launched at Plymouth 18.3.1813; by 1833 to Woolwich] |
| 1814-1828 | **Arrow** | Breakwater Department | Ex schooner; 1828 BU |
| 1815-___? | | Sheer vessel | Plymouth Dyd; ____ |
| | | 65', 53'4½" × 20'2" × 9'4". 115 44/94 tons | |
| | | PLANS: Lines & profile/frame/rig/deck/platforms | |
| NOTE: With the two Chatham sheer hulks, the only other vessel to be specifically built as a hulk | | | |
| 1815-1844? | **Fountain** | tank schooner | Plymouth Dyd; may have been BU 1844; but may alternatively have lasted until 1864?; this confusion is probably due to there being a vessel with the same name at Jamaica at this time. |
| | | ?66'6", ____ × 19' × ____. 106 59/94 tons | |
| | | PLANS: Specification? | |
| 1819-1869 | **Tavy** | sailing lighter (No 1) | Plymouth Dyd; 01.1869 BU |
| | | ?54'2", 43'5½" × 19'6" × 8'6". 87 84/94 tons | |
| | | PLANS: Lines, profile, midsection. (Presumably this vessel; described as a replacement for the *William & Ann*). | |
| 1821-1850 | **Devonport** | Anchor hoy & buoy boat | Plymouth Dyd; later to Bermuda |
| 1822-1831? | **John** | sailing lighter | Purchased from the Victualling Department; 07.1831 sold? |
| 1825-___? | | Powder barge | ____?, for the Powder Works |
| | | 31'7", 30' × 11'6" × 4'8". 16 23/94 tons | |
| | | PLANS: Lines, profile, deck | |
| 1828-after 1864 | **Arthur** | Victualling vessel | ____, Shaldon; still in service 1864 |
| 1830s?-after 1864 | | Mooring lighter No 3 | Ex Store vessel No 8; by 1854 Lump No 3; still in service 1864 |
| 1830s?-after 1864 | | Mooring lighter No 5? | Ex Store vessel No?; by 1854 Lump No 5; still in service 1864 |
| 1830s?-after 1864 | | Mooring lighter No 8 | Ex Store vessel No 1; by 1854 Lump No 8; 1864 to be BU |
| 1830s?-___? | | Store vessel No 2 | Origin & fate uncertain |
| 1830s?-after 1864 | | Store Vessel No 3 | Origin uncertain; still in service in 1864 |
| 1830s?-___? | | Store vessel No 4 | Origin & fate uncertain |
| 1830s?-___? | | Store (or mud) vessel No 5 | Origin & fate uncertain. [This may connect with the second plan mentioned under the 1813 mud vessel; but there is another plan dated 1838 (Profile, midsection) showing 'a vessel to carry away rocks fitted with scuttles in the bottom to let the rubble fall through' (ie. a hopper barge) which may refer to this vessel |
| 1831-1869 | **Coronation** | tank vesse | lPlymouth Dyd; sold 2.12.1869 to Marshall, Plymouth, to BU |
| | | 57'3½", 45'8¾" × 18'1½" × ____. 78 40/94 tons | |
| | | PLANS: As fitted 1841: Lines & profile/decks | |
| 1835-1868 | **Devon** | sailing lighter (Yard Craft No 4) | Chatham Dyd (Symonds design); 1868 wrecked |
| | | [As *Rochester* at Chatham, which see] | |
| 1836-1897? | **Cremill** | victualling hoy | (Royal William victualling craft No 2)Pembroke Dyd (Symonds design); 1897 BU |
| 1844-after 1900 | **Hamoaze** | sailing lighter (Yard Craft No 5) | Plymouth Dyd (Symonds design); 1884 Mooring lighter; 1885 Devonport Yard Craft No 10; still in service 1900; fate uncertain |
| 1844-1869 | **Plym** | | ____ Plymouth Dyd; 2.12.1869 sold |
| 1845-1870 | **Devon** | store carrier | Plymouth Dyd; 1870 BU |
| Between 1845 & 1854-after 1864 | **Diligence** | lighter | Ex transport 1814; still in service in 1864 |
| After 1845-1853 | **Adventure** | lighter | Ex transport 1809; 1853 sold |
| Between 1845 | | Luggage schooner | Origins obscure; still in service 1864 |

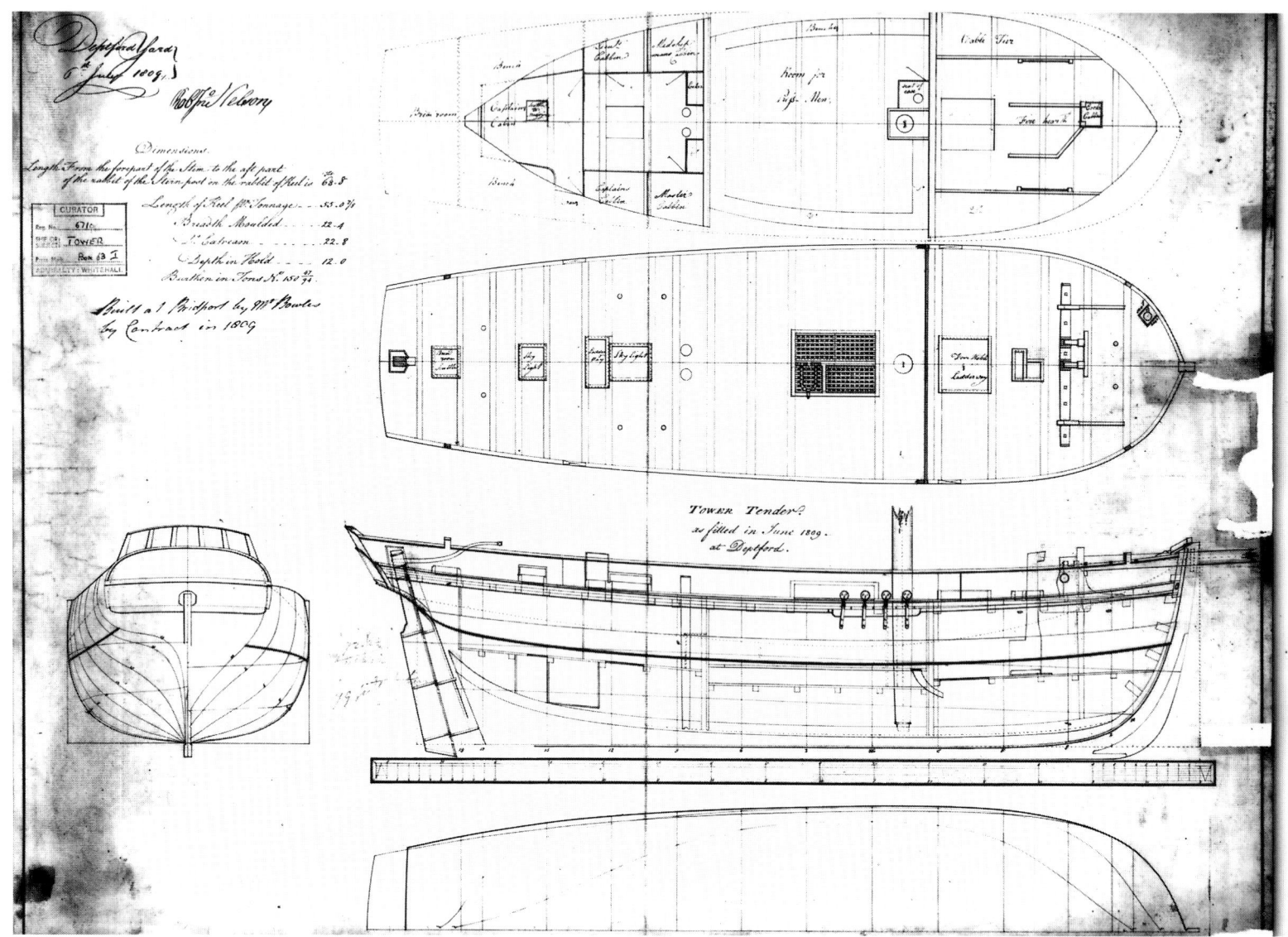

Hull plan (6710) 'as fitted' at Deptford after building in 1809

| | | | |
|---|---|---|---|
| & 1854-after 1864 | | | |
| Between 1845 & 1854-___? | *Confiance* | sailing vessel | Origin & fate uncertain |
| Between 1845 & 1854-___? | Mud vessel | | Origin & fate uncertain |
| 1852 | Mooring lighter | | Approved to build. [As mooring lighters built for Sheerness 1834.] |
| By 1860-1904 | *Medway* | Devonport Yard Craft No 6 | Ex cutter tender; 1876 renamed *Plymouth*; 1904 sold |
| 1860-___? | Mooring lighter No 8 | | Plymouth Dyd [Keel laid 21.4.56. Launched 12.11.60]; fate uncertain |
| ___?-___? | Y.C. No 27 | lighter | Origins uncertain; A.O. 17.11.1870 renamed *Fanny*; A.O. 4.11.1885 re-designated Y.C. No.D.17 [?] |
| 1864-1874 | Y.C. No 7 | sailing lighter | Allen & Warlow, Pembroke [Ordered 6.5.63. Keel laid 15.6.63. Launched 9.2.64] for £2,608 A.O.17.11.1870 renamed *Tortoise*; 1874 BU at Plymouth. 163 tons |

**Other vessels apparently connected with this Yard include:**

| | | |
|---|---|---|
| 1784 | Mooring lighter | Proposed. 66', 52'6" x 22'5" x 12'. 140 27/94 tons |
| | PLANS: Lines, profile, deck | |
| No date (Ca 1800?) | Mooring lighter | ___ |
| | 66', 52'10" x 22'4" x ___ 140 14/94 tons | |
| | PLANS: Lines, profile, deck | |
| No date | Mooring lighter | Proposed (Henslow design) |
| (Ca 1800?) | | |
| | 58'9", ___ x 20' x 10'7" ___ | |
| | PLANS: Lines, profile, deck | |
| 1813 | Stone carrying vessel for Breakwater | Proposed by Captain Huddert. Midships dropping well |
| | PLANS: Lines | |
| No date | Sailing lighter | Proposed |
| | 70', 55'6" x 22' x 11'. 142 88/94 tons | |
| | PLANS: Lines/profile & deck | |

## Pembroke Dockyard

| | | | |
|---|---|---|---|
| 1826-after 1864 | Mooring lighter | | Pembroke Dyd; still in service 1864 |
| 1831-after 1864 | Open barge | | Pembroke Dyd; still in service 1864 |
| By 1845-1850 | *Cracker* | pitch boat | ___ , 'near Portsmouth'; 1850 BU |
| By 1854-after 1864 | *Quail* | sailing lighter | Ex cutter 1830; still in service 1864 |

## Harwich Dockyard

| | | | |
|---|---|---|---|
| 1709-1713 | *Harwich* | hoy/lighter | Poulter, Harwich; 1713 to Deptford |

## At Other British Ports

**At the Tower of London:**

| | | | |
|---|---|---|---|
| 1790-1809 | *Deptford* | tender [brig] | Purchased whilst building by Woolcombe |

& Co, Thames as a mercantile brig for £1,400 on 13.1.1788 [Launched 13.2.88]; 1809/10 fitted for the Limerick station

____, 63'11½" x 21'6½" x 10'1". 158 tons. 14 men. 6 x 3pdrs + 6 swivels; later 4 x 12pdr carronades replaced the 3pdrs

| 1809-1817 | *Tower* | tender | Bools, Bridport; 1816? to Deptford as a hulk; 1817 to the Thames Police |

68'8", 55'0⅞" x 22'4" x 12'. 150 47/94 tons

PLANS: Lines, profile, decks

### At Deal:

| 1796 | | Anchor & cable boat (the largest) | |

____, 28' x 12' x 3'4". 15 88/94 tons

PLANS: Lines, profile, deck, midsection

### At Leith:

| By 1816–____? | | Sailing vessel | Origin & fate obscure |
| 1820-1828? | *Hope* | longboat | Ex buoy boat built at Deptford Dyd 1813; 1828 to Chatham for a tender? |

### At Falmouth:

| By 1816–after 1864 | *Security* | lighter | From Plymouth; still in service in 1864 |
| By 1845–____? | *Swale* (ex *Ceres*) ____ | | From ____ ; shortly afterwards to Portsmouth |

### At Portland:
See under Portsmouth Dockyard

### At Ireland:

| ____?–1712 | *Irish* | lighter | Became a yacht; 1712 sold |

### At Cork:

| 1808? | *Ceres* | lighter fitted to bring up anchors | |

____ 53'11", 40'0¾" x 17'3" x 9'1". 64 57/94 tons

PLANS: Profile & deck

| Ca 1816–____? | sailing vessel | | Origin & fate uncertain |
| 1823 | | Ordnance lighter | Pembroke Dyd |

____ 46'2", 37'3½" x 15'1" x 7'8". 45 11/94 tons

PLANS: Lines, profile, deck

| By 1830-1831 | *Camel* | lighter | From Plymouth; sold 1831 |
| By 1830-1832? | *Port Royal* | lighter | From Jamaica; out of service by 1832? |
| 1844–after 1864 | | Tank vessel | Pembroke Dyd (Symonds design); still in service in 1864 |

### At Limerick:

| 1810-1816 | *Deptford* | tender | From the Tower [fitted at Woolwich]; 1816 turned over, with her boats to the Hibernian Marine Society |

## Dockyard Vessels and Others, Location Uncertain

NOTE: Some of these vessels are probably not naval at all.

| 1716 | | Buoy boat | |

____, 31' x 9'6" x 4'3". 12 71/94 tons

| Early 18thC? | *Elizebeth* (&/or *Friends Goodwill*) | timber hoy(s) | Possibly hired vessels? |

PLANS: Lines & profile

| Early 18thC? | *Two Brothers* | sand barge | Mercantile? |

29'4", ____ x 16'9" x 5'6". About 30 tons

PLANS: Lines

| 1743–____? | | Lighter | Robert Carter, Limehouse [Launched 7.2.1743] |

63'2", 49'9¾" x 20'1¼" x 8'4½". 107 tons

| 1757–____? | | Small Lighter | Adams, Hamble [Launched 01.1757] About 42' long |

PLANS: (Slade Collection) Lines & profile

| Mid 18thC? | *Cupitt* | hoy/sloop | |

____ 34', 30'10½" x 11'4" x ____ 16 25/94 tons

PLANS: (Charnock Collection) Lines

| 1798–____? | | Naval transport; 2 masts | Pitcher, Northfleet |

86'1", 70'6¼" x 24'3" x 13'4½". 220 55/94 tons

PLANS: Lines, profile/decks (dated 1797)

| 1801 | | Longboat | |

____ 31' x 9'6" x 4'1" ____

PLANS: (Kennedy Collection) Lines & profile

| 1819 | | Sailing & mooring lighter | Design only (By Peake & Tucker) 'not built' |

63'9½", 46'8" x 22'5½" x 10'7¼". 125 51/94 tons

PLANS: Lines & profile

| 1829 | | Sailing lighter (*Falmouth*?) | Portsmouth Dyd |

PLANS: Specification

| No date | | Chain lighter ____ | |

45'4", 36'2¼" x 15'3" x 6'6". 44 71/94 tons

PLANS: Lines & profile

| No date | | Trinity (Trinity House?) Lighter No 1 (dredger/ 'ballast lighter'). | Presumably for the River Thames. Clinker built |

41'1", ____ x 20'2" x 6'10"

PLANS: Lines, profile, deck/dredging equipment

| No date | | Ballast lighter (dredger) | |

____ 40', ____ x 17'4" x 6'. About 35 tons

PLANS: Outline profile & deck

| No date | | Mooring lighter | |

____ 64'3½", 52'0⅛" x 20'9" x ____ 119 30/94 tons

PLANS: Lines & profile

| No date | | Pitch boat | |

____ 48', 31'4" x 14' x 6'6". 35 8/94 tons

PLANS: (Charnock Collection) Lines, profile, deck

| No date | | Chalk barge | |

____ 57', ____ x 16'4" x ____ 50 63/94 tons

PLANS: Outline profile & deck

### Ordnance vessels, location uncertain:

| Mid-late 18thC? | | Gunwharf boat No 21 | |

____ 26' x 6'1" x 2'7". ____

PLANS: Lines, profile, deck

| 1811 | | Two gun hoys | William Hayles, Poplar |

50'6½", 39'4⅝" x 17'10" x 7'3½". 66 58/94 tons

PLANS: Lines, midsection, profile/decks

| 1827 | | Vessel | Pitcher, Northfleet |

57'6", 46'2" x 18'2" x 9'. 81 3/94 tons

PLANS: Lines, profile/Specification.

| 1828 | | Powder vessel | |

____ 49'9", 39'7" x 16'10" x 8'2'". 59 61/94 tons

PLANS: Lines & profile

| 1837 | | Coasting vessel | Cowes (White?) |

57', 43'5¾" x 19' x 8'10". 82 tons

PLANS: Lines, profile, midsection/sails

| 1867 | *Emily* | powder vessel | D Robinson, Gosport [for War Dept] |

### Victualling department vessels, location uncertain:

| 1708 | | Victualling hoy | ____ |

PLANS: (Limekiln Collection) Lines

| 1810 | | Victualling hoy | Daniel List, Fishbourne |

50'6½", 39'4⅝" x 17'10" x 7'3½". 66 58/94 tons

PLANS: Lines, profile, midsection/deck

| 1815 | | Victualling hoy | Mark & John Richards & John Davidson, Heath [as 1810 hoy above] |
| 1827 | | Vessel | |

____ 57'6", 46'2" x 18'2" x 9'. 81 3/94 tons

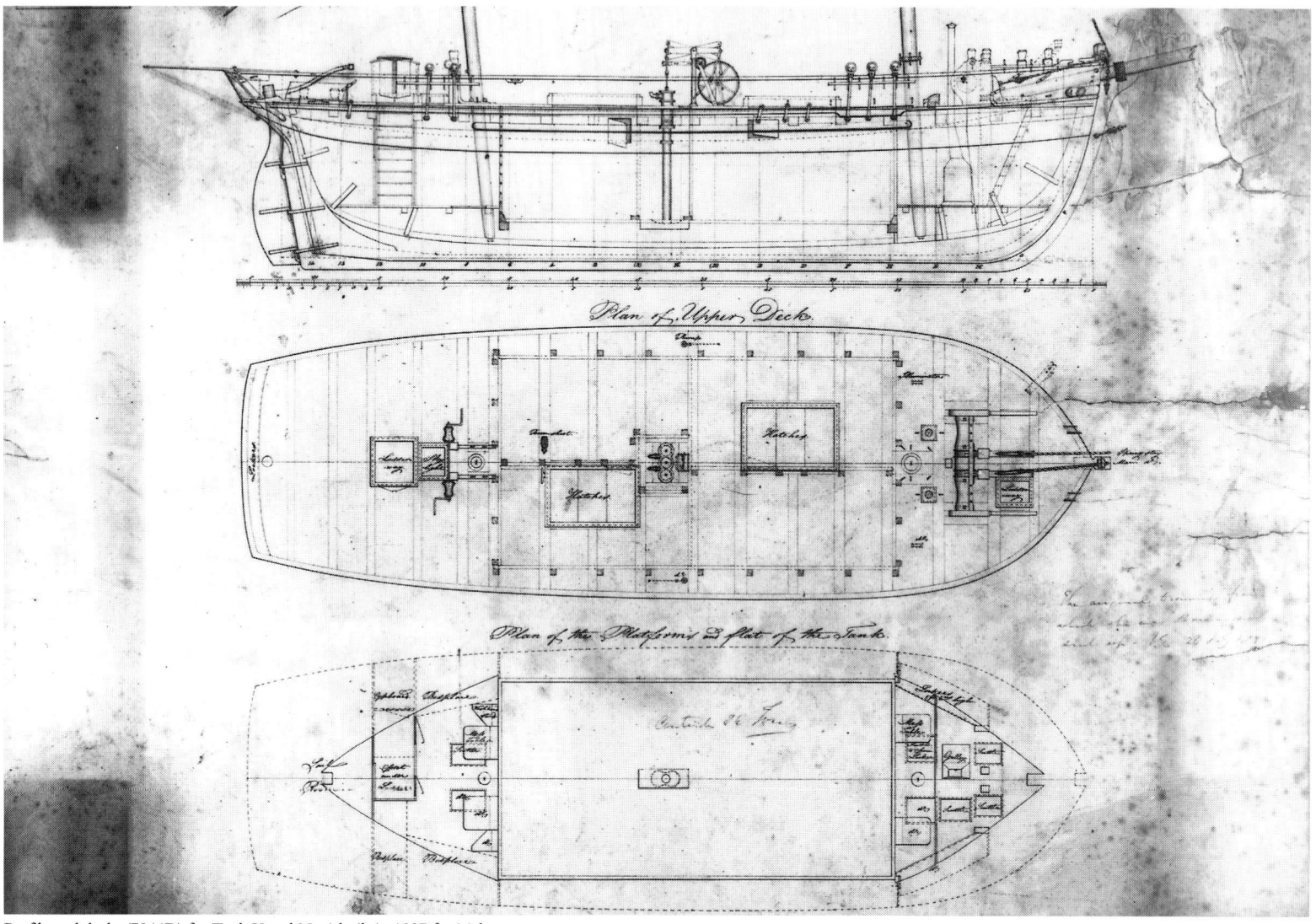

Profile and decks (7846D) for Tank Vessel No 1 built in 1837 for Malta

| | | |
|---|---|---|
| 1832 | Vessel | (Proposed, but 'not built') |

PLANS: Lines & profile/profile & decks

55', 44'4" x 17'3" x 8'8". 70 15/94 tons

PLANS: Lines, profile/deck

1832    Tank vessel    (Proposed)

56', 45'3" x 17'8" x 8'9". 75 4/94 tons

PLANS: Lines & profile

No date (late 18th or early 19thC?)    Open long boat for the use of the Victualling; built 'in the neighbourhood of Portsmouth'.

47', 38'6" x 12'2" x 7'6". 30 26/94 tons

PLANS: Lines & profile

### Impress service (press gang) tenders:

Ca 1800    *Marianne*    tender for impressed seamen, two masted (brig or schooner). Origins uncertain

72'2⅜", 60'5⅜" x 23' 0⅝" x 10'5". 170 93/94 tons

PLANS: Profile & upper deck/lower decks & midsection

# Abroad

### At Gibraltar:

| 1741-1750 | *Discovery* | pink | Purchased as a transport; 1750 sold |

74'6", 57'1¼" x 22'6½" x 8'0¾". 154 tons. 30/20/14 men. 10 or 12 guns + 6 or 4 swivels

| 1799-1801? | *New Adventure* | transport or tender | ?Purchased 1799; still in service 1801 |
| 1799-___ ? | | Two launches for weighing anchors | Woolwich Dyd |

38' x 11' x 4'6".

PLANS: Lines, profile, midsection. In 1804 two more were built for Malta

| 1815-___ ? | Tank vessel | | Plymouth Dyd |

66'6", 55'6" x 19' x 8'. 106 53/94 tons

PLANS: Lines/profile, decks, midsection

| 1860s-___ ? | *Kertch* | water tank vessel | Ex gunboat; fate uncertain (Yard Craft No 1) |

### At Port Mahon, Minorca:

| 1721-1724 | *Blast* | pitch boat | Ex bomb vessel; 1724 deleted |
| 1740-1756 | *Minorca* | lighter | Deptford Dyd (Surveyor: Stacey); 1750 put up for sale but no-one would buy her; 1756 lost at the taking of the Island by the French. [As *Royal Escape* at Deptford, which see] |

### At Malta:

| 1804-___ ? | Two launches for weighing anchors | Woolwich Dyd [As the two built in 1799 for Gibraltar, which see] |
| 1807-after 1864 | Tank vessel No 3 | Malta Dyd; still in service 1864 |
| 1810-after 1864 | Mud machine | Malta Dyd; still in service 1864 |
| 1811-after 1864 | Crane vessel | Malta Dyd; still in service 1864 |
| 1813-after 1864 | Square punt | Malta Dyd; still in service 1864 |
| 1813-after 1864 | Tank vessel No 4 | Malta Dyd; still in service 1864 |

# YARD CRAFT

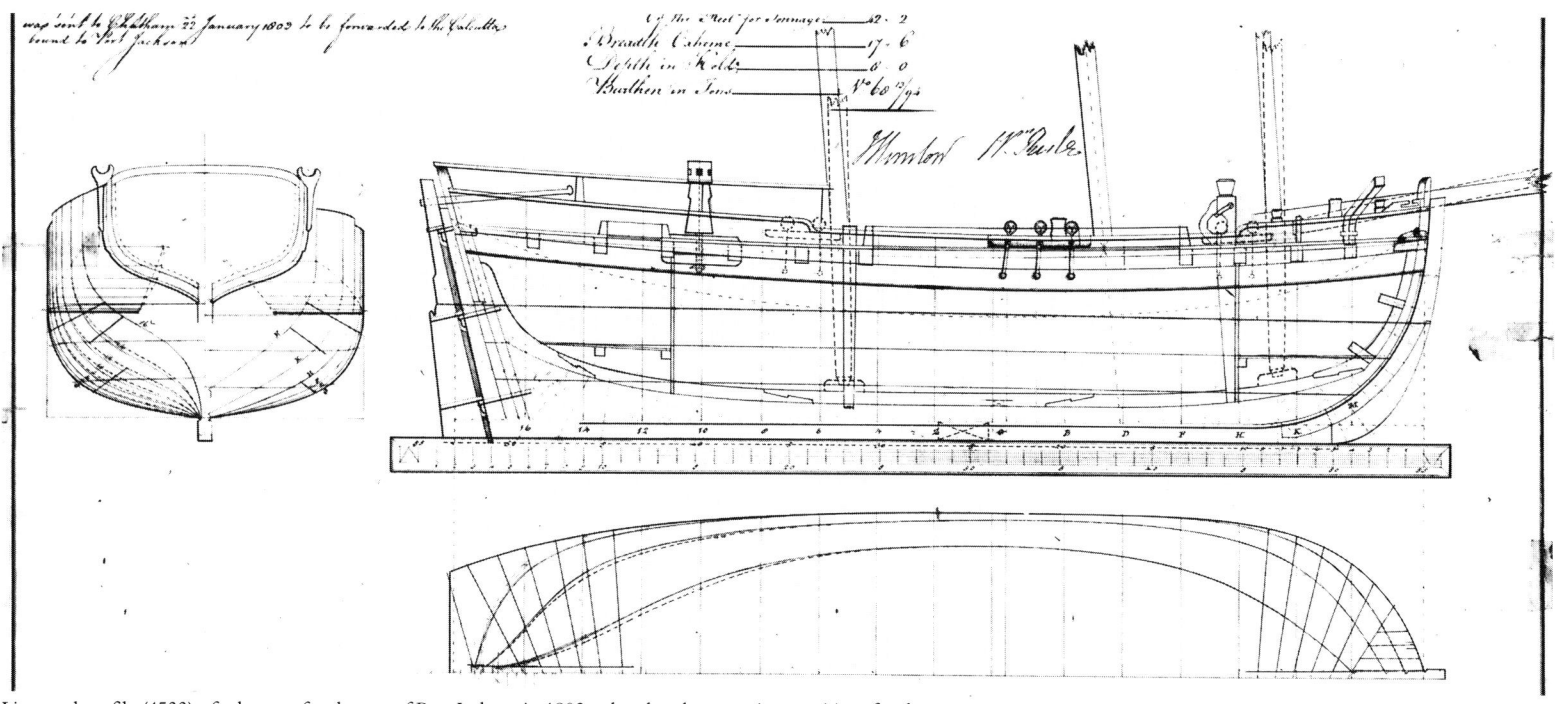

Lines and profile (4533) of schooner for the use of Port Jackson in 1803 - the plan shows various positions for the masts

| | | | |
|---|---|---|---|
| 1815-after 1864 | Mud punt No 2 | | Malta Dyd; still in service 1864 |
| 1822-after 1864 | Chain lighter | | Malta Dyd; still in service 1864 |
| By 1830-after 1864 | Mud punt No 1 | | Malta Dyd; still in service 1864 |
| By 1830-after 1864 | Anchor punt | | Malta Dyd; still in service 1864 |
| 1837-after 1864 | Tank vessel No 1 | | Chatham Dyd. (Symonds design); still in service in 1864. Ketch rig |

PLANS: Lines & profile/profile & decks/decks/sails/midsection

| | | | |
|---|---|---|---|
| 1853-after 1864 | Savage | chain lighter | Ex brig 1830; still in service 1864 |

### At Lisbon:

| | | | |
|---|---|---|---|
| 1708-1713? | Runner | transport, hoy | From Deptford 1708; 12.1713 sold? |
| 1708-1713 | Deptford | transport | From Deptford by 1708; 1713 sold |

### At Tangier:

| | | | |
|---|---|---|---|
| 1677-1683 | Pontoon ('Puntoone') | | Built at Tangier; destroyed during the evacuation of that port [267 tons] |

### At Madeira:

| | | | |
|---|---|---|---|
| 1808 | Eclipse | sloop (sloop rig) | Proposed to take up anchors at Madeira |

Ca 41'4", 31'2¼" x 14' x 6'5". ____

PLANS: Lines, profile

### At the Cape of Good Hope:

| | | | |
|---|---|---|---|
| 1818?-___? | Hardy | bullock vessel | Ex gunbrig; later convict hospital ship; fate uncertain |
| 1842-1847 | Fawn | tank vessel | Ex slaver schooner; 1847 sold |
| 1844-1866 | Progresso | tank vessel | 23.4.1844 purchased (ex slaver taken at Simons Bay 1843?); 1866 sold and/or 1869 BU |
| 1850-1857 | Dart | tender | Ex brig; 1857 back in Britain & to the Coastguard |
| 1860-1866 | Swift (Y.C. No 3) | mooring vessel | Ex brig; sold 1866 |
| 1864-___ | Y.C. No 4 | tank vessel | Deptford Dyd [Order 28.1.64, Keel laid 16.6.64, Launched 15.11.64] for Simons Bay |
| 1866?-1866 | Kent | barge | Purchased? for the light house works; not required; sold 05.1866 for £200 to the harbour works at Table Bay |

Also in 1819 plans (decks) were prepared for a sailing & mooring lighter for the Cape. In 1821 a transport was proposed but does not seem to have been built.

[79'6", 63'1⅜" x 24'5" x 16'7". 200 12/94 tons. Two masts;

PLANS: Lines, profile/decks/framing.]

### At Trincomalee:

| | | | |
|---|---|---|---|
| 1821- | Tank vessel No 2 | | Francis Schuyler, Cochin (Edye after 1864 design); still in service 1864 |

43', 32' x 13' x 7'. 31 64/94 tons

PLANS: Lines, profile, rig, deck, midsection/specification

| | | | |
|---|---|---|---|
| 1822-1850 | Cochin | schooner tender/tank vessel | Cochin (Edye design); Launched 23.4.1820?; in service 1822; sold 1850 |

53'6", 43'2" x 15'4" x 7'10". 53 92/94 tons

PLANS: Lines, profile, deck

| | | | |
|---|---|---|---|
| 1851-after 1864 | Teazer | tank vessel | Moulmein, Burma (Surveyor's Department design); still in service 1864 |

### At Botany Bay, Australia:

| | | | |
|---|---|---|---|
| 1786-___? | One 32' & one 28' cutter | | Burr; fate uncertain |

32' x 8'10" x 3'2"

PLANS: Lines, profile, deck

28' x 7'1" x 2'8"

PLANS: Lines, profile, deck

| | | | |
|---|---|---|---|
| 1803-___? | sailing barge | | Port Jackson; fate uncertain |

56'9", 36' x 17'6" x 5'6". 54 19/94 tons

PLANS: Profile, deck, midsection/pintles for rudder/scantlings

| | | | |
|---|---|---|---|
| 1803-___? | schooner | | Port Jackson; fate uncertain |

53', 42'2" x 17'6" x 8'. 68 13/94 tons

PLANS: Lines & profile/deck & frame/pintles

### Saint John's, Newfoundland:

| | | | |
|---|---|---|---|
| 1788-___? | Launch | | Portsmouth Dyd fate uncertain |

30'6" x 9' x 5'

PLANS: Lines, profile & deck

| | | | |
|---|---|---|---|
| 1798-___? | Launch | | Clark, Queenborough |

30' x 9' x 3'9"

PLANS: Lines, profile, deck

### At Halifax, Nova Scotia:

| | | | |
|---|---|---|---|
| 1809-1811 | Orange | 'petty auger' | Halifax Dyd; 1811 sold |

NOTE: Not certain that this was actually a yard craft; she is not listed with the yard craft and carried 20 men - but little else is known about her. Presumably she was of petiauger rig - a small version of schooner rig but with the fore mast raked forward and no headsails

| | | | |
|---|---|---|---|
| 1814-___? | riggers launch | | Halifax Dyd [As riggers launch |

| | | | |
|---|---|---|---|
| 1814-____? | Mooring lighter | | for Plymouth 1793, which see] Halifax Dyd |
| | 57'6", 43'5¾" x 19'11" x ____ 91 68/94 tons | | |
| | PLANS: Lines, profile, deck | | |
| 1817-after 1864 | Anchor lighter | | Halifax Dyd; still in service 1864 |
| 1823-after 1864 | Buoy boat | | Plymouth Dyd; still in service 1864 |
| 1839-after 1864 | Tank vessel | | Halifax Dyd; still in service 1864 |
| 1841-by 1854 | Buoy boat | | Halifax Dyd; deleted by 1854 |
| 1850-1859 | *Netley* | buoy boat | Ex cutter 1850/1852 at Bermuda and then Jamaica - but probably returned to Halifax; 1859 sold in Newfoundland |
| By 1854-after 1864 | *Hunter* | schooner | Origin uncertain; still in service 1864 |

## At Bermuda:

| | | | |
|---|---|---|---|
| 1812-after 1864 | Anchor hoy | | ____, Bermuda; still in service 1864 |
| 1819-after 1864 | *Hope* ____ | | ____, Bermuda; still in service 1864 |
| 1819-after 1864 | *Lord of the Isles* ____ | | ____, Bermuda; still in service 1864 |
| 1822-after 1864 | *Sea Breeze* ____ | | ____, Bermuda; still in service 1864 |
| 1826-after 1854 | *Kite* (ex *Aetna*) | sheer vessel | Humble & Hurry, Liverpool (purchased after completion?); still in service 1854 |
| 1826-by 1848 | *Porgy* ____ | | ____, Bermuda; deleted by 1848? |
| 1827-after 1864 | *Horn* | Boat? | ____ Purchased; still in service 1864 |
| 1828-after 1864 | *Albicore* ____ | | ____, Bermuda; still in service 1864 |
| 1829-after 1864 | Flat lumber boat No 1 | | ____, Bermuda; still in service 1864 |
| 1829-after 1864 | Tank schooner | | ____, Bermuda; still in service 1864 |
| 1839-after 1864 | *Hawk* | decked sailing vessel | Purchased (built ____, Bermuda); in service til 1864 |
| 1840-after 1864 | Flat lumber boat No 2 | | ____, Bermuda; still in service 1864 |
| 1840-after 1864 | Flat rubble boat No 1 | | ____, Bermuda; still in service 1864 |
| 1840-after 1864 | Flat rubble boat No 2 | | ____, Bermuda; still in service 1864 |
| 1842-after 1864 | Flat bell vessel (for diving bell) | | ____, Bermuda; still in service 1864 |
| By 1840s-after 1864 | Pitch boat | | Origin uncertain; still in service 1864 |
| 1844-after 1864 | *Pearl* | decked sailing vessel | ____, Bermuda; still in service 1864 |
| 1845-after 1864 | *Princess Alice* | decked sailing vessel | ____, Bermuda; still in service 1864 |
| 1848-after 1864 | *Clara* | decked sailing vessel | ____, Bermuda; still in service 1864 |
| 1848-after 1864 | *Porgy* | decked sailing vessel | ____, Bermuda; still in service 1864 |
| 1847- | *Cassius* | diving bell vessel | ____, Bermuda; purchased 7.9.1847; after 1864 still in service |
| 1849-after 1864 | Tank schooner | | ____, Bermuda; still in service 1864 |
| By 1850s- | *Hebe* | decked sailing | Origin uncertain; still in service 1864 |

| | | | |
|---|---|---|---|
| after 1864 | | vessel | |
| 1850-1852 | *Netley* | | See under Halifax |
| 1851-1870 | *Devonport* | anchor hoy & buoy boat | Plymouth Dyd; 1870 BU at Bermuda |

## At Antigua:

| | | | |
|---|---|---|---|
| 1794-____? | Launch | | Deptford Dyd fate uncertain |
| | 30' x 9'6" x 3'9" | | |
| | PLANS: Lines, profile, deck] | | |
| 1799-____? | Launch | | 'proposed by the Officers of Antigua Yard in lieu of their former Hawser Boat decayed ..... worn out and past repairs.' Antigua Dyd |
| | 39'6", 12' x 3'4" | | |
| | PLANS: Lines, profile, deck., midsection | | |
| 1800-____? | Schooner for use of the yard English Harbour, Antigua | | |
| | 40' x 12'6" x ____ 25 26/94 tons | | |
| | PLANS: Lines | | |
| No date (Ca 1800-1840?) | Dredger boat | | |
| | 43', 37' x 26' x 4'6". ____ | | |
| | PLANS: Elevation & midsection | | |

## At Jamaica:

| | | | |
|---|---|---|---|
| 1770-____? | Two launches | | Deptford Dyd; fate uncertain |
| | 33' x 11' x 3'10" | | |
| | PLANS: Lines | | |
| 1804-____? | Tank vessel, cedar built | | ____, Bermuda (contracted for and supervised by Goodrich & Sheddon, designed by Capt A.F. Evans 'under the auspices and by the direction of Rear Admiral J.T.Buckworth, K.G.') |
| | 60' x 23' x 9'6". 147 92/94 tons. 3 masts | | |
| | PLANS: Lines, profile, deck | | |
| 1808?-____? | Sailing lighter | | Jabez Bailey, Ipswich. [As *Saltash* 1807 at Plymouth and *Goodwill* 1810 at Sheerness.] |
| By 1816-1818 | *Port Royal* | sailing lighter | Origin unknown; 1818 sold. Possibly the vessel proposed to be built at Deptford Dyd for Jamaica in 1814 |
| | 41'5", 30'2⅝" x 15'10½" x 7'6". 40 48/91 | | |
| | PLANS: Lines & profile/deck/frame | | |
| 1816-after 1864 | *Bermuda* | tank vessel | ____, Bermuda; still in service 1864 |
| 1816-after 1864 | *Milford* | tank vessel | ____, Bermuda; still in service 1864 |
| 1820s?-by 1830 | *Port Royal* | lighter | Ex *Woolwich* from Woolwich; to Cork by 1830 |
| 1820s?-by 1830 | *Camel* | sailing vessel | Origin uncertain; by 1830 to Portsmouth |
| 1832-1832 | *Port Royal* | lighter ? | ex *Medway*?; from Sheerness; 1832 wrecked |
| 1833-____? | *Fountain* | tank vessel | Pembroke Dyd (Symonds design); sold 06.1843 or still in service 1864 (confusion probably caused by the existence of another *Fountain* at Plymouth) |
| 1845-after 1864 | Tank vessel | | ____, Bermuda; still in service 1864 |
| 1851-1851? | *Netley* | | see under Halifax |
| 1856-____? | Buoy boat | | Thomas Davis & J.J.Gutteridge, Bermuda (cost £3,299); fate uncertain |

## At Hong Kong:

| | | | |
|---|---|---|---|
| 1857-1869 | *Thistle* [YC 1] | | ____ Purchased in China 1857; 1869 sold |

## Chapter 5: Coastguard & Customs

There are three parts to this appendix. The first two lists are of old naval vessels (usually small) which were transferred to the Coastguard and to the Customs respectively, usually to act as accommodation hulks and 'watch vessels'. The third is of the 'cruisers', small vessels usually, though not invariably, rigged as cutters or ketches (often interchangeably, cutter rig being used in summer whilst the more manageable ketch rig was resorted to for winter cruising).

### Vessels to the Coastguard as Watch Vessels

These were old naval vessels, brigs, sloops, etc, which were turned into hulks manned by the Coastguard and anchored at strategic points along the coast as depot ships and for maintaining the watch for smugglers. They originated as part of the intensive system of the 'Coast Blockade' set up with major naval participation just after the ending of the Napoleonic War in order to curb smuggling. Watch vessel numbers (WV) seem to have been allocated in or before 1863.

| Year | Name | Type | Notes |
|---|---|---|---|
| 1824 | Snapper | gunbrig | Fate unknown |
|  | Bathurst | purchased brig | Sold 11.4.1858 to Castle to BU |
| 1825 | Griper | gunbrig | Fate unknown |
| 1826 | Pelter | gunbrig | Fate unknown |
|  | Adder | gunbrig | 12.1831 cast adrift when under tow and wrecked off Beachy Head; hull sold |
| 1831 | Clinker | gunbrig | At Yantlett Creek; WV 12; 1867 sold |
| 1833 | Chanticleer | brig | At the River Crouch; WV 5; 1871 BU |
|  | Shamrock | brig | From Woolwich; at Rochester; WV 18 [?]; by 1867 sold |
| 1834 | Emulous | brig | At 'Bugsley's Hole' then 'Haven Hole'; WV 13; 1865 sold |
|  | Havock | gunbrig | At Hamble River; 1859 BU |
| 1835 | Cadmus | brig | At Whitstable; WV 24; 1864 sold |
| 1836 | Eclipse | brig | At ____ ; WV 21; 1865 sold |
| 1838 | Icarus | brig | At Lymington; 1861 sold |
| After 1837 | Sylph | survey vessel | ____ ; 1888 sold |
| 1830s | Richmond | Purchased [?] brigantine | At West Mersea; 1865 to be disposed of |
| 1843 | Partridge | brig | For Southampton Water at Netley; 1864 sold |
| 1845 | Beagle | brig | ____ |
| After 1845 | Pelican | brig | At Rye Harbour; WV 29; 1865 sold |
| 1857 | Cygnet | brig | At Chichester; WV 30; 1877 BU |
|  | Penguin | brig | At the Hamble River; WV 31; 1871 sold |
|  | Star | brig | At Gravesend; WV 11; Ca 1899 BU |
|  | Linnet | brig | At ____ ; WV 36; 1867 sold |
|  | Dart | brig | At Beresford, Ireland; WV 26; 1875 BU |
|  | Amphitrite | 36 | At Portsmouth; 1860s lent to the contractor for the forts at Portsmouth (War Department); 1875 BU |
|  | Pandora | brig | At ____ ; 1862 sold |
|  | Southampton | 60 | Guard ship at Sheerness; 1868 to the Humber |
|  | Philomel | brig | WV 23; foundered in the Swale 1869 & sold to salvors in 1870, then BU |
|  | Eagle | 74/50 | At Falmouth; 1860 training ship in Southampton Water; 1862 to Liverpool; 1910 Mersey Division RNVR; 1918 renamed Egret; 1926 burnt; 1927 remains sold |
|  | Wellington | 74 | At Sheerness; 1862 to Liverpool |
|  | Melampus | 38 | At Southampton; 1858 to Portsmouth |
|  | Meander | 38 | At ____ ; 1859 to Ascension Island |
| By 1860 | Dove | revenue cruiser (1818) | At Foulness Island; WV 6; 1865 sold |
| 1861 | Dolphin | brig | At ____ ; WV 3; 1894 BU |
|  | Mortar Vessel No 21 |  | At the Tower of London; renamed Harpy; 1871 to be replaced by WV 52; 1872 BU |
| 1862 | Fanny | Admiralty yacht/ cutter tender | At Kingstown as tender to the Coastguard ship; 1879 still in service |

[See Yachts section; also PLANS: Outline profile/sails/lower deck/windlass.]

| Year | Name | Type | Notes |
|---|---|---|---|
| 1863 | Arab | brig | At Queenborough (the Swale); WV 18; 1879 BU |
|  | Britomart | brig | At Beresford, Ireland; WV 25; then at Burnham; 1874 BU |
|  | Despatch | brig | At the East Swale; WV 24; 1901 sold |
|  | Elk | brig | At ____ ; WV 13; 1871 [?] WV 28 at Tilbury; 1893 sold |
|  | Kangaroo | brig | At Stangate Creek; WV 20; 1870 at the River Crouch; 1897 sold |
|  | Mortar Vessel No 5 |  | At Castle Coote by 1871; WV 27; 1874 BU |
|  | Mortar Vessel No 7 |  | At Tilbury; WV 37; 1870 to Sheerness; 1874 BU |
|  | Mortar Vessel No 28 |  | By 1870 at Branksea; WV 39; 1876 sold |
|  | Mortar Vessel No 29 |  | At West Mersea; WV 21; 1875 Stangate; 1906 sold |
|  | Mortar Vessel No 31 |  | Replacing Kite at Cliffe Creek; WV 9; 1870 to War Office |
|  | Mortar Vessel No 41 |  | In the Hamble River; WV 32; 1871 defective & WV 14 to take her place; 1876 refit; 1881 sold |
|  | Mortar Vessel No 46 |  | Replaced Childers at Rochester; Cliffe Creek; WV 10; 1893 sold |
| 1864 | Mortar Vessel No 38 |  | At Stangate?; WV 4; 1873 to Hamford Water; 1874 Stangate; by 1886 at Walton-on-Naze; 1901 BU |
|  | Mortar Vessel No 39 |  | Wartering [?] Haven; WV 8; 1897 sold |
|  | Mortar Vessel No 43 |  | At Yantlet Creek; WV 12; 1898 sold |
|  | Mortar Vessel No 51 |  | At Haven Hole [Hole Haven?]; WV 13; sold |
| 1865 | Mortar Vessel No 50 |  | At Mucking Creek; WV 17; 1872 to Sheerness |
|  | Mortar Vessel No 55 |  | At Cowes, Isle of Wight; WV 6; replaced there by WV 17; 1876/7 in the River Crouch at Foulness; 1881 sold |
| 1866 | Tyrian |  | From Portsmouth as a Coastguard receiving hulk; may have eventually gone to Jamaica; 1892 sold |
|  | Mortar Vessel No 56 |  | WV 19; 1874 BU |
| 1867 | Mortar Vessel No 42 |  | To the Hamble River; WV 9; 1877 sold |
|  | Mortar Vessel No 49 |  | To Sheerness to replace Medusa at Colemouth [?] Creek; WV 22; 1898 sold |
|  | Mortar Vessel No 52 |  | WV 16; 1871 to replace Ready as Guard Ship at the Tower of London; 1894 sold |
| 1871 | Mortar Float No 107 |  | At Chichester Harbour; WV 42; 1891 sold |

### Vessels Transferred to the Customs

Mostly to take part in the great campaign in the 1820s against organised smuggling known as the 'Coast Blockade'

| Year | Name | Type | Notes |
|---|---|---|---|
| 1803? | Courser | gunbrig | Fate uncertain |

| 1806 | *Pandour* | small two decker | Fate uncertain |
|---|---|---|---|
| 1815 | *Escort* | gunbrig | Fate uncertain |
|  | *Redbreast* | gunbrig | Fate uncertain |
| 1824 | *Snapper* | brig | To the Coast Blockade; 1861 sold |
| 1825 | *Hyperion* | 32 | To the Coast Blockade; 1833 BU |
| 1826 | *Pelter* | brig | To the Coast Blockade near Folkestone; 1862 sold |
| 1828 | *Forrester* | Hoy | To the Coast Blockade; fate unknown |
|  | *Yarmouth* | Lighter | To the Coast Blockade in the Swale; fate unknown |

## Coastguard and Revenue (Customs) Cruisers of Naval Significance

All cutters unless otherwise stated. Customs vessels were briefly under Admiralty control in the period after 1816, and even outside this period many were recorded in the sources used for this list. Sometimes this was because the vessels were transferred temporarily or permanently to the Navy. In around the middle of the nineteenth century the Admiralty was supervising the construction of vessels for the Coastguard and plans of some of these vessels survive in the Admiralty Collection. Some manuscript naval lists include vessels from Coastguard or Customs, whilst the secondary sources used (Pitcairn Jones and Colledge) include some, usually sparse, details of such craft. The records used produce a partial and unsatisfactory record, but on the principle that a partial list is better than none at all this one is put forward in the hope that someone else will be goaded into producing a better one.

Coastguard vessels are asterisked [*]. Note that thanks to the partial and fragmentary records used this is a provisional list only; it is quite possible that in some cases there may be two entries for what is in fact only one vessel.

Details (when available) are given in the following order: *Date* of build or entry into service; the whole list is grouped in chronological order under the year of build. *Name. Type* (and *station* when available). *Builder and where built. Fate* (and any alterations when in service). *Dimensions* (Length on range of deck, length of keel for tonnage x breadth x depth in hold), followed by *tonnage*. Finally the existence of any *plans* is noted.

| 1775 | *Royal George* | Scots excise cutter 18 guns; 1818 out of service |
|---|---|---|
| 1776 | *Grayhound* | lugger for the Collector of St Mawes, built (John Parkey?) Cawsand 73'4", 56'4" 'x 23' x 8'6". 158 47/94 tons PLANS: (Mulgrave Collection) Lines & profile.] |
| 1782 | *Snapper* | built Whitstable; 1817 sold |
| 1788 | *Swan*; | 1790 to RN?; 26.5.1792 wrecked off Shoreham? |
| 1790 | *Hind* | transferred 1827 to the RN (see under cutters in main list) |
| Ca 1790? | *Fox* | in service till Ca 1792? |
| 1791 | *Hound* | built Cowes. 111 tons; still in service 1821 |
| 1792 | *Annesley* | built Bridport; 1818 sold |
|  | *Rose* | built Cowes; still in service 1828 |
| By 1793 | *Dolphin* | 139 tons?; 1797 still in service |
|  | *Falcon* | 14/16 guns; 1806 still in service |
|  | *Mermaid* | 1818 still in service |
|  | *Otter* | 1806 still in service |
|  | *Racer* | 1816 out of service |
| 1793 | *Eagle* | 12 guns 155 tons; 1818 out of service |
|  | *Fly* | 29 tons; 1806 out of service |
|  | *Greyhound* | 1806 out of service |
|  | *Liberty* | 4 guns; 1795 out of service |
|  | *Royal Charlotte* | no further note |
|  | *Stag* | 14 guns, 129 tons; 1796 still in service (same as 1794 *Stag*?) |
|  | *Swan* | [Ireland] 92 tons; (is this the *Swan* taken over by the RN in 1795 and then taken by the French in the same year?) |
|  | *Tartar* | 8 guns 90 tons; no further note |
| Ca 1793 | *Speedwell*? | 194 tons; 1816 sold (?the same vessel?) |
| 1794 | *Prince Augustus Frederick* * | built Cowes; 1817 fitted for the Preventive Service |
|  | *Stag* | built Cowes. 14 guns, 153 tons (same as 1793 *Stag*?) |
| 1795 | *Eagle* | 6 guns; excise vessel; 1798 out of service? |
|  | *Fly* | 18 tons; excise vessel; no further note |
|  | *Fox* | 6 guns, Ca 50 tons; 1805 hired by RN? |
|  | *Royal George* | Scots excise cutter 10 guns; 1818 out of service (?same as 1775 vessel of this name?) |
|  | *Prince of Wales* | Excise cutter 20 guns. 300 tons (and a separate revenue vessel of this name and date?) |
| Ca 1795 | *Ferret* | 4 guns, 45 tons; excise vessel; no further note |
|  | *Lark* | excise vessel; 1798 still in service? |
| 1796? | *Ranger* | built at Cowes (or ex Dutch?) 8 guns 97 tons; 10.1822 lost (or is this another vessel?) |
| 1796 | *Swallow* * | built Cowes; still in service 1816 |

Hull plan (6670) of the lugger *Alarm* 'as taken off' in 1811

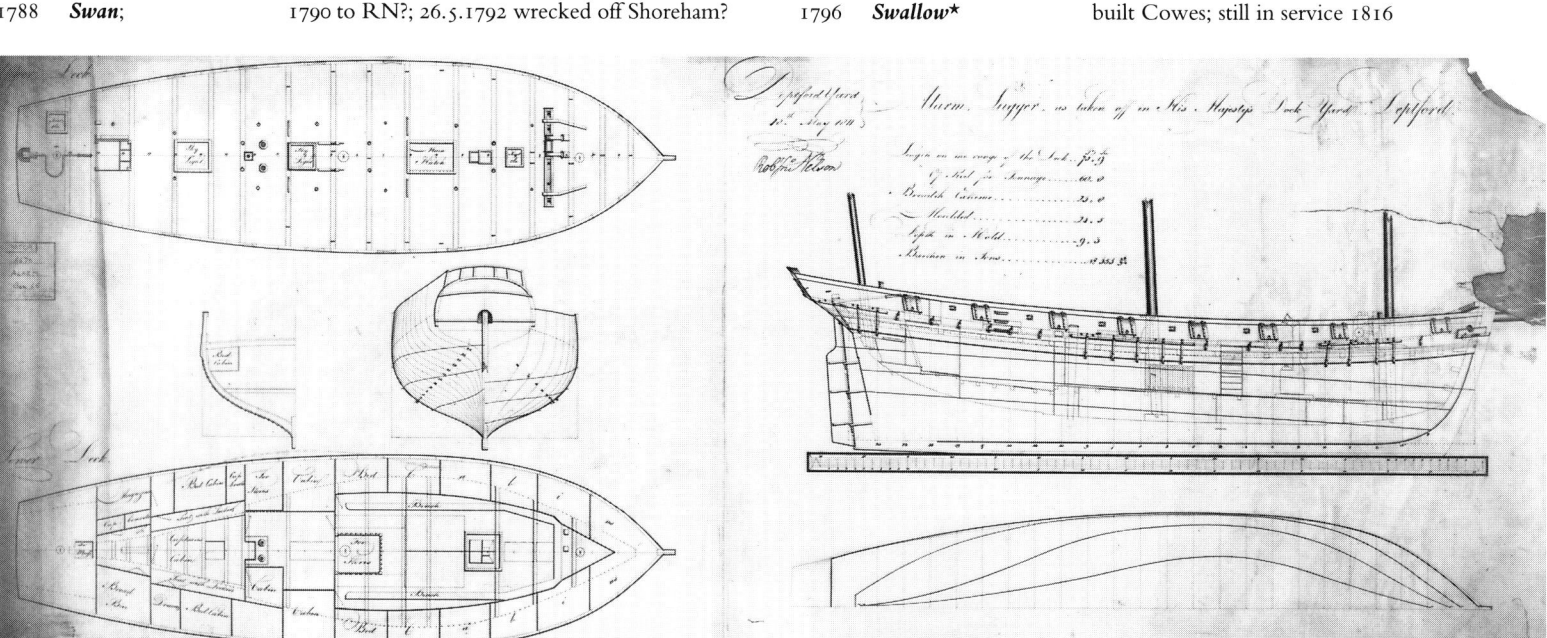

# COASTGUARD AND CUSTOMS VESSELS

|  |  |  |
|---|---|---|
|  | *Defence* | 8 guns, 76 tons; probably the same vessel is noted in 1802 |
|  | *Duke of York* | no further trace beyond 1796 |
| 1796 or 1797 | *Minerva* | built Plymouth; 1818 BU |
| By 1797 | *Beresford* | brig; no further trace beyond 1797 |
|  | *Diligence* | in service this year but no further note |
|  | *Lively* | 75 tons; excise vessel; still in service 1805 |
|  | *Viper* | 1813 still in service |
| 1797 | *Nancy* | out of service ca 1801 |
|  | *Repulse* | no further note |
| By 1798 | *Badger* | still in service 1803 |
| 1799 | *Dolphin* | 10 guns built at Cowes; 1821-1827 served in the RN as *Grecian* (which see) |

68'9", ___ x 26'3" x 9'6". 45 tons

| 1800 | *Antelope*★ | built Cowes; 1832 sold |
|---|---|---|
|  | *Resolution*★ | built Cowes; 1831 BU |
|  | *Yarmouth*★ | rebuilt at Yarmouth 1810; BU ca 1849 |

60', ___ x 20'5" x ___ . 105 tons

| 1801 | *Tartar* | 10 guns; no further note |
|---|---|---|
| 1802 | *Townsend* | built Bridport; 1823 sold |
| 1803 | *Prince of Wales* | built by Scotts, Greenock (Scots) |
|  | *Royal George* | built Cowes? 143 tons; 1843 still in service (if this is the same vessel?) |
| 1804 | *Providence* | built Dover; 1821 to the RN as *Quail* (which see) |
| 1805 | *Scout* | built Cowes; 1862 sold |
|  | *Eagle* | excise vessel; 10 guns; out of service 1827 |
| Ca 1806 | *Leopard* | 121 tons; out of service 1809 |
| 1806 | *Active*★ | built Cowes; in service 1816 |
|  | *Griper*★ | built Mevagissey; in service 1816 |
|  | *Prince of Wales* | brig built at Cowes; no further note |
|  | *Tiger* | (built Bridport 1805?); 1808 to RN as gun brig?; 1812 out of service? |
| By 1806 | *Violet* | lugger; transferred to the RN in 1806 |
| 1807 | *Industry* | built Bridport; 1832 (or 1817?) sold |
|  | *Stork* | built Cowes; 1826 still in service |
| 1808 | *Active* | 14/6 guns; still in service 1880 |

60'6", 47'2⅜" x 20'3" x 9'9". 101 23/94.
PLANS: Taken off 1839: Lines/profile/lower deck/upper deck.

|  | *Fox* | 10 guns excise vessel; built Mevagissey; still in service 1828 |
|---|---|---|
|  | *Lapwing*★ | built Mevagissey; still in service 1816 |

Dimensions as *Fancy*?
PLANS: Decks

|  | *Greyhound* | built Bridport; still in service 1844 |
|---|---|---|
|  | *Royal George* | brig; no further note |
|  | *Watchful*★ | built Mevagissey; 1822 sold |
| By 1809 | *Hawk* or *Hawke* | 6 guns; 1817 out of service? |
|  | *Hound* | 12 guns; 1862 still in service |
|  | *Vigilant* | 14 guns; 5.12.1819 wrecked in Torbay |
| 1809 | *Shark* | no further note |
| By 1810 | *Alarm* | lugger 8 guns; origins uncertain; to Navy 1810; given up to Customs 1811; no further information (see entry in Gunboats, luggers, etc section of Part II, Chapter 7). |

71'9", 60'8" x 21'8" x 9'3". 151. Men: 40. Guns: 8 x 12★
PLANS: Lines, profile, decks, midsection

| 1810 | *Dart*★ | purchased, built Mevagissey or Ramsgate 1809; 1831 still in service |
|---|---|---|

62'7", 47'8" x 22'4" x 10'. 127 tons

| 1813 | *Badger* | built Hastings; still in service 1845 |
|---|---|---|
|  | *Regent*★ | brig; taken on the stocks at Genoa where she was building for the French; probably given to the Coastguard; 1821 still in service; fate uncertain |

97'9", 80'4" x 29' x 13'. 359 44/94. Guns: 10 x 12★ + 2 x 9 + 2 x 6
PLANS: (Longstaff Collection) Lines

| 1814 | *Beresford* | brig; taken from the Americans and rebuilt at Greenock; 1820 to the Preventive Service; 1860 sold |
|---|---|---|
|  | *Lynx* | built Bridport; 1825 sold |
|  | *Richmond* | built Dublin; 1820 still in service |
|  | *Hawke*★ | built Bridport; 1875 still in service (possibly two vessels of this name in service between 1816 and 1822?) |
|  | *Seagull* | (uncertain whether for Revenue or not); 16.4.1817 renamed *Adder*; 1860 still in service |
| Ca 1814 | *Hart* | became a dockyard tender in 1822 at Woolwich |
| 1815 | *Wellington* | built Cowes; 1823 still in service |
|  | *Anacreon* | schooner [according to Colledge built for the Navy but then transferred to the Customs in 1816; no other note |
|  | *Harpy* | built Cowes, 138 tons; 1829 still in service |
| By 1816 | *Speedwell*★ | (?) ___ ; 1816 sold |

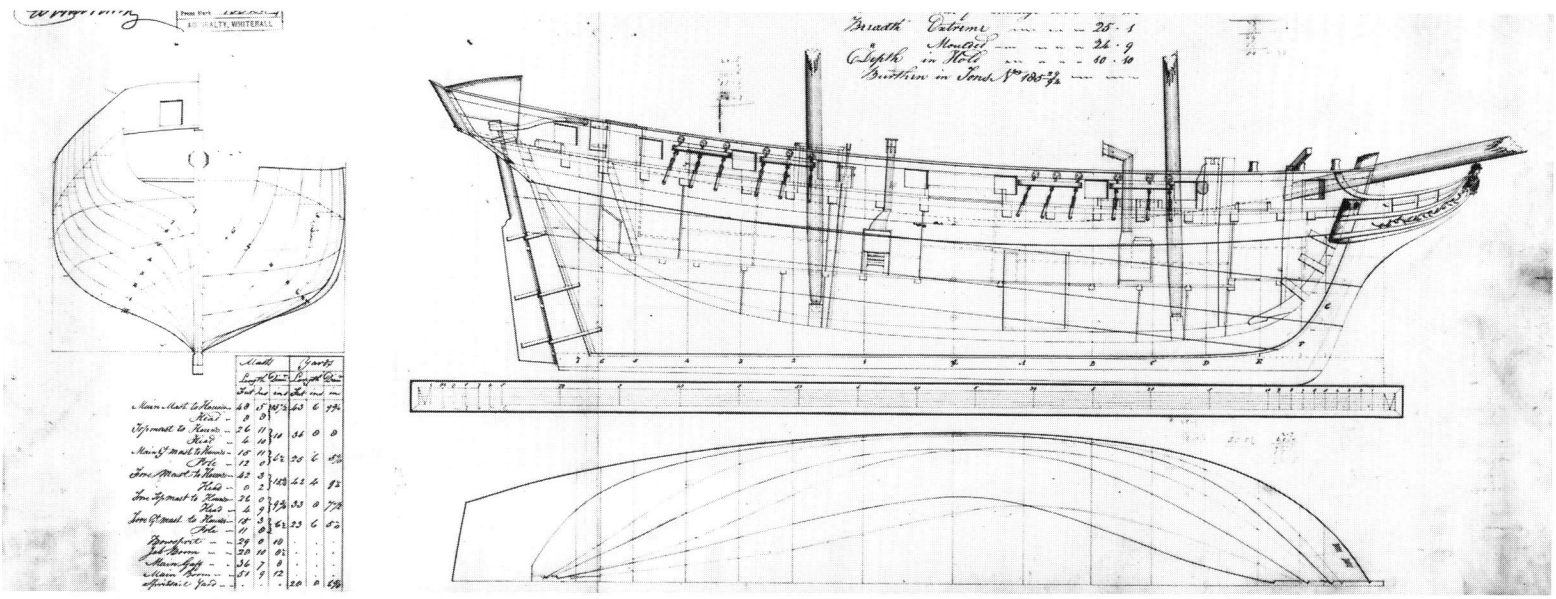

Lines and profile (3335) of the revenue brig *Shamrock*, plan drawn in 1817

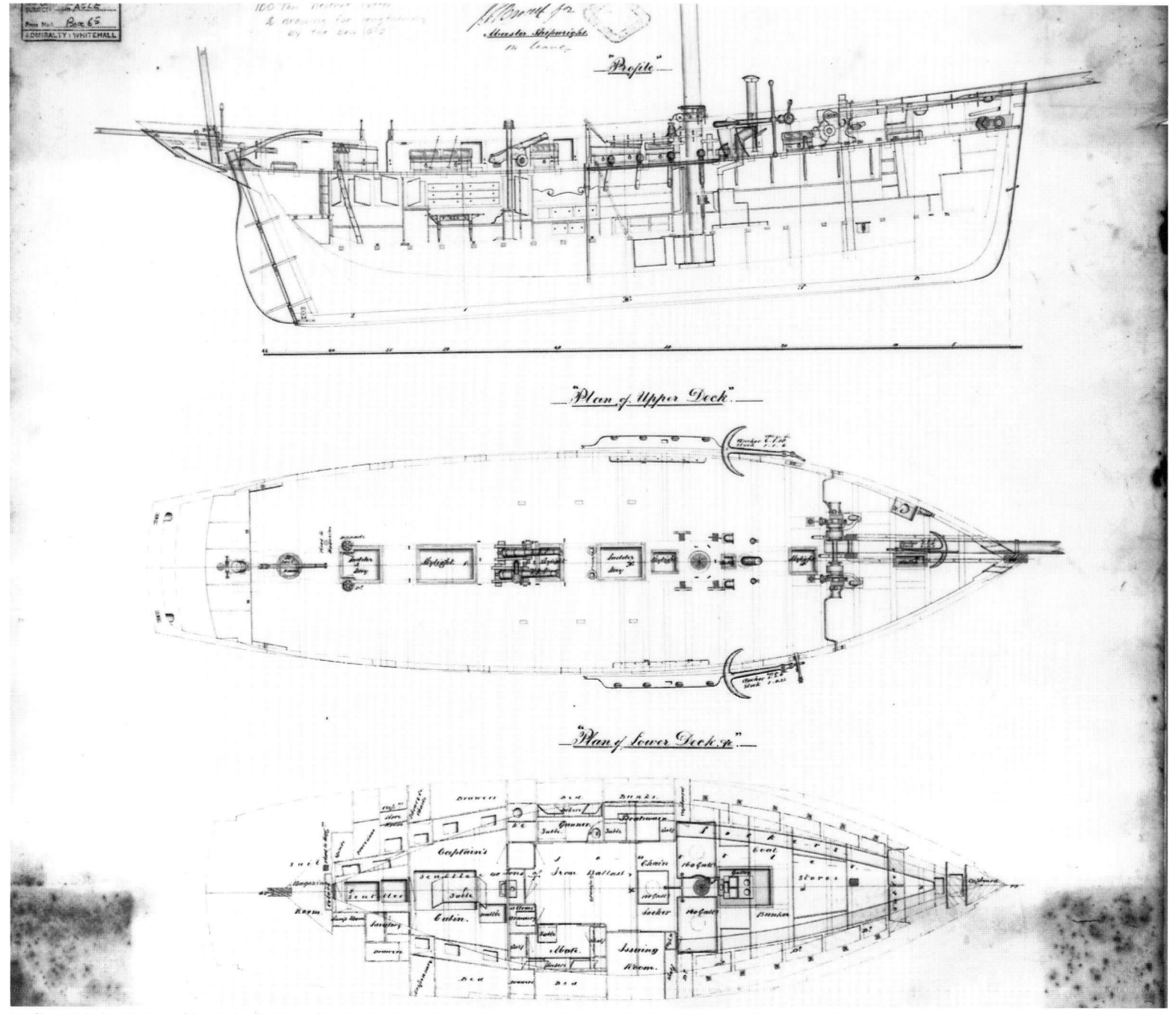

Profile and decks (7131A) of the *Eagle* of 1844 'as fitted' in 1873 – a plan which is typical of its period in the exquisite detailing shown of armament, furnishings and other fittings

|  |  |  |
|---|---|---|
|  | **Eagle** | 125 tons; 1840 still in service |
|  | **Lapwing** | 1844 still in service |
|  | **Tartar** | 1840 still in service |
| 1816? | **Princess Royal** | 1859 still in service (same vessel?) |
|  | 66'3½", 53'5½" x 22'8⅞" x 10'5⅛". 146 87/95 | |
|  | PLANS: Lines | |
|  | **Shamrock** | (ex *Resolution*) revenue brig on the Irish Station. 12 guns; 1838 still in service |
|  | 74'1", 55'4½" x 25'1" x 10'10". 185 29/94 | |
|  | PLANS: Lines & profile/decks | |
| 1816 | **Griper** | no further note |
| By 1817 | **Hardwicke** | Irish customs; 10.1820 wrecked in Dundrum Bay |
|  | **Whitworth** | disposed of by 1826 |
|  | **Wickham** | 1856 still in service |
| 1817 | **Fancy** | built Plymouth Dyd; fate uncertain |
|  | 67', 52'7" x 22'5" x ____, 140 51/94 | |
|  | PLANS: Lines/decks (as *Lapwing*) also for *Kite, Dove, Cheerful, Racer, Sprightly* | |
|  | **Kite** | built Plymouth Dyd; still in service 1838; fate uncertain. [As *Fancy*] |
|  | **Drake** | no further information |
|  | **Adder** | see *Seagull* 1814; still in service 1860 |
|  | **Defence** | built at Iwade; no further note |
|  | **Nimble** | 1843 still in service (?same vessel?) |
| By 1818 | **Scourge;** | 1826 still in service |
| 1818 | **Dove** | built Plymouth Dyd; 1865 sold. [as *Fancy*] |
|  | **Racer** | built Pembroke Dyd; 4.5.1830 ordered to be sold at Malta. [as *Fancy*] |
|  | **Sprightly*** | built Pembroke Dyd; 1819 to Revenue; 6.1.1821 wrecked off Portland. [as *Fancy*] |
|  | **Diligence** | built Pembroke Dyd; 18.9.1830 (or 1838?) wrecked on the coast of Ireland |
|  | 70'8⅜", 52'3¼" x 24' x 11'0⅜". 160 13/94 | |
|  | PLANS: Lines & profile/frame. Plans used for *Bramble* Class of naval cutters | |
| Ca 1818 | **Speedwell** | 1.2.1819 wrecked near Frazerburgh |

# COASTGUARD AND CUSTOMS VESSELS

|  | *Watchful* | no further note |
|---|---|---|
| By 1819 | *Swallow* | 6 x 6 + 2 small guns. 165 tons; [may have been the *Swallow* lost 30.11.1825, but probably not] |
| 1819 | *Cheerful* | built Plymouth Dyd; fate uncertain. [as *Fancy*] |
| 1820 | *Sylvia* | built Pembroke Dyd; fate uncertain |
| 1821 | *Skylark* | [see cutters in main list; *Vigilant* Class]; 1880 still in service |
|  | *Swift* | built Pembroke; 1846 still in service |
| By 1822 | *Hawke* | 1875 still in service |
| By 1823 | *Prince of Wales* | 1848 still in service |
| 1823 | *Hamilton* | 1860 still in service |
|  | *Scout* | 84 tons; 1859 still in service |
| 1825 | *Experiment* | built Cowes. 43 tons; 1843 still in service |
|  | ?*Viper* | (ex *Mermaid*?); 1845 still in service |
| By 1826 | *Defence** | 130 tons; 1847 sold |
|  | *Delight* | 70 tons?; 1890 still in service |
|  | *Hound* | 169 tons; 27.3.1835 wrecked at Weymouth |
| 1827 | *Stag* | built Cowes; 1880 still in service |

66'1", 52'6" x 21'10" x 9'10". 129 80/94
PLANS: Taken off 1839: Lines & midsection/profile/lower deck/upper deck

| By 1828 | *Chance* | 1880 still in service |
|---|---|---|
| 1828 | *Forester** | (dockyard craft listed 1827?) transfer?; 1862 sold |
| By 1829 | *Dolphin* | 1858 still in service |
|  | *Dove* | Ca 1840 renamed *Kangaroo* (could be the same *Dove* 1818-1865 listed under cutters in the main list) |
|  | *Hornet* | still in service 1839. |
|  | *Lion* | by 1869 out of service |
|  | *Lively* | 1860 out of service |
|  | *Mermaid* | 1880 still in service |
|  | *Ranger* | 1846 still in service |
|  | *Repulse* | 1836 still in service |
|  | *Royal Charlotte* | 1880 still in service |
|  | *Sprightly* | 1840 still in service |
|  | *Stork* | 1837 still in service |
| 1829 | *Fox* | built Cowes 1845 still in service |
|  | *Swallow* | 27.3.1835 wrecked in Weymouth Bay |
| 1830 | *Prince of Wales* | built Ransom & Ridley, ____ ; 1868 sold |
|  | *Speedwell* | (ex *Liverpool*) built Cowes; 28 tons; no further note |
| 1832 | *Adelaide* | 143 tons; still in service 1845 |
|  | *Squirrel* | built Cowes; 36 tons; no further note |
| By 1834 | *Camelion/Chameleon* | 27.8.1834 run down by *Castor* off Dover |
| 1834 | *Sylph* | 1860 still in service |

42'8", 36'4½" x 14' x 7'5". 35 tons
PLANS 1844?: Lines/lower deck/upper deck/sails

| By 1835 | *Victoria* | 1857 still in service |
|---|---|---|
| 1835 | *Lady Flora* | built ——— ; 1866 to be disposed of |
|  | *Camelion* | 1859 still in service |
| By 1836 | *Princess Royal* | 1859 still in service |
|  | *Racer* | 1860 still in service |
| 1836 | *Flying Fish* | built White, Cowes; 41 tons; out of service before 1852 |
|  | *Onyx* | built Cowes. 31 tons; 1860 still in service |
| By 1837 | *Redbreast* | 1838 still in service |
| 1837 | *Harpy* | built Cowes; 1869 sold as light vessel and served as such in Southampton Water till 1883 |
|  | ?*Squirrel* | built Pembroke; 16 guns?; no further note |
|  | *Vixen* | built Cowes; no further note |

| By 1838 | *Maria* | 1847 still in service (and until 1880?) |
|---|---|---|
| 1838 | *Adder* | built Cowes; 53 tons; no further information (is this the same as the 1817 *Adder*?) |
|  | *Neptune* | built Cowes. 42 tons; 1854 still in service |
|  | *Royal George* | ____ |

70'7", 55'6" x 22'6" x 10'1". 149 42/94
PLANS: Taken off 1839: lines/profile/lower deck/upper deck

| By 1840 | *Peterel* | ____ ; 1860 still in service |
|---|---|---|

53'2", 39'3½" x 17'5½" x 8'6". 62 45/94
PLANS: Lines: taken off 1840: Profile. 1841: Lines & midsection/profile/decks

|  | *Prince Albert* | 1856 still in service |
|---|---|---|
| 1840? | *Sealark* | 1844 still in service |

49'8", 39'0¾" x 17'10" x 8'2". 65
PLANS: Lines/upper deck. Also for *Skylark* 1841

1840 Cutter proposed for the service of the Herring Fishery in Scotland (Admiralty/Revenue?) by Admiralty Order 12.5.1840

62', 50'4¾" x 19'6" x 9'9". 100 16/94
PLANS: Lines & midsection/sails

|  | *Curlew* | (later *Providence*) Symonds design; 1857 still in service |
|---|---|---|

37', 28'7⅛" x 14'2" x 6'6". 30
PLANS: Lines. Also for *Despatch*

|  | *Lively** | 100 tons; 13.5.1870 sold at Harwich |
|---|---|---|
| By 1841 | *Gertrude* | 10 men; 1858 still in service |
| 1841 | *Skylark* | ____ [as *Sealark* + PLANS: Lines.] |
|  | *Dispatch* | ____ [as *Curlew*. PLANS: Lines.] |
| By 1842 | *Fairy* | 1855 still in service |
|  | *Frances* | 70 tons; 1890 still in service |
| By 1843 | *Egremont* | 1849 still in service |
|  | *Eliza* | 1860 still in service |
|  | *Governor* | 1860 still in service |
|  | *Harriet* | 1860 still in service |
| 1843 | *Nelson* | no further note |
| By 1844 | *Bat* | 1858 still in service |
|  | *Hind** | 41 tons; 1880 still in service |
|  | *King George** | tender; 1885 renamed *Flora*; 1901 wrecked at Kingstown |
|  | *Lady of the Lake* | 1860 still in service |
|  | *Margaret* | 70 tons; 1890 still in service |
| 1844 | *Eagle* | built ____ ; 1893 sold at Portsmouth. |

59', 47' x 20'3" x 10'1". 100
PLANS: Lines/Profile/lower deck/upper deck/midsection. 1870: Sails. 1873: Profile & deck.

|  | *Jane* | no further note |
|---|---|---|
| 1845 | *Rob Roy* | no further note |
| 1847 | Cutter of 160 tons | ____ |

70', 54'6½" x 23'6" x 11'1". 156 80/94
PLANS: Lines & midsection

|  | *Defence** | 60 tons; 1859 still in service |
|---|---|---|
| 1849 | *Safeguard* | brig; in service until 1862 |
| 1855 | *Mermaid** | purchased. 165 tons; 14.8.1890 sold to Tough and Henderson |
| By 1858 | *Fly** | 60 tons; Ca 1910-1918 hulked for use of pilots at Plymouth |
| 1858 | *Neptune** | 60 tons; 4.4.1905 sold at Chatham |
| 1859 | *Fairy** | tender; no further note |
| 1862 | *Fanny** | from RN; see chapter on Yachts; 31.10.1878 sunk off the Tuskar Rock in collision with SS *Helvetia* |
| 1880 | *Hind** | 130 tons; 27.11.1900 wrecked in the Shipwash |

# Chapter 6: Hulks

In the late Middle Ages a 'hulk' was a type of cargo vessel, the type that replaced the 'cog' (*kogge*) in the service of the Hanseatic League.[1] By the seventeenth century, however, it had transformed itself into a term for any old vessel which was reduced to a bare hull and used for 'harbour service', at anchor or beached. (Another use of the word 'hulk' is to describe a dismasted vessel, or any vessel deprived of its motive power can be described as being 'left a hulk'.) This was usually the fate of most warships that survived the hazards of the sea and the enemy to become too weak or too obsolete to be safely or usefully employed at sea.

**Lazaretto**: The name, and the concept, came from Italy where the practice of rigid quarantine to prevent, or at least limit, the spread of the Plague and other infectious diseases was developed. These were hulks adapted to be used as isolation hospitals and/or accommodation for men undergoing quarantine.

**Powder hulk**: Vessel for storing and issuing gunpowder, usually moored well away from the dockyard to which they were attached. Usually under the control of the Ordnance (which was a separate organisation to the Army or Navy, though later coming under the control of the former).

**Receiving ship**: Hulks intended to receive men, or stores, or both. Accommodation hulks and store hulks and vessels used for both purposes. The Royal Navy did not build barracks for its seamen in home ports until the beginning of the twentieth century; before this hulks were used for accommodating men between commissions or before they were assigned to ships (particularly pressed men who needed to be kept in some form of confinement to prevent desertion).

**Sheer hulk**: A vessel fitted with a pair of 'sheer legs' (two large spars forming an 'A frame') to hoist masts in and out of vessels. In effect, a floating crane. Some other navies used large cranes mounted on towers for this purpose (a fine example of one of these can be seen at Copenhagen Dockyard, for example) but the Royal Navy preferred to use sheer hulks, and in fact had three purpose built ones (the two Chatham sheer hulks and one at Plymouth), the only examples of hulks built as such the author has been able to find. As Brian Lavery points out,[2] the reason for the RN's preference must have come from the greater convenience and more rapid 'turn round' of being able to hoist masts in or out whilst a ship was in a roadstead rather than alongside the dockyard.

The entries in this list are given grouped by the dockyard or other place where the hulks were located. First come the Royal Dockyards, with the home ones listed before the overseas ones. They are followed by a listing of the hulks still in naval hands which were based elsewhere, and finally by those hulks which were transferred into the hands of various organisations (predominantly charitable). It is interesting to note that some of those which became training ships for boys (many of them reformatories, to hold what we would now call 'juvenile offenders') were destroyed by fire. One suspects that many of these fires were started deliberately by inmates; after all, setting fire to one's school is still a not infrequent form of adolescent protest.

## Hulks at Deptford Dockyard

| | | | |
|---|---|---|---|
| 1694-1707 | purchased as hulk | *Success* | [after being hired 1689-1692]; 1707 breakwater |
| 1706-1708 | hulk | *Kingfisher* | ex 40; 1708 to Harwich |
| 1708-1731? | hulk | *Gloucester* | ex 60; 1731? BU |
| 1725-1734 | mooring hulk? | *Mermaid* | ex 32; 1734 BU |
| 1731-1748 | hulk | *Success* | ex storeship; 1748 sold |
| 1739-1744 | hulk | *Rose* | ex 20; 1744 sold. |
| 1740-1741 | hulk | *Adventure* | ex 40; 1741 BU |
| 1745-1768 | hulk | *Panther* | ex 50; 1768 sold |
| 1767-1787 | hulk | *Bedford* | ex 70; 1787 sold |
| 1788-1816 | hulk | *Worcester* | ex 64; 1816 BU |
| 1794-1804 | purchased magazine | *Francis* | (ex mercantile *Francis*); barge purchased 1794 or 1795 apparently later to Blackwall; 1804 BU. [55'5", 44'5" x 15'10" x 4'9". 59 tons. 5 men. 4 x 'wall pieces' (musketoons/swivel mounted guns)] |
| 1798-1811 | receiving ship | *Janus* | ex Dutch 32; 1811 sold |
| 1805-1816 | receiving ship | *Sulphur* | ex purchased bomb; 1816 sold |
| 1806-1807 | receiving ship for pressed men | *Enterprize* | from the Tower; 1807 BU |
| 1810-1823 | receiving ship | *Dedaigneuse* | ex French 32; 1823 sold |
| 1814-1816 | receiving ship | *Pegasus* | ex frigate, 28; or at Chatham; 1816 sold |
| 1815-1828/ 1833-1834 | sheer hulk | *Monmouth* | ex purchased 64; 1828 to Woolwich; 1833 returned; 1834 BU |
| 1816-1818 | receiving ship for distressed seamen | *Perseus* | ex 22; 1818 to the Tower of London |
| 1816-1817 | hulk | *Tower* | ex tender at the Tower of London; 1817 to the Thames Police |
| 1819-1825 | convict ship | *Dromedary* | (ex *Howe*) ex storeship; 1825 to Bermuda |
| 1824-1834 | coal depot | *Driver* | ex sloop; 1834 BU |
| 1825-1836 | receiving ship | *Dover* | ex Franco-Italian frigate; 1831 quarantine service; 1836 sold |
| 1826-1829 | coal depot | *Bacchus* | ex brig; 1829 breakwater |
| 1826-1838 | convict ship | *Dasher* | ex sloop; from the Committee for Distressed Seamen via the Army; 1838 BU |
| After 1828-1834 | coal depot | *Portsmouth* | ex transport; from Woolwich; 1834 BU |
| 1841-1844 | convict ship | *Thames* | ex frigate, 38; 1844 to Bermuda |
| 1850-before 1861 | receiving ship | *Crocodile* | ex 28; before 1861 to London? |
| 1856-1858 | cholera ship | *Bacchante* | ex frigate, 38; from Sheerness; actually moored at Greenwich?; 1858 BU |

## Hulks at Woolwich Dockyard

| | | | |
|---|---|---|---|
| 1690?-1713 | hulk | *Saint David* | ex 54; from Kinsale?; 1713 sold |
| 1694-1715 | purchased as hulk | *Josiah* [*Joshua*?] | 1709 to Harwich for a short time?; 1715 breakwater |
| 1715-1740 | hulk | *Somerset* | ex 80; 1740 BU |
| 1740-1760 | hulk | *Devonshire* | ex 80; 1760 sold |
| 1742-1744 | moorings ship | *Phoenix* | ex 20; 1744 sold |
| 1760-1768 | hulk | *Tavistock* | ex 50; 1768 sold |
| 1769-1786 | hulk | *Guernsey* | ex 50; 1786 sold |
| 1770-1776 | receiving ship for impressed men | *Nightingale* | ex purchased 24; 1776 to the Tower of London |
| 1784-1802 | receiving ship | *Rainbow* | ex 44; 1802 sold |
| 1785-1815 | hulk | *Preston* | ex 50; 1815 BU |
| 1799-1815 | receiving ship | *Assurance* | ex 44; 1815 BU |
| 1799-1810 | hospital ship? | *Matilda* | ex French 28; 1810 BU |
| 1803-1815 | hospital ship for convicts | *Savage* | ex sloop; 1815 sold |
| 1803?-1810 | hulk? | *Triton* | ex frigate, 32; 1810 to Plymouth |
| 1803-1805 | hulk | *Quebec* | ex frigate, 32; 1805 back in service |
| 1812-1818 | receiving ship | *San Fiorenzo* | ex French Frigate; 1818 to Sheerness |

---

1 The hulk would appear to have been a vessel constructed in a particular way. For details see, for example, B Greenhill, *The Archaeology of the Boat* (London 1976).
2 In his chapter on harbour craft in *The Line of Battle*.

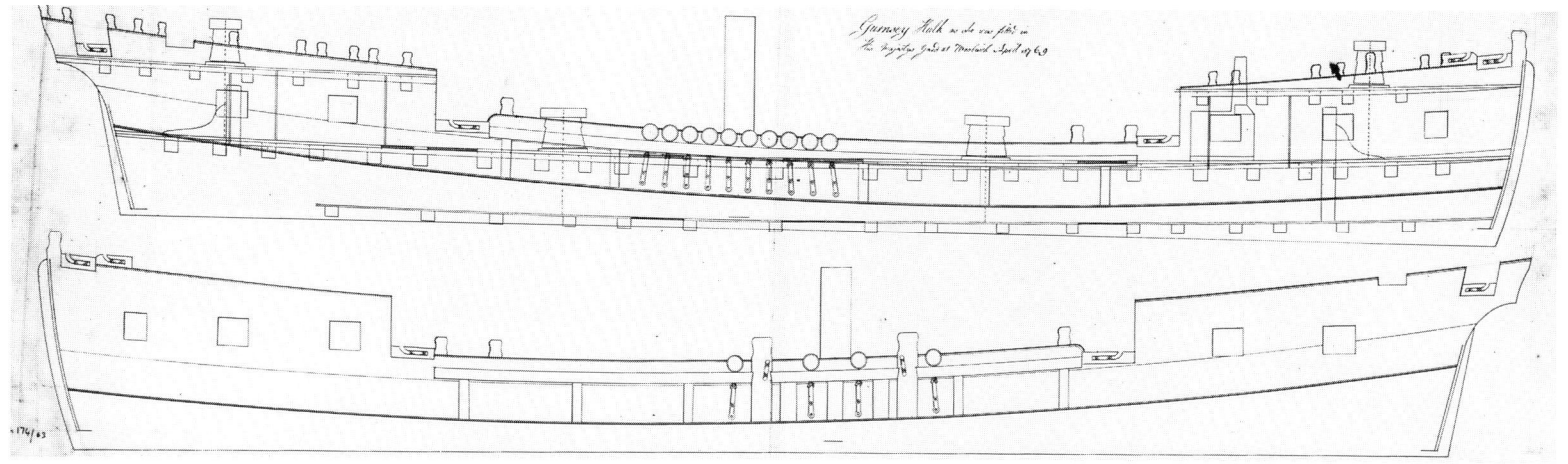

Internal and external above water profiles (Charnock collection Ch 174) of the old 50 gun ship *Guernsey* 'as fitted' cut down to a sheer hulk at Woolwich in 1769

| | | | |
|---|---|---|---|
| 1813-1817 | receiving ship | **Helder** | ex Dutch frigate; 1816 receiving ship for distressed seamen; 1817 BU |
| 1814-1832 | sheer hulk | **Sampson** | from Chatham; 1832 sold |
| 1814-1816 | hulk | **Porpoise** | ex storeship; 1816 sold |
| 1815-1828 | convict hospital ship | **Alonzo** | ex purchased sloop; 1828 to Portsmouth [PLANS: Decks & part profile] |
| 1818-1834 | hulk | **Discovery** | ex purchased sloop; 1834 BU |

NOTE: There is a well known print of this vessel as a hulk by E W Cooke

| | | | |
|---|---|---|---|
| 1823-1835 | convict ship | **Ethalion** | ex frigate, 36; 1824 temporary receiving ship; 1835 breakwater for Harwich |
| 1823-1837 | temporary convict ship | **Narcissus** | ex frigate, 32; 1824 convict hospital ship; 1837 sold |
| 1823-1825 | convict ship | **Sydney** | ex survey vessel; 1825 sold |
| 1824-1834 | receiving ship | **Eurydice** | ex 24; 1834 BU |
| 1824-1828 | temporary convict ship | **Heroine** | (ex *Venus*) ex frigate, 36/32; 1828 sold |
| 1824-1827 | temporary receiving ship | **Pheasant** | ex sloop; 1827 sold |
| 1827-1832 | floating magazine | **Snap** | ex gunbrig; 1832 sold |
| 1828-1833 | sheer hulk | **Monmouth** | from Deptford; 1833 to Deptford |
| 1828-? | coal depot | **Portsmouth** | ex transport; later to Deptford |
| 1830-1851 | purchased as coal depot | **Charger** | (ex mercantile paddle steamer *Hermes/Courier/George IV*?; built at Blackwall, 1824); 1851 BU |
| 1830-1862 | purchased as coal depot | **Messenger** | (ex mercantile paddle steamer *Duke of York* built at Blackwall, 1824); 1862 sold |
| 1830-1855 | convict ship | **Justitia** | (ex *Dolphin*, ex *Hindostan*); purchased 54; 1855 sold |
| 1831-1833 | quarantine service vessel | **Shamrock** | ex gunbrig; 1833 to the Customs |
| 1832-1857 | receiving ship | **Warrior** | ex 74; from Chatham; 1857 BU |
| 1833-1841 | convict and receiving ship | **Leven** | ex 28; 1841 to Limehouse |
| 1835-1874 | receiving ship | **Salsette** | ex frigate, 36; from Pembroke; 1874 BU |
| 1836-1858 | convict hospital ship | **Unite** | (ex *Imperieuse*) ex French frigate 38; 1858 BU |
| 1839-1872 | receiving ship | **Hebe** | ex frigate, 38; 1872 to Sheerness & BU |
| 1839-1841 | lazaretto? | **Plover** | ex brig; ? from being on loan to the Thames Tunnel (as accommodation ship?); 1841 sold |
| 1843-1857 | convict & receiving ship | **Sulphur** | ex bomb vessel; 1857 BU |
| 1847-1879 | 'flag ship' | **Fishguard** | ex frigate, 38; fitted for the Commodore; 1879 BU |
| 1848-1857 | convict ship | **Defence** | ex 74; 1857 burnt & BU |
| 1852-1869 | receiving ship & coal depot | **Aigle** | ex frigate, 36; 1869 to Sheerness |
| 1855-1896 | ordnance depot | **Talbot** | ex 28/sloop; 1896 sold |
| 1862-by 1873 | coal depot | **Mercury** | ex frigate, 38; by 1873 to Sheerness |

## Hulks at Chatham Dockyard

| | | | |
|---|---|---|---|
| 1673-1704 | hulk | **Arms of Horn** | ex Dutch prize (taken 1672); 1673 storeship at Sheerness; 1703 sank at her moorings (?at Sheerness?); 1704 BU |
| 1675-1703 | hulk | **Rotterdam/ Arms of Rotterdam** | ex Dutch prize taken 1673; 1675 hulked at Chatham; 1703 BU |
| 1694-1813 | sheer hulk | **Chatham** | see Yard Craft |

NOTE: This vessel her 1813 replacement, and one other built for Plymouth were the only sheer hulks, or hulks of any kind, built as such for the Royal Navy. All other hulks were conversions, purchases, or both

| | | | |
|---|---|---|---|
| 1713-1716 | hulk | **Saint George** | from Harwich; 1716 sunk as a foundation |
| 1715-1741 | hulk | **Sunderland** | ex 60; 1741 to Port Mahon |
| 1740-1742 | hulk | **Sterling Castle** | ex 60; 1742 to Sheerness |
| 1745-1749 | hospital ship | **Britannia** | ex 100; 1749 BU |
| 1748-1781 | hulk | **Winchester** | ex 50; from Sheerness; 1781 BU |
| 1749-1754 | church ship | **Sandwich** | ex 90; 1752 lazaretto; 1754 to Sheerness |
| 1755-1761 | prison ship (for the French) | **Cornwall** | ex 80; 1761 BU |
| 1758-1759 | hospital ship for soldiers | **Gloucester** | ex 50; 1759 to Sheerness |
| 1768 | To Commissioners of the Victualling | **Falkland** | ex 50; fate unknown |
| 1778-1779 | receiving ship | **Dunkirk** | ex 60; 1779 to the Downs, then to Plymouth |
| 1778-1783 | purchased as a prison ship | **Security** | ex mercantile?; classed as a Sixth Rate; 1783 sold [PLANS: 1778: decks. 1779: Profile & decks.] |
| 1780-1783 | convalescent ship | **Nabob** | ex storeship; later to Sheerness?; 1783 sold |
| 1782-1802 | receiving ship | **Prince Edward** | ex Dutch 62/56; 1802 sold |
| 1783-1802 | receiving ship | **Warwick** | ex 50; 1802 sold |
| 1785-1806 | receiving ship/ sheer hulk | **Rotterdam** | ex Dutch 50; 1806 sold [PLANS: Charnock Collection) Profile & upper deck/quarter deck.] |
| 1788-1816 | hospital ship | **Union** | ex 90; 1790 renamed *Sussex*; 1816 BU |
| 1788-1793 | receiving ship | **Hero** | ex 74; from Plymouth; 1793 prison ship; 1800 renamed *Rochester*; 1810 BU |
| 1794-1810 | prison ship | **Bristol** | ex 50; 1810 BU |
| 1794-1812 | lazaretto | **Eagle** | ex 64; 1800 renamed *Buckingham*; 1812 BU |
| 1794-1797 | lazaretto | **Director** | ex 64; 1797 reverted to active service as a 64 |
| 1797-1831 | hospital ship | **Argonaut** | ex French 64; 1831 BU |
| 1798-1811 | prison ship? | **Vryheid** | ex Dutch 72; 1811 sold |
| 1798-1814 | prison ship | **Gelykheid** | ex Dutch 64; 1814 sold (by then at Portsmouth?) |
| 1799-1801 | convalescent ship | **Standard** | ex 64; 1801 reverted to active service as a 64 |
| 1800-1814 | receiving ship | **Dordrecht** | ex Dutch 64; 1814 to Sheerness |
| 1800-1803 | prison ship | **Tromp** | ex Dutch 60; 1803 to Portsmouth |
| 1802-1817 | powder hulk/ prison ship | **Camperdown** | Dutch 64; 1817 sold |
| 1802-1818 | powder hulk | **Wassenaar** | ex Dutch 64; 1818 sold |

| | | | |
|---|---|---|---|
| 1802–1822 | powder ship | *Delft* | ex Dutch 64; 1822 breakwater at Harwich |
| 1803–1806 | powder hulk? | *Irresistible* | ex 74; 1806 BU |
| 1805–1814 | floating magazine | *Chatham* | from Falmouth; 1810 renamed *Tilbury*; 1814 BU |
| 1805–1816 | hulk | *Haerlem* | ex Dutch 64; at Stansgate Creek?; 1811 receiving ship; 1816 sold |
| 1806–1822 | 'stationary ship' | *Prince of Orange* | ex 74; initially at Sheerness?; 1811 powder magazine; 1822 sold |
| By 1807–1814 | prison ship | *Bahama* | ex Spanish 74; 1814 BU |
| 1807–1833 | receiving ship | *Namure* | ex 98/74; at the Nore?; 1833 BU |
| 1807–1814 | receiving ship? | *Brakel* | ex Dutch 54; 1814 sold |
| 1808–1814 | prison ship (for the Danes) | *Sampson* | ex 64; 1814 to Woolwich |
| By 1809–1809 | powder magazine | *Vlieter* | ex Dutch 44; 1809 to Sheerness |
| 1809–1814 | prison ship | *Fyen* | ex Danish 74; 1814 sold |
| 1809–1814 | prison ship | *Kronprincen* | ex Danish 74; 1814 sold |
| 1809–1815 | prison ship | *Trusty* | ex 50; 1815 BU |
| 1809–1814 | receiving ship | *Adamant* | ex 50; possibly to Sheerness?; 1814 BU |
| 1810–1817 | ballast ship | *Expedition* | ex 44; 1817 BU |
| 1810–1834 | prison ship | *Canada* | ex 74; 1814 powder magazine; 1826 convict ship; 1834 BU |
| 1811–1814 | receiving ship | *Frederick Swaern* | ex Danish frigate; 1814 sold |
| 1811–1830 | receiving ship | *Terpsichore* | ex frigate, 32; 1830 BU |
| 1811–1817 | receiving ship | *Guelderland* | ex Dutch 64; 1817 sold |
| 1812–1825 | prison ship | *Brunswick* | ex 74; 1814 powder magazine; 1825 to Sheerness |
| 1812–1830 | harbour flagship | *Ceres* | ex frigate, 32; from Sheerness; 1816 victualling depot; 1830 BU |
| 1813–1876 | sheer hulk | *Chatham* | built at Chatham to a design by Seppings of pitch pine; 1876 BU; see Yard Craft list. |

NOTE: Like her predecessor this vessel was actually purpose built, and with that predecessor and one other at Plymouth the only examples in the Royal Navy

| | | | |
|---|---|---|---|
| 1813–1817 | temporary prison hulk | *Defiance* | ex 74; 1817 BU |
| 1813–1835 | convict ship | *Edgar* | ex 74; 1815 renamed *Retribution*; 1835 BU |
| 1813–1827 | powder magazine | *Polyphemus* | 1826 for the Lieutenants of the Ordinary; 1827 BU ex 64; |
| 1814–1816 | receiving ship | *Pegasus* | ex frigate, 28; or at Deptford; 1816 sold |
| 1814–1816 | prison ship | *Belliqeux* | ex 64; 1816 BU |
| 1814–1832 | lazaretto | *Courageux* | ex 74; 1832 BU |
| 1819–1832 | receiving ship | *Warrior* ex 74; | 1831 temporary quarantine ship; 1832 to Woolwich |
| 1819–1840 | convict ship | *Ganymede* | ex corvette; 1840 capsized?; then BU |
| 1824–1838 | receiving ship | *Aboukir* | ex 74; 1831 hospital ship; 1838 sold |
| 1825–1847 | convict ship | *Euryalus* | ex frigate, 36; 1847 to Gibraltar |
| 1829–1838 | victualling depot | *Brune* | ex French frigate from Sheerness; 1838 sold |
| 1830–1859 | receiving ship/or hulk | *Tartar* | (or from 1827?); ex frigate, 36; 1859 BU |
| 1830–by 1854 | convict ship/coal depot | *Cumberland*; | ex 74; 1833 renamed *Fortitude*; by 1854 to Sheerness |
| 1831–1854 | quarantine service | *Dartmouth* | ex frigate, 36; 1854 BU at Deptford |
| 1833–1833 | harbour service, unspecified | *Creole* | ex frigate, 36; 1833 BU |
| 1833–1861 | receiving ship | *Hussar* | ex frigate, 38; 1861 target ship and burnt at Shoeburyness |
| By 1850–1852 | convict hospital ship | *Wye* | from Sheerness; 1852 BU |
| 1852–1875 | coal depot | *Ocean* | ex 110/80; from Sheerness; 1875 BU |
| 1850s–1866 | coal hulk | *Terpsichore* | ex 28; 1865 used for mining experiments; 1866 BU |
| 1855–1874 | powder depot | *Volage* | ex 28; 1855 to the War Department at Upnor (opposite side of the River Medway from Chatham); 1874 sold |
| 1860–1873 | hulk | *Unicorn* | ex frigate, 38; 1868 lent as a powder hulk to the War Department; 1873 to Dundee |
| 1861–1884 | receiving ship | *Gloucester* | ex 74/50; 1884 sold |
| 1862–1868 | flag & receiving ship | *Wellesley* | ex 74; 1868 to Purfleet |
| 1862–1864 | target ship | *Powerful* | ex 84; 1864 BU |
| 1873–1905 | base ship | *Pembroke* | ex 74/screw line of battle ship; 1890 renamed *Forte* as receiving hulk; 1905 sold |

NOTE: *Pembroke* gave her name to the subsequent 'stone frigate' (barracks) at Chatham

| | | | |
|---|---|---|---|
| 1880–1921 | chapel ship | *Helena* | from Ipswich; 1884 police accommodation ship; 1921 sold |
| 1885–1899 | training ship | *Clarence* | (ex *Royal William*); ex 120/screw line of battle ship; 1899 burnt by accident |
| 1891–1905 | receiving ship | *Royal Adelaide* | ex 104; from Plymouth; 1905 sold |
| 1895–1906 | ordnance store | *Melampus* | ex frigate, 38; from Portsmouth; 1906 sold |

## Hulks at Sheerness Dockyard

| | | | |
|---|---|---|---|
| 1687–1697 | hulk | *Saint George* | ex 70/66 built at Deptford 1622; 1697 breakwater |
| 1687–1699 | hulk | *Leopard* | ex 54 built at Deptford 1659; 1699 breakwater |
| 1698–1712 | hulk | *Medway Prize* | ex prize 50; 1709 to be sunk; 1709 to receive men; 1712 breakwater |
| 1709–1728 | hulk | *Kingfisher* | ex 40; from Harwich; 1728 BU |
| 1727–1745 | hulk | *Buckingham* | (ex *Revenge*) ex 70; 1745 breakwater |
| 1742–1771 | hulk | *Sterling Castle* | ex 70; from Chatham; 1771 BU |
| 1742–1749 | receiving ship | *Defiance* | ex 60; 1749 BU |
| 1745–1748 | hulk | *Winchester* | ex 50; from Harwich; 1748 to Chatham |
| 1750–1759 | quarantine hulk | *Looe* | ex 44; 1759 breakwater |
| 1752–1759 | quarantine hulk | *Launceston* | ex 44; 1759 reverted to service as a 44 |
| 1754–1770 | lazaretto | *Sandwich* | ex 90; from Chatham; 1770 BU |
| 1755–1764 | hospital ship | *Princess Caroline* | ex 80; 1764 BU |
| 1759–1764 | receiving ship | *Gloucester* | ex 50; from Chatham; 1764 BU |
| 1759–1762 | receiving ship | *Humber* | ex 44; 1762 to the Downs as a 44 again |
| 1764–1784 | temporary lazaretto | *Prince Frederick* | ex 70; 1784 sold |
| 1770–1787 | lazaretto | *Newark* | ex 80; 1787 BU |
| 1772–1810 | hulk | *Prince of Orange* | ex 70/60; 1810 sold |
| 1775–1782 | receiving ship (at the Nore) | *Conquestadore* | ex Spanish 60; 1782 BU. |
| 1777–1783 | hospital ship | *Orford* | ex 70; 1783 breakwater |
| 1779–1783 | receiving ship | *Greenwich* | ex storeship; 1783 sold |
| 1779–1779 | receiving ship | *Diligence* | ex sloop; 1779 returned to service as a fireship (never actually served as a hulk) |
| 1780?–1783 | slop ship | *Hinchinbrook* | ex American corvette; 1783 sold |
| 1783–1799 | receiving ship | *Princess Caroline* | ex Dutch 54; 1799 BU |
| 1784–1796 | receiving ship | *Prosperity* | ex storeship; 1796 BU |
| 1784–1819 | lazaretto | *Belleisle* | ex French 64; 1819 sold |
| 1786–1792 | purchased as slop ship | *Grantham* | (ex mercantile *Grantham*); 1792 BU |
| 1787–1810 | receiving ship | *Sandwich* | ex 90; 1794 prison ship (classed as a ship sloop); 1810 BU |
| 1788–1818 | lazaretto | *Princess Amelia* | ex 80; 1818 sold |
| 1789–1795 | ordnance hulk | *Merlin* | ex purchased corvette; 1795 sold |
| 1793–1806 | convalescent ship | *Grana* | ex Spanish corvette; 1806 sold |
| 1794–1794 | lazaretto | *Renown* | ex 50; 1794 BU |
| 1796–1801 | receiving ship | *Eolus* | ex frigate, 32; 1800 renamed *Guernsey*; 1801 BU |
| 1796–1818 | temporary receiving ship | *Belle Poule* | ex French frigate; 1818 sold |
| 1797–1806 | Powder hulk? | *Eclair* | ex French sloop; 1806 sold |
| 1797–1802 | slop ship | *Boreas* | ex frigate, 28; 1802 sold |
| 1798–1816 | Port Admiral's ship | *Vindictive* | ex Dutch 28; 1816 sold |
| 1799–1805 | (see under Cape of Good Hope) | *Broaderscarp* | ex Dutch 54/50 |
| 1799–1826 | lazaretto | *Valiant* | ex 74; 1826 BU |
| 1799–1843 | lazaretto | *Duke* | ex 90; from Portsmouth?; 1843 BU |
| 1800–1828 | hospital ship for lazarettos | *Lizard* | ex frigate, 28; 1828 sold |
| 1800–1815 | receiving ship | *Drochterland* | ex Dutch 44; 1815 BU |
| 1801–1816 | receiving ship | *Adventure* | ex 44; 1816 BU |
| 1801–1815 | receiving ship | *Reliance* | ex purchased storeship; 1815 sold |
| 1801–1802 | receiving ship | *Raison* | ex French corvette; 1802 sold |
| 1803–1830 | receiving ship | *Zealand* | ex Dutch 64; 1809 convict ship; 1815 *Justitia*; 1830 sold |
| 1803–1812 | receiving ship | *Ceres* | ex frigate, 32; 1812 to Chatham |

| | | | |
|---|---|---|---|
| 1803-1814 | mooring hulk? | *Winchelsea* | ex frigate, 32; 1814 sold |
| 1805-1809 | provision receiving ship | *Overyssel* | ex Dutch/Spanish 64; 1809 receiving ship breakwater at Harwich (1822 sold) |
| 1808-1818 | convict ship | *Discovery* | ex exploration vessel/sloop; 1818 to Woolwich |
| 1808-1816 | receiving ship | *Combatant* | ex sloop; 1816 BU |
| 1809-1825 | receiving ship | *Camilla* | ex 20; later floating barracks; 1825 breakwater (1831 sold) |
| 1809-1817 | sheer hulk | *Vlieter* | ex Dutch 44; from Chatham; 1817 BU |
| 1809-1816 | receiving ship | *Champion* | ex 24; 1816 sold |
| 1810-1815 | receiving ship | *Raisonable* | ex 64; 1815 BU |
| 1810-1816 | military depot & hospital ship | *Falcon* | ex purchased sloop; 1816 sold |
| 1813-1816 | receiving ship | *Quebec* | ex frigate, 32; 1816 BU |
| 1813-1864? | quarantine hulk | *Acute* | ex gunbrig; 1864? BU |
| 1814-1838 | lazaretto | *Christian 7th* | ex Danish 80; 1838 BU |
| 1814-1823 | receiving ship | *Dordrecht* | ex Dutch SLB; from Chatham; 1823 sold |
| 1816-1829 | victualling depot | *Brune* | ex French frigate; 1829 to Chatham |
| 1816-1836 | convict hulk | *Bellerophon* | ex 74; 1824 renamed *Captivity*; 1836 sold |
| 1816-1837 | sheer hulk | *Lion* | ex 64; from Plymouth; 1837 sold [PLANS: presumably for this vessel as sheer hulk: Sheer legs (2).] |
| 1817-1823 | convict ship | *Batavier* | ex Dutch 56; 1823 BU |
| 1817-1818 | receiving ship | *Alexandria* | ex frigate, 32; 1818 BU |
| 1818-1837 | lazaretto | *Saint Fiorenzo* | ex French 34; 1837 BU |
| 1818-1838 | lazaretto | *Imperieuse* | ex Spanish frigate; 1838 sold |
| 1819-1826 | hulk in the Swale? | *Congo* | ex steam vessel/survey ship; 1826 sold |
| 1820-1838 | receiving ship | *Temeraire* | ex 98; from Plymouth; 1838 sold for BU |

NOTE: Turner saw her at sunset being towed up the River Thames to be broken up, and the result was his painting 'The Fighting *Temeraire*'

| | | | |
|---|---|---|---|
| 1823-1836 | receiving ship | *Terrible* | ex 74; 1829 coal depot; 1836 BU |
| 1823-1832 | slop ship | *Desiree* | ex French frigate; 1832 sold |
| 1824-1843 | receiving ship | *Vengeur* | ex 74; 1843 BU |
| 1825-1826 | lazaretto | *Brunswick* | ex 74; from Chatham; 1826 BU |
| 1827-1850 | lazaretto | *Northumberland* | ex 74; 1850 BU |
| 1831-1859 | receiving ship/ temporary hulk | *Shannon* | ex frigate, 38; 1844 renamed *Saint Lawrence*; 1859 BU |
| 1831-1863 | lazaretto | *Duncan* | ex 74; from Portsmouth; 1863 BU |
| 1831-1836 | quarantine service | *Martial*; | ex gunbrig; 1836 sold |
| 1831-1850 | lazaretto | *Ramillies* | ex 74; 1850 BU |
| 1831-1852 | lazaretto | *Ocean* | ex 110/80; 1832? reverted to 80; later flag ship for the Captain of the Ordinary; 1852 coal depot then to Chatham |
| 1833-1848 | lazaretto | *Blossom* | ex sloop; 1848 BU |
| 1834-____? | convict hospital hulk & floating breakwater | *Wye* | ex 28; later to Chatham |
| 1836-1841 | lazaretto | *Menelaus* | ex frigate, 38; 1841 to Portsmouth |
| 1836-1838 | lazaretto | *Plover* | ex brig; 1838 lent to the Thames Tunnel |
| 1837-1858 | lazaretto | *Bacchante* | ex frigate, 38; briefly to Greenwich as a cholera ship; 1858 BU |
| 1837-1875 | receiving ship | *Nymphe* | ex frigate, 38; 1861 to the water police; 1863 Roman Catholic chapel ship; 1871 renamed *Handy*; 1875 BU |
| 1838-1871 | lazaretto | *Rhin* | ex French frigate; 1871 to the Thames as a cholera ship |
| 1839-1841 | target | *Raleigh* | ex brig; 1841 sold |
| 1841-1844 | storeship | *Tortoise* | from Pembroke; 1844 to Ascension Island |
| 1842-1869 | receiving ship | *Minotaur* | ex 74; 1859 guard ship; 1866 renamed *Hermes*; 1869 BU |
| 1846-1855 | store depot | *Eolus* | ex frigate, 38; 1855 to Portsmouth |
| 1848-1870 | coal depot? | *Fortitude* | (ex *Cumberland*) ex 74; from Chatham; 1870 sold |
| 1848-1857 | receiving & depot ship | *Wellington* | ex 74; 1857 to the Coastguard |
| 1848-1894 | marine barracks ship | *Benbow* | ex 74; 1854 prison ship for Russians; 1859 coal depot; 1894 sold for BU |
| 1851-1854 | storeship | *Atholl* | ex 26; 1854 to Greenock |
| 1854-1892 | coal depot | *Columbine* | ex brig; 1892 sold |
| 1854-1869 | prison ship for Russians | *Devonshire* | ex 74; from Greenwich; by 1860 school ship in 'Queenborough Swale'; 1869 BU |
| 1854-1872 | hospital ship | *Belleisle* | ex 74; (1866-1868 lent to the Seamens Hospital at Greenwich); 1872 BU |
| 1860-1903 | coal hulk | *Dido* | ex sloop; 1903 sold |
| 1863-1869 | sheer hulk | *Cumberland* | ex 70; 1869 training ship in the Clyde |
| 1865-1957 | jetty | *Cornwallis* | ex 74; 1916 renamed *Wildfire* as base ship; 1957 BU |
| 1868-1872 | mooring vessel | *Latona* | ex frigate, 38; 1872 to ____ |
| 1868-1874 | receiving ship | *Diana* | ex frigate, 38; 1874 BU |
| 1869-1870 | coal hulk | *Aigle* | from Woolwich; 1870 used for torpedo experiments; sunk during experiments and remains sold |
| 1872-1894 | powder hulk | *Leonidas* | ex frigate, 38; 1894 sold |
| By 1873-1906 | coal depot | *Mercury* | ex frigate, 38; 1906 sold |

## Hulks at Portsmouth Dockyard

| | | | |
|---|---|---|---|
| 1682-1712 | hulk | *French Ruby* | ex French prize taken 1666; hulked after storm damage; 1712 BU |
| 1690-1713 | hulk | *Saint David* | ex 54; hulked after salvage; 1713 sold |
| 1691-1717 | hulk | *Exeter* | ex 70; hulked after she was burnt; 1717 BU |
| 1703-____? | victualling storeship | *Bourbon Caro* | ex Spanish warship taken at Vigo in 1702; fate uncertain |
| 1703-____? | victualling storeship | *Santa Crux* | ex Spanish warship taken at Vigo in 1702; fate uncertain |
| By 1712-1713 | storeship | *Moderate* | ex French 56; 1713 sold |
| 1715-1742 | hulk | *Berwick* | ex 70; 1742 BU |
| 1740-1768 | hulk | *Yarmouth* | ex 70; 1768 BU |
| 1740-1745 | hulk | *Nonsuch* | ex 50; 1745 BU |
| 1740-1745 | hulk | *Hastings* | ex 40; 1745 sold |
| 1740-1763 | hospital ship | *Blenheim* | ex 90; 1763 BU |
| 1740 or 1743-1762 | hulk | *Captain* | ex 70; 1762 BU |
| 1742-1748 | receiving ship | *Dreadnought* | ex 60; 1748 BU |
| 1742-1749 | receiving ship | *Medway* | ex 60; 1749 BU |
| 1761-1784 | hulk | *Princessa* | ex Spanish 70; 1784 sold |
| 1770-1773 | temporary hulk | *Assistance* | ex 50; 1773 sold |
| 1777-1799 | receiving ship/hulk | *Essex* | ex 64; 1799 sold |
| 1778-1784 | prison ship/receiving ship for convalescent seamen | *Mars* | ex 74; 1784 sold |
| 1778-1789 | receiving ship? | *Lenox* | ex 74; 1781 guard ship; 1789 BU |
| 1778-1791 | hulk | *Firm* | ex 60; 1784 receiving ship; 1791 sold |
| 1778-1800 | receiving ship | *Modeste* | ex French 64; 1800 BU |
| 1778-1802 | receiving ship | *Warspite* | ex 74; 1800 renamed *Arundel*; 1802 sold |
| 1779-1779 | hulk | *Grasshopper* | ex sloop; 1779 renamed *Basilisk* and fitted as a fireship |
| 1780-1784 | receiving ship | *Diligent* | ex Spanish 70; 1784 at Spithead |
| 1780-1784 | receiving ship | *Dragon* | ex 74; 1784 sold |
| 1780-1783 | convalescent ship | *Lioness* | ex storeship; 1783 sold |
| 1780-1783 | slop ship | *Snake* | ex sloop; 1783 BU |
| 1784-1816 | sheer hulk | *Neptune* | ex 90; 1816 BU |
| 1790-1813 | receiving ship | *Royal William* | ex 100/84; 1813 BU |
| 1790?-1805 | receiving ship? | *Brisk* | ex sloop; 1805 sold |
| 1791 (or 1788)-1817 | sheer hulk | *Prince William* | ex Spanish 64; later fitted for the reception of guns; 1817 BU |
| 1792-1802 | slop ship | *Vulture* | ex sloop; 1802 sold |
| 1792-1816 | receiving ship | *Grafton* | ex 74; 1816 BU |
| 1794-1815 | prison ship | *Pegase* | ex French 74; 1801 prison hospital ship; 1815 BU [PLANS: 1799: Decks. 1804: decks. N.D. decks.] |
| 1794-1817 | receiving ship | *Alcide* | ex 74; 1817 BU |
| 1795-1816 | prison ship | *Vigilant* | ex 64; 1816 BU |
| 1795-1802 | receiving ship | *Gentille* | ex French 40; 1802 sold |
| 1795-1815 | prison ship | *Prothee* | ex French 64; 1815 BU |
| 1796-1821 | convict ship | *Laurel* | (ex *Daphne*) ex Dutch 24; 1821 sold |
| 1796-1815 | prison ship | *Royal Oak* | ex 74; 1805 renamed *Assistance*; 1815 BU |
| 1796-1818 | prison ship | *Monmouth* | ex 64; renamed *Captivity*; 1818 BU |
| 1796-1816 | receiving ship/hulk for Spithead | *Puissant* | ex 74; 1816 sold |
| 1796-1802/1803 | receiving ship | *Prudente* | ex French frigate; 1802 or 1803 sold |

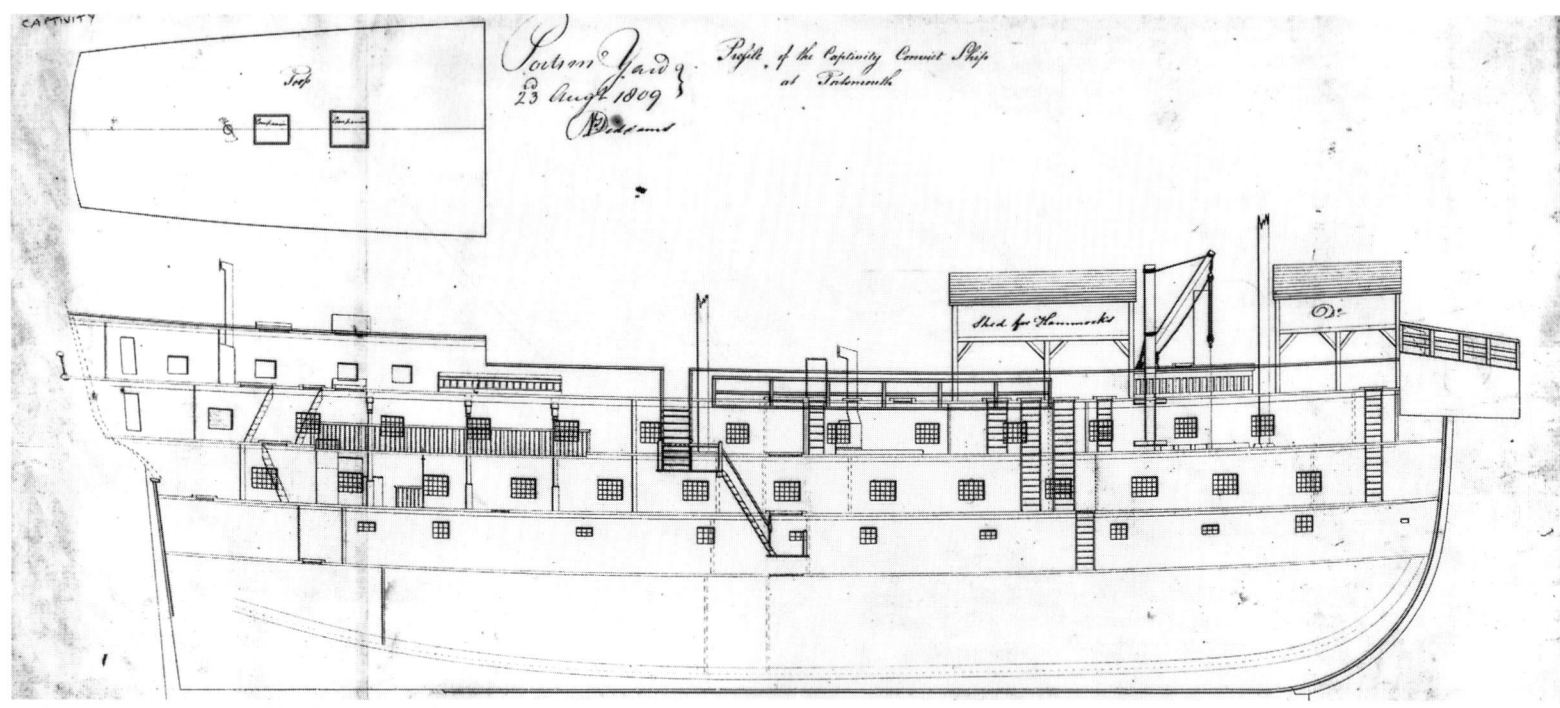

Profile and poop plan (8065) of the old 64 *Monmouth* converted to a convict ship and renamed *Captivity*. The plan shows her as she was in 1809

| | | | |
|---|---|---|---|
| 1796-1800 | slop ship/lighter | *Firebrand* | ex fire vessel; 1800 BU |
| 1790s?-1802 | temporary receiving ship | *Cygnet* | ex sloop; 1802 sold |
| 1797-1816 | prison ship | *Sultan* | ex 74; 1805 renamed *Suffolk*; 1816 BU |
| 1798-1816 | prison ship | *Crown* | ex 64; 1802 powder hulk; 1806 prison ship; 1816 BU |
| 1798-1799 | lazaretto? | *Duke* | ex 98; 1799 to Sheerness? |
| 1799-1814 | prison ship | *Fame* | ex 74; renamed *Guildford*; 1814 sold |
| 1800-1814 | prison ship? | *San Damaso* | ex Spanish 74; 1810 sold |
| 1801-1828 | prison ship | *San Antonio* | ex Spanish 74; 1810 powder magazine; 1827 sold; 1828 resold |
| 1801-1829 | powder hulk | *Bulldog* | sloop/bomb vessel; 1829 BU |
| 1801-1805 | hospital ship | *Medea* | ex frigate, 28; 1805 sold |
| 1802-1817 | convict ship | *Portland* | ex 50; 1817 sold. (Based in Langstone Harbour - the other side of Portsea Island from Portsmouth Harbour) |
| 1802?-1817 | receiving ship | *Robust* | ex 74; 1817 BU |
| 1803-1815 | prison ship? | *Tromp* | ex Dutch 60; from Chatham; 1811 receiving ship; 1815 sold |
| 1803-1805 | 'stationary service' | *Blonde* | ex frigate, 32; 1805 sold |
| 1803-1832 | slop ship | *Pearl* | ex frigate, 32; 1810 receiving ship; 1825 renamed *Prothee*; 1832 sold |
| 1804-1817 | Army prison ship | *Dido* | ex frigate, 28; 1817 sold |
| 1804-1823 | convict hospital ship | *Spiteful* | ex gunbrig; 1823 BU (Attending *Captivity*) |
| 1805?-1821 | convict hospital ship | *Sagesse* | ex French frigate; 1821 sold |
| 1805-1819 | lazaretto (at the Motherbank) | *Alexander* | ex 74; 1819 BU [PLANS: 1805: Outline decks.] |
| 1805-1814 | receiving ship | *Glenmore* | ex frigate, 36; 1814 sold |
| 1805-1810 | receiving ship | *Autumn* | ex purchased sloop; 1810 converted to bomb vessel and renamed *Strombolo* |
| 1806-1817 | medical depot | *Leander* | ex 50; renamed *Hygeia*; 1817 sold |
| 1806-1816 | temporary prison ship | *Marengo* | ex French 76; 1816 BU |
| 1806-1823 | receiving ship | *Perseverance* | ex frigate, 36; 1823 sold |
| 1807-1814 | receiving ship | *Cyclops* | ex frigate, 28; 1814 sold |
| Ca 1807-1817 | convalescent ship (later?) | *Gladiator* | ex 44; 1817 BU |
| 1808?-1825 | receiving ship | *Tre Kronen* | ex Danish 74; 1825 sold |
| 1808-1816 | victualler at Spithead | *Ildefonso* | ex Spanish 74; 1816 BU [PLANS: 1809: Profile & orlop/decks.] |
| 1808-1816 | prison ship | *Vengeance* | ex 74; 1816 BU |
| 1809-1816 | prison ship | *Veteran* | ex 64; 1816 BU |
| 1809-1817 | receiving ship | *Pluto* | ex fireship; 1817 sold |
| 1810-1836 | receiving ship | *Merlin* | ex purchased sloop; 1836 sold |
| 1811-1825 | receiving ship | *Skiold* | ex Danish 74; 1825 sold |
| 1811-1825 | receiving ship | *Odin* | ex Danish 74; 1825 sold |
| 1811-1825 | receiving ship | *Braave* | ex Dutch frigate; 1825 sold |
| 1811-1817 | receiving ship | *Cuba* | ex Spanish frigate; 1817 sold |
| 1811-1817 | prison ship | *Pompee* | ex French 80 1817 BU |

NOTE: Supposed to be the ship from which the British sailor's slang name for Portsmouth - 'Pompey' - comes

| | | | |
|---|---|---|---|
| 1811-1817 | receiving ship | *Speedy* | ex purchased sloop; 1817 sold |
| 1812-1817 | receiving ship | *Squirrel* | ex 24; 1817 sold |
| 1812-1816 | prison ship | *Waldemar* | ex Danish 80; 1816 BU |
| 1813-1815 | salvage vessel | *Caroline* | ex frigate, 36; 1815 BU (at Deptford by then) |
| 1814-1816 | temporary prison ship | *Blake* | (ex *Bombay*) ex 74; 1816 sold |
| 1814-1821 | temporary prison ship | *Atlas* | ex 98/74; 1815 powder magazine; 1821 BU |
| 1814-1823 | receiving ship | *Mars* | ex 74; 1823 BU |
| 1815-1817 | victualling depot | *Arve Princen* | (ex *Heir Apparent Frederick*) ex Danish 74; 1817 sold |
| 1815-1817 | hulk | *Bedford* | ex 74; 1817 BU |
| 1816-1837 | victualling & accommodation for Officers | *Prince* | ex 98 1837 BU |
| 1816-1848 | convict ship | *Leviathan* | ex 74; 1846 target ship; 1848 sold |
| 1817-1839 | sheer hulk | *Prince George* | ex 98; 1839 BU |
| 1819-1838 | convict hospital ship | *Racoon* | ex sloop; 1838 sold |
| 1819-1845 | receiving ship | *Swiftsure* | ex 74; 1845 BU |
| 1819-1852 | lazaretto | *Prometheus* | ex fireship/sloop; later? receiving ship?; 1839 renamed *Veteran*; 1852 BU |
| 1819-1840 | convict ship? | *Ganymede* | ex French corvette; 1840 BU |
| 1819-1853 | convict ship | *York* | ex 74; 1853 BU |
| 1822-1836 | receiving ship | *Emerald* | ex frigate, 36; 1836 BU |
| 1822-1837 | receiving ship | *Immortalite* | ex French frigate; 1837 sold |
| 1822-1834 | receiving ship | *Bucephalus* | ex frigate, 32; 1834 BU |
| 1823-1841 | slop ship | *Barrosa* | ex frigate, 36; later receiving ship/ordnance depot; 1841 sold |
| 1823-1851 | receiving ship | *Topaze* | ex French frigate; 1851 target then BU |
| 1823-1855 | receiving ship | *Blake* | (ex *Bombay*) ex 74; 1855 BU |
| 1824-date | flag & receiving ship | *Victory* | ex 100; 1920s rebuilt and preserved as national monument; still in existence and in naval service |
| 1825-1838 | church ship | *Venerable* | ex 74; 1838 BU |
| 1826-1861 | receiving ship | *Victorious* | ex 74; 1861 BU |
| 1826-1831 | lazaretto | *Duncan* | ex 74; 1831 to Sheerness |
| 1828-1835 | convict ship | *Alonzo* | from Woolwich; 1835 to Leith |

NMM photo negative X.1468. 'Vernon', a picture by W. Mackenzie Thomson showing the Navy's premier torpedo and mining establishment at some stage after 1886 when the old sailing frigate *Vernon* (in the foreground in this picture) was joined by the screw line of battle ship *Donegal* (which took her name whilst the frigate became *Actaeon*) and before these wooden vessels were retired in the early 1920s in favour of a shore establishment

| Years | Role | Name | Details |
|---|---|---|---|
| 1829-1853 | receiving ship | *Menai* | ex 28/sloop; 1852 target; 1853 BU |
| 1830-1835 | gunnery training ship | *Excellent* | ex 74; 1835 BU |

NOTE: This is the ship after which the Navy's gunnery school was named

| Years | Role | Name | Details |
|---|---|---|---|
| 1831-1852 | receiving ship | *Mersey* | ex 26; 1852 BU |
| 1831-1836 | lazaretto | *Albion* | ex 74; 1836 BU |
| 1831-1843 | temporary lazaretto | *Anson* | ex 74; by 1843 to Chatham, then to Tasmania |
| 1833-1849 | receiving ship | *Success* | ex 28; 1849 BU |
| 1833-1865 | receiving ship/ temporary hulk | *Blanche* | ex frigate, 38; 1865 sold to BU |
| 1834-1861 | gunnery training | *Excellent* | (ex *Boyne*) ex 104; 1859 renamed *Queen Charlotte*; 1861 BU |
| 1838-1867 | coal depot | *Maidstone* | ex frigate, 36; 1867 BU |
| 1838-1860 | receiving ship | *Dryad* | ex frigate, 36; 1860 BU |
| 1839-1846 | receiving ship | *Etna* | ex bomb vessel; possibly sent to Liverpool?; 1846 BU |
| 1841-1897 | lazaretto at the Motherbank | *Menelaus* | from Sheerness; 1897 sold |
| 1841-1860 | convict ship | *Briton* | ex frigate, 38; 1860 target, then BU |
| 1844-1861 | convict ship | *Stirling Castle* | ex 74; from Plymouth; 1861 BU |
| 1845-1866 | lazaretto at the Motherbank | *Tyrian* | ex brig; 1864 receiving ship; 1866 to the Coastguard |
| 1846-1906 | store depot | *Belvidera* | ex frigate, 36; 1852 receiving ship; 1906 sold |
| 1848-1905 | coal depot | *Malabar* | ex 74; 1883 renamed *Myrtle*; 1905 sold |
| 1848-1868 | Ordinary depot ship | *Illustrious* | ex 74; 1853 hospital ship; 1859 Ordinary guard ship; 1868 BU |
| 1850-1895 | receiving ship (& coal depot?) | *Blonde* | ex frigate, 38; 1866 temporary hospital ship; 1870 renamed *Calypso*; 1895 sold |
| 1852-by 1890 | coal depot | *Orestes* | ex sloop; later designated C.28; by 1890 to Plymouth |
| 1853-1860 | coal depot (& receiving ship?) | *Pitt* | ex 74; on to Portland |
| 1855-1859 | hospital ship | *Britannia* | ex 120; 1859 to Dartmouth |
| 1855-1867 | Roman Catholic chapel ship | *Thalia* | ex frigate, 38; 1867 BU |
| 1855-1886 | accommodation ship | *Eolus* | from Sheerness; 1861 lazaretto; 1886 BU |
| 1856-1892 | receiving ship | *Bellerophon* | (ex *Waterloo*) ex 80; 1892 sold |
| 1857-1866 | training ship | *Ganges* | ex 84; 1866 to Plymouth |
| 1858-1865 | gunnery trials ship | *Alfred* | (ex *Asia*) ex 74/50; 1865 BU |
| 1859-1908 | Ordinary guard ship | *Asia* | ex 84; 1908 sold for BU |
| 1859-1908 | coal depot | *Lily* | ex brig; Ca 1890 designated C.29; later (probably at Chatham) redesignated C.15; 1908 BU (at Chatham) |
| 1859-1867 | harbour police ship | *Champion* | ex sloop; 1864 lent to the Committee on floating obstructions and used for explosive experiments; 1867 BU |
| 1859-1892 | gunnery training ship | *Excellent* | (ex *Queen Charlotte*) ex 104; 1892 sold |
| 1860-1862 | target ship | *Sirius* | ex frigate, 38; 1862 BU |
| 1860-1860 | target ship | *Undaunted* | ex frigate, 38; 1860 BU |
| 1860-1893 | coal depot | *Carnatic* | ex 74; 1886 floating magazine for the War Office; 1891 to the Admiralty again; 1893 sold |
| 1860-1906 | coal depot | *Camperdown* | (ex *Trafalgar*) ex 104/110; 1882 renamed *Pitt*; 1906 sold |
| 1861-1864 | receiving ship | *Sultan* | ex 74; 1862 target; 1864 BU |
| 1861-1862 | target | *Java* | ex frigate, 50; 1862 BU |
| 1862?-1906 | training ship | *Saint Vincent* | ex L20; 1906 sold for BU |

NOTE: Gave her name to the training establishment for boys at Portsmouth

| Years | Role | Name | Details |
|---|---|---|---|
| 1862-1868 | target ship | *Siren* | ex brig; 1868 BU |
| 1862-1877 | water police ship | *Juno* | ex 26; 1877 seagoing training ship |
| 1863-1901 | target ship | *Thunderer* | ex 84; 1869 renamed *Comet*; 1870 renamed *Nettle*; 1901 sold |
| 1863-1908 | gunnery training ship | *Calcutta* | ex 84; 1908 sold |
| 1863-1909 | harbour service | *Duke of Wellington* | ex 110; 1909 sold for BU |
| 1864-1885 | 'lavatory' | *Laurel* | ex frigate, 38; 1885 sold to BU |
| 1864-1869 | target ship | *America* | ex 74; 1869 BU |
| 1866-1895 | Roman Catholic chapel ship | *Melampus* | ex frigate, 38; from the Coastguard at Southampton; 1886 to the War Office as an ordnance store; 1891 returned to the Admiralty; 1895 to Chatham |
| 1866-1889 | hospital ship | *Acteon* | ex 26; 1874 used for torpedo experiments and attached to *Vernon*; 1889 sold |
| 1866-1897 | powder depot | *Grampus* | (ex *Tremendous*) ex 74/50; 1883 War Department mine depot; 1897 sold |
| 1868-1879 | chapel ship | *Helena* | ex brig; 1879 to Ipswich |

| | | | |
|---|---|---|---|
| 1872-1875 | powder depot | *Latona* | ex frigate, 38; from Sheerness; 1874 to be training ship (but not done?); 1875 BU |
| 1873-1923 | torpedo instruction ship | *Vernon* | ex frigate, 50; from Portland; 1886 renamed *Actaeon*; 1923 sold |

NOTE: Gave her name to the Navy's torpedo and mining (and later diving) school at Portsmouth

| | | | |
|---|---|---|---|
| 1874-1904 | hulk? | *Hannibal* | ex 90; 1904 sold for BU |
| 1878-1924 | Instruction ship | *Marlborough* | ex 120/screw line of battle ship; 1904 renamed *Vernon II*; 1924 sold and foundered on her way to BU |
| 1905-1920 | training ship for boy artificers | *Hindostan* | ex 80; from Dartmouth; for boy artificers renamed *Fisguard III*; 1920 renamed *Hindostan*; 1921 sold for BU |

## Hulks at Plymouth Dockyard

| | | | |
|---|---|---|---|
| 1689-1689 | hulk | *Dover Prize* | ex French; 1689 wrecked |
| 1690-1730 | purchased as hulk | *Plymouth* | 1730 BU |
| 1699-1731 | hulk | *Thunderbolt Prize* | ex French 32; 1731 BU |
| 1702-1712 | tender to hulk | *Saudadoes Prize* | ex French 40/36; 1708 hulk; 1712 breakwater |
| 1731-1763 | hulk | *Jersey* | ex 50; 1763 breakwater |
| 1740-1749 | harbour hulk | *Enterprise* | ex 40; 1745 hospital ship; 1749 BU |
| 1740-1745 | hulk | *Saphire* | ex 40; 1745 sold |
| 1742-1783 | hulk | *Berwick* | ex 70; 1783 BU |
| 1764-1783 | hulk | *Barfleur* | ex 90; 1783 BU |
| 1775-1784 | receiving ship | *Saint Ann* | ex French 64; 1784 sold |
| 1778-1811 | receiving ship | *Warspite* | renamed *Arundel*; ex 74; 1811 BU |
| 1778-1782 | prison ship | *Cambridge* | ex 80; 1780 receiving ship; 1782 returned to service as an 80 (but see under 1790 below) |
| 1779-1792 | receiving ship for convicts | *Dunkirk* | ex 60; 1785 receiving ship; 1792 sold |
| 1779-1803 | temporary hulk | *Chichester* | ex 70; 1783 receiving ship; 1803 BU |
| 1781-1783 | slop ship | *Basilisk* | (ex *Grasshopper*) ex fireship (ex sloop); 1783 sold |
| 1780-1783 | for the reception of sick men | *Tiger* | ex 74; 1783 sold [PLANS: 1778: decks.] |
| 1783-1806 | receiving ship | *Yarmouth* | ex 70/74; 1806 sold |
| 1784-1809 | sheer hulk | *Princessa* | ex Spanish 70; 1809 BU [PLANS: 1809: Profile/main deck/quarter deck/forecastle.] |
| 1787-1788 | receiving ship | *Hero* | ex 74; 1788 to Chatham |
| 1787-1813 | temporary hulk | *Panther* | ex 60; 1813 BU |
| 1788-1808 | receiving ship | *Rippon* | ex 60; 1808 BU |
| 1789-1811 | receiving ship | *Medway* | ex 60; 1802 renamed *Arundel*; 1811 BU |
| 1790-1808 | receiving ship | *Cambridge* | ex 80; 1808 BU |
| 1793-1797 | receiving ship | *Medusa* | ex 50; 1797 seagoing hospital ship |
| 1793-1797 | convalescent ship | *Chatham* | ex 50; 1797 to Falmouth |
| 1794-1811 | hospital ship | *Engageante* | ex French frigate; 1811 BU |
| Ca 1794-1815 | hospital ship | *Caton* | ex French 64; 1815 sold |
| 1794-1814 | prison ship | *Bienfaisant* | ex French 64; 1814 BU [PLANS: 1794: Decks.] |
| 1794-1814 | prison ship | *Prudent* | ex 64; 1802 powder hulk; 1814 sold |
| 1795-1820 | prison ship or powder ship? | *Fortitude* | ex 74; 1820 BU |
| 1796-1814 | prison ship | *Europe* | ex 64; 1814 BU |
| 1796-1802 | prison ship? | *Commerce de Marseilles* | ex French 110; 1802 BU |
| 1797-1814 | prison ship | *San Ysidro* | ex Spanish 72; 1814 sold |
| 1798-1801 | temporary prison hospital ship | *Iphigenia* | ex frigate, 32; 1801 troopship |
| 1798-1811 | receiving ship | *Saldanha* | ex Dutch frigate; 1811 BU |
| 1798-1811 | hulk | *Resolue* | ex French frigate; 1811 BU |
| 1798?-1811 | harbour service? | *Myrmidon* | ex 24; 1811 BU |
| After | temporary | *Tholin* | ex Dutch 36; 1811 BU |
| 1798-1811 | receiving ship | | |
| 1799-1811 | hulk | *Proserpine* | ex French frigate; 1811 BU |
| 1800-1814 | prison ship? | *San Nicholas* | ex Spanish 80; 1814 sold |
| 1800-1804 | convalescent ship | *Prince Frederick* | ex Dutch 64; 1804 to Berehaven |
| 1801-1806 | receiving ship | *Musette* | ex French sloop; 1803 floating battery for the River Yare; 1806 sold |
| 1805-1817 | for the military medical staff | *Hornet* | ex sloop; 1817 sold |
| 1805-1825 | receiving ship | *Carnatic* | ex 74; 1815 renamed *Captain*; 1825 BU |
| 1806-1817 | receiving ship | *Egyptienne* | ex French 24 pdr frigate; 1817 sold |
| 1807-1811 | receiving ship | *Santa Gertruyda* | ex Spanish frigate; 1811 BU |
| 1808-1816 | prison ship | *Brave* | ex French 80/84; 1814 powder ship; 1816 BU |
| 1808-1816 | prison ship? | *Hector* | ex 74; 1816 BU |
| 1808-1822 | powder ship | *Alexandre* | ex French 74; 1822 sold |
| 1809-1813 | receiving ship | *Captain* | ex 74; 1813 burnt by accident & BU |
| 1810-1828 | receiving ship | *Intrepid* | ex 64; 1828 BU |
| 1810-1842 | sheer hulk | *Sans Pareil* | ex French 80; 1842 BU [PLANS: 1842: Profile/orlop/lower deck/lines/after part of upper deck/forward part of upper deck/scantlings.] |
| 1810-1814 | receiving ship | *Triton* | ex 32; from Woolwich; 1814 sold |
| 1811-1815 | receiving ship | *Clara* | ex Spanish 12 pdr frigate; 1815 sold |
| 1811-1815 | receiving ship | *Amsterdam* | ex Dutch frigate; from Cork; 1815 sold |
| 1811-1828 | receiving ship | *Virginie* | ex French 18 pdr frigate; 1827 sold; 1828 resold |
| 1811-1827 | receiving ship | *Peterell* | ex sloop; 1827 sold |
| 1811-1814 | receiving ship | *Rattlesnake* | ex sloop; 1814 sold |
| 1811-1814 | prison ship | *Ganges* | ex 74; 1814 sold |
| 1811-1819 | powder hulk | *Mont Blanc* | ex 74; 1819 sold |
| 1812-1821 | prison ship | *Vanguard* | ex 74; 1814 powder ship; 1821 BU |
| 1813-1820 | prison ship | *Temeraire* | ex 98; 1820 to Sheerness |
| 1813-1816 | temporary hospital ship | *Medusa* | ex frigate, 32; 1816 BU |
| 1813-1825 | prison ship | *Britannia* | ex 100; 1815 flag & receiving ship and renamed *Saint George*; 1825 renamed *Barfleur*; 1825 BU |
| 1813-1824 | powder hulk | *Gibraltar* | ex Spanish 80; 1824 to Pembroke |
| 1813-1818 | temporary prison ship | *Neptune* | ex 98; 1818 BU |
| 1811-1821 | army depot | *Caesar* | ex 80; 1821 BU |
| 1814-1835 | prison hospital ship | *Renown* | ex 74; 1835 BU (by this time at Deptford?) |
| 1814-1816 | sheer hulk | *Lion* | ex 64; 1816 to Sheerness |
| 1815-1832 | receiving ship (temporary hulk) | *Diadem* | ex 64; 1832 BU |
| 1816-1826 | diving bell ship | *Resolute* | ex gunbrig; 1826 to Bermuda |
| 1825-1841 | receiving ship | *Royal Sovereign* | ex 100; renamed *Captain*; 1841 BU |
| 1825-1834 | lazaretto | *Hannibal* | ex 74; ?to Pembroke?; 1834 BU |
| 1825-1860 | receiving ship | *Active* | ex frigate, 36; 1833 renamed *Argo*; 1860 BU |
| 1826-1841 | receiving ship/ slop ship | *Phoebe* | ex frigate, 36; 1841 sold |
| 1826-1865 | receiving ship | *Egeria* | ex sloop; 1843 breakwater department; later police ship; 1865 BU |
| 1826-1836 | breakwater department | *Carnation* | ex brig; 1836 sold |
| 1827-1865 | receiving ship | *Vigo* | ex 74; 1865 BU |
| 1831-1840 | coal depot ship | *Malta* | ex French 84; 1840 BU |
| 1831-1863 | receiving ship | *Lively* | ex frigate, 38; 1863 sold |
| 1831-1870 | lazaretto | *Jupiter* | ex 50; 1832 seagoing troopship; 1846 coal depot; 1870 BU |
| 1831-1837 | coal depot | *Tamar* | ex 26/sloop; 1837 sold |
| 1836-1862 | coal depot | *Pallas* | ex frigate, 36; 1862 sold |
| 1836- | ____ | *Medusa* | ex 38; |
| 1839-1844 | convict ship | *Stirling Castle* | ex 74; 1844 to Portsmouth |
| 1839-1842 | depot | *Nightingale* | ex packet; 1842 sold |
| 1839-1906 | receiving ship/ coal depot | *Impregnable* | ex 104; later flag ship; 1862 training ship; 1883 renamed *Kent*; 1891 renamed *Caledonia*; 1906 sold |
| 1842-1857 | temporary sheer hulk | *Spartiate* | ex French 80/76; 1857 BU |
| 1842-1868 | receiving ship | *Bellona* | (ex *Indus*) ex 74; 1868 BU |
| 1845-1864 | Keyham breakwater | *Lapwing* | ex brig; 1864 sold |

| | | | |
|---|---|---|---|
| 1840s?-1850 | lazaretto | *Redbreast* | ex tender from Sheerness; 1850 sold |
| Late 1840s?-1884 | training ship | *Agincourt* | ex 74; 1866 cholera hospital; 1870 receiving ship; 1884 sold to BU |
| 1846-1850 | provision depot | *Portland* | ex frigate, 50/60; 1850 fitted for service as a frigate again |
| 1846-1853 | provision depot | *Madagascar* | ex frigate, 38; 1853 to Rio de Janeiro |
| 1846-1875 | provision depot | *Andromache* | ex 28; 1854 powder depot; (later at Pembroke but then returned to Plymouth); 1875 BU |
| 1846-1863 | hospital ship | *Lancaster* | ex frigate, 50/60; later to the Ordinary; 1864 sold |
| 1846-____ | ____ | *Jupiter* | ex 50; ____ |
| 1846-____ | ____ | *Andromeda* | ex 38; ____ |
| 1847-1862 | provision depot | *Tyne* | ex 26; 1850 store ship; 1862 sold |
| After 1848-____ | ____ | *Agincourt* | ex 74; ____ |
| 1850-1883 | lazaretto | *Bacchus* | (ex *Arethusa*) from Liverpool; 1852 coal depot; 1883 sold to BU |
| 1852-1870 | lazaretto | *Lavinia* | from Liverpool; 1852 coal depot; 1870 sunk & wreck sold |
| 1853-1907 | coal depot (& receiving ship?) | *Nimrod* | ex 20; later yard craft C.1; later C.76?; 1907 sold |
| 1854-1861 | temporary hospital ship | *Inconstant* | ex 36; 1861 to Cork |
| 1862-1904 | coal depot | *Diligence* | ex transport; later to Portsmouth?; 1904 sold |
| 1863-1897 | powder depot | *Conquestador* | ex 74; from the War Office at Purfleet; 1897 sold |
| 1865-1906 | water police ship | *Leda* | ex frigate, 38; 1906 sold |
| 1866-1929 | training ship | *Ganges* | ex 84; from Portsmouth; 1906 renamed *Tenedos III*; 1910 renamed *Indus V*; 1922 renamed *Impregnable III*; 1929 sold for BU |
| 1866-1922 | accommodation hulk (tender to *Impregnable*) | *Circe* | ex frigate, 38; 1885 swimming bath; 1916 *Impregnable IV*; 1922 sold for BU |
| 1869-1908 | gunnery training ship | *Windsor Castle* | ex 110; renamed *Cambridge*; 1908 sold for BU |
| 1869-1891 | flag & receiving ship | *Royal Adelaide* | ex 110; 1891 to Chatham |
| 1869-1892 | floating factory | *Cambrian* | ex 36; 1892 sold |
| 1870-1902 | coal hulk | *Seahorse* | (became *Lavinia*) ex frigate, 38; 1902 sold |
| 1876-1921 | training ship for the Devon & Cornwall Training Ship Society | *Mount Edgcumbe* | (ex *Conway*, ex *Winchester*) ex frigate, 60; from Liverpool; 1921 sold |
| By 1890- | coal depot | *Orestes* | from Portsmouth; later yard craft Ca 1905 C.28; Ca 1905 sold |

## Hulks at Pembroke Dockyard

NOTE: All the lazarettos were based at Milford

| | | | |
|---|---|---|---|
| 1757-1762 | floating battery at Milford | *Chester* | ex 50; 1762 back in service as a 50 |
| 1805-1822 | lazaretto | *Siren* | ex frigate, 32; 1822 BU |
| 1813-1828 | accommodation for clerks, etc | *Lapwing* | ex frigate, 28; from Cork; 1828 BU |
| 1813-1821 | lazaretto | *Perlin* | ex Danish 18 pdr frigate; 1821 to Liverpool |
| 1813-1850 | lazaretto | *Triumph* | ex 74; 1850 BU |
| 1814-1828 | lazaretto | *Otter* | ex sloop; 1828 sold |
| 1814-1836 | lazaretto | *Santa Margarita* | ex Spanish 12 pdr frigate; 1836 sold [PLANS: 1815: Lower deck/upper deck/quarter deck & forecastle. ND: Decks.] |
| 1816-1817 | quarantine service | *Enchantress* | ex purchased brig; from Bristol?; 1817 to revenue service in the Blackwater. |
| 1824-1841 | coal depot | *Tortoise* | (ex *Sir Edward Hughes*) ex storeship (ex purchased frigate); 1841 to Sheerness |
| 1824-1827 | lazaretto | *Akbar* | (ex *Cornwallis*) ex purchased frigate; 1827 to Liverpool |
| 1824-1850 | lazaretto | *Dragon* | ex 74; 1832 marine barrack & receiving ship; 1842 *Fame*; 1850 BU |
| 1824-1827 | lazaretto | *Newcastle* | ex frigate, 60; 1827 to Liverpool |
| 1824-1836 | lazaretto | *Gibraltar* | ex Spanish 80; From Plymouth; 1836 BU |
| 1825-1831 | lazaretto | *Dreadnought* | ex 98/104; 1831 to Greenwich |
| 1825-1845 | lazaretto | *Ville de Paris* | ex 110; 1845 BU |
| 1825-1868 | lazaretto | *Saturn* | ex 74; 1845 flag & receiving ship; 1850 guard ship; 1868 BU |
| 1825-1846 | lazaretto | *Milford* | ex 74; 1846 BU |
| 1831-1835 | lazaretto | *Salsette* | ex frigate, 36; 1835 to Woolwich |
| 1836-1854 | lazaretto | *Mulgrave* | ex 74; 1844 powder ship; 1854 BU |
| 1846-1882 | lazaretto/guard ship | *Hope* | ex brig; 1863 to receive boilers of ships; 1882 BU |
| 1850s-1871 | powder ship | *Andromache* | ex 28; from Plymouth; 1871 to Plymouth |
| 1858-1863 | drill ship | *Active* | ex 36; 1863 to Sunderland |

## Hulks at Harwich Dockyard

| | | | |
|---|---|---|---|
| 1702-1713 | purchased as hulk | *Saint George* | 1713 to Chatham |
| 1708-1709 | hulk | *Kingfisher* | from Deptford; 1709 to Sheerness |
| 1709-1709? | hulk | *Josiah* | from Woolwich & returned there |
| 1744-1745 | hulk | *Winchester* | ex 50; 1745 to Sheerness |
| 1835 | breakwater | *Ethalion* | ex 36; from Woolwich |

## Naval Hulks, in Britain but not at the Royal Dockyards

NOTE: Listed anti-clockwise round the coast, starting from the Thames

### At the Tower of London:

| | | | |
|---|---|---|---|
| 1742-1748 | hospital ship | *Solebay* | ex 24; 1748 sold |
| 1757-1762 | hospital ship | *Phoenix* | ex 24; 1760 sold |
| 1775-1776 | hulk | *Lion/Lyon* | ex hoy from Portsmouth; 1776 survey ship |
| 1776-1783 | receiving ship/hospital ship | *Nightingale* | ex purchased 24; from Woolwich; 1783 breakwater |
| 1791-1806 | receiving ship for impressed men | *Enterprise* | ex frigate, 28; 1806 to Deptford (1807 BU) |
| 1803-1816 | receiving ship | *Resource* | ex frigate, 28; renamed *Enterprise*?; 1816 BU |
| 1818-1850 | receiving ship for seamen | *Perseus* | ex 22; from Deptford; 1850 BU |
| 1850-1861 | receiving ship | *Crocodile* | ex 28; from Deptford?; 1861 sold |

### In the Thames:

| | | | |
|---|---|---|---|
| 1841-1846 | hulk at Limehouse | *Leven* | from Woolwich; 1848 BU |
| 1844-1872 | hulk for coal | *Carron* | ex brig/paddle steamer ;1872 deleted |
| 1861-1903 | RNR | *President* | ex 60; first in the City Canal; training ship 1871 in the West India Docks at Poplar; 1903 renamed *Old President*; 1903 sold |

### Gravesend:

| | | | |
|---|---|---|---|
| 1803-1814 | receiving ship | *Corso* | ex Spanish sloop; 1814 sold |
| 1814-1814 | receiving ship | *Galgo* | ex purchased sloop; 1814 sold |
| 1815-1817 | receiving ship | *Port Mahon* | ex Spanish brig; 1817 to Thames Police |

### Yarmouth (Great Yarmouth):

| | | | |
|---|---|---|---|
| 1811-1815 | receiving ship | *Iris* | ex frigate, 32; renamed *Solebay?*; 1815 to the Marine Society |

### Hull (not sent?):

| | | | |
|---|---|---|---|
| 1818-1818 | tender | *Brev Drageren* | ex Danish brig; 1818 to the Thames for the Army |

### South Shields:

| | | | |
|---|---|---|---|
| 1860-1895 | training ship | *Castor* | ex frigate, 36; for Coast Volunteers; later for RNR; 1895 to Sheerness and into reserve (1902 sold) |
| 1863-1908 | RNR training ship | *Active* | ex 36; 1867 renamed *Tyne*; 1867 renamed *Barham*; 1908 sold. |

## In the Tyne:

| | | | | |
|---|---|---|---|---|
| 1868-1875 | hulk | *Cornwall* | | ex 74; from the London School Ship Society; renamed *Wellesley*; 1875 BU |
| 1874-1914 | hulk | *Boscawen* | | ex 70; from Southampton; renamed *Wellesley*; 1914 burnt and BU |

## Sunderland:

| | | | | |
|---|---|---|---|---|
| 1863-1906 | drill ship | *Active* | | ex 36; for Naval Volunteers; 1867 name changed to *Tyne*, then *Durham*; 1906 sold |
| 1861-1897 | training ship | *Trincomalee* | | ex frigate, 38/corvette 26; for the Royal Naval Volunteers; 1897 sold to Wheatley Cobb and to Falmouth as training ship '*Foudroyant*' |

## Leith:

| | | | | |
|---|---|---|---|---|
| 1813-1816 | receiving ship | *Latona* | | ex frigate, 38; 1816 sold |

## Dundee:

| | | | | |
|---|---|---|---|---|
| 1831-1833 | lazaretto | *Sybille* | | ex French 40/48; 1833 sold |
| 1860-1875 | training ship | *Brilliant* | | ex frigate, 36; for Coast Volunteers; 1875 to Inverness |
| 1873-date | drill ship | *Unicorn* | | ex frigate, 38; from Chatham; for RNR; 1941 renamed *Cressy*; 1959 renamed *Unicorn*; now being preserved at Dundee by a local association |

## Aberdeen:

| | | | | |
|---|---|---|---|---|
| 1861-1871 | training ship | *Conway* | | from Liverpool; renamed *Worcester*; for RNR; 1871 BU |
| 1870-1904 | training ship | *Clyde* | | ex frigate, 38; for RNR; 1904 sold |

## Inverness:

| | | | | |
|---|---|---|---|---|
| 1875-1908 | training ship | *Brilliant* | | ex frigate, 36; from Dundee; for RNR; 1889 renamed *Briton*; 1906 lent to the War Office as sleeping accommodation for militia; 1908 sold |

## Greenock:

| | | | | |
|---|---|---|---|---|
| 1814-1815 | receiving ship | *Ornen* | | ex Danish schooner; 1815 to Clyde Marine Society |
| 1854-1861 | depot ship ('rendezvous') | *Atholl* | | ex 28; from Sheerness; 1861 to Plymouth in reserve (BU 1863) |

## Liverpool:

| | | | | |
|---|---|---|---|---|
| 1795-1802 | receiving ship | *Acteon* | | ex 44; 1802 sold |
| 1815-1836 | lazaretto | *Experiment* | | from Falmouth; 1836 sold |
| 1821-1846 | lazaretto? | *Pearlen* | | ex Danish frigate; from Pembroke; 1846 sold |
| 1827-1864 | hulk | *Akbar* | | (ex *Cornwallis*) ex purchased frigate; from Pembroke; 1864 sold? |
| By 1830- | cutter tender | *Redbreast* | | ____ ? |
| 1836-1852 | lazaretto | *Lavinia* | | ex frigate; 1852 to Plymouth |
| 1836-1850 | lazaretto? | *Arethusa* | | ex frigate, 38; 1844 renamed *Bacchus*; 1850 to Plymouth |
| 1846-1863 | lazaretto | *Druid* | | ex frigate, 38; 1863 sold |
| 1863-1926 | RNR training ship | *Eagle* | | from Falmouth; 1918 renamed *Eaglet*; 1926 BU |

## Bristol:

| | | | | |
|---|---|---|---|---|
| 1813-1816 | receiving ship | *Enchantress* | | ex purchased brig; 1816 to Milford (Pembroke) |
| 1861-1911 | training ship | *Daedalus* | | ex frigate, 38; for the RNR; 1911 sold to BU |

## Falmouth:

| | | | | |
|---|---|---|---|---|
| 1797-1805 | convalescent ship | *Chatham* | | ex 50; from Plymouth; 1805 to Chatham |
| 1805-1815 | storeship | *Experiment* | | ex 44; 1815 to Liverpool |
| 1823-1851 | depot | *Astrea* | | ex frigate, 36; 1851 BU |
| 1832-1851 | coal depot | *Aurora* | | ex French frigate; 1851 BU |
| 1857-1863 | ____ | *Eagle* | | ex 74/50; 1863 to Liverpool RNR |

## Dartmouth:

| | | | | |
|---|---|---|---|---|
| 1859-1869 | training ship | *Britannia* | | from Portsmouth; for cadets; 1869 BU |
| 1864-1905 | training ship | *Hindostan* | | ex 80; 1905 to Portsmouth |
| 1869-1909 | training ship | *Prince of Wales* | | ex 110; renamed *Britannia*; for cadets; 1909 sold for BU |

## Portland:

| | | | | |
|---|---|---|---|---|
| 1860-by 1877 | coal depot | *Pitt* | | ex 74; from Portsmouth; later returned to Portsmouth; 1877 BU |
| 1860-1871 | hulk | *Hyacinth* | | ex sloop; 1871 BU |
| 1863-1873 | hulk | *Vernon* | | ex frigate, 50; 1873 to Portsmouth |

## Southampton:

| | | | | |
|---|---|---|---|---|
| 1862-1874 | drill ship | *Boscawen* | | ex 70; 1874 to the Tyne |

## Hulks in Ireland:

| | | | | |
|---|---|---|---|---|
| 1690-____ ? | hulk at Kinsale | *Saint David* | | ex 54; later to Woolwich |
| 1745-1749 | hulk at Kinsale | *Salisbury* | | ex 50; 1749 sold |
| 1797-____ | flagship/receiving ship at ____ ? | *Princess* | | ex Dutch storeship; ____ ? fate uncertain |
| 1804-1817 | hospital ship at Berehaven | *Prince Frederick* | | ex Dutch 64; from Plymouth; 1817 sold |
| 1806-1811 | storeship at Cork | *Amsterdam* | | ex Dutch frigate; 1811 to Plymouth |
| 1806-1834 | storeship at Haulbowline | *Eurus* | | ex Dutch frigate; 1834 BU |
| 1810-1813 | salvage vessel (for the wreck of *Britannia*) *Lapwing* | | | ex 28; 1813 to Pembroke |
| 1810-1816 | receiving ship | *Deptford* | | from the Tower of London; 1816 to the Hibernian Marine Society |
| 1810-by 1840s | depot at Waterford | *Trial* | | ex cutter/armed vessel; by 1840s to Callao, Peru |
| 1810-1814 | salvage vessel | *Alligator* | | ex frigate, 28; afterwards laid up at Plymouth; 1814 sold at Cork |
| 1815-1823 | hospital ship laid up at Cork | *Trent* | | ex frigate, 36; 1823 BU |
| 1822-1837 | convict ship at Cork | *Surprise* | | ex frigate, 38; 1837 sold |
| 1824-1837 | convict ship at Cork | *Essex* | | ex American frigate; 1837 sold |
| 1848-1878 | coal depot at Kingstown | *Wolf* | | ex sloop; 1878 BU |
| 1861-1867 | receiving ship & temporary flagship at Cork *Inconstant* | | | ex 36; 1863 sold; 1867 BU |
| 1863-1875 | powder depot | *Mermaid* | | ex frigate, 38; from War Department at Dublin at Purfleet; 1875 BU |
| 1877-1904 | hulk at Queenstown | *Alarm* | | from Plymouth; 1900 landing stage at Berehaven; 1904 sold |

# Hulks at Dockyards Abroad, etc

## Cadiz:

| | | | | |
|---|---|---|---|---|
| 1684-1690 | hulk | *Maria Prize* | | (Algerine, taken 1684); 1690 sold at Cadiz |
| 1694-1701 | hulk purchased at Cadiz | *Asia* | | 1701 lost [420 tons, 23 men] |
| 1694-1701 | hulk purchased at Cadiz | *Loyalty* | | 1701 lost |

## Lisbon:

| | | | | |
|---|---|---|---|---|
| 1703-Ca 1715 | hulk | *Content Prize* | | ex French 70/62; Ca 1715 sold? |

## Gibraltar:

| | | | | |
|---|---|---|---|---|
| 1757-Ca 1764 | hulk | *Ipswich* | | ex 70/64; Ca 1764 BU |
| 1795-____ ? | prison ship | *Aurore* | | ex French frigate; fate uncertain |
| 1798-1810 | hulk | *Guerier* | | ex French 74; 1810 BU |
| 1805-1818 | hulk classed as a sloop | *San Juan/ San Juan Nepomuceno* | | ex Spanish 74; 1818 sold |
| 1841-1857 | guard ship | *Reindeer* | | ex brig; 1857 sold |
| 1842-1884 | convict ship | *Owen Glendower* | | ex frigate, 36; 1884 sold |

| | | | |
|---|---|---|---|
| 1847-1883 | guard ship | *Samarang* | ex 28; 1883 BU |
| 1847-1860 | convict ship | *Euryalus* | from Chatham; 1859 renamed *Africa*; 1860 sold |

## Port Mahon:

| | | | |
|---|---|---|---|
| 1710-____ ? | hulk | *Elephant* | ex French storeship; fate uncertain |
| 1741-1744 | hospital ship | *Sunderland* | ex 60; from Chatham; 1744 condemned |
| 1741-1754 | hospital ship | *Sutherland* | (ex *Reserve*) ex 50; 1754 BU |
| 1742-1748 | hospital ship | *Rochester* | ex 50; 1744 renamed *Maidstone*; 1748 BU |

## Malta:

| | | | |
|---|---|---|---|
| 1800-1802 | prison ship | *Dego* | ex prize small line of battle ship; 1802 sold |
| 1807-1807 | receiving ship | *Lancaster* | purchased 64; not actually sent, and became a victualler instead |
| 1808-1816 | receiving ship | *Trident* | ex 64; 1816 sold |
| 1832-1857 | receiving ship | *Ceylon* | (ex *Bombay*) ex purchased frigate; 1857 sold |
| 1855-1902 | guard ship/ receiving ship | *Hibernia* | ex 120; 1902 sold |

## Alexandria, Egypt:

| | | | |
|---|---|---|---|
| 1837-1841 | coal depot | *Ariadne* | ex sloop; 1841 sold |

## Constantinople (Istanbul):

| | | | |
|---|---|---|---|
| 1855-1856 | coaling hulk purchased at Constantinople | ____ (no name?); 1856 sold | |
| 1855-1857 | store & receiving ship | *Melampus* | ex frigate, 38; ? never sent?; 1857 to Coastguard |

## Balaklava:

| | | | |
|---|---|---|---|
| 1855-1856 | coal depot purchased at Balaklava | *Medora* | ex mercantile barque; 1856 sold |

## Sierra Leone:

| | | | |
|---|---|---|---|
| 1830s-1841 | receiving ship | *Conflict* | ex brig; 1841 sold |
| 1861-1867 | coal depot | *Isis* | ex frigate 50; 1867 sold |

## Fernando Po ('Jallah Coffee'):

| | | | |
|---|---|---|---|
| 1862-1871 | depot ship | *Vindictive* | ex 74/frigate 50; 1871 sold |

## Ascension Island:

| | | | |
|---|---|---|---|
| 1811-1863 | receiving ship | *Tortoise* | from Chatham; 1863 BU |
| 1811-1846? | coal hulk | *Independencia* | ex slaver; 1847? sold |
| 1857-1870 | coal & store ship | *Meander* | ex frigate, 38; 1870 wrecked |
| 1865-1871 | store ship | *Flora* | 36; 1871 to Simonstown (Cape) |

## Saint Helena:

| | | | |
|---|---|---|---|
| 1819-by 1831 | hulk | *Camel* | ex storeship; by 1831 returned to Britain and sold |
| 1819-1821 | hulk (never sent?) | *Abundance* | ex storeship; 1821 laid up at Deptford; 1823 sold |

## Cape of Good Hope:

| | | | |
|---|---|---|---|
| 1800-1803 | hulk for Saldanha Bay (never sent?) | *Broaderscarp* | ex Dutch 54/50; 1803 at Sheerness & BU there in 1805 |
| 1834-1860 | receiving ship | *Badger* | ex brig; 1860 BU |
| 1845-1867 | Purchased light vessel & receiving ship | *Pacific* | ex mercantile?; 1867 BU |
| 1847-1873 | coal depot & receiving ship | *Seringapatam* | ex frigate, 38; 1873-1883 receiving ship then BU & remains sold |
| 1872-1891 | receiving ship | *Flora* | from Ascension Island; 1891 sold |

## Zanzibar:

| | | | |
|---|---|---|---|
| 1874-1884 | harbour storeship | *London* | ex 90; 1884 sold to BU |

## In the East Indies (Bombay?):

| | | | |
|---|---|---|---|
| 1757 or 1760-____ ? | hulk | *Kent* | ex 70; fate uncertain |

## Bombay:

| | | | |
|---|---|---|---|
| 1761-1765 | hulk | *Tiger* | ex 60; 1765 sold |
| By 1807-1810 | receiving ship | *Arrogant* | ex 74; 1810 sold for BU |
| 1809-1824 | Purchased as sheer hulk | *Arrogant* | ex mercantile *Ardasier*; 1824 to Trincomalee [PLANS: 1822: Waterline profile & decks.] |

## Trincomalee:

| | | | |
|---|---|---|---|
| 1748-1749 | hulk | *Preston* | ex 50; 1749 scuttled |
| 1819-1824 | to receive rice | *Challenger* | ex brig; 1824 sold |
| 1819-1824 | hospital ship | *Orlando* | ex frigate, 36; 1824 sold |
| 1824-1842 | sheer hulk | *Arrogant* | from Bombay; 1842 sold |
| 1847-1864 | receiving ship | *Saphire* | ex 28; 1864 sold |

## Hong Kong:

| | | | |
|---|---|---|---|
| 1842-1861 | seamens hospital | *Minden* | ex 74; 1861 sold |
| 1848-1865 | seamens hospital | *Alligator* | ex 28; 1854 for the use of the Canton Consulate; 1865 sold |
| after 1853-1865 | army depot | *Hercules* | ex 74; 1865 sold |
| 1857-1873 | hospital ship | *Melville* | ex 74; 1873 sold |
| 1857-1875 | floating barracks | *Princess Charlotte* | ex 110; 1875 sold |
| 1867-1906 | Army hospital ship | *Meeanee* | ex 80/screw; 1906 BU |

## Shanghai:

| | | | |
|---|---|---|---|
| 1867-1869 | hospital ship | *Acorn* | ex brig; 1869 sold |

## Tasmania ('Van Diemen's Land'):

| | | | |
|---|---|---|---|
| 1843-1851 | convict ship | *Anson* | from Portsmouth (via Chatham); 1851 BU |

## Halifax, Nova Scotia:

| | | | |
|---|---|---|---|
| 1776-1784 | hulk | *Boulogne* | ex French frigate; 1784 used for lengthening the wharf |
| 1776-1793 | hulk | *Pembroke* | ex 60; 1793 BU |
| 1794-1802 | prison ship | *Security* | purchased; sold per A.O.25.6.1802 |
| 1808-1824 | hospital ship | *Centurion* | ex 50; 1824 sank & BU |
| 1809-1820 | Floating magazine | *Inflexible* | ex 64; 1820 BU |
| 1813-1820 | prison hulk | *Success* | ex frigate, 32; 1820 BU |
| 1832-1879 | receiving ship & convict ship | *Pyramus* | ex frigate, 36; 1879 sold |
| 1840s-1859 | receiving ship & buoy boat | *Netley* | ex revenue cutter; 1859 sold |

## Newfoundland:

| | | | |
|---|---|---|---|
| 1779-1783 | prison ship | *Proteus* | ex French corvette; 1783 sold |

## New York:

| | | | |
|---|---|---|---|
| 1779-1780 | prison ship | *Hunter* | ex sloop; 1780 sold |
| 1780-1780 | prison ship | *Scorpion* | ex sloop; 1780 sold |
| 1780-1780 | prison ship | *Strombolo* | (ex *Grampus*) ex fireship (ex sloop); 1780 sold |
| Ca 1780-1783 | prison hospital ship | *Jersey* | ex 60; 1783 sold |

## Bermuda:

| | | | |
|---|---|---|---|
| 1811-1821 | depot ship | *Ruby* | ex 64; 1821 BU |
| 1813-1824 | prison ship | *Ardent* | ex purchased 64; 1824 BU |
| 1813-1816 | hospital ship | *Romulus* | ex frigate, 36; 1816 BU |
| 1823-1848 | convict ship | *Antelope* | ex 50; 1848 BU |
| 1825-1850 | receiving ship | *Royal Oak* | ex 74; 1850 BU |
| 1825-1864 | convict ship | *Dromedary* | (ex *Howe*); ex storeship; from Deptford; 1854 sold [PLANS: 1807: Profile & platforms. 1814: Profile & platforms. 1846: Profile/orlop/middle deck/main deck/upper deck.] |
| 1826-1852 | diving bell | *Resolute* | ex gunbrig; from Plymouth; 1844 |

| | | | | |
|---|---|---|---|---|
| | ship/ receiving ship | | convict hulk; 1852 BU | |
| 1826-1846 | receiving ship/ slop ship (& sheer hulk?) | Despatch | ex transport; 1846 wrecked & BU | |
| 1827-1848 | receiving ship/ accommodation hulk | Dotterel | ex brig sloop; 1848 BU | |
| 1827-1853 | convict ship | Coromandel | (ex Malabar); ex purchased 56; 1853 BU [PLANS: book of profile & deck plans.] | |
| 1828-1865 | receiving ship/ convict ship | Weymouth | ex purchased 44; 1865 sold | |
| Ca 1830-1838 | defective & laid up | Slaney | ex 28; 1838 BU | |
| 1843-1875 | convict ship | Tenedos | ex frigate, 38; 1863 accommodation ship; 1875 BU | |
| 1844-1863 | convict ship | Thames | ex frigate, 38; from Deptford; 1863 sunk & wreck sold for BU | |
| 1847-1865 | convict ship | Medway | ex 74; 1865 sold | |

**Barbadoes**:

| | | | |
|---|---|---|---|
| 1814-1816 | prison ship | Vestal | ex frigate, 28; 1816 sold |

**Havanna, Cuba**:

| | | | |
|---|---|---|---|
| 1837-1845 | receiving ship for negroes (freed slaver) | Romney | ex 50; 1845 sold |

**Martinique, West Indies**:

| | | | |
|---|---|---|---|
| 1796-1798 | prison ship | Superbe | ex French 22; 1798 sold |
| 1800-1802? | storeship | Vorsechterkite | (Dutch brig, built 1790 in America); purchased 1800 by order of Lord H Seymour; 1802 sold? [77'6", 56' x 23'9" x 6'11". 167.] |

**Tortola, West Indies**:

| | | | |
|---|---|---|---|
| 1802?-1879? | receiving ship | Eclair | French schooner; renamed Safety; BU 1875 or 1879? |

**Jamaica**:

| | | | |
|---|---|---|---|
| 1701-1707 | hulk | Lewis Prize | ex French 42/36; 1707 sold |
| 1728-1771 | hulk | Southampton | ex 40; 1771 sunk & BU after salvage? |
| 1742-1744 | hulk | Lark | ex 40; 1744 wrecked |
| 1742-___? | hulk | Alderney | ex bomb vessel; fate uncertain |
| 1743-___? | hulk | Vulcan | ex fireship; fate uncertain |
| 1782-1783 | purchased as hulk | Conception | ex mercantile? purchased by Rodney; 1783 sold [44 men] |
| Ca 1798-1806 | receiving ship | Admiral de Vries | ex Dutch 64; 1806 sold |
| By 1807-1818 | receiving ship | Shark | ex sloop; 1818 wrecked |
| 1807-1813 | army convalescent ship | Coromandel | ex purchased 56; 1813 sold |
| 1809-1815 | depot for gunpowder (at Port Royal) | Thrush | ex purchased sloop; 1815 wrecked & sold |
| 1813-1815 | powder ship | Barbadoes | ex American brig; 1815 wrecked |
| 1819-1826 | convalescent ship | Serapis | ex 44; 1826 sold |
| 1825-1843 | receiving ship | Magnificent | ex 74; 1843 sold |
| 1836-1849 | receiving ship & coal depot | Galatea | ex frigate, 36; 1849 BU |
| 1842-1862 | receiving ship | Imaum | ex donated 70; 1862/1866 BU |
| 1858-1867 | receiving ship | Marianne | ex slaver; 490 tons; purchased 1858; 1867 BU |

**Antigua**:

| | | | |
|---|---|---|---|
| 1744-1750 | hulk | Ludlow Castle | ex 40; 1750 sold |
| 1758-1763 | hulk | Kinsale | ex 44; 1763 sold |
| 1788-1797 | purchased as a hulk at Antigua (on 1.6.1797) Earl Denbigh | | ex mercantile?; 1797 sunk [73', 59' x 24' x 9'4". 181.] |
| 1801?-1804 | prison ship? | De Ruyter | ex Dutch 64; 1804 wrecked in a hurricane |
| 1803-1811 | purchased as a prison hospital ship Minerva | | ex mercantile?; 1811 BU |
| 1804-1816 | prison ship | Antigua | ex French L'Egyptienne captured by Hippomenes 1804; 1816 BU |
| 1813-1815 | station ship at Saint Johns Harbour | Spider | ex Spanish brig; 1815 BU |

**Rio de Janeiro, Brazil**:

| | | | |
|---|---|---|---|
| 1840-1854 | for freed slaves | Crescent | ex frigate, 38; 1854 sold |
| 1853-1863 | receiving ship | Madagascar | ex frigate 38; from Plymouth; 1863 sold |
| 1863-1875 | storeship | Egmont | ex 74; 1875 sold |

**Callao, Peru**:

| | | | |
|---|---|---|---|
| By 1840s?-1848 | coal depot | Trial | ex cutter/armed vessel; from Waterford; 1848 sold |
| 1847-1866 | coal depot | Naiad | ex frigate, 38; 1866 sold |

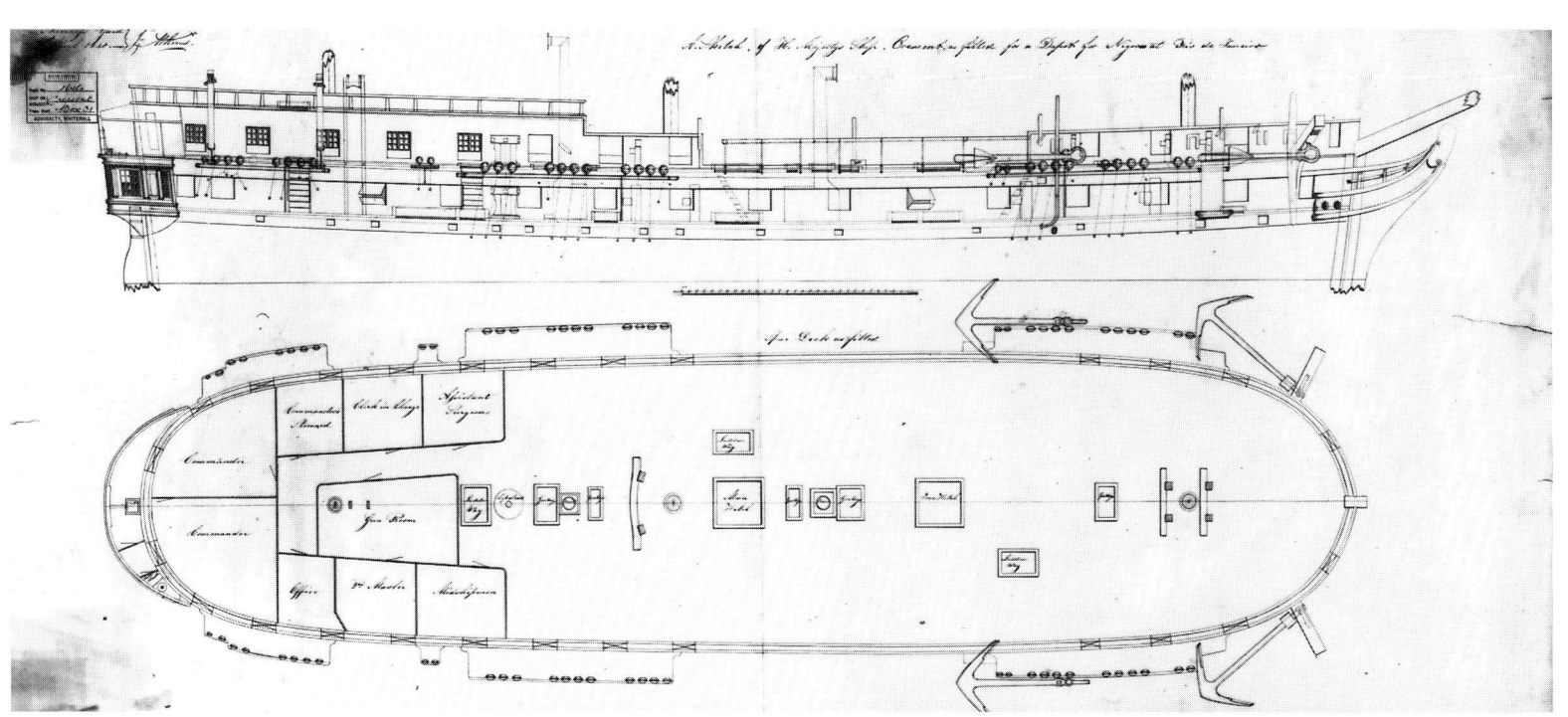

Above waterline profile and spar deck plan (1616) of the frigate Crescent as fitted in 1840 as an accommodation depot ship for freed negro slaves at Rio de Janeiro

HULKS

**Valparaiso, Chile:**
| | | | |
|---|---|---|---|
| 1843-1879 | coal depot | *Nereus* | ex frigate, 38; 1879 sold |

**Coquimbo:**
| | | | |
|---|---|---|---|
| 1877-1903 | ——— | *Liffey* | ex frigate, 50; 1903 sold? |

## Hulks, Location Uncertain

| | | | |
|---|---|---|---|
| 1689-1699 | ——— | *Magdalene* | captured by *Monmouth*; 24.8.1699 sold. [75', ____ x 28'9" x ____ . 285 tons] |
| 1696-___ | ——— | *Josiah* | ex purchased storeship; ____ [probably this is the *Joshua* at Woolwich; which see] |
| 1696-___ | ——— | *Suffolk* | ex purchased storeship; ____ |
| By 1699-___ | ——— | *Saint Nicholas* | ex French Sixth Rate; ____ |
| 1777-1794 | prison ship | *Justitia* | (see entry in Purchases and Prizes Napoleonic War chapter) |
| 1780?-1784 | ——— | *Achilles* | ex 60; 1784 sold |
| 1785-___ | ——— | *Fortunee* | ex French frigate (at Portsmouth?) struck off the list of the Navy |
| 1799-1803? | receiving ship | *Courageuse* | ex French frigate taken in the Mediterranean? 19.6.1799; 1803 still in service |
| 1800-___ | ——— | *America* | ex 64; ____ |
| 1802-1805 | prison ship | *Medee* | ex French frigate; 1805 sold |
| 1803-1806 | Trinity House | *Heroine* | purchased frigate; 1806 sold |
| 1869-1883 | coal hulk | *Forth* | ex 38; became *Jupiter*; 1883 sold |
| 1870-1875 | loaned training ship | *Goliath* | 80/screw; 1875 burnt by accident |
| 1870-1884 | coal hulk | *Hastings* | ex 74; 1884 sold |

## Vessels on Loan as Hulks to Other Organisations

**War Office (Army):**
| | | | |
|---|---|---|---|
| 1809-1817 | prison ship at Cowes, Isle of Wight [?] | | |
| | | *Buffalo* | ex purchased storeship; 1817 sold Cowes, Isle of Wight |
| 1814-1817 | depot at Cowes, Isle of Wight | | |
| | | *Valorous* | ex 22; 1817 sold |
| 1818-1825 | prison ship in Thames | *Brev Drageren* | ex Danish brig; from Hull; 1825 sold [PLANS: 1819: Decks.] |
| 1820-1822 | Intended for the Army in the West Indies but never sent there? | | |
| | | *Dasher* | ex sloop; from the Committee for Distressed Seamen (Thames); 1822 laid up at Deptford |
| 1855-1874 | powder depot at Upnor | *Volage* | ex 28; 1874 sold |
| 1856-1863 | powder ship at Purfleet | *Conquestador* | ex 74; 1863 returned to the RN at Portsmouth for Plymouth |
| 1858-1863 | powder ship at Purfleet | *Mermaid* | ex frigate, 38; 1863 to Dublin |
| 1867-1906 | Hospital ship at Hong Kong? | *Meeanee* | ex 80/screw; 1906 BU |
| 1886-1891 | floating magazine? at Portsmouth | ——— | ex 74; from Portsmouth; 1891 returned to Admiralty |
| 1906-1908 | sleeping accommodation for militia at Inverness | | |
| | | *Brilliant* | ex 36/22; 1908 returned & BU |

**Loans to Trinity House:**

In 1803 a number of vessels were handed over to Trinity House. These included: Frigates, 32: **Iris, Dedalus, Solebay, Retribution** (ex *Hermione*)★. Frigate, 28: **Vestal**. French 18 pdr frigates: **Unite** (ex *Imperieuse*), **Modeste**. All the above had reverted to the RN by 1805 except for the *Retributon* (★) which was BU in 1805, and *Vestal* which was not returned until 1810.

**Loans to the Atlantic Telegraph Co (Telegraph Construction & Maintenance Co):**
| | | | |
|---|---|---|---|
| 1864-1869 | | *Amethyst* | ex 26; 1869 sold to the Company |
| 1866-1869 | | *Iris* | ex 26; 1869 sold to the Company |

## River Thames

**London School Ship Society, Purfleet:**
| | | | |
|---|---|---|---|
| 1859-1868 | reformatory | *Cornwall* | ex 74; 1868 to the Tyne |
| 1868-1940 | reformatory | *Wellesley* | from Chatham; renamed *Cornwall*; 1940 sunk by German air attack |

**The Marine Society at Deptford, Greenwich or Woolwich:**
| | | | |
|---|---|---|---|
| 1799-1816 | training ship | *Thorn* | ex sloop; 1816 sold |
| 1815-1833 | training ship | *Solebay* | (ex *Iris*) ex frigate, 32; 1833 BU |
| 1833-1851 | training ship | *Iphigenia* | ex frigate, 36; 1851 BU |
| 1848-1865 | training ship | *Venus* | ex frigate, 38; 1865 BU |
| 1862-1876 | training ship | *Warspite* | ex 74/50; 1876 burnt |
| 1876-1918 | training ship | *Conqueror* | (ex *Waterloo*); ex 110; 1918 burnt |

**River Police:**
| | | | |
|---|---|---|---|
| 1817-1825 | tender | *Tower* | ex tender; from the Tower of London; 1825 sold |
| 1817-1837 | ——— | *Port Mahon* | ex Spanish brig; from Gravesend; 1837 sold |
| 1837-1857 | tender off Somerset House | *Investigator* | ex survey ship; 1857 BU |
| 1857-1895 | tender off Somerset House | *Royalist* | ex brig; 1895 sold |
| 1857-1874 | floating police station at Blackwall | *Scorpion* | ex brig; 1874 BU |

**Hospital at Greenwich (Later the Dreadnought Seamens Hospital):**
| | | | |
|---|---|---|---|
| 1831-1857 | hospital ship | *Dreadnought* | ex 104/98; from Pembroke; 1857 BU |
| 1849-1854 | temporary hospital ship | *Devonshire* | ex 74; 1854 to Sheerness |
| 1856-1871 | hospital ship | *Caledonia* | ex 120; renamed *Dreadnought* 1870 handed back to the RN and offered as a church ship for Limehouse, but not accepted; 1871 to the Metropolitan Asylums |
| 1866-1868 | temporary hospital ship | *Belleisle* | ex 74; from Sheerness; 1868 returned to Sheerness |

***Worcester* Training Ship, Greenhithe:**
| | | | |
|---|---|---|---|
| 1862-1885 | training ship | *Worcester* | frigate, 50; 1885 sold to BU |
| 1876-1948 | training ship | *Frederick William* | ex screw line of battle ship 110; renamed *Worcester*; 1948 sold, then foundered in the Thames, raised; 1953 BU |

**Committee of the Floating Church at the Tower of London:**
| | | | |
|---|---|---|---|
| 1828-1848 | floating church | *Brazen* | 26; 1848 BU |

**East India Company, Blackwall:**
| | | | |
|---|---|---|---|
| 1806-1838 | hulk | *Beschermer* | ex Dutch small line of battle ship; 1838 sold |

**West India Dock Company:**
| | | | |
|---|---|---|---|
| 1810-1815 | hulk | *Chichester* | ex 44; 1815 BU |
| 1815-1832 | receiving ship for boys | *Lancaster* | ex purchased 64; 1832 returned to the RN and sold |

**Seamens Home Society:**
| | | | |
|---|---|---|---|
| 1837-1874 | church ship | *Swan* | ex cutter; 1874 BU |

**Thames (Wapping) Tunnel Company:**
| | | | |
|---|---|---|---|
| 1838-1839 | accommodation ship | *Plover* | ex brig; from Sheerness; 1839 to Woolwich |

**National Refuge Society for the Destitute Boys of London, Greenhithe:**
| | | | |
|---|---|---|---|
| 1866-1889 | training ship | *Chichester* | ex frigate, 50; 1889 sold |

**Shaftesbury Homes:**
| | | | |
|---|---|---|---|
| 1874-1934 | training ship | *Arethusa* | ex frigate, 50; 1934 BU [NOTE: She was replaced by the steel barque *Pekin* (ex 'Flying P' Line) which was renamed |

'*Arethusa*', but is now part of the South Street Seaport display at New York under her original name.]

### Committee for Distressed Seamen:
| | | | |
|---|---|---|---|
| 1816-1819 | accommodation ship | *Abundance* | ex purchased storeship; 1819 storeship for Saint Helena |
| 1816-1818 at Deptford | | *Plover* | ex sloop; 1818 returned to the RN; 1819 sold |
| 1817-1817 at Blackwall | | *Batavier* | ex Dutch Small Line of Battle ship; 1817 to Sheerness |
| 1818-1820 | ____ | *Dasher* | ex sloop; 1820 to be fitted for the Army in the West Indies |
| 1820-1832 | hospital ship | *Grampus* | ex 50; 1832 sold |

### Peruvian Navy for the Crew of a Ship Building at Blackwall:
| | | | |
|---|---|---|---|
| 1857-1858 | accommodation ship | *Cleopatra* | ex 26; 1858 returned to the RN |

### Turkish Navy for the Crews of Ships Building on the Thames:
| | | | |
|---|---|---|---|
| 1864-1866 | accommodation ship | *Superb* | ex 80; in Ordinary again by 1866?; 1869 BU |

### Metropolitan Asylums:
1871-1875 hospital ship for male convalescent smallpox patients

| | | | |
|---|---|---|---|
| | | *Dreadnought* | (ex *Caledonia*) ex 120; from the Dreadnought Seamens Hospital; 1875 BU |
| 1877-1905 | training ship | *Exmouth* | ex 90; 1905 sold for BU |

### Cholera Ship:
| | | | |
|---|---|---|---|
| 1871-1884 | in the Thames? | *Rhin* | ex French frigate; from Sheerness; 1884 sold |

### Gravesend Alien Service:
| | | | |
|---|---|---|---|
| 1815-1858 | ____ | *Flamer* | ex gunbrig; 1858 sold for BU |

# East Anglia

### Church Ship at Ipswich:
| | | | |
|---|---|---|---|
| 1868-1880 | ____ | *Helena* | from Portsmouth; 1880 to Chatham |

# Humber

### Humber Training Ship Association:
| | | | |
|---|---|---|---|
| 1868-1912 | training ship | *Southampton* | ex frigate, 60; from the Coastguard at Sheerness; 1912 sold |

# North East Coast

### Seamens Mission North/South Shields:
| | | | |
|---|---|---|---|
| 1866-1885 | ____ | *Diamond* | ex 28; 1868 renamed *Joseph Straker*; 1885 sold to BU |

# Scotland

### Seamens Society, Leith:
| | | | |
|---|---|---|---|
| 1835-1842 | chapel ship | *Alonzo* | from Portsmouth; 1842 deliberately sunk |

### Dundee:
| | | | |
|---|---|---|---|
| 1869-1929 | loaned training ship | *Mars* | 80/screw; 1929 sold for BU |

### Northern Lights, Oban:
| | | | |
|---|---|---|---|
| 1860-1877 | coal depot | *Enterprise* | ex storeship; 1877 returned to Chatham?; 1903 sold |

### Clyde Marine Society:
| | | | |
|---|---|---|---|
| 1818- ____ | ____ | *Ornen* | ex Danish schooner; from Greenock; ____ |

### Clyde Industrial Training Ship Association:
| | | | |
|---|---|---|---|
| 1889-1889 | training ship for boys | *Cumberland* | ex 70; from Sheerness; 1889 destroyed by fire |

# Mersey

### Liverpool Juvenile Reformatory Association, Ltd:
| | | | |
|---|---|---|---|
| 1862-1908 | training ship | *Wellington* | ex 74; from the Coastguard; renamed *Akbar* 1908 sold for BU |

### Committee of the Liverpool Training Ship:
| | | | |
|---|---|---|---|
| 1864-1914 | training ship | *Indefatigable* | ex Frigate; 1914 sold |

### Church Society, Liverpool:
| | | | |
|---|---|---|---|
| 1826-1872 | church ship | *Tees* | ex 26; 1872 sold |

### Mercantile Marine Association, Liverpool:
| | | | |
|---|---|---|---|
| 1859-1861 | training ship | *Conway* | ex 28; 1861 to Aberdeen RNR |
| 1861-1876 | training ship | *Winchester* | ex Frigate, 50; renamed *Conway*; 1876 to Plymouth as *Mount Edgcombe* |
| 1876-1953 | training ship | *Nile* | ex 90; renamed *Conway*; later to the Menai Straits; 1953 stranded; 1956 wreck burnt |

### Liverpool Roman Catholic Reformatory Society:
| | | | |
|---|---|---|---|
| 1864-1884 | training ship | *Clarence* | ex 84; 1884 burnt & BU |
| 1885-1899 | training ship | *Royal William* | ex 120/screw; renamed *Clarence*; 1899 burnt by accident |

# Isle of Man

### Church ship:
| | | | |
|---|---|---|---|
| 1835-1846 | ____ | *Industry* | ex transport; 1846 BU |

# Wales

### Ragged School at Cardiff:
| | | | |
|---|---|---|---|
| 1860-1905 | school ship | *Havannah* | ex frigate, 36/19; 1905 sold |

### Hospital for Sick Seamen at Cardiff:
| | | | |
|---|---|---|---|
| 1866-1905 | hospital ship | *Hamadryad* | ex frigate, 38; 1905 sold |

### Church Ship at Cardiff:
| | | | |
|---|---|---|---|
| 1863-1892 | floating church | *Thisbe* | ex frigate, 38; 1892 sold |

# Bristol

### Bristol Training Ship Association:
| | | | |
|---|---|---|---|
| 1869-1906 | training ship | *Formidable* | ex 84; 1906 sold |

# Cornwall & Devon

### Wheatley Cobb's Training Ship at Falmouth:
| | | | |
|---|---|---|---|
| 1892-1897 | training ship | *Foudroyant* | ex 80 1897 wrecked on fund-raising cruise |
| 1897- ____ | training ship | *Trincomalee* | ex frigate, 38; from Sunderland; renamed '*Foudroyant*'; purchase (not loan); later to Portsmouth harbour as a youth training ship; recently (1992) to Hartlepool where she is being restored |
| 1912-1949 | training ship | *Implacable* | ex French 74; from Plymouth as loan; 1949 scuttled after an attempt to preserve her as a memorial to RN and Merchant Navy dead in both wars in a dock at Greenwich (now occupied by the *Cutty Sark*) had failed |

### Plymouth Guardians of the Poor:
| | | | |
|---|---|---|---|
| 1871-1879 | cholera hospital ship | *Pique* | ex 36; by 1879 returned |

**Plymouth Port Sanitary Authority:**
1879-1882   hospital ship   *Pique*   ex 36; 1882 see below

**Plymouth Town Council:**
1882-1910   hospital ship   *Pique*   ex 36; 1910 returned & sold

## Isle of Wight

**Bembridge Light Vessel:**
1821-1834   light vessel   *Havock*   ex gunbrig; 1834 to the Customs

**Guardians of the Poor, Cowes:**
1866-1866   as cholera hospital ship   *Acteon*   ex 26; 1866 returned to the RN

## Sussex

**Chapel Ship at Shoreham:**
1861-1862   chapel ship   *Herald*   ex 28; 1862 sold

## Ireland

**Kingstown Church Ship:**
1850s?-1867   church ship   *Ranger*   ex brig 1867 sold after sinking

**Cork Industrial Training Ship Organisation:**
1870-1873   school ship   *Creole*   ex 26; 1873 to BU; 1875 BU

**Hibernian Marine Society:**
1816-____   training ship?   *Deptford*   from Limerick; fate uncertain?

## Overseas

**Consulate at Canton, China:**
1854-1865   'used for the duties of HM Consulate'   *Alligator*   ex 28; from Hong Kong; 1865 sold

**New South Wales Government:**
1853-1876   diving bell vessel?   *Bramble*   ex cutter; later light ship; 1876 discarded
1867-1909   training ship   *Nelson*   ex 120/screw 1898 sold as store hulk; later coal hulk; 1928 BU

# INDEX

The index includes all the variants of names (including foreign and mercantile names) of all the vessels whose existence is noted in the three main parts of this work.

It is a nominal index given in alphabetical order of names (numbered vessels being listed at the end). The individual ships which bore each name are then listed in chronological order. Those ships which were the 'name ships' of classes (including the variants) are highlighted in bold print. The use of capital letters indicates that this was the name the ship served under in the Royal Navy whilst lower case print indicates a non-naval name (foreign or mercantile) when different from the naval name. The name is followed by two dates, the first being that of the year of launch or of acquisition by the Royal Navy or when the ship was given that name. The second is the year when the ship went out of service with the navy, or when the name was changed. The aim of giving both dates is to make it easier to look up which particular ship of the name is being referred to. Page references are only given for the 'main' names of ships and to the principal entry for that ship. In other words, entries for names which were changed before commissioning a ship, later renamings, or foreign and mercantile names (when different from the 'main' name entry which gives the relevant page. Page references are not given for listings in the 'Conversions' chapter (Part III Chapter 3) or the hulks (the vast majority) converted from active naval vessels (Part III Chapter 6) as these only repeat information given in the 'primary' entry on these ships.

## A


ABEILLE 1796-...? 260
Abénakise (French 1756)
see AURORA 1757
ABERCROMBIE 1809-1817 270
ABERGAVENNY 1795-1807 241
ABIGAIL 1812-1814 284
ABONDANCE 1781-1784 219
ABOUKIR 1798-1802 238
ABOUKIR 1807-1824 72
ABOUKIR 1848-1878 171
ABRAHAM'S OFFERING 1694-1694 188
ABRAM'S OFFERING
see ABRAHAM'S OFFERING
ABUNDANCE 1799-1816 260
**ACASTA** 1797-1821 118
ACERTIF 1808-1809 284
ACHATES 1808-1810 144
ACHATES 1809-1818 280
ACHERON 1803-1805 281
ACHILLE 1744-1745 201
ACHILLE 1794-1796 236
**ACHILLE** 1798-1865 110
ACHILLES 1747-1748 201
ACHILLES 1757-1784 76
ACHILLES 1781-1784 230
ACORN 1807-1819 132
ACORN 1826-1828 139
ACORN Cancelled 1831 139
**ACORN** 1838-1869 179
ACQUILON
see AQUILON
ACTAEON
see also ACTEON
ACTAEON 1757-1766 85
ACTEON 1775-1776 86
ACTEON 1778-1793 79
ACTEON 1805-1816 277
**ACTEON** 1831-1866 135
ACTAEON 1886-1923
see VERNON 1832
ACTIF 1794-1794 251
Actif (French)
see MORGIANA 1800
ACTIVE 1758-1778 85
ACTIVE 1776? 321
Active (mercantile)
see JACKALL 1778
ACTIVE 1779-1780 223
ACTIVE 1779? 231
Active
see SYLPH 1780
ACTIVE 1780-1782? 231
**ACTIVE** 1780-1796 84
ACTIVE 1794-1800 264
**ACTIVE** 1799-1825 119
ACTIVE 1800-...? 260
ACTIVE 1803-1803 288
ACTIVE 1803-1814 288
ACTIVE 1806-1816 333
ACTIVE 1808-1880 333
ACTIVE 1808-1816 333
ACTIVE 1845-1908 175
**ACUTE** 1797-1802 151
ACUTE 1804-1813 153
ADAMANT 1780-1809 77
Adamant (merchant 1798)
see BONETTA 1803
Adamant (merchant 1804)
see THRESHER 1804
Adams (American)
see DETROIT 1812
Adder (American)
see LEE 1780
ADDER 1782-1787 231
ADDER 1797-1805 151
ADDER 1805-1806 154
ADDER 1813-1832 154
ADDER 1817-1860 334
ADDER 1838? 335
ADELAIDE 1827-1833 292
ADELAIDE 1832?-1845 335
ADELAIDE 1848-1851 293
ADELPHI 1804-1805 288
ADELPHI 1810-1810 289
Admiraal De Ruijter (Dutch 1775)
see DE RUYTER 1799
Admiraal Tjerk Hiddes De Vries (Dutch 1783)
see ADMIRAL DE VRIES 1797
ADMIRAL BARRINGTON 1782-1783 232
ADMIRAL DE VRIES 1797-1806 240
ADMIRAL GARDNER 1808-1810 289
ADMIRAL MITCHELL 1800-1801 264
ADMIRAL MITCHELL 1803-1805 289
Admiral Parker (merchant)
see SUPPLY 1782
ADMIRAL PASLEY 1801-1802 264
Admiral Rainier (East Indiaman 1799)
see HINDOSTAN 1804
ADMIRAL RAINIER 1800-...? 260
ADMIRAL WHITSHED 1807-1808 289
ADMIRALTY YACHT 1814-1830 306
ADMIRALTY YACHT
see FANCY 1831
**ADONIS** 1806-1814 162
ADRIAN 1804-1810
see INDUSTRY 1804
ADVENTURE 1646-1692 13
ADVENTURE 1692-1709 25
ADVENTURE 1704-1705 194
ADVENTURE 1704-1705 194
ADVENTURE 1704-1711 194
ADVENTURE 1709?-1724 36
ADVENTURE 1726-1740 47
ADVENTURE 1741-1770 48
ADVENTURE 1756-1756 211
ADVENTURE 1757-1758 211
ADVENTURE 1763-1768 208
ADVENTURE 1771-1783 212
ADVENTURE 1776 302
ADVENTURE 1780 232
**ADVENTURE** 1784-1801 80
ADVENTURE 1813-1814 289
ADVENTURE 1821-1853
see AID 1809
ADVENTURE GALLEY 1696-1699 189
ADVENTURE PRIZE 1693-1695 186
ADVICE 1650-1701 13
ADVICE 1698-1711 24
ADVICE 1712-1744 35
ADVICE 1746-1756 47
ADVICE 1779-1793 226
ADVICE 1792-1793 235
ADVICE 1793-1793 235
ADVICE 1796-1807 255
ADVICE 1800-1805 160
ADVICE 1801-1802 264
ADVICE 1804-1804 289
ADVICE PRIZE 1693-1695 189
ADVICE PRIZE 1704-1712 193
AEOLUS
see EOLUS
AETNA 1691-1697 29
AETNA
see KITE 1826
AETNA
see also ETNA
AFFIANCE 1806-1810? 284
Affronteur (French)
see CAROLINE 1804
AFRICA 1692-1695 189
**AFRICA** 1761-1777 73
AFRICA 1781-1814 75
AFRICA 1803-1810 289
AFRICA 1859-1860
see EURYALUS 1809
AFRICAINE 1801-1816 244
AFRICAINE 1827-1867 123
AFRICAN 1825-1862 146
AGAMEMNON 1781-1809 74
AGAMEMNON not built 1840s 173
AGENORIA 1801-1801 263
AGRESSOR 1801-1815 153
AGINCOURT 1796-1812 239
AGINCOURT 1817-1848 114
AGLAIA 1782-1783 222
AGNES 1804-1806
see VENUS 1804
**AID** 1809-1821 168
AID 1838-1878 318
AIGLE 1782-1798 219
**AIGLE** 1801-1852 124
AIMABLE 1782-1814 219
AIMWELL 1794-1811 149
AJAX 1767-1785 70
AJAX 1798-1807 110
AJAX 1809-1864 114
AKBAR Cancelled 1809 114
AKBAR 1806-1862
see CORNWALLIS 1805
AKBAR 1862-1908
see WELLINGTON 1816
AKERS 1794-1801 254
Alaart (Danish)
see ALERT 1807
ALACRITY 1806-1811 141
ALACRITY 1818-1835 145
Aladdin (mercantile 1841)
see MUTINE 1825
ALARIC
see IRIS 1807
**ALARM** 1758-1812 83
ALARM 1763-1780 208
ALARM 1777-1778 228
ALARM 1799?
see HELDIN 1799
ALARM 1810-1811 283
ALARM 1812-1814 289
ALARM Cancelled 1820? 133
ALARM Cancelled 1832 134
ALARM 1845-1904 176
ALBACORE
see ALBECORE
ALBAN 1806-1812 162
ALBAN 1813-1822 283
**ALBAN** 1826-1860 146
ALBANAISE 1800-1800 260
ALBANY 1747-1763
see TAVISTOCK
ALBANY 1776-1782 224
**ALBATROSS** 1795-1807 139
ALBATROSS 1826-1833 295
ALBATROSS 1842-1860 179
ALBECORE 1781-1784 222
ALBECORE 1793-1802 129
ALBACORE 1804-1815 131
ALBECORE/ALBICORE 1828-1864 330
ALBEMARLE
see also ALBERMARLE
ALBERMARLE 1680-1701 11
ALBEMARLE 1704-1709 18
ALBEMARLE 1781-1784 219
**ALBION** 1763-1797 69
ALBION 1793-1794 264
ALBION 1798-1802? 249
Albion (merchant 1800)
see PROSPERO 1803
ALBION 1802-1831 112
ALBION 1803-1808 289
ALBION 1808-1812 289
**ALBION** 1842-1884 171
ALBOROUGH 1691-1696 30
ALBOROUGH 1706-1727 29
ALBOROUGH 1727-1742 50
ALBOROUGH 1743-1749 52
ALBOROUGH 1756-1777 89
ALBROUGH
see ALBOROUGH
ALCESTE 1799-1802 246
ALCESTE 1806-1817 271
ALCIDE 1755-1772 205
ALCIDE 1779-1817 69
Alcide (French privateer 1802)
see MUROS 1806
Alcion (French 1802)
see HALCYON 1803
ALCION
see HALCYON 1803
ALCMENE 1779-1784 219
ALCMENE 1794-1809 126
ALCMENE 1809-1816 272
Alcmène (French 1811)
see IMMORTALITE 1814
ALDBOROUGH
see ALBOROUGH
**ALDERNEY** 1735-1742 58
ALDERNEY 1743-1749 52
**ALDERNEY** 1757-1783 94
ALECTO 1781-1802 99
ALERT 1777-1778 101
ALERT 1778-1780 233
ALERT 1779-1792 98
ALERT 1790-1799 235
ALERT 1793-1794 129
ALERT 1798-1801 264
ALERT 1804-1804 289
ALERT 1804-1804 289
ALERT 1804-1812 276
Alert (merchant)
see MULGRAVE 1807
ALERT 1807-1809 278
ALERT 1808-1812 289
ALERT 1809-1809 289
ALERT 1812-1814 289
ALERT 1813-1832 142
ALERT 1835-1851 179
ALERT 1848-1850 295
Alerte (French)
see VIGILANTE 1793?
ALERTE 1793-1793 260
Alerte (French) 1789
see MINORCA 1799
ALEXANDER 1688?-1689 189
ALEXANDER 1778-1805 71
ALEXANDER/ALEXANDRIA 1796-1802 260
ALEXANDER 1817-1819 296
ALEXANDRE 1806-1808 267
ALEXANDRIA 1801-1804 247
ALEXANDRIA 1801-1803
see ALEXANDER 1796 tender
ALEXANDRIA 1806-1817 128
**ALFRED** 1778-1814 71
ALFRED 1778-1782 233
ALFRED 1793-1801 264
ALFRED 1819-1858
see ASIA 1811
ALGERINE 1814-1818
see TIGRESS 1808
ALGERINE 1810-1813 163
ALGERINE 1823-1826 145
ALGERINE 1829-1844 146
ALGIERS Cancelled 1834 171
ALGIERS
see CAESAR 1853
**ALGIERS** 1854-1870 171
ALKMAAR 1797-1815 239
ALLEGIANCE 1779-1782 224
ALLIANCE 1795-1802 245
Alliante (Dutch)
see ALLIANCE 1795
ALLIGATOR 1780-1782 96
ALLIGATOR 1787-1810 87
ALLIGATOR 1793-1801 263
ALLIGATOR 1821-1854 133
ALNWICK PACKET 1809-1809 289
ALONZO 1801-1815 250
ALPHEA 1806-1813 162
ALPHEUS 1814-1817 126
ALPHEUS 1839-1864 323
ALTHORP 1804-1805 289
Amarante (French)
see AMARANTHE 1796
AMARANTHE 1796-1800 251
AMARANTHE 1799-1804 247
AMARANTHE 1804-1815 140
AMAZON 1745-1763 198
**AMAZON** 1773-1794 84
**AMAZON** 1795-1797 123
**AMAZON** 1799-1817 121
AMAZON 1821-1863 120
AMBOYNA 1796-1802 260
AMBUSCADE 1745-1762 198
AMBUSCADE 1773-1798/1803-1810 84
AMBUSCADE 1798-1802 246
AMBUSCADE
see HELDER 1799
AMBUSCADE 1811-1812 272
AMBUSCADE
see AMPHION 1846
AMBUSH 1815-1815 284
AMELIA 1782 233
AMELIA 1796-1816 243
America (merchant)
see BLAZE 1739
AMERICA 1749-1756 79
AMERICA 1757-1771 76
AMERICA 1777-1800 74
AMERICA
see IMPETUEUX 1794
AMERICA 1810-1864 114
American Tartar (American)
see HINCHINBROOK 1777
Amérique (French)
see IMPETUEUX 1794
AMETHYST 1793-1795 242
AMETHYST 1799-1811 124
AMETHYST 1844-1869 176
Amethyste (Haitian)
see FELICITE 1809
AMITY 1703-1705 194
AMITY 1706-1707 194
AMITY 1794-1802 254
Amity (merchant 1801)
see INSPECTOR 1803
AMITY 1804-1804 289
AMPHION 1780-1796 84
**AMPHION** 1798-1820 126
AMPHION 1846-1864 126
AMPHITRITE 1778-1794 88
AMPHITRITE 1782 231
AMPHITRITE 1793-1794 264
AMPHITRITE 1795-1811
see POMONA 1778
AMPHITRITE 1799-1801 242
AMPHITRITE 1804-1805 271
AMPHITRITE 1816-1857 119
AMSTERDAM 1804-1806 273
AMY 1803-1804 289
ANACONDA 1813-1815? 280
ANACREON 1799-1802 253
ANACREON 1804-1805 284
ANACREON 1813-1814 130
ANACREON 1815-1816? 333
ANDROMACHE 1781-1811 84
ANDROMACHE 1812-1828
see PRINCESS CHARLOTTE 1799
ANDROMACHE 1832-1846 135
ANDROMEDA 1777-1780 86
**ANDROMEDA** 1784-1811 84
ANDROMEDA 1812-1816 275
ANDROMEDA
see NIMROD 1828
ANDROMEDA 1829-1846 123
Anfitrite (Spanish 1797)
see AMPHITRITE 1804
ANGEL 1694-1696 188
ANGELICA 1771-1783 301
ANGLESEA 1694-1719 23
ANGLESEA 1725-1742 47
ANGLESEA 1742-1745 48
ANGLESEA 1746-1764 49
ANHOLT 1811 284
Anibal (French)
see ACHILLE 1794
ANN 1703-1704 194
ANN/ANNE 1704-1708 194
ANN 1705 194
ANN 1705 194
ANN 1781-1855 316, 318, 324
ANN 1796-1801 264
ANN 1798-1802? 249
Ann (merchant)
see FIREDRAKE 1804
Ann (merchant? 1803?)
see ORESTES 1803
ANN 1803-1809 289
ANN 1804-1809 289
ANN 1807-1812 289
ANN 1809-1809 289
ANNA/ANNE 1739-1741 201
ANNA 1805-1809 284
Anna Maria (American)
see EXPRESS 1815
Ann and Amelia (East Indiaman 1799)
see MEDIATOR 1804
ANN AND CHRISTOPHER 1672-1692 15
ANN AND ELIZA 1855?-1864 317
ANN AND FRANCIS 1705-1707 194
Annapolis (merchant?)
see PORPOISE 1777
ANNA TERESA 1795-1801? 258
ANNE 1678-1690 12
Anne (merchant?)
see DEMARARA 1804
ANNE GALLEY 1739-1744 200
ANNE OF SANDWICH 1711-1711 194
ANNESLEY 1792-1818 332
ANSON 1747-1773 75
ANSON 1763-1774 208
ANSON 1781-1807 74
ANSON 1812-1831 114
ANT 1798-1815 260
ANT 1803-1806 289
ANT 1808-1810 289
ANT 1815-1857? 316, 317, 318, 322
ANT
see SINBAD 1834
ANTELOPE 1660-1693 13
ANTELOPE 1703-1738 25
ANTELOPE/ANTHELOPE 1702-1708 194
ANTELOPE 1742-1783 46
ANTELOPE 1783-1784? 231

352 INDEX


ANTELOPE 1800-1832 333
**ANTELOPE** 1802-1823 115
ANTELOPE 1805?-1832? 284
Antelope (French or Spanish) see FIREFLY 1807?
ANTELOPE 1812-1814 see FIREFLY 1807?
ANTHELOPE see ANTELOPE
Anthelope (merchant) see ROSE 1709
Anti-Briton (American) see TRIMMER 1782
ANTIGUA 1757-1763 207
ANTIGUA 1777-1781? 231
ANTIGUA 1779-1783 225
ANTIGUA 1804-1816 284, 346
Antilope (Dutch) see GIPSY 1805
APELLES 1808-1816 144
Apith (Russian) see BALTIC 1808
APOLLO 1747-1749 201
APOLLO 1774-1786 see GLORY 1762
APOLLO 1794-1799 118
**APOLLO** 1799-1804 124
APOLLO 1805-1856 121
Apollon (French 1716) see APOLLO 1747
Applewhite & Frere (merchant) see SALAMANDER 1757
AQUILON 1758-1776 85
Aquilon (Spanish 1750) see MORO 1762
AQUILON 1786-1816 84
Aquilon (French 1789) see ABOUKIR 1798
ARAB 1795-1796 249
ARAB 1797-1798 see ANT
ARAB 1798-1810 248
ARAB 1812-1823 142
ARAB 1847-1867 179
ARACHNE 1809-1837 142
**ARACHNE** 1847-1866 178
Araignée (French) see SPIDER 1782
ARAXES 1813-1817 120
ARBUTHNOT *Ca*1782 231
ARC-EN-CIEL 1756-1759 205
ARCHANGEL 1695-1695 189
ARCHANGEL 1774 301
**ARCHER** 1801-1815 153
Ardasier (merchant) see ARROGANT 1809
**ARDENT** 1764-1779 74
ARDENT 1782-1794 75
ARDENT 1796-1813 239
ARETHUSA 1759-1779 205
**ARETHUSA** 1781-1814 81
ARETHUSA 1817-1836 119
ARETHUSA 1849-1934 173
Aréthuse (French 1757) see ARETHUSA 1759
ARETHUSE 1793-1795 242
Aréthuse (French 1798) see RAVEN 1799
ARGO 1758-1776 85
ARGO 1780-1783 231
ARGO 1781-1816 80
Argo (Dutch) see JANUS 1796
ARGO hulk see ACTIVE 1799
ARGONAUT 1782-1797 216
ARGUS 1794-1799 264
ARGUS 1794-1798 260
ARGUS 1799-1811 250
ARGUS 1813-1828 142
Argus (French 1814-1820) see PLUMPER 1804
ARGUS Cancelled 1831 139
ARGYLE 1716-1719 see BONAVENTURE 1711
ARGYLE 1722-1748 45
ARIADNE 1776-1814 91
Ariadne (merchant 1803) see DOVE 1805
ARIADNE 1804-1814 289
ARIADNE 1806-1811 289
ARIADNE 1816-1837 137
ARIEL 1777-1779 91
ARIEL 1781-1802 97
'ARIEL' 1785 see note under ENTERPRISE Class 87
ARIEL 1806-1816 141
ARIEL 1820-1828 145
ARISTOCRAT 1794-1798 264
ARISTOCRAT 1799-1801 264
Aristotle (American 1806) see REDBRIDGE 1807
Arlequin (French) see PORTO
**ARMADA** 1810-1863 114
ARMIDE 1806-1815 271
ARMS OF HORN 1673-1704? 337
ARROGANT 1705-1709 191
**ARROGANT** 1761-1806 68
ARROGANT 1798-1801 253
Arrogante (French 1794) see ARROGANT 1798
ARROGANT 1809-1824 345
ARROGANT see PHAETON 1843
ARROW 1782-1783 231
ARROW 1796-1805 132
**ARROW** 1805-1828 161, 325
Arrow (American 1812) see WHITING 1812
**ARROW** 1823-1852 165
ARTHUR 1804-1805 see VENUS 1803
ARTHUR 1828-1864 325
ARTOIS 1780-1786 217
**ARTOIS** 1794-1797 118
ARUNDELL 1695-1713 26
ARUNDEL 1746-1765 87
ARUNDEL 1800-1802 see WARSPITE 1758
ARUNDEL 1802-1811 see MEDWAY 1753
ARVE PRINCEN 1811-1815 see HEIR APPARENT FREDERICK1807
Arve Princen Frederick (Danish 1782) see HEIR APPARENTFREDERICK 1807
ASIA 1694-1701 344
**ASIA** 1764-1804 73
Asia (East Indiaman) see SIR FRANCIS DRAKE 1805
ASIA 1811-1819 114
ASIA 1824-1859 107
ASP 1797-1816+ 151
ASP 1808-1814 279
**ASP** 1826-1826 165
ASSAULT 1797-1825 151
ASSIDUOUS 1823-1827 295
ASSISTANCE 1650-1699 13
ASSISTANCE 1699-1710 24
ASSISTANCE 1713-1720 35
ASSISTANCE 1725-1746 45
ASSISTANCE 1747-1770 76
ASSISTANCE 1771-1864 323
ASSISTANCE 1781-1802 77
ASSISTANCE 1805-1815 see ROYAL OAK 1769
ASSISTANCE 1809-1821 168
ASSISTANCE 1850-1854 294
ASSISTANT 1791-1802 235, 315
ASSOCIATION 1696-1707 18
ASSURANCE 1646-1696 13
ASSURANCE 1702-1712 191
ASSURANCE 1747-1753 79
ASSURANCE 1780-1799 80
Assuré (French 1696) see ASSURANCE 1702
ASTRAEA see ASTREA
ASTREA 1781-1801 84
ASTREA 1810-1823 124
ASTREA PRIZE 1740-1744 201
Astrée (French) see HINCHINBROOK 1778/9
Astrée (French 1808) see POMONE 1810
ATALANTA 1775-1781 96
ATALANTA 1797-1807 252
Atalanta (merchant) see CLYDE tender 1805
ATALANTA 1814-1814 284
ATALANTA 1816-1877 295
ATALANTA 1847-1868 180
ATALANTA 1878-1880 see JUNO 1844
Atalante (French 1784) see ESPION 1794
Atalante (French 1794) see ATALANTA 1797
Atalante (French) see HAWK 1803
ATALANTE 1808-1813 136
ATHENIAN 1800-1806 241
ATHENIEN see ATHENIAN 1800
ATHENIENNE 1796-1802 260
Athénienne (French/Maltese) see ATHENIAN 1800
ATHENS 1804-1805 289
**ATHOLL** 1820-1851 133
ATLANTIC 1793-1799 264
Atlas (French 1779-1781) see NORTHUMBERLAND 1743
ATLAS 1782-1814 65
ATLAS 1804-1804 289
Atlas (American) see SAINT LAWRENCE 1813
ATTACK 1794-1802 149
ATTACK 1804-1812 153
ATTENTIVE 1804-1812 153
ATTENTIVE 1812-1817 284
Audacieux (French) see VOLTIGEUR 1798
AUDACIOUS 1785-1815 69
AUDACIOUS not built 1840s 172
AUGUST 1705-1716 191
AUGUSTA 1736-1765 44
AUGUSTA 1771-1773 see CHARLOTTE 1711
AUGUSTA 1763-1777 73
AUGUSTA 1771-1773 see CHARLOTTE 1710
AUGUSTA 1795-1801? 258
AUGUSTA Cancelled 1810 115
AUGUSTA 1819-1823 295
Auguste (French 1704) see AUGUST 1705
Auguste (French 1740) see PORTLAND'S PRIZE 1746
AUGUSTUS 1782-1782 233
AUGUSTUS 1796-1801 260
AURORA 1757-1763 205
AURORA 1766-1769 83
AURORA 1777-1814 86
Aurora (American) see MENTOR 1781
AURORA 1804-1804 289
AURORA 1814-1832 273
AURORE 1793-1795 260
Aurore (French 1799) see CHARWELL 1801
Austerlitz (French 1795) see PULTUSK 1807
**AUSTRALIAN** 1846-...? 182
AUTUMN 1801-1805 251
AUXILIARY 1809-1809 289
AVENGER (1778) 1779-1783 see LUCIFER 1778
AVENGER 1794-1802 249
AVENGER 1803-1803 276
AVENGER 1804-1812 276
AVENGEUR see AVENGER 1794
AVENTURIER? 1793-1799 260
AVON 1805-1814 141
AXEMAN 1815 299
AZOFF see AZOV
**AZOV** 1855-1899 182


## B


BAB EL MANDEB 1801-1802 264
BABET 1794-1801 247
Baboo (mercantile 1835) see ASSISTANCE 1850
BACCHANTE 1803-1809 275
BACCHANTE 1811-1837 122
BACCHANTE Cancelled 1851 174
BACCHUS 1806-1807 162
BACCHUS 1807-1811 284
BACCHUS 1813-1826 142
BACCHUS 1844-1883 see ARETHUSA 1817
BADGER 1745-1762 55
BADGER 1776-1777 222
BADGER 1777-1783 222
BADGER 1794-1802 256
BADGER *Ca*1798-1803 333
BADGER 1808-1834 144
BADGER 1812-1814 289
BADGER 1813-1845? 333
BAHAMA 1805-1807 268
BALAHOU see BALLAHOO 1803
BALEINE 1760-1767 206
**BALLAHOO** 1803-1814 160
BALTIC 1808-1810 282
BALTIMORE 1691-1691 190
BALTIMORE 1742-1762 54
BALTIMORE 1745-1745 202
BANN 1814-1829 137
**BANTERER** 1807-1808 129
**BANTERER** 1810-1817 144
BARBADOES 1757-1763 207
BARBADOES 1778-1784 221
BARBADOES 1782-1786 231
BARBADOES 1800-1812 273
BARBADOES 1813-1817 279
BARBADOES see HIND 1814
BARBARA 1796-...? 260
BARBARA 1803-1803 289
BARBARA 1806-1815 162
BARBETTE 1805-1811 275
Barboude (French 1782) see BARBUDA
BARBUDA 1780-1782 231
BARDSEA 1802-1810 289
BARFLEUR 1697-1713 18
BARFLEUR 1716-1764 33
**BARFLEUR** 1768-1819 64
BARFLEUR 1825-1825 see BRITANNIA 1762
BARHAM 1811-1840 114
BARRACOUTA 1782-1792 228
BARRACOUTA 1804-1807 284
BARRACOUTA 1807-1816 141
BARRACOUTA 1820-1836 145
see DONEGAL 1798
Barras (French) see DONEGAL 1798
BARRINGTON 1748 201
BARROSA 1812-1823 124
BARWICK see BERWICK
BASILISK 1695-1729 30
BASILISK 1740-1750 58
BASILISK 1759-1762 99
BASILISK 1779-1781 see GRASSHOPPER 1777
BASILISK 1801-1818 152
BASILISK 1822-1846 165
BASING 1654-1660 see GUERNSEY 1660
BASING 1688-1698 189
BASING GALLY 1693-1694 189
Basque (French 1808) see FOXHOUND 1809
BAT 1836-1890 322
BAT *Ca*1844-1858 335
BATAVIER 1799-1817 241
BATHURST 1821-1858 292
BEACON 1804-1808 281
BEACON 1832-1846 see METEOR 1823
BEAGLE 1804-1814 142
BEAGLE 1820-1845 145
Béarnois (French 1808) see CURIEUX 1809
BEAULIEU 1791-1808 233
Beau Marseilles (French) see DEFENDER 1809
BEAUMONT 1780-1783 224
BEAUMONT 1804-1805 see ROSE 1804
BEAVER 1757-1761 205
**BEAVER** 1761-1783 94
BEAVER 1795-1808 249
BEAVER 1801-1801 264
BEAVER 1809-1829 145
BEAVER 1827-1854 318
BEAVERS PRIZE 1778-1780 see CONVERT 1777
Beckford see PORTSMOUTH 1756
Beckford see LAUREL 1759
BECKWITH 1816-1822 301
BEDFORD 1698-1736 20
BEDFORD 1741-1767 42
BEDFORD 1775-1815 70
BEDFORD GALLY 1697-1709 27
BEDFORD GALLY 1709-1725 36
BEE 1808-1849 316, 318, 319
BEE 1809-1809 289
BEELZEBUB see BELZEBUB
BEHEMOTH see SEIJEREN 1807
BELABOURER 1815 302
BELEM 1806-1806 284
Belerina (merchant 1809) see DART 1810
BELETTE 1793-1795 260
BELETTE 1798-1801 253
BELETTE 1806-1812 141
BELETTE 1814-1828 142
BELISARIUS see BELLISARIUS
BELLEISLE 1761-1784 205
BELLEISLE 1795-1814 258
BELLEISLE 1819-1854 113
Bella Josephine (1823) see ADELAIDE 1827
BELLE POULE 1780-1796 218
BELLE POULE 1806-1816 271
BELLEROPHON 1786-1816 69
BELLEROPHON 1824-1856 see WATERLOO 1818
BELLETTE see also BELETTE
BELLIQUEX 1758-1772 204
BELLIQUEX 1780-1814 74
Belliquex (French) see BELETTE 1798
BELLISARIUS 1781-1783 220
BELLONA 1747-1748 198
**BELLONA** 1760-1814 67
BELLONA 1778-1779/1780 234
BELLONA 1794-1805 256, 318
Bellona (Dutch 1794) see VINDICTIVE 1796
BELLONA 1804-1804 289
BELLONE 1806-1809 284
Bellona (Franco/Italian 1806) see DOVER 1811
BELLONA 1818-1842 see INDUS 1812
Bellone (French 1745) see BELLONA 1747
Bellone (French 1755) see REPULSE 1759
Bellone (French 1778) see PROSERPINE 1790
Bellone (French 1803) see JUNON 1810
Bellone (Franco/Italian 1806) see DOVER 1811
Bellotte (French 1781) see BELETTE 1793
Belmont (East Indiaman) see MONMOUTH 1796
BELVIDERA 1809-1846 124
BELZEBUB Cancelled 1812 148
BELZEBUB 1813-1820 148
BELZEBUB Cancelled 1832 149
BENBOW 1813-1848 114
BENGAL see FYEN 1807
BENJAMIN 1701-1703 194
BENJAMIN 1704-1704 194
BENJAMIN 1705-1705 194
BENJAMIN AND ANN 1794-1800 256
Bentinck (merchant) see PLOVER 1841
BERBICE 1780-1796 228
BERBICE 1782-1785? 231
BERBICE? 1793-1796 260
BERBICE 1805-1806 284
BERESFORD *Ca*1797 333
BERESFORD 1813-....? see NIAGARA, Lake Ontario
BERESFORD 1814-1860 333
BERGERE 1806-1811 276
BERMUDA 1795-1796 260
**BERMUDA** 1806-1808 136
BERMUDA 1808-1816 144
BERMUDA 1813-1841 285
BERMUDA 1816-1864 330
BERMUDA 1819-1821 295
**BERMUDA** 1848-1855 182
Bertin (French 1760) see BELLEISLE 1761
BERWICK 1679-1700 12
BERWICK 1700-1715 20
BERWICK 1723-1743 41
BERWICK 1743-1760 43
BERWICK 1775-1795/1805 70
BERWICK 1781-1786 230
BERWICK 1805-1816 see SAN JUAN
BERWICK 1809-1821 114
BERWICK PACKET 1809-1809 289
BESCHERMER 1799-1806 242
BESSIE/BESSY 1861-1870s 323
Bethia 1784 see BOUNTY 1787
Betsey (American) see YORK 1777
BETSEY 1790s 264
BETSEY 1803-1804 289
BETSEY 1804-1804 289
BETSEY 1804-1804 289
BETSEY 1804-1810 289
BETSEY 1806-1810 289
BETSEY GAINES [or GAINS] 1806-1810 289
BETSY *Ca*1770 298
BETSY 1782-1782 233
BETSY 1805?-1810 281
BETTY 1689-...? 189
BETTY 1728-1813 323
BETTY/BETTY PRIZE 1695-1695+ 185
BEULAH 1798-1802 264
BIDDEFORD 1695-1699 28
BIDDEFORD 1712-1726/1727 37
BIDDEFORD 1727-1736 50
BIDDEFORD 1740-1754 51
BIDDEFORD 1756-1761 89
BIEN AIME 1743-1746/1748 201
BIEN AIME 1793-1794 260
'Bien Aquise' see AURORA 1757
BIENFAISANT 1758-1783 204
Bien Venue (French 1788) see UNDAUNTED 1794
Birch (merchant) see RHINOCEROS 1781
BIRD 1764-1775 212
BISHOP 1797-1822 260
BITER 1797-1862 151
BITER 1804-1805 153
BITTERN 1796-1833 131
BITTERN 1840-1860 180
BLACK JOKE 1793-1800 264
BLACK JOKE 1806-1810 (1811) 289
Black Joke see DILIGENT 1813
BLACK JOKE 1827-1832 292
Black Prince (American 1774) see ALFRED 1778
**BLACK PRINCE** 1816-1855 115
BLACKSNAKE 1815 302
BLACKWALL 1696-1705 24
BLACKWELL HALL 1700-1701 194
BLADE OF WHEAT 1689-1689 186
**BLAKE** 1808-1814 113
BLAKE 1819-1823 see BOMBAY 1808
BLANCHE 1779-1780 231
BLANCHE 1786-1799 84
BLANCHE 1800-1805 124
BLANCHE 1805-1807 see AMPHITRITE 1804
BLANCHE 1809-1814 see BELLONE 1806
BLANCHE 1819-1833 119
BLANDFORD 1711-1719 37
BLANDFORD 1720-1742 49
BLANDFORD 1741-1767 51
BLANFORD see BLANDFORD
BLAST 1695-1724 30, 328
BLAST 1740-1745 60
BLAST 1759-1771 99
BLAST 1779-1783 see DRUID 1776
BLAZE 1691-1692 29
BLAZE 1694-1697 29
BLAZE 1739-1742 200
BLAZE 1745-1750? 201
BLAZE 1756-1757 207
BLAZER 1797-1803 151
BLAZER 1804-1814 153
BLENHEIM 1706-1709 see DUCHESS 1679
BLENHEIM 1709-1740 33
BLENHEIM 1761-1807 63
Blenheim (merchant 1803) see CORMORANT 1804
BLENHEIM see CHRISTIAN 7th. 1807
BLENHEIM 1813-1858 114
BLESSING 1694-1696 188
BLESSING 1804-1806 289
BLONDE 1760-1782 206
BLONDE 1787-1803 84
BLONDE 1793-1794* 260
BLONDE 1805-1811 see HEBE 1782
BLONDE see ISTER 1813
**BLONDE** 1819-1850 123
Bloodhound? (merchant) see DISCOVERY 1776
BLOODHOUND 1780s-1783 225
**BLOODHOUND** 1801-1816 152
BLOODLETTER 1815 302
BLOOM 1795-1797 260
BLOSSOM 1806-1833 130
BLUCHER/BLUTCHER 1814 299
**BOADICEA** 1797-1858 121
BOLD 1812-1813 155
BOLD see MANLY 1804
BOLINA 1803-1807 289
Bolivar (mercantile 1822) see NIMBLE 1826
BOLTON 1709-1817 303
BOLTON 1775-1776 231
BOMBAY see BOMBAY CASTLE 1782
BOMBAY 1790 235
BOMBAY 1805-1808 274
BOMBAY see BLAKE 1808
BOMBAY 1808-1819 72
BOMBAY 1828-1864 107
BOMBAY CASTLE 1782-1796 70
BONAVENTURE 1683-1699 13
BONAVENTURE 1689-1697 189
BONADVENTURE 1699-1711 24
BONADVENTURE 1705-1705 194
BONADVENTURE 1711-1716 35
Bonaparte [French]

INDEX 353


see SWAGGERER 1809
BONAVISTA 1778-1779 231
BONETTA 1699-1712 31
BONETTA 1718-1719 195
BONETTA 1721-1731 52
BONETTA 1732-1744 53
**BONETTA** 1756-1776 93
BONETTA 1779-1781/
1782-1797 96
BONETTA
see SWAN 1781
BONETTA 1798-1801 250
BONETTA 1803-1810 276
**BONETTA** 1836-1861 178
BONITA 1870-1887 317
BONNE CITOYENNE
1796-1819 249
Bonne Marseilles (French)
see DEFENDER 1809
Bonnet Rouge (French 1795)
see ABEILLE 1796
Bordelais (French)
see BOURDELOIS 1799
BOREAS 1757-1770 85
BOREAS 1774-1797 86
BOREAS 1806-1807 128
BOREAS
see HASFRUEN 1807
BORER 1794-1808 149
BORER 1812-1815 154
Borryan?
see SCORPION 1771
Boscawen (privateer 1745)
see MEDEA 1744
BOSCAWEN 1748,
1759-1760 202, 211
BOSCAWEN 1759 297
BOSCAWEN 1763-1773 208
**BOSCAWEN** Cancelled
1818 109
**BOSCAWEN** 1844-1914 173
BOSCAWEN 1873-1906
see TRAFALGAR 1841
BOSTON 1694-1695 189
BOSTON 1748-1752 87
BOSTON 1756-1757
see AMERICA 1749
BOSTON 1762-1811 83
BOSTON 1764-1768 301
Boston (American)
see CHARLESTOWN 1780
Boullogne (French)
see BOULOGNE 1762
BOULOGNE 1762-1776 206
Boulogne (French 1779-1781)
see JACKAL 1778
BOUNCER 1797-1802 151
BOUNCER 1804-1805 153
BOUNTIFUL 1782-1783 230
BOUNTY 1787-1789 233
BOUNTY 1809-1864+ 322
BOURBON CARO 1702-...? 339
BOURBONNAISE
1809-1817 272
BOURDELOIS 1799-1804 247
BOXER 1797-1809 151
BOXER 1812-1813 145
BOYNE 1692-1708 18
BOYNE 1708-1732 33
BOYNE 1739-1763 41
BOYNE 1766-1783 73
**BOYNE** 1790-1795 66
**BOYNE** 1810-1834 106
BRAAK 1795-1798 251
BRAAK 1799-1802 248
BRAAVE 1796-1811 246
BRADDICK 1804-1805
see SAMUEL BRADDICK
BRAGANZA
see PRINCESS CAROLINA 1807
Braganza (merchant)
see COCKBURN 1822
BRAKEL 1796-1807 241
BRAMBLE 1809-1815 163
**BRAMBLE** 1822-1853 165
BRAVE 1747-1748 202
Brave (French 1794)
see ARROGANT 1798
BRAVE 1798-1799 264
Brave (French)
see ARAB 1798
Brave (French privateer 1801)
see BARBADOES 1803
BRAVE 1805-1808 267
BRAVE 1806-1806 285
Brave Savage
see ROYAL SAVAGE 1775
BRAVO 1794-1803 151
BRAZEN 1781-1799 227
BRAZEN 1798-1800 250
**BRAZEN** Cancelled 1799 131
BRAZEN 1808-1827 131
BREAM 1807-1817? 161
BREDAH 1679-1690 12
BREDAH 1692-1730 20
BREDAH
see PRINCE OF ORANGE 1734
BREDALBANE 1849 296
BREV DRAGEREN
1807-1815 277
BRIDGEWATER 1698-1738 27
BRIDGWATER 1740-1743 91
BRIDGWATER 1744-1758 92
BRIGHTON 1795-1797 260
BRILLIANT 1729 195
BRILLIANT 1755-1756 212
BRILLIANT 1759-1776 82
BRILLIANT 1779-1811 86
Brilliant (merchant)
see ADVICE 1796
BRILLIANT
see ORONTES 1813
BRILLIANT 1814-1860 125
Brilliant (Franco-Italian)
see GENOA 1814
BRILLIANT PRIZE
1695?-1698 186
BRISEIS 1808-1816 144
BRISEIS 1829-1839 146
BRISK 1784-1805 98
BRISK 1805-1816 131
BRISK 1819-1843 145
BRISTOL 1653-1693 13
BRISTOL 1692-1697 190
BRISTOL 1693-1709 23
BRISTOL 1703-1704 194
BRISTOL 1711-1743 35
BRISTOL 1746-1768 47
BRISTOL 1775-1794 77
BRISTOL 1812-1814
see AGINCOURT 1796
BRITANNIA 1682-1716 11
BRITANNIA 1719-1745 39
BRITANNIA 1762-1813 62
BRITANNIA 1781-1782 230
BRITANNIA 1793-1796 264
BRITANNIA 1803-1811 289
BRITANNIA 1804-1809 289
BRITANNIA 1820-1855 104
BRITANNIA 1869-1909
see PRINCE OF WALES
BRITISH FAIR(E)
1793-1800 264
BRITISH FAIR 1803-1805 289
BRITISH FAIR 1807-1814 289
BRITOMART
see GLOMMEN 1807
BRITOMART 1808-1819 144
BRITOMART 1820-1843 145
**BRITOMART** 1847-1863 181
BRITON 1812-1841 119
BRITON 1869-1908
see BRILLIANT 1814
BROADERSCARP
1799-1805 242
Broaderschap [Dutch 1795]
see BROADERSCARP 1799
BROCHETTE 1759 297
BROCK *Ca*1815 302
BROCK 1817-1837 301
BRODERICK 1759-1760? 211
BROKE 1814-1814 289
BROKE
see FINCH 1813
BROOKE 1803-1809 285
BROTHERS 1795-1802 258
BRUISER 1797-1802 151
BRUIZER 1804-1815 153
BRUNE 1761-1792 206
BRUNE 1808-1816 272
BRUNSWICK 1765-1778 299
**BRUNSWICK** 1790-1812 109
**BRUNSWICK** 1855-1867 173
**BUCEPHALUS** 1808-1822 127
BUCKINGHAM 1711-1727
see REVENGE 1699
BUCKINGHAM 1731-1742 41
BUCKINGHAM 1751-1777 72
BUCKINGHAM 1800-1812
see EAGLE 1774
BUFFALO 1777-1783
see CAPTAIN 1743
BUFFALO 1797-1809 260
BUFFALO 1813-1840 284
BUFFALO *Ca*1815 301
BULLDOG 1782-1801 97
BULLDOG 1794-1794 256
BULLFROG 1838-1841 302
BULWARK Cancelled 1783 71
**BULWARK** 1807-1826 109
Bunkers Hill (American)
see SURPRIZE 1778
Buonaparte (French)
see PERT 1804
Buonaparte (French)
see SWAGGERER
BURCHETT 1708-1709 194
BURCHETT 1708, 1711 194
BURCHETT 1708-1709 194
BURFORD 1679-1698 11
BURFORD 1698-1719 20
BURFORD 1722-1752 41
**BURFORD** 1757-1785 73
BURLINGTON 1695-1733 23
BURLINGTON ...?-1833
see PRINCESS CHARLOTTE
BUSTARD 1806-1815 278
BUSTARD 1818-1829 145
BUSTLER 1782-1788 223
BUSTLER 1805-1808 154
BUSY 1778-1792 226
**BUSY** 1797-1806 140
Busy (merchant)
see ROCKET 1804
BUTE
see EARL OF BUTE
BUZZARD 1813?-1814
see HAWKE 1806
BUZZARD 1834-1843 146


# C


CABOT 1777-1783 222
CACAFUEGO 1799-1804 261
Cacique (Slaver)
see SNAP 1845
CADIZ MERCHANT
1688-1692 187
**CADMUS** 1808-1835 144
CAESAR Cancelled 1783 71
**CAESAR** 1793-1814 67
**CAESAR** 1853-1870 172
CA IRA 1795-1796 261
CALCUTTA 1795-1805 241
CALCUTTA 1831-1863 107
CALDWELL 1770 299
CALEDON 1808-1811 280
**CALEDONIA** 1808-1856 104
CALEDONIA 1812?-1815? 302
CALEDONIA 1891-1906
see IMPREGNABLE 1810
CALLIOPE 1808-1829 144
**CALLIOPE** 1837-1855 135
CALPE 1800-1802 253
CALYPSO 1783-1803 98
CALYPSO 1805-1821 140
CALYPSO
see HYENA cancelled 1828 146
CALYPSO 1826-1833 145
CALYPSO 1845-1866 177
CALYPSO 1870-1895
see BLONDE 1819
**CAMBRIAN** 1797-1828 116
CAMBRIAN 1841-1892 175
CAMBRIDGE 1666-1694 12
CAMBRIDGE 1695-1713 19
CAMBRIDGE 1715-1750 33
**CAMBRIDGE** 1755-1790 66
CAMBRIDGE (Danish)
see PRINCESS SOPHIA
**CAMBRIDGE** 1815-1856 109
CAMBRIDGE 1869-1898
see WINDSOR CASTLE 1858
CAMDEN 1797-1801 261
CAMEL 1776-1784 220
CAMEL *Ca*1779 298
CAMEL 1788-1784
see MEDIATOR 1782
CAMEL 1798-1831 324, 327
CAMEL 1813-1819 284
CAMEL 1820s-by 1830 330
CAMEL 1851?-1877 319
CAMELEON 1777-1780 96
CAMELEON 1781-1783 223
CAMELION 1795-1811 139
CAMELEON 1816-1849 145
CAMELION 18..-1834 335
CAMELION ?1855-1859? 335
CAMILLA 1776-1809 91
CAMILLA 1847-1861 180
CAMPBELL 1796-1803 261
CAMPERDOWN
1797-1802 240
CAMPERDOWN
1803-1804 289
CAMPERDOWN 1825-1860
see TRAFALGAR 1820
CAMPHAAN 1799-1801 253
CANADA 1711-1711 194
**CANADA** 1765-1810 69
CANADA 1775-1775/6 231
CANADA ?1777-1779? 231
CANADA 1805? 289
CANADA 1814-...? 298
CANADA Building 1815 301
CANCEAUX 1764-1782 212
Cannonière (French 1803-1809)
see MINERVE 1795
CANOPUS 1798-1862 237
CANSO
see CANCEAUX
CANSO 1813-1816 283
CANTERBURY 1692-1703 188
CANTERBURY 1693-1720 22
CANTERBURY 1711-1711 194
CANTERBURY 1722-1741 44
CANTERBURY 1745-1770 45
CANTERBURY 1756-1764 210,
318
CAPELIN 1804-1808 160
CAPTAIN 1678-1705 12
CAPTAIN 1708-1720 34
CAPTAIN 1722-1740 41
CAPTAIN 1743-1777 43
CAPTAIN 1787-1809 69
CAPTAIN 1815-1825
see CARNATIC 1783
CAPTAIN 1825-1841
see ROYAL SOVEREION 1786
CAPTIVITY 1796-1818
see MONMOUTH 1772
CAPTIVITY 1824-1836
see BELLEROPHON 1786
CARCASS 1695-1713 30
CARCASS 1740-1749 60
CARCASS 1759-1784 99
CARKASS
see CARCASS
CARLETON 1776 298
CARLISLE 1693-1696 22
CARLISLE 1699-1700 24
CARLISLE 1745-1745 202
CARLOTTA 1810-1815 280
CARMELA 1741-1742 201
CARMEN 1800-1802 246
CARMIN 1741-1742 201
CARNATIC 1783-1805 72
CARNATIC 1823-1860 115
CARNATION 1807-1808 141
CARNATION 1813-1826 142
CAROLINA 1716-1732
see PEREGRINE GALLEY 1700
and under yachts (1713)
CAROLINA
see ROYAL CAROLINE 1733
CAROLINA 1781-1784 225
CAROLINE 1798-1798 264
CAROLINE 1794-1802 256
CAROLINE 1795-1813 123
CAROLINE 1795-1802 258
Caroline (French)
see FLECHE 1798
Caroline (French 1802 or 1806)
see BOURBONNAISE 1809
CAROLINE 1804-1807 289
CAROLINE 1809-1809 289
CAROLINE 1809-1814? 285
Caroline (mercantile)
see FAWN 1839
CAROLINE 1859-1863 295
CARRERE 1801-1814 244
CARRIER 1805-1808 281
Carrière (Franco/Italian 1797)
see CARRERE 1801
Carron (East Indiaman 1804)
see DOVER 1810
CARRON 1813-1820 137
**CARRON** 1827-1844 146
CARTERET 1803-1805 289
CARYSFORT 1766-1813 85
CARYSFORT 1836-1861 175
**CASSANDRA** Cancelled 1782 82
CASSANDRA 1806-1807 162
CASSANDRA
see ALERT 1807
Cassard (French 1795)
see BRAVE 1806
CASSIUS 1848-after 1864 330
CASTILIAN 1809-1829 142
CASTLE 1672-1692 15
CASTLE OF MASTERLAND
1694-1695 187
CASTLEREAGH 1846-1848 295
CASTOR 1781-1781 231
CASTOR 1785-1819 84
Castor (Dutch 1780)
see SALDANHA 1796
**CASTOR** 1832-1902 126
Catalonia (Spanish)
see MUSTICO 1800
CATHERINE
see also KATHERINE
CATHERINE 1691-1694 189
CATHERINE 1794-1801 254
CATO 1782-1782 78
CATON 1782-1794 216
CAUSTIC 1815 299
CAWSAND 1795-1798 324
CECILIA 1806-1811 289
CELEBES 1806-1808 285
CENSEUR 1795-1795 261
CENSOR 1801-1816 152
CENTAUR 1746-1761 52
CENTAUR 1759-1782 203
CENTAUR 1797-1819 109
Centaure (French 1757)
see CENTAUR 1759
CENTINEL
see SENTINEL
CENTURION 1650-1689 13
CENTURION 1690-1729 23
CENTURION
1733-1744/1745-1769 44
CENTURION
see EAGLE 1745
CENTURION 1774-1808 77
CENTURION 1827-1828
see CLARENCE
CENTURION 1844-1870 172
CENTURION PRIZE
1743-1743 201
CEPHALUS 1807-1830 141
CERBERE 1800-1804 260
CERBERUS 1758-1778 85
CERBERUS 1779-1783 84
**CERBERUS** 1794-1814 126
Cerberus (Dutch 1784)
see TEXEL 1799
CERBERUS 1827-1860 120
**CERES** 1777-1778/
1779-1782 96
Cérès (French 1778-1779)
see CERES 1777
CERES 1781-1803 84
CERES 1793-1794 264
CERES 1801-1801 264
Ceres (East Indiaman)
see GRAMPUS 1795
CERES 1808-.... 327
Cérès (French 1810)
see SEINE 1814
CERES 1833-1877
see CERUS 1830
CERF 1803-1806 277
CERF 1805-1809
see CYANE 1796
CERUS 1830-1833 293
CESAR
see CAESAR 1793
CESAR 1806-1807 285
CEYLON 1808-1832
see BOMBAY 1805
CHALEUR 1764-1768
see CERBERE 1800
CHALLENGER Cancelled
1848 178
CHALLENGER 1806-1811 143
CHALLENGER 1813-1819 142
**CHALLENGER** 1826-1835 135
**CHAMPION** 1779-1809 88
CHAMPION 1783? 233
CHAMPION 1793-1800 264
CHAMPION 1803-1804 289
**CHAMPION** 1824-1859 138
CHAMPLAIN 1816?-*Ca*1832 299
CHANCE 1799-1800 261
CHANCE 1804-1806
see NANCY hired 1804
CHANCE 1815-1815 289
CHANCE ?1828-1880? 334
CHANTICLEER 1808-1833 144
CHAPMAN 1793-1798 264
CHAPMAN 1804-1806 289
Charente-Inférieure (French 1794)
see TRIBUNE 1796
CHARGER 1801-1814 153
CHARGER 1830-1851 337
CHARITY 1689-1692 189
CHARITY 1770 299
CHARLES 1668-1687
see SAINT GEORGE 1687
CHARLES 1688-1695 187
CHARLES 1745-1747 202
CHARLES 1804-1814 289
CHARLES 1811-1814 289
CHARLES GALLY
1676-1693 13
CHARLES GALLY
1693-1710 26
CHARLES GALLEY
1710-1726 36
Charles Jackson (merchant)
see TARTARUS 1797
CHARLES AND HENRY
1688-1689 187
CHARLES ROYAL
see ROYAL CHARLES 1673
CHARLESTOWN
1780-1783 219
CHARLOTTE 1677-1721 303
CHARLOTTE 1711-1771 303
CHARLOTTE 1763-1770 208
Charlotte
see ETNA 1756
CHARLOTTE
see ROYAL CHARLOTTE 1762
Charlotte
see ROYAL CHARLOTTE 1780
CHARLOTTE 1798-1799 261
CHARLOTTE 1800-1801 261
CHARLOTTE ?-1801 264
CHARLOTTE 1803-1806 289
CHARLOTTE 1805-1806 289
Charming Jenny (merchant)
see FIREBRAND 1739
Charming Molly (merchant)
see GRIFFIN 1778
CHARMING MOLLY
1796-1801 264
Charming Sally (American)
see BARBUDA 1780
CHARON 1778-1781 79
CHARON 1783-1805 79
CHARWELL 1801-1813 250
CHARWELL 1814-1837
see MOIRA, Lake Ontario
CHARWELL 1816 301
CHARYBDIS 1809-1819 142
CHARYBDIS 1831-1843 146
CHASER 1781-1782/
1783-1784 224
Chasseur (French)
see CHASER 1781
CHATHAM 1691-1718 23
CHATHAM 1694-1813 318
CHATHAM 1716-1743 304
CHATHAM 1721-1749 45
CHATHAM 1741-1807 304
**CHATHAM** 1758-1810 77
CHATHAM 1788-1830? 233
CHATHAM 1790-1794 233
CHATHAM 1793-1795 264
CHATHAM 1793-1867 305
CHATHAM 1811-1864 168
**CHATHAM** 1812-1817 115
CHATHAM 1813-1876 319
CHATHAM PRIZE
1704-1708 194
**CHEERFUL** 1806-1816 162
CHEERFUL 1819-...? 334
CHEERLY 1804-1815 153
**CHEROKEE** 1808-1828 144
CHEROKEE 1851-... 183
CHERUB 1806-1820 130
CHESAPEAKE 1813-1819 271
CHESAPEAKE Cancelled
1851 175
CHESTER 1691-1707 23
CHESTER 1708-1750 35
CHESTER 1744-1767 47
CHESTERFIELD 1745-1762 48
see POMONA 1761
Chevert (fRENCH)
see POMONA 1761
CHICHESTER 1695-1705 19
CHICHESTER 1706-1749 20
**CHICHESTER** 1753-1779 73
CHICHESTER 1785-1810 80
CHICHESTER 1809-1811 284
CHICHESTER 1843-1866 116
CHIFFONE 1801-1814 247
**CHILDERS** 1778-1811 98
CHILDERS 1812-1822 142
CHILDERS 1827-1865 135
CHILDS PLAY 1706-1707 192
CHIPPAWA 1769-1775 301
CHIPPEWA 1795 302
CHIPPEWAY 1812-1813 302
CHOMONDELY 1763-1771 208
CHOLMONDY 1704-1704 194
CHRISTIAN 7th.[VII]
1807-1814 267
CHUB 1807-1812 161
CHUB 1813-1814 298
CHURCHILL 1705-1708 194
CIRCE 1785-1810 87
CIRCE 1804-1814 128
CIRCE 1827-1866 120
City of Kingston (1821)
see UNION 1823
CLARA 1804-1811 273
CLARA 1848-after 1864 330
CLARENCE 1812-1827 114
CLARENCE 1827-1864 107
CLARENCE 1885-1899
see ROYAL WILLIAM 1833
CLAUDIA 1806-1809 162
CLEAVELAND 1671-1715 303
CLEEVLAND
see CLEAVELAND
CLEOPATRA 1779-1814 84
CLEOPATRA 1815 302
CLEOPATRA 1835-1862 175
Cléopatre (French 1791?)
see PROSELYTE 1793
Cléopatre (French 1791?)
see OISEAU 1793
CLEVELAND
see CLEAVELAND

CLEVELAND 1803-1806 289
CLEVELAND 1809-1810 289
CLINKER 1797-1802 151
CLINKER 1804-1806 153
CLINKER 1813-1831 155
CLINTON 1780-1784 219
CLINTON 1782 231
CLIO 1807-1845 141
CLORINDE 1803-1817 270
Clorinde (French 1808)
see AURORA 1814
CLYDE 1796-1805 118
CLYDE 1805-1826 285
CLYDE 1806-1814 118
CLYDE 1828-1870 120
Cocarde Nationale (French)
see NATIONAL COCKADE 1795
COCHIN 1822-1850 329
COCKATRICE 1781-1802 102
COCKATRICE 1804-1808 289
COCKATRICE
see BREV DRAGEREN 1807
COCKATRICE 1832-1858 166
COCKBURN 1822-1823 295
COCKBURN 1827-1837 302
COCKCHAFER 1793-1801 264
COCKCHAFER 1813-1815 283
Codrington
see CHEROKEE 1774
COLCHESTER 1694-1704 23
COLCHESTER 1707-1718 25
COLCHESTER 1721-1742 45
COLCHESTER 1744-1744 47
COLCHESTER 1746-1773 47
COLIBRI 1809-1813 285
**COLIBRI** Cancelled 1814 147
**COLLINGWOOD** 1841-1867 172
COLOMBE 1803-1811 277
COLONEL MURRAY 1815 299
COLOSSUS 1787-1798 72
**COLOSSUS** 1803-1826 112
COLOSSUS 1848-1867 172
COLPOYS 1804-1807 289
COLUMBIA 1812/13-1820 280
COLUMBIA 1829-1859 146
COLUMBINE 1806-1824 141
**COLUMBINE** 1826-1854 147
**COMBATANT** 1804-1816 136
COMBUSTION 1782-1784 226
COMEET 1795-1798 251
COMET 1695-1706 30
COMET 1742-1749 61
COMET 1756 211
COMET 1761-1763 211
COMET 1777-1778 224
COMET 1778-1782? 231
COMET 1779-1780 100
see DILIGENCE 1756
COMET 1783-1800 211
Comet (merchant 1800)
see SPY 1804
COMET 1807-1815 149
COMET 1828-1832 138
COMET 1869-1870 138
Comète (French)
see COMET 1761
COMMERCE 1803-1807 283
COMMERCE DE MARSEILLES 1793-1796 236
COMTE D'ESTAING 1782? 231
Comptesse de Maurepas (French)
see PORT ROYAL 1779
Comptesse de Provence (French)
see ROBECQUE 1782
Comte de Saint Florentine (French)
see SAINT FLORENTINE 1759
Comte de Toulouse (French 1703)
see THOULOUSE 1711
Comte de Vallance (French)
see SWIFT 1761
Comtesse d'Ayen (French)
see LURCHER 1761
COMUS 1806-1816 128
COMUS 1832-1862
see COMET 1828
CONCEPTION 1782-1783 346
CONCEPTION PRIZE 1690-1694 185
CONCORD 1697-...? 187
Concord
see PLUTO 1756
CONCORD 1803-1810 283
CONCORDE 1783-1811 219
Conde de Chincon (Spanish or French?)
see RIPPONS PRIZE
CONFEDERATE 1781-1782 218
Confederacy (American 1778)
see CONFEDERATE 1781
Confiance (Franco-American 1800)
see SKIPJACK 1807
CONFIANCE 1805-1810 275
Confiance (French 1809-1810)
see MINERVE 1795
CONFIANCE 1813-1822 142
CONFIANCE 1814-1814 299
CONFIANCE 1813-1814 360
CONFIANCE 1814 302
CONFIANCE 1816-1831 302
CONFIANCE 1827-by 1872 146
CONFIANCE 1840s 326
CONFLAGRATION 1781-1781 231
CONFLAGRATION 1783-1793 99
CONFLICT 1801-1804 153
CONFLICT 1805-1810 154
CONFLICT 1812-1830s 154
Conflit (French 1804-1814)
see CONFLICT 1801
**CONFOUNDER** 1805-1814 154
**CONGO** 1816-1826 169
CONGRESS 1781 232
Congress (American)
see GENERAL MONCK 1781
CONQUERANT 1798-1803 238
CONQUEROR 1746-1749 201
CONQUEROR 1758-1760 73
CONQUEROR 1773-1794 70
**CONQUEROR** 1801-1822 110
CONQUEROR 1862-1876
see WATERLOO 1833
**CONQUEST** 1794-1817 149
Conquistador (Spanish 1720-38)
see GLOUCESTER 1709
CONQUESTADORE 1762-1775 205
**CONQUESTADORE** 1810-1856 114
Conquistador (Spanish 1758)
see CONQUESTADORE 1762
CONSTANCE 1796-1801 264
CONSTANCE 1797-1806 248
CONSTANCE 1799-1804 264, 289
Constance (French)
see MARIA 1805
CONSTANCE Cancelled 1844? 175
**CONSTANCE** 1846-1875 173
CONSTANT 1801-1816 153
CONSTANT WARWICK 1666-1691 13
Constitution (French)
see MAIDA 1806
CONSTITUTION 1796-1801 264
CONSTITUTION 1804-1804 289
CONSTITUTION 1835 295
Content (French 1686)
see CONTENT PRIZE
CONTENT 1705-1706/1708-1715 195
CONTENT 1777-1780? 233
CONTENT 1799-1799 264
CONTENT PRIZE 1695-1702 185, 344
CONTEST 1797-1799 151
CONTEST 1799-1799 261
CONTEST 1799-1803 259
CONTEST 1804-1809 153
CONTEST 1812-1828 154
**CONTEST** 1846-1868 181
Convention Nationale (French)
see MARIE ANTOINETTE 1794
CONVERT
see LOYAL CONVERT 1776
CONVERT 1777-1778 221
CONVERT 1778-1782 217
CONVERT 1793-1794 261
**CONVULSION** 1804-1806 156
**CONWAY** 1814-1825 132
CONWAY 1832-1861 134
CONWAY
see WINCHESTER 1822
CONWAY 1876-1956
see NILE 1829
CONWAY CASTLE 1804-1809 289
COQUETTE 1783-1785? 231
**COQUETTE** 1807-1817 132
COQUETTE Cancelled 1840s 177
COQUILLE 1798-1798 261
CORDELIA 1808-1833 144
CORMORANT 1757-1762 206
CORMORANT 1776-1781 96
CORMORANT 1781-1783 221
**CORMORANT** 1794-1796 130
CORMORANT 1796-1800 248
CORMORANT 1804-1817 276
CORNELIA 1808-1814 127
CORNELIUS 1804-1805 285
CORNWALL 1692-1704 18
CORNWALL 1706-1722 19
CORNWALL 1726-1755 40
CORNWALL 1761-1780 68
Cornwall (mercantile 1796)
see VOLCANO 1797
CORNWALL
see HEIR APPARENT FREDERICK 1807
CORNWALL 1812-1859 114
CORNWALL *Ca*1815 302
CORNWALL 1868-1940
see WELLESLEY 1815
CORNWALLIS 1777-1782 228
CORNWALLIS 1781-1782 230
CORNWALLIS 1797?-1798 261
Cornwallis (1798)
see DESPATCH 1816
CORNWALLIS 1805-1812 270
**CORNWALLIS** 1813-1862 115
COROMANDEL 1795-1807 241
COROMANDEL 1815-1827
see MALABAR 1804
Corona (Franco-Italian 1809)
see DAEDALUS 1811
CORONACON
see CORONATION 1705
CORONATION 1685-1691 11
CORONATION 1705-1707 194
CORONATION 1831-1869 325
CORSO 1796-1803 254
Cosmopolite (French)
see DEMARARA 1806
COSSACK 1806-1816 129
Countess D'Ayen
see LURCHER 1761
COUNTESS OF ELGIN 1803-1814 289
COUNTESS OF SCARBOROUGH 1777-1779 234
Courageuse (French)
see COURAGEUX 1799
COURAGEUX 1761-1796 203
Courageux (French 1779?)
see LUTINE 1799
COURAGEUX 1799-1803 261
**COURAGEUX** 1800-1814 109
Coureer (Dutch)
see QUEEN MAB 1807
COUREUR 1778-1780 228
COUREUR 1798-1801 248
COUREUSE 1795-1799 255
COURIER 1798-1801 264
COURIER 1804-1806 289
COURIER
see QUEEN MAB 1807
Courier (merchant 1827)
see MONKEY 1831
Courier de l'Orient (1783)
see FORTUNE 1778
Courrier de New York (1783-*Ca*1794)
see ALLIGATOR
Couronne (French 1766)
see CA IRA 1793
**COURSER** 1797-1803 152
COVE 1835 295
COVENTRY 1694-1704/1709-1709 23
**COVENTRY** 1757-1783 85
CRACHE FEU 1795-1797 259
CRACKER 1797-1802 151
CRACKER 1797-1797 264
CRACKER 1804-1815 153
CRACKER 1826-1842 293, 322
CRACKER 1826-1826 165
CRACKER by 1845-1850 326
CRACKER Cancelled 1850 183
CRAFTY 1804-1807 285
CRANE 1777-1783 229
CRANE 1806-1808 161
CRANE 1809-1814 142
CRANE 1839-1862 179
CRASH 1797-1802 151
CRASH
see SCOURGE 1794
CREMILL 1836-1897? 325
CREOLE 1803-1804* 285
CREOLE 1813-1833 124
CREOLE 1845-1875 176
CRESCENT 1690-1698 187
CRESCENT 1758-1777 205
CRESCENT 1779-1781 86
CRESCENT 1784-1808 81
CRESCENT 1810-1840 122
CRESSY 1810-1832 114
**CRESSY** 1853-1867 172
CRESSY 1941-1959
see UNICORN 1824
CRETAN 1808-1814 279
CRIB 1804-1805 285
CRICKET 1808-1808 289
CROCODILE 1781-1784 88
CROCODILE 1806-1816 129
CROCODILE 1825-1850 133
**CROCUS** 1808-1815 144
CROWN 1654-1704 13
CROWN 1704-1719 25
CROWN 1747-1770 79
**CROWN** 1782-1798 75
CROWN 1801-1801 264
CROWN 1794-1800 256
CROWN HERRING 1694-...? 188
CROWN PRIZE 1691-1692 186
CRUELLE 1798-1801 261
CRUELLE 1800-1801 260
CRUISER 1705-1708 192
CRUISER
see also CRUIZER
CRUISER 1721-1732 52
Cruiser (merchant)
see KITE 1778
CRUISER 1797-1819 140
CRUISER 1828-1849 135
CRUIZER 1709-1712 194
CRUIZER 1721-1724 195
CRUIZER 1732-1745 35
CRUIZER 1746-1748 201
**CRUIZER** 1752-1770 92
CRUIZER 1759-1759 211
CRUIZER 1780-1791 227
CRUIZER 1804-1808 289
CRUIZER
see also CRUISER
CRUIZER PRIZE
see CRUISER 1705
CRYSLER by 1815 301
CUBA 1806-1811 274
CUBA 1857-1867 293
**CUCKOO** 1806-1810 161
CULLODEN
see PRINCE HENRY 1747
CULLODEN/CULLODEN SMACK 1746-1765 323
**CULLODEN** 1747-1770 67
**CULLODEN** 1776-1781 71
**CULLODEN** 1783-1813 72
CUMBERLAND 1695-1707 19
CUMBERLAND 1710-1732 33
CUMBERLAND 1739-1760 41
CUMBERLAND 1739-1742 200
CUMBERLAND? 1745? 201
CUMBERLAND 1774-1804? 70
Cumberland (American 1777)
see ROVER 1779
CUMBERLAND 1803-1804/1809-1810 285
CUMBERLAND 1807-1830 113
CUMBERLAND 1842-1889 173
CUPID 1777-1778 221
CUPID Not purchased 1781 227
CUPITT *Ca*1750s 327
CURACOA 1802?* 285
CURACOA 1809-1849 124
CURIEUX 1804-1809 277
CURIEUX 1809-1814 280
CURLEW 1795-1796 139
CURLEW 1803-1810 276
Curlew (American 1812)
see COLUMBIA 1812
CURLEW 1812-1822 142
CURLEW 1830-1849 146
CURLEW *Ca*1841-*Ca*1857 335
CUTTLE 1807-1813 161
Cuvera (East Indiaman 1798)
see MALABAR 1804
CYANE 1796-1805 131
CYANE 1806-1815 129
CYCLOPS 1779-1807 86
CYDNUS 1813-1816 120
CYGNET 1688-1693 187
CYGNET 1758-1768 207
CYGNET 1776-1790 96
CYGNET 1796-1799 264
CYGNET 1800-1801 264
CYGNET 1804-1815 131
CYGNET 1819-1835 145
CYGNET 1840-1877 179
**CYNTHIA** 1796-1809 131
CYNTHIA 1810-1816? 285
CYNTHIA 1826-1827 295
CYPRESS 1760-1760 211
CYRENE 1814-1828 137
CYRUS 1782-1786 231
**CYRUS** 1813-1823 137

# D

DAEDALUS
see DEDALUS
Damas Argentinas (Spanish)
see KANGAROO
DAME DE GRACE 1799-1799 261
Dame Ernouf (French)
see SEAFORTH 1805
DANAE 1759-1771 205
DANAE 1779-1797 219
DANAE 1798-1800 248
DANGEREUSE 1800-1800? 260
DANNEMARK 1807-1815 268
DANIEL AND CHRISTOPHER 1706-1706 194
DAPHNE 1776-1795/1797-1802 91
DAPHNE 1794-1796 264
DAPHNE 1796-1798 247
DAPHNE 1806-1816 129
DAPHNE Cancelled 1830 132
DAPHNE 1838-1865 177
DAPPER 1805-1814 154
**DARING** 1804-1813 153
DARING Cancelled 1843 179
**DARING** 1844-1865 180
DART 1782-1782 233
**DART** 1796-1809 132
DART 1803-1808 283
DART 1803-1805 289
DART 1805-1814 289
DART 1810-1813 285
DART 1810-1831 333
Dart (American 1813)
see PIKE 1813
DART 1814-...? 316
DART 1814-...? 316, 320
DART 1847-1857 179, 329
DARTMOUTH 1655-1690 13
DARTMOUTH 1693-1695/1702-1703 23
DARTMOUTH 1698-1714 24
DARTMOUTH 1716-1733 36
DARTMOUTH 1741-1747 46
DARTMOUTH Cancelled 1749 77
DARTMOUTH 1813-1831 124
**DASHER** 1797-1818 136
Dasher (merchant)
see THUNDER 1803
DAUNTLESS 1804-1807 136
DAUNTLESS 1808-1825 130
Dauphin (French 1743)
see RICHMOND 1746
Dauphin (Franco-American)
see SWIFT 1782
DAUPHINE 1780?-1784 230
DAUPHIN ROYAL 1796-1801 261
David (merchant)
see POUNCER 1797
DEAL 1812-1830 316
DEAL CASTLE 1697-1706 28
DEAL CASTLE 1706-1722 29
DEAL CASTLE 1727-1747 50
DEAL CASTLE 1746-1756 52
DEAL CASTLE 1756-1780 90
DE BRAAK
see BRAAK
DECADE 1798-1801 246
Décade Française (French 1794)
see DECADE 1798
DECOUVERTE 1803-1808 255
DECOUVERTE 1806-1816 285
**DECOY** 1810-1814 164
DEDAIGNEUSE 1801-1810 246
**DEE** 1814-1819 132
DEE
see AFRICAN 1825
**DEDALUS** 1780-1811 84
DAEDALUS 1790-1795 235
DEDALUS 1811-1813 270
DEDALUS 1826-1861 120
DEFENCE 1763-1811 68
DEFENCE 1796-1802? 334
DEFENCE 1804 285
Defence (merchant?)
see SQUIB 1804
DEFENCE 1815-1848 114
DEFENCE 1817 333
DEFENCE 1826-1847 334
DEFENCE 1847-1859? 335
DEFENDER 1797-1802 152
DEFENDER 1804-1809 153
DEFENDER 1809-1814 283
DEFIANCE 1676-1695 12
DEFIANCE 1695-1707 22
DEFIANCE 1707-1743 35
DEFIANCE 1730-1742 44
DEFIANCE 1744-1766 45
DEFIANCE 1772-1780 74
DEFIANCE 1781-1786 233
DEFIANCE 1782 234
DEFIANCE 1783-1813 70
DEFIANCE 1794-1797 256
DE GIER
see GIER 1799
DEGO 1800-1802 241
DELAVAL 1694 189
DELAWARE 1777-1783 219
DELFT 1797-1802 240
DELIGHT 1686?-1714 31
DELIGHT 1709-1713 37
DELIGHT 1778-1781 96
DELIGHT 1782 231
DELIGHT 1801-1805 250
DELIGHT 1806-1808 143
DELIGHT 1808-1814 278
DELIGHT 1819-1824 145
DELIGHT 1826?-1890? 334
DELIGHT 1829-1844 146
DELIGHT FRIGAT 1701-1702 194
DELPHINEN 1807-1808 278
DEMARARA 1804-1804 285
DEMERARA 1806-1813 285
De Meric
see CRUISER 1705
DENMARK 1785-1785
see PRINCE FREDERICK1807
DENMARK
see DANNEMARK 1807
DENNIS 1743-1833 325
DEPENDANCE 1776-1786 228
DEPTFORD 1665-1689 14
DEPTFORD 1687-1700 13
DEPTFORD 1700-1717 24
DEPTFORD 1702-1713 315, 329
DEPTFORD 1716-1811 315
DEPTFORD 1719-1724? 36
DEPTFORD 1732-1767 44
DEPTFORD 1735-1756 61
DEPTFORD 1781-1865 315, 319, 321
DEPTFORD 1784-1816 315, 324
DEPTFORD 1790-1816 326, 327
DEPTFORD'S PRIZE 1740-1744 200
DE RUYTER 1799-1804 241
DERWENT 1807-1817 141
DESIREE 1800-1823 244
DESPATCH
includes DISPATCH
DISPATCH 1692-1712 31
DISPATCH 1711-1711 195
DISPATCH 1745-1773 55
DISPATCH 1764-1780? 213
DISPATCH 1770-1771 211
DISPATCH 1776-1777 228
DISPATCH 1776? 231
DISPATCH 1777-1778 96
DESPATCH 1777-1783
see CHEROKEE 1774
DISPATCH 1783-1798
see ZEPHYR 1779
DISPATCH 1781-1795? 231
DISPATCH 1781-1782 231
DESPATCH 1795-1798 139
DISPATCH 1797-1801 264
DISPATCH 1797-1800 261
DESPATCH 1798-1820 317
DESPATCH 1799-1801 250
DESPATCH 1804-1811 140
DESPATCH 1812-1836 142
DISPATCH 1816-1846 293
DISPATCH 1841-.... 325
DESPATCH 1851-1863 180
DESPERATE 1799-1811 261
DESPERATE 1805-1814 153
Desperate (Bermuda 1797)
see FIERCE 1806
DESTRUCTION 1804-1806 156
DESTRUCTION 1815-1815 285
DETERMINEE 1799-1803 248
DETERMINEE
see NYMPHEN 1807
Déterminée (French)
see NETLEY 1807
DETROIT 1812-1812 302
DETROIT 1813-1813 302
DEUX AMIS 1796-1799 255
Deux Couronnes (French)
see RANGER 1747
Deux Frères (French 1784)
see JUSTE 1794
Deux Frères (French)
see TWO BROTHERS 1798/9
DEVASTATION 1803-1812 281
DEVASTATION Cancelled 1820 148
DEVON 1835-1868 325
DEVONPORT 1821-after 1864 325, 330
DEVONSHIRE 1692-1701 18
DEVONSHIRE 1704-1707 19
DEVONSHIRE 1710-1740 33
DEVONSHIRE 1745-1772 41
DEVONSHIRE 1804-1804 285
DEVONSHIRE 1812-1849 114
DEXTEROUS 1805-1816 153
D'Hautpoult (French)
see ABERCROMBIE 1809
Diable a Quatre (French 1792)
see IMOGEN 1800

INDEX 355

DIADEM 1782-1815 74
Diadem (merchant 1801)
see FALCON 1801
DIADEM 1804-1806 289
Diamant (French 1693-1697)
see DIAMOND 1651
Diamant (French)
see ISIS 1747
DIAMOND 1651-1693 13
DIAMOND 1708-1721 36
DIAMOND 1723-1744 47
DIAMOND 1741-1756 48
DIAMOND 1774-1784 84
DIAMOND 1793-1794 264
DIAMOND 1794-1812 118
DIAMOND 1816-1827 119
DIAMOND 1832-1833 296
**DIAMOND** 1848-1868 176
DIAMOND ROCK
1804-1805 285
DIANA 1757-1793 83
DIANA 1775-1775 213
DIANA 1794-1815 118
DIANA 1804-1805 288
DIANA 1807-1810 285
DIANA 1822-1868 120
Diane (French 1796)
see NIOBE 1800
DICTATOR 1783-1817 75
DIDO 1784-1804 87
**DIDO** 1836-1903 177
DIDON 1805 271
DILIGENCE 1692-1708 31
DILIGENCE 1709-1712 193
**DILIGENCE** 1756-1779 93
DILIGENCE 1759-1767 323
DILIGENCE 1759-1769 210
Diligence
see DISCOVERY 1776
DILIGENCE 1789 234
DILIGENCE 1800-1800 264
**DILIGENCE** 1795-1801 139
DILIGENCE *Ca*1797 333
DILIGENCE 1801-1812 251
Diligence (French)
see INCENDIARY 1804
DILIGENCE 1812-1814 285
DILIGENCE 1814-1862 168
DILIGENCE 1818-1830? 334
DILIGENCE GALLY
see DILIGENCE 1709
DILIGENT 1770-1775 213
Diligent (merchant)
see DISCOVERY 1776
DILIGENT 1776-1777 228
DILIGENT 1777-1779 229
DILIGENT/DILIGENTE
1780-1784 215
DILIGENT 1781-1790 231
DILIGENT 1790-1794 233
DILIGENT 1793-1800 264
DILIGENT 1799-1802
see PORPOISE 1798
DILIGENT 1800 261
Diligent (merchant?)
see SQUIB 1804
Diligent (French 1799)
see WOLF 1806
DILIGENT 1813-1814 284
Diligente (French 1691)
see DILIGENT PRIZE
Diligente (Spanish)
see DILIGENT 1780
Diligente (French 1809)
see SAINT PIERRE 1809
Diligente (French 1810)
see DILIGENT 1813
Diligentia (Spanish)
see LIGERA 1804
DILIGENT PRIZE
1694-1698 185
DIOMEDE 1781-1795 80
**DIOMEDE** 1798-1815 115
DIRECTOR 1784-1801 73
DISCOVERY 1692-1705 31
DISCOVERY 17..-1719 195
DISCOVERY 1741-1750 201, 328
DISCOVERY 1776-1797 229
DISCOVERY 1789-1808 234
DISCOVERY DOGGER
1711-1712 193
DISPATCH
see under DESPATCH
Dix Août (French 1795)
see BRAVE 1806
DIXMUND 1803-1807 285
DIXMUNDE 1795-1803 258
Dolfijn (Dutch 1774)
see DOLPHIN 1799
DOLLY 1796-1801 264
DOLLY 1803-1803 289
DOLORES 1807-1807? 285
DOLPHIN 1690-1709 26
DOLPHIN 1700s-1713 321
DOLPHIN 1711-1730 36
DOLPHIN 1732-1755 50
DOLPHIN 1751-1777 87
DOLPHIN 1781-1817 79
DOLPHIN 1793-1797? 332
DOLPHIN 1793-1796 264
DOLPHIN 1793-1801-
1802 264, 255
DOLPHIN 1799-1802 261
DOLPHIN 1799-1821 333
DOLPHIN 1813 285
DOLPHIN 1815-1830
see HINDOSTAN 1804
DOLPHIN 1829?-1858? 334
DOLPHIN 1836-1861 178
DOLPHINS PRIZE 1757-1760 207
DOMINICA 1805-1809 285
DOMINICA 1805-1806 285
DOMINICA 1809-1813/
1814-1814 283
DON CARLOS 1811
see INFANTE DON CARLOS
DONEGAL 1798-1845 236
DORDRECHT 1796-1823 239
DORIS 1795-1805 123
DORIS Cancelled 1806 127
DORIS 1808-1828 82
DOROTHEA 1817-1819 296
DOROTHY 1707-1708 194
DORSET 1753-1815 305
DORSET 1782 233
DORSET 1793-1802 264
DORSETSHIRE 1694-1706 19
DORSETSHIRE 1712-1749 33
DORSETSHIRE 1757-1775 73
Dos Amigos (Slaver)
see FAIR ROSAMOND 1832
DOTTEREL 1808-1827 142
DOURO 1831-1832 296
DOVE 1805-1805 282
DOVE 1805-1806 282
DOVE 1818-1865 334
DOVE 1823-1829 295
DOVE 1829?-1840? 335
DOVE 1823-1872 319
DOVE? Cancelled 1843? 179
DOVE
see KANGAROO 1852
DOVER 1654-1695 13
DOVER 1695-1730 24
DOVER 1741-1763 48
DOVER 1786-1806 80
DOVER 1793-1794 264
DOVER 1794-1801 264
DOVER 1794-...? 256
DOVER 1807?-1811 274
DOVER 1808-1810 289
DOVER 1811-1825 274
DOVER PRIZE
1689-1689 189, 342
DOVER PRIZE
1693-1698 185
DRAGON 1647-1690 13
DRAGON 1690-1707 23
DRAGON 1707-1712 35
DRAGON 1715-1733
see ORMOND 1711
DRAGON 1736-1757 44
Dragon (French 1758)
see EURUS 1758
DRAGON 1760-1780 68
DRAGON 1782-1785 226
**DRAGON** 1798-1824 110
DRAGON PRIZE
1689-1690 186
DRAGON'S PRIZE
1742?-1745 201
DRAKE 1653-1690/1 14
DRAKE 1694-1694 27
DRAKE 1705-1715? 32
DRAKE 1705-1727 303
DRAKE 1727-1749 304
DRAKE 1729-1740 52
DRAKE 1741-1742 53
DRAKE 1743-1748 55
DRAKE 1759-1760 211
DRAKE
see RESOLUTION 1770
DRAKE 1777-1778 224
DRAKE 1779 234
DRAKE 1779-1800 222
DRAKE 1798-1804* 261
DRAKE 1799-1801 264
DRAKE 1804-1808 276
DRAKE 1804-1805 289
DRAKE 1808-1822 145
DRAKE 1817 334
DRAKE 1834-1854 183, 322
DRAKE
see HART 1822
DREADFULL 1695-1695 30
DREADNOUGHT 1660-1690 12
DREADNOUGHT 1691-1705 22
DREADNOUGHT 1706-1721 23
DREADNOUGHT 1723-1742 44
DREADNOUGHT 1742-1784 45
**DREADNOUGHT**
1801-1825 106
DREADNOUGHT 1856-1875
see CALEDONIA 1808
DREADNOUGHT'S PRIZE
1748-1748 200
DRIVER 1797-1824 136
DROCHTERLAND
1799-1815 242
DROMEDARY 1777-1783 229
DROMEDARY
see JANUS 1778
DROMEDARY 1808-1819
see HOWE 1805
DROMEDARY
see LIVELY 1838
**DRUID** 1761-1773 94
DRUID 1776-1779 220
DRUID 1783-1813 84
DRUID 1825-1846 122
DRYAD 1795-1838 123
DUBLIN 1709-1752 303
DUBLIN
see DORSET 1753
**DUBLIN** 1757-1784 67
DUBLIN 1803-1810 289
DUBLIN 1812-1841 114
Duc D'Aiguillon (French)
see POSTILLION 1757
Duc D'Aquitaine (French 1754)
see DUKE OF AQUITAINE
1757
Duc De Chatres (French)
see INVERNESS 1746
DUC DE CHATRES
1781-1784 221
Duc De Chaulnes (French 1694-7)
see SCARBOROUGH 1694
DUC DE LA VAUGIGNON
1779-1779 226
DUC D'ESTISSAC
1781-1783 221
Duc De Wagram (French 1805)
see DOMINICA 1809
DUCHESS 1679-1701 11
DUCHESS OF BEDFORD
1804-1808 289
DUCHESS OF CUMBERLAND
1783 233
DUCHESS OF CUMBERLAND
1793-1801 264
DUCHESS OF CUMBERLAND
1803-1805 289
Duchess of Manchester (merchant)
see PETEREL 1777
DUCHESS OF YORK
1796-1800 264
DUCHESS OF YORK
1834-1836 295
DUCK 1833-after 1879 319
DUCKWORTH 1803-1803 289
Duff
see BEACON 1804
Du Guay Trouin (French)
see GUAY TROUIN 1780
Duguay-Trouin (French 1800)
see IMPLACABLE 1805
DUKE 1682-1701 11
DUKE 1728-1733
see VANGUARD 1710
DUKE 1739-1769 42
DUKE 1739-1742 200
DUKE 1746-1749 201
DUKE 1760-1761 211
**DUKE** 1777-1798 65
DUKE BRONTE 1803-1803 289
DUKE OF AQUITAINE
1757-1761 204
DUKE OF CLARENCE
1794-1801 264
DUKE OF CLARENCE
1803-1804 289
Duke of Cumberland (merchant)
see SHEERNESS
WATERBOAT 1747
DUKE OF CUMBERLAND
1759 297
DUKE OF CUMBERLAND
1803-1805 289
Duke of Cumberland (merchant)
see DROMEDARY 1777
DUKE OF CUMBERLAND
1803-1805 289
DUKE OF GLOUCESTER
1807-1813 299
DUKE OF KENT 1776 299
DUKE OF KENT 1801 299
DUKE OF KENT 1804-1805 289
DUKE OF KENT
Proposed 1809 285
DUKE OF RUTLAND 1784 234
**DUKE OF WELLINGTON**
1852-1909 170
DUKE OF YORK 1763-1766 208
DUKE OF YORK 1794-1799 264
DUKE OF YORK 1796 333
DUKE OF YORK 1803-1810 289
Duke of York (1824)
see MESSENGER 1830
DUKE WILLIAM 1746 202
DUKE WILLIAM 1763-1768 208
DUMBARTON 1685-1691 15
DUMBARTON CASTLE (1696)
1707-1708 192
DUNBARTON
see DUMBARTON
DUNCAN
see DOVER 1807
DUNCAN 1811-1826 114
DUNIRA
see IMMORTALITE 1814
DUNKIRK 1660-1704 12
DUNKIRK 1704-1729 23
DUNKIRK 1734-1749 44
**DUNKIRK** 1754-1778 76
DUNKIRK PRIZE
1705-1708 192
DUNMORE 1772-1796? 301
DUNWICH 1695-1714 28
DUQUESNE 1803-1805 267
Duquesne (French)
see UNIQUE 1808/9
DURHAM 1867-1908
see ACTIVE 1845
DURSLEY GALLEY
1719-1745 37
DUTCHESS
see DUCHESS
DWARF 1810-1824 164
DWARF 1826-1826 165

## E

EAGLE 1660-1694 14
EAGLE 1679-1699 12
EAGLE 1696-1703 32
EAGLE 1699-1707 20
EAGLE 1744-1745
see CENTURION 1733
EAGLE 1745-1767 45
EAGLE 1745-1757 202
EAGLE 1774-1794 74
EAGLE 1793-1818 332
EAGLE 1794-1804 256
EAGLE 1795-1798 332
EAGLE
see ECLIPSE 1803
EAGLE 1804-1918 113
EAGLE 1805-1827 333
Eagle (American)
see FINCH
EAGLE 1815-1815 286
EAGLE 1816?-1840? 333
EAGLE 1844-1893 335
EAGLET 1691-1693 30
EAGLET 1699 188
EAGLET 1918-1927
see EAGLE 1804
EARL/EARL GALLEY
1702-1707 195
EARL DENBIGH
see EARL OF DENBIGH
EARL MOIRA 1812 286
Earl Mornington
see DRAKE 1804
EARL OF BUTE 1776-1777 233
Earl of Chatham (merchant)
see CHATHAM 1790
EARL OF CHATHAM
1792/3-1799? 261
Earl of Denbigh (merchant)
see PELICAN 1777
EARL OF DENBIGH
1788-1797 346
EARL OF EGMONT
1770-1777 212
Earl of Mansfield (East Indiaman)
see WEYMOUTH 1795
EARL OF MOIRA
see MOIRA 1805
Earl of Mornington 1799
see DRAKE 1804
EARL OF NORTHAMPTON
1769-1774 212
Earl of Pembroke (merchant 1768)
see ENDEAVOUR 1768
EARL ST.VINCENT
1799-1801 264
EARL ST.VINCENT
1804-1808 289
EARL SPENCER
1799-1801 264
EARL SPENCER 1799-1799 264
EARL SPENCER 1803-1814 289
Earl Spencer (merchant)
see ALTHORP 1804
Earl Talbot (East Indiaman)
see AGINCOURT 1796
EARNEST 1805-1816 153
EBRINGTON 1805-1808 289
ECHO 1758-1770 206
ECHO 1780-1781 224
**ECHO** 1782-1797 97
**ECHO** 1797-1809 135
ECHO 1809-1817 142
ECHO 1827-1885 146
ECLAIR 1793-1797 248
ECLAIR 1795-1802 259
ECLAIR 1801-1807 261
ECLAIR 1807-1831 141
Eclair (French 1808)
see PICKLE 1808
ECLIPSE 1787-1787 233
ECLIPSE 1797-1802 152
ECLIPSE 1803-1807 277
Eclipse
see DECOUVERTE 1806
ECLIPSE 1807-1815 141
ECLIPSE 1808-.... 329
ECLIPSE 1819-1838 145
Ecureil (French 1706-1708)
see SQUIRREL 1704
Ecureil (French 1814-1819)
see BOUNCER 1804
EDEN 1814-1833 132
EDGAR 1668-1700 12
EDGAR 1702-1708 21
EDGAR 1709-1711 34
**EDGAR** 1758-1774 76
EDGAR 1779-1813 68
EDGAR
see HOOD 1859
EDINBURGH 1707-1709 192
EDINBURGH 1716-1717
see WARSPITE 1703
EDINBURGH 1721-1742 41
EDINBURGH 1744-1771 43
EDINBURGH 1811-1866 114
Edward (American 1812)
see VARIABLE 1814
EDWARD & SUSANNA
1689-1690 189
EDWINSTONE 1801-1802 264
EGERIA 1807-1826 130
EGMONT 1764-1776 211
**EGMONT** 1768-1799 70
EGMONT
see EARL OF EGMONT 1770
EGMONT 1810-1863 114
EGREMONT 1810
(American?) 286
EGREMONT 1843?-1849? 335
EGYPTIENNE 1801-1817 242
Egyptienne (French)
see ANTIGUA prison ship 1804
EIJDEREN/EIYDEREN
1807-1813 276
ELEANOR 1739-1742 200
ELECTRA 1806-1808 143
ELECTRA 1808-1816 279
ELECTRA Cancelled 1831 139
ELECTRA 1837-1862 177
ELEPHANT 1705-1709 194
ELEPHANT 1776-1779 228
ELEPHANT 1786-1830 69
ELEPHANT FLYBOAT
see ELEPHANT 1705
ELING 1795-1827 323
**ELING** 1796-1814 159
ELISABETH
see under ELIZABETH
ELIZA 1803-1812 289
ELIZA 1804-1814 289
ELIZA 1805-1811 289
ELIZA 1821-1822 296
ELIZA 1843?-1860? 335
ELIZA AND JANE 1803-1812 289
ELIZABETH 1679-1704 12
ELIZABETH 1704-1704 21
ELIZABETH 1705 194
ELIZABETH 1706-1737 21
ELIZABETH 1708 194
ELIZABETH early C18th 327
ELIZABETH 1721-1722 196
ELIZABETH 1737-1766 41
ELIZABETH 1741 201
**ELIZABETH** 1769-1797 70
Elizabeth (American)
see LUCIFER 1780?
Elizabeth
see SORLINGS 1789
ELIZABETH 1795-1801? 258
ELISABETH 1795-by 1830 324
Elizabeth (Merchant 1801)
see AVENGER 1803
ELIZABETH 1805-1812 289
Elizabeth (French)
see TRINIDAD 1805
ELIZABETH 1806-1807 286
ELIZABETH 1807-1820 113
ELIZABETH 1808-1814 286
ELIZABETH 1808-1809 289
ELIZABETH 1809-1809 289
ELIZABETH AND MARY
1804-1814 289
ELIZABETH AND SARAH
1688-1690 187
ELK 1804-1812 142
ELK 1813-1836 142
ELK 1847-1863 180
ELLEN 1804-1809 289
ELLENS 1803-1809 289
ELTHAM 1736-1763 48
ELVEN/ELVIN 1807-1814 277
Embuscade (French 1703)
see PLYMOUNT PRIZE 1709
Embuscade (French 1745)
see AMBUSCADE 1746
Embuscade (Dutch) 1795
see HELDER 1799
Embuscade (French 1798-1803)
see AMBUSCADE 1773
Embuscade (French 1789)
see AMBUSCADE 1798
EMERALD 1757-1761 206
EMERALD 1762-1793 83
EMERALD 1795-1822 123
EMERALD 1820-1847 293
**EMERALD** 1856-1869 174
Emeraude (French 1744)
see EMERALD 1757
EMILIA 1840 295
Emilien (French)
see TRINCOMALEE 1801
EMILY 1867 327
Empereur (French 1799)
see HART 1805
EMMA 1809-1811 289
EMPRESS MARY
1799-1804 260
Empress of Russia (merchant)
see VIGILANT 1777
Empress of Russia (merchant)
see VOLCANO 1780
EMS 1809-1815 286
EMULOUS 1806-1812 141
EMULOUS 1812-1817 280
EMULOUS 1819-1834 145
ENCHANTRESS 1804-1813 277
Encounter (French)
see DECOUVERTE 1806
ENCOUNTER 1805-1812 154
ENDEAVOUR 1694-1695 188
ENDEAVOUR
1695-1705 188, 317
ENDEAVOUR 1704-1704 194
ENDEAVOUR 1704-1706 194
ENDEAVOUR 1705-1707 194
ENDEAVOUR 1708-1713 323
ENDEAVOUR 1711-1711 194
ENDEAVOUR 1763-1771 208
ENDEAVOUR 1768-1778 212
ENDEAVOUR *Ca*1770? 213
ENDEAVOUR 1775-1782 212
ENDEAVOUR 1804-1812 289
ENDYMION 1779-1790 80
**ENDYMION** 1797-1868 116
Enfant Prodigue (French 1794)
see SAINT LUCIA 1803
ENGAGEANTE 1794-1811 245
ENGLAND 1693-1695 189
ENQUIRY 1691-1691 189
ENTERPRISE/ENTERPRIZE
1705-1707 192
ENTERPRISE/ENTERPRIZE
1709-1740 36
ENTERPRIZE
see LIVERPOOL 1741
ENTERPRIZE 1743?-1748 201
ENTERPRIZE 1744-1771
see NORWICH 1718
**ENTERPRIZE** 1774-1791 86
ENTERPRIZE 1801 262
ENTERPRIZE 1803-1816
see RESOURCE 1778
ENTERPRISE 1848-1903 294
ENTREPRENANTE
1798-1812 255
Envieuse (French 1769)
see AURORE 1793
Eolo (French 1799)
see NIMROD 1799
Eolo (Slaver)
see PICKLE 1852
EOLUS 1758-1796 83
EOLUS 1801-1817 126
EOLUS
see PIQUE 1801

356 INDEX


EOLUS 1825-1846 120
Epervier (French)
see EPERVOIR 1797
EPERVIER 1803-1811 277
EPERVIER 1812-1814 142
EPERVOIR 1797-1801 249
EPHIRA 1808-1811 144
EPHRAIM 1695-1695 189
EPREUVE/EPREVUE
1760-1764 207
EREBUS 1807-1819 149
EREBUS 1826-1848 148
ERIDANUS 1813-1818 125
ERNESTOWEN 1815 302
ERNE 1813-1819 137
ERUPTION 1804-1807 280
ERYNNIS 1758-1760 211
ESCORT 1801-1816 152
ESCORTE 1757-1768 206
ESDAILE 1808-1809 289
ESK 1813-1827 137
Espérance (French 1779)
see CLINTON 1780
ESPERANCE 1795-1798 254
Espérance (French 1808-1810)
see LAUREL 1806
Esperanza (French
see FLYING FISH 1793
ESPIEGLE 1793-1802 251
ESPIEGLE 1794-1802 262
ESPIEGLE 1804-1811 276
Espiègle (French 1804)
see ELECTRA 1808
ESPIEGLE 1812-1832 142
**ESPIEGLE** 1844-1861 181
ESPION 1782-1784 224
ESPION 1793-1794 248
ESPION 1794-1800 245
ESPION
see LITTLE BELT 1807
ESPOIR 1784 231
ESPOIR 1797-1804 252
ESPOIR 1804-1821 140
ESPOIR 1826-1853 145
ESSEX 1679-1699 12
ESSEX 1700-1736 20
ESSEX 1741-1757 42
ESSEX 1760-1777 73
ESSEX 1814-1824 273
ESSEX PRIZE 1694-1702 186
ESTHER 1760-1762 211
ESTHER 1763-1779 208
ETHALION 1797-1799 118
**ETHALION** 1802-1835 125
ETNA/AETNA 1691-1697 29
ETNA 1739-1746 200
ETNA 1756-1763 207
ETNA
see SCORPION 1771
ETNA 1776-1784 99
ETNA 1781 231
Etna (French 1794)
see CORMORANT 1796
ETNA 1803-1816 281
ETNA 1824-1839 148
AETNA
see KITE 1826
Etoile (French 1703)
see SWALLOW PRIZE 1705
Etoile (French 1814)
see TOPAZE 1814
ETRUSCO 1795?-1798 260
EUGENIA/EUGENIE
1797-1802? 252
Eugénie (French 1798)
see PANDOUR 1798
EUGENIE 1804-1810 276
EUPHRATES 1813-1818 125
EUPHRATES Cancelled 1831 123
EUPHROSYNE 1796-1802 260
EUROPA/EUROPE
1689-1690 190
EUROPA 1765-1796 73
EUROPA 1782 232
EUROPA 1783-1814 77
EUROPE
see EUROPA 1765
EUROTAS 1813-1817 120
EUROTAS 1829-1865 123
EURUS 1758-1760 206
EURUS 1796-1806 245
**EURYALUS** 1803-1825 124
EURYDICE 1781-1824 88
**EURYDICE** 1843-1878 176
EUSTATIA
see SAINT EUSTATIUS 1781
EVEILLE 1795-1796 262
EXCELLENT 1787-1830 68
EXCELLENT 1834-1859
see BOYNE 1810
EXCELLENT 1859-1892
see QUEEN CHARLOTTE 1810
EXERTION 1805-1812 154
EXETER 1680-1691 13
EXETER 1697-1740 22
EXETER 1744-1763 45
**EXETER** 1763-1784 73
EXMOUTH 1854-1905 171
EXPEDITION 1679-1699 13
EXPEDITION 1699-1709 20
Expédition (merchant)
see DEPTFORD 1702
EXPEDITION 1714-1715 34
EXPEDITION 1747-1764 79
EXPEDITION 1778-1816 101
EXPEDITION 1781-1783 224
EXPEDITION 1784-1810 80
EXPEDITION 1803-1804 289
EXPERIMENT 1689-1724 26
EXPERIMENT 1727-1738 50
EXPERIMENT 1740-1763 51
EXPERIMENT 1765-1768 212
**EXPERIMENT** 1774-1778 78
EXPERIMENT 1781-1785 225
EXPERIMENT 1784-1805 80
**EXPERIMENT** 1793-1796 159
EXPERIMENT 1794-1801 254
Experiment
see MANLY 1797
EXPERIMENT 1806-1807? 157
EXPERIMENT 1825-1843? 334
EXPLOSION 1779-1783
see SWAN 1767
EXPLOSION 1797-1807 254
EXPRESS 1695-1713 32
**EXPRESS** 1800-1813 160
EXPRESS 1815?-1827 283
EXPRESS 1835-1862 179
EXTRAVIGANT 1691-1692 187
EYDEREN
see EIJDEREN


## F


FAIR RHODIAN 1781-1782 231
FAIR ROSAMOND
1831-1845 293
FAIRY 1778-1811 96
FAIRY 1812-1821 142
FAIRY 1826-1840 145
FAIRY 1842?-1855? 335
FAIRY 1859 335
Faisan (French 1761)
see PHEASANT 1761
FAITH 1774 301
FALCON 1666-1694 13
FALCON 1694-1695 27
FALCON 1704-1709 27
**FALCON** 1744-1745 58
FALCON 1745-1759 57
FALCON 1771-1779 96
FALCON 1782-1800 98
FALCON ?1793-1806? 332
FALCON 1801-1816 251
FALCON 1807-1808 286
FALCON 1820-1838 145
FALKLAND 1696-1702 24
FALKLAND 1702-1718 24
FALKLAND 1720-1743 45
FALKLAND 1744-1768 47
FALKLAND PRIZE
1704-1705 191
FALMOUTH 1693-1702 23
FALMOUTH 1708-1724 35
FALMOUTH 1729-1747 45
FALMOUTH 1752-1765 76
FALMOUTH 1807-1883 318, 323
FALMOUTH 1814-1825 137
FALMOUTH 1829-... 327
FAMA 1804-1812 271
FAMA 1808-1808 286
FAME 1702-1703 194
FAME 1709-1710 193
FAME 1744-1745 199
FAME
see RENOWN 1747
**FAME** 1759-1799 67
Fame (merchant)
see PROSPERITY 1782
FAME 1806-1814 289
**FAME** 1805-1817 112
FAME 1842-1850
see DRAGON 1798
FANCY 1806-1811 154
FANCY 1808-1809 289
FANCY 1809-1811 289
FANCY 1811-1814 289
FANCY 1817 334
FANFAN 1666-1692 14
FANNY 1799-1801 264
Fanny (merchant 1801)
see PIGEON 1805
FANNY 1804-1804 290
FANNY 1805-1807 290
**FANNY** 1827-1838 181
FANNY 1831-1863 306, 331
FANNY 1845 295
FANTOME 1810-1814 280
FANTOME 1839-1865 180
FARMER'S GOODWILL
1699 189
FARMER'S INCREASE
1801-1801 264
FARQUER
see LIVELY 1755
Faucon/Faucon Anglais (French
1694)
see FALCON 1666
Faucon (French)
see HAWKE 1756
FAULCON
see FALCON
FAULKLAND
see FALKLAND
FAULKON
see FALCON
Faune (French)
see FAWN 1805
Fauvette (French 1743)
see MEDWAY'S PRIZE 1745
FAVERSHAM 1696-1711 27
FAVERSHAM 1712-1730 36
FAVERSHAM 1741-1749 48
Favori (French 1815-1824)
see MALLARD 1801
FAVORITE
see FAVOURITE
**FAVOURITE** 1757-1784 94
FAVOURITE 1794-1806 130
FAVOURITE 1803-1806 290
FAVOURITE 1803-1804 290
FAVOURITE 1805-1813? 286
FAVOURITE 1806-1821 130
FAVOURITE 1806-1807 282
FAVOURITE 1807-1811 290
**FAVOURITE** 1829-1890 138
Faune (French 1804)
see FAWN 1805
FAWN 1805-1806? 286
FAWN 1807-1818 132
FAWN 1840-1847 293, 329
FEARLESS 1794-1804 149
FEARLESS 1804-1812 153
FELICITE 1809-1809* 286
FELICITY 1775 301
FELICIDADE 1845-1845* 295
FELIX 1803-1807 282
Fenix (Spanish 1749)
see GIBRALTAR 1780
Ferme (French 1699)
see FIRME 1702
Ferme
see FIRME 1805
FERRET 1704-1706 32
FERRET 1711-1718 37
FERRET 1721-1731 52
FERRET 1743-1757 199
**FERRET** 1760-1775 94
**FERRET** 1763-1781 100
FERRET 1784-1801 98
FERRET 1794-1801 256
FERRET *Ca*1795 332
FERRET 1799-1799 262
FERRET 1808-1808 290
Ferret (American 1804)
see MOHAWK 1813
FERRET 1806-1813 140
FERRET 1813-1820
see NOVA SCOTIA 1812
FERRET 1821-1837 145
FERRET 1840-1869 179
FERRETER 1801-1807 152
FERVENT 1804-1864? 153, 322
FEVERSHAM
see FAVERSHAM
Fidèle (French 1794)
see TOURTERELLE 1795
Fidèlle (French 1809)
see LAUREL 1809
FIERCE 1806-1813
see BRAVE 1805
FIIEN
see FYEN
FIILA
see FYLLA
FINCH 1813-1814 298
FINDON 1801-1801 264
FINDON 1804-1810 290
FIREBRAND 1694-1707 29
FIREBRAND 1739-1743 200
FIREBRAND 1755-1757
see DOLPHIN 1732
FIREBRAND 1778-1781
see PORPOISE 1777
FIREBRAND 1794-1800 254, 322
FIREBRAND 1804-1804 281
FIREBRAND 1804-1807 281
FIREBRAND 1815 286
FIREDRAKE 1688-1689 30
FIREDRAKE 1693-1703 30
FIREDRAKE 1742-1763 61
FIREDRAKE 1804-1807 281
FIREFLY 1781-1782 231
Firefly (American?)
see REVENGE
FIREFLY 1803-1806? 286
FIREFLY 1807-1812 286
FIREFLY 1807-1807
see FLYING FISH 1803
FIREFLY 1828-1835 295
**FIRM** 1759-1791 76
**FIRM** 1794-1802 150
FIRM
see DIOMEDE 1798
FIRM 1804-1811 153
FIRME 1702-1713 191
FIRME 1805-1814 268
FISGARD 1797-1814 244
FISGARD 1819-1879 119
FISHGUARD
see FISGARD
FISGARD III 1905-1920
see HINDOSTAN 1841
FLAMBEAU 1804-1807 281
FLAMBOROUGH
1697-1705 28
FLAMBOROUGH
1707-1727 36
FLAMBOROUGH
1727-1749 50
FLAMBOROUGH
1756-1772 89
FLAMBOROUGH PRIZE
1757-1763 207
FLAME 1690-1697 29
FLAMER 1797-1802 152
FLAMER 1804-1858 153
FLASH 1804-1807 281
FLECHE 1794-1795 262
FLECHE 1798-1810 250
FLEMISH 1673-1713 318
FLEUR DE LA MER
1808-1811 286
Fleur de Lys (French 1785)
see PIQUE 1795
FLIGHT
see DOVE 1805
FLIRT 1778-1782 231
FLIRT 1782-1795 99
FLIRT 1794-1798 264
FLIRT 1798-1800 264
FLIRT 1801-1801 264
FLIRT 1803-1806 290
FLIRT 1812 286
FLORA Cancelled 1757 96
FLORA 1761-1778 206
**FLORA** 1780-1808 81
FLORA 1783 233
FLORA 1793-1795 264
FLORA 1794-1798 264
FLORA 1800-1801 264
FLORA 1809-1809 290
FLORA 1844-1891 175
FLORA 1885-1901
see KING GEORGE 1844
Flore Americaine (French)
see FLORA 1761
FLORENCE
see FAVOURITE 1803
FLORENTINA 1800-1803 246
FLORIDA 1764-1772 212
FLORIDA 1774-1778 213
FLORIDA 1778-1782 228
FLORIDA 1814-1819 276
FLY 1694-1695 32
FLY 1695-1712 31
FLY 1732-1751 53
**FLY** 1752-1772 92
FLY 1763-1771 208
FLY 1776-1802 96
FLY 1779 233
FLY 1780-1781 231
FLY 1795-1801 264
FLY 1793-1806 332
FLY 1795 332
FLY 1804-1808 290
FLY 1804-1805 131
**FLY** 1805-1812 143
Fly (American 1811)
see SEALARK 1812
FLY 1813-1828 142
FLY 1816 295
FLY 1825-1863 316, 318
**FLY** 1831-1903 139
FLY 1843-.... 319
FLY ?1858-1918? 335
FLY BY NIGHT 1804-1804 290
FLYE
see FLY
FLYING FISH 1778-1782 231
FLYING FISH 1782-1783 231, 233
FLYING FISH
1793-1795/1797-1799? 255
FLYING FISH 1803-1809 286
FLYING FISH 1804-1804 290
FLYING FISH 1804-1807? 160
FLYING FISH 1805-1808 282
FLYING FISH 1807-1811 278
FLYING FISH 1809-1814 290
FLYING FISH 1813-1814-
1817 290, 295
FLYING FISH 1836-by 1852 335
**FLYING FISH** 1844-1852 180
FLYVENDE FISK (Danish 1788)
see FLYING FISH
FOGO 1747 201
FOLKESTONE 1703-1728 25
FOLKESTONE 1741-1749 48
FOLKESTONE 1764-1778 100
FOLKESTONE 1804-1814 290
FORCE 1794-1802 149
FORDWICH 1704 194
FORESIGHT 1650-1698 13
FORESTER 1693-1752 321
FORESTER 1748-1828 321
FORESTER 1794-1800? 256
FORESTER 1806-1819 141
FORESTER Cancelled 1820s 146
FORESTER 1828-1862 334
FORESTER 1832-1843 146
FORMIDABLE 1759-1768 203
FORMIDABLE 1777-1813 64
Formidable (French 1793)
see BELLEISLE 1795
Formidable (French)
see BRAVE 1805
**FORMIDABLE** 1825-1906 107
FORRESTER
see FORESTER
FORT DIAMOND 1804 286
FORTE 1799-1801 262
**FORTE** 1814-1844 122
FORTE 1832 295
FORTE 1890-1905 hulk
see PEMBROKE 1812
FORTH
see TIGRIS 1813
FORTH 1813-1819 116
FORTH 1833- after 1869 123
FORTITUDE 1780-1820 69
FORTITUDE 1833-1870
see CUMBERLAND 1807
FORTUNE/FORTUNE PRIZE
1692-1698 188
FORTUNE 1699/
1700-1700 189, 194
FORTUNE 1706-1706 194
FORTUNE 1707-1707 194
FORTUNE 1709-1713 38
FORTUNE 1746-1770
see FALCON 1744
FORTUNE 1770-1772? 231
FORTUNE 1778-1780 96
FORTUNE 1779-1780 224
FORTUNE 1780-1797 223
FORTUNE 1782-1782 233
Fortune (merchant)
see WRANGLER 1797
FORTUNE 1797-1798 264
FORTUNE 1798-1799 262
Fortune (merchant)
see TORCH 1804
FORTUNEE 1779-1785 218
FORTUNEE 1800-1818 123
FORTUNEE 1801-1801 264
FORWARD 1805-1815 153
FOUDRE 1800-1801 262
FOUDROYANT 1758-1787 203
**FOUDROYANT** 1798-1890 109
Foudroyant (French)
see POULETTE 1799
'Foudroyant' 1897-date
see TRINCOMALEE 1817
FOUGEUX 1747-1759 198
FOUNTAIN 1815-1844? 325
FOUNTAIN 1833-after 1864 330
FOUR BROTHERS
1794-1801 256
FOWEY 1692-1693 189
FOWEY 1696-1704 26
FOWEY 1705-1709 27
FOWEY 1709-1746 36
FOWEY 1744-1748 48
FOWEY 1749-1780 87
FOWEY 1798-1800 264
FOWEY 1795-1801? 258
FOWEY 1806 286
FOWEY 1813
clerical error for TOWEY of
this date
FOWY/FOWYE
see FOWEY
FOX 1690-1692 29
FOX 1699-1699 31
FOX/FOX PRIZE 1705-1706 192
FOX 1708-1724
see NIGHTINGALE 1703
FOX 1727-1738 50
FOX 1740-1745 51
FOX 1746-1751 52
FOX 1773-1778 86
FOX 1780-1816 84
FOX 1783 233
FOX *Ca*1790-1792 332
FOX 1793/4-1797 264
FOX 1794-1801 264
FOX 1790s 264
FOX 1794-1797 264
FOX 1795-1803 332
FOX 1796-1801 264
FOX 1796-1802 264
FOX 1799-1799 262
FOX 1803-1804 290
FOX 1803-1804 290
FOX 1803-1805 290
FOX 1808-1828? 333
FOX 1812-1812 290
FOX 1829-1845? 335
FOX 1829-1882 120
FOXHOUND 1806-1809 141
FOXHOUND 1809-1816 280
FOXHOUND
Cancelled 1831 146
FRACOUR 1782 231
FRANCES 1804-1805
see NANCY 1804
FRANCES ?1842-1890? 332
FRANCHISE 1803-1815 273
FRANCIS 1795-... 336
FRANCIS 1796-....? 302
FRANCIS AND
LOUIS/FRANCIS AND
LUCY 1704-1706 194
Franklin (American)
see NESTOR
Franklin (French 1797)
see CANOPUS 1798
FRATERNITY 1739 202
Frederick (French)
see PELICAN 1781
FREDERICK
see DENMARK yacht
FREDERICK 1795-1799
or 1803? 262
FREDERICK 1804-1804
see LORD NELSON hired 1804
FREDERICKSTEIN
1807-1813 274
FREDERICKSWAERN
1807-1814 274
FREDERICK WILLIAM
1860-1876 170
FREIJA
see FREYA 1807
Fremantle (merchant)
see BUFFALO 1797
FRENCH RUBY 1666-1712 339
FREYA 1807-1816 271
Friedland (Franco/Italian 1808)
see DELIGHT 1808
FRIENDLY SOCIETY
1693-1696 189
Friends (merchant 1800)
see EUGENIE 1804
FRIENDS ADVENTURE
1705-1706 194
FRIENDS ADVENTURE
1711-1711 194
FRIENDS GOODWILL
1705-1705 194
FRIENDS GOODWILL
1707-1707 194
Friends Goodwill (merchant)
see LITTLE BELT 1812
FRIENDS GOODWILL
early C18th 327
FRIENDSHIP 1702-1704 194
FRIENDSHIP 1704-1705 194
FRIENDSHIP 1706-1708 194
FRIENDSHIP 1763-1771 208
FRIENDSHIP 1776-1777 233
FRIENDSHIP 1793-1794 264
FRIENDSHIP 1794-1801 254
FRIENDSHIP 1795-1801 258
FRIENDSHIP 1801-1801 264
Friendship (merchant 1800)
see SENTINEL 1804
FRIENDSHIP 1803-1807 290
Frisk (merchant 1803)
see CARRIER 1805
FRISK
see FOX hired 1803
FROG 1689-1689 189
FROLICK 1806-1813 141
Frolic (American 1813)
see FLORIDA 1814
FROLIC 1820-1838 145
**FROLIC** 1842-1865 180
FROLICK

INDEX 357

see FROLIC
FUBBS 1682-1701 303
FUBBS 1701-1723 303
FUBBS 1724-1781 304
FUERTE 1807-1812 275
FULMINANTE 1798-1801 255
Furie (Dutch)
see WILHELMINA 1798
Furie (Dutch)
see FURY 1799
FURIEUSE 1809-1816 272
FURIOUS 1797-1802 152
FURIOUS 1804-1815 153
FURNACE 1695-1725 30
FURNACE 1740-1763 60
FURNACE 1778-1783 225
FURNACE 1797-1802 152
FURNACE 1807-1811? 286
FURY 1779-1787 96
FURY 1782 231
FURY 1790-1811 129
FURY 1794-1802 256
FURY 1814-1825 148
FUZE 1804-1807 281
FUZEE 1692
probably clerical error for RUZEE (which see)
FYEN 1807-1814 268
FYLLA 1807-1814 275

## G

GABRIEL 1694 189
GAGE 1776 302
GAIETE/GAYETE 1797-1808 249
GAINSBOROUGH 1653-1660
see SWALLOW 1660
Gaité (French)
see GAIETE 1797
GALATEA 1776-1783 91
GALATEA 1794-1809 126
GALATEA 1795-1805? 264
GALATEA 1810-1849 124
GALATHEE 1799-1806? 254
GALGO/GALGO PRIZE 1742-1743 200
GALGO
see SWALLOW 1744
Galgo (Spanish)
see CHANCE 1799
GALGO 1801-1814 251
GALICIA 1741-1741 201
GALLANT 1797-1802 152
GALLANT 1804-1815 153
GAMBIER 1808-1814 290
**GANGES** 1782-1814 72
GANGES 1821-1906 107
GANNET 1796-1798 264
GANNET 1800-1814 253
GANNET 1814-1838 142
GANYMEDE 1809-1840 275
GARLAND 1660-1698 13
GARLAND 1703-1709 25
GARLAND 1712-1721
see SCARBOROUGH 1696
GARLAND 1724-1744 49
GARLAND 1748-1783 52
GARLAND 1762-...? 211
GARLAND 1795-1798
see SIBYL 1779
GARLAND 1798-1803 262
GARLAND 1800-1803 248
Garland
see GALGO 1801
GARLAND 1807-1817 128
GARLAND'S PRIZE 1741-1744 201
GASPEE 1764-1772 211
GASPEE 1774-1775*/1776-1777 231
GAYETE
see GAIETE 1797
Gelderland (Dutch 1781)
see GUELDERLAND 1799
GELYKHEID 1797-1814 240
GENERAL BERESFORD
see PRINCE REGENT 1812
GENERAL BROCK 1814-1815 299
GENERAL COOTE 1804-1804 290
GENERAL CONCLEUX 1808?-1811 283
GENERAL DRUMMOND 1815 299
GENERAL DUFF 1797-1802 262
GENERAL ELLIOT 1801-1801 264
Général Ernouf (French)
see LILLY 1795
GENERAL HUNTER
see HUNTER 1812
GENERAL LAKE 1797-1802 262
General Lally (French 1757)
see FLAMBOROUGH PRIZE 1757
GENERAL MONCK 1781-1782 231
GENERAL SIMCOE 1814-1815 299
GENERAL SMALL 1797-1801 264
General Washington (American privateer)
see GENERAL MONCK 1781
GENEREUX 1800-1816 237
Genie (French)
see VENGEUR 1794
GENOA 1814-1838 267
GENTILLE 1795-1802 245
GEORGE 1689-1690 189
GEORGE 1705-1708 194
GEORGE 1706-1706 194
GEORGE 1755-1756 299
GEORGE 1756-1762 211
GEORGE *Ca*1770-1776 231
GEORGE 1782-1782 233
GEORGE 1794-1798 256
GEORGE 1796?-1798 262
GEORGE 1798-1801 264
GEORGE AND ANN 1704-1705 194
GEORGE AND MARY 1704-1706 194
George Hibbert (merchant 1803)
see SPEEDY 1803
GEORGIANA 1804-1804 290
see KING GEORGE hired 1804
GERMAIN 1764? 213
GERMAINE 1779-1781 231
GERMAINE 1781?-1784 234
GERMOON/GERMOON PRIZE 1691-1700 186
GERTRUDE 1804-1804
see FLYING FISH hired 1804
GERTRUDE ?1841-1858? 335
GIBRALTAR 1711-1725 37
GIBRALTAR 1727-1749 50
**GIBRALTAR** 1754-1773 89
GIBRALTAR 1779-1781 223
GIBRALTAR 1780-1836 214
GIBRALTAR'S PRIZE 1759-1761 207
'GIBRALTAR'
see Hamilton design gunboat No 14
GIER/GIERE 1799-1803 253
GIPSY 1799-1801 262
Gipsy (American 1805/6)
see FLEUR DE LA MER
GIPSY 1805-1808 286
**GIPSY** 1836-1892 182
GIRONDE 1800-1801 253
GLADIATOR 1783-1817 80
GLADWIN 1764 301
Glaneur (French)
see GIBRALTAR'S PRIZE 1759
GLASGOW 1707-1719 193
GLASGOW 1745-1746 52
GLASGOW 1757-1779 90
GLASGOW 1814-1828 116
GLATTON 1795-1830 241
GLEANER 1808-1814 290
GLEANER 1809-1814 284
GLENGARRY 1815 302
GLENMORE 1796-1814 123
Gleneur (American)
see OTTER 1778
Gloire (French 1707)
see SWEEPSTAKES 1709
Gloire (French 1742)
see GLORY 1747
GLOIRE 1781-1783 228
GLOIRE 1795-1802 245
Gloire (French 1800)
see TRINCOMALEE 1801
GLOIRE 1806-1812 271
GLOIRE 1814-1817 273
GLOMMEN 1807-1809 278
GLORIEUX 1782-1782 215
GLORIOSO 1747-1749 201
GLORY 1747-1763 198
GLORY 1762-1774 83
GLORY 1788-1825 65
GLORY 1804-1809 290
GLOUCESTER 1695-1708 22
GLOUCESTER 1704-1707 194
GLOUCESTER 1709-1709 35
GLOUCESTER 1711-1725 35
GLOUCESTER 1737-1742 46
GLOUCESTER 1745-1764 47
GLOUCESTER
see FLORIDA 1764
Gloucester (merchant)
see EXPLOSION 1797
GLOUCESTER 1807-1813 299
GLOUCESTER 1812-1884 114
GLUCKSTADT 1807-1814 278
GLYKSTAT
see GLUCKSTADT 1807
GOELAN 1793-1794 251
GOELAN 1803-1810 277
Goeland (French 1787)
see GOELAN 1793
Goeland (French 1801)
see GOELAN 1803
GOLDFINCH 1808-1838 144
GOLIATH 1781-1815 68
GOLIATH
see CLARENCE 1827
GOLIATH 1842-after 1870? 172
GOOD DESIGN 1797-1802 264
Good Intent (merchant 1776)
see FLAMBEAU
Good Intent (merchant)
see RAMBLER 1778
GOOD INTENT 1792?-1793 262
GOOD INTENT 1801-1801 264
GOODLY 1740-1741 201
GOODWILL 1706-1791 320
GOODWILL 1705-1711 194
GOODWILL 1763-1786 208
GOODWILL 1791-*Ca*1816? 320
GOODWILL 1810-1858 319, 321
GOODWIN PRIZE 1691-1695 186
GOONONG ASSI? 1797-1801 262
GOREE 1759-1763 102
GOREE BRIG 1759-1763 207
GOREE 1800-1806 253
GOREE
see FAVOURITE 1794
GORGON 1785-1817 80
GOSHAWK 1805-1816 143
GOSHAWK 1814-1815 147
GOSHAWK
see NERBUDDA 1848
GOSPORT 1696-1706 27
GOSPORT 1707-1735 36
GOSPORT 1741-1768 48
Gosport Packet (merchant)
see ASSISTANT 1791
GOSSAMER 1823-1861 293
GOVERNOR ?1843-1860? 332
Governor Trumbull
see TOBAGO 1778
GOZO 1800-1804? 255
GRACE 1763-1772 208
GRACE 1781-1781 233
GRACE 1793-1801 265
GRACE 1794-1798 256
Gracieuse (French)
see ROCHESTER PRIZE 1702
Gracieuse (French 1787)
see UNITE 1796
GRACIEUSE 1804-1807 286
GRAFTON 1679-1699 13
GRAFTON 1694-1696 188
GRAFTON 1700-1707 20
GRAFTON 1709-1722 34
GRAFTON 1725-1744 41
GRAFTON 1750-1767 72
GRAFTON 1771-1816 69
GRAMONT 1757-1762 207
GRAMPUS 1707-1708 194
GRAMPUS 1731-1742 52
GRAMPUS 1743-1744 55
**GRAMPUS** 1746-1772 91
GRAMPUS 1777-1779
see BUCKINGHAM 1751
GRAMPUS 1777-1778 231
**GRAMPUS** 1782-1794 78
GRAMPUS 1795-1799 241
GRAMPUS 1802-1832 115
GRAMPUS 1845-1897
see TREMENDOUS 1784
GRANA 1781-1800 220
GRANADO 1693-1694 30
GRANADO 1695-1718 30
**GRANADO** 1742-1763 61
GRANADOE
see GRANADO
GRAND FALCONER 1796-1801 265
GRAND TURK 1745-1749 199
GRANICUS 1813-1817 126
GRANTHAM 1654-1660
see GARLAND 1660
GRANTHAM 1786-1792 338
GRAPPLER 1797-1803 152
GRASSHOPPER 1776-1779 221
GRASSHOPPER 1799-1812 316
GRASSHOPPER 1806-1811 141
GRASSHOPPER 1813-1832 142
GRAVESEND 1704-1705 195
GRECIAN 1814-1822 283
GRECIAN 1821-1827 295
GRECIAN 1838-1865 179
GREENFISH 1690-1705 32
GREEN LINNET 1803-1809 290
GREENWICH 1666-1699 13
GREENWICH 1699-1724 24
GREENWICH 1731-1744 45
GREENWICH 1748-1757 76
GREENWICH 1777-1779 230
GREENWICH 1778-1779 225
GREENWICH 1778-1782 231
GREENWICH 1827-1836
see RODNEY 1809
GRENADA 1804-1809 286
GRENADA 1807-1814? 287
GRENVILLE 1763-1775 209
Grenville (merchant)
see TORTOISE 1777
GREYHOUND 1672-1698 14
GREYHOUND 1694-1698 188
GREYHOUND 1703-1711 25
GREYHOUND 1712-1718 37
GREYHOUND 1720-1741 49
GREYHOUND 1741-1768 51
GREYHOUND 1763-1780 208
GREYHOUND 1773-1781 86
GREYHOUND 1776-... 332
GREYHOUND 1780-1781 231
GREYHOUND 1783-1808 84
GREYHOUND 1783-1783 233
GREYHOUND 1798-1799 265
GREYHOUND 1808-1844? 333
GREYHOUND
see EUPHRATES 1813
GRIFFIN 1690-1701 29
GRIFFIN 1702-1737 29
GRIFFIN 1713-1714 194
GRIFFIN 1758-1761 85
GRIFFIN 1778-1786 226
GRIFFIN 1794-1801 265
GRIFFIN 1803-1805 290
GRIFFON 1808-1819 279
GRIFFON
see GRIFFIN
GRIFFON 1832-1869 146
GRINDER ...?-1832 295
GRIPER 1797-1802 152
GRIPER 1804-1807 153
GRIPER *Ca*1806-1864 334
GRIPER 1813-1868? 155
GROS ISLE *Ca*1783 228
GROUPER 1804-1811 161
GROWLER 1797-1797/1799 152
GROWLER 1804-1815 153
Growler (American)
see CHUB 1813
GUACHAPIN 1801-1811 253
GUADELOUPE 1760-1762 211
GUADELUPE 1763-1781 85
GUADELOUPE 1809-1814 280
GUARDIAN 1784-1794 80
GUARDLAND/GUARLAND
see GARLAND
GUAY TROUIN 1780-1787 221
GUELDERLAND 1799-1817 240
Guelderland (Dutch)
see HELDER 1808
Guêpe (French 1798)
see WASP 1800
GUERIER 1798-1802 238
GUERNSEY 1654-1693 13
GUERNSEY 1696-1716 24
GUERNSEY 1717-1737 36
GUERNSEY 1740-1786 46
GUERNSEY 1800-1801
see EOLUS 1758
GUERNSEY'S PRIZE 1746-1747 201
GUERRIER
see GUERIER and GUERRIERE GUERRIERE 1806-1812
GUILDFORD 1799-1814
see FAME 1759
Guillaume Tell (French 1795)
see MALTA 1800
Guipuscano (Spanish)
see PRINCE WILLIAM 1780
Guirlande (French 1757)
see CYGNET 1758
Guirlande (French 1759)
see GARLAND 1762
GULNARE 1827/8-1842 295
GULNARE 1844-1850 296
GULNARE 1855-1863 295
Guter Erwatung (Dutch)
see THUNDER 1797

## H

Haarlem (Dutch 1791)
see HAERLEM 1797
Haarlem (Dutch)
see AMBOYNA 1796
Haasje (Dutch)
see PHOSPHORUS 1804
HADDOCK 1805-1809 161
HAERLEM 1778-1779 229
HAERLEM 1797-1816 240
HALCYON 1803-1812 277
HALCYON 1813-1814 142
HALCYON Cancelled 1831 146
HALDIMAND 1771 299
HALF MOON 1685-1692 15
HALIFAX 1756-1756 302
HALIFAX 1756-1756 299
HALIFAX 1756-1757 211
HALIFAX 1768-1775 212
HALIFAX 1775-1779 212
HALIFAX 1780-1781 221
HALIFAX 1800-1801
see MARY 1797
HALIFAX 1806-1814 131
HALSTARR 1807-1809 286
HAMADRYAD 1797-1798 246
HAMADRYAD 1804-1815 271
HAMADRYAD 1823-1905 120
HAMILTON ?1823-1860? 334
HAMMOND 1782-1783 231
HAMOAZE 1702-1713 323
HAMOAZE 1800-1843 324
HAMOAZE 1844-after 1900 183, 325
HAMPSHIRE 1653-1697 13
HAMPSHIRE 1698-1739 24
HAMPSHIRE 1741-1766 46
HAMPTON COURT 1678-1701 12
HAMPTON COURT 1701-1707 20
HAMPTON COURT 1709-1741 34
HAMPTON COURT 1744-1774 43
Hancock (American 1776)
see IRIS 1777
HANDY 1871-1875 hulk
see NYMPHE 1812
HANNAH 1798-1802 265
HANNAH 1803-1805 290
HANNIBAL 1689-1691 189
HANNIBAL 1779-1782 77
HANNIBAL 1780 (1782?)-1788? 224
HANNIBAL 1786-1801 71
HANNIBAL 1793-1793 265
Hannibal (French 1801-1823)
see HANNIBAL 1786
HANNIBAL 1804-1804 290
HANNIBAL 1810-1825 112
Hannibal (American 1810)
see ANDROMEDA 1812
HANIBAL Cancelled 1846 171
**HANNIBAL** 1854-1904 172
HANS FREUEN
see HASFRUEN
HAPPY 1711-1735 37
HAPPY 1725-1729 52
HAPPY 1754-1766 92
HAPPY JENNET 1745-1747 202
HAPPY LADD 1855-1856 296
HAPPY RETURN 1660-1691 13
HAPPY RETURN 1695-1695 189
HAPPY RETURN 1804-1805 290
Harbinger (merchant 1850)
see RECRUIT 1846
HARBOUR Ca.1801-after 1864 325
Hardi (French 1800)
see ROSARIO 1800
HARDWICK 1748-1748 202
HARDWICK ?1817-1820 334
HARDY 1797-1802 152
HARDY 1804-1822? 153
HARE 1703-1704 194
HARE 1709-1712 193
Harlequin (American)
see PORTO 1780
HARLEQUIN 1781-...? 231
HARLEQUIN 1796-1802 262
HARLEQUIN 1804-1809 290
HARLEQUIN
see ELVEN 1807
HARLEQUIN 1813-1829 142
HARLEQUIN 1815-1816 287
HARLEQUIN 1836-1889 178
Harlingen (Dutch)
see AMBOYNA 1796
HARMAN 1702-1705 194
HARMAN 1801-1801 265
Harmonie (French)
see GRENADA
Harmonie (French?)
see UNIQUE 1808
HARMONY Not approved 1797 317
Harmony (merchant 1804)
see VOLUNTEER 1804
HARP/HARPE 1691-1693 30
HARPIE
see HARPY
HARPY 1777-1783 221
HARPY 1796-1817 139
HARPY 1815-1829? 333
HARPY 1825-1841 145
HARPY 1837-1869 335
HARRIER 1804-1809 142
HARRIER 1813-1829 142
HARRIER 1831-1840 139
HARRIET 1804-1804
see ROSE, hired 1804
HARRIET 1805 287
HARRIET ?1843-1860? 335
HARRIOTT 1781-1784 230
HART 1691-1692 30
HART 1793-1817 262
HART 1796-1801 265
HART 1805-1810 287
HART *Ca*1814-1822-1875 306, 333
HARWICH 1684-1691 12
HARWICH 1695-1700 24
HARWICH 1709-1714 315, 326
HARWICH 1743-1760 47
Hasard
see HAZARDOUS
Hasard (French)
see HAZARD 1744
HASFRUEN 1807-1814 271
Hasselaar (Dutch)
see HALSTARR 1807
HASTINGS 1695-1697 26
HASTINGS 1698-1707 27
HASTINGS 1707-1745 36
HASTINGS 1741-1763 48
HASTINGS 1819-1866 292
HASTY 1797-1802 152
HASTY 1812-1827? 154
HAUGHTY 1797-1802 152
HAUGHTY 1804-1816 153
Haultpoult (French 1807)
see ABERCROMBIE 1809
HAVANNAH 1811-1905 124
Havfruen (Danish 1789)
see HASFRUEN 1807
HAVICK/HAVIK 1796-1800 249
HAVOCK 1805-1821 154
Havraise (French 1799)
see CLORINDE 1803
HAVRE DE GRACE 1800-...? 260
HAWK/HAWKE 1690-1712 29
HAWK/HAWKE 1721-1740 52
HAWK/HAWKE 1741-1747 53
**HAWK** 1756-1759/1761-1781 92
HAWK ?1775-1776 213
HAWK 1782-1790 233
**HAWK/HAWKE** 1793-1803 129
HAWK/HAWKE 1803-1804 276
HAWK 1806-1813? 277
HAWK/HAWKE ? 1809-1817? 333
HAWK 1839-after 1864 330
Hawke (merchant)
see CAMELEON 1781
HAWKE 1798-1802 265
HAWKE 1794-1796 256, 262
HAWKE 1803-1805 290
HAWKE 1822-1875? 335
HAWKE 1820-1865 115
HAYLING 1706-1729 321
HAYLING 1729-1759 321
HAYLING 1760-1782 321
HAZARD 1711-1714 37
HAZARD 1744-1745/1746-1749 55
**HAZARD** 1749-1783 94
HAZARD 1793-1801 265
HAZARD 1794-1817 130
HAZARD 1837-1866 138
HAZARDOUS 1703-1706 196
HAZARD PRIZE 1756-1759 207
HEART OF OAK 1777?-1782? 229
HEART OF OAK 1794-1796 254
HEART OF OAK 1801-1801 265
HEARTY 1805-1816 154
HEARTY 1824-1827 145
HEBE 1782-1805 217
HEBE 1798-1802 265
HEBE 1804-1812 290
HEBE 1804-1813 128
Hébé (French 1808)
see GANYMEDE 1809
HEBE 1826-1872 120
HEBE 1850s-after 1864 330
HEBRUS 1813-1817 126
HECATE 1797-1809? 152
HECATE 1809-1817 140

HECLA 1797-1813 254
**HECLA** 1815-1831 148
HECTOR 1703-1718 25
HECTOR 1721-1742 47
HECTOR 1743-1762 48
HECTOR 1763-1773 208
HECTOR 1774-1816 70
Hector (merchant)
see OSTRICH 1777
HECTOR 1782-1782 214
Hector (Dutch 1784)
see PANDOUR 1799
Hector (American)
see VULCAN 1796
HECTOR 1797-...? 262
HEIR APPARENT FREDERICK 1807-1811 268
Helden
see LONDON 1756
HELDER 1799-1807 246
HELDER
see HELDIN 1799
HELDER 1808-1813 274
HELDERENBURG 1686-1688 15
HELDIN 1799-1802 246
HELENA 1778-1778/1779-1798 222
HELENA 1781-1802
see ATALANTA 1775
HELENA 1804-1814 131
HELENA Cancelled 1831 146
HELENA 1843-1921 180
Hélène (French?)
see HELENA 1778
HELICON 1808-1829 144
Hell-Hound (Dutch)
see CONTEST 1799
Henri (French)
see CALEDON 1808
HENRIETTA 1660-1689 12
HENRIETTA 1682-1721 303
Henriquetta (Slaver)
see BLACK JOKE 1827
Henry 1804 (merchant)
see RAILLEUR 1804
HENRY AND JANE 1704-1704 194
HENRY AND RICHARD 1711-1711 194
HENRY AND SUSAN 1704-1704 194
HENRY PRIZE 1690-1698 185
HEPHAESTION
see NAYADEN 1807
Herald
see MASTIFF 1797
Herald (merchant 1801)
see SCOURGE 1803
HERALD 1806-1817 130
Herald (American 1811)
see BARBADOES 1813
HERALD 1822-1862 133
HERCULE 1798-1810 238
**HERCULES** 1759-1784 67
Hercules (Dutch)
see PYLADES 1781
Hercules (Dutch)
see DELFT 1797
Hercules (merchant 1801)
see MERLIN 1803
HERCULES 1815-1865 114
HERMES 1796-1797 251
HERMES 1798-1802 249
HERMES 1803-1810 276
HERMES 1811-1814 137
Hermes (1824)
see CHARGER 1830
HERMES 1866-1869
see MINOTAUR 1816
HERMIONE 1760-1763 211
**HERMIONE** 1782-1797 84
Hermione (French 1803)
see MILAN 1805
Hermosa Marianne (Spanish)
see PORT MORANT 1780
**HERO** 1759-1800 67
**HERO** 1803-1811 112
HERO 1804-1805 290
HERO 1809-1811 290
HERO
see WELLINGTON 1816
HEROINE 1783-1803 219
HEROINE 1809-1828
see VENUS 1758
HEROINE 1841-1878 179
HERON 1804-1816 276
HERON 1812-1831 142
HERON 1847-1859 179
HERRING 1807-1816? 160
Hertigennan [af Sodermanland]
see PORPOISE 1780
HESPER 1809-1817 132
HESTER
see ESTHER
HESTER ?1720-1721? 196
Heureuse (French)
see HEUREUX 1799
Heureuse Révolution (Haiti)
see FELICITE 1809
HEUREUX 1799-1806 248
HEUREUX 1807-1814 278
HEXHAM 1798-1802 265
HIBERNIA
see PRINCE OF WALES 1765
**HIBERNIA** 1804-1902 105
HIGHFLYER 1813-1813 283
HIGHFLYER 1822-1833 164
HINCHINBROOK 1745-1746 58
HINCHINBROOK 1775-1777 231
HINCHINBROOK 1777-1780? 221
HINCHINBROOK 1777-1778 222
HINCHINBROOK 1778/9-1782/3 231
HINCHINGBROKE
see HINCHINBROOK
HIND 1691-1697 31
HIND 1709-1709 37
HIND 1709-1711 194
HIND 1712-1721 37
HIND 1741-1743 201
HIND 1744-1747 57
HIND 1749-1794 87
HIND 1785-1811 85
HIND (1790) 1827-1844 332, 293
HIND 1793 262
HIND 1794-1801 265
HIND 1803-1804 290
HIND 1804-1805 290
HIND 1814-1829 137
HIND ?1844-1880? 335
HIND 1880 335
HINDOSTAN 1795-1804 241
HINDOSTAN 1804-1820 270
Hindostan (East Indiaman 1813)
see BUFFALO 1813
HINDUSTAN 1841-1905/1920-1921 109
HIPPOMENES 1803-1813 276
HIRONDELLE 1801-1801 265
HIRONDELLE 1804-1808 287
HOBART 1794-1803 249
Hoche (French)
see DONEGAL 1798
Hocquart (French 1704)
see DUNKIRK PRIZE 1705
HOGUE 1811-1865 114
HOLDERNESS 1779? 231
HOLLY 1809-1814 163
HOLSTEIN 1801-1805 241
HOLYGOLAND 1704-1706 194
HOOD 1859-1888 173
HOPE 1678-1695 12
HOPE 1704-1704 195
HOPE 1764-1776 211
HOPE 1771-1779 301
HOPE 1776-1779 224
HOPE 1780-1785 227
HOPE 1780-1781 223
HOPE 1794-1799? 256
HOPE 1795-1807 251
HOPE 1796-1797 265
Hope (merchant)
see RATTLER 1797
HOPE 1798-1800 265
HOPE 1799-1802 265
Hope (merchant 1800)
see HYENA 1804
HOPE 1803-1807 290
HOPE 1803-1811 290
HOPE 1803-1805 290
HOPE 1808-1819 144
HOPE 1809-1809 290
HOPE 1813-after 1879 319, 323, 327
HOPE 1819-after 1864 330
HOPE 1824-1888 145
HOPE 1826 296
HOPE 1830?-1860? 296
HOPE 1847-after 1854 318
HOPEWELL 1689 189
HOPEWELL 1690-1690 29
HOPEWELL 1690-1692 187
HOPEWELL 1694-1696 188
HOPEWELL 1704-1704 195
HOPEWELL 1704/1706 195
HORATIO 1807-1865 121
HORN BOAT 1827-after 1864 330
HORNET 1745-1770 55
HORNET 1763-1772 208
HORNET 1776-1791 96
HORNET 1779 231
HORNET
see HOUND 1790
HORNET 1794-1795 256
HORNET 1794-1817 130
HORNET ?1829-1839? 334
**HORNET** 1831-1845 165
HORNET? Cancelled 1849? 183
HOSIER 1801-1802 262
HOTSPUR 1810-1821 124
HOTSPUR 1828-1868 123
HOUND 1690-1692 29
HOUND 1700-1714 32
HOUND 1732-1745 53
HOUND 1745-1773 57
HOUND 1776-1784 96
**HOUND** 1790-1794 129
HOUND 1791-1821 332
HOUND 1796-1800 139
HOUND 1801-1812 251
HOUND ?1809-1862? 333
HOUND ?1826-1835 334
HOUND 1846-1887 179
HOWE 1805-1808 284
HOWE 1815-1854 104
HOWE
see WINDSOR CASTLE 1756
HUDSON 1793-1793 265
Huit Amis (French 1799)
see BONETTA 1798
HUMBER 1690 189
HUMBER 1693-1707 19
HUMBER 1708-1723 33
HUMBER 1726-1728 40
HUMBER 1748-1762 79
HUMBER 1804-1811 290
HUMBER 1806-1808? 287
HUNTER 1690-1710 29
Hunter (merchant)
see VULCAN 1739
**HUNTER** 1756-1780 92
HUNTER 1763-1771 209
HUNTER 1796-1797 249
Hunter
see MOSQUITO 1799
HUNTER 1801-1809 253
HUNTER 1812-1813 302
HUNTER By 1854-after 1864 330
HURD 1808-1809 287
HURLEY/HORLY 1704-1706 195
HURLEY 1798-1801 265
HURMON
see HARMON?
HURON 1763 301
HURON 1815-1831 302
HUSSAR 1757-1762 85
HUSSAR 1763-1780 86
HUSSAR 1778 231
HUSSAR 1780-1786 228
HUSSAR 1780-1783 220
HUSSAR 1784-1796 87
HUSSAR 1798-1800 262
HUSSAR 1799-1804 121
HUSSAR 1807-1866 122
HUSSAR 1812-...? 287
Hussard (French 1779)
see ECHO 1780
Hussard (French 1797)
see SURINAM 1799
Hussard (French)
see HUSSAR 1798
HYACINTH 1806-1820 130
HYACINTH 1829-1871 138
HYACINTHE/HYACINTHIA
see PRECIOUS STONE 1692
HYAENA
see HYENA
HYDRA 1778-1783 220
HYDRA 1791-1792 235
Hydra
see RESOLUE 1796
**HYDRA** 1797-1820 121
HYENA 1778-1793/1797-1802 88
HYENA
see HUSSAR 1799
HYENA 1804-1822 274
HYENA
see CALYPSO 1826
HYENA Cancelled 1831 146
Hyène (French 1793-1797)
see HYENA 1778
HYGEIA 1806-1817
see LEANDER 1780
**HYPERION** 1807-1825 127
HYPOLLITUS
see FREYA 1807
HYTHE 1806-1872? 325

## I

**ICARUS** 1814-1838 147
ICICLE *Ca*1812 298
IDAS 1808-1812 290
IDAS 1809-1810 290
Iéna (French 1805)
see VICTOR 1808
IGNITION 1804-1807 281
ILDEFONSO 1805-1816 268
ILLUSTRIOUS 1789-1795 69
ILLUSTRIOUS 1803-1868 112
Illyric/Illyrien (Franco/Italian 1808)
see DELIGHT 1808
IMAUM 1836-1862 292
IMMORTALITE 1798-1806 244
IMMORTALITE 1806-1811 271
IMMORTALITE 1814-1837 273
IMMORTALITE 1859-1883 174
IMOGEN 1800-1805 250
IMOGEN/IMOGENE 1805-1817 143
IMOGENE 1831-1840 134
Impériale (French)
see SUBTLE 1807
IMPERIEUSE 1793-1802 243
IMPERIEUSE 1801-1805
see AMPHITRITE 1799
IMPERIEUSE 1804-1838 271
IMPERIEUSE 1852-1867 174
Impétueux (French)
see VILLE DE PARIS 1782
IMPETUEUX 1794-1794 262
IMPETUEUX 1794-1813 236
IMPETUOUS
see IMPETUEUX
IMPLACABLE 1805-1949 267
IMPREGNABLE 1786-1799 64
**IMPREGNABLE** 1810-1883 106
IMPREGNABLE III 1922-1929
see GANGES 1821
IMPREGNABLE IV 1916-1922
Inabordable (French 1794)
see LEGORE 1796
**INCENDIARY** 1778-1781 99
INCENDIARY 1782-1801 99
INCENDIARY 1804-1812 280
**INCONSTANT** 1783-1817 82
INCONSTANT Cancelled 1832 123
**INCONSTANT** 1836-1866 175
Inconstante (French 1790)
see CONVERT 1793
INDEFATIGABLE 1784-1816 74
Indéfatigable (French)
see DESPATCH 1799
INDEFATIGABLE 1804-1805 287
INDEFATIGABLE Cancelled 1834? 173
**INDEFATIGABLE** 1848-1914 174
INDEPENDANCE 1780-...? 228
see RACER 1812
INDEPENDENCIA 1844-1847? 345
INDIAN 1805-1817 136
INDIAN 1855-1856 296
INDIGNANT 1805-1811 154
Indivisible (French 1799)
see ALEXANDRE 1806
INDUS 1790 235
INDUS 1812-1818 114
**INDUS** 1839-1898 109
INDUS V 1910-1922
see GANGES 1821
INDUSTRIOUS ANN 1801-1801 265
INDUSTRY 1740 202
INDUSTRY 1760?-1766? 211
INDUSTRY 1794-1795 254
INDUSTRY 1801-1801 265
INDUSTRY 1803-1809 290
INDUSTRY 1804-1804 290
INDUSTRY 1804-1804 290
INDUSTRY 1807 287
INDUSTRY 1807-1817 or 1831 333
INDUSTRY 1814-1846 169
INFANTA 1762-1775 204
Infanta (Spanish 1787)
see SALAMENE 1799
Infanta Amelia (Spanish)
see PORPOISE 1799
INFANTE DON CARLOS 1804-1810 287
Infatigable (French 1799)
see IMMORTALITE 1806
**INFERNAL** 1757-1774 99
INFERNAL 1778-1783 225
INFERNAL 1815-1831 148
INFLEXIBLE 1776-...? 298
**INFLEXIBLE** 1780-1820 75
INQUIRY 1691 188
Insolent (French 1794)
see ARROGANT 1798
INSOLENT 1811-1818
see ARROGANT 1798
**INSPECTOR** 1782-1802 97
INSPECTOR 1803-1810 276
INSPECTOR 1822 296
INTEGRITY 1805?-1809? 287
INTELLIGENCE 1695-1699 31
INTELIGENT 1805-1864? 154, 322
INTREPID 1747-1765 197
INTREPID 1747-1748 202
**INTREPID** 1770-1828 74
Intrepid
see DEVASTATION 1803
INVERNESS 1746-1750 99
INVESTIGATOR 1798-1810 250
**INVESTIGATOR** 1811-1867 169
INVESTIGATOR 1848-1853? 294
INVETERATE 1805-1807 155
INVINCIBLE 1747-1758 197
INVINCIBLE 1765-1801 69
INVINCIBLE 1808-1861 72
Invincible Général Buonaparte (French)
see BRAZEN 1798
IPHIGENIA 1780-1801 84
IPHIGENIA
see IMPERIEUSE 1804
IPHIGENIA 1808-1851 82
Iphigénie (French 1810)
see GLOIRE 1814
IPSWICH 1694-1727 20
IPSWICH 1730-*Ca*1764 41
Irene (Dutch)
see GRASSHOPPER 1806
IRIS 1777-1781 217
IRIS 1783-1811? 84
IRIS 1807?-1816 271
Iris (French)
see RAINBOW 1809
IRIS 1840-1870 176
IRISH *Ca*1700-1712 303, 327
IRRESISTIBLE 1782-1806 69
IRRESISTIBLE 1805-1816
see SWIFTSURE 1787
IRRESISTIBLE 1859-1894 172
ISABELLA 1683-1702 303
ISABELLA 1703-1775 303
ISABELLA 1817 296
ISIS 1747-1766 198
ISIS 1774-1810 77
**ISIS** 1819-1867 116
ISLE OF WIGHT 1673-1701 303
ISLE OF WIGHT 1701-1712 303
ISTER 1813-1819 125

## J

JACK 1780-1781 232
JACKAL 1778-1779/1781-1785 227
Jackal (merchant)
see MONKEY 1780?
JACKAL 1781-1782 233
JACKAL 1792-1794 235
JACKALL 1801-1807 152
Jackal (pirate)
see ASSIDUOUS 1823
JACKDAW 1806-1816 161
JACKDAW 1830-1835 167
JACK TAR 1794-1794 251
JACOB AND JOHN 1704 195
Jacobin (French 1794)
see MATILDA 1794
JAHDE 1809-1815 287
JALOUSE 1797-1807 252
JALOUSE 1809-1819 130
JAMAICA 1710-1715 37
**JAMAICA** 1744-1770 58
JAMAICA 1780-1783 231
JAMAICA 1798-1814 247
JAMAICA Cancelled 1829 118
James 1774 (merchant)
see FLASH 1804
JAMES 1803-1807 290
JAMES AND ELIZABETH 1796-1802 262
JAMES AND MARY 1703-1704 195
JAMES AND MARY 1744-1748 202
JAMES AND WILLIAM 1794-1799 257
JAMES GALLEY 1676-1694 13
JAMES WATT 1853-1875 172
JANE 1706-1708 195
JANE 1747 201
JANE 1781-1782 225
JANE 1798-1801 265
Jane (merchant 1795)
see WATCHFUL 1804
JANE 1816-1821 296
JANE 1844 335
JANISSARY 1801-1802 262
JANUS 1778-1800 79
JANUS 1796-1798 246
JASEUR 1807-1809 287
JASEUR 1813-1845 142
JASON 1747-1763 198
JASON 1763-1785 83
Jason (French 1779)
see ARGONAUT 1782
JASON 1794-1798 118
Jason (Dutch 1770)
see PROSELYTE 1796
JASON 1800-1801 124
JASON 1803-1803 290
JASON 1804-1815 128
Jason 1803
see HERON 1804
JASON 1813-1814 287
JASON Cancelled 1831 123
JASPER 1808-1817 144
JASPER 1820-1828 145
JAVA 1806-1807 287
JAVA 1811-1812 272
**JAVA** 1815-1862 116
JEAN BART 1793 262
Jean Bart (French 1793)
see ARAB 1795
Jean Bart (French 1795)
see LAUREL 1795
Jean Jaques Rousseau (French 1795)
see MARENGO 1806
Jean-Jaques Rousseau (French)
see HARLEQUIN 1806
Jeany (merchant 1803)
see IGNITION 1804
JEFFERIES 1701-1706 195
JEMMY 1662-1722 303
Jena (French)
see GRENADA 1807
Jena (French 1805)
see VICTOR 1808
JENNET
see BETSY Hired, 1804
JERSEY 1654-1691 13
Jersey (French 1691-1716)
see JERSEY 1654
JERSEY 1693-1698 27
JERSEY 1698-1731 24
JERSEY 1736-1783 44
JERSEY *Ca*1776 302
JERUSALEM 1689 189
Jeux (French 1689)
see PLAY PRIZE
Jeux (French 1689 - not as above)
see CHILDS PLAY
JEWEL
see ALCMENE 1809
JOHANNA 1711 195
JOHN 1705 195
John (merchant 1766)
see WILDFIRE 1804
JOHN 1804-1805 290
JOHN 1822-...? 325
JOHN AND ANN 1705-1706 195
JOHN AND ANN 1711 195
JOHN AND ANNE 1744-1746 202
JOHN AND BRIDGET 1704-1711 195
JOHN AND ELIZABETH 1705-1706 195
JOHN AND ELIZABETH 1711 195
JOHN AND FRANCIS 1711 195
JOHN AND JANE 1706-1708 195
JOHN AND JOSEPH 1705 195
JOHN AND MARTHA 1694-1698 188
JOHN AND MARY 1705 195
JOHN AND MARY 1705-1706 195
JOHN AND MARY 1706-1708 195
JOHN AND MARY 1707 195
JOHN AND MARY 1711 195
JOHN AND MARY 1803-1803 290
JOHN AND ROBERT 1707 195
JOHN AND SARAH 1706 195
JOHN AND SARAH 1707-1708 195
JOHN AND SUSAN 1705-1706 195
JOHN AND SUSANN 1703-1704 195
JOHN BULL 1804-1808 290
John Gordon (merchant 1801)
see FIREFLY 1803

INDEX 359

JOHN OF DUBLIN 1689-1690 189
JOHNSTON 1760-1764 299
Joli (French Fireship)
see JOLLY PRIZE 1693
JOLLY 1709-1714 193
JOLLY PRIZE 1693-1698 186
JOSEPH 1692-1698 188, 189
Joseph (merchant)
see JAMAICA 1780
JOSEPH 1796-1801 265
JOSEPH 1803-1805 290
JOSEPH 1807-1809 290
Joséphine (French ?)
see MORNE FORTUNEE 1804
Josephine (Portuguese)
see PROMPT 1804
Joseph Straker
see DIAMOND 1848
JOSIAH 1694-1715 189, 336
JOYFULL PRIZE 1694-by 1699 186
JUDITH 1798-...? 259
JULIA 1804-1805?
see LORD NELSON 1804
JULIA 1806-1817 143
Julia (American 1812)
see CONFIANCE 1813
JULIA 1814-...? 300
JULIA 1864-1871 318
JULIAN 1690-1698 188
JULIAN 1704 194
JULINES AND ELIZABETH 1758-1762 211
JULIUS Cancelled 1810 115
JUMNA 1848-1862 180
JUMPER 1705 195
JUNIPER 1809-1814 163
JUNO 1757-1778 83
JUNO 1780-1811 84
JUNO 1844-1878 176
Junon (French 1782)
see PRINCESS CHARLOTTE 1799
JUNON 1809-1809 287
JUNON 1810-1817 272
JUPITER 1707-1708 194
JUPITER 1778-1808 77
Jupiter (Dutch 1782)
see CAMPERDOWN 1797
Jupiter (French 1794)
see MAIDA 1806
**JUPITER** 1813-1870 115
JUPITER
see FORTH 1833
JUSTE 1794-1811 236
JUSTITIA 1777-1795? 232
JUSTITIA 1807-1817 268
JUSTITIA 1815-1830
see ZEALAND
JUSTITIA 1830-1855 269
see HINDOSTAN 1804

## K

Kaikusroo (East Indiaman 1799)
see HOWE 1805
KANGAROO 1795-1802 139
KANGAROO 1805-1815 131
KANGAROO 1818-1828 294
KANGAROO 1829-1834 295
KANGAROO *Ca*1840?
see DOVE revenue cutter 1829?
KANGAROO 1852-1863 180
KATE KARNEY 1808-1808 290
KATHERINE
see ROYAL KATHERINE 1664
KATHERINE 1671-1720 303
KATHERINE 1694-1701 189
KATHERINE 1721-1801 304
KEMPHAAN 1799-1801
see CAMPHAAN
Kenau Hasselaar
see HALSTARR
KENDAL 1809-1810 290
KENNINGTON 1736-1749 51
KENNINGTON 1756-1774 89
KENT 1679-1698 13
KENT 1698-1722 20
KENT 1724-1744 41
KENT 1746-1757/1760? 43
KENT 1762-1784 68
KENT 1776-1777 233
KENT 1798-1801 265
**KENT** 1798-1880 110
KENT 1798-1802 262
KENT 1804-1804 290
KENT 1811-1818 318
KENT 1866?-1866 329
KENT 1883-1891
see IMPREGNABLE 1810
KEPPEL 1778-1783 222
KEPPEL 1778-1779 232
Keppel (merchant)
see PROVIDENCE 1782
KERTCH 1855-after 1860 182, 328
KESTREL 1846-1852 294
KINGFISH 1806-1812 287
KINGFISHER 1675-1699 13
KINGFISHER 1684-1694 15
KINGFISHER 1699-1706 24
KINGFISHER 1745-1763 57
KINGFISHER/KINGSFISHER 1770-1778 96
KINGFISHER/KINGSFISHER 1782-1798 223
KINGFISHER 1804-1816 131
KINGFISHER 1823-1838 145
KINGFISHER 1845-1890 180
KINGFISHER 1890-1907
see MARTIN 1850
King George (American)
see ALLEGIANCE 1779
KING GEORGE 1796-1801 265
KING GEORGE 1803-1804 290
KING GEORGE ?1844-1885 335
King of Portugal
see VESUVIUS 1756
KING OF PRUSSIA 1763-1768 209
KINGSFISHER
see KINGFISHER
KINGSMILL 1797-1802 262
KINGSTON 1697-1716 22
KINGSTON 1719-1736 35
KINGSTON 1746-1748 202
KINGSTON 1748 202
KINGSTON 1740-1762 45
KINGSTON 1804-1805 290
KINGSTON
see PORTLAND 1822
KINGSTON 1814-1832
see PRINCE REGENT
KINGSTON 1815 302
KINGSTON 1858-1867 293
KINSALE 1700-1723 27
KINSALE 1724-1741 47
KINSALE 1741-1763 48
KITCHEN 1670-1698 30, 303
**KITE** 1764-1771 100
KITE 1778-1793 226
KITE 1795-1805 139
KITE 1805-1815 143
KITE 1817-1838? 334
KITE 1826-1854? 330
KITTY 1780-1788? 228
KITTY 1804-1805 290
Komeet (Dutch)
see COMEET 1795
KRONPRINCEN 1807-1809 269
Krondprincesse Marie (Danish)
see KRONPRINCESSEN
KRONPRINCESSEN 1807-1810 269
Krondprinds Frederick (Danish)
see KRONPRINCEN

## L

LABRADOR 1777-1779 232
LACEDEMONIAN 1796-1797 262
LACEDEMONIAN 1812-1822 119
Lacédémonienne (French)
see LACEDEMONIAN
LADY ANN 1799-1801 265
Lady Augusta
see FLYING FISH 1817
Lady Cathcart
see METEOR 1797
LADY CHARLOTTE 1799-1801 265
LADY DUNCAN 1798-1801 265
LADY FLORA 1835-1866 335
LADY JANE 1795-1800 265
LADY LOUISA 1829 296
LADY MACKWORTH 1779-1782 233
LADY MELVILLE 1804-1805 290
LADY NELSON 1800-1825 262, 265
LADY OF THE LAKE ? 1844-1860? 335
LADY PREVOST 1812-1813 302
LADY TAYLOR 1793-1794 265
LADY WARREN 1804-1807 290
Lady Washington (American?)
see HOPE 1780
LANCASTER 1694-1710 19
LANCASTER 1722-1743 40
LANCASTER 1749-1773 41, 44
LANCASTER 1797-1832 239
LANCASTER 1823-1864 116
LANDRAIL 1806-1814 161
LANDRAIL 1817-by 1822 296
LANGPORT 1654-1660
see HENRIETTA 1660
LAPWING 1764-1765 100
LAPWING 1785-1828 87
LAPWING 1808-1844? 334
LAPWING 1825-1864 145
LARK(E) 1675-1698 14
LARK 1703-1723 25
LARK 1726-1744 47
LARK 1744-1757 48
LARK 1762-1778 83
LARK 1780-1784 227
Lark
see MONKEY 1780
LARK 1799-1801 265
LARK 1801-1801 265
LARK *Ca*1795-1798 332
LARK 1794-1809 130
**LARK** 1830-1860 166
LARNE 1814-1820 137
LARNE 1832-1866
see LIGHTNING 1829
Larne and Sophie
see SATELLITE 1825
LASCELLES 1800-1801 265
Lascelles (East Indiaman)
see MADRAS 1795
LASTOFF
see LOWESTOFFE
**LATONA** 1781-1816 81
LATONA 1821-1875 120
LAUNCESTON 1711-1726 36
LAUNCESTON 1741-1784 48
LAURA 1806-1812 162
LAUREL 1759-1763 207
LAUREL 1763-1771 209
LAUREL 1779-1780 87
LAUREL Cancelled 1783 85
LAUREL 1795-1797 247
LAUREL 1798-1821
see DAPHNE 1796
LAUREL 1802-1802 265
**LAUREL** 1806-1808 128
LAUREL 1809-1812 272
LAUREL 1813-1885 122
LAURESTINUS 1810-1813
see LAUREL 1806
**LAVINIA** 1806-1870 118
LAVINIA 1870-1902
see SEAHORSE 1830
LAWRELL 1704-1705 195
LAWRENCE 1782? 232
LAYSTAFFE
see LOWESTOFT
Lazovski (French 1789)
see ESPOIR 1797
LEAK/LEAKE 1705-1707 195
LEAK/LEAKE 1709-1712 195
LEANDER 1780-1806 77
Leander (Merchant 1795)
see STROMBOLO 1797
Leander 1801
see CURLEW 1803
**LEANDER** 1813-1830 116
**LEANDER** 1848-1867 174
LEDA 1783-1795 82
**LEDA** 1800-1808 119
LEDA 1809-1817 124
LEDA 1828-1906 123
LEE 1776-1779? 298
LEE 1780-1784 228
LEE 1814-1822 137
LEGERE 1796-1807 299
LEGERE 1798-1801 262
LEGERE
see NID ELVIN 1807
LEGORE 1694? 186
LEIGHTON 1798-1803? 262
LEITH 1777-1782 233
LEITH 1798-1802 265
LEITH 1804-1805 290
LEMON 1812-1813 287
LEMON FLOWER 1695-1705 195
LENOX 1678-1701 12
LENOX 1701-1721 21
LENOX 1723-1756 41
LENOX 1758-1789 67
LEOCADIA 1781-1794 218
LEOCADIA 1805-1814 277
LEONIDAS 1807-1894 119
LEOPARD (1659) 1687-1699 338
LEOPARD 1703-1719 25
LEOPARD 1721-1740 45
LEOPARD 1741-1761 46
Léopard (French)
see MARGATE 1746
LEOPARD 1770-1772 213
LEOPARD 1790-1814 77
LEOPARD 1794-1795? 257
LEOPARD 1800s-1808 322
LEOPARD *Ca*1806-1809 333
LEVANT 1689-1690 189
LEVANT 1758-1780 85
LEVANT
see VENUS 1807
LEVANT 1813-1820 137
LEVEN 1813-1848 137
LEVERET 1806-1807 141
LEVERET 1808-1822 144
LEVERET 1825-1843 145
LEVIATHAN 1777-1780
SEE NORTHUMBERLAND 1750
**LEVIATHAN** 1790-1848 72
LEWIS PRIZE 1694-1707 185
LEYDEN 1799-1815 241
Leyson (American 1812)
see PICTOU/PICTON 1813
LIBERTY 1768-1769 213
LIBERTY 1779-1816 226
LIBERTY 1793-1801 265
LIBERTY 1793-1798 332
LIBERTY 1798-...? 259
Liberty
see LINNET 1835
LIBERTY 1850-1905 178
see LITCHFIELD
LICORNE 1778-1783 217
LICORNE
see UNICORN 1776
LIFFEY
see FYLLA 1807
LIFFEY
see ERIDANUS 1813
LIFFEY 1813-1827 116
LIFFEY 1856-1877 175
LIGERA 1804-1814 275
LIGHTNING 1691-1705 29
LIGHTNING 1740-1746 60
LIGHTNING 1755-1762
see VIPER 1746
LIGHTNING 1779-1783
see SYLPH 1776
LIGHTNING 1806-1816 149
LIGHTNING 1829-1832 138
Lille Belt (Danish)
see LITTLE BELT
LILLY 1801-1804
see SPENCER 1795
LILLY 1804-1811 276
LILY 1782 233
LILY 1837-1905 178
LIME
see LYME
LIMNADE 1781-.... 299
LINCOLN 1695-1703 23
LINNET 1801-1801 265
LINNET 1806-1813 282
LINNET 1814?-.... 298
LINNET 1817-1833 164
LINNET 1835-1857 179
LION/LYON 1640-1698 12
LION/LYON 1703-1709 317
LION/LYON 1709-1752 314
LION/LYON 1710-1735 35
LION/LYON 1738-1765 44
LION/LYON 1753-1786 102, 321
LION/LYON 1763-1771 209
LION/LYON 1777-1837 74
LION 1782-1785 228
LION 1793-1794 265
LION 1793-1801 265
LION 1794-1795 257
Lion (French)
see BELLEISLE 1795
LION 1804-1805 290
LION 1804-1805 290
LION 1828-by 1869 335
LION 1823-1826 296
LION 1847-1905 172
LION 1855?-1869 317
LIONESS 1777-1779 229
LISBORNE 1781-1783 225
LITCHFIELD 1694-1720 23
LITCHFIELD 1730-1744 45
LITCHFIELD 1746-1758 47
LITCHFIELD PRIZE 1703-1706 192
LITTLE BELT 1807-1811 275
LITTLE BELT 1812-1813 302
LITTLE LONDON 1672-1697 15
LITTLE LUCY 1777-1778 232
LITTLE SALLY 1770s or 80s 232
LITTLE SARAH 1708-1708 195
LIVELY 1709-1712 37
LIVELY 1713-1738 37
LIVELY 1740-1750 51
LIVELY 1755-1759? 299
LIVELY 1756-1778/1781-1784 89
LIVELY 1779-1782 222
LIVELY 1794-1798 126
LIVELY 1794-1798 254
LIVELY 1797-after 1830 322
LIVELY ?1797-1805? 333
LIVELY 1799-1800? 262
LIVELY 1805-1805 290
**LIVELY** 1804-1810 121
LIVELY 1813-1863 119
LIVELY ?18291860 335
LIVELY 1838-1894 323
LIVELY 1840-1870 335
Lively Lass (merchant)
see ANT 1795
LIVELY PRIZE 1689 185
LIVERPOOL 1741-1755 48
LIVERPOOL 1758-1778 85
LIVERPOOL 1814-1822 116
LIVERPOOL Cancelled 1829 116
Liverpool
see SPEEDWELL 1830
Liverpool (Omani 1826)
see IMAUM 1836
LIZARD 1693-1696 27
LIZARD 1697-1714 28
LIZARD 1744-1748 58
LIZARD 1757-1828 85
LIZARD 1782-1783 232
LIZARD 1782?-1785? 232
LOBERON 1812-...? 287
LOCHEGEON 1759-1778? 297
LOCUST 1801-1814 153
LOIRE 1798-1818 244
LONDON 1670-1706 11
LONDON 1704-1706, 1711 195
LONDON 1706-1747 18
LONDON 1756-1756 299
LONDON 1757-1758 211
LONDON 1756-1758 210
LONDON 1759-1764? 211
**LONDON** 1766-1811 64
LONDON 1793-1801 265
LONDON
see ROYAL ADELAIDE 1828
LONDON 1840-1884 107
LONDON PACKET
see LONDON 1793
LOO
see LOOE
LOOE 1696-1697 27
LOOE 1697-1705 27
LOOE 1707-1737 36
LOOE 1741-1744 48
LOOE 1745-1759 48
LOOE 1759-1763
see LIVERPOOL 1741
LORD COCHRANE 1808-1814 290
LORD COCHRANE 1815 299
LORD DUNCAN 1798-1801 265
Lord Duncan (merchant 1798)
see TICKLER 1808
LORD DUNCAN
see DUNCAN 1807
LORD ELDON 1804-1807 290
LORD HOOD 1797-1798 265
LORD HOWE 1763-1771 209
LORD HOWE 1777-1789? 232
LORD KEITH 1804-1808
see ACTIVE, hired 1803
Lord Lenox (merchant)
see FIREBRAND 1804
LORD MELVILLE 1804-1811 290
Lord Melville (merchant 1804)
see PORPOISE 1804
LORD MELVILLE
see MELVILLE 1813
LORD MULGRAVE 1793-1797 265
LORD NELSON 1798-1801 265
LORD NELSON 1801-1807? 262
LORD NELSON 1803-1804 290
LORD NELSON 1804-1804 290
LORD NELSON 1804-1808 290
LORD NELSON 1807-1809 290
LORD OF THE ISLES 1819-1864 330
LORD ST.VINCENT 1801-1802 265
LORD SIDMOUTH 1812-1812 290
Lord Star ? (yacht 1846)
see ROSE 1857
LORD WELLINGTON 1815 299
Lottery (American 1811)
see CANSO 1813
LOUDOUN 1756 302
LOUDON 1779-1781 233
LOUISA 1794-1798 257
LOUISA 1798-1802 262
LOUISA 1799-1801 265
LOUISA 1803-1811 290
LOUISA 1827-1876 316, 321
LOUISA 1835-1841 293
LOUISBOURG 1746-1747 201
LOUP CERVIER?
see PEACOCK 1812
LOVE AND FRIENDSHIP 1796-1803 262
Lovely Lass
see SYLPH 1776
LOWESTOFFE 1697-1722 37
LOWESTOFF 1723-1744 49
LOWESTOFF 1742-1749 51
**LOWESTOFF** 1756-1760 85
**LOWESTOFF** 1761-1801 84
LOWESTOFFE Cancelled 1805 82
LOWESTOFFE'S PRIZE 1777-1779 222
LOYAL CONVERT 1776 298
LOYALIST 1779-1781 224
LOYAL MERCHANT 1701-1703 195
Loyal Oak [merchant?]
see CONFLAGRATION 1781
LOYALTY 1694-1694 189
LOYALTY 1694-1701 344
LUCIFER 1778-1779 221
LUCIFER 1783?-1784 226
LUCIFER 1803-1811 281
LUCY 1777-1777 233
LUCY 1804-1805
see ALERT 1804
LUDLOW 1698-1703 27
LUDLOW CASTLE 1707-1722 36
LUDLOW CASTLE 1724-1750 47
LUDLOW CASTLE 1744-1771 48
LUMLEY CASTLE 1693-1693 189
LURCHER 1761-1762 208
LURCHER 1763-1778 100
LURCHER 1783-1783
see PIGMY 1781
LURCHER 1796-1801 265
LUTIN 1793-1796 251
Lutin (French 1804)
see HAWK 1806
LUTINE 1793-1799 244
LUTINE 1799-1802 250
LYDALL 1705-1705 195
LYME 1654-1660
see MONTAGUE 1660
LYME 1694-1720 26
LYME 1720-1739 49
**LYME** 1740-1747 51
**LYME** 1748-1760 85
LYNN 1696-1713 27
LYNN 1715-1732 36
LYNN 1741-1763 48
LYNX 1761-1777 94
LYNX 1777-1783 221
LYNX 1794-1813 130
Lynx (French 1804)
see HEUREUX 1807
Lynx (French 1814-1834)
see CONFLICT 1801
Lynx (American privateer 1812)
see MUSQUIDOBIT 1813
LYNX Cancelled 1814 142
LYNX 1814-1825? 333
LYNX 1833-1845 146
LYON
see LION
LYRA 1808-1818 144
LYRA 1821-1845 145
LYRE/LYRA 1803-1812 290
LYS 1745 1749 198
LYS 1755 211
LYSANDER 296

## M

MACASSAR 1806-1807 287
MACCLESFIELD 1755-1758 211
MACEDONIAN 1810-1812 122
Machault (French 1755)
see CORMORANT 1757
MACHINE 1692-1696? 188
MACKEREL 1804-1815 160
Macreuse (French 1794)
see DECADE 1798
MACKWORTH
See LADY MACKWORTH
MADAGASCAR 1811-1819 272
MADAGASCAR 1822-1863 123
MADELENA 1748-1748 201
MADRAS 1795-1807 241
MADRAS
see MEEANEE 1838
MAEANDER
see MEANDER
MAESTERLANDT
see CASTLE OF MASTERLAND
MAGDALEN/MAGDALENE 1764[1769?]-1777 212
MAGDALEN 1780 232
MAGDALEN 1804-1805 290
MAGDALENE

1689-1699 189, 347
MAGICIENNE 1781-1810 219
MAGICIENNE 1812-1845 124
MAGNANIME 1748-1775 197
**MAGNANIME** 1780-1813 74
MAGNET 1807-1809 142
MAGNET 1809-1812 279
Magnet (American)
see ATTENTIVE 1812
MAGNET 1814-1846 300
MAGNET 1823-1847 145
MAGNIFICENT 1766-1804 69
MAGNIFICENT 1806-1843 113
MAGPIE 1806-1807 161
**MAGPIE** 1826-1826 165
MAGPIE 1830-1846? 167, 311
MAHONESA 1796-1798 246
MAHONISA
see MAHONESA
MAIDA 1806-1814 267
MAIDSTONE 1654-1660
see MARY ROSE 1660
MAIDSTONE 1693-1714 27
MAIDSTONE 1705-1705 195
MAIDSTONE 1744-1748
see ROCHESTER 1716
MAIDSTONE 1744-1747 47
MAIDSTONE 1758-1794 85
MAIDSTONE 1795-1810 126
MAIDSTONE 1811-1867 124
MAIDSTONE'S PRIZE
1746-....? 201
MAJESTIC 1785-1816 69
Majestic
see HERMES 1803
MAJESTIC 1804-1805 290
MAJESTIC 1853-1868 172
MALABAR 1795-1796 241
MALABAR 1804-1815 270
MALABAR 1818-1905 113
MALACCA 1809-1816 124
MALACCA 1853-1869 176
MALLARD 1801-1804 153
MALTA 1800-1840 237
Malta (Spanish 1797)
see GOZO 1800
MANCHESTER 1814-1815 290
Manchester (packet 1805)
see DOVE 1823
MANDARIN 1810-1810 287
Mandurese (Dutch)
see MANDARIN
MANILLA 1760-1763 210
MANILLA 1780-1782 229
MANILLA 1809-1812 124
MANILLA Cancelled 1831 123
MANLY 1797-1802 259
MANLY 1804-1811/
1813-1814 153
MANLY 1812-1833 145
MANOEVERER
1845-1894 318, 321
MANSTICE
see MUSTICO 1800
Marat (French 1793)
see BELLEISLE 1795
MARATHON
see DEFENCE 1815
MARATHON
see DENMARK 1807
MARCHIONESS OF
QUEENSBURY 1829-1830 293
Marchioness of Salisbury (packet)
see NIGHTINGALE 1829
MARCIA 1804-1804
see MARY 1804
MARECHAL DE BRETAGNE?
1780 232
Maréchal de Richelieu (French
1756)
see ECHO 1758
MARECHAL DE COBOURG
1794-1801 265
MARENGO 1806-1816 267
MARGARET 1744
see MARGARETTA
MARGARET 1795-1798 262
MARGARET 1803-1809 290
MARGARET ?1844-1890? 335
MARGARET AND ANN
1802-1807 290
MARGARETTA 1744-1744 201
MARGARETTA 1775-1775 213
MARGATE 1698-1707
see JERSEY 1693
MARGATE 1706-1706 195
MARGATE 1709-1712 37
MARGATE 1746-1749 199
MARIA 1776-1779? 298
MARIA 1802-1802 265
MARIA 1803-1805 290
MARIA 1803-1812 290
MARIA 1805-1807 287
MARIA
see MOIRA 1805
MARIA 1808?-1808 287
MARIA 1809?-1814? 287
MARIA 1813-1814 290
MARIA 1813-1814 290
MARIA ?1838-1892? 335
MARIANA 1694-1698
see MARY ANN
MARIANA/MARIANNE
1805-1806 287
MARIANNA PRIZE
1693-1698 186
Marianne (French)
see MARIANNA PRIZE
Marianne (merchant)
see WOOLWICH 1788
MARIANNE 1799-1801 262
MARIANNE 1858-1867 346
MARIA PRIZE 1684-1690 344
Maria Reijgersbergen (Dutch)
see JAVA 1806
MARIA TERESA 1748 201
Marie (French)
see MARY 1797
MARIE
see IRIS 1807
MARIE ANTOINETTE
1795-1797 255
Marie de Pontchartrain (French)
see MARY DE
PONCHARTINE
MARIE ROSE
see MARY ROSE
Marie Victoire (French 1757)
see TARTAR'S PRIZE 1757
MARIGOLD 1655-1712 321
MARINER 1801-1814 153
MARINER 1846-1865 180
Marjory (Merchant)
MARLBOROUGH
1706-1725 18
MARLBOROUGH
1732-1763 40
MARLBOROUGH
1767-1800 69
MARLBOROUGH
1807-1835 112
MARLBOROUGH
1855-1904 170
MARQUIS DE BRETIGNY
1780-...? 228
MARQUIS DE SEIGNELLY
1780-1786 221
Marquis de Vandreuil (French 1754)
see RACEHORSE 1757
Marquis of Cornwallis (Bombay
Marine 1801)
see CORNWALLIS 1805
Marquis of Granby (Merchant)
see RESOLUTION 1771
Marquis of Rockingham (Merchant)
see ADVENTURE 1771
Marquis of Salisbury (Merchant)
see SWALLOW 1824
MARS 1746-1755 197
MARS 1759-1784 67
MARS 1781-1784 218
Mars (Dutch)
see PRINCE EDWARD 1781
Mars (Dutch)
see ORESTES 1781
**MARS** 1794-1823 109
Mars (Dutch 1780)
see VLIETER 1799
Mars (French 1799)
see GARLAND 1800
MARS 1804-1805 290
MARS 1848-1869 172
Marseillaise (French)
see AVENGER 1794
MARSHAL BERESFORD
1815 299
MARSHAL DE COBOURG
see MARECHAL DE
COBOURG
MARSOUIN 1798-1798 262
MARSTON MOOR 1654-1660
see YORK 1660
MARTEN
see MARTIN 1850
Marten Harpentzoon Tromp
(Dutch 1779)
see TROMP
MARTHA 1707-1710 195
MARTHA AND MARY
1795-1801? 258
MARTIAL 1805-1836 145
MARTIN 1694-1702 31
MARTIN 1744-1744 202
MARTIN 1761-1784 95
MARTIN 1790-1800 129
MARTIN 1805-1806 131
MARTIN 1809-1817 136
MARTIN 1817?-1817 296
MARTIN 1821-1826 137
MARTIN 1850-1890 180
MARY 1660-1703 12
MARY 1667-1726 303
MARY 1690-1690 189
MARY 1694 189
MARY 1704-1728 22
MARY 1704-1704 195
MARY 1704-1705 195
MARY 1705-1705 195
MARY 1706-1706 195
MARY 1706-1706 195
MARY 1706-1708 195
MARY 1707-1708 195
MARY 1711-1711 195
MARY 1712?-1728 323
MARY 1728-1816 304
MARY 1728-after 1771 323
Mary (merchant)
see VULCAN 1745
Mary (merchant)
see NIMBLE 1778
MARY 1790s 265
MARY 1793-1794 265
MARY 1794-1798 257
MARY 1797-1800 253
MARY 1797-1807? 262
MARY 1800-1801 265
MARY 1801-1839 316, 319
MARY 1803-1808 290
MARY 1803-1804 290
MARY 1804-1804 290
MARY 1804-1807 290
MARY 1805-1805 287
MARY 1809-1812 290
MARY 1812-1813 302
MARY 1838-.. 317
MARY AND BETSY
1795-1801? 258
MARY AND ELLEN
1804-1810 290
MARY AND LUCY
1796-1799 262
MARY AND MARTHA
see MARTHA AND MARY
MARY AND SARAH
1703-1704 195
MARY ANN 1694-1698 188
MARY ANN 1790s? 265
MARY ANN 1800-..? 262
MARY ANN 1802-1802 265
MARY ANNE ..-1795 262
MARY DE PONCHARTINE
1694 186
MARY GALLEY 1687-1708 13
MARY GALLEY 1708-1721 36
MARY GALLEY 1727-1743 47
MARY GALLEY 1744-1764 48
MARYGOLD
see MARIGOLD
Mary Gordon (mercantile 1839)
see ROYALIST 1841
Maryland Planter (mercantile)
see PROSERPINE 1757 and
hired 1755-6? 211
MARY ROSE 1660-1691 13
MARY ROSE 1799-1801 262
MARYS 1805-1812 290
MASTIFF 1797-1802? 259
MASTIFF 1813-1851 155
see MATILDA 1805
MATILDA 1794-1810 247
Matilda
see HAMADRYAD 1804
MATILDA 1805 287
MATHEWS 1706-1708 195
Maure (French 1696)
see MOOR 1711
MAYFLOWER 1694-1696 189
MAYFLOWER 1695-1695 189
MAYFLOWER 1705/
1709-1711 195
MAYFLOWER 1745-1745 203
MAYFLOWER
1794-after 1799 318
MEANDER 1813-1817 120
MEANDER 1840-1870 123
MECKLENBURGH
1763-1773 209
MEDEA 1744-1745 201
MEDEA 1778-1805 86
MEDEA Cancelled 1802? 128
MEDEA (Spanish 1797)
see IMPERIEUSE 1804
MEDEE 1800-1805 262
Medee
see MEDEA
MEDIATOR 1745-1745 201
MEDIATOR 1782-1788 80
MEDIATOR 1804-1809 270
MEDINA 1772-1832
see PORTSMOUTH 1703
MEDINA 1813-1832 137
MEDITERRANEAN
1757-...? 210
MEDITERRANEAN PACKET
1793-1794 265
MEDORA 1855-1856 345
MEDUSA 1785-1798 78
MEDUSA 1801-1816 126
MEDUSA Cancelled 1831 120
MEDWAY 1693-1716 22
MEDWAY 1718-1749 35
MEDWAY 1742-1748 45
**MEDWAY** 1753-1782 76
MEDWAY 1756-1764 210, 320
MEDWAY 1781-1785 318
MEDWAY 1796-by 1830 320
MEDWAY
see TREKRONEN 1807
MEDWAY 1812-1865 114
MEDWAY 1830-1832 321, 330
MEDWAY by 1860-1876 326
MEDWAY PRIZE
1697-1712 185
MEDWAY PRIZE
1704-1713 192
MEDWAYS PRIZE
1745-1750 198
MEEANEE 1848-after 1867 172
Meermin (Dutch)
see MIERMEN 1796
MEGAERA 1783-1817 100
MELAMPE 1757-1764 205
**MELAMPUS** 1785-1815 82
MELAMPUS 1820-1906 120
MELEAGER 1785-1801 84
MELEAGER 1806-1808 82
MELPOMENE 1794-1815 243
MELPOMENE 1815-1821 273
MELPOMENE 1857-1875 174
MELVILLE 1805-1808 277
MELVILLE 1813-1814 300
MELVILLE 1817-1873 115
Menagère (French 1776)
see ALBEMARLE 1781
MENAI 1814-1853 132
MENDOVI 1797-1811 253
MENELAUS 1810-1897 122
MENTOR 1780-1781 232
MENTOR 1781-1783 222
MENTOR 1793-1801 265
MERCATOR 1804-1805 290
MERCHANT 1777-1782 233
Mercure (French 1745-1757)
see MERCURY 1744
Mercure (French 1797)
see TROMPEUSE 1797
Mercurius (Dutch)
see HERMES 1796
MERCURIUS 1807-1816 278
MERCURY 1694-1697 32
MERCURY 1739-1744 200
MERCURY 1744-1745 201
MERCURY 1745-1753 52
MERCURY 1756-1777 90
MERCURY 1763- .... 299
MERCURY 1779-1814 86
MERCURY 1807-1838 287
MERCURY 1826-1906 120
MERCURY 1837-1876 182
MEREDITH 1763-1784 209
Meric (French)
see CRUISER 1705
MERLIN 1666-1698 303
MERLIN 1699-1712 31
MERLIN 1744-1748 55
MERLIN 1753 202
MERLIN 1756-1757 93
MERLIN 1757-1777 206
MERLIN 1780-1795 219
**MERLIN** 1796-1803 131
MERLIN 1803-1836 276
MERMAID 1655-1689 14
MERMAID 1689-1707 26
MERMAID 1692-1693 186
MERMAID 1707-1734 36
Mermaid
see ETNA 1739
MERMAID
see RUBY 1708
MERMAID
see KENNINGTON 1749
**MERMAID** 1749-1760 88
**MERMAID** 1761-1778 86
MERMAID 1784-1815 84
MERMAID 1793-1818 332
MERMAID 1799-1801 262
MERMAID 1817-1820 296
MERMAID
see VIPER 1825?
MERMAID 1826-1875 120
MERMAID ?1829-1880? 335
MERMAID 1853-1890 335
MEROPE 1808-1815 144
MERSEY 1782 232
MERSEY 1814-1852 132
MERTON 1744 202
MESSENGER 1694-1701 32
MESSENGER 1830-1862 337
METEOR 1797-1802 259
METEOR 1803-1811 281
METEOR 1812-1816
see STAR 1805
METEOR 1823-1832 148
MICHIGAN 1763-.... 301
MIERMEN 1796-1801 251
MIGNONNE 1794-1797 262
MIGNONNE 1803-1804 287
MIGNONNE (?)
see MUSETTE 1806
MILAN 1805-1815 271
Milan (French 1807)
see ACHATES 1809
**MILBROOK** 1797-1808 159
MILFORD 1689-1694 26
MILFORD 1695-1697 26
MILFORD 1697-1705
see SCARBOROUGH 1694
MILFORD 1705-1729 27
MILFORD 1744-1749 [44]
see ADVICE 1712
MILFORD 1759-1785 85
**MILFORD** 1809-1846 111
MILFORD 1816-after 1864 330
Mince (French)
see CHATHAM PRIZE 1704
MINDEN 1810-1861 72
MINERVA 1759-1778 82
MINERVA 1780-1803 81
MINERVA 1781-1783 230
Minerva (merchant)
see BOUNTIFUL 1782 or
TOBAGO 1782
MINERVA 1793-1794 265
MINERVA 1793-1801 265
MINERVA 1790s 265
MINERVA 1797-1818 333
Minerva
see READY 1797
Minerva (Dutch)
see BRAAK 1799
MINERVA/MINERVE
1803-1811 287
MINERVA 1803-1804 290
MINERVA 1804-1805 290
MINERVA 1805-1816 128
MINERVA 1820-1895 120
Minerve (French 1782)
see SAN FIORENZO 1794
MINERVE 1795-1803 243
MINERVE 1803
see MINERVA
Minerve (French 1804)
see ALCESTE 1806
MINION 1793-1794 265
MINORCA 1740-1756 328
**MINORCA** 1779-1781 102
MINORCA 1799-1802 253
MINORCA laid down 1799 262
MINORCA 1805-1814 140
MINOTAUR 1793-1810 72
MINOTAUR 1816-1866 72
MINSTREL 1807-1817 130
MINX 1797-1802 259
MINX 1801-1809 153
**MINX** 1829-1833 166
MISTLETOE 1809-1816 163
MISSISAUGA 1793 299
MISSASACO/MISSISAGA
1759-1765 299
MODENA 1690-1691 189
MODERATE 1702-1713 191
Modéré (French 1685)
see MODERATE
MODESTE 1759-1800 205
MODESTE 1793-1814 243
MODESTE 1814-1816 273
**MODESTE** 1837-1866 177
MOHAWK 1756-1756 299
MOHAWK 1759-1764 299
MOHAWK 1781-.... 299
MOHAWK 1782?-1783 222
MOHAWK 1795-1803 299
MOHAWK 1798-1799 262
Mohawk (French privateer)
see SAINT CHRISTOPHER
1807
MOHAWK 1813-1814 287
MOHAWK
see ONTARIO 1813
MOIRA 1805-1814 299
Mollineux (merchant)
see STROMBOLO 1739
MOLLY 1781-1781 or 1782 233
MOLLY 1781-1782 233
MONARCA 1780-1791 215
MONARCH 1747-1760 197
**MONARCH** 1765-1813 69
MONARCH 1832-1866 107
Monarch (Merchant)
see HOUND 1801
Monarque (French 1747)
see MONARCH 1747
MONCK 1659-1702 12
MONCK 1702-1720 22
MONCK PRIZE
see MONK PRIZE
MONGOOSE 1800-1803 262
MONK PRIZE 1709-1712 193
MONDOVI 1798
see MENDOVI
MONDOVI
see DELPHINEN 1807
MONKEY 1780-1786 227
MONKEY 1801-1810 152
MONKEY 1826-1831 165
MONKEY 1831-1833 293
MONMOUTH 1666-1698 303
MONMOUTH 1667-1700 12
MONMOUTH 1700-1716 22
MONMOUTH 1718-1744 35
MONMOUTH 1742-1767 42
MONMOUTH 1772-1798 74
MONMOUTH 1796-1834 239
MONMOUTH 1868-1902 hulk
see HOTSPUR 1828
MONSIEUR 1780-1783 218
MONTAGUE 1660/
1675-1698 12
MONTAGUE 1698-1714 22
MONTAGUE 1716-1749 35
**MONTAGUE** 1757-1774 76
MONTAGUE 1779-1818 71
MONT BLANC 1805-1811 268
MONTEGO BAY 1796-1800 262
MONTREAL 1761-1779 83
MONTREAL 1813-1832 300
MONTREAL 1814-1842
see WOLFE 1813 or ROYAL
GEORGE 1813?
MONTREAL 1836-1848 302
MOOR 1711-1716 192
MORDAUNT 1683-1693 15
MORGIANA 1800-1811 253
MORGIANA 1811-1825 136
MORNE FORTUNEE
1803/4-1804 277
MORNE FORTUNEE
1804-1809 287
MORNE FORTUNEE
1808-1813 279
MORNING STAR
1763-1773 209
MORNING STAR
1781-1782 225
MORO 1762-1770 203
MORRISTON 1804-1812 290
MORS AUT GLORIA
1810-...? 287
MORTAR 1693-1703 30
MORTAR 1742-1749 61
MORTAR 1759-1774 99
MORWELHAM 1785-1816 324
MORWELHAM 1795-1801? 258
MOSAMBIQUE 1804-1810 287
MOSELLE 1794-1802 249
MOSELLE 1804-1815 140
MOSQUITO 1799-1803 262
MOSQUITO
see also MUSQUITO
MOTHER RAMSAY
1803-.... 290
MOUCHERON 1801-1807 253
MOUNTAGUE
see MONTAGUE
MOUNT EDGCUMBE
1876-1921
ex WINCHESTER 1822
Mountford (Indiaman)
see ALARM 1777
MOZAMBIQUE
see MOSAMBIQUE
MUDD 1704-1705 195
Mulet (French 1784)
see MULETTE 1793
MULETTE 1795-1796 262
MULGRAVE 1807? 287
MULGRAVE 1812-1854 114
MULLET 1793-1796 161
MULLET
see MULETTE 1793
MULLET 1807-1814
MUROS 1806-1808 275
MUROS 1809-1822 144
MURRAY 1759-.... 299
MUSCOVIA MERCHANT
1701-1703 195
MUSCOVY MERCHANT
*Ca*1694 189

INDEX 361


MUSETTE 1796-1806 249
MUSETTE 1803-1808 287
MUSETTE 1806-1814 277
MUSQUENONGE 1759-1778? 297
MUSQUIDOBIT 1813-1820 283
MUSQUITO 1777 232
MUSQUITO 1793/ 1794-1799 255
**MUSQUITO** 1794-1795 150
MUSQUITO 1804-1822 140
MUSQUITO 1825-1843 145
MUSQUITO 1851-1867 180
MUSTICO 1800-1816 262
MUTIN/MUTINE 1779-1798 226
MUTINE 1797-1807 252
MUTINE 1806-1819 141
MUTINE 1825-1841 145
**MUTINE** 1844-1848 180
**MYRMIDON** 1781-1811 89
MYRMIDON 1813-1823 137
MYRTLE 1807-1818 132
MYRTLE 1825-1829 125
MYRTLE 1883-1905
see MALABAR 1818

## N

NABOB 1777-1783 229
**NAIAD** 1797-1866 121
NAIADE 1783-1784 220
NAJADEN (Danish 1795)
see NYADEN
NAMURE 1697-1722 18
NAMURE 1729-1749 40
**NAMURE** 1756-1833 63
NANCY 1789-1814 302
NANCY 1793-1794 265
NANCY 1794-1801 254
NANCY 1797-1801? 333
NANCY 1804-1804 290
NANCY 1804-1804 290
NANCY 1809-1813 287
**NANKIN** 1850-1895 174
Napoléon (French)
see DOMINICA 1805
NARCISSUS 1781-1796 91
NARCISSUS 1795-1801 265
**NARCISSUS** 1801-1837 127
NARCISSUS 1845 296
NARCISSUS Cancelled 1848 177
**NARCISSUS** 1859-1883 174
NASSAU 1698-1706 20
NASSAU 1707-1736 34
NASSAU 1740-1770 42
NASSAU 1785-1799 74
NASSAU 1805-1814
see HOLSTEIN 1801
NATHANIEL 1689-1692 187
NATIONAL COCKADE 1795-1796? 263
**NAUTILUS** 1762-1780 95
NAUTILUS 1784-1779 98
**NAUTILUS** 1804-1807 132
NAUTILUS 1807-1823 141
Nautilus (American 1803)
see EMULOUS 1812
NAUTILUS 1830-1878 146
NAVY 1673-1698 303
NAVY TRANSPORT 1705-1730 315
NAVY TRANSPORT 1730-1742 315
NAVY TRANSPORT 1752-1791 315
Nayade (French 1793)
see MELVILLE 1805
Nayade (French)
see NAUTILUS 1805
NAYADEN
see NYADEN
NEARQUE 1806-1814 277
NECKER 1781-1782 232
Necker (French 1794)
see RAISON 1795
NEGRO 1804-1814
see NIGER 1759
NEGRESSE 1799-1803 263
NELLY 1815 302
NELSON 1799-1801? 265
NELSON 1804-1805 290
**NELSON** 1814-1928 104
NELSON 1815 302
NELSON 1838-1840 318
NELSON 1843-.... 335
NEMESIS 1780-1814 86
NEMESIS 1798-1804? 265
NEMESIS 1826-1866 123
NEPEAN 1809-1809 290
NEPTUNE 1683-1710 11
NEPTUNE 1710-1724 33
NEPTUNE 1730-1750 40
NEPTUNE 1742 202
**NEPTUNE** 1756-1816 63
Neptune (merchant)
see LION 1782
NEPTUNE 1796-1798 265
**NEPTUNE** 1797-1818 106
NEPTUNE 1798-1801 265
NEPTUNE 1803-1812 290
NEPTUNE 1803-1812 290
NEPTUNE 1807-1814 290
NEPTUNE 1832-1875 105
NEPTUNE 1838-1854? 335
NEPTUNE 1858-1905 335
NERBUDDA 1848-1855 178
NEREIDE 1797-1816 246
NEREIDE 1810-1816 272
Neréide (French 1809)
see MADAGASCAR 1811
NEREIDE
see NYMPHE 1812
NEREUS 1809-1817 126
NEREUS 1821-1879 120
NESTOR 1781-1783 232
**NETLEY** 1798-1806 159, 259
NETLEY 1807-1808 287
NETLEY 1808-1810 279
NETLEY 1814-1837
see PRINCE REGENT
NETLEY 1823-1848 293, 330
NETTLE 1870-1901
see THUNDERER 1831
Nettuno (Franco/Italian 1807)
see CRETAN 1808
NEW ADVENTURE 1799-1801? 328
NEW AFRICA 1690-1695 189
NEWARK 1695-1713 19
NEWARK 1717-1741 34
NEWARK 1747-1787 41
NEWASH 1815-1832 302
NEW AUGUSTA 1763-1765 211
NEW BETSY 1794-1798 257
New Brunswick
see SUPPLY 1793
NEWCASTLE 1653-1703 13
NEWCASTLE 1704-1728 25
NEWCASTLE 1733-1746 45
NEWCASTLE 1750-1761 76
**NEWCASTLE** 1813-1827 116
New Concord
see PLUTO 1756
New Grove (merchant 1799)
see ACHERON 1803
NEWPORT 1694-1696 28
NEWPORT 1698-1714
see ORFORD 1695
NEW UNION 1790s 265
NIAGARA 1814-1842?
see ROYAL GEORGE
NIAGARA 1815 301
NIAGARA 1838-1843
see PRINCE REGENT
NID ELVIN 1807-1814 278
NIEMEN 1809-1815 272
NIEMEN 1820-1828 133
NIEUPORT 1795-1810? 259
**NIGER** 1759-1804 83
Niger (merchant 1809)
see VIPER 1809
NIGER 1813-1820 120
NIGHTINGALE 1703-1707 28
NIGHTINGALE 1707-1716 37
NIGHTINGALE 1746-1770 199
NIGHTINGALE 1805-1815 143
NIGHTINGALE 1825-1829 165
NIGHTINGALE 1829-1842 294
NILE 1799-1801 265
NILE 1799-1800 265
NILE 1804-1806 290
NILE 1804-1806 290
NILE 1806?-1811 282
NILE 1839-1876 107
NIMBLE 1778-1781 226
NIMBLE 1781-1808 227
NIMBLE 1803-1814 290
**NIMBLE** 1811-1812 164
NIMBLE 1812?-1816 282
NIMBLE 1817?-1843? 334
NIMBLE 1826-1826 165
NIMBLE 1826-1834 292
NIMBLE 1826-1838 322
NIMROD 1777 233
NIMROD 1794-1802 265
NIMROD 1799-1811 250
NIMROD 1803-1808 290
Nimrod (American 1804)
see NETLEY 1808
NIMROD 1808-1814 290
NIMROD 1812-1827 142
NIMROD 1828-1907 133
Ninfa (Spanish 1794)
see HAMADRYAD 1797
NIOBE 1800-1816 244
NIOBE 1849-1862 177
Nisus (French 1806)
see GUADELOUPE 1809
NISUS 1810-1822 122
NONPAREIL 1807-1813 283
NONSUCH 1668-1685 13
NONSUCH 1686-1714 321
NONSUCH 1696-1716 24
NONSUCH 1717-1745 36
NONSUCH 1741-1766 46
NONSUCH 1774-1802 74
NONSUCH
see NORGE 1807
NORFOLK 1693-1718 19
NORFOLK 1719-1749 40
NORFOLK
see PRINCESS AMELIA 1757
NORFOLK 1757-1774 67
NORFOLK 1804-1809 290
NORFOLK 1809-1814 290
NORFOLK
see KRONPRINCEN 1807
NORGE 1807-1816 269
NORTH 1778-1779 224
NORTH AMERICAN 1778?-1782? 232
NORTH STAR 1810-1817 132
NORTH STAR 1824-1860 133
NORTHUMBERLAND 1679-1701 13
NORTHUMBERLAND 1702-1703 21
NORTHUMBERLAND 1705-1719 21
NORTHUMBERLAND 1721-1739 41
NORTHUMBERLAND 1743-1744 43
NORTHUMBERLAND 1750-1777 72
NORTHUMBERLAND 1794-1795 237
**NORTHUMBERLAND** 1798-1850 110
NORWICH 1691-1692 23
NORWICH 1693-1712 23
NORWICH 1718-1744 36
NORWICH 1745-1769 47
NOTTINGHAM 1703-1716 23
NOTTINGHAM 1719-1739 35
NOTTINGHAM 1745-1773 45
NOTTINGHAM 1794-1800 257
NOTTINGHAM PRIZE 1704-1706 193
Nouvelle Entreprise (Franco-Italian 1808)
see THEODOSIA 1808
Nouvelle Eugénie (French)
see EUGENIA 1797
NOVA SCOTIA 1812-1813 280
NOX 1798-1801 265
Nuestra Senora de la Paz (Spanish 1778)
see GRANA 1781
Nuestra Senora del Carmen (Spanish 1770)
see CARMEN 1800
Nuestra Senora del Carmin (Spanish)
see CARMIN
Nuestra Senora del Carmen y San Antonio (Spanish 1715-8)
see HAMPTON COURT 1701
Nuestra Senora del Covadonga (Spanish)
see CENTURION PRIZE
Nuestra Senora del Monte Carmelo (Spanish)
see CARMELO
Nuestra Senora del Rosario (Spanish)
see ROSARIO 1797
NYADEN 1807-1812 272
NYMPH 1778-1783 96
NYMPHE 1780-1810 218
NYMPHE 1793-1794 265
NYMPHE/NYMPH 1803-1814 290
NYMPHE 1812-1871 122
NYMPHEN/NYMPHIN 1807-1816 271

## O

OAKE ROYAL
see ROYAL OAK 1674
OBERON 1805-1816 143
OBSERVATEUR 1806-1814 278
OBSERVER 1781?-1784 225
**OCEAN** 1761-1791 64
**OCEAN** 1805-1875 105
OCTAVIA 1849-1876 173
ODIN 1807-1825 269
Ohio (American)
see SAUK 1814
OISEAU 1762 211
OISEAU 1779-1783 219
OISEAU 1793-1816
Oiseau (French 1801)
see REDBRIDGE 1804
OLD PORTSMOUTH 1741-1772
see PORTSMOUTH 1703
OLD PRESIDENT 1903-1903
see PRESIDENT 1829
OLD TRUELOVE 1720-...? 316, 321
OLIVE BRANCH 1690-1690 189
OLIVE BRANCH 1794-1802 254
Oliver Cromwell (American)
see CONVERT 1777
OLYMPIA 1806-1815 162
ONONDAGA 1759-1764 299
ONONDAGA 1790-1793 299
ONONDAGA 1793?-.... 299
ONTARIO 1755-1756 299
ONTARIO 1780-1780 299
ONTARIO 1813-1832 142
ONYX 1808-1819 144
ONYX 1822-1837 145
ONYX 1836-1860? 335
Opiniâtre (French 1689)
see CHILDS PLAY
OPOSSUM 1808-1819 144
OPOSSUM 1821-1841 145
ORANGE 1809-1811 327
Oreste (French 1809)
see WELLINGTON 1810
ORESTES 1781-1799 223
ORESTES 1803-1805 276
ORESTES 1805-1817 143
ORESTES
see WELLINGTON 1810
**ORESTES** 1824-1905 137
Orestes (Slaver)
see KINGSTON 1858
ORFORD 1695-1698 28
ORFORD 1698-1709 20
ORFORD 1713-1745 34
ORFORD 1749-1783 72
ORFORD Cancelled 1809 114
ORFORD PRIZE 1703-...? 192
ORFORD PRIZE 1708-1709 193
ORIFLAMME 1761 211
ORINOCO
see ORONOOKO
ORION 1787-1814 69
**ORION** 1854-1867 173
ORLANDO 1811-1824 82
ORMOND 1711-1715 35
ORNEN 1807-1815 283
ORONOQUE 1782? 232
ORONOOKO 1805-1811 287
ORONTES 1813-1817 125
ORPHEUS 1774-1778 84
ORPHEUS 1780-1807 84
Orpheus (Dutch 1803)
see HELDER 1808
ORPHEUS 1809-1819 124
ORPHEUS Cancelled 1831 123
ORQUIJO/ORQUIXO
see URQUIJO
ORTENZIA/ORTENZA 1808-1812 287
OSBORNE 1804-1805 291
**OSPREY** 1797-1813 135
**OSPREY** 1844-1846 180
OSSORY 1682-1705 11
OSSORY 1711-1716 33
OSTEND 1795-1809? 259
OSTRICH 1777-1782 224
OSWEGO 1755-1756 299
OTTAWA Building 1779 302
OTTAWA 1794 302
OTTER 1701-1702 32
OTTER 1710-1713 37
OTTER 1721-1742 52
OTTER 1742-1763 53
OTTER 1767-1778 96
OTTER 1778-1783 224
OTTER 1782-1801 98
OTTER ?1793-1806? 332
OTTER 1805-1828 131
Ouataouaise (French 1759)
see ?JOHNSTON 1760
Overijssel (Dutch)
see OVERYSSEL
OVERYSSEL 1795-1822 239
OWEN GLENDOWER 1808-1884 124
OWNER'S ADVENT 1694-1698 188
OWNER'S ADVENTURE 1704-1705 194
OWNER'S ENDEAVOUR 1689 189
OWNER'S GOODWILL 1694-1705 188, 315
OWNER'S GOODWILL 1706-1708 194
OWNER'S GOODWILL 1706-1706 194
Owner's Goodwill
see STROMBOLO 1756
OWNER'S LOVE 1688-1697 187
OX 1807-1807 291
OXFORD 1674-1702 13
OXFORD 1702-1723 24
OXFORD 1727-1758 45
Oxford (merchant 1803)
see ALERT, 1804
OXFORD
see JUSTITIA 1807
OXFORD PRIZE 1701-1703 194

## P

PACAHUNTA 1780-1782 224
PACIFIC 1777-1781 229
Pacific (merchant 1802)
see SWIFT 1804
PACIFIC 1845-1867 345
Pacifique (French?)
see PACIFIC 1777
PACTOLUS 1813-1818 120
PAKENHAM 1797-1800? 263
see SNAP 1808
PALLAS 1757-1783 82
Pallas (French 1777)
see CONVERT 1778
**PALLAS** 1793-1798 126
Pallas (French 1800)
see PIQUE 1800
PALLAS
see MINERVA 1780
PALLAS
see SHANNON 1803
PALLAS 1804-1810 128
Pallas (Dutch)
see CELEBES 1806
PALLAS 1816-1862 124
PALMA
see GLOIRE 1814
PANCEY
see PANSY
PANDORA 1779-1791 88
PANDORA
see PANDOUR 1795
PANDORA 1806-1811 142
PANDORA [later LYNX] Cancelled [1814] 142
PANDORA 1813-1831 142
**PANDORA** 1833-1857 178
PANDOUR 1795-1797 251
PANDOUR 1798-1800 253
PANDOUR 1799-1799 242
PANDOUR
see COSSACK 1806
PANSY 1747-1748 202
PANTALOON 1831-1852 292
PANTHER 1703-1713 25
PANTHER 1716-1748 35
PANTHER 1746-1756 47
PANTHER 1758-1813 76
Panthère (French 1744)
see AMAZON 1745
PANZEY
see PANSY
PAPILLON 1803-1806 287
PAPILLON 1809-1815 280
PARAGON 1804-1805 291
PARAMOUR 1694-1706 31
PARTHIAN 1808-1828 144
PARTHIAN 1829-1830 296
PARTRIDGE 1809-1816 130
PARTRIDGE 1822-1824 145
PARTRIDGE 1829-1843 146
PATRIOT 1808-1813 287
PASLEY 1790s 265
PAULINA 1805-1816 143
PAZ 1807-1814? 283
PEACE 1703-1704 195
PEACOCK 1806-1813 141
PEACOCK 1812-1814 277
PEACOCK 1815? 287
PEARL 1651-1697 13
PEARL 1693-1694 186
PEARL 1708-1723 36
PEARL 1726-1744 47
PEARL 1744-1759 48
PEARL 1762-1825 83
PEARL 1780? 232
**PEARL** 1828-1851 138
PEARL
see IMOGENE 1831
PEARL 1844-after 1864 330
PEARLEN
see PERLIN 1807
PEDRO 1796?-1801 263
PEGASE 1782-1815 215
Pégase (French 1794)
see DONEGAL 1798
PEGASUS 1776-1777 96
PEGASUS 1779-1816 87
PEGASUS Cancelled 1831 120
**PEGGY** 1749-1769? 91
PEGGY 1779-1779 233
PEGGY 1796-1799 265
PEGGY 1804-1804 291
PEGGY 1804-1804 287
Pèlerine (French)
see ARETHUSA 1759
Pelham (merchant 1744)
see SALAMANDER 1745
PELICAN 1690-1692 187
PELICAN 1697 189
PELICAN 1706-1706 195
PELICAN 1757-1763 207
PELICAN 1777- (1787?) 228
PELICAN 1777-1781 88
PELICAN 1781-1783 222
PELICAN 1795-1806 139
PELICAN 1806-1812 277
PELICAN
see ECLAIR 1807
PELICAN 1812-1845? 142
PELORUS 1808-1841 142
PELTER 1794-1802 149
PELTER 1804-1809/1810 153
PELTER 1813-1826 155
PEMBROKE 1690-1694 26
PEMBROKE 1694-1709 22
PEMBROKE 1709-1709, 1713 195
PEMBROKE 1710-1726 35
PEMBROKE 1733-1749 44
**PEMBROKE** 1757-1793 76
PEMBROKE 1812-1890 114
PEMBROKES PRIZE 1740-1744 200
PENANG
see MALACCA 1809
PENDENNIS 1679-1689 13
PENDENNIS 1695-1705 24
PENELOPE 1778-1779 88
PENELOPE 1783-1797 84
PENELOPE 1795-1799 265
**PENELOPE** 1798-1815 124
PENELOPE 1829-1864 120
PENGUIN 1751-1760
see DOLPHIN 1732
PENGUIN 1777-1782? 232
**PENGUIN** Cancelled 1782 102
PENGUIN 1798-1809
see COMEET 1795
PENGUIN 1813-1815 142
PENGUIN 1838-1857 179
Pensée (French 1793)
see VIDETTE 1800
PENZANCE 1694/5-1713 28
PENZANCE 1747-1766 79
Péralty (French 1807-1808)
see BARBARA 1806
Perçante (French 1795)
see JAMAICA 1796
PERDRIX 1795-1799 247
Peregrina (Spanish)
see PEREGRINE 1742
PEREGRINE/PEREGRINE GALLEY 1700-1716 28
PEREGRINE 1749-1762
see ROYAL CAROLINE 1733
PEREGRINE/PEREGRINE PRIZE 1742-1743 200
Perle (French)
see PEARL 1780
Perle (French 1790)
see AMETHYST 1793
PERLEN/PERLIN 1807-1813 271
PERSEUS 1776-1805 91
PERSEUS 1804-1805 291
PERSEUS 1812-1850 128
**PERSEVERANCE** 1781-1823 82
PERSEVERANCE 1806-1808 291
PERSIAN 1809-1813 142
PERSIAN 1839-1866 179
PERT 1804-1807 287
PERT 1808-1813 287
PERUVIAN 1808-1813 142
PETER 1706-1706 195
PETEREL 1777-1788 230
PETERELL 1794-1827 129
PETEREL Cancelled 1831 168
PETEREL 1838-1862 179
PETEREL ?1843-1860? 335
PETIT BOSTON 1793-1793 263
PETITE VICTOIRE 1793-1793 263
Peuple (French)
see PORT ROYAL
Peuple Souverain (French 1757)
see GUERRIER 1798
PHAETON 1691-1692 29
PHAETON 1739-1743 200
PHAETON 1782-1820 81

Phaeton (French 1804) see MUSETTE 1806
**PHAETON** 1848-1875 174
PHEASANT 1761-1761 207
PHEASANT 1778-1781 226
PHEASANT 1798-1827 131
PHEASANT Cancelled 1832 139
PHENIX see PHOENIX
PHILADELPHIA 1778-1782 232
PHILOMEL 1806-1817 141
PHILOMEL 1823-1833 145
PHILOMEL 1842-1857 179
PHIPPS 1808-1812 283
**PHOEBE** 1795-1841 123
PHOEBE 1804-1805 see BETSEY Hired 1804
PHOEBE 1810-1814 291
PHOEBE 1854-1875 174
PHOENIX 1671-1692 13
PHOENIX 1692-1698 188
PHOENIX 1694-1708 29
PHENIX 1704-1704 194
PHOENIX 1709-1727 37, 38
PHOENIX 1728-1744 50
PHOENIX 1743-1760 52
**PHOENIX** 1759-1780 79
PHOENIX 1783-1816 82
PHOENIX 1794-1801 265
PHOENIX 1799-1801 265
PHOENIX 1803-1804 291
PHOENIX 1804-1804 291
PHOENIX 1810-1814 291
PHOSPHORUS 1804-1810 281
PICKLE 1800-1801 263
PICKLE 1802-1808 see STING 1800
PICKLE see ECLAIR 1801
PICKLE 1808-1818 287
**PICKLE** 1827-1847 166
PICKLE 1852-1854 292
PICTON see PICTOU
PICTOU 1813-1813/ 1814-1818 283
PIEDMONTAISE/PIEMONTAISE 1808-1815 272
PIEDMONTAISE 1815-1815 see PRESIDENT 1806
PIERCER 1794-1802 149
PIERCER 1804-1814 153
Pierre Czar (French, American built 1805) see ALGERINE
PIGEON 1805-1805 282
PIGEON 1806-1809 161
PIGEON 1827-1847 145
PIGMY see RANGER 1779
**PIGMY** 1781-1793 102
PIGMY 1798-1805 see MUTIN 1779
PIGMY 1806-1807 287
PIGMY 1807-1814 see RANGER 1806
**PIGMY** 1810-1823 163
PIGOT 1778-1778 228
Pigot see LANCASTER 1797
Pijl (Dutch 1784) see PYL 1796
PIKE 1806-1809 161
PIKE 1808?-1811 see FLEUR DE LA MER
PIKE 1813-1836 283
PILCHARD 1805-1813 161
PILOT 1807-1828 141
**PILOT** 1838-1862 179
PILOTE 1780-1799 227
PINCHER 1797-1802 259
PINCHER 1804-1816 153
PINCHER 1827-1838 166
PIONEER 1804-1807 287
PIONEER 1810-1849 163
PIQUE 1795-1798 245
PIQUE 1800-1819 244
PIQUE Cancelled 1832 or 1833 123
**PIQUE** 1834-1862 175
PITT 1763-1766 209
Pitt (merchant) see BEAVER 1776 or 1777
PITT 1809-1812 291
PITT 1805-1807 282
Pitt (E.I.Co.1805) see DORIS 1808
PITT 1816-1877 114
PITT 1882-1906 see TRAFALGAR 1820
PLACENTIA 1780-1782 223
**PLACENTIA** 1790-1794 158
**PLANTAGENET** 1801-1817 109
PLAY PRIZE 1689-1697 185
PLEASANT HILL 1790s 265
PLEASANT HILL 1804-1810 291
PLOVER 1796-1819 131
PLOVER 1821-1841 145
PLOVER 1842-1854 295
PLOW 1706-1706 196
PLUMPER 1795-1802 149
PLUMPER 1804-1805 153
PLUMPER 1807-1812 153
PLUMPER 1813-1833 154
PLUTO 1745-1747 200
PLUTO 1756-1762 207
PLUTO 1778-1780 see TAMAR 1758
**PLUTO** 1782-1817 100
PLYM 1795?-1802 263
PLYM 1844-1869 325
PLYMOUTH 1653-1703 12
PLYMOUTH 1690-1730 342
PLYMOUTH 1704-1742 323
PLYMOUTH 1705-1705 23
PLYMOUTH 1708-1720 35
PLYMOUTH 1722-1764 44
PLYMOUTH 1742-1806 323
PLYMOUTH 1755-1793 305
PLYMOUTH 1779-1798 324
PLYMOUTH 1786-1815 324
PLYMOUTH 1797-1801 265
PLYMOUTH 1796-1830 305
PLYMOUTH 1799-after 1864 324
PLYMOUTH 1806-1815 318
PLYMOUTH 1814-1870 306
PLYMOUTH 1876-1904 see MEDWAY by 1860
Plymouth Packet (merchant) see PLYM 1795
PLYMOUTH PRIZE 1709-1709 194
PLYMTON 1778-1830s 323
POCAHONTAS see PACAHUNTA
POCOCK 1808-1809 291
PODARGUS 1808-1833 144
POICTIERS 1809-1857 114
Poisson Volante? see FLYING FISH 1793
Poisson Volante (French) see FLYING FISH 1803
POLACCA 1756-1769 210
POLECAT 1782?-1782 232
Policy (merchant) see AUGUSTA 1819
POLL 1808-1809 291
Polly (American?) see RACEHORSE 1777
POLLY 1798-... ? 259
POLLY 1801-1801 265
POLYPHEMUS 1782-1827 74
POMONA 1761-1776 207
POMONA 1778-1795 86
POMONA 1797-1801 265
Pomona (Spanish 1794) see CUBA 1806
POMONA/POMONE 1810-1816 272
POMONE 1794-1802 242
POMONE 1805-1811 119
Pomone (French 1804) see AMBUSCADE 1811
POMPEE 1793-1817 236
POMPEY 1740-1741 202
PONDICHERRY 1780-1784 229
POOLE 1696-1737 27
Poole (merchant) see PHAETON 1739
POOLE 1745-1765 48
PORCUPINE 1746-1763 199
**PORCUPINE** 1777-1805 88
PORCUPINE 1777-1788 222
PORCUPINE see SERPENT 1789
PORCUPINE 1805-1807 291
PORCUPINE 1807-1816 129
PORCUPINE Cancelled 1832 133
PORGEY 1807-1810 161
PORGY 1826-by 1848? 330
PORGY 1848-after 1864 330
PORPOISE 1777-1778 220
PORPOISE 1780-1783 229
**PORPOISE** 1798-1799 168
PORPOISE 1799-1803 260
PORPOISE 1804-1805 291
PORPOISE 1804-1816 276
PORT ANTONIO 1757-after 1763 210
PORT ANTONIO 1779-1783 224
PORT D'ESPAGNE 1806-1811 287
PORTIA 1810-1817 144
PORTLAND 1653-1692 13
PORTLAND 1693-1719 23
PORTLAND 1723-1743 45
PORTLAND 1744-1763 47
**PORTLAND** 1770-1817 77
PORTLAND 1796-1802 263
PORTLAND 1822-1862 116
PORTLANDS PRIZE 1746-1749 198
PORT MAHON 1711-1740 37
PORT MAHON 1740-1763 51
PORT MAHON 1798-1837 253
PORT MORANT 1780?-1782? 230
PORTO 1780-1782 221
PORT ROYAL 1757-1763 210
PORT ROYAL 1779-1783 232
PORT ROYAL 1781 211
PORT ROYAL 1796-1797 263
PORT ROYAL By 1816-1818 330
PORT ROYAL 1820s-by 1830 see MEDWAY at Portsmouth
PORT ROYAL 1832-1832 see MEDWAY from Sheerness
PORTSEA see LIVELY 1838
PORTSMOUTH 1650-1689 13
PORTSMOUTH 1679-1703 30, 303
PORTSMOUTH 1690-1696 26
PORTSMOUTH 1703-1741 303
PORTSMOUTH 1707-1728 36
PORTSMOUTH see ELTHAM 1736
PORTSMOUTH 1742-1747 61
PORTSMOUTH 1742-1793 305
PORTSMOUTH 1747-1767 323
PORTSMOUTH 1756-1759 210
PORTSMOUTH 1794-1869 305
PORTSMOUTH 1811-1834 168
PORTSMOUTH PRIZE 1694-1696 185
POST BOY 1694-1694 32
POST BOY 1695-1695 32
POST BOY 1695-1702 31
Postillion (French 1691) see POSTILLION PRIZE
POSTILLION 1709-1711 195
POSTILLION 1745-1748 202
POSTILLION 1757-1763 205
POSTILLION PRIZE 1701-1709 193
Poulette (French) see SPITFIRE 1793
POULETTE 1793-1796 263
POULETTE 1799-1814 248
POUNCER 1797-1802 259
POWERFUL 1783-1812 70
POWERFUL 1826-1864 107
PRECIOUS STONE 1692-...? 186
Premier Consul (French 1788) see SOPHIE 1798
Premier Consul (French 1800) see SCOUT 1801
PRESIDENT 1650-1660 see BONADVENTURE 1660
PRESIDENT 1806-1815 271
PRESIDENT 1815-1818 270
**PRESIDENT** 1829-1903 117
PRESTON 1653-1660 see ANTELOPE 1660
PRESTON 1716-1739 see SALISBURY 1698
PRESTON 1742-1749 46
**PRESTON** 1757-1815 77
PRESTWOOD 1793-1801 265
PRETTY LASS 1804-1805 291
PREVOST 1804-1806 287
PREVOYANTE 1795-1819 245
PRIMROSE 1807-1809 141
**PRIMROSE** 1810-1832 147
PRINCE 1670-1692 11
PRINCE 1705-1707 see OSSORY 1682
PRINCE 1715-1738 see TRIUMPH 1698
PRINCE 1750-1776 40
PRINCE 1788-1837 64
PRINCE ALBERT ?1840-1856? 335
PRINCE ALBERT see PRINCESS ROYAL 1853
PRINCE ARTHUR 1808-1808 145
PRINCE AUGUSTUS FREDERICK 1794-1817? 332
PRINCE AUGUSTUS FREDERICK 1816-1821 296
Prince Charles (French) see HAZARD 1744
PRINCE COBURG 1795-1798 265
PRINCE D'AUVERGNE 1794-1801 265
PRINCE DE NEUFCHATEL 1814-1815 287
PRINCE EDWARD 1745-1766 49
PRINCE EDWARD 1772-1775 213
PRINCE EDWARD 1780-1782 223
PRINCE EDWARD 1781-1812 216
PRINCE EDWARD 1793-1794 265
PRINCE FREDERICK see EXPEDITION 1714
PRINCE FREDERICK 1740-1784 42
PRINCE FREDERICK 1796-1817 239
PRINCE FREDERICK 1807-1816 see 'DENMARK' yacht 1785
PRINCE GEORGE 1693-1695 189
PRINCE GEORGE 1701-1719 18
PRINCE GEORGE 1723-1758 40
PRINCE GEORGE 1763-1771 209
PRINCE GEORGE 1772-1839 64
PRINCE HENRY 1747-1764 79
PRINCE OF ORANGE 1692-1697 189
PRINCE OF ORANGE 1734-1810 41
Prince of Orange (merchant) see PHEASANT 1778
PRINCE OF ORANGE 1799-1822 239
PRINCE OF ORANGE 1804-1805 291
PRINCE OF WALES 1690s-by 1699 187
PRINCE OF WALES 1756-1758 211
PRINCE OF WALES 1765-1783 69
Prince of Wales see SUPPLY 1777
Prince of Wales see RELIANCE 1793
PRINCE OF WALES 1794-1822 64
Prince of Wales (Revenue service 1794) see THRUSH 1806
PRINCE OF WALES 1795 332
PRINCE OF WALES 1795-1801 or 1797-1803? 265
PRINCE OF WALES *Ca*1801 260
PRINCE OF WALES 1803 333
PRINCE OF WALES 1806 333
PRINCE OF WALES ? 1823-1848? 334
PRINCE OF WALES 1830 335
PRINCE OF WALES 1850-...... 306
**PRINCE OF WALES** 1860-1869 170
PRINCE REGENT 1812-1813 299
PRINCE REGENT 1814-1814 300
PRINCE REGENT 1820-1847 306
Prince Regent (Packet 1821) see CYNTHIA 1826
PRINCE REGENT 1823-1875 104
PRINCESS 1716-1728 see OSSORY 1711
PRINCESS 1795-after 1797 260
PRINCESSA/PRINCESA 1740-1784 198
PRINCESSA 1780-1784 216
PRINCESS ALICE 1845-after 1864 330
PRINCESS AMELIA 1728-1752 see HUMBER 1726
PRINCESS AMELIA see NORFOLK 1757
**PRINCESS AMELIA** 1757-1818 66
PRINCESS AMELIA (Cancelled 1800) 115
PRINCESS AMELIA 1816-1818 see DENMARK 1785
PRINCESS ANN 1692-1699 189
PRINCESS ANN 1701-1706 195
PRINCESS ANNE 1701-1702 see DUCHESS 1679
PRINCESS AUGUSTA 1773-1818 see CHARLOTTE 1711
PRINCESS AUGUSTA see 'DENMARK' 1785
PRINCESS AUGUSTA 1803-1814 291
PRINCESS CAROLINA/PRINCESS CAROLINE 1731-1764 40
PRINCESS CAROLINA/PRINCESS CAROLINE 1780-1783 216
PRINCESS CAROLINA 1807-1813 269
PRINCESS CHARLOTTE 1799-1812 244
PRINCESS CHARLOTTE 1804-1805 291
PRINCESS CHARLOTTE 1807? 287
Princess Charlotte (merchant) see VIPER 1807
PRINCESS CHARLOTTE 1814-1815 300
**PRINCESS CHARLOTTE** 1825-1875 106
PRINCESS LOUISA 1728-1736 48
PRINCESS LOUISA 1737-1742 see SWALLOW 1732
PRINCESS LOUISA 1744-1766 45
PRINCESS MARY 1728-1737 see MARY 1704
PRINCESS MARY 1742-1766 45
PRINCESS MARY 1808-1809 287
PRINCESS OF ORANGE see PRINCE OF ORANGE 1799
PRINCESS OF WALES 1777-1778 233
PRINCESS OF WALES 1797-1801 265
PRINCESS OF WALES 1803-1814 291
PRINCESS ROYAL 1728-1763 see OSSORY 1711
PRINCESS ROYAL 1739-1750 201
PRINCESS ROYAL 1773-1807 64
PRINCESS ROYAL 1795-1797 265
Princess Royal (East Indiaman) see ARDENT 1796
PRINCESS ROYAL 1804-1805 291
PRINCESS ROYAL ?1836-1859? 334
**PRINCESS ROYAL** 1853-1872 171
PRINCESS SOPHIA FREDERICA 1807-1816 269
Princessen Carolina (Dutch) see PRINCESS CAROLINA 1780
PRINCE WILLIAM 1780-1817 215
PRINCE WILLIAM 1797-1801 265
PRINCE WILLIAM 1804-1812 291
Prince William Henry (merchant) see PROVIDENCE 1798
Principe de Asturias (Spain) see CUMBERLAND 1695
Prinses Frederika Louise Wilhelmina [Dutch 1789] see BRAAVE 1796
Prinses Louisa [Dutch 1769] see BROADERSCARP 1799
Prinsesse Carolina (Danish 1805) see PRINCESS CAROLINA 1807
Prindsesse Sophie Frederica (Danish 1773) see PRINCESS SOPHIA FREDERICA 1807
Prins Frederick [Dutch 1779] see PRINCE FREDERICK
Prins Frederick Willem [Dutch 1788] see GELYKHEID 1797
PROCRIS 1806-1815 141
PROCRIS 1822-1837 145
PROGRESSO 1843-1866 329
PROHIBITION 1699-1702 31
**PROJECT** 1806-1810 156
PROMETHEUS 1807-1839 149
PROMPT 1702-1703 191
PROMPT 1840-1845 293
PROMPTE 1793-1813 247
**PROMPTE** 1814-1814 127
PROSELYTE 1780-1785 218
PROSELYTE 1793-1794 263
PROSELYTE 1796-1801 246
PROSELYTE 1804-1808 274
PROSELYTE 1804-1809 287
PROSERPINE 1756-1756 207
PROSERPINE 1757-1763 207
PROSERPINE 1777-1779 86
Proserpine (French 1785) see AMELIA 1796
PROSERPINE 1798-1799 263
Proserpine (Dutch 1801) see AMSTERDAM 1804
PROSERPINE 1807-1809 126
PROSERPINE 1830-1864 130
PROSPERITY 1782-1784 230
PROSPERO 1803-1807 281
PROSPERO 1809-1816 144
PROSPEROUS 1689 189
PROSPEROUS 1704-1706 195
PROSPEROUS 1711-1711 195
PROSPEROUS 1711-1711 195
PROSPEROUS 1790s 265
PROTECTOR 1756-1761 212
Protector (American 1776) see HUSSAR 1780
PROTECTOR 1805-1833 153
Protée (French 1772) see PROTHEE
PROTEUS 1777-1779 220
PROTHEE 1780-1815 215
PROTHEE 1825-1832 see PEARL 1762
PROVIDENCE 1689 189
PROVIDENCE 1691-1707 188
PROVIDENCE 1704-1706 195
PROVIDENCE 1780-1784 218
PROVIDENCE 1779? 232
PROVIDENCE 1782-1784 231
PROVIDENCE 1791-1797 235
PROVIDENCE 1798-1804 263
PROVIDENCE 1800-1801 265
PROVIDENCE 1801-1802 265
PROVIDENCE 1804-1821 333
PROVIDENCE 1804-1804 291
PROVIDENCE 1804-1812 291
PROVIDENCE by 1810 322
PROVIDENCE 1822-1829 see QUAIL 1817
PROVIDENCE see CURLEW
PROVO see PREVOST 1805
PRUDENCE 1693-1695 189
PRUDENCE 1800-...? 263
PRUDENT 1768-1814 73
PRUDENTE 1779-1802/3 217
PRUDENTE see WOLF 1806
Prussian Eagle (merchant) see TORTOISE 1782
PSYCHE 1805-1812 274
Psyche (Dutch) see SCIPIO 1807
PSYCHE 1814-1814 127
PSYCHE 1814-1830? 300
Ptarmigan (merchant 1849) see RESOLUTE 1850
PUCK 1855-1856 293
PUISSANT 1793-1816 237
PUTULSK 1807-1810 278
PYL 1796-1801 251
Pylade (French 1804) see VIMIERA 1808
Pylade (Franco-Italian 1807) see CARLOTTA 1810
PYLADES 1781-1790 223
**PYLADES** 1794-1796/ 1798-1815 129
Pylades (Dutch 1804) see SURINAM 1804
**PYLADES** 1824-1845 137
**PYRAMUS** 1810-1879 124
Pyrrhus (French 1791) see MONT BLANC 1811

## Q

QUAIL 1806-1816 161
**QUAIL** 1817-1829 164
QUAIL 1817-1822 293
QUAIL 1830-1859 166, 34, 326
QUAKER 1671-1698 15
Quatre Frères (French) see TRANSFER 1797
QUEBEC 1760-1779 83
QUEBEC 1775?-1782? 232
QUEBEC 1781-1816 84

INDEX 363


QUEBEC 1815 302
QUEEN 1693-1709 17
QUEEN
see ROYAL GEORGE 1715
**QUEEN** 1769-1821 65
QUEEN 1778-1782 233
QUEEN 1793-1794 265
QUEEN 1793-1794 265
QUEEN 1794-1801 265
QUEEN 1801-1801 265
QUEEN 1804-1805 291
**QUEEN** 1839-1871 170
Queen Anne Gally (merchant)
see DILIGENCE 1709
QUEENBOROUGH
1671-1718 303
QUEENBOROUGH
1694-1719 28
QUEENBOROUGH
1718-1777 304
QUEENBOROUGH
1744-1746
see FOWEY
QUEENBOROUGH
1748-1761 87
QUEENBOROUGH
1800-1801 265
QUEENBOROUGH
1804-1805 291
QUEEN CHARLOTTE
1790-1800 63
QUEEN CHARLOTTE
1800-1801 265
QUEEN CHARLOTTE
1803-1814 291
QUEEN CHARLOTTE
1803-1812 291
QUEEN CHARLOTTE
1804-1805 291
QUEEN CHARLOTTE
1809-1818 302
QUEEN CHARLOTTE
1810-1859 63
QUEEN CHARLOTTE
1859-1861
see UNION 1811
QUEEN MAB
see COQUETTE 1807
QUEEN MAB 1807-1812? 287
QUEEN PRIZE 1694-.... 186
QUEENSTON 1815 301
QUINBOROUGH
see QUEENBOROUGH

## R

Raadhuis van Haarlem (Dutch)
see STATHOUSE VAN HARLEM
RACEHORSE 1757-1775 207
RACEHORSE 1776-1776 224
RACEHORSE 1777-1778 224
RACEHORSE 1779-1781 228
RACEHORSE 1781-1799 221
RACEHORSE 1801-1802 265
RACEHORSE 1806-1822 140
RACEHORSE 1830-1901 138
RACER 1793?-1816 322
RACER 1810-1810 164
RACER 1812-1814 283
Racer (American privateer 1811)
see SHELBURNE 1813
RACER 1818-1830 334
RACER 1833-1852 178
RACER ?1836-1860? 335
RACOON 1782-1782 232
RACOON 1795-1806 139
RACOON 1808-1838 132
RAIKES 1780-1780 229
RAILLEUR 1783 232
RAILLEUR 1797-1800 249
RAILLEUR 1804-1810 276
RAINBOW 1747-1802 79
RAINBOW 1806-1807 288
RAINBOW 1809-1815 275
RAINBOW 1823-1838 133
RAINBOW PRIZE
1696-1698 185
RAINIER
see ADMIRAL RAINIER
RAISON 1795-1801 247
RAISON
see GLUCKSTADT 1807
RAISONABLE 1758-1762 204
RAISONABLE 1768-1815 74
RALEIGH
see ADVENTURE 1771
RALEIGH 1778-1783 217
RALEIGH 1806-1841 141
**RALEIGH** 1845-1857 173
RAMBLER 1778-1787 226
RAMBLER 1778-1779 233
RAMBLER 1796-1810 251
RAMILLIES 1706-1741
see ROYAL KATHERINE 1703
RAMILLIES 1749-1760 40
**RAMILLIES** 1763-1782 69
RAMILLIES 1785-1850 71
Ramillies (merchant 1804) 142
see PROSELYTE 1804
RANELAGH 1697-1723 19
RANELAGH
see PRINCESS CAROLINE 1731
RANGER 1747-1749 198
RANGER 1752-1783 92
Ranger (American 1777)
see HALIFAX 1780
RANGER 1777-1781? 233
RANGER 1779-1784 232
RANGER 1787-1794 233
RANGER 1794-1805 129
RANGER 1796? 332
RANGER 1810-1810 291
RANGER 1805-...? 284
RANGER 1806-1807 282
RANGER 1807-1814 130
RANGER By 1816-1821 316
RANGER 1820-1832 133
RANGER ?1829-1846? 334
RANGER 1835-1867 179
RAPID 1804-1808 153
**RAPID** 1808-1814 147
Rapid
see NOVA SCOTIA 1812
RAPID 1829-1838 145
RAPID 1840-1856 178
RAPIDE 1808-1814 288
RAPOSA/RAPOSO
1806-1808 288
RATION 1779-.... 298
Rattler (merchant 1780)
see SPARROW 1796
RATTLER 1783-1792 98
RATTLER 1793-1796 265
RATTLER 1795-1815 129
RATTLER 1797-1802? 259
Rattler (merchant 1798)
see WOLVERINE 1798
**RATTLESNAKE** 1777-1781 101
RATTLESNAKE 1779-1781 233
Rattlesnake (American)
see CORMORANT 1781
RATTLESNAKE 1783-1786
see CORMORANT 1781
RATTLESNAKE 1791-1814 129
RATTLESNAKE
see HERON 1812
RATTLESNAKE 1822-1860 133
RAVEN 1745-1763 55
RAVEN 1771-1780 211
RAVEN 1782-1783 80
see CERES 1777
RAVEN 1796-1798 139
RAVEN 1799-1800 253
RAVEN 1804-1805 142
RAVEN 1805-1816 143
RAVEN 1829-1859 167
READY 1797-1802 259
REBECCA 1691-1691 189
REBECCA 1804-1810 291
REBUFF 1805-1814 154
RECOVERY
see MINERVA 1759
RECOVERY 1781-1783
see HOPE 1776
RECOVERY 1782-1786 233
RECOVERY 1797-1801
see PORT ROYAL 1796
RECRUIT 1806-1822 141
RECRUIT 1829-1832 145
**RECRUIT** 1846-1849 181
REDBREAST 1805-1815 153
REDBREAST 1817-1850 164
REDBREAST ?1837-1838? 335
REDBRIDGE 1699 190
REDBRIDGE 1796-1803 159
REDBRIDGE 1804-1805 288
REDBRIDGE 1804-1806 288
REDBRIDGE 1807-1808? 288
REDESDALE 1809-1810 291
RED LION
see LION 1640
REDOUBT 1793-1802 255
REDOUBTABLE 1815-1841 114
REDPOLE 1808-1828 144
REDWING 1806-1827 141
REDWING Cancelled 1831 or 1832 139
REFUGE
see RESOLUTE 1850
REGARD 1798-1802 265
Régénérée (French 1793)
see ALEXANDRIA 1801
REGENT 1813-1821 333
REGULUS 1785-1816 80
REGULUS 1804-1806 288
Régulus (French privateer)
see MORNE FORTUNEE
Reina (Spanish 1743)
see REYNA 1762
REINDEER 1804-1814 142
REINDEER 1829 145
Relampago
see INDIAN 1855
RELIANCE 1793-1815 260
RELIANCE 1812-1814 291
RENARA 1806 288
RENARD 1781-1782 232
RENARD 1797-1805 249
RENARD 1803-1809 276
RENARD 1808-1818 145
Renard (French)
see CRAFTY
RENARD
see also REYNARD
RENEGADE 1823-1826 296
RENIRA 1847-1850 294
RENNIE 1813-1863
325, 318, 319, 321
Renommée (French 1744)
see RENOWN 1747
RENOMMEE 1796-1810 246
Renommée (French 1805 or 1807)
see JAVA 1811
RENOWN 1747-1771 198
RENOWN 1774-1794 77
RENOWN 1798-1835 110
REPRISE 1795-1797
see REVOLUTIONAIRE
REPRISAL 1794-1794 253
Républicain (French 1791)
see MONT BLANC 1805
REPUBLICAINE/REPUBLICAN
1795-1803 263
Républicaine Francaise (French 1793)
see RENOMMEE 1796
REPULSE 1759-1776 205
REPULSE 1779-1782 226
REPULSE 1780-1800 74
REPULSE 1781-1783 232
REPULSE 1794-1795 257
REPULSE 1797 333
**REPULSE** 1803-1820 113
REPULSE ?1829-1836? 335
REQUIN/REQUIEN
1795-1801 251
Requin (French 1806)
see SABINE 1808
RESEARCH 1846-1859 295
RESERVE 1650-1701 13
RESERVE 1701-1703 24
RESERVE 1704-1716 25
RESISTANCE 1782-1795 80
Résistance (French 1793)
see FISGUARD
RESISTANCE 1801-1803 123
RESISTANCE 1805-1858 121
RESOLUE 1795-1802 263
RESOLUE 1798-1811 246
RESOLUTE 1805-1852 154
RESOLUTE 1850-1879 294
RESOLUTION 1667-1698 12
RESOLUTION 1698-1703 20
RESOLUTION 1705-1707 21
RESOLUTION 1708-1711 34
RESOLUTION 1758-1759 67
RESOLUTION 1770-1813 70
RESOLUTION 1771-1782 212
Resolution (merchant)
see DRAKE 1777
RESOLUTION 1779-1797 226
Resolution
see LARK 1780
RESOLUTION 1780-1782 232
RESOLUTION 1784-1789
see ROMULUS 1777
RESOLUTION 1795-1801 265
RESOLUTION 1798 263
RESOLUTION 1800-1831 333
RESOLUTION 1807-1814 291
RESOLUTION
see SHAMROCK 1816
RESOURCE 1778-1803 86
RESTAURATION
see RESTORATION
RESTORATION 1678-1701 12
RESTORATION 1702-1703 21
RESTORATION 1706-1711 26
Restoration?
see LOYALIST 1779
RETALIATION 1799-1800
see HERMIONE 1782
RETALIATION 1815 302
RETRIBUTION 1800-1803
see HERMIONE 1782
RETRIBUTION 1813-1835
see EDGAR 1779
REUNION 1793-1796 245
Revanche (French)
see HOBART 1794
Revenant (French 1809)
see VICTOR 1808
REVENGE 1699-1711 20
REVENGE 1716-1716
see SWIFTSURE 1696
REVENGE 1718-1740 35
REVENGE 1740 202
REVENGE 1742-1787 42
REVENGE 1778-1779 232
REVENGE 1780-1782 228
REVENGE 1796-1798 263
Revenge (American)
see FLYING FISH 1805
**REVENGE** 1805-1849 110
REVENGE 1806 288
Revolutie (Dutch 1779)
see PRINCE FREDERICK 1796
Révolution (French)
see CENSEUR 1795
REVOLUTIONAIRE
1794-1822 243
REVOLUTIONAIRE
1794-1795 263
REWARD 1804-1806 291
REYNA 1762-1772 204
REYNARD
see also RENARD
REYNARD 1821-1857 145, 319
RHIN 1806-1884 271
RHINOCEROS 1781-1784 230
RHINOCEROS 1793 263
RHODA 1804-1805
see INDUSTRY 1804
Rhodes (American)
see BARBADOES 1782
RHODIAN 1809-1813 145
RICHARD AND JOHN
1688-1692 187
RICHARD AND JOHN
1704 195
RICHARD AND MARGATE
1705 195
RICHARD AND MARTHA
1689 189
RICHMOND 1656-1698 13
RICHMOND 1746-1749 199
RICHMOND 1757-1781 83
RICHMOND 1806-1814 154
RICHMOND 1814-1820? 333
RIFLEMAN 1804-1809 281
RIFLEMAN 1809-1836 142
RINALDO 1808-1835 144
RINGDOVE 1806-1829 141
RINGDOVE 1833-1850 178
RIPON
see RIPPON
RIPPON 1712-1729 35
RIPPON 1735-1751 44
RIPPON 1758-1808 76
RIPPON 1812-1821 114
RIPPON'S PRIZE
1744-1747 198
Rittenhouse (American)
see ALBANY 1776
Rivoli (French 1807)
see ASP 1808
Rivoli (French, American built 1807)
see PERT 1806
RIVOLI 1812-1819 270
ROAST BEEF 1760-1761 211
ROBECQUE 1782-1783 222
Robert (French)
see ESPION 1793
ROBERT 1794-1799 258
ROBERT 1801-1801 265
ROB ROY 1845 335
ROBUST 1758 211
ROBUST 1764-1817 69
Robuste (French 1793)
see SCOURGE 1796
ROCHESTER 1692-1714 23
ROCHESTER 1705-1705 195
ROCHESTER 1706-1706 195
ROCHESTER 1707-1707 195
ROCHESTER 1716-1744 36
ROCHESTER
see MAIDSTONE 1744
ROCHESTER 1746-1746 202
ROCHESTER 1749-1777 47
ROCHESTER 1800-1810
see HERO 1759
ROCHESTER 1833-1877 319
ROCHESTER PRIZE
1702-1712 193
**ROCHFORT** 1814-1826 115
ROCKET 1804-1807 281
RODNEY 1759 211
RODNEY 1781-1782 232
RODNEY 1809-1827 114
**RODNEY** 1833-1882 107
ROE 1691-1697 31
ROEBUCK 1688-1692 187
ROEBUCK 1690-1701 26
ROEBUCK 1704-1725 26
ROEBUCK 1733-1743 48
ROEBUCK 1743-1769 48
**ROEBUCK** 1774-1811 79
Roebuck (merchant)
see SWAN 1781
ROEHAMPTON 1755-1756 211
ROLLA 1806-1810 277
**ROLLA** 1808-1822 144
ROLLA 1829-1868 145
ROMAN 1808-1814 279
Roman Emperor (merchant)
see PLUTO 1745
ROMAN EMPEROR
1757-1763 207
ROMNEY 1694-1707 23
ROMNEY 1708-1723 35
ROMNEY 1726-1757 45
**ROMNEY** 1762-1804 77
ROMNEY 1815-1845 115
ROMULUS 1777-1781 79
ROMULUS 1785-1816 81
Ronco (Franco/Italian 1807)
see TUSCAN 1808
ROOK 1806-1808 161
ROSA 1802-1803 263
ROSA 1804-1812 291
**ROSAMOND** 1807-1815 132
ROSARIO 1797-1800 254
ROSARIO 1800-1809 250
ROSARIO 1808-1812 145
ROSE 1674-1698 13
ROSE
see SALLY ROSE 1684
ROSE 1709-1711 193
ROSE 1712-1722 37
ROSE 1724-1744 49
ROSE 1740-1755 51
ROSE 1757-1779 90
ROSE 1783-1794 82
Rose (Revenue cutter)
see RANGER 1787
ROSE 1792-1828? 333
ROSE 1793-1801 265
ROSE 1794-1800 265
ROSE 1803-1804 291
ROSE 1804-1804 291
ROSE 1804-1804 291
ROSE 1805-1806 288
ROSE 1805-1817 131
**ROSE** 1821-1851 137
ROSE 1838-1864 296
ROSE 1857-1862 293
Rostan (French)
see CRESCENT 1758
Rosebud
see ACTIVE 1780
ROSINA 1804-1805 291
ROTA 1807-1816 271
ROTTERDAM 1672-1703 337
ROTTERDAM 1781-1806 217
ROVER 1779-1780 224
ROVER 1781-1781 222
Rover
see REDOUBT 1793
ROVER 1798-1801 265
ROVER 1796-1798 249
ROVER 1804-1804 291
ROVER 1808-1828 141
ROVER Cancelled 1831 139
**ROVER** 1832-1845 177
ROVER 1853-1862 180
ROWCLIFFE 1799-1801 265
ROWENA 1808-1809 291
Royal (French)
see TRITON PRIZE 1705
ROYAL ADELAIDE
1828-1905 106
ROYAL ADELAIDE
1834-1877 306
Royal Admiral (East Indiaman)
see YORK 1796
**ROYAL ALBERT** 1854-1883 170
ROYAL ANN/ROYAL
ANNE 1704-1729 18
ROYAL ANN 1756-1767
see ROYAL GEORGE 1715
ROYAL ANN
see ROYAL GEORGE 1756
ROYAL ANNE GALLEY
1709-1721 36
ROYAL CAROLINE
1716-1732 304
see PEREGRINE 1700
ROYAL CAROLINE
1734-1749 304
ROYAL CAROLINE
1750-1761 305
ROYAL CHARLES
1673-1693 11
ROYAL CHARLOTTE
see CHARLOTTE 1749
ROYAL CHARLOTTE
1761-1820
see ROYAL CAROLINE 1750
ROYAL CHARLOTTE
1764-1770 301
ROYAL CHARLOTTE
1780-1783 230
ROYAL CHARLOTTE 1793 332
Royal Charlotte (East Indiaman)
see MALABAR 1795
ROYAL CHARLOTTE
1824-1832 306
ROYAL CHARLOTTE
?1829-1880? 335
ROYAL ESCAPE 1660-1714 315
ROYAL ESCAPE 1714-1736 315
ROYAL ESCAPE 1736-1750 315
ROYAL ESCAPE 1749-1791 315
ROYAL ESCAPE 1792-1877
315, 319, 321
ROYAL EXCHANGE 1759 211
ROYAL FREDERICK
see QUEEN 1839
ROYAL FREDERICK
see FREDERICK WILLIAM 1860
ROYAL GEORGE 1714-1715
see VICTORY 1695
ROYAL GEORGE 1715-1756 33
ROYAL GEORGE 1746 202
**ROYAL GEORGE**
1756-1782 62
ROYAL GEORGE
1775-1818 332
ROYAL GEORGE
1777-1779? 298
ROYAL GEORGE
1788-1822 63
ROYAL GEORGE
1795-1818? 332
Royal George (Revenue 1803)
see BUSTARD 1806
ROYAL GEORGE
1803-1843? 333
ROYAL GEORGE 1808 333
ROYAL GEORGE
1809-1814 299
ROYAL GEORGE
1813-1814 299
Royal George (Merchant?)
see SPEEDWELL 1815
ROYAL GEORGE
1817-1905 306
ROYAL GEORGE
1827-1875 104
ROYAL GEORGE
1838-....? 335
Royal Hollondais (Franco/Dutch)
see CHATHAM 1809
ROYALIST 1798-1800 265
ROYALIST
see FLIRT 1794-1798
ROYALIST 1797-1801 263
ROYALIST 1804-1805 291
ROYALIST 1807-1819 141
ROYALIST 1825-1838 145
ROYALIST 1841-1857 292
Royaliste (French 1788)
see UNDAUNTED 1794
ROYAL JAMES 1675-1695 11
ROYAL KATHERINE
1664-1702 11
ROYAL KATHERINE
1703-1706 18
Royal Louis (American)
see ALBECORE 1781
Royal Mary (Scots 1698)
see GLASGOW 1707
ROYAL OAK 1674-1690 12
ROYAL OAK 1690-1710 20
ROYAL OAK 1713-1737 34
ROYAL OAK 1741-1764 42
**ROYAL OAK** 1769-1805 70
ROYAL OAK
see RENOWN 1798
ROYAL OAK 1809-1850 112
ROYAL SAVAGE
1775-1775 298
ROYAL SOVEREIGN
1637-1696 11
ROYAL SOVEREIGN
1701-1766 17
**ROYAL SOVEREIGN**
1786-1825 63
ROYAL SOVEREIGN
1804-1850 305
ROYAL SOVEREIGN
Cancelled 1838 171
ROYAL SOVEREIGN
1857-1885 170
ROYAL TRANSPORT

1695-1698 28
ROYALTY 1797-...? 259
ROYAL WILLIAM 1692-1714 17
Royal William (Scots 1696) see EDINBURGH 1707
ROYAL WILLIAM 1719-1813 39
ROYAL WILLIAM 1833-1885 105
Rubis (French 1707-8) see RUBY 1706
RUBIS 1747-1748 202
RUBY 1651-1706 13
RUBY 1706-1707 25
RUBY 1708-1748 35
RUBY 1745-1765 47
RUBY 1776-1821 74
RUBY PRIZE 1694-1698 185
RUMNEY see ROMNEY
RUNNER TRANSPORT 1706-1708? 315, 329
RUPERT 1666-1701 12
RUPERT 1703-1736 22
RUPERT 1740-1769 45
RUPERT PRIZE 1692-1700 186
RUPERTS PRIZE 1741-1743 200
RUSSELL 1692-1707 18
RUSSELL 1709-1726 33
RUSSELL 1735-1761 40
RUSSELL 1764-1811 69
RUSSELL 1822-1865 114
Russian Eagle (merchant) see TORTOISE 1782
RUYTER see ADMIRAL DE RUYTER
RUZEE 1692-1692 188
RYE 1696-1727 27
RYE 1727-1735 50
RYE 1740-1744 51
RYE 1746-1763 52

## S

SABINA 1804-1804 see CHAMPION 1803
SABINE 1808-1818 279
SABRINA 1806-1816 130
SABRINA 1838 296
SAFEGUARD 1797-1802 259
SAFEGUARD 1804-1811 153
SAFEGUARD 1849-1862 335
SAFETY 1802-1879? hulk see ECLAIR 1795
SAGESSE 1803-1821 275
SAINT ALBANS 1687-1693 13
SAINT ALBANS 1706-1717 25
SAINT ALBANS 1718-1734 36
SAINT ALBANS 1737-1744 46
SAINT ALBANS 1747-1765 75
**SAINT ALBANS** 1764-1814 73
SAINT ALBANS PRIZE 1690?-1698 186
SAINT ANDREW 1670-1701 11
SAINT ANN 1693-1703 189
SAINT ANNE 1756-1758 211
SAINT ANN 1761-1784 205
SAINT ANTHONY 1694-1694 189
Saint Antoine (French) see SAN ANTONIO 1801
SAINT ANTONIO 1700-1707 194
SAINT ANTONIO 1762-1775 205
SAINT ANTONIO see SAN ANTONIO 1801
SAINT CHRISTOPHER 1807-1811 288
SAINT CROIX 1793-1794 263
SAINT DAVID 1667-1713 13
SAINT DOMINGO 1809-1816 113
SAINT EUSTATIUS 1781-1783 221
SAINT FERMIN see SAN FERMIN
SAINT FIORENZO see SAN FIORENZO 1794
SAINT FLORENTINE 1759-1771 204
SAINT GEORGE 1687-1701 11
SAINT GEORGE (1622) 1687-1697 338
SAINT GEORGE 1701-1741 18
SAINT GEORGE 1702-1716 343, 337
SAINT GEORGE 1740-1774 40
Saint George 1754 see PELICAN 1757
SAINT GEORGE 1785-1811 65
SAINT GEORGE 1802-1802 265
SAINT GEORGE 1815-1825 see BRITANNIA 1762
SAINT GEORGE 1840-1883 104
Saint Jean (Maltese) see ATHENIAN 1800
SAINT JEAN D'ACRE Cancelled 1845 171
SAINT JOHN 1695-1698 188
SAINT JOHN 1764-1777 212
SAINT JOHN 1782-1783 232
SAINT JOSEPH 1704-1710 194
Saint Joseph (French 1808) see MAGNET 1809
SAINT LAWRENCE 1764-1766 212
SAINT LAWRENCE 1767-1775 213
SAINT LAWRENCE 1775-1776 212
SAINT LAWRENCE 1782? 232
SAINT LAWRENCE 1813-1815 283
SAINT LAWRENCE 1814-1832 300
SAINT LAWRENCE 1817? 296
SAINT LAWRENCE 1844-1859 see SHANNON 1806
SAINT LEWIS PRIZE 1696 185
SAINT LOE/SAINT LO 1701-1716 303
Saint Lewis see SAINT LEWIS PRIZE
Saint Louis (French 1696) see SAINT LEWIS PRIZE
SAINT LUCIA 1803-1807 288
SAINT MALOES see SAINT MALO MERCHANT/PRIZE
SAINT MALO MERCHANT 1689-1690 189
SAINT MALO PRIZE 1695-1696 189
SAINT MARTIN 1690-1695 186
SAINT MICHAEL 1669-1706 11
SAINT MICHAEL see MARLBOROUGH 1706
SAINT MICHAEL 1748-1748 202
SAINT MIGUEL see SAN MIGUEL 1782
SAINT NICHOLAS 1694-1694 188
SAINT NICHOLAS see SAN NICHOLAS 1797
SAINT NICHOLAS (OF DUNKIRK) 1690s-by 1699 187
SAINT PAUL 1679-1698 15
SAINT PHILLIPS CASTLE 1780s-1783 225
Saint Pierre (French, then Papal) see SPEEDY 1782
SAINT PIERRE 1796 263
SAINT PIERRE 1809-1814 277
SAINT QUINTIN 1744-1748 202
SAINT THOME 1800-1801? 263
SAINT VINCENT 1692-1698 188
SAINT VINCENT 1798-1802 265
SAINT VINCENT 1803-1806 291
SAINT VINCENT 1815-1906 104
SAINT VINCENTE see SAN VINCENTE
SALAMANDER 1687-1703 30
SALAMANDER 1703-1713 30
SALAMANDER 1730-1744 58
SALAMANDER 1745-1748 200
SALAMANDER 1757-1762 207
SALAMANDER 1778-1783 see SHARK 1776
SALAMANDER 1787-1799 234
SALAMANDER 1804-1807 281
SALAMINE/SALAMENE/SALEMENE 1799-1802 253
SALDANHA 1796-1811 243
SALDANHA 1809-1811 124
SALISBURY 1698-1703 24
SALISBURY 1707-1716 35
SALISBURY 1717-1724 36
SALISBURY 1726-1749 45
SALISBURY 1746-1761 47
SALISBURY 1769-1796 77
SALISBURY 1814-1837 115
SALISBURY PRIZE 1708-1716 see SALISBURY 1698
Sally (American 1754) see GRENVILLE 1763
Sally see SAINT LAWRENCE 1767
SALLY 1781-1783 230
SALLY 1797-1801 265
SALLY 1800-1800 265
SALLY 1801-1801 265
SALLY 1804-1807 291
SALLY 1809-1809 291
SALLY ROSE 1684-1696 15
SALOMAN/SALORMAN 1808-1808 288
SALSETTE 1806-1874 82
SALTASH 1732-1741 53
SALTASH 1741-1742 199
SALTASH 1742-1746 54
SALTASH 1746-1763 91
SALTASH 1748-1752 323
SALTASH 1756-1775 323
SALTASH 1808-1831 325
Salvador (Spanish 1781) see GIBRALTAR 1779
SALVADOR DEL MUNDO 1797-1815 236
SAMARANG 1808?-1814 see SCIPIO 1807
SAMARANG Cancelled 1820 139
SAMARANG 1822-1883 133
SAMPSON 1678-1689 15
SAMPSON 1689-1689 189
SAMPSON 1781-1832 74
SAMUEL & HENRY 1689-1695 190
SAMUEL BRADDICK 1804-1804 291
SAMUELL 1704-1704 195
San Alejandro (Spanish 1749) see GIBRALTAR 1780
San Antonio (Spanish 1761) see SAINT ANTONIO 1762
SAN ANTONIO 1801-1828 239
SAN BRUNO 1780 233
San Carlos (Spanish 1717-8) see CUMBERLAND 1695
San Carlos (Naples/Austria 1720-1773) see CUMBERLAND 1695
SAN CARLOS 1779-1784 230
San Cristobal (Spanish 1784) see BAHAMA 1805
San Damaso (Spanish 1750) see MORO 1762
SAN DAMASO 1797-1814 238
SANDFLY 1794-1803 150
SAN DOMINGO see SAINT DOMINGO
SANDWICH 1679-1709 11
SANDWICH 1711-1711 195
SANDWICH 1715-1770 33
**SANDWICH** 1759-1810 63
SANDWICH 1780-1781 230
SANDWICH 1782-1783 233
SANDWICH 1798-1801 265
SANDWICH 1798-1799 265
SANDWICH 1804-1814 291
SANDWICH 1803-1805 288
SANDWICH 1807-1809 111
see PITT 1809
SANDWICH Cancelled 1811
San Felipe Apostol (Spanish 1781) see OVERYSSEL 1795
SAN FIORENZO 1794-1837 243
SAN FIORENZO Cancelled 1856 174
SAN FIRMIN 1780-1780 223
SAN GENARO 1762-1763 211
San Giovanni (Maltese 1790s) see ATHENIAN 1800
San Gregorio (Spanish 1755) see SOVERANO 1762
San Ildefonso (Spanish) see ILDEFONSO
San Isidro (Spanish 1768) see SAN YSIDRO
SAN JOSEPH/JOSEF 1797-1849 236
San Josef (Spanish, Greek-built 1796) see CALPE 1800
San Josepho (French 1808?) see MAGNET 1809
SAN JUAN 1805-1808 268
San Juan Nepomuceno (Spanish 1776) see SAN JUAN
SAN LEON 1798-1800 253
San Lorenzo (Spanish 1747) see TIGER 1762
San Luis Gonzaga (Spanish 1750) see INFANTA 1762
SAN MIGUEL 1782-1791 215
SAN NICHOLAS/SAN NICOLAS 1797-1814 237
SAN RAFAEL 1780 232
SAN RAFAEL 1805-1810 267
Sans Pareil (French 1695-7) see NONSUCH 1668
SANS PAREIL 1794-1842 237
**SANS PAREIL** 1851-1867 173
Sans Pareille (French 1798) see DELIGHT 1801
SANTA AMMONICA see SANTA MONICA 1779
Santa Barbara (Spanish 1750) see PRINCESSA 1780
Santa Clara (Spanish 1784) see CLARA 1804
SANTA CRUX 1702-...? 339
SANTA DOROTHEA 1798-1814 246
Santa Florentina (Spanish 1786) see FLORENTINA 1800
SANTA GERTRUYDA 1804-1811 273
Santa Leocadia (Spanish 1777) see LEOCADIA 1781
SANTA LUCIA 1780-1783 232
SANTA MARGARITA 1779-1836 218
Santa Matilda (Spanish 1778) see DRYAD 1804
SANTA MONICA 1779-1782 217
SANTA TERESA 1780 232
SANTA TERESA 1799-1802 246
Santa Ursula (Spanish 1797) see AMPHITRITE 1804
SAN VINCENTE 1780-1783 224
SAN YSIDRO 1797-1813 239
SAPPHIRE 1675-1696 13
SAPPHIRE 1689-1690 190
SAPPHIRE 1708-1745 36
SAPPHIRE 1741-1784 48
SAPPHIRE 1806-1822 130
**SAPPHIRE** 1827-1864 135
SAPPHIRE'S PRIZE 1745-1745 200
SAPPHO 1806-1830 141
SAPPHO 1837-1859 178
SARACEN 1804-1812 142
SARACEN 1812-1819 142
SARACEN 1831-1862 146
SARAH 1704-1705 195
SARAH 1704-1711 195
SARAH 1756-1758 211
SARAH AND BETTY 1701-1703 195
Sarah Ann (merchant 1800) see METEOR 1803
SARDINE 1796-1806 249
SARDOINE 1761-1768 207
SARPEDON 1809-1812 145
SARPEN 1807-1811 278
SARTINE 1778-1780 219
SATELLITE 1806-1811 143
SATELLITE 1812-1824 142
SATELLITE 1825 296
**SATELLITE** 1826-1849 138
SATISFACTION 1777-1782 233
SATISFACTION 1794-1802 254
SATURN 1786-1868 68
SAUDADOES 1670/1673-1696 14
SAUDADOES PRIZE 1692-1692 185
SAUK 1814-1831? 302
SAUMEREZ 1804-1805 291
SAVAGE 1748-1748 202
**SAVAGE** 1750-1756 92
SAVAGE 1778-1815 96
SAVAGE 1805-1819 143
SAVAGE 1830-1864 146, 329
SAVANNAH 1779-1779 232
SAXON 1747-1748 202
SCAMANDER see LIVELY 1813
**SCAMANDER** 1813-1819 125
SCARBOROUGH 1691-1694 31
SCARBOROUGH 1694-1694 26
SCARBOROUGH 1695-1710 26
SCARBOROUGH 1711-1720 36
SCARBOROUGH 1722-1737 49
SCARBOROUGH 1739-1744 201
SCARBOROUGH 1740-1749 51
SCARBOROUGH 1756-1780 90
SCARBOROUGH 1812-1836 114
SCEPTRE 1689 190
SCEPTRE 1781-1799 75
SCEPTRE 1802-1821 113
SCEPTRE 1805-1806 291
SCIPIO 1739-1747 200
SCIPIO 1782-1798 75
Scipio see HECLA 1797
SCIPIO 1807-1808? 277
SCIPIO see BULWARK 1807
SCIPION 1793-1793 263
SCIPION 1805-1819 268
SCORPION 1742 202
SCORPION 1746-1762 57
SCORPION 1771-1780 211
SCORPION 1785-1802 97
SCORPION 1794-1804 258
SCORPION 1803-1819 140
Scorpion (American) see CONFIANCE 1814
SCORPION 1827 296
SCORPION 1832-1874 146
SCOURGE 1779-1795 223
SCOURGE 1779-1783 232
SCOURGE 1794-1803 258
SCOURGE 1796-1802 249
SCOURGE 1803-1816 276
SCOURGE 1804-1806 288
SCOURGE? 1818-1826 334
SCOUT 1695-1703 303
SCOUT 1780-1794 223
SCOUT 1800-1801 250
SCOUT 1801-1802 251
SCOUT 1804-1827 140
SCOUT 1805-1862 333
SCOUT 1823-1859? 335
**SCOUT** 1832-1852 139
SCOUT BOAT 1695-1703 32
SCRUB 1815-1827 284
SCRUB 1827-1828 296
SCYLLA 1809-1846 142
SEA BREEZE 1822-after 1864 330
SEAFLOWER 1704-1706 196
SEAFLOWER 1782?-1814 228
**SEAFLOWER** 1830-1866 181
SEAFORD 1695-1697 28
SEAFORD 1697-1722 28
SEAFORD 1724-1740 49
SEAFORD 1741-1754 51
**SEAFORD** 1754-1784 90
SEAFORD PRIZE 1708-1712 193
SEAFORTH 1805-1806 288
SEAGULL 1795-1804/5 139
**SEAGULL** 1805-1808 143
SEAGULL 1808-1814 279
SEAGULL 1831-1852 168
SEAHORSE 1694-1704 28
SEAHORSE 1694-1698 188, 319
SEAHORSE 1710-1711 37
SEAHORSE 1712-1727 37
SEAHORSE 1727-1748 50
**SEAHORSE** 1748-1784 88
SEAHORSE 1794- 1819 118
SEAHORSE 1830- 1870 123
SEALARK 1806-1809 161
SEALARK 1812-1820 283
SEALARK Cancelled 1831 146
SEALARK 1843-1898 179
SEALARK 1840?-1844? 335
SEA NYMPH 1780-1782 232
SEA NYMPH 1780-1780 233
SEA NYMPH 1800-1800 265
SECURITY 1778-1783 337
SECURITY 1785-after 1864 324, 327
SECURITY 1794-1802 345
SEDGEMOOR 1687-1689 13
Seijeren (Danish 1795) see SYEREN
Seine (French 1688) see FALKLAND PRIZE 1704
Seine (French?) see SNAKE 1777
SEINE 1798-1803 244
SEINE 1802-1813 see AMBUSCADE 1798
SEINE 1814-1823 273
SELBY 1654-1660 see EAGLE 1660
SELBY 1781-1783 232
SELBY 1798-1801 250
SELF 1801-1801 263
SEMIRAMIS 1808-1844 124
SENECA 1771-1788 299
SENEGAL 1760-1778/1780-1780 95
SENEGAL see RACEHORSE 1777
SENSIBLE 1798-1802 246
SENSIBLE see ROTA 1807
SENTINEL 1804-1812 277
SERAPIS 1779-1779 79
SERAPIS 1782-1819 80
Sérieux (French 1740) see INTREPID 1747
SERIN 1794-1796 254
**SERINGAPATAM** 1819-1883 122
SERPENT 1693-1694 30
SERPENT 1695-1703 30
SERPENT 1742-1748 61
SERPENT Cancelled 1783 97
SERPENT 1789-1806 129
SERPENT 1793-1793 263
SERPENT 1794-1802? 258
Serpent (French) see ASP 1808
Serpent (French, American built 1807) see PERT 1808
SERPENT Cancelled 1810 131
SERPENT 1832-1861 178
SEVEN SISTERS 1794-1800 258
SEVENTH HOY or SEVERN HOY 1689-1690 190
SEVERN 1693-1734 23
SEVERN 1739-1746/1747 46
SEVERN 1747-1759 76
SEVERN 1786-1804 80
Severn see SULPHUR 1797
SEVERN 1801-1802 265
Severn (East Indiaman 1812) see CAMEL 1813
SEVERN 1813-1825 116
SEVERN see TAGUS 1813
SEVERN Cancelled 1831 123
SEVERN 1856-1876 173
SHAMROCK 1808-1811 163
SHAMROCK 1812-1833 154
SHAMROCK 1816?-1838? 334
SHANNON 1757-1765 85
SHANNON 1796-1802 126
SHANNON 1797-1802 263
SHANNON 1803-1803 82
SHANNON 1806-1844 119
SHANNON see CHUB 1813
SHANNON 1855-1871 174
SHARK(E) 1691-1698 31
SHARK(E) 1699-1703 31
SHARK(E) 1711-1722 37
SHARK(E) 1723-1732 52
SHARK(E) 1732-1755 53
SHARK 1776-1778 p 220
SHARK 1779-1818 96
SHARK 1780-1780 220
SHARK 1794-1795 258
SHARK 1809-.... 333
SHARPSHOOTER 1805-1816 153
SHEARWATER 1808-1832 144
SHEERNESS 1690-1730 26
SHEERNESS 1691-*Ca*1700 319
SHEERNESS 1732-1744 50
SHEERNESS 1743-1768 52
SHEERNESS 1758-1811 319, 320
SHEERNESS 1787-1805 80
SHEERNESS 1788-1810? 324
SHEERNESS 1800-1801 265
SHEERNESS 1803-1805 291
SHEERNESS WATERBOAT 1698-1723 319
SHEERNESS WATERBOAT 1747-1772 320
SHELANAGIG 1781-1781 232
SHELBURNE 1813-1817 283
SHELBY 1781-1783 232
SHELDRAKE 1806-1817 143
SHELDRAKE 1825-1853 145
SHEPPEY 1813-1853? 317, 318, 320
SHEPPEY 1856-...? see DRAKE 1834
**SHERBOURNE** 1763-1784 100
SHIRLEY 1747-1748 202
SHORAM see SHOREHAM
SHOREHAM 1693-1719 26
SHOREHAM 1720-1744 49
SHOREHAM 1744-1758 52
SHOREHAM PRIZE 1709-1712 193
SHOREHAM'S PRIZE 1746-1747 200
SHREWSBURY 1695-1711 19
SHREWSBURY 1713-1750 33
SHREWSBURY 1758-1783 67
SIAM [SYAM] 1694? 189
SIAM [SYAM] 1701-1703 195
SIBYL 1779-1795 86
Sibylle (French) see SYBILLE 1794
SIDNEY SMITH 1799-1801 265
SIGNET see CYGNET
SINBAD 1834-1866 183, 318, 321
SINCERE 1793-1797 263
SINCERITY 1793-1794 265
SIR ANDREW MITCHELL 1804-1805 291

INDEX 365

SIR ANDREW MITCHELL 1806-1807 288
Sir Charles Grey (merchant) see LILLY 1795
SIR EDWARD HAWKE 1768-1773 212
SIR EDWARD HUGHES 1806?-1809 274
SIREN 1745-1764 52
SIREN 1773-1777 86
SIREN 1779-1781 88
SIREN 1782-1822 84
SIREN/SYREN Cancelled 1806 127
SIREN 1814-1815 288
**SIREN** 1841-1868 180
SIRENE 1794-1796 see SERIN
Sirene (Dutch) see DAPHNE 1796
SIR FRANCIS DRAKE 1805-1809 274
SIR GEORGE PREVOST see ROYAL GEORGE 1813
SIR GEORGE PREVOST 1814-...? 299
SIR HOME POPHAM 1814-....? 299
SIR ISAAC BROCK 1813 300
SIRIUS 1786-1790 see BERWICK 1781
**SIRIUS** 1797-1810 123
SIRIUS 1813-1816 122
SIR JAMES YEO 1814-...? 299
Siro (American) see ATALANTA 1814
SIR SIDNEY BECKWITH 1815 299
SIR SIDNEY SMITH see SIDNEY SMITH
SIR SIDNEY SMITH see MAGNET 1814?
SIR THOMAS PASLEY 1800-1800 265
SIR THOMAS PASLEY 1801-1802 265
SIR THOMAS TROUBRIDGE 1804-1806 291
SISTERS 1812-1863 322
SKIOLD 1807-1825 269
SKIPJACK 1807-1812 288
SKIPJACK 1827-1841 166
SKYLARK 1806-1812 143
SKYLARK 1812 288
SKYLARK 1821-after 1880 165
SKYLARK 1826-1843 145
SKYLARK 1841 335
SKYLARK see SKIPJACK 1827
SLANEY 1813-1838 137
SLY 1798-1800 266
SMYRNA FACTOR 1693-1695 190
SMIRNA [SMYRNA] FACTOR 1701-1705 195
SMYRNA MERCHANT 1689-1697 190
SNAKE 1711-1714 194
SNAKE 1776-1781 232
SNAKE 1777-1780 224
**SNAKE** 1797-1816 135
**SNAKE** 1832-1847 178
SNAP 1808-1811 279
SNAP 1812-1832 154
SNAP 1845-1847 296
SNAPPER 1782 332
SNAPPER 1804-1811 160
SNAPPER 1813-1824 155
SNIPE 1801-1846 152
SNIPE 1828-1860 165
Soberano (Spanish 1755) see SOVEREIGN 1762
Société de Grenze (French) see SOCIETY hired 1690
SOCIETY 1690-1697 190
SOCIETY 1694-1698 188
SOCIETY 1709-1711 195
SOESDYKE 1692-1713 303
SOLEBAY 1694-1709 24
SOLEBAY 1711-1748 37
SOLEBAY 1742-1763 51
SOLEBAY 1763-1782 86
SOLEBAY 1785-1809 84
SOLEBAY 1811-1833 see IRIS 1783
Solitaire (French privateer) see AMPHITRITE 1782
SOLITAIRE 1782-1786 216
SOMBRERO 1811 288
SOMERS 1807-1812 288
Somers (American) see SAUK 1814
SOMERSET 1698-1740 19
SOMERSET 1731-1746 40
SOMERSET 1748-1778 72
SOMERSET see SKIOLD 1807
SOPHIA 1685-1712/3 15, 315, 321
SOPHIA 1792 or 1795 299
SOPHIA 1804-1807 241
SOPHIA 1850-1853 294
SOPHIE 1782-1784 219
SOPHIE 1798-1809 250
SOPHIE 1809-1825 142
SORLINGS 1694-1705 26
SORLINGS 1706-1717 26
SORLINGS 1789-1809 234
SORLINGS by 1816-...? 316
SOUTHAMPTON 1693-1699 23
SOUTHAMPTON 1699-1771 24
SOUTHAMPTON 1704-1704 195
SOUTHAMPTON 1708-1708 195
SOUTHAMPTON 1757-1812 82
SOUTHAMPTON 1820-1857 116
SOUTH CAROLINA 1782 232
SOUTHSEA CASTLE 1696-1697 27
SOUTHSEA CASTLE 1697-1699 27
SOUTHSEA CASTLE 1708-1723 36
SOUTHSEA CASTLE 1724-1744 36
SOUTHSEA CASTLE 1745-1762 48
Souverain (French 1757) see GUERIER 1798
SOVERANO 1762-1770 204
SOVEREIGN see ROYAL SOVEREIGN 1637
SOVEREIGN OF THE SEAS see ROYAL SOVEREIGN 1637
**SPANKER** 1794-1810 151
SPARKLER 1797-1802 151
SPARKLER 1804-1808 153
SPARROW 1796-1805 263
SPARROW 1805-1816 143
SPARROW 1828-1860 165
SPARROWHAWK 1807-1841 141
SPARTAN 1806-1822 121
SPARTAN Cancelled 1831 123
**SPARTAN** 1841-1862 176
SPARTIATE 1798-1857 237
SPEAKER 1650-1660 see MARY 1660
SPEAKER 1756-1758 211
SPEAKER 1798-1802 263
SPECULATOR 1794-1801 266
SPECULATOR 1803-1813 291
SPEEDWELL 1688-1692 187
SPEEDWELL 1689-1689 189
SPEEDWELL 1690-1702 26
SPEEDWELL 1702-1715 27
SPEEDWELL 1706-1707 195
SPEEDWELL 1716-1720 37
SPEEDWELL 1744-1750 55
SPEEDWELL 1752-1779 92
SPEEDWELL 1761-1761 211
SPEEDWELL 1780-1807? 223
SPEEDWELL 1793?-1816 332
SPEEDWELL 1797-1801 266
SPEEDWELL 1801-1802 266
SPEEDWELL 1804-1804 291
Speedwell (1797) see LINNET 1806
SPEEDWELL 1815-1834 281
SPEEDWELL 1818-1819 334
SPEEDWELL 1830-.... 335
SPEEDWELL 1841-1855 295
SPEEDWELL PRIZE 1708-1712 194
**SPEEDY** 1782-1801 98
SPEEDY 1796?-1804 299
SPEEDY 1803-1817 276
SPEEDY 1828-1866 165
SPENCE 1723-1729 52
SPENCE 1730-1749 52
SPENCER 1691 190
SPENCER 1795-1804 249
**SPENCER** 1800-1822 110
Spencer (American privateer 1811) see COCKCHAFER 1813
SPEY 1814-1822 137
SPEY 1827-1840 145
SPHINX 1748-1770 87
**SPHINX** 1775-1811 90
SPHINX 1815-1835 145
SPIDER 1782-1806 228
SPIDER 1794-1796 266
SPIDER 1803-1804 291
SPIDER 1806-1815 278
SPIDER 1835-1861 165
SPITEFUL 1794-1800? 263
SPITEFUL 1797-1823 152
SPITFIRE 1778-1778 228
SPITFIRE 1779-1780 see SPEEDWELL 1752
SPITFIRE 1782-1825 99
SPITFIRE 1793-1794 255
SPITFIRE 1798-1801 263
SPITFIRE 1815 302
SPRAGGE see YOUNG SPRAG 1673
**SPRIGHTLY** 1777-1777 101
SPRIGHTLY 1778-1801 101
SPRIGHTLY 1779-1782 228
Sprightly (merchant) see SULTANA 1780
SPRIGHTLY 1803-1809 291
SPRIGHTLY 1805-1815 282
SPRIGHTLY 1818-1821 334
SPRIGHTLY ?1829-1840? 335
Spring (merchant) see LUCIFER 1803
Springfield (American) see CAROLINA 1781
SPRY 1742-1742 202
SPY 1690-1693 29
SPY 1693-1706 31
SPY 1721-1731 52
SPY 1732-1745 53
SPY 1756-1773 93
SPY 1763-1773 210
SPY 1776-1778 96
SPY see ESPION 1793
SPY 1804-1813 276
SPY 1841-1862 178
SPYE see SPY
SQUIB 1804-1805 281
SQUERRELL see SQUIRREL
SQUIRREL 1694-1714 303
SQUIRREL 1703-1703 28
SQUIRREL 1704-1706 28
SQUIRREL 1707-1727 37
SQUIRREL 1727-1749 50
SQUIRREL see ALDERNEY 1743
**SQUIRREL** 1755-1763 90
**SQUIRREL** 1785-1817 89
SQUIRREL 1832 &/or 1837 335
SQUIRREL 1853-1879 178
STAG 1758-1783 83
STAG 1793-1798 (or 1794?) 332
STAG 1794-1800 126
STAG 1794-1801 266
STAG 1804-1804 291
STAG 1812-1821 124
STAG 1827-1880? 334
STAG 1830-1866 123
STANDARD 1782-1816 74
Stanislaus (French 1780) see PROSELYTE 1780
STANLEY 1778 232
STAR 1694-1712 30
STAR 1778-1783 224
STAR 1795-1802 139
STAR 1805-1812 131
STAR 1808-1828 318
STAR 1814-1830? see MELVILLE 1813
**STAR** 1835-1857 179
STARLING 1801-1804 152
STARLING 1805-1814 154
**STARLING** 1817-1828 165
STARLING 1829-1844 167
STATELY 1784-1814 74
STATIRA 1807-1815 121
STATIRA Cancelled 1832 123
STAUNCH 1797-1803 259
STAUNCH 1804-1811 153
STAUNCH 1854-1870 317
STAVERENS ....?-1811 284
STEADY 1782-1784 230
STEADY 1797-1803 152
STEADY 1804-1815 153
Steece (Danish 1798) see WARNING 1807
Ster (Dutch) see HOPE 1795
STERLING CASTLE 1679-1699 13
STERLING CASTLE 1699-1703 20
STERLING CASTLE 1705-1720 21
STERLING CASTLE 1723-1771 41
STERLING CASTLE 1742-1762 43
STERLING CASTLE see also STIRLING CASTLE
STING 1800-1802 244
STIRLING CASTLE see also STERLING CASTLE
STIRLING CASTLE 1775-1780 74
STIRLING CASTLE 1811-1861 114
STORK 1746-1747 202
STORK 1756-1758 94
STORK 1778-1780 232
STORK 1796-1816 130
STORK 1807-1826? 333
STORK ?1829-1837? 335
STORMONT 1781-1784 225
STRAFFORD 1714-1726 35
STRAFFORD 1735-1756 44
STRENUOUS 1805-1814 155
STROMBOLO 1691-1713 29
STROMBOLO 1739-1744 200
STROMBOLO 1756-1768 207
STROMBOLO 1772-1780 see GRAMPUS 1746
STROMBOLO 1797-1809 254
STROMBOLO 1799-1799 263
STROMBOLO 1810-1815 see AUTUMN 1801
SUBSTITUTE 1782-1783 229
SUBTLE 1807-1807 288
SUBTLE 1808-1812 283
SUCCESS 1689-1692 190
SUCCESS 1694-1707 190, 336
SUCCESS 1704-1708 195
SUCCESS 1709-1709/1710 193
SUCCESS 1709-1748 37
SUCCESS 1711-1711 195
SUCCESS 1712-1743 37
SUCCESS 1740-1749 51
SUCCESS 1781-1820 84
Success (French 1801) see SUCCESS 1781
SUCCESS 1795-1802 258
SUCCESS see ETNA 1803
SUCCESS 1825-1849 133
SUFFISANTE 1795-1803 251
SUFFISANTE 1803-1807 277
SUFFOLK 1680-1698 13
SUFFOLK/SUFFOLK HAGBOAT 1694-1713 189
SUFFOLK 1699-1736 20
SUFFOLK 1740-1770 42
**SUFFOLK** 1765-1803 70
SUFFOLK 1804-1814 291
SUFFOLK 1805-1816 see SOLTAN 1775
SULPHUR 1778-1783 225
SULPHUR 1797-1816 255
SULPHUR 1826-1857 148
SULTAN 1775-1805 70
SULTAN 1802?-1804 288
SULTAN 1807-1864 112
SULTANA 1768-1773 212
SULTANA 1776-1783 232
SULTANA 1780-1799 227
SULTANE 1814-1819 273
SUNDERLAND 1693-1744 22
SUNDERLAND 1724-1742 44
SUNDERLAND 1744-1761 45
SUN PRIZE 1692-1701 185
SUN PRIZE 1704-1713 192
SUPERB see SUPERBE 1710
SUPERB 1736-1757 44
SUPERB 1760-1783 68
SUPERB 1798-1826 110
SUPERB see FOUDROYANT 1798
SUPERB 1842-1869 172
SUPERBE 1710-1733 192
SUPERBE 1795-1798 247
SUPERIEURE 1803-1814 277
SUPPLY 1688-1690 190
SUPPLY 1691-1747 318, 319
SUPPLY 1705-1705 195
SUPPLY 1725-1757 315, 318
SUPPLY 1759-1792 169, 315
SUPPLY 1777-1779 229
SUPPLY 1782-1784 230
SUPPLY 1793-1802? 260
SUPPLY 1798-1834 317
SUPPLY 1801-1801 266
SUPPLY 1816-1843 322
SURINAM 1799-1803 (1807) 250
SURINAM 1804-1808 288
SURINAM 1805-1825 140
SURLY 1806-1837 162
SURPRIZE 1746-1770 52
SURPRIZE 1774-1783 86
SURPRIZE 1778-1783 224
SURPRIZE 1779/1780-1786 226
SURPRIZE 1786-1792 234
SURPRIZE 1796-1802 247
SURPRISE 1799-1800 263
SURPRISE 1812-1837 119
SURPRISE 1814-1832 302
SURVEILLANTE 1803-1814 270
SURVEYOR 1813-1814 288
SUSANNAH 1704-1707 195
SUSANNAH 1804-1804 291
SUSSEX 1693-1694 19
SUSSEX 1790-1816 see UNION 1756
SUSSEX 1790s 266
SUSSEX OAK 1804-1808 291
SUTHERLAND 1716-1754 see RESERVE 1704
SUTHERLAND 1741-1770 46
**SUTLEJ** 1855-1869 174
SUWOROW see BLACK JOKE 1793
SWAGGERER 1809-1815 288
SWALE By 1845-...? 327
SWALLOW 1660-1692 13
SWALLOW 1699-1703 31
SWALLOW 1703-1717 24
SWALLOW 1719-1729 36
SWALLOW 1732-1737 44
**SWALLOW** 1744-1744 55
SWALLOW 1745-1769 58
SWALLOW 1769-1777 96
SWALLOW 1779-1781 222
Swallow (1779) see LILLY 1804
SWALLOW 1781-1793 223
SWALLOW 1795-1802 140
SWALLOW 1796-1816 332
SWALLOW 1805-1815 140
SWALLOW 1811-1830? 316
SWALLOW ?1819-1825 334
SWALLOW ....-1825 296
SWALLOW 1824-1836 296
SWALLOW 1829-1835 335
SWALLOW PRIZE 1692-1696 186
SWALLOW PRIZE 1705-1711 192
SWAN 1673-1689 15
SWAN 1694-1707 28
SWAN 1709/10-1713 37
SWAN 1745-1763
SWAN 1767-1779/1783-1814 55
**SWAN** 1781-1782 96
SWAN 1788/1790-1792 223
SWAN 1793-1795? 235, 332
SWAN 1799-1801 332, 263
SWAN 1803-1811 266
SWAN 1811-1874 291
SWEEPSTAKES 1666-1698 164
SWEEPSTAKES 1708-1709 13
SWEEPSTAKES 1709-1716 36
SWIFT 1695-1696 31
SWIFT 1696-1698 32
SWIFT 1699-1702 31
SWIFT 1704-1719 32
SWIFT 1721-1741 52
SWIFT 1741-1756 53
SWIFT 1748 202
SWIFT 1761-1762 208
SWIFT 1763-1770 96
SWIFT 1763-1773 209
SWIFT 1773-1784 213
SWIFT 1777-1778 96
SWIFT 1778-1781 222
SWIFT 1783?-1784 224
SWIFT 1793-1799 129
SWIFT 1793-1794 266
SWIFT 1794-1802 255
SWIFT 1796-1801 266
SWIFT 1799-1801 266
SWIFT 1803-1806 291
SWIFT 1803-1804 291
SWIFT 1804-1814 276
SWIFT 1817-1821 164
SWIFT 1821-1846? 334
SWIFT 1835-1866 179, 329
Swift Gally (Merchant) see SUCCESS 1709
SWIFT PRIZE 1689-1695 185
SWIFTS PRIZE 1717 196
SWIFTSURE 1673-1696 12
SWIFTSURE 1696-1716 20
SWIFTSURE 1750-1773 72
Swiftsure? (Merchant) see SEAFLOWER 1782
SWIFTSURE 1787-1805 71
**SWIFTSURE** 1804-1845 113
SWINGER 1794-1802 149
SWINGER 1798-1799 263
SWINGER 1804-1812 153
SWINGER 1813-1864? 154, 322
SYAM see SIAM
SYBILLE 1794-1833 243
SYBILLE 1847-1866 175
SYDNEY 1813-1825 284
SYEREN 1807-1815 270
SYLPH 1776-1779 220
SYLPH 1780-1782 227
SYLPH 1795-1811 140
SYLPH 1803-1809 291
SYLPH 1812-1815 136
**SYLPH** 1821- 1888 169
SYLPH 1834-1860? 335
Sylphe (French 1804) see SEAGULL 1808
SYLVIA 1806-1816 162
SYLVIA 1820-.... 334
**SYLVIA** 1827-1859 166, 322
SYMETRY 1776-1776 233
SYREN see SIREN
Syren (American 1803) see SIREN 1814

## T

TAGUS 1813-1822 125
TALAVERA see WATERLOO 1818
TALAVERA 1818-1840 113
TALBOT 1691-1691/1693-1694 31
Talbot see PROTEUS 1777
TALBOT 1807-1815 132
TALBOT 1824-1896 133
TAMAR 1758-1778 94
TAMAR 1795-1798 324
TAMAR/TAMER 1796-1810 118
TAMAR 1814-1837 132
Tamise (French 1793-6) See TAMISE 1758
TANAIS 1813-1819 120
TANG 1807-1811 161
TAPAGEUR 1779-1780 226
TAPAGEUR/TAPIGEUR 1781? 232
TAPAGEUSE 1806-? 288
Tape de l'Oeuil see DOMINICA 1805
TARLETON 1782/1793-1794 233, 263
TARTAN PRIZE 1692-1696 186
TARTAR 1702-1732 27
TARTAR 1734-1755 51
**TARTAR** 1756-1797 85
Tartar (merchant) see TRUE BRITON 1778
TARTAR 1794-1801 266
TARTAR 1793 332
TARTAR 1801-1811 127
TARTAR 1801 334
TARTAR 1814-1859 125
TARTAR ?1816-1840? 333
Tartare (French 1780-1782) see TRUE BRITON 1778
TARTAR'S PRIZE 1757-1760 206
TARTARUS 1797-1804 255
TARTARUS 1806-1816 149
Tartu (French 1788) see URANIA 1797
TARTUFFE 1763-1770 209
TAUNTON 1654-1660 see CROWN 1660
TAVISTOCK 1745-1747 55
TAVISTOCK 1746-1747 202
TAVISTOCK 1747-1768 76
TAVISTOCK 1748-1767 323
TAVY 1797-after 1864 324
TAVY 1819-1869 325
TAY 1813-1816 137
TAYLOR 1758-1758 211
TEASER 1801-1801 266
TEASER/TEAZER 1794-1802 149
TEASER/TEAZER 1798-1799 263
TEASER 1804-1815 153
TEAZER 1851-after 1864 329
TECUMSETH 1814 299
TECUMSETH 1814-1833 302
TEES 1813-1872 132
TELEGRAPH 1798-1801 266
Telegraph (merchant 1799) see RIFLEMAN 1804
TELEGRAPH 1812?-1817 283
TELEMACHUS 1796-1801 266
TELEMACHUS 1803-1804 291
TEMERAIRE 1759-1784 203
TEMERAIRE 1798-1838 106
Tempête (French 1689-1726) see FIREDRAKE 1688
**TEMPLE** 1758-1762 73
TENEDOS 1812-1875 119
TENEDOS III see GANGES 1821

366 INDEX

TERESA
see FREDERICKSTEIN
TERMAGANT 1780-1795 220
**TERMAGANT** 1796-1819 131
TERMAGANT 1822-1824
see HERALD
TERMAGANT 1838-1845 146
TERPSICHORE 1760-1766 206
TERPSICHORE 1785-1830 84
Terpsichore (French 1812)
see MODESTE 1814
TERPSICHORE 1847-1866 178
TERRIBLE 1694-1710 26
TERRIBLE 1730-1749 58
TERRIBLE 1747-1763 197
TERRIBLE 1762-1781 69
TERRIBLE 1785-1836 71
TERRIER 1782-1783 232
TERRIER 1798-1800 266
TERROR 1696-1705 30
TERROR 1742-1754 61
TERROR 1759-1774 99
TERROR 1779-1812 99
TERROR 1794-1804 263
TERROR 1813-1848 147
Teulié (Franco/Italian 1808)
see ROMAN 1808
TEXEL 1799-1808 241
**THAIS** 1806-1818 149
THAIS 1829-1833 146
THALIA 1782-1814 81
THALIA 1830-1887 120
THAMES 1758-1793/1796-1803 83
Thames (collier 1803)
see AVENGER 1804
THAMES 1804-1806 291
**THAMES** 1805-1816 127
THAMES 1805-1872 316, 318, 321
THAMES 1823-1863 120
Thamise (French)
see THAMES 1758
THEBAN
see PERLIN 1807
THEBAN 1809-1817 124
THEBAN Cancelled 1831 123
Thebe (merchant)
see DILIGENCE 1812
THEODOCIA
see VENTURER 1807
THEODOSIA 1808-1814 283
THEOPHILUS 1705-1705 195
Theresa (French?)
see PORT ANTONIO 1779
THESEUS 1786-1814 71
THETIS 1717 196
THETIS 1747-1767 49
**THETIS** 1773-1781 84
**THETIS** 1782-1814 81
Thétis (French 1787)
see BRUNE 1808
THETIS 1793-1794 266
THETIS 1793-1794 266
THETIS 1796-1802 263
THETIS 1798-1800 266
THETIS 1817-1830 119
**THETIS** 1846-1855 175
THISBE 1783-1815 86
THISBE 1824-1892 120
THISTLE 1808-1811 163
THISTLE 1812-1823 154
THISTLE 1857-1869 330
THOLIN 1796-1811 243
THOMAS 1704-1705 195
THOMAS 1796-1798 263
THOMAS 1808-1809 288
THOMAS AND DANIEL 1689-1689 190
THOMAS AND ELEANOR 1804-1805 291
THOMAS AND ELIZABETH 1688-1692 187
THOMAS AND ELIZABETH 1711-1711 195
THOMAS AND HESTER 1704-1706 195
THOMAS AND JAMES 1691-1697 189
THOMAS AND KATHERINE 1705-1705 195
THOMAS AND MARY 1705-1708 195
THOMAS AND MARY 1708-1708 195
THOMAS AND MARY 1721-1721 196
THORN 1779-1779/1782-1816 96
THOULOUSE 1711-1712 194
THRACIAN 1809-1841 142
THRASHER 1804-1814 277
THREE BROTHERS 1689-1689 190
THREE BROTHERS 1704-1706 195
THREE BROTHERS 1777-1782 233
THREE SISTERS 1777-1780 233
THRUSH 1806-1819 276
Thulen (Dutch)
see THOLIN 1796
THUNDER 1695-1696 30
THUNDER 1718-1734 195
THUNDER 1740-1744 60
THUNDER 1759-1774 99
THUNDER 1775-1778 bomb
see RACEHORSE 1757
THUNDER 1779-1781 99
THUNDER 1797-1802 255
THUNDER 1803-1814 281
THUNDER Cancelled 1812 148
THUNDER 1815 302
THUNDER 1829-1851 148
THUNDERBOLT PRIZE 1696-1698 185
THUNDERER 1760-1780 67
THUNDERER 1776-....? 298
THUNDERER 1783-1814 71
THUNDERER
see TALAVERA 1818
THUNDERER 1831-1869 107
TIBER 1813-1820 120
TIBER Cancelled 1831 123
TICKLER 1781-1783 225
TICKLER 1794-1802 149
TICKLER 1798-1800 263
TICKLER 1804-1808 153
TICKLER 1808-1816 282
TIGER/TYGER 1647-1681-1702 13
TIGER 1695-1696 190
TIGER/TYGER 1702-1718 24
TIGER/TYGER 1722-1743 36
TIGER/TYGER
see HARWICH 1743
TIGER/TYGER 1747-1765 75
TIGER/TYGER 1762-1783 204
TIGER/TYGER
see ARDENT 1764
TIGER 1794-1798 258
TIGER
see GRAMPUS 1802
TIGER 1805-1808 or 1808-1812 288
TIGER 1806-1808? 333
TIGER PRIZE 1678-1696 15
TIGRE 1795-1817 238
Tigre (Spanish 1747)
see TIGER 1762
Tigre (French 1793 or 1798?)
see DRAKE 1798
TIGRE 1809-1809 291
TIGRESS 1797-1802 152
TIGRESS 1804-1808 153
TIGRESS 1808-1814 282
Tigress (American)
see SURPRISE
TIGRIS 1813-1819 125
TIGRIS Cancelled 1832 123
TILBURY 1699-1726 24
TILBURY 1733-1742 44
TILBURY 1745-1757 45
TILBURY 1810-1814
see CHATHAM 1738
**TISIPHONE** 1781-1816 99
TOBAGO 1777?-1781 232
TOBAGO 1782-1783 225
TOBAGO 1805-1806 288
TOBAGO 1807? 288
TOLIAPIS ....? 319
Tom
see MINX 1797
TONNANT 1798-1821 237
TOPAZE 1793-1814 244
Topaze (French 1804)
see ALCMENE 1809
TOPAZE 1814-1851 273
TOPAZE 1858-1884 175
TORBAY 1693-1716 19
TORBAY 1719-1750 34
TORBAY 1750-1816 [74]
see NEPTUNE 1730
TORBAY
see KRON PRINCESSEN 1807
TORCH 1804-1811 281
TORCH 1805-1811 288
TORMENTOR 1795-.... 263
TORONTO 1817-.... 301
Torride (French 1775)
see TORRIDGE below
TORRIDGE 1799-1801 263
TORRINGTON 1654-1660
see DREADNOUGHT 1660
TORRINGTON 1729-1744 48
TORRINGTON 1743-1783 48
TORTOISE 1777-1780 229
TORTOISE 1780-1787 324
TORTOISE 1782-1785 231
TORTOISE 1789-1863 324
TORTOISE 1809-1865
see SIR EDWARD HUGHES 1807
TOURTERELLE 1795-1816? 247
TOWER 1809-1825 327
TOWEY 1814-1822 132
TOWNSEND 1802-1823 333
TOWZER 1798-1799 263
**TRAFALGAR** 1820-1825 106
TRAFALGAR 1841-1873 105
TRANSFER 1797-1802 252
TRANSFER
see MERCURIUS 1807
Transport No 45
see MEDORA 1855
TRANSIT 1809-1815 163
TRANSPORTER 1677-1702 318
TRANSPORTER 1704-1713 318
TRAVE 1813-1821 273
TRAVELLER 1804-1805 291
TREE CRONEN/TREKRONEN 1807-1825 269
TRELAWNY 1744-1747 201
TREMENDOUS 1784-1845 72
TRENT 1757-1764 85
TRENT 1796-1823 123
TRENT 1817-1818 296
Trente-et-un-Mai (French 1791)
see MONT BLANC 1805
TREPASSEY 1779-1781/1782-1784 224
TREPASSEY 1790-1803 158
TRIAL
see also TRYALL
TRIAL/TRYALL 1744-1746 58
TRIAL/TRYALL 1776-1776 232
TRIAL/TRYALL 1780s 228
**TRIAL** 1790-1848 158
TRIAL 1807-1815 288
TRIAL 1810-1818? 282
TRIBUNE 1796-1797 246
TRIBUNE 1803-1839 82
TRIBUNE Cancelled 1840s 177
TRIDENT/TRYDENT PRIZE 1695-1702 185
TRIDENT/TRYDENT 1747-1763 198
TRIDENT 1768-1816 73
TRIDENT PRIZE 1705-1713 194
TRIMMER 1782-1801 225
TRINCOMALEE 1799-1799 263
TRINCOMALEE 1801-1803/1806-1808? 263
TRINCOMALEE 1817-1897 119
TRINCULO 1809-1841 142
TRINIDAD 1805-1806 288
TRINIDAD 1808-1809 288
TRITON/TRYTON 1702-1709 192
TRITON/TRYTON 1745-1758 52
TRITON/TRYTON 1741-1741 202
TRITON 1773-1796 86
Triton
see NABOB 1777
**TRITON** 1796-1814 127
TRITONA 1811-1811 288
TRITON PRIZE/TRYTON PRIZE 1705-1713 192
Triumfo (Spain)
see TRIUMPH 1739
TRIUMPH/TRYUMPH 1698-1714 18
TRIUMPH 1739-1740 202
TRIUMPH/TRYUMPH 1764-1850 68
TROMP/TROMPE 1796-1800 239
TROMPEUSE 1794-1796 251
TROMPEUSE 1797-1800 249
TROMPEUSE 1800-1811 250
TROUBRIDGE 1804-1810 291
Trudaine (French 1754?)
see BEAVER 1757
TRUE BLUE 1807-1809 291
TRUE BRITON 1778-1780/1782-1785 226
TRUE BRITON 1812-1814 291
TRUELOVE 1694-1698 188
TRUELOVE 1694-1714 321
TRUELOVE 1704-1708 195
TRUELOVE 1707-*Ca*1770 321
TRUELOVE 1720-...?
see OLD TRUELOVE
TRUMPETT 1694-1695 188
**TRUSTY** 1782-1815 78
TRYALL 1703-1713 323
TRYALL 1710-1719 37
TRYALL 1719-1731 37
TRYALL 1732-1741 53
TRYALL 1803-1807 291
TRYALL
see also TRIAL
TRYAL PRIZE 1741-1742 202
TRYDENT
see TRIDENT
TRYTON
see TRITON
TRYUMPH
see TRIUMPH
Tsukuba/Tsukuba Kan (Japan)
see MALACCA 1853
TURBULENT 1805-1808 154
TUSCAN 1808-1818 278
TWEED 1759-1776 83
**TWEED** 1807-1813 130
TWEED
see GLENMORE 1796
TWEED 1823-1852 133
TWO BROTHERS 1706-1706 195
TWO BROTHERS early C18th? 327
TWO BROTHERS 1798-1802? 259
Two Crowns (merchant)
see RANGER 1747
TWO FRIENDS 1803-1807 291
Two Sisters (merchant)
see PINCHER 1797
TYGER
see TIGER
TYNE 1814-1825 131
**TYNE** 1826-1862 134
TYNE 1867-1867
see ACTIVE 1845
TYRIAN 1808-1819 145
TYRIAN 1826-1866 146
TYRONE 1798-1802 263

## U

ULYSSES 1779-1816 79
**UMPIRE**
see ROYAL GEORGE 1788 63
UNDAUNTED 1794-1795 247
UNDAUNTED 1795-1796
see ARETHUSE 1795
UNDAUNTED 1799-1800 263
UNDAUNTED 1807-1860 122
UNICORN 1748-1771 85
UNICORN 1776-1787 91
UNICORN
see THALIA 1782
UNICORN 1794-1815 126
UNICORN 1824-date 120
Unie (Dutch 1795)
see DROCHTERLAND 1799
UNION 1709-1718
see ALBERMARLE 1704
UNION 1726-1749 40
UNION 1756-1790 63
UNION 1759-1760 211
Union
see ELEPHANT 1776
Union (merchant)
see PORCUPINE 1777
UNION 1782-1782 233
UNION 1794-1800 258
UNION 1793-1793 266
UNION 1793-1794 266
UNION 1799-1801 266
UNION 1801-1801 266
Union
see DILIGENCE 1801
UNION 1803-1809 291
UNION 1805-1810 281
UNION 1809-1811 291
UNION 1811-1833 106
UNION 1823-1828 296
UNIQUE 1803?-1806 288
UNIQUE 1808-1809 288
Unité (French 1794)
see SURPRISE 1796
UNITE 1796-1802 246
UNITE 1803-1806
see IMPERIEUSE 1793
United (merchant 1777)
see SALAMANDER 1804
UNITED BROTHERS 1803-1807 291
UNITY 1665?-1712? 318
UNITY 1688-1707 187, 321
UNITY 1689 190
UNITY 1692-1695 190
UNITY 1693-1713 318
UNITY 1704-1706 195
UNITY 1704-1706 195
UNITY 1707-1773 323
UNITY 1720-1788 323
UNITY 1721-1722 196
Unity (merchant 1770)
see ERRUPTION 1804
UNITY 1788-1878 324
UNITY 1804-1810 290
Unity Gally
see CRUIZER 1709
URANIA 1797-1807 243
URANIE
see URANIA
URCHIN 1797-1800 263
URGENT 1804-1816 153
URQUIJO 1805-1805 288
URSULA 1745-1745 202
UTILE 1796-1798 248
UTILE 1799-1801 253
UTILE 1804-1814 276
UTILE
see EIJDEREN 1807
UTRECHT 1799-1815 241

## V

Vadette?
see VALETTE
Vaillant (French 1691-1716)
see MARY ROSE 1660
Vaillante (French 1795)
see DANAE 1798
Vaillante (French 1801)
see BARBETTE 1805
VAINQUER/VANQUER 1748-1748 202
VALETTE 1800-1801 266
VALEUR 1705-1718 192
VALEUR 1759-1764 206
**VALIANT/VALLIANT** 1759-1828 68
VALIANT 1794-1801 266
VALIANT 1807-1823 113
VALIANT Cancelled 1832 110
VALOROUS 1804-1817 136
VALOROUS 1816-1829 137
VALURE
see VALEUR 1705
VAN
see DOVE?
VAN 1824-1866 317, 318
VANGUARD 1678-1705 11
VANGUARD 1710-1728 33
VANGUARD 1748-1774 72
VANGUARD 1781-1782 232
VANGUARD 1787-1821 68
**VANGUARD** 1835-1875 172
VANNEAU 1793-1796 263
VANSITTART 1821-1829 296
Var (French 1806 )
see CHICHESTER 1809
VARIABLE 1811?-1814
see REDBRIDGE 1807
VARIABLE 1814?-1817 280
VARIABLE
see PIGEON 1827
Variante (French 1785)
see UNITE 1796
VARIATION 1794-1808 317
VAUGHAN 1781-1783 225
VAUGHAN 1782-1783 232
VAUTOUR 1809/1810-1813 280
Vedette (French 1793)
see VIDETTE 1800
VENERABLE 1784-1804 71
VENERABLE 1808-1838 113
VENGEANCE 1758-1766 206
VENGEANCE 1774-1816 70
VENGEANCE 1793-1804 256
VENGEANCE 1800-1801 244
Vengeance (French 1793 or 1799)
see SCOUT 1800
Vengeance (American 1812 privateer)
see TELEGRAPH 1812
VENGEANCE 1824-1897 107
Vengeur
see CHARLOTTE 1798
VENGEUR
see TOBAGO 1805
**VENGEUR** 1810-1843 114
VENSEJO (+ Spanish 1799)
see VINCEJO
VENOM 1794-1799? 263
Ventueux (French 1795)
see ECLIPSE 1803
VENTURA 1806-1806 288
VENTURE 1798-1802
see RANGER 1787
VENTURE
see FLYVENDE FISK 1807
VENTURER 1807-1814? 288
**VENUS** 1759-1809 82
Venus (Dutch 1778)
see AMARANTHE 1796
Vénus (French)
see MUSQUITO 1793
VENUS 1793-1797 266
VENUS 1794-1801 266
VENUS 1796-1801 266
VENUS 1798-1802 266
VENUS 1803-1806 291
VENUS 1803-1804 291
VENUS 1804-1804 291
Vénus (French 1806)
see NEREIDE 1810
VENUS 1807-1815 272
VENUS 1820-1865 119
VENUS PACKET 1815-1815 291
VERNON 1781-1782 233
**VERNON** 1832-1886 173
VERNON II 1904-1924
Vertu (French 1794)
see VIRTU 1803
VESTA 1806-1816 162
VESTAL 1757-1775 82
VESTAL 1777-1777 91
VESTAL 1779-1816 87
VESTAL Cancelled 1831 135
**VESTAL** 1833-1862 175
Vestale (French 1757)
see FLORA 1761
VESUVE 1795-1802 259
VESUVIUS 1691-1692/1693 29
VESUVIUS 1694-1705 188
VESUVIUS 1739-1742 200
VESUVIUS 1756-1763 207
VESUVIUS 1771
see RAVEN 1771
VESUVIUS 1775-1812 99
VESUVIUS
see VULCAN 1777
VESUVIUS 1797-1812 263
**VESUVIUS** 1813-1819 147
VESUVIUS Cancelled 1831 148
VETERAN 1787-1816 75
VETERAN 1839-1852
see PROMETHEUS 1807
Viala (French 1794)
see MAIDA 1806
Victoire (French)
see ? SPIDER 1782
VICTOIRE 1795-1800 263
VICTOIRE 1797-1801 254
VICTOIRE
see ORNEN 1807
VICTOR 1779-1780? (or 1783?) 224
VICTOR 1798-1806 135
VICTOR 1808-1809/1810-1810? 288
VICTOR 1814-1842 142
VICTORIA
see ROYAL SOVEREIGN cancelled 1838? 335
VICTORIA ?1835-1857?
VICTORIA
see WINDSOR CASTLE 1858
VICTORIEUSE 1795-1805 251
VICTORIOUS 1785-1803 71
VICTORIOUS 1808-1861 113
VICTORY 1620-1692 11
VICTORY 1691-1695
see ROYAL JAMES 1675
VICTORY 1695-1714/1715-1716 17
VICTORY 1738-1744 39
VICTORY 1763-... 301
VICTORY 1765-date 62
VIDETTE 1800-1802 263
Vierge (French)
see VIRGIN 1760
VIGILANT/VIGILANTE 1745-1759 197
VIGILANT 1755-1756 299
VIGILANT 1760-1778? 297
VIGILANT 1774-1816 74
VIGILANT 1777-1780 229
VIGILANT 1793-1801 266
VIGILANT 1798-1800 263
VIGILANT 1803-1807 288
VIGILANT 1806-1807
see SUBTLE 1807
VIGILANT ?1809-1819 333
VIGILANT 1821-1832 165
VIGILANTE 1793-1793 263
Vigilante (French 1802)
see SUFFISANTE 1803
Vigilante (Spanish)
see SPIDER 1806
VIGO
see DARTMOUTH 1693
VIGO 1810-1865 114
Vilaret (French 1803?)
see RAPIDE 1808
Ville de Milan (French 1803)
see MILAN 1805
VILLE DE PARIS 1782-1782 214
**VILLE DE PARIS** 1795-1845 105
VIMIERA 1808-1814 279
VINCEJO/VINCEGO 1799-1804 253

INDEX 367

VINDICTIVE 1779-1783 232
VINDICTIVE 1796-1798 247
VINDICTIVE 1813-1871 114
VIOLET 1806-1812 283, 333
VIOLET 1812-1814? 288
VIOLET 1835-1848 294
VIPER 1746-1746 202
VIPER 1746-1755 58
**VIPER** 1756-1779 93
VIPER 1777-1789 228
VIPER 1779-1783 232
VIPER 1781-1803?
see GREYHOUND 1780
VIPER 1794-1797 251
VIPER 1794-1801 258
VIPER ?1797-1813? 333
VIPER 1807-1809 282
VIPER 1809-1814 282
Viper (American 1804)
see MOHAWK 1813
VIPER 1825?-1845? 335
VIPER 1831-1851 165
Vipère (French 1793)
see VIPER 1794
VIPERE 1793-1793 263
VIRAGO 1793-1800 263
VIRAGO 1805-1816 154
VIRGIN? 1760-1764 210
VIRGINIA 1778-1782 219
Virginia (American)
see GIBRALTAR 1779
VIRGINIA 1796-1800 263
VIRGINIE 1796-1828 243
VIRGINIE 1808-1811 278
VIRGIN PRIZE 1690-1698 186
VIRTU 1803-1810 271
Visitation
see EXTRAVIGANT 1691
VITTORIA
see PRINCESS CHARLOTTE
VIVO 1800-1801 253
VIXEN 1801-1815 145
VIXEN 1837-...? 335
VLIETER 1799-1817 242
VOLADOR 1807-1807 288
Volage (French 1741)
see VIPER 1746
VOLAGE 1798-1804 248
see ECLIPSE 1803
VOLAGE 1807-1818 128
**VOLAGE** 1825-1874 134
VOLCANO 1778-1781 225
VOLCANO 1780s-1784 226
VOLCANO/VULCANO 1797-1810 255
VOLCANO 1804-1810 288
VOLCANO 1810-1816
see HERON 1804
VOLCANO Cancelled 1832 148
VOLONTAIRE 1806-1816 271
Voltaire (French 1794)
see MAIDA 1806
VOLTIGEUR 1798-1802 249
Voltigeur (French 1805)
see PELICAN 1806
VOLTIGEUR
see SARPEN 1807
Volunteer (merchant 1803)
see UTILE 1804
VOLUNTEER 1804-1812 277
Vrijheid (Dutch 1782)
see VRYHEID
VRYHEID 1797-1811 240
VULCAN 1691-1708 29
VULCAN 1739-1743 200
VULCAN 1745-1749 200
VULCAN 1777-1781 225
VULCAN 1780-1780 232
VULCAN 1783-1793 100
VULCAN 1796-1801
see VOLCANO 264
VULTURE 1690-1708 29
VULTURE 1744-1761 57
VULTURE 1763-1771 96
VULTURE 1776-1802 96
VULTURE 1803-1814 276

# W

WAAKSAMHEIDT/WAAKZAMHEID 1798-1802 248
WAGER 1739-1741 201
WAGER 1744-1763 52
WAGGON 1760-1778? 297
WAGTAIL 1806-1807 161
WAKEFIELD 1656-1660
see RICHMOND 1660
WALDEMAR 1807-1816 267
Waldemar Melgounoff (Russian?)
see STEADY 1782
WALKER 1804-1805 291
Waller/Walter (French merchant?)
see FIREBRAND 1804
WANDERER 1806-1817 130
WANDERER 1835-1850 178
Warley
see CALCUTTA 1795
WARNING 1807-1814 278
WARRIOR 1781-1857 71
Warrior (merchant 1801)
see VULTURE 1803
WARRINGTON 1692-1693 190
WARSPIGHT
see WARSPITE
WARSPITE 1666-1702 12
WARSPITE 1703-1716 21
WARSPITE/WARSPIGHT 1758-1800 67
WARSPITE/WARSPIGHT 1807-1876 112
WARSPITE
see WATERLOO 1833
WARWICK 1696-1709 24
WARWICK 1711-1726 35
WARWICK 1733-1756 44
WARWICK 1767-1802 77
Washington (Dutch 1795)
see PRINCE OF ORANGE 1799
WASHINGTON 1776-17.. 298
**WASP** 1749-1781 92
WASP 1782-1798 223
WASP 1793 264
WASP 1794-1801 258
WASP 1800-1811 250
WASP 1800-1878 324
**WASP** 1809-1809 288
WASP 1812-1847 142
Wasp (American 1806)
see PEACOCK 1812
WASP 1822-1829 296
WASSANAAR/WASSENAAR 1797-1802 240
WATCHFUL 1804-1814 277
WATCHFUL 1808-1822 333
**WATERLOO** 1818-1824 109
WATERLOO 1833-1862 105
WATERWITCH 1781-1781 233
WATERWITCH 1834-1861 292
Waterwitch (merchant )
see FALCON 1820
WEASELL/WEAZLE 1704-1712 32
WEAZEL
see WEAZLE
WEAZLE 1721-1732 52
WEAZLE 1745-1779 199
WEAZLE 1783-1799 98
WEAZLE/WEAZEL 253 1799-1804 253
WEAZLE 1808-1811 291
WEAZLE 1805-1815 140
WEAZLE 1808-1812 288
WEAZLE 1822-1844 145
WEE PET 18.. -1849 295
WELCOME 1689-1690 190
WELCOME 1704-1704 195
WELCOME 1704-1708 195
WELCOME 1705-1706 195
WELCOME 1775
Wellesley (1797)
see WEYMOUTH 1804
WELLESLEY
see CORNWALL 1812
WELLESLEY 1815-1868 115
WELLESLEY 1894-1914
see BOSCAWEN 1844
WELLINGTON 1804-1805 291
WELLINGTON 1810-1812 280
WELLINGTON 1815-1823 333
WELLINGTON 1816-1862 114
**WELLS** 1764-1780 100
WESER 1813-1817 273
WEST FLORIDA
see FLORIDA
WEYMOUTH 1693-1717 23
WEYMOUTH 1719-1733 36
WEYMOUTH 1736-1745 44
WEYMOUTH 1752-1772 75
WEYMOUTH 1795-1800 241
WEYMOUTH 1804-1865 270
WHITBY 1781-1785 231
WHITEHAVEN 1747-1747 120
WHITING 1711-1712 194
WHITING 1805-1812 161
WHITING 1812-1816 283
WHITWORTH ?1817- 1826? 334
WICKHAM ?1817-1856? 334
WIGEON 1806-1808 161
WIGHT 1835-1864 336
WILD 1692-1694 189
WILD BOAR 1808-1810 144
WILDFIRE 1804-1807 281
WILDFIRE 1916-1957
see CORNWALLIS 1813
WILD PRIZE
see WILD
WILHELMINA 1798-1813 246
Willemstadt /Willemstadtler Boetzelaar (Dutch)
see PRINCESS 1795
WILLIAM 1694 189
William (merchant)
see CANCEAUX 1764
WILLIAM 1778-1783 233
WILLIAM 1794-1801 258
WILLIAM 1795-1802 258
WILLIAM 1798-1810 249
WILLIAM 1804-1852 325
WILLIAM 1808-1810 288
William Adventure (merchant 1763)
see FUZE 1804
WILLIAM AND ANN 1704-1706 195
WILLIAM AND ANNE 1757-1757 211
WILLIAM AND ANN 1795-1819 324
WILLIAM AND ANN 1795-1795 264
WILLIAM AND ELIZABETH 1695-1695 189
WILLIAM AND ELIZABETH 1707-1711 195
WILLIAM AND JAMES 1705-1705 195
William and Jean
see PORTSMOUTH TRANSPORT
WILLIAM AND LUCY 1795-1797? 258
WILLIAM AND MARGARET/MARGATE 1706-1706 195
WILLIAM AND MARY 1689-1689 190
WILLIAM AND MARY 1694-1801 303
WILLIAM AND MARY 1694-1696 188
WILLIAM AND MARY 1704-1711 195
WILLIAM AND MARY 1705-1711 195
WILLIAM AND MARY 1706-1706 195
William and Mary (Merchant )
see SANDWICH 1803/PITT 1805
WILLIAM AND MARY 1807-1849 306
WILLIAM AND REBECCA (WILLIAM AND ROBERT?) 1689-1689 190
WILLIAM AND THOMAS 1689-1689 190
William Bayard (American 1812)
see ALBAN 1813
WILLIAM PITT 1796-1799 266
WILLIAMSON 1760-1761 299
WILLINGTON 1801-1802 266
WILSON 1794-1797? 258
Wimbury (1803)
see ESPIEGLE 1804
WINCHELSEA/WINCHELSEY 1694-1706 26
WINCHELSEA 1706-1707 27
WINCHELSEA/WINCHELSEY 1708-1735 192
WINCHELSEA 1740-1758/1761 51
WINCHELSEA 1763-1774 209
WINCHELSEA 1764-1814 83
WINCHESTER 1693-1695 22
WINCHESTER 1698-1716 24
WINCHESTER 1717-1781 36
WINCHESTER 1744-1769 47
WINCHESTER 1822-1876 116
WINCHESTER
WINDSOR 1695-1725 22
WINDSOR 1729-1742 44
WINDSOR 1745-1717 45
WINDSOR CASTLE 1678-1693 11
WINDSOR CASTLE 1702-1706
see DUCHESS 1679
WINDSOR CASTLE 1790-1839 64
WINDSOR CASTLE
see DUKE OF WELLINGTON 1852
WINDSOR CASTLE 1858-1908 170
WINSBY 1654-1660
see HAPPY RETURN 1660
Winterton (East Indiaman)
see COROMANDEL 1795
WIZARD 1805-1816 143
WIZARD 1824-1826 296
WIZARD 1830-1859 146
WOLF 1690-1692 29
WOLF 1699-1704/1708-1712 31
WOLF 1731-1741 52
WOLF 1742-1749 54
WOLF 1754-1781 92
WOLF 1780-1780 232
WOLF 1794-1803 258
WOLF 1800-1802
see PANDOUR 1798
WOLF 1801-1829 264
WOLF 1804-1806 131
WOLF 1806-1811 278
WOLF 1814-1825 144
**WOLF** 1826-1878 138
WOLFE
see also WOLF
WOLFE
see ROYAL GEORGE 1809
WOLFE
see MONTREAL 1813
WOLFE 1813-1814 300
WOLFE Building 1815 (Cancelled) 301
WOLVERENE 1798-1804 253
WOLVERENE 1805-1816 140
WOLVERENE 1836-1855 178
WOODCOCK 1806-1807 161
WOODLARK 1805-1805 153
WOODLARK 1808-1818 144
WOODLARK 1821-1864 164
WOOLF
see WOLF
WOOLWICH 1675-1702 13
WOOLWICH 1702-1736 24
WOOLWICH 1704-1704 195
WOOLWICH 1705-1711 195
WOOLWICH 1705-1726 317
WOOLWICH 1726-1767 317
WOOLWICH 1739 1813 317
WOOLWICH 1741-1747 46
WOOLWICH Cancelled 1748 76
WOOLWICH 1749-1762 79
WOOLWICH 1785-1813 80
WOOLWICH 1788-1800 234
WOOLWICH 1815-1832? 318, 327, 330
WORCESTER 1651-1660
see DUNKIRK 1660
WORCESTER 1698-1713 24
WORCESTER 1714-1733 35
WORCESTER 1735-1765 44
Worcester (merchant)
see VESUVIUS 1739
**WORCESTER** 1769-1816 74
WORCESTER 1843-1885 116
WORCESTER 1861-1871
see CONWAY 1832
WORCESTER 1876-1953
see FREDERICK WILLIAM 1860
WORCESTER PRIZE 1705-1708 193
WRANGLER 1797-1802 259
WRANGLER 1804-1815 153
WRENN 1694-1697 31
WRIGHT 1797-1801 266
WYE 1811-1852 132
WYNDATT building 1779 302

# X

XENOPHON 1798-1798 250

# Y

Yankee (American)
see ROYAL SAVAGE 1775
YARMOUTH 1695-1707 20
YARMOUTH 1709-1768 34
YARMOUTH 1745-1801 43
YARMOUTH 1800-1828 316, 318
YARMOUTH 1800-1810-1849? 333
YARMOUTH
see WALDEMAR 1807
YORK 1660-1703 12
YORK/YORKE 1706-1737 23
YORK 1739-1754 44
YORK 1744-1745 202
YORK 1748-1748 202
YORK 1753-1772 76
YORK 1777-1779 224
YORK 1779-1781? 229
YORK 1796-1804 239
YORK 1807-1853 112
YORK 1815 302
Yorkshire
see CAMEL 1776
YOUNG HAZARD 1779-1779 233
YOUNG HEBE 1839-1842 296
YOUNG HEBE 1843-1850 293
YOUNG LADY 1694-1695? 188
YOUNG SPRAG 1673-1693 15
YTHAN 1809-1809 291

# Z

Zachari (Maltese)
see DEGO
Zaragozana (Pirate/slaver 1820)
see RENEGADE 1823
ZEALAND/ZEELAND 1796-1813 239
ZEALOUS 1785-1816 69
ZEBRA 1777-1778 96
**ZEBRA** 1780-1812 97
ZEBRA 1815-1840 142
ZEBRA
see JUMNA 1848
Zefir (Dutch 1786)
see EURUS 1796
ZENOBIA 1806-1806 162
ZENOBIA 1807-1835 141
Zéphir (French?)
see BEAUMONT 1780
ZEPHYR 1757-1778
see MERLIN 1756
ZEPHYR 1779-1783 222
ZEPHYR 1795-1808 251
ZEPHYR ...-1809 288
ZEPHYR 1809-1816 144
ZEPHYR 1823-1836 145

## Numbered Vessels

1 & 2 Advice boats
see EXPRESS and ADVICE 1800
1 to 3 Gunboats 1793-1799? 264
1 to 3 Gunboats 183, 302
1 to 85 Gunboats (Hamilton design) built 1808-1809 157
1 Gunboat
see TOWZER 1798
1 Row Galley Gunboat 1801 157
1 to 2 Mortar Vessels 1854 183
2 Advice boat
see ADVICE 1800
2 Gunboat
see TIGER 1798
3 Gunboat
see MERMAID 1799
4 Gunboat
see ASSAULT 1797
4 Gunboat
see SWINGER 1798
5 Gunboat
see ASP 1797
5 Gunboat
see TEAZER 1798
6 Gunboat
see ACUTE 1797
7 Gunboat
see SPARKLER 1797
8 Gunboat
see BOUNCER 1797
9 Gunboat
see BOXER 1797
10 Gunboat
see BITER 1797
11 Gunboat
see BRUIZER 1797
12 Gunboat
see BLAZER 1797
13 Gunboat
see CRACKER 1797
14 Gunboat?
15 Gunboat
see CRASH 1797
16 Gunboat
see CONTEST 1797
17 Gunboat
see ADDER 1797
18 Gunboat
see SPITEFUL 1797
19 Gunboat
see STEADY 1797
20 Gunboat
see COURSER 1797
21 Gunboat
see DEFENDER 1797
22 Gunboat
see ECLIPSE 1797
23 Gunboat
see FURIOUS 1797
24 Gunboat
see FLAMER 1797
25 Gunboat
see FURNACE 1797
26 Gunboat
see GROWLER 1797
27 Gunboat
see GRIPER 1797
28 Gunboat
see GRAPPLER 1797
29 Gunboat
see GALLANT 1797
30 Gunboat
see HARDY 1797
31 Gunboat
see HAUGHTY 1797
32 Gunboat
see HECATE 1797
33 Gunboat
see HASTY 1797
34 Gunboat
see METEOR 1797
35 Gunboat
see MASTIFF 1797
36 Gunboat
see MINX 1797
37 Gunboat
see MANLY 1797
38 Gunboat
see POUNCER 1797
39 Gunboat
see PINCHER 1797
40 Gunboat
see WRANGLER 1797
41 Gunboat
see RATTLER 1797
42 Gunboat
see READY 1797
43 Gunboat
see SAFEGUARD 1797
44 Gunboat
see STAUNCH 1797
45 Gunboat
see TIGRESS 1797